Lecture Notes in Artificial Intelligence 4251

Edited by J. G. Carbonell and J. Siekmann

Subseries of Lecture Notes in Computer Science

AF698370

Bogdan Gabrys   Robert J. Howlett
Lakhmi C. Jain (Eds.)

Knowledge-Based Intelligent Information and Engineering Systems

10th International Conference, KES 2006
Bournemouth, UK, October 9-11, 2006
Proceedings, Part I

Springer

Series Editors

Jaime G. Carbonell, Carnegie Mellon University, Pittsburgh, PA, USA
Jörg Siekmann, University of Saarland, Saarbrücken, Germany

Volume Editors

Bogdan Gabrys
Bournemouth University
School of Design, Engineering and Computing
Computational Intelligence Research Group
Talbot Campus, Fern Barrow, Poole, GH12 5BB, UK
E-mail: bgabrys@bournemouth.ac.uk

Robert J. Howlett
University of Brighton
School of Engineering
Centre for SMART Systems, Brighton BN2 4GJ, UK
E-mail: r.j.howlett@brighton.ac.uk

Lakhmi C. Jain
University of South Australia
School of Electrical and Information Engineering
Knowledge-Based Intelligent Information and Engineering Systems Centre
Adelaide, Mawson Lakes Campus, South Australia SA 5095, Australia
E-mail: Lakhmi.Jain@unisa.edu.au

Library of Congress Control Number: 2006933827

CR Subject Classification (1998): I.2, H.4, H.3, J.1, H.5, K.6, K.4

LNCS Sublibrary: SL 7 – Artificial Intelligence

ISSN 0302-9743
ISBN-10 3-540-46535-9 Springer Berlin Heidelberg New York
ISBN-13 978-3-540-46535-5 Springer Berlin Heidelberg New York

This work is subject to copyright. All rights are reserved, whether the whole or part of the material is concerned, specifically the rights of translation, reprinting, re-use of illustrations, recitation, broadcasting, reproduction on microfilms or in any other way, and storage in data banks. Duplication of this publication or parts thereof is permitted only under the provisions of the German Copyright Law of September 9, 1965, in its current version, and permission for use must always be obtained from Springer. Violations are liable to prosecution under the German Copyright Law.

Springer is a part of Springer Science+Business Media

springer.com

© Springer-Verlag Berlin Heidelberg 2006
Printed in Germany

Typesetting: Camera-ready by author, data conversion by Scientific Publishing Services, Chennai, India
Printed on acid-free paper SPIN: 11892960 06/3142 5 4 3 2 1 0

Preface

Delegates and friends, we are very pleased to extend to you the sincerest of welcomes to this, the 10th International Conference on Knowledge Based and Intelligent Information and Engineering Systems at the Bournemouth International Centre in Bournemouth, UK, brought to you by KES International.

This is a special KES conference, as it is the 10th in the series, and as such, it represents an occasion for celebration and an opportunity for reflection. The first KES conference was held in 1997 and was organised by the KES conference founder, Lakhmi Jain. In 1997, 1998 and 1999 the KES conferences were held in Adelaide, Australia. In 2000 the conference moved out of Australia to be held in Brighton, UK; in 2001 it was in Osaka, Japan; in 2002, Crema near Milan, Italy; in 2003, Oxford, UK; in 2004, Wellington, New Zealand; and in 2005, Melbourne, Australia. The next two conferences are planned to be in Italy and Croatia. Delegate numbers have grown from about 100 in 1997, to a regular figure in excess of 500. The conference attracts delegates from many different countries, in Europe, Australasia, the Pacific Rim, Asia and the Americas, and may truly be said to be 'International'. Formed in 2001, KES International has developed into a worldwide organisation that provides a professional community for researchers in the discipline of knowledge-based and intelligent engineering and information systems, and through this, opportunities for publication, networking and interaction. Published by IOS Press in the Netherlands, the KES Journal is organised by joint Editors-in-Chief R.J. Howlett and B. Gabrys. There are Associate Editors in the UK, the US, Poland, Australia, Japan, Germany and the Czech Republic. The Journal accepts academic papers from authors in many countries of the world and has approximately 600 subscribers in about 50 countries. KES produces a book series, also published by IOS Press, and there are plans for the development of focus groups, each with associated publications and symposia, and other ventures such as thematic summer schools.

The KES conference continues to be a major feature of the KES organisation, and KES 2006 looks set to continue the tradition of excellence in KES conferences. This year a policy decision was made not to seek to grow the conference beyond its current size, and to aim for about 450-500 papers based on their high quality. The papers for KES 2006 were either submitted to Invited Sessions, chaired and organised by respected experts in their fields, or to General Sessions, managed by Track Chairs and an extensive International Programme Committee. Whichever route they came through, all papers for KES 2006 were thoroughly reviewed. There were 1395 submissions for KES 2006 of which 480 were published, an acceptance rate of 34%.

Thanks are due to the very many people who have given their time and goodwill freely to make the conference a success.

We would like to thank the KES 2006 International Programme Committee who were essential in providing their reviews of the papers. We are very grateful for this service, without which the conference would not have been possible. We thank the high-profile keynote speakers for providing interesting and informed talks to catalyse subsequent discussions.

An important distinction of KES conferences over others is the Invited Session Programme. Invited Sessions give new and established researchers an opportunity to present a "mini-conference" of their own. By this means they can bring to public view a topic at the leading edge of intelligent systems. This mechanism for feeding new blood into the research is very valuable. For this reason we must thank the Invited Session Chairs who have contributed in this way.

The conference administrators, Maria Booth and Jo Sawyer at the Universities of Brighton and Bournemouth respectively, and the local committees, have all worked extremely hard to bring the conference to a high level of organisation, and our special thanks go to them too.

In some ways, the most important contributors to KES 2006 were the authors, presenters and delegates without whom the conference could not have taken place. So we thank them for their contributions.

In less than a decade, KES has grown from a small conference in a single country, to an international organisation involving over 1000 researchers drawn from universities and companies across the world. This is a remarkable achievement. KES is now in an excellent position to continue its mission to facilitate international research and co-operation in the area of applied intelligent systems, and to do this in new and innovative ways. We hope you all find KES 2006 a worthwhile, informative and enjoyable experience. We also hope that you will participate in some of the new activities we have planned for KES, for the next 10 years, to benefit the intelligent systems community.

August 2006

Bob Howlett
Bogdan Gabrys
Lakhmi Jain

KES 2006 Conference Organization

Joint KES 2006 General Chairs

Bogdan Gabrys
Computational Intelligence Research Group
School of Design, Engineering and Computing
University of Bournemouth, UK

Robert J. Howlett
Centre for SMART Systems
School of Engineering
University of Brighton, UK

Conference Founder and Honorary Programme Committee Chair

Lakhmi C. Jain
Knowledge-Based Intelligent Information and Engineering Systems Centre
University of South Australia, Australia

Local Organising Committee (University of Brighton)

Maria Booth, KES Administrator
Shaun Lee, Simon Walters, Anna Howlett
Nigel Shippam, PROSE Software Support
Anthony Wood, Web Site Design

Local Organising Committee (University of Bournemouth)

Jo Sawyer, KES 2006 Local Committee Coordinator
Michael Haddrell, Mark Eastwood, Zoheir Sahel, Paul Rogers

International Programme Committee and KES 2006 Reviewers

Abbass, Hussein	University of New South Wales, Australia
Abe, Akinori	ATR Intelligent Robotics and Communication Labs, Japan
Adachi, Yoshinori	Chubu University, Japan
Alpaslan, Ferda	Middle East Technical University, Turkey
Ang, Marcello	National University of Singapore, Singapore
Angelov, Plamen	Lancaster University, UK
Anguita, Davide	DIBE - University of Genoa, Italy
Anogeianakis, Giorgos	Aristotle University of Thessaloniki, Greece
Arotaritei, Dragos	Polytechnic Institute of Iasi, Romania

Arroyo-Figueroa, Gustavo	Electrical Research Institute of Mexico, Mexico
Asano, Akira	Hiroshima University, Japan
Asari, Vijayan	Old Dominion University, USA
Augusto, Juan	University of Ulster at Jordanstown, UK
Baba, Norio	Osaka Kyoiku University, Japan
Bajaj, Preeti	G.H. Raisoni College of Engineering, Nagpur, India
Bargiela, Andrzej	Nottingham Trent University, UK
Berthold, Michael	University of Konstanz, Germany
Berthouze, Nadia	University of Aizu, Japan
Binachi-Berthouze, Nadia	University of Aizu, Japan
Bod, Rens	University of St Andrews, UK
Bosc, Patrick	IRISA/ENSSAT, France
Bouvry, Pascal	University of Applied Sciences, Luxembourg
Brown, David	University of Portsmouth, UK
Burrell, Phillip	London South Bank University, UK
Cangelosi, Angelo	University of Plymouth, UK
Ceravolo, Paolo	University of Milan, Italy
Chakraborty, Basabi	Iwate Prefectural University, Japan
Chen, Yen-Wei	Ritsumeikan University, Japan
Chen-Burger, Yun-Heh	University of Edinburgh, UK
Chung, Jaehak	Inha University, Korea
Cios, Krzysztof	University of Colorado at Denver and Health Sciences Center, USA
Coello, Carlos A.	LANIA, Mexico
Coghill, George	University of Auckland, New Zealand
Corbett, Dan	SAIC, USA
Corchado, Emilio	University of Burgos, Spain
Correa da Silva, Flavio	University of Sao Paulo, Brazil
Cuzzocrea, Alfredo	University of Calabria, Italy
Damiani, Ernesto	University of Milan, Italy
Deep, Kusum	Indian Institute of Technology, Roorkee, India
Deng, Da	University of Otago, Dunedin, New Zealand
Dubey, Venky	University of Bournemouth, UK
Dubois, Didier	University Paul Sabatier, France
Duch, Wlodzislaw	Nicolaus Copernicus University, Poland
Eldin, Amr Ali	Delft University of Technology, The Netherlands
Far, Behrouz	University of Calgary, Canada
Finn, Anthony	DSTO, Australia
Flórez-López, Raquel	University of Leon, Spain
Fortino, Giancarlo	Università della Calabria, Italy
Fuchino, Tetsuo	Tokyo Institute of Technology, Japan
Fyfe, Colin	University of Paisley, UK
Gabrys, Bogdan	University of Bournemouth, UK

Galitsky, Boris	Birkbeck College University of London, UK
Ghosh, Ashish	Indian Statistical Institute, Kolkata, India
Girolami, Mark	University of Glasgow, UK
Gorodetski, Vladimir	St. Petersburg Institute of Informatics, Russia
Grana, Manuel	Universidad Pais Vasco, Spain
Grana Romay, Manuel	Universidad Pais Vasco, Spain
Grzech, Adam	Wroclaw University of Technology, Poland
Grzymala-Busse, Jerzy	University of Kansas, USA
Gu, Dongbing	University of Essex, UK
Håkansson, Anne	Uppsala University, Sweden
Hanh H., Phan	State University of New York, USA
Hansen, Lars Kai	Technical University of Denmark, Denmark
Harrison, Robert	The University of Sheffield, UK
Hasebrook, Joachim. P	University of Luebeck, Germany
Hatem, Ahriz	The Robert Gordon University, Aberdeen, UK
Hatzilygeroudis, Ioannis	University of Patras, Greece
Helic, Denis	Technical University of Graz, Austria
Hildebrand, Lars	University of Dortmund, Germany
Hirai, Yuzo	Institute of Information Sciences and Electronics, Japan
Hong, Tzung-Pei	National University of Kaohsiung, Taiwan
Honghai, Liu	The University of Aberdeen, UK
Hori, Satoshi	Institute of Technologists, Japan
Horio, Keiichi	Kyushu Institute of Technology, Japan
Howlett, Robert J.	University of Brighton, UK
Huaglory, Tianfield	Glasgow Caledonian University, UK
Iluminada, Baturone	University of Seville, Spain
Imada, Akira	Brest State Technical University, Belasus
Ishibuchi, Hisao	Osaka Prefecture University, Japan
Ishida, Yoshiteru	Toyohashi University of Technology, Japan
Ishida, Yoshiteru	Toyohashi University, Japan
Ishii, Naohiro	Aichi Institute of Technology, Japan
Jacquenet, François	University of Saint-Etienne, France
Jadranka, Sunde	DSTO, Australia
Jain, Lakhmi C.	University of South Australia, Australia
Jarvis, Dennis	Agent Oriented Software Pty. Ltd., Australia
Jesse, Norbert	University of Dortmund, Germany
Kacprzyk, Janusz	Polish Academy of Sciences, Poland
Karacapilidis, Nikos	University of Patras, Greece
Karny, Miroslav	Institute of Information Theory and Automation, Czech Republic
Kasabov, Nik	Auckland University of Technology, New Zealand
Katarzyniak , Radoslaw	Wroclaw University of Technology, Poland

Kazuhiko, Tsuda	University of Tsukuba, Japan
Keskar, Avinash	Visvesvaraya National Institute of Technology, India
Kim, Dong-Hwa	Hanbat National University, Korea
Kim, Jong Tae	SungKyunKwan University, Republic of Korea
Kim, Sangkyun	Yonsei University, South Korea
Kim,Tai-hoon	Korea
Kittler, Josef	University of Surrey, UK
Kóczy, Tamás, László	Budapest University of Technology and Economics, Hungary
Koenig, Andreas	Technical University of Kaiserslautern, Germany
Kojiri, Tomoko	Nagoya University, Japan
Konar, Amit	Jadavpur University, India
Koshizen, Takamasa	Honda Research Institute Japan Co., Ltd., Japan
Kunifuji, Susumu	School of Knowledge Science, Japan
Kurgan, Lukasz	University of Alberta, Canada
Kusiak, Andrew	The University of Iowa, USA
Lanzi, Pier Luca	Politecnico di Milano, Italy
Lee, Dong Chun	Howon University, Korea
Lee, Geuk	Hannam Howon University, Korea
Lee, Hong Joo	Dankook University, South Korea
Lee, Hsuan-Shih	National Taiwan Ocean University, Taiwan
Lee, Raymond	Hong Kong Polytechnic University, Hong Kong, China
Liu, Yubao	Sun Yat-Sen University, China
Lovrek, Ignac	University of Zagreb, Croatia
Lu, Hongen	La Trobe University, Australia
Luccini, Marco	University of Pavia, Italy
Mackin, Kenneth J.	Tokyo University of Information Sciences, Japan
Main, J.	La Trobe University, Australia
Mandic, Danilo	Imperial College London, UK
Maojo, Victor	Universidad Politécnica de Madrid
Martin, Trevor	Bristol University, UK
Masulli, Francesco	University of Pisa, Italy
Mattila, Jorma	Lappeenranta Univeristy of Technology, Finland
Mazumdar, Jagannath	University of South Australia, USA
McKay, Bob	University of New South Wales, Australia
Mera, Kazuya	University of Hiroshima, Japan
Mesiar,Radko	STU Bratislava, Slovakia
Mira, Jose	ETS de Ingeneria Informática (UNED), Spain
Monekosso, Dorothy	University of Kingston, UK
Montani, Stefania	Universitá del Piemonte Orientale, Italy
Morch, Anders	University of Oslo, Norway
Munemori, Jun	Wakayama University, Japan

Munemorim Jun	Wakayama University, Japan
Murthy, Venu K.	RMIT University, Australia
Nakamatsu, Kazumi	University of Hyogo, Japan
Nakano, Ryohei	Nagoya Institute of Technology, Japan
Nakao, Zensho	University of Ryukyus, Japan
Nakashima, Tomoharu	Osaka Prefecture University, Japan
Narasimhan, Lakshmi	University of Newcastle, Australia
Nauck, Detlef	BT, UK
Navia-Vázquez, Angel	Univ. Carlos III de Madrid, Spain
Nayak, Richi	Queensland University of Technology, Brisbane, Australia
Neagu, Ciprian	University of Bradford, UK
Negoita, Mircea	KES, New Zealand
Nguyen, Ngoc Thanh	Wroclaw University of Technology, Poland
Nishida, Toyoaki	University of Kyoto, Japan
Niskanen, Vesa A.	University of Helsinki, Finland
O'Connell, Robert	University of Missouri-Columbia, USA
Ong, Kok-Leong	Deakin University, Australia
Palade, Vasile	University of Oxford, UK
Palaniswami, Marimuthu	The University of Melbourne, Australia
Pant, Millie	BITS-Pilani, India
Papis, Costas	University of Piraeus, Greece
Paprzycki, Macin	Warsaw School of Social Psychology, Poland
Park, Gwi-Tae	Korea University, Korea
Pedrycz, Witold	University of Alberta, Canada
Peña-Reyes, Carlos-Andrés	Novartis Institutes for Biomedical Research, USA
Piedad, Brox	University of Seville, Spain
Polani, Daniel	University of Hertfordshire, UK
Popescu, Theodor	National Institute for Research and Development Informatics, Romania
Rajesh, R.	Bharathiar University, India
Reusch, Bernd	University of Dortmund, Germany
Rhee, Phill	Inha University, Korea
Rose, John	Ritsumeikan Asia Pacific University, Japan
Ruan, Da	The Belgian Nuclear Research Centre, Belgium
Ruta, Dymitr	BT Exact, UK
Rutkowski, Leszek	Technical University of Czestochowa, Poland
Sato-llic, Mika	University of Tsukuba, Japan
Sawada, Hideyuki	Kagawa University, Japan
Seiffert, Udo	Leibniz-Institute of Plant Genetics, Gatersleben, Germany
Semeraro, Giovanni	Università degli Studi di Bari, Italy
Sharma, Dharmendra	University of Canberra, Australia

Sirlantzis, Konstantinos	University of Kent, UK
Skabar, A.	La Trobe University, Australia
Sobecki, Janusz	Wroclaw University of Technology, Poland
Soo, Von-Wun	National University of Kaohsiung, Taiwan
Sordo, Margarita	Harvard Medical School, USA
Stumptner, Markus	University of South Australia, Australia
Stytz, Martin	Institute for Defense Analyses, USA
Suetake, Noriaki	Yamaguchi University, Japan
Sujitjorn, Sarawut	Suranaree University of Technology
Sun, Zhahao	University of Wollongong, Australia
Szczerbicki, Edward	University of Newcastle, Australia
Takahash, Masakazu	Simane University, Japan
Taki, Hirokazu	Wakayama University, Japan
Tanaka, Takushi	Fukuoka Institute of Technology, Japan
Tanaka-Yamawaki, Mieko	Tottori University, Japan
Teodorescu, Horia-Nicolai	Romanian Academy, Romania
Thalmann, Daniel	EPFL, Switzerland
Thatcher, Steven	University of South Australia, Australia
Tolk, Andreas	Virginia Modeling Analysis & Simulation center, USA
Torresen, Jim	University of Oslo, Norway
Treur, Jan	Vrije Universiteit Amsterdam, Netherlands
Turchetti, Claudio	Università Politecnica delle Marche, Italy
Tweedale, J.	DSTO, Australia
Uchino, Eiji	Yamaguchi University, Japan
Unland, Rainer	University of Duisburg-Essen, Germany
Verdegay, JoseLuis	University of Granada, Spain
Virvou, Maria	University of Piraeus, Greece
Walters, Simon	University of Brighton, UK
Wang, Dianhui	La Trobe University, Australia
Wang, Lipo	Nanyang Tech University, Singapore
Wang, Pei	Temple University, USA
Watada, Junzo	Waseda University, Japan
Watanabe, Keigo	Saga University, Japan
Watanabe, Toyohide	Nagoya University, Japan
Wermter, Stefan	University of Sunderland, UK
Wren, Gloria	Loyola College in Maryland, USA
Yamashita, Yoshiyuko	Tohoku University, Sedai, Japan
Yoo, Seong-Joon	Sejong University, Korea
Zahlmann, Gudrun	Siemens Medical Solutions; Med Projekt CTB, Germany
Zambarbieri, Daniela	University of Pavia, Italy
Zha, Xuan	NIST, USA

Zharkova, Valentina	Bradford University, Bradford, UK
Zhiwen, Yu	Northwestern Polytechnical University, China
Zurada, Jacek	University of Louisville, USA

General Track Chairs

Generic Intelligent Systems Topics

Track Title	Track Chair
Artificial Neural Networks and Connectionists Systems	Ryohei Nakano, Nagoya Institute of Technology, Japan
Fuzzy and Neuro-Fuzzy Systems	Detlef Nauck, BT, UK
Evolutionary Computation	Zensho Nakao, University of Ryukyus, Japan
Machine Learning and Classical AI	Mark Girolami, University of Glasgow, UK
Agent Systems	Ngoc Thanh Nguyen, Wroclaw University of Technology, Poland
Knowledge Based and Expert Systems	Anne Hakansson, Uppsala University, Sweden
Hybrid Intelligent Systems	Vasile Palade, Oxford University, UK
Miscellaneous Intelligent Algorithms	Honghai Liu, University of Portsmouth, UK

Applications of Intelligent Systems

Track Title	Track Chair
Intelligent Vision and Image Processing	Tuan Pham, James Cook University, Australia
Intelligent Data Mining	Michael Berthold, University of Konstanz, Germany
Knowledge Management and Ontologies	Edward Szczerbicki, University of Newcastle, Australia
Web Intelligence, Multimedia, e-Learning and Teaching	Andreas Nuernberger, University of Magdeburg, Germany
Intelligent Signal Processing, Control and Robotics	Miroslav Karny, Academy of Science, Czech Republic
Other Intelligent Systems Applications	Anthony Finn, Defence Science & Technology Organisation, Australia

Invited Session Chairs

Zhaohao Sun, University of Wollongong, Australia
Gavin Finnie, Bond University, Queensland, Australia
R.J. Howlett, University of Brighton, UK
Naohiro Ishii, Aichi Institute of Technology, Japan
Yuji Iwahori, Chubu University, Japan
Sangkyun Kim, Yonsei University, South Korea
Hong Joo Lee, Dankook University, South Korea
Yen-Wei Chen, Ritsumeikan University, Japan
Mika Sato-llic, University of Tsukuba, Japan
Yoshiteru Ishida, Toyohashi University of Technology, Japan
Dorothy Monekosso, Kingston University, UK
Toyoaki Nishida, The University of Kyoto, Japan
Ngoc Thanh Nguyen, Wroclaw University of Technology, Poland
Rainer Unland, University of Duisburg-Essen, Germany
Tsuda Kazuhiko, The University of Tsukuba, Japan
Masakazu Takahash, Simane University, Japan
Daniela Zambarbieriini, Università di Pavia, Italy
Angelo Marco Luccini, Giunti Interactive Labs, Italy
Baturone Iluminada, Institute de Microelectronica de Sevilla, University of Seville, Spain
Brox Poedad, Institute de Microelectronica de Sevilla, University of Seville, Spain
David Brown, University of Portsmouth, UK
Bogdan Gabrys, University of Bournemouth, UK
Davide Anguita, DIBE - University of Genoa, Italy
P. Urlings, DSTO, Australia
J. Tweedale, DSTO, Australia
C. Sioutis, DSTO, Australia
Gloria Wren, Loyola College in Maryland, USA
Nikhil Ichalkaranje, UNISA, Australia
Yoshiyuki Yamashita, Tohoku University, Sendai, Japan
Tetsuo Fuchino, Tokyo Institute of Technology, Japan
Hirokazu Taki, Wakayama University, Japan
Satoshi Hori, Institute of Technologists, Japan
Raymond Lee, Hong Kong Polytechnic University, China
Tai-hoon Kim, Korea University of Technology and Education, Republic of Korea
Kazumi Nakamatsu, University of Hyogo, Japan
Hsuan-Shih Lee, National Taiwan Ocean University, Taiwan
Ryohei Nakano, Nagoya Institute of Technology, Japan
Kazumi Saito, NTT Communication Science Laboratories, Japan
Giorgos Anogeianakis, Aristotle University of Thessaloniki, Greece
Toyohide Watanabe, Nagoya University, Japan

Tomoko Kojiri, Nagoya University, Japan
Naoto Mukai, Nagoya University, Japan
Maria Virvou, University of Piraeus, Greece
Yoshinori Adachi, Chubu University, Japan
Nobuhiro Inuzuka, Nagoya Institute of Technology, Japan
Jun Feng, Hohai University, China
Ioannis Hatzilygeroudis, University of Patras, Greece
Constantinos Koutsojannis, University of Patras, Greece
Akinori Abe, ATR Intelligent Robotics & Communication Labs, Japan
Shoji, ATR Intelligent Robotics & Communication Labs, Japan
Ohsawa, ATR Intelligent Robotics & Communication Labs, Japan
Phill Kyu Rhee, Inha University, Korea
Rezaul Bashar, Inha University, Korea
Jun Munemori, Wakayama University, Japan
Takashi Yoshino, Wakayama University, Japan
Takaya Yuizono, Shimane University, Japan
Gwi-Tae Park, Korea University, South Korea
Manuel Grana, Universidad Pais Vasco, Spain
Richard Duro, Universidad de A Coruna, Spain
Daniel Polani, University of Hertfordshire, UK
Mikhail Prokopenko, CSIRO, Australia
Dong Hwa Kim, Hanbat National University, Korea
Vesa A. Niskanen, University of Helsinki, Finland
Emilio Corchado, University of Burgos, Spain
Hujun Yun, University of Manchester, UK
Jaehak Chung, Inha University Korea, South Korea
Da Deng, University of Otago, New Zealand
Mengjie Zhang, University of Wellington, New Zealand
Valentina Zharkova, University of Bradford, UK
Jie Zhang, George Mason University, USA
Richi Nayak, Queensland University of Technology, Australia
Lakhmi Jain, University of South Australia, Australia
Dong Chun Lee, Howon University, Korea
Giovanni Semeraro, Universitá degli Studi di Bari, Italy
Eugenio Di Sciascio, Politecnico di Bari, Italy
Tommaso Di Noia, Politecnico di Bari, Italy
Norio Baba, Osaka Kyoiku University, Japan
Takumi Ichimura, Hiroshima City University, Japan
Kazuya Mera, Hiroshima City University, Japan
Janusz Sobecki, Wroclaw University of Technology, Poland
Przemyslaw Kazienko, Wroclaw University of Technology, Poland
Dariusz Król, Wroclaw University of Technology, Poland

Kenneth J. Mackin, Tokyo University of Information Sciences, Japan
Susumu Kunifuji, Japan Advanced Institute of Science and Technology, Japan
Motoki Miura, Japan Advanced Institute of Science and Technology, Japan
Jong Tae Kim, SungKyunKwan University, Republic of Korea
Junzo Watada, Waseda University, Japan
Radoslaw Katarzyniak, Wroclaw University of Technology, Poland
Geuk Lee, Hannam Howon University, Korea
Il Seok Ko, Chungbuk Provincial Univ., Korea
Ernesto Damiani, University of Milan, Italy
Paolo Ceravolo, University of Milan, Italy
Dharmendra Sharma, University of Canberra, Australia
Bala Balachandran, University of Canberra, Australia
Wanla Ma, University of Canberra, Australia
Danilo P. Mandic, Imperial College London, UK
Tomasz Rutkowski, RIKEN, Japan
Toshihisa Tanaka, TUAT, Tokyo, Japan
Martin Golz, Schmalkalden, Germany

Keynote Lectures

Evolving Intelligent Systems: Methods and Applications
Nikola Kasabov, Knowledge Engineering and Discovery Research Institute (KEDRI), Auckland University of Technology, New Zealand

Feature Selection in Pattern Recognition
Josef Kittler, University of Surrey, UK

Ant Colony Optimisation
Luca Maria Gambardella, Istituto Dalle Molle di Studi sull'Intelligenza Artificiale (IDSIA), Switzerland

Evolvable Genetic Robots in a Ubiquitous World
Jong-Hwan Kim, Korea Advanced Institute of Science and Technology (KAIST), Republic of Korea

Industrial Applications of Intelligent Systems
Ben Azvine, BT, UK

Towards Multimodal Computing: Extracting Information from Signal Nonlinearity and Determinism
Danilo Mandic, Imperial College London, UK

Table of Contents – Part I

Generic Intelligent Systems

Artificial Neural Networks and Connectionists Systems

Fuzzy and Neuro-fuzzy Systems

Evolutionary Computation

Machine Learning and Classical AI

Agent Systems

Knowledge Based and Expert Systems

Hybrid Intelligent Systems

Miscellaneous Intelligent Algorithms

Knowledge Management and Ontologies

Web Intelligence, Multimedia, e-Learning and Teaching

Intelligent Signal Processing, Control and Robotics

Intelligent Systems Applications

Invited Session Papers

Experience Management and Engineering Workshop

Invited Sessions

Industrial Applications of Intelligent Systems

Information Engineering and Applications in Ubiquotous Computing Environments

Table of Contents – Part II

Computational Intelligence for Signal and Image Processing

Soft Data Analysis

Immunity-Based Systems: Immunoinformatics

Ambient Intelligence: Algorithms, Methods and Applications

Communicative Intelligence

Knowledge-Based Systems for e-Business

Neuro-fuzzy Techniques for Image Processing Applications

Knowledge-Based Interface Systems (1)

Nature Inspired Data Mining

Intelligent Agents and Their Applications

AI for Decision Making

Intelligent Data Processing in Process Systems and Plants

Skill Acquisition and Ubiquitous Human Computer Interaction

IATA – Intelligent Agent Technology and Applications

Computational Intelligence Approaches and Methods for Security Engineering

Hybrid Information Technology Using Computational Intelligence

Logic Based Intelligent Information Systems

Knowledge-Based Mult-criteria Decision Support

Neural Information Processing for Data Mining

Sharing of Learning Knowledge in an Information Age

Intelligent Information Management in a Knowledge-Based Society

Knowledge/Software Engineering Aspects of Intelligent Systems Applications

Knowledge-Based Interface Systems (2)

Intelligent Databases in the Virtual Information Community

Hybrid Intelligent Systems in Medicine and Health Care

Table of Contents – Part III

Chance Discovery

Context Aware Evolvable and Adaptable Systems and Their Applications

Advanced Groupware and Network Services

Computational Methods for Intelligent Neuro-fuzzy Applications

Intelligent Information Processing for Remote Sensing

Evolutionary and Self-organising Sensors, Actuators and Processing Hardware

Human Intelligent Technology

Connectionist Models and Their Applications

Intelligent and Cognitive Communication Systems

Innovations in Intelligent Agents and Applications

Intelligent Pattern Recognition and Classification in Astrophysical and Medical Images

Intelligent Multimedia Solutions and Security in the Next Generation Mobile Information Systems

Engineered Applications of Semantic Web – SWEA

Intelligent Techniques in the Stock Market

Soft Computing Techniques and Their Applications

Human Computer Intelligent Systems

Recommender Agents and Adaptive Web-Based Systems

Intelligent Data Analysis for Complex Environments

Creativity Support Systems

Artificial Intelligence Applications in Power Electronics

Soft Computing Approaches to Management Engineering

Knowledge Processing in Intelligent and Cognitive Communicative Systems

Intelligent and Secure Digital Content Management

Business Knowledge Modelling and Maintenance

Innovations in Intelligent Systems and Their Applications

Intelligent Agents and Their Applications

Signal Processing Techniques for Knowledge Extraction and Information Fusion

Integration of Ant Colony SOM and K-Means for Clustering Analysis

Sheng-Chai Chi[1] and Chih Chieh Yang[2]

Department of Industrial Engineering and Management Information
Huafan University
Taipei County, Taiwan 223
[1] scchi@huafan.hfu.edu.tw,
[2] m9223020@cat.hfu.edu.tw

Abstract. In data analysis techniques, the capability of SOM and K-means for clustering large-scale databases has already been confirmed. The most remarkable advantage of SOM-based two-stage methods is in saving time without the considerable computations required by conventional clustering methods for large and complicated data sets. In this research, we propose and evaluate a two-stage clustering method, which combines an ant-based SOM and K-means. The ant-based SOM clustering model, ABSOM, embeds the exploitation and exploration rules of state transition into the conventional SOM algorithm to avoid falling into local minima. After application to four practical data sets, the ABSOM itself not only performs better than Kohonen's SOM but also it works very well in the two-stage clustering analysis when it is taken as the preprocessing technique for the ABSOM+K-means method.

1 Introduction

With the substantial decline of data storage cost, rapid improvement of computer performance and the popularization of computer networks, great volumes of various types of information are being produced in our surrounding environment. The resulting large-scale databases have led to the development of data storage and analysis techniques such as data warehousing and data mining to manage them.

Of these data analysis techniques, the capability of K-means and Kohonen's SOM for clustering large-scale databases has been confirmed [11]. Even though SOM or conventional clustering methods have their superior features for clustering analysis, their combinations into two-stage methods are generally much more powerful than individual methods. Not only do two-stage methods conquer the drawbacks of conventional hierarchical and partitive clustering methods, but their clustering results are more accurate [13]. A two-stage method usually combines either a conventional hierarchical method or SOM with a partitive method. The function of the hierarchical method or SOM is commonly used to determine the number of groups in the first stage, and the K-means method is used to implement the clustering process in the second stage.

Generally, a two-stage method including a SOM-based algorithm has two ways of functioning. In the first way, SOM is initially adopted to determine the number of

B. Gabrys, R.J. Howlett, and L.C. Jain (Eds.): KES 2006, Part I, LNAI 4251, pp. 1 – 8, 2006.
© Springer-Verlag Berlin Heidelberg 2006

groups and the initial group centers of K-means, i.e. the respective weight vectors of the groups, from the network topology. The merit of SOM is that it faithfully presents the structure of the original data set through a simple two or three-dimensional network topology [9,15]. This advantage overcomes the problems of K-means in appropriately determining the number of groups and their initial group centers because these parameters substantially affect the final result [10,12]. Based on the results from SOM, the K-means method then further clusters the data set unambiguously so that its is clear why each data sample is assigned to its group according to the mapping results on the network topology, especially since some data samples are located near the border between two groups [14].

In the second way, SOM initially maps a large-scale data set upon the network topology and generates the topological coordinates of prototypes for further clustering in the second stage. The method employed in the second stage may be SOM, a hierarchical agglomerative clustering method, or a partitive clustering method like K-means. The main advantage of the SOM-based two-stage method is savings of time without the considerable computations required by conventional clustering methods for large and complicated data sets [16]. Therefore, this research proposes an effective and efficient method for preprocessing a large-scale data set under the limited time conditions to complete the purpose of data mining and knowledge discovery.

This paper has five sections. Section 1 is the introduction to the research motivation and objectives. In Section 2, the proposed SOM algorithm based on an ant colony optimization is described in detail. In Section 3, the two-stage method of ant colony SOM and K-means using four public datasets is verified by comparison with the two-stage method of Kohonen's SOM with K-means. Finally, conclusions and future study are presented.

2 ABSOM Algorithm

In this study, we propose an Ant-Based Self-Organization Map (ABSOM), which embeds the concept of ACS (Ant Colony System) in the SOM algorithm, in order to avoid falling into local minimum during the solution search process [1,2]. The detailed algorithm is presented below.

Step 1: Parameter setting

(a) Topological network setting

The topological network $M = \{c_1, c_2 \ldots c_j \ldots c_m\}$ is defined so that M consists of $m = m_1 \times m_2$ map units where m_1 and m_2 represent the number of units located on the two sides of the network, respectively. According to the probability distribution $P(\xi_i)$ where ξ_i stands for an input vector, the n-dimensional weight vector $w_{c_j} \in R^n$ of every map unit c_j is initialized. Set the start time $t = 0$ and the initial neighborhood radius $\lambda(0)$.

(b) Pheromone matrix setting

Construct a pheromone matrix $T = \{\tau_{c_1}, \tau_{c_2}, \ldots, \tau_{c_j}, \ldots, \tau_{c_m}\}$ and give each element of the matrix an initial pheromone $\tau_{c_j}(0) = 0$. Define q_0 as the pheromone threshold for choosing the exploitation or exploration updating rule. The value range of q_0 is limited to the interval [0, 1].

(c) Other parameter settings

N_{dlen} and $N_{trainlen}$ represent the number of total data samples and training epochs, respectively. D_{c_j} is defined to be the set of data samples mapped on the map unit c_j.

Step 2: Read in and normalize the training data set.

Step 3: Compute all the neighborhood radii that will be used during the training processes.

Step 4: Select a sample randomly. Calculate the Euclidean distance $\Delta\delta_{ji}$ between map unit c_j on the topological network and input vector ξ_i, and compute pheromone τ_{c_j} for map unit c_j.

$$\Delta\delta_{ji}(t) = \sum_{k=1}^{n} \left\| \xi_{ik} - \varpi_{jk} \right\| \tag{1}$$

Step 5: Conduct the updating processes for the weight vectors

(a) After the set of data samples are mapped onto the map units, find the Best Match Unit (BMU) s for each data sample according to the following formula based on the exploitation and exploration updating rules of the ant-based algorithm:

$$s = \begin{cases} \arg\min \ \Delta\delta_{ji}(t), & \text{if } q \le q_0 \qquad \text{(exploitation)} \\ p_{c_j} = \dfrac{\tau_{c_j}(t)}{\sum_{c_j=c_1}^{c_j=c_N} \tau_{c_j}(t)}, & \text{otherwise} \qquad \text{(exploration)} \end{cases} \tag{2}$$

where q : a real number generated randomly from 0 to 1;

q_0 : the threshold to choose the exploitation updating rule when q is no greater than q_0;

p_{c_j} : the probability that map unit c_j is the BMU when the exploration rule is chosen.

To record the data samples that have been mapped on unit c_j, the accumulated data set can be expressed as

$$D_{c_j}(t+1) = \begin{cases} \{D_{c_j}(t), \xi_i\}, & \text{if } c_j = s \\ \{D_{c_j}(t)\}, & \textit{otherwise} \end{cases} \tag{3}$$

(b) Update the pheromone for each map unit

$$\Delta\tau_{c_j}(t+1) = 1/\Delta\delta_{ji}(t+1), \qquad \text{if } c_j = s \tag{4}$$

$$\tau_{c_j}(t+1) = \begin{cases} \beta\tau_{c_j}(t) + \alpha\Delta\tau_{c_j}(t+1), & \text{if } c_j = s \\ \beta\tau_{c_j}(t), & \text{otherwise} \end{cases} \tag{5}$$

where α is the accumulating rate of pheromone because of Euclidean distance and β is the decay rate of pheromone.

(c) Update the weight vector for each map unit

$$\Delta w_{c_j}(t) = lr(t) \cdot (\xi_i - w_s) \cdot h_{rs}(t) \tag{6}$$

where $lr(t)$ is the learning rate of connection weights at time t; $h_{rs}(t)$ is the neighborhood radius at time t.

$$w_{c_j}(t+1) = w_{c_j}(t) + \Delta w_{c_j}(t) \tag{7}$$

Step 6: Decrease the neighborhood radius and update the neighborhood coefficient

$$\lambda(t+1) = \varepsilon\lambda(t) \tag{8}$$

$$h_{rs}(t+1) = f(r_{c_j}, \lambda(t)) = \exp(-r_{c_j} / \lambda(t)) \tag{9}$$

where $\lambda(t)$: the neighborhood radius at time t;

ε: the decreasing learning rate of neighborhood radius $\varepsilon \in (0,1)$;

r_{c_j}: the topological distance between the winner (or the BMU) and unit c_j on the map topology.

Step 7: Check the stopping criteria N_{dlen}, $N_{trainlen}$.

3 Evaluation of ABSOM + K-means

Although an artificial data set with given solution might be created to examine the feasibility of a developed model, we consider that it will be more meaningful to test the model directly on practical data sets. Thus, four public data sets [15] are employed below. Those are Fisher's iris data (Iris), Forina's wine recognition data (Wine), optical recognition of handwritten digits (Optical) and pen-based recognition of handwritten digits (Pen-Based).

As mentioned, the visualization results from the ABSOM and SOM can be used to determine the number of groups and initial center of each group for clustering in the following stage, K-means. To verify the usefulness of visualization, the DB (Davies-Bouldin) index values from K-means for these two algorithms are measured and compared [5]. The verification procedure has two stages. In the first stage, the ABSOM and SOM are conducted for clustering the four data sets. The results of their QE (Quantization Error) and TE (Topological Error) values [6,8] are shown in Table 1. Then, the U-Matrix method, contained in the SOM Toolbox 2.0, is adopted to depict the similarity of the mapped data by the ABSOM and SOM in terms of the whole variables and individual variable.

Comparing grey level and shape of data distribution for each variable with those for the overall variables, we can easily identify where the key (principal) variables come from and eliminate those unimportant variables from original data set if any. If the conclusion can be confirmed by using Principal Component Analysis (PCA) [7] or general regression method, the less important variables may be eliminated to reflect a positive influence on the clustering purpose. Then, the U-Matrix method is combined with the block diagram method to generate the topological maps from the ABSOM and the SOM for the Optical data set shown in Fig. 1 and 2, respectively. The blocks in Fig. 1 and 2 represent the mapping results from the ABSOM and the SOM.

Table 1. QE and TE values of the four data sets from the SOM and ABSOM before visualization process

	Data set I (Iris)		Data set II (Wine)	
Algorithm	SOM	ABSOM	SOM	ABSOM
Q.E.	0.08451	**0.08367**	0.5001	**0.4971**
T.E.	0.2889	**0.2753**	0.4398	**0.3901**
	Data set III (Pen-Based)		Data set IV (Optical)	
Algorithm	SOM	ABSOM	SOM	ABSOM
Q.E.	0.4108	**0.3756**	0.9471	**0.8777**
T.E.	**0.0799**	0.0938	**0.0624**	0.1059

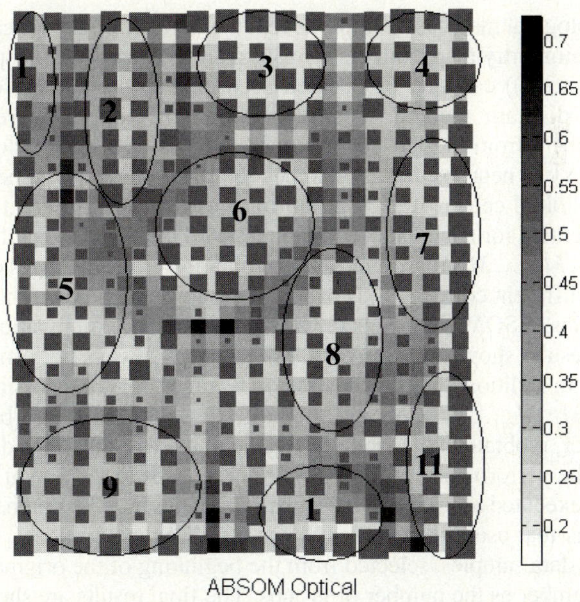

Fig. 1. Topological map produced by the U-Matrix and block diagram methods for the clustering result of Optical data set from the ABSOM

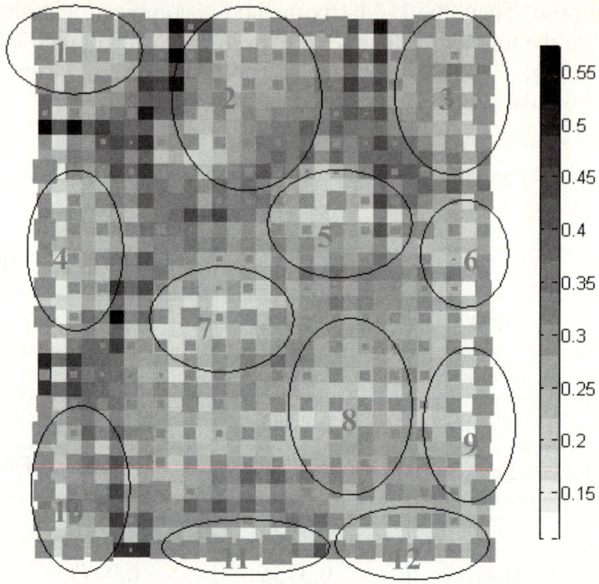

Fig. 2. Topological map produced by the U-Matrix and block diagram methods for the clustering result of Optical data set from the SOM

After the topological map has been produced, the number of groups can be determined according to the similarity map from the U-matrix, and the groups corresponding center(s) (or prototype vector(s)) can also be decided according to the number of times mapped from the block diagram method. The neurons on the topology are encoded using incremental numbers from the left-and-top corner to the right-and-bottom corner. There may be more than two neurons that are mapped the same number of times.

To determine initial center of each group for the K-means, different combinations of possible group centers for each data set obtained from the ABSOM and SOM are tested. The tuple set (M_1, M_2,…, M_n) represents the combination of the averages of all the possible group centers if different combinations of initial group centers exist. The best result from ABSOM+K-means or SOM+K-means for each data set is the combination with the lowest DB value. The results show that the ABSOM+K-means always performs better than the SOM+K-means. In addition, the combination of averages of all the possible group centers, represented by (M_1, M_2,…, M_n), obtains the best DB value in the Pen-based and Optical data sets; however, it obtained the worst DB value in the Iris and Wine data sets.

To compare the performance of the ABSOM+K-means to the one of SOM+K-means, K-means is also executed alone using the same four data sets. The number of groups is set to be the same as that used in the ABSOM or the SOM experiments. The initial group centers are those data samples, selected from the beginning of the original data set, which have the same number as the number of groups. The final results are shown in Table 2 in which we found that the ABSOM+K-means performs better than the SOM+K-means while the SOM+K-means performs better than the K-means alone.

Table 2. Comparison of K-means, SOM+K-means and ABSOM+K-means using the four data sets

Data set	I(Iris)	II(Wine)	III(Pen-Based)		IV(Optical)	
Cluster	3	3	12	13	11	12
K-means	0.7337	**1.3627**	1.6688	1.7288	1.9616	1.9092
SOM + K-means	0.7990	**1.3627**	1.4013	N/A	N/A	1.4746
ABSOM + K-means	**0.7189**	**1.3627**	N/A	**1.3650**	**1.2975**	N/A

Thus, we can confirm that the ABSOM performs better than the conventional SOM on the evaluation criterion QE. In addition, after comparing the ABSOM+K-means and SOM+K-means, we conclude that the ABSOM in all respects performs better than the conventional SOM in clustering analysis for the four data sets.

4 Conclusions and Future Study

This research proposes and evaluates a two-stage clustering method that combines an ant colony SOM and K-means. The SOM clustering model, ABSOM, embeds the exploitation and exploration rules of state transition into the conventional SOM algorithm. As shown by applying the method to four practical data sets, the ABSOM itself not only performs better than Kohonen's SOM but also it works very well in the two-stage clustering analysis when it is used as the preprocessing technique for the ABSOM+K-means method.

In the evaluation process, the U-Matrix method is combined with the block diagram method to generate a U-Matrix block diagram on the network topology, which facilitates the visualization ability in determining initial prototypes (or group centers) for the K-means in the following clustering analysis. The results show that the proposed ABSOM+K-means method performs better than the SOM+K-means method for the clustering analysis.

As indicated by other recent studies, the fixed structure of the conventional SOM neural networks does not afford to respond to actual data distribution in the original data set, if the data set is constituted by a number of groups containing vague boundaries and complicatedly hierarchical relations. Therefore, our future research work will extend the proposed ABSOM approach to create a "growable" ant colony SOM such as growing grid in [3] to deal with this type of problem.

Moreover, as mentioned in the literature review, there are many indices in addition to the Davis-Bouldin index, such as Dunn's index, or RMSSTD, that can be utilized to conduct the performance evaluation of a clustering method after/during the clustering process. The use of such performance indices will assist direct and real-time evaluations for clustering quality.

Acknowledgements

The authors would like to thank the National Science Council of the Republic of China for supporting this research under grant no.: NSC 94-2213-E-211-009. Additionally, the invaluable recommendations from the anonymous reviewers are very appreciated.

References

1. Blum, C. and Dorigo, M.: Hyper-cube Framework for Ant Colony Optimization. IEEE Transaction on System, Man and Cybernetics Part B. 4:2 (2004) 1161–1172.
2. Bonabeau, E., Dorigo, M. and Théraulaz, G.: Swarm Intelligence: from Natural to Artificial Systems. Oxford University Press, New York (1999).
3. Fritzke, B.: Growing Grid: a Self-organizing Network with Constant Neighborhood Range and Adaptation Strength. Neural Processing Letters, 2:5 (1995) 9-13.
4. Günter, S., Bunk, H.: Validation Indices for Graph Clustering. Pattern Recognition Letters, 24 (2003) 1107-1113.
5. Halkidi, M., Batistakis, Y. and Vazirgiannis, M.: On Clustering Validation Techniques. Journal of Intelligent Information Systems, 17 (2001) 107-145.
6. Heskes, T. Self-organizing Maps, Vector Quantization, and Mixture Modeling. IEEE Transactions on Neural Networks, 12:6 (2001) 1299-1305.
7. Jollife, I.: Principal Component Analysis. Springer-Verlag, New York (1986).
8. Kiviluoto, K.: Topology Preservation in Self-Organizing Maps, Proceedings of International Conference on Neural Networks (ICNN) (1996) 294-299.
9. Kohonen, T.: Self-organizing Map, 2nd ed., Springer-Verlag, Berlin (1995).
10. Kuo, R.J., Chi, S.C. and Teng, P.W.: Generalized Part Family Formation through Fuzzy Self-organizing Feature Map Neural Network. Computers and Industrial Engineering, 4 (2001) 79-100.
11. Mao, J. and Jain, A.K.: A Self-organizing Network for Hyperellipsoidal Clustering (HEC). IEEE Transactions Neural Networks, 7 (1996) 16–29.
12. Peña, J.M., Lozano, J.A. and Larrañaga, P.: An Empirical Comparison of Four Initialization Methods for the K-means Algorithm. Pattern Recognition Letters, 20 (1999) 1027-1040.
13. Sharma, S.: Applied Multivariate Techniques. John Wiley & Sons (1996).
14. Sugiyama, A. and Kotani, M.: Analysis of Gene Expression Data by Using Self-organizing Maps and K-means Clustering. Proc. of Int. J. Conf. on Neural Networks (2002) 1342-1345.
15. UCI machine learning repository, *http://www.ics.uci.edu/~mlearn/MLRepository.html*
16. Vesanto, J. and Alhoniemi, E.: Clustering of the Self-organizing Map. IEEE Transactions on Neural Networks, 11:3 (2000) 586-600.

An Improved Text Categorization Methodology Based on Second and Third Order Probabilistic Feature Extraction and Neural Network Classifiers

D.A. Karras

Chalkis Institute of Technology, Automation Department & Hellenic Open University,
Rodu 2, Ano Iliupolis, Athens 16342, Greece
dakarras@teihal.gr, dakarras@ieee.org, dakarras@usa.net

Abstract. In this paper we deal with the main aspect of the problem of extracting meaning from documents, namely, with the problem of text categorization, outlining a novel and systematic approach to its solution. We present a text categorization system for non-domain specific full-text documents. The main contribution of this paper lies on the feature extraction methodology which, first, involves word semantic categories and not raw words as other rival approaches. As a consequence of coping with the problem of dimensionality reduction, the proposed approach introduces a novel second and third order feature extraction approach for text categorization by considering word semantic categories cooccurrence analysis. The suggested methodology compares favorably to widely accepted, raw word frequency based techniques in a collection of documents concerning the Dewey Decimal Classification (DDC) system. In these comparisons different Multilayer Perceptrons (MLP) algorithms, the LVQ and the conventional k-NN technique are involved.

Keywords: Text Classification, Neural Networks, higher order feature extraction.

1 Introduction

Numerous text categorization systems exist in the literature. They can be grouped into clustering or mapping systems where the categories are not defined by humans and into supervised text categorization systems, where the categories are predefined. Regarding the first group, several domain-specific methods have been developed, such as NLP (Natural Language Processing) systems for Software Libraries and Software Reuse [1], NLP architectures for collaborative computing systems solving the vocabulary problem by clustering semantically similar terms [2], through statistics and Hopfield type of neural networks, etc. Moreover, non-domain specific text categorization methods are included in this group, such as the recently proposed system for self-organization and mapping of general document collections [3], by applying a modification of the SOFM neural net algorithm.

Concerning the second group, a number of statistical classification and machine learning techniques have been applied to automatic text categorisation, including

B. Gabrys, R.J. Howlett, and L.C. Jain (Eds.): KES 2006, Part I, LNAI 4251, pp. 9 – 20, 2006.
© Springer-Verlag Berlin Heidelberg 2006

domain-specific rule learning based systems [4], regression models [5], nearest neighbour classifiers [5], decision trees [6], Bayesian classifiers [6], Support Vector Machines [7] and neural networks [8]. On the other hand, several feature extraction methodologies have already been applied to this field. The basic methodology is the frequency of word occurrence technique representing a document through the vector space model (SMART) [9]. A multitude of feature extraction methods stem from the SMART representation, like word frequency weighting schemes (Boolean, tf x idf (inverse document frequency), tfc, entropy, etc.) [10] as well as feature extraction techniques involving dimensionality reduction through feature selection schemes (document frequency thresholding, information gain, hypothesis-testing, etc.) [10] and through feature combination or transformation (For instance, Latent Semantic Indexing (LSI) related methodologies) [10]. For an excellent survey of the state-of-the-art in supervised text categorization we should point out [10].

In the herein introduced approach we present a supervised text categorization system for non-specific domain full-text documents, aimed at extracting meaning according to well defined and systematically derived subject categories used by librarians. This system is based on a novel feature extraction scheme based on semantics processing through first, second and third order statistical procedures. More specifically, the proposed feature extraction technique has the following attributes.

- It involves the extraction of first, second as well as third order characteristics of the input documents, based on processing individual and pairs of word semantic categories (Word semantic category Affinity - Mean distance between two word semantic categories) found in these documents.
- Both the text categorization indexing scheme and the word semantic category extraction scheme are based on widely accepted state of the art approaches. More specifically, concerning the former, the widely adopted by librarians DDC methodology [11] is considered. Concerning the latter, the use of the WorldNet [11] as a tool for processing the documents and associating semantic categories to their words has been adopted due to its high credibility and acceptance among NLP researchers. WordNet is a thesaurus that helps us to put the words we extract into a fixed number of semantic categories. These basic features will be analyzed with more detail in the next paragraphs.

To be able to measure progress in this field it is recommended to use standardized collection of documents for analysis and testing. One such data set is the Reuters-21578 collection [10]. Its categorization scheme, however, is domain specific and not scientifically derived and widely accepted by librarians as that defined by DDC. Therefore, since the final goal is to design a text categorization system for realistic applications and not for benchmarking purposes, the DDC document collection is herein exclusively involved. In the following section 2, the suggested supervised text categorization approach is described in detail. In section 3 the experimental study is outlined. And finally, section 4 presents the conclusions and prospects of this research effort.

2 The Proposed Feature Extraction Methodology Involving Second and Third Order Probabilistic Features for Text Categorization

The suggested procedure is mainly divided into two phases. In the first phase we extract some first and second order characteristics, that is, frequency of word semantic categories appearance (extraction of keywords) and word semantic categories affinity (extraction of pairs of words semantic categories), from a set of documents collected from the Internet and labeled using DDC methodology [11]. We then use the widely accepted NLP semantics extraction program, the thesaurus called WordNet, in order to put the words we have extracted into a fixed number of semantic categories. As illustrated in the next paragraphs, the application of WordNet results in substituting a word by the vector of its semantic categories. After this stage, we use these semantic categories vectors instead of the initial words for subsequently extracting the first and second order document features of the proposed supervised text categorization system. In the second phase, employing these features, MLP Neural Networks are trained so that they can classify documents into a set of DDC text categories (note that these document categories are not related in any way with the categories of the WordNet). In the following paragraphs of this section we give a more detailed description of our novel methodology for extracting the first and second order characteristics, advancing from words to WordNet categories, creating the patterns and finally training the MLP neural networks.

The collection of documents used in this research effort, concerning the development of the proposed system, comes from the Internet site (link.bubll.ac.uk/isc2), which contains documents classified and accessed according to DDC (Dewey Decimal Classification). DDC is a document content classification system, internationally known and used in libraries and other organizations where a subject classification (catalogs) of documents is needed.

DDC defines a set of 10 main categories of documents that cover all possible subjects a document could refer to. Each one of them is divided into 10 Divisions and each division into 10 Sections. The classes of documents in DDC are defined through a systematic and objective way and not arbitrarily and subjectively. This is the reason we use DDC instead of defining our own classes of documents or following the classification schemes found in large domain-specific document collections found in the Internet, like the Reuters database. These document collections can only serve as test beds for evaluation of the preliminary design and not of the final design of text categorization real world systems. The DDC classes used in this paper are the following:

(1) Generalities,(2) Philosophy & related disciplines, (3) Religion, (4) Social sciences, (5) Language, (6) Pure sciences, (7) Technology (Applied sciences), (8) The arts, (9) Literature & rhetoric and (10 General geography, history, etc.

For each of these 10 classes a collection of 150 documents was made. The average length of these 1500 files was 256 words. Out of them, 1000 (10x100) documents (67%) comprise the training set and the rest 500 (10x50) documents (33%) comprise the test set. Following the process defined next, each document will be represented by a training (or test) pattern. So, there are 1000 training and 500 test patterns for training and testing the neural networks. We must note here that all collected documents are in English.

The first step is to remove a certain set of words, which are of less importance for meaning extraction. Specifically, it is assumed that nouns and verbs that can be found in a document are more relevant to meaning extraction, while certain adverbs, propositions and especially articles and adjectives are less relevant. This assumption is supported by the psycholinguistic theories underlying WordNet [11]. The set of such words excluded from further processing in this rsearch effort are: **"I","you", "he","she","it", "us", "them", "the","of","end","to", "in","that","is", "was","for", "with", "as","on","be", "at","by", "this","had","not","are","but",“from","or","have","an","they","which","th ere","all", "would","their", "we","him", "been","has", "when", "will", "more",“no","if","out",“so","said","what","up","its","about","into","them" ,"than","can","only","other", "onto"**

WordNet [11] is actually something more than a simple thesaurus. It is based on the latest psycholinguistic theories (a combination of psychological and linguistic theories) about the human lexical memory. The initial idea was to provide an aid to use in searching dictionaries conceptually, rather than merely alphabetically. In WordNet, the English nouns and verbs are organized in sets of synonyms. Each one of these sets describes a lexical concept. These sets are related in many ways to each other.

The most important of these relations between sets of word semantic categories (lexical concepts) considered in this paper, is the Hyponymy/Hyperonymy or Superset/Subset relation. A concept described by a set of synonyms {x,x',...} is a hyponym of a concept described by another set of synonyms {y,y',...}, if people whose native language is English accept phrases such as "An x is a (kind of) y". In this way, concepts, and eventually words, are organized in a tree structure. The root (or a set of roots) of this tree describes the most general concept. For example, there is a set of 25 concepts (e.g. Process, Natural Object, Food, Communication, Knowledge and so on), which are the most general with regards to nouns.

The most important reason for using WordNet in this work is the fact that we can achieve dimensionality reduction of the space vectors representing documents and used in training the neural network subject classifiers. In particular, we use WordNet to represent each word we have extracted from the texts with a set of WordNet concepts (in our case there are 229 such concepts), or semantic categories of words. Thus, after this step, all texts are represented by space vectors with 229 dimensions of WordNet semantic categories, instead of the words extracted from them. The following list shows the 229 WordNet categories used in this work:

Another reason for using WordNet is that in this way we try to give a solution to the Vocabulary Problem [2]. The vocabulary problem refers to the problem that rises when different people, with different knowledge background, training and experience, use different sets of terms when dealing with a certain subject. By using WordNet, different words with the same meaning (synonyms) are grouped under the same concept (set of synonyms). Although WordNet is not able to classify every given word, and especially scientific terms, into one of the available concepts, it can, however, classify a large number of the words extracted and therefore, give a quite good solution to the vocabulary problem.

Table 1. The subset of WorldNet word Semantic Categories used in the proposed methodology

Ability,Could,Kill,Prose, Abstraction,Cover,knowledge_domain,Province Accept,Covering,land, Psychological_feature,Accomplishment,Create,language,quality Act,Creation,leader,quantity, Activity,Creator,line,record Affirm,currency,linguistic_relation,reject,African,decelerate,liquid,relation Agree,decide,location,relationship,American,decoration,look,religious_ceremony Analyze,device,lose,religious_person,Animal,differ,make,remove Area,disease,matter,resprire,Art,displease,means,rock Artifact,document,measure,rule,artificial_language,enter,medicine,science Asian,entertainer,message,scientist,Athlete,entity,military_action,show Attribute,equal,military_man,social_class,Attitude,equipment,military_unit,social_event Attribute,european,mineral,social_group,basic_cognitive_process,event,misc,social_relation Be,excel,money,social_science,Become,expert,motion,solid Belief,express,move,speech,biological_group,feel,munber,speech_act body_part,feeling,music,spiritual_being,Calculation,fight,name,spiritual_leader Calculus,fluid,names,state,Capitalist,food,nation,statement Care,form,native,structure,Categorize,genealogy,natural_language,succeed Category,generality,natural_object,support,Caucasian,genitor,natural_phenomenon,surface Cause,geological_time,natural_process,symbol,Ceremony,geology,natural_science,symptom Change,get,nature,take,Characteristic,give,nonreligious_person,think chemical_compound,go,number,time,chemical_element,goods,occupation,time_period chemical_process,group,organization,time_unit,City,group_action,part,touch Classification,grouping,permit,trait,Collection,hair,person,transfer Color,have,phenomenon,travel,Commerce,higher_cognitive_process,physical,treat Communicate,holy_day,physical_structure,understand,Communication,ideal,planet,undo Communicator,imaginary_being,plant,union,Concept,imaginary_place,point,unit Confirm,implement,possession,use,Connect,inform,power,vehicle Content,information,print_medium,want,Continent,institution,printing,wish Continue,interact,process,word,Control,judge,procession,work, Convey,keep,property,worker, Writing

It has been previously mentioned that a novel and significant feature of this work is, actually, the use of pairs of word semantic categories apart from extracting keywords, that is, individual word (nouns and verbs) lexical concepts, in order to extract content information. Involving correlations of word semantic categories equals incorporating second order information in the document space vector representation.

The previous step involved the use of WordNet for passing from words and pairs of words to semantic categories of words and pairs of semantic categories. The result for each text is finally one vector of 229 elements (the number of WordNet categories) containing information about the frequency of occurrence of WordNet semantic categories in the text, instead of words occurrence frequency as in the simple word frequency weighting indexing [10] method. In addition, two matrices of 229x229 elements are derived, each containing information about the pairs of semantic categories cooccurrence. We then use cooccurrence matrices analysis, involved in texture processing of images, to transform each of the two matrices into a vector of six elements. All this methodology is described in more detail in the following.

It should be pointed out that in the proposed methodology, after preprocessing, each document could be represented (using a program in C language) in a tree structure. Each node of this tree stores information for each new word found, comprised of,

- The word itself, The number of appearances of this word in the certain text, A list with the positions of the word in the text, The DDC class in which the document belongs

It must be mentioned here that the following characters are the ones considered as separators between two words in a text:

- Line feed (\n), Tab (\t), One or more spaces (SPACE), (\r), Page feed (\f)

Having this information we create at first a vector containing each different word and the frequency of its appearance in the text. This completes the second step in the proposed feature extraction technique (words selection is the first step).

The third step in the suggested feature extraction methodology is the creation of the first out of the two matrices mentioned above, the Full Affinity Matrix. This matrix contains information about word affinity, and more specifically information about the mean distance between any pair of words found in a certain text. The number stored in position [i,j] of this matrix is the mean distance of the words i and j in the text.
The mean distance between two words i and j is calculated as follows:

1. SUM[i,j] is the total sum of distances between the two words i and j for all the possible pairs (i,j) of these two words in the text.
2. N[i,j] is the number of all the possible pairs (i,j) of the words i and j found in the text.
3. $AVN[i,j] = \frac{SUM[i,\ j]}{N[i,\ j]}$ is the mean distance between the two words.

The next step is to create the second of the matrices, called 2-D Adjoining Affinity Matrix. This matrix also contains information about word affinity in texts, that is, about mean distance between words as in the previous case of the full affinity matrix. The difference, however, is that not all possible pairs of words are examined as before. This matrix contains information about neighboring or adjoining words in text. The following example is quite explanatory:

"Time slips by like grains of sand". In the above phrase, adjoining words are Time (position 1) with slips (position 2), Slips (position 2) with like (position 4), Like (position 4) with grains (position 5), Grains (position 5) with sand (position 7)

Whereas, in creating the full affinity matrix we would also consider the pairs (time, grains), (like, time), (time, like), and so on.
After creating the word frequency of occurrence vector and the two affinity matrices, the fourth step is to involve WordNet in use. The idea, as shown before, is to associate each word with a vector of semantic categories by giving the word as input to WordNet. For example, giving "car" as input to WordNet, the result is (after automatic processing of the output) the following set:
(vehicle, artifact, entity, area, structure)

WORD	WORDNET CATEGORIES
Like	judge, think
Time	event, measure, abstraction, time, property, attribute, information, know-ledge, psychological_feature

The total number of word semantic categories is 229. Therefore each word frequency vector is transformed into word semantic categories frequency vector of 229 elements, while each word affinity matrix (full or adjoining) is transformed into a 229x229 matrix of word semantic categories affinity matrix. The process for these transformations is easy to understand.

First, in the case of the frequency vectors, each word is taken and its corresponding set of n semantic categories is found, word_i = (semantic_word_i_1, semantic_word_i_2, …. semantic_word_i_n). The element in position [i] of the semantic categories frequency vector is increased by one if the semantic category in position [i] of that vector is found in the set of semantic categories associated to that particular word and produced by WordNet. Otherwise the element remains as before. Finally, each element of the semantic categories vector shows how many times the corresponding category is found in the text (by counting the matching semantic categories in the whole set of words extracted from the text).

Second, in the case of the two affinity matrices, each pair of words [i,j] encountered in the text is considered, along with the corresponding information about their associated total sum of distances (SUM[i,j]) and the number of their pairs (N[i,j]) occurred in the document. The set of semantic categories for both words is then found. The element [k,l] of the semantic categories affinity matrix is changed if category k is found in the categories set of word i and category l is found in the categories set of word j. More specifically, the corresponding total sum in the semantic categories affinity matrix is increased by SUM[i,j] and the number of pairs is increased by N[i,j].

Thus, after finishing the fourth step, each text is represented by:

- A 229-element WordNet categories frequency vector
- A 229x229-element semantic categories affinity matrix (Full Affinity Matrix)
- A 229x229-element semantic categories affinity matrix (2-D Adjoining Affinity Matrix) – semantic categories pairs frequencies.

The above four steps lead to the extraction of first and second order features.

The fifth step involves the extraction of third order features. To this end we form the 3-D 229x229x229-element semantic categories affinity matrix (3-D Adjoining Affinity Matrix). The (i,j,k) element of such a matrix represents the frequency of the triple (i,j,k) of semantic categories inside the text. (i,j) are neighboring semantic categories in the text with distance 1. The same holds for the pair (j,k). That is, when calculating (i,j,k) triple's frequency inside the text, triple components should exist in the given order (i,j,k) and be neighboring semantic word categories to count.

The sixth step in the proposed feature extraction approach is to transform the above described affinity matrices into vectors of 6 elements by applying Cooccurrence Matrices Analysis to each one of them. The final result is that each text is represented by a vector of 229+6+6+6=247 elements by joining the semantic categories frequency vector and the three vectors obtained by applying the cooccurrence matrices analysis measures (of 6 dimensions each) to each of the two 2-D affinity matrices as well as to the 3-D affinity matrix.

In this sixth step, concerning the feature extraction approach herein adopted, some ideas taken from Texture Analysis are utilized. Texture analysis as a process is part of what is known as Image Processing. The purpose of texture analysis in image processing is to find relations between the pixel intensities that create an image. Such a relation is for example the mean and the variance of the intensity of the pixels. A common tool for texture analysis is the Cooccurrence Matrix [12]. The purpose of the cooccurrence matrix is to describe the relation between the current pixel intensity and the intensity (gray level) of the neighboring pixels. The creation of the cooccurrence matrix is the first step of the texture analysis process. The second step is to extract a number of statistical measures from the matrix.

After cooccurrence matrices formation in image texture analysis, several measures are extracted from these cooccurrence matrices in order to best describe the texture of the source image in terms of classification accuracy. The following 6 measures in table 2 are an example of texture measures [12]. Cooccurrence matrices analysis in texture processing could be associated with word cooccurrence matrices for text understanding, in a straightforward way, where, each semantic word category (from 1 to 229) is associated with a pixel intensity. The rationale underlying such an association is explained below. It is clear that similar measures could be extracted for both word affinity matrices. Thus, the six measures shown in table 2 are, also, the ones used in this work. In every equation, f(r,c) represents the position [r,c] of the associated matrix. Therefore, texture analysis process ends up with a small set of numbers describing an image area in terms of texture information, which results in information compression. Instead of the source image (a matrix of pixels) there is a set of numbers (the results of applying the texture measures) that represent textural information in a condensed way. This is the reason why it is herein attempted to associate texture analysis and text understanding analysis, namely, dimensionality reduction of the input vectors produced in text processing. The same ideas emerging from texture analysis in image processing are applied to the document text categorization problem.

Table 2. The cooccurrence 2-D matrices analysis associated measures

1.	**Energy:**	M_1 **= Sum Sum f(r,c)**^{2}
2.	**Entropy:**	M_2 **= Sum Sum log(f(r,c) * f(r,c)**
3.	**Contrast:**	M_3 **= Sum Sum (r-c)**^{2} *** f(r,c)**
4.	**Homogeneity:**	M_4 **= Sum Sum f(r,c)/(1+|r-c|)**
5.	**Correlation:**	M_5 **= Sum Sum {(r*c) f(r,c) –**m_r m_c**}/**s_r s_c
	where, $\mathbf{m_i}$ is the mean value and $\mathbf{s_i}$ is the variance of row (column) i.	
6.	**Inverse Difference Moment:**	M_6 **= Sum Sum f(r,c)/((1+|(r-c)|)**
	where, the double summation symbol {Sum Sum} ranges for all rows and columns respectively.	

The above ideas about 2-D cooccurrence matrices analysis could be easily generalized to 3-D cooccurrence matrix analysis for handling the triples (i,j,k) of the 3-D adjoining affinity matrix above defined in step 5. The relevantr measures are proposed herein in the next table 3.

Table 3. The cooccurrence 3-D matrices analysis proposed generalized associated measures

1. Energy: M_1 = **Sum Sum Sum** $f(r,c,w)^2$
2. Entropy: M_2 = **Sum Sum Sum** $\log(f(r,c,w) * f(r,c,w)$
3. Contrast: M_3 = **Sum Sum Sum** $\{(r-c)^2 * f(r,c) + (c-w)^2 * f(c,w) + (r-w)^2 * f(r,w)\}$
4. Homogeneity: M_4 = **Sum Sum Sum** $f(r,c,w) / (1+\lvert r-c\rvert+\lvert c-w\rvert+\lvert r-w\rvert)$
5. Correlation: M_5 = **Sum Sum Sum** $\{(r*c*w)\, f(r,c,w) - m_r\, m_c\, m_w\} / s_r\, s_c\, s_w$
where, m_i is the mean value and s_i is the variance of row (column) i.
6. Inverse Difference Moment: M_6 = **Sum Sum Sum** $f(r,c,w)/(1+\lvert r-c\rvert+\lvert c-w\rvert+\lvert r-w\rvert)$
where, the double summation symbol {Sum Sum Sum} ranges for all rows and columns respectively.

While cooccurrence matrices in texture analysis contain information about the correlation of neighboring pixel intensities in an image, the proposed affinity matrices contain information about the correlations of WordNet semantic categories associated with a text. Autocorrelation matrices are involved in both definitions and therefore, both processes bear some similarities. Thus, the measures defined in tables 2,3 are applied to the word affinity matrices too. After the application of cooccurrence based matrices analysis, there will be a total of 1500 vectors of 247 elements each. The next thing to discuss is how neural networks are trained by utilizing these vectors so as to correctly classify documents in the 10 DDC classes.

An important aspect concerning neural networks training is the normalization of their input vectors. Normalization of the set of features presented in this paper is achieved by substituting each feature value x with $\frac{x - Min}{Max - Min}$ where Max and Min are the maximum and the minimum values of the set where this feature value belongs in, respectively. Actually, thirteen such sets of values exist for the document space vectors under processing. Namely, one for the word semantic categories frequencies, six for the cooccurrence analysis measures with respect to the full affinity matrix and six for the cooccurrence analysis measures with respect to the adjoining affinity matrix. This normalization procedure constitutes the sixth and final step of the feature extraction stage of the proposed text categorization methodology.

There are several reasons for involving supervised neural networks instead of traditional statistical classifiers:

- Their capability to learn difficult tasks and their satisfactory generalization even in the case of classification problems with multiple input features and classes [13].
- Their capability to generalize well even with noisy data and missing information [13].
- They are universal approximators whose generalization capabilities are not affected, at least to the first order, from the curse of dimensionality [13].

3 Experimental Study

In order to evaluate the proposed text categorization methodology an extensive experimental study has been conducted and herein is presented. Two different sets of experiments have been organized. The first one involves the collection of documents based on the DDC classification approach and the herein proposed feature extraction methodology. However, apart from the proposed methodology of combining first, second as well as third order features we compare feature extraction using only first order features and only first and second order features as well. In the second set of experiments, in order to validate our approach, we have applied a standard feature extraction method as the **tf x idf** word frequency weighting scheme [10] to the DDC collection of 1500 documents. Therefore, a set of 1500 document space vectors, different from the ones of the first set, has been obtained. The total number of different words encountered in all the 1500 documents, after removing the words specified in section 2 (words selection procedure), are 6533. Therefore, all the word frequencies based document space vectors used in this set of experiments are of 6533 dimensions.

In both sets of experiments the classifiers reported in the next paragraphs define the second stage of the text categorization system. While, in the first set they are applied to the proposed document space vectors, analyzed in section 2, in the second set of experiments they have been applied to the document space vectors derived using the **tf x idf** technique.

Concerning both sets of experiments, a cross-validation methodology [13] has been applied to all the text categorization methods herein involved, in order to ensure statistical validity of the results, due to the relatively not large number of documents included in this DDC collection. As mentioned above, the 1500 document space vectors constructed from the DDC collection are divided in the training set of patterns (1000 of them) and in the test set of patterns (500 of them). In order to apply the cross-validation ethodology [13], 500 different pairs of such training and test sets have been created from the set of 1500 document space vectors by randomly assigning each of them to a training or test set.

A variety of MLP training models has been applied to both sets of experiments, including standard on-line Backpropagation, RPROP and SCG (Scaled Conjugate Gradients). Several architectures have been investigated of the types 247-x-10, 241-x-10, 229-x-10 and 6533-x-10 (input neurons- hidden layer neurons- output neurons). Only the best of these MLP model architectures and training parameters, out of the many ones investigated, are reported in tables 4 and 5. The MLP simulation environment is the SNNS (Stuttgart Neural Network) Simulator [14]. For comparison the LVQ and the k-NN classifier, reported to be one of the best algorithms for text categorization [10] are, also, involved in these experiments.

Concerning the performance measures utilized in this study, mean and variance of the classification accuracy are reported in tables 4 and 5. To this end, if one classifier results in $G_1\%$, $G_2\%$, $G_3\%$, ..., $G_{500}\%$ over all the 10 classes classification accuracies with respect to each one of the 500 different testing sets of patterns associated with the cross-validation procedure, then,

- Mean Accuracy = $(G_1\%+G_2\%+G_3\%+ \ldots.+ G_{500}\%)/500$
- Accuracy Variance = VARIANCE $(G_1\%, G_2\%, G_3\%, \ldots., G_{500}\%)$

where, each Gi% refers to the over all categories classification accuracy (number of correctly classified patterns/ total number of patterns) for the patterns encountered in the ith test set and not to each category classification accuracy separately. These performance measures are the usually involved statistical measures in cross-validation experiments [13]. The following tables 4 and 5 show the results obtained by conducting the above defined experimental study. Table 4 illustrates the results obtained by applying the proposed feature extraction methodology and its variants, while table 5 presents the results obtained by applying the **tf x idf** technique. These favorable text classification performance results show the validity of our approach concerning the novel document space vector extraction.

Table 4. Text Categorization accuracy obtained involving the proposed, in section 2, document space vectors extraction methodology. The statistics obtained after 500 runs of all algorithms. The relevant results in each cell of the above table are in the form (results involving first, second and third order features/ results involving only first and second order features/ results involving only first order features). The same form holds for the architectures.

Performance Measures / Text Classifier	Mean Accuracy (%)	Accuracy Variance (%)
LVQ, 40 code book vectors	70.5/69.8/69.3	2.5/2.4/2.2
LVQ, 28 code book vectors	71.1/70.6/70.2	2.5/2.5/2.4
Vanilla BP, 247/241/229-12-10, (0.6, 0.2)	61.2/60.8/60.7	2.8/2.8/2.6
RPROP, 247/241/229-12-10, (1,100,10)	76.0/75.5/75.2	3.6/3.5/3.2
RPROP, 247/241/229-12-10, (0.1,50,4)	74.3/74.1/73.6	3.1/3.1/3.0
RPROP, 247/241/229-14-10, (0.1,50,4)	62.4/62.3/61.7	2.9/2.7/2.6
SCG, 247/241/229-15-10, (0,0,0,0)	63.9/63.4/63.2	1.9/1.8/1.5
Vanilla BP, 247/241/229-18-10, (0.6,0.2)	62.1/61.5/61.1	1.1/1.1/1.0
RPROP, 247/241/229-18-10, (1,100,10)	72.4/72.2/71.3	0.8/0.8/0.6
Vanilla BP, 247/241/229-24-10, (0.4,0.1)	70.0/69.6/69.0	3.0/3.0/2.7
K-NN (nearest neighbor)-(K= 30)	69.3/69.2/69.0	2.7/2.6/2.3

Table 5. Text Categorization accuracy obtained involving the tf x idf word frequency based document space vectors extraction methodology. The statistics obtained after 500 runs of all the algorithms.

Performance Measures / Text Classifier	Mean Accuracy (%)	Accuracy Variance (%)
LVQ, 28 code book vectors	67.7	2.2
Vanilla BP (On-line BP), 6533-320-10, (0.4,0.5)	57.3	4.1
RPROP, 6533-320-10, (1,100,10)	68.4	5.2
SCG, 6533-320-10, (0,0,0,0)	61.7	2.3
K-NN (nearest neighbor)-(K= 30)	67.6	2.1

4 Conclusions and Prospects

This paper outlines a feasible solution to the problem of text categorization. The collection of documents and their content categorization is based on a widely accepted and systematic methodology, namely, the DDC system. The techniques suggested involves neural networks and novel feature extraction methods based on first, second and third order characteristics of word concept frequencies and affinities estimated by application of a widely accepted NLP thesaurus tool, namely, the WordNet as well as by statistical techniques stemmed from Texture processing in image analysis. The promising results obtained are favorably compared to other well established text categorization document space vectors formation techniques. Future aspects of our work include, also, the designing of a complete system based on Computational Intelligence and Artificial Intelligence Techniques for dealing with the full DDC text categorization problem, involving 1000 human understandable categories.

References

[1] Merkl D., "Text Classification with Self-Organizing Maps: Some lessons learned", Neurocomputing, vol. 21, pp. 61-77, 1998.

[2] Chen H., et al, "Information Visualization for Collaborative Computing", IEEE Computer, pp. 75-82, Aug. 1998

[3] Kohonen T., et al, "Self-Organization of a Massive Document Collection", IEEE Trans. Neural Networks, vol. 11, no 3, pp. 574-585, 2000.

[4] Cohen W. J. and Singer Y.,"Context-sensitive learning methods for text categorization", In SIGIR'96: Proc. 19th Annual Int. ACM SIGIR Conf. on Research and Development in Information Retrieval, pp. 307-315, 1996.

[5] Yang Y. and Pedersen J. P, "Feature selection in statistical learning of text categorization", In the 14th Int. Conf. on Machine Learning, pp. 412-420, 1997.

[6] Lewis D. and Ringuette M., "A comparison of two learning algorithms for text classification", 3rd Annual Symposium on Document Analysis and Information Retrieval, pp. 81-93, 1994.

[7] Joachims T., "Text categorization with support vector machines: Learning with many relevant features", In Proc. 10th European Conference on Machine Learning (ECML), Springer Verlag, 1998.

[8] Wiener E., Pedersen J. O., and Weigend A. S., "A neural network approach to topic spotting", 4th annual symposium on document analysis and Information retrieval, pp. 22-34, 1993.

[9] Salton G. and McGill M. J., "An Introduction to Modern Information Retrieval", McGraw-Hill, 1983.

[10] Aas K. and Eikvil L., "Text Categorisation: A Survey", Technical Report, Norwegian Computing Center, P.B. 114 Blindern, N-0314 Oslo, Norway, 1999

[11] Chan, Mai L., Comaromi J. P. and Satija M. P.,. 1994. "Dewey Decimal Classification: a practical guide". Albany, N.Y.: Forest Press.

[12] Haralick, R. M., Shanmugam, K. and Dinstein, I. "Textural Features for Image Classification", IEEE Trans. Syst., Man and Cybern., Vol. SMC-3, 6, pp. 610-621, 1973.

[13] Haykin S., "Neural Networks. A comprehensive foundation", Prentice Hall, 1999.

[14] Zell A., Mamier G., Vogt M., et al. , "SNNS Stuttgart Neural Network Simulator, User Manual Version 4.1", Report No 6/95, University of Stuttgart, http://www-ra.informatik.uni-tuebingen.de/SNNS/UserManual/UserManual.html

An Improved OIF Elman Neural Network and Its Applications to Stock Market

Limin Wang[1,2], Yanchun Liang[1], Xiaohu Shi[1], Ming Li[2], and Xuming Han[1,3]

[1] College of Computer Science and Technology, Jilin University, Key Laboratory of Symbol Computation and Knowledge Engineering of Ministry of Education, Changchun 130012, China
ycliang@jlu.edu.cn

[2] Department of Computer Science and Technology, Changchun Taxation College, Changchun 130117, China

[3] Institute of Information and Spreading Engineering, Changchun University of Technology, Changchun 130012, China

Abstract. An improved model is proposed based on the OIF Elman neural network by introducing direction and time profit factors and applied to the prediction of the composite index of stock. Simulation results show that the proposed model is feasible and effective. Comparisons are also made when the stock exchange is performed using prediction results from different models. It shows that the proposed model could improve the prediction precision evidently and realize the main purpose for investors to obtain more profits.

1 Introduction

Stock market could indicate the status of economy, the high venture and high profit are its important characters. Therefore the prediction of the stock index is the interesting problem of people. The stock operation can be considered as a huge and most complicated nonlinear dynamic system. The variation of stock indexes has strong nonlinear feature. Though many methods have been proposed for the prediction of stock indexes, such as the moving average method, point drawing method, K drawing method, season change and judge analysis, etc., their prediction results are not very well for the strong nonlinear feature of the stock market. It was proved by Hornik that the artificial neural network (ANN) could approximate any nonlinear functions at any precision if the network structure could be selected appropriately. Therefore ANN has the capability to uncover the hidden relations among the sample data [1]. Accordingly, the rationality and applicability are the particular advantages of ANN when to construct stock models. Recently, there are some researchers who have developed some methods based on ANN to the field of the finance forecasting and have obtained satisfactory results. There are three kinds of common stock forecasting models, namely, Multi-layer Perception (MP), Radial Basis Function (RBF), and Support Vector Machine (SVM). But there are obvious limitations when using these methods, such as the slow convergence speed, being easy to get into local optimal values and bad generalization [2]. In this paper, in order

B. Gabrys, R.J. Howlett, and L.C. Jain (Eds.): KES 2006, Part I, LNAI 4251, pp. 21 – 28, 2006.
© Springer-Verlag Berlin Heidelberg 2006

to obtain more accurate forecasting precision, the dynamic feedback neural network, namely, the output-input feedback Elman neural network (OIF Elman NN) is used to forecast the stock indexes. Considering that the main purpose of investors is to obtain more profits, this paper proposed improved OIF Elman neural network model by introducing profit factors to OIF Elman neural network, including direction profit factor and time profit factor. Numerical results show that the proposed model can provide more accurate forecasting precision for stock indexes. Moreover, when the stock exchange is performed using the prediction results from different models, the proposed model can improve the profit evidently compared with the OIF Elman neural network.

2 A Brief Introduction of OIF Elman Neural Network

The OIF Elman neural network (OIF Elman NN) based on Elman neural network (Elman NN) is proposed in Ref [3], which has two particular layers called context layer and context II layer besides the conventional input, hidden and output layers. The context layer and context II layer could memorize the former values of the hidden and output layer nodes, respectively, so they could be regarded as time-delay operators. Therefore the feedforward part of the network could be modified in the learning process while the recurrent part is fixed. The structure of OIF Elman NN is depicted as Fig.1.The mathematic model of OIF Elman neural network is:

$$x(k) = f(w^{I1}x_c(k) + w^{I2}u(k-1) + w^{I4}y_c(k)), \tag{1}$$

$$x_C(k) = \alpha x_C(k-1) + x(k-1), \tag{2}$$

$$y_C(k) = \gamma y_C(k-1) + y(k-1), \tag{3}$$

$$y(k) = w^{I3}x(k). \tag{4}$$

Where w^{I1}, w^{I2}, w^{I3}, w^{I4} are, respectively, the weights which are link the hidden and context layers, the hidden and input layers, the output and hidden layers and the hidden and context II layers, $x_c(k)$ and $x(k)$ are the outputs of context layer and hidden layer, $y_c(k)$ and $y(k)$are the outputs of context II layer and output layer, α, γ are the feedback gains of the self-connections of context and context II layers, respectively. Where $f(x)$ is usually selected as *sigmoid* function, namely that:

$$f(x) = \frac{1}{1+e^{-x}}. \tag{5}$$

Let the given real output of the kth sample be denoted by $y_d(k)$, then the objective function is selected as follows:

$$E(k) = \frac{1}{2}(y_d(k) - y(k))^T(y_d(k) - y(k)). \tag{6}$$

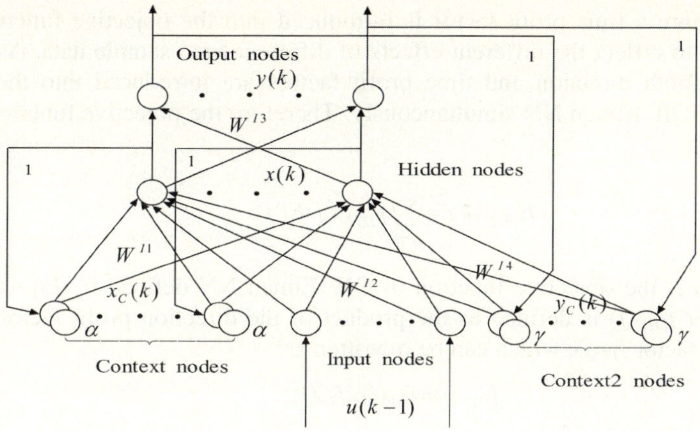

Fig. 1. The structure of OIF Elman NN structure

3 OIF Elman Neural Network with Profit Factors

Just as OIF Elman NN does, the errors between the forecasting values and the objective ones are generally regarded as the evaluation criterion of the forecasting models. While considering that the main purpose of investors is to obtain more profits, it seems that the profit should be paid more attention rather than forecasting precision when the stock indexes' prediction is performed. By experimenting, it could be found that the profits of stock exchange dealt according to the prediction results are not always accordant with the forecasting precision but accordant with the correctness of the fluctuation trend forecasting [4]. While by further observing, it was found that if the fluctuation trend of stock indexes is predicted correctly by the calculating model, the profits are always high; and vice versa. Therefore considering this point, we propose an improved OIF Elman neural network by introducing direction profit factor and time profit factor into OIF Elman model.

For convenient calculation, the stock exchange fee is not considered, and we assume that there is only one stock in the market. Therefore we perform the stock exchange according to the following rule: if we predict that the stock price will rise next working day, then we buy the stock as much as money permitted; otherwise we trade off all the stock in hand. The profit could be defined by follows:

$$Profit=((Remaining-Seed)/\ Seed)\times 100\%. \tag{7}$$

Where *Profit* is the profit by exchanging stock, *Seed* the value of the initial fund, and *Remaining* the value of the final fund, respectively.

In order to get better forecasting and assessment results, a direction profit factor is introduced to the objective function of OIF Elman NN which could represent the punishment trend that is not accord with the real value. In addition, since the inputs of the neural network are the former composite indexes of stock, which can be considered as a time series, so the data which are near to the current time should play the more important role compared with that of the ones which are far to the current time, therefore a time profit factor is introduced into the objective function of OIF Elman NN to reflect the different effects of different time sample data. As results of the above, both direction and time profit factors are introduced into the objective function of OIF Elman NN simultaneously. Therefore the objective function could be rewritten as:

$$E_{DTP}(k)=\sum_{j=1}^{k} f_{DTP}(j)E(j)\,, \tag{8}$$

Where $E(j)$ is the objective function of OIF Elman NN defined by Eq. (6), and the profit factor $f_{DTP}(j)$ is defined as the product of the direction profit factor $f_{DP}(j)$ and time profit factor $f_{TP}(j)$, which can be rewritten as:

$$f_{DTP}(j)=f_{DP}(j)\times f_{TP}(j), \tag{9}$$

Here $f_{DP}(j)$ and $f_{TP}(j)$ are defined as follows [5-6]:

$$f_{DP}(j)=\begin{cases} g & if\ (y_d(j)-y_d(j-1))(y(j)-y(j-1))\le 0 \\ h & otherwise \end{cases}, \tag{10}$$

$$f_{TP}(j)=1/(1+\exp(a-2a\cdot j\cdot k^{-1}))\,. \tag{11}$$

Here g is a big value compared with h, therefore $f_{DP}(j)$ could represent the punishment, and a is the discount rate, $f_{TP}(j)$ could strengthen the effects of sample data near the current time and weaken the effects of those far from the current time.

Calculating the partial derivatives of $E_{DTP}(k)$ with respect to weights and let them be zeros, we have the learning algorithm of OIFENNPF as follows:

$$\Delta w_{jl}^{I1}(k)=\sum_{t=1}^{k}\eta_1 f_{DTP}(t)\sum_{i=1}^{m}(\delta_i^0 w_{ij}^{I3})\frac{\partial x_j(t)}{\partial w_{jl}^{I1}}\quad (j=1,2,\cdots,n;\ l=1,2,\cdots,n)\,, \tag{12}$$

$$\Delta w_{jq}^{I2}(k)=\sum_{t=1}^{k}\eta_2 f_{DTP}(t)\delta_j^h u_q(t-1)\quad (j=1,2,\cdots,n;\ q=1,2,\cdots,r)\,, \tag{13}$$

$$\Delta w_{ij}^{I3}(k)=\sum_{t=1}^{k}\eta_3 f_{DTP}(t)\delta_i^0 x_j(t)\quad (i=1,2,\cdots,m;\ j=1,2,\cdots,n)\,, \tag{14}$$

$$\Delta w_{jl}^{I4}(k)=\sum_{t=1}^{k}\eta_4 f_{DTP}(t)\sum_{i=1}^{m}(\delta_i^0 w_{ij}^{I3})\frac{\partial x_j(t)}{\partial w_{js}^{I4}}\quad (j=1,2,\cdots,n;\ s=1,2,\cdots,m)\,, \tag{15}$$

$$\delta_i^0=(y_{d,i}(k)-y_i(k))\,, \tag{16}$$

$$\delta_j^h=\sum_{i=1}^{m}(\delta_i^0 w_{ij}^{I3})f_j'(\cdot)\,, \tag{17}$$

$$\frac{\partial x_j(k)}{\partial w_{jl}^{I1}}=f_j'(\cdot)x_l(k-1)+\alpha\frac{\partial x_j(k-1)}{\partial w_{jl}^{I1}}\quad (j=1,2,\cdots,n;\ l=1,2,\cdots,n)\,, \tag{18}$$

$$\frac{\partial x_j(k)}{\partial w_{js}^{I4}} = f_j'(\cdot) y_s(k-1) + \gamma \frac{\partial x_j(k-1)}{\partial w_{js}^{I4}} \quad (j = 1,2,\cdots,n; s = 1,2,\cdots,m). \tag{19}$$

Where η_1, η_2, η_3 and η_4 are the learning steps of the w^{I1}, w^{I2}, w^{I3} and w^{I4}, respectively.

4 Experimental Results

The experiment data are selected from the composite indexes in Shanghai Stock Exchange (CISSE) from 4 January 1999 to 29 June 2000. The network structures are selected as 3-10-1, namely with 3 input nodes, 10 hidden and 1 output nodes, respectively. The inputs of the networks are the composite indexes of the stock of three consecutive days, and the output is the forecasting value of the fourth day. Before learning, the input data should be normalized into the unit interval [0, 1] by using the equation:

$$\hat{x}_i = (x_i - \min(x_i)) / (\max(x_i) - \min(x_i)) \quad (i = 1,2,\cdots,N). \tag{20}$$

Where x_i is the composite index of the ith day, $\min(x_i)$ is the minimal value of all x_i, $\max(x_i)$ is the maximal value of all x_i, and $\hat{x}_i$ is the normalized value of x_i, respectively.

To evaluate the effectiveness of the prediction models, we use the absolute average error (AAE) to represent the forecasting precision, the absolute average error (AAE):

$$AAE = \frac{1}{N} \sum_{i=1}^{N} |y_d(i) - y(i)| \quad (i = 1,\cdots,N). \tag{21}$$

4.1 Selection of the Parameter Values of the Direction and Time Profit Factors

4.1.1 Selection of the Parameter Value *a* for the Time Profit Factor

The parameters, in the example, are selected by trial and error as: γ=0.1, g=2.0, h=0.1, η_1=0.2, η_2=0.2, η_3=0.01, η_4=0.01, respectively. According to the minimum average value of the absolute average error (AAE), the discount rate a can be obtained is 0.1 by using OIFENNPF network, all the results obtained are listed in Table 1.

Table 1. Selection for the value of *a*

α \ a	0.1	0.3	0.5	0.8
0.1	0.000291	0.000951	0.002988	0.038963
0.3	0.000473	0.001469	0.004389	0.051165
0.5	0.000769	0.002268	0.006447	0.067188
0.8	0.001251	0.003503	0.009472	0.088232
average	0.000696	0.002048	0.005824	0.061387

4.1.2 Selection of the Parameter Values of *g* and *h* for the Direction Profit Factor

In order to explain the results effectively, in the example, the learning steps are selected as: η_1=0.2, η_2=0.2, η_3=0.01, η_4=0.01, respectively. The other parameters are selected as: a=0.1, γ=0.1. The direction factor in Eq. (10) is selected by using OIFENNPF network. According to the minimum average values of the absolute average error (AAE), the direction factors h and g obtained by Table 2 and Table 3 are 0.1 and 2.0 respectively.

Table 2. Selection for the value of h (g=2.0)

α \ h	0.1	0.3	0.5
0.1	0.024920	0.049704	0.180344
0.2	0.039486	0.078757	0.227014
0.3	0.062566	0.124792	0.285759
0.5	0.099138	0.197735	0.313314
0.8	0.157088	0.248905	0.394390
average	0.076640	0.139979	0.280164

Table 3. Selection for the value of g (h=0.1)

α \ g	1.0	1.5	2.0
0.1	0.525853	0.031589	0.016613
0.2	0.417752	0.050053	0.026324
0.3	0.331872	0.079310	0.041711
0.5	0.263646	0.125668	0.066092
0.8	0.209446	0.199124	0.104725
average	0.349714	0.097149	0.051093

4.2 Prediction of the Composite Index of Stock

Firstly, the OIFENNPF is applied to do the prediction of the composite indexes of stock, the comparison of the OIFENNPF with OIF Elman NN is also performed. The

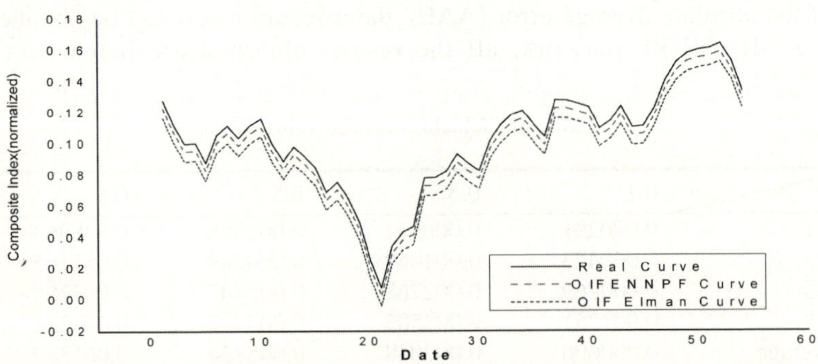

Fig. 2. Comparison results of OIF Elman and OIFENNPF models for the stock

parameters are selected by trial and error as: α=0.1, γ=0.1, η_1=0.2, η_2=0.2, η_3=0.01, η_4=0.01, a=0.1, g=2, h=0.1. The prediction results are shown in Fig. 2. It could be found clearly that the forecasting curve obtained from OIFENNPF is closer to the real curve than that obtained from the OIF Elman NN.

4.3 Comparisons of Stock Profit

In this section, OIF Elman and OIFENNPF models are applied to the prediction of the stock composite indexes. The parameters are selected by trial and error as: α=0.1, η_1=0.1, η_2=0.2, η_3=0.02, η_4=0.01, a=0.1, g=2, h=0.1. According to their prediction results, the stock exchanges are performed respectively in the following rule: if the stock price is predicted rising next working day, then we buy the stock as much as money permitted; otherwise trade off all the stock in hand. The initial money funds are set as 1000 RMB. Table 4 and Table 5 show the results after 50 and 100 iterations, respectively. In the tables, the second column represents the absolute average error, the third column the final fund, and the fourth column the profit, respectively.

Table 4. Errors, remaining and profits using different models (after 50 iterations)

Model	AAE	Remaining(RMB)	Profit Rate (%)
OIF Elman	0.000850	1149.9182	17.1079
OIFENNPF	0.000051	1395.0054	39.5005

Table 5. Errors, remaining and profits using different models (after 100 iterations)

Model	AAE	Remaining(RMB)	Profit Rate (%)
OIF Elman	0.000297	1182.9043	18.2904
OIFENNPF	0.000021	1699.9914	69.9991

From the tables it can be seen that the profits of the stock exchange are not always accordant with the forecasting precision. In other words, the high forecasting precision does not mean the high profit. While the profits obtained according to the results of the proposed OIFENNPF model is obviously higher than that according to OIF Elman model. For example, after 50 and 100 iterations, the profit results of OIFENNPF increase 2.3089 and 3.8271 times than that of OIF Elman model, respectively. In conclusion, the obtained profit of the stock exchange according to the prediction result of OIFENNPF model is higher than that of OIF Elman model evidently.

5 Conclusions

In order to obtain more accurate forecasting results and realize the main purpose for investors to get more profits, an improved OIFENNPF model is proposed by introducing direction and time profit factors to OIF Elman neural network. Numerical results show that the improved model proposed in this paper is feasible and effective for stock investing. The proposed models could improve the forecasting precision and profits obviously and possess the characteristic of quick convergence, which could be useful in the field of the finance.

Acknowledgment

The authors are grateful to the support of the National Natural Science Foundation of China under Grant No. 60433020, the science-technology development project for international collaboration of Jilin Province of China under Grant No. 20050705-2, the doctoral funds of the National Education Ministry of China under Grant No. 20030183060, the Graduate Innovation Lab of Jilin University under Grant No. 503041, and "985" Project of Jilin University.

References

1. Hornik, K., Stinchcombe, M., White, H., Universal approximation of an unknown mapping and its derivatives using multilayer feedforward networks. Neural Networks, Vol. 3. (1990) 551-560.
2. Liu, X. Y., He, J. F., Stock market prediction based on neural networks. Acta Scientiarum Naturalium Universitatis Nankaiensis, 1998, 31: 39-44.
3. Shi, X. H., Liang, Y. C., Lee, H. P., Lin, W. Z., Xu, X., Lim, S. P., Improved Elman networks and applications for controlling ultrasonic motors. Applied Artificial Intelligence, Vol. 18. (2004) 603-629.
4. Yao, J. T., Tan, C. L., Time dependent Directional Profit Model for Financial Time Series Forecasting[A]. International Joint Conference on Neural Networks 2000. Como Italy: Los Alamitos Calif., IEEE Computer Society, (2000) 291-296.
5. Caldwell, R. B., Performances metrics for neural network-based trading system development. Neuro Ve$t Journal, Vol. 3. (1995) 22-26.
6. Refenes, A. N., Brntz, Y., Bunn, D. W., Burgess, A. N., Zapranis, A. D., Financial time series modeling with discounted least squares back propagation. Neural Computing, Vol. 14. (1997) 123-138.

Fuzzy Logic Controller for Turbojet Engine of Unmanned Aircraft

Min Seok Jie[1], Eun Jong Mo[1], Gyo Young Hong[2], and Kang Woong Lee[1]

[1] School of Electronics, Telecommunication and Computer Engineering, Hankuk Aviation University, 200-1, Hwajeon-dong, Deokyang-gu, Koyang-city, Kyonggi-do, 412-791, Korea
tomsey@korea.com, moe@pweh.com, kwlee@mail.hangkong.ac.kr
[2] Department of Avionics and Computer Simulation, Hanseo University, TaeAn-Gun, ChungNam, 357-953, Korea
kiathgy@hanseo.ac.kr

Abstract. This paper proposes a turbojet engine controller for unmanned aircraft using Fuzzy-PI(Proportional Integral) algorithm. To prevent surge or flame out which could occur during engine acceleration or deceleration operation, the PI type fuzzy system controls the fuel flow input value. The fuzzy logic with logarithm function is designed to obtain the fast acceleration and deceleration of the engine under the condition that the operating point should stay between the surge and flame out control lines during the transient. This paper presents a fuzzy inference rule made by the logarithm function can improve the characteristics of tracking error in steady state. The proposed control scheme is proved by computer simulation using a linear engine model. Simulation results on the state space model of a turbo jet engine illustrate that the proposed controller achieves the desired performance.

Keywords: fuzzy control, logarithm function, turbojet engine, unmanned aircraft, fuel flow control.

1 Introduction

It is obvious that different control approach of aircraft engine is applied depending upon the purpose of the flight or flight condition such as starting, take-off, or landing operation. Depending on the external factors such as air temperature or pressure, the engine control scheme may be different as well.

For military aircraft, sharp maneuvering is prerequisite requirement of flight control, but for the commercial aircraft, smooth maneuvering is required. The application of different control approaches has been widely studied to achieve flight safety and reliability.

The method to control aircraft engine in the past was very limited to basic operations such as engine start or thrust control. With the introduction of automatic engine control system as well as high performance applications, the development of engine control technology is becoming apparent. For applying automatic engine control system into both military and commercial aircraft engine, it is required to quick response characteristics to any change of speed or thrust during the engine acceleration or deceleration. Automatic control of fuel flow or compressor rotating speed is applied to prevent surge which is occurred in high thrust operation

B. Gabrys, R.J. Howlett, and L.C. Jain (Eds.): KES 2006, Part I, LNAI 4251, pp. 29 – 36, 2006.
© Springer-Verlag Berlin Heidelberg 2006

environment such as take off or landing. In addition, the automatic engine control system controls rotor speed by measuring the inlet and outlet temperature. As a result of that, it minimizes the distress of the parent material of the engine so that it can achieve engine life cycle extension as well as fuel efficiency.

As mentioned above, to automatically control an aircraft engine, each system should be realized into digital format. Fuel flow control is proposed to control rotor speed automatically. The desired thrust can be achieved by varying the fuel flow or exhaust nozzle dimension, which is recently proved by applying multivariable control approach [1]. Application of multivariable control approach is used to design for linear models using LQG/LTR [2], [3] approach or H-infinity control [4] synthesis formulation. This technique is a robust control to reduce disturbance or parameter variations, however, it may weaken control performance because of uncertainty of each engine model which is linearized at each operating point.

However, small engine such as unmanned aerial vehicle is required to system simplicity by limiting the dimension of exhaust nozzle so that the engine control system suited to SISO(Single Input Single Output) system [5] can be designed to achieve desired engine performance by controlling fuel flow effectively.

One important factor to be considered to design engine control system is to design a protection system which controls the engine not to occur surge or flame out during engine acceleration or deceleration. When high thrust is applied, like take-off operation, surge can happen and it is one of the major causes to lead flight accident or incident. Therefore it is important to consider implementing a surge prevention system in the aircraft engine. To prevent surge effectively, engine acceleration should be controlled to run under the surge control line. In this case, if the reference command is set as a compressor rotating speed in the initial stage, the engine may run exceed the surge margin by sudden acceleration, however through surge control system, the fuel flow is controlled to reduce the engine rotating speed to drop it under the surge limit.

This paper presents a design approach applying fuzzy inference method to a small turbo jet engine for unmanned aircraft in which fuel flow is used as only control input. To prevent surge which would be occurred during engine acceleration, this system includes 1) setting a reference acceleration speed for surge prevention, and then 2) improving error characteristics in steady state by log scale of residual errors between the reference acceleration speed and the actual compressor rotating speed, and 3) deciding fuel flow control input using fuzzy inference method which control the actual acceleration speed follow the reference acceleration speed. The proposed control capabilities are proved by simulation using linear engine model.

2 Turbo Jet Engine System and Fuzzy Inference Control

2.1 Turbo Jet Engine System

Simplified turbo jet engine consists of the three main stages as shown in Fig. 1. In the model considered, the air is compressed by compressors. The compressed air then enters a combustion chamber where it undergoes combustion with injected fuel. The subsequent decompression takes place in turbines, where part of the thermodynamic energy of the gas is transferred through the engine shaft back to other mechanical parts. An important energy balance is that the energy needed for air compression comes from the energy transferred to the shafts in the turbines. If the engine is in steady-state, then the energy needed in the compressors is equal to the energy produced in the turbines. Since the engine thrust is in proportion to the rotor speed of compressor, the engine thrust can be estimated by measuring the rotor speed of compressor and the estimated thrust value can be applied to engine control system.

The engine acceleration and deceleration control for generating the desired thrust is designed to control fuel flow so that the desired thrust can be controlled under the command of speed which is transferred into the rotor speed of compressor.

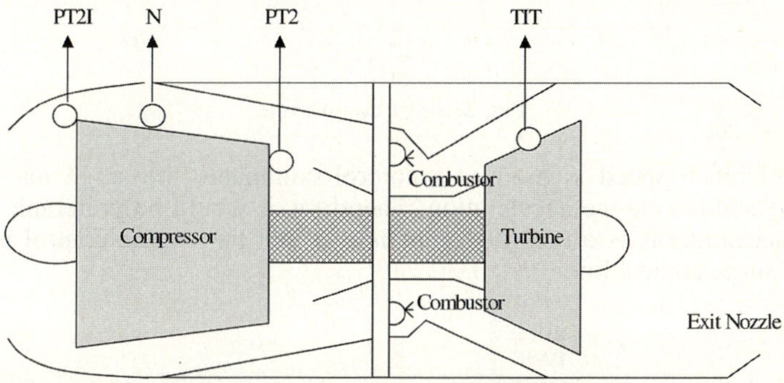

Fig. 1. Schematic of a turbojet engine

2.1 Fuzzy Inference Control

During engine acceleration or starting, pressure of the compressor is decreased by increasing of air mass. As a result of this, air may turn back and flow from the smooth to turbulent. This disturbed state in which the energy supplied to the compressor is simply wasted in churning the air around is known as surging. To control surge effectively, the engine acceleration should be controlled to follow the surge control line in which the surge margin line is closely positioned.

As shown in Fig. 2., acceleration control is achieved by setting a surge margin line in which sufficient surge margin is secured, and having engine control line to be positioned near the surge control line. In this case, if the reference command is set up as a compressor rotating speed, then in the initial stage the engine may run over the surge margin by sudden acceleration, however through the surge control, the fuel flow is controlled. As a result of that maintaining of slow engine acceleration is achieved.

For surge control, it is required to establish surge limit about 90% of the surge margin and select an engine acceleration speed as a reference command which controls the engine operate under the operation line not to run over the surge margin line, and select a flue flow control system which controls the compressor acceleration speed to control the compressor rotating speed follow the reference command.

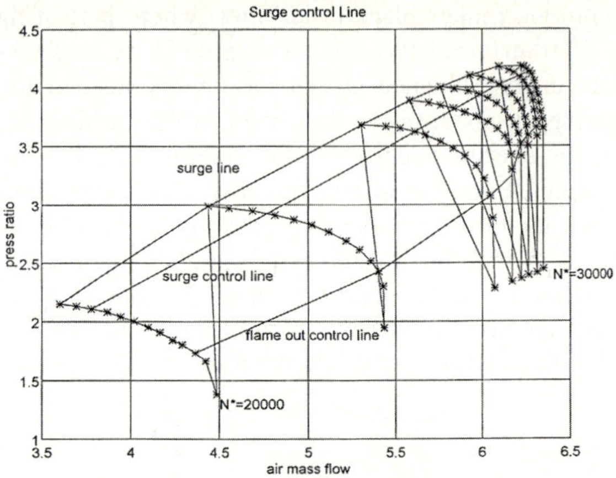

Fig. 2. Surge control line

If the reference speed is used as a control command, the surge may be easily occurred by sudden engine acceleration. Therefore, it would be preferable to use the reference acceleration as a control command, so that the engine control is achieved around the surge control line.

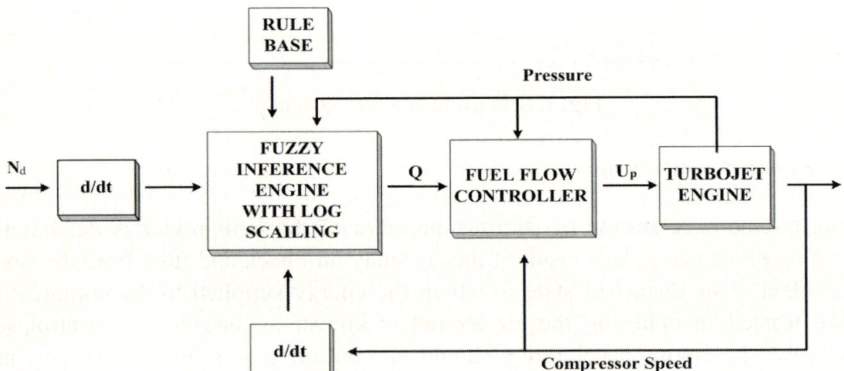

Fig. 3. Structure of the engine control system using fuzzy-inference method

Fig. 3., shows the structure of the engine control system using fuzzy inference method where $\dot{N}_d$ is reference acceleration on which the surge margin is reflected. We define $e_N = \log_{10}(\dot{N}_d - \dot{N})$ as the error of reference acceleration speed to the actual acceleration speed. We present the method to improve steady state error of compressor rotating speed by log scaling the error of compressor rotating acceleration speed.

And, the fuzzy inference fuel flow control input is described using PI control structure as follows;

$$u_p(t) = K_p(e_N, e_p)e_N + K_I(e_N, e_p)\int_0^t e_N(\tau)d\tau \tag{1}$$

where, e_p is an error of compression ratio at inlet of compressor. The control gain K_p and K_I are chosen as fuzzy inference method using e_N, e_p.

Fuzzy inference input variable defines e_N, e_p, the output variable defines PI control gain weight value Q.

The Table 1. defines the fuzzy rule of the input fuzzy variables and output fuzzy variables, and divides language value of input variable into 7 steps and the output fuzzy variables into 5 steps and the membership function is defined as Fig. 4.

The output fuzzy variables obtained from applying fuzzy rule are used to control input through de-fuzzification.

Table 1. Fuzzy rules

e_P \ e_N	NB	NM	NS	ZE	PS	PM	PB
NB	VB	VB	VB	VB	VB	VB	VB
NM	VB	VB	B	M	B	VB	VB
NS	VB	B	M	S	M	B	VB
ZE	B	M	S	VS	S	M	B
PS	VB	B	M	S	M	B	VB
PM	VB	VB	B	M	B	VB	VB
PB	VB	VB	VB	VB	VB	VB	VB

If the language values of fuzzy input variables are $e_N = x_1^0, e_p = x_2^0$ and the comparability of the condition "x is A" is $A(x^0)$, then the compatibility of the two inputs are defined using Mamdani method.

$$W_i = A_{i1}(x_1^0) \times A_{i2}(x_2^0), \qquad i = 1,2 \cdots n \tag{2}$$

Once the result of inference of fuzzy rule is generated by using Mamdani method, the general inference Q is generated using the center of area method as follows

$$Q = \frac{\sum_{i=1}^{n} W_i Q_i}{\sum_{i=1}^{n} W_i} \tag{3}$$

We define the PI control gain using inference result Q as follows

$$K_p = Q \times k_1, \; K_I = Q \times k_2 \qquad k_1, k_2 \text{ :constant} \tag{4}$$

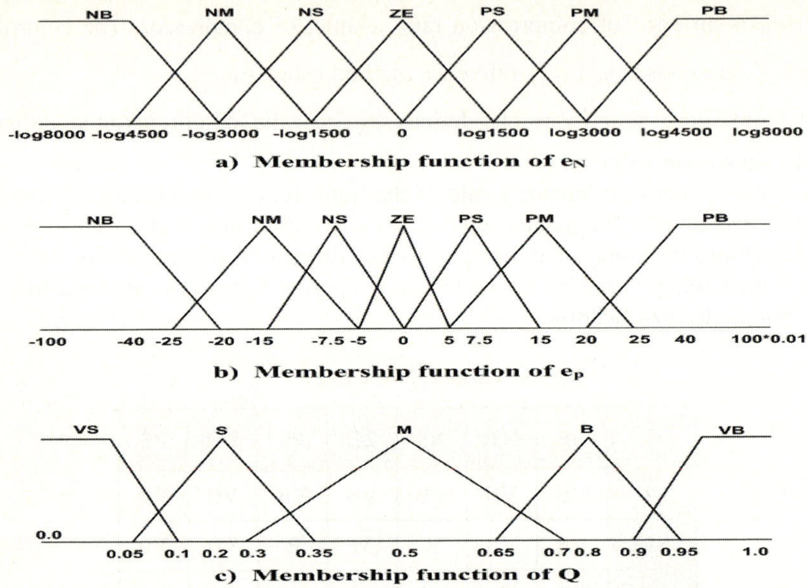

Fig. 4. Membership functions for input and output variables

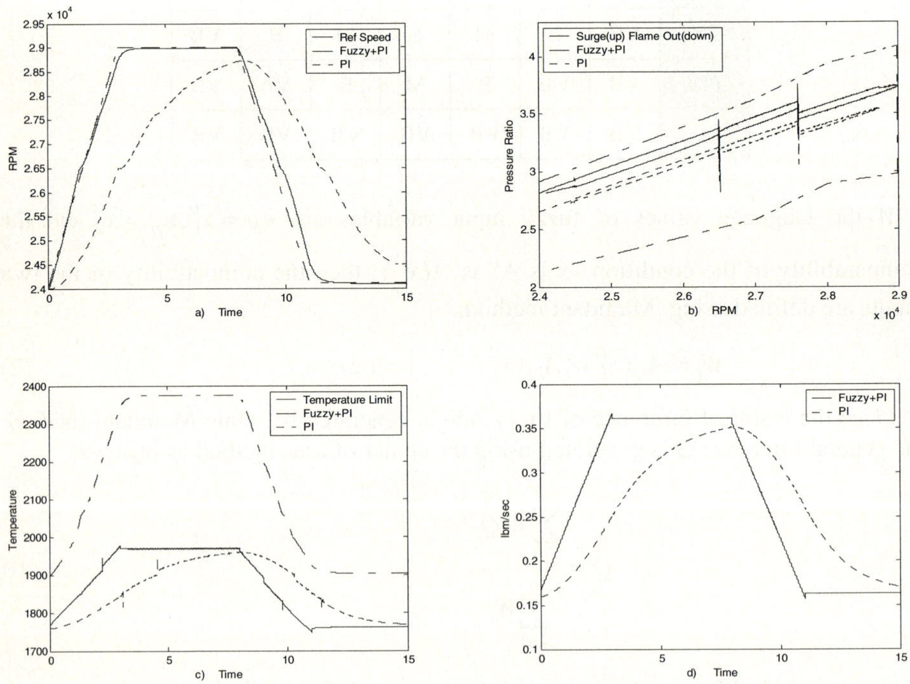

Fig. 5. Simulation results (a) compressor rotation speed (b) compressor outlet pressure ratio(c) turbine inlet temperature (d) fuel flow rate

3 Simulation

The performance of proposed fuzzy inference engine control method presented in this paper is shown by simulation using MATLAB. We use a linear model expressed with state space equation.

$$\dot{x}_p(t) = A_p x_p(t) + B_p u_p(t) \tag{5}$$

Where, $x_p = [x_{p1}\ x_{p2}\ x_{p3}]^T$

x_{p1} : compressor rotation speed x_{p2} : turbine inlet temperature

x_{p3} : compressor outlet pressure u_p : fuel flow

As shown in Fig. 5 the engine adopting the fuzzy PI controller operates successfully without occurring any surge or flame out event. The rotor rotating speed at the initial state set to 24000 rpm, and accelerates by 1700rps until reaching to 29000rpm, and maintains this state for 5 seconds, and then reduces the speed by 1600rps to 24100 rpm.

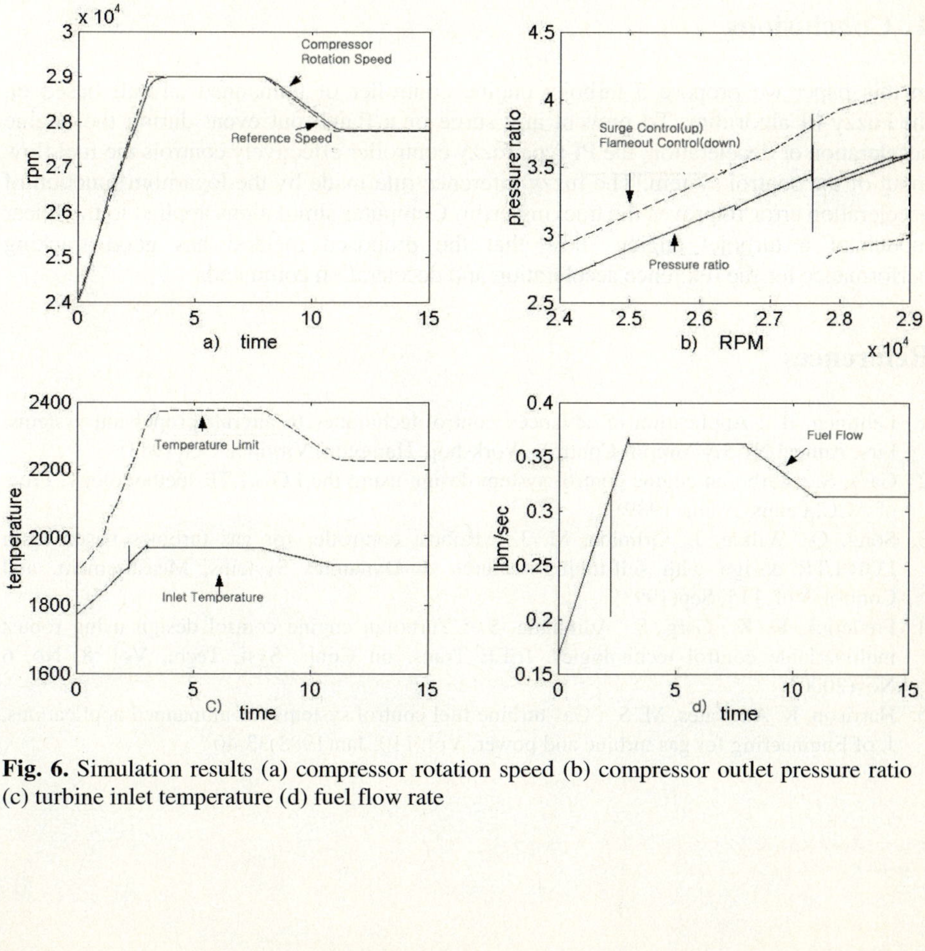

Fig. 6. Simulation results (a) compressor rotation speed (b) compressor outlet pressure ratio (c) turbine inlet temperature (d) fuel flow rate

The control gains of the proposed fuzzy inference PI controller are $K_p = Q \times 0.0000016$, $K_I = Q \times 0.000018$. The weigh value Q is determined depending on the control state of the fuzzy inference. As shown through computer simulation in Fig. 5.(a), (b) the engine run to follow the reference speed line indicated by a dashed line without exceeding both the surge control line and the flame out control line. In the Fig. 5.(a), (b), the pickings are found at 26500 rpm & 27500 rpm, however, those are one of phenomena which can occur during the rotating speed changes in the engine linear model.

Comparing its outstanding performance with PI controller shown by dotted line in Fig.5, the proposed fuzzy inference PI controller achieves much improved transient and steady state response characteristic as well as RPM & pressure ratio than the general PI controller. It also shows that the proposed Fuzzy PI controller makes the engine run to follow close to the reference speed line, and surge line.

As shown in Fig 6 (b) if the compressor rotating speed or pressure ratio exceeds the surge control line, the surge protection system using fuzzy PI controller activates by forcefully reducing the fuel flow and dropping the rotating speed under the surge control limit line. Through this simulation we show the surge protection system works as expected.

4 Conclusions

In this paper we propose a turbojet engine controller of unmanned aircraft based on the Fuzzy-PI algorithm. To prevent any surge or a flame out event during the engine acceleration or deceleration, the PI-type fuzzy controller effectively controls the fuel flow input of the control system. The fuzzy inference rule made by the logarithm function of acceleration error improves the tracking error. Computer simulations applied to the linear model of a turbojet engine show that the proposed method has good tracking performance for the reference acceleration and deceleration commands.

References

1. Lehtinen, B. : Application of advanced control techniques to aircraft propulsion systems. First Annual NASA Aircraft Controls Workshop, Hampton, Virginia, Oct(1983)
2. Garg, S. : Turbofan engine control system design using the LQG/LTR methodology. Proc. of ACC, Pennsylvania(1989)
3. Song, Q., Wilkie, J., Grimlbe, M. J. : Robust controller for gas turbines based upon LQR/LTR design with self-tuning features. J. Dynamics Systems, Measurement, and Control, Vol. 115, Sep(1993)
4. Frederick, D. K., Garg, S., Adibhatla, S. : Turbofan engine control design using robust multivariable control technologies. IEEE Trans. on Contr. Syst. Tech., Vol. 8, No. 6 Nov(2000)
5. Harrison, R. A., Yates, M. S. : Gas turbine fuel control systems for unmanned applications. J. of Engineering for gas turbine and power, Vol. 110, Jan(1988)33-40

Fuzzy Linear Programming Approach to Multi-mode Distribution Planning Problem

Bilge Bilgen and Irem Ozkarahan

Dokuz Eylul University
Department of Industrial Engineering, 35100, Izmir/Turkey
bilge.bilgen@deu.edu.tr, irem.ozkarahan@deu.edu.tr

Abstract. In this study we address the multi-product, multi-period, multi-mode distribution planning problem. The objective of this paper is to present a real distribution planning problem in which rail/road transportation is integrated within a whole focus on supply chain management. However, in real world problems, practical situations are often not well-defined and thus can not be described precisely. Therefore fuzzy mathematical programming becomes a valuable extension of traditional crisp optimization models. This paper also illustrates how a fuzzy linear programming approach be used to model and solve the multi-mode transportation problem.

1 Introduction

Transportation models play an important role in logistics and supply chain management (SCM) for reducing costs and improving service. This paper attempts to optimize a multi-product, multi-period distribution planning problem with consideration of multi-mode transportation within a supply chain. The distribution system considered here derived from current operations of a major organization that manages the supply chain from the time a grower delivers to a storage facility until the time that product is delivered to an end use customer.

In real problems, it is supposed that parameters involved in objective function and constraints are forecasted by expert's judgments, therefore such values are not always precise. Under many conditions, crisp data are inadequate to model real-life situations. Fuzzy set theory was proposed by Zadeh [18] and has been found extensive applications in various fields. It has been widely used to represent uncertain or preference information in many types of applications. Fuzzy sets theory has been implemented in mathematical programming since 1970 when Bellman and Zadeh [1] introduced the basic concepts of fuzzy goals, fuzzy constraints, and fuzzy decisions. Many of the developments in the area of fuzzy mathematical programming (FMP) are based on this seminal paper. Tanaka et al. [13] and Zimmerman [19] were among the first researchers to address FMP. In FMP, the standard problem varies depending upon what elements of the standard MP are to be considered fuzzy (resources and/or objective and/or parameters). Many FMP approaches have been developed for these combinations and can be classified into different categories according to different criteria. A detailed discussion of the FMP procedures can be found in [8, 20].

B. Gabrys, R.J. Howlett, and L.C. Jain (Eds.): KES 2006, Part I, LNAI 4251, pp. 37 – 45, 2006.
© Springer-Verlag Berlin Heidelberg 2006

Our model is considered as a transportation problem with some side constraints. Most studies in the literature have focused on using traditional analytical and simulation based methods to model the transportation problem; a few studies have attempted to model the problem in a fuzzy environment, but their scope has been limited. Since the transportation problem is essentially a linear program, one straightforward idea is to apply the existing fuzzy linear programming techniques. Related studies on the use of fuzzy programming method to solve fuzzy transportation problems include [3,4,7,9,12]. These researchers emphasize on investigating the theory and algorithm. These models are either classical two-index transportation problem or solid transportation problem [15]. To the best of the author's knowledge, no study has attempted to deal with multi-period, multi-commodity, multi-mode distribution planning problem with fuzzy goal, and constraints. The novel aspect of the model is that it combines fuzzy linear programming and multi-index distribution planning problem within supply chain to solve an actual case.

The purposes of this study are two fold: first to develop linear programming model to find the amount of products transported by mode and carrier to minimize the total cost, secondly to model the problem in a fuzzy environment.

In this paper, a linear programming model is first presented which determines the amount of products transported by route, by transportation mode, by carrier, and by period. Then using Zimmermann's approach we fuzzify the corresponding fuzzy elements, and transform LP model into a fuzzy model, which is solved using an auxiliary model. Two kinds of aggregation schemes are applied. A real-life example is used to illustrate the potential savings which can be attained by using fuzzy model. The rest of the paper is organized as follows. Next section describes the real life distribution planning problem. Section 3 presents a mathematical formulation of the problem and Section 4 describes the formulation of the problem in a fuzzy environment. Computational experiments are reported in Section 5, followed by our conclusions and extensions discussed in the final section.

2 Problem Description

The supply chain network considered in the model consists of multiple upcountry storage sites, and customers. Due to page restriction, in this paper we consider a restricted supply chain that excludes decisions concerning distribution to overseas customers. A complete description of the problem, data and the model can be found in Bilgen [2]. The planning horizon, in keeping with the midterm nature of the model, ranges from 1 month to 3 months. Different transportation modes are used between sites. Each storage site has limited out-loading capacity for each transportation mode. Products stored in storage sites are then transported to customers by rail and/or road transport. Destination sites also have limited in-loading capacity for each transportation mode. The following details are added:

- Possible transportation links are defined between storage sites and customers, including the transport modes and carriers. While for some transportation links only rail or only road transportation is available, for the others both transportation modes are available.

- Road and rail transportation are contracted out with long term contracts. Contracted freight rate per route, per mode and carrier is given.
- It is assumed that, customer's demand for a product and the shipment of that product must take place at the same period.

The overall problem can be stated as follows:

Given: cost parameters (out-load, in-loading cost and transportation cost per unit), transportation data, supply data, demand data, and capacity parameters.

Determine: the amount of products transported from each storage site to each customer by transport mode, by route, by carrier, by product and by period and the amount of products remaining at storage sites by product, and period.

Although SCM has seen great interest from the researchers, very little research has been done that attempts to address the SCM problems integrating with transportation related decisions. Transportation problems within SCM are ignored even though it is the most important field that would benefit from the modeling approach.

3 Mathematical Formulation

To describe the problem mathematically, we present a linear programming model. Before presenting a linear programming formulation for the distribution planning problem described in section 2, we first introduce the notation that will be used throughout the paper. Using this notation, a mathematical programming model can be formulated to solve the problem studied in this paper.
The notation used is described in Table 1.

Table 1. Notation

Notation	Remark
Sets	
S	set of storage sites $\{1,2,\ldots,S\}$
K	set of customers $\{1,2,\ldots,K\}$
M	set of transportation modes $\{RL,RD\}$
C	set of carriers $\{1,2,\ldots,C\}$
Parameters	
$SSupply_{js}$	the amount of original product j supplied to storage site s
$OutCap_{sm}$	out-loading capacity at storage site s, by using transportation m.
$InCap_{km}$	in-loading capacity at customer k, by using transportation mode m.
$ComCap_s$	combined out-loading capacity for storage site s.
$ComCap_k$	combined in-loading capacity for customer k.
$CCap_{mc}$	carrier capacity on transportation mode m and carrier c.
$OutLCost_s$	out-loading cost for storage site s
$InLCost_k$	in-loading cost for customer k
$FCost_{skmc}$	freight cost for route from storage site s to customer k, by using transportation mode m, and carrier c
D_{jkt}	demand for product j, customer k, in period t
Variables	
z_{jskmct}	the amount of product j transported from storage site s, to customer k, using transportation mode m and carrier c, in time period t.
SI_{jst}	the amount of product j that remains at storage site s, in the end of period t.

The LP formulation of our model is as follows:
Min

$$\sum_j\sum_s\sum_k\sum_m\sum_c\sum_t FCost_{skmc} z_{jskmct} + \sum_j\sum_s\sum_k\sum_m\sum_c\sum_t (OutLCost_s + InLCost_k) z_{jskmct} \quad (1)$$

$$SI_{js0} = SSupply_{js} \qquad \forall j,s \quad (2)$$

$$\sum_k\sum_m\sum_c z_{jskmct} - SI_{jst-1} + SI_{jst} = 0 \qquad \forall j,s,t \quad (3)$$

$$\sum_j\sum_k\sum_c z_{jskmct} \le OutCap_{sm} \qquad \forall s,m,t \quad (4)$$

$$\sum_j\sum_s\sum_c z_{jskmct} \le InLoad_{km} \qquad \forall k,m,t \quad (5)$$

$$\sum_j\sum_k\sum_m\sum_c z_{jskmct} \le ComCap_s \qquad \forall s,t \quad (6)$$

$$\sum_j\sum_s\sum_m\sum_c z_{jskmct} \le ComCap_k \qquad \forall k,t \quad (7)$$

$$\sum_j z_{jskmct} < FCap_{mct} \qquad \forall s,k,m,c,t \quad (8)$$

$$\sum_s\sum_m\sum_c z_{jskmct} = D_{jkt} \qquad \forall j,k,t \quad (9)$$

The total cost is computed as the sum of the freight cost from the storage sites to the customers, out-loading cost at storage sites, in-loading cost at customers. The transportation between storage sites and customers is done by haulage companies using different types of transportation modes. Constraints (2) assure that the initial amount of product *j* at storage site *s* is equal to the inventory for that product *j,* at storage site *s* at the end of the time period zero. Constraints (3) represent whether there is sufficient product *j* available to transport from storage site *s*, in time period *t*. Constraints (4) assure that, the amount of products transported to customers by different carriers must be smaller than the out-loading capacity for storage site *s* and transportation mode *m*, and in time period *t*. Constraints (5) enforce that the amount of products transported from all storage sites to each customer and mode is bounded by the in-loading capacity of that customer. Constraints (6)-(7) embodied that the combined out-loading capacity for each storage site and combined in-loading capacity for each customers, respectively. Constraints (8) assure that the amount of products loaded to each carrier type on that route must be smaller than the capacity of that carrier type. Finally demand constraints are imposed by equations (9).

Applying mathematical programming models to solve real world problems is a challenging task, because decision makers find it difficult to specify goals and constraints exactly and because the parameters used in these models can not be estimated exactly. The next section describes the problem in a fuzzy environment.

4 Fuzzy Distribution Planning Model

The distribution planning formulation presented in the preceding section is a deterministic model. It assumes certainty in all aspects of the problem. In this section, we formulate a FMP model which overcomes the rigidity of the deterministic model. In a fuzzy environment mathematical models must take into account fuzzy constraints, vague goals and ambiguous parameters. Membership functions are used to incorporate fuzziness or to represent the linguistic variables for applications of fuzzy set theory.

In the deterministic formulation, the total cost consists of shipping, out-loading, and in-loading costs. Most of these costs are not easily measured and hence a great deal of human perception is needed in their assessments. The objective function becomes fuzzy in character because of these fuzzy costs. On the other hand, in the original study it is assumed that customer's demand for a product and the shipment of that product must take place at the same period. But it is not realistic. A decision maker would want the goal to reach some aspiration level of the objective function or allow some violations on the constraints instead of actually minimizing the objective function and strictly satisfying the constraints. This leads to study of fuzzy linear programming. Also, supply, and market demand are considered as fuzzy data represented by intervals of tolerance. FMP models can provide a much more flexible depiction of the way a real-world system actually behaves.

Among various techniques in the literature, Zimmermann's [19], Werner's [16], Chanas's [3], and Verdegay's [17] approaches and concepts are the most typical ones for the class of FMP problems. Distribution planning system considered within this paper is modeled according to Zimmermann's approach [19], where the objective function as well as the constraints are described by fuzzy inequalities and are explicitly specified by their corresponding membership functions. An advantage of Zimmermann's approach over other methods is that different combination of membership functions and aggregating operators can be used which result in the equivalent linear models [14].

In this section, the strict requirements in the distribution planning problem are relaxed and the problem is reformulated as a fuzzy model. For constraints involving the supply and demand, we incorporate fuzzy constraints and a fuzzy goal and employ linear membership functions for them.

The objective function (1) and constraints (3) and (9) of the traditional model can be easily modified according to fuzzy set theory, to a fuzzy model. Each one of the constraints of the fuzzy model, (3) and (9), will be presented by a fuzzy set whose membership function is defined as $\mu_i(x)$:

$$\mu_i(x) = \begin{cases} 1 & if\ (Ax)_i \le b_i \\ 1 - \dfrac{(Ax)_i - b_i}{p_i} & if b_i < (Ax)_i \le b_i + p_i \\ 0 & if\ (Ax)_i > b_i + p_i \end{cases} \quad (10)$$

Where $(Ax)_i$ represents the left hand side of the fuzzy constraints, b_i represents the right hand side of the fuzzy constraints, and p_i is the tolerance level which a decision maker can tolerate in the accomplishment of the *i-th* constraint of the fuzzy inequality.

The membership function of the objective is denoted by $\mu_z(z)$ with the following definition:

$$\mu_z(z) = \begin{cases} 1 & z < z_0, \\ 1 + \dfrac{z_0 - z}{p_0} & z_0 \le z \le z_0 + p_0, \\ 0 & z_0 + p_0 < z \end{cases} \quad (11)$$

Where, p_0 is an aspiration level of objective value. According to Bellman and Zadeh [1] approach, the optimal solution of problem is such a solution of problem which simultaneously fulfills the constraints and the goal to a maximal degree. In FMP, fuzzy operators are used to transform FMP to traditional mathematical programming. In this research two types of aggregators have been applied; namely the "max-min operator" [1], and the "convex combination of min operator and max operator" [20].

Case1: Max-min operator and linear non-increasing membership function

The equivalent LP formulation can be expressed as (2), (4)-(8), (12)-(15).

If the memberships of fuzzy relationship are non-increasing linear functions, then the equivalent linear programming formulations for the three fuzzy inequalities can be expressed as:

$$\text{Max } \lambda_1 \quad (12)$$

Subject to:

$$\sum_j\sum_s\sum_k\sum_m\sum_c\sum_t FCost_{skmc} z_{jskmct} + \sum_j\sum_s\sum_k\sum_m\sum_c\sum_t (OutLCost_s + InLCost_k) z_{jskmct} + \lambda_1 P_0 \le Z_0 + P_0 \quad (13)$$

$$\sum_k\sum_m\sum_c z_{jskmct} - SI_{jst-1} + SI_{jst} + \lambda_1 P_q \le P_q \quad , \forall j,s,t \quad (14)$$

$$\sum_s\sum_m\sum_c z_{jskmct} + \lambda_1 P_r = D_{jkt} + P_r \,, \quad \forall j,k,t \quad (15)$$

Where, Z_0 is a goal of the fuzzy objective function, P_0 is an aspiration level of objective value. For specifying the parameter Z_0, it is useful to consult the computational result of crisp linear programming problem given in the previous section. P_q and P_r are the tolerance levels for the corresponding fuzzy relationships. The non-fuzzy constraints (2) and (4)-(8) are incorporated to the fuzzy model remaining unalterable. The objective function maximizes the satisfaction level of the least satisfied constraint (minimum operator). If λ is one, then all the constraints are satisfied. If they are between 0 and 1; they lie in fuzzy region. For this min-operator, the $\leq$ constraint structure and maximizing criterion are such that the maximum of the minimum of all possible values of λ will be selected [10].

Case 2: Convex combination of the min-operator and max-operator, and linear non-increasing membership function.

The main difference between this model and the previous one is given by the use as operator, for aggregation of the fuzzy objective and fuzzy constraints of the convex combination of min-operator and max-operator [20].

Aggregating the fuzzy sets using the convex combination of the min-operator and max-operator, the fuzzy set decision is defined as:

$$\mu_D(x) = \gamma \min \mu_i(x) + (1-\gamma) \max \mu_i(x) \tag{16}$$

Where $\mu_i(x)$ is the membership functions of the fuzzy sets which are aggregated using the convex combination of the min-operator and the max-operator, γ is the degree of compensation, and i denotes the i-*th* fuzzy constraint. γ is set at 0.6, an effective value in most circumstances [20]. Miller et al. [10], and Mula et al. [11] adopt this approach to solve their problems. The equivalent LP formulation consists of (2), (4)-(8), (13)-(15), and (17)-(21).

$$\text{Max} \qquad \gamma\lambda_1 + (1-\gamma)\lambda_2 \tag{17}$$

Subject to:

$$\sum_j\sum_s\sum_k\sum_m\sum_c\sum_t FCost_{skmc} z_{jskmct} + \sum_j\sum_s\sum_k\sum_m\sum_c\sum_t (OutLCost_s + InLCost_k) z_{jskmct} \lambda_2 P_0 \leq Z_0 + P_0 + M\pi_1 \tag{18}$$

$$\sum_k\sum_m\sum_c z_{jskmct} - SI_{jst-1} + SI_{jst} + \lambda_2 P_q \leq P_q + M\pi_2 \;, \quad \forall j,s,t \tag{19}$$

$$\sum_s\sum_m\sum_c z_{jskmct} + \lambda_2 P_r = D_{jkt} + P_r + M\pi_3 \;, \qquad \forall j,k,t \tag{20}$$

$$\sum_{i=1}^{n} \pi_i \leq n-1, \qquad \pi_k \in (0,1) \tag{21}$$

The objective function and fuzzy constraints are represented by two groups of constraints, one for the min operator (13), (14), (15); the other one for the max operator (18), (19), (20) and (21). λ_1 is the degree of satisfaction of the least satisfied constraint, λ_2 is the degree of satisfaction of the most satisfied constraint. The objective function maximizes the linear combination of the least satisfied constraint and the satisfaction level of the most satisfied constraint. M is a very large number.

5 Computational Results

In this section, we present computational experiments that were carried out on a real application. The illustrative case study is provided to test the model. The proposed multi-period, multi-commodity distribution planning model described above were tested on a real life project of a company that exports wheat to the customers. We choose an MS-ACCESS database to store all necessary data. The company has eight grain types, eight storage sites, four customers, two types transportation mode, four carrier types for the road transportation and three time periods.

The illustrated problem has been solved by ILOG OPL Studio Version 3.7 [6]. We made all test runs on a personal computer based on a 1.4 GHz Pentium IV processor equipped with 1 GB RAM. Cumulative indexing, compact data structure are used. Since many combinations are physically impossible, all summations run over the allowable combinations of the indices.

Table 2 summarizes the results from three different models. Model I is based upon the traditional model. Model II represents the fuzzy model with the non-increasing linear membership function and uses max-min operator. Model III represents the fuzzy model with the non-increasing linear membership function and uses convex combination of the min-operator and the max-operator. The total costs according to fuzzy LP approaches are less than that of the crisp approach. The higher cost for the crisp approach is a result of the rigidity of bounds in the linear programming model.

Table 2. Summary of Computational Results

Model	Total Cost
Model I	1751800
Model II	1477800
Model III	1469500

6 Concluding Remarks

In this paper, we proposed an efficient mathematical programming formulation to form the distribution network to minimize the total cost. We believe that the rigid requirements in this model do not accurately characterize the real life wheat distribution planning. A fuzzy linear programming model was then developed to deal with fuzziness in the light of an obscure estimation of parameters. This study provides insights that resulted in savings to the wheat company while increasing customers' satisfaction. The fuzziness considered in this paper is limited to fuzzy objective, and fuzzy constraints. Several parameters such as, in/out-load capacity, demand, supply values and carrier capacity can also be considered as a fuzzy set or fuzzy equation. The impact of different membership functions and operators can be examined on computational performance. New version of the proposed model will be formulated by considering the formulation and solution methods proposed by Delgado et al. in [5].

Acknowledgements. The authors would like to thank Dr. Koray Dogan for providing us with all data and information in the test case.

References

1. Bellman, R.E., Zadeh, L.A.: Decision making in a fuzzy environment. Management Science, 17 (1970) 141-164.
2. Bilgen, B.: Modeling of various distribution planning problems within supply chain. Unpublished Ph.D. dissertation, Dokuz Eylul University, Izmir, 2005.
3. Chanas, S., Kolodziejczyk, W., Machaj, A.: A fuzzy approach to the transportation problem. Fuzzy Sets and Systems, 13 (1984) 211-221.
4. Chanas, S., Kuchta, D.: Fuzzy integer transportation problem. Fuzzy Sets and Systems, 98 (1998) 291-298.
5. Delgado, M., Verdegay J.L., Vila, M.A.: A general model for fuzzy linear programming. Fuzzy Sets and Systems, 29 (1989) 21-29.
6. ILOG OPL Studio 3.7.: Language Manual. Gentilly, France: (2003), ILOG SA.
7. Jimenez, F., Verdegay, J.L.: Solving fuzzy solid transportation problems by an evolutionary algorithm based parametric approach. European Journal of Operational Research, 117 (1999) 485-510.
8. Lai Y.J., Hwang, C.L.: Fuzzy mathematical programming- methods and applications. Lecture Notes in Economics and Mathematical Systems, 394 (1992).
9. Liu, S-T., Kao, C.: Solving fuzzy transportation problems based on extension principle. European Journal of Operational Research, 153 (2004) 661-674.
10. Miller, W.A., Leung, L.C., Azhar, T.M., Sargent, S.: Fuzzy production planning model for fresh tomato packing. International Journal of Production Economics, 53 (1997) 227-238.
11. Mula, J., Poler, R., Garcia J.P.: MRP with flexible constraints: A fuzzy mathematical programming approach. Fuzzy Sets and Systems, 157 (2006) 74-97.
12. Sakawa, M., Nishizaki, I., Uemura, Y.: Fuzzy programming and profit and cost allocation for a production and transportation problem. European Journal of Operational Research, 131 (2001) 1-15.
13. Tanaka, H., Okuda, T., Asai, K.: On fuzzy mathematical programming, Journal of Cybernetics, 3 (1974) 37-46.
14. Tsai, C.-C., Chu, C.-H., Barta, T. A.: Modeling and analysis of a manufacturing cell formation problem with fuzzy mixed-integer programming, IIE Transactions, 29 (1997) 533-547.
15. Tzeng, G.H., Teodorovic, D., Hwang, M.J.: Fuzzy bicriteria multi-index transportation problems for coal allocation planning of Taipower. European Journal of Operational Research, 95 (1996) 62-72.
16. Werner, B.: An interactive fuzzy programming system. Fuzzy Sets and Systems, 23 (1987) 131-147.
17. Verdegay, J.L.: Applications of fuzzy optimization in operational research. Control Cybernetics, 13 (1984) 229-239.
18. Zadeh, L.A.: Fuzzy sets. Information and Control, 8 (1965) 338-353.
19. Zimmermann H.-J.: Description and optimization of fuzzy systems. International Journal of General System, 2 (1976) 209-215.
20. Zimmermann H.-J.: Fuzzy Set Theory and Its Applications. 3rd edn. Kluwer Academic Publishers, Boston (1996).

Hardware Implementation of Fuzzy Controllers-Approach and Constraints

Amol Deshmukh[1], Preeti Bajaj[2], and A.G. Keskar[3]

[1] Asst.Professor & Research Associate, ETRX Dept, G.H.Raisoni College of Engineering, Nagpur, India
amolydeshmukh@yahoo.com
[2] Professor & Head, ETRX Dept, G.H.Raisoni College of Engineering, Nagpur, India
preetib123@yahoo.com
[3] Professor & Head, ETRX & CSE Dept, VNIT, Nagpur, India

Abstract. The design approaches & constraints for hardware implementation of fuzzy logic controllers are discussed here. The digital, analog & mixed analog domains are compared. Apart from these the Memory based approach & MFC based approach are discussed. Silicon area & operational speed are the major constraints for fuzzy controllers. Most hardware implementations of fuzzy controllers use piece-wise linear functions to represent membership functions. The existing analog approach for the design of fuzzy controllers offer parallel computing with lower hardware resources but they implement much lesser rules than the fully digital ones. Compared to its analog counterpart, the digital approach has greater flexibility, easier design automation, and good compatibility with other digital systems but they are more complex and need larger chip area. Analog approach can be faster & enough flexible to compete digital fuzzy approaches. Working with digital signal limits the application to low-speed problems due to sequential calculation of fuzzy logic quantities.

1 Introduction

The increased popularity of hybrid intelligent systems in the recent times lies to the extensive success of these systems in many real world complex problems. The main reason for the success is the synergy derived by the computational intelligent components such as fuzzy logic. Fuzzy inference techniques are becoming an attractive approach to solve control and decision-making problems. This is mainly due to its inherent ability to manage the intuitive and ambiguous behavioral rules given by a human operator to describe a complex system. The application of fuzzy technologies to real-time control problems implies that hardware realizations be adapted to the fuzzy paradigm. Many microelectronics implementations of fuzzy controllers have been proposed recently. However, if fuzzy controllers are to be massively adopted in consumer products, they must fulfill some additional characteristics. First, they must be flexible, that is, suitable for adapting their functionality to different applications. This implies the capability to program the knowledge base and select different inference mechanisms. On the other hand,

B. Gabrys, R.J. Howlett, and L.C. Jain (Eds.): KES 2006, Part I, LNAI 4251, pp. 46 – 51, 2006.
© Springer-Verlag Berlin Heidelberg 2006

considering fuzzy controllers as integrated circuits, they must be efficient in terms of silicon area and operational speed. More recently, there has been considerable interest in the development of dedicated hardware implementations, which facilitate high speed processing[1,2]. However, the main drawback of such an approach is that it is only cost effective for high-volume applications. A more feasible methodology for lower volume problems demands an application of general-purpose or programmable hardware such as the FPGAs[3].

2 Methodology

Fuzzy logic controller (FLC) is composed by a knowledge base, that comprises the information given by the process operator in the form of linguistic control rules, a Fuzzification interface, which has the effect of transforming crisp data into fuzzy sets, an Inference system, that uses them joined with the knowledge base to make inference by means of a reasoning method and a Defuzzification interface, that translates the fuzzy control action so obtained to a real control action as shown in Fig.1.

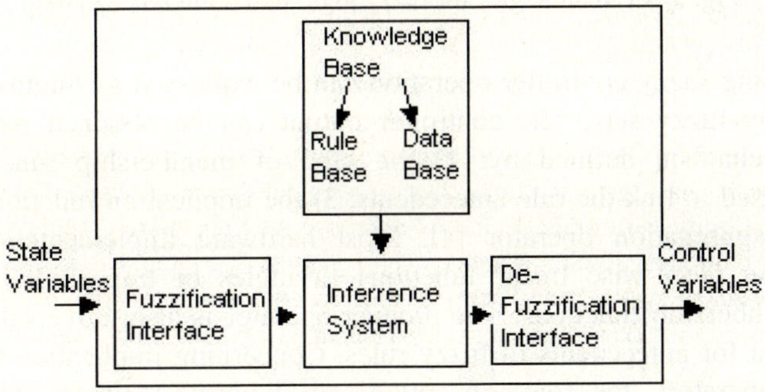

Fig. 1. Generic structure of FLC

A relevant feature of fuzzy systems is that they are universal approximators and, hence, potentially suitable for any application. The problem is how to design them. On one side, the numbers of inputs, outputs, and corresponding linguistic labels covering them have to be selected. On the other side, the types of membership functions, antecedents' connective and fuzzy implication as well as the methods of rule aggregation and defuzzification have to be chosen. VLSI implementations of fuzzy systems offer the advantages of high inference speed with low area and power consumption, but they lack of flexibility, that is, the design variables are fixed in a fuzzy chip. Since flexibility can be increased by programmability of several parameters, it is important to select efficiently these parameters to achieve a good trade-off between hardware simplicity and approximation capability.

Fig.2 explains the two basic VLSI design strategies for implementing the fuzzy logic-based inference system.

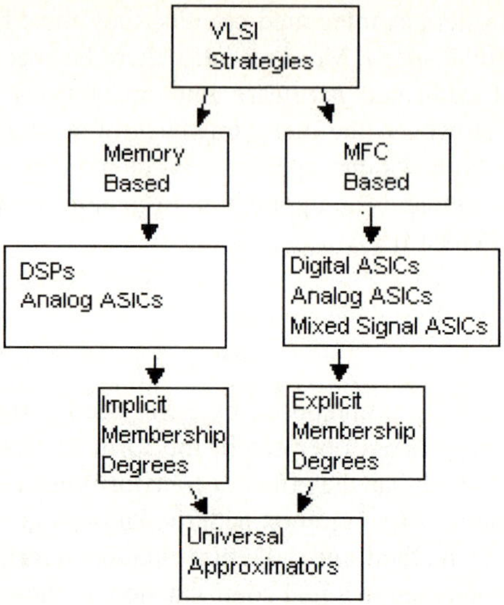

Fig. 2. VLSI strategies for fuzzy logic-based inference system

Rules describing fuzzy controller operation can be expressed as linguistics variables represented by fuzzy sets. The controller output can be obtained by applying an inference mechanism defined by: 1) the kind of membership functions; 2) the connectives used to link the rule antecedents; 3) the implication function chosen; and 4) the rule aggregation operator [4]. Most hardware implementations of fuzzy controllers use piece-wise linear functions (triangles or trapezoids, generally) to represent membership functions. The product operator is also universally adopted as the connective for antecedents in fuzzy rules. Concerning implication functions and aggregation operators, the common options are T-norms (Min or Product) and T-conorms (Max or Sum), respectively. A set of well-known inference mechanisms can be obtained by combining these operators (Min- Max, Dot-Sum, etc.).

3 Design Approach

Three different levels can be considered in the design flow of fuzzy logic controllers: algorithmic, architectural and circuit [5]. The algorithmic level specifies the functional behavior of the controller, defining the shape of the membership functions, the implication mechanism, the defuzzification method, etc. Concerning the physical realization, a particular architecture must be selected, and the required building blocks must be identified. Finally, these building blocks can be implemented as an integrated circuit according to the implementation technique & more suitable to the specific application. The design tasks at each level can be supported by different CAD tools. Traditionally, specific fuzzy tools can be used to simulate at the algorithmic level, while architectural and circuit simulations can be embedded in conventional circuit design environments.

Approaches:

1) FPGA Implementation of Fuzzy Controllers is carried out by Lago et.al [3]. Two different approaches suggested for logic synthesis of Boolean equation describing the controller input-output relations, dedicated hardware to implement the fuzzy algorithm according to a specific architecture based on VHDL cell library. Synthesis process accelerated by means of CAD tools, which translates the high level description of controller.
2) Design and application of Analog Fuzzy logic controller is being carried out by Shuwei Guo et. Al[6]. It includes the analog fuzzy logic hardware, its implementation & application to autonomous mobile system. An adjustable membership function, reconfigurable inference engine & small area defuzzification block were the features of Novel Analog controller.
3) A VHDL package for description of Fuzzy logic controllers is developed by D. Galan et.al [7]. It stressed on use of VHDL for modeling & simulation which provides a formal description of the system and allows the use of specific description styles to cover the different abstraction levels.
4) FPGA Implementation of Fuzzy systems, Neural Networks & Fuzzy Neural Networks was undertaken by J.J.Blake et.al [8]. It stressed on development of dedicated hardware implementation. Zero order Sugeno fuzzy model was considered and resulted in reduction in number of interconnects between input & hidden layers.
5) I.Baturone et.al. carried out Optimization of Adaptive Fuzzy Processor Design. Mixed signal blocks based on digitally programmed current mirrors are used & Error-descent learning algorithm is used for tuning [9].
6) Evolutionary design of Fuzzy Logic Controllers was carried out by Carlos Cotta et.al [10]. It used a class of genetic algorithms to provide suboptimal fuzzy rule bases & applied to cart centering problem. Fuzzy interpreter was used to test the resulting fuzzy rules.

Generally a digital fuzzy system is either a fuzzy (co-)processor or a digital ASIC, which contains logic circuits to compute the fuzzy algorithm, memories to store fuzzy rules, and generators or look-up tables for membership functions of the input and output variables. The digital systems are more complex and need larger chip area. Analog approach can be faster & enough flexible to compete digital fuzzy approaches. Working with digital signals limits the application to low-speed problems due to the sequential calculation of fuzzy logic quantities.

4 Constraints

The most direct strategy to approximate a given surface is to implement a look-up table that contains all the output values corresponding to every possible combination of the input signals. If there are u inputs, x=(x1, ..., xu), coded by N bits, a look-up table divides the input space into 2N•u cells and associates to each cell a constant output value. Hence, look-up tables implement piecewise constant interpolators.

The microelectronic circuits suitable to carry out this strategy are standard digital memories and PLDs. One advantage of this realization is that the response time is very short because the only operation is to retrieve data from a memory and this can be performed in a few nanoseconds. As a drawback, look-up table implementations are practical only when the number of inputs and their quantization degrees are low, otherwise the total number of bits to store is very large.

A significant weak point in analog implementations is the lack of suitable memory elements. However, in the recent times several dynamic and non-volatile analog memory circuits have been developed. This enables the implementation of compact analog architectures. Fig.3. compares the Analog & Digital approaches

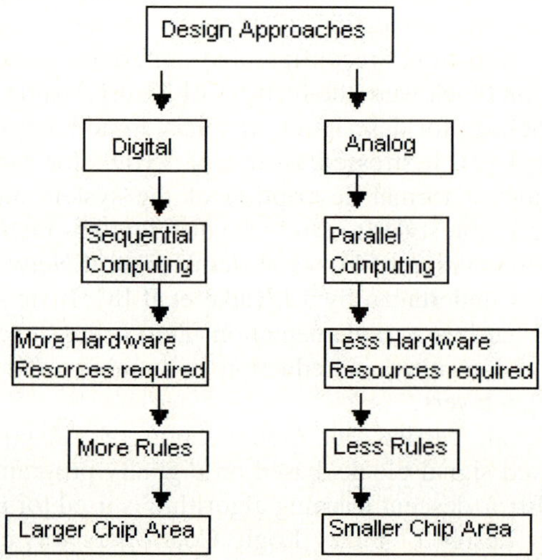

Fig. 3. Comparison of Analog & Digital Implementations

However, most of the digital systems require A/D and D/A converters to communicate with sensors and/or actuators. The hybrid or mixed signal parallel/sequential architectures can match the analog processor cost, speed and power consumption to the application constraints. Furthermore, in general, sequentialization exhibits flexibility to emulate a different number of network structures and models with much better efficiency than a fully parallel architecture [11]. Thus there is a need for mixed signal design as a complete solution for design & development of fuzzy controllers.

5 Conclusion

Generally a digital fuzzy system is either a fuzzy (co-)processor or a digital ASIC, which contains logic circuits to compute the fuzzy algorithm, memories to store fuzzy rules, and generators or look-up tables for membership functions of the input and output variables. The digital systems are more complex and need larger chip area. Analog approach can be faster & enough flexible to compete digital fuzzy approaches. Working with digital signals limits the application to low-speed problems due to the sequential calculation of fuzzy logic quantities.

References

[1] Yamakawa, A fuzzy inference engine in non-linear analog mode and its application to a fuzzy logic control, IEEE Trans. on Neural Networks,pp. 496-522, Vol. 4, No. 3, May 1993.

[2] Watanabe, et.al., A VLSI fuzzy logic controller with reconfigurable, cascadable architecture, IEEE Journal of Solid State circuits, pp.376-382, Vol. 25, No. 2, April 1990.

[3] E. Lago, M. A. Hinojosa, C. J. Jiménez, A. Barriga, S. Sánchez-Solano FPGA Implementation Of Fuzzy Controllers, XII Conference on Design of Circuits and Integrated Systems (DCIS'97), pp. 715-720, Sevilla, November 18-21. 1997

[4] C. J. Jiménez, S. Sánchez Solano, A. Barriga, Hardware Implementation Of A General Purpose Fuzzy Controller, Sixth International Fuzzy Systems Association World Congress (IFSA'95),Vol. 2, pp. 185-188, Sao Paulo - Brazil, July 21-28, 1995.

[5] A. Barriga, et.al. Automatic Synthesis Of Fuzzy Logic Controller, Mathware & Soft Computing, Vol. III, no 3, pp. 425-434, September 1996

[6] S. Guo, et.al, Design and Application of an Analog Fuzzy Logic Controller, IEEE Transactions on Fuzzy Systems, Vol. 4, No. 4, p-p. 429-438, November 1996.

[7] D. Galán, C. J. Jiménez, A. Barriga, S. Sánchez Solano VHDL Package For Description Of Fuzzy Logic Controllers, European Design Automation Conference (EURO-VHDL'95), pp. 528-533, Brighton - Great Britain, September 18-22, 1995.

[8] J.J.Blake et.al. The Implementation of Fuzzy systems, Neural Networks & Fuzzy Neural Networks using FPGAs, Information Sciences, Vol. 112, p-p151-168, 1998

[9] I. Baturone, S. Sánchez Solano, A. Barriga, J. L. Huertas, Optimization Of Adaptive Fuzzy Processor Design, XIII Conference on Design of Circuits and Integrated Systems (DCIS'98), pp. 316-321, Madrid, November 17-20. 1998

[10] Carlos Cotta, Enrique Alba and José M Troya, Evolutionary Design Of Fuzzy Logic Controllers, XIII Conference on Design of Circuits and Integrated Systems (DCIS'98),pp. 316-321, Madrid, November 17-20. 1998

[11] J. Madrenas, E. Alarcón, J. Cosp, J.M. Moreno, A. Poveda and J. Cabestany, Mixed-Signal VLSI for Neural and Fuzzy Sequential Processors, *Proceedings of the 2000 IEEE* International Symposium on Circuits and Systems - ISCAS'00, Geneva, Switzerland, June 2000

The Design of Self-tuning Strategy of Genetically Optimized Fuzzy-PI Controller for HVDC System

Zhong-xian Wang, Jueng-je Yang, and Tae-chon Ahn

Department of Electrical Electronic and Information Engineering, Wonkwang University, 344-2, ShinYong-Dong, Ik-San, Jeonbuk-Do, Republic of Korea
{tcahn, zhxwang, yjjksm}@wonkwang.ac.kr

Abstract. In this paper, we study an approach to design a self-tuning Fuzzy-PI controller in HVDC(High Voltage Direct Current) system. In the rectifier of traditional HVDC system, turning on, turning off, triggering and protections of thyristors have lots of problems that can make the dynamic instability and cannot damp the dynamic disturbance efficiently. In order to solve the above problems, we adapt Fuzzy-PI controller with and without self-tuning for the fire angle control of rectifier. The performance of the Fuzzy-PI controller is sensitive to the variety of scaling factors. The design procedure dwells on the use of evolutionary computing(Genetic Algorithms, GAs). Then we can obtain the optimal scaling factors of the FIS controller by Genetic Algorithms. Then, we adopt fuzzy logic controller to tune the scaling factors on line. A comparative study has been performed among traditional PI controller and Fuzzy-PI controller with and without self-tuning, to prove the superiority of the proposed scheme.

Keywords: Fuzzy-PI controller, Self-tuning controller, HVDC, Genetic Algorithms(GAs).

1 Introduction

From the beginning of electric power history, DC transmission lines and cables have less expensive and more advantageous than those for three-phase AC transmission. As power generation and demand are increasing, in order to handle large bulk of power, we need utilize the savings that DC transmission offers. Not only it is used for long distance power transmission, also it is being used as a part of the AC network to enhance the stability of the system[1]. But the operation and control of HVDC links pose a challenge for the designers to choose the proper control strategy under various operation conditions[2]. The HVDC system traditionally uses PI controllers to control the DC current thereby keeping the current order at the required level. However, in controlling a nonlinear plant such as the fire angle of the rectifier side, the model controls such as fuzzy controllers show better performance to the dynamic disturbances than traditional PI controllers[3].

In this paper, we study an approach to design a Fuzzy-PI controller based on Genetic Algorithms(GAs) and self tuning Fuzzy-PI controller in HVDC system. The paper also includes the experimental study dealing with the rectifier side current controller and deriving the optimal control parameters through GAs. The performance of

B. Gabrys, R.J. Howlett, and L.C. Jain (Eds.): KES 2006, Part I, LNAI 4251, pp. 52 – 59, 2006.
© Springer-Verlag Berlin Heidelberg 2006

systems under control is evaluated by the method of IAE(Integral of the Absolute value of Error)[4].

A Node Circuit Analysis simulation program was used in this study. The program has the capability of detailed modeling of transmission lines. In addition, the program is very similar with the real HVDC system[5].

2 HVDC System Model

A two pole point-to-point HVDC system has been simulated under the environment of MATLAB[5]. Each element on either side of the DC link and the transmission lines is represented in detail.

Part 1: HVDC model system

Generally speaking, the HVDC system can be divided into four parts - generator side, rectifier side, inverter side and the load bus. In this paper, we only discuss about rectifier side. It's not necessary to illustrate each part in detail in our model. So, we assume the inverter side and the load to be voltage resource. The system shown in Figure 1 is re-divided into four segments - generator, transformer including one Y-Y connection type and one Y-△connection type, the 12-pulse rectifier consisting of two 6-pulse bridges in series and voltage resource containing inverter side and the load bus[6].

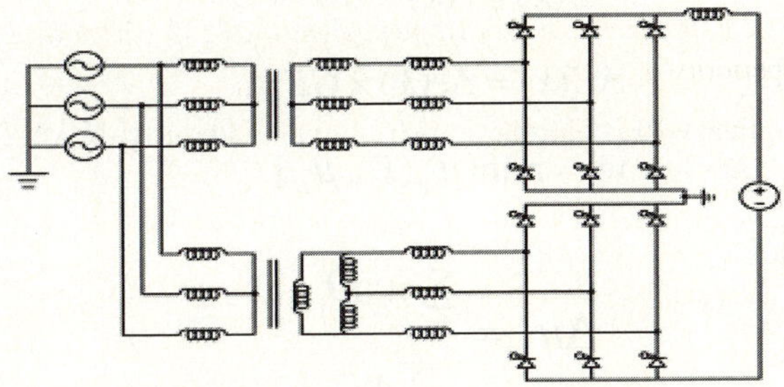

Fig. 1. HVDC real model circuit

Part 2: Rectifier control system

The characteristic curve of voltage current control is operated through the constant voltage current control of a rectifier and a constant extinction angle control of an inverter in steady state.

3 Proposed Methods of Controller Design

In order to improve the dynamic performance for rectifier current controller in the HVDC system, designed strategies of fuzzy controller with and without self-tuning part based on Genetic Algorithms(GAs) are as follows:

3.1 Fuzzy-PI Controller

The block diagram of Fuzzy-PI controller is shown in Figure 2[8][11].

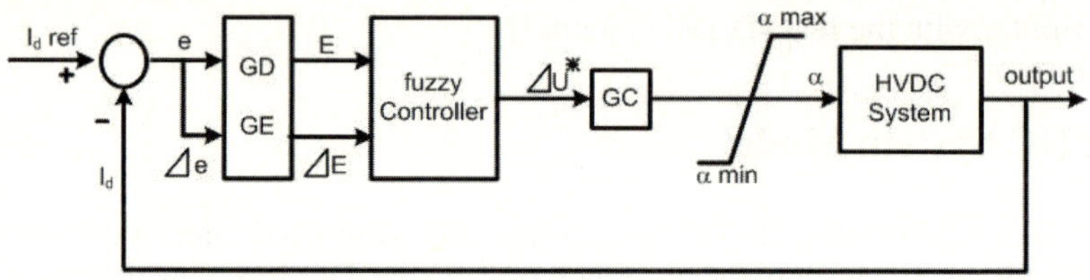

Fig. 2. Block diagram of Fuzzy-PI controller

Here, the current error(e) and its derivative error(△e) are used to adjust the input variables(E, △E) by the scaling factors(GD,GE) that we can derive from the Genetic Algorithms(GAs).

$$e(k) = I_{dref}(k) - I_d(k) \tag{1}$$

$$\Delta e(k) = \frac{e(k) - e(k-1)}{T} \tag{2}$$

$$E(k) = e(k) \times GD \tag{3}$$

$$\Delta E(k) = \Delta e(k) \times GE \tag{4}$$

$$w_i = \min[\mu_A(E), \mu_B] \tag{5}$$

$$\Delta u^* = \frac{\sum_{i=1}^{n} w_i D_i}{\sum_{i=1}^{n} w_i} \tag{6}$$

$$\alpha(k) = \Delta u^*(k) \times GC \tag{7}$$

$$\alpha(k) = \alpha(k-1) - \Delta\alpha(k) \tag{8}$$

From equations (1)~(8) and Figure 2, we know if we want to control the fire angle, we have to determine the parameters(GC,GD,GE). So, we adapted the GAs to find the optimal parameters. We will discuss about the GAs and its estimation in detail in next segment. Now, we need make a FIS(fuzzy inference system).

The above Fuzzy PI controller consists of rules of the form[7].

IF E is A^l and △E is B^l, THEN $\triangle U^*$ is C^l Where A^l, B^l and is C^l are fuzzy sets, and l=1,2,…,m.

Suppose that the domains of interval of the two input variables(E, △E) and the output variable(△U) are [-1,+1] and [-1,+1] respectively. The inputs E and △E are fuzzified into 7 sets, as shown in Figure 3(a),

NB: Negative Big, NM: Negative Medium, NS: Negative small, ZO: Zero, Ps: Positive small, PM: Positive Medium and PB: Positive Big.

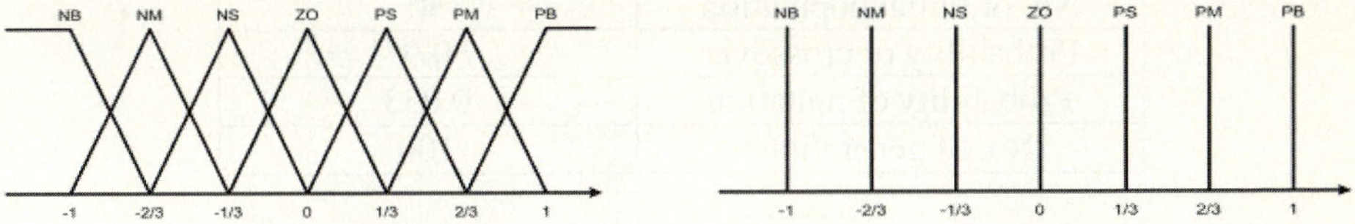

(a) E and ΔE (b) ΔU

Fig. 3. Membership functions for E, ΔE and ΔU

The membership functions are as shown in Figure 3(a). Thus, a complete fuzzy rule base consists of 49 rules. For simplicity, assume that C^l is the fuzzy sets NB(m3), NM(m2), NS(m1), ZO(0), PS(-m1), PM(-m2) and PB(-m3),whose membership functions are shown in Figure 3(b).

The collection of the rules is shown in Table 1.The min-max product by equation(5) is used for the compositional rule of inference and the defuzzification method is the center of gravity as expressed by equation(6).

Table 1. FIS rules

		ΔE						
		NB	NM	NS	ZO	PS	PM	PB
E	NB	-m3	-m3	-m3	-m3	-m2	-m1	0
	NM	-m3	-m3	-m3	-m2	-m1	0	m1
	NS	-m3	-m3	-m2	-m1	0	m1	m2
	ZO	-m3	-m2	-m1	0	m1	m2	m3
	PS	-m2	-m1	0	m1	m2	m3	m3
	PM	-m1	0	m1	m2	m3	m3	m3
	PB	0	m1	m2	m3	m3	m3	m3

3.2 Genetic Algorithms(GAs)

In this study, we use GAs to find the parameters of the scaling factors of Fuzzy-PI controller[8]. To use a GA, we need represent a solution to our problem as a genome. To find the best one, the GAs create a population of solutions and apply genetic operators such as mutation and crossover to evolve the solutions[9]. To obtain the satisfactory dynamic performance of transient process, we adapt IAE to be the smallest objective function. Select the equation(9) to be the optimal fitness of the parameter determination[4].

$$J = \int_0^t |e(\tau)| d\tau \tag{9}$$

The simulation coefficients and architectures of populations are described in Table 2

Table 2. The coefficients of GAs simulation

Items	Fuzzy-PI controller
No. of initial population	30
Probability of crossover	0.9
Probability of mutation	0.033
No. of generations	100

3.3 Self-tuning Fuzzy-PI Controller

We proposed a method of self-tuning Fuzzy-PI controller whose block diagram is shown in Figure 4.

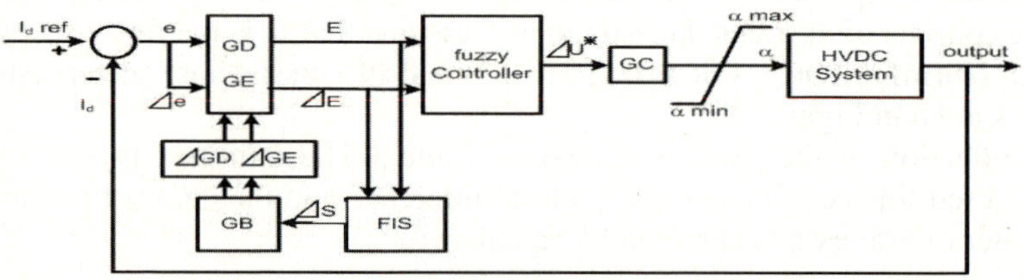

Fig. 4. Block diagram of self-tuning Fuzzy-PI controller

Here, we add a self-tuning controller base on Figure 2. We can adjust the GD and GE on-line using FIS. From equations(1)-(6), we can get,

$$GD(new)=GD(old)+\triangle GD \tag{10}$$

$$GE(new)=GE(old)+\triangle GE \tag{11}$$

$$\triangle GD=\triangle s*GB*0.001 \tag{12}$$

$$\triangle GE=\triangle s*GB*0.00001 \tag{13}$$

Here, the scaling factors(GD,GE) are derived by GAs. GB is coefficient that adjusts the $\triangle$GD and $\triangle$GE. when $\triangle S>=0$, GB is 24.5; when $\triangle S<0$ GB is /24.5.

The above FIS consists of rules of the form[7].

IF E is A^l and $\triangle$E is B^l, THEN $\triangle U^*$ is C^l

Where A^l, B^l and is C^l are fuzzy sets, and l=1,2,…,m.

Suppose that the domains of interval of the two input variables(E, $\triangle$E) and the output variable($\triangle$S) are [-1,+1] and [-1,+1] respectively. The inputs E and $\triangle$E are fuzzified into 3 sets, as show in Figure 5(a), NB: Negative Big, ZO: Zero and PB: Positive Big.

The membership functions are as shown in Figure 5(a). Thus, a complete fuzzy rule consists of 9 rules. For simplicity, assume that Cl is the fuzzy sets NB(m3), NM(m2), ZO(0), PM(-m2) and PB(-m3), whose membership functions are shown in Figure5(b).

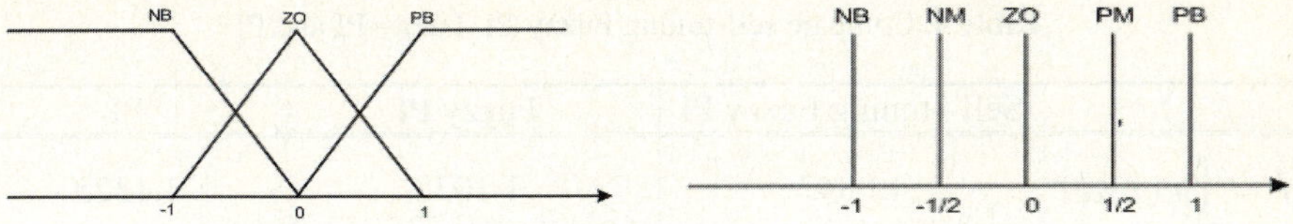

(a) E and ΔE (b) ΔS

Fig. 5. Membership functions for E, ΔE and ΔS

Table 3. The collection of the rules

E\EC	PB	ZO	NB
PB	PB	PM	ZO
ZO	PM	ZO	NM
NB	ZO	NM	NB

The collection of the rules is shown in Table 3.The min-max product by equation(5) is used for the compositional rule of inference and the defuzzification method is the center of gravity as expressed by equation(6).

4 Simulation and Studies

In last chapter, we used GAs to find the optimal parameters. Also we used FIS to tune the parameters. Then, we get the fined optimal parameters for Fuzzy-PI(GC, GD, GE), PI(Kp, Ki) controllers and self-tuning Fuzzy-PI controller(GB) using the same procedure.

GC: 0.04, GD: 0.00005, GE: 0.008 / Kp: -0.1492, Ki: -16.00 / GB: 24.5

In this part, we compare self-tuning Fuzzy-PI controller with Fuzzy-PI controller.

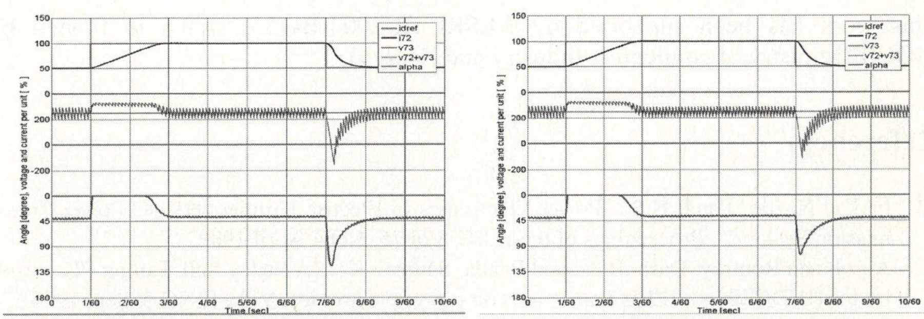

(a) Self-tuning Fuzzy-PI controller (b) Fuzzy-PI controller

Fig. 6. (a) self-tuning Fuzzy-PI controller (b) Fuzzy-PI controller based on GAs

Table 4. Compare self-tuning Fuzzy-PI, Fuzzy-PI and PI

	Self –tuning Fuzzy PI	Fuzzy PI	PI
$J = \int_0^t \lvert e(\tau) \rvert d\tau$	1,0847	1.1038	1.1323
rising time	802× Δt	1401××Δt	1403×Δt
overshoot	0.2307%	0.5545%	3.7330%
undershoot	1.3001	1.3016%	21.661%

*note: Δt = 0.25[sec]/15000[sample]

After comparison of self-tuning Fuzzy-PI, Fuzzy-PI and PI controllers, we know the self-tuning Fuzzy-PI has better performance in terms of overshoot and undershoot, as shown in Table 4. Not only there are little overshoot and undershoot in self-tuning Fuzzy-PI controller, and also IAE and rising time in self-tuning Fuzzy-PI controller is smaller. When the reference direct current changed, the voltage and alpha vary small.

5 Conclusions

In the paper, we study the Fuzzy-PI controller, self-tuning Fuzzy-PI controller and Genetic Algorithms(GAs). We use GAs to find the optimal parameters of Fuzzy and PI controller. Then, we compare the fuzzy controller with traditional PI and self-tuning Fuzzy-PI controller. Through the comparison of three types of controller in different case, we know the self-tuning Fuzzy-PI controller shows the better performances in terms of overshoot and undershoot And rising time. Therefore, the proposed design method of self-tuning Fuzzy-PI controller can be useful tools for system stability and fast damping the system disturbance in HVDC system.

Acknowledgement

This work has been supported by KESRI (R-2004-B-133), which in funded by MOCIE(Ministry of commerce, Industry and Energy).

References

1. Farhad Nozari, Patel, H.S.: "Power Electronics in Electric Utilities: HVDC Power Transmission Systems", Proceedings of the IEEE, Vol.76, NO.4 April 1988.
2. Aurobinda Routray, Dash ,P.K., and Panda, Sanjeev K. : "A Fuzzy Self-Tuning PI controller for HVDC links", IEEE Transctions on power electronics, Vol.11, NO.5, Sep, 1996.
3. Yoon, J.Y., Hwang, Gi-Hyun. and Park, June-Ho. : "A Genetic Algorithm Approach to Design the Optimal Fuzzy Controller for Rectifier Current Contol in HVDC System" Journal of Electrical Engineering and Information Science Vol.3, No.6, Page 792 ~ 797, Dec, 1998.
4. Liu, Jin-Kun Advanced PID Control with MATLAB Simulation, Electronic Industry Publishing Company, October,2002.

5. Ahn, Tae-Chon. and so on. : "Development of Adaptive Granular Control Technique for HVDC Transmission System" report, Feb, 2005.
6. Dash, P.K., Liew, A.C. and Routray, A.: "HIGH - performance controllers for HVDC transmission links "IEE Pro-Gener.Transm.Distrib., Vol.141, No.5, Sep, 1994.
7. Wang, Li-Xin. "A Course of Fuzzy System and Control" PRENTICE HALL 1997.
8. Oh, Sung-Kwun., Pedrycz, Witold., Rho, Seok-Beom. and Ahn, Tae-Chon. "Parameter Estimation of Fuzzy Controller and Its Application to Inverted Pendulum" Engineering Applications of Artificial Intelligence Volume 17, Issue 1 , February 2004, Pages 37-60.
9. Zhou, Xiao-Xin. "The Analysis and Control of AC/DC Power System", 2004.
10. Dash, P.K., Routray, Aurobinda., Panda, S.K. and Liew, A.C.:"FUZZY TUNING OF DC LINK CONTROLLERS" IEEE Catalogue No. 95TH8130
11. Wang, Zhongxian., Yang, Juengje., Rho, Seokbeom. and Ahn, Taechon., "A New Design of Fuzzy controller for HVDC system with the aid of GAs", Journal of Control, Automation, and Systems Engineering, ISSN 1225-9845, Vol.12, No.3, Page 221 ~ 226, March, 2006.

A Hybrid Evolutionary Algorithm for the Euclidean Steiner Tree Problem Using Local Searches*

Byounghak Yang

Industrial Engineering, Kyungwon University, Bockjung-dong San 65, Sujung-gu, Seongnam-si, Kyunggi-do, Korea
byang@kyungwon.ac.kr

Abstract. In this paper we introduce a hybrid evolutionary algorithm for the Euclidean Steiner tree problem (ESTP) which is to find a minimum-length Euclidean interconnection of a set of terminals in the plane. The individual is an assembly of locations of a non-fixed number of Steiner points and number of Steiner points. We use the operators crossover, mutation and selection. A Steiner points pool is introduced base on the optimal 3-terminal Steiner points generation and the minimum spanning tree. An initial population is generated by random selecting from the Steiner points pool. To improve the solution quality, some Steiner points in individual are deleted and rearranged. Randomly generated Steiner points are inserted in selected individual to search a new solution. Experimental results show that the quality of solution is improved by the hybrid operator. The gap between the optimal solution and the solution of hybrid evolutionary algorithm is less than 0.3%.

1 Introduction

Given a set V of n terminals in the plane, the Euclidean Steiner tree problem (ESTP) in the plane is to find a shortest network interconnecting V while allowing for addition of auxiliary points called Steiner points. In ESTP, Steiner points are unknown, therefore we should find the optimal number of Steiner points and their location on Euclidean plane. There are three kinds of variants of Steiner tree problem. First one is ESTP, second one is the rectilinear STP (RSTP) and last one is the STP in graph

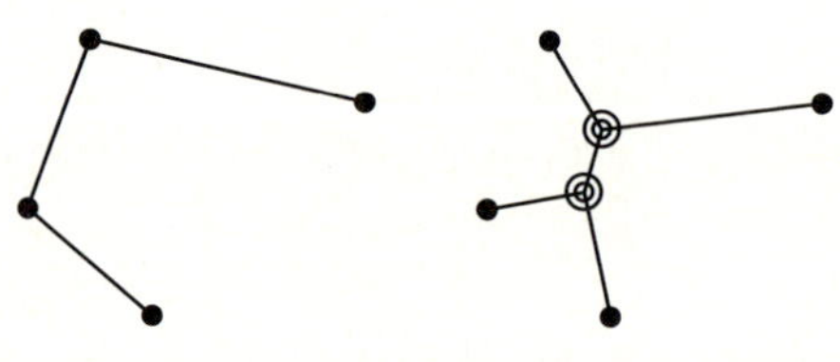

(a) (b)

Fig. 1. A minimum Euclidean spanning tree(a) and a minimum Euclidean Steiner tree(b)

* This research was supported by the Kyungwon University Research Fund in 2006.

B. Gabrys, R.J. Howlett, and L.C. Jain (Eds.): KES 2006, Part I, LNAI 4251, pp. 60 – 67, 2006.
© Springer-Verlag Berlin Heidelberg 2006

(GSTP). In GSTP, the set S and the location of Steiner points are given. It is well-known that the solution to ESTP will be the minimal spanning tree (MST) on some set of points V∪S while a set S is given Steiner points set. Let MST(V) be the cost of MST on set of terminals V. Then the reduction rate R(V) = {MST(V∪S) - MST(V) } / MST(V) is used as performance measure for algorithms on the Steiner tree problem. Du and Hwang [5] showed the upper limit of reduction rate for the Euclidean Steiner tree is at most 13.4%. The ESTP is known to be NP-complete [6]. Warme, Winter and Zachariasen [16] presented an exact algorithm for ESTP and showed that the average reduction rate of optimal solutions was about 3.3% for the OR-library instances. Since comprehensive survey of work relating to Steiner tree problems has been given by Hwang, Richards and Winter [8] and Zachariasen [19], we shall not give a complete literature. Beasley [3] presented standard instances for Steiner tree problem and heuristics algorithms [4] for the ESTP with Goffinet. Wakabayashi [15], Julstrom [11] Yang [18] introduced genetic algorithms (GA) for RSTP. Sock and Ahn [13] used genetic algorithm to solve the degree constrained minimum spanning tree. Some GA for ESTP were presented. Hesser [7] used an individual as an assembly of the (x,y) positions of a fixed number of Steiner points. Joseph and Harris [10] introduced an individual which is an array of the (x,y) position of Steiner points and some extra flags used in Steiner point generation. They mentioned that the solution quality of their GA was not so good and the problem lies in how they generate Steiner points, not in how the GA operators work. Barreiros [2] assumed the number of Steiner point is equal to the number of the problem terminals. Two consecutive Steiner points are always considered to be connected and these points form a 'backbone' of the Steiner tree. Every terminal is always considered to be connected to the nearest Steiner points in backbone. He used this backbone as individual in GA for small sized problem. For solving the normal sized problem, he first divided terminals to small sub set to solve small sized problem and then connected small tree to original tree. Main motivation of this research is to develop a hybrid evolutionary algorithm to find near optimal solution for the ESTP.

Hwang et al[8] showed that the fact that all of the terminal will be of degree 1,2, or 3, the Steiner points are all of degree 3, any two edges meet at an angle of al least 120° in the minimal Steiner tree, and at most n-2 Steiner points will be added to the tree. Torricelli [9] introduced an optimal solution for the 3 terminals case with points A, B, and C. His solution involved forming a triangle out of the three points and on each newly formed edge an equilateral triangle was created. Each newly formed equilateral triangle (ABX) was then circumscribed with a circle. The resulting intersection of the circles and line CX is known as an optimal Steiner point as Torricelli point [9]. Simpson [9] devised an alternate method of constructing a Steiner point. As with Torricelli, Simpson created equilateral triangles on each edge of triangle ABC. Then a Simpson line is drawn between the vertex of each equilateral triangle that is not part of the original triangle ABC and the opposite vertex of triangle ABC. The intersection of these three Simpson lines is the Steiner point [9].

2 Evolutionary Algorithm

Evolutionary algorithms are based on models of organic evolution. They model the collective learning process within a population of individuals. The starting population is initialized by randomization or some heuristics method, and evolves toward successively better regions of search space by means of randomized process of recombination, mutation and selection. Each individual is evaluated as the fitness value, and the selection process favors those individuals of higher quality to reproduce more often than worse individuals. The recombination mechanism allows for mixing of parental information while passing it to their dependents, and mutation introduces innovation into the population. Although simplistic from a biologist's viewpoint, these algorithms are sufficiently complex to provide robust and powerful adaptive search mechanisms [1]. In this research, the set of locations of Steiner points in Euclidean space become an individual and the individual is evaluated by minimum spanning tree with terminals and Steiner points in the individual.

2.1 General Evolutionary Algorithm

In this research, the individual is an assembly of (x_i,y_i) of a non-fixed number of Steiner points and number of Steiner points as follows ; $\{m, (x_1,y_1)$, (x_2,y_2) ,……, (x_m,y_m) } where m is the number of Steiner points. Each individual has different length depending on its number of Steiner points, and represents one Steiner Tree. The fitness of the individual corresponds to the length of the MST that can be constructed by using the original terminals and the Steiner points in the individual by the Prim algorithm [12].

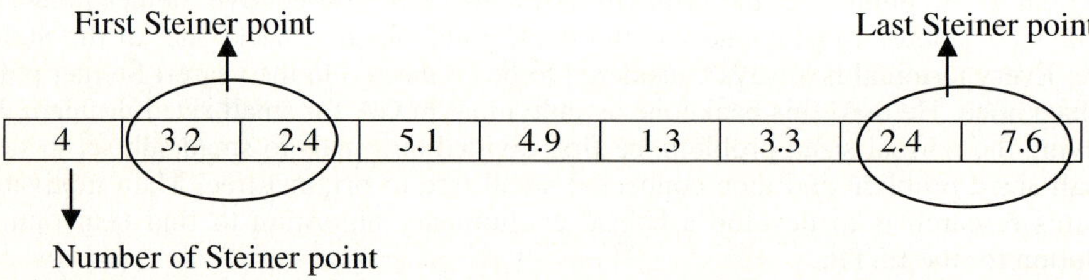

Fig. 2. The individual for evolutionary algorithm

The recombination or crossover operator exchanges a part of an individual between two individual. We choose some Steiner points in each individual and perform arithmetic crossover with other Steiner points in the other individual. We have two parents to crossover as P1={3, (1.2,3.6), (2.5,4.7), (3.6,2.4)} and P2={2, (6.4,4.1), (7.1,8.5)}. By the random number, we choose the number of exchanging which is less than the number of Steiner points in both parents. Then by the random number, some Steiner points are choose and exchanged. In this case, we choose the 2 as exchanging number. First and third Steiner points are selected in P1, and two Steiner points are selected in P2. Therefore we have two children by performing arithmetic crossover as follows: O1={3, (2.7,3.2), (2.5,4.7), (2.8,2.1)} and O2={2, (5.3,3.8), (8.5,9.1)}.

The mutation operator changes the locations of some selected Steiner points. Let (x_i,y_i) be the selected location of Steiner point. Then we can have new location (x_i+v,y_i+h) where v and h are some random number.

We use the tournament selection. The k-tournament selection method select a single individual by choosing some number k of individuals randomly from the population and selecting the best individual from this group to survive to the next generation.

The process is repeated as often as necessary to fill the new population. A common tournament size is k=2.

For the 3-terminal ESTP, the Torricelli point is the optimal Steiner point. We assume that the Torricelli point of some adjacent 3-terminal has high probability to be survived in the optimal Steiner tree. For the convenience to find adjacent 3-terminal, we make a minimum spanning tree for the V and find every directly connected 3 terminals in the minimum spanning tree as like Fig. 3(b). And the Torricelli point for these connected 3 terminals is calculated. We make Steiner point for every connected 3 terminals in minimum spanning tree, and put in Steiner points pool as like Fig. 3(c) and use one of candidate Steiner points. We make initial population for the evolutionary algorithm as following procedure:

Step 1: Make a minimum spanning tree (MST) for the terminals.
Step 2: Choose every connected three terminals in MST. Let them be 3-neighbor terminals (3NT).
Step 3: Make a Steiner point for each 3NT and add it in the Steiner points pool.
Step 4: Choose some Steiner points from Steiner points pool by random function and make one individual. Repeat Step 4 until all individuals are decided.

We make an initial population with 200 individuals.

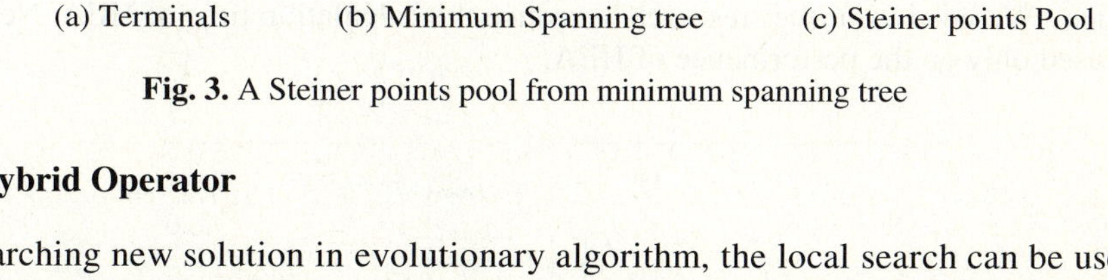

(a) Terminals (b) Minimum Spanning tree (c) Steiner points Pool

Fig. 3. A Steiner points pool from minimum spanning tree

2.2 Hybrid Operator

For searching new solution in evolutionary algorithm, the local search can be used. In this research, an insertion operator, a deletion operator and a location rearrangement operator are introduced. For some selected individual, randomly generated Steiner points in Euclidean space are inserted in current tree. Some Steiner points in tree are connected with only one or two other node. Those kinds of Steiner points in tree are obviously poor to reduce the tree cost. We introduce a deletion operator to delete poor Steiner points which are connected less than 2 other node in Steiner tree. For the Steiner point connected with 3 other node, the optimal location of Steiner point is found by the Torricelli point for the 3-terminal case. We trace the Steiner point which is connected with exactly 3 other node(A,B,C) and calculate the optimal location (S) of Steiner point for those 3 node(A,B,C) and move the location of Steiner point to S*

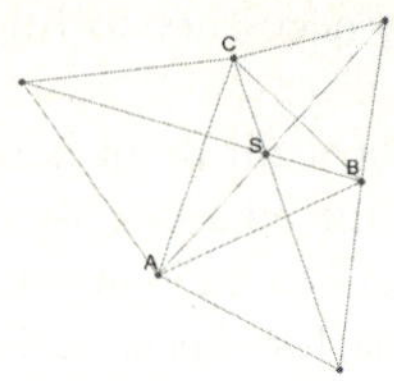

Fig. 4. The location rearrangement based on 3-terminal optimal Steiner points. S is the optimal Steiner point for node (A,B,C)

2.3 Alternative Algorithms

We introduced two evolutionary algorithms. First one is the general evolutionary algorithm (GEA) without the hybrid operator. Second one is the evolutionary algorithm (HEA) with the hybrid operator which is proposed in 2.2.

3 Computational Experiment

The computational study was made on Pentium IV processor. The evolutionary algorithm was programmed in Visual Studio. Problem instances are from OR-Library [3], 15 instances for each problem size 10, 20… 100, 250, 500. First computational experiment is to compare the reduction rate and the calculation time of the GEA and those of HEA for the 10 to 100 terminal instances. The results are shown in Fig. 5. For the reduction rate, HEA is very close to the optimal solution and GEA worse than the other heuristics. For the calculation time, GEA is a little bit better than HEA. Even the calculation time of HEA is loner than that of GEA, the solution quality of HEA is quite good for us to accept HEA as candidate method to compare with the other heuristics. We need to further research to reduce the calculation time of HEA. Next tests focused only on the performance of HEA.

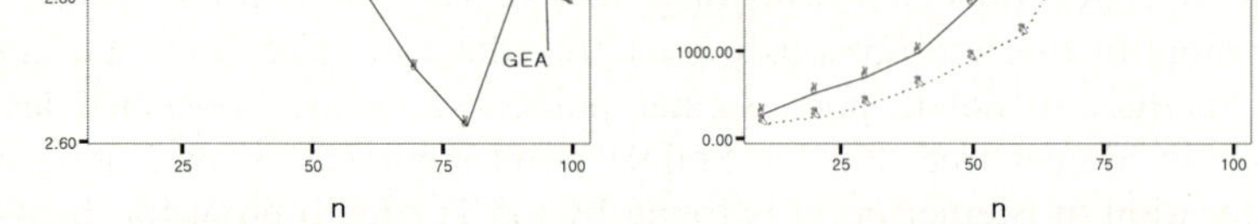

(a) Reduction rate (b) Calculation time

Fig. 5. The reduction rate and the calculation time of HEA and GEA

We compare four other heuristics with HEA. First one heuristic is introduced by-Beaseley and Goffinet [4], the second one is repeated descent (RD) heuristics, the third one is simulated annealing (SA) and the last one is tabu search (TS). Zachariasen showed the results of those four heuristics [19]. In Table 1, the gray area shows relatively better reduction rate in algorithms. The HEA gives larger percentage reduction in most instances than any other heuristic. Table 2 shows the Steiner tree cost and the % gap between the solution of HEA and the optimal solution. The Steiner tree cost of HEA is 99.98% of the optimal solution. And the worst case of % gap is 0.31%. Therefore even the worst case solution of HEA is 99.7% of the optimal solution. The 46 test problems for optimal solution were given by Soukup and Chow [14]. Warme [17] showed the optimal solutions of these problems. The 46 test problems were solved by HEA and GEA to compare with the optimal solutions. GEA failed to find optimal solution for 11 instances, but HEA found every optimal solution in 46 instances.

Table 1. The comparison of reduction rate of Beasley's, RD,SA,TS, HEA and the optimal solution

n	Beasley and Goffinet	RD	SA	TS	HEA	OPT
10	3.22	3.23	3.23	3.23	3.25	3.25
20	3.12	3.15	3.16	3.14	3.16	3.16
30	2.95	3.06	3.06	3.02	3.07	3.07
40	2.97	3.12	3.12	3.07	3.14	3.14
50	2.92	3.03	3.02	3.00	3.03	3.03
60	3.18	3.27	3.27	3.21	3.27	3.27
70	2.95	3.11	3.10	3.03	3.11	3.11
80	2.92	3.03	3.03	2.98	3.03	3.04
90	2.95	3.11	3.10	3.02	3.11	3.12
100	3.07	3.25	3.24	3.15	3.26	3.27
250	-	3.17	3.17	3.08	3.17	3.21
500	-	3.27	3.30	3.20	3.13	3.33

Table 2. The avearge tree cost of HEA and the optimal solution

n	HEA	Optimal	%gap of HEA
10	1.942215349	1.942215349	0
20	2.87810175	2.87810175	0
30	3.593265915	3.593073382	0.005358441
40	4.311809942	4.311739882	0.001624864
50	4.698299923	4.698293543	0.000135788
60	5.016830578	5.016418499	0.008214592
70	5.479710322	5.479394647	0.005761123
80	5.923445092	5.922725505	0.012149592
90	6.217807579	6.217427048	0.006120401
100	6.624427992	6.623445917	0.01482725
250	10.21870438	10.21579968	0.028433365
500	14.2563492	14.22732965	0.203970433
mean	5.930080668	5.927163738	0.023882987

4 Conclusion

In this paper we have introduced a hybrid evolutionary algorithm for the Euclidean Steiner tree problem. There were a couple of genetic approaches for ESTP or RSTP. They focused on the representation of Steiner tree in genetic algorithm. On the other hand, we focused on how we generate or search better Steiner points. For generating new Steiner point and improving the solution quality in evolutionary strategy, we introduced the hybrid operator which consisted of the insertion, the deletion and the Steiner points rearrangement. Computational results showed that the hybrid evolutionary strategy gives better Steiner tree solution than the other heuristics. As comparing with optimal solution, our hybrid evolutionary strategy found every optimal solution among 46 instances. And the worst gab between the values of optimal solution and that of evolutionary strategy is less than 0.31%. Our evolutionary strategy should solve the minimum spanning tree problems for the all individual in population; therefore the calculation time of evolutionary strategy is much bigger than that of the non-genetic heuristic algorithm. We are trying to modify our methods to reduce the calculation time to adapt our algorithm to ESTP with side constraint.

References

1. Bäck, T. : Evolutionary Algorithm in Theory and Practice, Oxford University Press (1996)
2. Barreiros, J. : An Hierarchic Genetic Algorithm for Computing (near) Optimal Euclidean Stein Steiner Trees. *Workshop on Application of hybrid Evolutionary Algorithms to NP-Complete Problems*, Chicago, 2003
3. Beasley, J. E.: OR-Library: Distributing Test Problems by Electronic Mail. *Journal of the Operational Research Society* 41 (1990) 1069-1072
4. Beasley, J. E, Goffinet, F. : A Delaunay Triangulation-based Heuristic for the Euclidean Steiner problem. *Networks* 24 (1994) 215-224
5. Du, D. Z., Hwang, F. K. : A Proof of Gilbert and Pollak's Conjecture on the Steiner Ratio. *Algorithmica* 7 (1992) 121-135
6. Garey, M. R., Graham, L. R. , Johnson, D. S. : The Complexity of Computing Steiner Minimal Trees. *SIAM Journal on Applied Mathematics 32* (1977)835-859
7. Hesser, J., Manner R., Stucky, O. : Optimization of Steiner Trees using Genetic Algorithms, *Proceedings of the Third International Conference on Genetic Algorithm* (1989) 231-236
8. Hwang, F. K., Richards, D. S., Winter, P.: The Steiner Tree problem. *Annals of Discrete Mathematics* 53(1992) .
9. Ivanov, A. O., Tuzhilin, A. A. : Minimal Networks: The Steiner Problem and its Geberalization, CRC Press, (1994)
10. Joseph, J., Harris, F. C. : A Genetic Algorithm for the Steiner Minimal Tree Problem. *Proceedings of International Conference on Intelligent Systems* (1996)
11. Julstrom, B.A. : Encoding Rectilinear Trees as Lists of Edges. *Proceedings of the 16th ACM Symposium on Applied Computing* (2001) 356-360
12. Prim, R. C. : Shortest Connection Networks and Some Generalizations. *Bell System Technical Journal* 56 (1957) 1389-1401
13. Sock, S.M., Ahn, B.H. : A New Tree Representation for Evolutionary Algorithms. *Journal of the Korean Institute of Industrial Engineers* 31(2005) 10-19

14. Soukup, J., Chow, W.F. : Set of Test Problems for the Minimum Length Connection Networks. *ACM/SIGMAP Newsletter* 15(1973) 48-51
15. Wakabayashi, S. : A Genetic Algorithm for Generating a Set of Rectilinear Steiner Trees in VLSI Interconnection Layout. *Information processing Society of Japan Journal* 43(5), 2002
16. Warme, D.M., Winter, P., Zachariasen, M. : Exact Algorithms for Plane Steiner Tree Problems : A Computational Study In: D.Z. Du, J.M. Smith and J.H. Rubinstein (eds.): Advances in Steiner Tree, Kluser Academic Publishers (1998)
17. Warme, D.M. : Http://www.group-w-inc.com/~warme/research
18. Yang, B.H. : An Evolution Algorithm for the Rectilinear Steiner Tree Problem. *LNCS* 3483(2005) 241-249
19. Zachariasen, M. : Local search for the Steiner tree problem in the Euclidean plane. *European Journal of Operational Research* 119 (1999) 282-300

Use of Cluster Validity in Designing Adaptive Gabor Wavelet Based Face Recognition

Eun Sung Jung and Phill Kyu Rhee

Dept. of Computer Science & Engineering Inha University
Yong-Hyun Dong , Incheon, South Korea
eunsung@im.inha.ac.kr, pkrhee@inha.ac.kr

Abstract. Face images in various situations due to facial expression, view point, illumination conditions, noise, etc. make identification process difficult. In this paper, the situation information of face images, what we call image context, is used to improve performance of a face recognition system. The proposed system partitions face images into several image contexts (groups) based on cluster validity, and takes adaptation to individual partitioned groups. In Gabor wavelet based face recognition, we apply weights to individual elements of facial feature, and those weights are trained by Genetic algorithm. We tried to use several unsupervised learning methods, clustering algorithms here, to partition face images into proper image contexts. There exists no formal way to decide the suitability of clustering algorithms for aiming at high recognition rate. We discuss about the process of cluster evaluation using the proposed cluster validity measure in designing adaptive face recognition. We achieved encouraging results though extensive experiments.

1 Introduction

Face recognition is a difficult problem because there are numerous variations in images for same person. It is caused by facial expression, view point, illumination conditions, noise, etc. There are two ways to solve this problem. One is to find features independent from variations of situations, or preprocessing to make normalized image [1, 2, 3]. And the other is to define situation specific actions to recognize a face properly [4, 5]. We take the second way. We applied face recognition process to concept of the *Context awareness* [6]. We use situation information of face images as a *context*, and use the context information to select a proper action method for face recognition.

Context awareness is an emerging application area with the aim of easing man-machine interaction [6]. For example in the case of a mobile device the man-machine interaction can be given. Mobile device recognize owners status, like location, health, etc. And it takes action proper for recognized situation. Crowley suggested a framework for context aware observation of human activity [6]. In context awareness, situation models are used to specify an architecture in which reflexive processes are dynamically composed to form federations for observing and predicting situations.

We applied context awareness to Gabor feature based face recognition process. In the proposed approach, variations of images are defined as image contexts. Image

B. Gabrys, R.J. Howlett, and L.C. Jain (Eds.): KES 2006, Part I, LNAI 4251, pp. 68 – 77, 2006.
© Springer-Verlag Berlin Heidelberg 2006

contexts are tried to be recognized by a context recognition method in order to select proper actions on face recognition. We construct different Gabor feature space for each image context (cluster). We found that the result of the context recognition representing image contexts affects the performance of the system. In the process of Gabor feature based face recognition, we assigned weights to each facial feature point that is derived from the cluster validity step. Every image context has its own optimized Gabor feature space that is produced by Genetic algorithm. It corresponds to action for context adaptation. Each weight sets are learned by Genetic algorithm. It corresponds to context adaptation. Through this processes, a face recognition system achieves robustness against various situations like illumination, facial expression, face color, etc.

For *context recognition*, we use unsupervised learning method, k-means algorithm here. There are many possible clustering results that can be constructed. It is, however, not easy to decide which clustering result is suitable for high recognition rate. Furthermore, if *context adaptation* process takes long time, it is impossible to test all possible clustering results. Hence, it is needed to predict performance of the process after adaptation process. We'll attack this problem with cluster evaluation using cluster validity measure, by analyzing with results after adaptation.

At section 2, 3, we'll show an unsupervised learning algorithm as context recognition, and analyze the result by cluster validity approach. At section 4, we'll show the context aware Gabor feature based face recognition process in which feature weights are trained by Genetic Algorithm, and at section 5, analyze the results with the cluster analysis results.

2 Context Recognition by Unsupervised Learning

Context recognition, the first step of context awareness plays an important role in context awareness. Thus, selecting algorithms, features, parameters for context recognition is very important for high performance of the entire system. It must have a relation with action methods which will be applied at the last step of context aware adaptation.

We adopt unsupervised learning clustering algorithm for context recognition process. Especially we use k-means clustering algorithm. Its goal is to find the *k* mean vectors $\mu_1, \mu_2, \ldots, \mu_k$, which are related to each cluster members. Let $\{x_i, i = 1,2,\ldots,n\}$ be the set of *n* patters. Let denote *j*th feature of x_i. Define for $i = 1,2,\ldots,n$ and $k = 1,2,\ldots,K$,

$$w_{ik} = \begin{cases} 1, & \text{if } i\text{th pattern belongs to } k\text{th cluster,} \\ 0, & \text{otherwise.} \end{cases} \tag{1}$$

Then, the matrix $W = [w_{ij}]$ has the properties that

$$w_{ij} \in \{0,1\} \text{ and, } \sum_{j=1}^{K} w_{ij} = 1. \tag{2}$$

All objects are divided into K clusters such that some metric relative to the mean vectors of the clusters is minimized. The algorithm is:

Algorithm: *k*-Means Clustering.

```
begin initialize n, c, μ_1, μ_2, ..., μ_k
        do classify n samples according to nearest μ_i
              recompute μ_i
        until no change in μ_i
   return μ_1, μ_2, ..., μ_k
end
```

We used coordinates of facial feature points for feature set of context recognition. These are obtained by chaining coordinates of facial feature points.

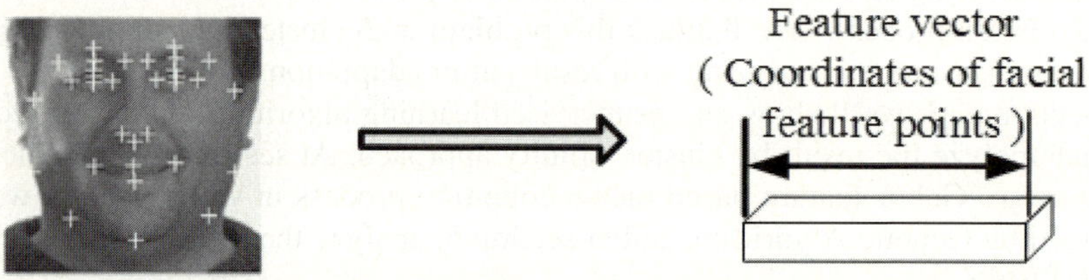

Fig. 1. Feature set for clustering (Coordinates of facial feature points)

We expect that we can get good clustering results which are reflecting variation of situations, like illumination, facial expression, by using these three feature sets. Whether or not we are satisfied to clustering result, it is not important problem for our primary goal - improving performance of the system, or raising recognition rate of face recognition. In next section we'll show the process of analyzing cluster results. And that will be compared with results of context adaptable face recognition.

3 Measurement of Cluster Validity

We use unsupervised learning algorithm for context recognition to recognize situation information of face images. The face recognition system is adapted to each situation. So, the roll of context recognition process is very important to improve performance of the system. But, because performance of the system can be decided after adaptation process, we can't decide the suitability of a clustering algorithm. We tried to solve this problem by cluster validity approach.

The main subject of cluster validity is evaluating and assessing of clustering results. There are two approaches to investigate cluster validity [7]. The first approach, *external criteria* based approach takes advantage of pre-specified structure, which is imposed on a data set and reflects our intuition onto the clustering structure of the data set. In the second approach, *internal criteria*, the clustering results are evaluated in terms of quantities that involve the vectors of the data set themselves. Since we don't have pre-specified, we adopted the second approach, internal criteria. Especially, we investigated Dunn's method [8, 9, 10] which is commonly used.

Dunn's Measure
Let $C=\{C_1,...,C_k\}$ be a clustering of a set of objects D, $\delta : C\times C \rightarrow \mathrm{R}$ be a cluster to cluster distance measure, and $\Delta : C \rightarrow \mathrm{R}$ be a cluster diameter measure. Then all measures I of the form

$$I(C)=\frac{\min_{i\neq j}\{\delta(C_i,C_j)\}}{\max_{1\leq l\leq k}\{\Delta(C_l)\}} \tag{3}$$

are called *Dunn indices*.

Originally, Dunn used

$$\begin{aligned}\delta(C_i,C_j)&=\min_{x\in C_i,y\in C_j} d(x,y) \quad \text{and}\\ \Delta(C_i)&=\max_{x,y\in C_i} d(x,y)\end{aligned} \tag{4}$$

Where $d : D\times D \rightarrow \mathrm{R}$ is a function that measures the distance between objects of D. With this settings the measure yields high values for clustering with compact and very well separated clusters.

Enhanced Dunn's Measure
Bezdek recognized that the index is very noise sensitive. Bezdek experienced that the combination of

$$\begin{aligned}\Delta(C_i,C_j)&=\frac{1}{|C_i||C_j|}\sum_{x\in C_i,y\in C_j} d(x,y) \quad \text{and}\\ \Delta(C_i)&=2\left(\frac{\sum_{x\in C_i} d(x,c_i)}{|C_i|}\right)\end{aligned} \tag{5}$$

gave reliable results for several data sets from different domains [11]. Here, c_i denotes the centroid of cluster C_i.

But, with these general cluster validity measures which are just measuring structural properties of clustering result, the link between cluster validity and performance of the system cannot be found. We will devise a new measure which is suitable for our face recognition system.

Proposed Cluster Validity Measure for Adaptive Face Recognition
In our face recognition system, distances of Gabor vectors extracted from each facial feature are used to find the nearest face image. We have found that there are feature points which are more important for right face recognition, and in different situations (illumination condition, facial expression, noise, etc.), these important feature sets are different. On the process of adaptation, high values of weights are assigned to important facial feature points of each situation. In our cluster validity measure, the importance of a

facial feature set is decided with the variance of distances between enrolled face images and test images in the adaptation data set. A feature with low value of variance is regarded as more important.

At first, for each clusters, and each facial feature points, standard deviations of distances between Gabor vectors extracted from enrolled image and test image in adaptation set are computed. The equation below shows the standard deviation *S* of cluster *c*, feature point *p*:

$$S_{cp} = \sqrt{\frac{\sum_{i=1}^{N_c} \left(d(x_{ip}, y_{ip}) - \overline{d(x_p, y_p)} \right)^2}{N_c}} \tag{6}$$

In each cluster, all *S* for features are summed, and multiplied by number of cluster. It is summed together for all clusters, and divided by number of all vectors. So cluster validity *V* is:

$$V = \frac{1}{\left(\frac{\sum_{i=1}^{C} \sum_{j=1}^{P} |C_i| S_{ij}}{N} \right)} \tag{7}$$

C denotes number of clusters, *P* denotes number of facial features, and *N* denotes number of all vectors.

4 Design of Adaptable Face Recognition Using Cluster Validity

We applied the concept of context-aware adaptable system to Gabor feature based face recognition. Image Context is recognized in the way described in the section 3 for the context adaptation. We construct different feature weights for each image context. Thus, clustering result representing image contexts affects the performance of the system. Since the process of finding the weights of Gabor feature space using Genetic algorithm requires too much time, it is impossible to test all possible clustering methods. We try to solve this problem through the evaluation of clustering results using cluster validity measures. In the process of selecting feature weights for Gabor feature based face recognition, we use optimal clustering result that is derived from the cluster validity step.

Proposed Scheme face Recognition using Cluster Validity Measure

Fig.2 shows the whole process of our system roughly. In the process of context adaptation, contexts are recognized by clustering algorithms, i.e. k-means algorithm here. Clustering result is used to find the weights of Gabor feature space for each image context. Because the process of finding feature weights using Genetic algorithm requires too much time, it is impossible to test every clustering result. We try to solve

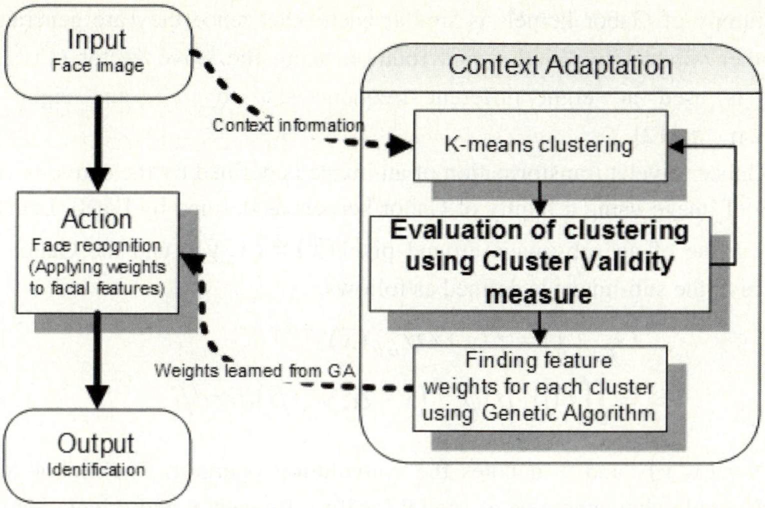

Fig. 2. The diagram which shows the whole process of context adaptation in our face recognition system

this problem through evaluation of clustering results using cluster validity measures. Every image context has its own weight set that are learned by Genetic algorithm. Fitness function of Genetic algorithm is recognition rate and scattering function testing on every cluster members using learned weights. It corresponds to context adaptation. Through this processes, a face recognition system robust against various situations like illumination, facial expression, face color, etc., can be produced.

Gabor Wavelet based Feature Representation

The proposed face recognition employs Gabor feature vector, which is generated using the Gabor wavelet transform. The kernels of the Gabor wavelets show biological relevance to 2-D receptive field profiles of mammalian cortical cells [12]. The kernels show desirable characteristics of spatial locality and orientation selectivity, a suitable choice for face image feature extraction for classification. The Gabor kernels for a facial feature point are defined as follows [12]:

$$\psi_{\mu,\nu}(\vec{x})=\frac{\left\|(\vec{k})_{\mu,\nu}\right\|^2}{\sigma^2}\exp\left(-\frac{\left\|(\vec{k})_{\mu,\nu}\right\|^2\left\|(\vec{x})\right\|^2}{2\sigma^2}\right)\left[\exp\left(i(\vec{k})_{\mu,\nu}(\vec{x})\right)-\exp\left(-\frac{\sigma^2}{2}\right)\right] \tag{6}$$

where μ and ν denote the orientation and dilation of the Gabor kernels, $(\vec{x})=(x,y)$, $\|\bullet\|$ denotes the norm operator, and the wave vector $(\vec{k})_{\mu,\nu}$ is defined as follows:

$$(\vec{k})_{\mu,\nu}=\begin{pmatrix}k_\nu\cos\phi_\mu\\ k_\nu\sin\phi_\mu\end{pmatrix},\ k_\nu=2^{\frac{-\nu+2}{2}},\ \phi_\mu=\pi\frac{\mu}{8}. \tag{7}$$

The family of Gabor kernels is similar each other since they are generated from one mother wavelet by dilation and rotation using the wave vector $(\vec{k})_{\mu,\nu}$. Gabor wavelet is used at eight different frequencies, $\nu = 0,\ldots,7$, and eight orientations, $\mu = 0,\ldots,7$ [12].

The Gabor wavelet transformation of an image is defined by the convolution of the sub-area of image using a family of Gabor kernels as defined by Eq(8). Let $f(\vec{x})$ be the gray value of an sub-image around pixel $(\vec{x}) = (x, y)$, and the Gabor wavelet transform of the sub-image is defined as follows:

$$\begin{aligned} G_{\mu,\nu}(\vec{x}) &= f(\vec{x}) * \psi_{\mu,\nu}(\vec{x}) \\ &= \iint f(\alpha,\beta)\psi_{\mu,\nu}(x-\alpha, y-\beta)\,d\alpha\, d\beta \end{aligned} \tag{8}$$

where $\vec{x} = (x, y)$, and $*$ denotes the convolution operator. The Gabor transform shows desirable characteristics of spatial locality, frequency and orientation selectivity similar to those of the Gabor kernels. Each $G_{\mu,\nu}(\vec{x})$ consists of different local, frequency and orientation characteristics at the facial feature point $\vec{x}$. Those features are concatenated together to generate a feature vector F for the facial feature point $\vec{x}$. F, the feature vector of sub-image around $\vec{x}$ is defined as the concatenation of 64 complex coefficients as follows:

$$F(\vec{x}) = \left(G_{0,0}(\vec{x})^t G_{0,1}(\vec{x})^t \ldots G_{7,7}(\vec{x})^t\right)^t \tag{9}$$

The feature vector thus includes all the Gabor transform at the facial feature point $\vec{x}$, $G_{\mu,\nu}(\vec{x})$, $\mu = 0,\ldots,7$, $\nu = 0,\ldots,7$.

$$V = (F(\vec{x}_1)F(\vec{x}_2)\ldots F(\vec{x}_n)) \tag{10}$$

Chromosome Encoding and Fitness function

The GA is employed to search among the different combinations of feature representations. The optimality of the chromosome is defined by classification accuracy and generalization capability. Fig.3 shows the encoding of chromosome description.

As the GA searches the genospace, the GA makes its choices via genetic operators as a function of probability distribution driven by fitness function. The genetic operators used here are selection, crossover, and mutation [13].

Weight of facial feature point-1	Weight of facial feature point-2	Weight of facial feature point-3	……	Weight of facial feature point-*n*

Fig. 3. The chromosome description of the proposed scheme

The GA needs a salient fitness function to evaluate current population and chooses offspring for the next generation. Evolution or adaptation will be guided by a fitness function defined in terms of the system accuracy and the class scattering criterion as follows:

$$\eta(V) = \lambda_1 \eta_\varsigma(V) + \lambda_2 \eta_g(V), \tag{11}$$

where $\eta_\varsigma(V)$ is the term for the system correctness, i.e., successful recognition rate and $\eta_g(V)$ is the term for class generalization [14]. λ_1 and λ_2 are positive parameters that indicate the weight of each term, respectively.

5 Experimental Results

In this section, we'll discuss the effectiveness of the proposed context-ware adaptable face recognition system taking advantage of the cluster validity measure. We'll observe how cluster validity is related to context adaptation performance.

Our face recognition system is consisted of two parts. One is adaptation procedure, and the other is actual recognition procedure. We divide FERET database [15] into two groups. The first group, which contains 500 pairs of images, is for adaptation. In this group, face images are clustered by k-means algorithm for context recognition. Using that clustering result, weights for each facial feature point are selected by Genetic Algorithm. And the other group, which contains 1200 pairs of images, is for evaluation. In this group, face recognition test is done using the clustering result, feature weights which is gotten from adaptation process. Of course, each group set involves images for enrollment and test.

Table 1. The face recognition rates on each clustering result and the measurements of cluster validity using Dunn's measure and our cluster validity measure

No.	Recognition rate (Before adaptation)	Recognition rate (After adaptation)	Cluster validity (Dunn measure)	Cluster validity (Our measure)
1	91.91 %	92.24 %	0.767	29.018
2		93.56 %	0.798	29.056
3		92.57 %	0.803	29.020
4		92.57 %	0.825	29.058
5		92.74 %	0.787	29.104
6		91.75 %	0.777	28.974
7		91.58 %	0.812	28.961
8		92.57 %	0.795	29.021
9		92.08 %	0.780	29.034
10		93.03 %	0.816	29.072

For the comparison of cluster validity with system performance, ten cases of clustering result are produced and used. Dunn measure and our cluster validity measure are used for checking cluster validity. Table 1 is showing the results.

Fig. 4 is showing the relationship between cluster validity and performance of face recognition. In the case using Dunn's measure, we cannot find clear relation between cluster validity and recognition rate. Using our cluster validity measure, we can find the good relationship between cluster validity and recognition rate.

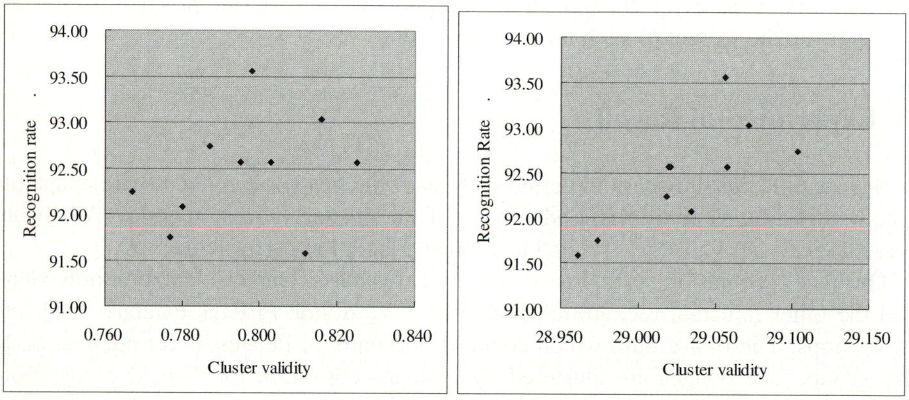

(a) (b)

Fig. 4. The relationship between cluster validity and recognition rate after adaptation((a): Using Dunn measure, (b): Using Our cluster validity measure)

We can see that system performance and our cluster validity measure have a close relation. At the most case, clustering with high cluster validity gets high recognition rate after adaptation. With our cluster validity measure which is designed for our face recognition procedure, we can choose optimal clustering for high performance of our context adaptable face recognition considering top three clustering methods.

6 Conclusion

In this paper, the situation information of face images, called image context, is used to improve performance of a face recognition system. Image contexts in various situations are distinguished each other by facial expression, view point, illumination conditions, noise, etc. By using cluster validity measure, we can expect performance of the adaptation using some context recognition method. We can choose a best clustering method for the system to be able to attain better performance. The proposed context aware adaptable system use unsupervised learning algorithms for context recognition. The proposed system partitions face images into several image contexts (groups) based on cluster validity. It assigns weights to each facial feature points represented by Gabor wavelet. We presented the process of cluster evaluation using the proposed cluster validity measure, and achieved encouraging experimental results though extensive experiments. The proposed method can play an important roll in designing systems which can adapt to uneven situations automatically.

References

1. Du, Bo, Shiguang, Qing, Shan Laiyun, Gao, Wen: Empirical comparisons of several preprocessing methods for illumination insensitive face recognition. Proceedings of the ICASSP '05, IEEE International Conference on Acoustic, Speec, and Signal Processing, Vol. 2, ii/981 - ii/984. (2005)
2. Sawides, M., Kumar, B.V.K.V., Khosla, P.K.: "Corefaces" - robust shift invariant PCA based correlation filter for illumination tolerant face recognition. Proceedings of the 2004 IEEE Computer Society Conference on Computer Vision and Pattern Recognition, CVPR 2004, Volume 2, 27, II-834 - II-841. (2004)
3. Zhou, Shaohua, Chellappa, R.: Illuminating light field: image-based face recognition across illuminations and poses. Proceedings of the Sixth IEEE International Conference on Automatic Face and Gesture Recognition, 229 – 234, 17-19 May 2004.
4. Lanitis, A., Taylor, C.J.: Towards automatic face identification robust to ageing variation. Proceedings of the Fourth IEEE International Conference on Automatic Face and Gesture Recognition. (2000) 391 – 396
5. Demir, E., Akarun, L., Alpaydin, E.: Two-stage approach for pose invariant face recognition. Proceedings of the IEEE International Conference on Acoustics, Speech, and Signal Processing, the ICASSP '00, Volume 6. (2000) 2342 – 2344
6. Crowley, J.L.: Context aware observation of human activities. Proceedings of the 2002 IEEE International Conference on Context aware observation of human activities. Multimedia and Expo 2002, ICME '02, Volume 1. (2002) 909 – 912
7. Theodoridis, S., Koutroubas, K.: Pattern recognition. Academic Press. (1999)
8. Bezdek, J.C., Pal., N.R.: Some new indexes of cluster validity. IEEE Transactions, Systems, Man and Cybernetics, Part B, Volume 28, Issue 3. (1998) 301 – 315
9. Halkidi, M., Batistakis, Y., Vazirgiannis, M.: Cluster Validity Methods: Part I. SIGMOD Record. (2002)
10. Stein, Benno, Meyer zu Eissen, Sven, Wissbrock, Frank: On Cluster Validity and the Information Need of Users. Proceedings of the 3rd IASTED International Conference on Artificial Intelligence and Applications (AIA'03). Benalmadena, Spain. (2003)
11. Bezdek, J. C., Li, W.Q., Attikiouzel, Y., Windham, M.: A Geometric Approach to Cluster Validity for Normal Mixtures. Soft Computing 1. (1997)
12. Wiskott., L., Fellous, J.M., Kruger, N., von der Marsburg, C.: Face Recognition by Elastic Bunch Graph Matching, Technical Report, IR-INI 96-08. Institut fur Neuroinformatik, Ruhr Universitat Bochum, D-44780, Bochum, Germany. (1996)
13. Goldberg, D.E.: Genetic Algorithm in Search, Optimization and Machine Learning. Addison-Wesley Publishing Company, Inc. Reading, Massachusetts. (1989)
14. Jeon, In Ja, Kwon, Ki Sang, Rhee, Phill Kyu: Optimal Gabor Encoding Scheme for Face Recognition Using Genetic Algorithm. KES 2004. (2004) 227-236
15. Phillips, P.J., Wechsler, H., Huang, J., Rauss, P.: The FERET database and evaluation procedure for face recognition algorithms. Image and Vision Computing, 16(5) (1998) 295-306

Self-adaptive Classifier Fusion for Expression-Insensitive Face Recognition

Eun Sung Jung, Soon Woong Lee, and Phill Kyu Rhee

Dept. Of Computer Science & Engineering
Inha University, Yong-Hyun Dong,
Incheon, South Korea
{eunsung, lsw}@im.inha.ac.kr, pkrhee@inha.ac.kr

Abstract. We address a self-adaptive face recognition scheme which is insensitive to facial expression variations. The proposed method takes advantage of self-adaptive classifier fusion based on facial geometry and RBF warping technology. Most previous face recognition schemes usually show vulnerability under changing facial expressions. The proposed scheme discriminates input face images into one of several context categories. The context categories are decided by unsupervised learning method based on the facial geometries that are derived from either scanned mosaic face images and/or coordinates of facial feature points. The proposed method provides a self-adaptive preprocessing and feature representation in accordance with the identified context category using the genetic algorithm and knowledge accumulation mechanism. The superiority of the proposed method is shown using FERET database where face images are relatively exposed to wide range of facial expression variation.

1 Introduction

In this paper, we discuss about the framework of self-adaptive classifier fusion and its application to robust face recognition that can behave in an adaptive and robust manner under dynamic variations. The framework can be thought as an augmented evolvable system framework which provides not only the capability of self adaptation taking advantage of context-awareness and context knowledge accumulation. The context knowledge of individual environmental contexts and their associated chromosomes is stored in the context knowledge base. The most effective action configuration of the system is determined for a current context/environment by employing the popular learning computing method, e.g., genetic algorithm (GA). GA explores a most effective chromosome that describes the action configuration for each identified environmental context. The context-awareness consists of context modeling and identification. Context modeling can be performed by an unsupervised learning method [1, 2] such as Fuzzy Art, SOM, K-means, etc. Context identification can be implemented by a normal classification method such as NN, k-NN, etc.

Even though many algorithms and techniques are invented, face recognition still remains a difficult problem yet. One of the most crucial problems in designing

B. Gabrys, R.J. Howlett, and L.C. Jain (Eds.): KES 2006, Part I, LNAI 4251, pp. 78 – 85, 2006.
© Springer-Verlag Berlin Heidelberg 2006

efficient face recognition is to eliminate or bypass the effect of changing facial expression [5, 15]. Recently, Bronstein proposes 3D face recognition method by trying to compute facial expression invariant representation [3]. However, the construction of 3D image itself either is much expensive or time consuming to be used in real world applications. Martinez tried to recognize a face from a single image including facial expression variation [4]. He argued that image matching with weighted Euclidean distance outperforms the non-weighted Euclidean distance based matching. Wavelet based face recognition system has been tested in the presence of varying facial expressions using JFEE database in [5]. However, the database they used is too small, i.e. only containing 10 person images. Adaptive Principal Component Analysis (APCA) is proposed by [15]. It first applies PCA to construct a subspace for image representation, rotate the subspace to provide insensitivity to facial expression, and then warps the subspace.

In general, classifier fusion schemes are expected to produce superior performance to a single classifier in terms of accuracy and reliability. Similar approaches can be found in the literature of classifier fusion/combination, classifier ensemble, expert mixture, etc [7, 8, 9, 10]. Traditional system fusion schemes can not be applicable in dynamically changing situations or environments since they usually require too much computation time or resource to achieve an optimal performance. The context-aware classifier fusion method [6] is applied to the adaptive preprocessing and feature vector selection for reconstructing multiple classifiers for a robust face recognition system. An adaptive preprocessing and classifier fusion method using context-awareness is tested for efficient recognition of faces with varying facial expression. The role of adaptive preprocessing is to produce well-preprocessed image for finding an optimal Gabor feature vectors so that the system could achieve best performance with face images of varying facial expression. This paper is organized as follows.

In the section 2, we discuss about the architecture for context-aware classifier fusion applying to robust recognition of faces with varying facial expression. We give the experimental results and the concluding remarks in the section 3 and 4, respectively.

2 Facial Expression-Insensitive Recognition Using Self-adaptive Classifier Fusion

In this section, we briefly discuss about a novel model of classifier fusion with the capability of autonomous adaptation, called self-adaptive classifier fusion [6]. Classifier fusion is to combine a set of classifier primitives to configure an efficient classifier system. A classifier system is modeled as having multiple stages each of which consists of several competitive action primitives. The action primitives are basic functional elements that can be heterogeneous, homogeneous, or hybrid operational entities. The same functional elements with different behaviors due to different parameters, threshold, etc. are treated as different action primitives. Classifier fusion can be thought to combine action primitive at each and produces an efficient classifier system for a given operational environment. Self-adaptive classifier fusion is an augmented classifier fusion with the capability of autonomous learning and self-growing knowledge base.

2.1 The Architecture of Self-adaptive Classifier Fusion

The realization of the model of self-adaptive classifier fusion will be discussed in this session. We devise an architecture that consists of the context-aware unit (CAU), the control unit (CU), the action unit (AU), the genetic algorithm unit(GAU) and the context knowledge base(CKB) (see Fig. 1).

Varying facial geometry, which is modeled as context information, is observed and categorized by the CAU. The CAU is trained and implemented by K-means and K-NN algorithms, respectively. The adaptive action of the AU for individual facial geometry context is learned or evolved by the GAU and stored in the CKB. The AU consists of the module of preprocessing, feature extraction, and class decision. The CU searches the CKB for a proper action of each identified context, i.e. a currently identified facial geometry. The preprocessing module includes image normalization, image warping and histogram equalization. The preprocessed image is transformed into the Gabor feature vector. The Gabor feature vector is restructured using the identified context information, i.e., category. Finally, the face class is decided by the class decision module. The class decision module performs the task of face recognition using Cosine distance. In this way, the proposed system can achieve the optimal performance under uneven facial expression. The details will be discussed in the following.

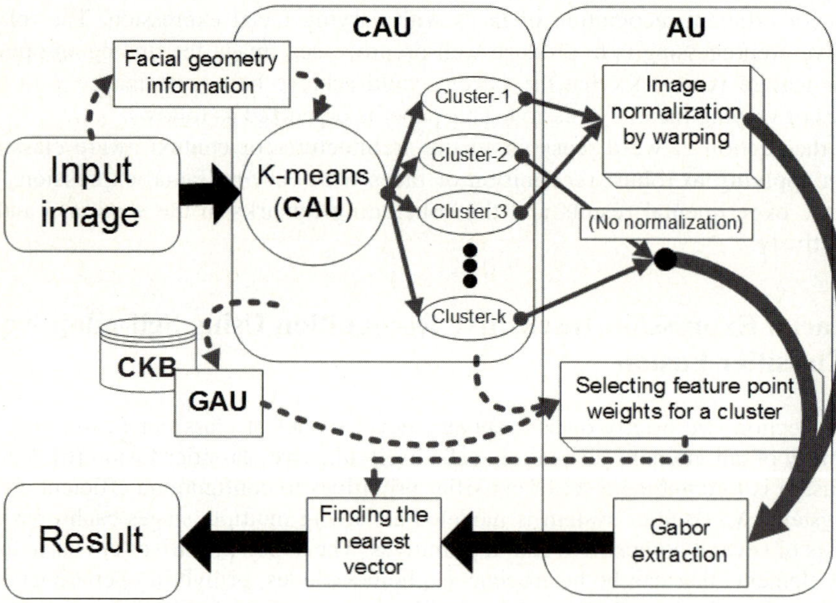

Fig. 1. The face recognition scheme using facial geometry based self-adaptive classifier fusion in the Gabor feature space

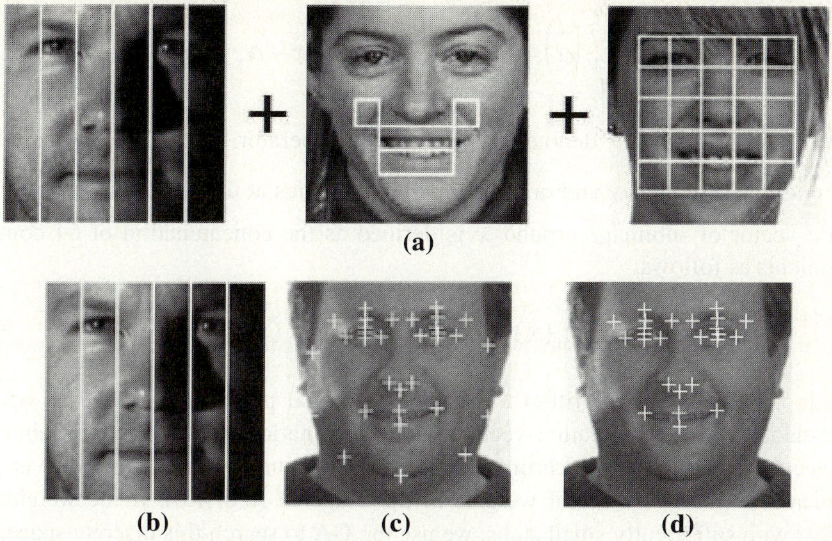

Fig. 2. Facial geometry representation by hybrid scanning (a), vertical scanning (b), coordinates of facial feature points (c) and facial feature points excepting face edges (d)

2.2 Context-Awareness Using Facial Geometry for Self-adaptive Classifier Fusion

The CAU tries to determine the facial expression context of each image by identifying the geometrical characteristics of facial images, especially, coming from facial expression variations. Context data is defined as any observable and relevant attributes, its interaction with other entities and/or surrounding environment at an instance of time. Context data can be various configurations, computing resource availability, dynamic task requirement, application condition, environmental condition, etc [11]. Derived context feature is derived or extracted from context data, and is used to decided context category. The CAU is implemented by K-means algorithm. The $1\times s$ input vector of the K-means is generated by resizing the $n\times m$ input image at time, where $s << n\times m$. We use $n\times m$ pixels as the context information. We used face image of 128×128 spatial resolution and 256 gray levels as input context data here. Four methods of derived context feature were investigated here as shown in Fig. 2.

2.3 Gabor Wavelet and Feature Space Using Genetic Algorithm

The feature space of the proposed face recognition system is constructed by the Gabor wavelet transform [12]. The Gabor transform shows desirable characteristics of spatial locality, frequency and orientation selectivity similar to those of the Gabor kernels. The convolution coefficient for kernels of different frequencies and orientations starting at a particular fiducial point is calculated. Gabor wavelet used at five different frequencies, ν= 0, . . . , 7, and eight orientations, μ= 0, . . . , 7 [13]. The Gabor wavelet transformation of an image is defined by the convolution of the sub-area of image using a family of Gabor kernels as defined by Eq(9).

$$G_{\mu,\nu}(\vec{x}) = f(\vec{\alpha}) * \psi_{\mu,\nu}(\vec{\alpha}) = \iint f(\alpha,\beta)\psi_{\mu,\nu}(x-\alpha, y-\beta)\,d\alpha\,d\beta \quad (9)$$

where $(\vec{x}) = (x, y)$, and * denotes the convolution operator. Each $G_{\mu,\nu}(\vec{x})$ consists of different local, frequency and orientation characteristics at the fiducial point $\vec{x}$. F, the feature vector of subimage around $\vec{x}$ is defined as the concatenation of 64 complex coefficients as follows:

$$F(\vec{x}) = (G_{0,0}(\vec{x})^t\, G_{0,1}(\vec{x})^t \;\cdots\; G_{7,7}(\vec{x})^t\,)^t \quad (10)$$

We adopt the Genetic algorithm to search an optimal preprocessing and an optimal representation of Gabor feature vector. Each combination of weights for Gabor feature vector is encoded by a chromosome. The contribution of fiducial point is in n-dimensional space by a set of weights in the range of (0.0, 1.0). If the weights are discrete with sufficiently small steps, we use the GA to search this discrete space. The control module derives the classifier being balanced between successful recognition and generalization capabilities as follows:

$$\mu(V) = \lambda_1\eta_s(V) + \lambda_2\eta_g(V) \quad (11)$$

Where $\eta_s(V)$ the term for the system correctness, i.e., successful recognition is rate and $\eta_g(V)$ is the term for class generalization. λ_1 and λ_2 are positive parameters that indicate the weight of each term, respectively.

2.4 The Facial Geometry Based Self-adaptive Recognition Insensitive to Facial Expression Changes

The self-adaptive classifier fusion method is applied to the preprocessing for relieving facial expression variations and the feature representation for expression-insensitive face recognition. The CKB stores the identified contexts and their matched actions those will be performed by the AU in the form of chromosomes. The detail of constructing the CKB for self-adaptive classifier fusion is given in the following.

1. Partition face images into several context categories according to their facial geometry.
2. Select a type of preprocessing for individual context categories. (RBF warping is carried out in this step, if necessary.)
3. Derive the Gabor feature $F(x_i)$ for each fiducial point, and normalize it. Concatenate the Gabor vectors for fiducial points to generate the entire Gabor feature space.
4. Begin the Gabor feature representation optimization until a predefined criterion is met.

1) Generate a new Gabor feature representation population.
2) Evaluate the fitness function $\mu(\mathrm{V}) = \lambda_1 \eta_s(V) + \lambda_2 \eta_g(V)$ of the classifiers using the newly derived Gabor feature representation population. If the criterion is met, go to Step 5.
3) Search for the Gabor feature representation in the Gabor feature space population that maximizes the fitness function and keep those as the best chromosomes.
4) Applying GA's genetic operators to generate new population of Gabor feature space. Go to step 4.2).

5. Steps 2 - 4 are repeated until a best feature vector is reached to the predefined criterion, or a generation exceeds a desired criterion value.
6. Update the CKB for the identified context category of facial geometry as the chromosome and reconstructed Gabor feature space.

The recognition task is performed by constructing self-adaptive classifiers using the generated CKB as follows:

1) Identify the context category of input face image using the CAU.
2) Search a chromosome representing optimal preprocessing and Gabor feature representation for the identified context category.
3) Perform the self-adaptive restructuring Gabor representation (multiple classifier) using the matched chromosome.
4) Derive the Gabor feature $F(x_i)$ for each fiducial point of the enhanced image, and normalize it.
5) Concatenate the Gabor features for fiducial points to generate total Gabor feature space.
6) Perform the task of recognition using the restructured Gabor feature vector.

3 Experimental Results

The feasibility of the proposed face recognition scheme has been tested using FERET database [14]. The training of the proposed face recognition scheme consists of two steps: the first step is training CAU by k-means algorithm for determining facial geometry based context categories, and the second step is training weights of each Gabor vectors for feature points. Finally, the evaluation is performed. We divided FERET database into two groups, training set (500 pairs of face images) and evaluation set (700 pairs of face images). Each group has face images for enrollment from FERET fafa set and those for test from fafb set.. Table 1 is showing the result of evaluation of proposed face recognition scheme for each case – not applying any methods, using feature points only, using warping only, using feature point weights and warping. Appliance of feature point weights improved performance at most clusters (so, total recognition rate is raised). And using warping process on some clusters, we got more improvement in performance of our system. We can make a face recognition system which is robust against the variance of facial expression by situation clustering and adaptation through using selective feature weights and warping.

Table 1. The result of face recognition test on FERET database using situation recognition scheme by clustering algorithm (tested on evaluation set (700 images in FERET database) by the system which is adapted in training set (500 images in FERET database))

Used method	Cluster 0	Cluster 1	Cluster 2	Cluster 3	Cluster 4	Cluster 5	Total
No feature point weights, no warping	91.73%	84.75%	100.00%	94.20%	92.71%	85.71%	**91.91%**
Using feature point weights on whole data							**92.90%**
Using warping on whole data							**92.41%**
Using adaptive feature point weights and warping on whole data							**92.74%**
Using feature point weights for each cluster	91.73%	89.83%	88.89%	91.30%	95.44%	85.71%	**93.40%**
Using warping on selected clusters	91.73%	84.75%	100.00%	94.20%	94.53%	85.71%	**92.90%**
Adaptive feature point weights for each cluster + Adaptive warping on selected clusters	91.73%	88.14%	88.89%	91.30%	96.35%	85.71%	**93.73%**

4 Conclusion

In this paper, we propose a novel adaptive and evolutionary technique combining context-awareness and classifier fusion for robust face recognition under varying facial expression. The performance of face recognition is closely related to the amount of variation observed in face image. In our self-adaptive classifier-fusion scheme input images are divided into several groups which indicate each situation. And doing proper action to each situation group, it maximizes the performance of the face recognition system. Out experimental results are showing the reliability and usability of our idea. Our scheme can be adapted to other applications by defining proper context domain and action methods. More robustness and applicability of our scheme can be accomplished by further approaches for other domains.

References

1. Nam, M.Y., Rhee, P.K.: A Novel Image Preprocessing by Evolvable Neural Network. LNAI3214, vol. 3 (2004) 843-854
2. Nam, M.Y., Rhee, P.K.: An Efficient Face Recognition for Variant Illumination Condition. ISPACS2005, vol. 1 (2004) 111-115
3. Bronstein, A.M., Bronstein, M.M., Kimmel, R.: Expression-invariant 3D face recognition. Proc. Audio- and Video-based Biometric Person Authentication (AVBPA), Lecture Notes in Comp. Science No. 2688. Springer (2003) 62-69
4. Martinez, A.M.: Recognizing Expression Variant Faces from a Single Sample Image per Class. Proc. IEEE Computer Vision and Pattern Recognition (CVPR), Madison (WI) (2003)
5. Sellahewa, H., Jassim, S.: Face Recognition in the presence of expression and/or illumination variation. Fourth IEEE Workshop on Automatic Identification Advanced Technologies. (2005) 144 - 148
6. Jeon, In Ja, Kwon, Ki Sang, Rhee, Phill Kyu: Optimal Gabor Encoding Scheme for Face Recognition Using Genetic Algorithm. KES 2004 (2004) 227-236
7. Kuncheva, L.I., Jain, L.C.: Designing classifier fusion systems by genetic algorithms. IEEE Transactions on Evolutionary Computation, vol.4, no. 4. (2000) 327-336
8. Kim, Sang-Woon, Oommen, B.J.: On using prototype reduction schemes and classifier fusion strategies to optimize kernel-based nonlinear subspace methods. IEEE Trans. Pattern Analysis and Machine Intelligence, vol. 27. (2005) 455-460
9. Ludmila, I., Kuncheva: Switching Between Selection and Fusion in Combining Classifiers: An Experiment. IEEE Transactions on Systems, Man, and Cybernetics - part B: cybernetics, vol.32, no.2. (2002) 146-156
10. Liu, C., Wechsler, H.: Gabor Feature Classifier for Face Recognition, Computer Vision. 8th IEEE International Conference (2001) 270-275
11. Yau, S., Karim, F., Wang, Y., Wang, B., Gupta, S.: Reconfigurable Context-Sensitive Middleware for Pervasive Computing. IEEE Pervasive Computing. (2002) 33-40
12. Faugman, J.: Uncertainty relation for resolution in space, spatial frequency, and orientation optimization by two-dimensional cortical filters. Journal Opt. Soc. Amer. 2(7). (1985) 675-676
13. Potzsch, M., Kruger, N., Von der Malsburg, C.: Improving Object recognition by Transforming Gabor Filter responses. Network: Computation in Neural Systems, vol.7, no.2. (1996) 341-347
14. Phillips, P.: The FERET database and evaluation procedure for face recognition algorithms. Image and Vision Computing, vol. 16, no. 5. (1999) 295-306
15. Shaokang, Chen, Brian, C.Lovell: Illumination and Expression Invariant Face Recognition with One Sample Image. Proceedings of the 17th International Conference on Pattern Recognition (ICPR'04), vol. 1. (2004) 300-303

Improved Harmony Search from Ensemble of Music Players

Zong Woo Geem

Johns Hopkins University, Environmental Planning and Management Program,
729 Fallsgrove Drive #6133, Rockville, Maryland 20850, USA
geem@jhu.edu

Abstract. A music phenomenon-inspired algorithm, harmony search was further developed by considering ensemble among music players. The harmony search algorithm conceptualizes a group of musicians together trying to find better state of harmony, where each player produces a sound based on one of three operations (random selection, memory consideration, and pitch adjustment). In this study, one more operation (ensemble consideration) was added to the original algorithm structure. The new operation considers relationship among decision variables, and the value of each decision variable can be determined from the strong relationship with other variables. The improved harmony search algorithm was applied to the design of water distribution network. Results showed that the improved algorithm found better solution than those of genetic algorithm, simulated annealing, and original HS algorithm.

1 Introduction

Human beings have developed new technology by imitating natural or behavioral phenomena on the earth. Likewise, optimization researchers in various fields have developed better algorithms by analogizing solution-searching processes to biological and behavioral phenomena [1-4]. Genetic algorithm (GA), which is one of the most popular phenomenon-inspired algorithms, has performed good job in various optimization problems, sometimes finding better solutions than those of mathematics-based algorithms although it does not completely mimics every mechanism and structure of DNA or RNA [5].

Recently, another phenomenon-based algorithm, harmony search (HS), which mimics music improvisation process has been proposed [6]. The HS algorithm has been successfully applied to various real-world optimization problems such as structural design [7-8], water network design [9-11], traffic routing [12-13], fluid leakage detection [14], and model parameter calibration [15-16]. HS even found better solution within much less time for water pump switching problem than branch and bound method originally coded in IBM subroutine or Microsoft Excel Solver [17].

While the original HS algorithm uses three mechanisms (random selection, memory consideration, and pitch adjustment) as each musician's performing behavior, there is still a possibility to include more mechanisms by imitating real music performance. In this study, one more mechanism, named ensemble consideration, is devised to consider the relationship among musicians. Just like certain musician has

B. Gabrys, R.J. Howlett, and L.C. Jain (Eds.): KES 2006, Part I, LNAI 4251, pp. 86 – 93, 2006.
© Springer-Verlag Berlin Heidelberg 2006

stronger relationship with specific musician in a group, this mechanism enables the algorithm to combine closely related variables together.

2 Harmony Search Algorithm

HS algorithm conceptualized a behavioral phenomenon of music players in improvisation process, where each player continues to polish its tune in order to produce better harmony.

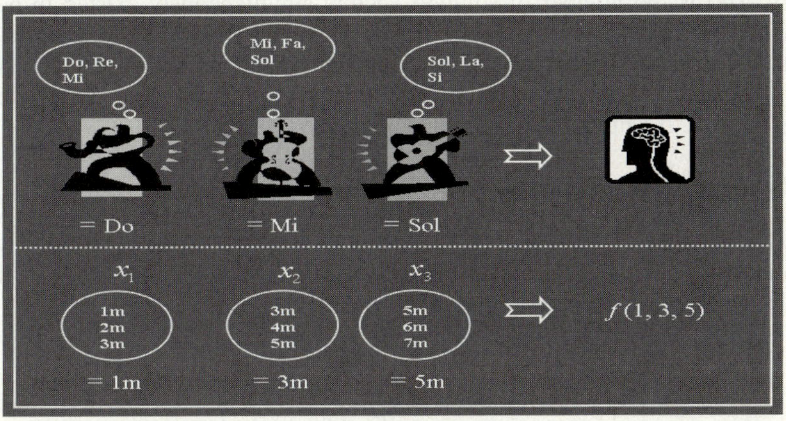

Fig. 1. Analogy between Improvisation and Optimization

Figure 1 illustrates the analogy between improvisation and optimization. Each music player (saxophonist, double bassist, and guitarist) in the figure can be matched with each decision variable (x_1 , x_2 , and x_3), and the candidate range of each music instrument (saxophone = {Do, Re, Mi}; double bass = {Mi, Fa, Sol}; and guitar = {Sol, La, Si}) can be matched with the candidate range of each variable value (x_1 = {1, 2, 3}; x_2 = {3, 4, 5}; and x_3 = {5, 6, 7}). If the saxophonist sounds the note Do, the double bassist sounds Mi, and the guitarist sounds Sol, these notes together becomes a new harmony (Do, Mi, Sol). If this new harmony has good quality, it is kept in each musician's memory. Likewise, if a new solution vector (1m, 3m, 5m) is generated in optimization process and it has good fitness in terms of objective function, it is kept in computer memory. This activity is repeated until (near-) optimal solution is reached. The HS algorithm is explained in detail in the following steps.

2.1 Problem Initialization

At first, the problem is formulated as the following optimization form:

$$\text{Optimize } f(\boldsymbol{x}) \tag{1}$$

$$\text{Subject to } x_i \in \mathbf{X}_i, i = 1,2,\ldots, N . \tag{2}$$

where $f(\cdot)$ is an objective (or fitness) function (= aesthetic standard); $\boldsymbol{x}$ is a solution vector (= harmony) composed of each decision variable x_i (= each music instrument); $\mathbf{X}_i$ is the set of candidate values (= set of candidate pitches) for each decision variable x_i, which becomes $\mathbf{X}_i = \{x_i(1), x_i(2), ..., x_i(K)\}$ ($x_i(1) < x_i(2) < ... < x_i(K)$) for discrete decision variables; N is the number of decision variables (= number of music instruments); and K is the number of candidate values for the discrete decision variables (= number of candidate pitches for each instrument).

The HS algorithm's parameters are also specified in this step: HMS (harmony memory size) is the number of solution vectors simultaneously existed in harmony memory; HMCR (harmony memory considering rate; $0 \le \text{HMCR} \le 1$) is the rate of memory consideration; PAR (pitch adjusting rate; $0 \le \text{PAR} \le 1$) is the rate of pitch adjustment; and MaxImp (maximum number of improvisations) is the maximum number of objective function evaluations.

2.2 Harmony Memory Initialization

Harmony Memory (HM), a matrix shown in Equation 3, is initially filled with as many randomly generated solution vectors as HMS. Corresponding function value of each random vector is also stored in HM.

$$\left[\begin{array}{cccc|c} x_1^1 & x_2^1 & \cdots & x_N^1 & f(\boldsymbol{x}^1) \\ x_1^2 & x_2^2 & \cdots & x_N^2 & f(\boldsymbol{x}^2) \\ \vdots & \cdots & \cdots & \cdots & \vdots \\ x_1^{HMS} & x_2^{HMS} & \cdots & x_N^{HMS} & f(\boldsymbol{x}^{HMS}) \end{array}\right] \tag{3}$$

2.3 New Harmony Improvisation

Next, a new harmony, $\boldsymbol{x}' = (x_1', x_2', ..., x_N')$ is improvised by the following three mechanisms: random selection, memory consideration, and pitch adjustment.

Random Selection. For the value of each decision variable x_i' in the new harmony vector, as a musician plays any possible pitch within the candidate range of musical instrument (for example, {Do, Re, Mi, Fa, Sol, La, Si, Do+} in Figure 2), the value can be randomly chosen within the value range $\mathbf{X}_i$ with a probability of (1-HMCR).

$$x_i' \leftarrow x_i' \in \boldsymbol{X}_i = \{x_i(1), x_i(2), ..., x_i(K)\} \quad \text{w.p.} \quad \mathit{(1 - HMCR)} \tag{4}$$

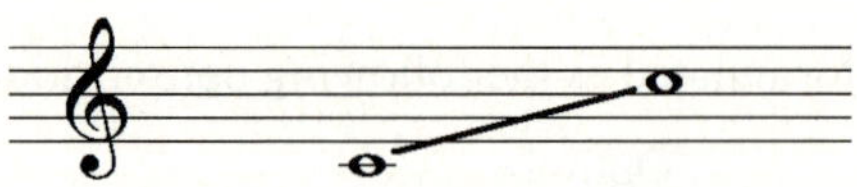

Fig. 2. Playable Range of Musical Instrument

Memory Consideration. Instead of the above-mentioned random selection, as a musician plays any pitch out of the preferred pitches stored in his/her memory (for example, {Do, Mi, Do, Sol, Do} in Figure 3), the value of decision variable x_i' can be chosen from any pitches stored in HM with a probability of HMCR.

$$x_i' \leftarrow x_i' \in \{x_i^1, x_i^2, ..., x_i^{HMS}\} \quad \text{w.p.} \quad HMCR \tag{5}$$

Fig. 3. Preferred Pitches Stored in Harmony Memory

Pitch Adjustment. Once one pitch is obtained in memory consideration as described in the above section, a musician can further adjust the pitch to neighboring pitches (for example, the note Sol can be adjusted to Fa or La as shown in Figure 4) with a probability of HMCR × PAR while the original pitch obtained in memory consideration is just kept with a probability of HMCR × (1-PAR).

$$x_i' \leftarrow \begin{cases} x_i(k+m) & \text{w.p.} \quad HMCR \times PAR \times 0.5 \\ x_i(k-m) & \text{w.p.} \quad HMCR \times PAR \times 0.5 \\ x_i(k) & \text{w.p.} \quad HMCR \times (1-PAR) \end{cases} \tag{6}$$

where $x_i(k)$ (the kth element in $\mathbf{X}_i$) in right hand side is identical to x_i' which is originally obtained in memory consideration; m is a neighboring index for discrete decision variables ($m \in \{1, 2, ...\}$; m usually has a value of 1).

Fig. 4. Pitch Adjustment

Ensemble Consideration. After the new harmony $\boldsymbol{x}' = (x_1', x_2', ..., x_N')$ is obtained using the above-mentioned three rules, one more operation can be considered from the relationship among decision variables. Just as a player has even stronger relationship with specific player in a music group (for example, the first and second players have stronger relationship in Figure 5), the new operation, ensemble consideration, enables the algorithm to combine closely related variables together.

Fig. 5. Relationship between Music Players

The value of decision variable x_i' can be chosen based on the value of other variable x_j' when x_i and x_j have strongest relationship among decision variables:

$$x_i' \leftarrow fn(x_j') \quad \text{where} \quad \max_{i \neq j}\left\{\left[Corr(\boldsymbol{x}_i, \boldsymbol{x}_j)\right]^2\right\} \tag{7}$$

where $Corr(\cdot)$ is statistical correlation function between $\boldsymbol{x}_i = (x_i^1, x_i^2, ..., x_i^{HMS})$ and $\boldsymbol{x}_j = (x_j^1, x_j^2, ..., x_j^{HMS})$, producing r (correlation coefficient); $\max(\cdot)$ is maximum function which finds the strongest correlated variable x_j for the variable x_i based on r^2 (determination coefficient); and $fn(\cdot)$ is relationship function which produces the value of x_i' based on the value of x_j'. The relationship function $fn(\cdot)$ finds the most frequent value of x_i from the pairs of x_i and x_j in HM where x_j is equal to x_j'.

In order to operate the ensemble consideration with certain amount of chance, a new parameter ECR (ensemble consideration rate; $0 \le \text{ECR} \le 1$) is introduced. While GA preserves highly fitted structure (building blocks) implicitly [1], HS can explicitly preserve highly fitted structure using this ensemble consideration.

Violated Harmony Consideration. Once the new harmony $\boldsymbol{x}' = (x_1', x_2', ..., x_N')$ is obtained using the above-mentioned four rules, it is then checked whether it violates problem constraints. Although the new harmony violates the constraints, it has still chance to be included in HM, just as rule-violated harmony (for example, parallel fifth violation as shown in Figure 6) was still used by musicians such as famous composer Ludwig van Beethoven.

Violated harmony can be considered by taxing a penalty. The penalty technique has been already used for considering constraints in many mathematical or phenomenon-inspired optimization techniques.

Fig. 6. Harmony Violating Parallel Fifth

2.4 Harmony Memory Update

If the new harmony vector $\boldsymbol{x}'$ is better than the worst harmony in the HM in terms of objective function, the new harmony is included in the HM and the existing worst harmony is excluded from the HM. If the termination criterion (MaxImp) is reached, computation is ended. Otherwise, section 2.3 and 2.4 are repeated.

In order to prevent premature HM where identical harmony vectors are predominant, the number of identical vectors in HM can be limited (Normally, up to two identical vectors).

3 Application to the Water Network Design

The improved HS algorithm including new operation of ensemble consideration is applied to the design of water distribution network, which is a combinatorial optimization problem to find minimal pipe diameters while satisfying water demand and pressure at each node. The real world network is Hanoi network as shown in Figure 7.

The following parameter values were used for the optimization: number of decision variables = 34; number of candidate values (=diameters) for the decision variables = 6; HMS = {30, 50, 100}; HMCR = {0.7, 0.9, 0.95}; PAR = {0.05, 0.1}; ECR = {0.0, 0.005, 0.01}; and MaxImp = 10,000. All solutions are hydraulically verified using a standard simulator, EPANET 2.0 [18].

The improved HS algorithm could find less costs ($6,197,960 for ECR = 0.005; and $6,116,506 for ECR = 0.01) than original HS algorithm ($6,221,158 for ECR = 0.0), where each ECR case performed 18 runs with different algorithm parameters (3 HMS's × 2 HMCR's × 3 PAR's). Each run took 9 seconds on Intel 1.8GHz CPU, evaluating the objective function 10,000 times, which is $3.5 \times 10^{-21}\%$ (= 10^4 / 6^{34}) of total solution space! The cost ranges of the proposed HS are [6.198E6, 7.689E6] (ECR = 0.005) and [6.117E6, 7.659E6] (ECR = 0.01) while that of the original HS is [6.221E6, 7.466E6]. From these cost ranges, it appears that the solutions of the ensemble HS fluctuates more dynamically than those of the original HS.

When further compared with other nature-inspired algorithms for this problem, improved HS (ECR = 0.01) could find better solution ($6,081,087) after 30,000 evaluations while GA found $6,187,320 after 1,000,000 evaluations [19] and simulated annealing (SA) found $6,093,469 after 53,000 evaluations [20].

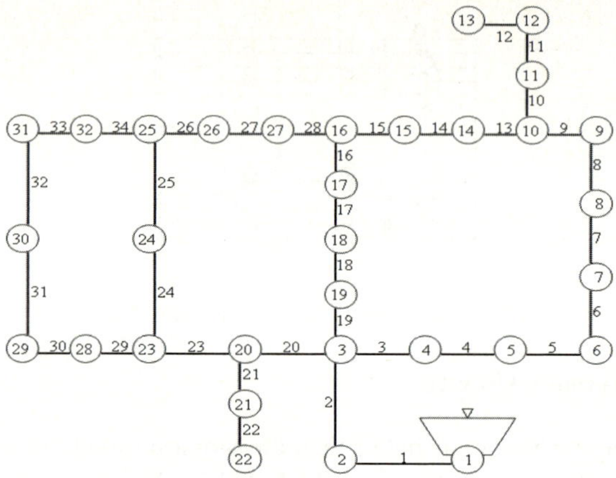

Fig. 7. Hanoi Water Network

4 Conclusions

A music phenomenon-based HS algorithm was further developed by mimicking musicians' interdependence. The value of one variable can be chosen from the value of another variable when they are strongly correlated in terms of statistical coefficient. The improved HS algorithm was then applied to a water network design. The ensemble-considered HS algorithm could find less-cost design than the original HS and other nature-inspired algorithms such as GA and SA.

Although GA keeps short-length good structures (building blocks) implicitly, this mechanism works well only when adjacent variables in a chromosome are strongly correlated. However, in real world problems (e.g., Hanoi network in Figure 7), this assumption does not seem always true; for example, pipe 28 and pipe 29 do not appear to have strong relationship because they are not located in neighboring place. Nevertheless, in a vector (or chromosome) in GA, these two variables have more chance not to be separated just because they are neighboring variables each other. That is the reason why this study developed the new operation, ensemble consideration, which can explicitly combine closely-related variables together, regardless of how far the two variables are separated in a solution vector.

References

1. Goldberg, D. E.: Genetic Algorithms in Search Optimization and Machine Learning. Addison Wesley, MA, USA (1989)
2. Kirkpatrick, S., Gelatt, C., and Vecchi, M.: Optimization by Simulated Annealing. Science. 220(4598) (1983) 671-680
3. Glover, F.: Heuristic for Integer Programming using Surrogate Constraints. Decision Sciences. 8(1) (1977) 156-166

4. Dorigo, M., Maniezzo, V., and Colorni, A.: The Ant System: Optimization by a Colony of Cooperating Agents. IEEE Transactions on Systems, Man, and Cybernetics-Part B. 26(1) (1996) 29-41
5. Ha, S., Lowenhaupt, K., Rich, A., Kim, Y., and Kim, K.: Crystal Structure of a Junction between B-DNA and Z-DNA Reveals Two Extruded Bases. Nature. 437(20) (2005) 1183-1186
6. Geem, Z. W., Kim, J. H., and Loganathan, G. V.: A New Heuristic Optimization Algorithm: Harmony Search. Simulation. 76(2) (2001) 60-68
7. Lee, K. S. and Geem, Z. W.: A New Structural Optimization Method Based on the Harmony Search Algorithm. Computers and Structures. 82(9-10) (2004) 781-798
8. Lee, K. S., Geem, Z. W., Lee, S. –H., and Bae, K. –W.: The Harmony Search Heuristic Algorithm for Discrete Structural Optimization. Engineering Optimization. 37(7) (2005) 663-684
9. Geem, Z. W. and Park, Y.: Harmony Search for Layout of Rectilinear Branched Networks. WSEAS Transactions on Systems. 5(6) (2006) 1349-1354
10. Afshar, M. H. and Marino, M. A.: Application of an Ant Algorithm for Layout Optimization of Tree Networks. Engineering Optimization. 38(3) (2006) 353-369
11. Geem, Z. W.: Optimal Cost Design of Water Distribution Networks using Harmony Search. Engineering Optimization. 38(3) (2006) 259-280
12. Geem, Z. W., Tseng, C. -L., and Park, Y.: Harmony Search for Generalized Orienteering Problem: Best Touring in China. Lecture Notes in Computer Science. 3612 (2005) 741-750
13. Geem, Z. W., Lee, K. S., and Park, Y.: Application of Harmony Search to Vehicle Routing. American Journal of Applied Sciences. 2(12) (2005) 1552-1557
14. Kim, S. H., Yoo, W. S., Oh, K. J., Hwang, I. S., and Oh, J. E.: Transient Analysis and Leakage Detection Algorithm Using GA and HS Algorithm for a Pipeline System. Journal of Mechanical Science and Technology. 20(3) (2006) 426-434
15. Kim, J. H., Geem, Z. W., and Kim, E. S.: Parameter Estimation of the Nonlinear Muskingum Model using Harmony Search. Journal of the American Water Resources Association. 37(5) (2001) 1131-1138
16. Paik, K., Kim, J. H., Kim, H. S., and Lee, D. R.: A Conceptual Rainfall-Runoff Model Considering Seasonal Variation. Hydrological Processes. 19 (2005) 3837-3850
17. Geem, Z. W.: Harmony Search in Water Pump Switching Problem. Lecture Notes in Computer Science. 3612 (2005) 751-760
18. Rossman, L. A.: EPANET2 Users Manual. US Environmental Protection Agency, Cincinnati, OH, USA (2000)
19. Savic, D. A. and Walters, G. A.: Genetic Algorithms for Least-Cost Design of Water Distribution Networks. Journal of Water Resources Planning and Management, ASCE. 123(2) (1997) 67-77
20. Cunha, M. C. and Sousa, J.: Hydraulic Infrastructures Design Using Simulated Annealing. Journal of Infrastructure Systems, ASCE. 7(1) (2001) 32-39

Performance Comparison of Genetic and Tabu Search Algorithms for System Identification

Aytekin Bagis

Erciyes University, Dept. of Electronic Eng., 38039, Kayseri, Turkey
bagis@erciyes.edu.tr

Abstract. This paper presents a performance comparison of the genetic and tabu search algorithm for system identification operations of different processes. The identification procedure is based on open-loop step response analysis of the processes. Each of the two algorithms is applied to determine the optimal parameter values of the processes to be modelled. Simulation results demonstrated that the presented algorithms can be efficiently used in the identification problems.

1 Introduction

One of the main problems in a design of a control system is the accurate identification of the system to be controlled. For tuning the controller, there are several parameters that reflect the control objectives, and the controller parameters are affected by the parameters of the identified process model. If a mathematical model of the plant can be derived, then it is possible to apply various design techniques for determining parameters of the controller that will meet the transient and steady-state specifications of the closed-loop systems. Therefore, in order to provide an accurate and efficient solution for linear control problems, a good model of the process should be obtained.

Identification of linear systems from closed-loop data has been addressed in the literature [1-5]. Ljung discussed and implemented many of the approaches to process identification using parametric and nonparametric models [1]. Solbrand et al. presented recursive methods for off-line identification [4]. Pintelon et al. gave an extensive survey on parametric identification of transfer functions in the frequency domain [6]. Schoukens et al. discussed use of a piece wise constant excitation method for linear dynamic system identification [7]. Rolain et al. used frequency domain identification methods to estimate the system order [8].

Since the average process engineer does not have the skill or the time to do a full system identification of each subprocess in a plant, proper or exact system identification is normally not done in the process industry. There is thus need for computer tools for obtaining a process model without too much effort. Heuristic optimization algorithms based on artificial intelligence can provide a fast and efficient solution to system identification problems. Genetic algorithms (GAs) and tabu search algorithms (TSAs) have been widely used as most successful optimization methods in areas such as optimization of the objective function, training of neural networks, tuning of fuzzy membership functions, machine learning, system identification and control [9]. These problems are difficult to optimize using traditional methods.

B. Gabrys, R.J. Howlett, and L.C. Jain (Eds.): KES 2006, Part I, LNAI 4251, pp. 94 – 101, 2006.
© Springer-Verlag Berlin Heidelberg 2006

In this work, the performance comparison of the TSA and GA is presented for system identification operations of different processes. The identification procedure is based on open-loop step response analysis. The main aim of each algorithm is to determine the optimal parameter values of the processes to be modelled. The paper is organized as follows. In Section 2, a brief review of the TSA and GA is presented. Section 3 presents the application of the algorithms for identification procedure. Simulation results are given in Section 4. Conclusions are presented in the last section.

2 Genetic Algorithms and Tabu Search Algorithm

2.1 Genetic Algorithms (GAs)

GAs are general purpose search algorithms which use principles inspired by natural genetics to evolve solutions to problems [9]. They work with a population of chromosomes, each one representing a possible solution to a given problem. The fitness values of all chromosomes are evaluated by calculating an objective function called as fitness function. The fitness function is the performance index of a GA and it describes the quality of each individual or chromosome. Based on the fitness of each individual, a group of the best chromosomes is selected through the selection process. The genetic operators, crossover and mutation, are applied to this selected population in order to improve the next generation solution. By exchanging information from each individual in a population, GAs preserve a better individual and yield higher fitness value by generation such that the performance can be improved. A GA requires the following major steps in order to solve a particular optimization problem: *1) Genetic representation of candidate solutions 2) Definition of a fitness function 3) Generation of an initial population 4) Description of the genetic operators 5) Selection of the GA parameters* (population size, number of generations, number of bits per variable, and probabilities of applying genetic operators).

2.2 Tabu Search Algorithm (TSA)

Tabu search was first introduced by Glover as an intelligent search technique to overcome local optimality [9]. The search moves from one solution to another, in order to improve the quality of the solutions visited. The simple tabu search starts with a feasible solution and chooses the best move according to an evaluation function. The move is a process that transforms the search from the current solution to its neighbouring solution. The neighbourhood is defined as the set of points reachable with a suitable sequence of local perturbations, starting from the current solution. New solutions are searched in the neighbourhood of the current one.

To avoid performing a move returning to a recently visited region, the reverses of the last moves are forbidden. These moves are then considered as tabu. They are recorded in a list called as tabu list, which is updated at each step. The objective of this list is to exclude moves which would bring the algorithm back to where it was at some previous iteration and keep it trapped in a local optimum. An admissible move is a move that is not on the tabu list. A move remains as a tabu move only for a given number of iterations.

During the iterations of the optimization process, candidate solutions are checked with respect to the tabu conditions and the next solutions is determined. The tabu conditions employed are based on the recency and frequency memories storing the information about the past steps of the search. The recency based memory prevents cycles of length less than or equal to a predetermined number of iterations from occurring in the trajectory. The frequency based memory keeps the number of change of the solution vector elements. If an element k of the solution vector does not satisfy the conditions (i) recency(k)>r.K, (ii) frequency(k)<f.avgfreq, then it is accepted as tabu and not used to create a neighbour. Here, r and f are recency and frequency factors, K is the number of the elements in solution vector, and avgfreq is the average frequency value. If a move on the tabu list leads to a solution with an objective function value strictly better than the best obtained so far, it is possible to allow this move. This operation called as aspiration criterion is a tool used to release the tabu status of a move if the move has the potential to lead the search to appealing solution regions.

The performance of TSA may be affected by the several factors: initial solution, type of move, size of neighbourhood, tabu list size, aspiration criterion, and stopping criterion.

3 Application of Algorithms to the Identification Problem

For the identification problem, four different processes with different orders are considered as follows:

$$\mathbf{G1(s)} = \frac{e^{-0.5S}}{(s+1)^2}, \qquad \mathbf{G2(s)} = \frac{4.228}{(s+0.5)(s^2+1.64s+8.456)},$$
$$\mathbf{G3(s)} = \frac{27}{(s+1)(s+3)^3} \qquad \mathbf{G4(s)} = \frac{2.1299}{0.356s^4+0.8116s^3+8.5057s^2+3.4020s+2.1299} \tag{1}$$

,

Fig.1 shows the structure of the controller self tuner and plant modelling scheme. A time-domain self tuning controller scheme consists of two main structures: (1) system identification mechanism, (2) controller tuning mechanism. Optimum model parameters are determined by the optimization algorithms. Our strategy to handle several different dynamics is to employ an appropriate reduced order model to capture major characteristics of their dynamics and yet have a common format on which a unified control design can be conducted. It is common and well-accepted practice to approximate high-order processes by low-order plus dead time models. Using model reduction method, a dynamical system can be represented by the first or second-order transfer functions as follows:

$$\mathbf{Gm1(s)} = \frac{e^{-LS}}{as+b}, \qquad \mathbf{Gm2(s)} = \frac{e^{-Ls}}{as^2+bs+c}. \tag{2.a, b}$$

where L, a, b, and c are the unknown (estimated) model parameters. These unknown parameters of the process model are found by the optimization algorithms according to the open-loop step response of the process. In order to have a good agreement of the model and process responses, the following performance function needs to be considered during the optimization procedure:

$$J(L,a,b,c) = \int_0^{\infty} (y_p(t+1) - y_m(t+1))^2 .dt \tag{3}$$

where $y_p(t+1)$ is the process step response, and $y_m(t+1)$ is the model step response which may be produced by the transfer function Gm1 or Gm2. During the optimization procedure, this function called as the integral square error (ISE) is minimized by the GA and TSA for better model parameters.

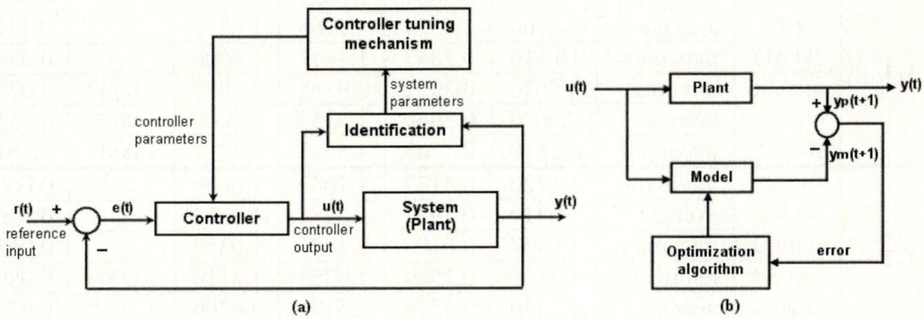

Fig. 1. (a) Block diagram of the controller self tuner, (b) Plant modelling scheme

4 Simulation Results

For the identification study, the model structures given in (2) are used. The main objective of each algorithm is to determine the unknown model parameters L, a, b, and c such that the model achieves high agreement with the process dynamics for the open-loop step response. The control parameter values of the GA and TS algorithms are given in Table 1. For all of the simulations, the iteration or generation number is taken as 50. The number of bits per variable is 8. Thus, the total number of bits in the string describing each member of the solutions is 32 (number of the variables x number of bits per variable=4x8).

Table 1. The values of the control parameters for GA and TSA

Genetic Algorithm	Tabu Search Algorithm
♦ population size = 30	♦ recency factor (r) = 0.5
♦ crossover rate = 0.8	♦ frequency factor (f) = 2.0
♦ mutation rate = 0.2	♦ iteration number = 50
♦ generation number = 50	♦ tabu conditions: **(1)** recency(k)>r.K, **(2)** frequency(k)<f.avgfreq
♦ type of the crossover = single point	
♦ selection operator = roulette wheel	♦ tabu list size = 4
♦ number of the variables = 4 (L, a, b, c)	♦ number of variables = 4 (L, a, b, c)
♦ number of bits per variable = 8	♦ number of neighbourhood per variable = 8
♦ variable limits: $L \in [0,5]$, $a \in [1,10]$ $b \in [1,10]$, $c \in [1,10]$	♦ variable limits: $L \in [0,5]$, $a \in [1,10]$ $b \in [1,10]$, $c \in [1,10]$

To observe the performance of the algorithms, three different starting points are taken. These are minimum (L=0, a=b=c=1), average (L=2.5, a=b=c=5), and maximum (L=5, a=b=c=10) starting points. After the optimization process, the final state of the model parameters is as shown in Table 2a and Table 2b.

Table 2a. Model parameters optimized by the GA

Process	Model	Starting Point	Time (sec)	L	a	b	c	J (ISE)
G1	first order (Gm1)	minimum	1.3750	0.9804	1.6353	1.0000	_	0.0041
		average	3.0160	1.4314	1.7765	1.0353	_	0.1274
		maximum	10.110	0.7843	1.8471	1.0000	_	0.0840
	second order (Gm2)	minimum	3.5620	0.3922	2.9765	2.2000	1.0000	0.0772
		average	8.0620	0.2941	1.0353	2.2000	1.0000	0.0530
		maximum	8.2810	0.1765	4.4941	2.5229	1.0000	0.1516
G2	first order (Gm1)	minimum	1.1720	0.3137	1.7059	1.0000	_	0.0061
		average	14.310	0.3922	1.3882	1.0353	_	0.0367
		maximum	13.437	0.0196	2.4471	1.0353	_	0.0633
	second order (Gm2)	minimum	1.1100	0.1569	1.6353	1.8118	1.0000	0.0802
		average	8.3440	0.4314	2.5176	3.4706	1.1059	0.6741
		maximum	6.5320	0.9412	1.0353	2.2000	1.0353	0.2969
G3	first order (Gm1)	minimum	0.9060	0.6275	1.5647	1.0000	_	0.0104
		average	6.8600	0.5490	2.3765	1.0000	_	0.1173
		maximum	6.0620	0.2157	1.4588	1.0353	_	0.0954
	second order (Gm2)	minimum	1.1250	0.0392	1.2824	1.8471	1.0000	0.0038
		average	9.7500	0.4706	1.3882	1.7765	1.0353	0.0721
		maximum	11.719	0.2549	1.3176	2.2706	1.0000	0.0542
G4	first order (Gm1)	minimum	0.8750	1.5686	1.0000	1.0000	_	0.3012
		average	8.3900	1.4510	2.1647	1.0000	_	0.6150
		maximum	12.859	1.1373	1.1765	1.1059	_	0.7106
	second order (Gm2)	minimum	0.8590	0.0000	3.4706	1.2824	1.0000	0.0541
		average	6.2660	0.0980	2.0941	1.1412	1.0353	0.4002
		maximum	4.1410	0.4706	3.8941	2.0235	1.0000	0.1895

Step responses of the models optimized by the algorithms for different processes are given in Figs.2-3. These graphics are drawn by using the parameter values with the minimum error. For example, for the graphic given in Fig.2a, the values of the first-order model parameters determined by the GA are L=0.9804, a=1.6353, and b=1.0000 (Table 2a). These values for the TSA are L=1.2549, a=1.3176, and b=1.0000 (Table 2b). On the other hand, the values of the second-order model parameters determined by the GA are L=0.2941, a=1.0353, b=2.2000, and c=1.0000 (Table 2a). Similarly, these parameter values in the second-order model optimized by the TSA are used as L=0.3529, a=1.2824, b=2.1294, and c=1.0000 (Table 2b). Fig.3b shows the second-order model parameters determined by the TSA for the process G4. The variation of the performance function J is also given in this figure.

Simulation results show that the TSA gives better results than the GA (Table 2a,b). The error of the performance function J in the approximation of the model parameters is very small in any condition (Table 2b). For the G1, G2, G3, and G4 processes, when the GA is used, the minimum values of the modelling errors are 0.0041, 0.0061, 0.0038, and 0.0541, respectively (Table 2a). On the other hand, if TSA is used in the identification procedure, these values are obtained as 0.0006, 0.0086, 0.0017, and 0.0014, respectively (Table 2b). The effectiveness of the TSA is obtained especially in the identification of the second-order plus dead time model. The minimum error values of the second-order models for G1, G2, G3, and G4 processes in the use of the GA are 0.053, 0.0802, 0.0038, and 0.0541, respectively (Table 2a). When the TSA is employed in the optimization, the values of 0.0006, 0.0112, 0.0017, and 0.0014 are obtained, respectively (Table 2b). These numerical results and simulation plots clearly show that the models identified by the TSA can successfully represent the processes considered (Table 2b, Figs.2-3).

Table 2b. Model parameters optimized by the TSA

Process	Model	Starting Point	Time (sec)	L	a	b	c	J (ISE)
G1	first order (Gm1)	minimum	1.6720	1.2549	1.3176	1.0000	_	0.0112
		average	7.7190	0.6078	2.1294	1.0000	_	0.0224
		maximum	6.2500	0.6078	1.9529	1.0000	_	0.0181
	second order (Gm2)	minimum	2.0620	0.3529	1.2824	2.1294	1.0000	0.0006
		average	6.2810	0.1569	2.1294	2.2706	1.0000	0.0071
		maximum	9.6090	0.2941	1.3882	2.0941	1.0000	0.0019
G2	first order (Gm1)	minimum	1.2030	0.1569	2.1294	1.0000	_	0.0086
		average	3.7810	0.1569	2.1294	1.0000	_	0.0086
		maximum	6.9530	0.4510	1.5294	1.0000	_	0.0095
	second order (Gm2)	minimum	1.1720	0.0000	1.0000	2.1647	1.0000	0.0112
		average	5.4680	0.0000	1.0000	2.1647	1.0000	0.0112
		maximum	11.9850	0.0000	1.0000	2.0941	1.0000	0.0119
G3	first order (Gm1)	minimum	1.2340	1.2549	1.0000	1.0000	_	0.0379
		average	3.7820	0.3137	2.1294	1.0000	_	0.0515
		maximum	6.9530	0.6078	1.4588	1.0000	_	0.0080
	second order (Gm2)	minimum	1.5780	0.0000	1.3176	2.1294	1.0000	0.0112
		average	4.8280	0.0000	1.5647	2.1294	1.0000	0.0132
		maximum	9.6720	0.1373	1.1059	1.8118	1.0000	0.0017
G4	first order (Gm1)	minimum	1.2340	1.5686	1.0000	1.0000	_	0.3012
		average	4.1100	0.6275	2.1294	1.0000	_	0.4423
		maximum	8.5310	1.2353	1.2118	1.0000	_	0.3115
	second order (Gm2)	minimum	2.3910	0.0196	3.9647	1.5647	1.0000	0.0014
		average	3.7350	0.0392	3.8588	1.5647	1.0000	0.0014
		maximum	8.3900	0.2941	3.2235	1.4235	1.0000	0.0107

In TSA, the quality of the starting solution is an important factor that plays an essential role in the search process. The diversification of candidate solutions to be evaluated in iterations of the search widely relies on the quality of the initial solution. This ensures that large areas of the space are searched and solutions do not get stuck in local minima. It is clearly seen that the optimum values of the model parameters can be obtained by using TSA in any starting solution (Table 2b). Depending on the values of a, b, and c, the model may have real or complex poles. Hence, second-order models are more suitable for representing both monotonic and oscillatory processes. Thus, for G4 process with the oscillations, the error value of the second-order model is smaller than the first-order model (Table 2a,b). The modelling error in TSA based optimization procedure has a remarkable reduction. This result also exhibits the computation power of the TSA with the minimum time (Table 2b, Fig.3bi,ii).

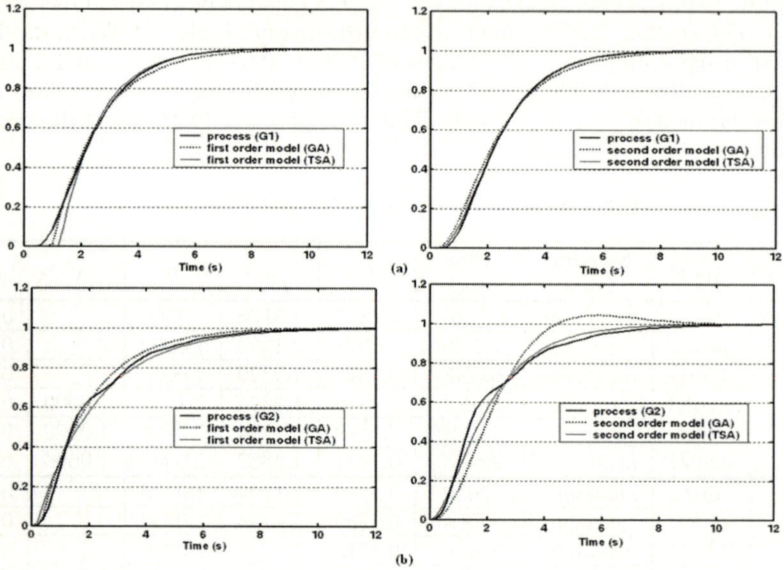

Fig. 2. Step responses of the models optimized by the algorithms for the processes G1 and G2

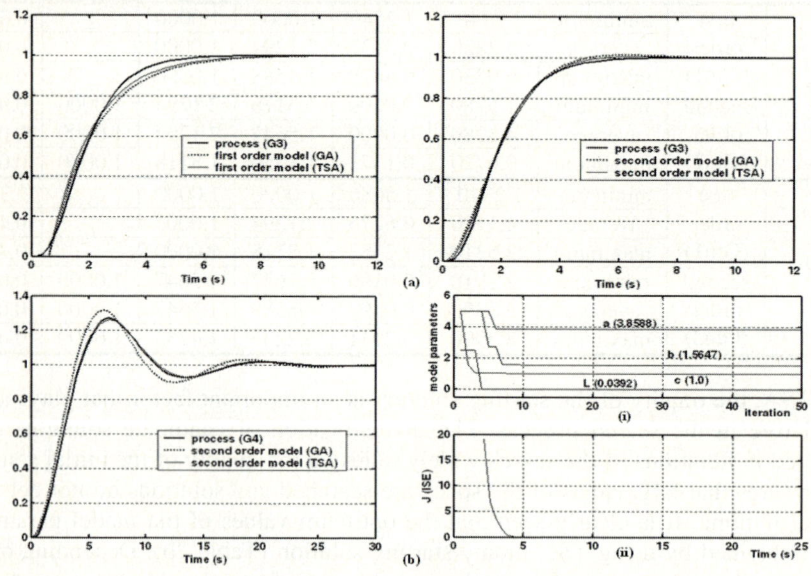

Fig. 3. (a),(b) Step responses of the models optimized by the algorithms for the processes G3 and G4. (i) Second-order model parameters determined by the TSA for the process G4, (ii) Performance function J versus time.

Although GAs use a group of the solution called as the population during the optimization procedure, the quality of the final solutions obtained by the algorithm is not very good (Table 2a, Fig.2b). However, in general, the GA has the satisfactory performance for all of the processes. It is noted that, by carefully adjusting the population size and the other control parameters of the algorithm, the desirable results can be obtained. Consequently, it is demonstrated that the GA and TSA constitute an efficient and fast optimizing alternatives for the system identification problems. These methods can be successfully employed for considerably increasing of the system performance in a control mechanism.

5 Conclusion

TSA and GAs have been applied to the system identification operations of different processes. The open-loop step responses obtained by the algorithms are compared to those obtained by the processes to be identified. Simulation results show that the TSA demonstrates a better performance in terms of the performance function considered. It is concluded that both GA and TSA can be successfully used in system identification problems.

References

1. Ljung, L.: System Identification. Prentice-Hall Englewood Cliffs, NJ. (1987).
2. Pintelon, R., Guillaume, P., Rolain, Y., Verbeyst, F.: Identification of Linear Systems Captured in a Feedback Loop. IEEE Trans. on Inst. and Measurement 41(6) (1992) 747-754.
3. Schoukens, J., Pintelon, R., Vandersteen, G., Guillaume, P.: Frequency Domain System Identification using Nonparametric Noise Models Estimated from A Small Number of Data Sets. Automatica 33(6) (1997) 1073-1086.
4. Solbrand, G., Ahlen, A., Ljung, L.: Recursive Methods for Off-Line Identification. Int.Journal of Control 41(1) (1985) 177-191.
5. Söderström, T., Stoica, P.: System Identification. Prentice-Hall Englewood Cliffs NJ. (1989).
6. Pintelon, R., Guillaume, P., Rolain, Y., Schoukens, J., VanHamme, H.: Parametric Identification of Transfer Functions in the Frequency Domain-A Survey. IEEE Trans. on Automatic Control 39(11) (1994) 2245-2260.
7. Schoukens, J., Pintelon, R., VanHamme, H.: Identification of Linear Dynamic Systems using Piecewise Constant Excitations: Use, Misuse and Alternatives. Automatica 30(7) (1994) 1153-1169.
8. Rolain, Y., Schoukens, J., Pintelon, R.: Order Estimation for Linear Time-Invariant Systems using Frequency Domain Identification Methods. IEEE Trans. on Automatic Control 42(10) (1997).
9. Pham, D.T., Karaboga, D.: Intelligent Optimisation Techniques: Genetic Algorithms, Tabu Search, Simulated Annealing and Neural Networks. Springer-Verlag. (2000).

Towards Self-configuring Evolvable FPGA Using Feedback Cross-Checking

Robert Ross and Richard Hall

Dept. Computer Science & Engineering,
La Trobe University, Melbourne,
3086, Victoria, Australia
rj3ross@students.latrobe.edu.au

Abstract. Evolvable hardware supposedly avoids single point of failure in circuitry. However, its control circuit is currently separate, thus a single point of failure exists in their interface. In contrast, we implement self-maintaining circuits in a faulty field programmable gate array simulation. We evaluate these circuits with respect to diagnosis reliability and reconfiguration performance. An implementation of our model could increase the survivability of critical engineering applications.

1 Introduction

In electronic circuits, *system failure* occurs when a system deviates from its requirements [1], which may be catastrophic in critical applications (eg. aerospace, defense and medical) [2,3,4]. Such systems have redundant hardware so that operational reliability is uncompromised by isolated failure [5]. Evolvable hardware (EHW) has logic and interconnections which are reconfigured to re-create fault-free circuits (via evolutionary algorithms) [6] Field Programmable Gate Arrays (FPGA) are programmable logic devices that are popular for EHW [7]. Current applications include microwave filters for cellular phones [8], power management systems [9], and critical path circuitry in NASA space-craft [10].

Control circuits for diagnosis and reconfiguration are currently separated from FPGA EHW for two reasons. Firstly, the majority of FPGAs can only be externally reconfigured; only recently Xilinx released their Virtex FPGAs with internal reconfiguration support [11,12]. Secondly, few tools support EHW implementation [7,12]. In contrast, in our FPGA simulation, both diagnostic and reconfiguration circuitry are built-in. Redundant diagnosis modules check a target circuit and also each other via feedback (ala [1,13]). Our simulation sidesteps hardware issues (eg. fault generation and propagation delays).

The rest of this paper is as follows. In section 2 we discuss requirements for modelling FPGA faults and self-configuration, and describe their implementation in section 3. In section 4 we make predictions of our fault-tolerant circuitry design in FPGA hardware and describe future work in section 5.

B. Gabrys, R.J. Howlett, and L.C. Jain (Eds.): KES 2006, Part I, LNAI 4251, pp. 102–109, 2006.
© Springer-Verlag Berlin Heidelberg 2006

2 Requirements

To simulate self-configuring evolvable FPGA we require three components:

1. *A representation of FPGA.* FPGA's contain millions of *logic* blocks and *control* blocks arranged in a lattice [14,15]. We model the (recent) Xilinx Virtex II FPGA architecture, whose *logic blocks* (Figure 1) contain a Lookup Table (LUT) which returns a selected logic gate output, a D-Latch (memory) and an output multiplexer. Logic blocks can be represented by a four inputs and one output (a second inverted output can be taken from the flip-flop) [16]. Control (switch) blocks route interconnections between logic blocks (allowing complete control of signal paths) and clock signals are distributed throughout the board by a separate set of wires [17].

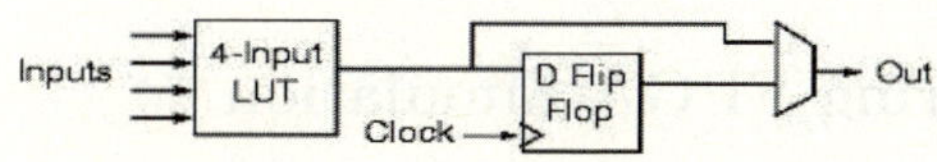

Fig. 1. FPGA logic block

2. *A representation of faults in FPGA.* We will represent a common basic type: single stuck-at faults (SSA), whereby a single node is connected (permanently) either to voltage supply rails or to ground. SSA is a good model for circuits damaged through electro-static discharge and radiation [18].
3. *Circuit models.* We required four modules: functional, output, management and reconfiguration (see Figure 2). The *functional module* is the target circuitry maintained by the EHW which is triplicated (we chose to implement a simple full adder).

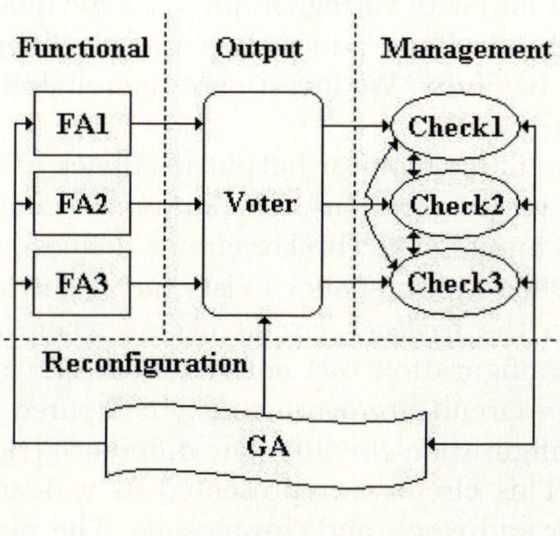

Fig. 2. Circuit Layout

The *output module* will detect faults in the functional modules (for our circuitry it will have inputs of 3 sums & 3 carries) and will trigger reconfiguration accordingly. This module can develop faults which are diagnosed by the management module.

The *management module* will contain three checking circuits. Each of these circuits will diagnose faults in the output module, reconfiguration module, and other checking circuits.

The *reconfiguration module* will re-create circuits by manipulating their logical structure and inter-connectedness (represented by a bit-stream). Following other EHW efforts, we will use genetic algorithms to directly manipulate this bit-stream [19,20]. We only implement the *mutation* operator (ala evolutionary programming [19]) which is amenable to hardware implementation via random bit inversion.

These requirements guided our simulation design.

3 Self-configuring FPGA Simulation

Our simulator consisted of an FPGA fault emulator and fault-tolerant circuit logic implementation. The fault emulator required three bit-stream representations: logic blocks, control blocks, and faults. *Logic blocks* are represented as a 5-tuple: its 2D coordinate in the FPGA lattice; LUT type; MUX type, and two output connections. *Control blocks* are represented as a bitstream tuple; its 2D coordinate in the FPGA lattice; and the on-off status of each of its required 56 different switches. *Faults* (assumed to occur on buses) are represented as a five-tuple; fault id; its 2D coordinate in the FPGA lattice; direction (vertical or horizontal); bus line (0..3); and stuck-at value (0 or 1). Our simulated circuits had their logic and control blocks represented as four aggregated bit-streams.

1. *Functional.* We chose the the simple full-adder (5 logic blocks) for its simplicity of design and widespread use. A total of three full adders were implemented in the circuit; results were propagated to the output circuit.
2. *Output.* We used a majority voting circuit (17 logic blocks), as it is simple to implement in hardware. If one adder gives a result different to the other two, it was assumed to be faulty. We investigate the reliability of this assumption later in this paper.
3. *Feedback checkers.* These monitor output, feedback and reconfiguration circuits (and trigger reconfiguration). The *output* voter circuit should report a discrepancy in its inputs. The checker circuit defines an error in the output circuit to occur when a discrepancy exists but no error is reported, or vice versa. An error in the *feedback* circuit occurs when it triggers a reconfiguration but a reconfiguration was not required, or vice versa. All parts of the *reconfiguration* circuit are continuously compared to the parts of other (redundant) reconfiguration circuits. Any difference triggers reconfiguration.
4. *Reconfiguration.* This circuit is represented as a 3-tuple; random number generator, circular addressor, and chromosone. The *random number generator* is implemented as a 5-bit Linear shift register which continuously cycles.

This value is sampled when faults are introduced into the system, thus the randomness of the values is correlated with the randomness of samples. The *circular addressor* provides a memory pointer into the chromosone bitsteam. During simulation this pointer cycles through the chromosone, ANDing bits with the random number generator output (see Pseudocode).

Mutation PseudoCode:

```
FOR (i = 0; i < numCycles; ++i){
    randNum = getRandomNumber();  \\This value will be either 0 or 1
    FOR (j = 0; j < numBlocks; ++j){
        \\Each logic block is XORed with (randomNumber AND AddressLocation)
        LogicBlockStoredValue[j] ^= (blockAddresser AND randNum);
    }
    blockAddresser = blockAddresser >> 1; //Shift addresser position by 1
}
```

Chromosones must be chosen judiciously; mutating working blocks easily causes errors. We found the repair time of the genetic algorithm (our largest circuit) particularly poor when treating it as an entire chromosone (results not shown). We were able to improve this performance by focussing on subchromosonal blocks of size four (this size was chosen after hand tuning) by subtracting the faulty chromosone bitstream from a working chromosone stream, significantly reducing the search space. Such an approach is in keeping with the spirit of genetic algorithms which use mutations primarily to avoid local optimums [19,20]. We diverged from genetic algorithms in the sense that we have a population of one (asexual reproduction) for the sake of simplicity, and we use a boolean fitness function (the proposed mutation either passes or fails the checker (triggering further reconfiguration)). Envisaging a fitness function for partially-correct circuits was beyond the scope of this paper.

We now evaluate our self-configuring FPGA EHW.

4 Evaluation

We evaluate our circuits with respect to diagnosis reliability and reconfiguration performance. With respect to *diagnosis reliability*, we mentioned previously that majority voting can be incorrect when the majority of input modules output the same erroneous result, and we want to estimate how often that could occur, demonstrating the effect of timely reconfiguration on stability. Calculations are based on a scenario of using full adder circuits connected to a voting circuit as implemented in the simulator with the parameters given in table 1.

Following the full adder logic, there are 5 independent bus lines which if stuck low would result in the sum value to be 0 (which is erroneous). Likewise there are 8 different bus lines which if held low would result in the carry value being 0. Using these values and knowing that the number of bus lines within the full

Table 1. Reliability Parameters

Pr(line stuck low)	10^{-6}/cycle (pessimistically)
Number of bus lines for 1 full adder	68
Circuit inputs	
CarryIn	1
A	1
B	1
Correct Outputs	
Sum	1
Carry	1

adder is 68 we can compute the probability of errors per cycle, given that a stuck low fault has occurred in the full adder for the inputs listed above:

$$Pr(SumError|StuckLowFaultOccured) = 5/68$$
$$Pr(CarryError|StuckLowFaultOccured) = 8/68$$
$$Pr(Error|StuckLowFaultOccured) = 13/68$$
$$Pr(StuckLowFaultInFA) = 68 \times 10^{-6}$$

Since we are calculating out the worst case scenario, calculations will continue for the overall output and will ignore any additional faults which may cancel out the original fault. Now the probability that the output will be faulty given the above calculations and three available full adders is:

$$Pr(CarryError) = (13/68 \times 10^{-6}) \times 3 = 39 \times 10^{-6}$$

The probability of a second full adder developing the same erroneous output and therefore causing system failure per cycle is:

$$Pr(CarryError|CarryError) = 39 \times 10^{-6} \times (13/68 \times 68 \times 10^{-6} \times 2) = 1.014 \times 10^{-9}$$

Note that the second full adder would have to fail before the first faulty full adder had been repaired to cause system failure.

With respect to *reconfiguration performance*, we introduced crippling SSA faults into each module separately in order to isolate the mean reconfiguration cycles and mean time to repair. We are particularly interested in knowing if the system could repair itself in a timely manner so that faults can be repaired fast enough (in general) to avoid system failure. Table 2 lists the reconfiguration performance leading to repair of each module.

Since each reconfiguration takes 20 master clock cycles, our simulation runs at approximately 23Hz (causing the slow repair times). In contrast, the Xilinx Virtex II FPGA (on which our simulation is based) has a 420MHz clock, thus full adder repair times (1133 seconds simulated) would complete in $\approx 62\ \mu$ s.

Table 2. Reconfiguration Performance

	Number of Tests Performed	Mean reconfiguration Cycles to fix circuit	Mean Time Taken (Seconds)
Full Adder	20	1303	1133
Voting Circuit	20	1197	1041
Checker Circuit	20	564	491
Rand Number Gen	20	126.55	110
Reconfig Circuit	20	611.25	531
Averages	20	760.36	661.2

The best case reconfiguration occurred on the checker circuit, taking only 32 reconfiguration cycles and completing the repair in a scaled hardware time of 150ns. The worst case reconfiguration repair also occurred on the checker circuit, taking 3578 reconfiguration cycles at a scaled time of 170μs. Note that the checker circuit was treated as a chromosone in its entirety.

5 Conclusion

In this paper we introduced a model of self-configuring evolvable FPGA that uses feedback cross-checking to avoid infinite regress. We used a minimalist hybrid of genetic algorithms and evolutionary programming to represent and mutate sub-chromosomal faults in circuits. There are many avenues for future work that we could explore:

- An effective means of generating faults in hardware to allow implementation.
- Other fault types, voting schemes, genetic algorithms and complex circuitry.
- Techniques to select minimal circuit chromosones that requires mutation.
- An address decoding scheme for blocks that improves resource management and performance in large lattices.

Our simulation performs self-diagnosis and self-repair so a direct self-configurable FPGA hardware implemention would require no external control circuitry. Such an implementation would maintain a full adder with a diagnostic reliability of 99% and a mean time to repair of 36μs (from our simulated mean of 661s). This self-configurable FPGA could be used in conjunction with external circuitry to provide even greater reliability for critical applications.

Acknowledgement. We would like to thank Adrian Stoica and Ricardo Zebulum (from the Biologically Inspired Technology and Systems (BITS) Group, Flight Systems Electronics Section, Jet Propulsion Laboratory, California Institute of Technology), for inspiring this research at KES2004, and for their assistance. We would also like to thank the Department of Computer Science and Computer Engineering at La Trobe University for supporting this research.

References

1. Anderson, T., Lee, P.A.: Fault Tolerance - Principles and Practice. Springer–Verlag, New York (1992)
2. Jafari, R., Dabiri, F., Brisk, P., Sarrafzadeh, M.: Adaptive and fault tolerant medical vest for life-critical medical monitoring. In: ACM symposium on Applied computing, New York, NY, USA, ACM Press (2005) 272–279
3. Dingman, C.P., Marshall, J.: Measuring robustness of a fault-tolerant aerospace system. In: FTCS 95: Twenty-Fifth International Symposium on Fault-Tolerant Computing, Washington DC USA, IEEE Computer Society (1995) 522
4. Moser, L., Melliar-Smith, M.: Demonstration of fault tolerance for corba applications. In: ARPA Information Survivability Conference and Exposition. Volume 2., Washington, DC, USA, IEEE Computer Society (2003) 87
5. Ellisona, R., Linger, R., Lipson, H., Mead, N., Moore, A.: Foundations for survivable systems engineering. Technical report, CERT Coordination Center, Software Engineering Institute Carnegie Mellon University (2001)
6. Greenwood, G.W., Hunter, D., Ramsden, E.: Fault recovery in linear systems via intrinsic evolution. In: Proceedings of the 2003 NASA/DoD Conference on Evolvable Hardware. Volume 1. (2003) 59–65
7. Sekanina, L.: Towards evolvable IP cores for FPGA's. In: NASA/DoD Conference on Evolvable Hardware. Volume 1. (2003) 145–154
8. Kasai, Y., Sakanashi, H., Murakawa, M., Kiryu, S., Marston, N., Higuchi, T.: Initial evaluation of an evolvable microwave circuit. In Miller, J., Thompson, A., Thomson, P., Fogarty, T.C., eds.: Evolvable Systems: From Biology to Hardware, Third International Conference on Evolvable Systems (ICES-2000). Volume 1801 of LNCS., Edinburgh, Scotland, UK, Springer Verlag (2000) 103–112
9. Tian, L., Arslan, T.: An evolutionary power management algorithm for SoC based EHW systems. In: NASA/DoD Conference on Evolvable Hardware. Volume 1. (2003) 117–124
10. Vigander, S.: Evolutionary fault repair of electronics in space applications. Dept. of Computer & Information Science, Norwegian University of Science and Technology (NTNU), Trondheim. (2001)
11. Eck, V., Kalra, P., LeBlanc, R., McManus, J.: In-circuit partial reconfiguration of rocketIO attributes. Technical report, XILINX (2004)
12. Sekanina, L.: On routine implementation of virtual evolvable devices using COMBO6. NASA/DoD Conference on Evolvable Hardware (2004)
13. Eberbach, E.: On expressiveness of evolutionary computation: Is EC algorithmic? In: World Congress on Computational Intelligence WCCI'2002. (2002) 564–569
14. Rangasayee, K.: The spartan-II family: The complete package. Technical report, Xilinx (2000)
15. Shanthi, A.P., Parthasarathi, R.: Exploring FPGA structures for evolving fault tolerant hardware. In: NASA/DoD Conference on Evolvable Hardware. Volume 1. (2003) 174–181
16. Goldstein, S., Budiu, M., Mishra, M., Venkataramani, G.: Reconfigurable computing and electronic nanotechnology. ASAP '03, The Hague, Netherlands (2003)
17. Trahan, J.L.: FPGA background. EE7700: Course on Run-Time Reconfiguration, Dept. Electrical & Computer Engineering, Louisiana State University (2005)

18. Garvie, M., Thompson, A.: Evolution of combinatorial and sequential on-line self-diagnosing hardware. In: 2003 NASA/DoD Conference on Evolvable Hardware. Volume 1. (2003) 167–173
19. Bäck, T. In: Evolutionary Algorithms in Theory and Practice. Oxford University Press, New York (1996) 80 – 121
20. Michalewicz, Z. In: Genetic Algorithms + Data Structures = Evolution Programs. Springer–Verlag, Berlin (1992) 55 – 62

A Self-adjusting Load Sharing Mechanism Including an Improved Response Time Using Evolutionary Information

Seonghoon Lee[1], Donghee Shim[2], and Dongyoung Cho[2]

[1] Department of Computer Science, Cheonan University
115, Anseo-dong, Cheonan, Choongnam, Korea
shlee@cheonan.ac.kr
[2] School of Information Technology & Engineering, Jeonju University
1200, 3rd Street Hyoja-dong Wansan-Koo, Jeon-Ju, Chonbuk, Republic of Korea
{dhshim, chody}@jj.ac.kr

Abstract. A load sharing mechanism is an important factor in computer system. In sender-initiated load sharing algorithms, when a distributed system becomes to heavy system load, it is difficult to find a suitable receiver because most processors have additional tasks to send. The sender continues to send unnecessary request messages for load transfer until a receiver is found while the system load is heavy. Because of these unnecessary request messages it results in inefficient communications, low cpu utilization, and low system throughput. To solve these problems, we propose a self-adjusting evolutionary algorithm approach for improved sender-initiated load sharing in distributed systems. This algorithm decreases response time and increases acceptance rate. Compared with the conventional sender-initiated load sharing algorithms, we show that the proposed algorithm performs better.

1 Introduction

Distributed systems consist of a collection of autonomous computers connected network. The primary advantages of these systems are high performance, availability, and extensibility at low cost. To improve a performance of distributed systems, it is essential to keep the system load to each processor equally.

An objective of load sharing in distributed systems is to allocate tasks among the processors to maximize the utilization of processors and to minimize the mean response time. Load sharing algorithms can be largely classified into three classes: static, dynamic, adaptive. Our approach is based on the dynamic load sharing algorithm. In dynamic scheme, an overloaded processor(sender) sends excess tasks to an underloaded processor(receiver) during execution.

Dynamic load sharing algorithms are specialized into three methods: sender-initiated, receiver-initiated, symmetrically-initiated. Basically our approach is a sender-initiated algorithm.

Under sender-initiated algorithms, load sharing activity is initiated by a sender trying to send a task to a receiver[1, 2]. In sender-initiated algorithm, decision of task transfer is made in each processor independently. A request message for the task

B. Gabrys, R.J. Howlett, and L.C. Jain (Eds.): KES 2006, Part I, LNAI 4251, pp. 110 – 121, 2006.
© Springer-Verlag Berlin Heidelberg 2006

transfer is initially issued from a sender to an another processor randomly selected. If the selected processor is receiver, it returns an accept message. And the receiver is ready for receiving an additional task from sender. Otherwise, it returns a reject message, and the sender tries for others until receiving an accept message. If all the request messages are rejected, no task transfer takes place. While distributed systems remain to light system load, a sender-initiated algorithm performs well. But when a distributed system becomes to heavy system load, it is difficult to find a suitable receiver because most processors have additional tasks to send. So, many request and reject messages are repeatedly sent back and forth, and a lot of time is consumed before execution. Therefore, much of the task processing time is consumed, and causes low system throughput, low cpu utilization

To solve these problems in sender-initiated algorithm, we use a new evolutionary algorithm. A new evolutionary algorithm evolves strategy for determining a destination processor to receive a task in sender-initiated algorithm. In this scheme, a number of request messages issued before accepting a task are determined by proposed evolutionary algorithm. The proposed evolutionary algorithm applies to a population of binary strings. Each gene in the string stands for a number of processors which request messages should be sent off.

The rest of the paper is organized as follows. Section 2 presents the Evolutionary Algorithm-based sender-initiated approach. Section 3 presents several experiments to compare with conventional method. Finally the conclusions are presented in Section 4.

2 Evolutionary Algorithm-Based Approach

In this section, we describe various factors to be needed for EA-based load sharing. That is, load measure, representation method, fitness function and algorithm.

2.1 Load Measure

We employ the CPU queue length as a suitable load index because this measure is known the most suitable index[5]. This measure means a number of tasks in CPU queue residing in a processor.

We use a 3-level scheme to represent a load state on its own CPU queue length of a processor. Table 1 shows the 3-level load measurement scheme. T_{up} and T_{low} are algorithm design parameters and are called *upper* and *lower thresholds* respectively.

Table 1. 3-level load measurement scheme

Load state	Meaning	Criteria
L-load	light-load	$CQL \leq T_{low}$
N-load	normal-load	$T_{low} < CQL \leq T_{up}$
H-load	heavy-load	$CQL > T_{up}$

(CQL : CPU Queue Length)

The transfer policy use the threshold policy that makes decisions based on the CPU queue length. The transfer policy is triggered when a task arrives. A node identifies as a ***sender*** if a new task originating at the node makes the CPU queue length exceed T_{up}. A node identifies itself as a suitable ***receiver*** for a task acquisition if the node's CPU queue length will not cause to exceed T_{low}.

2.2 Representation

Each processor in distributed systems has its own population which evolutionary operators are applied to. There are many encoding methods; Binary encoding, Character and real-valued encoding and tree encoding[12]. We use binary encoding method in this paper. So, a string in population can be defined as a binary-coded vector $<v_0, v_1, ..., v_{n-1}>$ which indicates a set of processors to which the request messages are sent off. If the request message is transferred to the processor P_i(where $0 \leq i \leq n$-1, n is the total number of processors), then v_i=1, otherwise v_i=0. Each string has its own fitness value. We select a string by a probability proportional to its fitness value, and transfer the request messages to the processors indicated by the string. When ten processors exist in distributed system, the representation is displayed as fig 1.

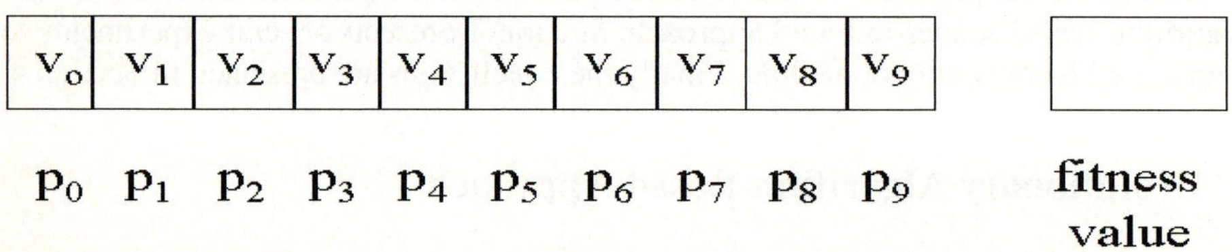

Fig. 1. Representation for processors

2.3 Load Sharing Approach

2.3.1 Overview

In sender-based load sharing approach using evolutionary algorithm, Processors received the request message from the sender send accept message or reject message depending on its own CPU queue length. In the case of more than two accept messages returned, one is selected at random.

Suppose that there are 10 processors in distributed systems, and the processor P_0 is a sender. Then, evolutionary algorithm is performed to decide a suitable receiver. It is selected a string by a probability proportional to its fitness value. Suppose a selected string is <-, 1, 0, 1, 0, 0, 1, 1, 0, 0>, then the sender P_0 sends request messages to the processors (P_1, P_3, P_6, P_7). After each processor(P_1, P_3, P_6, P_7) receives a request message from the processor P_0, each processor checks its load state. If the processor P_3 is a light load state, the processor P_3 sends back an accept message to the processor P_0. Then the processor P_0 transfers a task to the processor P_3.

2.3.2 Fitness Function

Each string included in a population is evaluated by the fitness function using following formula in sender-initiated approach.

$$F_i = \left(\frac{1}{\alpha \times TMP + \beta \times TMT + \gamma \times TTP} \right)$$

α, β, γ used above formula mean the weights for parameters such as *TMP*, *TMT*, *TTP*. The purpose of the weights is to be operated equally for each parameter to fitness function F_i.

Firstly, *TMP*(Total Message Processing time) is the summation of the processing times for request messages to be transferred. This parameter is defined by the following formula. The *ReMN* is the number of messages to be transferred. It means the number of bits set '1' in selected string. The objective of this parameter is to select a string with the fewest number of messages to be transferred.

$$TMP = \sum_{k \in x} \left(ReMN_k \times TimeUnit \right)$$

(where, x={i | v_i=1 for $0 \leq i \leq n$-1})

Secondly, *TMT*(Total Message Transfer time) means the summation of each message transfer times(*EMTT*) from the sender to processors corresponding to bits set '1' in selected string. The objective of this parameter is to select a string with the shortest distance eventually. So, we define the *TMT* as the following formula.

$$TMT = \sum_{k \in x} EMTT_k$$

(where x={i | v_i=1 for $0 \leq i \leq n$-1})

Last, *TTP*(Total Task Processing time) is the summation of the times needed to perform a task at each processor corresponding to bits set '1' in selected string. This parameter is defined by the following formula. The objective of this parameter is to select a string with the fewest loads. Load in parameter *TTP* is the volume of CPU queue length in the processor.

$$TTP = \sum_{k \in x} \left(Load_k \times TimeUnit \right)$$

(where x={i | v_i=1 for $0 \leq i \leq n$-1})

So, in order to have a largest fitness value, each parameter such as *TMP*, *TMT*, *TTP* must have small values as possible as. That is, *TMP* must have the fewer number of request messages, and *TMT* must have the shortest distance, and *TTP* should have the fewer number of tasks.

Eventually, a string with the largest fitness value in population is selected. And after evolutionary_operation is performed, the request messages are transferred to processors corresponding to bits set '1' in selected string.

2.3.3 Algorithm

This algorithm consists of five modules such as Initialization, Check_load, String_evaluation, Evolutionary_operation and Message_evaluation. Evolutionary_ operation module consists of three sub-modules such as Local_improvement_operation, Reproduction, Crossover. These modules are executed at each processor in distributed systems.

The algorithm of the proposed method for sender-initiated load sharing is presented as fig 2.

Algorithm : EA-based sender-initiated load sharing algorithm

Procedure Evolutionary_Algorithm Approach
{ Initialization();
while (Check_load())
if ($Load_i > T_{up}$) {
Individual_evaluation();
Evolutionary_operation();
Message_evaluation(); }
Process a task in local processor;
}
Procedure Evolutionary_operation()
{Local_improvement_operation();
Reproduction();
Crossover();
}

Fig. 2. Proposed algorithm

An Initialization module is executed in each processor. A population of strings is randomly generated without duplication.

A Check_load module is used to observe its own processor's load by checking the CPU queue length, whenever a task is arrived in a processor. If the observed load is heavy, the load sharing algorithm performs the following modules.

A Individual_evaluation module calculates the fitness value of strings in the population.

A Evolutionary_operation module such as Local_improvement_operation, Reproduction, Crossover is executed on the population in such a way as follows. Distributed systems consist of groups with autonomous computers. When each group consists of many processors, we can suppose that there are p parts in a string corresponding to the groups. The following evolutionary operations are applied to each string, and new population of strings is generated:

(1) Local_Improvement_Operation

String 1 is chosen. A copy version of the string 1 is generated and part 1 of the newly generated string is mutated. This new string is evaluated by proposed fitness function. If the evaluated value of the new string is higher than that of the original string, replace the original string with the new string. After this, the local improvement of part 2 of string 1 is done repeatedly. This local improvement is applied to each part one by one. When the local improvement of all the parts is finished, new string 1 is generated. String 2 is then chosen, and the above-mentioned local improvement is done. This local_improvement_operation is applied to all the strings in population.

```
/* Algorithms for local_improvement_operation */
for (i=1; i<=total_string_number; i++)
{
  select string[i];
  generate copy version of the selected string[i];
  for (j=1; j<=total_part_number; j++)
   /* total_part_number = p */
   {
    select a part[j] of the copy version;
    apply mutation operator to part[j];
    evaluate the mutated new string;
    if (fitness of new string > fitness of original string)
        original string ← new string;
   }
}
```

(2) Reproduction

The reproduction operation is applied to the newly generated strings. We use the "*wheel of fortune*" technique[4].

(3) Crossover

The crossover operation is applied to the newly generated strings. These newly generated strings are evaluated. We applied to the "one- point" crossover operator in this paper[4].

One-point crossover used in this paper differs from the pure one-point crossover operator. In pure one-point crossover, crossover activity generates based on randomly selected crossover point in the string. But boundaries between parts(p) are used as an alternative of crossover points in this paper. So we select a boundary among many boundaries at random. And a selected boundary is used as a crossover point. This purpose is to preserve an effect of the Local_improvement_operation of the previous phase. Therefore, the crossover activity in this paper is represented as fig 3.

Suppose that there are 5 parts in distributed systems. A boundary among the many boundaries(B_1, B_2, B_3, B_4) is determined at random as a crossover point. If a boundary B_3 is selected as a crossover point, crossover activity generate based on the B_3. So, the effect of the local_improvement_operation in the previous phase is preserved through crossover activity.

The Evolutionary_operation selects a string from the population at the probability proportional to its fitness, and then sends off the request messages according to the contents of the selected string.

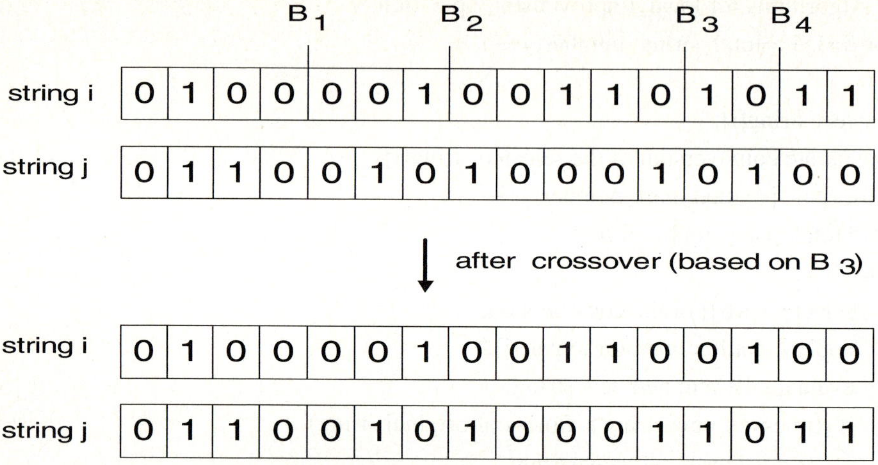

Fig. 3. Crossover Activity

A Message_evaluation module is used whenever a processor receives a message from other processors. When a processor P_i receives a request message, it sends back an accept or reject message depending on its CPU queue length.

3 Experiments

We executed several experiments on the proposed evolutionary algorithm approach to compare with a conventional sender-initiated algorithm.

Our experiments have the following assumptions. Firstly, each task size and task type are the same. Secondly, the number of parts(p) in a string is four. In evolutionary algorithm, crossover probability(P_c) is 0.7, mutation probability(P_m) is 0.1. The values of these parameters P_c, P_m were known as the most suitable values in various applications[3]. Table 2 shows the detailed contents of parameters used in our experiments.

Table 2. Contents of parameter

number of processor	24
P_c	0.7
P_m	0.1
number of strings	50
number of tasks to be performed	5000

The parameters and values for fitness value of sender-initiated load sharing algorithm are the same as the table 3. The load rating over systems supposed about 60 percent.

Table 3. Weight values for *TMP*, *TMT* and *TTP*

Weights for *TMP*	0.025
Weights for *TMT*	0.01
Weights for *TTP*	0.02

[Experiment 1]. We compared the performance of proposed method with a conventional method in this experiment by using the parameters on the table 2 and table 3. The experiment is to observe change of response time when the number of tasks to be performed is 5000.

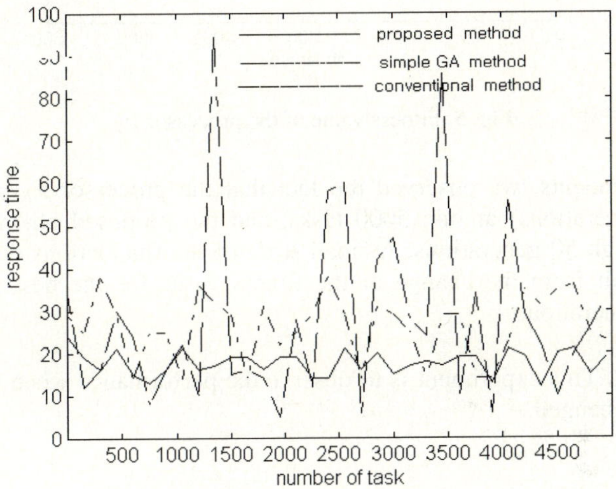

Fig. 4. Result of response time

Fig 4 shows result of the experiment 1. In conventional methods, when the sender determines a suitable receiver, it select a processor in distributed systems randomly, and receive the load state information from the selected processor. The algorithm determines the selected processor as receiver if the load of randomly selected processor is T_{low}(light-load). These processes are repeated until a suitable receiver is searched. So, the result of response time shows the severe fluctuation. In the proposed algorithm, the algorithm shows the low response time because the load sharing activity performs the proposed evolutionary_operation considering load states when it determines a receiver.

[Experiment 2]. This experiment is to observe the convergence of the fitness function for the best string in the population corresponding to a specific processor in distributed systems.

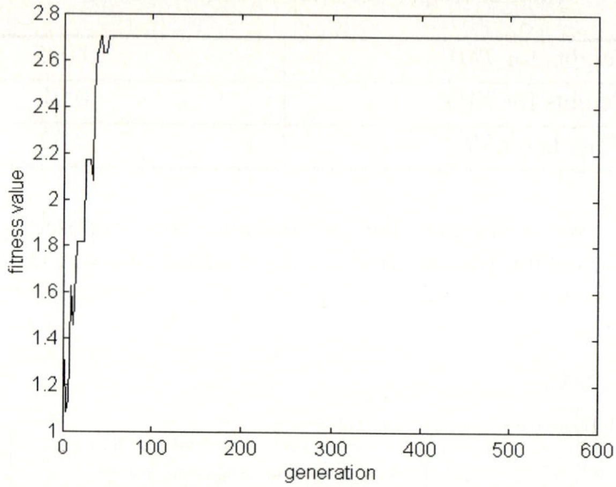

Fig. 5. Fitness value of the processor P_6

In this experiments, we observed the fact that the processor P_6 performs about 550tasks(550generations) among 5000 tasks, and the proposed algorithm generally converges through 50 generations. A small scale of the fluctuations displayed in this experiment result from the change of the fitness value for the best string selected through each generation.

[Experiment 3]. This experiment is to observe the performance when the probability of crossover is changed.

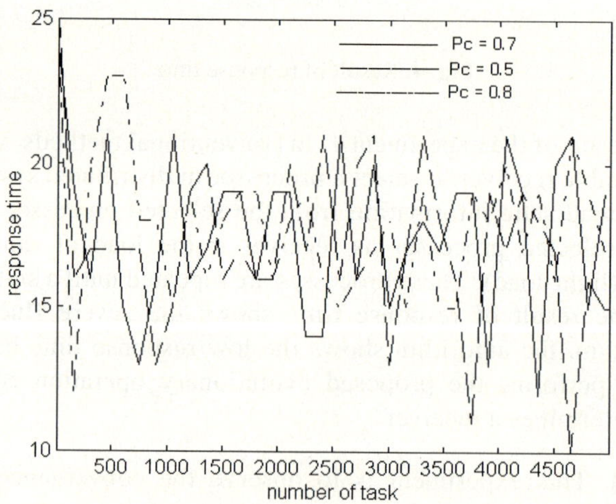

Fig. 6. Result depending on the changes of P_c

Fig 6 shows the result of response time depending on the changes of P_c when P_m is 0.1. In accordance with value of P_c, It shows a different performance. But the proposed algorithm shows better performance than that of conventional algorithm and simple evolutionary algorithm approach.

[Experiment 4]. This experiment is to observe the performance when the probability of mutation is changed

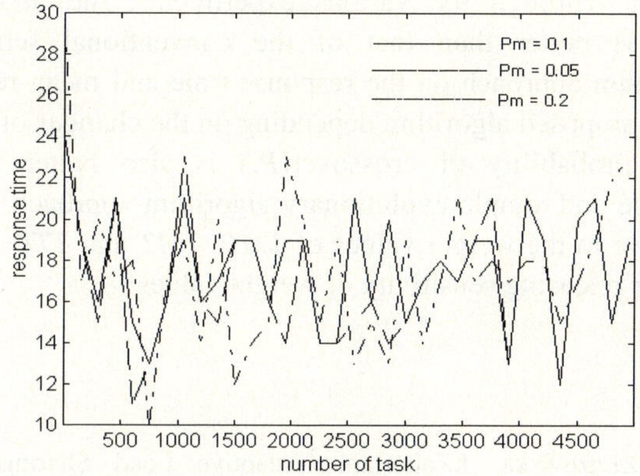

Fig. 7. Result depending on the changes of P_m

Fig 7 shows the result of the response time depending on the changes of P_m when P_c is 0.7. In accordance with value of P_m, It shows a different performance. But the proposed algorithm shows better performance than that of conventional algorithm and simple evolutionary algorithm approach.

[Experiment 5]. This experiment is to observe the response time when the system load is 80percentage

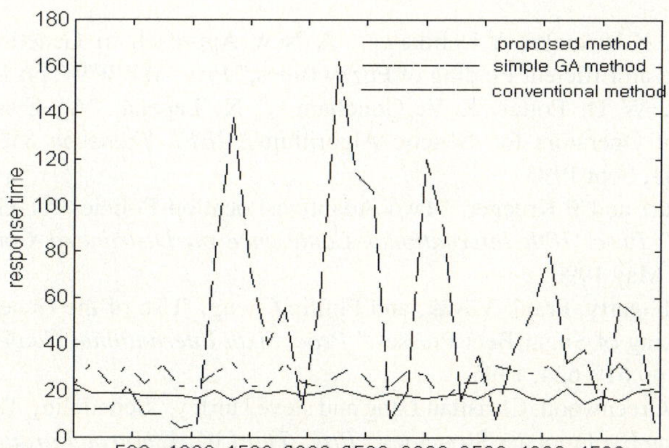

Fig. 8. Response time when system load is 80%

4 Conclusions

We propose new dynamic load sharing scheme in distributed system that is based on the new evolutionary algorithm with a local improvement operation. The proposed evolutionary algorithm is used to decide to suitable candidate receivers which task transfer request messages should be sent off. Several experiments have been done to compare the proposed scheme with a conventional algorithm and simple evolutionary algorithm approach. Through the various experiments, the performances of the proposed scheme is better than that of the conventional scheme and simple evolutionary algorithm approach on the response time and mean response time. The performance of the proposed algorithm depending on the changes of the probability of mutation(P_m) and probability of crossover(P_c) is also better than that of the conventional scheme and simple evolutionary algorithm approach. But the proposed algorithm is sensitive to the weight values of *TMP*, *TMT* and *TTP*. In future, we will study on method for releasing sensitivity of weight values

References

[1] D.L.Eager, E.D.Lazowska, J.Zahorjan, "Adaptive Load Sharing in Homogeneous Distributed Systems," *IEEE Trans on Software Engineering,* vol.12, no.5, pp.662-675, May 1986.

[2] N. G.Shivaratri, P.Krueger, and M.Singhal, "Load Distributing for Locally Distributed Systems," *IEEE COMPUTER*, vol.25, no.12, pp.33-44, December 1992.

[3] J.Grefenstette, "Optimization of Control Parameters for Genetic Algorithms," *IEEE Trans on SMC*, vol.SMC-16, no.1, pp.122-128, January 1986.

[4] J.R. Filho and P. C. Treleaven, "Genetic-Algorithm Programming Environments," *IEEE COMPUTER*, pp.28-43, June 1994.

[5] T. Kunz, "The Influence of Different Workload Descriptions on a Heuristic Load Balancing Scheme," *IEEE Trans on Software Engineering,* vol.17, No.7, pp.725-730, July 1991.

[6] T.Furuhashi, K.Nakaoka, Y.Uchikawa, "A New Approach to Genetic Based Machine Learning and an Efficient Finding of Fuzzy Rules," *Proc. WWW'94*, pp.114-122, 1994.

[7] J A. Miller, W D. Potter, R V. Gondham, C N. Lapena, "An Evaluation of Local Improvement Operators for Genetic Algorithms," *IEEE Trans on SMC*, vol.23, No 5, pp.1340-1351, Sept 1993.

[8] N.G.Shivaratri and P.Krueger, "Two Adaptive Location Policies for Global Scheduling Algorithms," *Proc. 10th International Conference on Distributed Computing Systems*, pp.502-509, May 1990.

[9] Terence C. Fogarty, Frank Vavak, and Phillip Cheng, "Use of the Genetic Algorithm for Load Balancing of Sugar Beet Presses," *Proc. Sixth International Conference on Genetic Algorithms*, pp.617-624, 1995.

[10] Garrism W. Greenwood, Christian Lang and steve Hurley, "Scheduling Tasks in Real-Time Systems using Evolutionary Strategies," *Proc. Third Workshop on Parallel and Distributed Real-Time Systems*, pp.195-196, 1995.

[11] Gilbert Syswerda, Jeff Palmucci, "The application of Genetic Algorithms to Resource Scheduling," *Proc. Fourth International Conference on Genetic Algorithms*, pp.502-508, 1991.

[12] Melanie Mitchell, *An Introduction to Genetic Algorithms*, MIT Press, 1996.

[13] M.Srinivas, and L.M.Patnait, "Adaptive Probabilities of Crossover and Mutation in Genetic Algorithms," *IEEE Trans on SMC*, vol.24, no.4, pp.656-667, April 1994.

A New Memetic Algorithm Using Particle Swarm Optimization and Genetic Algorithm

Sang-Moon Soak, Sang-Wook Lee, N.P. Mahalik, and Byung-Ha Ahn

Dept. of Mechatronics, Gwangju Institute of Science and Technology, South Korea
{soakbong, yashin96, nmahalik, bayhay}@gist.ac.kr

Abstract. We describe a new memetic algorithm scheme combining a genetic algorithm (GA) and a particle swarm optimization algorithm (PSO). This memetic scheme uses the basic dynamics of PSO instead of the concept of the survival of fittest (selection strategy) in GA. Even though the scheme does not use a selection strategy, it shows that the algorithm can find good results and can be an alternative approach for network based optimization problems.

We test it in the context of a memetic algorithm applied to well known spanning tree based optimization problem, the degree constrained minimum spanning tree problem (DCMST). We compare with existing evolutionary algorithms (EAs), including *EA using edge window decoder* and *EA using edge-set encoding*, which represent the current state of the art on the DCMST. The new memetic algorithm demonstrates superior performance on the smaller and lower degree instances of the well-used 'Structured Hard' DCMST problems, and similar performance on the larger and higher degree instances.

1 Introduction

For a long time, various population based and nature inspired search algorithms have been developed based on principles existing in the nature. The most fundamental example is *Evolutionary Algorithms (EAs)*. The most outstanding feature distinguished EAs from general optimization algorithms is to use a set of solutions, *population* and to use *the survival of fittest*. Therefore, if a solution finds a better result than the others in population, it can generate more offspring for the next generation.

Recently, a new population based and nature inspired search algorithm was introduced, which is called *the particle swarm optimization (PSO)* [1], [2]. The motivation for the development of this method was based on the simulation of simplified animal social behaviors such as fish schooling, bird flocking, etc.

Many literatures showed that PSO can give better results than other optimization algorithms in continuous function optimization problems [15], [16]. But, unfortunately until now, it has been mainly applied to continuous optimization problems even though there are a few attempts for combinatorial optimization problems like scheduling problems [1], [3].

PSO has a strong local search (exploitation) ability to find the most optimistic result and converges relatively rapidly. Meanwhile it has a disadvantage of falling

B. Gabrys, R.J. Howlett, and L.C. Jain (Eds.): KES 2006, Part I, LNAI 4251, pp. 122–129, 2006.
© Springer-Verlag Berlin Heidelberg 2006

into a local minimum because of the loss of diversity [5], [6]. On the other hand, GAs have an efficient exploration ability to search a wide range of search space. But, GAs also have a disadvantage that does not have an effective local search mechanism for accurately searching near a good solution. So, we want to combine an exploitation ability of PSO and an exploration ability of GA in an algorithm (a *memetic algorithm : MA*).

Even though from the optimization point of view, it is still in dispute, the success of an MA depends on the tradeoff between the exploration abilities of the underlying EA and the exploitation abilities of the local searchers used. Anyway, MA also has a disadvantage. Sometimes, MA falls into local optimums so that it can not achieve a goal.

Therefore, we propose a new memetic algorithm to overcome this weakness. The main idea is to replace a selection strategy in GA with the concept of trajectory in PSP (the use of *pBest* (private best solution) and *gBest* (global best solution)) in a simple way and then performs a crossover-based hill climbing. Section 2 provides full details of the degree constrained minimum spanning tree problem (DCMST), while section 3 describes the proposed memetic algorithm hybridizing the Genetic Algorithms and Particle Swarm Optimization. In section 4 we compare the new memetic algorithm with the previous ESs using different encodings and the state of the art Edge Window Decoder (EWD) [12], [13]. We conclude in section 5, summarizing our findings and considering future work.

2 The Degree Constrained Minimum Spanning Tree Problem: DCMST

Research on the *degree constrained minimum spanning tree* (DCMST) problem seems to have begun with Narula and Ho [10].

Consider an undirected complete graph $G = (V,\ E)$, where $V = \{1,\ 2, ..., N\}$ is the set of N nodes and $E = \{1, 2, ..., M\}$ is the set of M edges with given distance (or cost). The DCMST on a graph is the problem of generating a minimum spanning tree T satisfying with degree constraints on the number of edges that can be incident to nodes of the graph. Unlike the MST, the DCMST cannot, in general, be found using a polynomial time algorithm, and in fact it is a well-known NP-hard problem of importance in communications network design, road network design and other network-related problems [13].

More formally, the DCMST can be described as follows.

$$\text{Minimize} \quad f(x) = \left[\sum_{i=1}^{M} x_i \cdot C_i \mid d_j \ \leq \ b_j, x \in T \right]. \tag{1}$$

where, d_j is the degree of node j and b_j is the degree constraint on node j and C_i is the cost of an edge i. x_i indicates the decision variable and if an edge e_i is included in a tree T, it is 1, otherwise 0.

3 The Proposed Memetic Algorithm

The Encoding and Decoding Method: We use the *Edge Window Decoder (EWD)* encoding [12], [13] for the proposed memetic algorithm because it is very simple to encode and decode a spanning tree.

First, EWD generates initial solutions following two conditions.

- If there are N nodes in a network, the length of string is $L = 2 \times (N-1)$.
- All nodes should appear at least a time and at most $d-1$ times in the string over the whole generations. Here, d indicates *the degree constraint.*

Second, a tree construction routine (TCR) is applied to the solution generated. Soak. et.al [12], [13]proposed three TCRs, the Cycle-Free TCR (CF-TCR), the Cycle-Breaking TCR (CB-TCR) and the Degree Constrained Kruskal TCR (dK-TCR). Among them, in their experiments at DCMST, dK-TCR was superior to the previous EAs [12]. So, we will use dK-TCR as a decoding method.

To apply dK-TCR to solutions, edges in the solution are extracted from left to right, considering each two genes. For example, let the generated solution be $[1, 3, 2, 5, 1, ...]$. Considering each two genes (edge window), the extracted edges are (1, 3), (3, 2), (2, 5), (5, 1) ... etc. and they are sorted from least cost edge and included in the edge set $E\{\cdot\}$. In here, note that an edge (i, i) is not considered and an edge (i, j) and (j, i) are the same so that only one edge of them is included in $E\{\cdot\}$. And then, dK-TCR generates a tree using edges in $E\{\cdot\}$ in accordance with the order as a Kruskal algorithm, avoiding a cycle and satisfying a degree constraint. After all edges in $E\{\cdot\}$ are used, the generated tree may not include all nodes in the network because of a degree constraint. In that case, a random edge is considered. If the edge does not result in a cycle and satisfies a degree constraint, the edge is included in the tree. Otherwise, other edge is considered arbitrarily. This process is continued until all nodes are included in the tree.

The Genetic Operators: For the new memetic algorithm, we use two different crossover operators, Common Gene Preservation Crossover (CGPX) [12] and Adjacent Node Crossover (ANX) [13]. These crossover operators showed a good performance at different application problems because of different features. CGPX has a strong locality and ANX is known as it can preserve population diversity for a long number of generations. So, we can expect a synergy using these operators together.

The *Common Gene Preservation Crossover* (CGPX) is designed to focus on the inheritability of common genetic information, as well as exploit edge-cost information. CPGX produces two children, O^1 and O^2, from two parents, P^1 and P^2, as follows:

We use P_i^k (or O_i^k) to denote the node at position i in individual k. To produce O^1, we first set $O_1^1 = P_1^1$. Then, working from left to right (i.e. from $i = 2$ to $i = 2(N-1)$), we compare P_i^1 and P_i^2. Whenever these are the same, i.e. $P_i^1 = P_i^2 = n$ we make $O_i^1 = n$, unless this would mean more than $d-1$ ns appearing in O^1. If that is the case, then a random choice of node is made from those with fewer than $d-1$ occurrences in the string so far. When $P_i^1 \neq P_i^2$, we compare the costs of the edges (O_{i-1}^1, P_i^1) and (O_{i-1}^1, P_i^2), and make $O_i^1 = P_i^l$, where $l = 1$ or $l = 2$, whichever corresponds to the lower cost of these two edges. However, if one of $l = 1$ or $l = 2$ violates the degree constraint, then we use the other; if both would violate the degree constraint, then we make a random valid choice for O_i^1. We use precisely the same process to generate O^2 except that we begin with $O_1^2 = P_1^2$.

ANX generates offspring following the procedure below.

First a node-adjacency map is built for each parent. For example, the set A_4^1 lists all nodes which are adjacent to node 4 in parent A^1. So, if $A^1 = (1, 4, 2, 1, 5, 2, 3, 1)$, then $A_4^1 = \{1, 2\}$. A random node is then chosen to be the starting node of the offspring and becomes the first "current" node. ANX then iterates the following process until the child contains $2(N-1)$ nodes: where c is the current node, we first find $E = A_c^1 \cap A_c^2$. If E is not empty, then the node which be the least distance from c in this intersection is selected (breaking ties randomly); this becomes the next node in the child, and now becomes the current node c. If E is empty, then the next gene is simply selected from $S = A_c^1 \cup A_c^2$ the same way. Finally, if both A_c^1 and A_c^2 are empty, then a random valid node is chosen. In each step, if a node d is selected as the next node in the child, d is removed from the adjacency map A_c^1 and A_c^2 and c is also removed from the adjacency map A_d^1 and A_d^2. For more details, see [13].

We use a simple *swap* mutation for experiments. First, it selects two positions randomly and then swap the genes at those positions.

The Local Search Operators: In the context of evolutionary computation, the local search method can be divided into two ways; *crossover based hill-climbing algorithm* and *mutation based hill-climbing algorithm*. The most simple hill-climbing algorithm will be a mutation operator itself because it can result in a small change in phenotype in comparison with a crossover operator. So, the majority of hill-climbing algorithms are a mutation based hill-climbing [7], [9]. Anyway, a local search method depends on the encoding used. In most of the mutation based local search algorithms, the used encodings were a *nonredundant representation*; genotype is mapped to phenotype one by one. On the other hand, if a mutation based local search algorithm is applied to a redundant representation like EWD, sometimes it may not give any change to parent's phenotype [14].

Anyway, some research papers based on a crossover based hill-climbing algorithm are found in literature [4], [8], [11]. If a crossover operator can have high heritability, it can play a role in a memetic algorithm as a local search operator. We showed that CGPX has high locality while ANX has slightly high locality and at the same time it can maintain population diversity for the whole genetic search process [14]. So, the proposed algorithm first searches the narrow area using CGPX and then widens the search space using ANX a little.

Repair Process: We note that the CGPX and ANX operator does not ensure that the child maintains degree constraints (i.e. that each node occurs at lease one times and at most *d - 1* times in the string). We note that it would be straightforward to make appropriate modifications to CGPX and ANX to be applied to DCMST problems.

In these cases the offspring will need to be repaired, and we adopt a simple repair process as follows, which takes advantage of maintaining a check on the number of occurrences $C(k)$ of each node k as the child is developed. If there is some node k with $C(k) = 0$, we first select a gene r randomly, but such that $2 \leq C(r) \leq d-1$, and replace the node r in the child with k. This is repeated until every node has at least one occurrence.

Removing the Survival of Fittest in GA: In PSO, each solution maintains its own best solution ($pBest$) and it has a relationship to the process (*a trajectory*) that the solution has been evolved for several generations. Moreover, solutions do not share information with other particles (solutions) but share with only the best particles ($gBest$). On the other hand, in GA all solutions share information with other solutions in population through a crossover operator and the trajectory of a solution is not maintained because of *a selection strategy*.

We think that keeping the past history of a solution and guiding toward $pBest$ or $gBest$ may be a good alternative to do an efficient search because PSO have already proved its performance in several fields [1], [2], [3]. Therefore, to employ PSO into GA, we replace a *selection strategy* with the concept of a *trajectory* in PSO. Figure. 1 shows the psudocode of the proposed memetic algorithm.

4 Experiments

In these experiments, we evaluate the new MA by comparing it with the state of the art Edge Window Decoder [12], Edge Set encoding, Network Random Keys (NetKeys) encoding and the well known Prüfer number encoding. Space limitations prevent us from explaining the comparative encoding schemes, however the associated references at the end of this article include URLs from which interested readers can freely obtain such details.

We note that our benchmark problems are the structured hard DCMST problems (SHRD problems) [12], [13]. We used the same benchmark instances used in the literature [12] and final results on the papers for the comparison. Such structured hard DCMST instances were generated by using Krishnamoorthy's method and problems with 30 to 100 nodes respectively were produced, each leading to three DCMSTs, one for each of the degree constraints 3, 4 and 5.

Experiments continue for 50,000 evaluations because Soak et.al [12] mentioned that all methods appeared to converge well before this limitation, and each trial is repeated 10 times with different random seeds. The control parameters of the MA, Max_Iter1 and Max_Iter2, are set 1 and $(N \times 0.2)$ respectively.

Table. 1 summarizes all our experimental results on the test SHRD instances. In it, a table entry gives the relative difference between the mean (over the ten trials) solution C for the algorithm, and the mean solution for ten trial runs of d-Prim (C_{d-Prim}) as follows:

$$Gains = (C_{d-Prim} - C)/C_{d-Prim} \times 100\% \quad (2)$$

```
The Proposed Memetic Algorithm
The Encoding Method : EWD, The Decoding Method : dK-TCR,
Crossover (Pc = 100%) : CGPX and ANX,
Mutation (Pm = 100%) : Reciprocal Exchange Mutation
1. Initialization;
2. Evaluation;
3. Crossover-Based Local Search (CBLS);
        for Max_Iter1 then
                Offspring_Pool{} ← CGPX(P1, P2);
            Best_Offspring ← Best(Offspring_Pool{});
        if ((Best_Offspring < P1) || (Best_Offspring < P2))
                if (P1 < P2)
                        P1 ← Best_Offspring;
                else
                        P2 ← Best_Offspring;
        else        // * PSO Concept * //
                RA ← random(2);        // * return 0 or 1 * //
                if RA == 0
                        for Max_Iter2 then
                                Offspring_Pool{} ← ANX(pBEST_P1, gBEST);
                else
                        for Max_Iter2 then
                                Offspring_Pool{} ← ANX(pBEST_P2, gBEST);
                Best_Offspring ← Best(Offspring_Pool{});
                if ((Best_Offspring < pBest) || (Best_Offspring < gBest))
                        pBest or gBest ← Best_Offspring;
                else
                        if ((Best_Offspring < P1) || (Best_Offspring < P2))
                                if (P1 < P2)
                                        P1 ← Best_Offspring;
                                else
                                        P2 ← Best_Offspring;
                        else
                                P1 or P2 ← Best_Offspring;
4. Mutation;
```

Fig. 1. Pseudocode of the proposed memetic algorithm

Bold indicates the best result over all algorithms for a particular problem. In this table, the other results except GAPSO referred to the best results in [12] and [13].

In the comparison to prüfer, NetKey and Edge Set (ES) encoding, GAPSO shows superior results at all test instances regardless a network size and a degree constraint. In the comparison to EWD, it shows slightly better results at the most of the test instances and worse results at some test instances. But, the difference is very small. Taking the worst and the best results into account, GAPSO shows slightly better results. Unfortunately, we did not compare the computation time of each algorithm because of the difference of computer system used for experiments. Anyway for fair comparison, we used the same number of evaluation.

Table. 2 indicates t-test restuls between GAPSO vs. EWD. Although it is very difficult to distinguish the difference between the two algorithms at larger size and higher degree network problems, this table shows that most results of the statistical analysis are valid with 99% confidence interval and over 99.9% confidence interval at some instances.

Table 1. Experiment Results (Gains) on SHRD instances

Instances		Prüfer	NetKey	ES	EWD				GAPSO		
					CF-TCR	*d*K-TCR					
N	*d*	Avg.	Avg.	Avg.	Avg.	Worst	Avg.	Best.	Worst	Avg.	Best.
SHRD30	3	2.25	3.49	3.84	3.14	3.80	3.86	3.89	3.89	**3.89**	3.89
	4	4.85	4.93	6.04	5.62	6.05	6.06	6.09	6.05	**6.08**	6.09
	5	1.81	3.76	4.75	4.43	4.74	**4.76**	4.78	4.74	4.75	4.78
SHRD40	3	5.58	6.23	6.59	6.16	6.64	6.71	6.76	6.74	**6.76**	6.78
	4	2.79	3.67	5.10	4.71	5.13	5.17	5.21	5.19	**5.21**	5.21
	5	2.95	2.51	4.80	4.42	4.73	4.89	4.95	4.87	**4.92**	4.95
SHRD50	3	18.92	18.41	19.71	19.40	19.75	19.80	19.85	19.83	**19.84**	19.85
	4	13.03	12.80	14.53	14.26	14.62	14.67	14.72	14.68	**14.72**	14.75
	5	10.31	8.98	12.13	11.90	12.21	12.28	12.32	12.32	**12.35**	12.36
SHRD60	3	15.43	14.80	15.88	15.68	16.21	16.23	16.25	16.23	**16.26**	16.27
	4	6.92	5.64	8.54	8.29	8.83	8.86	8.91	8.80	**8.87**	8.91
	5	3.52	0.92	4.89	4.32	4.98	**5.05**	5.08	4.86	5.04	5.09
SHRD70	3	11.51	11.28	12.39	12.00	12.51	12.54	12.56	12.56	**12.57**	12.58
	4	7.38	6.62	8.67	8.29	8.81	**8.83**	8.87	8.81	**8.83**	8.85
	5	7.00	5.46	8.12	7.99	8.40	8.36	8.42	8.24	**8.36**	8.43
SHRD100	3	23.64	22.37	24.02	23.86	24.26	24.28	24.29	24.30	**24.31**	24.32
	4	12.40	10.40	13.40	13.38	13.73	**13.76**	13.78	13.72	**13.76**	13.79
	5	6.21	2.45	7.56	7.39	7.86	**7.87**	7.89	7.84	**7.87**	7.91

Note: *N* is the number of nodes; *d* is the degree constraint.

Table 2. Statistical Analysis : t-Test results

Instances		GAPSO vs. EWD		Instances		GAPSO vs. EWD	
		t	*p*-value			*t*	*p*-value
SHRD30	3	3.354	0.003	SHRD60	3	4.609	<0.001
	4	3.182	0.005		4	0.65	0.523
	5	-0.884	0.388		5	-0.25	0.8
SHRD40	3	3.81	<0.001	SHRD70	3	5.59	<0.001
	4	3.556	0.002		4	0	1
	5	1.345	0.195		5	-1.93	0.069
SHRD50	3	3.897	<0.001	SHRD100	3	7.31	<0.001
	4	4.457	<0.001		4	0.444	0.662
	5	5.891	<0.001		5	0.74	0.468

5 Conclusions

The DCMST is an NP-hard combinatorial optimization problem which is important in practice. In this paper we have introduced a new memetic algorithm which combines simply a genetic algorithm with a particle swarm optimization algorithm, and which uses a crossover-based hill-climbing as a local searcher. And we showed that it outperforms (at least on the standard testbed studied) EA using the edge window decoder and EA using the edge set encoding, which are currently the state of the art in the published literature on the DCMST. The statistical results also showed that the new concept (replacing the concept of *the survival of fittest* with the concept of *the trajectory*) can be a good alternative. Especially, we performed the *t*-test and the results showed the very high confidence of 99% at most of the cases, over 99.9% at some cases.

References

1. L. Cagnina, S. Esquivel and R. Gallard, Particle Swarm Optimization for Sequenceing Problem: A Case Study, in proc. of IEEE EC, (2004) 536–541.
2. R. C. Eberhart and J. Kennedy, Particles swarm optimization, in Proc. of IEEE Int. Conf. on Neural Network, (1995) 1942–1948.
3. X. Hu and R. C. Eberhart, Swam Intelligence for Permutation Optimization: A Case Study of n-Queens Problem, in Proc. of IEEE Swarm Intelligence Symposium, (2003) 243–246.
4. T. Jones, Crossover, Macromutation, and Population-based Search, Proc. of the 6th Int. Conf. on Genetic Algorithms, (1995).
5. T. Krink, J.S. Vesterstroem and J. Riget, Particle Swarm Optimization with Spatial Particle Extension, in Proc. of the IEEE Congress on EC, (2002) 1474–1479.
6. T. Krink and M. Lovbjerg, The life cycle model: combining particle swarm optimization, genetic algorithms and hill climbers, in Proc. of Parallel Problem Solving from Nature VII, (2002) 621–630.
7. T.L. Lau and E.P.K. Tsang, Applying a Mutation-Based Genetic Algorithm to Processor Configuration Problems, Proc., 8th IEEE Conf. on Tools with AI, (1996).
8. M. Lozano, F. Herrera, N. Krasnogor and D. Molina, Real-Coded Memetic Algorithms with Crossover Hill-Climbing, Evolutionary Computation, vol. 12, no. 3, (2004) 273–302.
9. P. Merz and B. Freisleben, Fitness Landscapes, Memetic Algorithms, and Greedy Operators for Graph Bipartitioning, Evolutionary Computation, vol. 8, no. 1, (2000) 61–91.
10. S.C. Narula and C.A. Ho, Degree-constrained minimum spanning tree, *Computer and Operations Research*,**7**:239–249, (1980).
11. U.M. O'reilly and F. Oppacher, Hybridized Crossover-Based Search Techniques for Program Discovery, Proc. of the 1995 World Conf. on EC, (1995) 573-578.
12. S. M. Soak, D. Corne and B. H. Ahn, A Powerful New Encoding for Tree-Based Combinatorial Optimisation Problems, in Proc. of PPSN VII, Lecture Notes in Computer Science, vol. 3242, (2004) 430–439.
13. S.M, Soak, D. Corne and B.H. Ahn, The Edge-Window-Decoder Representation for Tree-Based Problems, appear to IEEE Trans. on Evolutionary Computation, April, (2006).
14. S.M, Soak, D. Corne and B.H. Ahn, On a property analysis of representations for constrained spanning tree problems, 7th Int. Conf. on Artificial Evolution (2005).
15. X.H. Wang and J.J. Li, Hybrid Particle Swarm Optimization With Simulated Annealing, in Proc. of the 3rd Int. Conf. on Machine Learning and Cybernetics, (2004) 2402–2405.
16. W.J. Zhang and X.F. Xie, DEPSO: Hybrid Particle Swarm with Differential Evolution Operator, in Proc. of IEEE Systems, Man and Cybernetics, (2003) 3816–3821.

Mathematical and Empirical Analysis of the Real World Tournament Selection

S.W. Lee, S.M. Soak, N.P. Mahalik, B.H. Ahn, and M.G. Jeon

Dept. of Mechatronics, Gwangju Institute of Science and Technology, South Korea
{yashin96, soakbong, nmahalik, bayhay, mgjeon}@gist.ac.kr

Abstract. This paper provides a theoretical and empirical analysis of a recently proposed selection method, which is called 'Real World Tournament Selection' (RWTS). First of all, a characteristic of the selection probability of RWTS is analyzed. Secondly, the performance of RWTS is more widely tested at three benchmark problems - function optimization problems, the traveling salesman problem and the multdimensional knapsack problem. Finally, RWTS is compared with five well-known selection strategies. The experimental results show that it can achieve an encouraging performance, not only solution quality but also computation time.

1 Introduction

When we apply evolutionary algorithms (EAs) to a problem, it decides the choice of an appropriate selection strategy is one of the very important issues because the survival of the fittest, which is the basic concept of EAs. Nevertheless, selection methods have long remained overlooked, in comparison with mutation and crossover operations, because it is not concerned with searching for a good solution.

Specifically, a mathematical analysis of the selection methods has generally been overlooked. Bäck [1] analyzed the selective pressure of most selection methods used in evolutionary algorithms. Blickle et. al [3] analyzed tournament selection and found uncovered some of its mathematical characteristics. Julstrom [4] investigated the same properties between different selection methods. An overview of selection methods is well described in [2].

In this paper, we analyze Real World Tournament Selection (RWTS) [5], [6]. The RWTS is a new way of producing an entire population. The population is produced by a series of two-chromosome tournament competitions. The difference between RWTS and tournament selection (TS) is that in RWTS, all chromosomes in the population are paired according to the index number so that all solution are considered for the competition.

Soak et. al [5], [6] already applied RWTS to tree-based optimization problems, and their experimental results showed that RWTS can give superior performance to the general tournament selection in all test instances except the evolutionary algorithm using the edge set encoding (in this case, RWTS showed similar with TS or a little worse performance than TS). However, Soak et. al did not give a

B. Gabrys, R.J. Howlett, and L.C. Jain (Eds.): KES 2006, Part I, LNAI 4251, pp. 130–137, 2006.
© Springer-Verlag Berlin Heidelberg 2006

specific mathematical model for RWTS, only compared it with TS and applied it to only tree-based optimization problems. So, in this paper, first we want to give a mathematical rationale of RWTS. Second we want to compare it with various selection strategies on many different problems: the function optimization problem, the traveling salesman problem and the multidimensional knapsack problem. Finally, we show that it can be used at the wide range of application and give good performance.

This paper is organized as follows. Overview of selection methods is described in Section 2. Section 3 presents the proposed selection method, RWTS. Experimental results are presented in Section 4, and finally, some concluding remarks are given in Section 5.

2 Overview of Selection Method

In this section, five selection methods (Tournament Selection with Replacement, Tournament Selection without Replacement, Probabilistic 2-tournament selection, Linear Ranking Selection and Exponential Ranking Selection) are briefly reviewed, and then compared with RWTS. These five selection methods are widely used by researchers and can be analyzed mathematically. It is assumed that all individuals have a different fitness, and the population is ordered in a way such that the lowest rank 1 is assigned to the worst individual, and the highest rank N is assigned to the best groups among N individuals.

2.1 Tournament Selection

The selection probabilities for tournament selection with replacement (TSwR) and without replacement (TSwoR) derived in [2] are expressed as

$$p_i^{TSwR} = \frac{i^t - (i-1)^t}{N^t} \tag{1}$$

$$p_i^{TSwoR} = \frac{\binom{i-1}{t-1}}{\binom{N}{t}} , \tag{2}$$

where, t is tournament size and $\binom{a}{b} = \frac{a!}{b!(a-b)!}$.

The Probabilistic 2-tournament selection (PTS) is presented by Julstrom [4] as

$$p_i^{PTS} = \frac{2(i-1)}{N(N-1)}q + \frac{2(N-i)}{N(N-1)}(1-q) , \tag{3}$$

where $0.5 < q < 1.0$.

2.2 Ranking Selection

The selection probabilities of the linear (LRS) and the exponential (ERS) ranking selection are well described in [4] as

$$p_i^{LRS} = \frac{1}{N}\left((2-s) + \frac{2(i-1)(s-1)}{N-1}\right) \tag{4}$$

$$p_i^{ERS} = \frac{r^{N-i}(1-r)}{1-r^N} \quad , \tag{5}$$

where, $1 < s < 2$ and $0 < r < 1.0 (r \simeq 1.0)$.

3 Real World Tournament Selection (RWTS)

RWTS is derived from the real world tournament competition. In RWTS, each individual in the population is paired with neighboring individual without replacement or omission. If the last individual has no neighboring individual (population is *odd*), it competes with a randomly selected individual among previous winners in the same tournament level. When all competitions are over in present tournament level, only the winners are inserted in the mating pool and move on the next tournament level. The process is repeated until the mating pool is full. Fig. 1 shows how RWTS works with a population of size 7 and population of size 8. The numbers inside circle mean a fitness value (higher value is fitter) and a dotted circle represents the individual which is selected randomly from the previous winner at the same tournament level for mating with the lastly remaining single individual. As the result of the tournament competition, all checked individuals (total is population size) are inserted in the mating pool. In case (b) ($N = 2^n$), the last level is dummy level for filling the population size.

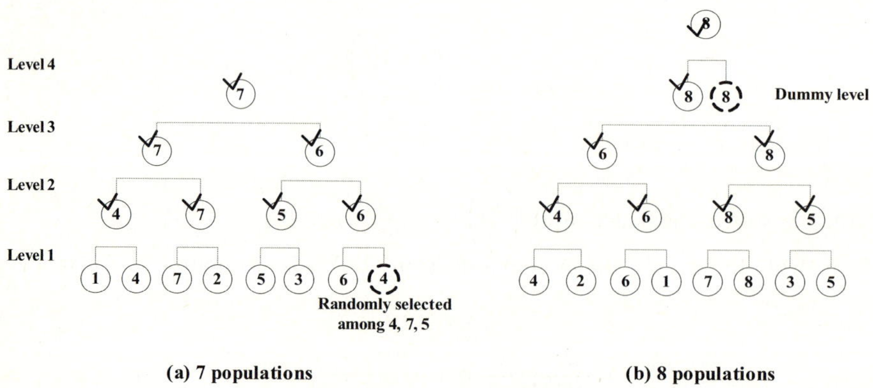

Fig. 1. The description of RWTS

For mathematical analysis of RWTS, we assume that the population size is $N = 2^n$ like (b) of fig. 1, and each individual has a different fitness, and the population is ordered in a way such that the worst individual is assigned the lowest rank 1, and the best individual is assigned the highest rank N. If $N = 2^n$, RWTS performs tournament competitions of $(n+1)$ level (the last level is dummy level).

In tournament level 1, the probability that the i-th ranked individual wins is $\frac{i-1}{N-1}$ because among $N-1$ individuals, i-th ranked individual must be paired with individuals below i rank. If i-th ranked individual survives in the level 1, it can go the tournament level 2 and competes with a neighbor, one of other winners at level 1. In level 2, the condition that i-th ranked individual wins is that two neighbor individuals must be below the i rank among the remained $i-2$ competitors (in level 1, there is already 1 individual below i rank). So, the probability which i-th ranked individual can survive in level 2 is $\frac{(i-2)(i-3)}{(N-2)(N-3)}$ under the assumption that i-th ranked individual survives in level 1. As we integrate two results, the total probability that i-th ranked individual in level 1 and 2 can survive, is given by $\frac{i-1}{N-1}\left(1+\frac{(i-2)(i-3)}{(N-2)(N-3)}\right) = \frac{i-1}{N-1} + \frac{(i-1)(i-2)(i-3)}{(N-1)(N-2)(N-3)}$.

Generalizing this equation as any level k, the i ranking individual must be paired with $2^k - 1$ individual below i ranking for surviving in level k. Now, the question is that which level i ranking individual can go to. This is easily given by calculating $log_2 i$. The upper-limit of a tournament level that i ranking individual can go is $|log_2 i|$, where $|x|$ means that discards down to decimal place. Therefore, from all of the above result, the total probability which i ranking individual from level 1 to level $|log_2 i|$ can survive is formulated by $p_{i,total}^{RWTS} = \sum_{x=1}^{|log_2 i|}\left(\prod_{n=1}^{2^m-1}\frac{i-n}{N-n}\right)$.

$\sum_{i=1}^{N} p_{i,total}^{RWTS}$ is $(N-1)$. The N ranking individual from dummy level is selected as the last individual which will go to the mating pool. Thus, the selection probability of i ranking individual is $\frac{p_{i,total}^{RWTS}}{N}$ (in case of N ranking which means best individual, $\frac{(p_{i,total}^{RWTS}+1)}{N} = \frac{n+1}{N}$).

$$p_i^{RWTS} = \frac{1}{N}\sum_{x=1}^{|log_2 i|}\left(\prod_{n=1}^{2^m-1}\frac{i-n}{N-n}\right) \qquad \text{for } i = 1, \cdots, N-1$$

$$= \frac{n+1}{N} \qquad \text{for } i = N \tag{6}$$

Actually, the selection probability of the best individual in RWTS is meaninglessness. It is deterministic because the best individual is inserted $(n+1)$ times in the mating pool; the best individual never loses in competitions. Therefore, RWTS implies *Elitism strategy* with elite size $(n+1)$ (*pop_size* is even) or n (*pop_size* is odd).

4 Experimental Results

In this experiment, we compare RWTS with other selection methods. First, the probability distributions are compared mathematically. And then their performances are compared through the three kinds of benchmark problems, function optimization problems, the traveling salesman problem (TSP) and the multidimensional knapsack problem (MKP).

4.1 Probability Distribution Comparison

Fig. 2 shows the distribution of the selection probabilities of six selection methods with $pop\_size = 128$. RWTS has almost similar values to other selection methods below 85% ranking; gentle slope. However, above 85% ranking, RWTS shows much higher selection probability than the other selection methods. That means RWTS has a strong selection pressure. Although it is difficult to say that what property of selection methods can help to improve the performance of EA, we can observe RWTS has significantly different selection probability distribution in comparison with other selection strategies.

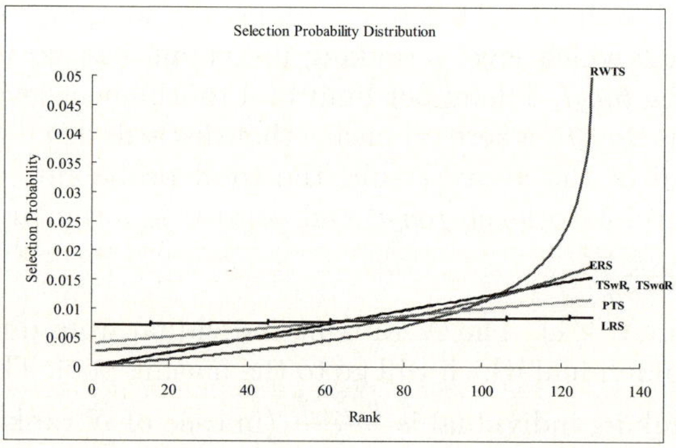

Fig. 2. The distributions of the selection probability

4.2 Experimental Comparison

In this paper, we tested several selection methods on three different problems, the function optimization problem (FOP), the traveling salesman problem (TSP) and the multiple knapsack problem (MKP).

For FOP, we selected two instances, the Grewangk Function and the Rastrigin Function.

Rastirgin Function:

$$f\left(x_i|_{i=1,20}\right) = 200 + \sum_{i=1}^{20} x_i^2 - 10 \cdot cos2\pi x_i, \quad x_i \in [-5.12, 5.12] \tag{7}$$

Grewangk Function:

$$f\left(x_i|_{i=1,10}\right) = \sum_{i=1}^{10} \frac{x_i^2}{4000} - \prod_{i=1}^{10} cos\left(\frac{x_i}{\sqrt{i}}\right) + 1, \quad x_i \in [-512, 512] \tag{8}$$

For TSP and MKP, we selected 5 instances. These two benchmark problems are available from OR-Library [7].

We use a simple GA with a binary encoding for FOP and a random key (RK) encoding for TSP and MKP in each case for fair comparison of the varying selection methods; maximum evaluation number $\simeq 10^5$; uniform crossover with probability = 0.6. And in case of FOP, we use varying mutation probability from 0.001/ bit to 0.009/ bit by step size = 0.002/ bit; and run each experiment 30 times. In case of TSP and MKP, we use one point swap mutation with probability = 0.4; and run each experiment 10 times.

The parameters concerned with each selection method are set with tournament size $t = 2$ in TSwR and TSwoR, probability $q = 0.75$ in PTS, constant $s = 1.1$ in LRS and constant $r = 0.985$ in ERS.

Table 1. The experimental result of FOP (*pop_size* = 128, *generation* = 1000)

	Rastrigin function														
	RWTS					TSwR					TSwoR				
Mut.	Min.	Ave.	Max.	STD.	CPU	Min.	Ave.	Max.	STD.	CPU	Min.	Ave.	Max.	STD.	CPU
0.001	21.93	36.82	55.77	8.17	7.85	21.99	38.05	49.00	6.30	8.03	23.43	35.95	51.49	7.79	8.03
0.003	5.73*	18.05	32.58	6.46	7.84	14.51	**27.53**	36.82	6.55	8.02	14.16*	**25.83**	39.45	6.48	8.03
0.005	5.75	**15.50**	25.16	5.74	7.84	13.65*	28.37	41.88	6.07	8.02	17.43	28.06	37.15	5.67	8.02
0.007	11.19	17.84	29.52	5.85	7.85	20.05	28.17	43.29	5.34	8.00	17.98	27.93	42.77	4.80	8.02
0.009	11.66	20.47	42.91	6.65	7.84	23.54	36.62	52.33	5.44	8.00	28.99	38.29	52.00	5.90	8.00
	PTS					LRS					ERS				
Mut.	Min.	Ave.	Max.	STD.	CPU	Min.	Ave.	Max.	STD.	CPU	Min.	Ave.	Max.	STD.	CPU
0.001	18.56	**29.48**	39.99	6.58	8.04	24.44	39.57	63.75	9.71	8.77	14.29	37.89	68.31	10.12	8.80
0.003	18.16*	30.86	42.13	6.36	8.03	12.10	22.87	43.05	7.32	8.76	12.81	23.35	35.64	6.49	8.79
0.005	30.68	40.69	54.34	6.35	8.02	5.63*	17.66	31.77	7.23	8.75	7.89	19.53	34.86	6.08	8.78
0.007	40.66	64.17	80.01	7.36	8.01	8.55	18.55	33.44	6.49	8.76	8.14	17.68	28.94	5.96	8.78
0.009	70.76	86.86	101.59	9.60	8.00	6.20	**14.42**	24.01	5.52	8.75	6.51*	**16.95**	25.43	5.24	8.78
	Grewangk function														
	RWTS					TSwR					TSwoR				
Mut.	Min.	Ave.	Max.	STD.	CPU	Min.	Ave.	Max.	STD.	CPU	Min.	Ave.	Max.	STD.	CPU
0.001	0.13	0.45	0.68	0.15	3.99	0.31	0.46	0.79	0.11	4.08	0.17	0.45	0.79	0.16	4.07
0.003	0.18	0.30	0.47	0.08	4.00	0.15	0.28	0.42	0.08	4.07	0.12*	**0.28**	0.46	0.10	4.07
0.005	0.10	**0.25**	0.43	0.08	3.99	0.12*	**0.25**	0.37	0.07	4.07	0.12	0.30	0.55	0.10	4.07
0.007	0.10	0.26	0.44	0.09	3.99	0.14	0.39	0.51	0.10	4.06	0.27	0.41	0.54	0.09	4.06
0.009	0.09*	**0.25**	0.35	0.06	3.99	0.42	0.58	0.80	0.09	4.06	0.39	0.57	0.74	0.09	4.06
	PTS					LRS					ERS				
Mut.	Min.	Ave.	Max.	STD.	CPU	Min.	Ave.	Max.	STD.	CPU	Min.	Ave.	Max.	STD.	CPU
0.001	0.20	0.35	0.65	0.12	4.09	0.27	0.51	0.81	0.15	5.02	0.27	0.52	0.81	0.15	5.02
0.003	0.10*	**0.28**	0.43	0.09	4.08	0.20	0.36	0.63	0.12	5.01	0.14	0.33	0.63	0.12	5.01
0.005	0.38	0.65	0.88	0.11	4.07	0.09	0.33	0.62	0.12	5.01	0.09	0.27	0.62	0.12	5.01
0.007	0.59	0.99	1.23	0.13	4.07	0.06*	0.26	0.40	0.08	5.01	0.11	0.29	0.40	0.08	5.01
0.009	1.14	1.31	1.70	0.12	4.08	0.07	**0.25**	0.41	0.09	5.01	0.07*	**0.25**	0.46	0.09	5.01

Table 1 shows the experimental result of the Rastrigin function and the Grewangk function. Bold indicates the best average value and * indicates the best Min value of each selection method. Because of space limitations we present only the result of 128 population, 1000 generation case, but three different experiments showed almost the same results.

Table 2. The experimental result of TSP (*pop_size* = 128, *generation* = 1000)

	eil51					eil76					eil101				
Sel.	Min.	Ave.	Max.	STD.	CPU	Min.	Ave.	Max.	STD.	CPU	Min.	Ave.	Max.	STD.	CPU
RWTS	**17.53**	**22.52**	28.76	4.19	2.40	**25.91**	**33.52**	41.73	3.84	3.67	**38.60**	**42.22**	46.06	2.39	5.01
TSwR	20.68	25.76	32.08	3.49	2.73	30.43	37.32	43.17	3.70	3.94	41.76	45.14	48.32	2.02	5.27
TSwoR	19.74	25.14	31.80	3.78	2.68	33.16	36.99	39.75	2.27	3.94	45.57	47.48	49.02	1.14	5.32
PTS	37.01	39.85	44.07	2.04	2.85	47.33	50.18	53.13	1.83	4.13	54.44	57.14	58.77	1.43	5.48
LRS	19.67	24.83	33.65	4.58	2.09	34.40	41.42	45.07	3.04	3.08	42.77	48.50	56.43	3.77	4.05
ERS	20.83	24.99	33.51	4.64	2.02	33.73	38.31	41.78	3.04	3.00	42.93	49.58	54.82	3.28	4.04

	gr202					pa561				
Sel.	Min.	Ave.	Max.	STD.	CPU	Min.	Ave.	Max.	STD.	CPU
RWTS	**48.72**	**52.49**	56.27	2.17	10.70	**79.93**	**80.51**	81.21	0.35	29.81
TSwR	54.97	57.21	60.62	1.64	10.94	81.22	82.04	82.54	0.45	32.08
TSwoR	54.75	56.31	58.43	1.08	11.60	80.90	82.00	82.75	0.51	31.31
PTS	62.79	64.91	66.11	0.92	11.44	83.67	84.11	84.55	0.26	31.30
LRS	59.06	61.01	65.26	1.83	9.16	82.77	84.12	84.90	0.51	26.65
ERS	59.24	61.83	66.25	2.24	8.45	80.36	83.61	85.61	1.40	26.09

Table 3. The experimental result of MKP (*pop_size* = 128, *generation* = 1000)

	(5, 100)					(5, 100)					(10, 100)				
Sel.	Min.	Ave.	Max.	STD.	CPU	Min.	Ave.	Max.	STD.	CPU	Min.	Ave.	Max.	STD.	CPU
RWTS	**1.01**	2.09	2.64	0.53	40.13	1.78	2.20	3.54	0.75	40.12	2.86	3.20	4.12	0.54	64.08
TSwR	1.39	2.14	3.43	0.68	40.37	2.09	2.88	4.05	0.65	42.97	3.06	3.71	4.65	0.54	62.10
TSwoR	1.72	2.29	2.77	0.48	40.22	1.66	2.15	2.62	0.30	40.13	3.12	3.96	4.84	0.63	59.96
PTS	1.28	**1.82**	2.44	0.50	41.02	**1.41**	**2.03**	3.06	0.63	42.45	**1.83**	**2.62**	3.47	0.53	65.62
LRS	3.42	5.44	8.20	1.95	41.63	3.12	4.22	5.18	0.82	39.70	4.05	4.82	5.79	0.59	61.60
ERS	2.89	3.44	4.05	0.47	40.99	3.49	5.32	9.50	2.20	42.59	4.32	5.35	7.51	1.12	65.99

	(10, 100)					(10, 100)				
Sel.	Min.	Ave.	Max.	STD.	CPU	Min.	Ave.	Max.	STD.	CPU
RWTS	2.86	3.39	3.85	0.36	64.86	3.87	3.99	4.17	0.11	63.75
TSwR	**2.76**	3.89	4.94	0.74	61.81	3.96	4.33	5.04	0.37	61.54
TSwoR	2.98	4.03	5.62	1.03	62.40	4.38	4.89	5.89	0.53	61.30
PTS	2.80	**3.36**	4.23	0.66	64.70	**2.37**	**2.88**	3.73	0.49	67.90
LRS	4.10	5.91	9.80	2.14	65.66	3.19	4.50	5.31	0.75	61.43
ERS	4.13	6.38	7.88	1.51	64.90	3.58	5.48	8.30	1.71	66.70

Clearly, RWTS was superior to the tournament selections (TSwR, TSwoR and PTS) and ERS. However, RWTS showed slightly poor performance than LRS for Rastrigin func. On the other hand, RWTS showed the best average performance at the Grewangk func. Moreover, RWTS takes less computation time than the others for both test functions. We could observe that the difference is bigger as the population size increases. Another encouraging fact is that the standard deviation of RWTS is relatively smaller than the others. This implies an EA using RWTS can give very stable results on different environments and problems.

Table 2 and 3 show another experimental results on TSP and MKP. In TSP table, the name of title means problem name in [7] and in MKP table, the name of title '(number1, number2)' means '(total number of resources in MKP, total number of projects in MKP)'. In here, each element indicates the solution quality. In case of TSP, it is measured by the percentage gap between the best solution found and the optimal solution i.e $(Best - OPT)/Best \times 100$. And in case of MKP, it is measured by the percentage gap between the best solution found and the optimal solution of the LP relaxation i.e $(OPT_{LP} - Best)/OPT_{LP} \times 100$.

RWTS indicated better performance at all test instances for TSP instances, while RWTS showed second best performance at all test instances for MKP; the best performance was returned by PTS, but the difference was very small. In case of computation time, RWTS took more computation time than LRS and ERS, while it took less computation time than TSwR, TSwoR and PTS at TSP. On the other hand, in case of MKP, it is very difficult to say which method takes less computation time. STD was relatively less than the others. So, we can say that if we use RWTS for solving some optimization problem, we can obtain at least a stable and good result.

5 Conclusion

Based on the selection probability distribution, we have analyzed a real world tournament selection (RWTS) and also showed that RWTS can give an encouraging performance in a wide range of problems from a function optimization problem to a combinatorial optimization problem.

In this paper, we have focussed on only the selection probability distribution for the mathematical analysis but we found the fact that RWTS has higher selection pressure and higher selection probability in high ranking and implies *Elitism.* That means it does not need a strategy for preserving the best solution in this generation.

Even though RWTS dose not give better performance at all test problems, one definite thing is that it can give a sure performance at FOP and TSP. So, we recommend that RWTS is employed to EAs for solving various kind of optimization problems.

Acknowledgement

This research is supported by the UCN Project, the MIC 21st Century Frontier R & D Program in Korea.

References

1. Bäck, T.: Selective pressure in evolutionary algorithms: A Characteriszation of selection mechanisms, Proc. of 1st IEEE Conf. on Evolutionary Computation. (1994) 57–62
2. Bäck, T.: Evolutionary Algorithms in Theory and Practice. Oxford University Press, Oxford, UK (1996)
3. Blickle, T. and Thiele, L.: A Mathematical Analysis of Tournament Selection, Genetic Algorithms (1995) 9–16
4. Julstrom, B.A: It's all the same to me: Revisiting rank-based probabilities and tournaments, Proc. of The Cong. on Evolutionary Computation, (1999) 1501–1505
5. Soak, S.M, Corne, D. and Ahn, B.H.: A Powerful New Encoding for Tree-Based Combinatorial Optimisation Problems, Lecture notes in computer science, no.3242 (2004) 430–439
6. Soak, S.M, Corne, D. and Ahn, B.H.: A New Encoding for the Degree Constrained Minimum Spanning Tree Problem, Lecture notes in Comuputer Science, no.3213, (2004) 952–958
7. http://people.brunel.ac.uk/ mastjjb/jeb/info.html

Link Mass Optimization of Serial Robot Manipulators Using Genetic Algorithm

Serdar Kucuk[1] and Zafer Bingul[2]

[1] Kocaeli University, Electronics and Computer Education,
Umuttepe Campus, Kocaeli, Turkey
skucuk@kou.edu.tr
[2] Kocaeli University, Mechatronics Engineering,
Vinsan Campus, Kocaeli, Turkey
zaferb@kou.edu.tr

Abstract. This paper presents the link mass optimization of a serial robot manipulator based on minimum joint torque requirements that is primary concern in the industrial robot applications. The optimization of link mass globally minimizes joint torques weighted by inverse of inertia matrix. Genetic algorithm (GA) was used to optimize energy produced by robot manipulator. The influences of GA parameters (population sizes and mutation rates) on the solution of this problem were examined by varying these parameters. The rigid body dynamics of a cylindrical three-link serial robot manipulator is used as an optimization model. A fifth order polynomial used for actuating the joints from initial position to the goal position in a smooth manner. The link masses are used as design variables limited to upper, and lower bounds.

1 Introduction

The energy minimization has been studied for a long time in the robotics and automated manufacturing systems. For example, Garg et al. [1] have developed a strategy for force balance and energy optimization of cooperating manipulators. Hirakawa and Kawamura [2] have presented an electrical energy minimization approach to solve the trajectory generation problem in redundant robot manipulators. Liu and Wang [3] used a dual neural network for the bi-criteria torque optimization of kinematically redundant manipulators. Minimization of joint torque of robot manipulators is one of the most important performance criteria. This balances between the total energy consumption and the torque distribution among the joints. Optimal performance based on energy consumption can be obtained from a robotic manipulator with optimum link masses. Having better link mass distribution and minimizing joint torques are the ways of improving the energy efficiency in robot manipulators. The joint torques can be locally minimized using inverse of inertia matrix [4].

A lot of optimization techniques such as genetic algorithm [10], neural network [5] and minimax algorithms [6] have been used in robotics application problems. Recently, Tang et al. [5] have used neural network approaches for joint torque optimization. Stocco et al. [6] have proposed a new discrete global minimax algorithm for optimization of robot parameters. In this paper, GA is applied for optimizing the link

B. Gabrys, R.J. Howlett, and L.C. Jain (Eds.): KES 2006, Part I, LNAI 4251, pp. 138 – 144, 2006.
© Springer-Verlag Berlin Heidelberg 2006

masses of three link robot manipulator in order to obtain minimum energy consumed by the joints. GA optimization possesses some unique procedures that separate from the other optimization techniques given as follows:

i. GA uses objective function instead of derivatives.
ii. It searches over the entire population instead of a single point.
iii. It deals with parameter coding instead of parameters themselves.

These features provide GA to be a robust and useful optimization technique over the other search techniques [7].

In recent years, GA optimization techniques have been widely used in many other engineering applications, such as machine learning, scheduling, patterns recognition and robot motion planning [8], [9], [10]. GA that provides a robust search procedure for solving difficult problems is an adaptive heuristic search algorithm based on the evolutionary ideas of natural selection and genetic. The basic concept of GA is designed to simulate processes in natural system necessary for evolution, specifically those that follow the principles first laid down by Charles Darwin of survival of the fittest. As such they represent an intelligent exploitation of a random search within a defined search space to solve a problem. The obtained optima are an end product containing the best elements of previous generations where the attributes of a stronger individual tend to be carried forward into following generation. The rule is survival of the fittest will.

A GA begins with a set of solutions (represented by chromosomes) called population that is created randomly. Solutions from one population are taken and used to form a new population. Basically, the new population is generated by using following fundamental genetic evolution processes which are reproduction, crossover and mutation.

At the reproduction process, chromosomes are chosen based on natural selections. The chromosomes that comprise the new population are selected according to their fitness values defined with respect to some specific criteria such as roulette wheel selection, rank selection or steady state selection. The fittest chromosomes have a higher probability of reproducing one or more offspring in the next generation in proportion to the value of their fitness.

At the crossover stage, two members of population exchange their genes. Crossover can be implemented in many ways such as having a single crossover point or many crossover points which are chosen randomly. A simple crossover can be implemented as follows. In the first step, the new reproduced members in the mating pool are mated randomly. In the second step, two new members are generated by swapping all characteristics from a randomly selected crossover point.

At the mutation process, value of a particular gene in a particular chromosome is changed. This means changing a gene value from 1 to 0 or vice versa. These changes are implemented rarely. Therefore, the probability of mutation in a chromosome is generally kept very small.

2 Problem Formulation

Complete dynamic model of a robotic manipulator can be derived by using relatively simple approach called Lagrangian formulation that is based on the generalized coordinates, energy and generalized forces. The generalized coordinate q is the distance d for the prismatic joint or the joint angle, Θ for the revolute joint. If the manipulator is moving freely in the workspace, then the dynamic equation of motion of n DOF robot manipulator can be given in the following matrix form.

$$D(q)\ddot{q}+C(q,\dot{q})+G(q)=\tau \quad (1)$$

where, τ is the nx1 generalized torque vector applied at joints. q, $\dot{q}$ and $\ddot{q}$ are the nx1 joint variables vector of the robot manipulator, its first and second differential forms, respectively. The first term $D(q)\ddot{q}$ is nxn inertial coefficient matrix that represents the inertial forces and torques. The second term $C(q,\dot{q})$ is the nx1 Coriolis and centrifugal force vector. The last term $G(q)$ is the nx1 gravitational force vector.

The local optimization of the joint torgue weigheted by inertia results in resolutions with global characteristics. That is the solution to the local minimization problem [3]. The formulation of the problem is

$$\text{Minimize } \tau^{T}D^{-1}\tau \ (m_1,\ m_2,\ m_3)$$
$$\text{subjected to}$$
$$J(q)\ddot{\theta}+\dot{J}(q)\dot{\theta}-\ddot{r}=0 \quad (2)$$

where, $r=f(q)$ is the position vector of the end-effector. $J(q)$ is the Jacobian matrix and defined as $J(q)=\partial f(q)/\partial q$. The objective function is also minimize the kinetic energy given in equation 3.

$$\int_0^{t_f}\dot{\theta}^{T}\frac{D\cdot\dot{\theta}}{2} \quad (3)$$

3 Example

The rigid body dynamics model of three-link robot manipulator whose link parameters assigned according to the DH method [11] is given in Fig. 1. For simplification, each link of the robot manipulator is modelled as a homogeneous cylindrical or prismatic beam of mass m_i which is located at the centre of each link.

Generalized torque vector applied at joints are

$$\tau=[\tau_1 \quad \tau_2 \quad \tau_3]^{T} \quad (4)$$

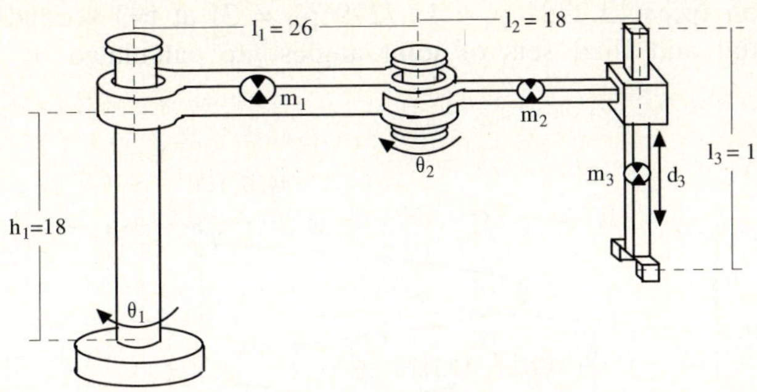

Fig. 1. The rigid body dynamics model of the three-link robot manipulator

where

$$\tau_1 = [\frac{1}{4}m_1 l_1^2 + I_{zz_1} + m_2(\frac{1}{4}l_2^2 + l_1 l_2 \cos\theta_2 + l_1^2) + I_{zz_2} + m_3(l_2^2 + 2l_1 l_2 \cos\theta_2 + l_1^2)$$
$$+ I_{zz_3}]\ddot{q}_1 + [m_2(\frac{1}{4}l_2^2 + \frac{1}{2}l_1 l_2 \cos\theta_2) + I_{zz_2} + m_3(l_2^2 + l_1 l_2 \cos\theta_2) + I_{zz_3}]\ddot{q}_2$$
$$-[m_2 l_1 l_2 s\theta_2 + 2m_3 l_1 l_2 s\theta_2]\dot{q}_1\dot{q}_2 - [\frac{1}{2}m_2 l_1 l_2 s\theta_2 + m_3 l_1 l_2 s\theta_2]\dot{q}_2^2$$

$$\tau_2 = [m_2(\frac{1}{4}l_2^2 + \frac{1}{2}l_1 l_2 \cos\theta_2) + I_{zz_2} + m_3(l_2^2 + l_1 l_2 \cos\theta_2) + I_{zz_3}]\ddot{q}_1$$
$$+[\frac{1}{4}m_2 l_2^2 + I_{zz_2} + l_2^2 m_3 + I_{zz_3}]\ddot{q}_2 + [\frac{1}{2}l_1 l_2 m_2 s\theta_2 + l_1 l_2 m_3 s\theta_2]\dot{q}_1^2$$

$$\tau_3 = m_3\ddot{q}_3 + g_0 m_3$$

The minimization problem is

$$\tau^T H^{-1}\tau = \left(-\tau_1 \frac{l_2^2 m_3 + Izz_3}{m_3 l_1^2(m_3 l_2^2 c^2\theta_2 - l_2^2 m_3 - Izz_3)} + \tau_2 \frac{l_2^2 m_3 + m_3 l_1 l_2 c\theta_2 + Izz_3}{m_3 l_1^2(m_3 l_2^2 c^2\theta_2 - l_2^2 m_3 - Izz_3)}\right)\tau_1$$
$$+\left(\tau_1 \frac{l_2^2 m_3 + m_3 l_1 l_2 c\theta_2 + Izz_3}{m_3 l_1^2(m_3 l_2^2 c^2\theta_2 - l_2^2 m_3 - Izz_3)} - \tau_2 \frac{l_2^2 m_3 + 2m_3 l_1 l_2 c\theta_2 + m_3 l_1^2 + Izz_3}{m_3 l_1^2(m_3 l_2^2 c^2\theta_2 - l_2^2 m_3 - Izz_3)}\right)\tau_2$$
$$+m_3(\ddot{q}_3 + g_0) \quad (5)$$

3.1 The Solution Procedure with Genetic Algorithm Approach

Fig. 2 shows the displacement of the end-effector varying in time. In Fig. 2, the end-effector starts from the initial position (p_{xi}=44 p_{yi}=0 p_{zi} =18) at t=0 second, and reach the final position (p_{xf}= 13.2721 p_{yf}= 12.7279 p_{zf} = 2) at t=3 seconds in Cartesian space. The initial and final sets of joint angles are calculated by using inverse kinematics.

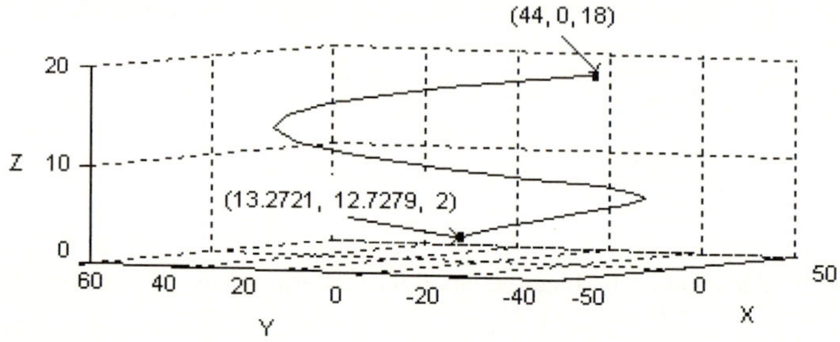

Fig. 2. The displacement of the end-effector (Fig. 1) varying in time

Ten samples obtained from Fig. 2 at time intervals between 0 and 3 seconds for the optimization problem are given in Table 1. Pos.-i, Vel.-i and Acc.-i represent position, velocity and acceleration of ith joint of the robot manipulator, respectively ($1 \leq i \leq 3$).

Table 1. Position, velocity and acceleration samples of first, second and third joints

Sample	Pos.-1	Vel.-1	Acc.-1	Pos.-2	Vel.-2	Acc.-2	Pos.-3	Vel.-2	Acc.-3
1	0.126	3.737	72.177	0.047	1.401	27.066	0.005	0.166	3.207
2	6.917	48.071	203.37	2.594	18.02	76.266	0.307	2.136	9.039
3	31.220	115.20	228.97	11.707	43.20	85.866	1.387	5.120	10.176
4	75.555	177.77	177.77	28.333	66.66	66.666	3.358	7.901	7.901
5	135.53	217.07	78.577	50.823	81.40	29.466	6.023	9.647	3.492
6	202.43	223.00	-39.822	75.912	83.62	-14.933	8.997	9.911	-1.769
7	265.82	194.13	-48.622	99.684	72.80	-55.733	11.814	8.628	-6.605
8	316.12	137.67	-19.022	118.54	51.62	-82.133	14.050	6.118	-9.734
9	347.22	69.444	-22.222	130.20	26.04	-83.333	15.432	3.086	-9.876
10	359.03	13.937	-29.422	134.63	5.226	-48.533	15.957	0.619	-5.752

In robot design optimization problem, the link masses m_1, m_2 and m_3 are limited to an upper bound of 10 and to a lower bound of 0. The objective function does not only depend on the design variables but also the joint variables (q_1, q_2, q_3) which have lower and upper limits ($0<q_1<360$, $0<q_2<135$ and $0<q_3<16$), on the joint velocities ($\dot{q}_1, \dot{q}_2, \dot{q}_3$) and joint accelerations ($\ddot{q}_1, \ddot{q}_2, \ddot{q}_3$). The initial and final velocities of each joint are defined as zero. In order to optimize link masses, the objective function should be as small as possible at all working positions, velocities and accelerations. The following relationship was adapted to specify the corresponding fitness function:

$$\tau^T D^{-1} \tau = \tau^T D^{-1} \tau(1) + \tau^T D^{-1} \tau(2) + \ldots + \tau^T D^{-1} \tau(9) + \tau^T D^{-1} \tau(10) \quad (6)$$

In the GA solution approach, the influences of different population sizes and mutation rates were examined to find the best GA parameters for the mass optimization problem where the minimizing total kinetics energy. The GA parameters, population sizes and mutation rates were changed between 20-60 and 0.005-0.1 where the number of iterations was taken 50 and 100, respectively. The GA parameters used in this study were summarized as follows:

Population size : 20, 40, 60
Mutation rate : 0.005, 0.01, 0.1
Number of iteration : 50, 100

All results related these GA parameters are given in Table 2. As can be seen from Table 2, population sizes of 20, mutation rate of 0.01 and iteration number of 50 produce minimum kinetic energy of 0. 892 10^{-14} where the optimum link masses are m_1=9.257, m_2=1.642 and m_3=4.222.

Table 2. The GA parameters, fitness function and optimized link masses

Population size	20	20	20	20	20	20	40	40	40
Mutation rate	0.005	0.01	0.1	0.005	0.01	0.1	0.005	0.01	0.1
Number of iteration	50	50	50	100	100	100	50	50	50
Fitness Value (10^{-14})	1.255	0.892	5.27	1.619	3.389	554.1	1.850	8.517	6.520
m_1	0.977	9.257	6.559	3.421	3.167	0.146	7.507	4.271	9.462
m_2	1.915	1.642	0.508	1.661	0.508	0.039	0.185	0.254	0.156
m_3	1.877	4.222	0.205	0.889	1.446	0.009	2.717	0.332	0.664

Population size	40	40	40	60	60	60	60	60	60
Mutation rate	0.005	0.01	0.1	0.005	0.01	0.1	0.005	0.01	0.1
Number of iteration	100	100	100	50	50	50	100	100	100
Fitness Value (10^{-14})	1.339	2.994	13.67	4.191	1.876	12.35	2.456	4.362	19.42
m_1	3.040	3.773	9.677	5.347	3.949	0.078	6.432	6.041	1.349
m_2	0.048	0.889	0.117	0.449	1.397	0.078	0.713	0.381	0.058
m_3	3.958	0.351	0.263	0.772	0.498	0.351	1.388	0.811	0.215

6 Conclusion

In this paper, GA optimization technique is applied to find the optimum link masses for the three-link serial robot manipulator. The influences of different population sizes and mutation rates were searched to achive the best GA parameters for the mass optimization. The optimum link masses obtained at minimum torque requirements show that the GA optimization technique is very consistant in robotic design applications. Mathematically simple and easy coding features of GA also provide convinient solution approach in robotic optimization problems.

References

1. Garg, D., Ruengcharungpong, C.: Force Balance and Energy Optimization in Cooperating Manipulation. Proceeding of the 23th Annual Pittsburgh Modeling and Simulation Conference, Vol. 23. 2017-2024, (1992).
2. Hirakawa, H. A., Kawamura, A.: Trajectory Generation for Redundant Manipulators Using Variational Approach Applied to Minimization of Consumed Electrical Energy. Proceeding of the 40th International Workshop on Advanced Motion Control, 687-692, (1996).

3. Lui, S., Wang J. : A Dual Neural Network for Bi-criteria Torque Optimization of Redundant Robot Manipulators. Lecture Notes in Computer Science, Vol. 3316, (2004).
4. Nedungadi, A., Kazerouinian K.: A Local Solution with Global Characteristics for Joint Torque Optimization of Redundant Manipulator. J. Robot. Syst. Vol. 6, (1989).
5. Tang W., Wang J.: Two Recurrent Neural Networks for Local Joint Torque Optimization of Kinematically Redundant Manipulators. IEEE Trans. Syst. Man Cyber, Vol. 32(5), (2002).
6. Stocco, L., Salcudean, S.E., Sassani, F.: Fast Constraint Global Minimax Optimization of Robot Parameters. Robotica, Vol. 16. 595-605 (1998).
7. Garg, D., Kumar, M.: Optimization Techniques Applied to Multiple Manipulators for Path Planning and Torque Minimization. Journal for Engineering Applications of Artificial Intelligence, Vol. 15. 241-252 (2002).
8. Chen, M., Zalzasa, A.: A Genetic Approach to Motion Planning of Redundant Mobile Manipulator Systems Considering Safety and Configuration. Journal of Robotic Systems, Vol. 14. 529-544 (1997).
9. Pires, E., Machado, J.:A Trajectory Planner for Manipulators using Genetic Algorithms. Proceeding of the IEEE Int. Symposium on Assembly and Task Planning, 163-168 (1999).
10. Choi, S., Choi, Y., Kim, J.: Optimal Working Trajectory Generation for a Biped Robot Using Genetic Algorithm. Proceedings of IEEE/RSJ International Conference on Intelligent Robots and Systems, Vol. 3. 1456-1461 (1999).
11. Denavit, J., Hartenberg R. S. : A Kinematic Notation for Lower-Pair Mechanisms Based on Matrices. Journal of Applied Mechanics, Vol. 22(2), (1955).

An Algorithm for Eliminating the Inconsistencies Caused During Discretization

Feng Honghai[1,2], Liu Baoyan[3], He LiYun[3], Yang Bingru[2], Chen Yumei[4], and Zhao Shuo[5]

[1] Urban & Rural Construction School, Hebei Agricultural University, 071001 Baoding, China
[2] Information Engineering School, University of Science and Technology Beijing, 100083 Beijing, China
[3] China Academy of Traditional Chinese Medicine, 100700 Beijing, China
liuby@tcmcec.com
[4] Tian'e Chemical Fiber Company of Hebei Baoding, 071000 Baoding, China
[5] Hebei Agricultural University, 071001 Baoding, China

Abstract. Rough set based rule generation methods need discretization of the continuous values. However, most existing discretization methods cause inconsistencies. In this paper, we propose an algorithm that can eliminate the inconsistencies caused during the course of discretization. The algorithm can be integrated into the discaretization algorithms that cannot avoid causing inconsistencies to eliminate the inconsistencies. Three data experimental results show that the algorithm is available.

1 Introduction

Data sets frequently have attributes with numerical domains that make them unsuitable for certain data mining algorithms that deal mainly with nominal attributes such as decision trees and rough set theory. To use such algorithms we need to replace numerical attributes with nominal attributes that represent intervals of numerical domains with discrete values. This process, known as "discretization" has received a great deal of attention in data mining literature and includes a variety of ideas ranging from fixed k-interval discretization [1], fuzzy discretization [2,3], Shannon-entropy discretization due to Fayyad and Irani presented in [4,5].

However, discretization may cause some inconsistencies. In this paper, we propose an algorithm that can eliminate the inconsistencies caused during the course of discretization. The algorithm can be integrated into the discretization algorithms that cannot avoid causing inconsistencies to eliminate the inconsistencies. Three data sets experimental results show that the algorithm is available.

2 Basic Concepts of Discretization and Rough Set Theory

2.1 Basic Concepts of Discretization

The discretization process can be described generically as follows. Let B be a numerical attribute of a set of objects. The set of values of the components of these

B. Gabrys, R.J. Howlett, and L.C. Jain (Eds.): KES 2006, Part I, LNAI 4251, pp. 145 – 152, 2006.
© Springer-Verlag Berlin Heidelberg 2006

objects that correspond to the B attribute is the active domain of B and is denoted by adom(B).

To discretize B we select a sequence of numbers $t_1 < t_2 < \cdots < t_l$ in adom(B). Next, the attribute B is replaced by the nominal attribute B that has l + 1 distinct values in its active domain { $k_0, k_1, \cdots k_l$ }. Each B-component b of an object o is replaced by the discretized B-component k defined by

$$k = \begin{cases} k_0, & if \quad b \le t_1 \\ k_1, & if \quad t_i < b \le t_{i+1} \quad for \quad 1 \le i \le l - 1 \\ k_l, & if \quad t_l < b \end{cases}$$

The numbers $t_1 < t_2 < \cdots < t_l$ define the discretization process and they will be referred to as class separators.

2.2 Indiscernibility Relation

In a decision table DT=$\langle U, A, V, f \rangle$, to every subset of attributes $B \subseteq A$, a binary relation, denoted by $IND(B)$, called the B -indiscernibility relation, is associated and defined as follows:

$$IND(B) = \left\{ \begin{aligned} &(x_i, x_j) \in U \times U : a \in B, f(x_i, a) \\ &= f(x_j, a) \end{aligned} \right\} \tag{1}$$

2.3 Lower and Upper Approximations

The rough sets approach to data analysis hinges on two basic concepts, namely the lower and the upper approximations of a set, referring to:

-The elements that doubtlessly belong to the set, and
-The elements that possibly belong to the set.

Let X denotes the subset of elements of the universe U ($X \subset U$). The lower approximation of X in B ($B \subseteq A$), denoted as $B\_(X)$, is defined as the union of all these elementary sets that are contained in X . More formally:

$$POS_B(X) = B\_(X) = \{x_i \in U \| [x_i]_{IND(B)} \subset X\}$$

The upper approximation of the set X , denoted as $B^{-}(X)$, is the union of these elementary sets, which have a non-empty intersection with X :

$$B^{-}(X) = \{x_i \in U \| [x_i]_{IND(B)} \cap X \ne 0\}$$

The difference:

$$\mathrm{BN}(X) = B^{-}(X) - B\_(X)$$

is called a boundary of X in U.

3 The Numeric Examples That Cause Inconsistencies

Proposition (1): If the positive regions are held during the course of discretization, after discretization the inconsistent discrete examples are caused by the numeric examples that are in the boundary region of the corresponding decision class.

Proof

Given the condition attribute C, and decision attribute D, suppose IND $(C) = \{C_1, C_2 \ldots C_i\}$, IND $(D) = \{D_1, D_2 \ldots D_j\}$, and $C_i \subseteq D_j$

$\because C_i \subseteq D_j$ $\therefore C_i \rightarrow D_j$ and C_i is the positive region of D_j, namely, the numeric examples in C_i can only belong to the class D_j. If the positive region of the D_j has been held during the course of discretization, the examples discretized can only belong to the class D_j too.

Suppose C_{i+1} is the boundary region of D_j, so the numeric examples in C_{i+1} may belong to class D_j and $\neg D_j$, after discretization the discrete examples may belong to D_j and $\neg D_j$ too.

Proposition (2): The inconsistent examples are first caused by the numeric examples that are in the borderline of its interval, i.e., the numeric examples that have the least distance to the cuts cause the inconsistent discrete examples.

4 Algorithm

(1) Input the inconsistent examples that are in boundary region of any decision class and their cuts.

(2) For (i=0, i<n, i++)// i is the number of the inconsistent examples caused during //the course of the discretization, and n is the amount of the //examples

For (j=0, j<m, j++)// j is the number of the attribute values of the inconsistent //examples, and m is the amount of the attributes.

Calculate the differences of the continuous attribute values to their cuts

Select the cuts and the values that have the minimum distance to their cuts

For (k=0, k<l, k++)// k is the number of the cuts, and the l is amount //of cuts

Calculate the amount of the examples whose values have the minimum distance to their cuts

If two group of inconsistent examples are in different sides of one cut

Select the group that has more examples (if the amount of examples in two groups is equal, select any one of the two group)

Put the values into the opposite interval that is on the other side of the cut.

(6) If the inconsistent examples become consistent, adjust the intervals and cuts. Else select the inconsistent examples goto (2).

5 Micronutrients in SARS Patients' Body Data Set Experiments

The whole SARS data set includes 60 cases among which examples 31~60 are SARS patients and 61~90 are healthy human beings. Table 1 gives partial experimental results of micronutrients that are essential in minute amounts for the proper growth and metabolism of human beings.

Table 1. Partial examples of whole SARS data set

U	1	2	3	4	5	6	7	C	U	1	2	3	4	5	6	7	C
31	66	15.8	24.5	700	112	179	513	0	68	158	14.4	37.0	*1025*	101	44.6	72.5	1
32	185	15.7	31.5	701	125	184	427	0	69	*133*	22.8	31.0	1633	401	180	*899*	1
43	186	9.26	37.1	958	*233*	73.0	347	0	70	156	135	322	6747	1090	*228*	810	1
52	154	13.8	*53.3*	621	105	160	723	0	72	247	17.3	*8.65*	2554	241	77.9	373	1
53	179	12.2	17.9	1139	150	*45.2*	218	0	74	209	*6.43*	86.9	2157	288	74.0	219	1
55	175	5.84	24.9	807	123	55.6	*126*	0									
60	178	*28.8*	32.4	992	112	70.2	169	0									

Attribute "1", "2", "3", "4", "5", "6", "7" denote micronutrient Zn, Cu, Fe, Ca, Mg, K, Na respectively, and the decision attribute "C" denotes the class "SARS" and "healthy". $V_C = \{0,1\}$, where "0" denotes "SARS", "1" denotes "healthy".

For attribute "1" (Zn), [13.5, 132] = $POS_{Zn}(C=0)$, [133, 226] = $BN_{Zn}(C=0)$ = $BN_{Zn}(C=1)$ and [235, 247] = $POS_{Zn}(C=1)$. So the cuts are (132+133)/2=132.5, and (226+235)/2=230.5.

For attribute "2" (Cu), [3.27, 5.84] = $POS_{Cu}(C=0)$, [6.43, 28.8] = $BN_{Cu}(C=0)$ = $BN_{Cu}(C=1)$ and [65.4, 135] = $POS_{Cu}(C=1)$. So the cuts are (5.84+6.43)/2=6.135, and 47.1.

For attribute "3" (Fe), [4.04, 8.17] = $POS_{Fe}(C=0)$, [8.65, 53.3] = $BN_{Fe}(C=0)$ = $BN_{Fe}(C=1)$ and [61.7, 322] = $POS_{Fe}(C=1)$. So the cuts are 8.41, and 57.5.

For attribute "4" (Ca), [135, 992] = $POS_{Ca}(C=0)$, [1025, 1437] = $BN_{Ca}(C=0)$ = $BN_{Ca}(C=1)$ and [1521, 6747] = $POS_{Ca}(C=1)$. So the cuts are 1008.5, and 1479.

For attribute "5" (Mg), [32.6, 95.5] = $POS_{Mg}(C=0)$, [99.1, 233] = $BN_{Mg}(C=0)$ = $BN_{Mg}(C=1)$ and [241, 1090] = $POS_{Mg}(C=1)$. So the cuts are 97.3, and 237.

For attribute "6" (K), [31, 44.6] = $POS_K(C=1)$, [45.2, 228] = $BN_K(C=0)$ = $BN_K(C=1)$ and [239, 1314] = $POS_K(C=0)$. So the cuts are 44.9, and 233.5.

For attribute "7" (Na), [67.3, 109] = $POS_{Na}(C=1)$, [126, 899] = $BN_{Na}(C=0)$ = $BN_{Na}(C=1)$ and [1092, 1372] = $POS_{Na}(C=0)$. So the cuts are 117.5, and 995.5.

For every attribute, 0, 1, and 2 denote the three intervals respectively in ascending order.

However, example 23 and example 32 become inconsistent ones. The values of attribute K are 45.2 and 47.9, whereas the cut is 44.9, so the value that has the least distance to the cut is 45.2, additionally, 45.2 is a borderline value in [45.2, 228]. Using the algorithm proposed above the value 45.2 should fall under the opposite interval (anterior interval), that is, it is discretized into 0 not the 1, the value 45.2 is represented by 0 not 1. Accordingly, the cut is changed into 45.2.

6 Iris Plants Database Experiment

In the data set, there are 150 instances (50 in each of three classes), and there are 4 numeric condition attributes named sepal length (SL), sepal width (SW), petal length (PL), and petal width (PW) respectively, the only decision attribute is the class (C) whose values are Setosa, Versicolor, and Virginica.

For attribute PW, [0.1, 0.6] = POS_{PW}(C= Setosa), [1, 1.3] = POS_{PW}(C= Versicolor), [1.4, 1.8] = POS_{PW}(C= Versicolor and C= Virginica) and [1.9, 2.5] = POS_{PW}(C=Virginica). So the cuts are 0.8, 1.35, and 1.85.

For attribute PL, [1, 1.9] = POS_{PL}(C= Setosa), [3, 4.4] = POS_{PL} (C= Versicolor), [4.5, 4.7] = POS_{PL} (C= Versicolor and C=Virginica), where the amount of the examples that belong to class Versicolor is more than the amount of the examples that belong to class Virginica. [4.8, 5.1] = POS_{PL} (C= Versicolor and C=Virginica), where the amount of the examples that belong to class Versicolor is less than the amount of the examples that belong to class Virginica. and [5.2, 6.9] = POS_{PL}(C=Virginica). So the cuts are 2.45, 4.45, 4.75, and 5.15.

For attribute SW, [2, 2.9] = BN_{SW}(C= Versicolor) = BN_{SW}(C=Virginica)= BN_{SW}(C= Setosa), where the amount of the examples that belong to the classes Virginica and Versicolor is more than the amount of the examples that belong to the class Setosa , [3, 3.3] = BN_{SW}(C= Versicolor) = BN_{SW}(C=Virginica)= BN_{SW}(C= Setosa). [3.4, 4.4] = BN_{SW}(C= Versicolor) = BN_{SW}(C=Virginica) = BN_{SW}(C= Setosa), where the amount of the examples that belong to the class Versicolor is more than the amount of the examples that belong to the any one of the other two classes. So the cuts are 2.95, 3.35.

For attribute SL, [4.3, 4.8] = POS_{SL}(C= Setosa), [4.9, 5.4] = BN_{SL}(C= Versicolor) = BN_{SL}((C=Virginica)= BN_{SL}((C= Setosa), where the amount of the examples that belong to the class Setosa is more than the amount of the examples that belong to the classes Virginica and Versicolor, [5.5, 6.2] = BN_{SL}(C= Versicolor) = BN_{SL}(C=Virginica)= BN_{SL}(C= Setosa), where the amount of the examples that belong to the class Versicolor is more than the amount of the examples that belong to the classes Virginica and Setosa. [6.3, 7.9] = BN_{SL} (C= Versicolor) = BN_{SL} (C=Virginica) = BN_{SL} (C= Setosa), where the amount of the examples that belong to the class Virginica is more than the amount of the examples that belong to the classes Versicolor and Setosa. So the cuts are 4.85, 5.45, and 6.25.

Furthermore, examples 84, 120, and 127 become inconsistent ones; examples 71, 128, 139, and 150 become inconsistent ones; examples 73, 77, 124, and 134 become inconsistent ones.

Table 2. Inconsistent examples 84, 120, and 127

U	PW	cut	PL	cut	SW	cut	SL	cut	class
84	1.6		5.1	5.15	2.7		6.0		Versicolorlor
120	1.5		5.0		2.2		6.0		virginica
127	1.8	1.85	4.8		2.8		6.2	6.25	virginica

Table 3. Inconsistent examples 73, 77, 124, and 134

U	PW	cut	PL	cut	SW	cut	SL	cut	class
73	1.5		4.9		2.5		6.3	6.25	Versicolorlor
77	1.4	1.35	4.8		2.8		6.8		Versicolorlor
134	1.5		5.1	5.15	2.8		6.3	6.25	Virginica
124	1.8	1.85	4.9		2.7		6.3	6.25	Virginica

Table 4. Inconsistent examples 71,128, 139, and 150

U	PW	cut	PL	cut	SW	cut	SL	cut	class
71	1.8	1.85	4.8		3.2		5.9		Versicolorlor
128	1.8	1.85	4.9		3.0	2.95	6.1		virginica
139	1.8	1.85	4.8		3.0	2.95	6.0		virginica
150	1.8	1.85	5.1	5.15	3.0	2.95	5.9		virginica

For attribute PW, 1.4 and 1.8 are the borderline values, and they have the least distance to their corresponding cuts. Since [1.4, 1.8] is a boundary region of "class=vesicolor and virginica", so we put 1.8 into its posterior interval [1.9, 2.5] and put the value 1.4 into its anterior interval [1, 1.3], the three intervals [1, 1.3], [1.4, 1.8], and [1.9, 2.5] turn into [1, 1.4], [1.5, 1.7], [1.8, 2.5].

For attribute PL, 5.1 is the borderline value, and it has the least distance to its corresponding cut. Since [4.5, 5.1] is a boundary region of "class=vesicolor or virginica", so we put 5.1 into its posterior interval [5.2, 6.9] and so we turn the boundary regions [4.5, 5.1] and [5.2, 6.9] into [4.5, 5.0] and [5.1, 6.9].

For attribute SW, 3.0 is the borderline value, and it has the least distance to its cut, so we put 3.0 into its anterior interval [2, 2.9], so the intervals [2, 2.9] and [3, 3.3] are turned into [2, 3] and [3.1, 3.3].

For attribute SL, since [6.3, 7.9] is a positive region of class viginica, value 6.3 cannot be put into [5.5, 6.2], contrarily, [5.5, 6.2] is a boundary region of class versicolor, virginica and setosa, value 6.2 can be put into [6.3, 7.9]. So the two intervals become [5.5, 6.1] and [6.2, 7.9].

After the above adjustment works, again we get two inconsistent examples that are 85 and 107.

Table 5. Inconsistent examples 85 and 107

U	PW	cut	PL	cut	SW	cut	SL	cut	class
85	1.5		4.5		3.0		5.4	5.45	Versicolorlor
107	1.7		4.5		2.5		4.9	4.85	Virginica

In the two inconsistent examples, values 1.5 and 1.7 (attribute PW), 4.5 (attribute PL), 3.0 (attribute SW) are all the borderline values, but their intervals have been adjusted in the first step, so in the second step, these intervals should not be adjusted. For values 5.4 and 4.9 (attribute SL), they are all the borderline values, and have the least distance to their cuts, so they will be put into the opposite intervals. However, the two values are on different sides of one cut, so we only adjust one of them, for example, we adjust the value 4.9, so the intervals [4.3, 4.8], [4.9, 5.4] are turned into [4.3, 4.9], [5.0, 5.3].

After one step of discretization and two steps of adjustments, we get the consistent discretization results.

7 Clinical SARS Data Experiment

We obtained the clinical SARS data of 524 SARS patients from Beijing in 2003. For condition attributes "the nth day of the patient having been ill", "age", "the highest body temperature" and the decision attributes "state of illness" 4309 records were gotten out of 8722 records. After discretization we get only two inconsistent records. Additionally, and after using the above algorithm the two inconsistent examples become consistent ones. The discretization rules for the attribute "the nth day of the patient having been ill" are:

If "state of an illness"=1 then the values of "the nth day of the patient having been ill" are discretized into 0.

If "state of an illness"=1 or 2 then the values of "the nth day of the patient having been ill" are discretized into 1.

If "state of an illness"=1, 2, or 3 then the values of "the nth day of the patient having been ill" are discretized into 2.

The discretization rules for the attributes "age" and "the highest body temperature" are the same as the rules for the attribute "the nth day of the patient having been ill".

From the above experimental results, using our discritization algorithm and the corresponding algorithm for eliminating inconsistencies only three continuous attributes are needed for us to get the consistent discretization results, which shows the higher discretization efficiency.

8 Conclusions

(1) In most cases, discretization causes inconsistencies.
(2) After discretization, we should not delete the repeated examples straightway. On the contrary, we should check whether there are any inconsistent examples. If there are some inconsistent examples, use the algorithm proposed above to adjust the discretization results.
(3) If the positive regions are held during the course of discretization, after discretization the inconsistent discrete examples are caused by the numeric examples that are in the boundary region of the corresponding decision class.

(4) The inconsistent examples are firstly caused by the numeric examples that are in the borderline of its interval. When adjusting the intervals the borderline values should be taken into account first.
(5) The algorithm proposed in this paper has higher discretization efficiency.

Acknowledgements

The paper is supported by the Special Foundation of SARS of the National High-Tech Research and Development Major Program of China (863)- "Clinical Research on the Treatment of Severe Acute Respiratory Syndrome with Integrated Traditional and Western (No.2003AA208101)".

References

1. Dougherty J, Kohavi R, Sahami M.: Supervised and unsupervised discretization of continuous features. Proc. of the 12th International Conference on Machine Learning. (1995) 194–202
2. Kononenko I.: Naive Bayes classifier and continuous attributes Informatica Vol 16 (1992) 1–8.
3. Kononenko I.: Inductive and Bayesian learning in medical diagnosis. Appl Artif Intell Vol 7 (1993) 317–37.
4. Fayyad UM. On the Induction of Decision Trees for Multiple Concept Learning. Ph.D. thesis, University of Michigan (1991).
5. Fayyad UM, Irani K.: Multiinterval discretization of continuous-valued attributes for classification learning. Proc. of the 12th International Joint Conference on Artificial Intelligence (1993) 1022–27.

A Combinational Clustering Method Based on Artificial Immune System and Support Vector Machine

Zhonghua Li and Hong-Zhou Tan

Department of Electronics and Communication Engineering,
Sun Yat-sen University, Guangzhou 510275, China
{lizhongh,issthz}@mail.sysu.edu.cn

Abstract. Clustering is one branch of unsupervised machine learning theory, which has a wide variety of applications in pattern recognition, image processing, economics, document categorization, web mining, etc. Today, we constantly face how to handle a large number of similar data items, which drives many researchers to contribute themselves to this field. Support vector machine provides a new pathway for clustering, however, it behaves bad in handling massive data. As an emergent theory, artificial immune system can effectively recognize antigens and produce the memory antibodies. This mechanism is constantly used to achieve representative or feature data from raw data. A combinational clustering method is proposed in this paper based on artificial immune system and support vector machine. Experimentation in functionality and performance is done in detail. Finally a more challenging application in elevator industry is conducted. The results strongly indicate that this combinational clustering in this paper is of feasibility and of practice.

Indexed Terms: Data clustering, combinational clustering, artificial immune system, support vector machine.

1 Introduction

For a simpler approach to identifying similar data items, *clustering analysis* may be used. It refers to a large number of algorithms and methods for grouping together similar data items with the groupings relationships between the items. It is a useful first step in the data analysis process because it can guide the efforts of any further study that might required. Clustering analysis has been widely applied to such fields as pattern recognition, image processing, economics, document categorization, web mining, etc [1]. A common question facing researchers in many areas is how to organize observed data into meaningful structures. The general clustering methods include Tree Clustering, Block Clustering, K-means Clustering, Evolutionary Maximization Clustering, and so on [3].

Since 2000, *support vector machine* (SVM) have gathered numerous eyeballs around the machine-learning world, which almost includes three branches: classification, regression and clustering [2]. Unlike support vector classification, support vector clustering is unsupervised learning. Data points are mapped from data

B. Gabrys, R.J. Howlett, and L.C. Jain (Eds.): KES 2006, Part I, LNAI 4251, pp. 153 – 162, 2006.
© Springer-Verlag Berlin Heidelberg 2006

space to a high dimensional feature space by means of a Gaussian kernel. In feature space, the smallest sphere is searched which can enclose the image of the data. This sphere is then mapped back into data space. Several contours are produced to enclose the data points. Points enclosed by each separate contour are of the same cluster. However, large-scale data items from data space are directly mapped to a high dimensional feature space, which increases the time complexity of SVM. To overcome this demerit, it is essential to extract the feature data from the raw dataset, which will be feed to SVM for clustering analysis.

In the recent years, artificial immune system interests any researches due to its rich immune mechanisms and powerful information processing capability [4-7], where de Castro and von Zuben creatively applied *artificial immune system* (AIS) to data clustering [6]. A classical *artificial immune network* called aiNet was constructed under the spirit of biological immune system, here the raw data was considered as the antigens of immune system, and the feature data was outputted as memory cells through the evolutionary interaction between the given antigens and a number of antibody candidates. Later an improved clustering method based on AIS was presented, which explores a new pathway for clustering large-scale data items. As long as the related parameters are chose properly, AIS is no longer time-consuming for any type of data items.

In the past, we attempted to combine AIS and SVM for data clustering, where AIS is responsible for extracting feature data from the raw data and SVM focuses on clustering the extracted feature data [12]. Based on the previous work, this paper will propose an improved combinational clustering method (ICCM) for further study. Our goal is advance ICCM more efficient and more stable. In the entry of this combined method, we will present a tactful policy that adds an extra block for data processing from the viewpoint of systems solutions.

The remainder of this paper is organized as follows. Some basic features about biological immune system and SVM are firstly reviewed in Section 2. Section 3 proposes a modified combined clustering method based on AIS and SVM, and gives some technical details. Simulation experiments and discusses are listed in Section 4. Finally conclusions are drawn in Section 5.

2 Features of Related Theories

A. Support Vector Machine

SVMs have been widely adopted for classification, regression and novelty detection, recently used for clustering analysis too. The basis of this support vector clustering is density estimation through SVM training. Support vector clustering is a boundary-based clustering method, where the support information is used to construct cluster boundaries. It is able to detect arbitrary shape clusters with a hierarchical structure in high dimensional data. It provides a way to deal with outliers and by using kernel methods; explicit calculations in feature space are not necessary [2].

Given $\{x_i\} \in \chi$ be a dataset with N points, Φ is a non-linear transformation from χ to some high dimension feature space. The smallest enclosing sphere of radius R is looked for. When soft constraints are incorporated by adding a slack variable ξ_j, R should satisfy the following constraint.

$$\|\Phi(x_j) - a\| \le R^2 + \xi_j, \forall j, \tag{1}$$

where $\|\bullet\|$ is the Euclidean norm, a is the center of the sphere, and $\xi_j \ge 0$. To solve this problem, the Lagrangian theorem is employed.

$$L = R^2 - \sum_j (R^2 + \xi_j - \|\Phi(x) - a\|)\beta_j - \sum_j \xi_j \mu_j + C\sum_j \xi_j \tag{2}$$

where $\beta_j \ge 0$ and $\mu_j \ge 0$ are both Lagrange multipliers, C is a constant, $C\sum_j \xi_j$ is a penalty term. Then Equation (2) can be transformed into its dual problem below,

$$W = \sum_j \Phi(x_j)^2 \beta_j - \sum_{i,j} \beta_i \beta_j \Phi(x_i) \bullet \Phi(x_j), \quad 0 \le \beta_j \le C \quad j = 1,\dots,N \tag{3}$$

The dot products $\Phi(x_i) \bullet \Phi(x_j)$ can be represented by a Mercer kernel $K(x_i, x_j)$. A Gaussian kernel is typically chose,

$$K(x_i, x_j) = e^{-q\|x_i - x_j\|^2} \tag{4}$$

At each point x, the distance of its image in feature space from the center of the sphere is defined as follows.

$$R^2(x) = \|\Phi(x) - a\|^2 \tag{5}$$

Finally, the contour enclosing the points in data space could be defined by:

$$\{x \mid R(x) = R\} \tag{6}$$

Support vectors lie on the boundary of each cluster. Bounded support vectors are outside, and all other points are inside the clusters.

B. Biological Immune system

Biological immune system can detect the evasion of external pathogens and destroy them in order to keep us healthy and secure. As immune system is being understood better and better, it's some biological mechanisms are helpful to handle some complex problems in theoretical and practical fields. For the sake of similarity, all models inspired from immune system developed based on biological immune system are called artificial immune system or AIS. Since the first AIS model births, its many excellent features interest a number of researchers around the world [4-10].

There are two classes of responses within immune system: primary immune response and secondary immune response. Primary immune response is provoked when the immune system encounters an antigen for the first time. After the infection has been cleared, a number of these antibodies will retain in the body and contribute to the secondary immune response. The secondary immune response is characterized by more rapid and more abundant production of the antibody. In addition, the secondary immune response can be fired from an antigen that is a similar, although not identical, antigen to the one that originally established the memory.

Identification of the antigen is achieved by complementary matching between the paratope and the epitope, comparable to a lock and a key. The strength of the bond will depend on how closely the two match. The closer the match between the antibody and the antigen, the stronger the molecular binding and the better the recognition.

The immune system is made up of numerous B cells and T cells which are constantly produced in the bone marrow and thymus, respectively. The level of B cell stimulation depends not only on the success of the match to the antigen, but also on how well it matches other B cells in the immune system. For similarity, the effect of T cells will be ignored in this paper.

If the stimulation of a B-cell reaches a certain threshold, the B-cell is transformed into a blast cell, which begins to divide rapidly, producing clones of itself. In order to allow the immune system adaptive, the clones that are produced turn on a mutation mechanism which generates mutations in the gene coding for the antibody molecule, which is called somatic hyper-mutation. However, if the stimulation level falls below the threshold, the B-cell will not replicate and will die timely. About 5% of the B cells in the biological immune system die each day in this manner.

Immunological memory is achieved by maintaining a population of B cells after the primary response. This allows for an area of a B-cell population to cluster similar B cells and produce a self-supporting structure to aid the immune response. By repeating the process of mutation and selection a number of times, the immune system learns to produce better matches for the antigen.

3 An Improved Combinational Clustering Method

On basis of our previous work [11][12], an improved combined clustering method will be presented in this section. Followed a clear theoretical framework, the corresponding technical details will be explained later.

A. ICCM proposed in this paper

Fig.1 illustrates the theoretical framework of ICCM. The procedure of the whole algorithm is divided into two phases. In the figure, each rectangular block denotes a key step, and the corresponding data shapes are given in the brackets. One phase includes data input, immune suppression, aiNet and feature data output, which is prepared for the next phase. The other consists of feature data input, SVM and support vectors output.

It is specially pointed out that a step about immune suppression is added in ICCM, compared to the earlier combined clustering model (CCM) in the literature

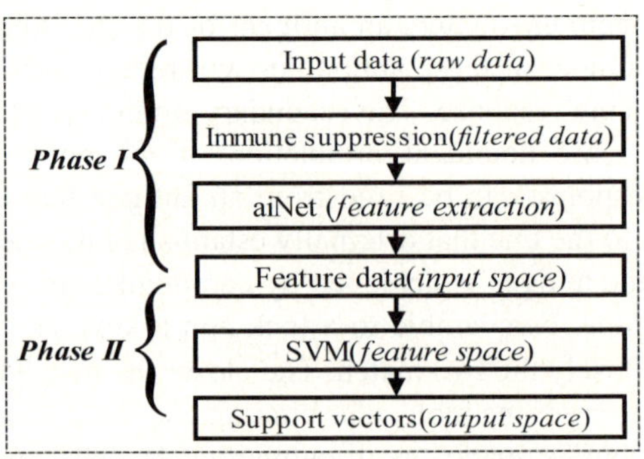

Fig. 1. Theoretical framework of ICCM

[12]. Inspired by the stimulation mechanism between any two B-cells, immune suppression operator is introduced to roughly remove the redundancy of data items before aiNet. This helps to decrease the computational load of aiNet, because the time complexity of aiNet is approximately proportional to the number of antigens.

For large-scale dataset, we consider the following two scenarios as describe in literature [11]. One scenario (*scenario_a*) is to regard the raw dataset with large-scale data as an entity, and the other (*scenatio_b*) is to partition the raw dataset into several data blocks. Whatever *scenerio_a* or *scenario_b* is employed, aiNet always has large increase in *CPU time* of execution.

In order to further increase the computational efficiency, we directly sample some data from the antigens as the initial antibodies and feed them to both previous CCM algorithm and ICCM algorithm in this paper.

B. aiNet in Phase I

As the core procedure in the first phase of ICCM, aiNet is responsible for feature extraction from the filtered data items. In other word, we employ a classical aiNet, proposed by de Castro and von Zuben, to conduct data compression.

According to the aiNet, the basis procedures can be constructed as follows.

Step 1. *Input* data items as *Ab*s and *standardize* them if necessary.

Step 2. *Produce n* random *Ab*s in the interval (0,1).

Step 3. While the termination condition is reached, go to *Step 7*; else *continue*;
For each antigen,

- *Calculate* the *affinity* between itself and candidate antibodies.
- *Select m* antibodies with the largest affinity to be network cells.
- *Clone* network cells. The bigger *affinity* is, the more the cloned cells are.
- *Mutate* each cloned cell. The mutation rate of each cloned cell is inversely exponentially proportional to its *affinity*.
- *Re-calculate* the affinity between the antigen and each mutated cell, and *choose* mutated cells with the largest affinity to enter memory cell subset M_p.
- *Eliminate* individuals in M_p whose similarity between each other is below the suppression threshold σ_s , and *retain* the representative individuals.

Step 4. *Incorporate* M_p into memory cell set *M*.

Step 5. *Eliminate* individuals in *M* whose similarity is less than σ_s ;

Step 6. Randomly *recruit d%* new individuals as next-generation *Ab*s together with *M*. Then return to *Step 3*.

Step 7. Output *M* as feature data items.

The termination condition of aiNet is either the average similarity of memory cell set *M* keeps unchanged or changes enough little between two adjacent generations, or its iteration reaches the maximum value *Gen* predefined before aiNet runs.

C. SVM in Phase II

In the second phase of ICCM, we can exploit support vector machine to cluster the feature dataset output from the first phase. Support vector clustering [2] is reviewed as follows.

1) Determine the width parameter q of Gaussian kernel and the constant C;
2) Calculate the kernel matrix of reduced data points;
3) Solve the Lagrangian multipliers β_j, finding support vectors and outliers;
4) Calculate the radius of super-plane and $C\sum_j \xi_j$ of distance equation;
5) Compute the adjacent matrix of reduce data points;
6) Search the cluster assignments;
7) Calculate the performance index of clustering approach.
8) Map back to data space to show the cluster results.

The termination status is that the clusters don't vary within a wide range over q.

4 Experiments and Results

In this section, we will firstly arrange a series of experimentations to evaluate our ICCM method and to compare it with the previous CCM approach. Then this section will end with a more challenging application in elevator traffic pattern recognition.

A. Experimental Preparation

For the sake of simplicity, we choose the test dataset in the literature [6] as *basis data*, on which some new test datasets are constructed. This basis data is visually drawn in Fig.2. There are five subgroups of data, and each subgroup denotes one pattern, marked in different colors and shapes.

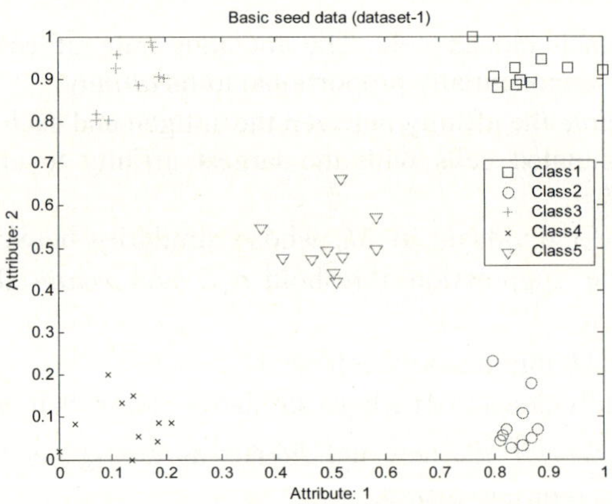

Fig. 2. Training patterns of the basis data

In the experimentations, we concentrate on the noise tolerance and the time complexity of ICCM. Therefore we evaluate ICCM on three kinds of test datasets: *dataset_*1 (the first dataset is the basis data), *dataset_*2(the second dataset is acquired by simple replication) and *dataset_*3 (the three dataset is produced by means of clone and mutation). Here, *dataset_*2 is utilized to evaluate its ability to clustering large-scale data, and *dataset_*3 is exploited for examining its noise tolerance.

Unless there is extra emphasis, we simply define some parameters of aiNet as follows: the aiNet suppression threshold value t_s =0.1; the rate of natural death p_r=1.0; the number of best-matching cells taken for each *Ag* *m*=4; the clone number multiplier *Nc*=20; the maximum generations of immune evolutionary computation *Gen*=15; the parameter for controlling average similarity of memory cells *sc*=0.01.

For *scenario_a* and *scenario_b*, the threshold value δ_s should be less than σ_s of aiNet, because there is just a single data filter operation on the raw data before aiNet. Here we make $\delta_s = 0.05$ without special emphasis.

All experiments are conducted in Matlab 6.5 and the main configuration of the computer is Intel Pentium (R) 4 CPU 2.40GHz and 512MB Memory, and Windows XP.

B. Experimental Results

To keep this paper concise, we will do some experiments only on performance evaluation of ICCM, which is compared to CCM in the work [12]. After a number of tests, we select the parameters q =30 and C =100 for SVM, which have preferable adaptability to a large variety of datasets.

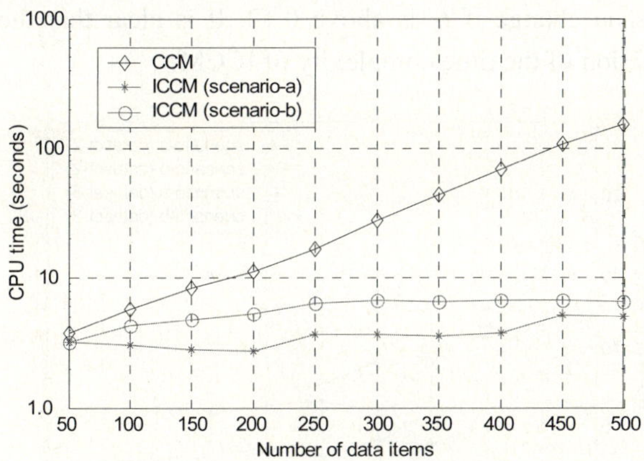

Fig. 3. CPU time over number of data items in *dataset_2*

Firstly we will study how the CPU time changes over number of data items. Here two datasets (*dataset_2* and *dataset_2*) are considered. For *dataset_2*, the CPU time of ICCM over number of datasets is illustrated in Fig.3. As the number of data items increases, CPU time of CCM ascends steeply like an exponential line, while our ICCM always has a lower and more moderate CPU time, less than 10 seconds. For *dataset_3*, the similar phenomenon occurs in Fig.4.

In the first phase of ICCM, aiNet is an iterative process whose time complexity is heavily affected by the number of antigens. If the number of antigens is reduced, the CPU time of ICCM will be depressed remarkably. From Fig.3 and Fig.4, therefore, we can reasonably conclude that the data filter step contributes much to a decrease in CPU time of ICCM.

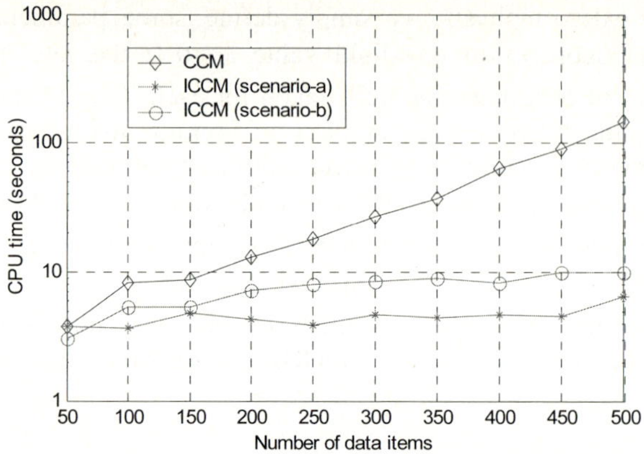

Fig. 4. CPU time over number of data items in dataset_3

Fig.5 illustrates how the immune suppression threshold t_s influences CPU time of ICCM in this paper. As t_s increments, the CPU time of ICCM decreases gradually. If t_s is below 0.08, the CPU time of ICCM has a sharp decrease, while the CPU time of ICCM almost has no change if t_s is above 0.12. It is clear that the increase of t_s results in the reduction of the time complexity of ICCM.

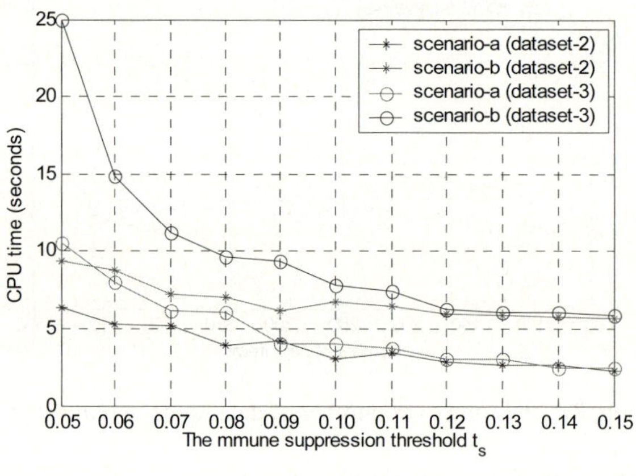

Fig. 5. CPU time over the immune suppression threshold t_s

In Fig.5, *scenario_a* has lower time than *scenario_b* for the same datasets, which is because *scenario_a* compresses the data more than *scenario_b*. At the same time, *dataset*_2 cost less time than *dataset*_3 for the same scenarios, because *dataset*_3 has more impurity than *dataset*_2. As long as t_s is defined below 0.10, ICCM has a CPU time, no more than 10 seconds.

C. An Application in Elevator Traffic Pattern Analysis

Elevator traffic analysis is crucial for developing a well-performed vertical transportation system. For different elevator traffic pattern, it is necessary to design a specific control strategy, which has been the usual road for elevator engineers. For this sake, we will apply ICCM proposed in this paper to analyze the elevator traffic patterns under a typical office building. The practical data is collected from the elevator traffic within some building in Guangzhou of China in 2003.

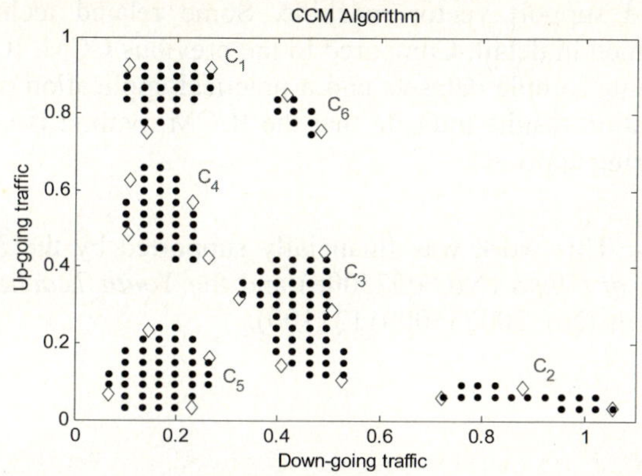

Fig. 6. Elevator traffic patterns from CCM algorithm

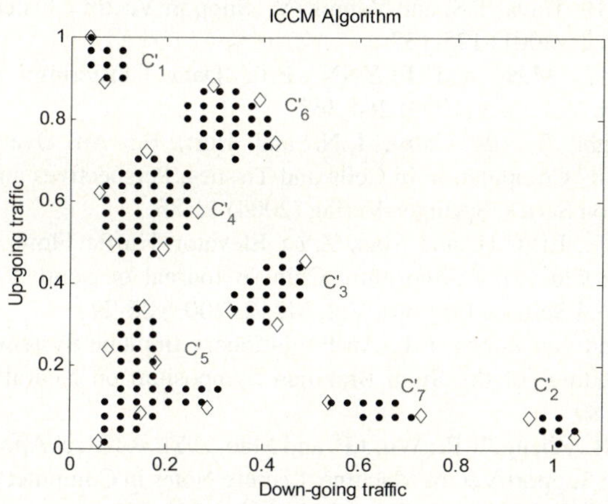

Fig. 7. Elevator traffic patterns from ICCM algorithm

For the sake of comparison, Elevator traffic pattern analysis would be conducted by CCM and ICCM, respectively. CCM algorithm and ICCM algorithm have the same parameter settings. Fig.6 illustrates the clustering results from CCM algorithm, where six key traffic patterns are displayed clearly, while Fig.7 shows seven typical traffic modes, as a result of ICCM algorithm. Each pattern or mode is denoted by a block of clustering area. Although the number of traffic patterns which two clustering approaches reach is not equal, their clustering results take on similar looks, which indicates that both CCM and ICCM effectively perform elevator traffic analysis. For the same task, elevator traffic analysis, CCM algorithm requires almost 9.2 seconds while ICCM algorithm costs only 6.3 seconds. It is clear that ICCM algorithm in this paper outperforms CCM algorithm in computational efficiency.

5 Conclusions and Future Work

This paper proposes an improved combinational clustering method based on artificial immune system and support vector machine. Some related technical spirits are described and explained in detail. Compared to the previous CCM, ICCM has a lower CPU time for the same sample datasets and a practical application of elevator traffic analysis. The simulation results indicate that the ICCM method is an applicable and effective data clustering approach.

Acknowledgements. This work was financially supported by the *National Natural Science Foundation of China* (No.60575006) and the *Youth Teacher Foundation of Sun Yat-sen University* (No. 2005350001131100).

References

1. Jajuga, K., Sokolowski, A. and Bock, H.H.: Classification, Clustering and Data Analysis,Springer-Verlag, Berlin Heidelberg New York (2002)
2. Asa, B.H., David, H., Hava, T.S. and Vapnik, V.: Support Vector Clustering, J. of Machine Learning Research 2, (2001) 125-137
3. Jain, A.K., Murty, M.N. and FLYNN, P.J.: Data Clustering: A Review, *ACM ComputingSurveys*, Vol.31, 3 (1999) 264-363
4. Timmis, J., Knight, T., de Catro, L.N. and Hart, E.: An Overview of artificial immunesystems, In “Computation in Cells and Tissues: Perspectives and Tools Thought”, NaturalComputation Series, Springer-Verlag (2004) 51-86
5. Li, Z.H., Zhu, Y.F., Li, C.H. and Mao, Z.Y.: Elevator Traffic Flow Analysis Based on Artificial Immune Clustering Algorithm. Chinese Journal of South China University of Technology (Natural Science Edition), Vol. 31, 12 (2003) 26-29.
6. de Castro, L.N. and von Zuben, F.J.: An Evolutionary Immune System Network for Data Clustering. Proceedings of the Sixth Brazilian Symposium on Neural Networks, Rio de Janeiro (2000) 84-89
7. Li, Z.H., Chen, S.B., Zheng, R.R., Wu, J.P. and Mao, Z.Y.: A Novel Approach to Clustering Analysis Based on Support Vector Machine. Lecture Notes in Computer Science, Vol.3173, Springer-Verlag Berlin Heidelberg New York (2004) 565-571
8. Li, J., Gao, X.B. and Jiao, L.C.: A Novel Clustering Algorithm Based on Immune Network with Limited Resource, Lecture Notes in Artificial Intelligence, Vol.3339, Springer-Verlag, Berlin Heidelberg New York (2004) 319-331
9. Younsi, R. and Wang, W.J.: A New Artificial Immune System Algorithm for Clustering,Lecture Notes in Computer Science, Vol.3177, Springer-Verlag, Berlin Heidelberg NewYork (2004) 58-64
10. de Catro, L.N. and Timmis, J.: Artificial Immune systems: A New Computational Intelligence Approach, Springer-Verlag, London (2002)
11. Li, Z.H. and Tan, H.-Z.: An Improved Clustering Method for Large-scale Data Based on artificial Immune systems, Dynamics of Continuous, Discrete and Impulsive Systems, Series B: Applications and Algorithms (2006) in press
12. Li, Z.H. and Tan, H.Z.: Combining Artificial Immune System with Support Vector Machine for Clustering Analysis, Dynamics of ontinuous, Discrete and Impulsive ystems, Series B: Applications and Algorithms (2006) in pressZ. Li and H.-Z. Tan 162.

User Preference Through Learning User Profile for Ubiquitous Recommendation Systems

Kyung-Yong Jung

School of Computer Information Engineering,
Sangji University, Korea
kyjung@sangji.ac.kr

Abstract. As ubiquitous commerce is coming, the ubiquitous recommendation systems utilize collaborative filtering to help users with fast searches for the best suitable items by analyzing the similar preference. However, collaborative filtering may not provide high quality recommendation because it does not consider user's preference on the attribute, the first rater problem, and the sparsity problem. This paper proposes the user preference through learning user profile for ubiquitous recommendation systems to solve the current problems. In addition, to determine the similarity between the users belonging to particular categories and new users, we assign different statistical values to the preference through learning user profile. We evaluated the proposed method on the EachMovie dataset and it was found to significantly outperform the previously proposed method.

1 Introduction

In ubiquitous commerce, users spend much time and effort in finding the best suitable items since more and more information is placed on-line. To save their time and effort in searching the items they want, the ubiquitous recommendation system is required. The ubiquitous recommendation system should predict the most suitable items for users by retrieving and analyzing their preferences. The ubiquitous recommendation system utilizes in general an information filtering technique called collaborative filtering [18]. Collaborative filtering is based on the ratings of other users who have similar preferences and is widely used for such recommendation as Amazon.com and IMDb.com. On the other hand, collaborative filtering have the first rater problem because they recommend based on the interest of users in items, not considering the items' contents. Also, in case users estimate preference on a lot of items, systems have a shortcoming that the accuracy of prediction is fallen due to the sparsity of {user-item} matrix [1,4]. The other reason of sparsity for {user-item} matrix in the ubiquitous recommendation system is the missing value caused from partial rating on items. Methods used in [2,10] solve only the first rater problem not the sparsity problem. As the method of LSI [2,12], SVD [14] decreases the dimension of {user-item} matrix, it solves the sparsity problem of the collaborative filtering. However, it does not solve the first rater problem. In addition, a method used in [14], it just tried to solve both the first rater problem and the sparsity problem.

B. Gabrys, R.J. Howlett, and L.C. Jain (Eds.): KES 2006, Part I, LNAI 4251, pp. 163 – 170, 2006.
© Springer-Verlag Berlin Heidelberg 2006

The sparsity of matrix and first rater problem, and missing value are solved by reconstructing user preferences, which are elements of a {user-item} matrix, based on the discovery of user preference through Bayesian categorization. The proposed method was tested in database which user rated preference rating on items, and proved its effectiveness compared with existent methods.

The rest of this paper is organized as follows. Section 2 describes briefly a {user-item} matrix in ubiquitous recommendation systems. Section 3 illustrates the proposed the user preference though learning user profile in detail. In Section 4, the experimental results are presented the conclusions are given in Section 5.

2 {User-Item} Matrix in Ubiquitous Recommendation Systems

The ubiquitous recommendation system using collaborative filtering is based on the ratings of the users who have similar preferences according to a {user-item} matrix. The user in ubiquitous recommendation systems does not provide preferences on all items. Therefore, a missing value is created in a {user-item} matrix. The missing value results in the sparsity of a {user-item} matrix. In this paper, the user preference generation is presented, in order to reduce the sparsity of a {user-item} matrix caused by the missing value. The proposed approach uses a user profile, in order to convert a sparse rating {user-item} matrix into a {user-item} full matrix.

The input data is composed of a user and a two dimension matrix. The row in the matrix represents the list of items, the column represents the users, and the value of matrix represents the user's preference of various items. If m items are defined, which are composed of p feature vectors and a group of n users, a user group is expressed as $U=\{user_i\}$ $(i=1,2,...,n)$, the item group is expressed as $I=\{item_j\}$ $(j=1,2,...,m)$. $R=\{r_{ij}\}$ $(i=1,2,...,n\ j=1,2,...,m)$ representing a matrix of {user-item}. The element in matrix r_{ij} means $user_i$'s preference to $item_j$ [6]. Table 1 is a {user-item} matrix in ubiquitous recommendation systems.

Table 1. {User-item} matrix in ubiquitous recommendation systems

	$item_1$	$item_2$	$item_3$	$item_4$	...	$item_j$	...	$item_m$
$user_1$	r_{11}	r_{12}	r_{13}	r_{14}	...	r_{1j}	...	r_{1m}
$user_2$	r_{21}	r_{22}	r_{23}	r_{24}	...	r_{2j}	...	r_{2m}
...	...	...	...	...	...	...	...	...
$user_i$	r_{i1}	r_{i2}	r_{i3}	r_{i4}	...	r_{ij}	...	r_{im}
...	...	...	...	...	...	...	...	...
$user_n$	r_{n1}	r_{n2}	r_{n3}	r_{n4}	...	r_{nj}	...	r_{nm}

The ubiquitous recommendation systems use the preference information, for specific items. The preference levels are represented on a scale of 0~1.0 in increments of 0.2, a total of 6 degrees, only when the value is higher than 0.5 is the user classified as showing interest. The items used are movie-related items gleaned by the EachMovie. r_{ij} in Table 1 is defined as $r_{ij} \in \{\phi,0,0.2,0.4,0.6,0.8,1\}$ $(i=1,2..,n)$ $(j=1,2,..,m)$. Namely, the element of matrix r_{ij} is in one of 6 degrees or no evaluation case. ϕ means that $user_i$ does not rate $item_j$. The ubiquitous recommendation systems then use this rating matrix in order to derive predictions [3,5,17].

3 User Preferences Though Learning User Profile

Previous studies of automatic learning of profile include use of probability [4,8], use of statistics [13] and use of vector similarity [7,8]. Among them, learning user profile through Bayesian probability is effective method [15]. Since learning the ontological user profile through use of simple Naïve Bayesian classifier extracts all word appeared in web pages, it is hard to reflect characteristic of web pages accurately. Mistaken learning user profile caused by this reduces accuracy of classification. Because of this, Bayesian classification method [11] that uses TF·IDF to make it more accurate was suggested. The suggested method extracts characteristic of web pages through use of TF·IDF from web pages. It also gives weight to characteristic extracted from web pages and so mistaken classification caused by noises is reduced more than simple Naïve Bayesian classifier [12]. However, since characteristic of extracted web pages does not reflect semantic relation, it could not resolve the problem of mistaken classification caused by ambiguity of words [6]. To resolve this problem, this paper suggests the preference through learning user profile for ubiquitous recommendation systems.

3.1 Learning User Profile Using Naïve Bayesian Classifier

We use a multinomial text model [14,15,16], in which a web page is modeled as an ordered sequence of ordered sequence of word events drawn from the same vocabulary, *V*. The Naïve Bayesian assumption states that the probability of each word is dependent on the web page classes but independent of the word's context and position [10]. For each class c_j, and word, $w_k \in V$, the probability, $P(c_j)$ and $P(w_k|c_j)$ must be estimated from the training data. Then the posterior probability of each class given a web page, *D*, is computed using Bayes rule by Equation (1).

$$p(c_j \mid D) = \frac{P(c_j)}{P(D)} \prod_{i=1}^{|D|} P(a_i \mid c_j) \tag{1}$$

Where a_i is the *i*th word in the web page, and $|D|$ is the length of the page in words. Since for any given page, the prior $P(D)$ is a constant, this factor can be ignored if all that is desired is a rating rather than a probability estimate. A ranking is produced by sorting pages by their odds ratio, $P(c_1|D)/P(c_0|D)$, where c_1 represents the positive class and c_0 represents the negative class. An item is classified as positive if the odds are greater than 1, and negative otherwise. In case, since items are represented as a vector of web pages, d_m, one for each slot s_m, the probability of each word given the category and the slot $P(w_k|c_j, s_m)$, must be estimated and the posterior category probability for a film, *F*, computed using Equation (2).

$$p(c_j \mid F) = \frac{P(c_j)}{P(F)} \prod_{m=1}^{|S|} \prod_{i=1}^{|d_m|} f_{nk} \cdot [Log_2 \frac{n}{DF} + 1] \cdot P(a_{m,i} \mid c_j,\ s_m) \tag{2}$$

Where *S* is the number of slots and $a_{m,i}$ is the *i*th word in the *m*th slot. The class with the highest probability determines the predicted rating. The *Laplace* smoothing [10] is used to avoid zero probability estimates for words that do not appear in the limited training set. Finally, calculation with logarithms of probabilities is used to avoid underflow. Naïve Bayesian classifier through use of TF·IDF makes morphological analysis to extract characteristic of web pages and extracts only nouns from its outcome [13]. f_{nk} is relative frequency of word n_k against all words within all web pages and n is the number of web pages and DF is the number of learning web pages where word n_k appeared. If characteristic of web pages is $\{a_1, a_{2,\dots,} a_{i,\dots,} a_D\}$, Naïve Bayesian classifier classifies web pages into one class among $\{c_1, c_{2,\dots,} c_{i,\dots,} c_j\}$.

$$p((\mathrm{w}_{k1}, \mathrm{w}_{k2}, \mathrm{w}_{k(r-1)}, w_{k(r-1)}) \mid c_j, s_m) = \sum_{e=1}^{N} f_{nk} \cdot [Log_2 \frac{n}{DF} + 1] \cdot a_{ej} \cdot n_{kem} / (\sum_{e=1}^{N} a_{ej} \mid d_m \mid) \quad (3)$$

Equation (3) is used to give probability of word $(w_{k1}, w_{k2} \dots, w_{k(r-1)}, w_{kr})$ within c_j, s_m is expressed as $p((w_{k1}, w_{k2} \dots, w_{k(r-1)}, w_{kr}) \mid c_j, s_m)$. If a word appears n times in a item F_e, it is counted as occurring $a_{e1}n$ times in a positive item and $a_{e0}n$ times in a negative item. Where n_{kem} is the count of the number of times word w_k, appears in item F_e, in slot s_m, and denotes the total weighted length of web pages in category c_j and slot s_m. Figure 1 shows the learning user profile with the extracted information.

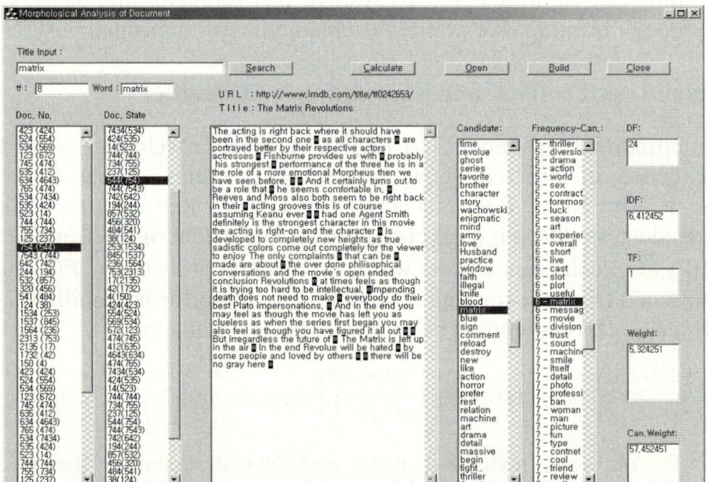

Fig. 1. Learning user profile with extracted information

3.2 User Preferences for Recommendations

To produce recommendations, the ubiquitous recommendation system learns the user profile, predicts the ratings for the un-rated items, and finally ranks the items by their ratings [2,13,15]. It treats items rated 1-3 as negative instances, and those rated 4-6 as positive instances. The scores are ranked based on the natural log of the posterior odds of positive. When predicting for a target item, the ubiquitous recommendation system computes the posterior probability of each category given the target item. Then the predicted rating of the posterior probability distribution for the categories is computed and used as the predicted ratings by Equation (4). Here, $P(i)$ is the posterior probability for category i.

$$predicted\ score = \sum_{i=1}^{6} i \cdot P(i) \quad (4)$$

We use the predicted ratings rather than choosing the most probable category to better represent the continuity of the ratings. As with relevance feedback in information retrieval [19], this cycle can be repeated several times to recommend the good results.

4 Evaluations

In this section, we describe the experimental methodology and metrics we use to compare different prediction algorithm; and present the results of our experiment. We use {user-item} matrix provided by the EachMovie dataset [9] and the movie details from the Internet Movie Database (IMDb). The data set contains 72,916 users who entered a total of 2,811,983 numeric ratings for 1,628 different items. Each rating comes from the set {0.0, 0.2, 0.4, 0.6, 0.8, and 1.0} where 0.0 is the lowest score and 1.0 is the highest. In this paper, to have a quick turn-around time for our experiments, we only used a subset of the EachMovie dataset. This dataset contains 7,893 randomly selected users and 1,461 items for which content were available from IMDb. The reduced dataset has 299,997 ratings for 1,289 items. The dataset provides optional demographic data such as age, gender, and zip code supplied by each person. For each item, information such as the name, categories, release data and IMDb URL are provided. Finally the dataset provides the actual rating data provided by each user for various items. User ratings range from one-to-six stars. One stars (★) indicate extreme dislike for an item and six stars (★★★★★★) indicate high praise. This dataset contained 0.5562 negative user ratings (i.e. ≤3). The distribution over the 1-6 ratings is given in Table 2.

Table 2. Distribution of user ratings

User rating	★	★★	★★★	★★★★	★★★★★	★★★★★★
No. of items	291	237	189	198	179	195

The textual data obtained from IMDb has many real-world aspects. The users reported that while rating the items, some pages retrieved directly from IMDb.com contained synopses that were clearly intended for other items. There were also spelling errors that a simple approach such as ours would treat as separate instances. In addition, the amount and quality of description of the items tended to vary across a wide range.

The EachMovie dataset was processed for a total of 20,864 users, such that each user rated at least 80 items and 1,612 kinds of item through the data integrity. This constituted the training set for the Naïve Bayes learning phase. 20,862 users were selected for the systematic sample base on the preprocessed EachMovie dataset. The items are categorized into 10 categories: action, animation, art foreign, classic, comedy, drama, family, horror, romance, and thriller, where an item may belong to more than one category. The training set is constructed from the items for which the preference is determined based on the attribute. Table 3 shows the classification of the items into 10 categories. The learning set is obtained from the training set by assigning the estimated value. Many users selected the action and drama categories as the representative attributes, because most users like these two types of items. The item of the training set is learned, in order to grant the estimated value by means of the Naïve Bayes algorithm.

Table 3. Training set

Categories		Item rated the preference
1	Action	Golden eye, Clueless, Monkeys, Star gate, Star Wars, Drop Zone
2	Animation	Toy Story, Eden, Heavy Metal, Pocahontas, Space Jam, Robin
3	Art/Foreign	Four Rooms, Birdcage, Antonia's Line, Birdcage, Stalker, Diva
4	Classic	Jumanji, Balto, Happy Gilmore, Foreign Student, Alien, Annie
5	Comedy	Ace Ventura, Bronx Tale, Fatal Instinct, Four Rooms, Heather
6	Drama	Sabrina, Nixon, Ventura, Clerks, Get Shorty, Noon, Cape Fear
7	Family	Casper, Apollo13, Bad Boys, Batman Forever, Fly Away Home
8	Horror	Copycat, Screamers, Mary Reilly, Babe, Clueless, M, Braindead
9	Romance	American President, Swiss Family Robinson, Benny & Joon
10	Thriller	Die Had, Taxi Driver, Crimson Tide, The Net, Breakdown, Water

Table 4. {user-item} matrix given to Bayesain estimated value

	$item_5$	$item_6$	$item_{10}$	$item_{18}$	......	$item_{21}$	......	$item_{127}$
$user_1$	0.0025	0.0038	0.0098	0.0035	......	0.0097	......	0.0084
$user_{19}$	0.0012	0.0015	0.0024	0.0018	......	0.0075	......	0.0027
$user_{21}$	0.0020	0.0037	0.0034	0.0023	......	0.0037	......	0.0037
$user_{35}$	0.0008	0.0013	0.0006	0.0019	......	0.0026	......	0.0008
$user_{45}$	0.0012	0.0020	0.0017	0.0026	......	0.0047	......	0.0007
$user_{55}$	0.0014	0.0029	0.0045	0.0038	......	0.0055	......	0.0002

Table 4 shows the information about the missing values, for which the weight value of the user preference for a particular item is assigned differently depending on the classified categories in the learning set. If a new user is classified according to the Naïve Bayes classifier, the new user is differently granted the estimated value to item, in order to include similarity between the users in the classified categories and the new user. The shaded boxes represent the predictive preference obtained by granting the posterior probability to the missing value using Equation (4).

In this paper, and Mean Absolute Error (MAE) and Rank Score Measure both suggested by paper [3] are used to gauge performance. MAE is used, in order to evaluate single item ubiquitous recommendation systems. Rank scoring is used to evaluate the performance of systems that recommend items from ranked lists.

We considered the proposed method using the user preference through learning user profile (UP_LUP) and hypothesized that is would outperform the stand-alone collaborative filtering approach. However, it is also important to compare our approach with that obtained using a combination of content based filtering and collaborative filtering. By comparing the method proposed and those proposed by Soboroff [16], Pazzani [12], and Fab [1], we are able to compare with the predictive accuracy values, such as the MAE and Rank scoring.

Figure 2 shows the MAE and Rank scoring as the frequency with which the user evaluates the *n*th rating number is increased. Figure 2 shows that the system proposed by Soboroff [16] that solves the first rater problem exhibits lower performance when the number of evaluations is lower. The other methods show higher performance than that of Soboroff. As a result, the method developed by Pazzani and the UP_LUP method that solve both the sparsity and first rater problem show the highest accuracy rates.

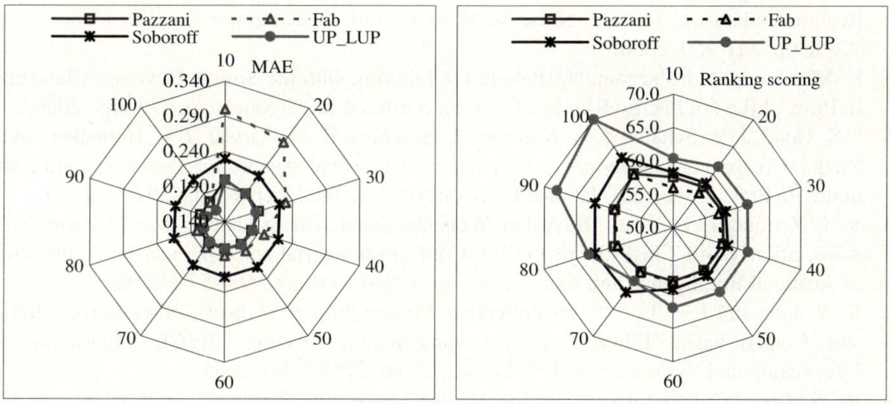

Fig. 2. MAE, Rank scoring of *n*th rating number

5 Conclusions

It is crucial for the ubiquitous recommendation system to have a capability of making accurate prediction by learning user profile. Although collaborative filtering is widely used for recommendation, some efforts to overcome its drawbacks should be made to improve the prediction quality. In this paper, the user preference through learning user profile for ubiquitous recommendation systems is solve the sparsity problem and the first rater problem. We calculate the weight value of the preference using a weighted binary classifier that map the actual one-to-six rating. So reflect the information of item to the statistical value. The proposed method was compared with the existent methods that combine collaborative filtering and content based filtering. As a result, the proposed method shows higher performance than existent method in both comparisons. The ubiquitous recommendation system with the proposal method can be a choice to resolve the first rater problem and the sparsity problem while it gives high prediction quality. In a future study, the proposed method will be further verified with a larger number of users and in a combined approach, involving the proposed method and memory-based method.

Acknowledgements

We thank the DES systems research center from permitting us to use the EachMovie dataset. This research was supported by Sangji University Research Fund, 2006.

References

1. M. Balabanovic and Y. Shoham, "Fab: Content-based, Collaborative Recommendation," Communication of the Association of Computing Machinery, Vol. 40, No. 3, pp. 66-72, 1997.
2. J. L. Herlocker, J. A. Konstan, L. G. Terveen, and J. T. Riedl, "Evaluating Collaborative Filtering Recommender Systems," ACM Transactions on Information Systems, Vol. 22, No. 1, pp. 5-53, 2004.
3. J. L. Herlocker, J. A. Konstan, and J. Riedl, "Explaining Collaborative Filtering Recommendations," In Proc. of the ACM 2000 Conf. on Computer Supported Cooperative Work, pp. 241-250, 2000.
4. K. Miyahara, M. J. Pazzani, "Collaborative Filtering with the Simple Bayesian Classifier," In Proc. of the 6th Pacific Rim Int. Conf. on Artificial Intelligence, pp. 679-689, 2000.
5. 5 N. Good, J. B. Schafer, J. A. Konstan, A. Borchers, B. M. Sarwar, J. L. Herlocker, and J. Riedl, "Combining Collaborative Filtering with Personal Agents for Better Recommendations," In Proc. of the 16th National Conf. on Artificial Intelligence, pp. 439-446, 1999.
6. S. J. Ko and J. H. Lee, "Bayesian Web Document Classification through Optimizing Association Word," In Proc. of the Int. Conf. on Industrial and Engineering Applications of Artificial Intelligence and Expert Systems, LNAI 2718, pp. 565-574, 2003.
7. K. Y. Jung and J. H. Lee, "User Preference Mining through Hybrid Collaborative Filtering and Content-based Filtering in Recommendation System," IEICE Transaction on Information and Systems, Vol. E87-D, No. 12, pp. 2781-2790, 2004.
8. W. S. Lee, "Collaborative Learning for Recommender Systems," In Proc. of the 18th International Conference on Machine Learning, pp. 314-321, 1997.
9. P. McJones, EachMovie dataset, URL: http://www.research.digital.com/SRC/eachmovie
10. T. Mitchell, *Machine Learning*, McGraw-hill, New York, pp. 154-200, 1997.
11. Introduction to Rainbow URL: http://www.cs.cmu.edu/afs/cs/project/theo-11/www/naïve-bayes.html.
12. K. Miyahara and M. J. Pazzani, "Collaborative Filtering with the Simple Bayesian Classifier," In Proc. of the 6th Pacific Rim International Conf. on Artificial Intelligence, pp. 679-689, 2000.
13. T. Joachims, "A Probabilistic Analysis of the Rocchio Algorithm with TF·IDF for Text Categorization," In Proc. of the 14th Int. Conf. on Machine Learning, pp. 143-151, 1997.
14. P. Melville, R. J. Mooney, and R. Nagarajan, "Content-Boosted Collaborative Filtering for Improved Recommendations," In Proc. of the National Conf. on Artificial Intelligence, pp. 187-192, 2002.
15. R. Raymond, P. Bennett, and L. Roy, "Book Recommending Using Text Categorization with Extracted Information," In Proc. of the AAAI'98/ICML Workshop on Learning Categorization, pp. 49-54, Madison, WI, 1998.
16. I. Soboroff and C. K. Nicholas, "Related, but not Relevant: Content-Based Collaborative Filtering in TREC-8," Information Retrieval, Vol. 5, No. 2-3, pp. 189-208, 2002.
17. K. Yu, Z. Wen, X.W Xu, and M. Ester, "Feature Weighting and Instance Selection for Collaborative Filtering: An Information-Theoretic Approach," Knowledge and Information Systems, Vol. 5, No. 2, pp. 201-224, Springer-Verlag, 2003.
18. J. H. Hwang, M. S. Gu, and K. H. Ryu, "Context-Based Recommendation Service in Ubiquitous Commerce," In Proc. of the Int. Conf. on Computational Science and Its Applications, LNCS 3481, pp. 966-976, 2005.
19. J. Wang, A. P. de Vries, and M. J.T. Reinders, "A User-Item Relevance Model for Log-based Collaborative Filtering," In Proc. of the European Conf. on Information Retrieval, LNCS 3936, pp. 37-48, 2006.

Automated Recognition of Cellular Phenotypes by Support Vector Machines with Feature Reduction

Y. Mao[1], Z. Xia[1], D. Pi[1], X. Zhou[2], Y. Sun[1], and S.T.C. Wong[2]

[1] Zhejiang University, National Laboratory of Industrial Control Technology, Institute of Industrial Process Control, Hangzhou 310027, P.R. China
[2] Harvard University, Harvard Center for Neurodegeneration and Repair, Harvard Medical School and Brigham and Women's Hospital, Harvard Medical School, 220 Longwood Avenue, Goldenson Building 524, Boston, MA 02115, USA
ymao@iipc.zju.edu.cn

Abstract. In this paper, wrapper based feature selection by support vector machine is used for cellular multi-phenotypic mitotic analysis (MMA) in high content screening (HCS). Haralick texture feature subset and Zernike polynomial moment subset are used respectively or combined together as extracted digital feature set for original cellular images. Feature reduction is done by support vector machine based recursive feature elimination algorithm on these feature sets. With optimal feature subset selected, fuzzy support vector machine are adopted to judge the cellular phenotype. The results indicate Haralick texture feature subset is complementary with Zernike polynomial moment subset, when these two feature subsets are combined together; the cellular phase identification system achieved 99.17% accuracy, which is better than only one feature subset of them is used. The recognition accuracy with feature reduction is better than that achieved when no feature reduction done or using PCA as feature recombination tool on these datasets.

1 Introduction

Estimating cell phenotype in cellular mitosis is very important to study drug effects on cancer cells. Cell phenotype in cell cycle progress (including inter-phase, prophase, metaphase, anaphase, etc.) may be imaged and tracked by automated microscope combined with the availability of fluorescence probe techniques. High content pharmaceutical screening based on these techniques to monitor the drug effects on interested cells or samples becomes more and more popular in biomedical domain [1]. This kind of technologies tries to massively parallelize analyses currently done in single pipeline (e.g. by multi-well plates), and enable researchers to study intracellular events on many different phenotypic cells [2]. A main challenge for designing such a system lies on how to automate extraction of single cellular images and recognition of phenotype of these single cells. The latter is considered even more difficult for unstable factor in cellular image segmentation (caused by uneven lightness or overlapping of cells) and vague difference between many cellular phenotypes. A good approach to deal with this problem is to introduce artificial intelligence or learning machine [3, 4].

B. Gabrys, R.J. Howlett, and L.C. Jain (Eds.): KES 2006, Part I, LNAI 4251, pp. 171 – 178, 2006.
© Springer-Verlag Berlin Heidelberg 2006

In [2], a three-channel imaging system is used to analyze three different components localized in the same cells. The three kinds of components are three cellular organisms, including actin, microtubules and DNA. According to the biological mechanism analyses of cellular mitosis, a simple computational cellular phase identification system is constructed by using some simple cellular image region properties and logical operation based on the cellular components extracted from the three-channel images. In the image dataset obtained for screening Monastrol suppressors, the method achieved a success recognition rate of cellular phase between 90% and 95% on 30 images from 10 compounds treated samples in three channels.

In this work, an improved cellular phenotype recognition method based on texture features of three-channel cellular images is proposed. Key cellular pattern characteristics are selected respectively from 3×78 Haralick texture features or 3×49 Zernike polynomial moments or these two feature subsets together (3×127 features) by support vector machine based recursive feature elimination. The size of optimal feature subset is determined by 5-fold cross-validation, and then fuzzy support vector machine combined with the features selected are employed to classify the three-class cellular phenotypes, which are monoasters, metaphase and anaphase. On the unicellular image dataset extracted on 30 images from 10 compounds treated samples in three channels, our algorithms achieved 99.17% classification success rate when Zernike feature subset and Haralick feature set are combined together as initialized feature set, which is better than the methods used in [2]. The experimental results indicated 1) Haralick texture feature subset is complementary with Zernike polynomial moment subset; when these two feature subsets are combined together; the cellular phase identification system is better than only one feature subset of them is used; 2) the recognition accuracy with feature reduction is better than that achieved when no feature reduction done or using PCA as feature recombination tool on these datasets.

2 Systems and Methods

2.1 Three-Channel Imaging System

Cellular mitosis is often divided into six phases: inter-phase, prophase, metaphase, anaphase, telophase, and cytokinesis. These phases are grouped into three categories: inter-phase and prophase as monoasters, metaphase, and the other three original phases are composed as anaphase according to the morphology of condensing DNA for nuclei, moving microtubule for mitosis and moving actin for cytoplasm. These three components are scanned by a digital microscope using three different channels combined with reactive organic dyes. The representative three components of one cell in three different phases are described in Fig.1.

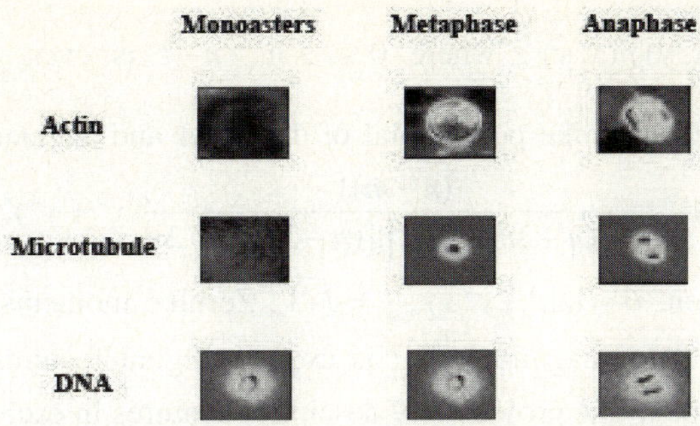

Fig. 1. Three representative cell patterns in cellular mitosis are presented in three channels

2.2 Cellular Image Segmentation and Unicellular Sample Localization

In order to extract accurate information for cellular pattern analyses, the images from three channels are integrated.

Firstly, these images are segmented to localize the single cells by the methods used in [2]: edge detecting is done by Laplacian of Gaussian operator; line filter is used to remove fragmented lines of background noise; eccentricity thresholding is used to find out region edge line with big curvature; and then dilation, fill, erosion, open and area-open operators are used sequentially to formulate cell masks; finally, watershed algorithm is used to separate touching objects and all objects on the four borders of images are removed.

Then unicellular image sample is located by cell masks from segmentation results of the three channels. Masks from DNA channel are employed on original images from all channels. But two near-neighboring regions in DNA channel masks may belong to a unique cell in cellular anaphase. To avoid identifying such regions as two cellular samples, cellular masks from other two channels are used to do assistant decision. If two cellular masks in DNA channel map to a same region mask in other two channels, such two regions are considered belonging to a same cell. By this method, all cells could be localized and described by information from three channels.

2.3 Texture Features Extraction for Unicellular Image

Two types of numerical features are employed to describe the texture of single cell patterns.

The first feature set is the Zernike polynomial moments [5] through order 12. After unicellular sample is localized, the smaller rectangular region containing the cell is used to describe its image area. In Zernike moments, the rectangular region is converted to a unit circle by locating the center of cell image and normalizing all sample pixel coordinates, $f(x, y)$ is used as resulting normalized image pixels. The Zernike moments for unicellular sample were calculated as

$Z_{nl} = \frac{n+1}{\pi}\sum_x\sum_y V_{nl}^*(x,y)f(x,y)$, where $0 \le l \le n$, $n-l$ is even, $V_{nl}^*(x,y)$ is the complex conjugate of a Zernike polynomial of degree n and angular dependence l: $V_{nl}(x,y) = \sum_{m=0}^{(n-1)/2}(-1)^m \frac{(n-m)!}{m![(n-2m+l)/2]![(n-2m-l)/2]!}(x^2+y^2)^{n/2-m}e^{il\theta}$, where $0 \le l \le n$, $n-l$ is even, $\theta = \tan^{-1}(y/x)$, $i = \sqrt{-1}$. Zernike moments through degree 12 is calculated and their magnitute $|Z_{nl}|$ is used as digital features for unicellular sample [5]. For each cell, this provided 49 desciptive features in each channel, totally 3×49 features available.

The other feature set is Haralick texture features, which contain 14 features related to intuition descriptions of image texture in each direction (also named gray-level co-occurrence matrix) [6]. Six directional ($0^o, 30^o, 60^o, 90^o, 120^o, 150^o$) Haralick texture features are calculated in each unicellular image of three channels. 3×6×14 Haralick features are calculated for each cell. Maximum correlation coefficient was abnegated for its computational instability. Finally, 3×6×13 descriptive features are available.

The two feature sets are both invariant to position and rotation of cells, and insensitive to changes in the scale of intensity values [7]. For each single cell image, there are 381 numerical texture features from three channels totally. Toolbox for extracting these two feature sets are downloaded from http://murphylab.web.cmu.edu/software.

2.4 Feature Reduction and Cellular Phase Identification by Fuzzy Support Vector Machine

Many literatures suggest feature reduction to be done before decision-making [8, 9, 10]. In this work, support vector machine based recursive feature elimination (SVM-RFE) is used to select optimal subsets for datasets containing Zernike feature set or Haralick feature set or these two feature sets together.

Support vector machine based recursive feature elimination (SVM-RFE) is first proposed by Guyon et al. [10], and has been applied in several gene-expression studies [9, 10] for selecting important features. Recursive feature elimination is a circulation procedure for eliminating features by a criterion. It consists of three steps: 1) train the classifier; 2) compute the ranking criterion; 3) remove the features with smallest ranking score. When support vector machine is used as classifier, the ranking criterion is relative to the realization of SVM. In support vector machine [11], a classification function is realized as $f(x) = \sum_{i=1}^N \alpha_i y_i K(\mathbf{x}_i, \mathbf{x}) + b$, where the coefficients $\boldsymbol{\alpha} = (\alpha_i)$ and b are obtained by training over a set of examples $S = \{(\boldsymbol{x}_i, y_i)\}_{i=1,\ldots,N}$, $\boldsymbol{x}_i \in \boldsymbol{R}^n, y_i \in \{-1,1\}$. In the linear case, the SVM expansion defines the hyper-plane $f(\boldsymbol{x}) = \langle \boldsymbol{w}, \boldsymbol{x} \rangle + b$, with $\boldsymbol{w} = (w_i) = \sum_{i=1}^N \alpha_i y_i \boldsymbol{x}_i$. The idea is to define the importance of a feature of the SVM in terms of its contribution to a cost function $J(\boldsymbol{\alpha})$. At each step of the RFE procedure, the SVM is trained on the given dataset, J is computed and the feature with smallest contribution to J is discarded.

The SVM must be retrained after each elimination operation, because the importance of a feature with medium-low importance may be promoted by removing a correlated feature. By SVM-RFE, a ranked list of features is thus obtained, from which the top several features may be selected as optimal feature subset according to the predictive accuracy of *n*-fold cross-validation. In this work, *n* is set to 5.

Based on the optimal feature subsets chosed by SVM-RFE between every two classes, fuzzy support vector machine is employed to classify the three-class cellular phenotypes: monoasters, metaphase and anaphase.

Fuzzy support vector machine (FSVM) is a relatively new method first proposed by Abe and Inoue [12]. It is proposed to deal with unclassifiable regions when using one vs. rest or pairwise classification method. FSVM is an improved pairwise classification method with SVM. A fuzzy membership function is introduced into the decision function based on pairwise classification. For the data in the classifiable regions, FSVM gives out the same classification results as pairwise SVM method; and for the data in the unclassifiable regions, FSVM generates better classification results than the pairwise SVM. In the process of being trained, FSVM is the same as pairwise SVM [12]. If there are K classes in training dataset, there are $K(K-1)/2$ SVMs trained in FSVM. The decision function of FSVM is constructed based on these $K(K-1)/2$ SVM classification functions. Define $f_{st}(\mathbf{x}) = -f_{ts}(\mathbf{x}), (s \neq t; s,t = 1,\cdots,K)$, the fuzzy membership function $m_{st}(\mathbf{x})$ is introduced on the directions orthogonal to $f_{st}(\mathbf{x}) = 0$ as $m_{st}(\mathbf{x}) = \begin{cases} 1 & \text{for } f_{st}(\mathbf{x}) \geq 1 \\ f_{st}(\mathbf{x}) & otherwise \end{cases}$. Using $m_{st}(\mathbf{x}), (s \neq t; s,t = 1,\cdots,K)$, the class $i, i \in \{1,\cdots,K\}$ membership of $\mathbf{x}$ is defined as $m_s(\mathbf{x}) = \min_{t=1,\cdots,K} m_{st}(\mathbf{x})$, which is equivalent to $m_s(\mathbf{x}) = \min\left(1, \min_{s \neq t, t=1,\cdots,K} f_{st}(\mathbf{x})\right)$. An unknown sample $\mathbf{x}$ is classified by $\arg\max_{s=1,\cdots,n} m_s(\mathbf{x})$.

In our work, 5-fold cross-validation is used to test the performance of our cellular multi-phenotypic mitotic analysis method. To prove the effectiveness of feature reduction, the recognition algorithms without feature reduction and feature recombination by PCA are compared with our method. When PCA is used for feature recombination, numbers of principle components are decided by 5-fold cross validation, the same as SVM-RFE used.

3 Experimental Results

600 cells are extracted as our original dataset on 30 images from 10 compounds treated cases in three channels, and their phenotypes are recognized and labelled manully, in which status of 300 cells are monoasters, 200 are metaphase and 100 are anaphase. By Zernike polynomial moments, 147 features are calculated, and dataset 1 is constructed containing these features. By Haralick texture method, 234 features are extracted, which constructed dataset 2. Dataset 1 and 2 are combined together to construct dataset 3.

On these three datasets, performance indices of our methods are listed in Table 1-4. In Table 1, prediction accuracy of FSVM combined with SVM-RFE, PCA respectively or with all features on three datasets is described. FSVM combined with SVM-RFE achieved 99.17% accuracy on dataset 3. Using the same method, 96.33% accuracy is achieved on dataset1 and 90.83% on dataset 2, which are lower than that on dataset 3. Using FSVM combined with PCA or FSVM with all features, the similar conclusion is also drawn. It means when Zernike moment features combined with Haralick texture features, a higher performance will be achieved, these two feature subsets are complementary with each other, this relation can be revealed from comparison among rows in Table 1. When SVM-RFE combined with cross-validation is used as feature reduction tool, it performs better than PCA used as feature recombination tool or no treatment on original digital features, which can be revealed from comparison among columns in Table 1.

Table 1. Prediction accuracy of FSVM combined with SVM-RFE, PCA respectively or with all features on three datasets

Dataset used	Prediction accuracy of FSVM with following methods		
	SVM-RFE	PCA	All features
Dataset 1 (Zernike)	96.33%	90.50%	91.50%
Dataset 2 (Haralick)	90.83%	90.00%	88.67%
Dataset 3 (Both)	99.17%	91.50%	92.83%

In Table 2-4, more detailed prediction accuracy of FSVM combined SVM-RFE on dataset 1-3 is described by confusion matrix. When Zernike moment features are combined with Haralick texture features, prediction accuracy of every single class is higher than that when only Zernike moment feature subset or Haralick texture feature subset used. The cellular phenotypic recognition accuracy reported here is better than that in [2], which is below 97.64%, and recognition accuracy of cells in every phenotype is also promoted. In [2], 97.64% cells in monoasters are correctly classified, 95.43% as cells in metaphase and 96.88% as cells in anaphase. Here, they are 99.7%, 99.0% and 98.0%.

Table 2. Prediction accuracy of FSVM combined with SVM-RFE on dataset 1 containing 147 Zernike moment features described by confusion matrix

True phenotypes	Prediction accuracy by FSVM & SVM-RFE on dataset 1		
	Monoasters	Metaphase	Anaphase
Monoasters	98.3%	0.7%	1.0%
Metaphase	1.5%	93.5%	5.0%
Anaphase	2.0%.	2.0%	96.0%

Table 3. Prediction accuracy of FSVM combined with SVM-RFE on dataset 2 containing 234 Haralick texture features described by confusion matrix

True phenotypes	Prediction accuracy by FSVM & SVM-RFE on dataset 2		
	Monoasters	Metaphase	Anaphase
Monoasters	96.7%	2.7%	0.6%
Metaphase	6.0%	83.0%	11.0%
Anaphase	3.0%.	8.0%	89.0%

Table 4. Prediction accuracy of FSVM combined with SVM-RFE on dataset 3 containing 147 Zernike moment features and 234 Haralick texture features described by confusion matrix

True phenotypes	Prediction accuracy by FSVM & SVM-RFE on dataset 3		
	Monoasters	Metaphase	Anaphase
Monoasters	99.7%	0.0%	0.3%
Metaphase	0.0%	99.0%	1.0%
Anaphase	2.0%.	0.0%	98.0%

4 Conclusion

In this work, an improved cellular phenotype recognition method based on texture features of three-channel cellular images is proposed. Haralick texture feature subset and Zernike polynomial moment subset are used respectively or combined together as extracted digital feature set for original cellular images. Feature reduction is done by support vector machine based recursive feature elimination algorithm on these feature sets. With optimal feature subsets selected, fuzzy support vector machine are adopted to judge the cellular phenotype. The results indicate Haralick texture feature subset is complementary with Zernike polynomial moment subset, when these two feature subsets are combined together; the cellular phase identification system achieved 99.17% accuracy, which is better than only one feature subset of them is used. The recognition accuracy with feature reduction is better than that achieved when no feature reduction done or using PCA as feature recombination tool on these datasets.

References

1. Perlman, Z., Slack, M., Feng, Y., Mitchison, T., Wu, L., Altsschule, S.: Multidimensional drug profiling by automated microscopy. Science 2004; **306**: 1194-8.
2. Zhou, X., Cao, X., Perlman, Z., Wong, S.: A computerized cellular imaging system for high content analysis in Monastrol suppressor screens. Journal of Biomedical informatics, In press, 2005.
3. Huang, K., Velliste, M., Murphy, R.: Feature reduction for improved recognition of subcellular location patterns in fluorescence microscope images. Proc. SPIE, 2003, 4962:307-318.

4. Boland, M., Murphy, R.: A neural network classifier capable of recognizing the patterns of all major subcellular structures in fluorescence microscope images of hela cells. Bioinformatics, 2001, **17**:1213-1223.
5. Teague, M.: Image analysis via the general theory of moments. Journal of the optical society of America, 1980, **70**, 920-930.
6. Haralick, R., Shapiro, L.: Computer and robot vision. Addison-Wesley, Reading, MA, 1992.
7. Khotanzad, A., Hong, Y.: Rotation invariant image recognition using features selected via a systematic method. Pattern recognition, 1990, **23**:1089-1101.
8. Boland, M., Markey, M., Murphy, R.: Automated recognition of patterns characteristic of subcellular structures in fluorescence microscopy images. Cytometry, 1998, **22**:366-375.
9. Mao, Y., Zhou, X., Pi D., Wong, S., Sun, X.: Multi-class cancer classification by using fuzzy support vector machine and binary decision tree with gene selection. Journal of Biomedicine and Biotechnology, 2005. **2**: 160-171.
10. Guyon, I., Weston, J., Barnhill, S., Vapnik, V.: Gene selection for cancer classification using support vector machines. Machine learning 2002; **46**: 389-422.
11. Vapnik, V.: The nature of statistical learning theory. New York, Springer, 2000.
12. Abe, S., Inoue, T.: Fuzzy support vector machines for multiclass problems. in European Symposium on Artificial Neural Networks. 2002. Bruges.Belgium.

Power Transformer Fault Diagnosis Using Support Vector Machines and Artificial Neural Networks with Clonal Selection Algorithms Optimization

Ming-Yuan Cho, Tsair-Fwu Lee, Shih-Wei Gau, and Ching-Nan Shih

Department of Electrical Engineering, National Kaohsiung University of Applied Science, Kaohsiung, Taiwan 807, ROC
mycho@mail.ee.kuas.edu.tw, tflee@bit.kuas.edu.tw, stone@cc.kuas.edu.tw

Abstract. This paper presents an innovative method based on Artificial Neural Network (ANN) and multi-layer Support Vector Machine (SVM) for the purpose of fault diagnosis of power transformers. A clonal selection algorithm (CSA) based encoding technique is applied to improve the accuracy of classification, which demonstrated in the literature for the first time. With features and RBF kernel parameters selection to predict incipient fault of power transformer improve the accuracy of classification systems and the generalization performance. The proposed approach is distinguished by removing redundant input features that may be confusing the classifier and optimizing the selection of kernel parameters. Simulation results of practice data demonstrate the effectiveness and high efficiency of the proposed approach, which makes operation faster and also increases the accuracy of the classification.

1 Introduction

This Paper proposes a combination method to improve support vector machines (SVM) based fault diagnosis for power transformers through the operation of clonal selection algorithm (CSA). The SVM is a relatively new kind of learning machine and soft computing method based on the structural risk minimization (SRM) principle and statistical machine learning theory (SLT) that performs structural risk minimization on a nested set structure of separating hyperplanes which was proposed by Vapnik and his group at AT&T Bell Laboratories [1]. SVMs have been introduced as a new methodology for solving classification and functional regression with many successful applications [2]. It is powerful for the problem with small sampling, nonlinear and high dimension. It provides a unique solution and is a strongly regularized method appropriate for ill-defined problems.

Power transformers are definitely vital devices in a transmission and distribution system, as expensive items; need to be carefully monitored throughout their operation. Fault diagnosis of it is important for safety of the device and relevant power system. To predict incipient fault is gaining importance in industry because of the failure of a power transformer may cause an interruption in power supply and loss of revenue. Therefore, it is of great importance to detect incipient failures in power transformers

B. Gabrys, R.J. Howlett, and L.C. Jain (Eds.): KES 2006, Part I, LNAI 4251, pp. 179 – 186, 2006.
© Springer-Verlag Berlin Heidelberg 2006

as early as possible, so that we can switch them safely and improve the reliability of power systems [3]. The most widely applied method for incipient faults in power transformers can be detected and monitored based on dissolved gas analysis (DGA) which has been proved that are related closely to transformer's internal faults [3]. Fault gases are produced by degradation of the transformer oil and solid insulating materials, such as paper, pressboard and transformer board, which are all made of cellulose. According to the IEC 599 [4], the fault related gases include hydrogen (H_2), oxygen (O_2), nitrogen (N_2), methane (CH_4), ethylene (C_2H_4), acetylene (C_2H_2), ethane (C_2H_6), carbon dioxide (CO_2), and carbon monoxide (CO) that can be determined from a DGA test. As study results report, corona or partial discharge, thermal heating and arcing are the three main causes for insulation degradation in a transformer [5]. The energy dissipation is least in corona or partial discharge, medium in thermal heating, and highest in arcing [6].

Various fault diagnosis techniques have been proposed in the literature, including the conventional key gas method, ratio-based method [7] and recently artificial intelligence methods, such as expert system [8], fuzzy logic [9] and artificial neural networks (ANNs) [10], and the hybrid combinations of these methods have given promising results [11]. The conventional key gas and ratio methods are based on experience in fault diagnosis using DGA data, which may vary from utility to utility due to the heuristic nature of the methods and the fact that no general mathematical formulation can be utilized. The expert system and fuzzy logic approaches can take DGA standards and other human expertise to form decision-making system. Information, such as influence of transformer size, manufacturer, volume of oil, gassing rates and history of the diagnosis result can be utilized. However, there are some intrinsic shortcomings, for example, the difficulty of acquiring knowledge and maintaining a database. Both methods need a large knowledge base that must be constructed manually. In order to overcome the shortcomings, one solution is realized to obtain its optimization result through evolutionary algorithm [12]. And a novel evolution computation enhanced method was proposed to ameliorate the fuzzy diagnosis capabilities [13]. The traditional ANN method can directly acquire experience from the training data, and overcome some of the shortcomings of the expert system. However, it suffers from a number of weaknesses, including the need for a large number of controlling parameters, difficulty in obtaining a stable solution and the danger of over-fitting. To overcome the drawbacks of traditional neural networks, a new type of neutral network is proposed for incipient fault diagnosis of power based on extension theory [7]. As ANNs, expert system and fuzzy logic approaches have their advantages and disadvantages, hybrid artificial intelligence approaches are also under consideration. Their disadvantages could be overcome by connecting expert system, fuzzy logic, ANNs and evolutionary algorithms as a whole [14].

Recently, SVM, a novel network algorithm, originally developed by Vapnik and Cortes [15] has emerged as one powerful tool for data analysis. It is powerful for the problem with small sampling, nonlinear and high dimension. SVM has been widely used for many applications, such as face recognition [11], time series forecasting [16], fault detection [17] and modeling of nonlinear dynamic systems [18].

In this study, we explored the feasibility of applying an ANN and multi-layer SVMs with features and RBF kernel parameters selection to predict incipient fault of power transformer by means of CSA-based encoding techniques for the first time. It is capable of improving the accuracy of classification systems by removing redundant input features that may be confusing the classifier, and hence improving the generalization performance. Simulation results of practice data demonstrate the effectiveness and high efficiency of the proposed approach which makes operation faster and also increases the accuracy of the classification.

2 Proposed Method of Fault Diagnosis

In many practical classification tasks we are encountered with the problem, which we have a very high dimensional input space and we want to find out the combination of the original input features which contribute most to the task. The overall structure and main components of proposed method are depicted in Fig.1. CSA performs the optimal features and parameters searching for SVM classifier. In SVM classifier, we are taking existing Vapnik-Chervonenkis (VC) theoretical bounds on the generalization error for SVMs instead of performing cross-validation [15]. This is computationally much faster than *k*-fold cross-validation, since we do not have to retrain times on each feature subset but just once. Additionally in general the error bounds have a higher bias than cross-validation in practical situations they often have a lower variance and can thus reduce the overfitting of the wrapper algorithm. The feature selection process is done in an evolutionary way by the CSA-based encoding which also allows us to optimize different parameters of the SVM, like the regularization parameter in parallel.

In general, there is no superior kernel function for machine learning, and the performance of a kernel function rather depends on the specific application. Also, the parameters in a kernel function play the important role of representing the structure of a sample space. The optimal set of the parameters maximizing the classification performance can be selected by a machine learning method. In a learning phase, the structure of a sample space is learned by a kernel function, and the knowledge of a sample space is contained in the set of parameters. Furthermore, the optimal set of features also should be chosen in the learning phase. In our method, CSA technique is exploited to obtain the optimal set of features as well as the optimal parameters for a kernel function. Simulating an immune system procedure, CSA creates improved diagnosis classification containing a set of features and parameters by the iterative process of reproduction, evaluation, and selection process. At the end of learning stage, the CSA-based SVM consists of a set of features and parameters for a kernel function. The CSA-based SVM classifier is obtained after learning phase to be used to classify new DGA pattern samples in classification phase [18].

As suggested above, DGA approach is inherently imprecise and requires other techniques to obtain better results. Hence, in this section we shall apply the proposed method of CSA-based SVMs optimization algorithm to conquer the DGA test for power transformer.

Up to now, most software cannot do effective automatic features and parameters selection. For SVMs, that is naturally a very complex non-linear and non-quadratic global optimization problem. In general, Based on minimize the number of the support vectors, the bigger margin is chose. Lee and Lin [19] make full use of training samples to adopt the exhaustive search for the optimal hyper parameter, which is minimize the expectation test of the leave-one-out (loo) error rate. But we are taking existing VC theoretical bounds on the generalization error for SVMs instead of performing loo cross-validation.

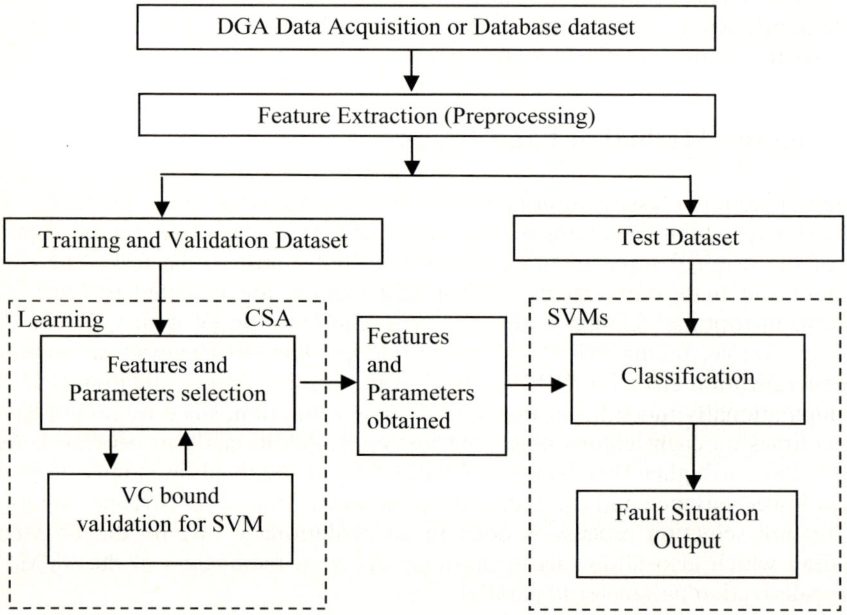

Fig. 1. Overall structure of proposed method

We briefly review the VC theoretical bounds over here; we only deal with the problem of classification problem. Consider the following loss function

$$L(y, f(\mathbf{x},\alpha)) = \begin{cases} 0 & \text{if } y = f(\mathbf{x},\alpha) \\ 1 & \text{if } y \neq f(\mathbf{x},\alpha) \end{cases} \tag{1}$$

Due to the probability distribution of the expected risk functional is unknown; we may replace it by an empirical risk function.

$$R_{emp}(\alpha) = \frac{1}{l}\sum_{i=1}^{l} L(y_i, f(\mathbf{x}_i,\alpha)) \tag{2}$$

In order to know the quality of the empirical risk $R_{emp}(\alpha)$ to approximate the expected risk, Vapnik presented the following bound theorem [20]. With probability at least 1-η (0≤η≤1), the inequality must hold true simultaneously for all functions.

$$R(\alpha) \leq R_{emp}(\alpha) + \frac{\varepsilon}{2}\left[1+\sqrt{1+\frac{4R_{emp}(\alpha)}{\varepsilon}}\right] \tag{3}$$

Where $\varepsilon = (4/l)(h(\ln(2l/h)+1-\ln\eta)$ and h is the VC dimension of the set of functions $f(\mathbf{x},\alpha)$, $\alpha\in\Lambda$.

From Eq. (3), we can see that the minimization of the expected risk $R(\alpha)$ is equal to the minimization of the two terms on the right-hand side of Eq. (3) at the same time. The first term on the right of Eq. (3) $R_{emp}(\alpha)$ is minimized by learning process. The second term varies with the VC dimension h and the number of examples l. The smaller the VC dimension h and the larger the number of examples l, the smaller the value of the second term.

3 Experimental Results

Two sets of DGA data were compiled for the feasibility study of our proposed method. The first data set consisted of 134 gas records from the IEC TC10 Database that had been discriminated into 6 classes based on IEC 60599 described values corresponding to different actual conditions [21]. The second data set contained 635 historical dissolved gas records from different transformers of the Taipower Company, which is the sole electricity supplier in Taiwan. These data were divided into three data sets: the training data set, the validation data set and the testing data set. This study employed the DGA data to show the classification performances of CSA-based SVM models as compared with those of the ANN model.

The features we concerned not only three basic gas ratios of DGA data, but also including their mean (μ); root mean square (rms), variance (σ^2); skewness (normalized 3rd central moment, γ_3), kurtosis (normalized fourth central moment, γ_4), normalized fifth to tenth central moments (γ_5-γ_{10}) as follows:

$$\gamma_n = \frac{E\{[y_i - \mu]^n\}}{\sigma^n}, \; n = 3,10 \tag{4}$$

where E represents the expected value of the function. Total numbers of derived features are 36. We randomly extracted instances for learning set, validation set and test set for the corresponding phase

3.1 Performance Comparison of ANNs and SVMs for the IEC TC 10 Dataset

In this section classification results are presented for stand alone ANNs and SVMs without features and parameters selection. For each case of stand alone ANN, number of neurons in the hidden layer was kept at 24 and for stand alone SVMs, a constant width (σ) of 0.5 and tradeoff regularization parameter (C) of 100 was used. These values were chosen on the basis of initial trials of training.

Table 1, 2 shown the effects of signal processing on the classification results for straight ANNs and SVMs with the first IEC TC 10 dataset. In each case all the features from the signals without and with signal preprocessing were used. The test success of stand alone ANNs was 78.7% whereas it was 81.8% for SVMs. In each case, the training time of ANNs was much higher than SVM. And the performances are improved by CSA optimization to 95.3% and 99.4% respectively.

The computation time (on a PC with Pentium IV processor of 1.7 GHz and 512 MB RAM) for training the classifiers are also shown, though direct comparison is difficult due to difference in code efficiency. These values are mentioned for rough comparison only. However, it should be mentioned that the difference in computation time should not be very important if the training is done off-line.

Table 1. Performance with 36 features for the IEC TC 10 dataset

Classifier	Stand alone			
	No. of inputs	Training success(%)	Test success(%)	Training time(s)
ANN	36	78.7	83.6	16.385
SVM	36	81.8	90.4	0.937

Table 2. Performance with CSA optimization for the IEC TC 10 dataset

Classifier	With CSA Optimization			
	No. of inputs	Training success(%)	Test success(%)	Training time(s)
ANN	12	89.9	95.3	3.212
SVM	4	97.9	99.4	0.435

3.2 Performance Comparison of ANNs and SVMs for the Taipower Company Dataset

In this section, classification results are presented for ANNs and SVMs with features and parameters selection based on CSAs optimization. In each case, all 36 features were auto-selected from the corresponding range of input features by the proposed method. In the case of ANNs, the number of neurons in the hidden layer was selected n the range of 10–30 whereas for SVMs, the RBF kernel width was auto-selected in the range of 0.1–2.0 with CSA optimization.

Table 3,4 shown the classification results along with and without the selected parameters for each DGA data. In all cases, the input features were selected by CSAs from the entire range. The test success improved substantially with feature selection for both ANNs and SVMs. Test success was nearly 100% in case of ANNs whereas it was 100% for all signals in the case of SVMs for training and testing set. After features and parameters selection, it can be seen that the performances on the unseen test sets have all increased.

Table 3. Performance with 36 features for the Taipower Company dataset

Classifier	Stand alone			
	No. of inputs	Training success(%)	Test success(%)	Training time(s)
ANN	36	77.6	82.9	20.221
SVM	36	79.0	89.7	1.104

Table 4. Performance with CSA optimization for the Taipower Company dataset

Classifier	With CSA Optimization			
	No. of inputs	Training success(%)	Test success(%)	Training time(s)
ANN	4	100	99.98	5.335
SVM	4	100	100	0.676

4 Conclusions

In this paper, the selection of input features and the appropriate classifier parameters have been optimized using a CSA-based approach. A procedure is presented for detection of power transformer faulty situation using two classifiers, namely, ANNs and SVMs with CSA-based features and parameters selection from DGA inherently imprecise signals. The appropriate radius-margin bound is selected as the objective function for automatic parameters selection for SVM with experiment test. Experiment results of practice data demonstrate the effectiveness and high efficiency of the proposed approach. The classification accuracy of SVMs was better than of ANNs, without CSA in this application. With CSA-based selection, the performances of both classifiers were comparable at nearly 100%. The results show the potential application of CSAs for off-line features and parameters selection. This also opens up the potential use of optimized features and classifier parameters for real-time implementation leading to possible development of an automated DGA condition monitoring and diagnostic system. We conclude that our method is not only able to figure out optimal parameters but also minimize the number of features that SVM classifier should process and consequently maximize the diagnosis rate of DGA.

References

1. Vapnik V, Golowich S, Smola A. "Support Vector Method for Function Approximation, Regression Estimation, And Signal Processing [A]." In: Mozer M, Jordan M, Petsche T (eds). Neural Information Processing Systems [M]. MIT Press, 1997, 9.
2. O. Chapelle, V. Vapnik, O. Bousqet, and S. Mukherjee. Choosing Multiple Parameters for Support Vector Machines. *Machine Learning*, 46(1):131 – 159, 2002.
3. IEC Publication 599, "Interpretation of The Analysis of Gases in Transformers And Other Oil-Filled Electrical Equipment In Service," Fisrt Edition, 1978.
4. "IEEE Guide for The Interpretation of Gases Generated In Oil-Immersed Transformers," IEEE Std.C57.104-1991.
5. R.R. Rogers, "IEEE And IEC Codes to Interpret Faults in Transformers Using Gas in Oil Analysis," IEEE Trans Electron. Insulat. 13 (5), pp.349–354,1978.
6. M.H. Wang, "A Novel Extension Method for Transformer Fault Diagnosis," IEEE Trans. Power Deliv. 18 (1),pp.164–169,2003.
7. N.C.Wang,"Development of Monitoring and On-line Diagnosing System for Large Power Transformers," (Power Reserch Institute,TPC) .
8. Yuanping.Ni, "One Intelligent Method for Fault Diagnosis and Its Application," IEEE Trans. Power Deliv, pp755-758,2000.

9. Ganyun Lv., Haozhong Cheng, Haibao Zhai, Lixin Dong, “Electric Power Systems Research 75,” 9–15, 2005 Elsevier B.V. All rights reserved.
10. L.B. Jack, A.K. Nandi, “Fault Detection Using Support Vector Machines and Artificial Neural Networks: Augmented by Genetic Algorithms,”Mech. Syst. Signal Process. 16 (2–3), pp.373–390, 2002.
11. W.C. Chan, C.W. Chan, K.C. Cheung, C.J. Harris, “On The Modeling of Nonlinear Dynamic Systems Using Support Vector Neural Networks,” Eng. Appl. Artif. Intell. 14, pp.105–113,2001.
12. H. Frohlich. Feature Selection for Support Vector Machines by Means of Genetic Algorithms. Master’ s thesis, University of Marburg, 2002.
13. M,Dong, ”Fault Diagnosis Model Power Transformer Based on Statistical Learning Theory and Dissolved Gas Analysis,” IEEE 2004 Internation Symposium on Electrical Insulation,p85-88.
14. Y.C. Huang, H.T. Yang, C.L. Huang, Developing a new transformer fault diagnosis system through evolutionary fuzzy logic, IEEE Trans. Power Deliv. 12 (2) (1997) 761–767.
15. C. Cortes, V. Vapnik, Support-vector Networks, Machine Learn. 20 (3) (1995) 273–295.
16. E.H. Tay Francis, L.J. Cao, Application of support vector machines in financial time series forecasting, Omega. 29 (4) (2001) 309– 317.
17. ”BECTA”,Power Station Magazine in Russia,No.6,1998.
18. W.W. Yan, H.H. Shao, “Application of Support Vector Machine Nonlinear Classifier to Fault Diagnoses,” Proceedings of the Fourth World Congress Intelligent Control and Automation, 10–14 Shanghai, China, pp. 2697–2670, June 2002.
19. J. H. Lee, and C. J. Lin, Automatic model selection for support vector machines, Technical Report, Dept. of Computer Science and Information Engineering, Taipei, Taiwan, November 2000.
20. V.N. Vapnik, “The Nature of Statistical Learning Theory,” Springer-Verlag, New York, 1995.
21. Michel Duval, ”Interpretation of Gas-In-Oil Analysis Using New IEC Publication 60599 and IEC TC 10 Databases,” IEEE Electrical Insulation Magazine, 2001 — Vol. 17, No. 2

A Semi-naive Bayesian Learning Method for Utilizing Unlabeled Data

Chang-Hwan Lee

Dept. of Info. & Comm.,
Dong Guk University
3-26 Pil-Dong, Chung-Gu, Seoul, Korea
chlee@dgu.ac.kr

Abstract. In many applications, an enormous amount of unlabeled data is available with little cost. Therefore, it is natural to ask whether we can take advantage of these unlabeled data in classification learning. In this paper, we analyzed the role of unlabeled data in the context of naive Bayesian learning. Experimental results show that including unlabeled data as part of training data can significantly improve the performance of classification accuracy.

1 Introduction

In current supervised learning methods, the algorithms need significant amount of labeled data to obtain satisfactory classification performance. Pure supervised learning from limited training data set will have poor classification performance. However, in many domain applications, labeling is often done by a person, which is a time-consuming process. For example, in the task of classifying text documents, most of users of a practical system would not have the patience to label thousand articles. In fact, in many instances, the labeling process is harder or more expensive than the sampling step required to obtain the observations.

Most computational models of supervised learning rely on labeled training data, and ignore the possible role of unlabeled data. Therefore, situations in which both labeled and unlabeled samples are available arise naturally, and the investigation of the simultaneous use of both kinds of observations in learning leads immediately to questions of the relative value of labeled and unlabeled samples.

It is known that unlabeled data alone are generally insufficient to yield better-than-random classification because there is no information about the class label [1]. However, unlabeled data do contain information about the joint distribution over features other than the class label. Because of this, they can sometimes be used, together with a sample of labeled data, to increase classification accuracy in certain problem settings [7]. It is important to notice that this result depends on the critical assumption that the data indeed have been generated using the same parametric model as used in classification. Many applications require classifiers built by machine learning to categorize incoming data automatically. Since in many applications, an enormous amount of unlabeled data is available with little

B. Gabrys, R.J. Howlett, and L.C. Jain (Eds.): KES 2006, Part I, LNAI 4251, pp. 187–194, 2006.
© Springer-Verlag Berlin Heidelberg 2006

cost, it is natural to ask whether, in addition to human labeled data, one can also take advantage of the unlabeled data.

The purpose of this paper is to explore and study some techniques for improving the accuracy of classification algorithms by utilizing unlabeled data that may be available in large numbers and with no extra cost. We are to show how unlabeled data can improve the performance of classification learning in the context of naive Bayesian classifier. We present an algorithm and experimental results demonstrating that unlabeled data can significantly improve learning accuracy in certain learning problems. We identify the abstract problem structure that enables the algorithm to successfully utilize this unlabeled data, and show that unlabeled data will boost learning accuracy for problems in this class.

Related Work

Most classic methods of learning with unlabeled data use a generative model for the classifier and use Expectation-Maximization(EM) method. The EM algorithm was first introduced in [4] as a procedure for estimating the parameter values that maximize the likelihood function when there are missing data. There are two steps in EM algorithm, expectation(E-step) and maximization(M-step). The E-step calculates the expected values of the sufficient statistics given the current parameter estimates. The M-step makes maximum likelihood estimates of the parameters given the estimated values of the sufficient statistics.

In co-training method [10] [12], two classifiers are defined using two distinct feature sets. Co-training then consists of iteratively using the output of one classifier to train the other. Variations of co-training have been applied to many application areas. Co-training is a learning algorithm in which the redundancy of the learning task is captured by training two classifiers using separate views of the same data. This enables bootstrapping from a small set of labeled training data via a large set of unlabeled data. In co-training, two classifiers are defined using distinct feature sets. Co-training then consists of iteratively using the output of one classifier to train the other. Variations of co-training have been applied to many applications areas.

Another related area of research that has very different goal is that of active learning [2] [8] where the algorithm repeatedly selects an unlabeled example, asks an expert to provide the correct label, and then rebuilds its hypothesis. An important component of active learning is in the selection of the unlabeled example.

2 Problem Formulation

This section presents a probabilistic framework for characterizing the nature of classifiers. This paper employs naive Bayes–a well-known probabilistic classifier [9], as the foundation upon which we will later build in order to incorporate unlabeled data.

The naive Bayes method applies to learning tasks where each instance x is described by a conjunction of attribute values and where the target function can take on any value from some finite set V. A set of training examples of the target function is provided, and a new instance is presented, described by the tuple of attribute values $(a_1, a_2, \ldots, a_n)$. The learner is asked to predict the target value, or classification, for this new instance. The Bayesian approach to classifying the new instance is to assign the most probable target value, given the attributes $(a_1, a_2, \ldots, a_n)$ that describe the instance. Suppose a_i denote the i-th attribute. The naive Bayes expression for the probability of a new instance given its class is given as

$$v = argmax_{v_j \in V} \; P(v_j) \prod_i P(a_i|v_j)$$

Thus the parameters of an individual mixture components are a multinomial distribution. The only other parameters of the model are mixture weights(class prior probabilities), which indicate the probabilities of selecting the different mixture components. Thus the complete collection of model parameters is a set of multinomials and prior probabilities over those multinomials. Learning a naive Bayes consists of estimating these parameters of the generative model by using a set of labeled training data.

Suppose the estimate of $P(a_i|v_j)$ and $P(v_j)$ are denoted as $\hat{p_{ij}}$ and α_j, respectively, then $\hat{p_{ij}}$ is defined as

$$\hat{p_{ij}} = \frac{P(a_i \wedge v_j)}{P(v_j)} = \frac{N(a_i \wedge v_j)}{N(v_j)}, \qquad \hat{\alpha_j} = \frac{N(v_j)}{N} \tag{1}$$

where $N(t)$ means the number of data satisfying the condition t.

The parameter estimation formulae in Equation 1 that result from this maximization are the familiar ratios of empirical counts. Furthermore, the counts in both the numerator and denominator are augmented with pseudo-counts. The class prior probabilities are estimated in the same manner, and also involve a ratio of counts with smoothing: The use of this type of prior is sometimes referred to as Laplace smoothing. Smoothing is necessary to prevent zero probabilities for infrequently occurring. Therefore, in this paper, we use Laplace smoothing and, thus, the final probability estimates are given as:

$$\hat{p_{ij}} = \frac{N(a_i \wedge v_j)}{N(v_j) + N}, \qquad \hat{\alpha_j} = \frac{N(v_j)}{N + C} \tag{2}$$

where N represents the total number of data and C represents the number of class values, respectively.

Given estimates of these parameters calculated from the training data according to Equation 2, it is possible to turn the generative model backwards and calculate the probability that a particular mixture component generated a given data.

Incorporating Unlabeled Data

We have modified the naive Bayesian algorithm in order to process the unlabeled data. The basic mechanism of processing unlabeled datais as follows. The algorithm first calculates the estimates of parameters using only labeled data. After acquiring estimated values of parameters, the algorithm classifies unlabeled data. It calculates the weights of each value of the target feature, and the target value with the highest weight is assigned to the value of the target attribute. After the class values of each unlabeled data are calculated, the algorithm is trained again using both originally labeled data and formerly unlabeled data. The algorithm iterates this process until there is no or very little changes in the estimated target values of the unlabeled data. Figure 1 represents the brief pseudo code of the algorithm.

```
Given : L : labeled data, U : unlabeled data

/* process labeled data */
calculate P̂_ij and α̂_j and using all data in L
while (no change in the values of P̂_ij and α̂_j) do
    /* process unlabeled data */
    for each data (a_1, ..., a_t) ∈ U do
        calculate v satisfying v = max_j α̂_j ∏_i P̂_ij
        assign label v to (a_1, ..., a_t)
    end-for
    calculate P̂_ij and α̂_j using both L and U
end-do
```

Fig. 1. Pseudo code of the algorithm for incorporating unlabeled data

3 Empirical Studies

In this section, we present experimental results of the influence of unlabeled data using the proposed algorithm. We have tested the effectiveness of unlabeled data using a set of synthetic data to precisely analyze the effect of estimating parameters.

The data is generated according to multiplicative distribution. For each attribute i, two parameters, p_i and q_i are associated with the attribute. The parameter p_i represents the probability that attribute i becomes 1 given the class value is 0. Similarly, q_i represents the probability that attribute i becomes 1 given the class value is 1. There is also another parameter, α, which represents the probability that class 0 is selected. These parameters can be defined as follows: $\alpha = P(v = 0), \quad p_i = P(a_i = 1|v = 0), \quad q_i = P(a_i = 1|v = 1)$

The values of these parameters are determined as given in Figure 3. Using these values, we generated a set of labeled data as well we unlabeled data.

Figure 3 shows the pseudo code of the algorithm for generating data set. We could also calculate the accuracy of optimal classifier in case test data are generated based on the algorithm in Figure 4.

$\alpha = 0.5$
$p_1 = 0.7,\ p_2 = 0.3,\ p_3 = 0.8,\ p_4 = 0.3,\ p_5 = 0.2,\ p_6 = 0.1,\ p_7 = 0.8$
$q_1 = 0.2,\ q_2 = 0.6,\ q_3 = 0.3,\ q_4 = 0.3,\ q_5 = 0.6,\ q_6 = 0.7,\ q_7 = 0.8$

Fig. 2. Parameter values for generating sample data

```
Given : α : prob. that class 0 is selected
p_i : prob. that i-th attribute = 1 when target value =0
q_i : prob. that i-th attribute = 1 when target value =1
t : number of attributes, n : number of data, D : the data set

D := {}
repeat n times
    determine the class value v (v = 0 with probability α)
    for j=1 to t
        if v=0 then
            determine the value of α_j (α_j = 1 with probability p_i)
    else if v=1 then
            determine the value of α_j (α_j = 1 with probability q_i) end-if
D := D ∪ (a_1, ..., a_t, v);
return D
```

Fig. 3. Algorithm for generating sample data

```
Given : α : prob. that class 0 is selected
p_i : prob. that i-th attribute = 1 when target value =0
q_i : prob. that i-th attribute = 1 when target value =1
t : number of attributes, D : the data set

correct := incorrect := 0
for each tuple (a_1, ..., a_t, v) ∈ D do
    Sum_0 := α; Sum_1 := 1 − α;
    for i=1 to t
        if v=0 then
            if α_i = 1 then Δ_i = p_i else Δ_i = 1 − p_i end-if
            Sum_0 := Sum_0 * Δ_i;
        else if v=1 then
            if α_i = 1 then δ_i = q_i else δ_i = 1 − q_i end-if
            Sum_1 := Sum_1 * δ_i;
        end-if
    end-for
    if Sum_0 > Sum_1 then output := 0 else output := 1 end-if
    if output = v then correct++ else incorrect++ end-if
    end-for
return correct/(correct+incorrect);
```

Fig. 4. Algorithm for calculating optimal classification accuracy

The size of test data is given as 1,000, disjoint with training data, and the optimal Bayes accuracy using the values in Figure 3 is calculated as 86.6%.

The proposed algorithm first trains a classifier with only the available labeled data, and uses the classifier to assign labels of each unlabeled data by calculating the expectation of the missing class labels. It then trains a new classifier using all the data–both the originally labeled and the formerly unlabeled–and iterates. Over the course of several experimental comparisons, we show that unlabeled data can significantly increase performance.

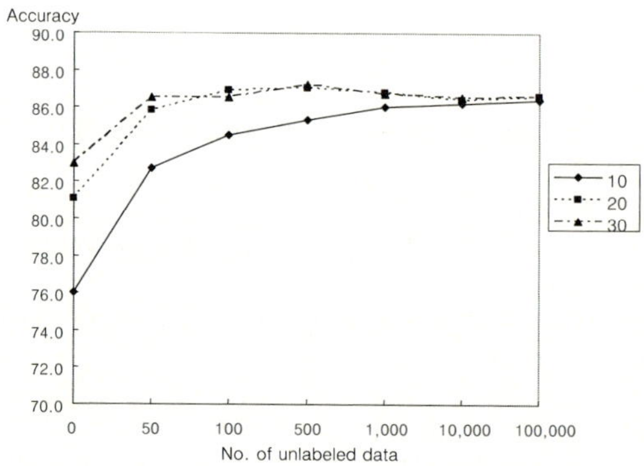

Fig. 5. Accuracies based on the number of unlabeled data (1)

Figure 5 shows the effect of using unlabeled data in case the number of labeled data is given as 10, 20 and 30, respectively. As shown in Figure 5, when the number of labeled data are small, the classification accuracy could be greatly improved by incorporating unlabeled data. We vary the amount of unlabeled data from 0 to 100,000 and see how the effect of using unlabeled data. When only 10 labeled data is given, the classification accuracy is initially given as 76%. However, as the amount of unlabeled data increases, the accuracy increases as well. The accuracy reaches 86.6%, the optimal accuracy for given data set, when sufficient amount(100,000 in this experiment) of unlabeled data is given.

Figure 6 shows the effect of using unlabeled data in case the number of labeled data is moderate(e.g., 50, 100 and 500, respectively). The classification accuracy increases as more unlabeled data are available. However, in Figure 6, we see that initial accuracies using labeled data are higher than that of Figure 5 since larger number of labeled data are available in Figure 6. Furthermore, the system required less number of unlabeled data in order to approach the optimal accuracy(86.6%), and the effect of unlabeled data in the case of Figure result-2 is less significant than that of Figure 5.

Figure 7 shows the effect of using unlabeled data in case the number of labeled data is large(e.g., 1,000, 5,000 and 10,000, respectively). In these cases, since the system already haveenough number of labeled data, the use of unlabeled data does not show any significant improvement in terms of classification accuracy. When 10,000 cases of labeled data is given, the system already reached the optimal accuracy.

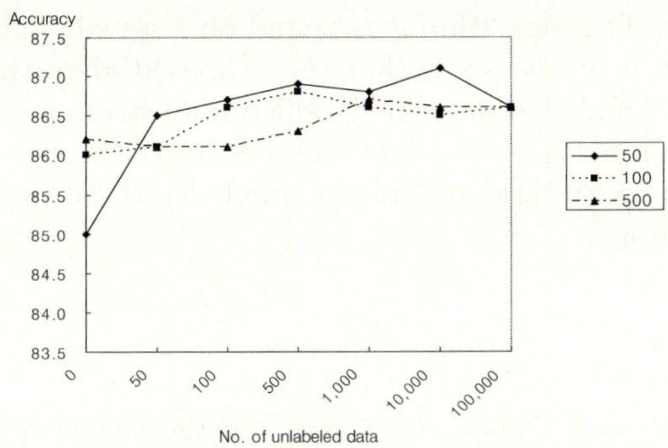

Fig. 6. Accuracies based on the number of unlabeled data (2)

As we can see in Figure 5, the reduction in the number of labeled data needed can be significant. In order to identify the data with the optimal accuracy(86.6%), a traditional learner, with no unlabeled data, requires more than 5,000 labeled data; alternatively our algorithm takes advantage of unlabeled data and requires only 10 labeled data along with 10,000 unlabeled data to achieve the same accuracy. Thus, in this task, the proposed method in this paper significantly reduces the need for labeled training data. With only 10,000 labeled data, accuracy is improved from 76.0% to 86.6% by adding unlabeled data. These findings illustrate the power of unlabeled data, and also demonstrate the strength of the algorithms proposed here.

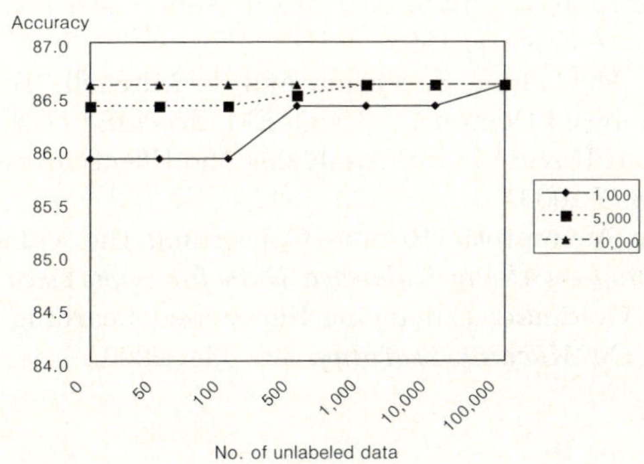

Fig. 7. Accuracies based on the number of unlabeled data (3)

4 Conclusion

In this paper, we demonstrate the effectiveness of using unlabeled data in classification learning in the context of naive Bayesian method. We proposed a semi-naive Bayesian learning method, which could incorporate unlabeled data in naive Bayesian learning. The algorithm was tested on a set of synthesized data.

The experimental results show that the proposed algorithm could effectively take advantage of unlabeled data in classification learning. Compared with pure supervised learning method, the algorithm needs significantly smaller amount of labeled data, along with large number of unlabeled data, to achieve the optimal classification accuracy.

References

1. Vittorio Castelli and Thomas M. Cover. On the Exponential Value of Labeled Samples *Pattern Recognition Letters*, Vol. 16, pp. 105-111, 1995
2. D. Cohn et al. Active Learning with Statistical Models, *Journal of Artificial Intelligence Research*, Vol. 4, 129-145, 1996
3. F. De Comite et al. Positive and Unlabeled Examples Help Learning, *Tenth Int'l Conf. on Algorithmic Learning Theory*, 219-230, 1999
4. A. P. Dempster, N. M. Laird, and D. B. Rubin. Maximum Likelihood from Incomplete Data via the EM Algorithm, *Journal of Royal Statistical Society*, Vol. 39, pp. 1-38, 1977
5. R. Duda, et al. *Pattern Classification*, 2nd edition, John Wiley & Sons, 2001
6. Sally Goldman and Yan Zhou. Enhancing Supervised Learning with Unlabeled Data *Int'l Conf. on Machine Leaning*, 2000
7. T. Hofmann. Text Categorization with Labeled and Unlabeled Data: A Generative Model Approach, *NIPS 99 Workshop on Using Unlabeled Data for Supervised Learning*, 1999
8. R. Liere and P. Tadepalli. Active Learning with Committees for Text Categorization, *14th National Conf. on Artificial Intelligence*, 591-596, 1997
9. Tom Mitchell. *Machine Learning*, McGraw Hill, 1997
10. T. Mitchell. The Role of Unlabeled Data in Supervised Learning, *6th Int'l Colloquium on Cognitive Science*, 1999
11. K. Nigam, A.K. McCallum, S. Thrun, and T. Mitchell. Text Classification from Labeled and Unlabeled Documents Using EM, *Machine Learning*, 39, 103-134, 2000
12. Kamal Nigam and Rayid Ghani. Analyzing the Effectiveness and Applicability of Co-training, *CIKM* 2000
13. T. Zhang. Some Asymptotic Results Concerning the Value of Unlabeled Data, *NIPS 99 Workshop on Using Unlabeled Data for Supervised Learning*, 1999
14. Y. Zhou and S. Goldman. Enhancing Supervised Learning with Unlabeled Data *17th Int'l Conf. On Machine Learning*, 327-334, 2000

Quantitative and Ordinal Association Rules Mining (QAR Mining)

Filip Karel

Department of cybernetics, Czech Technical University in Prague,
Technická 2, Praha 6, 166 27
karelf1@fel.cvut.cz

Abstract. Association rules have exhibited an excellent ability to identify interesting association relationships among a set of binary variables describing huge amount of transactions. Although the rules can be relatively easily generalized to other variable types, the generalization can result in a computationally expensive algorithm generating a prohibitive number of redundant rules of little significance. This danger especially applies to quantitative and ordinal variables. This paper presents and verifies an alternative approach to the quantitative and ordinal association rule mining. In this approach, quantitative or ordinal variables are not immediately transformed into a set of binary variables. Instead, it applies simple arithmetic operations in order to construct the cedents and searches for areas of increased association which are finally decomposed into conjunctions of literals. This scenario outputs rules that do not syntactically differentiate from classical association rules.

Keywords: association rules, quantitative attributes, ordinal attributes.

1 Introduction

The discovery of quantitative and ordinal association rules in large databases is considered as an interesting and important research problem. Recently, different aspects of the problem have been studied, and several approaches have been presented in the literature. Authors mostly agree that standard algorithms are ineffective and often provide us with illogical or unusable results. Quantitative or ordinal attributes bring many specifics. One of them is the optimalization of discretization of quantitative attributes. Another question is how to measure the quality of generated rules. Some authors use classical measures like support and confidence [3],[4],[8], other authors point out that in the case of quantitative it is better to use different measures [5],[9],[10]. Principle of proposed approaches is almost always very similar. It is based on finding suitable intervals of attributes which are then used in some kind of binary tests which are the basic principle of classical association rules mining. Distinctions among authors are in the methodology of selecting intervals and in measures used to evaluate these intervals. In general the authors show better results of their algorithms and more intuitive concept, but they are not primarily focused on time consumption and also they

B. Gabrys, R.J. Howlett, and L.C. Jain (Eds.): KES 2006, Part I, LNAI 4251, pp. 195–202, 2006.
© Springer-Verlag Berlin Heidelberg 2006

often do not provide a complex algorithm. Authors in [11] discus this problem, but do not provide any algorithm which saves time during QAR mining. Authors also do not address the problem of attributes combination in detail.

In this paper an innovative algorithm for QAR mining is proposed. The main advantages of this algorithm are significantly lower time consumption and a reduction of the number of redundant rules. On the other hand, as this algorithm is not based on a complete search through the state space it cannot be guaranteed that all rules are found.

Association rule is an implication $X \rightarrow Y$. The left hand side of this implication is called *antecedent*, the right hand side is *succedent*. Generally antecedent and succedent can be called *cedents*. Cedents are constructed of *attributes*, which are included in the database. Cedent can consist of one attribute - *trivial cedent*, or more attributes - *non-trivial cedent*.

The rest of paper is organized as follows. Section 2 provides an overview of different types of attributes and describes the preprocessing process. Section 3 is focused on combination of attributes into non-trivial cedents. The following section 4 examines areas of interest and describes how to decompose the non-trivial cedents and how to select valid rules. Section 5 gives examples of experimental results, summarizes the paper and discusses open problems.

2 Attributes' Preparation and Preprocessing

The data mining literature contains a variety of terms describing different types of data or attributes. In this paper the following division is used:

Qualitative attributes - they are only divided in categories, but not numerical measures

- nominal - attributes that are exhaustively divided into mutually exclusive categories with no rankings that can be applied to these categories (names, colors)
- ordinal - the categories into which they are classified can be ordered (evaluation of an action = {very good, good, bad, very bad})

Quantitative attributes - attributes that are measured on a numerical scale and to which arithmetic operations can be applied

- discrete - have a measurement of scale composed of distinct numbers with gap in between (number of cars)
- continuous - can ideally take any value (height, distance)

The proposed algorithm deals with qualitative ordinal, and quantitative discrete and continuous attributes.

Attributes can take different values and different number of values. These differences among attributes can very heavily influence the results of association rules mining, so it is necessary to preprocess them before combining. The common methods for attributes adaptation can be used - **normalization** and **discretization**.

Using normalization we reach all attributes take values in the same range. All attributes are normalized to values between 0 and *max_value*.

There are many advantages of using discrete values over continuous ones. Discrete features are closer to a knowledge-level representation than continuous ones. For both users and experts, discrete features are easier to understand, use, and explain. As reported in [12], discretization makes learning more accurate and faster. Discussing advantages and disadvantages of different types of discretization is out of the scope of this paper. Generally speaking, the higher degree of discretization (number of discrete values), the higher the computational costs. The lower degree of discretization, the higher information loss. One of the main advantages of algorithm proposed in this paper is its low time consumption. We can therefore afford higher degree of discretization.

3 Combining Attributes

To create non-trivial cedents, several attributes have to be combined together and they have to be represented by one number.

Combination can be realized by basic operations like adding and subtraction, which are simple enough and they are not increasing time consumption of the whole algorithm. Moreover, the following decomposition of the non-trivial cedent is simple when these operations are used. We can see on figure 1 there can be quite big differences among non-trivial cedents' histograms depending on the way how cedents are created. If there are only two operations (addition and subtraction) then we have 2^{n-1} of combinations how to create a non-trivial cedent, where n is the number of attributes in a non-trivial cedent. We have $\binom{m}{n}$ combinations of attributes which can create the non-trivial cedent, where m is total number of attributes in the database. Both antecedent and succedent have to be created so overall time consumption is $2^{n_a+n_s-2} * \binom{m_a}{n_a} * \binom{m_s}{n_s}$. Testing all possibilities could therefore be very time demanding.

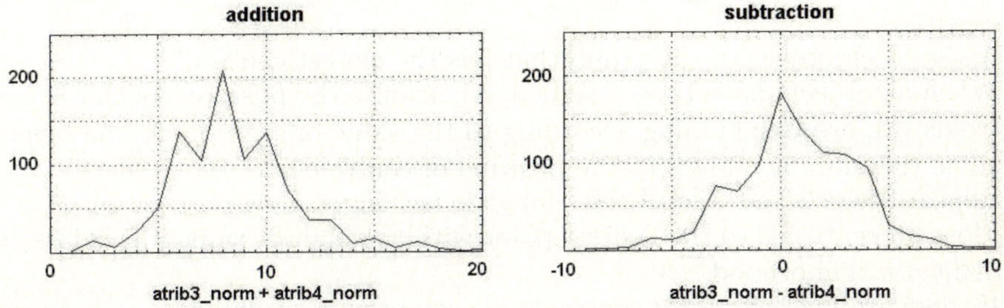

Fig. 1. Cedents histograms

Experiments have demonstrated it is more suitable to assess the sign to an attribute on the antecedent side in order to how it influences attribute(s) on the succedent side (positively or negatively). For this decision a non-parametric *spearman rank correlation coefficient* is used as follows. The succedent is created by addition of all attributes. Then, spearman rank correlation coefficient is computed with each of the antecedent attributes. If the coefficient is positive (negative) then + (-) is assigned to the antecedent attribute.

If we combine attributes without normalization and discretization, the histogram of the resulting cedent is influenced more by attributes with high number of high values. This fact can heavily influence further rules generation. If we combine normalized attributes, all of them have equal weight and resulting cedent is influenced by all attributes equally. In following discretization, numbers of records in particular discretization bins are more balanced.

4 Areas of Interest

This section describes how to identify areas of interest and how to mine rules from these areas of interests. After constructing the non-trivial cedents, the contingency table of differences between real and expected values is used. The size of the table is $\sum_{i=1}^{n} D_A(i)$ x $\sum_{i=1}^{m} D_S(i)$, where n is the number of attributes creating antecedent, m is the number of attributes creating succedent, D_A is the degree of discretization of the i-th antecedent attribute, D_S is the degree of discretization of the i-th succedent attribute.

The interesting areas to identify are the areas with positive values. In these areas there are more records than expected in case of cedents independence. In most cases, all these areas cannot be examined because of huge time consumption. So the task is to identify the most interesting areas, this means the largest areas with the highest positive values.

For this task the standard clustering algorithm is followed. To increase the quality of generated rules and to reduce the number of redundant and similar rules, points 5 and 6 are added.

1. Place K points into the space represented by the objects (points with positive values) that are being clustered. These points represent initial groups' centroids.
2. Assign each object to the group that has the closest centroid.
3. When all objects have been assigned, recalculate the positions of the K centroids (we use weightening according to the value of each point, the biggest value the more it influences the position of centroid).
4. Repeat Steps 2 and 3 until the centroids no longer move.
5. Move all centroids to the nearest point with the highest value if it is in some defined neighborhood.
6. If there are two (or more) centroids in each others' neighborhood, keep just one with the highest value.

After the clustering algorithm finishes, we have the position of final centroids in the matrix (contingency table). First point's coordinate represents the value of an antecedent, second represents the value of a succedent. We receive K points coordinates, where K is number of interesting areas we want to gain.

Decomposition points' coordinates into the concrete values of antecedent's and succedent's attributes has to follow now. I demonstrate it on the example of a non-trivial antecedent made of two attributes and non-trivial succedent made of 2 attributes. The degree of discretization is 10. Therefore the contingency table of differences has size of 20 x 20. Let us decide to gain just one interesting area and suppose that we received the centroid coordinates [13, 5], where 13 represents antecedent axis and 5 succedent axis. We now have to decompose these coordinate's values into the attributes' values.

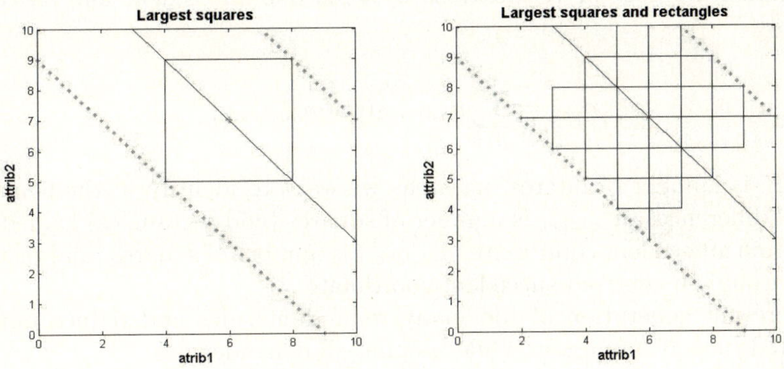

Fig. 2. Antecedent decomposition

We can imagine the situation in 2D space - Figure 2 (for more attributes, more dimensional space is used). Coordinates X and Y represents the two antecedent attributes. The full line represents points which satisfies the condition $attrib_1 + attrib_2 = 13$. Some tolerance area have to be set up to gain rules with sufficient support and to use the positive points in the centroid surrounding. The tolerance area is represented by dotted lines. All points in this area satisfies condition $10 \leq (attrib_1 + attrib_2) \leq 16$.

There are 3 options how to go through the space bordered by the lines

1. **Largest squares** - we take into account only the largest squares, we achieve lowest time consumption - $n_{ver} \approx (X_{max} - X_{min})^{D-1}$, where n_{ver} is number of verifications. We have of course limited resolution, The example square on the left side of figure 2 illustrate $attrib_1 = 4..8$ & $attrib_2 = 5..9$.
2. **Largest squares and rectangles** - we take into account all largest squares and rectangles, time consumption is higher, $n_{ver} \approx ((X_{max} - X_{min} * tol\_width))^{D-1}$. The example square and rectangles on the right side of figure 2 illustrate possibilities from $attrib_1 = 5$ & $attrib_2 = 3..10$ to $attrib_1 = 2..10$ & $attrib_2 = 7$.
3. **All squares and rectangles** - we consider all squares and rectangles in the area, time consumption of this possibility is the highest of all possibilities, $n_{ver} \approx (X_{max} - X_{min})^{tol\_width+(D-1)}$.

Obviously, the highest time consumption, the better resolution is obtained. But even the most time consuming option has lower time consumption in comparison with classical approaches.

All the combinations of antecedent and succedent squares and rectangles have to be verified for the valid rules. There are many measures for quality of association rules described in literature. I'm using the basic measures of *support* and *confidence* [1] ensembled by the *lift* measure [2]. One rule is described by one square (rectangle) of antecedent side and one square (rectangle) from succedent side of a particular centroid.

Maximum number of verifications between one antecedent and succedent is given by

$$n_{ver} = \sum_{i=1}^{K} n_{antrect}(i) * n_{succrect}(i),$$

where K is number of interesting areas we want to identify in the contingency table of differences, $n_{antrect}$ is number of squares (and rectangles) gained from i-th centroid antecedent coordinate, $n_{succrect}$ is number of squares (and rectangles) gained from i-th centroid succedent coordinate.

To prevent generation of too many redundant rules and reduce number of generated rules to minimum while keeping all dependencies

- minimum thresholds for support, confidence and lift (*minsupp, minconf* and *minlift*) are set up,
- from each area represented by one centroid only two best rules are selected - one with the highest confidence (lift), which also satisfies condition *minsupp* and one with highest support, which also satisfies conditions *minconf* and *minlift*.

The number of rules is limited by $2 * K$ for each atecedent-succedent combination. For the whole algorithm there are

$$n_{rules} \leq 2 * K * n_{comb},$$

where n_{rules} is number of rules, n_{comb} is number of possible antecedent-succedent combinations, $n_{comb} \leq \binom{m_a}{n_a} * \binom{m_s}{n_s}$.

So the number of generated rules can be influenced through parameter K. If we want to have just a few rules and gain the main dependencies in data, we set K lower, if we want to gain all valid rules we set up K higher.

5 Experimental Results and Conclusions

An innovative algorithm for QAR mining was introduced in this paper. This algorithm consists of four basic steps.

1. Preprocessing of attributes (normalization and discretization).
2. Non-trivial cedents creation (addition and subtraction).

3. Identification of interesting areas in contingency table.
4. Cedents decomposition into rules and best rules selection.

The comparison of classical mining and QAR mining was made over the STULONG data domain [13], [14]. It concerns a dataset describing the data collected during a longitudinal study of atherosclerosis prevention on around 1400 middle-aged men. I worked with data representing entry examination of men, ten basic attributes from the database were selected and I tried to find association among selected attributes.

Time consumption is represented by number of verifications, i.e. number of candidate rules tested for the *minsupp*, *minconf* and *minlift* thresholds. On figure 3 we can see how number of verifications depends on number of attributes in database and how number of valid rules depends on interval length. Interval length represents maximum range of one attribute in the rule. For example if we have interval length three, attribute in rule can take values 1..4, 2..5, ... and it can't take values 1..5, 2..6, ... Non-trivial antecedent made of two attributes and trivial succedent were used in this experiment.

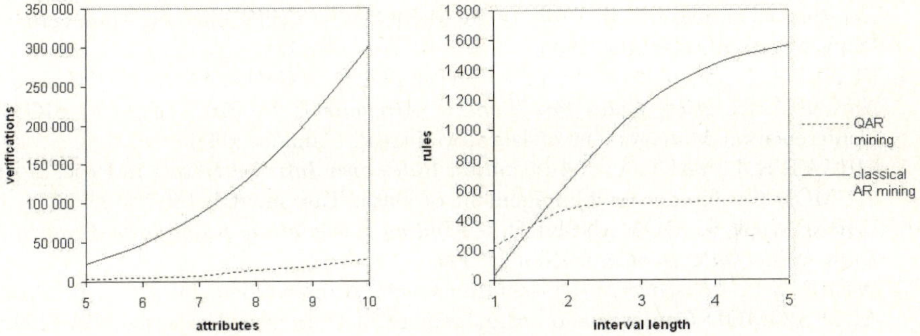

Fig. 3. Experimental results

The number of verifications grows very quickly with the number of attributes. For ten attributes we have over 300 000 verifications using classical approach, while the number of verifications using QAR mining approach is approximately ten times lower. The time needed for non-trivial cedent construction, interesting areas identifying and cedents decomposition have to be added to the QAR approach time consumption. Still the time consumption reaches approximately 15% of the classical approach time consumption.

The number of rules is higher using classical approach but number of dependencies described by these rules is almost the same. This means that QAR mining approach reduces the number of redundant rules and contributes to the transparency of generated rules.

The main disadvantage of the proposed algorithm is that it is not based on complete searching. It cannot be guaranteed that all rules which satisfy conditions of *minsupp*, *minconf* and *minlift* are found. Practical experiments demonstrated that in most cases 90-95% of all dependencies are identified. The number of rules to describe the dependecies is significantly lower. Positive characteristics of proposed algorithm can be more valued in working with large databases with high number of attributes or when looking for more complicated dependencies with more attributes on antecedent or succedent side.

Acknowledgement. This work was supported by the Universities Development Fund CR, project nr. 877 /2006.

References

1. AGRAWAL R., IMELISKI T., SWAMI A., *"Mining Association Rules Between Sets of Items in Large Databases"*, In Proc. of ACM SIGMOD Conference on Management of Data, pages 207-216, Washington, D.C., 1993.
2. PIATETSKY-SHAPIRO G., *"Discovery, analysis, and presentation of strong rules"*, in Knowledge Discovery in Databases, Cambridge, MA: AAAI/MIT, 1991, pp. 229-248.
3. SRIKANT R., AGRAWAL R., *"Mining Quantitative Association Rules in Large Relational Databases"*, In Proc. of ACM SIGMOD Conference on Management of Data, Montreal, Canada, 1996.
4. FUKUDA T., MORIMOTO Y., MORISHITA S., TOKUYAMA T., *"Mining Optimized Association Rules for Numeric Attributes"*, In Proc. of ACM SIGMOD Conference on Management of Data, Montreal, Canada, 1996.
5. MILLER R.J., YANG Y., *"Association Rules over Interval Data"*, In Proc. of ACM SIGMOD Conference on Management of Data, Tuscon, AZ, 1997.
6. IMBERMAN S., DOMANSKI B., *"Finding Association Rules from Quantitative Data Using Data Booleanization"*, 1999.
7. WEBB G.I., *"Discovering Associations with Numeric Variables"*, ", In Proc. of ACM SIGMOD Conference on Management of Data, San Francisco, CA, 2001.
8. RASTOGI R. SHIM K., *"Mining Optimized Association Rules with Categorical and Numeric Attributes"*, IEEE Transactions on Knowledge and Data Engineering, vol. 14, no. 1, 2002.
9. GUILLAUME S., *"Discovery of Ordinal Association Rules"*, In Proceedings of the Sixth Pacific-Asia Conference on Knowledge Discovery and Data Mining (PAKDD'02), 322-327, Taipei, Taiwan, 6-8 May 2002.
10. AUMANN Y., LINDELL Y., *"A Statistical Theory for Quantitative Association Rules"*, Journal of Intelligent Information Systems, vol. 20, 255-283, 2003.
11. WIJSEN J., MEERSMAN R., *"On the Complexity of Mining Quantitative Association Rules"*, Data Mining and Knowledge Discovery, 2, 263–281, 1998.
12. DOUGHERTY J., KOHAVI R., SAHAMI M., *Supervised and unsupervised discretization of continuous features*, Proceedings of the Twelfth International Conference on Machine Learning (pp. 194–202), Tahoe City, CA, 1995.
13. STULONG project, WWW page, http://euromise.vse.cz/stulong.
14. KLEMA J., NOVAKOVA L., KAREL M., STEPANKOVA O., *Trend Analysis in Stulong Data*, In Proceedings of ECML/PKDD'04 Discovery Challenge, 2004.

Extraction of Spatial Rules Using a Decision Tree Method: A Case Study in Urban Growth Modeling

Jungyeop Kim[1], Yunhee Kang[2], Sungeon Hong[3], and Soohong Park[1]

[1] Dept. of Geoinformatic Engineering, Inha University, Incheon, Korea
jyfloo@hotmail.com, shpark@inha.ac.kr
[2] Dept. of Computer Science & Information Engineering, Inha University, Incheon, Korea
yjfloo@datamining.inha.ac.kr
[3] Dept. of Welfare-Land Information, Cheongju University, Cheongju, Korea
hongsu2005@cju.ac.kr

Abstract. Data mining refers to the extraction of knowledge from large amounts of data using pattern recognition, statistical methods, and artificial intelligence. The decision tree method, a well-known data mining technique, is used to extract decision rules because using it, we can understand the grounds of classification or prediction. The decision tree method thus can be used to analyze spatial data. There is an enormous amount of spatial data in GIS (Geographic Information System), which has been studied in modeling with these data. Spatial modeling as applied to a decision tree method can be carried out more effectively. In this paper, we extracted spatial rules using the decision tree method, and then applied them to urban growth modeling based on Cellular Automata (CA). An evaluation comparing the model using the decision tree method (the proposed model) with the standard UGM showed that the proposed model is more accurate.

1 Introduction

Simulation modeling is one way of predicting future changes under a plausible scenario using constraints and a relevant data set. Simulation models generally have low prediction accuracy due to the incompleteness of both (simulation) model rules and input data sets. These problems may be partly solved using data mining techniques. There are many data mining techniques available. Some of these include association rule algorithms, genetic algorithms, neural networks, decision tree methods, link analyses, clustering, and fuzzy sets. The decision tree method is one of the techniques that has been used for extracting patterns and discovering features efficiently in large database sets. Accordingly, many research communities have used decision tree methods. Most of the resultant studies have focused on non-spatial data; however, considering the features of the decision tree methods, it is possible to use one of them to model spatial data, particularly in simulation modeling. This paper applies a decision tree method to a (simulation) modeling application, more specifically Cellular Automata (CA)-based urban growth modeling, and evaluates and demonstrates the

B. Gabrys, R.J. Howlett, and L.C. Jain (Eds.): KES 2006, Part I, LNAI 4251, pp. 203–211, 2006.
© Springer-Verlag Berlin Heidelberg 2006

feasibility of the decision tree method to simulation modeling by comparing it to a well known modeling example. CA-based urban growth modeling was chosen for this study because it has recently received much attention from the GIS research community as a new way of modeling urban growth. In addition, the CA-based urban growth models entail universal transition rules that predict or simulate future trends. Transition rules are basically spatial rules and are generally derived from previous research or developed by the researcher in a subjective manner. Thus, constructing transition rules for the CA-based model is one of the most important steps affecting modeling results.

Combining a CA-based urban growth model and decision tree methods also has very significant practical implications. When considering the impacts of rapid urbanization, unemployment, disease and other factors, it is vital for us to analyze how cities have been developed and to predict how cities will be expanded in the near future, in order to establish a plan. As a result, analysis and prediction of urban growth yields important data for urban policies (Kang, Park 2000). Accordingly, there have been many studies conducted on urban growth modeling (Jeong, Lee, Kim 2002; Chen, Gong, He 2002; Choi 2001; Zhang, Wang 2001). However, there is no study on modeling through spatial-rule extraction. The present study aimed both to extract spatial rules using a decision tree method and to show that an urban growth model applying the decision tree type of method is more effective than other models.

2 Cellular Automata (CA) and Decision Tree Algorithm

2.1 CA

Cellular Automata (CA) are both a body of knowledge and set of techniques for solving complex dynamic-systems problems. CA evolve in discrete time steps by updating their local state according to a universal rule that is applied to each cell synchronously at each time step. The value (local state) of each cell is determined by a geometrical configuration of neighbor cells, and is specified in the transition rule. Updated values of individual cells then become the inputs for the next iteration. A Cellular Automaton include four components: a cellular space, local states, neighborhoods and a transition rule. The cellular space is a multidimensional space commonly consisting of triangular, square, and hexagonal shapes. Local states represent the status of each cell encoded by numerical values at a given time step. A neighborhood is defined as a set of cells located around a cell of interest. Transition Rules specify how each automaton is updated according to the defined neighborhood and current states of cells (Park 1997). In Fig 1, the lighter color gray cell within the 3 x 3 window (i.e, the neighborhood) is changed to a darker gray cell in a discrete time step according to the transition rule.

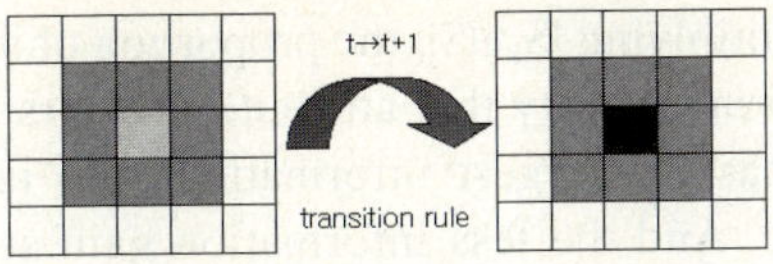

Fig. 1. Concept of Cellular Automata (CA)

2.2 Decision Tree Method

A decision tree method can not only extract satisfactory results under imperfect conditions but also shorten the work time required for establishing a model. Moreover, it helps us to reduce much effort in the data-conversion stage because it can directly deal with continuous values. After all, using a decision tree method, we can easily understand what sorts of attributes have a decisive effect on each category value. We can convert those results into urban growth rules and express them as "If - then" sentences for establishing a model (Jang, Hong, Jang 2002; Han, Kamber 2001).

Algorithm C5.0 is generally used for building a decision tree. The steps are as follows. First, choose one attribute and make a 'root node', and then make trees per each attribute value. Second, under the tree, set the branches that have attribute values among the training set. Third, recursively apply this course to subsets of the training set and stop the application if one class includes all of the training sets. The algorithm thus proceeds. Finally, divide the training set into subsets in accordance with the attribute values and then proceed with division of the subsets according to the other attribute values.

In the decision tree using Algorithm C5.0, each node acquires an information gain of attribute values and then makes a decision tree by using those values. In other words, the attribute that has the biggest information gain becomes the node standing at the highest place. In information theory, we usually use entropy to obtain this information gain. In order to explain this entropy, let us suppose that gathering S consists of examples that belong to n different classes each other. In this case, when we consider p_i as the i times proportion of the classes, the entropy that represents the standard of S' impurity is defined as follows:

$$E(S) \equiv \sum_{i=1}^{n} - p_i \log_2 p_i \tag{1}$$

Using this entropy value, we can find the attribute with which the entropy is the lowest for the classification. In short, it allows us to obtain information gain. We can obtain information gain by dividing S into n in accordance with attribute value A. The formula follows:

$$Gain(A) \equiv E(S) - \sum_{k=1}^{n} E(S_k) \frac{|S_k|}{|S|} \tag{2}$$

In the above formula, E(S) is the S entropy before the division and $E(S_k)$ is the average entropy value of each subset after the division. It is not an arithmetic mean but it is obtained by adding $E(S_k)$, which represents the entropy value of each subset to the computed value when multiplying $|S_k|/|S|$, the proportion of subset in S, by the weight. Gain (A) is the information quantity that attribute A offers for classification. Therefore, the attribute that has the largest information gain is selected as the highest node in the decision tree. And the less information gain an attribute has, the lower position it is in. Fig 2 shows the segment of the decision tree produced by Algorithm C5.0, with the established data. We can apply this decision tree to the proposed model. Land use is the highest node because it has the largest attribute value; the second-highest ranking node is the urbCnt. After all, attributes are assigned their position, from highest to lowest, in accordance with the quantity of the information gain (Fig. 2).

```
landuse =<1 [mode: no] (43888)
 urbCnt =< 4 [mode: no] (37422, 0.872) -> no
 urbCnt > 4 [mode: no] (6466)
  urCnt =< 9 [mode: no] (5068)
   slopeDeg =<44 [mode: no] (4449)
    urbCnt =< 6 [mode: no] (2602)
     distRoadM =< 0 [mode: yes] (550)
      urbCnt =< 5 [mode: no] (331, 0.526) -> no
      urbCnt > 5 [mode: yes] (219)
       slopeDeg =< 12 [mode: no] (70, 0.614) -> no
       slopeDeg > 12 [mode: yes] (149, 0.638) -> yes
     distRoadM > 0 [mode: no] (2052, 0.682) -> no
```

Fig. 2. Decision tree by Algorithm C5.0

3 Urban Growth Modeling Based on CA and Decision Tree

3.1 Model Framework

The model components for this study were constructed in accordance with CA. First, the cellular space that is latticed could have diverse forms but it is usually square and two-dimensional. In our study, when establishing the data, we made the cellular space 100m x 100m because the space must be at least 100m in order to distinguish the urban block.

Second, the specific values we graded represent the local state. In this study, we made layers that represent the following characteristics: urban/non-urban, water bodies (river, lake, and reservoir), distance from transportation (distance-to-road), neighbor urban cell count, slope, and green-belt. And we used them as the values of the local state. We realize that other factors (e.g., cultural, social and politic factors, etc.) have important effects on urban growth, but it is very difficult to establish quantitative data for those. And to quantitatively analyze them is also impossible. Thus, in the present study, we only applied the physical data.

Third, as a method to establish the range of the neighborhood cells, we carried out an experiment on cellular spaces of 3 x 3, 5 x 5, 7 x 7, and 9 x 9 size. As a result, we found that the 5 x 5 size cell offered the results most similar to the real data in the experiment.

Finally, the transition rule becomes the important CA factor that determines how the center of the neighbor will change according to the condition of the neighborhood cells and constraints as time passes from t to t+l.. In this study, we applied the decision tree method by using Algorithm C5.0 to apply the transition rule.

3.2 Data Set

We established data for settlement area, urban area, agricultural area, rest area, road network, water bodies, green-belt, DEM, contour, administrative district and others through four periods (1960s, 1970s, 1980s and 1990s).

We preprocessed the established data and selected that which was suitable to the UGM and proposed model. First, we combined the settlement area and urban area into one urban area. The agricultural area and the rest areas were unified as a non-urban area. This data was used to count the number of urban cells. Second, because the road network data has a direct effect on urban growth, for the experiment, the road network data was used to compute the distance from the road. Finally, the water bodies, green-belt and DEM were used to limit the urban growth. In Korea, the green-belt system has been established since 1971, and we collected green-belt data from that time to the present. After creating the DEM by using a contoured digital topographical map, we used it as slope for the study. Because the elevation value was less changeable over time, we used the DEM values in the 1990s as substitutes for the 1960s, 1970s and 1980s data. Table 1 shows the data that we used in our experiment.

Table 1. Data types and Time

Thematic Data	Scale	Time
urban area	1:50,000	1960s, 1970s, 1980s, 1990s
non-urban area	1:50,000	1960s, 1970s, 1980s, 1990s
road network	1:50,000	1960s, 1970s, 1980s, 1990s
water bodies	1:50,000	1960s, 1970s, 1980s, 1990s
green-belt	1:50,000	1970s, 1980s, 1990s
Administrative district	1:50,000	1960s, 1970s, 1980s, 1990s
DEM	1:25,000	1998

4 Simulation and Results

4.1 Embodiment of Urban Growth Model

In applying the extracted rules to the cell, we covered the whole experimental area. This course is one cycle of a model. Until now, there has been no study that has established the length of a cycle. We tried to determine the correct cycle by comparison with the real data, and as a result, we confirmed that one cycle is similar to one year of urban growth. One cycle indicates one year, and 30 cycles reflect urban growth over 30 years. We used the 1960s data as the input data of the simulation models. After 30 cycles, we could represent the accuracy of our model through experiment results (Table 2). The denominator is the number of urban cells in the actual data, and the numerator is the number of urban cells in the model.

Table 2. Model Comparison

<table>
<tr><th colspan="2">Test \ Time</th><th>1970s</th><th>1980s</th><th>1990s</th></tr>
<tr><td rowspan="2">α</td><td>Proposed Model</td><td>4693/3324 = 1.41</td><td>8916/5002 = 1.78</td><td>10813/8391 = 1.28</td></tr>
<tr><td>UGM</td><td>8257/3324 = 2.48</td><td>9968/5002 = 1.99</td><td>10827/8391 = 1.29</td></tr>
<tr><td rowspan="2">β</td><td>Proposed Model</td><td>2481/3324 = 0.75</td><td>4390/5002 = 0.87</td><td>7276/8391 = 0.87</td></tr>
<tr><td>UGM</td><td>2879/3324 = 0.86</td><td>4365/5002 = 0.87</td><td>7072/8391 = 0.84</td></tr>
<tr><td rowspan="2">γ</td><td>Proposed Model</td><td>2481/5536 = 0.45</td><td>4390/9528 = 0.46</td><td>7276/11928 = 0.61</td></tr>
<tr><td>UGM</td><td>2879/8702 = 0.33</td><td>4365/10605 = 0.41</td><td>7072/12146 = 0.58</td></tr>
</table>

In the proposed model, spatial rules can be applied to a necessary transition rule as components of CA. We used the decision tree method with the attributes land use (landuse), urban count of neighbor (urbCnt), slope (slopeDeg), and distance from road (distRoadM). The decision tree (Fig. 2) constructed through this method was transformed into spatial rules. In Fige 2, 'no' indicates the non-urban area and 'yes' indicates the urban area. We only used 'yes' because the purpose of our study was to extract those cells that are able to grow into an urban area.

For the experiment, we used Clementine 6.5. Clementine 6.5 allocates a level of confidence to each of the rules of the constructed decision tree. Decimal points represent the confidence level (Fig 2). The confidence, as the conditional probability, may be expressed as follows:

$$confidence(A \Rightarrow B) = \frac{A \ and \ B \ transaction}{A \ transaction} \tag{3}$$

In the above formula, the denominator is the number of cells and the numerator is the number of cells that qualified to be an urban area. First, imposed the random numbers were imposed on each of the cells, and then the confidence was compared with the spatial rules. If the value is lower than the confidence, the cell can grow into an urban area. But if higher, it should remain in the present situation. This method improves the realism of the model by adding random factors.

4.2 Urban Growth Simulation

The range of our experiment was from the 1960s to the 1990s. We analyzed the accuracy of the results of the modeling by using the following formulae:

$$\alpha = \frac{urban\ cell\ count\ of\ simulation}{urban\ cell\ count\ of\ actual\ data},\quad \beta = \frac{A \cap B}{B},\quad \gamma = \frac{A \cap B}{A \cup B} \tag{4}$$

where A is the modeled urban area and B is the actual urban area, α is the comparison between the result of the simulation and the actual urban cell, and β is the rate of coincidence between the result of the simulation and the actual urban cell. Lee-Sallee (γ) shows how correctly coincident are the results of modeling with the spatial shape of the actual urban area. Then, we compared the accuracy value with that of the UGM.

According to the results of the UGM, there was a boom of urban growth in the 1970s. As a result, β in the UGM is more correct than in our experiment. But the Lee-Sallee coefficient in our model is more similar to reality than it is that in the UGM (Table 2).

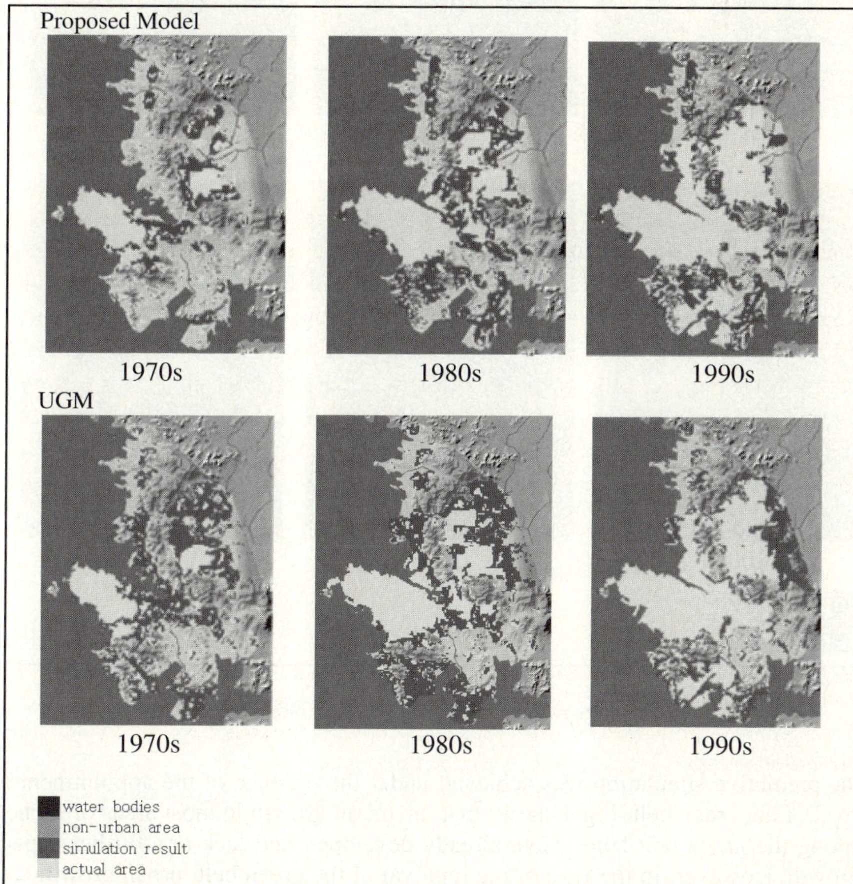

Fig. 3. Comparison of the real data with the simulation results

In the 1980s, both UGM and our proposed model have similar β values, however Lee-Sallee coefficient of our proposed model is slightly lower than that of UGM. This result also repeats in the 1990s. In comparison, our model represents the better results, and thus it is more suitable for estimation of urban growth (Fig 3; Table 2). We found that our model is effective when extracting spatial rules in areas that cannot reveal the characteristic of urban growth.

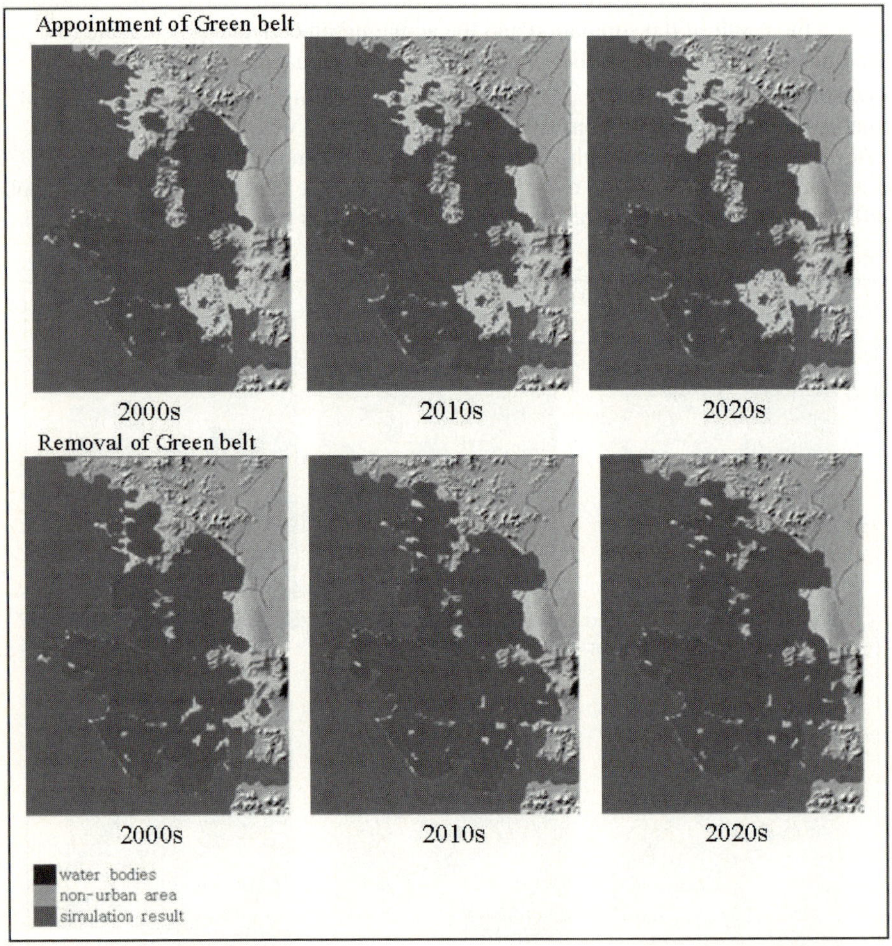

Fig. 4. Urban growth simulation from the 1990s to the 2020s

The predictive simulation was achieved under the premise of the appointment and removal of the green belt. Fig 4 shows that the urban growth in most areas of Incheon, excepting the green-belt zone, have already developed and lack any further capacity for growth. However, in the case of the removal of the green belt, urban growth could be continuous through 2010, excepting some mountainous areas.

5 Conclusions

This study aimed to create a more realistic urban growth model by directly extracting spatial rules using a decision tree method and adapting them to an urban growth model. To accomplish this, first we selected Incheon for the experiment area and established data for 30 years. Second, we extracted the spatial rules by using the decision tree method in order to understand the patterns that Incheon has exhibited in urban development over those 30 years. Finally, we adapted the rules to the our model and proved that our experimental model coincides with the data of the pattern of past urban growth.

When we compared our experimental result with the present UGM method by statistical analysis, we found that our experimental result was more correct than that of the present UGM. In other words, we could verify that our experimental result presents the more realistic urban growth pattern in comparison with the real data and other statistics.

Urban growth is affected by different factors (e.g. social and political factors.) Because it is impossible to establish data for those factors and to evaluate them, this study employed only a physical-data model. Despite this, our model could produce results similar to the real data by extracting spatial rules, using the decision tree method, suitable for the experimental areas. Therefore, we expect that our new model, using decision tree, will be adaptable to other areas in which data has not been completely established.

References

1. Chang-young Choi: Urban Growth Modelling by Cellular Automata. M.A. dissertation. Graduate School Gyeongsang National University (2001).
2. Jae-Joon Jeong, Chang-Moo Lee, Yong-Il Kim: Developments of Cellular Automata Model for the Urban Growth. Korean Planner Association. Vol 37. No 1 (2002).
3. Jiawei Han, Micheline Kamber: Data Mining Concepts and Techniques. MORGAN KAUFMANN PUBLISHERS (2001).
4. Jin Chen, peng Gong, Chunyang He, Wei Luo, Masayuki Tamura, and Peijun Sho: Assessment of the Urban Development Plan of Beijing by Using a CA-Based Urban Growth Model. Photogrammetric Engineering & Remote Sensing. Vol 68. NO 10 (2002) 1063-1071.
5. Nam-Sik Jang, Sung-Wann Hong, Jae-Ho Jang: DataMining. Daechung media (2002).
6. Soohong Park: Design and implementation of An Integrated CA-GIS System. The Journal of Geographic Information System Association of Korea. Vol 9. No 2 (2001) 185-206.
7. Soohong Park, Yong-Gin Joo, and Yun-Ho Shin: Design and Development of a Spatio-Temporal GIS Database for Urban Growth Modeling and Prediction. The Geographical Journal of Korea. Vol 36. No 4 (2002) 313-326.
8. Xinsheng Zhang and Yeqiao Wang: Spatial Dynamic Modeling for Urban Development. Photogrammetric Engineering & Remote Sensing. Vol 67. NO 9 (2001) 1049-1057.
9. Youngok kang,Soohong Park: A Study on the Urban Growth Forecasting for the Seoul Metropolitan Area. Journal of the Korean Geographical Society (2000) 621-639.
10. http://www.ncgia.ucsb.edu/projects/gig/.

Classification Using Multiple and Negative Target Rules

Jiuyong Li and Jason Jones

Department of Mathematics and Computing,
University of Southern Queensland,
Australia, 4350
jiuyong@usq.edu.au, jonesj@usq.edu.au

Abstract. Rules are a type of human-understandable knowledge, and rule-based methods are very popular in building decision support systems. However, most current rule based classification systems build small classifiers where no rules account for exceptional instances and a default prediction plays a major role in the prediction. In this paper, we discuss two schemes to build rule based classifiers using multiple and negative target rules. In such schemes, negative rules pick up exceptional instances and multiple rules provide alternative predictions. The default prediction is removed and hence all predictions relate to rules providing explanations for the predictions. One risk for building a large rule based classifier is that it may overfit training data and results in low predictive accuracy. We show experimentally that one classifier is more accurate than a benchmark rule based classifier, C4.5rules.

Keywords: classification, association rule, negative and multiple rule.

1 Introduction

Rules are a type of human-understandable knowledge, and therefore rule-based methods are very popular in building decision support systems. Last twenty years saw a lot of rule classification systems, such as, the covering algorithm based systems, e.g. AQ15 [8], CN2 [2,3], decision tree based systems, e.g. C4.5rules [9], and association rule based systems, e.g. CBA [6] and CMAR [5].

Apart from some routine processes, such as data preprocessing and result presentation, rule based classification usually involves two stages, training and test. Consider a relational data set where each record is assigned a category (class), called a training data set. In the training stage, a rule set is generated from the training data set. Each rule associates a pattern with a class. In the test stage, rules are used to predict classes of records that have no class information. If the predictive class is the class that a record is supposed to belong to, then the prediction is correct. Otherwise, the prediction is wrong. The proportional of correct predictions on the test data is the accuracy of a classifier.

A classifier refers to a rule set and the mechanism to make predictions. Highly accurate classifiers are generally preferred. One commonly used rule based classifier is ordered rule based classifiers. Rules are organised as a sequence, e.g. in the

B. Gabrys, R.J. Howlett, and L.C. Jain (Eds.): KES 2006, Part I, LNAI 4251, pp. 212–219, 2006.
© Springer-Verlag Berlin Heidelberg 2006

descending accuracy order. When it classifies a coming record, the first matching rule in the sequence makes the prediction. This sequence is usually tailed by a default class. When there is no rules in the sequence matching the coming record, the class of the record is predicted as the default one. C4.5rules [9] and CBA [6] employ this model.

An ordered rule based classifier is simple and effective. Its predictions are easy to be interpreted and its wrong predictions are easy to be traced down since only one rule is used. It makes a prediction based on the most likelihood. This is because that a rule with higher accuracy usually precede a rule with lower accuracy and the accuracy approximates the conditional probability when a data set is big.

Traditional classification problems normally involve two classes and hence a prediction is either one class or the other class. When the number of classes is big, it is difficult to predict all instances to one class. One reason is that some patterns are association with multiple classes. If we restrict predictive class to one, this results in low accuracy of some rules. These low accurate rules make predictions unreliable and lose their utility significantly. It is desirable to have some multiple target rules to overcome this drawback of single target rules. Practically, predicting an instance to belong to two classes out of ten possible classes with 90% accuracy is more useful than predicting it to belong to one class with 50% accuracy.

In additional, predictions made by the default class may be misleading. For example, in data set Hypothyroid, 95.2% records belong to class Negative and only 4.8 % records belong to class Hypothyroid. So, if we set the default prediction as Negative, then a classifier that has no rule will give 95.2% accuracy. You can see that how accuracy is floated by the default prediction. Further, this distribution knowledge is too general to be useful. For example, a doctor uses his patient data to build a rule based diagnosis system. 95% patients coming to see him are healthy, and hence the system sets the default as healthy. Though the default easily picks up 95% accuracy, this accuracy is meaningless for the doctor.

In this paper, we propose to use negative and multiple rules to build rule based classifiers. Negative rules summarise exceptional instances of low accurate rules and multiple target rules provide alternative predictions. We also drop the misleading default predictions in the classifiers. A risk of building a classifier from a large rule set is that it may reduce the accuracy of the classifier because a large model usually tends to overfit data. We experimentally demonstrate that the new schemes do not overfit data, and can improve accuracy of classifiers over a benchmark rule based classifier, C4.5rules.

2 Multiple and Negative Rule Based Classifier

2.1 Multiple and Negative Rules

Let us start with an example. We observer that 60% customers buying product a also buy product b, and hence summarise this phenomenon as rule $a \rightarrow b$.

Afterwards, when a customer put product a in his/her shopping trolley, he/she is recommended to buy product b. Should store mangers promote product b targeting this group of customers buying product a? We look at some possible consequences. There may be 10% customers buying product a hate the product b, and therefore this promotion angers these customers. Further, other 30% customers may be annoyed since they are not interested in product b at all.

A solution is to find the 10% customers and exclude them from the the promotion list; and to find another product that is of interest to the 30% of customers and bind two products in one promotion. This solution minimises the probability of annoying customers. This is the basic idea for negative and multiple target rules.

Many real world problems have a number of classes. In some cases, a pattern is association with two or three classes, and it is impossible to have an accurate rule to associate an instance to one class. Therefore it is necessary to extend the traditional classification rules to target multiple classes. Practically, these rules are interesting if they restrict the choices to two or three among ten or twenty classes. In e-commence applications the number of classes may be thousands.

For example, there are ten classes in a data set, and we have two rules $A \rightarrow c_1$ (conf $= 48\%$) and $A \rightarrow c_2$ (conf $= 42\%$). Either rule is accurate enough to be a rule alone, but the joint rule $a \rightarrow c_1 \vee c_2$ has the confidence of 90% and is an accurate rule. Predictively, eliminating eight of the potential ten classes as outcomes means that the rule is still informative.

Multiple target rules are different from general association rules. The consequent of a general association rule can be a number of conjunctive items, but the consequent of a multiple target rule is a number of disjunctive items.

Almost all statements have exceptions, and rules have exceptions too. Rule $AX \rightarrow \neg c_1$ is an exceptional rule of rule $A \rightarrow c_1$. This means that pattern A generally associates with class c_1 but does not when it occurs with pattern X. Using negative rules to filter out exceptional instances of a regular rule can increase the accuracy of the regular rule.

For example, $A \rightarrow c_1$ $(\text{supp} = 15\%, \text{conf} = 50\%)$ and $AX \rightarrow \neg c_1$ $(\text{supp} = 5\%, \text{conf} = 100\%)$. The combination of two rules will produce the confidence of 75% since 5% exceptional instances are removed from the former rule.

2.2 Two MTNT Classifiers

We will present two classifier models with multiple and negative rules in this section.

The first model (MTNT1) does not directly use multiple and negative target rules for making prediction, but simply provides some backup explanations for low accurate rules. A low accurate rule is coupled by some negative and multiple target rules, and as a result users will be aware of possible adverse effects of the prediction made by the rule and possible alternative prediction.

In MTNT1, all regular rules are sorted by their estimated predictive accuracy and there is no default class at the end of this rule sequence. Confidence is accuracy on the training data, but accuracy on test data is required for prediction. We use method in [4] to estimate predictive accuracy of a rule.

In classification, the first matching regular rule classifies a record. Multiple and negative target rules do not participate in the classification, but provide some backup explanations. An example of MTNT1 classifier is shown in Figure 1.

Rule 1: $A \rightarrow c_1$		acc = 95%
Rule 2: $B \rightarrow c_2$		acc = 92%
...	...	...
...	...	...
Rule 11: $D \rightarrow c_5$		acc = 70%
$DE \rightarrow \neg c_5$		
$DF \rightarrow \neg c_5$		
...	...	...
$D \rightarrow c_2 \vee c_5$		
Rule 12: ...		...
...	...	...
...	...	...
No default class		

Fig. 1. An example for MTNT1 classifier. Rules are sorted by their estimated accuracy and low accurate rules are coupled by multiple and negative rules for better understanding. For example, c_2 is an alternative prediction for Rule 11 and some exceptional instances of Rule 11 are explained by the following negative rules.

The second model (MTNT2) makes use of multiple and negative target rules in predictions. We use a covering algorithm based method to sort regular rules in classifier as C4.5rules [9] and CBA [6]. The rule with the fewest false positive errors is put the first. Then all covered records by the selected rule are removed, and false positive errors are recomputed for the remaining rules. This procedure repeats until there is no records or rules left. Unlike C4.5rules [9] and CBA [6], there is no default class. Remaining rules are appended to this sequence in the accuracy decreasing order.

Multiple and negative rules are inserted in the classifier in the following way. Negative target rules are put before their corresponding regular rule to filter exceptional instances. All multiple target rules are appended to the end in the accuracy decreasing order. An example of MTNT2 is shown in Figure 2.

In classification, for a record to be classified, it is compared with rules from the top to the bottom. If a matching rule is a single target rule without any negative rules, then the rule classifies it. If the matching rule has negative rules, we move into the negative rule subset and match each of them. If no negative rule matches the record, then the rule classifies it. Otherwise, no prediction is made. We move on to the next rule and repeat the process.

Exceptional instances of a regular rule are removed by the negative rules. As a result, the rule group, including a regular rule and some negative rules,

Rule 1: $A \rightarrow c_1$		acc = %95
Rule 2: $B \rightarrow c_2$		acc = %92
...	...	...
...	...	...
Rule 11: $DE \rightarrow \neg c_5$		
$DF \rightarrow \neg c_5$		
$D \rightarrow c_5$		acc = %70
...	...	...
...	...	...
Rule 49: $E \rightarrow c_2 \vee c_3$		acc = %92
Rule 50: $D \rightarrow c_2 \vee c_5$		acc = %90
...	...	...
...	...	...
No default class		

Fig. 2. An example for MTNT2 classifier. Regular rules are ordered by a covering algorithm based method to minimise false positive errors. Negative target rules filter the exceptional instances for the following regular rule, and as a result the regular rule makes more accurate predictions. All multiple target rules reside at the end in the accuracy decreasing order to make alternative predictions when an accurate prediction is impossible.

is more accurate than the single regular rule. Alternative predictions are made by multiple target rules. In some cases, we may not define a coming record in a single class. It is still useful to classify it into two or three possible classes when the number of all possible classes is big. Therefore, we put a set of multiple target rules at the end of classifier to for this purpose.

This classifier sometimes results in the classification of two or more classes simultaneously, however, we do not consider this as the conflicting prediction. When the number of classes is big, this prediction narrows the choice to a smaller number, e.g. 2. This prediction is useful since it excludes many other choices. This prediction is not as accurate as the prediction of one class, but in some cases it is impossible to predict one class accurately. We scale down the predictive accuracy of multiple target rules and details are presented in the following section.

3 Experimental Results

In the previous section, we mainly discuss how to incorporate multiple and negative target rules in a classifier to improve the explanatory ability of a rule based classifier. We also drop the default prediction in the classifier so that every prediction relates to rules. However, a major concern is that a large classifier may overfit data and reduce its predictive accuracy. In this section, we will show that our classifiers do not sacrifice the accuracy of classifiers.

We carried out experiments by using 10-fold cross validation on 7 data sets from the UCI Machine Learning Repository [1]. We chose them since they contain 4 or more classes each. Multiple and negative target rules are not suitable for two-class data sets.

We compare the MTNT classifiers with c4.5rules [9]. We chose c4.5rules since we are able to modify the code to drop its default prediction. In addition, nearly all new classifiers have been compared with C4.5, and hence interesting readers can compare the MTNT classifiers with other classifiers indirectly.

In the experiments, we used the local support. The *local support* of rule $A \rightarrow c$ is $\text{supp}(Ac)/\text{supp}(c)$. It avoids too many rules in the large distributed classes and too few rules in the small distributed classes. We explored negative rules to depth 3, and therefore the maximum length of any negative rule is the length of regular + 3. The maximum classification rule length was set as 6, the minimum accuracy threshold was set as 50%, the high accuracy threshold was set as 90%, and the number of multiple targets was set as 2.

There is no default prediction for MTNT classifiers. If no one rule matches a record, an error is counted.

A brief description of data sets and classifiers is given in Table 1. MTNT classifiers are four times larger than C4.5rules on average. This is because the default prediction is a vital part in a C4.5rules classifier. The construction of C4.5rules classifiers has a post-pruning process. All rules and the default prediction are considered as a whole. Any rule that does not contribute to increase the accuracy is eliminated. Removed rules may not be as good as the default in terms of accuracy, but they provide direct reasons for correct or wrong predictions. In other words, they make a classifier more understandable. Therefore, we keep larger classifiers.

On average, a regular rule in MTNT classifiers has 2 negative target rules and four regular rules have 1 multiple target rule. We did not set the minimum support requirement for negative target rules and therefore their number is comparatively big. In contrast,the number of the multiple target rules is small since the they have to reach the high accuracy threshold.

Table 1. A brief description of data sets and classifiers. In classifier size columns, (M) means #multiple rules, (N) means #negative rules, and no symbol means #regular rules.

Data sets			classifier size	
Name	#records	#classes	MTNT	C4.5rules
anneal	898	5	42 + 30(M) + 51(N)	22
auto	204	7	50 + 6(M) + 244(N)	27
glass	214	7	29 + 2(M) + 21(N)	12
led7	3200	10	161 + 32(M) + 76(N)	32
lymph	148	4	33 + 10(M) + 123(N)	11
vehicle	846	4	242 + 42(M) + 510(N)	47
zoo	101	7	8 + 6(M) + 7(N)	9
Average	n/a	n/a	80 + 18(M) + 147(N)	23

Table 2. Accuracy of different classifiers (in %)

data set name	MTNT1 no default	MTNT2 no default	C4.5rules no default	C4.5rules default
anneal	95.6	96.9	90.3	93.5
auto	70.3	77.5	75.1	78.0
glass	71.6	74.9	63.6	72.5
led7	72.0	73.9	73.2	73.2
lymph	80.5	83.1	73.1	78.4
vehicle	69.8	70.6	67.3	71.9
zoo	93.1	94.8	92.1	92.1
Average	79.0	81.6	76.4	79.9

To have a fair comparison, predictions made by multiple target rules are scaled down so that they scored $1 - \frac{Num\ Target}{Max\ Class}$ for a correct prediction instead of 1. For example, for a data set with 4 classes, a multiple target rule with two classes makes a correct prediction. We consider this as a 0.5 correct prediction instead of 1. Therefore, we do not expect the multiple target rules to increase the accuracy of classifiers, but to improve the understandability of the predictions.

MTNT2 is more accurate than C4.5rules (with default) and MTNT1 is nearly as accurate as C4.5rules (with default), as shown in Table 2. Consider both MTNT classifiers do not include the default prediction. The MTNT classifiers have successfully dropped the default prediction while maintaining accuracy. The default is replaced by some additional rules, which make correct or wrong predictions directly associate with rules.

We are also aware that a number of new rule based classifiers, e.g. CBA [6] and its enhancement [7], have been proposed. They are more accurate than the C4.5rules. However, they make use of the default prediction, a factor that reduces the explanatory ability of a rule based classifier. We did not compare with them since we are unable to drop their default predictions.

4 Conclusions

In this paper, We proposed to incorporate multiple and negative rules in rule based classifiers. Negative rules are used to pick up exceptional instances and multiple rules are used to provide alternative predictions. They improve explanatory ability of predictions made by low accurate rules by characterising their exceptional instances and providing alternatives. We proposed two schemes to bind the multiple and negative rules to classifiers and remove the default prediction. We experimentally shows that one proposed classifier is more accurate than a benchmark rule based classifier, C4.5rules.

The utility of multiple and negative rules has strong practical implication in eCommerce applications, e.g. target-commercial and personalization. In these applications, the number of possible targets is very big, and rules are usually low accurate. The utility of multiple and negative rules is an approach to obtain more certain information in these data.

Acknowledgement

This project has been partially supported by Australian Research Council Discovery Grant DP0559090.

References

1. E. K. C. Blake and C. J. Merz. UCI repository of machine learning databases, http://www.ics.uci.edu/~mlearn/MLRepository.html, 1998.
2. P. Clark and R. Boswell. Rule induction with CN2: Some recent improvements. In *Machine Learning - EWSL-91*, pages 151–163, 1991.
3. P. Clark and T. Niblett. The CN2 induction algorithm. *Machine Learning*, 3(4):261–283, 1989.
4. J. Li, H. Shen, and R. Topor. Mining the optimal class association rule set. *Knowledge-Based System*, 15(7):399–405, 2002.
5. W. Li, J. Han, and J. Pei. CMAR: Accurate and efficient classification based on multiple class-association rules. In *Proceedings 2001 IEEE International Conference on Data Mining (ICDM 2001)*, pages 369–376. IEEE Computer Society Press, 2001.
6. B. Liu, W. Hsu, and Y. Ma. Integrating classification and association rule mining. In *Proceedings of the Fourth International Conference on Knowledge Discovery and Data Mining (KDD-98)*, pages 27–31, 1998.
7. B. Liu, Y. Ma, and C. Wong. Improving an association rule based classifier. In *4th European Conference on Principles and Practice of Knowledge Discovery in Databases PKDD*, pages 504–509, 2000.
8. R. Michalski, I. Mozetic, J. Hong, and N. Lavrac. The AQ15 inductive learning system: an overview and experiments. In *Proceedings of IMAL 1986*, Orsay, 1986. Université de Paris-Sud.
9. J. R. Quinlan. *C4.5: Programs for Machine Learning*. Morgan Kaufmann, San Mateo, CA, 1993.

Coherent Formation for Agents Using Flocking with Cellular Automata

Leo Bi and Richard Hall

Dept. Computer Science & Engineering,
La Trobe University, Melbourne,
3086, Victoria, Australia
leobi@msn.com

Abstract. While flocking assists an agent to coordinate its movement locally, it cannnot steer agents towards a particular destination in a coherent formation. We augment flocking by situating each agent in a cellular automata with local relations to neighbouring agents (which impact steering); coherent formation is thus constructed bottom-up. We implement our agents in the massive multiplayer online role playing game Lineage II in Java, and evaluate our formations with respect to coherence under different terrains. Applications include computer games, computer-generated movies and robots in hazardous environments.

1 Introduction

Several distributed and centralised approaches compute the simultaneous coherent motion of multiple entities; typical applications include crowd and army simulation in virtual or robotic environments [1]. *Distributed* approaches (eg. flocking [2,3], grid searching [4], and social potential field [5]) achieve local collision avoidance, but groups can easily break up and become stuck in certain terrain. On the other hand, *Centralised* approaches (eg. randomised path planner [6] and probabilistic road map [7]) scale poorly to high numbers of agents and cease functioning in specific types of terrain [8].

In our distributed coherent formation approach an agent is positioned in a virtual cellular automata (links communicate location); next internal state is determined by current internal state and the state of neighbouring nodes [9]. These cellular automata have been designed to interact in such a way that a coherent formation of agents is produced bottom-up (ala amorphous computing [10]); coherent formation is not an emergent property of our cellular automata (ala artificial life [11,12]). We represent both square and triangle formations.

The rest of this paper is as follows. In section 2 we discuss requirements of each agent, and describe their implementation in section 3. In section 4 we evaluate our formation's coherence with respect to different leader movements and terrain. Finally, we draw conclusions and describe future work in section 5.

B. Gabrys, R.J. Howlett, and L.C. Jain (Eds.): KES 2006, Part I, LNAI 4251, pp. 220–227, 2006.
© Springer-Verlag Berlin Heidelberg 2006

2 Requirements

Our agents had eight requirements:

1. *Collision avoidance.* In the real world (and in realistic virtual worlds), object collisions causes damage. In order to avoiding damaging themselves, agents will attempt to move around (as opposed to into) other objects.
2. *Movement towards a target destination.* The target destination of agents will depend upon their role. Our agents will have one of two destinations, depending on whether they are leader or follower agents. A *leader* agent will select a point in the world map and move towards that point. A *follower* agent will keep a distance and angle from the agent which it is following.
3. *Relationships with other agents.* Most agents in formation (see Figure 1) have a *parent* agent (the agent they follow) and *children* agent(s), the agents that are following them. There are two exceptions: leader agents (no parent); and leaf agents (right at the back of the formation thus cannot have children).

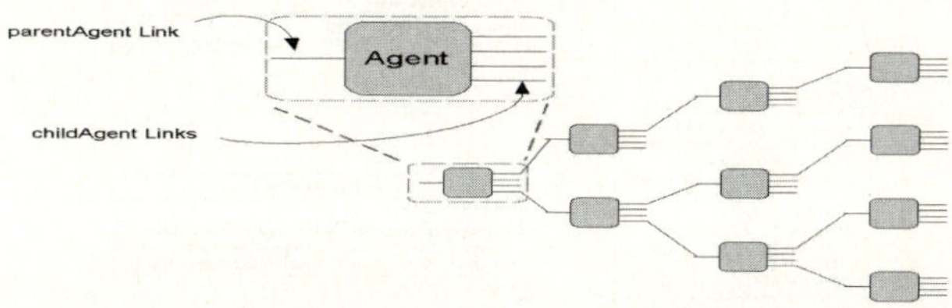

Fig. 1. Agent relations

4. *Formation flexibility.* Followers will use the leader's formation specification, and will have a minimal number of parameters to permit both square and triangle formations.
5. *Maintain coherency through obstacles.* The formation will maintain coherency amidst obstacles as much as possible and regroup as soon as possible.
6. *Maintain coherency despite decimation.* The formation will regroup and continue despite agents being killed off (ala military simulation).
7. *Asynchronous communication.* Regardless of when agents are allocated cycles, agents will be able to (depending on the situation) associate (request to adopt free agents) and disassociate (turn their children into free agents).
8. *Small overhead.* In a virtual environment, the more memory and compute required per agent means less agents can be supported by the system as massive multiplayer online game represent agents in real-time.

These requirements approximate those of several previous distributed approaches to coherent formation: agents are asynchronous, identical, memoryless, anonymous, and non-communicative [13,14,15] - we maintain the asynchronous requirement. Our agents will be identical except for the leader agent and will have no memory of past events. They will remember their relationships which are facilitated by simple communication.

3 Coherent Formation Simulaton

Our simulation was developed bottom–up in three stages (in light of our requirements) which are represented in our simulation class diagram (see Figure 2). Firstly, we made agents follow a player-directed agent while avoiding obstacles (see subsection 3.1). Secondly, we gave agents the ability to make and break relationships with each other and follow each other in a simple chain (see subsection 3.2). Finally, we created the leader agent, as distinct from the follower agent, and gave it the ability to impose a formation specification on its following agents (see subsection 3.3).

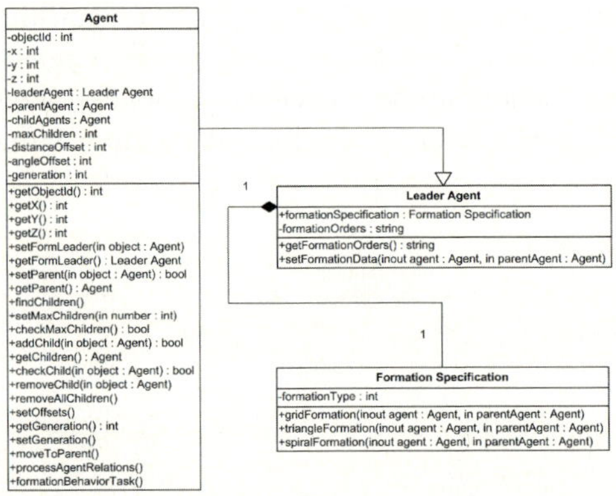

Fig. 2. Simulation Class Diagram

We implemented these classes in the free open-source server *L2J* which works with the massive multiplayer online role playing game Lineage II by NCsoft. Agents in formation achieve communication by accessing relevant agent's attributes at least once per second, scheduled randomly by the game engine.

3.1 Agent Navigation

For each agent, other objects in the world act as either movement repulsors or attractors (ala flocking [3]). There can be a large number of *repulsors* for an agent, defined to be any non-parent object within a local (hand-tuned) radius. For each of these objects the agent constructs a euclidean distance vector, weighted by closeness, then sum these vectors into a repulsor vector.

For each agent there is one *attractor*; for follower agents it is a specific locationwith respect to their parent's location in terms of distance and angular relation to the parent's bearing. Agents orient their bearing identically with their parent's bearing so that in our 3D virtual environment agents are 'looking' the same way. The main reason for matching orientation is because when both agents are moving at the same speed, the distance between parent and children can be maintained (its also aesthetic). With *leaders*, their attractor is defined to be any place or object othen than agents in their formation. We felt it was against the spirit of the formation to create a loop whereby leaders follow their own leaf agents. However, leaders can follow a leaf agent in another formation, thus formations of formations can be constructed bottom-up.

Once both attractors and repulsors have been calculated, their values are normalised and summed into an agent's movement vector. They are normalised because the repulsors collectively are as important as an agent's attractor. This movement vector is discretised to our virtual environment's coordinate space.

3.2 Agent Relationships

Agents have several criteria that they need to meet in order to become either parents or children. To become a *parent*, all agents (bar the leader) must have a parent. In addition, an agent must have empty child slots (all agents are limited to the same small number of children to maintain regular order in the formation). Conversely, existing parents must cease being parents if they lose their parent. Where both these conditions are satisfied, an agent can query all other agents within a certain radius and adopt free agents as children.

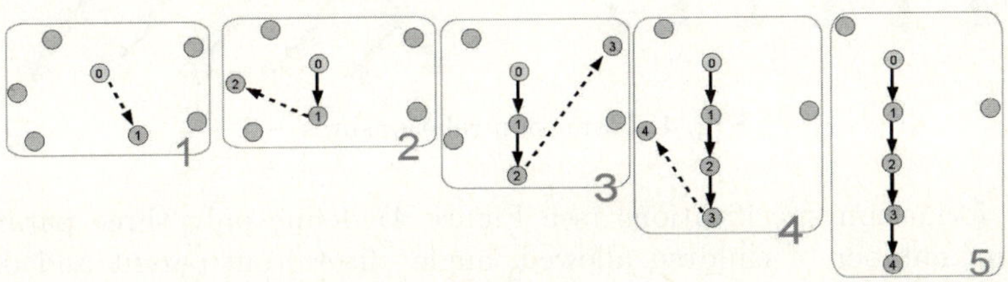

Fig. 3. Single-file forming

In order to become a *child*, an agent has to satisfy five criteria: it must be an agent which can join this (or any) formation; the parent must have a child slot available for it; it must not have a parent already; it cannot be a child of itself; and it must not share the same leader and have a lower generation. A 'generation' describes the rank of an agent (a leader is generation 0, the rank of agents behind the leader is generation 1 etc.) Thus if an agent with a lower generation was to become a child of a higher generation, it could cause a loop. Such loops can occur (without the latter criteria) when a lower generation agent is decimated (eg. killed in battle) and there are suddenly a number of agents asynchronously associating and disassociating with children.

Once an agent satisfies the criteria to become a child it needs to establish itself in five ways. Firstly, it needs to move at the same speed as its parent, to maintain its distance. Secondly, it needs to set its generation equal to its parents generation+1 to assist formation changes. Thirdly, it needs to be able to recognise its parent so it knows which agent it is to follow. Fourthly, it needs to recognise its leader so that it can identify which formation it previously belonged to if it becomes separated (loyalty). With no other information, a group of agents can form a simple single-file formation (see Figure 3).

3.3 Leader Relationships

While all followers know their leader's identity, a leader does more than simply standing at the front of the formation. Leaders contain the formation specification which they communicate directly to new children. Our representation thus diverges minimally from (true) cellular automata as it should request this information from its parent directly. It could be stored in each agent, but it would be identical for all agents in the one formation and a waste of memory for the many agents (eg. vendors and quest-givers) who would never participate in formations. Since formations need to continue if a leader is decimated, the leader is controlled by a virtual agent (immortal and invisible).

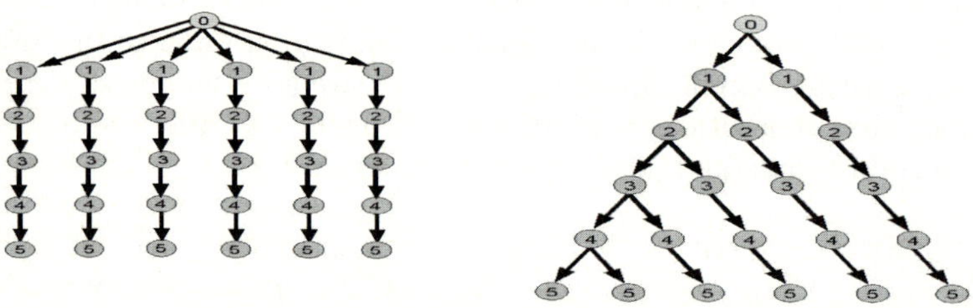

Fig. 4. Formation relationships

Our formation specifications (see Figure 4) define only three parameters: maximum number of children allowed, angle offset from parent and distance to parent. With respect to *number of children*, regardless of formation, agents with a generation equal to the maximum formation depth receive a zero child allowance. With the grid formation, the leader is allowed six children, whereas all other (non-leaf) parents are allowed one child. With the triangle formation, one child of each generation is allowed two children while the rest have one.

Angle offsets from parents are defined differently depending on the formation. With the grid formation, all agents of generation 1 have a specific angle with respect to the leader, but all other agents follow directly behind their parents. Agents of generation 1 are told what their angle should be; thus leaders keep track of column leaders. With the triangle formation, the angle depends on the number of existing children. The first child follows behind and to the right, the second child behind and to the left. Agents simply calculate their angle depending on whether they are the first or second child.

4 Evaluation

We evaluated our grid formation with respect to coherence in free-space and cluttered terrains. With *free-space* terrain we calculate (averaged over three runs), how long it takes the formation to come to a complete halt behind the leader after the leader has halted (stop times). We tested grids of width 6 and depths 4,6, and 9 agents respectively. The leader either runs (in a straight line) for a certain number of paces or rotates on the spot. With rotations, we measured the time taken for the agents to have completely reformed behind the leader if the leader turns either 90° or 180°. We took three samples for each rotation, reporting the average and the interval between the maximum and minimum stop times. Our samples in Figure 5 were manually with a stop watch as system timing information was unavailable while using Lineage II.

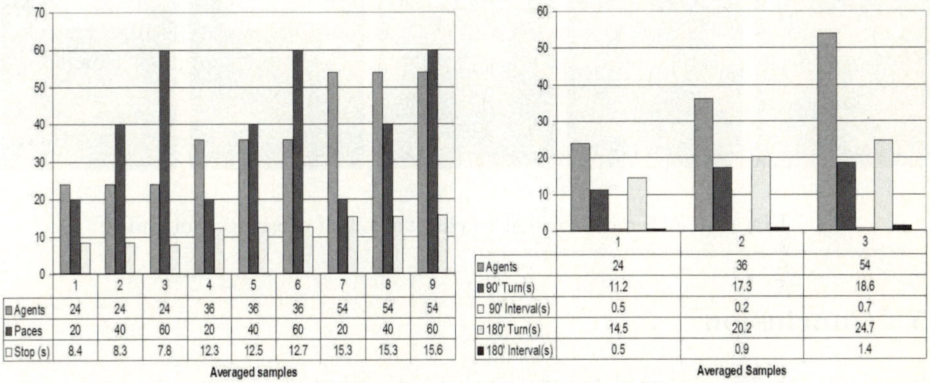

Fig. 5. Stop times for straight line (left) and rotation (right)

We also evaluated whether our agents stayed in formation despite *cluttered terrain*. Figure 6 shows a grid formation moving through rocks (top set) and through itself (bottom set). Leftmost images show where the leader is about to go; rightmost images show regrouped agents; middle images are snapshots of follower agents interpolated between these endpoint images.

Our evaluation provides evidence that our agent formations largely satisfy our requirements. With respect to *free-space* straight-line stop times, it appears that for a specific number of followers, stop times are the same, independent of distance. With respect to rotation stop times, it appears that there is very little variation for a specific number of followers. For both stop times, there appears to be a non-polynomial relationship to the number of agents. On the other hand, with respect to *cluttered terrains*, our agents were successful in maintaining formation despite static or dynamic obstacles; we lost no agents from our formations on any of the terrains we tested.

We also wanted to evaluate the scalability of our architecture, either locally or online, but circumstances were against us. The Windows task manager showed no noticeable difference with our largest group of agents (54), but we were unable to determine how many agents were required to hit computational limits; agents are spawned randomly on a massive map. In addition, we were unable to evaluate our architecture online because a new client version of the game was released during our development with which our server was incompatible. Coding the L2J server is complex; there are more than 20,000 source files.

Fig. 6. Cluttered terrains: ro cks (top), self-crossing (bottom)

5 Conclusion

In this paper we presented an approach to distributed motion planning where agents group together in a coherent formation and implemented this approach in the virtual environment L2J. We ran several real-time tests to evaluate coherency; we investigated different numbers of agents following the leader in a straight line, following a rotating leader; and navigating through difficult terrain. We also tested the formations ability to repair after an agent was deleted (data not shown); in all of our tests agents robustly maintained formation.

Our approach was based on a hybrid of flocking with cellular automata. Navigation was achieved via summing attractors and repulsors, whereas the internal state of agents was modified by nearest-neighbour interactions. Out approach provided several features: collision avoidance; steering; following relationships; multiple formation types; and coherency despite obstacles and decimation. It seems to be quite suitable for asynchronous virtual environments due to its processing and memory overheads (at least for small formations). Using cellular automata provided several advantages: it allows our formations to temporarily deform around obstacles as necessary; no global path for the group is planned; coherency is maintained irrespective of group size; and we can construct formations of formations. Future work will include creating formations of different shapes, looking for optimisations, and evaluating the success of our approach in other environments.

Acknowledgement This research was inspired by Professor Harold Abelson (MIT) who gave a talk on amorphous computing at the Department of Computer Science and Computer Engineering at La Trobe University in December 2004.

References

1. Fujimura, K.: Motion Planning in Dynamic Environments. Springer Verlag (1991)
2. Reynolds, C.W.: Flocks, herds and schools: A distributed behavioral model. Proceeding of SIGGRAPH '87 (1987)
3. Reynolds, C.W.: Steering behaviors for autonomous characters. GDC 1999 (1999)
4. Russell, S., Norvig, P.: Artificial Intelligence: A Modern Approach. Prentice-Hall, Inc (2003)
5. Reif, J., Wang, H.: Social potential fields: A distributed behavioral control for autonoomous robots (1995)
6. Barraquand, J., Latombe, J.: Robot motion planning: A distributed representation approach. International Journal of Robotics Research **10** (1991) 628–649
7. Kavraki, L., Latombe, J.: Probabilistic roadmaps for robot path planning (1998)
8. Kamphuis, A., Overmars, M.: Motion planning for coherent groups of entities. IEEE Int. Conf. on Robotics and Automation (2004)
9. Langdon, W.: Genetic programming and data structures. University College, London (1996)
10. Abelson, H., Allen, D., Coore, D., Hanson, C., Homsy, G., Knight, T.F., Nagpal, R., Rauch, E., Sussman, G.J., Weiss, R.: Amorphous computing. Communications of the ACM **43**(5) (2000) 74–82
11. Langton, C.G.: Studying artificial life with cellular automata. Physica D **22** (1986) 120–149
12. Langton, C.: Artificial Life. In C. Langton. (ed) Artificial Life, Santa Fe Institute Studies in the Sciences of Complexity, Addison-Wesley (1989)
13. Gervasi, V., Prencipe, G.: Flocking by a set of autonomous mobile robots. Technical Report TR-01-24 (2001)
14. Flocchini, P., Prencipe, G., Santoro, N., Widmayer, P.: Hard tasks for weak robots: The role of common knowledge in pattern formation by autonomous mobile robots. In: ISAAC. (1999) 93–102
15. Prencipe., G.: Corda: Distributed coordination of a set of autonomous mobile robots. European Research Seminar on Advances in Distributed Systems (2001) 185–190

Managing Emergent Processes

John Debenham and Simeon Simoff

Faculty of IT, University of Technology, Sydney, Australia
{debenham, simeon}@it.uts.edu.au
http://e-markets.org.au

Abstract. Emergent processes are non-routine business processes whose execution is guided by the knowledge that emerges during a process instance. Managing emergent processes needs an intelligent agent that is guided not by a process goal, but by a continual in-flow of information, where the integrity of each chunk of information may be uncertain.

1 Introduction

Emergent processes are business processes that are not predefined and are ad hoc. These processes typically take place at the higher levels of organisations [1], and are distinct from production workflows [2]. Emergent processes are opportunistic in nature whereas production workflows are routine [3]. The tasks involved in an emergent process may be carried out by collaborative groups as well as by individuals [4]. Further, the goal of an emergent process instance may mutate as the instance matures. From the management perspective, emergent processes are "knowledge-driven". A *knowledge-driven process* is guided by its "process knowledge" and "performance knowledge". *Process knowledge* is information either generated by the individual users or is extracted from the environment, and includes background information. *Performance knowledge* describes how the other agents together with their 'owners' perform, including how reliable they are. Plan-based agent architectures such as BDI [5] are not directly suitable for managing knowledge-driven processes — this would require machinery to manage the mutations of the process goal. The agent architecture described here is based on information theory, and uses two forms of entropy-based inference [6] which are based on random worlds [7].

A multiagent system for emergent process management consists of the set of agents $\{X_i\}_{i=0}^{n}$ — the description following is written from the point of view of agent X_0 that interacts with the other n agents. In the text, the agent X_ω is "an other" agent — i.e. $\omega \neq 0$. X_0 decides what to do — such as what message to send — on the basis of its information that may be qualified by expressions of degrees of belief. X_0 uses this information to calculate, and continually recalculate, probability distributions for that which it does not know.

2 Process Management

Following [2] a *business process* is "a set of one or more linked procedures or activities which collectively realise a business objective or policy goal, normally

B. Gabrys, R.J. Howlett, and L.C. Jain (Eds.): KES 2006, Part I, LNAI 4251, pp. 228–235, 2006.
© Springer-Verlag Berlin Heidelberg 2006

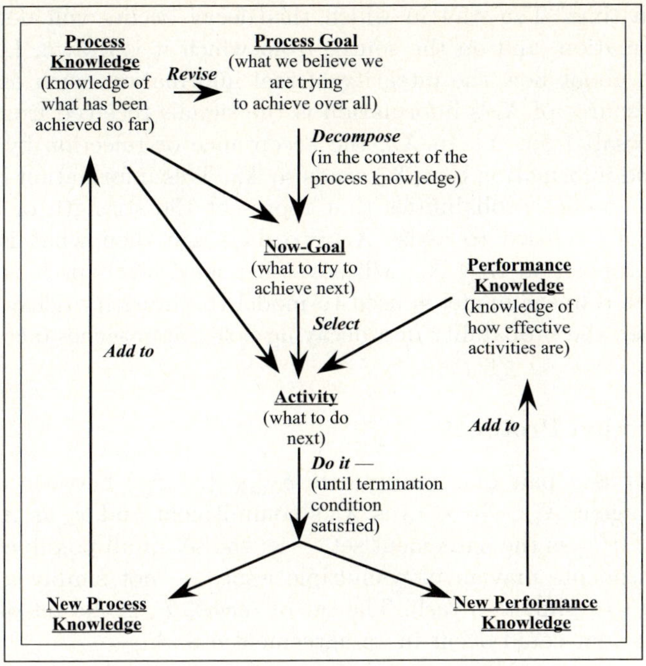

Fig. 1. Knowledge-driven processes

within the context of an organisational structure defining functional roles and relationships". An *activity* is a description of a piece of work that forms one logical step within a process. A particular process is called a (process) *instance.* Three classes of business process are:

- A *task-driven process* has a unique decomposition into a — possibly conditional — sequence of activities. Each of these activities has a goal and is associated with a task that "always" achieves this goal.
- A *goal-driven process* has a process goal, and achievement of that goal is the termination condition for the process. The process goal may have various decompositions into sequences of sub-goals where some of these sequences of tasks may work better than others [4]. The possibility of task failure is a feature of goal-driven processes.
- A *knowledge-driven process* may have a process goal, but the goal may be vague and may mutate [3]. Mutations are determined by the knowledge generated during the process. So the process knowledge gives direction to knowledge-driven processes — see Fig. 1.

3 Emergent Process Agent X_0

X_0 operates in an information-rich environment that includes the Internet. The integrity of X_0's information, including information extracted from the Internet, will decay in time. The way in which this decay occurs will depend on the type of information, and on the source from which it is drawn. Little appears to be known about how the integrity of real information, such as news-feeds, decays. One source of X_0's information is the signals received from X_ω. These include proposals from X_ω to X_0, the acceptance or rejection by X_ω of X_0's proposals, and information that X_ω sends to X_0. This information is augmented by X_0 with sentence probabilities that represent the strength of her belief in its truth. If X_ω refused to assist X_0 two days ago then what is X_0's belief now in the proposition that X_ω will assist her now? Perhaps it is around 0.1. For simplicity, a linear model is used to model the integrity decay of these beliefs, and when the probability of a decaying belief approaches 0.5[1] the belief is discarded.

3.1 Interaction Protocol

An *agreement* is a pair of commitments $\delta_{X_0:X_\omega}(x_0, x_\omega)$ between an agent X_0 and another agent X_ω, where x_0 is X_0's commitment and x_ω is X_ω's commitment. $\mathcal{A} = \{\delta_i\}_{i=1}^{D}$ is the agreement set — ie: the set of all possible agreements. These commitments may involve multiple issues — not simply a single issue such as time to complete a task. The set of *terms*, $\mathcal{T}$, is the set of all possible commitments that could occur in an agreement $a \in \mathcal{A}$.

The agents communicate using sentences in a first-order language $\mathcal{C}$. Predicates in $\mathcal{C}$ include: Delegate($\cdot$), Join($\cdot$), Accept($\cdot$), Reject($\cdot$), Inform($\cdot$) and Quit($\cdot$). Delegate$((X_0, \rho), (X_\omega, G_i))$ means "X_0 proposes to recompense X_ω with ρ if X_ω agrees to take responsibility for an individual goal G_i". Join$((X_0, \rho), (X_\omega, G_i))$ means "X_0 proposes to recompense X_ω with ρ if X_ω agrees to contribute to cooperative goal G_j". Accept(δ) means "the sender accepts your proposed agreement δ". Reject(δ) means "the sender rejects your proposed agreement δ". Inform$((X_0, \mathcal{I}_k), X_\omega)$ means "X_0 offers information $\mathcal{I}_k$ to X_ω". Quit($\cdot$) means "the sender quits — the interaction ends". So for these predicates, and in this discussion, an agreement δ has the form $((X_0, \rho), (X_\omega, G_i))$.

3.2 Agent Architecture

X_0 uses the language $\mathcal{C}$ for external communication, and the language $\mathcal{L}$ for internal representation. One predicate in $\mathcal{L}$ is: $\text{Acc}_\text{d}(((X_0, \rho), (X_\omega, G_i)))$. The proposition $(\text{Acc}_\text{d}(\delta) \mid \mathcal{I}_t)$ means: "X_0 will be comfortable accepting the delegation agreement δ with agent X_ω given that X_0 knows information $\mathcal{I}_t$ at time t". The idea is that X_0 will accept delegation agreement δ if $\mathbf{P}(\text{Acc}_\text{d}(\delta) \mid \mathcal{I}_t) \geq \alpha$ for some threshold constant α. Similarly Acc_j for Join($\cdot$) agreements. The probability distribution $\mathbf{P}(\text{Agg}_\text{d}((X_0, \rho), (X_\omega, G_i)))$ is agent X_0's estimate of the probability that agent X_ω will agree to the Delegate agreement δ [or $\text{Agg}_\text{j}(\cdot)$ for Join($\cdot$) agreements] — it is estimated in Sec. 4.

Each incoming message M from source S received at time t is time-stamped and source-stamped, $M_{[S,t]}$, and placed in an *in box*, $\mathcal{X}$, as it arrives. X_0 has an

[1] A sentence probability of 0.5 represents null information, ie: "maybe, maybe not".

information repository $\mathcal{I}$, a *knowledge base* $\mathcal{K}$ and a *belief set* $\mathcal{B}$. Each of these three sets contains statements in a first-order language $\mathcal{L}$. $\mathcal{I}$ contains statements in $\mathcal{L}$ together with sentence probability functions of time. $\mathcal{I}_t$ is the state of $\mathcal{I}$ at time t and may be inconsistent. At some particular time t, $\mathcal{K}_t$ contains statements that X_0 believes are true at time t, such as $\forall x(\text{Accept}(x) \leftrightarrow \neg\text{Reject}(x))$. The belief set $\mathcal{B}_t = \{\beta_i\}$ contains statements that are each qualified with a *given sentence probability*, $\mathbf{B}(\beta_i)$, that represents X_0's belief in the truth of the statement at time t. The distinction between the knowledge base $\mathcal{K}$ and the belief set $\mathcal{B}$ is simply that $\mathcal{K}$ contains unqualified statements and $\mathcal{B}$ contains statements that are qualified with sentence probabilities. $\mathcal{K}$ and $\mathcal{B}$ play different roles in the method described in Sec. 3.3; $\mathcal{K}_t \cup \mathcal{B}_t$ is required by that method to be consistent.

X_0's actions are determined by its "strategy". A *strategy* is a function $\mathbf{S} : \mathcal{K} \times \mathcal{B} \to \mathcal{A}$ where $\mathcal{A}$ is the set of actions. At certain distinct times the function $\mathbf{S}$ is applied to $\mathcal{K}$ and $\mathcal{B}$ and the agent does something. The set of actions, $\mathcal{A}$, includes sending Delegate(·), Join(·), Accept(·), Reject(·), Inform(·) and Quit(·) messages to X_ω. The way in which $\mathbf{S}$ works is described in Secs. 4. Two "instants of time" before the $\mathbf{S}$ function is activated, an "import function" and a "revision function" are activated. The import function $\mathbf{I} : (\mathcal{X} \times \mathcal{I}_{t^-}) \to \mathcal{I}_t$ clears the in-box, using its "import rules". An import rule takes a message M, written in language $\mathcal{C}$, and from it derives sentences written in language $\mathcal{L}$ to which it attaches decay functions, and adds these sentences together with their decay functions to $\mathcal{I}_{t^-}$ to form $\mathcal{I}_t$. An *import rule* has the form: $\mathbf{P}(S \mid M_{[X_\omega,t]}) = f(M, X_\omega, t) \in [0,1]$, where S is a statement, M is a message and f is the decay function. Then the belief revision function $\mathbf{R} : \mathcal{I}_{t^-} \to (\mathcal{I}_t \times \mathcal{K}_t \times \mathcal{B}_t)$ deletes any statements in $\mathcal{I}_{t^-}$ whose sentence probability functions have a value that is ≈ 0.5 at time t.

X_0 uses three things to construct proposals: an estimate of the likelihood that X_ω will accept any agreement [Sec. 4], an estimate of the likelihood that X_0 will, in hindsight, feel comfortable accepting any particular agreement, and an estimate of when X_ω may quit and leave the interaction — see [8].

3.3 Inference

The agent X_0 employs two forms of inference: maximum entropy inference and minimum relative entropy inference. Let $\mathcal{G}$ be the set of all positive ground literals that can be constructed using the predicate, function and constant symbols in $\mathcal{L}$. A *possible world* is a valuation function $\mathcal{V} : \mathcal{G} \to \{\top, \bot\}$. $\mathcal{V}$ denotes the set of all possible worlds, and $\mathcal{V}_\mathcal{K}$ denotes the set of possible worlds that are consistent with a knowledge base $\mathcal{K}$ [7].

A *random world* for $\mathcal{K}$ is a probability distribution $\mathcal{W}_\mathcal{K} = (p_i)$ over $\mathcal{V}_\mathcal{K} = (\mathcal{V}_i)$, where $\mathcal{W}_\mathcal{K}$ expresses an agent's degree of belief that each of the possible worlds is the actual world. The *derived sentence probability* of any $\sigma \in \mathcal{L}$, *with respect to* a random world $\mathcal{W}_\mathcal{K}$ is:

$$(\forall \sigma \in \mathcal{L})\mathbf{P}_{\mathcal{W}_\mathcal{K}}(\sigma) \triangleq \sum_n \{\, p_n \;:\; \sigma \textit{ is } \top \textit{ in } \mathcal{V}_n \,\} \tag{1}$$

A random world $\mathcal{W}_\mathcal{K}$ is *consistent* with the agent's beliefs $\mathcal{B}$ if: $(\forall\beta \in \mathcal{B})(\mathbf{B}(\beta) = \mathbf{P}_{\mathcal{W}_\mathcal{K}}(\beta))$. That is, for each belief its derived sentence probability as calculated using Eqn. 1 is equal to its given sentence probability.

The *entropy* of a discrete random variable X with probability mass function $\{p_i\}$ is [6]: $\mathbf{H}(X) = -\sum_n p_n \log p_n$ where: $p_n \geq 0$ and $\sum_n p_n = 1$. Let $\mathcal{W}_{\{\mathcal{K},\mathcal{B}\}}$ be the "maximum entropy probability distribution over $\mathcal{V}_\mathcal{K}$ that is consistent with $\mathcal{B}$". Given an agent with $\mathcal{K}$ and $\mathcal{B}$, *maximum entropy inference* states that for *any* sentence, $\sigma \in \mathcal{L}$, its *derived sentence probability* is:

$$(\forall\sigma \in \mathcal{L})\mathbf{P}(\sigma) \triangleq \mathbf{P}_{\mathcal{W}_{\{\mathcal{K},\mathcal{B}\}}}(\sigma) \tag{2}$$

Using Eqn. 2, the derived sentence probability for any belief, β_i, is equal to its given sentence probability. So the term *sentence probability* is used without ambiguity.

If X is a discrete random variable taking a finite number of possible values $\{x_i\}$ with probabilities $\{p_i\}$ then the *entropy* is the average uncertainty removed by discovering the true value of X, and is given by $\mathbf{H}(X) = -\sum_n p_n \log p_n$. The direct optimization of $\mathbf{H}(X)$ subject to a number, θ, of linear constraints of the form $\sum_n p_n g_k(x_n) = \overline{g}_k$ for given constants $\overline{g}_k$, where $k = 1, \ldots, \theta$, is a difficult problem. Fortunately this problem has the same unique solution as the *maximum likelihood problem* for the Gibbs distribution.

Given a prior probability distribution $\underline{q} = (q_i)_{i=1}^n$ and a set of constraints, the *principle of minimum relative entropy* chooses the posterior probability distribution $\underline{p} = (p_i)_{i=1}^n$ that has the least relative entropy with respect to $\underline{q}$:

$$\arg\min_{\underline{p}} \sum_{i=1}^{n} p_i \log \frac{p_i}{q_i} \tag{3}$$

and that satisfies the given constraints. The principle of minimum relative entropy is a generalization of the principle of maximum entropy. If the prior distribution $\underline{q}$ is uniform, the relative entropy of $\underline{p}$ with respect to $\underline{q}$ differs from $-\mathbf{H}(\underline{p})$ only by a constant. So the principle of maximum entropy is equivalent to the principle of minimum relative entropy with a uniform prior distribution.

4 Interaction

X_0 interacts with its opponents $\{X_i\}_{i=1}^n$. It is assumed that goals are initially triggered externally to the system. For example, X_0's 'owner' may have an idea that she believes has value, and triggers an emergent process to explore the idea's worth. The interaction protocol is simple, if X_0 sends a Delegate(·) or a Join(·) message to X_ω then an interaction has commenced and continues until one agent sends an Accept(·) or a Quit(·) message. This assumes that agents respond in reasonable time which is fair in an essentially cooperative system.

To support the agreement-exchange process, X_0 has do two different things. First, it must respond to proposals received from X_ω. Second, it must construct proposals, and possibly information, to send to X_ω — that is described now. Maximum entropy inference is used to 'fill in' missing values with the "maximally noncommittal" probability distribution. To illustrate this suppose that X_0 proposes to delegate a process to X_ω. This process involves: X_ω delivering — using an Inform(·) message — u chapters for a report in so-many days v. This section describes machinery for estimating the probabilities $\mathbf{P}(\text{Agg}_\text{d}((X_0, u), (X_\omega, G_v))$ where the predicate $\text{Agg}_\text{d}((X_0, u), (X_\omega, G_v))$ means "X_ω will accept X_0's delegation proposal $((X_0, u), (X_\omega, G_v))$".

X_0 assumes the following two preference relations for X_ω, and $\mathcal{K}$ contains:

$$\kappa_{11}: \quad \forall x, y, z((x < y) \rightarrow (\text{Agg}_\text{d}((X_0, y), (X_\omega, G_z)) \rightarrow \text{Agg}_\text{d}((X_0, x), (X_\omega, G_z))))$$
$$\kappa_{12}: \quad \forall x, y, z((x < y) \rightarrow (\text{Agg}_\text{d}((X_0, z), (X_\omega, G_x)) \rightarrow \text{Agg}_\text{d}((X_0, z), (X_\omega, G_y))))$$

As noted in Sec. 3.3, these sentences conveniently reduce the number of possible worlds. The two preference relations κ_{11} and κ_{12} induce a partial ordering on the sentence probabilities in the $\mathbf{P}(\text{Agg}_\text{d}((X_0, u), (X_\omega, G_v)))$ array.

Suppose that X_0 has the following historical data on similar dealings with X_ω. Three months ago X_ω asked for ten days to deliver four chapters. Two months ago X_0 proposed one day to deliver three chapters and X_ω refused. One month ago X_ω asked for eight days to deliver two chapters. $\mathcal{B}$ contains:
$\beta_{11}: \text{Agg}_\text{d}((X_0, 4), (X_\omega, G_{10}))$; $\beta_{12}: \text{Agg}_\text{d}((X_0, 3), (X_\omega, G_1))$ and
$\beta_{13}: \text{Agg}_\text{d}((X_0, 2), (X_\omega, G_8))$, and assuming a 10% decay in integrity for each month: $\mathbf{P}(\beta_{11}) = 0.7$, $\mathbf{P}(\beta_{12}) = 0.2$ and $\mathbf{P}(\beta_{13}) = 0.9$

The distribution $\mathbf{W}_{\{\mathcal{K},\mathcal{B}\}}$ has just five different probabilities in it. The probability matrix for the proposition $\text{Agg}_\text{d}((X_0, u), (X_\omega, G_v))$ is:

$v \setminus u$	1	2	3	4	5
11	0.9967	0.9607	0.8428	0.7066	0.3533
10	0.9803	0.9476	0.8330	**0.7000**	0.3500
9	0.9533	0.9238	0.8125	0.6828	0.3414
8	0.9262	**0.9000**	0.7920	0.6655	0.3328
7	0.8249	0.8019	0.7074	0.5945	0.2972
6	0.7235	0.7039	0.6228	0.5234	0.2617
5	0.6222	0.6058	0.5383	0.4523	0.2262
4	0.5208	0.5077	0.4537	0.3813	0.1906
3	0.4195	0.4096	0.3691	0.3102	0.1551
2	0.3181	0.3116	0.2846	0.2391	0.1196
1	0.2168	0.2135	**0.2000**	0.1681	0.0840

In this array, the derived sentence probabilities for the three sentences in $\mathcal{B}$ are shown in bold type; they are exactly their given values. X_0's *interaction strategy* is a function $\mathbf{S}: \mathcal{K} \times \mathcal{B} \rightarrow \mathcal{A}$ where $\mathcal{A}$ is the set of actions that send Delegate(·), Join(·), Accept(·), Reject(·), Inform(·) and Quit(·) messages to X_ω. If X_0 sends any message to X_ω then she is giving X_ω information about herself.

5 Delegation

The mechanism that X_0 uses for managing process delegation is described in full. Join($\cdot$) messages are managed similarly. This next section discusses the sorts of payoff measures and estimates that are available, and that are combined to give a value for the expected payoff vector $\underline{\nu}_i$ for each agent. Let $\mathbf{P}(A \gg)$ denote A is the 'best choice' in terms of some combination of the parameter estimates described following. These measurements are then used by agent X_0 to determine $\mathbf{P}(X_i \gg)$, and then in turn to determine the delegation strategy $(p_i)_{i=1}^n$.

5.1 Performance Parameters — Choosing the 'Best' Collaborator

Agent X_0 continually measures the performance of itself and of other agents in the system using four measures. Three are: *time*, *cost* and *likelihood of success*. The last one is a *value* parameter that is attached to other agents. Time is the total time taken to termination. Cost is the actual cost of the of resources allocated. For example, the time that the agent — possibly with a human 'assistant' – actually spent working on that process. The likelihood of success is the probability that an agent will deliver its response within its constraints. The value parameter is the value added to a process by an agent. It is often very difficult to measure — it is treated here by a subjective estimate.

The three parameters *time*, *cost* and likelihood of *success* are observed and recorded every time an agent, including X_0, delivers, or fails to deliver, its commitments. This generates a large amount of data whose significance can reasonably be expected to degrade over time. So a cumulative estimate only is retained. The integrity of information 'evaporates' as time goes by. If we have the set of observable outcomes as $O = \{o_1, o_2, \ldots, o_m\}$ then complete ignorance of the expected outcome means that our expectation over these outcomes is $\frac{1}{m}$ — i.e. the unconstrained maximum entropy distribution. This natural decay of information integrity is offset by new observations.

Given one of the parameters, u, with m possible outcomes[2], suppose that $P^t(u' \mid \delta)$ is the estimate at time t of the probability that the actual outcome u' will be observed given that the agent being observed has committed to δ. Suppose that X_0 observes the actual outcome r, on the basis of this outcome X_0 believes that the probability of r being observed at the next time is g_r. Then let $P^t_{g_r}(u' \mid \delta)$ be the posterior minimum relative entropy distribution calculated using Eqn. 3 with prior distribution $P^t(u' \mid \delta)$ and satisfying the constraint that $P^t_{g_r}(r \mid \delta) = g_r$. Then update $P^t(u' \mid \delta)$ with:

$$P^{t+1}(u' \mid \delta) = \frac{1-\rho}{n} + \rho \cdot P^t_{g_r}(u' \mid \delta) \tag{4}$$

This equation determines the development of $P^t(u' \mid \delta)$ for some large $\rho \in [0, 1]$.

The method determined by Eqn. 4 is used by X_0 to update its estimates of probability distributions for all the agents that it deals with. For example, if $P^t(\cdot)$

[2] The *success* parameter has only two possible outcomes 'succeed' and 'fail'.

is X_0's estimate of the time that X_ω will take to deliver on a particular type of agreement. Suppose that at time t, X_ω delivers her response after having taken time u. Then X_0 attaches a belief (i.e. a sentence probability) to the proposition that this is how X_ω will behave at time $t+1$. This becomes the constraint in the minimum relative entropy calculation and then Eqn. 4 gives $P^{t+1}(\cdot)$. This method may be extended to estimate the probability that one agent is a better choice than a number of other agents. For example, if there are three agents to choose from, A, B, and C, then: $\mathbf{P}(A \gg) = \mathbf{P}((A \gg B) \wedge (A \gg C))$, and Eqn. 1 estimates the probability of the two propositions on the right-hand side.

5.2 Delegation Strategy

A delegation strategy is a probability distribution $\{p_i\}_{i=1}^n$ that determines who from $\{X_i\}_{i=1}^n$ to offer responsibility to for doing what. The delegation strategy achieves this stochastically by determining instead n probabilities $(p_1, \ldots, p_n)$ where p_i is the probability that the i'th agent will be selected, and $\sum_i p_i = 1$. The choice of the agent to delegate to is then made with these probabilities. The expression of the delegation strategy in terms of probabilities enables the strategy to balance conflicting goals, such as achieving process quality and process efficiency.

$\mathbf{P}(X_i \gg)$ is the probability that X_i is the 'best' choice. A strategy that continually chose the 'best' on the basis of historic data is flawed because an agent who "goes through a bad patch" may never be chosen — this means that if an agent wants the quiet life all is would have to do is make a series of mistakes. The delegation strategy chooses agents with probability $p_i = \mathbf{P}(X_i \gg)$. So the probability that X_0 will attempt to delegate a process to X_ω is equal to the probability that X_0 estimates X_ω to be the 'best' choice for the job.

References

1. Jain, A., Aparicio, M., Singh, M.: Agents for process coherence in virtual enterprises. Communications of the ACM 42 (1999) 62–69
2. Fischer, L.: The Workflow Handbook 2003. Future Strategies Inc. (2003)
3. Dourish, P.: Using metalevel techniques in a flexible toolkit for CSCW applications. ACM Transactions on Computer-Human Interaction (TOCHI) 5 (1998) 109–155
4. Smith, H., Fingar, P.: Business Process Management (BPM): The Third Wave. Meghan-Kiffer Press (2003)
5. Wooldridge, M.: Multiagent Systems. Wiley (2002)
6. MacKay, D.: Information Theory, Inference and Learning Algorithms. Cambridge University Press (2003)
7. Halpern, J.:Reasoning about Uncertainty. MIT Press (2003)
8. Debenham, J.: Bargaining with information. In Jennings, N., Sierra, C., Sonenberg, L., Tambe, M., eds.: Proceedings Third International Conference on Autonomous Agents and Multi Agent Systems AAMAS-2004, ACM (2004) 664–671
9. Jaynes, E.: Probability Theory The Logic of Science. Cambridge University Press (2003)

Teamwork Coordination in Large-Scale Mobile Agent Networks

Gordan Jezic, Mario Kusek, and Vjekoslav Sinkovic

University of Zagreb
Faculty of Electrical Engineering and Computing
Department of Telecommunications
Unska 3, HR-10000 Zagreb, Croatia
{gordan.jezic, mario.kusek, vjekoslav.sinkovic}@fer.hr

Abstract. This paper deals with multi-agent systems coordinated via teamwork in large-scale networks. A formal model based on the Fragmented Mobile Agent Network is proposed. An agent team is described as a knowledge-based system that includes subteams consisting of coordination agents and mobile subteam members organised according to a predefined subteam coordination plan. A case study considering a scenario in which multi-operation teamwork agents upgrade software employed in a Grid network was conducted. A performance analysis based on simulations of networks with different fragment sizes and subteam coordination plans is presented.

Keywords: mobile agents, teamwork, coordination, large-scale network, Grid.

1 Introduction

Service management and software configuration operations in distributed systems become very demanding tasks as the number of computers and/or geographical distances between them grow. The situation gets worse with an increase in the complexity of the network and the number of nodes. In dynamic and large-scale environments with a large number of nodes, more complex operations, such as remote software testing or tracing on each of the systems, are required. In [1] we presented a system called the Remote Maintenance Shell (RMS) developed for Grid service management. It enables the service management and maintenance of several versions of service without suspending their regular operation. It is based on remote operations performed by mobile software agents.

A software agent is a program which represents a user in a network and performs tasks on behalf of this user. Mobile software agents offer advantages for service management in networks since they provide full user and service mobility. In [2], the authors consider mobility management for personal agents which represent users in a network and support terminal, user and service mobility. Groups of agents organised into teams and coordinated via teamwork can be applied to the problem of distributed service/software management in large-scale networks. An approach in which

B. Gabrys, R.J. Howlett, and L.C. Jain (Eds.): KES 2006, Part I, LNAI 4251, pp. 236–243, 2006.
© Springer-Verlag Berlin Heidelberg 2006

cooperative agents are organised into teams with dynamically evolving subteams is presented in [3].

In this paper, we present a model of a multi-agent system consisting of stationary and mobile agents coordinated via teamwork. They can flexibly achieve joint goals in large-scale distributed systems by applying knowledge sharing. Knowledge is shared between subteam coordination agents (SCAs) and mobile subteam members in order to perform software management operations in a network. Each SCA manages one part of the coordination of the agents. This paper continues the work presented in [4] where a multi-agent system is organised as an agent team and coordinated according to a shared plan. We extend this model to include a large number of nodes and introduce the idea of a Fragmented Mobile Agent Network (F-MAN). An F-MAN divides the network into network segments operated by agent subteams. It represents a team coordination infrastructure in which an agent team is divided into subteams responsible for executing operations on network fragments. Subteam member agents directly communicate with each other and share knowledge needed for operation execution. In [5] we presented a multi-agent system as a hybrid knowledge-based system in which knowledge is shared between a master agent and team agents. Knowledge-based extensions which enable agent communication are defined in [6]. This paper focuses on the effects of dividing an agent network into an F-MAN and the impact of organising agent teams in order to improve execution time.

Different approaches regarding organisations in multi-agent systems have been considered [7]. In this paper, joint intentions [8] and BDI (Beliefs-Desires-Intentions) agents [9] are not used because most of the agents considered are not intelligent. In our approach, agents use a shared plan such as that described in [10]. The agents are coordinated only with respect to action execution but not planning. A single agent performs centralised planning and then distributes actions to the other agents.

The paper is organised as follows: Section 2 deals with team organisation in a fragmented mobile agent network and subteam coordination plans considered as the basis for identification of agent knowledge sharing. A case study describing software upgrading in a Grid network is given in Section 3. Performance analysis based on simulation is described in Section 4, and Section 5 concludes the paper.

2 Building Agent Teams in the Fragmented Mobile Agent Network

The Mobile Agent Network (MAN) is used for modelling agent organisation and coordination in an agent team. The MAN is represented by the triple: $\{A, S, N\}$, where A is a multi-agent system $A = \{agent_1, agent_2, ..., agent_k, ..., agent_p\}$ consisting of co-operating and communicating mobile agents, S represents a set of processing nodes $S = \{S_1, S_2, ..., S_i, ..., S_m\}$ at which the agents perform services, and N is a network that connects processing nodes and allows agent mobility. Each processing node S_i has a unique addressi from the set of addresses, $address = \{address_1, address_2, ..., address_i, ..., address_m\}$. An agent is defined by the triple a $agent_k = \{name_k, address_k, task_k\}$, where namek defines the agent's unique identification, $address_k \in address$ represents the list of nodes to be visited by the agent and $task_k$ denotes the functionality it provides in the form of task$_k$ = $\{s_1, s_2, ..., s_i, ..., s_q\}$ representing a set of assigned elementary operations si. When hosted by node $S_i \in address_k$, $agent_k$ performs operation $s_i \in task_k$. If an operation requires specific data, the agent carries this data during migration [4]. This model is implemented in Grid networks where agent teams participate in software upgrading operations.

In a large-scale network, such as Grid, with a large number of processing nodes and complex user requests with many and/or dependent operations, it is beneficial to divide the network into fragments and to divide a team of agents into subteams. We define a fragmented mobile agent network (F-MAN) as an extension of MAN. The F-MAN is represented by a triple {SA, FS, N}, where SA is a multi-agent system organised as a team and divided into subteams $SA = \{SA_1, SA_2, ..., SA_j, ..., SA_n\}$, FS represents a set of network fragments $FS = \{f_1, f_2, ..., f_j, ..., f_n\}$ on which subteams perform services, and N is a network. Each subteam SA_j performs services in a specified fragment, $f_j \Rightarrow SA_j$. Each fragment f_j includes a set of processing nodes $f_j = \{S_{j1}, S_{j2}, ..., S_{jq}\}$.

A subteam is organised as a knowledge-based system which includes one subteam coordination agent (SCA) and subteam members which share plan and knowledge. An initial request is first submitted to a management agent (MA). The MA then defines fragments and divides the initial request into fragment requests which are sent to the SCAs of each fragment, $CA = \{SCA_1, SCA_2, ..., SCA_j, ..., SCA_n\}$. The number of SCAs is equal to the number of the fragments which in turn depends on the network size. After receiving a request, each SCA creates a subteam, a shared plan, distributes operations among them and starts their executions.

The majority of knowledge is concentrated at the SCAs of the subteams. Each SCA possesses knowledge of its capabilities, subteam capabilities and situation-specific knowledge [11]. An SCA must be able to, on the basis of a user request and operations, create a shared subteam coordination plan (SCP), form a subteam of member agents, and send them to perform the corresponding operations [12]. Situation–specific knowledge represents the capability of the SCA to make a decision in any unforeseen situation while executing the SCP. It can modify subteam agents' tasks during execution. Each SCA manages coordination in one fragment, and thus manages part of the whole coordination.

2.1 Organising Subteams

The F-MAN divides a team of agents *A* into subteams *SA* assigned to network fragments *FS*. Figure 1 shows an example of an F-MAN organisation with three fragments and three subteams. Each subteam is organised by an SCA according to a shared subteam coordination plan (SCP). The SCP defines subteam behaviour, the complexity of subteam members and the coordination mechanism required.

An SCP defines the complexity of subteam agents (i.e. the number of elementary operations per agent) and the size of a subteam. We chose two SCPs for organising agents within subteams according to the results from [4]: SCP1 - a single subteam agent can execute multiple operations on all nodes in the fragment, and SCP2 - a subteam agent can execute multiple operations on only one node in the fragment.

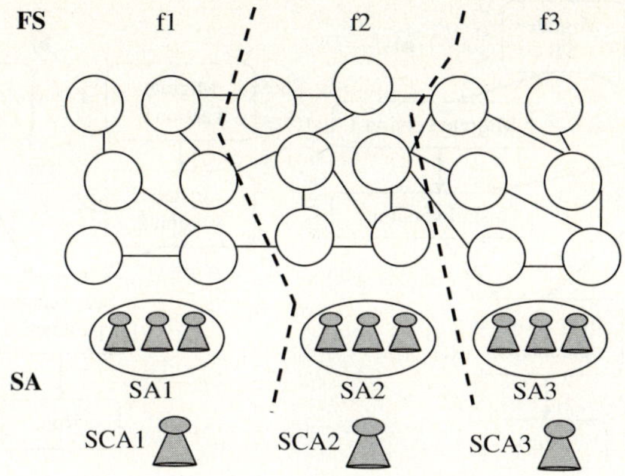

Fig. 1. A Fragmented Mobile Agent Network

3 Case Study: Software Upgrading in a Grid Network

The MA-RMS (Multi-Agent Remote Maintenance Shell) is a distributed system comprised of two main components: the RMS Console and the RMS Maintenance Environment. The RMS Console is the client part of the RMS which offers a GUI enabling the user to define management operations (request) on remote nodes. The computer on which it is installed and run on is referred to as the management station. The management station represents a centralized location from where management is performed. The RMS Maintenance Environment, also called the RMS Core, is the server part of RMS which must be preinstalled on remote network nodes in order for them to be managed by the MA-RMS. A more detailed description of those components and their internal architecture is given in [1, 5].

The MA-RMS is used to manage services employed in test Grid environments at CERN [1]. A Grid infrastructure is composed of a large-scale distributed system which includes a set of virtual organisations consisting of sets of individuals and/or institutions (hosting environments) where Grid services are run [13]. The division of a Grid network into smaller parts (fragments) enables the simultaneous management of multiple hosting environments without handling each individually [14].

Software operations considered in this case study are the following: software migration (s1), installation (s2), setting execution parameters (s3), running (s4), and stopping (s5). Each remote software operation is represented by an elementary operation with corresponding input and output parameters. User requests for remote software operations are presented by a directed acyclic graph G = (T, L) where T denotes a list of elementary operations, and L represents a set of directed edges which define precedence relations between tasks [5] (Figure 2).

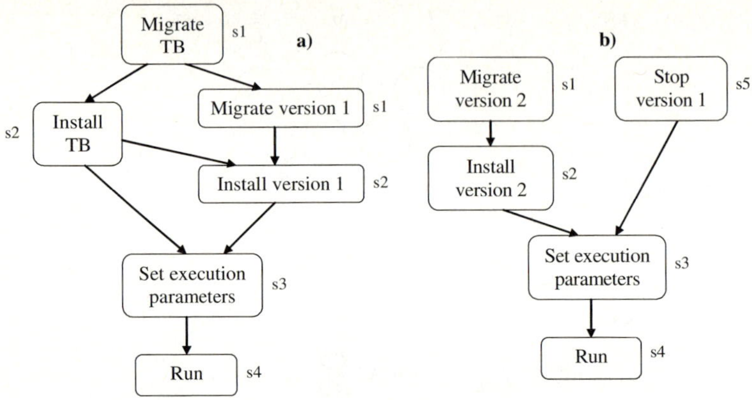

Fig. 2. Scenario graph for software starting and upgrading

3.1 MA-RMS Multi-operation Teamwork Agents

An MA-RMS user inputs its desired end state via the GUI. That end state is passed to the MA which defines fragments and their corresponding requests which are then sent to the SCAs of each fragment. The SCAs then generate the necessary operations, assign them to their subteam, and create and start multi-operation teamwork agent(s). Finally, to each multi-operation agent is assigned one or several operations, according to one of the SCPs defined in the previous Section.

Each operation contains input and output variables. The input variables specify the necessary preconditions that need to be satisfied in order for operation execution to begin. The output variables of a certain operation specify all the other operations which depend on it, i.e. define the operations that need to be notified upon its completion. That way, a network of interdependencies is created between operations. The operations do not communicate directly, but through their enclosing agents.

In this case study, which consists of two parts, the desired end state is to "introduce software at a single node and upgrade existing software with a newer version at that node". Eleven operations are used in this scenario to introduce, run and upgrade software. Figure 2a shows an operations graph for one Grid network node. The first operation is testbed migration (migrate TB). Testbed is an adaptation layer between the MA-RMS and the software version [1]. Testbed migration is followed by testbed installation (install TB) and version (i.e. service) migration (migrate version), which in turn depend on testbed migration. Version installation (install version) depends on version migration, while the operation responsible for setting execution parameters depends on both tesbed and version installation. The last operation executed is responsible for running the software (run software) and can be executed after setting the execution parameters. The second part of the scenario is concerned with upgrading software with a newer version (Figure 2b). The following operations must be executed: migration and installation of the new version, termination of the old version, setting the new version to be the active one and starting the new version.

4 Performance Analysis

We simulated the execution of software management operations in a large-scale mobile agent network. The simulation included 100 network nodes divided into different number of network fragments, from 1 to 10. The performance analysis presented in this section is based on the time execution assessment defined in [4, 5]. The agent team performance evaluation is based on the following assumptions:

a) each agent belongs to only one subteam and agents can not change subteams;
b) all subteams are organised with the same SCP;
c) agent migration and global dialog time are the same;
d) migration time for a loaded agent is multiplied by 10 because it carries data;
e) agent migration and dialog time depend on the available network transmission rate;
f) operation execution time depends on the node capacity and is the same for all operations and all nodes in the network;
g) agent creation and death consumes time equal to operation execution time;
h) time is measured in discrete time units Δt.

A team includes SCAs and MA-RMS multi-operation agents coordinated via teamwork according to predefined SCPs (SCP1 and SCP2). Since all subteams have the same SCP, subteams are cloned and placed in other network fragments. A clone agent is defined as an agent member that performs the same task in any fragment.

Figure 3 shows total execution time for SCP1 depending on the number of fragments (F1 - F10), network delay (represented by agent migration and dialog time, i.e. communication between agents), and node capacity (represented by operation execution time at a node). The previous graph shows that the total execution time for a network without fragments (F0) and with one fragment (F1) is practically the same, which means that introducing a fragmentation mechanism is insignificant. However, increasing the number of fragments decreases the total execution time.

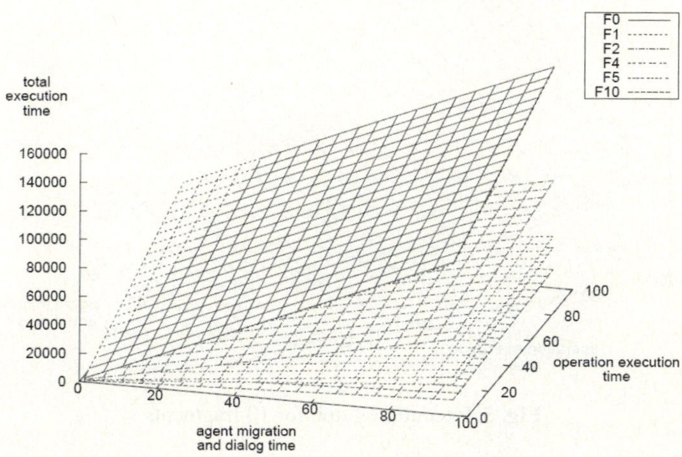

Fig. 3. Execution graph for SCP1

Figure 4 shows the simulation results for SCP2. From both graphs it can be concluded that a larger number of fragments gives better results. The graph in Figure 5 shows the total execution time for both SCP1 and SCP2 with 10 fragments. It shows that SCP2 gives much better results than SCP1.

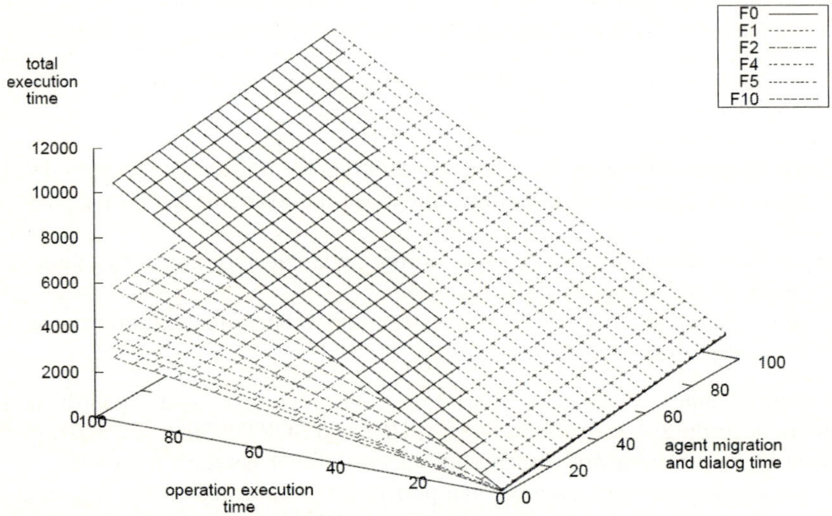

Fig. 4. Execution graph for SCP2

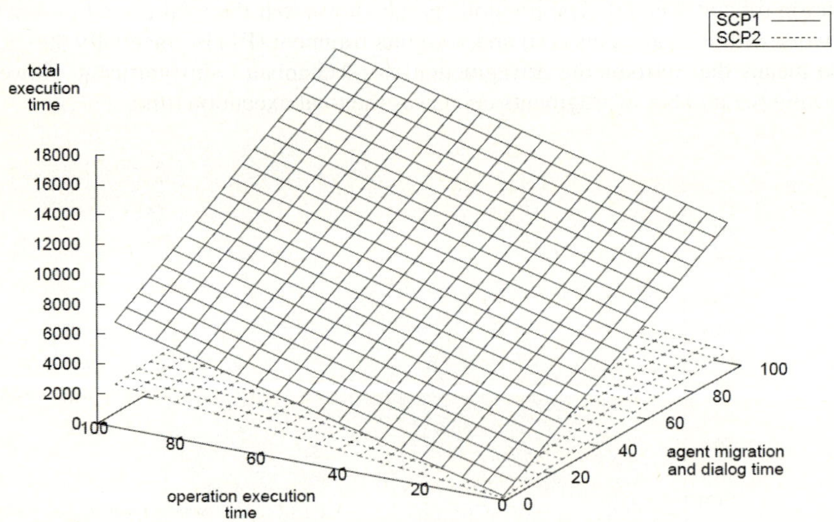

Fig. 5. Execution graph for 10 fragments

5 Conclusion

In this paper, we discuss multi-agent systems coordinated via teamwork in large-scale networks. The Fragmented Mobile Agent Network organised as a knowledge-based system divided into subteams is proposed as an extension of the MAN. A case study describing software upgrading in a Grid network using multi-operation teamwork agents is used to analyse the proposed approach. From the performance analysis it can be concluded that introducing fragments improves results with respect to agent systems without fragments. Namely, with an increase in the number of fragments, the total execution time decreases. Regarding the subteam coordination plan, SCP2 is better than SCP1.

Future work will include multi-agent planning and eliminating certain simplifications considered in the paper. Namely, member agents will be migrated from one fragment to another and will thus be able to perform operations in more fragments. Furthermore, the F-MAN will be implemented in the MA-RMS system and simulation results verified.

References

1. Jezic G., Kusek M., Marenic T., Lovrek I., Desic S., Trzec K., Dellas B.: Grid Service Management by Using Remote Maintenance Shell, Grid Services Engineering and Management, Springer-Verlag, LNAI 3270 (2004) 136-149.
2. Lovrek, I., V. Sinkovic, Knowledge-based Mobility Management in All-Mobile Network, KES 03, LNAI 2774, Springer-Verlag, (2003) 661-667.
3. Scerri, P., Z. Xu, E. Liao, J. Lai, K. Szcara, Scaling Teamwork to Very Large Teams, In Proceedings of AAMAS-04, (2004) 888-895.
4. Kusek, M., I. Lovrek, V. Sinkovic: Agent team coordination in the mobile agent network. KES 05, LNAI 3681 Springer-Verlag, (2005) 240-246.
5. Jezic, G., Kusek, M., Desic, S., Caric, A., Huljenic, D.: Multi-Agent System for Remote Software Operation, KES 03, LNAI 2774, Springer-Verlag, (2003) 675-682.
6. Lovrek I., V. Sinkovic, G. Jezic: Communicating Agents in Mobile Agent Network, KES 02, Frontiers in Artificial Intelligence and Applications 82 (2002) pp. 126-130.
7. Weiss, G. (ed.): Multiagent Systems, MIT Press, Cambridge (1999)
8. Smith, I., P. Cohen, J. Bradshaw, M. Greaves, and H. Holmback.: Designing conversation policies using joint intention theory, In Proceedings of Third International Conference on Multi-Agent Systems (ICMAS98), (1998) 269-276.
9. JACK TM Intelligent Agents Teams Manual, Agent Oriented Software (2004) http://www.agent-software.com
10. Grosz, B. and S. Kraus, The Evolution of SharedPlans, in A. Rao, M. Wooldridge, Eds., Foundations and Theories of Rational Agencies (1998) 227-262.
11. Giampapa, J.A., Sycara, K.: Team-oriented agent coordination in the RETSINA multi-agent system, Tech. rep. CMU-RI-TR-02-34, Robotics Institute, Carnegie Mellon Univ. (2002).
12. Scerri P. et al., Challenges in Building Very Large Teams, MMAS 2004, LNCS 3446, Springer-Verlag, (2005) 86-103.
13. Foster, I, Kesselman, C., Nick, J., Tuecke, S.: Grid Services for Distributed System Integration. Computer, 35(6), (2002) 37–46.
14. Foster, I., Kesselman, C. (ed.): The Grid 2: Blueprint for a New Computing Infrastructure and the Grid, (Chapter 29), Morgan Kaufman (2004).

Real-Time Search Algorithms for Exploration and Mapping

In-Cheol Kim

Department of Computer Science, Kyonggi University
Suwon-si, Kyonggi-do, 442-760, South Korea
kic@kyonggi.ac.kr

Abstract. In this paper, we consider the problem of exploring unknown environment with an intelligent virtual agent. Traditionally research efforts to address the exploration and mapping problem have focused on the graph-based space representations and the graph search algorithms. In this paper, we propose DFS-RTA* and DFS-PHA*, two real-time graph search algorithms for exploring and mapping an unknown environment. Both algorithms are based upon the simple depth-first search strategy. However, they adopt different real-time shortest path-finding methods for fast backtracking to the last unexhausted node. Through some experiments with a virtual agent deploying in a 3D interactive computer game environment, we confirm the completeness and efficiency of two algorithms.

1 Introduction

Suppose that an agent has to construct a complete map of an unknown environment using a path that is as short as possible. An agent has to explore all nodes and edges of an unknown, strongly connected directed graph. The agent visits an edge when it traverses the edge. A node or edge is explored when it is visited for the first time. The goal is to determine a map of the graph using the minimum number R of edge traversals. At any point in time the robot knows (1) all visited nodes and edges and can recognize them when encountered again; and (2) the number of visited edges leaving any visited node. The agent does not know the head of unvisited edges leaving a visited node or the unvisited edges leading into a visited node. At each point in time, the agent visits a current node and has the choice of leaving the current node by traversing a specific known or an arbitrary unvisited outgoing edge. An edge can only be traversed from tail to head, not vice versa.

If the graph is Eulerian, $2m$ edge traversals suffice, where m is the number of edges. This immediately implies that undirected graphs can be explored with at most $4m$ traversals [2]. For a non-Eulerian graph, let the deficiency d be the minimum number of edges that have to be added to make the graph Eulerian. Recently Kwek [7] proposed an efficient depth-first search strategy for exploring an unknown strongly connected graph G with m edges and n vertices by traversing at most $\min(mn, dn^2+m)$ edges. In this paper, we propose DFS-RTA* and DFS-PHA*, two

B. Gabrys, R.J. Howlett, and L.C. Jain (Eds.): KES 2006, Part I, LNAI 4251, pp. 244 – 251, 2006.
© Springer-Verlag Berlin Heidelberg 2006

real-time graph search algorithms for exploring and mapping an unknown environment. Both algorithms are based upon the simple depth-first search strategy. However, they adopt different real-time shortest path-finding methods for fast backtracking to the last unexhausted node. Through some experiments with a virtual agent deploying in a 3D interactive computer game environment, we confirm the completeness and efficiency of two algorithms.

Table 1. RTA* Algorithm

```
Function : RTA*(S_NODE, G_NODE)
C_NODE = S_NODE ; best = 0
do
  for each neighbor Y_NODE of C_NODE
    f = h(Y_NODE,G_NODE) + k(C_NODE, Y_NODE)
    if f < best, B_NODE = Y_NODE ; best = f
  advanceTo(B_NODE)
  C_NODE = B_NODE
until  C_NODE = G_NODE
```

2 Real-Time Search for Shortest Path Finding

State-space search algorithms for finding the shortest path can be divided into two groups: *off-line* and *real-time*. Off-line algorithms, such as the A* algorithm [8], compute an entire solution path before executing the first step in the path. Real-time algorithms, such as the RTA*(Real-Time A*)[6], perform sufficient computation to determine a plausible next move, execute that move, then perform further computation to determine the following move, and so on, until the goal state is reached. These real-time or on-line algorithms direct an agent to interleave planning and actions in the real world. These algorithms can not guarantee to find the optimal solution, but usually find a suboptimal solution more rapidly than off-line algorithms. The RTA* algorithm shown in Table 1 calculates $f(x')=h(x')+k(x,x')$ for each neighbor x' of the current state x, where $h(x')$ is the current heuristic estimate of the distance from x' to the goal state, and $k(x,x')$ is the distance between x and x'. And then the algorithm moves to a neighbor with the minimum $f(x')$ value. The RTA* revises a table of heuristic estimates of the distances from each state to the goal state during the search process. Therefore, the algorithm is guaranteed to be complete in the sense that it will eventually reach the goal, if certain conditions are satisfied. Fig. 2 illustrates the search process of the RTA* algorithm on the example graph of Fig. 1.

The PHA* algorithm [5], which is another real-time search algorithm for finding the shortest path, is a 2-level algorithm. As summarized in Table 2, the upper level is a regular A* algorithm [8], which chooses at each cycle which node from the open list to expand. It usually chooses to expand the node with the smallest *f*-value in the open list, regardless of whether the agent has visited that node before or not. On the other hand, the lower level directs the agent to that node in order to explore it.

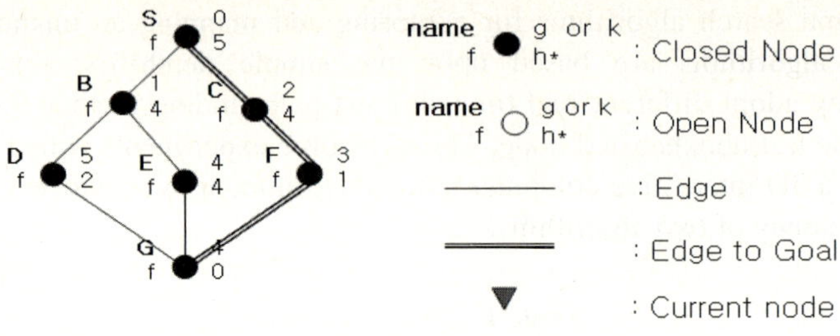

Fig. 1. An Example Graph

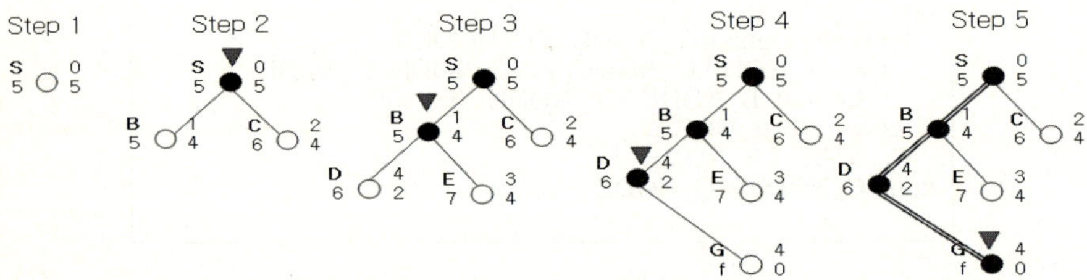

Fig. 2. Search Directed by RTA*

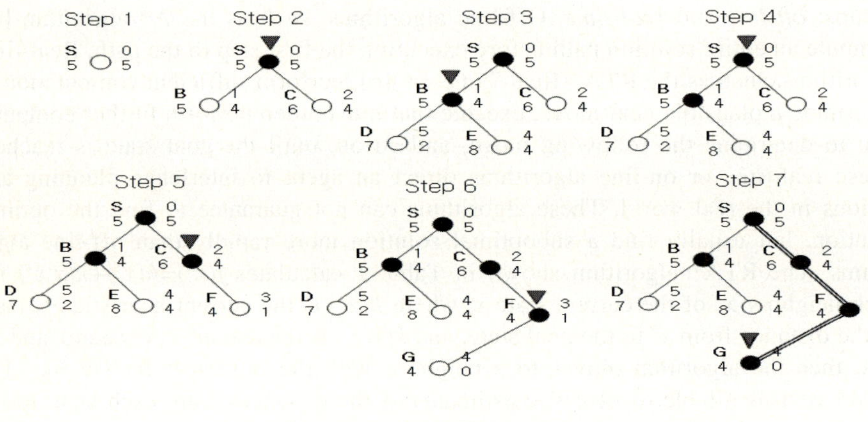

Fig. 3. Search Directed by PHA*

The lower level must use some sort of a navigational algorithm such as Positional DFS (P-DFS), Directional DFS (D-DFS), or A*DFS. Due to the A* algorithm used in the upper level, this PHA* algorithm can guarantee to find the optimal solution. Fig. 3 illustrates the search process of the PHA* algorithm on the example graph of Fig. 1.

Table 2. PHA* Algorithm

```
Function : PHA*(S_NODE, G_NODE)
C_NODE = S_NODE ; best = 0
do
  N_NODE = best node from OPEN_LIST
  if N_NODE = explored
    lowerLevel(C_NODE , N_NODE)
    C_NODE = N_NODE
until  C_NODE = G_NODE

Function : lowerLevel(C_NODE, G_NODE)
S_NODE = C_NODE
f = 0
do
  for each neighbor Y_NODE of C_NODE
    f = g(Y_NODE) + h(Y_NODE,G_NODE)
     if f < best, B_NODE = Y_NODE ; best = f
  moveTo(B_NODE)
  C_NODE = B_NODE
until  C_NODE =  G_NODE
```

3 Search Algorithms for Exploration and Mapping

We consider real-time search algorithms based upon a simple depth-first strategy like Kwek's. Table 3 summarizes the simple depth-first search strategy for exploring an unknown environment. It simply traverses an unvisited edge when there is one until the agent is stuck. As the agent traverses the edges in this greedily manner, we push the current node into a stack. When the agent is stuck, we simply pop the stack until either the stack is empty or the popped node *y* is not exhausted. In the former case, the agent has already traversed all the edges of G. In the latter case, Kwek claimed that there is a path from *u* that leads to *y* in the graph G' obtained by the agent's exploration so far. In Kwek's depth-first search strategy, the agent therefore traverses this path to *y* and repeats the greedy traversal of the unvisited edges. In our depth-first search strategy for exploring and mapping, however, the agent tries to find an optimal path to *y* by applying a real-time shortest path-finding procedure such as RTA* or PHA*. In other words, the *moveShortestPath* function in the definition of our simpleDFS algorithm can be instantiated with the RTA* or PHA* function defined above. We call the real-time depth-first exploration algorithm using RTA* (or PHA) as *DFS-RTA** (or *DFS-PHA**).

Following our search strategy, the agent does not try to find any path to *y* in the graph G' obtained so far, but find an optimal path by traversing even unexplored nodes and edges outside of G'. Fig. 4 shows an example search directed by our simpleDFS strategy. Suppose the agent has already traversed the cyclic path A-E-H-I-G-D-C-B-A and then it is stuck. It pops the nodes A, B, C, and D from the stack until it encounters the unexhausted node G. And then it tries to move to the node G through the shortest path and traverses the unexplored edge to F.

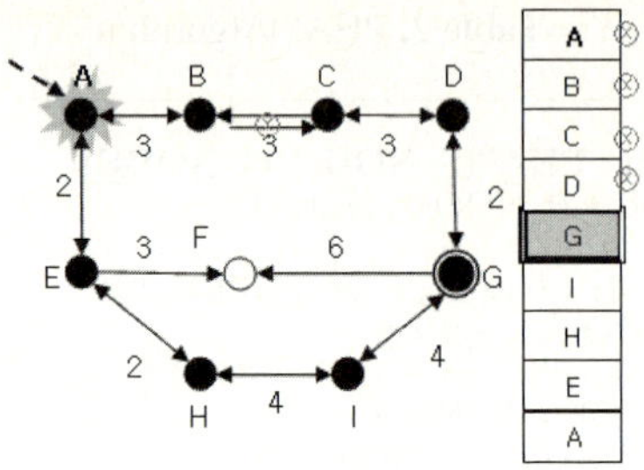

Fig. 4. Search Directed by simpleDFS

Table 3. SimpleDFS Algorithm

```
Function : simpleDFS(S_NODE)
C_NODE = S_NODE   /* start node */
N_NODE, Y_NODE = null
do
  while(exhausted(C_NODE) != true)
    do
    N_NODE = selectNextNode(C_NODE)
    advanceTo(N_NODE)
    push( HISTORY, C_NODE ) /* history stack */
    C_NODE = N_NODE
    end
  do
    Y_NODE = pop(HISTORY)
  until((exhausted(Y_NODE) != true) or
      (empty(HISTORY) = true))
  if (exhausted(Y_NODE) != true),
   moveShortestPath(C_NODE, Y_NODE)
   C_NODE = Y_NODE
until (empty(HISTORY) = true)

Function : exhausted(C_NODE )
for each neighbor Y_NODE of C_NODE
  if unvisited(Y_NODE), return false
return true

Function : selectNextNode(C_NODE)
f = 10000   /* a large value */
for each neighbor Y_NODE of C_NODE
    if distance(C_NODE, Y_NODE) <  f,
     B_NODE = Y_NODE
     f = distance(C_NODE, Y_NODE)
return B_NODE
```

4 Implementation

In order to test our exploration algorithms, we implemented an intelligent virtual agent called UTBot. The UTBot is a bot client of the Gamebots system. The Gamebots [1] is a multi-agent system infrastructure derived from Unreal Tournament (UT). Unreal

Tournament (UT) is a category of video games known as first-person shooters, where all real time players exist in a 3D virtual world with simulated physics. The Gamebots allows UT characters to be controlled over client-server network connections by feeding sensory information to bot clients and delivering action commands issued from bot clients back to the game server. In a dynamic virtual environment built on the Gamebots system, the UTBot must display human-level capabilities to play successfully, such as learning a map of their 3D environment, planning paths, and coordinating team activities under considering their adversaries. In order to assist UT characters' move in a 3D virtual world, the UT game maps contains many waypoints perceivable by characters on itself. These waypoints and links connecting two adjacent waypoints constitute a directional graph model for UTBot's exploration. The UTBot was built upon a generic behavior-based agent architecture, CAA (Context-sensitive Agent Architecture). Our CAA consists of (1) a world model; (2) an internal model; (3) a behavior library; (4) a set of sensors and effectors; and (5) an interpreter. Based on the proposed DFS-RTA* and DFS-PHA* algorithms, we implemented the UTBot's exploring behaviors. The UTBot provides a visualization tool shown in Fig. 5. This visualization tool can help us keep track and analyze the UTBot's exploring behaviors.

5 Experiments

In order to compare with our DFS-RTA* and DFS-PHA*, we implemented another depth-first search algorithm for exploration. The DFS-DFS algorithm uses a simple depth-first search strategy not only for traversing the entire graph systematically before stuck, but also for finding a backtracking path to the unexhausted node from the obtained graph G'. Four different UT game maps of 90, 120, 150, and 180 nodes were prepared for experiments. On each game map, we tried individual exploration algorithms (DFS-DFS, DFS-RTA*, DFS-PHA*) five times with different starting point. Fig. 6, 7, and 8 represent the result of experiments. Fig. 6 shows the average number of visited nodes, Fig. 7 shows the average traveled distance, and Fig. 8 shows the average consumed time of individual exploration algorithms, respectively.

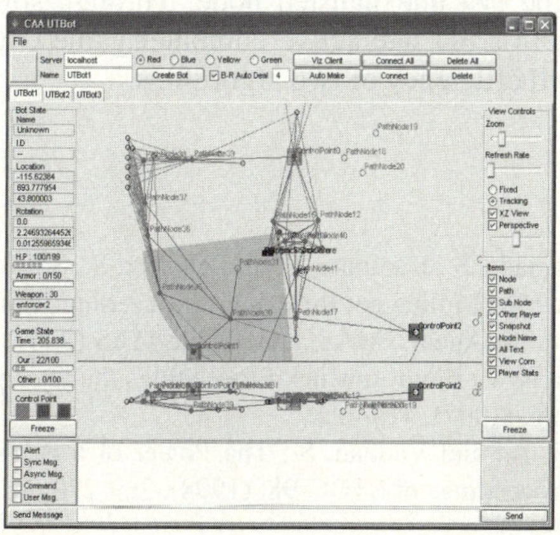

Fig. 5. Visualization Tool

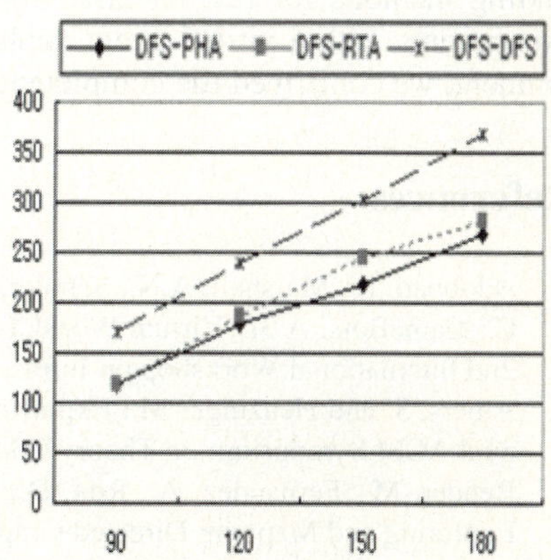

Fig. 6. Visited Nodes

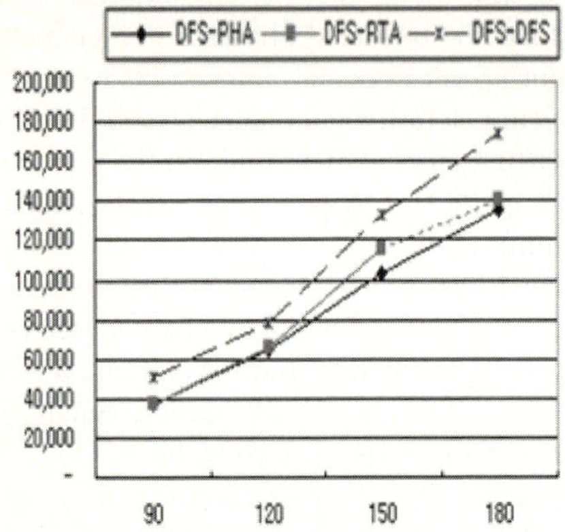

Fig. 7. Traveled Distance

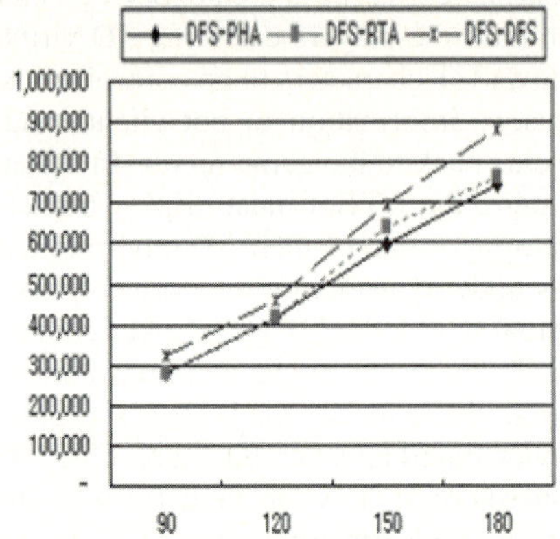

Fig. 8. Consumed Time

We can find out that our DFS-RTA* and DFS-PHA* algorithms outperforms the simple DFS-DFS algorithm in all three different performance measures. In particular, two algorithms show high efficiency in terms of the visited nodes and the traveled distance. Moreover, the more complex and larger world maps are used, the clearer superiority of our algorithms is. We also notice that DFA-PHA* slightly outperforms DFS-RTA* among two proposed algorithms. The reason is in part that the algorithm PHA* can guarantee to find an optimal path to the destination, but the RTA* can guarantee just a suboptimal path.

6 Conclusions

We presented DFS-RTA* and DFS-PHA*, two real-time graph search algorithms for exploring and mapping an unknown environment. Both algorithms are based upon the simple depth-first search strategy, but they use different real-time shortest path-finding methods for fast backtracking to the last unexhausted node. Through some experiments with a virtual agent deploying in a 3D interactive computer game environment, we confirmed the completeness and efficiency of two algorithms.

References

1. Adobbati, R., Marshall, A.N., Scholer, A., Tejada, S., Kaminka, G.A., Schaffer, S., Sollitto, C.: GameBots: A 3D Virtual World Test Bed for Multiagent Research. Proceedings of the 2nd International Workshop on Infrastructure for Agents, MAS, and Scable MAS (2001)
2. Albers, S. and Henzinger M.: Exploring Unknown Environments. Proceedings the 29th Annual ACM Symposium on Theory Computing, (1997), 416-425.
3. Bender, M., Fernandez, A., Ron, D., Sahai, A., and Vadhan, S.: The Power of a Pebble: Exploring and Mapping Directed Graphs. Proceedings of STOC-98, (1998), 269-278

4. Deng, X., Kameda, T., and Papadimitriou, C.: How to Learn in an Unknown Environment. Proceedings of the 32nd Symposium on the Foundations of Computer Science, (1991), 298-303.
5. Felner A., Stern R., Kraus S., Ben-Yair A., Netanyahu N.S.: PHA*: Finding the Shortest Path with A* in An Unknown Physical Environment. Journal of Artificial Intelligence Research (JAIR), Vol.21 (2004), 631-670
6. Korf, R.E.: Real-time Heuristic Search. Artificial Intelligence, Vol.42, No.3, (1990), 189-211
7. Kwek, S.: On a simple Depth-First Search Strategy for exploring Unknown Graphs. Proceedings of the 5th International Workshop on Algorithms and Data Structures, LNCS 1272, (1997), 345-353
8. Pearl, J. and Kim, J.E.: Studies in Semi-Admissible Heuristics. IEEE Transaction on PAMI, Vol.4, No.201, (1982), 392-400

Real-Time Control of Decentralized Autonomous Flexible Manufacturing Systems by Using Memory and Oblivion

Hidehiko Yamamoto[1], Jaber Abu Qudeiri[2], and M.A. Jamali[3]

[1] Dept. of Human and Information Systems, Gifu University,
1-1, Yanagido, Gifu-shi, 501-1193, Japan
yam-h@cc.gifu-u.ac.jp
http://www.gifu-u.ac.jp/~yamlab/
[2] Graduate School of Engineering, Gifu University
[3] Industrial Engineering Dept., Technology University of Troyes, France

Abstract. This paper describes a method that uses memory to determine a priority ranking for competing hypotheses. The aim is to increase the reasoning efficiency of a system the author calls reasoning to anticipate the future (RAF), which controls automatic guided vehicles (AGVs) in autonomous decentralized flexible manufacturing systems (AD-FMSs). The system includes memory data of past production conditions and AGV actions. Using these memory data, the system reorders hypotheses by giving the highest priority ranking to the hypothesis that is most likely to be true. The system was applied to an AD-FMS that was constructed on a computer. The results showed that, compared with conventional reasoning, this reasoning system reduced the number of hypothesis replacements until a true hypothesis was reached.

1 Introduction

Autonomous decentralized flexible manufacturing systems (AD-FMSs) are one approach to the operation of production plants in the 21st century [1,2]. In a study on AD-FMS, the author examined the movements of automatic guided vehicles (AGVs) in terms of controlling which part should be taken next to which machining center during FMS operation [3]. In this study, a system called reasoning to anticipate the future (RAF) was developed that can determine the next action of the AGV at a given point in time, by anticipating the next actions available to the AGV (next action) for several steps into the future and anticipating in advance the FMS operating status that will be produced in the near future. This RAF applies hypothetical reasoning to the number of next actions that can be considered for the AGV (competing hypotheses). However, if the number of agents included in the hypothetical reasoning process in the RAF is increased, the number of next actions that are considered as competing hypotheses also increases. As a result, the replacement of true and false hypotheses and number of repetitions of discrete production simulations produced by these replacements are increased, giving rise to the problem of decreased reasoning efficiency of the RAF. The present article reports a method to solve these problems.

B. Gabrys, R.J. Howlett, and L.C. Jain (Eds.): KES 2006, Part I, LNAI 4251, pp. 252 – 259, 2006.
© Springer-Verlag Berlin Heidelberg 2006

Reported methods to increase the efficiency of hypothetical reasoning include avoiding multiple searches of the same search trees, compiling knowledge from previously conducted reasoning, and applying learning functions using previous reasoning results. The approach in the present study is to express characteristic information of an AD-FMS as memory data that are added and deleted, in order to establish a priority order among hypotheses competing to be selected as true. This method differs from conventional methods, and seeks to make optimum use of the characteristics of AD-FMSs.

Research on predicting future situations has included statistical prediction[4] and a system in which the weight expressing the goodness of the expert predicted value is constantly updated[5]. The present method differs from conventional prediction methods in that it makes predictions based on memory, similarity, and discrete production simulations.

2 Decentralized Autonomous FMS

The AD-FMS dealt with in this paper consists of internal agents including a parts warehouse for the supply of parts into the AD-FMS, a product warehouse where completed parts are kept, AGVs that transport single parts, and several machining centers (MCs), as shown in Fig. 1. There are several units of the same type of MC that do the same work. Groups of the same type of MC are called MC groups, and are designated as MC_1, MC_2, and so on. To distinguish between individual MCs in a group, a further designation is made with a hyphenated number after the group name, as MC_{1-1}, MC_{1-2}, and so on.

In the AD-FMS in this study, production is carried out with each AGV determining its own actions through RAF. RAF can anticipate up to several future steps available to the AGV. Then, through predictions based on discrete production simulations of phenomena that may occur within these several future steps, RAF works backward to the present to decide which of the options the AGV should choose at the present point in time[3]. This process is controlled by the application of hypothetical reasoning to the several competing options (hypotheses) available to the AGV.

Fig. 2 shows the flow of RAF. RAF includes two parallel reasoning processes: hypothetical reasoning for moving decisions (HR-MD), which decides the destination of an AGV, and hypothetical reasoning for parts decisions, which decides which parts to input. Occurrence of a contradiction causes the reasoning to backtrack. HR-MD that was applied in this study is described in the following.

To decide the next action of an AGV, HR-MD (1) sets a priority ranking for destinations of MC groups that perform the same machining process, (2) sets a priority ranking for parts warehouse and product warehouse destinations, and (3) sets a priority ranking for each MC within a group. It then sets a priority ranking for destinations available to the AGV in each of the competing hypothesis, in the order of likelihood of being adopted as a true hypothesis. This ranking is called a moving priority ranking (MP-ranking). Then, from among the competing hypotheses ranked by priority, one is selected as a true hypothesis from among the hypotheses with high priority rankings, and the remaining are considered false hypotheses. The hypothetical reasoning then begins.

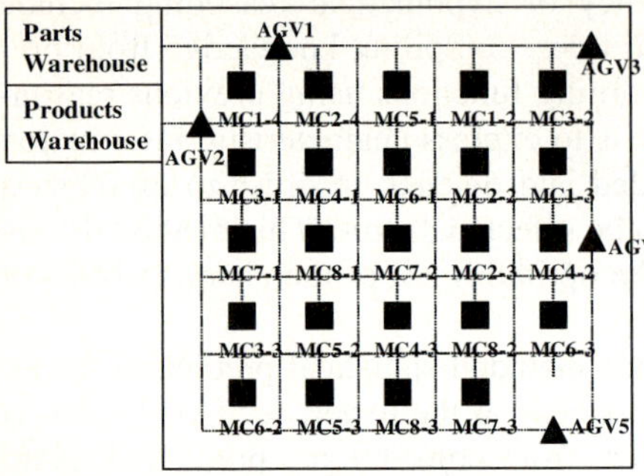

Fig. 1. AD-FMS example

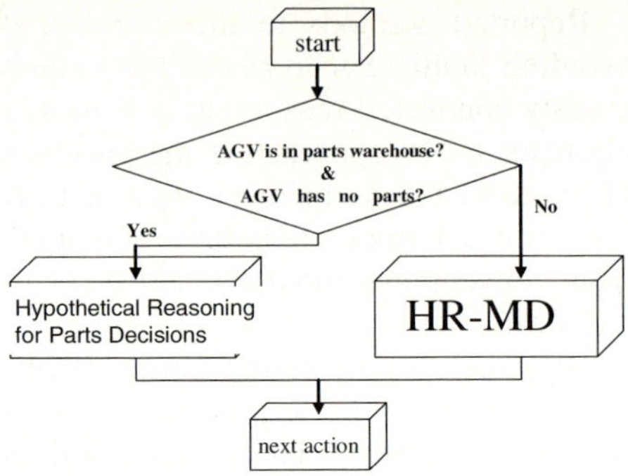

Fig. 2. Flow of RAF

3 Moving Priority Decision Based on Memory and Oblivion

3.1 Contents of Memory

In RAF the reasoning and simulations are repeated continuously as true and false hypotheses are exchanged, until contradictions are no longer found. Because of this, a problem occurs in that the number of exchanges of true and false hypotheses, and number of production simulations that must be re-executed because of these exchanges, increase with greater numbers of MCs or AGVs. In the present study, therefore, good reasoning efficiency was defined as fewer exchanges or production simulations. To solve this problem of excessive exchanges, a method to rank competing hypotheses by oblivion and memory (ranking by oblivion and memory; ROM) was proposed to improve the reasoning efficiency of RAF.

An important point in ROM is which memories are retained and how these data are used. In the following, therefore, we will first consider the contents of memory.

The memory expresses 4 items: the 3 production situations of (a) moving priority ranking, (b) the location of the AGV, and (c) the location of the parts, together with (d) the AGV destination decided at that time (determined action). Next, the contents of these 4 items, as well as 3 terms using these 4 items, are defined as follows.

[Definition] (a) Moving priority ranking: the priority ranking determined for each destination available to the AGVs as reasoned by HR-MD.

[Definition] (b) AGV positions: the current position of each AGV or the destination it is moving toward. This corresponds to the parts warehouse, Product warehouse and each MC.

[Definition] (c) Parts positions: the place where each part existing in the AD-FMS is located. This corresponds to the parts warehouse, product warehouse, each MC and each AGV.

[Definition] (d) Decided action: the final destination selected for the AGV under the production situations of (a)~(c).

[Definition] Single memory: the memory set of (a), (b), (c) and (d).

[Definition] Whole memory: the set of all single memories, M_{number}.

[Definition] Oblivion degree: the values attached to all elements of whole memory, from 1 to M_{number}. The values are decided using a carry up method. For example, if a new memory corresponding to a single memory is assigned oblivion degree M_{number}, the single memory whose oblivion degree had been M_{number} is given a new oblivion degree of $M_{number} - 1$. This carry up method is carried out with other single memories in whole memory. As a result, the single memory whose oblivion degree had been 1 up to that point will be deleted from whole memory. The smaller the oblivion degree is, the sooner the memory is forgotten.

According to the above definitions, ROM can have many memories. Like humans, ROM has memories that are both easier and more difficult to forget.

The relation between whole memory and single memory is shown in Fig. 3. Examples of ROM memory contents (a) ~ (c) are shown in Table 1 to Table 3.

Table 1 shows the contents of memory (a). The table indicates that the MP-rankings, $P'(x)$, for the destinations of the seven competing hypotheses, $MC_{1\text{-}1}$, $MC_{1\text{-}2}$, $MC_{1\text{-}3}$, $MC_{2\text{-}1}$, $MC_{2\text{-}2}$, parts warehouse and products warehouse are 4, 1, 3, 2, 5, 7, 6, respectively. Table 2 shows the contents of memory (b). It indicates that ${}_{AGV}L(m)$, the location of each AGV, is parts warehouse, $MC_{1\text{-}2}$, $MC_{2\text{-}2}$ and $MC_{2\text{-}1}$ respectively. Table 3 shows the contents of memory (c). It indicates that ${}_{Parts}L(s)$, the location of each part, is $MC_{1\text{-}2}$, AGV_1, $MC_{2\text{-}1}$ and $MC_{2\text{-}2}$ respectively.

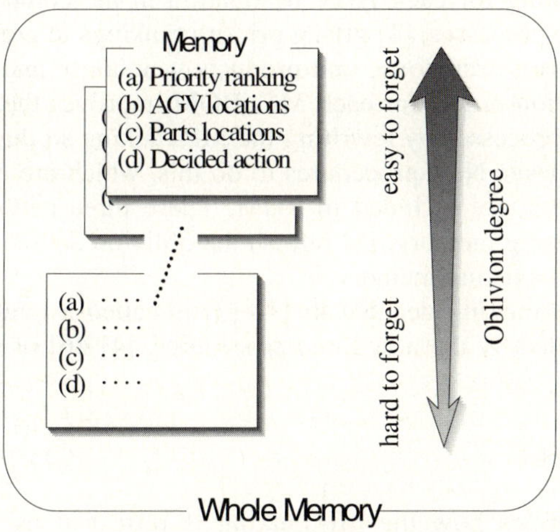

Fig. 3. Memory construction

Table 1. Example of memory (a)

x	1	2	3	4	5	6	7
$G(x)$	MC1-1	MC1-2	MC1-3	MC2-1	MC2-2	Parts Warehouse	Products Warehouse
$P'(x)$	4	1	3	2	5	7	6

Table 2. Example of memory (b)

n	1	2	3	4	5	6	7
Gmemory(n)	MC1-1	MC1-2	MC1-3	MC2-1	MC2-2	Parts Warehouse	Products Warehouse
P (n)	**4**	3	6	**2**	**5**	1	7

Table 3. Example of memory (c)

m	1	2	3	4
AGVs	AGV1	AGV2	AGV3	AGV4
AGVL(m)	Parts Warehouse	MC1-2	MC2-2	MC2-1

3.2 How to Use ROM

ROM is included in the process of HR-MD and works to improve MP-ranking. In the HR-MD decision which is one of the hypothetical reasoning tasks in RAF described in section 2, MP-ranking for each AGV destination in the competing hypotheses is decided through three processes, [1] giving priority rankings to group MCs, [2] giving priority rankings to parts warehouse and products warehouse and [3] giving priority rankings to the selection orders for each MC. ROM improves the MP-ranking determined through three processes by rewriting the MP-ranking so that it is likely to be a realizable true hypothesis. New procedures to do this, which are carried out after the above three procedures, are included in ROM. These three new procedures are [4] refresh MP-ranking using memory, [5] refresh the oblivion degree, and [6] add a new memory and delete an existing memory.

Therefore, the MP-ranking decided in [1]~[3] is called a temporary MP-ranking and the ranking decided by the new three procedures, [4]~[6] of ROM, is called the final MP-ranking.

3.3 How ROM Works

This section first defines how the MP-ranking is refreshed using memory in [4], which is one of the new three procedures of ROM.

[Definition] Refresh MP-ranking using memory: carry out the following three procedures.

[Procedure 4-1] Find the degree of similarity between the current production situation and the content of each M_{number} single memory in whole memory.

[Procedure 4-2] Find the single memory whose priority ranking is the highest and insert the determined action into the top position in the temporary MP-ranking.

[Procedure 4-3] Move the priority rankings in the second and lower positions down in the order. □

In order to carry out Procedure 4-2, it is necessary to check out the similarity between the current production situation and the content of each M_{number} single memory in the whole memory. The similarity is obtained using the following three similarity degrees.

<similarity degree $_{\text{move}}$A>: the similarity in the MP-ranking
<similarity degree $_{\text{agv}}$A>: the similarity in the current positions of each AGV
<similarity degree $_{\text{parts}}$A>: the similarity in the current positions of each part

The similarity degree $_{\text{move}}$A is the value expressing the difference between MP-ranking (a), which is in the memory, and the current temporary MP-ranking which is acquired during HR-MD□The similarity degree $_{\text{agv}}$A is the value expressing the difference between AGV positions (b) and the current positions of each AGV in AD-FMS. The similarity degree $_{\text{parts}}$A is the value expressing the difference between parts positions (c) and the current positions of each part in AD-FMS. The numerical values of the similarity degrees are calculated by adding 1 if the current production situation is the same as a situation in memory, and adding 0 if the current production situation is not the same as any in memory.

By [Procedure 4-1], the similarity degree between the current production situation and the single memory for M_{number} units can be calculated. Next, [Procedure 4-2] is carried out. It selects the single memory with the highest similarity degree, and insert the determined action (d) of the single memory into the top position in the temporary MP-ranking. Third, [Procedure 4-3] moves the existing priority rankings down one place in order and, as a result, a new MP-ranking or the final MP-ranking can be decided.

Next, the followings definitions correspond to [5], refreshing the oblivion degree.

[Definition] [5] Refresh the oblivion degree: Insert the single memory with a highest similarity degree into the lowest place, M_{number}, of the oblivion degree in the whole memory. □

Due to refreshing a single memory once selected is not easily removed from whole memory.

Finally, the function of [6], which adds a new memory and deletes an existing memory, is defined as below.

[Definition] [6] Add a new memory and deletes an existing memory: The following two procedures are carried out.

[Procedure 6-1] The AGVs next action that is finally decided without contradictions is considered to be a new (d) determined action. The (a) MP-ranking, (b) AGV positions and (d) parts locations at this time including the new (d) determined action are memorized in the middle range of the oblivion degree among the whole memory as a new single memory.

[Procedure 6-2] Single memories whose oblivion degrees are higher than the new single memory are in turn moved up and, as a result, the single memory that has had the highest oblivion degree is deleted from the whole memory. □

Through this procedure [6], whenever a real AGV's destination is decided, a new memory is added and at the same time a memory with a low frequency of selection as an AGV destination is deleted.

4 Simulation Experiments

The ROM proposed in the last paper was applied to the AD-FMS created on computer in simulation experiments. The AD-FMS had 9 types of parts, 8 MC groups, 3 of each MC, and 5 AGVs as shown in Fig.1.

On the assumption that unanticipated troubles occur during actual FMS operation, trouble conditions were set in which each AGV randomly broke down 3 times a day, each time for 5 minutes, and part processing time on each MC was randomly extended 10%. Ten different random number sequences were used as the random conditions for these troubles. The number of simple memories stored in the total memory of ROM was set at 20.

Operating results are summarized in Tables 4. The table shows the number of hypotheses until the true hypothesis was selected (hypothesis selection number) during operation when the AD-FMS was operated for 24 hours with each of the 10 types of trouble. For comparison, the table also shows the hypothesis selection number when using HR-MD with a temporary MP-ranking (conventional method [3,6]). From the table, we see that the hypothesis selection number with ROM was reduced to less than half that with the conventional method for all of the 10 types of trouble. For example, the mean hypothesis selection number was 899,943.5 with the conventional method and 445,763.2 with ROM. This indicates that the MP-ranking of ROM is better than the temporary MP-ranking at ordering hypothesis in order of possibility of being selected as true, confirming that there is a reduction in the numbers of wasted true/false hypothesis replacements and production simulations that are executed.

Table 4. Results of Experiments

Hypotheses Selection Number	Conventional method	**ROM**
Condition 1	893,532	452,539
Condition 2	905,398	445,126
Condition 3	921,744	441,059
Condition 4	902,827	446,284
Condition 5	886,551	446,901
Condition 6	898,857	446,815
Condition 7	899,596	444,296
Condition 8	900,342	445,106
Condition 9	891,872	444,910
Condition 10	898,716	444,599
Average	899,943.5	445,763.5

5 Conclusions

The present study proposes a system called ROM that determines a hypothesis priority ranking using memory, to improve the reasoning efficiency of RAF. In this system, the AGV itself decides its movements in AD-FMS. ROM stores previous production situations and the action of the AGV decided at those times as memories. Using these memories, hypotheses are ranked according to possibility of being selected from among competing hypotheses as true, with the aim of improving the efficiency of the RAF. This ROM repeatedly adds new memories and deletes existing ones, so that the memory contents are constantly refreshed.

ROM was applied to AD-FMSs constructed on a computer and the number of hypothesis replacements until a true hypothesis was determined was compared with that from a conventional system. It was found that under all conditions ROM reduced the number of hypothesis replacements to half that with a conventional system, demonstrating the validity of this system.

References

1. Moriwaki, T. and Hino, R., Decentralized Job Shop Scheduling by Recursive Propagation Method, International Journal of JSME, Series C, Vol. 45, No. 2, (2002), 551.
2. Laengle, Th. et al., A Distributed Control Architecture for Autonomous Robot Systems. Workshop on Environment Modelling and Motion Planning for Autonomous Robots , World Scientific, Series in Machine Perception and Artificial Intelligence, Vol. 21, (1994), 384.
3. Yamamoto, H.: Decentralized Autonomous FMS Control by Hypothetical Reasoning including Discrete Simulator, Proceedings of The Fourteenth International Conference on Industrial & Engineering Applications of Artificial Intelligence & Expert Systems (IEA/AIE-2001), Lecture Notes in the Computer Science/Lecture Notes in Artificial Intelligence series, Springer-Verlag (eds. L. Monostori, J. Vancza and M. Ali), Budapest, (2001)571-581.
4. Berger, J.O.: Statistic Decision Theory and Bayesian Analysis (2nd Ed.), 1985, Springer-Verlag.
5. Variant, L.: A Theory of the Learnable, Communication of the ACM, Vol. 27, No.11, (1984)1134.
6. Yamamoto, H. and Marui, E., Intelligent Real-time Production Control To Realize Autonomous Decentralized FMS, Proceedings of International Conference on Industrial Engineering and Production Management (IEPM2003), Porto, (CD-ROM) (2003).

Adaptive QoS Control Mechanism in Flexible Videoconference System

Sungdoke Lee[1], Dongsoo Han[1], and Sanggil Kang[2]

[1] School of Engineering, Information and Communication University
P. O. Box 77, Yuseong, Daejeon, 305-600 Korea
{sdlee, dshan}@icu.ac.kr
[2] Computer Science & Engineering
Inha University
253 Younghyun-dong, Nam-gu, Incheon, 402-751 Korea
sgkang@inha.ac.kr

Abstract. In this paper, we propose an adaptive QoS (Quality of Service) control mechanism and its corresponding architecture using multi-agent framework to improve flexibility of a videoconference system (FVCS). The proposed mechanism realizes more flexible by changing their QoS control strategies dynamically than a conventional videoconference system. We modelled the prototype of our modified FVCS and implemented the proposed mechanism by varying QoS parameters. The prototype system shows its capability of flexible problem solving against QoS degradation, along with other possible problems within the given time limitation.

1 Introduction

To use videoconference system (VCS) on heterogeneous computers and network environments, users have to consider the status of system resources, situations of other site over network, and working condition of videoconference processes on machines in order to maintain a comfortable system operation [1]-[4]. Usually, these tasks of Quality of Service (QoS) burden novice users. To reduce the users' burden, Flexible Videoconference System (FVCS) [5]-[7] has been designed by adding some flexible features to traditional VCS. FVCS can change its functions and performances autonomously in accordance with changes of user requirements or system/network environments. In the research area of adaptive QoS control based on the network environments, an FVCS has been developed in order to control its outgoing data rate for considering congestion condition of the network. However, most FVCSs control the QoS in a static fashion, which makes their problem solving capability monolithic. Because of that, flexible behavior considering importance or emergence of given problems during conference is difficult to achieve.

In this paper, we propose an adaptive QoS control mechanism to overcome the limitation explained above. The adaptive QoS control mechanism can deal with the problems such as exhaustion of resources, changing the QoS control dynamically based on the characteristics of the problem, status of problem solving

B. Gabrys, R.J. Howlett, and L.C. Jain (Eds.): KES 2006, Part I, LNAI 4251, pp. 260–267, 2006.
© Springer-Verlag Berlin Heidelberg 2006

process, and user requirements. The QoS control is done by a new architecture of knowledge processing called M-INTER to switching problem solving strategy. In the experimental section, we develop our prototype FVCS and show its capability of solving the problems of QoS decreasing, along with other possible problems within the given time limit.

The remainder of this paper is organized as follows: Section 2 briefly explains basic concept of FVCS. Section 3 then presents adaptive QoS control mechanism and its architecture. Section 4 shows experiment result and its analysis using the prototype system. In Section 5, we conclude our paper.

2 Flexible Videoconference System

To lighten users' burdens of VCS, FVCS [5]-[7] is attained by embedding the following functionality to the existing VCS, i.e., (F1) service configuration function at the start of a session and (F2) service tuning function during the session. Here, (F1) composes the most suitable service configuration of VCS automatically by selecting best software modules and deciding their set-up parameters under the given conditions of environments and users. The function reduces users' burden at the start up of videoconference session. (F2) adjusts the QoS autonomously according to the changes of network/computational environments or user requirements against the QoS. The function is realized by two phase tuning operations, i.e., parameter operation tuning for small-scale changes and reconfiguration of videoconference service for large-scale changes. In this paper we focus on (F2) and propose some improvement in (F2).

Due to the static QoS control, most existing FVCSs have several problems as follows: (P1) Acceptable range of environmental changes is small: Most FVCSs are designed and customized to use on a specific network environment, i.e. the Internet environment. For an unexpected environment, QoS control ability is extremely limited. (P2) Load balancing capability is very limited: Since each user terminal plays roles of both a sender and a receiver for videoconferencing, heavy load on CPU and I/O of each terminal node can be imposed, which causes limited capability to deal with various types of changes. (P3) Meta level requirements of QoS are not considered: To achieve an effective and delicate QoS control, meta level monitoring capability for detecting classes of problem and progress status of problem resolving process is needed.

3 Strategy-Centric Adaptive QoS Control

3.1 Adaptive QoS Control with M-INTER Model

To overcome the problems of traditional QoS control mechanisms described in the previous section, especially (P3), we embed an adaptive QoS control mechanism to VCM agents in FVCS and realizes the service tuning function (F2). The adaptive QoS control mechanism is designed along with the following policies: (1) Introduce the knowledge representation scheme of meta level knowledge

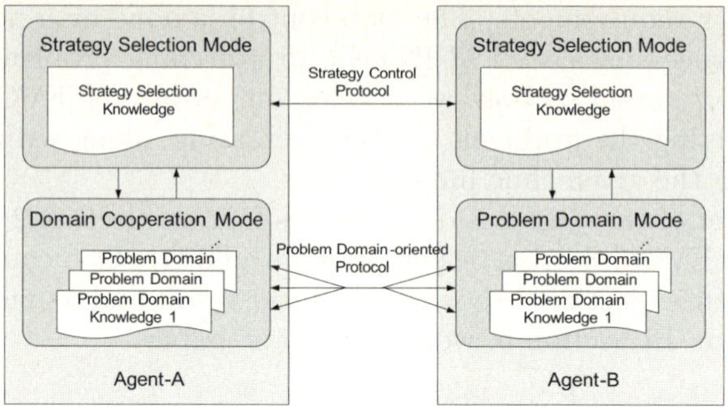

Fig. 1. Conceptual scheme

to control the problem solving processes. (2) Give a design of function to incorporate multiple QoS control strategies. (3) Design a knowledge processing mechanism to switch the QoS control strategies using meta level knowledge. (4) Realize the proposed mechanism as software modules to promote re-usability and maintenance during agent design and implementation.

Fig. 1 represents the concept of adaptive QoS control mechanism. In the mechanism, the knowledge processing of QoS control will be performed in the following two different modes of agents in FVCS, namely Strategy Selection Mode and Domain Cooperation Mode. In Strategy Selection Mode agents monitor the meta-level conditions of cooperative behavior such as a class of given problem, a level of improvement during problem solving process, the rest period until deadline, etc. With the conditions, agents select the most adequate strategy by using Strategy Selection Knowledge. Moreover, each agent has to negotiate to select the best strategy and exchange information of each cooperation status and problem domain knowledge of them. Hence, we defined a strategy control protocol between agents in order to promote the meta level cooperation. In Domain Cooperation Mode, the problem domain specific cooperation is performed between agents. A problem domain means a class of problems to be resolved, such as video QoS control problem domain and audio QoS control problem domain. In a problem domain, several strategies are prepared to be activated. Each strategy corresponds to a different method to resolve a specific problem, and the strategy is realized by a Problem Domain Knowledge in an agent and Problem Domain-oriented Protocol (DoP) between agents. In this mode, one strategy is selected and activated during cooperation of agents with respect to situation of both the external environment and the cooperation status. The selection of strategy is in charge of Strategy Selection Mode.

The adaptive QoS control is realized by the alternative transition of these two modes. When a problem occurs, firstly, an agent begins to negotiate with other agents of FVCS to decide the most proper strategy on the given conditions in Strategy Selection Mode. Next, the agents transit to Domain Cooperation Mode, and they begin to perform the problem domain-oriented cooperation using specified Problem Domain Knowledge and DoP.

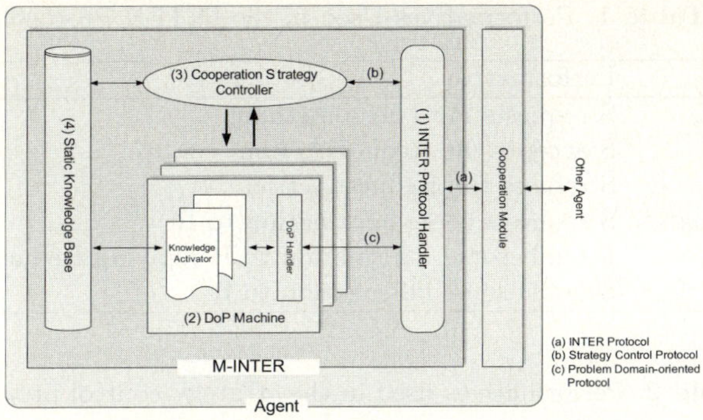

Fig. 2. M-INTER architecture

3.2 Meta-Interaction Architecture

To accomplish the strategy-centric adaptive QoS control, we propose Meta-Interaction (M-INTER) Architecture which consists of a new architecture of knowledge in VCM agents and a new protocol between agents as seen in Fig. 2.

From Fig. 2, INTER Protocol Handler is a simple message handling module to cope with inter-agent communication messages driven by INTER Protocol. The primary protocol used for cooperating between agents is represented in communication primitives as in Table 1. In the table, "S" stands for a sender of a message, while "R" stands for a recipient of a message. When an agent A asks an agent B to do something, the agent A sends a RequestAction message to the agent B. After receiving the message, the agent B decides whether it can accept the request or not, and replies an Acceptance or a Refusal message to the agent A [6].

Problem Domain-oriented Protocol Machine (DoP Machine) is a protocol handling module to achieve the problem domain-oriented cooperation in Domain Cooperation Mode. The DoP Machine is an implementation model of the Problem Domain Knowledge in Fig. 1. The DoP Machine consists of a DoP Handler and several Knowledge Activators(KAs). DoP Handler is a simple parser of DoP, while Knowledge Activator decides actions of an agent based on the static knowledge when it receives a message from other agent. There can be several KAs in a DoP machine. Each KA can be realized by different knowledge processing implementations. For instance, a KA can be implemented by a rule-based inference engine, on other hand, other KAs can be implemented by a simple procedural-type inference module. Cooperation Strategy Controller is a strategy control module activated in Strategy Selection Mode. This module is charged with selecting DoP Machine and Knowledge Activator and negotiating with other agents using Strategy Control Protocol as seen in Table 2.

The selection is done in accordance with the meta level requirements of QoS and cooperation status. Make-Coop related messages (Request, Acceptance, Refusal-Make-Coop messages) are used when two agents begin cooperative problem solving. In the messages some kinds of information about required

Table 1. Performatives Used in the INTER Protocol

Performative	Performative
RequestAction	S requests R to do something
Acceptance	S accepts the RequestAction
Refusal	S refuses the RequestAction
RequestInformation	S requests some information to R
Information	S sends some information to R replying RequestInformation
Report	S sends some information to R

Table 2. Performatives used in the strategy control protocol

Performative	Performative
Request-Make-Coop	S requests R to start cooperation
Acceptance-Make-Coop	S accepts a request from R to start cooperation
Refusal-Make-Coop	S refuses a request from R to start cooperation
Request-Close-Coop	S requests R to terminate cooperation
Acceptance-Close-Coop	S accepts a request from R to terminate cooperation
Refusal-Close-Coop	S refuses a request from R to terminate cooperation
Request-Change-Protocol	S requests R to change protocol
Acceptance-Change-Protocol	S accepts a request from R to change protocol
Refusal-Change-Protocol	S refuses a request from R to change protocol
Request-Change-Coop-Status	S requests R to change cooperation status
Acceptance-Change-Coop-Status	S accepts a request from R to change cooperation status
Refuse-Change-Coop-Status	S refuses a request from R to change cooperation status

cooperation such as goal and deadline are exchanged. Close-Coop related messages are used to close the cooperation. Change-Protocol related messages are used in the case that change of DoP machine is required during a cooperation. Moreover Change-Coop-Status related messages are used to change cooperation status such as goal and deadline. Static Knowledge Base is a container of expert knowledge that is used by Cooperation Strategy Controller and Knowledge Activators.

By applying the architecture described above, it can change cooperation strategies flexibly switching DoP machines and Knowledge Activators. The mechanism can improve dynamic range against the changes on environments and user requirements, and it can also realize meta level requirements of QoS.

3.3 Applying M-INTER Model to FVCS

To apply M-INTER model to FVCS, we define four types of DoP Machines for QoS control of video process in FVCS as follows: (1) Basic Protocol Machine: a simple protocol machine to control QoS of video in both sites. By the protocol, VCM agents direct videoconference processes repeatedly to increase/decrease the values of QoS parameters in a fixed range until resource conditions are recovered. (2) Compromise Level Protocol Machine: a deliberative type protocol machine to adjust QoS by trial and error strategy. With this protocol, VCM agents have each meta state on limitations of degradation of QoS parameters, namely compromise level. VCM agents perform negotiation to find the compromise point each other

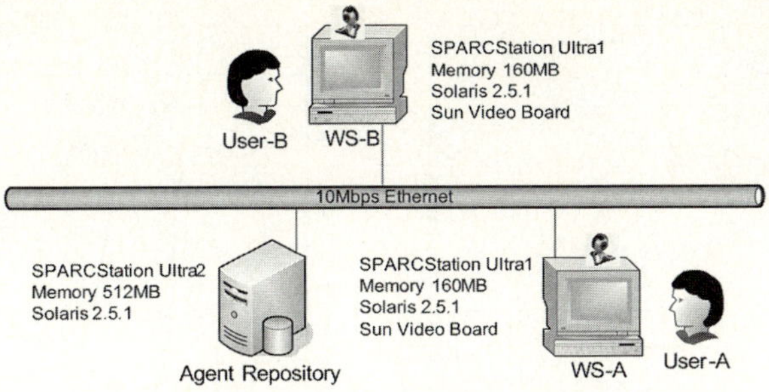

Fig. 3. Experiment environment

and to change their compromise level dynamically. (3) Time Restricted Protocol Machine: a protocol machine that can cooperate considering time restrictions such as deadline in the first priority. With this protocol, cooperation between agents is terminated forcedly regardless of the degree of problem solvency. (4) Reactive Protocol Machine: a reactive type protocol machine to reduce communication overhead between agents. Although accuracy of parameter tuning is not guaranteed, quick response against the changes is enabled. This strategy is used on the unstable environment where resources are expected to be changed at very short time interval.

4 Experiments and Evaluation

The proposed architecture based on M-INTER model is embedded to the VCM agents of FVCS, described in section 3. We have used ADIPS Framework [8] as an agent-based computing infrastructure. The proposed architecture of M-INTER model is written in Tcl/Tk programming language [9] extending the agent's knowledge architecture provided by original ADIPS Framework.

The FVCS based on M-INTER architecture has been implemented and experimented under the environment shown in Fig. 3. To evaluate the flexibility of agents provided by our model, we have changed the CPU resources forcedly, and have monitored the system's behavior. The result clearly shows that the experiment with the proposed architecture acts much more flexibly than the existing systems. Firstly, some extra load on CPU of WS-B has been added externally and observed the changes of QoS parameters of video process, i.e., frame rate, encoding quality, and resolution. At that time, on WS-A in Fig. 3, User-A represents his requirement of smoothness in movement of video to the highest priority, second highest priority to video quality and lowest priority to video resolution. On the other hand, User-B on WS-B represents its requirements with the highest priority to video quality, second highest priority to smoothness, and lowest priority to resolution. When the problem of CPU resources insufficiency is occurred, we provided a limited time to solve the problem to the system. The given time limit was 120 seconds, respectively.

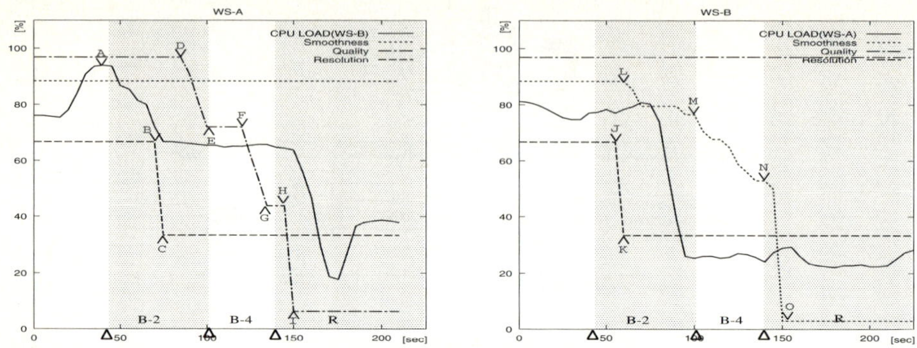

Fig. 4. Behavior of FVCS against CPU variation (time limit 120s): (a) Change of QoS at User-A, (b) Change of QoS at User-B

Fig. 4 represents the transition of the parameters' values controlled by the agents. In the graph, x-axis represents the time (second) and y-axis represents each parameter values observed at the recipient site. The parameter values are expressed in percentage when the following values are regarded as 100%, CPU load: 100%, Smoothness in movement: 35-fps, Quality: 32-level, Resolution: 3-level. In the graph, symbol triangle indicates the switching time of DoP machines or Knowledge Activators (KAs). 'B-1(the KA 1 of Basic Protocol Machine)', 'R(the Reactive Protocol Machines)' etc. represent the types of DoP machines and the KAs used in the time slot. Basically, when the CPU resources are found insufficient (maybe CPU load of WS-A or WS-B increases), the agents aim to maintain the stability of the system by considering the user's requirements. In this case, it might reduce the QoS and consequently release some CPU resources. When the CPU load of WS-B increases (at point A or after 40 seconds), agents of FVCS began cooperative actions. At first, the KA 2 of the Basic Protocol Machine (B-2) is selected. There exist five types of KAs in Basic Protocol Machine. If this numerical value of this KA becomes big, the slope becomes sharper. In the area of 'B-2'of Fig. 4 (a), the resolution of video provided to User-A at WS-A was reduced at point B-C according to user priority. Secondly, the video quality was reduced at point D-E. While the resolution of video provided to User-B at WS-B was reduced at point J-K in 'B-2' shown in Fig. 4 (b). In the next instance, smoothness was reduced at point L-M. In this stage, the Cooperation Strategy Controller starts activating. It calculates the remaining time and the degree of problem solution (in this case, the release of CPU resource). By considering these results it selects the KA without changing the protocol machines. As a result, the parameter value has a sharp declination between (M-N) points. When the remaining time becomes very small (at the warning stage), the Cooperation Strategy Controller starts activating again and changes its protocol from Basic to Reactive Protocol Machine, 'R'. During a Reactive Protocol session, only the highest priority QoS parameter remain unchanged, but other QoS parameters are decreased to the minimum, without considering any further conditions. Therefore, CPU resources are released (points H-I, N-O). In this given time limit (120 seconds), the agents try different strategies to satisfy the user requirements as much as possible and finally the system succeeds in releasing the CPU resources.

5 Conclusion

In this paper, we have proposed a mechanism and it's architecture called M-INTER to accomplish adaptive QoS control. This model extends the functions of the QoS control mechanism by sophisticated cooperation among agents in FVCS. The proposed architecture analyzes the property of the problem occurred on QoS, considers every step during a session, changes the strategies dynamically and solves the problem even more flexibly. We have implemented the proposed architecture and carried out experiments by applying it to FVCS. The experimental results clearly show that the flexibility is improved.

For the future work of our system, we need to develop an algorithm for improving the efficiency of the Cooperative Strategy Control.

References

1. MaCanne, S., Jacobson, V.: Vic: a flexible framework for packet video. ACM Multimedia, Nov. (1995) 511-522
2. Turletti, T., Huitema, C.: Videoconferencing on the Internet. IEEE/ACM Trans. on Networking, Vol. 4, No. 3. (1996) 340-351
3. Bolliger, J., Gross, T.: A Framework-Based Approach to the Development of Network-Aware Applications. IEEE Trans. on Software Engineering, Vol. 24, No. 5. (1998)
4. Jacobson, V., McCanne, S.: Visual Audio Tool. Lawrence Berkeley Laboratory. ftp://ftp.ee.lbl.gov/conferencing/vat
5. Suganuma, T., Kinoshita, T., Sugawara, K., Shiratori, N.: Flexible Videoconference System based on ADIPS Framework. Proc. of the 3rd Int. Conf. and Exhibition on the Practical Application of Intelligent Agents and Multi-Agent Technology (1998) 83-100
6. Lee, S. D., Karahashi, T., Suganuma, T., Kinoshita, T., Shiratori, N.: Construction and Evaluation of Agent Domain Knowledge for Flexible Videoconference System. Trans. IEICEJ, Vol. J83-B, No. 2. (2000) 195-206
7. Shiratori, N., Sugawara, K., Kinoshita, T., Chakraborty, G.: Flexible Networks: Basic Concepts and Architecture. IEICE Trans. Commun., Vol. E77-B, No. 11. (1994) 1287-1294
8. Kinoshita, T., Sugawara, K.: ADIPS Framework for Flexible Distributed Systems. Lecture Notes in Computer Science, Vol. 1599. Springer-Verlag, Berlin Heidelberg New York (1998) 18-32
9. Ousterhout, J. K.: Tcl and the Tk Toolkit. Addison-Wesley (1994)

Multi-agent Architecture for Automating Satellite Control Operations Planning and Execution

Adriana C. Biancho, Andreia C. de Aquino, Mauricio G.V. Ferreira, José Demisio S. da Silva, and Luciana S. Cardoso

National Institute for Space Research (INPE). Applied Computing Postgraduate Program. Av. dos Astronautas, 1758. CEP 12227-010. São José dos Campos, Brazil
{adcarnie, ancarnie, demisio}@lac.inpe.br, mauricio@ccs.inpe.br, luciana@dss.inpe.br

Abstract. Reducing the costs of space operations is an increasing demand which may be achieved by automating the ground segment operations. This work proposes a Multi-Agent Ground-operation Automation architecture named MAGA that aggregates agents responsible for automating the ground segment operation planning and execution. MAGA architecture manages ground resource allocation for multi-satellite tracking, plans the satellite control operations and automates plan execution. For managing the sharing of ground resources, it identifies satellites with time-conflicting visibility periods and reduces the least priority satellite tracking. For the planning process, MAGA architecture analyzes whether a satellite tracking period is sufficient to achieve all the tracking goals and allows the elimination of lesser priority goals in case of insufficient time. The satellite control planning problem is specified in PDDL 2.2 and the satellite control plans generated according to the temporal planning paradigm.

1 Introduction

In the field of satellite control, there has been a general interest in automating the space operation planning and execution. The automation of space operations represents a way of reducing the in-orbit satellite maintenance costs as well as increasing the satellite control system reliability.

The automation described in this work is strongly concerned with the tasks of controlling multiple satellites, generating satellite control plans and automating plan execution. In order to realize these tasks, there is a set of restrictions to be managed, such as the sharing of ground stations for multi-satellite tracking and time-critical operations to be executed during the limited period of time a satellite becomes visible to a ground station.

Ground stations are where the antennas which capture satellite signals are located at. The time window in which a satellite becomes visible to a ground station is named satellite visibility period or satellite pass [1].

Concerning the task of controlling multiple satellites that share the use of ground stations, the visibility period of one satellite may conflict with the visibility period of another one. When this situation takes place, the visibility period conflicting part of the least priority satellite must be cancelled. Canceling a satellite conflicting visibility

B. Gabrys, R.J. Howlett, and L.C. Jain (Eds.): KES 2006, Part I, LNAI 4251, pp. 268 – 275, 2006.
© Springer-Verlag Berlin Heidelberg 2006

period means to reduce the time window in which the satellite will be tracked by the ground station. This time window reduction generates a time restriction that must be considered while generating the satellite control plan.

Satellite control plans, named Flight Operation Plans (FOP), contain all the operations related to the control of in-orbit satellites. FOP generation requires that these operations be inserted at specific instants of time in order to achieve the satellite tracking goals. These goals comprise, for instance, calibration measure execution, telemetry reception, telecommand sending, and distance and speed measure execution.

MAGA architecture presents a solution by planning the ground resource utilization in order to meet multiple satellite trackings, by generating a FOP for each satellite and by automating FOP execution.

For planning the ground resource utilization, it considers the satellite priority rankings in order to reduce their tracking periods when conflicts occur. For plan generation, it reasons whether or not there is sufficient time to achieve all the tracking goals and allows disconsidering goals in case of insufficient time. For plan execution, MAGA architecture restricts the human satellite operators to the execution monitoring and allows their intervention in case of anomalies.

The following section relates some fundamental concepts of the Artificial Intelligence planning area with the satellite control problem. Section 3 presents MAGA architecture and describes how its agents actuate to provide the solution. Section 4 presents an overview of the architecture behavior. Following a discussion of related research in Section 5, the paper concludes in Section 6 with a summary of this work contributions.

2 Planner Agent and the Satellite Control Problem

Planning problems involve a set of initial states, a set of goals and the corresponding actions that contribute to achieve these goals. A planner agent is an agent responsible for solving planning problems. This type of agent uses a planning algorithm called planner to plan a sequence of actions whose execution will lead to the desired goals.

The representation of planning problems has been a concern since 1971 when Fikes and Nilsson developed the STRIPS language [2]. From this time on, other researchers have proposed planning problem representation languages based on STRIPS aiming at developing a more expressive language for real planning problems.

PDDL (Planning Domain Description Language) was the language adopted in this work to model the satellite control planning problem [3] [4]. The choice of using PDDL is due to the fact that PDDL 2.2 allows planning problem modellers to specify actions with duration and deterministic unconditional exogenous events, which are facts that will become true or false at time points that are known to the planner in advance, independently of the actions that the planner chooses to execute. These two features are very important for the generation of Flight Operation Plans due to the fact that space operations have duration and must be adjusted to the well-defined time windows of the satellite tracking periods.

PDDL separates the planning problem description into two files. The Planning Domain Description file contains the domain types, functions, predicates and actions.

An action is associated to a precondition and an effect, and may be assigned an execution duration. The Problem Description file contains the objects present in the problem instance, the initial states and the goals.

In the satellite control problem, a planning action consists of a space operation that is concerned either with the control of satellite on-board equipments status or with obtaining the satellite exploitation expected products. Planning goals are represented by tracking goals to which we propose the assignment of priorities according to their execution relevance in the domain.

3 MAGA Architecture

MAGA architecture agents are specialists in executing specific tasks which result in the generation of a final plan to be executed. Figure 1 illustrates MAGA architecture.

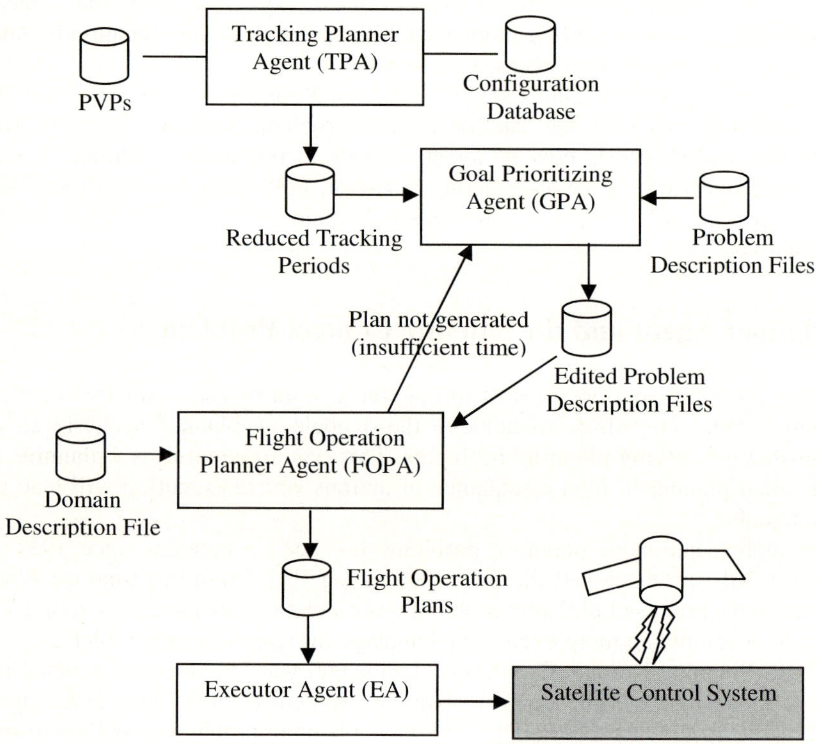

Fig. 1. Multi-Agent Space Operation Automation architecture (MAGA architecture)

The Configuration Database contains all the satellites and ground stations configuration parameters including the sattelite and tracking goal priorities. A Pass Visibility Prevision (PVP) file contains data about the future passes of one satellite over a specific ground station. This file provides, for instance, predictions about the ground antenna pointing data.

The following sub-sections describe the MAGA architecture agent responsibilities.

3.1 Tracking Planner Agent (TPA)

This agent generates Tracking Plans (TP) defining which satellites may be tracked by a ground station, the order they can be tracked, and the tracking duration.

The time window a satellite can be really tracked is named the satellite tracking period which may comprise its whole visibility period or just a portion of it. A satellite tracking period comprises just a portion of its visibility period when the visibility period of a second satellite conflicts with that of the first one. In this case, the tracking period of the satellite with less priority is reduced to avoid time conflict.

TPA thus manages the problem of multi-satellite tracking with conflicting visibility periods (concerning the same ground station) by cancelling or shortening the tracking of the satellite with less priority.

TPA also applies other criteria to define the sequence of satellites to be really tracked. The restrictions imposed by these criteria are: (i) when a satellite is visible to two or more ground stations simultaneously, it must be tracked by the ground station with the highest priority; (ii) the tracking of satellites with visibility periods shorter than a pre-defined duration are not considered; (iii) there must be a minimal time interval between the end of a satellite tracking and the beginning of the tracking of another one; and (iv) a satellite calibration operation may be cancelled whenever a time conflict occurs, independently of the satellite priority.

Figure 2 illustrates how the Tracking Planner Agent solves the problem of multi-satellite conflicting tracking periods. In Figure 2(a) it is shown a two satellite typical tracking by a ground station and three important variables that concerns this problem.

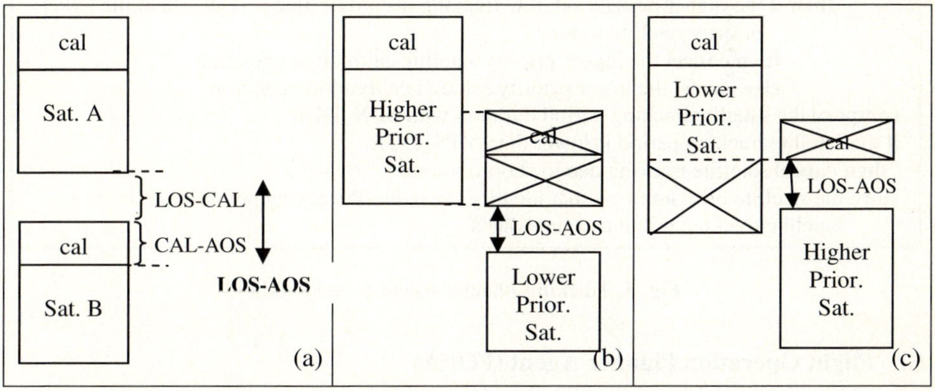

Fig. 2. Solution to the problem of multi-satellite conflicting tracking periods

Variable LOS-AOS (Loss of Signal-Acquisition of Signal) represents the minimal time interval between the end of satellite A tracking period and the beginning of satellite B tracking period. Variable LOS-CAL (Loss of Signal-Calibration) represents the minimal time interval between the end of satellite A tracking period and the beginning of satellite B calibration operation. Variable CAL-AOS (Calibration-Acquisition of

Signal) represents the minimal time interval between the beginning of satellite B calibration and its acquisition of signal (tracking beginning).

The calibration operation consists of sending a signal from the Satellite Control Center to test the ground station equipments before the satellite tracking beginning. Figures 2(b) and 2(c) illustrate the canceling of a lower priority satellite tracking period portion at the satellite tracking beginning and at its end, respectively.

A simplified pseudo-code for the Tracking Planner Agent problem is presented in Figure 3.

```
Function TRACKING-PLANNER-AGENT
  (perception: PVPs files, Configuration Database)
  returns new satellite tracking initial and final dates, orbit numbers that identify the
            reduced tracking periods

  select PVPs files (different satellites) for the same ground station that contain the same]
          prediction initial and final dates
      compare the satellite visibility period duration with the minimal duration value
              (variable MIN_DUR) for a satellite to be tracked
  if the satellite visibility period is lower than MIN_DUR
    then cancel satellite tracking due to short duration
  retrieve the satellite priorities from the Configuration Database
  compare a higher priority satellite tracking instant of time with the tracking instant of
              time of a lower priority satellite
  if the difference between these instants of time are equal or lower than LOS-AOS
    then cancel the lower priority satellite tracking instant of time
    else if the difference between these instants of time are equal or lower than LOS-CAL
           + CAL-AOS
         then if the higher priority satellite tracking instant of time is higher than the lower
                 priority satellite time
                then cancel the higher priority satellite calibration operation
                else cancel the lower priority satellite calibration operation
compare the satellite tracking period duration with MIN_DUR
if the satellite tracking period is lower than MIN_DUR
  then cancel satellite tracking due to short duration
store the satellite orbit numbers that identify the reduced tracking periods and the new
       satellite tracking initial and final dates
```

Fig. 3. Tracking Planner Agent pseudo-code

3.2 Flight Operation Planner Agent (FOPA)

This agent is responsible for generating the Flight Operation Plan which contains satellite control operations (planning actions) to be executed by an Executor Agent during the satellite tracking period. FOPA generates a Flight Operation Plan for each satellite to be tracked by a specific ground station antenna.

FOPA has as input the Planning Domain file and the Problem Description files, written in PDDL 2.2. For plan generation, it uses the temporal planner LPG-TD [5] as its reasoning mechanism.

3.3 Goal Prioritizing Agent (GPA)

This agent acts when a satellite tracking period is reduced in order to avoid time conflict with another satellite. In this case, the satellite tracking period is generally not enough to execute all the original goals previously defined in the PDDL Problem Description file. The Goal Prioritizing Agent (GPA) by assigning priorities to goals, makes it possible to consider solely the most relevant goals for planning. The aim is to fit the planning actions to the short-time tracking period.

For achieving that, GPA iteratively selects the lesser priority goals from the PDDL Problem Description file and eliminates them until it is possible to generate a plan to a subset of the original goals. These two tasks are performed by the GPA subcomponents named Goal Selector and Goal Editor, respectively.

In order to implement this solution, each goal must be annotated with a priority which comprises a symbolic value that might change from one satellite tracking to another, to better specify the need for the goal execution in the next tracking period.

3.4 Executor Agent (EA)

This agent is responsible for the Flight Operation Plan automated execution at the specified instants of time. Obeying the sequence of operations (planning actions) specified in the Flight Operation Plan, the Executor Agent calls the Satellite Control System functions related to each plan operation.

4 MAGA Architecture Behavior

The Tracking Planner Agent (TPA) obtains from the PVPs files all the satellites visibility period initial and ending times for a specific ground station. Once this information was obtained, this agent compares the several satellites initial and ending times, reducing the least priority satellite tracking period when some intersection occurs.

After performing its task, the Tracking Planner Agent (TPA) communicates with the Goal Prioritizing Agent (GPA) to provide a list containing the orbit numbers that identify the satellite tracking periods reduced by TPA and the new satellite tracking initial and final dates.

For each reduced tracking period, GPA selects the least priority goal from the PDDL Problem Description file and eliminates it. These two tasks are performed by the Goal Prioritizing Agent sub-components Goal Selector and Goal Editor, respectively. Once the least priority goal was eliminated from the Problem Description file, the Flight Operation Planner Agent (FOPA) attempts to replan by considering as input the new Problem Description file.

The loop – select goal by priority, remove goal from the planner agent (FOPA) input, and attempt to replan – is iteratively executed until FOPA can generate a Flight Operation Plan that achieves a subset of the original goals, i.e. the remainder most relevant goals, within the reduced tracking period restriction of time. The generated FOP is hence ready for being automatically executed by the Executor Agent.

5 Related Works

MAGA architecture addresses two main problems that are separately treated by the Space Operation and the Artificial Intelligence Planning communities.

The first problem consists of scheduling the ground resource utilization for multi-satellite tracking. Zheng'an et al. [6] propose some criteria aiming at supporting the maximum number of satellite trackings by a network of ground stations in a certain period. MAGA architecture does not address this issue but addresses the multi-satellite tracking conflict resolution itself.

The second problem consists of having a large number of possible goals of differing values and of having to choose a subset that can be accomplished within a limited time. Unfortunately, the existing planner algorithms are generally designed to accept a well-defined set of goals and terminate in failure unless the entire set of goals can be achieved.

MAGA architecture (Goal Prioritizing Agent) provides a solution to this problem by assisting the classical planner (Flight Operation Planner Agent) in choosing which goals to achieve. Smith [7] also provides advice on the goals to the planner. Nevertheless, he addresses planning problems to which the cost of achieving a set of goals depends on the order in which the goals are achieved.

For the problem of multi-satellite control planning we identified that a simpler solution could be adopted as the order in which goals are achieved does not influence on the solution. Therefore, MAGA architecture simply adopts a descending-ordered ranking of goals (priorities) and suggests the removal of the lesser priority goals when satellite tracking time conflicts occur.

Coddington [8] also reasons about goals with priorities but does that in a distinct way from ours. Coddington edits the plan when there is insufficient time to achieve the entire set of goals. It removes from the plan a goal and all of its associated actions and constraints. But for doing that, there is the associated cost of maintaining the dependencies between actions and goals during the planning process.

MAGA architecture (Goal Prioritizing Agent) prefers editing the input for the planner rather than the already generated plan. By solely editing the input for the planner, we avoid having to interfere on the planner activity.

Therefore, in MAGA architecture the planner is considered as a black-box subcomponent with its input being adjusted to portray the varying context of the satellite control problem. The advantage of treating the planner as a black-box subcomponent is the possibility of replacing it as the Artificial Intelligence planning area evolves.

6 Conclusions

In this paper, a new multi-agent automation planning and execution architecture is proposed to the context of multi-satellite control problem.

MAGA architecture solves the problem of controlling satellites with conflicting time tracking periods. When a set of satellites share the same ground station and an intersection occurs between the visibility periods of two satellites, the Tracking Planner Agent reduces the tracking period of the least priority one. This time period reduction means the MAGA architecure planner agent (FOPA) will not have sufficient time to achieve the entire set of original goals (tracking goals).

At this point, the Goal Prioritizing Agent selects and removes goals so that FOPA can generate a plan that achieves the set of remainder goals within the reduced satellite tracking period. Goal priorities are defined by the satellite operator and are configurable in order to portrait the next satellite tracking requirements. So, goals disconsidered for the generation of a satellite control plan (FOP) may have their priorities augmented in the next plan generation. This enables to configure the planning task to reflect the actual satellite tracking requirements.

The main contributions of this work are related to the preparation of an adequate input for the Flight Operation Planner Agent which has an A.I. planner algorithm as its reasoning mechanism. As the planner algorithms are not currently prepared to abandon the achieving of some goals in favor of others (according to their priorities to the problem) during the planning process, MAGA architecture proposes a way of dealing with this restriction by scheduling in advance the ground resource utilization according to the satellite priority ranking and by iteratively editing the planner input goals.

Furthermore, the MAGA architecture aspect of operation execution automation facilitates the work of human satellite operators by performing the most repetitive tasks. This automation allows a single human operator to control several satellite trackings thus reducing the in-orbit satellite maintenance costs as well as increasing the satellite control system reliability.

References

1. ECSS-E-70: Part 1A: Space Engineering – Ground Systems and Operations. Part 1 – Principles and Requirements. ESA Publications Division, Netherlands (2000)
2. Fikes, R., Nilsson, N.: STRIPS – A New Approach to the Application of Theorem Proving to Problem-Solving. Artificial Intelligence, Vol. 3-4, N. 2 (1971) 189-208
3. Fox, M., Long, D.: PDDL 2.1 – An Extension to PDDL for Expressing Temporal Planning Domains. Journal of Artificial Intelligence Research, Vol. 20 (2003) 61-124
4. Edelkamp, S., Hoffmann, J.: PDDL 2.2 – The Language for the Classical Part of the 4th International Planning Competition. Technical Report 195. Albert Ludwigs Universität, Institüt fur Informatik, Freiburg, Germany (2004)
5. Gerevini, A., Saetti, A., Serina, I., Toninelli, P.: Planning in PDDL2.2 Domains with LPG-TD. In International Planning Competition booklet, 14th International Conference on Automated Planning and Scheduling. Whistler, Canada, June (2004)
6. Zheng'an, Z., Tongjie, Y., Lun, L.: Planning and Scheduling of Operations of Small Satellites. Proceedings of the Eight International Conference on Space Operations. Montreal, Canada, May (2004)
7. Smith, D.: Choosing Objectives in Over-Subscription Planning. Proceedings of the 14th International Conference on Automated Planning and Scheduling. AAAI Publisher. Whistler, Canada, June (2004) 393-401
8. Coddington, A.: A Continuous Planning Framework with Durative Actions. Proceedings of the Ninth International Symposium on Temporal Representation and Reasoning. IEEE Computer Society. Manchester, UK, July (2002) 108-114

Web-Based Agent System for Interworking Between Wired Network Devices and Mobile Devices

Sang Tae Kim[1], KeeHyun Park[2], and Hyun Deok Kim[1]

[1]Dept. of Electronic Engineering, Kyungpook National University, Daegu 702-701, South Korea
[2]College of Information and Communication, Keimyung University, Daegu 704-701, South Korea
hyundkim@ee.knu.ac.kr

Abstract. A web-based agent system has been proposed and demonstrated to realize the interworking services between wired network devices and mobile devices. The agent system enables the resource-limited mobile device to support Jini network services without any additional client program installation in it. It also provides the list of services available on the network in real time, which facilitates the users to utilize various services through the web browser of the mobile device.

1 Introduction

As the mobile devices become more powerful enough to run much more sophisticated applications, the interworking services between the mobile devices and other electronic devices become more important and feasible. The mobile device should be able to interact with other mobile devices or even with non-mobile devices anytime and anywhere in the world to support the interworking services. The agent system plays a key role to realize the interworking services since the devices may employ different operating systems, different platforms and different protocols [1].

Though various middlewares are available at this moment, it is not easy to realize the interworking services between wired network devices and mobile devices since the resources of the mobile devices may not be sufficient to operate the middlewares required for the interworking services. For example, Jini is one of the widely used middlewares for the interworking between the home network devices and the home appliances. However, Jini network technology cannot be applied directly to the mobile devices since the resources of the mobile device such as a processing power and a memory size are not sufficient to run Java Virtual Machine (JVM) [2]. Most of the middlewares have similar problems since they have been developed originally as a part of the desktop applications and thus they are too heavy to be applied to the mobile devices [3]-[4].

It is also important to provide a user-friendly interface on the mobile device to facilitate the utilization of the services. The web browser has become one of the

B. Gabrys, R.J. Howlett, and L.C. Jain (Eds.): KES 2006, Part I, LNAI 4251, pp. 276–283, 2006.
© Springer-Verlag Berlin Heidelberg 2006

most widely used user interfaces even in the mobile devices. The users can utilize any service on the network through the web browser. Thus the web browser is a very promising user interface to implement the interworking services between wired network devices and mobile devices.

In this paper, we proposed and demonstrated a web-based agent system employing Jini network technology to realize the seamless interworking services between the wired network elements and the mobile devices. The web-based agent system enables the resource-limited mobile device to participate in Jini network without any additional client program installation in it. The users utilize all the services on the Jini network through a user-friendly interface of the mobile device since the interworking services are provided through the web browser of the mobile device. In the section 2, we will briefly review the Jini network technology. After that we will describe the web-based agent system in the section 3 and the experimental results in the section 4. The conclusion will be given in the section 5.

2 Jini Network Technology

The Jini network technology is built around the Java programming language and environment. It realizes an object-oriented distributed network where the devices and the applications are treated as services. The users can easily utilize the services through the network without any change of the specific configuration of the client. In addition, there is no need to install any additional programs such as device drivers into the client. It makes the dynamic network management easy and simple since the Jini network is intrinsically a distributed network. The Jini network supports a plug and play operation when a device or an application is added to the network. The Jini network components are composed of a service provider, a service client and a lookup service as shown in Fig. 1.

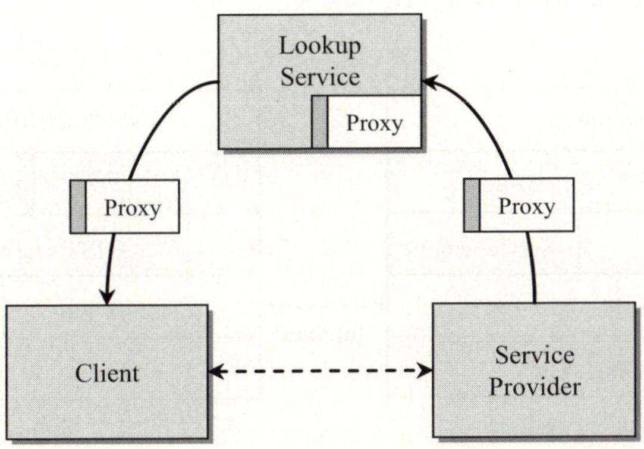

Fig. 1. Jini network components

The characteristics of the Jini network released by Sun Microsystems are as follows [5].

1. Conciseness: In Jini network, the services communicate each other and thus it provides a simple method to add and to utilize the network nodes such as a scanner and a printer.
2. Reliability: Jini informs automatically when the list of services available changes. It also provides a structure to recover the system with a problem and supports the client to find other service when a problem occurs in a specific service.
3. Expandability: Jini network technology can organize, expand and combine community because it can form a group by sharing each services. It also provides the environment the user utilize all the services within the group.

The Jini network technology is originally developed for a desktop environment and implemented by using a JVM. However, the mobile devices such as cellular phones and portable digital assistants (PDAs) have insufficient resources and thus have several problems to perform the operations related floating points, reflections and error processing. Namely, it is difficult to apply the Jini technology to the mobile device since the mobile device may not support the JVM. Thus, it is important to develop an agent system which enables the mobile device to participate in the Jini network even when the resources of the device are insufficient.

3 Configuration of the Web-Based Agent System

Recently, the web browsers are widespread and used in various devices such as workstations, personal computers (PCs) and even mobile devices. Thus we have implemented a web-based agent system to support the Jini network technology on the mobile device, which facilitates the users to utilize the services through a user-friendly interface. The configuration of the web-based agent system with a mobile device is shown in Fig. 2.

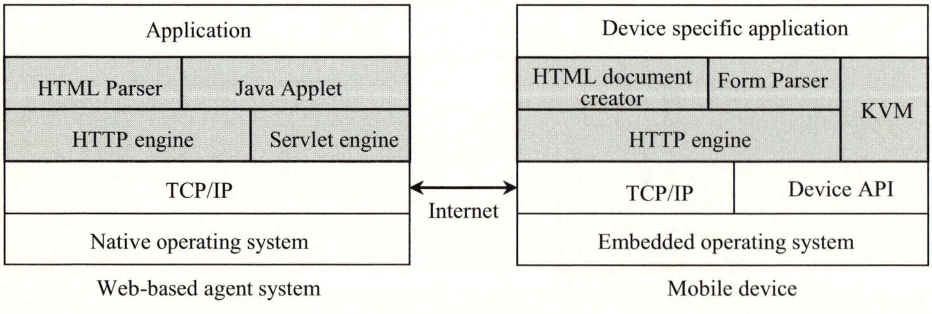

Fig. 2. Configuration of the web-based agent system with a mobile device

The web-based agent system has the JVM to execute the Servlet engine on which the Java applet is executed. Since the Java applet is executed in the web-based agent system, the web-based agent system can participate in the Jini network and can utilize Jini network services. The web-based agent system also has a HTTP engine and a HTML Parser to support web-based services.

The mobile device is implemented by using the Kilobyte Virtual Machine (KVM) which is widely used to support Java environments in mobile devices. The mobile device has a HTTP engine and a HTML document creator to support a web browser. The web browser of the mobile device is used to access the Jini agent system by using a HTTP protocol. If the mobile device is accessed to the web-based agent system, Java applet of the web-based agent system is downloaded to the mobile devices. Thus, the users can access the Servlet engine of the web-based agent system through the web browser of the mobile device. The Form Parser of the mobile device converts the Java message transmitted from the agent system to a proper one readable via the HTML. The connection between the web-based agent system and the mobile devices is established by using a TCP/IP interface. Other devices are also connected to the web-based agent system through the TCP/IP interface.

The web-based agent system provides a plug and play operation in the distributed network, which is a very important feature since it facilitates the interworking between devices. Namely, the user can easily add or remove devices without any information on the device in Jini network. The web-based agent system is designed to support a plug and play operation. Thus, the agent system provides a dynamic service list and the users can explore the services available in a real time through a web browser of the mobile device.

4 Demonstration Results

To evaluate the validity of the proposed web-based agent system, we implemented a Jini network with the web-based agent system, a mobile device and wired network elements. The network configuration is shown in Fig. 3.

The client (mobile device) was connected directly to a public internet service provider (ISP) network though a 2.4 GHz wireless access point and the web-based agent system located on the local access network (LAN) which was connected to the public ISP network. The wired network elements were a personal computer (PC) with a file server, a printer and a scanner. The client device was a PDA (*Compaq iPAQ PocketPC H*4700). We used a KVM and a Connected Limited Device Configuration (CLDC) on the mobile device and the operating system of the PDA was Microsoft Windows for Consumer Electronics (CE). The version of the Java was J2sdk 1.3.1 [6].

Both the browsers of the mobile device and the web-based agent system were web browsers, Internet Explorer. The user interface of the mobile device was implemented by using a Java applet and the web server of the web-based agent system was implemented by using a Servlet, which enabled the client to participate in the Jini network through the web browser. It is notable that we installed

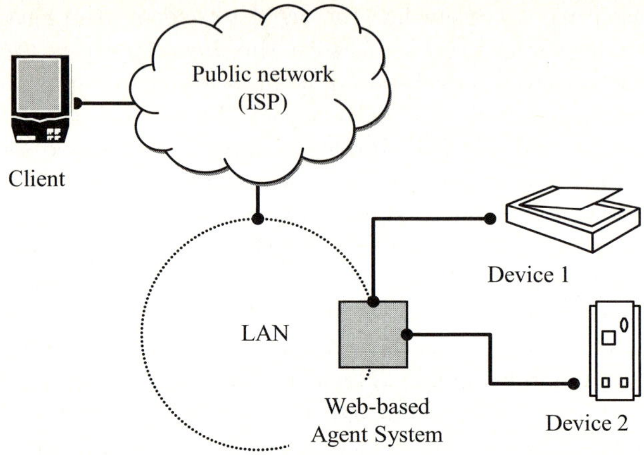

Fig. 3. Jini network with web-based agent system

Jeode virtual machine into the client to support a Java applet since the Microsoft Internet Explorer included in the Windows CE does not support a Java applet [7]. The web-based agent system was implemented to provide a service list available to the client and thus the users can easily search the services available in the Jini network. The servlet of the web-based agent system receives the proxy ID and the service ID of the specific service registered to the Jini network from the Jini Lookup Service. The servlet lists up the available service list form the proxy ID and the service ID and provides the service list available to the mobile device. The process of the service list provision is as follows. The constructor of the service list initializes the user interface and sets up all the objects required for the client to participate in the Jini network. It also creates a `Discoverer`, which tries to discover the services available through a `DiscoveryListener` on a `LookupDiscovery` instance. Finally, the constructor display `ServiceItem` to the user. An example of the service list of the mobile device (client) is shown in Fig. 4.

The source code to implement a dynamic service list is as follows:

```
import java.util.ArrayList;
public class Client implements Runnable
 {
  private java.util.List vec = new ArrayList();
   class Discoverer implements DiscoveryListener
    {
      vec.add(newregs[i].getLocator().getHost()+" ");
        for(int j = 0; j < vec.size(); j++) {
           dic += (String)vec.get(j);
         }
    }
 }
```

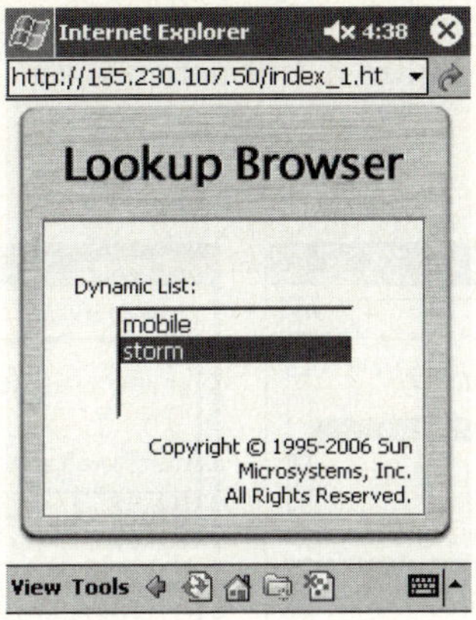

Fig. 4. The service list of the devices

Jini network technology provides a security policy by using Java2 security mechanisms, which grants the client the right to access a specific service provider. Each client should have the security policy file denoting the right of access and the client should submit the policy file through a `java.security.policy property` to the look server to utilize a service. The lookup server examines the policy file submitted from the client and determines whether to allow or to deny the access requested by the client. We have created two policy files, one for web-based agent system and the other for the mobile devices. Two files are similar but the policy file of the web-based agent system denotes the right to access Jini network, while the policy file of the mobile device denotes the right to access only the web-based agent system.

The policy files have includes a security file as a subset. Following is the context of the security file of the mobile device.

```
grant { permission java.security.AllPermission; };
```

The policy file was registered in `java.security.policy` of mobile device and the context of the policy file as follows.

```
policy.url.3=file:/Windows/lib/security/mypolicy
```

We have demonstrated two exemplary services, a file download service and a printing service. The selection process of the file to be downloaded from the file server on the Jini network is shown in Fig. 5(a) and the downloading process is shown in Fig. 5(b). If the user select a file server from the service list through the web browser of the client, the client sends the proxy ID and the service ID of the file server to the web-based agent system. The agent system searches the file server and informs the client of the usability of the file server. After receiving the message, the client accesses the file server and the user can download files in the file server. Since the Java applet is downloaded from the web-based agent system the user can download the file by executing the Java applet through a web browser. Thus there is no need to install the Java applet in the mobile device, which makes it possible to utilize a Jini network service with a resource-limited mobile device.

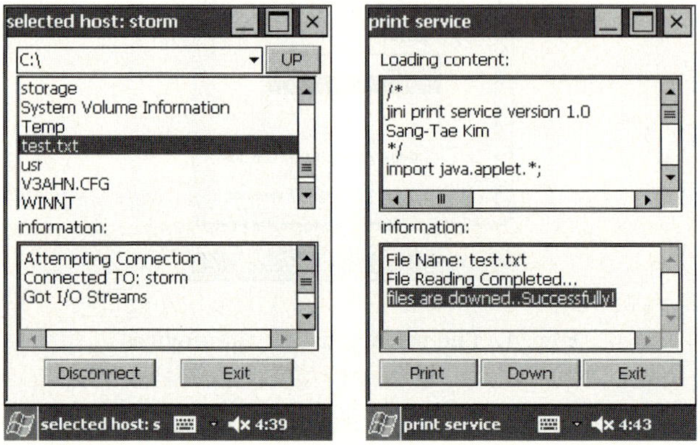

(a) (b)

Fig. 5. File download service: File search process (a), File download processing (b)

After reading a downloaded file (test.txt), we have the file to be printed by using another service provider (a printer). The file selection process to be printed and the print process are shown in Fig. 6 (a) and (b), respectively. If the user execute the printing service through the web browser of the mobile device, the client sends the proxy ID and the service ID of the printer to the web-based agent system.

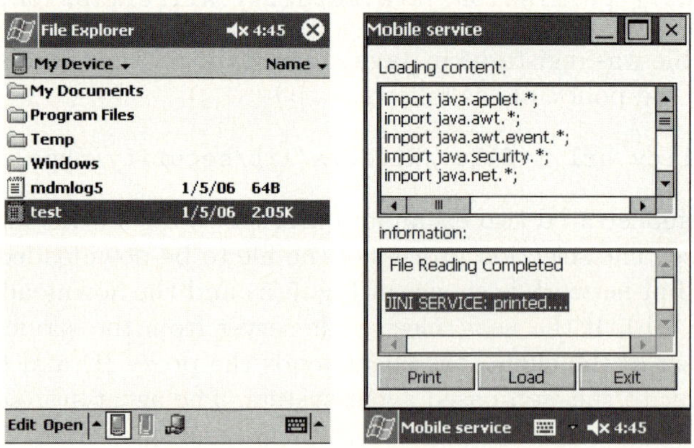

(a) (b)

Fig. 6. Printing service: File search process (a), print processing (b)

The agent system searches the printer and informs the client of the usability of the printer. After receiving the message, the client transmits data to the printer to execute a file printing. When the file printing service has been completed the printer server sends a message to the Jini Lookup service to inform the client has completed the use of the service. It is notable that there is no need to install any printer driver in the client to use a file printing service.

5 Conclusion

A web-based agent system has been implemented to support the interworking services between wired network devices and mobile devices. It enables the mobile device to participate in a Jini network without any additional client program installation in it. It also provides a service list through a web browser of the mobile device, which facilitates the users to search and to utilize various services available on a Jini network in real time. Since the web-based agent system enables an internetworking between the mobile device and other electric devices over a Jini network, it realizes seamless services over mobile networks and wired networks.

Acknowledgement. This work was partially supported by the Mobile Technology Commercialization Center(MTCC) and the Brain Korea 21(BK21) Project.

References

1. Sang Tae, Kim., Byoung Ju, Yun., Hyun Deok, Kim.: Mobile Agent System for Jini Networks Employing Remote Method Invocation Technology. KES (2005)
2. Landis, S., Vasudevan, V.: Reaching out to the cell phone with Jini. System Sciences (2002)
3. Sing, Li.: Professional Jini. Wrox Press Ltd (2000)
4. Gupta, R., Talwar, S., Agrawal, D.P.: Jini home networking: a step toward pervasive computing. IEEE (2002)
5. Sun Microsystems.: Java Object Serialization (1998)
6. Sun Microsystems.: Java 2 Platform Micro Edition. http://java.sun.com/j2me/index.jsp (2002)
7. Insignia.: Jeode Platform White Paper (2001)
8. John W., Muchow.: Core J2ME. Sun Microsystems Press (2002)
9. Hinkmond Wong.: Developing Jini Applications Using J2ME Technology. Addison Wesley (2002)
10. W. Keith Edwards.: Core JINI. Sun Microsystems Press (2002)
11. Michael Jernimo., Jack Weast.: UPnP Design by Example. Intel Press (2003)
12. McDowell, C., Shankari, K.: Connecting Non-Java Devices to a Jini Network. Proceedings of the Technology of Object-Oriented Languages and System (2000)

Load Protection Model Based on Intelligent Agent Regulation

Dragan Jevtic[1], Marijan Kunstic[1], and Stjepan Matijasevic[2]

[1] University of Zagreb, Faculty of Electrical Engineering and Computing, Unska 3, 10000 Zagreb, Croatia
{dragan.jevtic, marijan.kunstic}@fer.hr
[2] Helix, IX. Juzna obala 18, 10020 Zagreb, Croatia
stjepan.matijasevic@helix.hr

Abstract. Paralleling rapid advancement in the telecommunication network expansion is necessary for advanced network traffic management surveillance. The increasing number and variety of services being offered by networks have emphasized the demand for optimized load management strategies. The paper deals with regulation of a mobile agent moving toward service processing resource in the part of the agent network. We have constructed the agent architecture for the control of service processing load. The goals of the controlling system were both to protect the processing load and to predict the arrival rate of client's requests. Self-adaptive property is implemented by reinforcement Q-learning. The analysis is based on experimentation through simulations.

Keywords: intelligent agent, load protection, reinforcement learning.

1 Introduction

The selection and routing process in the agent environments is crucial for optimal reduction of time spent in service and for optimal resource usage inside the communication network. Most commonly the approaches and methods are designed to support the communication service directly. This means that the resources requested by the communication service are connected or reserved only when needed. Similarly, the protective mechanism activates when collision or failure is detected. Early identification of requirements, irregular states and failures is very complex due to the network's dynamics. There are two classes of communication services that play dominant role in communication. These are the real-time and non-real time services with main distinction based on their sensitivity to delay.

The paper discusses an agent system modeled for processing of the server load protection. Maximum benefits from the intelligent agent-based approach for this application were predicted because of the agent's ability to act toward the assigned goal using the on-line and self-adaptive learning techniques. The goal had been set to continuous saving a certain amount of resources for the services requiring on-line

B. Gabrys, R.J. Howlett, and L.C. Jain (Eds.): KES 2006, Part I, LNAI 4251, pp. 284 – 291, 2006.
© Springer-Verlag Berlin Heidelberg 2006

execution, to minimize rejection and to predict request arrival rate. By strict follow-up of the goal the protection mechanism could be automatically expanded, whereas the other, such as the on-line or higher priority services, was beyond the prediction. These other objects, agents or services were supported indirectly by monitoring the processor's power only, and by dynamically assigning available power to the lower priority resources [1, 3].

We assumed the agent system with one intelligent agent in the network where the agents migrated from one node to another node and performed the operations on behalf of their users [2]. The intelligent agent represented an autonomous and self-adaptive entity [1, 3, 6], which normally collaborated with other agents within the agent's environment. We designed a heterogeneous agent system with four types of agents. These were an intelligent agent (signed as a load manager agent - *LMA*), selector agents (*SA*), client agents (*CA*) and service provider agents (*SPA*). *LMA*, *SA* and *SPA* agents were immobile and *CA* agents were assumed to be mobile. We assumed that the *CA* agent was the entity acting on behalf of its user by moving through the network. The immobile triple *LMA*, *SA* and *SPA* were used to regulate movement of *CA*.

This paper is organized as follows: first, the reinforcement learning and internal representation of the intelligent agent is described and general description of Q-algorithm is specified in Section 2. The agent structure for load protection, agent collaboration model and reward settings is introduced and described in Section 3. The outcomes of simulations are given by diagrams and in table, and are described in section 4, followed by Conclusion.

2 Reinforcement Learning

Reinforcement learning (RL) is a problem faced by an agent that learns the function through trial and error interactions within the environment. The RL algorithm pertains directly to the agent's experience interacting within the defined environment and varying the policy in real-time. It can be efficiently used for solving sequential decision-making problems. Basic idea is to use the experience for gradual discovery of optimal *value function*. This is the function that finds the best long-term return an agent can get from a given state, when a particular action following optimal policy is undertaken. Having detected the outcome of such an action the agent uses the reinforcement learning algorithm Q-learning (Watkins 1989), or Sarsa, to improve the long-term estimate of the value function associated with the pair state and action [4, 5].

2.1 Intelligent Agent Internal Representation

Instead of dealing with the environment directly, the intelligent agent can work together with other agents to solve a given task. Separation from the environment enables the intelligent agent to assign specialized tasks to specialized agents and to manage their activity. Internal representation of the intelligent agent consists of the action, state and monitoring devices. Management rules for the action-state alteration were implemented in the *change* mechanism as shown in fig. 1. Various structures of the agent's internal representation could be elaborated, but these elements were recognized as common.

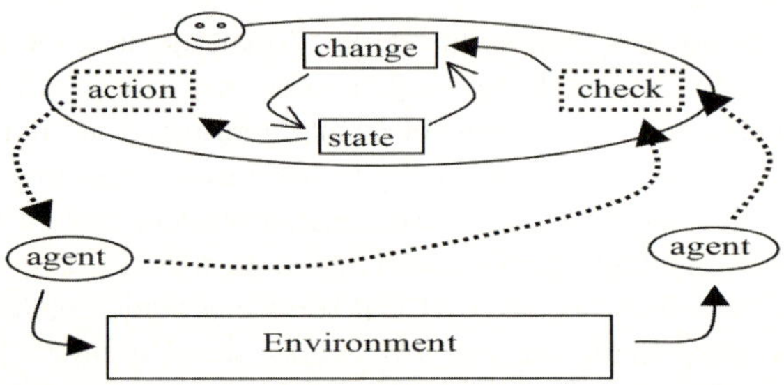

Fig. 1. The intelligent agent operates with specialized agents. Information flow consists of the orders from the intelligent agent and notifications from the specialized agents. Four components represent the intelligent agent internal structure: action, check, change and state.

Relevant to the agent's internal structure the intelligent agent could be in the function shown in fig. 1. The agent received a scalar value or a reward for every executed action. The potential problem was in defining the appropriate contribution of different parameters and in expressing them as a unique scalar value presented to the agent.

2.2 Q-Learning

The main step in any learning is the update phase at which the result of an action changes the values of agent's internal representations.

Q-learning operates with the experience of each state transition to update one element of a table. This table has an entry, *Q*(*s*,*a*), for each pair of state *s* and action *a*. After transition $S_t \rightarrow S_{t+1}$, having taken action a_t and having received a reward r_{t+1}, this algorithm performs the update,

$$Q_{t+1}(s_t, a_t) = (1-\alpha)Q_t(s_t, a_t) + \alpha\,[REWARD(s_t, a_t) + \gamma \max_{a_{t+1}} Q_t(s_{t+1}, a_{t+1})], \quad (1)$$

where $0 < \alpha < 1$ is learning rate and γ is a discount factor. After an adequate number of actions this process converges so that the *greedy policy* with respect to *Q* is optimal. The greedy policy is to select in each state *s* an action *a* for which *Q*(*s*, *a*) is the largest. Thus, this algorithm enables finding optimal policy just from the experience and without a model of environmental dynamics.

3 Agent Model and Load Protection

We considered the agent's supported communication services in which we assumed that the *CA* agents moved through the network on behalf of their users. We presumed that these *CA* agents supported the non-real time services i.e. that their movement could be delayed without significant impairment of the service quality. Nevertheless, we believed that the general rule for the non-real time services was soonest possible completion of all service requests, irrespective of whether delay was a critical execution parameter.

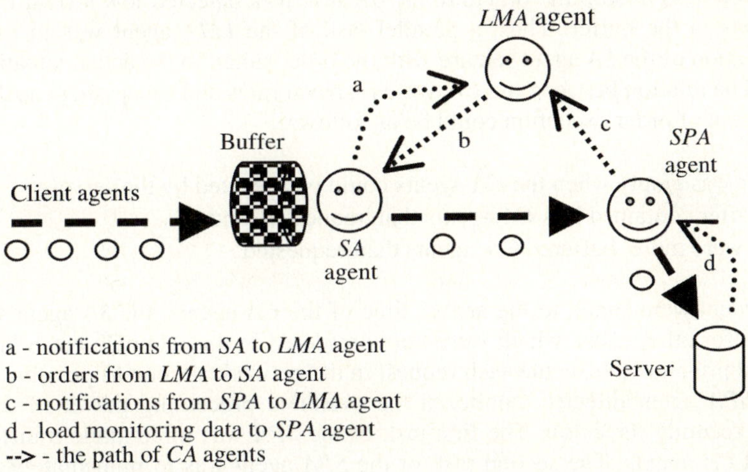

Fig. 2. The model of agent environment used to control the processing load. Triple *LMA*, *SA* and *SPA* were used to control transfer of *CA* agents to the server.

Furthermore, we assumed the existence of the real time services in the same agent environment. That class of services are known to require prompt support i.e. they allow slight delay that is usually acceptable for the user. The agent model is given in fig. 2. Tasks of the *LMA* agent were to protect the server load and to self-adapt its own orders to the input arrival rate of the *CA* agents.

3.1 Agent's Triple for Input-Output Adaptation

As already mentioned, the intelligent *LMA* agent was used to manage the flow of *CA* agents on their way to the service processing resource i.e. server. The *LMA* was responsible for decisions about individual transfer for the *CA* agents brought by multiple selection choice. The *CA* agents were received and accumulated in the buffer, at the entry to our specifically created agent environment (Fig. 2). The *LMA* was sending the transfer order to *SA* agent at each time step and was waiting for notification. The *SA* agent was responsible for taking over the requested number of the *CA* agents from the buffer and for sending them to the *SPA* agent. If the buffered number of the *CA* agents were below of the requested, the *SA* agent would send the existing *CA* and wait for other *CA* agents until the whole request was completed. The *SA* would then immediately notify the *LMA* of the completion of the requested action. Then the *LMA* would send another request.

As already mentioned, the task of the *LMA* agent was to order the *SA* agent to transfer the *CA* agents from the buffer to the *SPA* agent (server). These orders specified the exact number of the *CA* agents to be transferred toward for the current time step. In our simulations the *LMA* received 4 different orders. These were the requests for 0 *CA* agents, 1 *CA* agent, 2 *CA* agents and 3 *CA* agents.
The request for zero *CA* agents was a simple "no action" and meant two things. The first one was the *LMA* agent's estimation to stop the increase of servers' load and the second one was to stop the orders to *SA* because of a detected low arrival rate of the *CA* agents in the buffer. Thus, a parallel task of the *LMA* agent was to care about minimization of the *SA* agent seizure with the order suited to the actual situation in the buffer. The relation between the *CA* agents' arrival rates and processing capabilities at the moment of order execution could be as follows:

- buffer was empty when the *CA* agents could be accepted by the server,
- the buffer contained less *CA* agents than requested, and
- there were more buffered CA agents than requested.

If the orders were suited to the arrival time of the *CA* agents, the *SA* agent would be liberated for other tasks, which were not considered in this example but followed the proposed principle to execute each request in the network as soon as possible.

The SPA agent directly monitored the available processing power of the server before executing its action. The first task of the *SPA* was to continue transfer of the received *CA* agents. The second task of the *SPA* agent was to immediately reject all *CA* agents that, if transferred, might cause excess of the estimated value. Rejection of the *CA* agents implicated their death [2]. Strictly applied, the *CA* agent death at that moment denoted inability to further process the service requests carried by the rejected *CA* agents, and cancellation of service.

3.2 Reward Settings

A reward signal was calculated relevant to the selected action. The reward was calculated with equation 2 when the requested actions were one or two or three *CA* agents chosen to transmit for processing.

$$\mathrm{Re}\,ward = 1 - \left(\frac{achieved\_load}{estimated\_load} + waiting\_time \right) \quad (2)$$

The parameter *achieved_load* indicated actual amount of the processor load expressed in percentage. *Achieved_load* was real processor load achieved after getting the *CA* agent. Occupancy of the processing power was restricted by the parameter *estimated_load*. That value was a boundary for dividing the processing power into two parts for two different service classes. The parameter *waiting_time* was used to minimize the *SA's* waiting for execution of the requested action. That parameter was used to decrease the reward in case of prolonged waiting for the *CA* agents. For example, the arrival rate of the *CA* agents was repeatedly shorter than the rate of the requests.

$$\mathrm{Re}\,ward = 1 - \frac{achieved\_load}{estimated\_load} \tag{3}$$

In case of the *LMA* request for zero the *CA* agent's equation 3 was applied. The parameter *waiting_time* was excluded because the *LMA* agent request was not the order for transport and, therefore, waiting for the *CA* agents was not included.

4 Experimental Results

The source of the *CA* agents was modeled as a discrete agent generator with the following characteristics. One agent was turned out in 5 seconds and then each 5 seconds frequency of the agent creation was increased by 5%. That progress was going on until the frequency of 5 agents per second was achieved. The agents were accumulated in the buffer and then transferred to the server under the *LMA* control. The task of the *LMA* was to adjust its orders to the actual amount of load on the server and to follow actual frequency of the *CA* agent's arrival.

The server's capability was described as a processing power with discrete values between 1 and 100 percentages. The estimated processing load was set to 33%. As a boundary between two classes of services, that value ensured 67% of the processor's power for higher priority services. That boundary could not be exceeded by processing of the *CA* agents, but could be automatically lowered in case the bursting of higher class services exceed the estimated 67% [1, 3].

As stated before, the task of the *SPA* agent was to reject every *CA* agent tending to increase the processing power of the server over the estimated value. On the other hand, the task of the *LMA* was to minimize the *CA* agent's death. To elaborate on these performances two circumstances were depicted and described. Learning rate and a discount factor were set to 1 and 0 respectively. In the first case during simulation the input arrival rate of the *CA* agents was higher than the processing rate. The diagram in fig. 3 shows cumulative presentation of the rejected *CA* agents and the achieved processing load for 275 time steps. The estimated load limit was achieved quickly, always after 15 time steps, and then the rejection of agents began and ended after approximately 130 time steps (fig. 3). In that case processing of every *CA* agent was set to occupy 3% of the processor power and 3000 msec of the processing time.

In the second case, during simulation the input arrival rate was changed periodically, as described at the beginning of this section. The diagram in fig. 4 shows periodical achievement of the estimated load, followed immediately by rejection of the *CA* agents. These rejections were reduced with every next attempt to achieve the estimated load. However, the recorded number of *CA* agents in the input buffer was zero as a consequence of the *LMA* orders. In that case processing of every *CA* agent occupied 4% of the processor power and 2000 msec of the processor time. Simulation results were completed for 1600 time steps.

Table 1. Total number of orders for the *CA* agent's transfer

A)		*B)*	
LMA Orders for:	Total:	*LMA* Orders for:	Total:
0 *CA* agents	278	0 *CA* agents	26
1 *CA* agent	91	1 *CA* agent	149
2 *CA* agents	37	2 *CA* agents	131
3 *CA* agents	32	3 *CA* agents	516

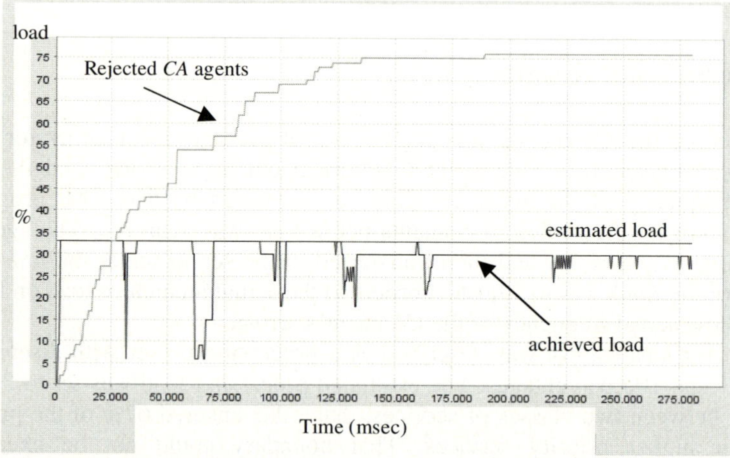

Fig. 3. The diagram shows the effects of the load protection and cumulative amount of the rejected *CA* agents. After approximately 130 time steps the *LMA* agent stopped rejecting the *CA* agents and kept the current load under the estimated.

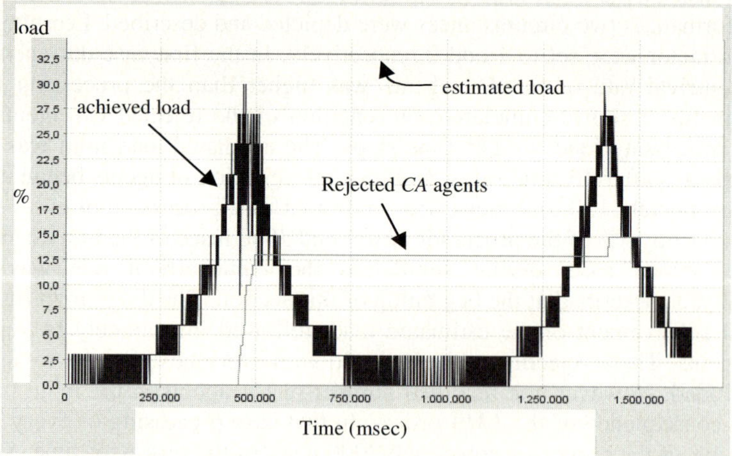

Fig. 4. The *LMA* agent follows the input arrival rate keeping the *CA* agents in the input buffer at zero except if the boundary is reached, when the *SPA* starts with rejection

The arrangement of *LMA* orders is illustrated in table 1. Table shows the distribution of *LMA* orders for two simulated circumstances. Part *A* of the table relates to the circumstance in fig. 3, and part *B* of the table relates to the circumstance in fig. 4. In the case of load protection the continuously activated *LMA* agent frequently selected the request for 0 *CA* agents and then more frequently requested for 1 *CA* agent. Conversely, when the *CA* agents' arrival rate varied considerably different distribution of the *LMA* requests was detected (see table 1).

5 Conclusion

This work applies the idea of intelligent agent learning to control the processor load. The processing load was dedicated for two classes of services separated by the boundary controlled by the intelligent agent. That boundary was the estimated maximal processing power that could be used for client agents processing. The mission of the intelligent agent was to select a fitting order from the set of four orders to follow the input properties when the agent arrival rate was lower than the processing rate, and to minimize agents' death when the arrival rate exceed the server's allowed processing power. Shortly, the task of the intelligent agent was to handle the flow of client's agents toward processing resources. Simulation results showed capability of the proposed agent system to follow the given goal under the circumstances selected and described here. Because of the influence of unrelated parameters and constraints the problem remains how to express them as a unique scalar value and then to follow global objective.

References

1. Jevtic D., Kunstic M., Ouzecki D.: The Effect of Alteration in Service Environments with Distributed Intelligent Agents. Lecture Notes in Computer Science/LNAI 3683, Part 3. Springer-Verlag, Berlin Heidelberg New York (2005) 16-22
2. Kusek, M., Lovrek, I., Sinkovic V.: Agent Team Coordination in the Mobile Agent Network. Lecture Notes in Computer Science/LNAI 3681, Part 1. Springer-Verlag, Berlin Heidelberg New York (2005), 240-246
3. Jevtic D., Kunstic M., Jerkovic N.: The Intelligent Agent-Based Control of Service Processing Capacity. Lecture Notes in Computer Science/LNAI 2774, Part 2. Springer-Verlag, Berlin Heidelberg New York (2003) 668-674
4. Bertsekas, D. P., Tsitsiklis, J. N.: Neuro-Dynamic Programming. Athena Scientific, MIT, Belmont, Massachusetts (1996)
5. Watkins, C.J.C.H., Dayan, P.: Q-learning. Machine Learning, Vol. 8, Kluwer Academic Publishers, (1992) 55-68
6. Luck, M., d'Inverno, M.: A Conceptual Framework for Agent Definition and Development. The Computer Journal Vol. 44 (1), (2001) 1-20

Reserve Price Recommendation by Similarity-Based Time Series Analysis for Internet Auction Systems

Min Jung Ko[1] and Yong Kyu Lee[2]

[1] Institute of Computer Science, Dongguk University
[2] Corresponding Author, Department of Computer Engineering, Dongguk University Pil-dong, Jung-gu, Seoul 100-715, Rep. of Korea
{mjgo, yklee}@dongguk.edu

Abstract. It is very important that sellers provide reasonable reserve prices for auction items in internet auction systems. We recently have proposed the agent based system in order to generate reserve prices automatically, based on the case similarity of information retrieval theory and the moving average of time series analysis. However, the approaches have some problems that they can not reflect the recent trend of auction prices on the generated reserve prices properly, because they only provide the bid prices or average prices of items, by using the past auction data for sellers. In this paper, in order to overcome the problems, we propose Similarity-Based Time Series Analysis which search the past bid price through case similarity and then generate reserve prices by using time series analysis. Through performance experiments, we show that the successful bid rate of the new method can be improved by preventing sellers from making unreasonable reserve prices compared with the past methods.

1 Introduction

The number of purchases by internet auctions has been increasing as electronic commerce has been generalized. In those systems, seller's proposing reserve prices are very important [4][5]. But at the same time, sellers have trouble with deciding prices of items because sellers get the past reserve prices of auction items [3][11]. Therefore the success bid rate may drop due to the large gap between the reserve price-price decided by seller and the real bid price. So sellers sometimes may fail in an internet auction, and then participate in auction again or sell their items in off-line.

In this paper, we also present a method to generate reserve prices and bid prices based on the similarity method [6][9] in an internet auction system. It generates reserve prices for sellers based on the most similar case. The case base was constructed using the past auction records and we searched results from the electronic commerce sites on the Web. However, it can't reflect the recent trend of auction prices well on the generated reserve prices as it simply provides the winning bid price of the most similar item. In order to overcome this problem, we propose a new method that generates reserve prices based on the time-series analysis method [7], which has been used to analyze the moving trends of the prices of an item. We have chosen the moving average method, the exponential smoothing method, and the least square method

B. Gabrys, R.J. Howlett, and L.C. Jain (Eds.): KES 2006, Part I, LNAI 4251, pp. 292 – 299, 2006.
© Springer-Verlag Berlin Heidelberg 2006

because they are most widely used [9][10]. However, we can't find out the studies that are used for recommending reserve prices of auction items in internet auction systems.

These methods are applied to the prices history data searched from the case base using the previously presented similarity method. By using Similarity-Based Time series Analysis, we can generate the most similar reserve prices, which well reflect the current trends of the item prices. Through performance experiments, we show that the successful bid rate of the new method can be increased by preventing sellers from making unreasonable reserve prices compared with the previous methods.

This paper is structured as follows. Section 2 explains the method that recommends the reserve price. In Section 3 we will prove the effectiveness of our system through a series of experiments. Finally, in the last section, we will be discuss conclusions and further researches.

2 Reserve Price Recommendation Methods

Fig. 1 explains the procedure that an appropriate reserve price for an auction item is generated. The case base stores the records consisting of the reserve price, bidding prices, winning bid price with detailed item information obtained from the past auctions.

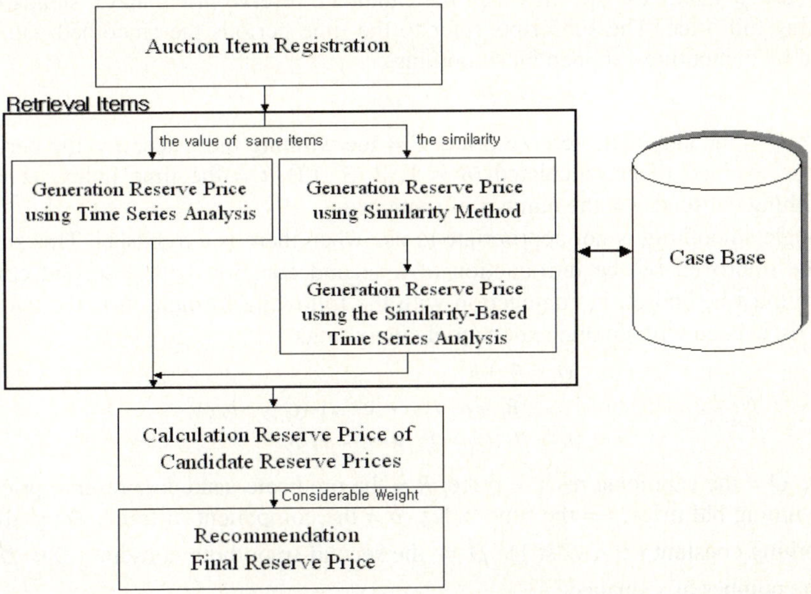

Fig. 1. Reserve Pricing Procedure

2.1 Recommendation the Reserve Price Using Time Series Analysis

We provide methods that automatically generate the candidate reserve price as applying the time series analysis to internet auctions systems.

Moving Average. We can obtain the candidate reserve price by applying winning bid price of auction items to their formula. The method is the same as the following.

$$Q_t = \frac{1}{n}\sum_{i=1}^{n}(P_{t-i}) \tag{1}$$

where Q = the candidate reserve price, P = the winning bid price, n = the period for moving average to be calculated (n=1, 2, 3,12), t = the time index, i = the number of sequence.

Another method of computing the average is by adding each value divided by the number of values. It is called the weighted moving average and the candidate reserve price is generated using the following formula.

$$Q_t = \sum_{i=1}^{n} w_{t-i} \cdot P_{t-i} \tag{2}$$

where Q = the candidate reserve price, P = the winning bid price, n = the period for moving average to be calculated (n = 1, 2, 3,12), t = the time index, w = the weight of the price, i = the number of sequence.

Exponential Smoothing. The single exponential smoothing scheme begins by setting Q, where Q stands for the smoothed the candidate reserve price, and P stands for the winning bid price. The subscripts refer to the time periods the smoothed value Q is found by computing exponential smoothing.

$$Q_t = \alpha \cdot P_{t-i} + (1-\alpha) \cdot V_{t-i} \tag{3}$$

where Q = the candidate reserve price, P = the winning bid price, n = the period for moving average to be calculated (n = 1, 2, 3, 12), t = the time index, α = the smoothing constant, i = the number of sequence.

Single smoothing is not appropriate to use when there is a trends[1]. This situation can be improved by the introduction of a second equation with a second constant, which must be chosen in conjunction with the following formula. It is the two equations associated with double exponential smoothing.

$$Q_t = R_t + b_i \tag{4}$$

$$R_t = \alpha \cdot P_{t-i} + (1-\alpha) \cdot (Q_{t-i} + b_{t-i}) \tag{5}$$

$$b_t = \beta \cdot (Q_t - Q_{t-i}) + (1-\beta) \cdot b_{t-i} \tag{6}$$

where Q = the candidate reserve price, R = the predicate candidate reserve price, P = the winning bid price, t = the time index, b = the component of trend, α = the first smoothing constant ($0 < \alpha < 1$), β = the second smoothing constant ($0 < \beta < 1$), i = the number of sequence.

Least Square. The candidate reserve price is generated using the following formula.

$$Q_t = X + Y \cdot D_t \tag{7}$$

$$Y = (\sum_{i=1}^{n} (D_t - \overline{D}) \cdot (P_t - \overline{P})) / (\sum_{i=1}^{n} (D_t - \overline{D})) \tag{8}$$

$$X = \overline{P} - Y \cdot \overline{D} \tag{9}$$

where Q = the candidate reserve price, P = the winning bid price, D = the selling month of auction items, $\overline{P}$ = the mean of the winning bid price, $\overline{D}$ = the mean of the selling month, X, Y = the least square const, i = the number of sequence.

We recommend the final reserve price based on the calculation the candidate reserve prices for seller to internet auction systems. The method is the same as the following.

$$V_t = \frac{1}{n} \sum_{i=1}^{n} Q_t \tag{10}$$

where V = the final reserve price, Q = the candidate reserve price, n = the number of the candidate reserve price, t = the time index, i = the number of sequence.

2.2 Recommendation the Reserve Price Using Similarity Method

We can obtain the final reserve prices by applying similarity of auction items to their formula. The method is the same as the following.

$$V_t = \frac{1}{n} \sum_{i=1}^{n} w_i \cdot P_t \tag{11}$$

where V = the final reserve price, P = the winning bid price, n = the number of the candidate reserve price, w = weight (*1/similarity*), t = the time index, i = the number of sequence.

The vector space model [2] is used to retrieve ranked similar cases from the history database. The record consists of the attributes describing a case. The similarity measure is as follows:

$$similarity(r_j, q_k) = \frac{\sum_{i=1}^{n} (tr_{ij} \times tq_{ik})}{\sqrt{\sum_{i=1}^{n} tr_{ij}^2 \times \sum_{i=1}^{n} tq_{ik}^2}} \tag{12}$$

where tr_{ij} = *i*-th term in the vector for record *j*, tq_{ik} = *i*-th term in the vector for query *k*, and n = number of fields in the record.

2.3 Recommendation the Reserve Price Using Similarity-Based Time Series Analysis

We can obtain the final reserve prices by applying candidate reserve price of auction items to their formula. The method is the same as the following.

$$V_t = \frac{1}{n} \sum_{i=1}^{n} w_i \cdot Q_i \tag{13}$$

where V = the final reserve price, Q = the candidate reserve price, n = the number of the candidate reserve price, w = weight (*1/similarity*), t = the time index, i = the number of sequence.

3 Experimental Evaluation

Performance experiments have been performed using an auction history database containing the past auction data of a real commercial auction system. The database contains the 2,854 auction data used the same model cars [12] and 1,512 cell phone for the 2 years [14]. The past winning bid prices of each auction item was stored by month. Also, collecting the data of 413 used cars [13] and 103 used cell phone [15], we experiment the performance of their reserve price recommendations.

For the comparison, we use the normalized absolute error as follows:

$$E = \frac{1}{n}\sum_{i=1}^{n}(|b_i - r_i|)/b_i \tag{14}$$

where n = the number of items, b_i = the winning bid price, r_i = the reserve price given by the seller (the lowest bid price for an unsuccessful bid).

Electronic commerce Agent is a kind of software that performs automatic process instead of buyer, seller and coordinator[6]. It can be divided into the search agent, negotiation agent and shopping agent. In this paper, the reserve price is not decided by the seller but it is automatically generated by the system. We define the systems without an agent as "no agent" and the systems with an agent as "agent".

In Fig. 2, it means that the moving average is especially more effective than other methods in the car, namely decreasing weight is better than giving weight equally to the past 1 year data. Also, the exponential smoothing is especially the most effective in the cellphone, because experimental data are less influenced by seasonal trend. So this result can be changeable according to experimental data.

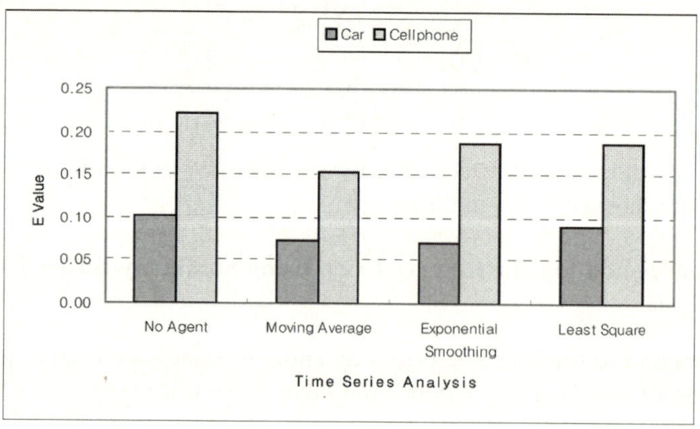

Fig. 2. Comparison of *E* value in Time Series Analysis

In Fig. 3, we get the value of E, 0.055, 0.119 through the similarity method and 0.103, 0.221 through the no agent. This means that the similarity method is more effective than the no agent, because experimental data are influenced by weighted similarity.

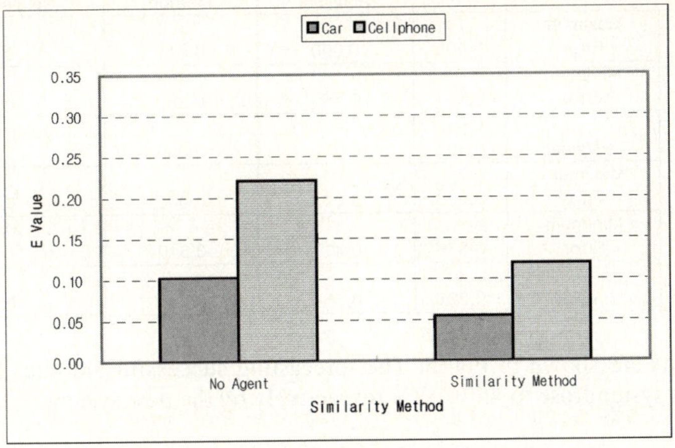

Fig. 3. Comparison of E value in Similarity Method

Fig. 4 shows the result of the E in obtained by the formula (13). As you can see, we get the result that the least square is more effective than in the previous system.

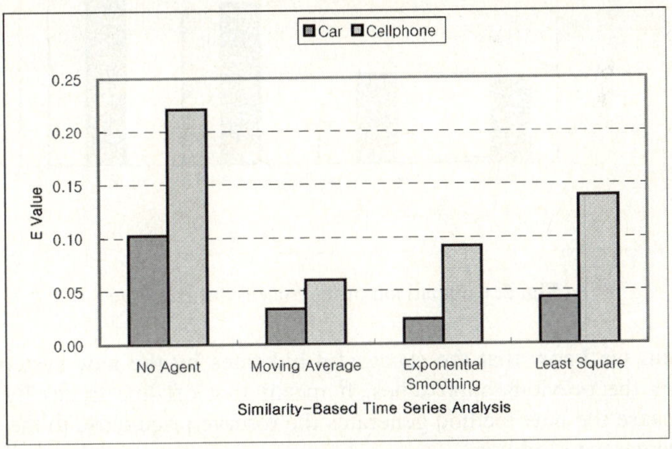

Fig. 4. Comparison of E value in Similarity-Based Time Series Analysis

Table 1 means that the forecasting errors of the new system are smaller than those of the previous system. Especially the range from the maximum to the minimum in error and the value of E through Time series analysis are smaller and better than those through No Agent.

Table 1. Difference of the *E* Values

		No Agent	Agent		
			Time Series Analysis	Similarity Method	Similarity-Based Time Series Analysis
Car	Maximum Error	0.000	0.000	0.011	0.000
	Minimum Error	0.714	0.565	0.282	0.192
	E Value	0.103	0.070	0.055	0.023
Cellphone	Maximum Error	0.038	0.012	0.000	0.000
	Minimum Error	0.578	0.417	0.318	0.285
	E Value	0.221	0.154	0.119	0.060

The results are shown in Fig. 5. The forecasting successful bid rate, 15%, 1% by the previous system rose to 45%, 37% respectively by the new system.

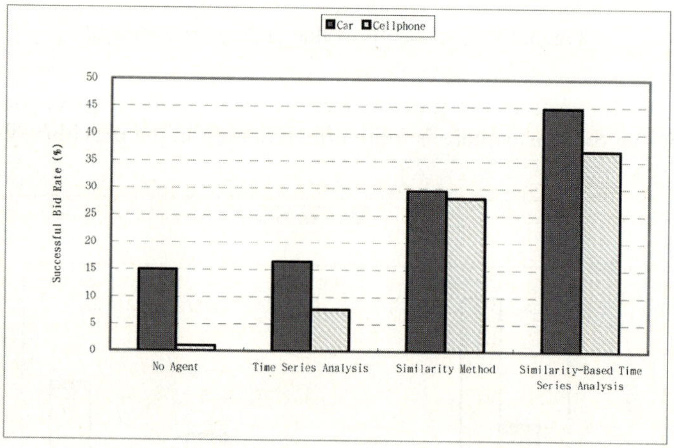

Fig. 5. Comparison of the Successful Bid Rate

As a result, we know that the successful bid rates by the new system are higher than those by the previous approaches. It means that off-line trades by coordinator decrease because the new method generates the reserve price close to the winning bid price and decreases re-bidding.

4 Conclusions

In this paper, we have proposed a pricing agent that automatically recommends reserve prices to sellers for internet auction systems. It uses the Similarity-Based Time

Series Analysis method to get appropriate reserve prices from the past auction data and the prices of items are gathered from electronic commerce sites on the web.

According to the results of performance experiments, the successful bid rate can be improved by preventing sellers from making unreasonable reserve prices. Through the experiments, we showed that the successful bid rate of the reserve pricing agent improved from 15% (used cars), 1% (used cellphone) by no agent, 16%, 4% by the time series analysis and 30%, 28% by the similarity method to 45%, 37% respectively by new system.

It means that the unsuccessful bid rate go down by preventing auctions from becoming unsuccessful due to higher reserve price than the real bid price. Also our system can decreases cases of successful auctions with too low bid price. As a result, we know that the Similarity-Based Time Series Analysis is the most suitable for recommending the reserve prices in auction systems through the experiments. In the future, we have to study the pricing agent with a large database of various auction items.

References

1. Carbone, H. C.: Forecasting Trends and Seasonal by Exponentially Weighted Moving Averages. Carnegie Institute of Technology, Vol. 2, No. 52 (1957)
2. Frakes, W. B.: Information Retrieval- Data Structures & Algorithms. Prentice Hall (1992)
3. Drucker, P. F., et. al.: Accurate Business Forecasting. Harvard Business School Press (1991)
4. Heck, E. V., Vervest, P.: How Should CIO's Deal with Web-Based Auctions?. Communications of the ACM, Vol. 41, No. 7, July (1998) 99-100
5. Jones, C. P. Investments-Analysis and Management. 8th Edition, Wiley (2001)
6. Ko, M. J., Lee, Y. K.: Automatic Reserve Price Generation for an Internet Auction System Using Moving Average. Journal of Society for e-Business Studies, Vol. 9, No. 2, May (2004) 17-31
7. LeBaron, B.: Do Moving Average Tradiule Imply Nonlinearities in Foreign Exchange Markets?. Technical Report 9222, Social Systems Research Institute, Univ. of Wisconsin, Madison (1992)
8. Lee, Y. K., Kim, S. W., Ko, M. J., Park S. E.: Pricing Agents for a Group Buying System. EurAsia ICT 2002, Lecture Notes in Computer Science, vol. 2510 (2002) 693-700
9. Shumway, R. H., Stoffer D. S.: Time Series Analysis and Its Application. Springer (2002)
10. Tashman, L. J.: Out-of-sample Tests of Forecasting Accuracy: an Analysis and Review. Int'l Journal of Forecasting, Vol. 16, No. 4, October (2000) 437-450
11. eBay Auction Guideline: http://webhelp.ebay.com/cgi-bin/eHNC (2003)
12. Korea Encar List: http://www.encar.co.kr/proc23/23000.jsp (2004)
13. Korea Glovis Secondhand Trend: http://www.hkaa.co.kr/Sise/SerCond01.aspx (2004)
14. Korea Cellphone Site: http://used4989.com/home2 (2004)
15. Korea Phonemore Site: http://www.phonemoa.co.kr (2004)

Use of an Intelligent Agent for an E-Commerce Bargaining System

Pao-Hua Chou[1], Kuang-Ku Chen[2], and Menq-Jiun Wu[1]

[1] Department of Mechatronics Engineering, National Chunghua University of Education
[2] College of Business Administration, National Chunghua University of Education
No.2, Shi-da Road, Chunghua City, Taiwan, R.O.C.
S92631001@mail.ncue.edu.tw, Kungku83@ms47.hinet.tw,
mwu@cc.ncue.edu.tw

Abstract. As the Internet has become mainstream, so has Electronic commerce (E-Commerce) become an important part of conducting business. Electronic auctions assist merchants and consumers in reducing costs and enable customized delivery of goods and services. Our goal is to develop a bargaining server prototype and to test its feasibility in the E-Marketplaces. The present auction system is the first and currently the only Internet auction system that supports complex auctions and bids via price-quantity graphs and mobile agents. The agent system allows users to use the system as a third-party auction site, which exploits the bargaining opportunities and market development in the Internet. The most significant achievement of this research is the development of a new e-commerce bargain architecture that adopts an agent's properties and a multi-agent framework. It also creates a user friendly interface for Graphic User's Interface (GUI). The implementation prototype system demonstrates excellent performance and the system can be easily implemented in a large bargain database of the e-commerce marketplace for wide accessibility.

1 Introduction

In recent years technological advances of the Internet and Web have continuously boosted the prosperity of electronic commerce. Through the Internet, different business entities, including suppliers, retailers and consumers can now easily interact with each other and conduct transactions within a minimum amount of time. Although the business-to-business (B2B) model is the main focus in the current electronic commerce development, the business-to-consumer (B2C) model has shown immense potential in the electronic marketplace through the encouragement of on-line trading due to continuing technological advances of the Internet. Nowadays, many companies have been digitalizing their enterprise to enhance their operating performance.

However, possessing advanced hardware facilities that can take advantage of the Internet infrastructure and maintaining captivating Web sites are not the only factors in guaranteeing success in the electronic market. In order to facilitate transactions, the problems associated with complex activities in electronic commerce must be resolved. The abundance of information available on the Internet allows consumers to communicate

B. Gabrys, R.J. Howlett, and L.C. Jain (Eds.): KES 2006, Part I, LNAI 4251, pp. 300–309, 2006.
© Springer-Verlag Berlin Heidelberg 2006

with sellers for a bargain. Sellers may have their own individual considerations for their corresponding marketing strategies. Therefore, the traditional commerce negotiation process, similar to human-based life bargaining between buyers and sellers, will also arise in the electronic market in order for both parties to reach an agreement that is satisfactory to both. However, the time consuming nature of human-based negotiation makes it unsuitable for the new electronic trading in which potential customers can easily access services through the Internet enterprises.

As the Internet moves further into the mainstream of everyday life, e-commerce is becoming an important mechanism for conducting business. E-commerce helps merchants and consumers to reduce costs and enables customized delivery of goods and services. Among current business models, electronic auctions are emerging as one of the most successful e-commerce technologies. Our goal is to employ the multi-agent system and its capabilities which integrate the e-commerce framework and the bargain platform. In addition, we also developed the e-commerce multi-agent bargain system for graphic user interface architectures. The remainder of this paper is organized as follows. Section 2 describes the background and development of multi-agent architecture and reviews the e-commerce marketplace theories. Section 3 demonstrates and explains in detail how to construct the multi-agent bargaining system architecture. Implementation results are presented in Section 4. Finally, conclusions are drawn in Section 5.

2 Preliminary

In the development of the e-commerce bargaining system in collaboration with multi-agent application some technologies are required, such as the agent system, multi-agent system, recommendation and negotiation process, and the marketplace function. These will be discussed in the following sub-sections.

2.1 Agent

The aim of this section is to provide an understanding of what agents are, and some of the issues associated with building them. A definition is in order for the term not to lose its meaning. The definition presented here is adapted from Wooldridge and Jennings [6]. “An agent is a computer system that is situated in some environment, and is capable of autonomous action in this environment in order to meet its design objectives”.

Figure 1 provides an abstract view of an agent. The diagram delineates an agent generating action outputs in order to affect its environment. In most domains of reasonable complexity an agent will not have complete control over its environment. Agents will have at best partial control in which it can exert influence. Thus, from the standpoint of the agent, the same action executed twice in apparent identical circumstances might appear to have entirely different effects, and in particular, it may fail to have the desired effect.

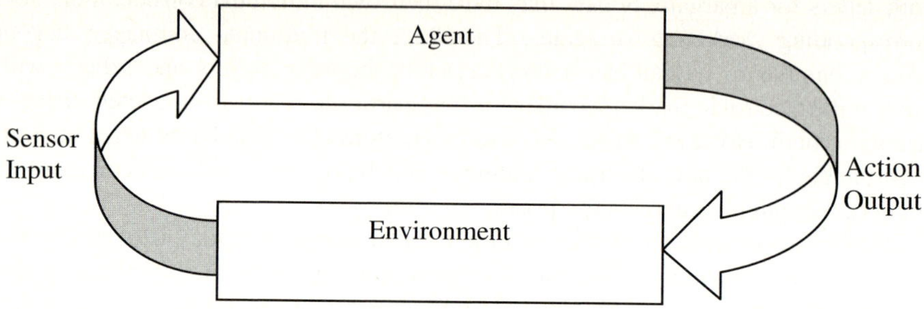

Fig. 1. An Agent in its Environment

An agent possesses the following minimal characteristics [4]:

1. Delegation: An agent performs a set of tasks on behalf of a user (or other agents) that are explicitly approved by the user.
2. Communication skills: An agent must be able to interact with the user to receive task delegation instructions, and inform the user of the task status and completion through an agent-user interface or through an agent communication language.
3. Autonomy: An agent operates without direct intervention (e.g., in the background) to the extent of the user's specified delegation. The level of autonomy of an agent can range from the authority to initiate a nightly backup to negotiating the best price of a product for the user.
4. Monitoring: An agent must possess the ability to monitor its environment in order to perform tasks autonomously.
5. Actuation: An agent needs to be able to affect its environment via an actuation mechanism for autonomous operations.
6. Intelligence: An agent must be capable of interpreting the monitored events to make appropriate actuation decisions for autonomous operations.

2.2 Single-Agent Systems

Single-agent systems are based on the centralized process model. On the other hand, multi-agent systems are based on distributed systems. Centralized single-agent systems employ a single agent which makes all the decisions while other agents act as remote slaves. Single-agent systems may have several entities such as actuators or even robots. However, if each entity sends its perceptions to and receives its actions from a single centralized process, the system remains a single-agent system.

In a single-agent system in which the agent models itself as part of the environment, some extra-environmental components such as actuators, sensors, and even other agents exist. However, other agents are considered to behave like an actuator or sensor in the sense that they do not possess any goals. Figure 2 shows the single-agent framework [5].

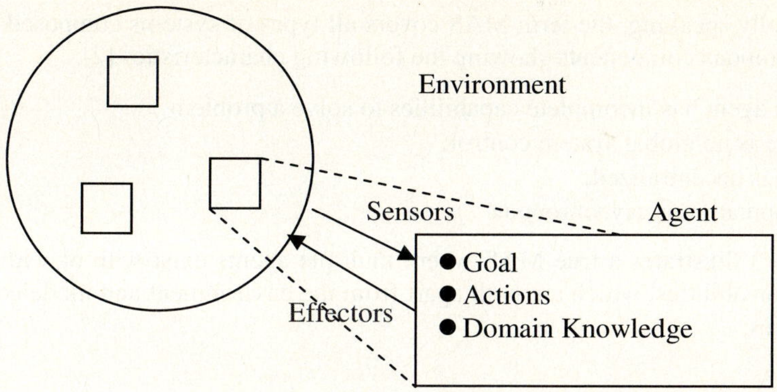

Fig. 2. A General Single-Agent Framework

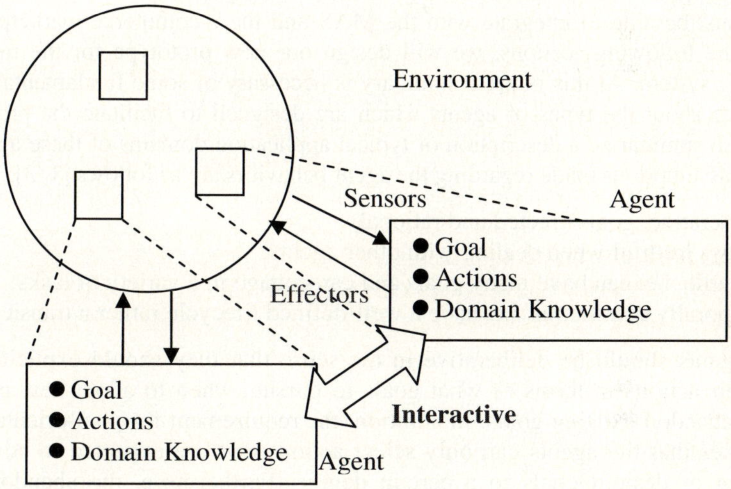

Fig. 3. A General Single-Agent Framework

2.3 Multi-Agent Systems

Various definitions from different disciplines have been proposed for the multi-agent system (MAS). As seen from Distributed Artificial Intelligence (DAI), MAS is defined as a loosely coupled network of problem-solving entities that work together to find answers to problems that are beyond the individual capabilities or knowledge of each entity. MAS differs from single-agent systems in that there are several agents, and these agents are aware of each other's goals and actions. In addition to the awareness of each other's intentions and behaviors, these agents can communicate with each other in a fully general MAS, either to help an individual agent in achieving its goal or (in very rare, circumstances) preventing it. This particular difference places the system at a higher level. Various benefits from communication between agents are discussed in the following chapters.

Generally speaking, the term MAS covers all types of systems composed of multiple autonomous components showing the following characteristics [2]:

- Each agent has incomplete capabilities to solve a problem;
- There is no global system control;
- Data is decentralized;
- Computation is asynchronous.

Figure 3 illustrates a true MAS where multiple agents exist with or without communication abilities, which are both apart from the environment and modeled as separate entities.

3 System Architecture

Our primary goal is to design a multi-agent bargaining system. Basically, the bargaining system must be able to integrate with the MAS and the e-commerce marketplace concept. In the following sections, we will design our new prototype for the multi-agent bargaining system. At this point, a summary is necessary of some fundamental assumptions made about the types of agents which are designed to facilitate the process. We should also summarize a description of typical application domains of these agents. The principal assumptions made regarding the agent behaviors are as follows [1, 3]:

(1) Deliberative, goal-directed and rational;
(2) Always truthful when dealing with other agents;
(3) Versatile, i.e. can have many goals and can engage in a variety of tasks;
(4) Temporally continuous, i.e. have a well-defined lifecycle rather a transient one.

The agents should be deliberative in the sense that they should explicitly reason about their actions in terms of what goals to pursue, when to adopt new goals, and when to abandon existing goals. In addition, the requirement for goal-oriented behaviors implies that the agents can only select actions that are expected to advance the attainment of desired goals to a certain degree. Furthermore, the abandonment of goals is only appropriate when achievement of the goals is remote or the purpose for achieving the goal is no longer upheld. Figure 4 shows the architecture of the agent that is not too dissimilar from other collaborative agent architectures in the literature.

As shown in Figure 4, the agent includes the following components:

(1) a Mailbox that handles communications between the agent and other agents.
(2) a Message Handler that processes incoming messages from the Mailbox, dispatching them to the relevant components of the agent.
(3) a Coordination Engine that makes decisions concerning the agent's goals, e.g. how they should be pursued, when to abandon them, etc. It is also responsible for coordinating the agent's interactions with other agents using its known coordination protocols and strategies, e.g. the various auction protocols or the contract net protocol.
(4) an Acquaintance Database that describes the agent's relationships with other agents in the society, and its beliefs about the capabilities of those agents. The Coordination Engine uses information in this database when making collaborative arrangements with other agents.

(5) a Planner and Scheduler that plans the agent's tasks based on decisions taken by the Coordination Engine and the resources and task specifications available to the agent.
(6) a Resource Database that maintains a list of resources that is owned by and available to the agent. The Resource Database also supports a direct interface to external systems, which allows it to dynamically link to and utilizes proprietary databases.
(7) an Ontology Database that stores the logical definition of each fact type: its legal attributes, the range of legal values for each attribute, any constraints between attribute values, and any relationships between the attributes of the fact and other facts.
(8) a Task/Plan Database that provides logical descriptions of planning operators known to the agent.
(9) an Execution Monitor that maintains the agent's internal clock, and starts, stops and monitors tasks that have been scheduled for execution or termination by the Planner/Scheduler.

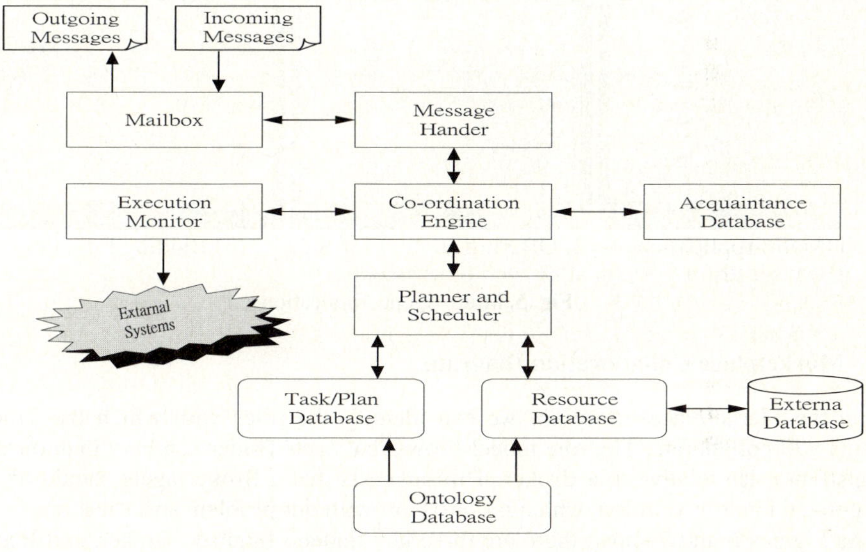

Fig. 4. The Architecture of the Agent

4 System Implementation

The decentralized nature of this type of marketplace implies that potential buyers require a means of finding vendors, and that prospective vendors require a means of advertising. In static marketplaces (i.e. where membership remains unchanged), it may be sufficient to furnish each agent a priority with knowledge of its peers. A more flexible alternative, and one used by dynamic marketplaces, is to designate the role of broker to at least one agent. The broker maintains an up-to-date registry of prospective buyers and sellers, providing an efficient means for agents to locate appropriate transaction partners.

Figure 5 shows the prototype of our multi-agent application. There are two major parts to the whole system: The Bargain Browser and the Price Configuration

Pane. The Bargain Browser serves as the transaction browser and represents the transaction process in the browser. The Bargain Browser appears in the "A" area of Figure 5(1). In addition, Price Configuration Pane serves as a pricing tool for sale and buys items in the "B" area of Figure 5(1). We will explain each component in the following sections.

(1)Main Application (2) Offering an item for Sale (3) Bidding for a item

Fig. 5. Multi-Agent Application

4.1 Marketplace Collaboration Diagram

From the Marketplace viewpoint we can identify the roles with which the Trader agents will collaborate. The role model shows that each Trader can play Inquirer and Registrant roles relative to a Broker. This suggests that a Broker agent should be incorporated into our solution, which is consistent with our problem specification.

As Figures 6 and 7 show, there are three key traders: Inquirer, Broker, and Registrant. Table 1 describes the detailed interactions between the marketplaces and the involvement of three key traders: Inquirer (Buyer), Broker, and Registrant (Seller).

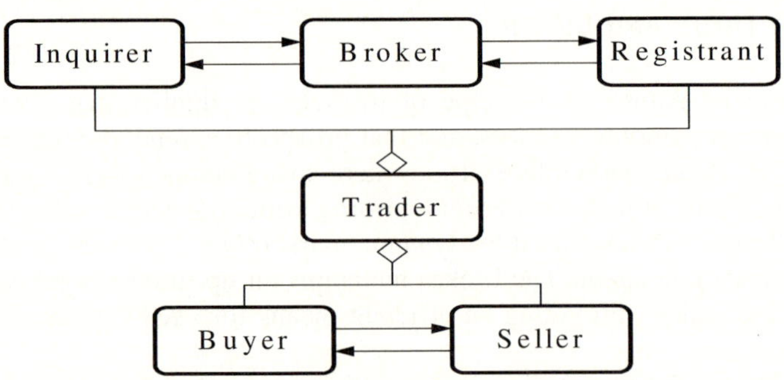

Fig. 6. The Marketplace Application

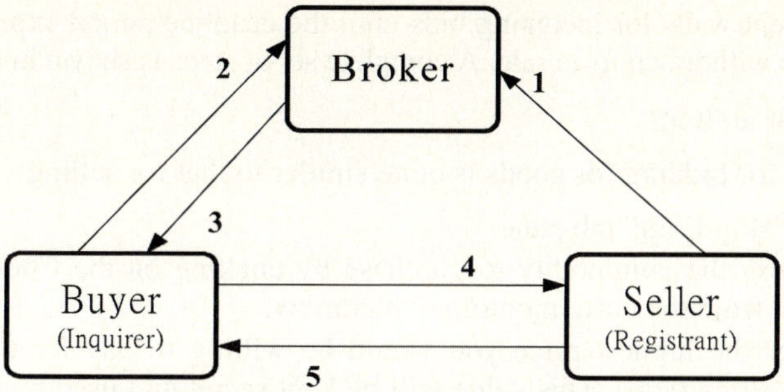

Fig. 7. The Interactions between the Marketplaces

Table 1. Interaction Process

	Collaboration	Explanation
1	Registration	Sellers register or de-register their items for sale
2	Find Request	Asks for agents selling items of interest
3	Find Response	A list of agents matching the desired criteria
4	Buyer Offer	Message containing buyer's proposal
5	Seller Response	Seller's reply to a previously submitted offer

4.2 Running the E-Commerce Bargain Application

Assuming that all the agents will run on the same machine, launching the application will involve the following processes. Once launched, the Fruit Market agents will display their front-end GUIs and await user instructions. Other than replying to register themselves with the Name Server, agents will not perform any autonomous actions. Now users have two options, either offer an item for sale or formulate a new bid.

(1) Offering an Item for Sale

The procedure for tendering goods is relatively simple:

- Go to the "Sell Fruit" tab pane.
- Now, choose the commodity to be sold either by typing its name into the Sell text field or clicking on the Choose button and selecting it from the current ontology hierarchy.
- Next, enter the reserve price of the item (this is the lowest price you are willing to accept) in the Reserve Price field. This value will be kept secret and used by the negotiation strategy to create an asking price, which will be the price quoted to potential buyers.
- The elapsed time before the expiration of the sale offer can be entered in the Deadline field. This sets the period of time that potential buyers have to submit bids and negotiate.

Finally, once the selling parameters have been entered, click on the Trade button. This will notify the broker of the item for sale, and thus any interested parties. Once offered for sale, the agent waits for incoming bids until the deadline period expires, whereupon the item will be withdrawn from sale. A complete set of steps is shown in Figure 5(2).

(2) Bidding for an Item

The procedure for bidding for goods is quite similar to that for selling:

- Go to the "Buy Fruit" tab pane.
- Now, choose the commodity to purchase by clicking on the Choose button and selecting it from the current ontology hierarchy.
- Next, enter the highest price you would be willing to pay for this item in the Maximum Price field. This value will be kept secret and used by the negotiation strategy to create a bid price, which will be offered to the seller.
- The elapsed time before the completion of the purchase can be entered in the Deadline field.
- Finally, once the bidding parameters have been entered, click on the Trade button. The agent will then consult the broker for the presence of matching items for sale.

The complete set of steps is shown in Figure 5(3).

(3) Negotiation Process

As demonstrated in Figure 8, the negotiation process functions by altering the attributes of the current goal and determines the next state of the agent's coordination engine by returning one of the following three types of messages:

- OK - i.e. the bid was acceptable, agents now move into the settlement stage.
- MESSAGE - i.e. the bid was not acceptable but close enough to warrant a counter-proposal.
- FAIL - this ends the negotiation, because the bid was too lo v and deadline has expired

If the negotiation results in a price acceptable to both parties, the coordination engine of each agent will move into the settlement stage.

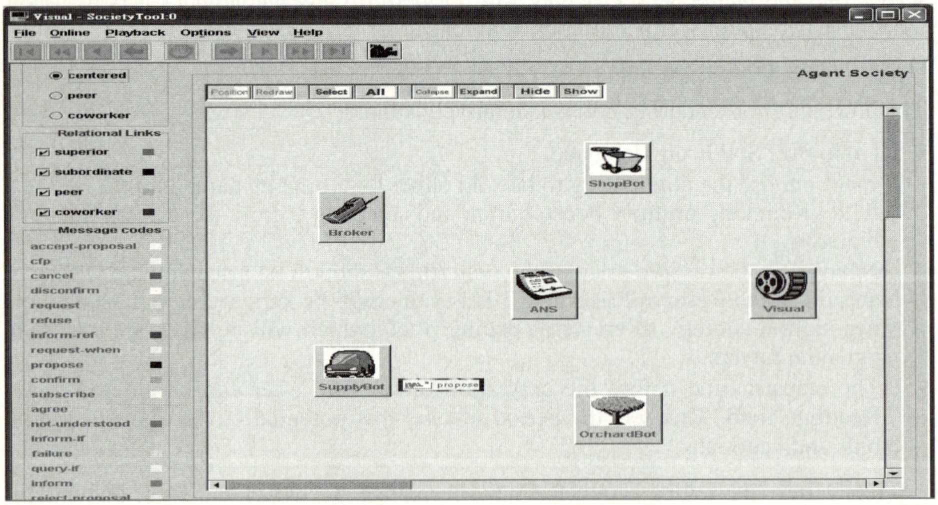

Fig. 8. Negotiation Process

5 Conclusion

In this research, we introduced an e-commerce bargaining system through the multi-agent framework. The user can use the bidding agent to sell/buy items and simultaneously browse the transaction information on the e-marketplace platform through the application. The prototype system demonstrated an excellent performance from the experiments and the system can be easily implemented in a large bargain transaction database of the e-commerce marketplace for wide accessibility. The techniques and methodology we developed in this research can be applied to many other applications, by taking advantage of the friendly graphic user interface and Java-based application, to significantly improve the traditional e-commerce bidding interaction and communication.

References

1. D'Hulst, R., and Rodgers, G.J.: Bid Distributions of Competing Agents in Simple Models of Auctions, Physica *A*, Vol. 294, (2001) 447-464
2. Flores-Mendez R.A.: Towards the Standardization of Multi-Agent System Architecture: *An Overview*, (1999) 18-24
3. Lee, K.Y., Yun, J.S., and Jo, G. S.: MoCAAS: Auction Agent system Using a Collaborative Mobile Agent in Electronic Commerce, Expert Systems with Applications, Vol. 24, (2003) 183–187
4. Löthman, H.: Intelligent Agent Platform in Distributed Software Architecture, Oulu Polytechnic, Raahe Institute of Computer Engineering, Bachelor's thesis, (2001) 134-136
5. Stone, P., and Veloso, M.: Multi-agent Systems: A Survey from a Machine Learning Perspective, Autonomous Robots, Vol. 8, No. 3, (2000) 345-383
6. Wooldridge, M.: This is MyWorld: the logic of an agent-oriented testbed for DAI, In Intelligent Agents: Theories, Architectures and Languages, *LNAI*, Vol. 890, (1995) 160-178

Automatic Classification for Grouping Designs in Fashion Design Recommendation Agent System

Kyung-Yong Jung

School of Computer Information Engineering
Sangji University, Korea
kyjung@sangji.ac.kr

Abstract. Nowadays, users spend much time and effort in finding the best suitable designs since more and more information is placed on-line. To save their time and effort in searching the designs they want, the user-adapting recommendation system is required. In this paper, Automatic Classification for Grouping designs (ACG) in a Fashion Design Recommendation Agent System (FDRAS) is proposed. The ACG algorithm groups designs into clusters based on these classified designs. It is possible that if the design requires simultaneous regrouping in all other groups, the ACG algorithm can be used to improve efficiency of information retrieval and sorting, in the FDRAS datasets. The proposed method is evaluated on a large database, significantly outperforming the nearest-neighbor model and k-mean clustering in the prototype user-adapting FDRAS. This method can solve the large-scale dataset problem without deteriorating accuracy quality.

1 Introduction

Information filtering techniques can be used to create the user-adapting recommendation systems, adapting their presentation to the user's preference [3,7]. The use of information filtering is evaluated in the prototype Fashion Design Recommendation Agent System (FDRAS). The techniques proposed for information filtering, fall into two classes: Collaborative filtering and Content based filtering. In recent research, the combination of both approaches in the context of the FDRAS [5] was studied. In this paper, the grouping method is focused on for the user-adapting recommendation system. EM algorithm [2] is an obvious method for grouping users, but does not work because it cannot efficiently construct two users that must be in the same class each time. The k-mean algorithm [1] is fast but ad hoc. The nearest-neighbor model [2] works well and has the virtue of being easily extended to much more complex models, however, this model is computationally expensive. For the purpose of solving this question, the Automatic Classification for Grouping designs (ACG algorithm), is proposed. This allows accurate recommendation of new designs due to grouping items into clusters based on retrieved designs. The ACG algorithm groups designs into clusters based on these classified designs. The FDRAS, is first described in more detail, in Section 2, which implements acquisitions of content models. In Section 3,

B. Gabrys, R.J. Howlett, and L.C. Jain (Eds.): KES 2006, Part I, LNAI 4251, pp. 310–317, 2006.
© Springer-Verlag Berlin Heidelberg 2006

automatic classification for grouping designs, is introduced. The proposed method is evaluated on a large database, significantly outperforming the nearest-neighbor model and k-mean clustering, in Section 4. The conclusions are given in Section 5.

2 FDRAS: Fashion Design Recommendation Agent System

Figure 1 shows the system overview of the Fashion Design Recommendation Agent System (FDRAS) using the extensive prediction method through collaborative filtering and content based filtering [4,5]. The purpose of the FDRAS project is to investigate how the combination of content based filtering and collaborative filtering can be used to build the user-adapting recommendation systems, in which the navigation through information is adapted to each user. The chosen application is a recommendation system that allows the user to browse through designs.

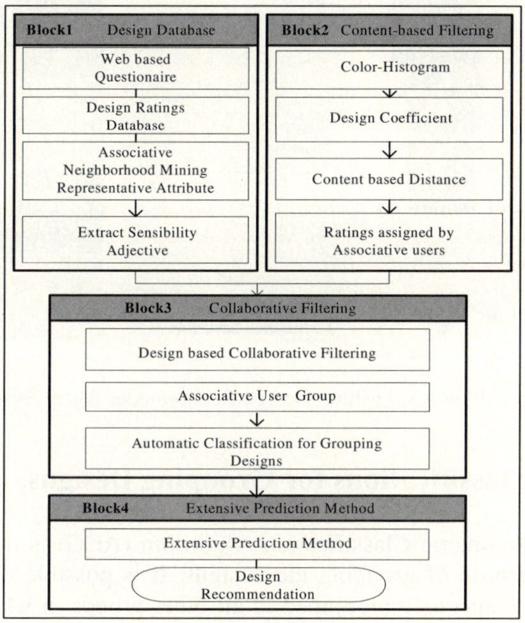

Fig. 1. System overview of the Fashion Design Recommendation Agent System

In the FDRAS project, we used filtering techniques to develop the user-adapting recommendation system, in which the hypertext structure is created for each specific user, based on predictions of what this particular user is likely to prefer. The basic idea is that the user is asked to provide ratings for the designs that he or she views during his or her visit. The system then selects designs similar to the designs with high ratings using content based filtering. Collaborative filtering is also used to make predictions based on the ratings that other users have provided during previous visits [9,10]. These predictions are then used to present appropriate designs to the user, so that the more relevant designs are seen first.

The FDRAS system has a topology that adapts itself to the user's taste and choices. This topology is achieved by the use of classifications, which are virtual classifications that contain designs of a chosen classification sorted according to personalized predictions produced by collaborative filtering. The user may choose from several classifications that are interconnected, so that he retains the ability to choose from among different designs and classifications while at the same time benefiting from the system's recommendations. Figure 2 shows the menu used for browsing the classification in the FDRAS system. When the user enters a classification, he is presented with designs sorted according to his preference. The user can then continues in the classification, choose to see more details of a particular design, or enter a new classification.

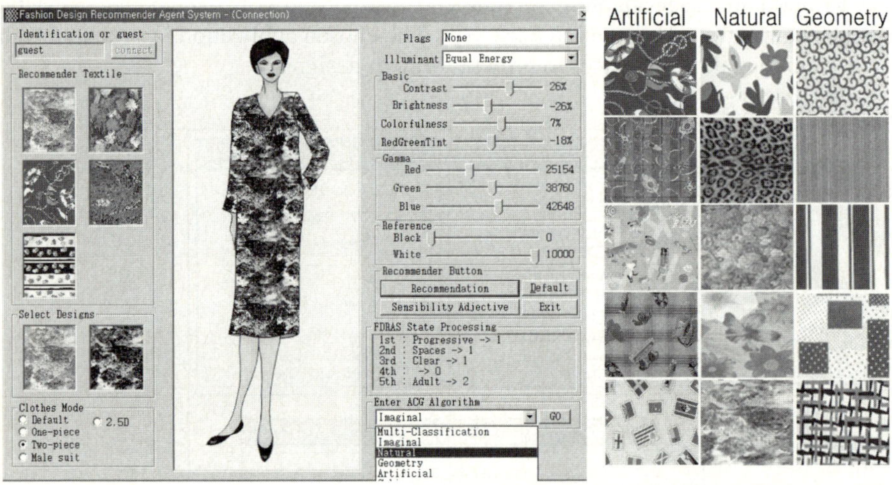

Fig. 2. FDRAS: Fashion Design Recommeder Agent System

3 Automatic Classifications for Grouping Designs

In this paper, an Automatic Classification Algorithm (ACG) is designed and implemented, for the purpose of grouping the designs. It is possible that if the design is required to be simultaneously regrouped in all other groups in which they occur, the ACG algorithm regrouped the designs into the relevant group. In addition, the ACG algorithm can result in improved predictive performance, under cold-start circumstance [8] that starts from small designs. The main steps of the ACG algorithm are presented in Table 1. The ACG algorithm will assign $Group_j$(i) to the design. In the first case, if the new design comes, the ACG algorithm must assign $Group_j$(i) to new design. To start with $Group_1$(i) is defined to be value 1. Then, in order to decide $Group_j$(i), the ACG algorithm calculates the average similarity between the new design and designs in the group. It is a second case where the $Group_j$(i) of existing designs is assigned newly. When design retrieves newly, design's $Group_j$(i) can be changed. In this case, before the design goes to another group, average similarity is calculated between the design, and the others in the current group, by Equation (1). If similarity between the design and the others in current group is less than the similarity_threshold, the design's $Group_j$(i) is changeless, otherwise the design goes to the next group.

$$Average\_Similarity(Group_j(i)\phi Pdesign)$$
$$= \frac{\sum_{i=1}^{classnum} Pdesign \cdot Group_j(i)}{\sum_{i=1}^{classnum} Pdesign^2 + \sum_{i=1}^{classnum} Group_j(i)^2 + \sum_{i=1}^{classnum} Pdesign \cdot Group_j(i)} \quad (1)$$

The process of calculating the average similarity in Equation (1) is presented. Let {*Pdesign*} be a design that newly enters in the ACG algorithm, or that $Group_j$(i) needs to be changed. Let {$Group_1$(i),$Group_2$(i),…,$Group_{designnum}$(i)} be designs that exist in a group. The element of the matrix bears the probability of designs in a class for designs in all class. ∅ at Equation (1) means that similarity between *Pdesign* and designs in $Group_j$(i) is calculated using Jaccard's method [9].

Table 1. Automatic classification algorithms for grouping designs

```
Algorithm Automatic classification algorithm for grouping designs
Input: classified designs group → Group_j(i), new design or update design
Output: modified designs group → Group_j(i)
Max_Sim ← 0 // the biggest similarity value
Index ← 0 // the biggest similarity index
if updated existent design then
   J_Sim ← Similarity() // existent designs
   if J_Sim ≥ Similarity_threshold then
      UpdateGroup(j) // assign existent group
      return
   endif
endif
for j ← 1 to all number of group do
   J_Sim ← Similarity()  // calculate similarity
   if J_Sim < Similarity_threshold then
      continue
   else
      if Max_Sim < J_Sim then
         Max_Sim ← j
      endif
   endif
endfor
InsertGroup(index) // automatic classification for grouping designs
return
```

3.2 Example of Grouping Designs

It is shown that an example of designs categorization and grouping designs. 140 designs are selected from design books. Attempts were made to evenly choose the type of design sources, motif size, hue contrast, value contrast, and the type of motif arrangement, with/without outline and motif variation if possible. For example, the class of design source included natural, man-made, imagination, and symbolisms and designs were selected from each source. All designs classified each 20-design with 7 categorized of the imaginal, the natural, the oriental, the western, the geometry, the artificial, and the modern. In this example, the target categorization for designs is 7 classes. Table 2 presents an example of designs that categorize with 7 classes.

Table 2 can be recomposed through the probability that designs occur in classes. In the result, Table 3 demonstrates that the number of the designs is transformed.

The ACG algorithm can assign designs to $Group_j$(i). Initially, $Group_j$(i) can be made by grouping designs into clusters of similar interest. In order to retrieve the optical cluster of design in Table 3, the ACG algorithm calculates similarity among designs. The result of calculating is presented in Table 4. When similarity between a design and the others is not less than 0.5, the ACG algorithm can group designs into clusters with the same $Group_j$(i). Here, it is decided that the similarity_threshold is 0.5.

When pairs are found where the similarity_threshold is not less than 0.5 in Table 4, four pairs of $\{(design_1, design_2), (design_1, design_4), (design_1, design_6), (design_3, design_5)\}$ can be found in Table 5. Hence, the ACG algorithm makes 2 groups, of which one group is composed of design $\{design_1, design_2, design_4, design_6\}$ and the other group is composed of design $\{design_3, design_5\}$.

Table 2. Categorized designs with 7 classes

	imaginal	*natural*	*oriental*	*western*	*geometry*	*artificial*	*modern*
$design_1$	1	3	0	4	0	2	0
$design_2$	1	2	2	3	0	0	0
$design_3$	2	0	1	0	3	0	0
$design_4$	0	1	0	1	0	1	1
$design_5$	1	0	2	0	2	2	3
$design_6$	0	2	1	0	1	2	2

Table 3. Probability of designs

	imaginal	*natural*	*oriental*	*western*	*geometry*	*artificial*	*modern*
$design_1$	0.1000	0.3000	0.0000	0.4000	0.0000	0.2000	0.0000
$design_2$	0.1250	0.2500	0.2500	0.3750	0.0000	0.0000	0.0000
$design_3$	0.3333	0.0000	1.6666	0.0000	0.5000	0.0000	0.0000
$design_4$	0.0000	0.3333	0.0000	0.3333	0.0000	0.3333	0.3333
$design_5$	0.1000	0.0000	0.2000	0.0000	0.2000	0.2000	0.3000
$design_6$	0.0000	0.2500	0.1250	0.0000	0.1250	0.2500	0.2500

Table 4. Similarity among designs

	$design_1$	$design_2$	$design_3$	$design_4$	$design_5$	$design_6$
$design_1$	1.0000	**0.9130**	0.0434	**0.6720**	0.0233	**0.6690**
$design_2$	**0.9130**	1.0000	0.0941	0.0149	0.0667	0.2000
$design_3$	0.0434	0.0941	1.0000	0.3147	**0.6702**	0.6702
$design_4$	**0.6720**	0.0149	0.3147	1.0000	0.3194	0.1319
$design_5$	0.0233	0.0667	**0.6702**	0.3194	1.0000	0.4667
$design_6$	**0.6690**	0.2000	0.4417	0.1319	0.4667	1.0000

Table 5. Grouping designs by ACG algorithm

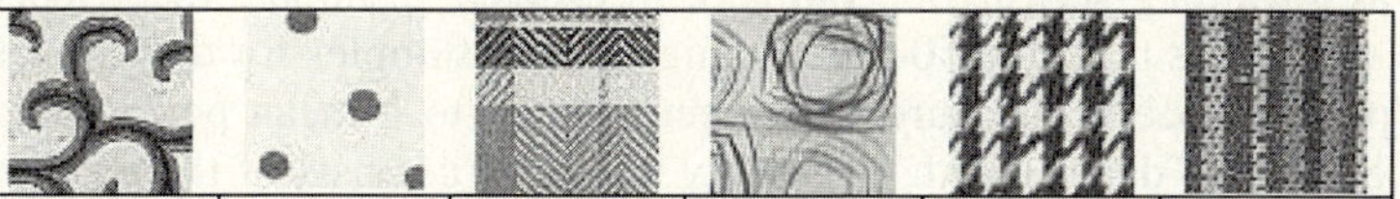

$design_1$	$design_2$	$design_4$	$design_6$	$design_3$	$design_5$
$Group_1$				$Group_2$	

After the ACG algorithm creates the design groups, the design groups can change frequently. The design groups change frequently, as a result of two factors. The first factor is that a new design enters the FDRAS. The second factor is that the interest of existent designs in a group can change frequently. In Table 6, a case where a new design enters the FDRAS is described.

Table 6. Desgins retrieved by design $\{design_7\}$

	imaginal	*natural*	*oriental*	*western*	*geometry*	*artificial*	*modern*
$design_7$	3	0	2	1	2	0	1

The ACG algorithm can calculate the similarity between design $\{design_7\}$ and designs in all groups. Then, the average similarity is 0.2548 at $Group_1$ and is 0.5670 at $Group_2$ in Table 7. In addition, if there are a few groups that average similarity is not less than 0.5, the ACG algorithm assigns $Group_j$(i) with the largest average similarity to new design.

Table 7. Similarity between design $\{design_7\}$ and designs in $Group_1$, $Group_2$

$design_7$	$design_1$	$design_2$	$design_3$	$design_4$	$design_5$	$design_6$	*average*
$Group_1$	0.1975	0.3934		0.1800		0.3185	0.2548
$Group_2$			0.4976		0.6364		0.5670

Finally, design $\{design_7\}$ is categorized into $Group_2$. However, if the average similarity is less than similarity_threshold 0.5 in all groups, design $\{design_7\}$ is categorized into new $Group_3$, and has not been until this time. In addition, if design $\{design_4\}$ is interested in new categorization, $Group_j$(i) must be changed. In this case, $Group_j$(i) of design $\{design_4\}$ can be changed in the above process.

4 Evaluation

In order to evaluate the performance of the Automatic Classifications for Grouping designs (ACG), grouping methods for collaborative filtering are compared with the Nearest-Neighbor Model (NNM) and K-Means Clustering (KMC). Experiments are performed on a subset of design rating data, collected from the FDRAS datasets [5]. The data represents an experiment of 512 users (259 male and 253 female), with each user rating 140 kinds of design, through data integrity. We ran several sets of off-line experiment in order to resolve questions concerning the application of the automatic classification for grouping designs in the FDRAS. For the experiment the user ratings are split in a test-set and training-set. For 300 users with more than 100 ratings a subset of 80 ratings is sampled in the test-set (6,152 ratings). The remaining ratings are used as a training-set (35,080 ratings). In order to eliminate the effects of biased splits the splitting is repeated 50 times using random samples for the test-set. For each split the performance is measured and accumulated to a mean performance over all splits. The FDRAS dataset with sensibility adjective consists of the evaluation rating values, sensibility adjectives, user-attributes, user-profile, and the design information.

In addition to the above, it is important to evaluate accuracy and recall. In order to quantify this with a single measurement, the F-measure in Equation (2) is used, which is a weighted combination of accuracy and recall frequency used in information retrieval [6,8].

$$F\_measure = \frac{(\beta^2+1)\times P\times R}{\beta^2\times P+R} \qquad P=\frac{x}{x+y}100\% \qquad R=\frac{x}{x+z}100\% \tag{2}$$

P and R in Equation (2) represent the accuracy and recall. “x” is the number of items, which appear in both classes. “y” is the number of items, which appear in classes categorized by the first method but not in the class categorized by second method. “z” is the number of items, which appears in the class categorized by the second method but not in the class categorized by the first method. The larger F-measure is, the better the classification performance is.

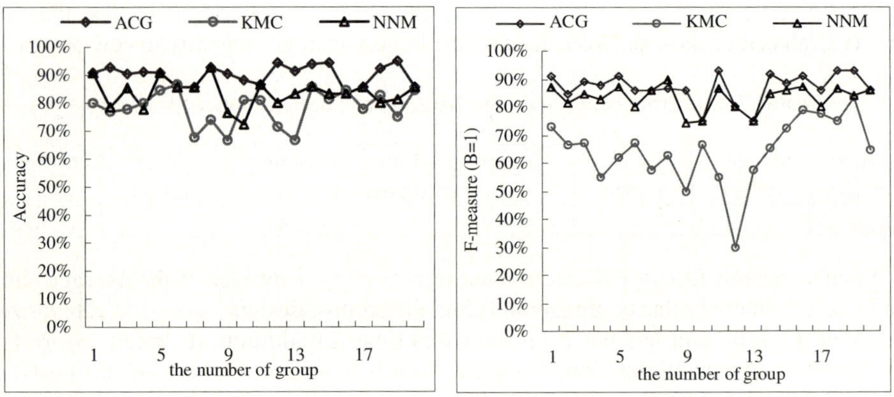

(a) Accuracy of group (b) F-measure of group

Fig. 3. Performance of the ACG method compared to KMC method and NNM method

Figure 3 summarizes the performance of the three methods. In Figure 3(a), it is seen that ACG is much more accurate than the other methods (average 92.23 for ACG vs. 76.36 for KMC vs. 84.62 for NNM). In Figure 3(b), both ACG, as well as NNM, can be seen to have a significant advantage in recall (average 89.96 for ACG vs. 88.43 for NNM) but KMC is low in recall (average 61.35 for KMC). In addition, ACG has a substantial advantage over KMC. This method can solve the large scale dataset problem without deteriorating accuracy quality. These results are encouraging and provide empirical evidence that the use of ACG can lead to improved performance of the FDRAS.

5 Conclusion

In this paper, the shortcomings of algorithms for grouping items based collaborative filtering techniques are identified, and the method of solving these problems through automatic classification for grouping designs is demonstrated. The contributions of this paper are twofold. First, it is demonstrated that, if a user requires to be simultaneously regrouped in all other groups in which they occur, it is possible to regroup the user into the relevant group. Second, it is demonstrated that it is possible to lead to improved predictive performance under cold-start circumstance. In experimenting for grouping designs, the automatic classification for grouping designs has been proven, and is much more efficient than the nearest-neighbor model and k-means clustering. Furthermore, the integration of categorizations and prediction algorithm will be explored, with the hope that prediction results in general improve and become increasingly reliable.

References

1. K. Alsabti, S. Ranka, and V. Singh, “An Efficient K-Means Clustering Algorithm,” In Proc. of the Workshop on High-Performance Data Mining, 1998.
2. J. L. Herlocker, J. A. Konstan, L. G. Terveen, and J. T. Riedl, "Evaluating Collaborative Filtering Recommender Systems," ACM Transactions on Information Systems (TOIS) archive, Vol. 22, No. 1, pp. 5-53, 2004.
3. J. H. Hwang, M. S. Gu, and K. H. Ryu, “Context-Based Recommendation Service in Ubiquitous Commerce,” In Proc. of the Int. Conf. on Computational Science and Its Applications, LNCS 3481, pp. 966-976, 2005.
4. K. Y. Jung and J. H. Lee, “User Preference Mining through Hybrid Collaborative Filtering and Content-based Filtering in Recommendation System," IEICE Transaction on Information and Systems, Vol. E87-D, No. 12, pp. 2781-2790, 2004.
5. K. Y. Jung, Y. J. Na, and J. H. Lee, "FDRAS: Fashion Design Recommender Agent System using the Extraction of Representative Sensibility and the Two-Way Filtering on Textile," In Proc. of Int. Conf. on Database and Expert Systems Applications, LNCS 2736, pp. 631-640, 2003.
6. S. J. Ko and J. H. Lee, “Feature Selection Using Association Word Mining for Classification,” In Proc of Int. Conf. Database and Expert Systems Applications, LNCS 2113, pp. 211-220, 2001.
7. A. Kohrs and B. Merialdo, “Improving Collaborative Filtering with Multimedia Indexing Techniques to Create User-Adapting Web Sites,” In Proc. of the ACM Int. Conf. on Multimedia, pp. 27-36, 1999.
8. R. Torres, S.M. McNee, M. Abel, J.A. Konstan, and J. Riedl, "Enhancing Digital Libraries with TechLens+," In Proc. of the 4th ACM/IEEE Joint Conf. on Digital Libraries, pp. 228-237, 2004.
9. J. Wang, A. P. de Vries, and M. J. T. Reinders, "A User-Item Relevance Model for Log-based Collaborative Filtering," In Proc. of the European Conf. on Information Retrieval, LNCS 3936, pp. 37-48, 2006.
10. K. Yu, X. Xu, M. Ester, and H. P. Kriegel, "Feature Weighting and Instance Selection for Collaborative Filtering: An Information-Theoretic Approach," Knowledge and Information Systems, Vol. 5, No. 2, pp. 201-224, Springer-Verlag, 2003.

Agent Behaviour in Double Auction Electronic Market for Communication Resources

Krunoslav Trzec[1], Ignac Lovrek[2], and Branko Mikac[2]

[1] Ericsson Nikola Tesla, R&D Centre,
Krapinska 45, HR-10000 Zagreb, Croatia
krunoslav.trzec@ericsson.com
[2] University of Zagreb,
Faculty of Electrical Engineering and Computing,
Department of Telecommunications
Unska 3, HR-10000 Zagreb, Croatia
{ignac.lovrek, branko.mikac}@fer.hr

Abstract. The paper deals with behaviour of trading agents with the lowest level of bounded rationality on an electronic market where the double auction is applied as a trading mechanism. Two types of double auction have been considered: the clearing house and the continuous double auction. The agent behaviour is modelled using business strategies based on adaptive learning algorithms and imitation which is governed by replicator dynamics from evolutionary game theory. A case study describing optical bandwidth trading on the double auction electronic market is analyzed.

1 Introduction

Electronic communications market with various information and communication services, i.e. voice, data and multimedia, fixed or mobile, is a highly dynamic environment. Provisioning of global and complex services requires new business relationships among service providers in order to fulfil different users' needs. Traditional leasing enhanced with dynamic pricing of communication resources' surpluses according to actual requirements, is a promising solution for evolving telecommunications market.

Software agent-based electronic market (e-market) for communication resources defined as homogeneous commodities, i.e. goods that are evaluated only by their price attribute, is elaborated in the paper. The trading agents act as buyers or sellers on behalf of service providers in order to accomplish given tasks autonomously. Double auction (DA) electronic market with a relatively small number of trading agents, which are relevant for trading of communication resources, such as optical bandwidth (i.e. units of bit rate), is studied. The double auction mechanism is considered because it enables good profits for the traders, as well as efficient market outcomes [1]. Models of agent behaviour based on adaptive learning, and suitable for agents with the lowest level of bounded rationality, are considered.

B. Gabrys, R.J. Howlett, and L.C. Jain (Eds.): KES 2006, Part I, LNAI 4251, pp. 318–325, 2006.
© Springer-Verlag Berlin Heidelberg 2006

The rest of the paper is organized as follows: Section 2 deals with adaptive agent behaviour in the double auction and elaborates the applied business strategies that learn from experience. Evolutionary selection process of business strategies is analyzed in Section 3. A case study describing agent-based optical bandwidth trading is given in Section 4, while section 5 concludes the paper.

2 Business Strategies for Double Auction

An auction is a trading mechanism with an explicit set of rules determining resource allocation and prices on the basis of traders' messages which include an offered price (called a bid for an offer to buy and an ask for an offer to sell). This competitive process serves to aggregate the scattered information about traders' declared valuations and to dynamically set a transaction price. An auction is one-sided if only bids or only asks are permitted, and two-sided if both are permitted. Moreover, an auction can be either one-shot or repeated. A repeated auction consists of several rounds.

The double auction is a two-sided repeated auction that allows the strategic behaviour of trading agents on the e-market [1]. Applying strategic behaviour, the agents form beliefs based on an analysis of what others might do [2]. The consequence of strategic behaviour is that agents' declared valuations (i.e. offered prices) are intentionally set to be different from their real valuations (i.e. limit prices). A standard approach to analyzing strategic behaviour is based on game theory [2, 3]. In particular, a double auction is usually modelled as a static game with incomplete information in which agent actions (i.e. offered prices) represents game-theoretic pure strategies and agent types determine their real valuations. In case of a DA, the methodology based on game theory searches for the Bayes-Nash equilibrium [3] in which the selected agent pure strategies best respond to each other. The main drawback of the game-theoretic approach is the hypothesis that trading agents are capable of calculating the Bayes-Nash equilibrium, i.e. they maximize the expected payoffs. In other words, perfectly rational agents are assumed.

A double auction requires more realistic models of agent behaviour. In this paper, strategies based on learning algorithms that are adaptive and backward-looking (i.e. agents only respond to their previous payoffs) are proposed for agents with the lowest level of bounded rationality. The level of bounded rationality is measured by agents' steps of thinking proposed by Camerer et al. [2]. Using zero steps of thinking, agents do not reason strategically at all. In other words, trading agents with the lowest level of bounded rationality are more willing to make a random guess in the first round of a DA and learn from subsequent experience, than to think hard before learning.

Consequently, two business strategies are used to model the adaptive behaviour of trading agents. The first one is the strategy proposed by Priest and van Tol [4], denoted by PvT. It is based on the "Zero-Intelligence Plus" (ZIP) strategy introduced by Cliff and Bruten [5]. An agent that applies the PvT strategy is initially assigned a random profit margin in the appropriate range. The agent then monitors bids and asks as well as transaction prices on the e-market, and uses the Widrow-Hoff learning rule with momentum [6] to modify its offered price in order to minimize the mean square root error with respect to the information regarding current transaction price.

The second business strategy used to model adaptive agent behaviour is based on a modified version of the Roth-Erev myopic stimuli-response learning algorithm [7], denoted by RE. It is more generic than the PvT strategy as it uses a three-parameter reinforcement learning algorithm which forms the basis of a model of human behaviour in games with multiple interacting players. In contrast to the PvT strategy, the RE strategy facilitate sensitivity of the learning rule to agents' activities on the e-market by capturing the psychological assumptions associated with human experimentation and memory through experimentation and recency parameters. The former takes into account not only actions that were successful in the past, but also reinforce actions similar to (i.e. near) successful one more strongly than actions that are further away, while the later is used to model the agent behaviour in which recent experience plays a larger role than past experience.

It is supposed that the agents' selection of a business strategy at the beginning of a DA is based on imitation that can be captured by the replicator dynamics (RD) [8]. The RD models an evolutionary selection process in which agents imitate a behaviour that appears to be more successful, with a probability to the expected payoff. Two types of double auctions have been considered: the clearing house (CH) and the continuous double auction (CDA) [1]. In a CH, an auctioneer waits for all the buyers and sellers to place bids and asks before clearing the market. The transaction prices set by the auctioneer are either discriminatory (set individually for each matched buyer-seller pair) or uniform (set to an equal value for all matched buyer-seller pairs) [1, 9]. On the other hand, a CDA continually accepts bids, which either immediately match pending asks, or remain standing until matched by latter asks.

3 Evolutionary Selection Process of Business Strategies

At the start of its first double auction, each of the M agents which participate on the electronic market chooses randomly to play one of the N available strategies according to a game-theoretic mixed strategy $m_i = (b_{i,1}, \ldots, b_{i,N})$. Here, $b_{i,j}$ indicates the probability that agent i plays strategy j subject to the constraints that $b_{i,j} \in [0, 1]$ and $\Sigma b_{i,j} = 1$. The agent selection of a mixed strategy for subsequent double auctions is done by imitating the behaviour of those trading agents that are more successful, i.e. earn higher payoffs, in previous auctions. The payoff to agent i is a real-valued function u of the strategy played by i and the strategies played by all other agents. Since DA symmetry is supposed, all agents that play the same strategy have the same payoff which is, in fact, a function of the number of agents playing each strategy. The profile of all the agents' mixed strategies is denoted with $\mathbf{m}$, whereas the profile of the mixed strategies for all agents except i is indicated by $\mathbf{m}_{-i}$. The case when trading agent i plays pure strategy j with a probability of one is marked with e^j. Formally, profile $\mathbf{m}^*$ constitutes a Nash equilibrium if for all agents i and all m_i

$$u(\mathbf{m}^*) \geq u(m_i, \mathbf{m}_{-i}^*). \tag{1}$$

For a symmetrical normal form game, $\mathbf{m}^*$ is a symmetric Nash equilibrium if it is the best replay against itself, i.e. if in Eq. 1, all mixed strategies in profile $\mathbf{m}_{-i}^*$ are equal to mixed strategy m_i*. Consequently, in the further analysis, an arbitrary, symmetric mixed strategy will be denoted by m and the probability that a given trading agent plays business strategy j by m_j.

In order to find symmetric Nash equilibriums that are more likely to be discovered by imitating trading agents, the evolutionary model based on replicator dynamics has been used. The RD equation is given by

$$\frac{dm_j}{dt} = \left[u(e^j, t) - \langle u(t) \rangle\right] m_j, \tag{2}$$

where $u(e^j, t)$ is the average payoff to business strategy j when all players play mixed strategy m, and $\langle u(t) \rangle$ is the mean payoff when all players play m. The calculation of average and mean payoffs for CH and CDA is performed by using methodology proposed by Walsh et al. [10] which transforms DA and its huge number of available pure strategies and types into a one-shot game with a limited number of high-level strategies. High-level strategies represent business strategies that are treated as simple pure strategies in a normal form game. For a small number of trading agents and strategies, this gives a relatively small normal form game payoff matrix amenable to the game-theoretic solution. The payoff matrix is calibrated by running simulations of the DA game. The variations in payoffs due to different agent types are averaged over many samples of type information resulting in a single mean payoff for each agent in each entry in the payoff matrix.

For any initial mixed strategy, we can discover the eventual outcome of the evolutionary process by finding the stationary points of Eq. 2, i.e. by solving $dm_j/dt = 0$ for all j. Since the replicator dynamics has the property that all symmetric Nash equilibriums of the game are stationary points of its phase space (i.e. direction field), by overlaying the direction field on the identified equilibriums we can see which of them are more likely to be discovered by trading agents. Those symmetric Nash equilibriums that represent stationary points, at which a larger range of initial mixed strategies will end up, are equilibriums that are more likely to be reached (assuming a uniform initial distribution). When mixed strategy trajectories governed by Eq. 2 converge to a symmetric Nash equilibrium, the equilibrium is asymptotically stable and represents an attractor. The basin of attraction for a stationary point is the region of mixed strategy space which, governed by the replicator dynamics, leads to the stationary point (i.e. attractor) that in our case denotes the long-term (evolutionary) stable agent behaviour on the double auction electronic market.

4 Case Study: Agent-Based Optical Bandwidth Trading

Optical bandwidth trading on the telecommunication market is chosen as a case study [11], taken from the broader research of agents and their application in a new generation networks [12]. In order to capture strategy payoffs of agents and enable an analysis of their behaviours using the replicator dynamics, each entry in the payoff matrix was computed by averaging the payoff of each business strategy across 2000 DA simulations. All double auction simulations (CH and CDA) consisted of 100 rounds and were performed using the JASA auction simulator [13], in which a discriminatory midpoint pricing rule was employed. At the beginning of each simulation, half the agents were randomly assigned as buyers and the remainder as sellers. For each type of double auction, limit prices were chosen from the same uniform distribution.

The simulations were restricted to private values scenarios [1, 9] in which trading agents have well-defined valuations for the unit of bit rate (Gbit/s) being negotiated and where one agent's valuation does not affect the valuation of another. This assumption implies that traders have no knowledge regarding the real supply and demand on the e-market, and thus face an inherently complex situation. Moreover, it is assumed that buyers and sellers are risk neutral – i.e. they seek to maximize their expected profits. This assumption is realistic for markets in which buyers and sellers do not resell their units of bit rate. The limit prices for both buyers and sellers define the supply and demand curves. They are drawn independently from a uniform random distribution in range (0, 50). The limit prices that did not yield a competitive equilibrium price [1] were discarded.

The trading agents applied either business strategies based on learning algorithms (RE or PvT) or the truth-telling strategy (denoted by TT) that simply declares the agent's limit price (i.e. real valuation of bit rate unit) and represents a non-strategic irrational behaviour. The parameters of the learning algorithm in the RE business strategy were set to the values obtained by Erev and Roth [7]. Consequently, the experimentation parameter was set to 0.2, the recency parameter to 0.1, and the scaling parameter to 9. In each auction round, the agents chose among 25 feasible actions (bid or ask prices). The Widrow-Hoff learning rule is characterized by two parameters: learning rate and momentum, which were set to 0.3 and 0.05, respectively.

The imitation of trading agents on the e-market is analyzed using replicator dynamics. At the start of first double auction, each trading agent was randomly assigned a probabilistic combination of business strategies, i.e. mixed strategy $m(b_1, b_2, b_3)$, where b_1, b_2, and b_3 denote the PvT, RE, and TT strategy choice probabilities, respectively. The initial mixed strategy was progressively adjusted at the start of subsequent double auctions as a result of the dynamics described by Eq. 2. In Fig. 1 and 2, which show the agent adjustments for the CH and CDA when 6 and 10 traders were on the e-market, respectively, the mixed strategies are indicated by points of the unit simplex (triangle). The choice when an agent selects a business strategy with a probability that equals one is marked by the one of the three vertices of the simplex.

A sequence of agent adjustments in Fig. 1 and 2 is depicted by a trajectory. Each simplex contains the trajectories generated from 250 randomly sampled initial mixed strategies. Since a majority of the trajectories start inside the simplex and head outwards, i.e. towards the edges and the three vertices, arrows are not placed on the trajectories in order to make the figures as clear as possible. The direction of the trajectories for both the CH and CDA indicates that the vertices of the simplex act as attractors, i.e. as asymptotically stable symmetric Nash equilibriums. The grey shading, which is proportional to $|dm/dt|$, denotes the adjustment rate of the agent behaviour. The set of trajectories leading to each vertex indicates the basin of attraction of the corresponding vertex. The likelihood of one vertex relative to another can be assessed by comparing the sizes of their respective basins of attraction.

From Fig. 1 and 2 it can be deduced that for a CH mechanism an increase in the number of trading agents from 6 to 10 results in adjustment of agent behaviour from the RE business strategy towards the PvT business strategy.

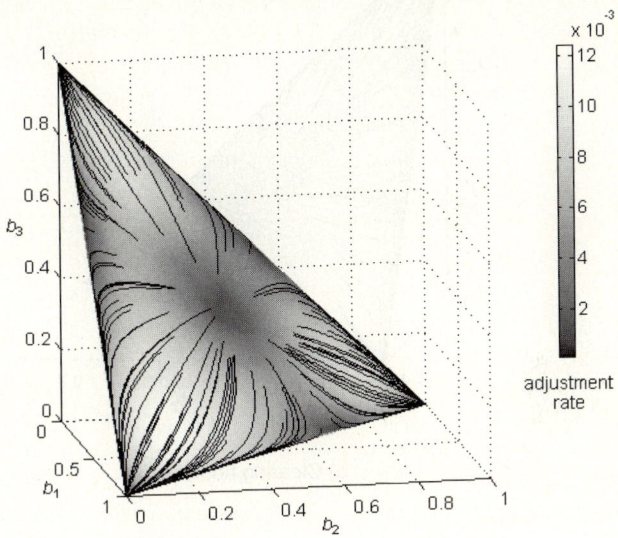

a) Clearing house

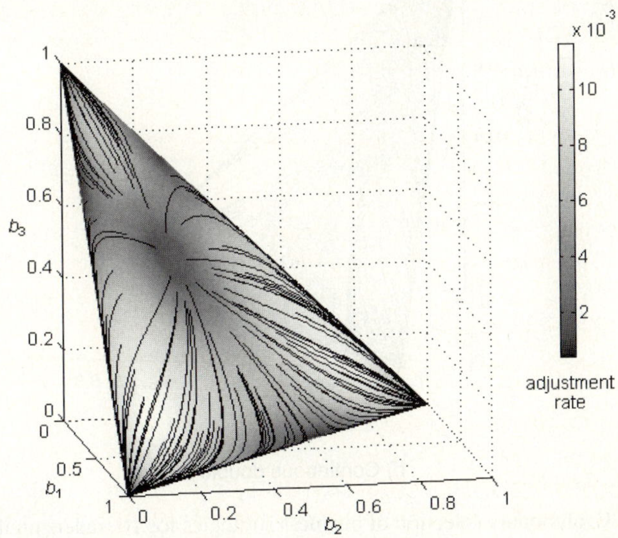

b) Continuous double auction

Fig. 1. Evolutionary selection of business strategies for 6 traders on the DA e-market

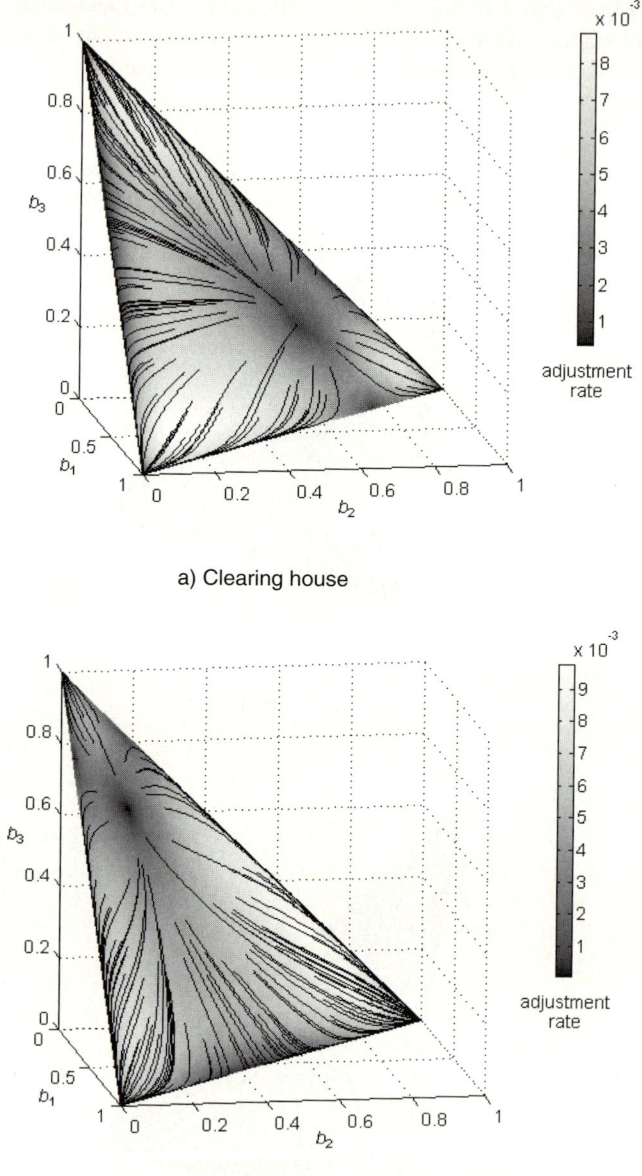

a) Clearing house

b) Continuous double auction

Fig. 2. Evolutionary selection of business strategies for 10 traders on the DA e-market

On the other hand, Fig. 1 and 2 shows that for a CDA the increase in the number of trading agents does not have significant influence on agent behaviour as in both cases the most likely strategy that traders choose is the RE business strategy.

Moreover, truth-telling strategy becomes less probable in a CDA as the number of trading agents increases.

4 Conclusion

An analysis of the evolutionary selection process regarding business strategies of the agents with the lowest level of bounded rationality on the double auction e-market was conducted. It identified the asymptotically stable symmetric Nash equilibriums (i.e. long-term stable agent behaviours), their relative likelihood, and the adjustment rate of agent behaviour towards them for six and ten traders in an optical bandwidth trading scenario. Future work will include the analysis of trading agents with higher level of bounded rationality that besides adaptive learning have strategic foresights.

References

1. Friedman, D., Rust, J. (eds.): The Double Auction Market: Institutions, Theories, and Evidence. Perseus Publishing, Cambridge (1993)
2. Huck, S. (ed.): Advances in Understanding Strategic Behaviour: Game Theory, Experiments and Bounded Rationality. Palgrave Macmillian, Hampshire (2004)
3. Vega-Redondo, F.: Economics and the Theory of Games. Cambridge University Press, Cambridge (2003)
4. Priest, C., van Tol, M.: Adaptive Agents in a Persistent Shout Double Auction. In: Proceedings of the 1st International Conference on Information and Computation Economies, Charleston, USA, (1998) 11–18
5. Cliff, D., Bruten, J.: Minimal-Intelligence Agents for Bargaining Behaviors in Market-Based Environments. Technical Report HPL 97-91, HP Labs, Bristol (1997)
6. Konar, A.: Artificial Intelligence and Soft Computing: Behavioral and Cognitive Modeling of the Human Brain. CRC Press, Boca Raton (1999)
7. Erev, I., Roth, A. E.: Predicting How People Play Games: Reinforcement Learning in Experimental Games with Unique, Mixed Strategy Equilibria. American Economic Review, 4 (1998) 848–881
8. Weibull, J. W.: Evolutionary Game Theory. MIT Press, Cambridge (1997)
9. Krishna, V.: Auction Theory. Academic Press, San Diego (2002)
10. Walsh, W. E., Das, R., Tesauro, G., Kephart, J. O.: Analyzing Complex Strategic Interactions in Multi-Agent Systems. In: Proceeding of the AAAI 2002 Workshop on Game Theoretic and Decision Theoretic Agents, Edmonton, Canada, (2002) 109–118
11. Trzec, K., Mikac, B.: Agent-Based Electronic Market for Optical Bandwidth Trading. In: Proceedings of the 12th International Conference on Software, Telecommunications and Computer Networks (SoftCOM 2004), Split, Croatia, Venice, Italy, (2004) 463–467
12. Lovrek, I., Sinkovic, V.: Mobility Management for Personal Agents in the All-Mobile Network. In: Negoita, M. G., Howlett, R. J., Jain, L. C. (eds.): Knowledge-Based Intelligent Information and Engineering Systems, 8th International Conference, KES 2004, Wellington, New Zealand, 2004. Proceedings. Vol. 1. Lecture Notes in Computer Science, Vol. 3213. Springer-Verlag, Berlin Heidelberg New York (2004) 1143–1149
13. Phelps, S., Tamma, V., Wooldridge, M., Dickinson, I.: Toward Open Negotiation. IEEE Internet Computing, 2 (2004) 70–75

Using Prototypical Cases and Prototypicality Measures for Diagnosis of Dysmorphic Syndromes

Rainer Schmidt and Tina Waligora

Institute for Medical Informatics and Biometry, University of Rostock, Germany
rainer.schmidt@medizin.uni-rostock.de

Abstract. Since diagnosis of dysmorphic syndromes is a domain with incomplete knowledge and where even experts have seen only few syndromes themselves during their lifetime, documentation of cases and the use of case-oriented techniques are popular. In dysmorphic systems, diagnosis usually is performed as a classification task, where a prototypicality measure is applied to determine the most probable syndrome. These measures differ from the usual Case-Based Reasoning similarity measures, because here cases and syndromes are not represented as attribute value pairs but as long lists of symptoms, and because query cases are not compared with cases but with prototypes. In contrast to these dysmorphic systems our approach additionally applies adaptation rules. These rules do not only consider single symptoms but combinations of them, which indicate high or low probabilities of specific syndromes.

1 Introduction

When a child is born with dysmorphic features or with multiple congenital malformations or if mental retardation is observed at a later stage, finding the correct diagnosis is extremely important. Knowledge of the nature and the etiology of the disease enables the pediatrician to predict the patient's future course. So, an initial goal for medical specialists is to diagnose a patient to a recognised syndrome. Genetic counselling and a course of treatments may then be established.

A dysmorphic syndrome describes a morphological disorder and it is characterised by a combination of various symptoms, which form a pattern of morphologic defects. An example is Down Syndrome which can be described in terms of characteristic clinical and radiographic manifestations such as mental retardation, sloping forehead, a flat nose, short broad hands and generally dwarfed physique [1].

The main problems of diagnosing dysmorphic syndromes are as follows [2]:

- more than 200 syndromes are known,
- many cases remain undiagnosed with respect to known syndromes,
- usually many symptoms are used to describe a case (between 40 and 130),
- every dysmorphic syndrome is characterised by nearly as many symptoms.

Furthermore, knowledge about dysmorphic disorders is continuously modified, new cases are observed that cannot be diagnosed (it exists even a journal that only

B. Gabrys, R.J. Howlett, and L.C. Jain (Eds.): KES 2006, Part I, LNAI 4251, pp. 326 – 333, 2006.
© Springer-Verlag Berlin Heidelberg 2006

publishes reports of observed interesting cases [3]), and sometimes even new syndromes are discovered. Usually, even experts of paediatric genetics only see a small count of dysmorphic syndromes during their lifetime.

So, we have developed a diagnostic system that uses a large case base. Starting point to build the case base was a large case collection of the paediatric genetics of the University of Munich, which consists of nearly 2.000 cases and 229 prototypes. A prototype (prototypical case) represents a dysmorphic syndrome by its typical symptoms. Most of the dysmorphic syndromes are already known and have been defined in literature. And nearly one third of our entire case base have been determined by semiautomatic knowledge acquisition, where an expert selected cases that should belong to same syndrome and subsequently a prototype, characterised by the most frequent symptoms of his cases, was generated. To this database we have added rare dysmorphic syndromes, namely from "clinical dysmorphology" [3] and from the London dysmorphic database [4].

1.1 Diagnostic Systems for Dysmorphic Syndromes

Systems to support diagnosis of dysmorphic syndromes have already been developed in the early 80s. The simple ones perform just information retrieval for rare syndromes, namely the London dysmorphic database [3], where syndromes a described by symptoms, and the Australian POSSUM, where syndromes are visualised [5]. Diagnosis by classification is done in a system developed by Wiener and Anneren [6]. They use more than 200 syndromes as database and apply Bayesian probability to determine the most probable syndromes. Another diagnostic system, which uses data from the London dysmorphic database was developed by Evans [7]. Though he claims to apply Case-Based Reasoning, in fact it is again just a classification, this time performed by Tversky's measure of dissimilarity [8]. The most interesting aspect of his approach is the use of weights for the symptoms. That means the symptoms are categorised in three groups – independent of the specific syndromes, instead only according to their intensity of expressing retardation or malformation. However, Evans admits that even features, that are usually unimportant or occur in very many syndromes sometimes play a vital role for discrimination between specific syndromes.

In our system the user can choose between two measures of dissimilarities between concepts, namely of Tversky [8] and the other one of Rosch and Mervis [9]. However, the novelty of our approach is that we do not only perform classification but subsequently apply adaptation rules. These rules do not only consider single symptoms but specific combinations of them, which indicate high or low probabilities of specific syndromes.

1.2 Case-Based Reasoning and Prototypicality Measures

Since the idea of Case-Based Reasoning (CBR) is to use former, already solved solutions (represented in form of cases) for current problems [10], CBR seems to be appropriate for diagnosis of dysmorphic syndromes. CBR consists of two main tasks [11], namely retrieval, that means searching for similar cases, and adaptation, that means adapting solutions of similar cases to the query case. For retrieval usually explicit similarity measure or, especially for large case bases, faster retrieval algorithms like Nearest Neighbour Matching [12] are applied. For adaptation only few general techniques exist [13], usually domain specific adaptation rules have to be acquired.

In CBR usually cases are represented as attribute-value pairs. In medicine, especially in diagnostic applications, this is not always the case, instead often a list of symptoms describes a patient's disease. Sometimes these lists can be very long, and often their lengths are not fixed but vary with the patient. For dysmorphic syndromes usually between 40 and 130 symptoms are used to characterise a patient.

Furthermore, for dysmorphic syndromes it is unreasonable to search for single similar patients (and of course none of the systems mentioned above does so) but for more general prototypes that contain the typical features of a syndrome. Prototypes are a generalisation from single cases. They fill the knowledge gap between the specificity of single cases and abstract knowledge in form of cases. Though the use of prototypes had been early introduced in the CBR community [14, 15], their use is still rather seldom. However, since doctors reason with typical cases anyway, in medical CBR systems prototypes are a rather common knowledge form (e.g. for antibiotics therapy advice in ICONS [16], for diabetes [17], and for eating disorders [18]).

So, to determine the most similar prototype for a given query patient instead of a similarity measure a prototypicality measure is required. One speciality is that for prototypes the list of symptoms is usually much shorter than for single cases.

The result should not be just the one and only most similar prototype, but a list of them – sorted according to their similarity. So, the usual CBR methods like indexing or nearest neighbour search are inappropriate. Instead, rather old measures for dissimilarities between concepts [8, 9] are applied and explained in the next section.

2 Diagnosis of Dysmorphic Syndromes

Our system consists of four steps (fig.1). At first the user has to select the symptoms that characterise a new patient. This selection is a long and very time consuming process, because we consider more than 800 symptoms. However, diagnosis of dysmorphic syndromes is not a task where the result is very urgent, but it usually requires thorough reasoning and afterwards a long-term therapy has to be started. Since our system is still in the evaluation phase, secondly the user can select a prototypicality measure. In routine use, this step shall be dropped and instead the measure with best evaluation results shall be used automatically. At present there are three choices. As humans look upon cases as more typical for a query case as more features they have in common [9], distances between prototypes and cases usually mainly consider the shared features. The first, rather simple measure just counts the number of matching symptoms of the query patient (X) and a prototype (Y) and normalises the result by dividing it by the number of symptoms characterising the syndrome.

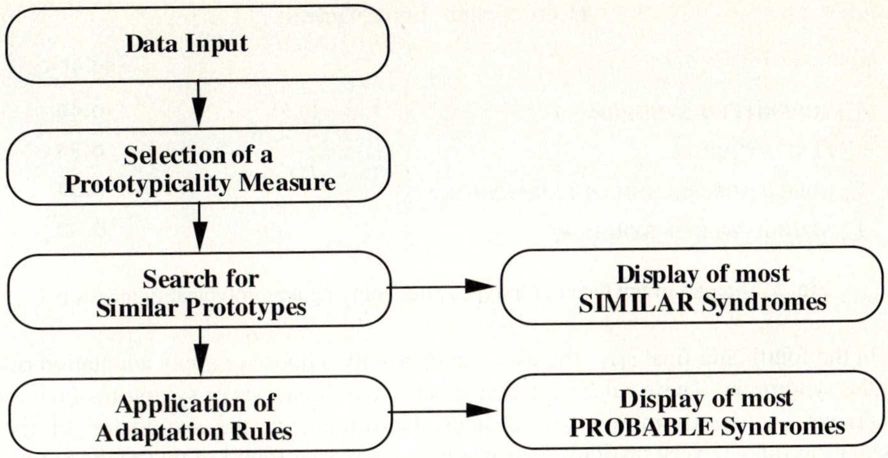

Fig. 1. Steps to diagnose dysmorphic syndromes

This normalisation is done, because the lengths of the lists of symptoms of the various prototypes vary very much. It is performed by the two other measures too.

$$\mathbf{D\,(X,Y) = \frac{F\,(X+Y)}{F\,(Y)}}$$

The second measure was developed by Tversky [8]. It is a measure of dissimilarity for concepts. In contrast to the first measure, additionally two numbers are subtracted from the number of matching symptoms. Firstly, the number of symptoms that are observed for the patient but are not used to characterise the prototype (X-Y), and secondly the number of symptoms used for the prototype but are not observed for the patient (Y-X) is subtracted.

$$\mathbf{D\,(X,Y) = \frac{f\,(X+Y) - f\,(X-Y) - f\,(Y-X)}{f\,(Y)}}$$

The third prototypicality measure was proposed by Rosch and Mervis [9]. It differs from Tversky's measure only in one point: the factor X-Y is not considered:

$$\mathbf{D\,(X,Y) = \frac{f\,(X+Y) - f\,(Y-X)}{f\,(Y)}}$$

In the third step to diagnose dysmorphoic syndromes, the chosen measure is sequentially applied on all prototypes (syndromes). Since the syndrome with maximal similarity is not always the right diagnosis, the 20 syndromes with best similarities are listed in a menu (fig.2).

Most similar prototypes:

Names of the prototypes	Similarities
☐ SHPRINTZEN-SYNDROM	0.49
☐ LENZ-SYNDROM	0.36
☐ BOERJESON-FORSSMAN-LEHMANN-S.	0.34
☐ STURGE-WEBER-SYNDROM	0.32

Fig. 2. Top part of the listed prototypes after applying a prototypicality measure

In the fourth and final step, the user can optionally choose to apply adaptation rules on the syndromes. These rules state that specific combinations of symptoms favour or disfavour specific dysmorphic syndromes. Unfortunately, the acquisition of these adaptation rules is very difficult, because they cannot be found in textbooks but have to be defined by experts of paediatric genetics. So far, we have got only 10 of them and so far, it is not possible that a syndrome can be favoured by one adaptation rule and disfavoured by another one at the same time. When we, hopefully, acquire more rules such a situation should in principle be possible but would indicate some sort of inconsistency of the rule set.

How shall the adaptation rules alter the results? Our first idea was that the adaptation rules should increase or decrease the similarity scores for favoured and disfavoured syndromes. But the question is how. Of course no medical expert can determine values to manipulate the similarities by adaptation rules and any general value for favoured or disfavoured syndromes would be arbitrary.

So, instead the result after applying adaptation rules is a menu that contains up to three lists (fig. 3). On top the favoured syndromes are depicted, then those neither favoured nor disfavoured, and at the bottom the disfavoured ones. Additionally, the user can get information about the specific rules that have been applied on a particular syndrome (e.g. fig. 4).

Names of the prototypes	Similarities	Applied rule
PROBABLE prototypes after application of the adaptation rules:		
☐ LENZ-SYNDROM	0.36	☐ REGEL-6
☐ DUBOWITZ-SYNDROM	0.24	☐ REGEL-9
Prototypes, no adaptation rules could be applied:		
☐ SHPRINTZEN-SYNDROM	0.49	
☐ BOERJESON-FORSSMAN-LEHMANN-S.	0.34	
☐ STURGE-WEBER-SYNDROM	0.32	
☐ LEOPARD-SYNDROM	0.31	

Fig. 3. Top part of the listed prototypes after additionally applying adaptation rules

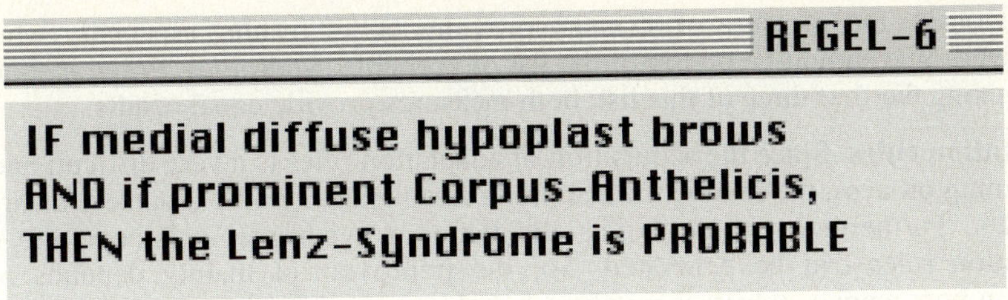

Fig. 4. Presented information about the applied adaptation rule

In the example presented by the figures 2, 3, and 4, the correct diagnosis is Lenz-syndrome. The computation of the prototypicality measure of Rosch and Mervis Lenz-syndrome was the most similar but one syndrome (here Tversky's measure provides a similar result, only the differences between the similarities are smaller). After application of adaptation rules, the ranking is not obvious. Two syndromes have been favoured, the more similar one is the right one. However, Dubowitz-syndrome is favoured too (by a completely different rule), because a specific combination of symptoms makes it probable, while other observed symptoms indicate a rather low similarity.

3 Results

Cases are difficult to diagnose when patients suffer from a very rare dymorphic syndrome for which neither detailed information can be found in literature nor many cases are stored in our case base. This makes evaluation difficult. If test cases are randomly chosen, frequently observed cases resp. syndromes are frequently selected and the results will probably be fine, because these syndromes are well-known. However, the main idea of the system is to support diagnosis of rare syndromes. So, we have chosen our test cases randomly but under the condition that every syndrome can be chosen only once.

For 100 cases we have compared the results obtained by both prototypicality measures (table 1).

Table 1. Comparison of prototypicality measures

Right Syndrome	Rosch and Mervis	Tversky
on Top	29	40
among top 3	57	57
among top 10	76	69

The results may seem to be rather poor. However, diagnosis of dysmorphic syndromes is very difficult and usually needs further investigation, because often a couple of syndromes are very similar. The first step is to provide the doctor with information about probable syndromes, so that he gets an idea which further investigations are appropriate. That means, the right diagnose among the three most probable syndromes is a already a good result.

Obviously, the measure of Tversky provides better results, especially when the right syndrome should be on top of the list of probable syndromes. When it should be only among the first three of this list, both measures provide equal results.

Adaptation rules. Since the acquisition of adaptation rules is a very difficult and time consuming process, the number of acquired rules is rather limited, namely at first just 10 rules. Furthermore, again holds: the better a syndrome is known, the easier adaptation rules can be generated. So, the improvement mainly depends on the question how many syndromes involved by adaptation rules are among the test set. In our experiment this was the case only for 5 syndromes. Since some had been already diagnose correctly without adaptation, there was just a small improvement (table 2).

Table 2. Results after applying adaptation rules

Right Syndrome	Rosch and Mervis	Tversky
on Top	32	42
among top 3	59	59
among top 10	77	71

Some more adaptation rules. Later on we acquired eight further adaptation rules and repeated the tests with the same test cases. The new adaptation rules again improved the results (table 3). It is obvious that with the number of acquired adaptation rules the quality of the program increases too. Unfortunately, the acquisition of these rules is very difficult and especially for very rare syndromes probably nearly impossible.

Table 3. Results after applying some more adaptation rules

Right Syndrome	Rosch and Mervis	Tversky
on Top	36	44
among top 3	65	64
among top 10	77	73

4 Conclusion

Diagnosis of dysmorphic syndromes is a very difficult task, because many syndromes exist, the syndromes can be described by various symptoms, many rare syndromes are still not well investigated, and from time to time new syndromes are discovered.

We have compared two prototypicality measures, where the one by Tversky provides slightly better results. Since the results were rather pure, we additionally have applied adaptation rules. We have shown that these rules can improve the results. Unfortunately, the acquisition of them is very difficult and time consuming. Furthermore, the main problem is to diagnose rare and not well investigated syndromes and for such syndromes it is nearly impossible to acquire adaptation rules.

However, since adaptation rules do not only favour specific syndromes but can be used to disfavour specific syndromes, the chance to diagnose even rare syndromes also increases by the count of disfavouring rules for well-known syndromes. So, the best way to improve the results seems to be to acuire more adaptation rules, however difficult this task may be.

References

1. Taybi, H., Lachman, R.S.: Radiology of Syndromes, Metabolic Disorders, and Skeletal Dysplasia. Year Book Medical Publishers, Chicago (1990)
2. Gierl, L., Stengel-Rutkowski, S.: Integrating Consultation and Semi-automatic Knowledge Acquisition in a Prototype-based Architecture: Experiences with Dysmorphic Syndromes. *Artificial Intelligence in Medicine 6* (1994) 29-49
3. Clinical Dysmorphology. www.clindysmorphol.com
4. Winter R.M., Baraitser M., Douglas J.M.: A computerised data base for the diagnosis of rare dysmorphic syndromes. Journal of medical genetics 21 (2) (1984) 121-123
5. Stromme P.: The diagnosis of syndromes by use of a dysmorphology database. Acta Paeditr Scand 80 (1) (1991) 106-109
6. Weiner F., Anneren G.: PC-based system for classifying dysmorphic syndromes in children. *Computer Methods and Programs in Biomedicine* 28 (1989) 111-117
7. Evans C.D.: A case-based assistant for diagnosis and analysis of dysmorphic syndromes. International Journal of Medical Informatics 20 (1995) 121-131
8. Tversky, A.: Features of Similarity. *Psychological Review 84* (4) (1977) 327-352
9. Rosch E., Mervis CB: Family Resemblance: Studies in the Internal Structures of Categories. *Cognitive Psychology 7* (1975) 573-605
10. Kolodner J: Case-Based Reasoning. Morgan Kaufmann, San Mateo (1993)
11. Aamodt A, Plaza E: Case-Based Reasoning: Foundation issues, methodological variation, and system approaches. *AICOM* 7 (1994) 39-59
12. Broder A: Strategies for efficient incremental nearest neighbor search. Pattern Recognition 23 (1990) 171-178
13. Wilke W, Smyth B, Cunningham P: Using configuration techniques for adaptation. In: Lenz M et al. (eds.): Case-Based Reasoning technology, from foundations to applications. Springer, Berlin (1998) 139-168
14. Schank RC: Dynamic Memory: a theory of learning in computer and people. Cambridge University Press, New York (1982)
15. Bareiss R: Exemplar-based knowledge acquisition. Academic Press, San Diego (1989)
16. Schmidt R, Gierl L: Case-based Reasoning for antibiotics therapy advice: an investigation of retrieval algorithms and prototypes. *Artificial Intelligence in Medicine* 23 (2001) 171-186
17. Bellazzi R, Montani S, Portinale: Retrieval in a prototype-based case library: a case study in diabetes therapy revision. In: Smyth B, Cunningham P (eds.): Proc European Workshop on Case-Based Reasoning, Springer, Berlin (1998) 64-75
18. Bichindaritz I: From cases to classes: focusing on abstraction in case-based reasoning. In: Burkhard H-D, Lenz M: (eds.): Proc German Workshop on Case-Based Reasoning, University Press, Berlin (1996) 62-69

ISOR: An Expert-System for Investigations of Therapy Inefficacy

Rainer Schmidt and Olga Vorobieva

Institute for Medical Informatics and Biometry, University of Rostock, D-18055 Rostock, Germany

Abstract. ISOR is a Case-Based system for long-term therapy support in the endocrine domain and in psychiatry. It performs typical therapeutic tasks, namely computing initial therapies, initial dose recommendations, and dose updates. Furthermore, ISOR deals especially with situations where therapies become ineffective. Causes for inefficacy have to be found and better therapy recommendations should be computed. In addition to the typical Case-Based Reasoning knowledge, namely former already solved cases, ISOR uses further knowledge forms, especially medical histories of query patients themselves, prototypical cases (prototypes) and therapies, conflicts, instructions etc. So, different forms and steps of retrieval are performed, while adaptation occurs as an interactive dialog with the user. Since therapy inefficacy can be caused by various circumstances, ISOR searches for former similar cases to get ideas about probable reasons that subsequently should be carefully investigated.

1 Introduction

In medical practice, therapies prescribed according to a certain diagnosis sometimes do not give desired results. Sometimes therapies are effective for some time but suddenly stop helping any more. There are many different reasons. A diagnosis might be erroneous, the state of a patient might have changed completely or the state might have changed just slightly but with important implications for an existing therapy. Furthermore, a patient might have caught an additional disease, some other complication might have occurred, or a patient might have changed his/her lifestyle (e.g. started a diet) etc.

For long-term therapy support in the endocrine domain and in psychiatry, we have developed a Case-Based Reasoning system, named ISOR, that not only performs typical therapeutic tasks but also especially deals with situations where therapies become ineffective. Therefore, it first attempts to find causes for inefficacy and subsequently computes new therapy recommendations that should perform better than those administered before.

ISOR is a medical Case-Based Reasoning system that deals with the following tasks:

- choose appropriate (initial) therapies,
- compute doses for chosen therapies,
- update dose recommendations according to laboratory test results,
- establish new doses of prescribed medicine according to changes in a patient's medical status or lifestyle,

B. Gabrys, R.J. Howlett, and L.C. Jain (Eds.): KES 2006, Part I, LNAI 4251, pp. 334–341, 2006.
© Springer-Verlag Berlin Heidelberg 2006

- find probable reasons why administered therapies are not as efficient as they should be,
- test obtained reasons for inefficacy and make sure that they are the real cause, and
- suggest recommendations to avoid inefficacy of prescribed therapies.

For psychiatric diseases some Case-Based Reasoning systems have been developed, which deal with specific diseases or problems, e.g. with Alzheimer's disease [1] or with eating disorders [2]. Since we do not want to discuss various psychiatric problems but intend to illustrate ISOR by understandable examples, in this paper we focus mainly on some endocrine and psychiatric disorders, namely on hypothyroidism and depressive symptoms. Inefficacy of pharmacological therapy for depression is a widely known problem (e.g. [3, 4, 5]). There are many approaches to solve this problem. Guidelines and algorithms have been created (e.g. [6, 7]). ISOR gives reference to a psychopharmacology algorithm [7] that is available on the website htp://mhc.com/Algorithms/Depression.

2 System Architecture

ISOR is designed to solve typical problems, especially inefficacy of prescribed therapies that can arise in different medical domains. Therefore most algorithms and functions are domain independent. Another goal is to cope with situations where important patient data is missing and/or where theoretical domain knowledge is controversial.

ISOR does not generate solutions itself. Its task is to help users by providing all available information and to support them to find optimal solutions. Users shall be doctors, maybe together with a patient. In addition to the typical Case-Based Reasoning knowledge, namely former already solved cases, ISOR uses further knowledge components, namely medical histories of query patients themselves and prototypical cases (prototypes). Furthermore, ISOR's knowledge base consists of therapies, conflicts, instructions etc. The architecture is shown in figure 1.

Technically, ISOR is implemented in Delphi 7, the format for the case and knowledge bases is Paradox 7, and retrieval is performed by SQL.

In this section we explain the components and in the next chapter we present examples to show how the main knowledge components work together.

2.1 Medical Case Histories

Ma and Knight [8] have introduced a concept of case history in Case-Based Reasoning. Such an approach is very useful when we deal with chronic patients, because often the same complications occur again, former successful solutions can be helpful again, while former unsuccessful solutions should be avoided.

The case history is written in the patient's individual base as a sequence of records. A patient's base contains his/her whole medical history, all medical information that is available: diseases, complications, therapies, circumstances of his/her life etc. Each record describes an episode in a patient's medical history. Episodes often characterise a specific problem. Since the case base is problem oriented, it contains just episodes and the same patient can be mentioned in the case base a few times, even concerning different problems.

Information from the patient's individual base can be useful for a current situation, because for patients with chronic diseases very similar problems often occur again. If a similar situation is found in the patient's history, it is up to the user to decide whether to start retrieval in the general case base or not.

In endocrinology, case histories are designed according to a standard scheme, one record per visit. Every record contains the results of laboratory tests and of an interrogatory about symptoms, complaints and physiological conditions of a patient.

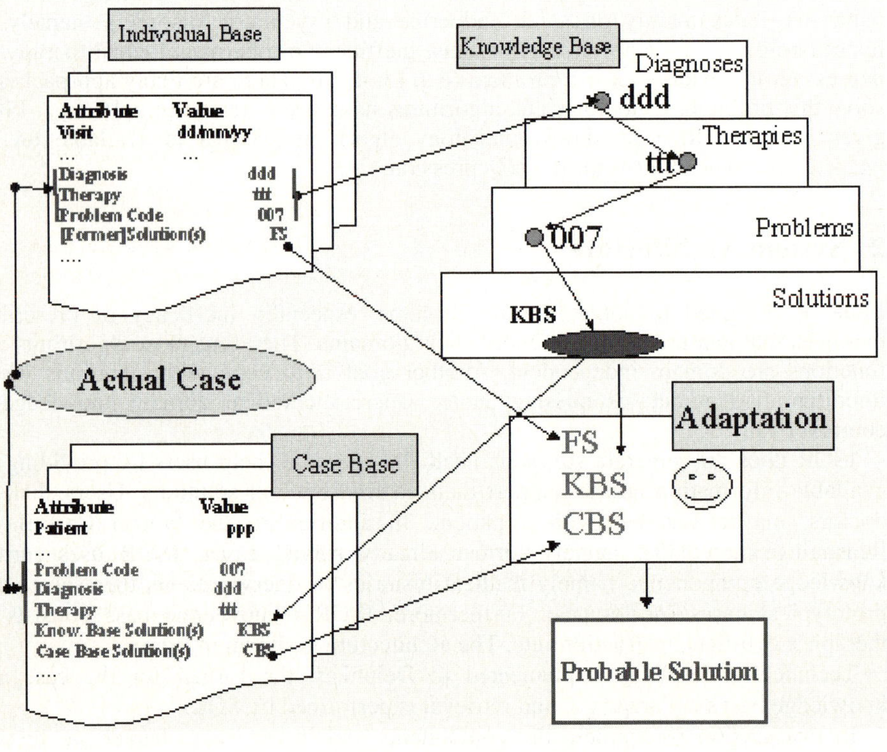

Fig. 1. System architecture

Therefore the retrieval of former similar situations from the individual base of an endocrine patient is easy to organise.

For psychiatric patients, case histories are often unsystematic and they can be structured in various forms. A general formalisation of psychiatric cases and their histories is not achieved yet. The design of case histories is problem dependent.

In both domains, we first search in the query patient's history for similar problems and for similar diagnoses.

2.2 Knowledge Base, Case Base, and Prototypes

The knowledge base contains information about problems and their solutions that are possible according to the domain theory. It has a tree structure and it consists of lists of diagnoses, corresponding therapies, conflicts, instructions, and medical problems (including solutions) that can arise from specific therapies. The knowledge base also contains links to guidelines, algorithms and references to correspondent publications [6, 7].

The case base is problem oriented. Thus a case in the case base is just a part of a patient's history, namely an episode that describes a specific problem that usually has a solution too. So, the case base represents decisions of doctors (diagnosis, therapies) for specific problems, and their generalisations and their theoretical foundations (see the examples). A solution is called "case solution", abbreviated "CS". Every case solution has (usually two) generalisations, which are formulated by doctors. The first one is expressed in terms of the knowledge base and it is used as a keyword for searching in the knowledge base. Such a generalisation is called "knowledge base solution", abbreviated "KBS". The second generalisation of a solution is expressed in common words and it is mainly used for dialogues. It is called "prompt solution", abbreviated "PS".

Former cases (attribute value pairs) in the case base are indexed by keywords. Each case contains keywords that have been explicitly placed by an expert. For retrieval three main keys are used: a code of the problem, a diagnosis, and a therapy. Further keys such as age, sex etc. can be used optionally.

Prototypes (generalized cases) play a particular role. Prototypes help to select a proper solution from the list of probable or available solutions. A prototype may help to point out a reason of inefficacy of a therapy or it may support the doctor's choice of a drug.

2.3 Retrieval, Adaptation, and Dialogue

For retrieval keywords are used. Since our system is problem oriented, the first one is a code that implies a specific problem. The second keyword is the diagnosis and the other ones are retrieved from the knowledge base.

Adaptation takes place as a dialogue between the doctor, the patient, and the system. The system presents different solutions, versions of them, and asks questions to manifest them. The doctor answers and selects suggestions, while the patient himself or herself suggests possible solutions that can be considered by the doctor and by the system.

We differentiate between two steps of adaptation. The first one occurs as a dialogue between ISOR and a user. Usually, doctors are the users. However, sometimes even a patient may take part in this dialogue. The goal of these dialogues is to select probable solutions from all information sources mentioned in sections 3.1 and 3.2. Pieces of information are retrieved by the use of keywords. Specific menus support the retrieval process. The first step of adaptation can be regarded as partly user based: ISOR presents lists of probable solutions and menus of keywords, the user selects the most adequate ones. The second adaptation means proving obtained solutions. This proving is rule based and it includes further dialogues, laboratory test results, and consultations with medical experts. While the procedures supporting the first adaptation step are domain independent, the adaptation rules of the second step are mainly domain dependent.

3 Examples

By three examples we illustrate how ISOR works. The first and the second one are from the endocrine domain, the third one deals with a psychiatric problem. Details about initial dose calculations and dose updates can be found in [9].

3.1 Hypothyroidism: Inefficacy of Levothyroxine Therapy

Every morning a mother gives her 10 year-old boy not only the prescribed Levothyroxine dose but also vitamin pills. These pills have not been prescribed but they are healthy and have lately been advertised on TV. Part of this medication is Sodium Hydrocarbonate (cooking soda) that causes problems with Levothyroxine.

Individual base. The same problem, inefficacy of Levothyroxine therapy, is retrieved from the patient's history. The solution of the former problem was that the boy did not take the drug regularly. This time it must be a different cause, because the mother controls the intake.

Knowledge base. It has a tree structure that is organised according to keys. One main key is *therapy* and the keyword is *Levothyroxine*. Another keyword is *instructions*. These instructions are represented in form of rules that concern the intake of Levothyroxine. For Levothyroxine a rather long list of instructions exists. Since the idea is that the boy may break an instruction, this list is sorted according to the observed frequency of offences against them in the case base.

Concerning these instructions a couple of questions are asked, e.g. whether the boy takes Sodium Hydrocarbonate together with Levothyroxine. Since the mother is not aware of the fact that Sodium Hydrocarbonate is contained in the vitamin pills, she gives a negative answer and no possible solution can be established by the knowledge base. However, *soda* is generated as one keyword for retrieval in the case base.

So, the following solutions are retrieved from the knowledge base, the third one does not fit for the boy.

Knowledge base solution 1: Sodium Hydrocarbon
Knowledge base solution 2: Soy
Knowledge base solution 3: Estrogene

Case base. Using the keyword *soda* eight cases with the following seven solutions are retrieved (case solution 4 occurs twice).

Case solution 1: Aspirin Upsa
Case solution 2: Cooking Soda
Case solution 3: Soluble juice
Case solution 4: Alka Seltzer
Case solution 5: "Invite"
Case solution 6: Vitamin "Teddy"
Case solution 7: Lime Pills"

Thus we get a list of drugs and beverages that contain sodium Hydrocarbonate, all of them belong to the generalised solution "soluble".

Solution. The boy admits to take Levothyroxine together with an instantiation of the generalised solution "soluble", namely soluble vitamin.

Recommendation. The boy is told to take vitamin four hours later than Levothyroxine. Additionally, further interactions between vitamin and Levothyroxine must be checked, because it might be necessary to adjust the Levothyroxine dose.

3.2 Inefficacy of Psychiatric Drugs

Originally, ISOR was developed for the endocrine domain, especially for hypothyroidism, but later on it has been generalised. Now it can solve the same types of problems in different medical domains. Now we present an example from psychiatry.

A man, 55 years of age, has been treated for depression for 15 years. Symptoms of depression appeared after he was cruelly beaten near his house. Since he did not see any connection between his depression and the violence, he did not tell it to his doctor. At first, the antidepressant Tofranil for intake in the morning and the sedative Tisercin for better sleep at bedtime were prescribed. Later on, during another depression stage the antidepressant Fluoxetine helped. Now, his problem is that neither Fluoxetine, nor any other proposed antidepressant helps any more.

Retrieval. Keywords are used to retrieve information from all data sources. Since optional keywords about a patient's feeling (e.g. *feeling worse*) are used for retrieval of the patient's medical history, even biographical events of a patient can be retrieved.

Individual base. Since the problem of inefficacy of an antidepressant never arose in the patient's past, no solution can be found. However, indirect information was retrieved. The keyword *feeling better* provided a trip to Switzerland, while the result of the keyword *feeling worse* provided a trip to Israel, where the latest very severe depression began.

Feeling better: A trip to Switzerland

Feeling worse: A trip to Israel

The knowledge base contains information about depression, anxiety and other psychiatric diseases, possible complications and references to their theoretical grounds [10, 11, 12]. References to similar problems are retrieved, the most remarkable one is a link to the algorithm for psychopharmacology of depression [7, 13]. Though the idea of the algorithm is to solve the problem of non-response to an antidepressant, it does not really fit here, because it does not cover the situation that a therapy helped for some time and then stopped having an effect.

Case base. Eleven cases with similar depression problems are retrieved. Three of them are characterised by the general idea *depression is secondary to anxiety resulting from a psychical trauma*.

Case solution 1: A severe stress during the World War 2 (a woman)

Case solution 2: A bad experience in a Jail (a young man)

Case solution 3: Sexual assault in childhood (a woman)

The other cases have solutions that are generalised to *changes in therapy*.

Adaptation. ISOR displays retrieved information pieces. In this case, two strategies are offered. The first one suggests trying some other therapy. This strategy is supported by the majority of the retrieved cases and partly by theoretical recommendations. The second strategy means to check the diagnosis. This strategy is supported by three retrieved cases and by the patient's medical history. The choice between both strategies is up to the user. In this example the doctor chooses to attempt the second strategy at first. The doctor is especially led by the patient's medical history, because Switzerland is usually associated with a safe life (especially in comparison to life in Russia), while living in Israel is considered as unsafe. Furthermore, this strategy is supported by the general situation that some sedative drugs (e.g. Tisercin at the beginning) had helped for some time.

ISOR offers a list of questions for the favoured strategy and as a result the doctor concludes that in this case depression is in fact only second to anxiety. The man is permanently afraid of possible violence and anxiety is based on strong fear that occurred long ago.

Explaining remarks. Diagnosing anxiety needs good medical skills, because patients try to suppress traumatic events from their memory [14]. In this example depression even served as a mechanism of suppression. The accepted case-based solution spared the patient unnecessary experiments with other psychopharmacological drugs.

So, the first problem is solved, a new diagnosis is ascertained.

The next problem is prescription of a therapy. According to the domain theory and to our knowledge base anxiety implies Neuroleptics [12, 15]. Many of them are available but a good choice is not trivial.

Individual base. From the patient's history those sedatives (Neuroleptics) are retrieved that he took in his lifetime and that had positive effects on his psychical condition: Tisercin and Paxil, which is a drug that has both sedative and antidepressive effects.

Prototype. Among those prototypes that have been defined by doctors (based on their long experience with cases) the prototypical solution Paxil is retrieved.

Adaptation. Before described, every drug must be checked for conflicts with the patient's additional diseases and already existing therapy. Though the query patient has already taken Paxil in the past, our system checks all possible conflicts. If necessary, adaptation has to be performed. In this case no conflicts are discovered and Paxil is prescribed.

4 Conclusion

We have presented a CBR system that helps doctors to solve medical problems, particularly to investigate causes of inefficacy of therapies. It includes different knowledge containers, namely a case base, a knowledge base, prototypes, and individual bases of patients that reflect their medical histories. Information retrieved from these containers is arranged in form of dialogues.

The case base plays a central role in the dialogue forming process. It serves as a kind of filter when the knowledge base suggests too many possible solutions for the problem (as in the first example). In this situation the most typical cases are retrieved from the case base. When a solution from the knowledge base is not convincing or when it is hardly adaptable, the case base may provide better alternatives (as in the third example).

Generalisations, keywords and references to other knowledge components belong to the case base. The adaptation program uses them to create dialogues. In the part that concerns the case base and the dialogues ISOR can be considered as domain independent.

The design of the case base and our implementation allow solving problems from different medical domains. Specific, domain dependant features are attributed mostly to the individual base, because every domain requires a special design of case histories. The knowledge base in ISOR is domain-oriented, but all algorithms and functions are completely domain independent.

Acknowledgement

We thank Dr. Monika Mix, Children's Hospital of the University Clinic of Rostock, and Prof. Nikolai Nikolaenko, Sechenov Institute of Evolutionary Physiology and Biochemistry in St.Petersburg, for their data and for their help and time during our consultations.

References

1. Marling, C., Whitehouse, P.: Case-Based Reasoning in the care of Alzheimer's disease patients. In: Aha, D.W., Watson, I. (eds.): Case-Based Reasoning Research and Development, Springer Berlin (2001) 702-715
2. Bichindaritz, I.: A case-based assistant for clinical psychiatry expertise. Journal of the American Medical Informatics Association, Symposium Supplement (1994) 673-677
3. Hirschfeld, R.M., et al.: Partial response and nonresponse to antidepressant therapy: current approaches and treatment options. J Clin Psychiatry 63 (9) (2002) 826-37
4. Lam, R.W., Wan, D.D., Cohen, N.L., Kennedy, S.H.: Combining antidepressants for treatment-resistant depression: a review. J Clin Psychiatry 63 (8) (2002) 685-93.
5. Cuffel, B.J., et al.: Remission, residual symptoms, and nonresponse in the usual treatment of major depression in managed clinical practice. J Clin Psychiatry 64 (4) (2003) 397-402
6. Expert Consensus Guideline Series Treatment of Posttraumatic Stress Dosorder. J Clin Psychiatry 60 (suppl 16) (2000) 1-76.
7. Osser, D.N., Patterson, R.D.: Algorithms for the pharmacotherapy of depression, parts one and two. Directions in Psychiatry 18 (1998) 303-334
8. Ma, J., Knight, B. A.: Framework for Historical Case-Based Reasoning. 5th International Conference on Case-Based Reasoning, Springer Berlin (2003) 246-260
9. Schmidt, R., Vorobieva, O.: Adaptation and Medical Case-Based Reasoning Focusing on Endocrine Therapy Support. In: Miksch, S., Hunter, J., Keravnou, E. (eds.): Artificial Intelligence in Medicine, Proceedings AIME-2005, Springer Berlin (2005) 308-317
10. Davidson, R.J.: Cerebral asymmetry and affective disorders: A developmental perspective. In: Cicchetti, D., Toth, S.L. (eds.) Internalizing and externalizing expressions of dysfunction. Rochester Symp. on Developmental Psychopathology 2, Hillsdale (1991) 123-133
11. Tucker, D.M., Liotti, M.: Neuropsychological mechanisms of anxiety and depression. In: Boller, F., Grafman, J. (eds.): Handbook of Neuropsychology, Vol. 3. Elsevier, Amsterdam (1989) 443-456
12. Gelder, M.G., Lopez-Ibor, U., Andeasen, N.C. (eds.): New Oxford Textbook of Psychiatry. Oxford University Press, Oxford (2000)
13. htp://mhc.com/Algorithms/Depression
14. Stein, M.B.: Attending to anxiety disorders in primary care. J Clin Psychiatry 64 (suppl 15) (2003) 35-39
15. Kalinowsky, L., Hippius, H.: Pharmacolological, convulsive and other somatic treatments in psychiatry. Grunee&Stratton, New York London (1969).

Reengineering for Knowledge in Knowledge Based Systems

Anne Håkansson[1] and Ronald L. Hartung[2]

[1] Department of Information Science, Computer Science, Uppsala University, Box 513, SE-751 20, Uppsala, Sweden
Anne.Hakansson@dis.uu.se
[2] Department of Computer Science, Franklin University, 201 S. Grant Avenue, Columbus, Ohio 43215, USA
Hartung@franklin.edu

Abstract. This paper presents an approach to reengineering knowledge-based systems. Commonly, reengineering is used to modify systems that have functioned for many years, but are no longer able to accomplish the tasks required, and therefore need to be updated. Reengineering can also be used to modify and extend the knowledge contained in these systems. This is an intricate task if the systems are large, complex and poorly documented. The rules in the knowledge base must be gathered, analyzed and understood. In this paper, we apply reengineering to the knowledge and the functionality of knowledge-based systems. The outcome of the reengineering process is presented in graphic representations using Unified Modeling Language diagrams.

1 Introduction

A wide variety of knowledge-based systems (KBS) have been developed to support decision-making across a great number of areas, with the oldest and best established being the field of medical services. Developing a system takes a considerable number of man months and, during this period, the requirements for the system may change. But the understanding of the problem may also change or discoveries may be made. Thus, as a result of change and invention, even a system that has been newly implemented for an organization can be out of date instantaneously [3].

Changing the system at this point may not be difficult or time-consuming, if, for example, the alteration required is only a minor modification in the source code and if extensive documentation is available for the system. However, it can present a considerable problem if a profound alteration is required in a large and complex system with complicated and highly interrelated source code. In such a situation, problems such as inconsistency, incompleteness and redundancy can easily arise.

Commonly, reverse engineering (also known as reengineering) is usually used to maintain and reuse source code [4]. Reengineering can also be used to transform the architecture of expert systems to obtain object-oriented systems' architecture [1]. During the process, the reengineering can avoid incorrectness, check for

B. Gabrys, R.J. Howlett, and L.C. Jain (Eds.): KES 2006, Part I, LNAI 4251, pp. 342 – 351, 2006.
© Springer-Verlag Berlin Heidelberg 2006

inconsistencies and detect incompleteness. However, since reengineering needs to be able to deal with a large amount of code, it is a cumbersome process [4]. Because of this, automated assistance is needed for reengineering of the domain knowledge and the functionality of a system.

In this paper, we present an approach to KBS-development based on reengineering principles. The reverse engineering for KBSs has to be as general as possible and should work with as many kinds of knowledge presentation as possible. Common representations for these kinds of systems are rule-based, fuzzy rules, frames, neural networks and hybrid systems that use a mix of these representations. In our approach, we utilize reengineering to handle systems developed with rule-based representation techniques.

If it is to work successfully, reengineering needs to gather up not only all the rules in the knowledge base, but also encapture all of the relationships between each rule and all others. It also collects up all the pre-stored facts and the user-given facts received in response to the questions posed to the end users, and the conclusions that can be presented to the end users. In addition, it is important to encapsulate the relationships between the rules and the facts and, also, between the rules and the conclusions. Taken together, these relationships constitute the reasoning strategy of the system.

During rule gathering, the reengineering process must be able to ensure that correctness, consistency and completeness are maintained. The definitions of these terms differ somewhat, but in this work, correctness is a match of the system's solution with respect to the opinions of human experts. Consistency is used in the sense that a knowledge base contains no rules that are conflicting, redundant, subsumed or circular. Completeness indicates that the rule base has no dead-end rules, i.e. no rules that have a premise that can never be satisfied [14]. The outcome from this reengineering process is presented as a graphic representation using Unified Modeling Language (UML) [9] diagrams.

2 Related Work

Several reengineering models have been developed during the years. Some operate on the source code level, and others recover the design and specifications [4]. These frequently use a knowledge-based approach to transform the architecture of a system from one form to another. An example of such a transformation is a model developed for the reengineering of expert systems to obtain an object-oriented architecture [1]. This model reengineers non-object-oriented systems into object-oriented architectures by using a knowledge-based approach. Other examples are have been used to change conventional architectures to object-oriented architectures [4] and transforming procedural programs to object-oriented programs, such as CORET (object-oriented reverse engineering) [10].

In our approach, we are not transforming architectures, but instead, using reengineering in KBSs to collect and represent the knowledge contained in the systems. This approach allows us to modify the knowledge and to extend the knowledge base with new knowledge. The collected knowledge is presented in a graphic form in UML diagrams to support modification by the developers.

3 Reengineering in Knowledge Based Systems

The existence of large quantities of undocumented code causes re-engineering to be a common and painful problem in industry. This code needs to be understood when it needs to be fixed, modified or transported to a new system. There are sets of techniques that have been developed to gain an understanding of the code [2]. The most common of these techniques is the ad hoc approach of a person reading the code and piecing together a picture of the system. There are also some tool-based techniques that have been applied and can certainly help [12]. However, the understanding still requires intense work by experienced software professionals.

In the case of KBSs and rule-based systems, this is also a severe problem. One advantage in conventional KBSs is modularity; where the modular structure of the code, whether object oriented classes or older procedural code, helps guide the discovery process. However it is typical for some of the older KBS to lack a modular system and to be driven by data. These KBSs lack the explicit control structures of conventional programs where the practitioner is faced with a tangle of propositions, facts and rules.

Three issues confronting the reengineering tool are to find:

- the part of the rules set that drives a particular conclusion
- the conclusions that are possible from a set of rules or inputs
- the difference set between the sets of rules or inputs that result in two different conclusions.

3.1 Reengineering for Knowledge in the Knowledge Base

Reengineering involves collecting all kinds of knowledge in a system and taking into account any special functionality, e.g., procedural code, utilized by the system to make use of the knowledge, in, for example, an inference mechanism. The knowledge is often presented as productions and heuristics rules and facts. The reengineering must collect different kinds of rules together and the relationships to all other rules to cover all of the knowledge contained in the knowledge base. Moreover, as pre-stored facts are utilized within some of the rules, these must not be omitted since the execution of the rules depends on these facts. In addition to the rules and facts, as the reengineering must collect all the information that is involved in the reasoning in the system, it is necessary to track down the user-given facts, in the form of answers given or the alternative responses associated with the questions posed, and the conclusions that will be presented for the different responses.

The reengineering uses different search methods to locate the knowledge in the system. Different search criteria are used to search for knowledge with each set of criteria depending on the system's representation and syntax. Consequently, we use different approaches to find rules and facts.

The first and simplest approach is to search for rules as predicates in the representation. The reengineering can easily pick up the rules using the keywords, i.e. "rule", if that is the word used to denote rules. See the example, rule number 105:

```
rule(11, "allergic purpura object", "Find doctor
immediately text", 1000):-
     check('general disease rule', 'general disease
     text', 1000),
     fact('fever object', '>', 'normal'),
     reply('one rash with strong red spot object','Yes'),
     reply('several symptoms: Vomit / Headache /
     Oversensitive to light / Pain when you bend your
     head forward ', 'No').
```

Fig. 1. Example of source code for a rule

The reengineering procedure continues by collecting the rules by using a search for all predicates referenced in the rule. Many of the associated rules can be found by using the content of the rules and relationships to other rules. One rule will usually reference several others and can, therefore, be found by investigating the content of the predicate, i.e., using "`check`". In general, the relationships between the rules result in a complex graph of dependencies.

By searching the internal contents of these rules, many of the pre-stored facts can be found, i.e., those labeled "`fact`", see Figure 1. These pre-stored facts are often stored in a database and are present even though the system does not execute commands involving them.

User-given facts are knowledge that is required by the system, but which, unfortunately, can only be found in the database during a session, i.e., "`reply`", as shown in Figure 1. Since, the user-given facts are not collected from the rules, the reengineering must reconstruct the user-given hypothetically. This is accomplished by inspecting the rules and determining what possible values will be given. This step is required because the facts inserted into the system will depend on consultation, and the values obtained will vary from one session to another.

Despite conducting a comprehensive search of all of the rules, there is no guarantee that all kinds of rules will be identified and picked up because, for example, some rules take on different structures depending on the purpose of the rule. These rules are different because they will accept different numbers of arguments depending upon the call, e.g., "`rule(105, 'sign of allergic purpura', 'Find a doctor immediately text', 1000)`" which has four arguments and "`check('general disease rule', 'general disease text', 1000)`" which has three arguments, and where the second alternative omits the rule number.

The second, and more complicated, approach is to search the syntactic form of the rules used to identify when the rules have a different appearance to that mentioned above, and use the presumed word "`rules`". Rules frequently have a predetermined pattern that can be used by the interpretation engine when reengineering grasps the rules in a knowledge base. This is referred to as the pattern of antecedent-consequent rules. Thus, determining the structure of the rules is the starting point for a search of this type, for which it is necessary to identify the first rule and then use the pattern revealed to find the rest of the rules.

The engine for interpretation can use the in-built predicate for interpreting antecedent-consequent rules. The predicate available in Prolog, called Clause, separates the predicate head from the body. These parts are then utilized by the clause predicate to check each part separately — the head, and the body. The reengineering utilizes this clause predicate by tracking the interpreter while forcing the system to execute the engine as though someone were consulting the system. This clause predicate uses the predicate's internal structure to interpret the rules, pre-stored facts and user-given facts to reach a conclusion or conclusions. This search works for a couple of KBSs, namely, those with predetermined patterns. However some systems use other kinds of self-developed interpreters.

In the third approach the reengineering tool looks for the rules without using any presumed word or pattern. If the relevant word or the pattern it should seek is unknown, the rules can be found by counting the number of occurrences of a predicate with the same name. The search will find predicates that occur frequently in the code. In a system, rules are, usually, the most common predicates and the number of occurrences of that predicate will generally exceed the number of times the other predicates occur. Again, the reengineering uses the same method as described in first approach to determine the pattern intrinsic to the rule and to look for other rules with the same pattern using the pattern or patterns to find all the other rules in the knowledge base. However, it is not certain that the reengineering procedure will find the rules of the system. Instead, the reengineering tool may find a predicate that is the code for functionality of the system rather than rules or facts. The knowledge engineer must analyze whether the predicate belongs to the knowledge or the functionality.

In addition to obtaining the rules, the other predicates, in the form of facts, questions used to obtain the user-given facts and conclusions all need to be found by the reengineering using the same kind of searching procedures as mentioned above. To find the questions and the conclusion, the rules are investigated. However, the connections between the rules, questions and the conclusions must also be collected.

3.2 Relationships Within the Set of Rules

The outcome of the reengineering is the knowledge, partitioned into rules and facts, which can be either pre-stored or user-given, and conclusions in different sets. Each of these can be considered to be distinct sets, thus there will be one set of rules, one set of pre-stored facts, and so on. Since, a knowledge base can contain several thousand rules, the set of rules will be huge and will involve a complex network of relationships. As a result of this, collecting the rules into a rule set will have an impact upon the comprehensibility of the rules. The rule set also contains the relationships between rules and to facts, which the reengineering must reveal.

Reengineering to determine the relationships is accomplished by traversing the network of rules. Almost every rule utilizes another rule because of the need to simplify the complex knowledge in the system and, hence, a complex structure of rules is created during the development of a knowledge base. Rules are built from the domain expert's expertise, but these rules usually, utilize facts and not other rules. The knowledge engineer transfers the domain knowledge from an expert to a system and, in so doing, learns how problem-specific rules are used. During the development of the knowledge base, different rules are generated to facilitate the handling of the production or heuristic rules, e.g., meta-rules, inferred rules and derived rules. These rules are highly interrelated to other rules.

Some rules are not really related to the domain problem [11] at all, e.g., meta-rules. Meta-rules are used to direct the problem solving and to determine how best to solve problems. These meta-rules are used when deciding which rules to apply to avoid a conflict within the reasoning strategy [8]. A meta-rule devises a strategy for the use of task-specific rules in the system [13] and is, therefore, dependent on other rules.

Some rules are specially constructed to enable them to be utilized by many other rules. Most commonly constructed rules are inferred rules. Inferred rules are derived from the initial rules and represent the process of derive new information from old [11]. An inferred rule links several different rules (e.g., A->B, B->C) by concluding that rule A implies rule C. In the inferred rules, the antecedent of the initial rule appears in the body of the rule. If the relationships between inferred rules and initial rules are found, it is important that they are preserved since they are affected by changes to the rule base.

Derived rules, as their name suggests, are rules that are derived from other rules. They are similar to inferred rules since they involve the process of linking new information, however, instead of inferring the rules, the rules are derived from other rules, making a derived rule a short form for several lines of primitive rules. The derived rules may have direct relationships to the rules that produced them, for which they are short cuts. These relationships, too, are important and must be retained.

As mention above, in the rule set, the rules also have a relationship to the facts, both pre-stored and user-given. Moreover, the majority of the rules are connected to the conclusions presented to the end users. The connections to the facts and conclusions must be taken care into consideration.

3.3 Reengineering for Source Code

Reengineering must operate on the source code, which implements the functionality, e.g. the functions and the predicates, used to handle the knowledge of the system. The functionality is essential for working with the knowledge, e.g., for using the interpretation engine, performing calculations and for consideration of factors relating to uncertainty.

Reengineering collects the functionality by searching in the source code. It follows the code and searches for the first occurrence the engine invokes rules. Other functionality is evident when the rules themselves need to calculate a result, e.g., to find averages. In these cases, the reengineering searches for the functionality from the rules, i.e., it searches for the rules that invoke this functionality. An additional functionality is the certainty factor. The interpreter usually calculates these certainty factors at the end of a session. To collect this functionality, the reengineering must again search in the source code. Hence, the engineering operates on the code for the interpretation engine.

3.4 Reengineering to Handle Problems Related to Consistency and Completeness

Consistency is a knowledge base with no rules that are conflicting, redundant, subsumed or circular. If the knowledge base has circular rules, the reengineering tool will run into an infinite loop, which must be taken care of during the reengineering. The structure of circular rules must be recognized to be able to avoid the potential loops.

Completeness means having a rule base without dead-end rules, i.e., avoiding the problems caused by having a premise that can never be reached or which does not exist. The reengineering would try the find the rule and run into a search problem. The tool works through the code until it finds the rule corresponding to a premise, but if the premise is lacking, the reengineering tool is searching for a non-existent rule. The reengineering tool keeps track of all the rules it has encountered, until, when a certain time limit is exceeded, concludes that the rule is missing.

The definition of correctness corresponds to the match of the conclusion of the system with the opinions of domain experts. Since correctness is human based, reengineering cannot make this kind of check, because such a check would require the system to reflect a domain expert's knowledge and reasoning to reach the same conclusions.

Graphic modeling is used to make the knowledge more accessible and the experts can use the models to check the rules and facts that are involved in a conclusion. If the models are easy to follow and understand, then it will be more straightforward for the users to judge consistency, completeness and correctness.

4 Presenting the Sets in UML Diagrams

The outcome of the reengineering tool has to be presented in a manner that is comprehensible for the engineer, enabling to maintain the system and reuse the knowledge. When a graphic representation is required, UML diagrams can used to present knowledge such as rules, facts, conclusions and reasoning strategies in KBSs as has been shown in the references [5, 6]. However in this paper, the UML diagrams are used to present the outcome of the reengineering.

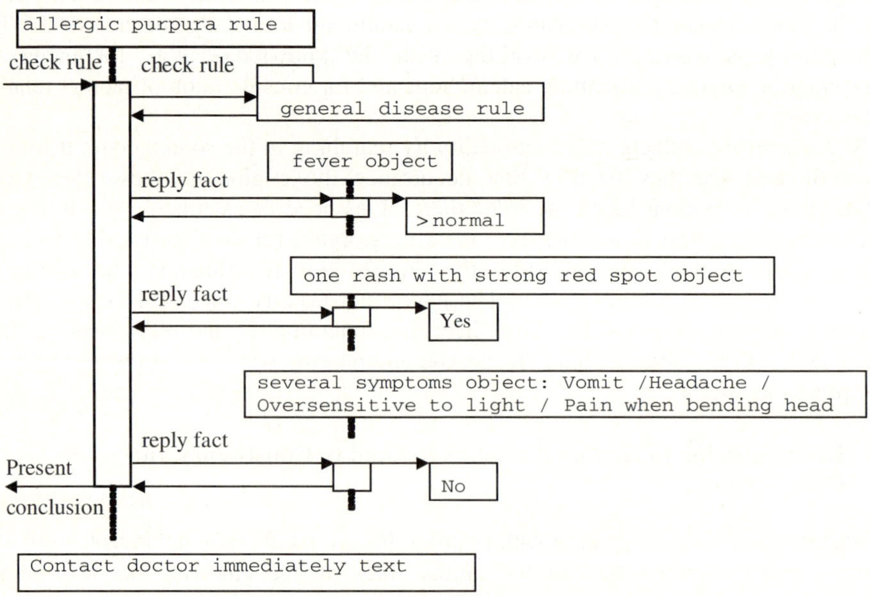

Fig. 2. Sequence diagram for the rule

The example presented in Figure 2 illustrates the engineering operation on the rule "allergic purpura rule", previously presented in Figure 1. Within the rule, another rule is implemented, the "general disease object". This incorporates a fact "fever object", which corresponds to a high body temperature, the fact "one rash with strong red spot object" with the user-given answer "Yes" and the fact "Several symptoms object: Vomit / Headache / Oversensitive to light / Pain when bending head forward" with the user-given answer "No". The conclusion "Find doctor immediately text" is connected to the rule. When a rule has been investigated, it can be folded into packages of UML, as demonstrated in the "general disease object" rule in Figure 2. The internal content of a rule is specified inside a package.

If a user is to be able to grasp the outcome of the reengineering, the tool must generate diagrams to answer specific questions posed by the user. For example, how a specified conclusion can be obtained from the rules. The diagram provides a graphic view of the answers.

The relationships between the rules together with the order of the system invoking the rules are presented along a lifeline. This lifeline illustrates the order with arrows presenting the rules, the pre-stored facts and the user-given facts.

At the end of the lifeline the conclusion is presented, in the "Contact the doctor immediately text" in Figure 2. In this example, the two conclusions that are attainable are: in the package "general disease rule", which cannot be seen in this example, and that at the end of the lifeline.

In addition, all kinds of relationships between the sets of rules or inputs that produce two conclusions will be presented in sequence diagram. However, to illustrate this more clearly we use a UML object diagram.

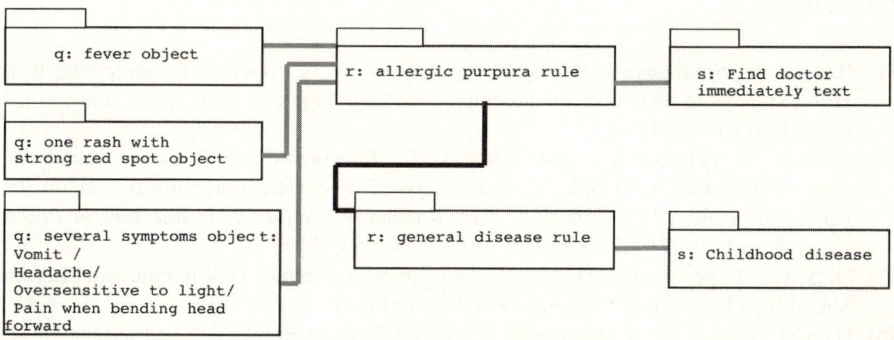

Fig. 3. Object diagram for the rule

Figure 3 presents the connections between questions, rules and conclusions using object diagrams. This example illustrates the same rule as was presented in Figure 2, revealing how the object diagram presents the connections between different parts more clearly than the sequence diagram.

5 Conclusions and Further Work

The problem with the current generation of KBSs and expert systems is the poor documentation and the deficiency of written information. This is complicated by the fact that not all expert systems have a modular structure. Some systems are developed without applying an architecture, and the code is not structured into different bases, such as knowledge bases.

In this paper, we have shown that reengineering can be used to gather knowledge. The reengineering has been applied to problems related to knowledge, relationships and consistency. The first and simplest approach, which is to search for rules, has been tested in the KANAL-system [7]. The system is an intelligent rule editor system, which is used to support a domain expert during the development of knowledge-based systems [7].

To comprehend the outcome of the reengineering and thereby to modify the knowledge, we present it through graphical representations, such as UML sequential diagrams. However, we have not worked with the application of reengineering to the source code handling the interpretation engine. It has previously been shown that the UML activity diagrams and state chart diagrams can used to illustrate source code in KBS, why they can be suitable for illustrating the outcome of the reengineering of the functionality.

Further work is needed to extend our reengineering tool to enable it to handle all kinds of knowledge representations. Reengineering operates on rule-based systems but it can easily be expanded to handle knowledge representation such as fuzzy rules and frames. Moreover, the principle of reengineering can be extended to handle neural networks and hybrid representation.

References

[1] Babiker, E., Simmons, D., Shannon, R., & Ellis, N.: A model for reengineering legacy expert systems to object-oriented architecture, Expert systems with applications, vol. 12, no. 3 (1997) 363-371.

[2] Carriere, J, O'Brian, L., and Verhoef, C. Reconstruction Software Architecture. Appearing Carriere J., O'Brian L., and Verhoef C. Reconstruction Software Architecture. Appearing in Bass L., Clements P., and Kazman R., *Software Architecture in Practice* (2nd edition), ISBN 0-321-15495-9, Addison-Wesley, 2003.

[3] Durkin, J.: Expert System Design and Development. Prentice Hall International Editions. MacMillian Publishing Company, New Jersey (1994).

[4] Gall, H., Klösch, R. & Mittermeir, R.: Using Domain Knowledge to Improve Reverse Engineering. *International Journal on Software Engineering and Knowledge Engineering*, World-Scientific Publishing, Vol. 6, No. 3, (1996) 477-505.

[5] Håkansson, A.: UML as an approach to Modelling Knowledge in Rule-based Systems. (ES2001) *The Twenty-first SGES International Conference on Knowledge Based Systems and Applied Artificial Intelligence*. Peterhouse College, Cambridge, UK; December 10th-12th (2001).

[6] Håkansson, A.: Supporting Illustration and Modification of the Reasoning Strategy by Visualisation. (SCAI'03) *The Eighth Scandinavian Conference on Artificial Intelligence,* Bergen, Norway, November 2th-4th (2003).

[7] Håkansson, A. Widmark, A. and Edman, 2000. A. The KANAL-system, an intelligent rule editor supporting the knowledge acquisition process. *Swedish Artificial Intelligence* Society Annual Workshop 2000 (SAIS 2000).

[8] Jackson, P.: *Introduction to Expert Systems*. Addison-Wesley Longman Limited. ISBN 0-201-87686-8 (1999).

[9] Jacobson, I. Rumbaugh, J. & Booch, G.: The Unified Modeling Language User Guide. Addison-Wesley, USA (1998).

[10] Mittermeir, R., Rauner-Reithmayer, D., Taschwer, M., Gall, H. & Weidl, J.: CORET Object-Oriented Re-Structuring. CORET Methodology and Tool. In: Mittermeir et al (Hrsg.): *CORET Methodology and Tool 1 Klagenfurt*, Wien: Inst. für Informatik-Systeme 26. November (1998).

[11] Merritt, D.: *Building Expert Systems in Prolog*. Springer-Verlag, New York Inc. ISBN 0-387-97016-9 (1989).

[12] Müller H., Orgun M., Tilley S. and Uhl J.: A Reverse Engineering Approach To Subsystem Structure Identification. In *Journal of Software Maintenance Research and Practice*, vol. 5, no.4, (1993)

[13] Negnevitsky, M.: Artificial Intelligence – A Guide to Intelligent Systems. Addison-Wesley, Pearson Education ISBN 0201-71159-1 (2002).

[14] Polat F. & Guvenir H.A.: UVT: A Unification-Based Tool for Knowledge Base Verification, *IEEE Expert: Intelligent Systems and Their Applications*, vol. 8, no. 3, June, (1993) 69-75.

Knowledge Representation for Knowledge Based Leadership System

Ronald L. Hartung[1] and Anne Håkansson[2]

[1] Department of Computer Science, Franklin University, 201 S. Grant Avenue, Columbus, Ohio 43215, USA
Hartung@franklin.edu

[2] Department of Information Science, Computer Science, Uppsala University, Box 513, SE-751 20, Uppsala, Sweden
Anne.Hakansson@dis.uu.se

Abstract. Leaders make decisions in different and sometimes difficult situations. These decisions are often converted into valuable knowledge needed by other people in decision positions. For example, students becoming leaders need knowledge about leadership to make good judgments. When the situation is sensitive, perhaps concerning personal changes or an unfamiliar business strategy, leaders need opinions and not just those from people involved in the company. An artificial system allows us to replicate a system of opinions and observations from outside advisors, which is why we base our research on this kind of system. In this paper we present a knowledge representation for a knowledge-based leadership system used in the experimental system we are developing.

1 Introduction

A useful way to spread leadership knowledge is to construct systems that help leaders access ideas and strategies to solve problems. Since this requires knowledge from other leaders, the ideal system would be knowledge-based, which would interact with the user, acting as a coach, a helper, and an information source. While systems like this are just beginning to exist, we have started to work on experimental systems, see Hartung and Håkansson [5], that will grow to reach the goal of an effective coach.

A necessary first step in the construction of these systems is to deal with representing knowledge in the domain of leadership. It is our thesis that knowledge in this domain has some challenges including: complexity of the knowledge, applying the wide range of domains where leadership, and separating leadership knowledge from the problem domain. It is important to characterize the knowledge that we associate with leadership, in order to represent it in a knowledge-based system.

Earlier work has been to develop an experimental prototype of a knowledge-based leadership system. From the work, we have found that there are three modes in a leadership support system: it may act as a guide by providing suggestions; it may evaluate the user's proposed ideas for the solution; or it may make the decision for the user. For the moment, we will not consider the decision-making task, because it seems questionable to remove the leadership from human control. It is difficult to decide what degree of control the system should assume.

B. Gabrys, R.J. Howlett, and L.C. Jain (Eds.): KES 2006, Part I, LNAI 4251, pp. 352 – 359, 2006.
© Springer-Verlag Berlin Heidelberg 2006

2 Related Work

An interesting result which supports the belief that knowledge-based leadership systems are possible, comes from the work in believable agents. To make one willing to use a leadership support system, it must present results in a believable way. The chosen knowledge and presentation of leadership knowledge will require that the system is acceptable to the user. One reason for this is that leadership is viewed as a human skill and the user must accept and adopt the suggestions of the system. One system that shows some of the potential for believable agents is Façade [7]. Another very good analysis of believable agents is found in believable social and emotional agents [9].

Façade illustrates the use of shallow understanding. The Façade system uses a very simple NLP front-end and reduces the input into speech acts. Speech acts are very high-level representations of what has been said, often focusing on the emotional content of the message. The representation scheme developed here uses speech acts and takes a high level view of the actions.

Another central idea, taken from "The leader as Martial Artist" [8], is to build a simplified emotional model for actors in a situation. This too is exploited in the representational scheme.

3 Types of Support Systems

There are several approaches to build systems supporting leadership decision-making. The simplest approach presents leadership stories to help the user formulate a solution. One good argument for this position can be found in Roger Schank's "Tell me a Story" [10]. In this case, the system can be as simple as a database of possible stories in which the user's problem acts as a query to find possible stories to match. The query function will require some use of natural language. Lest this appear too simple, this is what some people do when asked to help in decision-making. This can act as a guide, perhaps a weak one, but a guide, nonetheless. There are two reasons we have not chosen this approach. First of all the natural language query capability is not strong enough alone. Second, a richer representation allows us to apply computational methods and reasoning.

Up one level in complexity, the user proposes a solution and the system matches the existing story to the user by similarity measures. Compared to the simple guide level, this can present the stories to the user, allowing evaluation of relevance and usefulness. Also, given the goal, the system can show alternative stories that have different outcomes from those of the user may desire. This is stronger than the previous approach because the user provides more input to the system so there is more to use for matching up the stories.

The last level we consider requires the system to evaluate the story. The system will acquire the assumptions about other actors from the user. Because the stories depend on the actions of the actors, the assumptions about the motivations of the actors are critical to evaluating the stories. Validating the assumptions can be done by working through the rules in the actor's emotional/motivational model, and at each choice, asking the user if the rule being fired is valid, given the user's assessment of the other actor.

4 Leadership Knowledge

It is important to use leadership to accomplish specified goals. This can be concluded from Tannenbaum, Weschler & Massarik [12] in their statement "leadership is interpersonal influence, exercised in a situation, and directed, through the communication process, toward the attainment of a specified goal or goals". The leadership may be learned by following the earlier leaders' decisions but, from these decisions make there own active decision. Walter Lippman says that: "The final test of a leader is that he leaves behind in others the conviction and will to carry on." [2]. We learn by earlier leaders whose words and actions also teach new leaders as noted by their leadership. This is also described by Ralph Nader: "The function of leadership is to produce more leaders, not more followers." [2]. Conclusively, the goal of leadership is to accomplish leading other people by good judgments in different situations.

Leadership and management are often intertwined. One critical aspect of leadership is the ability to inspire and motivate people, whereas management often relies on authority alone. However, the ability to inspire and motivate is often not enough. Leaders need to make good decisions too and therefore, problem-solving skills are a useful part of leadership. In spite of good problem-solving, the plans may fail to achieve success, i.e., if the leader fails to motivate the people to implement them. So far, we have brought into play three domains: authority, inspiration/motivation, and problem solving in a domain. In this paper, the view adopted is the most traditional of leadership, i.e., the ability to inspire and motivate. The following is a good illustration of leadership principles.

4.1 Actions Taken for Leadership

To reach goals, the leadership needs to be successful. The following are actions that need to be taken to make the group work well [3]:

- Be able to embrace problems as possibilities.
- Letting the group attempt something new. This depends on the willingness of people to try new ways irrespective of the uncertainty of the ways.
- Presenting a positive attitude to all kinds of employees. This is especially important if the employee has a more subordinate position. "People become as they are considered" by the leader.
- Be flexible enough to be able to cooperate with dissidents. In settings in which people are too similar, the number of creative solutions frequently diminishes.
- Be willing to listen, answer and question.
- Organizing and prioritizing key factors in an organization. It is not the fastest group that has the effective organization, but rather the one that can prioritize the important key factors for success.
- Respecting the knowledge of others, as well as for one's own knowledge.
- When these actions are examined, it is reasonable to agree with each of the tenets.

What follows is the problem of applying them. Indeed, this is like reading Carl von Clauswitz or Sun Tzu: The principles make complete sense, but one can spend a lifetime learning to apply them. However, the actions are used as an approach to build a system for teaching and supporting leadership.

5 Representation Problems

To illustrate the representation, we have pulled two example stories from separate sources, Whitney and Packer [13] and Midell [8]. Indeed they were chosen almost randomly as examples to illustrate the flexibility of our approach. It is clear that two examples are not enough to build an argument for the validity of this approach. But they are very typical of what is commonly found in the literature. Please note, these are summarized and abridged.

The first example is found in Whitney and Packer [13]. Usually, Whitney gives some examples of "supporting relationships," and indeed his examples are both positive and negative. In this example, Whitney illustrates a negative example with a story, describing when he was working on a marketing campaign and he offered the boss the chance to do the public statements and interviews, "to take the bows". The boss deferred, and he went ahead. His boss cooled to him and only latter did he realize the boss did not like his taking the center stage and began to feel that Whitney was after his job.

In the second example, the staff of a facility approaches the leader of a group using the facility and complains that the group is making too much of a mess and not cleaning up their cups and dishes. The ensuing discussion becomes hostile, and even after an apology the leader of the staff continues to berate the group leader. Mindell [8] goes on to describe taking several approaches to defuse the situation, including attacking the staff leader as unfair, of sympathizing with the staff leader, and finally successfully bringing out the staff leaders distress at attacking the group over the mess.

These stories are typical of those found through out the literature. The first example will be used to illustrate our representation scheme, below. At the surface level, a story is a sequence of interactions between the characters in the story.

6 Representation

A number of story representation systems have been built. For example, in Communicating Everyday Experiences [1] and in Minimal Structures for Stories [11] the authors use representational schemes, which are aimed at narrative interactive stories. The previous references look at ways of representing the underlying goals and interactions, in order to generate a story. The leadership representation problem is the opposite. It needs to generate the representation from the stories.

The representation used has three parts: the goals, the interactions, and the emotional/motivational state. The basic idea of the interactions came from the references above, [1] and [11]. The emotional/motivational and the interaction model are inspired by the work in Façade [7]. These will be discussed in turn.

6.1 Goals, the Meta Level

Leadership stories are used to illustrate a purpose or goal. Goals must be captured in a natural language form and can be classified into related groups by using taxonomy. Similarity measures are applied to select suitable stories from the story repository of the system. The similarity of the user's goal entered into the system is compared to the goal of the story. The taxonomy is used to identify related stories.

The story may have a result that is different from the goal. This is especially true when the story is written as a counter example. In this case, the result is also recorded together with the goal.

6.2 Interactions

The story has a cast of actors and a sequence of interactions. The actors can be represented as objects, as they exchange messages in response to each other's messages and their internal motivations. The representation can be in either a logical form in predicates or graphically in diagram. These are fully equivalent. For the story representation, UML sequence diagram is chosen. The diagram is a helpful tool for human analysis because they can actually see the pattern of the knowledge and the result of the execution. The logical representation is used for reasoning and search in the system. The objects in the diagram are the actors (characters) in the story. The interactions are the arcs of the diagram, labeled with the actions. While the actors are obvious, the interactions require some elaboration. The interactions are the speech acts between the actors. They are all the spoken sentences in an abbreviated form using speech acts. The inspiration for the choice of interactions comes from the Facade system [7]. Speech acts are course grained where for example, speaking to an actor in a supportive manor is a speech act. In addition to the speech act label on an arc in the sequence diagram, the actual text used can be stored in the system, so it may be accessed as needed.

In order to represent the interactions at the proper level, the story is transformed to remove the non-leadership part of the story. Many leadership stories include details of the problem being solved as well as the leadership issues. It is extraneous to the application of leadership. Once this is removed, the story can be converted into a UML diagram with actors exchanging speech acts. The text of the full story is maintained and the non-leadership information is suppressed only in the representation. The story can still be presented completely because the speech acts in the story representation are linked to the corresponding text in the story. The following diagram shows the interactions in the first example.

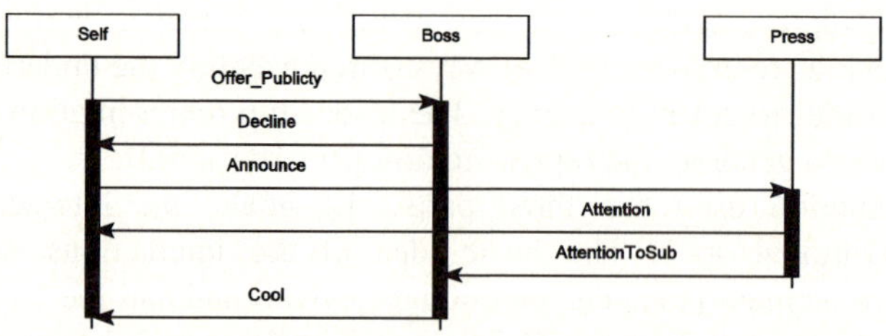

Fig. 1. The interactions corresponding to the first example

The diagram illustrates the different actors i.e., Self, Boss, and Press, and the interactions between the actors, e.g., Cool (meaning the expression of coldness toward the recipient actor). The input to the program is parameters invoking the actors and the actions. From these parameters the system automatically draws the diagram. The diagram in Figure 1 is drawn from the input of the actors and the actions between them, which correspond to the first example presented earlier.

From our tests, where only one is presented here, the representation has been robust enough to capture the critical aspects of the stories. Therefore, it seems many stories can be represented in this form. Some of the interactions are direct actions of the actors in the story. Some interactions are observations by the actors and, therefore, do not correspond to dialog or actions. This is the hardest part of the story to capture and is likely to be very hard to capture by a fully automated system.

6.3 Emotional/Motivation Representation

This is a representation of the goals and behaviors of the character. These are the things that drive the characters in the story. Sometimes this is presented explicitly as part of the story; sometimes it will need to be inferred. The goals and behaviors can be easily handled using rules where the rules capture the behaviors in the story.

This too will fit into the scheme proposed by the story representation. As interactions form a set of messages/method calls to the objects, i.e., actors, in the representation and the rules form the internal logic of the actors. In addition to the rules, some amount of internal state information is required. A measure of emotional response especially needs to be recorded in the actor's state, why an emotional state model is needed. A good example of this is found in [8]. The emotional state is really a vector of human emotional state and can be a combination of several facets, for example distress and anger together.

Emotional responses are complex in nature, but we do not need to capture a complete model of human behavior for story representation. Leadership stories generally contain a very shallow representation of human response. As more complex tasks are attempted, a more complex form may be needed. But rules have shown enough power for the present time.

The rules for each actor, for the first example, are given below. For illustration, text is used.

```
Self:
Starting emotional state is neural

If press announcement is needed, then offer the boss publicity
If offered to the boss and it is declined, then announce, and emotional state is happy
If press offers attention, then emotional state is happy
If boss is cool, then emotional state is worried
```

```
Boss:
Starting emotional state is neural

If offered publicity, then decline
If Employee gets attention, then emotional state is unhappy employee
If emotional state is unhappy with employee, then display coldness towards employee
```

```
Press:
If Announcement, then show attention to announcer and to Boss
```

In the system, we define a set of speech acts for the literal terms in the rules and we make it easier to deal with inference by reducing the variation and complexity in the acts. However, the text from the story is preserved to allow natural language processing in the system and also a readable version of the story. The text from the story is linked to the speech act it invokes. Like the speech acts, the rules can be linked to story text that describes the elements of emotional/motivational details.

6.4 Expressive Power

One of the interesting aspects of the representation is the separation into levels of the three characteristics. As more leadership knowledge is placed in to this representation, the levels of the representation can be used compare and contrast the stories with the needs of the users. This can be done by similarity measures. There will be several types of similarity. The coarsest type of similarity is the sequence of interactions from the sequence diagram. This can be applied without looking at the semantics of the interactions. The next type is the similarity of the speech acts, which are applied by using semantic tagging. And lastly, the similarity is the rules, which are being used to fire the actions that can define a similarity between stories. Each of these types, i.e., interactions, speech acts and rules, will reveal relations between leadership decision-making situations. These relations can be used to present alternative explanations and to contrast approaches. The expressive power of the representation is the use of the similarity measures to select stories to help the user of the system.

The users stated goal is used to locate sets of applicable sets of stories. The selected stories can be analyzed and compared, based on the similarity types describe above. The system in Hartung and Håkansson [5] collects information about the problem situation that allows the matching of stories. This includes issues like the number of participants, and facts about the participant's possible attitudes and motivations.

A second power from the representation in the form of speech acts is natural language generation. The work in Façade has demonstrated an approach to natural language generation [4]. The speech acts can form a template that can be used to generate a story, which is translated into the domain of the problem the user is trying to solve with leadership. This will require some work to achieve, but we view it as a possible step in the future.

7 Conclusions and Further Work

In this paper we have presented a knowledge representation for a knowledge-based leadership system that allows us to replicate a system of opinions and observations. The system makes the knowledge about leadership accessible for others that need the knowledge in different situations. It is knowledge about goals, interactions, and emotional/motivational representation presented as actors, actions and rules.

The current project is prototyping the representation to analyze leadership examples. The preliminary work is very promising. The next step is to implement searching to locate useful examples for the decision maker. This will be implemented in the framework of our experimental system. We also will attempt to use NLP systems for automated encoding of examples and develop a system that extracts assumptions about behavior from the rules in various stories but also present risk analysis.

References

1. Appian, P., Sundaram, H., Birchfield, D.: Communicating Everyday Experiences. SRMC October 2004, New York, USA, (2004)
2. Bolden,R.: Leadership Definitions. The University of Exeter, The Queen's Drive, Exeter, Devon, UK EX4 4QJ, http://www.leadership-studies.com/lsw/definitions.htm 2006-02-03, (2006)
3. Ekstam, K.: The Leadership's cornerstones – four factors for success. (Ledarskapets hörnstenar- fyra framgångsfaktorer). Kristianstads Boktryckeri, Sweden. ISBN 91-47-06560-5, (2002)
4. Kantrowitz, M.: GLINDA: Natural Language Text Generation in the Oz Interactive Fiction Project, School of Computer Science, Carnegie Mellon University report. CMU-CS-90-158 (1990)
5. Hartung, R., Håkansson, A.: A Knowledge Based Interactive Agent Assistant for Leadership" The 2nd European Conference on IS Management, Leadership and Governance. July 2006. Paris, France, (2006)
6. Liu, H.: Articulation, the Letter and the Spirit in the Aesthetics of Narrative. SRMC, Proceedings of the 1st ACM workshop on Story representation, mechanism and context, October 2004, New York, USA, (2004)
7. Mateas, M., Stern, A.: Façade: An Experiment in Building a Fully-Realized Interactive Drama. http://www.interactivestory.net/papers/MateasSternGDC03.pdf 2006-03-03
8. Mindell, A.: The leader as Martial Artist. Lao Tse Press, Portland, Oregon, USA, (2000)
9. Reilly, W.S.N.: Believable Social and Emotional Agents. Dissertation CMU-CS-96-138, Carnegie Mellon University, (1996)
10. Schank, R.C.: Tell Me A Story: A New Look at Real and Artificial Memory. Macmillan, New York. (1990).
11. Szilias, N., Rety, J.: Minimal Structures for Stories. SRMC, Proceedings of the 1st ACM workshop on Story representation, mechanism and context, October 2004, New York, USA, (2004)
12. Tannenbaum, R., Weschler, I., Massarik, F.: Leadership and Organization: A Behavioral Science Approach. McGraw-Hill book company, INC. New York Toronto London, (1961)
13. Whitney, J.O., Packer, T.: Power Plays: Shakespeare's Lessons in Leadership and Management. Simon and Schuster Inc., New York USA, (2001)

Modeling of the Vocational Suggestion with Restrictions (Graph Theory Approach)

Gregor Molan[1] and Marija Molan[2]

[1] HERMES SoftLab Research Group, Litijska 51, SI-1000 Ljubljana, Slovenia
gregor.molan@hermes.si
[2] Institute for Occupational Health, Poljanski nasip 58, SI-1000 Ljubljana, Slovenia
marija@molan.ws

Abstract. Support of decision making in the process of vocational counseling initiated development of an expert model. Modeling of the vocational suggestion is formalized on the basis of the graph theory. The construction of the model is presented with the graph. Local optimization with edge contraction of the given graph is based on minimization of graph's diameter. With further construction of the model graph the result is vocational suggestion with the restriction set. The model has been developed from the real data collected in Slovenian elementary school. Implementation of the graph theory, not only for presentation, but also interpretation of the model, gives the formalized base for development of user-friendly expert model.

1 Introduction

To reduce the risk of wrong child's decision in the vocational choosing process, vocational counselors have been included in the process of vocational counseling. Their support in the decision making process should help to avoid wrong decisions and increase the reliability of a particular child's decision. To achieve a higher level of reliability, counselors are looking for support tools that should help them in supporting of children in their decision making. The application of a standardized tool should reduce the perceived level of stress on the side of vocational counselors due to their individual responsibility for supporting the process of decision making. Previous positive experiences with development and implementation of the models have initiated development of the model for decision making support in the process of vocational counseling [1]. The expert model should be a tool for shaping the most adequate decisions on the basis of available data. The counselor advice in the process of vocational counseling should be supported with the expert model. The expert opinion and knowledge of experts have been incorporated in the model as the learned model's output.

Predictive validity of expert model for vocational counseling should be on long term if the suggestion will be accepted. Output of presented work is the expert model, which should be a valid tool for practical use in the process of vocational counseling. Long term validity will be proved in the future [2].

2 Problem

The main research goal of the project has been development of the expert model for support of decision making in the process of vocational counseling. The model has to

B. Gabrys, R.J. Howlett, and L.C. Jain (Eds.): KES 2006, Part I, LNAI 4251, pp. 360–367, 2006.
© Springer-Verlag Berlin Heidelberg 2006

reflect influence of (a) Child's health, (b) School results, (c) Interests, and (d) Social environment. For this purpose the problem has to be formalized and based on theory. In this paper the formalization and theoretical foundations of the model is presented. Theoretical foundations of the expert model with the mathematical logic and graph theory is the basement for the second step: computer programming. Calculations will be on a real data collected in the sample of 155 children at the age of 13 in north of Slovenia.

2.1 Input Data

Description of the data is organized in the subchapters with the adequate vertex level. Vertex level is the same as the database variable and the vertices in the same level are values of that variable. Graph, arcs, vertices, and weights are defined in the chapter Method. The first level is child's gender. Vertices on the levels 2, 3, and 4 describe physical status of general health estimated (by the specialist of school medicine). Vertices on the levels 5, 6, 7, 8, 9, 10, 11, and 12 describe psychical status of general health described by the specialist of school medicine.

Vertices on the level 13 and 14 describe health status of the child. Vertices on the level 15 describe school results and the level 16 describes influential factor. Vertices on the level 17 describe desired profession. There are three different meanings of the level 17:

[i] Child desired profession (*unproved validity of the value*),
[ii] Expert's advice for the child (*assumed as the valid value*),
[iii] Suggested profession as the output of the model (*modeled value*).

3 Method

To solve the special problem of finding adequate profession, general Vocational Counseling Model (VCM) is developed (Fig. 1).

The problem of choosing an adequate vocation in the process of vocational counseling is modeled with the theory. It means that the database query from the space of data about children and professions is transformed into the abstract space with the mathematical logic and the graph theory [3].

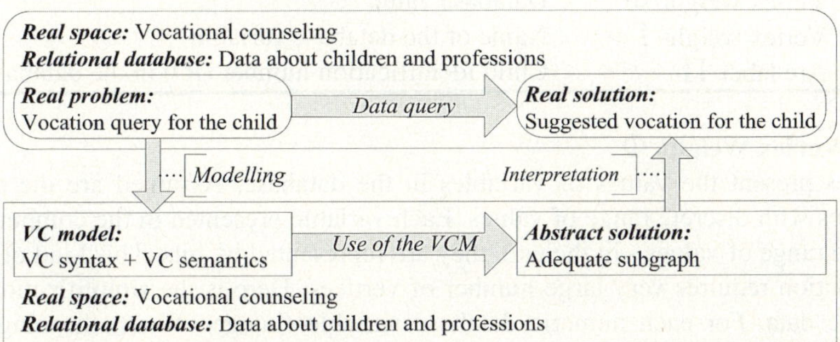

Fig. 1. High level design of the Vocational Counseling Model (VCM)

3.1 Formal Definition of the Abstract Space

The graph theory is used for presentation and interpretation of all the data that solve the real world problem.

Definition 1. A <u>graph</u> is a construction ***G=G(V,A,E)*** of finite set of ***<u>vertices</u>*** $V(G)=\{v_1, v_2,..., v_n\}$, finite set of ***<u>arcs</u>*** $A(G) \subseteq V(G) \times V(G)$

(i) $a \in A(G): \quad a = \overline{uv}, u,v \in V(G), u \neq v$

(ii) $u = init(a), v = ter(a)$

and finite set of *<u>edges</u>* $E(G) \subseteq V(G) \times V(G)$

(i) $e \in E(G): \quad e = uv, u,v \in V(G), u \neq v$

(ii) $\forall u,v \in V(G): \quad uv = vu$

An arc a is directed from the ***<u>initial vertex</u>*** $u = init(a)$ to the ***<u>terminal vertex</u>*** $v = ter(a)$.

<u>Vertex (arc, edge) weight (label)</u> on the graph is a function

$$W_{\Xi} : U \rightarrow \Xi, \quad U \in \{V, A, E\}, \Xi \in \{\Gamma, \Phi, \Pi\}$$

Each element from the set Ξ is an adequate range of values. To abbreviate the notation, instead of ***<u>vertex (arc, edge) weight (label)</u>*** W_{Ξ} the notation ***<u>vertex (arc, edge) weight (label)</u>*** Ξ can be used.

The data about vocational counseling are presented with graph *G=G(V,A,E)*. The graph in this article is weighted (labeled) directed multi graph. In the graph there are arcs (directed connections) and edges (undirected connections). There are weights for vertices, weights and labels for arcs, and weights for edges. Other definitions from graph theory (*edge contraction*, *graph minor*, etc.) are from basic literature [4].

Table 1. Weights and label on the graph *G=G(V,A,E)*

Weight, label	***Meaning for vocational counseling***
Vertex weight Φ	Database value
Vertex weight Γ	Name of the database variable
Arc label Π	Child identification number (ID) in the database

3.1.1 Vertex Weight Φ

Vertices present the values of variables in the database. Assumed are the database variables with discrete range of values. Each variable presented in the computer has a discrete range of values - in the end, they are represented by bits. This kind of a graph construction requires very large number of vertices. Here is the simplification of the numeric data. For each numeric database variable (integer, float), the range of its values has to be split into intervals. The name of each interval is unique and it is the weight of a vertex in the graph.

3.1.2 Vertex Weight Γ

The vertex weight Γ is the natural number of the database variable *level*. The vertex weight Γ is also identification of the variable in the database. Variables are arranged (sorted) from variables for attributes whose values are hard to change to the variables with modeled values. Vertices with weight 1 represent database variable for the most stable child attribute. Vertices with weight k are on the *k-th level* - represent database variables on the *k-th level* - they represent *k-th level child attribute*. Let the vertex weight be $\Gamma = \{1, 2, \ldots, L\}$. Vertices with weights from 1 to L-1 are called <u>*inner vertices*</u>.

$$\textit{Inner vertex } v \in V(G): \quad 1 < W_\Gamma(v) < L$$

The database variables for the vocational counseling are arranged by expert from the gender with weight 1 to the final vocational decision:

(i) $v \in V(G): \quad W_\Phi(v) \in \{\text{male, female}\} \quad \Rightarrow \quad W_\Gamma(v) = 1$

(ii) $v \in V(G): \quad W_\Phi(v) \in \text{"set of vocational decisions"} \quad \Rightarrow \quad W_\Gamma(v) = L$

Let x be the name of an arbitrary database variable with no required value. It means that it is possible for the variable x to have no value in the database. For such database variable x a special vertex named "no value for x" is assigned. The consequence is such: for each arc value there is exactly one arc with init vertex with weight i-1 and terminate vertex with weight i. This means that for each inner vertex there is the same number of terminated arcs and initial arcs:

$$1 < W_\Gamma(v) < L : \left|\{a \in A(G) \mid ter(a) = v\}\right| = \left|\{a \in A(G) \mid init(a) = v\}\right| \qquad (1)$$

Arcs are connections between vertices with neighboring labels (levels). For each arc $a = uv$, the corresponding initial vertex label $W_\Gamma(u)$ and the corresponding terminal vertex label $W_\Gamma(v)$ differ by exactly 1:

$$a = \overline{uv} \in A(G) \Rightarrow W_\Gamma(v) - W_\Gamma(u) = 1 \qquad (2)$$

3.1.3 Arc in Graph and Arc Label Π

Arcs are labeled with $\Pi = \{1, \ldots, C\}$; the label represents child identification.

Definition 2. Let it be $c \in \Pi$. The <u>***path k with label c***</u> P_c^k starts in a vertex on the level 1 (database variable with the most stable attribute) and terminates in a vertex on the level k. The <u>***path with label c***</u> P_c is the path P_c^k with $k = L$.

$$P_c^k = \left(\overline{v_0 v_1}, \overline{v_1 v_2}, \ldots, \overline{v_{k-1} v_k}\right) \quad \text{where}$$
$$\forall j = 1, \ldots, k : W_\Pi\left(\overline{v_{j-1} v_j}\right) = c, \; W_\Gamma(v_j) = j \qquad (3)$$

The path P_c^k with the chosen label c represents path of arcs for an adequate child c. It is the path from the first level vertex called "gender level" to the final vocation decision.

3.1.4 Edges in Graph

Edges are connections between the vertices with equal weight Γ - the vertices on the same level. The vertices are connected with edge if they are neighbors by means of their names. Names of vertices are the representation of variable values in the database. It means that edges are connections between variables with in the database.

3.2 Graph for Vocational Counseling

In this section are informal definitions needed for graph for Vocational Counseling. The set of **<u>*arc values for vertex*</u>** *v* is the set of all arc values that are connected with vertex *v*:

$$arc_{val}(v) = \begin{cases} arc_{val}^{init}(v); W_{\Gamma}(v) < L \\ arc_{val}^{ter}(v); W_{\Gamma}(v) = L \end{cases} \tag{4}$$

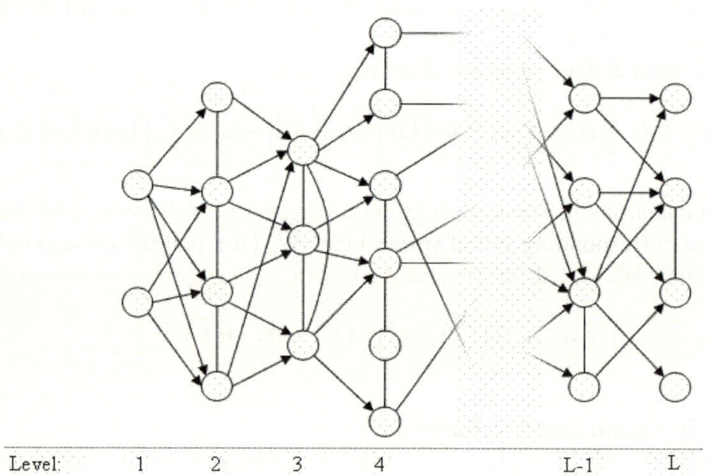

Fig. 2. Example of graph *G=G(V,A,E)* for vocational counseling. Names of vertices are not presented. Levels of vertices are presented at the bottom of the figure. Arcs can connect only vertices from the neighboring level, from lower to higher. Edges can connect only vertices from the same level.

<u>*Difference*</u> between vertex u and vertex v is natural number d(u,v):

$$d(u,v) = 1 - \frac{\left|(arc_{val}(u)) \cap (arc_{val}(v))\right|}{\left|(arc_{val}(u)) \cup (arc_{val}(v))\right|} + \left|W_{\Gamma}(u) - W_{\Gamma}(v)\right|$$

Vertex *u* and vertex *v* have the difference *d(u,v)* The difference *d(u,v)* is the strength of intersection set between sets $arc_{val}(v)$ and $arc_{val}(u)$. The difference *d(u,v)* between vertex *u* and vertex *v* can be interpreted as:

d(u,v)="number of different arc values"+ "the difference between their levels"

Function *d(u,v)* in the definition of the difference is the *metric* over vertices *V(G)*. The proof of *d(x,y)* being the metric is beyond the focus of this article (see [3],[5],[6]).

Assume graph $G = G(V, A, E)$ and path P_c for chosen label $c \in \Pi$. ***Difference $d(P_c)$ on the path*** P_c is the sum of differences between all vertices on the path P_c:

$$d(P_c) = \sum_{u_{j-1}v_j \in P} d(u_{j-1}, v_j)$$

Definition of the difference on the path is the basis for definition of the difference on the graph. The ***graph diameter d(G)*** is the biggest difference on all paths P_c:

$$d(G) = \max\{d(P_c) | c = 1,\ldots,C\}$$

3.3 Minimizing the Graph Diameter

The graph $G = G(V, A, E)$ is defined with allowance of contraction of its edges. Here is the algorithm for local optimization which is applied to modify a given graph and minimize its diameter.

Algorithm 1 Minimizing graph diameter (local optimization)

```
With graph G=G(V,A,E) do {
  Find edges that maximally minimizes graph diameter;
  Contract all found edges;
}
```

Note that the number of iterations is limited by the number of graph edges. The result of Algorithm 1 is the ***model graph*** $G' = G(V', A, E')$. Some of edges form E are contracted and consequently there are new vertices in V'. Let assume the level $l \in \Gamma$. In the case that there is a single vertex on the level l, then the level l has no influence on the decision process. The interpretation is that the database variable adequate to the level l is not important and such database variable is redundant. The result from model graph G' from Algorithm 1 is reduced set of database variables.

3.4 Modeling with the Model Graph

The model graph $G' = G(V', A, E')$ is constructed to be used in vocational counseling. Let it be $c \notin \Pi$ the new child with data presented with path P_c and let it be this path the adequate path in the model graph G'. Note that for this new child c there is no data about the profession - no data about the last level in the graph G'.

The path for child c is $P_c = \left(\overline{v_0 v_1}, \overline{v_1 v_2}, ..., \overline{v_{L-2} v_{L-1}}\right)$ and let the set $\{\Pi_1, \Pi_2, ..., \Pi_{L-1}\}$ be the set of computed sets for child c.

The set Π_1 is the set of children from model graph that have the same value $W_\Phi(\ )$ on the first level. It is the set of children with the same value of the first level of database variable as it is for the child c. The set Π_k is the set of children from the model that has the same paths from level 1 to level k as it is the path for child c. The set X is the ***restriction set*** and it is the set of pairs (e,l) of contracted edges and the level of the edge that is contracted. If there is no contraction needed, than the set X is empty.

Algorithm 2 Modeling of the vocational suggestion to child c with restrictions on the contracted levels

```
G'' := G' ;
X := { };
For level k ∈ {1,2,..., L} do {
  Π_k := { j | P_j^k ∈ G''; P_j^k = P_c^k };
  Until Π_k = { } repeat {
    Determine l ∈ {1,..,k} ∋ |Π_{l-1}| − |Π_l| ≥ |Π_{i-1}| − |Π_i|; ∀i ∈ {1,..,k};
    Contract the edge e on the level l ;
    X := X ∪ {(e,l)};
    Update G'' according to contraction of the edge e ;
    Update sets Π_1, Π_2,..., Π_k ;
  }
}
```

The interpretation of the empty set $\Pi_k = \{\ \}$ on the level k is that there is no adequate child from the model graph for the child c. In that case the Algorithm 2 determines the level $l \in \{1,2,...,k\}$ with the maximum reduction of power of sets $\{\Pi_1, \Pi_2, ..., \Pi_k\}$ and contract the edge e on the level l. Update the set X of contracted edges with the element (e,l). Algorithm continues with determination of the level with maximum power reduction and contraction of the edges until there is at least 1 element in the set Π_k. The level with contracted edges is called the ***contracted level***.

The result of Algorithm 2 is non empty set Π_L. The non emptiness of set Π_L is obvious from the corner case. In the corner case all edges on all levels are contracted and the set Π_L is the same as the set al all professions from the model. In such corner case the suggestion is the suggestion with restrictions on all levels.

4 Conclusion

Formalization of the expert model (VCM) with the graph theory clarified connections in the model and the vertex values. The graph defines the skeleton of the model. The semantics of the model is achieved with contraction of edges. On the basis of graph theory it is possible to present and interpret the model on the formal level. Experts have defined the data that should be incorporated into the model. They have defined the frame and the power of influence for each particular attribute. The graph theory is in the great advantage because of its power of presentation. It is much easier to clarify all the possible connections and to identify all the connection of attributes. Operations on the graph, corresponding to the nature of influence, lead to the output of the model - the vocational advice to the child. Proposed model should support decision making process in the real situation. Formalization is defined so far that the corresponding program code in Java is now in the process of implementation. Development of expert model in the social and preventive health care sciences should be much more effective and user-friendly with approach presented in this paper.

References

1. Molan M., Molan G.: Formalization of expert AH model for machine learning, Frant. artf. int. ell. appl.vol. 82, 2002.
2. Molan M., Molan G.: Validation of a method of vocational counselling, GIGA, Sabir (ed.). *Occupational health psychology : flexibility, quality of working life and health*, (European Academy of Occupational Health Psychology Conference Proceedings Series). Nottingham: The Institute of Work, Health & Organisations, 2003
3. Lloyd, J.W.: Foundations of Logic Programming, Springer-Verlag, Berlin, (Second edition), 1987
4. Diestel, R.: Graph theory (Graduate texts in mathematics), Springer, cop., New York, 2000
5. Molan, G.: Poizvedba po podatkih s teorijo grafov in matematično logiko, (Data Query with Graph Theory and Mathematical Logic), University in Ljubljana, Master Thesis, 1999
6. Rudin, W.: Functional analysis (International series in pure and applied mathematics), McGraw-Hill, New York, 1991

Robust Eye Detection Method for Varying Environment Using Illuminant Context-Awareness

Mi young Nam, Eun Jin Koh, and Phill Kyu Rhee

Dept. of Computer Science & Engineering, Inha University
253, Yong-Hyun Dong, Incheon, Korea
rera@im.inha.ac.kr, pkrhee@inha.ac.kr

Abstract. In this paper, we propose the efficient face and eye detection system using context based detector. The face detection system architecture use cascade method by illuminant face model. Also, we detect eye region after face detection. We construct nine classes to eye detector. It is enhanced eye detection ratio for varying illuminant face images. We define context to illumination class and distinguish class back propagation. Also, we made in context model using face illuminant. The multiple classifiers consist of face illuminant information. Context based Bayesian classifiers are employed for selection of face and eye detection windows. Face detection system is enhanced for face detection form multiple face class and non-face class. Proposed method is high performance more than single classifier.

1 Introduction

Face detection is one of the hardest problem in the research area of face recognition, and an important task in many computer vision applications such as human-computer interface, smart environment, ubiquitous computing, and multimedia retrieval. But, captured environment is different time and location. Face images are varying brightness value, therefore we are discrimination face image to illumination group. Many approaches have been research robust face detection for varying illuminant [1, 2, 3,4].

We propose the context – based face detector and eye detector which are multiple classifiers for face and eye class. Face and eye detection is constructed Bayesian classifier from face, non-face, eye and non-eye class. Context modeling is used supervised learning to six classes and context awareness is used back propagation.

The outline of this paper is as follows. In section 2, we present the architecture of the proposed face detection system. In section 3, we describe the method that search eye region using illuminant context awareness method. We give experimental results in section 4. Finally, we give concluding remarks in section 5.

2 The Face Detection Scheme

In this paper, we proposed the method that is multiple Bayesian classifiers for selecting eye location and face detection. The multiple Bayesian is defined from illumination and light. Because captured image form Camera is vary different external condition.

B. Gabrys, R.J. Howlett, and L.C. Jain (Eds.): KES 2006, Part I, LNAI 4251, pp. 368 – 375, 2006.
© Springer-Verlag Berlin Heidelberg 2006

2.1 Illuminant Context Modeling

We have tested three methods for illumination discrimination back propagation neural network based discrimination. We use FERET images of artificial facial images for context awareness modeling [5, 6, 7]. The artificial images are generated sine and cosine function to general illumination face images. We generated six classes for illuminant face images. Face and eye images are distinguish six contexts for robust face and eye detection in dynamic environment. Fig.1 shows face images clustering results.

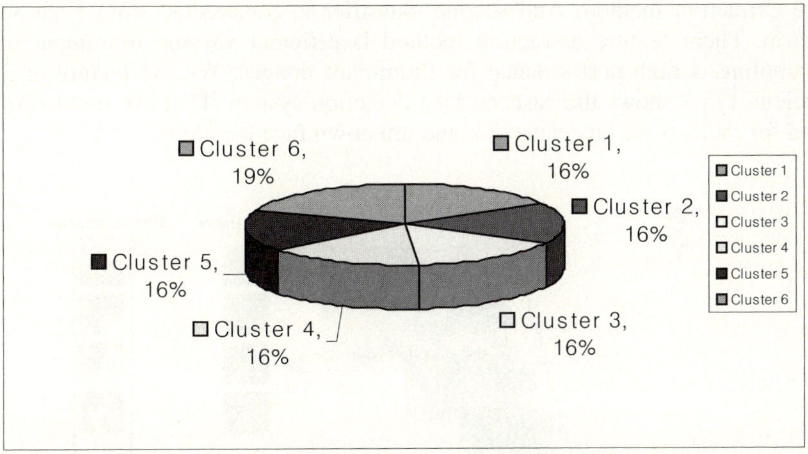

Fig. 1. Face image context modeling

Face data eye coordination is next figure. In figure, we used (a) face model because (a) face model is better (b) face model. Face image's eye position is important for face detection performance enhancement. Therefore we selected Fig.2 (b) because that is high performance face detection over 5% than Fig.2(a).

Example based face detection is important training images and vectorization method. Our proposed method is efficient for varying illuminant face and eye detection architecture.

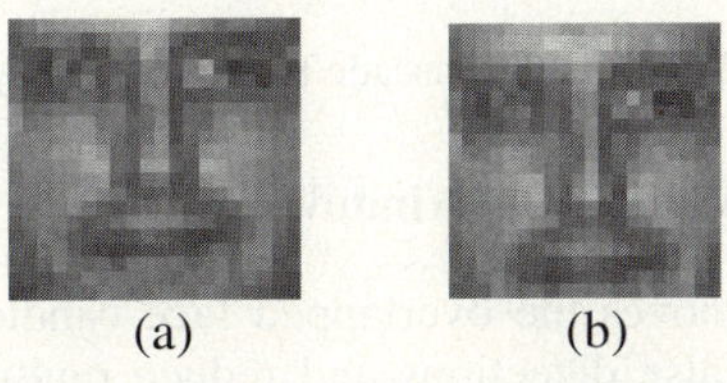

Fig. 2. Face modeling to training

Feature extraction method has much affect in face detection performance. The feature extraction method that was optimized by an experiment, as a result, raises good performance. Therefore, we can improve performance through efficient feature extraction and dimension decrease. We use two types of feature extraction method. One uses the 54 sub-regions. Other is the method using Haar wavelet transform.

2.2 Face Detection

A face detection system use multi-resolution methods and cascade face detection. The cascade face detector is generated two classifiers and these are classifiers of sub sampling and Haar wavelet feature extraction. Face detection system is two face detectors for each face group in Fig.2. Fist classifier is constructed from sub-sampling feature extraction method. And second classifier is constructed from Haar wavelet transform. There feature extraction method is efficient varying illuminant images. Sub-sampling is high performance for illuminant images. We use feature of 3 level coefficient. Fig.3 shows the cascade face detection system. That use multi-resolution method for multi-scale face detection and unknown face location.

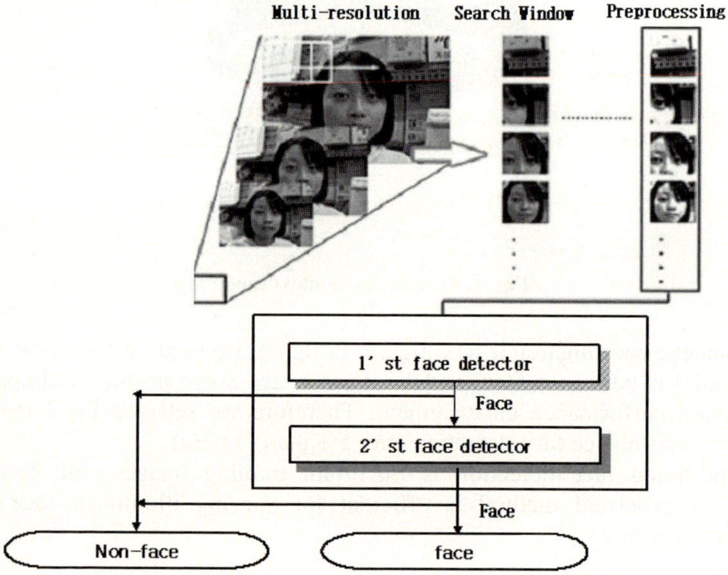

Fig. 3. This is the cascade face detection system

2.3 Merging Process of Candidate Windows

Merging heuristic that removes the overlapped face candidate windows is powerful process to remove most false detections and reduce multiple detections of a single face to one [7, 8, 9]. The merging heuristic used in this paper is as follows:

- Count the neighborhoods of each face candidate windows.
- Classify the candidate as a face if the number of its neighborhoods is above a threshold.

- And eliminate overlapped face windows as the number of neighborhoods. We can preserver the location with more number of detection, and remove locations with less detections.

3 The Illuminant Eye Detector

The outline of the proposed eye detection method is shown in Fig.4. Eye regions are searched by multiple Bayesian classifiers using integrated feature space to determine whether the 8×8 sub-window is eye region. We extract the eye feature is same face detection feature extraction.

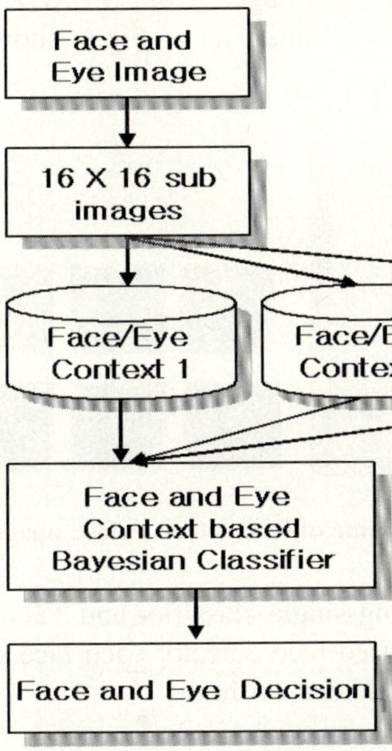

Fig. 4. The proposed scheme used for eye region detection

We generate Bayesian classifier under varying illuminant images. These classifiers are constructed in cascade form of multi steps. The 54-dimensional feature from illuminant 1'st classifier is used as classifier in first step. The illuminant 2'st classifier is used in second step.

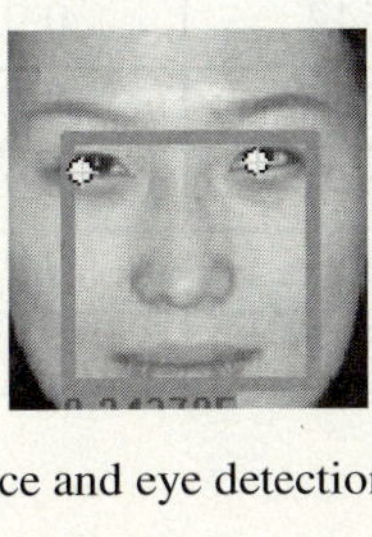

Fig. 5. Face and eye detection example

The locations which output through previous methods are good at being precision detect. But because eyes are very important feature at face recognition, we want to make robust precision position of eye. We regard pupil is robust precision point of eye and make an effort to point pupil. We use Gaussian filter and projection function to point center of pupil. In this case, center of detected window is moved far away from earlier position. It is a critical error and a primary factor of decrease of detect rate. Therefore, we need to restrict center of window to move for away.

4 Experiment

Face training images are generated 5588 images from FERET database and generate non-face images form images without face. Fig.6 show the training images example.

Fig. 6. Eye detection method is same of face detection face images

Table 1 shows face detection ratio using single classifier and Table 2 shows face detection performance using clustering based face detector each face six cluster. We can know that context based multiple face detector is high.

Table 1. Frontal face detection result using single classifier

Test Set	Face Detection	Face Detection Rate	False Detections
Context1	570	0.98	4
Context2	542	0.93	6
Context3	560	0.96	9
Context4	530	0.91	8
Context5	543	0.93	2
Context6	599	0.88	3

Table 2. Frontal face detection result using multi-classifier using illuminant context-awareness

Test Set	Face Detection	Face Detection Rate	False Detections
Context1	578	0.99	2
Context2	568	0.98	3
Context3	569	0.98	5
Context4	540	0.93	5
Context5	562	0.97	3
Context6	635	0.93	1

In Table 2, we can know that face detection ratio is different for context. We can reduce false detection ratio using multiple face detector. In the face detection system, false detection ratio is important measurement for system performance measurement.

Table 3 shows eye detection ratio using context based eye detector each eye cluster. Eye cluster is distinguished nine clusters and nine eye detectors. When eye detection ratio is 0.18, eye detection ratio is high other error ratio. Table 3 shows that we achieved very encouraging experimental results in the error rate 0.18. The first set has the some faces that are not frontal.

Table 3. Eye location results using illuminant context-awareness. (Err = 0.18)

		Context-Awareness Eye Detection		
FERET	Data Number	AR	FA	AR%
Context0	10	8	2	80.00%
Context 1	77	73	4	94.81%
Context 2	18	17	1	94.44%
Context 3	413	402	11	97.34%
Context 4	16	16	0	100.00%
Context 5	78	76	2	97.44%
Context 6	305	301	4	98.69%
Context 7	81	79	2	97.53%
Context 8	95	90	5	94.74%
Total	1093	1062	31	97.16%

Fig .7 shows face detection ratio for CMU dataset. CMU dataset include varying illuminant, complex images and background. The multiple face detectors are high performance more than single face detector.

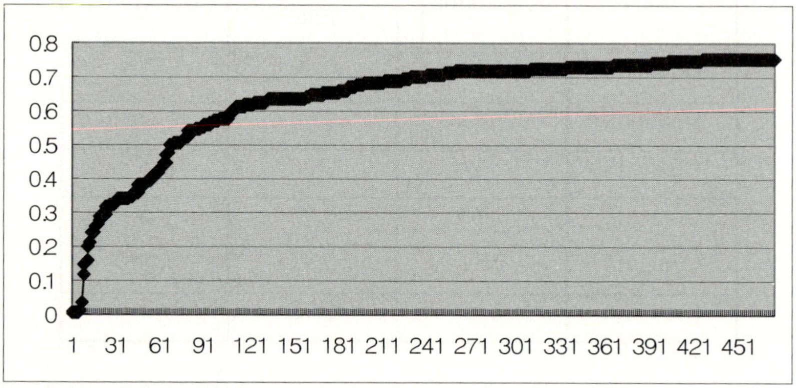

Fig. 7. Face detection ROC curve using rotated face images

5 Concluding Remarks

We proposed efficient object detection method for varying environment. The face detector is constructed multiple classifier of cascade method include two face detector, and eye detector is generated nine Bayesian classifiers according illuminant. The cascade face detector is efficient for face detection in illuminant face images, sub-sampling method is good more than other vectorization method. Multiple eye detection classifiers must be improved eye detection ratio. Bayesian classifiers using features from intensity and texture information and it results face candidate regions. This proposed method can detect 96% of faces and 97.16% of eyes in artificial FERET images. There are some works that can improve the proposed method.

References

1. SMITH, P., SHAH, M., AND LOBO, N. D. V. , "Monitoring, head/eye motion for driver alertness with one camera," In *Proceedings of the 2000 International Conference on Pattern Recognition*, Session P4.3A, (2000)
2. A. Haro, M. F. : Detecting and tracking eyes by using their physiological properties, dynamics, and appearance. Proc. Of IEEE Conf. on CVPR (2000)
3. S.Baskan, M. M. B., V.Atalay : Projection based method for segmentation of human face and its evaluation.. *Pattern Recognition Letters* 23 (2002) pp.1623-1629
4. T. V. Pham, et. Al.,: Face detection by aggregated Bayesian network classifiers. Pattern Recognition Letters 23, (2002) pp. 451-461
5. Ludmila I. Kuncheva, James C. Bezdek, and Robert P. W. Duin, "Decision templates for multiple classifier fusion: An experimental comparison," Pattern Recognition, (2001) pp. 299–314

6. Mi Young Nam, "An Efficient Face Location using Integrated Feature Space," LECTURE NOTES IN ARTIFICIAL INTELLIGENCE 3682, (2005) pp.327-335
7. Mi Young Namand Phill Kyu Rhee, "Human Face Detection using Skin Color Context Awareness and Context-based Bayesian Classifier,", LECTURE NOTES IN ARTIFICIAL INTELLIGENCE 3682, (2005) pp.298-307
8. S.Lucey, S.Sridharan, V.Chandran : Improved facial feature detection for AVSP via unsupervised clustering and dicriminant analysis. *EURASIP Journal on Applied Signal Processing*, vol 3 (2003) pp.264-275
9. C. Liu. : A Bayesian Discriminating Features Method for Face Detection. IEEE Trans. Pattern Analysis and Machine Intelligence, vol.25 (2003) pp.725-740

INMA: A Knowledge-Based Authoring Tool for Music Education

Maria Virvou, Aristomenis S. Lampropoulos, and George A. Tsihrintzis

Department of Informatics, University of Piraeus,
80 Karaoli and Dimitriou St., Piraeus 185 34, Greece
{mvirvou, arislamp, geoatsi}@unipi.gr

Abstract. A knowledge-based authoring tool and its domain-knowledge acquisition capabilities are described. The authoring tool is called INMA and assists human tutors to create their own Intelligent Tutoring Systems (ITSs) for music education. Learner modeling in the resulting ITSs is primarily based on a cognitive theory that models human reasoning and is called Human Plausible Reasoning. The particular theory makes use of a very detailed knowledge representation of the domain of the part of music to be taught in the form of hierarchies, so that similarities, dissimilarities, generalizations and specializations among the tutoring concepts may be inferred. INMA incorporates a special knowledge acquisition component that can assist instructional designers on the task of constructing the music hierarchies required. The operation of this component is based on content-based retrieval and semantic meta-data.

1 Introduction

Sophisticated approaches to computer-based education attempt to incorporate adaptivity to the features of tutoring systems. Adaptivity aims at dynamically customizing the tutoring advice of the educational application to the individual needs of the learner. Adaptivity to individual learners' needs may be achieved using the technology of Intelligent Tutoring Systems (ITSs). As Self [7] points out, ITS research is the only part of the general information technology and education that has as its scientific goal to make computationally precise and explicit forms of educational, psychological and social knowledge, which are often left implicit. The adaptivity of ITSs is primarily based on their user modeling component. User modeling components constantly collect information about users through their interaction with the system and create an internal representation of the user's mental state and needs. This representation is then used to generate advice tailored to the individual users. In this way, an ITS may provide individualized tutoring to learners.

However, ITSs have often been criticized that they represent immensely complex approaches to building learning environments (e.g. [1]). A solution to the problem of construction of ITSs is knowledge-based authoring tools that provide the facility of building multiple cost-effective ITSs. Knowledge-based authoring tools are meant to be used by instructors, who are not necessarily computer experts and wish to author their own ITSs on a particular tutoring domain. At the same time, knowledge-based authoring tools should be based on domain-independent methods that can be used for

B. Gabrys, R.J. Howlett, and L.C. Jain (Eds.): KES 2006, Part I, LNAI 4251, pp. 376–383, 2006.
© Springer-Verlag Berlin Heidelberg 2006

the semi-automatic creation of sophisticated ITSs by instructional designers. Indeed, Murray [6] highlights the potential of ITS authoring tools in giving the instructional designer a combination of facilities to produce visually appealing, interactive screens and a deep representation of content and pedagogy.

In this paper, we describe a knowledge-based authoring tool that can be used by instructional designers to create ITSs for music education. The authoring tool is called *INMA* (*IN*telligent *M*usic-education *A*uthor). In the context of INMA's reasoning facilities, a domain-independent method [10] has been used for providing user modeling capabilities to the ITSs that can be generated using INMA. The domain-independent method is primarily based on an adaptation of a cognitive theory that aims to model human reasoning and is called Human Plausible Reasoning [2] henceforth referred to as HPR. The human reasoning that is modeled by the cognitive theory is used in INMA to model the reasoning of learners. One major problem in the use of HPR in the context of the authoring tool is that HPR assumes the existence of detailed domain hierarchies in the knowledge representation so that its inference mechanism may be used efficiently. In order to resolve this problem and alleviate the cognitive burden of human tutors (prospective authors) who have to create the domain hierarchies, there is a special knowledge acquisition component in the authoring tool that assists human tutors in the creation and maintenance of these hierarchies. The knowledge acquisition component of INMA combines two types of meta-information in the description and quantification of the relations between data and hierarchies in a music collection. These types are: (1) content-dependent meta-data and (2) content-independent meta-data [3, 4, 9].

2 Overall Architecture and Operation of INMA

INMA is a knowledge–based authoring tool for music. The incorporation and adaptation of HPR in INMA provides intelligent tutoring environments where students will be able to answer questions concerning the tutoring domain and the tutoring system will be able to perform error diagnosis, so that the students' weaknesses and possible misconceptions are revealed. Then appropriate tutoring and advice is given to the students. HPR formalizes the plausible inferences based on similarities, dissimilarities, generalizations and specializations that people often use to make plausible guesses about matters that they know partially. These inferences may lead to either correct or incorrect guesses; in any case these guesses are plausible. INMA uses HPR for error diagnosis. We assume that a student who has not mastered the domain to be taught may give wrong answers to questions based on erroneous plausible guesses that s/he makes using his/her plausible reasoning on other parts of the domain that s/he knows.

Instructors who may wish to use the authoring tool to produce their own music- ITS, have to provide information to the system. The information required relates to the domain knowledge, the construction of tests and specification of the way that students are going to be marked in tests. Moreover, the use of HPR pre-supposes that the domain representation concerning music has been constructed in the form of hierarchies. In the case of an authoring tool, such as INMA, the knowledge representation would have to be constructed by the instructional designer. Thus the initial input to the authoring tool would involve a manual process in which a human instructional designer (in this case, a music tutor) acts as an author whose responsibility is to input data to the system via a dialog box. This manual process is quite difficult for instructors.

In view of the above problem, there is a component in INMA which is called Domain Knowledge Representation Assistant (DKRA), as in Figure 1. The DKRA provides the mechanisms that help the tutor construct and maintain the domain knowledge hierarchies. The DKRA realizes a semi-automatic process in which the human tutor does not have to create the knowledge representation manually, but instead uses a set of tools that allow him/her to interact with the data. This mechanism is very important given the fact that there is a huge increase in the availability of digital music, making it ncreasingly difficult to perform the task of querying and browsing a music database. At present, music data are usually organized on the basis of textual meta-information, such as the ID3 format. This is an extension to the popular MP3 format which allows the user to add tags to the beginning or end of the file, containing such information as song title, artist, album name, genre, etc. Despite its extended capabilities, ID3-based systems still suffer from serious drawbacks, as the textual description of audio content is subjective and the relevant meta-data have to be entered and updated manually.

The tutor explores through the data set to acquire knowledge, taking into account both qualitative (subjective) and quantitative (objective) aspects of the data. Thus, he/she is not only able to create hierarchies, but can also quantify and adjust these hierarchies to both the data and his/her expertise. In other words, the direct manual process is converted into a semi-automatic process, in which the teacher combines his/her primary knowledge with the knowledge extracted from the data, and the teacher plays the role of a supervisor of the knowledge representation creation process.

The overall architecture of INMA is illustrated in Figure 1. The human tutor (author) interacts with the tutor interface in order to construct the knowledge representation of the music syllabus to be taught in the form of hierarchies and in order to create a database of domain problems and exercises for the students to solve. The hierarchies of music are used by the HPR student modeler for error diagnosis. The results of the student modeling process are taken into account by the Tutoring Module that forms individualised advice that is presented to the student.

3 Description of the Domain Knowledge

The initial input to INMA is a description of knowledge concerning a specific domain given by a human teacher who is acting as an author. At first, the domain has to be described in terms of hierarchies, which constitute the knowledge representation of HPR. Therefore the author has to decide what the main concepts of the lesson are that may be represented in hierarchies. Then s/he may create hierarchies by using the Domain Knowledge Representation Assistant. Then the author inserts facts that s/he wishes to be taught to students and which are relevant to the main concepts of the hierarchies. Finally, the authoring tool may assist the author to construct tests that consist of questions relating to the domain factual knowledge.

For example, an author wishes to create a lesson in music about artists in the genres of Greek music. S/he may create the main hierarchy that represents the relations of genres-artists as the author perceives them by using the dialogue box shown in the example of Figure 2.

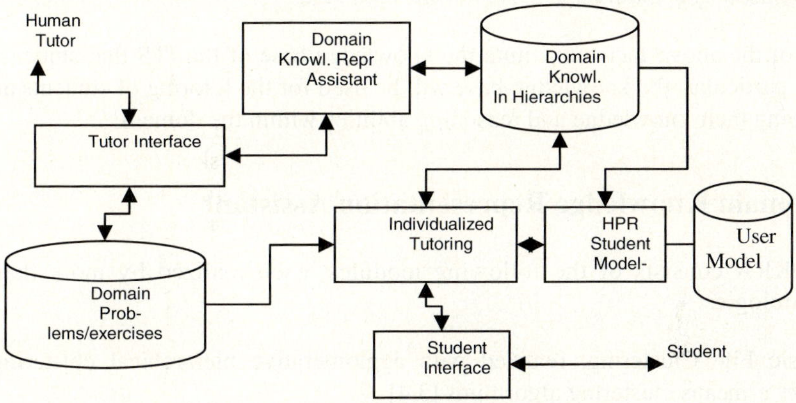

Fig. 1. INMA Architecture

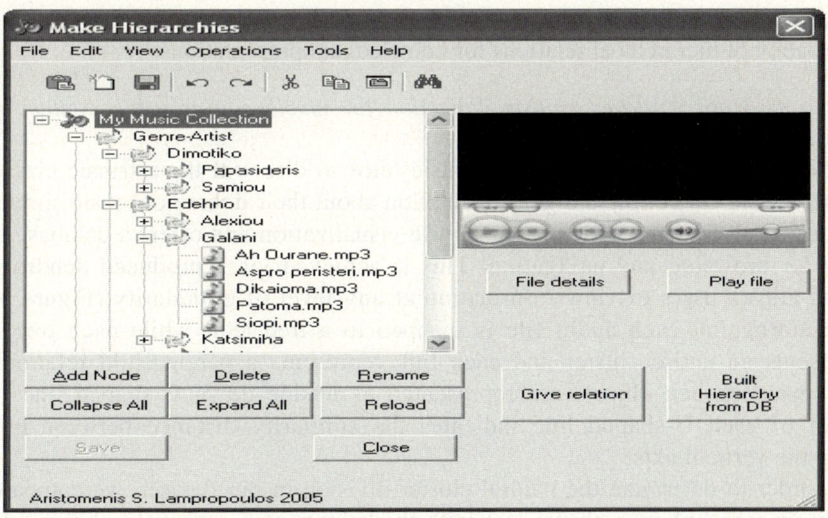

Fig. 2. Creation of domain knowledge

The author enters the facts in the relation that s/he wishes to teach to students in the particular lesson and the system produces the hierarchies. Examples of facts may be the following:

- The four main genres in Greek music are named *Rebetico*, *Dimotiko*, *Laiko*, and *Entechno*.
- Representative artists of genre *Entechno* are the following artists: *Alexiou*, *Galani* etc.
- Representative songs of genre *Entechno* are the songs entitled: "*Aspro Peristeri*", "*Koimisou aggeloudi mou*", "*Ah ourane mou*" etc.

All of the above facts constitute the knowledge-base of the ITS that students may use. In particular, the knowledge-base will be used for the tutoring of students and for examining their knowledge and reasoning abilities within the domain.

4 Domain Knowledge Representation Assistant

The DKRA consists of the following modules, each realized by more than one methodologies:

- Music File Clustering, realized with agglomerative hierarchical clustering and fuzzy c-means clustering algorithms [3,4]
- Genre Classification, realized with multilayer perceptron and adaptive boosting algorithms [9,11]
- Relevance feedback, which utilizes both types of meta-data and creates quantitative relations in the form of semantic networks [4,5]
- Creation of hierarchical relations for knowledge representations.

Step 1: Tutor explores music collection for music genres using content-based meta-data.

Music file clustering allows the music tutor to discover the intrinsic classes in his/her music collection and get information about their degree of music similarity. Clustering also allows efficient music file visualization and offers a database overview for browsing and navigation. This is achieved via a produced dendrogram, which allows users to view information at any level of granularity (Figure 3). In the dendrogram, each audio file is mapped to a tree leaf, while each tree node represents an entire cluster and each link represents a parent-child relationship. The links between objects are represented as upside down U-shaped lines. The height of each U-shaped line indicates the similarity distance between objects along the vertical axis.

In order to determine the natural cluster division in our dataset, we compare the height of each link in a cluster tree with the heights of neighbouring links at lower levels in the tree. On the one hand, if a link is approximately at the same height with neighbouring links, this indicates that there exist high degrees of similarity between objects joined at this level of the hierarchy. If the height of a link differs significantly from the heights of neighbouring links, this indicates that the degrees of similarity between objects at this level of the cluster tree are low and the link is said to be inconsistent with the links around it. Highly similar nodes or sub-trees have joining points that are far from the root.

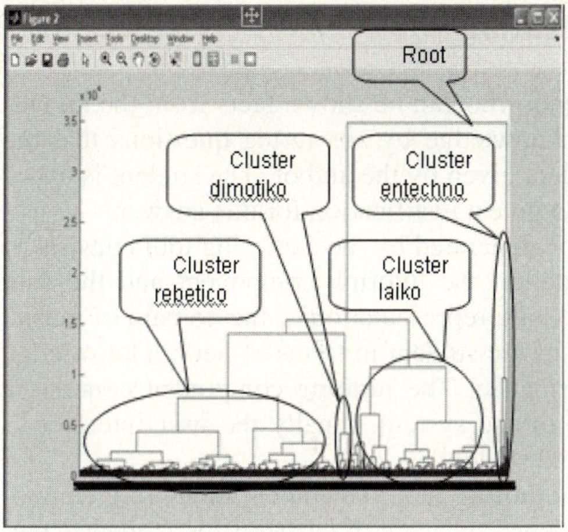

Fig. 3. Agglomerative hierarchical clustering

Table 1. Experimental results from classifiers

Classifier Name	Correctly Classified Instances (%)
Weak.classifiers.bayes.NaiveBayes	61.86
3-nearest neighbours classifier	60.72
Nearest-neighbour classifier	67.58
Sequential minimal optimization	68.35
Weak.classifiers.lazy.KStar	69.39
ClassificationViaRegression (M5 pruned tree)	73.11
4 hidden layer Perceptron	73.21
10 hidden layer Percetron	73.97
Weak.classifiers.trees.RandomForest	74.16
Weak.classifiers.meta.AdaBoostM1	75.88

Genre classification allows the music tutor to build classifiers for his/her music collection in order to organize the music data in an automatic way and make queries by examples or compare new musical pieces. For example, our data set consists of 1100 music pieces from Greek songs of 4 genres (Rebetico, Dimotiko, Laiko, and Entechno). In Table 1, we present some results of our trial and evaluation of different classifiers contained in the machine-learning tool called WEKA [11]. Classification results were calculated using 10-fold cross-validation evaluation where the dataset to be evaluated was randomly partitioned so that 10% be used for testing and 90% be used for training. This process was iterated with different random partitions and the results were averaged.

Step 2: The instructional designer organises the music collection and creates hierarchical relations for knowledge representations.

Relevance feedback bridges the gap between the two levels of meta-data. The music tutor may provide relevance feedback during a session of retrieval to tell the system which files he/she considers "relevant" or "irrelevant". During this process, a unified metric function $G(J,Q) = a \cdot \underset{low-level}{sim(J,Q)} + (1-a) \cdot \underset{semantic}{sim(J,Q)} \quad a \in [0,1]$ measures the relevance between query music files (Q) and any music file (J) in the database. The parameter a is the weight which the system attached to the links between keywords (ID3 tags) and files that form a semantic network as in [4,5]. In a semantic network, each music file (or the corresponding content-dependent meta-data) can be complemented with a set of semantically relevant keywords (content-independent meta-data) and, vice versa, the same keyword (eg., the name of a genre class : "rebetico") can be associated with more than one files. Finally, the music tutor can build hierarchies automatically, which contain the facts that he/she (as an individual) wishes to teach.

5 Resulting ITSs for Music Education

The resulting ITSs may be used by students who can be shown facts from the knowledge base. Students may also test their knowledge by answering questions that the authoring tool has formed, based on the data given by the author. The student is asked to give an answer and then is also asked to give a justification for this answer.

The architecture of the ITSs that will be generated by the authoring tool consists of the domain knowledge, the student modeller, the tutoring component and the user interface. The domain knowledge consists of a representation of the domain of music. The student modeller constructs a model of the student in terms of her/his knowledge level and her/his problem-solving performance. The tutoring component contains a representation of the teaching strategies of the system. Finally the user interface is responsible for the communication with the student.

In the ITSs that are generated by the authoring tool, the student modelling component examines the correctness of the students' answers in terms of the students' factual knowledge and reasoning that they have used. The reasoning of the student is examined by the HPR Modeller, which is the component that is based on an adaptation of Human Plausible Reasoning. The state of the student's knowledge of the domain is examined by the Overlay Modeller, which is the sub-component of the student modelling that operates based on the overlay technique [8]. Information about each student concerning his/her knowledge and reasoning ability is recorded in his/her long term student model. The adaptive presentation is performed by the Tutoring Component, which is responsible for showing to the student the parts of the theory that s/he does not know or that s/he wants to read about.

For example, in an ITS for music a student may be asked the following question: "Who is a representative artist of genre Entechno?" If the student does not have a ready answer because s/he may not possess the appropriate piece of factual knowledge for this answer, s/he may answer: "My guess is "Alexiou Haris". I know that "Alexiou" performs the song with title "Aspro Peristeri"; this song is similar to the song with title "Ah ourane mou" which is representative to genre with label "Entechno". Therefore, it is likely that "Alexiou" would be representative of genre "Entechno" too." In this case, the data that is going to be used for the overlay model consists of the following correct facts that the user has shown evidence of knowing:

- The artist "Alexiou" performs the song with title "Aspro Peristeri".
- The song with title "Aspro Peristeri" is similar to the song with title "Ah ourane mou".
- The song with title "Ah ourane mou" is representative to genre with label "Entechno".

6 Conclusions and Future Work

INMA is an authoring tool designed to assist human tutors to create their own ITSs on music. Student modelling of the resulting ITSs is based on HPR, a cognitive theory with inference mechanisms that simulate plausible human reasoning (in this case students' reasoning). However, HPR assumes the existence of a detailed domain representation in the form of hierarchies. In order to reduce the cognitive load of prospective authors in the creation of these hierarchies, we have developed a module that assists tutors in organising their databases in the form of hierarchies. This module uses techniques from content-based retrieval and semantic meta-data.

INMA has been fully implemented and has been partly evaluated. However, the full evaluation of an authoring tool needs many experiments that can take place during different phases of its operation (use of the Knowledge Representation Assistant by human tutors, use of the rest of the components by human tutors, use of resulting ITSs by the respective students, diagnosis of the students' errors), and concern different aspects of the authoring tool. The full evaluation of INMA is within the immediate future plans of this research and will be reported shortly.

References

[1] T. Boyle, *Design for Multimedia Learning,* Prentice Hall, 1997

[2] A. Collins, and R. Michalski, "The logic of plausible reasoning: A core theory", *Cognitive Science* 13, pp. 1-49, 1989

[3] A.S. Lampropoulos, D.N. Sotiropoulos and G.A. Tsihrintzis, "Individualization of Music Similarity Perception via Feature Subset Selection", *IEEE, International Conference on Systems, Man & Cybernetics 2004*, The Hague, Netherlands, October 10-13, 2004.

[4] A.S. Lampropoulos, and G.A. Tsihrintzis, "Semantically Meaningful Music Retrieval with Content-Based Features and Fuzzy Clustering," *5th International Workshop on Image Analysis for Multimedia Interactive Services*, Lisbon, Portugal, 2004.

[5] Y. Lu, et al, "A Unified Framework for Semantics and Feature Based Relevance Feedback in Image Retrieval Systems", In *Proceedings of ACM MULTIMEDIA 2000*, Los Angeles California, pp 31-38, 2000.

[6] T. Murray, "Authoring intelligent tutoring systems: an analysis of the state of the art", *International Journal of Artificial Intelligence in Education*, 10, pp. 98-129, 1999

[7] J. Self, 'The defining characteristics of Intelligent Tutoring Systems research: ITSs care, precisely", *International Journal of Artificial Intelligence in Education*, Vol. 10, pp. 350-364, 1999

[8] J.C. Stansfield, B. Carr, and I.P. Goldstein, "Wumpus advisor I: a first implementation of a program that tutors logical and probabilistic reasoning skills" At Lab Memo 381. Massachusetts Institute of Technology, Cambridge, Massachusetts, 1976

[9] G. Tzanetakis and P. Cook "Musical Genre Classification of Audio Signals" *IEEE Transactions on Speech and Audio Processing*, 10(5), July 2002

[10] M. Virvou, "A Cognitive Theory in an Authoring Tool for Intelligent Tutoring Systems" *IEEE International Conference on Systems Man and Cybernetics 2002 (SMC 02)*, Vol. 2, pp. 410-415, Hammamet, Tunisia, October 2002

[11] WEKA: http://www.cs.waikato.ac.nz/ml/weka

ALIMOS: A Middleware System for *A*ccessing Digital Music *LI*braries in *MO*bile *S*ervices

P.S. Lampropoulou, A.S. Lampropoulos, and G.A. Tsihrintzis

University of Piraeus, Department of Informatics,
80 Karaoli and Dimitriou St, Piraeus 18534, Greece
{vlamp, arislamp, geoatsi}@unipi.gr

Abstract. We present a software tool, called ALIMOS, for accessing digital music libraries in mobile telecommunication services. The tool provides a semi-automatic interface that allows users to interact with their digital music library in a flexible way based on a combination of content and semantic information. The system architecture is based on a multi-tier application with a *front-end* and a *back-end* level. ALIMOS is fully implemented and relevant evaluation results are given.

1 Introduction

The term *mobile services* refers to services requested by a user of a mobile network through his/her mobile device. Early stage mobile services were centered on voice communication (i.e., telephony) and were available only in telecommunication networks. However, modern mobile phones offer more than just voice communication and provide access to a vast array of data and services over the Internet, which include multimedia (music, video, etc.) data accessing and browsing. This of course gives rise to the need for developing appropriate layers of software (*middleware*) as interfaces between a mobile device and data depositories.

In this paper, we present a semi-automatic interface that allows users to interact with their digital music libraries in mobile services. Specifically, we have created a software tool, called ALIMOS (for *A*ccessing *Li*braries in *Mo*bile *S*ervices), which automatically organizes a collection of music files in a flexible way. ALIMOS relies extensively on artificial intelligence techniques, such as clustering, machine learning, and semantic networks, to combine content and semantic information to retrieve information from music databases.

Specifically, the paper is organized as follows: Section 2 reviews previous related works on audio content analysis and existing digital music library systems, while Section 3 presents the ALIMOS architecture and related mobile technologies used in it. Section 4 describes the operation of ALIMOS and presents results from its evaluation by 100 users . Finally, conclusions and future research directions are given in Section 5.

B. Gabrys, R.J. Howlett, and L.C. Jain (Eds.): KES 2006, Part I, LNAI 4251, pp. 384–391, 2006.
© Springer-Verlag Berlin Heidelberg 2006

2 Related Work

Audio content analysis and related meta-data: In many recent works, various features and methods of audio content analysis have focused on the problem of *automatic musical genre classification* to provide techniques to organize digital music into categorical labels created by human experts [1,2]. Even though this is a difficult process, same genre music signals have been observed to share certain common characteristics. This latter fact is promising for the extraction of reliable audio meta-data in digital music libraries.

There exist two kinds of audio meta-data, namely *content-independent* and *content-dependent* meta-data. The information in the former, typically contained in id3 tags, is subjective and textual and offers a semantic (high-level) description of music files. On the other hand, content-dependent meta-data are objectively and automatically extracted from raw audio data and assign a low-level feature vector to each music file [3,4].

Considerable performance boost in music retrieval can be achieved by combining the audio (meta-)data organization with a relevance feedback scheme, that is an iterative process at each stage of which retrievals are updated and refined on the basis of user-supplied feedback. Relevance feedback schemes are generally classified into two approaches, namely query point movement (query refinement) and re-weighting (similarity measure refinement) and usually result in updated retrieval sets which are (perceived by the specific user as) more similar to the query.

Review of existing digital music libraries: A number of digital music library implementations have recently appeared in the literature, which include VARIATIONS [5], JUKEBOX [6], PATRON [7], The New Zealand digital library of music and audio data and the SMILE system [8], more recent systems for music delivery on the Web, like NAPSTER [9] and GNUTELLA [10], and MUSESCAPE [11]. These systems are characterized by limitations, such as platform-specificity or requirements for specific hardware-implemented decoders. NAPSTER and GNUTELLA are even accused of copyright law violation. Finally, MUSESCAPE is limited to desktop applications and does not allow access to existing libraries. Moreover, its music file organization relies only on objective features and does not allow the use of knowledge engineering-based techniques, such as construction of user models via semantic networks and user-supplied relevance feedback.

Some of the limitations of these systems were alleviated in a middleware tool for *web*-based digital music libraries developed in the recent past by Lampropoulos, Lampropoulou and Tsihrintzis [12]. The core of this tool was a content-based music retrieval system, earlier developed by Lampropoulos and Tsihrintzis as a desktop [13] and Web [14] application. During the development of the various components of this tool, a number of challenges arose, which were addressed successfully. Some of these challenges included : (1) developing new approaches to music genre classification based on the features extracted from signals that correspond to distinct musical instrument sources [16,15], (2) modelling the subjective perception of similarity between music files by incorporating into the system innovative relevance feedback processes [17], and (3) using artificial immune system-based clustering and classification algorithms to provide a clearer and more compact revelation of the intrinsic similarities in the music database [18].

3 ALIMOS Architecture

The ALIMOS architecture is illustrated in Fig. 1 and was developed as a multi-tier application with a *front-end* and a *back-end* level. Specifically, the front-end level includes all those modules that implement the communication between the user, i.e., the mobile network, and the application, while the back-end level refers to all those modules that implement the content- and semantics-based retrieval mechanism.

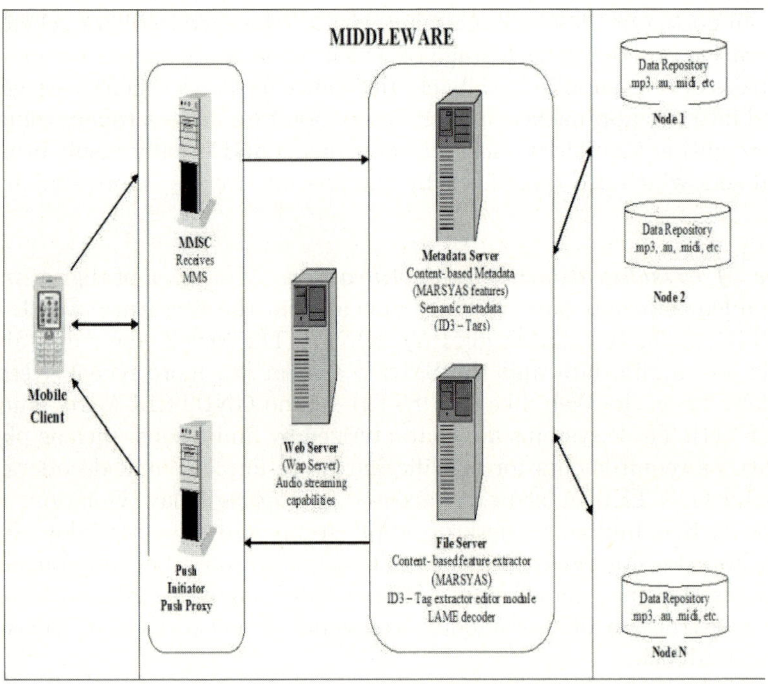

Fig. 1. ALIMOS Architecture

3.1 Mobile Technologies for the Front-End Level

The user submits a Multimedia Messaging Service (MMS) containing the query audio file and the server extracts it and stores the necessary user information (e.g., sender's telephone number). Next, the server submits the file to the

Content-Based-Retrieval (CBR) sub-system of the back-end level, which returns to the Push Proxy Gateway (PPG) a list of the five most relevant results. The PPG forwards this list to the user in the form of a Push Message. The results are not the actual audio files but their names, that is links to the WAP server. The user may download from the WAP server the audio file that he/she wants.

Push proxy protocol: In the "normal" client/server model, a client requests a service or information from a server, which in turn responds with transmitting information to the client. This is known as "pull" technology, as the client "pulls" information from the server [19]. Browsing the World Wide Web is a typical example of pull technology, where a user enters a URL (i.e., the request) which is sent to a server and the server returns with sending a Web page (the response) to the user. On the other hand, in the "push" technology, which is also based on the client/server model, there is no explicit request from the client before the server transmits its content. The WAP push framework introduces a means to transmit information to a device without a previous user request. Another way to say this is that whereas "pull" transactions of information are always initiated by the client, "push" transactions are server-initiated [20].

The push framework: A push operation in WAP is accomplished by allowing a Push Initiator (PI) to transmit push content and delivery instructions to a Push Proxy Gateway (PPG), which in turn delivers the push content to the WAP client according to the delivery instructions. The PI is typically an application that runs on an ordinary web server and communicates with the PPG using the Push Access Protocol (PAP). The PPG uses the Push Over-The-Air (OTA) Protocol to deliver the push content to the client [21].

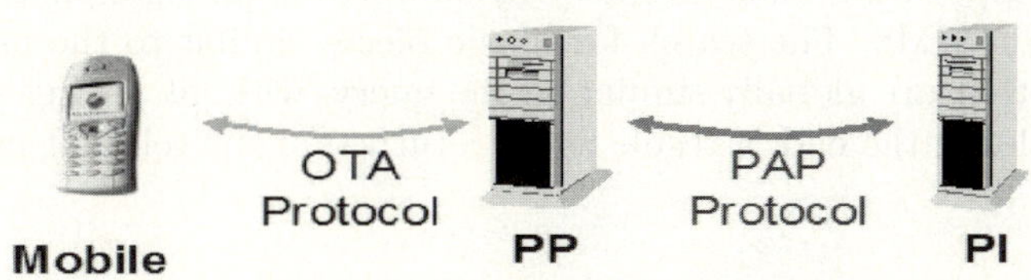

Fig. 2. Front-End Architecture

The push proxy gateway: The PAP is based on standard Internet protocols: XML is used to express the delivery instructions and the push content can be any of the Multipurpose Internet Mail Extension (MIME) type. These standards help make WAP Push flexible and extensible. In delivering the push content to the client, the PPG may potentially need to translate the client address provided by the PI into a format understood by the mobile network, transform the push content to adapt it to the client's capabilities, store the content if the client is currently unavailable, etc.

Realization of the front-end level: The front-end level of ALIMOS consists of three modules, as follows. The first module is a server that receives the MMS. This server is the implementation of a Multimedia Messaging Service Center (MMSC). This module is also responsible for extracting the audio clip from the user's MMS and sending it to the back-end level for the search to be processed. The second module is a Push initiator. This module receives the search results and submits it via the push framework to the third module, which is a PPG (Push Proxy Gateway) and its main duty is to push contents, (i.e., search results) from the Internet to the mobile network.

3.2 Back-End Level

The back-end level of ALIMOS includes a *feature extractor* that captures low-level audio features [13] and a Lame decoder that normalizes the format of all music files into a database-wide uniform sampling rate. The second module is an *id3 tag extractor* and an interface for the user to update the extracted semantic meta-data and add other possible semantic keywords. This module follows the framework developed for *image* retrieval in [23] and unifies semantics and content-based features. It links music files into a semantic network and allows user-supplied relevance feedback on both music semantic keywords (herein artist name, genre, etc.) and the music content-based features. The third module realizes the process of *clustering*, which is used for the initialization of ALIMOS and applied on low-level content-based features. Clustering reveals the predominant classes in the database and allows the user to annotate clusters with keywords of semantic knowledge such as albums, artists, etc. and to create semantic information even for those files to which no initial semantic meta-data are annotated (e.g., as in the case of empty id3 tags). The fourth module realizes the *retrieval process*. The application makes a query by submitting an example that the user has sent with a MMS. The search for music pieces similar to the user example is is for data that are globally similar to the query, without paying attention to the query scale. In the end, a result list is returned of the relevant music files in the database.

4 ALIMOS Operation and Evaluation

In ALIMOS, the user chooses first a query audio file from files stored in his/her device and sends it as a MMS to the service provider number. ALIMOS retrieves relevant files and returns a message to the user, which contains the names of the audio files as links (Fig. 3). The user may choose to download desired audio files in the result list through WAP.

To assess the performance of ALIMOS, we conducted and evaluation with 100 users belonging to different age groups and having various music style preferences and related music education levels. The evaluation process consisted of two stages: At first, each participant was asked to complete a questionnaire in which user background information was collected for statistical purposes, as summarized in Table 1. Next, each participant was asked to use ALIMOS to access

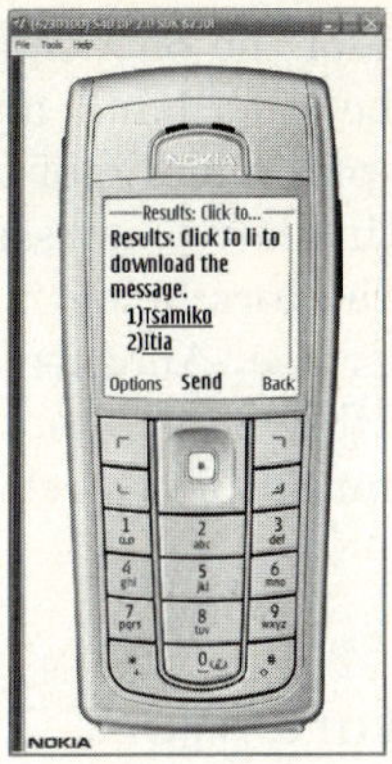

Fig. 3. Typical Receive Result

Table 1. ALIMOS Evaluation Participant Statistics

Favorite Music Genre	Dimotiko (7%), Laiko (43%), Rebetico (11%), Entechno (39%)
Overall Favorite Music Genre	Pop (42%), Laiko (35%), Entechno (23%)
Age Range	21 to 57
CDs Owned	10 to 600
Hours Spent per Week Listening to Music	1 to 70
Where Get Music *Usually* From	Radio/TV broadcasts (33%), MP3 (31%), CD/DVDs (27%), live concerts (6%), vinyl disks (3%)
Play Musical Instrument	39%, mostly the guitar or the piano
Professionally Involved in Music	3%
Rank Effectiveness	see Table 2 below
Previous Participation in Evaluation	47%

Table 2. Overall ALIMOS Performance Assessment

Ranking	**Excellent**	**Very Good**	**Good**	**Not Helpful**	**Misleading**
Percentage	22%	32%	27%	17%	2%

a data set (music library) of 1100 music pieces from Greek songs of 4 genres (Rebetico, Dimotiko, Laiko, and Entechno).

When asked to rank the overall ALIMOS performance, the participants responded as in Table 2. As seen, ALIMOS was ranked as "good", "very good" or "excellent" cumulatively by 81% of the participants, which is a clear indication of the success of ALIMOS. On the other hand, a percentage of 17% of the participants did not find ALIMOS helpful and 2% of the participants found ALIMOS misleading. The percentage of this latter assessment is low and may be due to reasons such as the specific evaluators being in a hurry or lacking motivation and/or familiarity with computer use. Another reason could lie with the *music database content* rather than ALIMOS (that is, the *CBMR tool*), which may not contain a sufficient number of music pieces relevant to the specific evaluators' interests.

5 Conclusions and Future Work

We presented ALIMOS, a software tool for accessing digital music libraries in mobile services. ALIMOS provides a semi-automatic interface for the user to automatically organize a collection of music files in a flexible way. ALIMOS relies extensively on artificial intelligence techniques, such as clustering, machine learning, and semantic networks, to combine content and semantic information to retrieve information from music databases. ALIMOS has been fully implemented and its operation was evaluated by 100 users.

Future related work will follow the avenue of improving further the classification/retrieval accuracy and efficiency of ALIMOS via incorporation of additional low-level music features (MPEG 7 low-level audio descriptors), alternative clustering algorithms and additional relevance feedback modules. We will also pursue a second round of evaluation, in which we will link ALIMOS to extensive libraries of Western music. The results of this and other related work will be announced shortly.

References

1. Aucoutier, J.J., Pachet F.: Representing Musical Genre : A State of the Art. Journal of New Music Research, 32(1) (2003), 83-93
2. Kosina, K.: Music Genre Recognition. PhD thesis. Hagenberg (2002)
3. Foote, J.: An overview of audio information retrieval. Multimedia Systems, **7(1)** (1999) 2–10
4. Tzanetakis, G.: Manipulation, Analysis and Retrieval Systems for Audio Signals. PhD thesis. Princeton University (2002)
5. Dunn, J.V, Mayer, C.A. : VARIATIONS: A Digital Library system at Indiana University, DL'99, Proceedings of the fourth ACM Conference on Digital Libraries, pp. 12-19, 1999.
6. Fonss-Jorgensen, E. : JUKEBOX: Final report, LIB-JUKEBOX/4-1049, Edited report no. 2, Aarhaus, Denmark, State and University Library, available at: http://www.sb.aau.dk/Jukebox/finalrep.html
7. Lyon, E., Maslin, J. : Audio and video on-demand for the performing arts, Project PATRON, International Journal of Electronic Library Research, Vol. 1 No 2, pp. 119-131, 1997.

8. Mcnab, R.J.,Smith, L.A.,Witten, I.H.,Henderson, C.L., Cunningham, S.J. : Toward the digital music library: tune retrieval from acoustic input, DL '96, Proceedings of the first ACM Conference on Digital Libraries, pp. 11-18, 1996.
9. http://www.napster.com.
10. http://www.gnutella.com.
11. Tzanetakis, G.: Musescape: A Tool for Changing Music Collections into Libraries.European Conference on Digital Libraries, (2003) 412–421.
12. Lampropoulos, A.S., Lampropoulou, P.S., Tsihrintzis, G.A.: A Middleware System for Web-based Digital Music Libraries. Procs.2005 IEEE/WIC/ACM International Joint Conference on Web Intelligence,Compiegne University of Technology, France, (September 19-22, 2005)
13. Lampropoulos, A.S., Tsihrintzis, G.A.: Semantically Meaningful Music Retrieval with Content-Based Features and Fuzzy Clustering. Procs.5th International Workshop on Image Analysis for Multimedia Interactive Services, Lisboa, Portugal (April 21-23, 2004)
14. Lampropoulos, A.S., Tsihrintzis, G.A.: Web-Based Instruction of Music Students through Internet Sources. Procs.5th International Conference on Information Communication Technologies in Education, Samos, Greece, (July 1-3, 2004)
15. Lampropoulos, A.S., Lampropoulou, P.S., Tsihrintzis, G.A.: Musical Genre Classification Enhanced by Source Separation Techniques. Procs.6th International Conference on Music Information Retrieval, London, UK, (September 11-15, 2005)
16. Lampropoulou, P.S., Lampropoulos, A.S., Tsihrintzis, G.A.: Musical Genre Classification of Audio Data Using Source Separation Techniques. Procs.5th EURASIP Conference on Speech and Image Processing, Multimedia Communications and Services,Smolenice, Slovak Republic, (June 29- July 2, 2005)
17. Lampropoulos, A.S., Sotiropoulos, D.N., and Tsihrintzis, G.A.: Individualization of Music Similarity Perception via Feature Subset Selection, IEEE International Conference on Systems, Man, and Cybernetics 2004. The Hague, The Netherlands (October 10-13, 2004)
18. Sotiropoulos, D.N., and Tsihrintzis, G.A.: Artificial Immune System-Based Music Piece Similarity Measures and Database Organization. Procs.5th EURASIP Conference on Speech and Image Processing, Multimedia Communications and Services,Smolenice, Slovak Republic, (June 29- July 2, 2005)
19. Push Access Protocol Specification, WAP-247-PAP-20010429-a, WAP Forum from http://www.wapforum.org.
20. Push Architectural Overview, WAP-250-PushArchOverview-20010703-p, WAP Forum from http://www.wapforum.org.
21. Push OTA Specification, WAP-235-PushOTA-20010308-a, WAP Forum from http://www.wapforum.org.
22. The Lame Project , Retrieved May 11, 2004 from http://lame.sourceforge.net.
23. Lu, Y., Hu, C., Zhu, X., Zhang, H., Yang, Q.: A unified framework for semantics and feature based relevance feedback in image retrieval systems. Procs. ACM MULTIMEDIA. Los Angeles, California. (2000) 31–38

Personalization Method for Tourist Point of Interest (POI) Recommendation

Eui-young Kang, Hanil Kim, and Jungwon Cho

Department of Computer Education, Cheju National University,
66 Jejudaehakno, Jeju-si, Jeju-do, 690-756 Korea
euiyoung1@hanmail.net, hikim@cheju.ac.kr, jwcho@cheju.ac.kr

Abstract. Today, travelers are provided large amount information which includes Web sites and tourist magazines about introduction of tourist spot. However, it is not easy for users to process the information in a short time. Therefore travelers prefer to receive pertinent information easier and have that information presented in a clear and concise manner. This paper proposes a personalization method for tourist Point of Interest (POI) Recommendation.

1 Introduction

Today, travelers are provided large amount of information which includes Web sites and magazines about introduction of tourist spot. However, it is not easy for users to process the information in a short time. For these reasons if users are asked to input too much information or, conversely, if they receive too much information, their ability to properly plan and to enjoy tour could be adversely affected. Therefore, users need to be able to obtain useful information with a minimum of effort and have that information presented to them in a clear and concise manner. There is a requirement for travelers to receive properly selected information in line with their situation and preference rather than large amounts of unnecessary information. For this reason, the tourist information service needs to introduce personalized technologies.

In this paper, we suggest the data possible to provide personalized tourist information, and the recommendation method for its application. In section 2, a related study is examined. In section 3, organized data elements are out-lined for the tourist POI (hereafter POI) recommendation. Section 4 describes the POI recommendation method using elements as stated above. Finally, the study is summarized advocating the direction of future development based on this system.

2 Related Work

2.1 The Metadata for POI Contents

In order to construct and offer specific contents economically, standardized con-tents are required. In many professional fields, the study of suitable metadata has been conducted.

B. Gabrys, R.J. Howlett, and L.C. Jain (Eds.): KES 2006, Part I, LNAI 4251, pp. 392 – 400, 2006.
© Springer-Verlag Berlin Heidelberg 2006

In the metadata of media, there are MPEG-7, FRBR of IFLA, and the ECHO project [1]. Additionally, Learning Object Metadata (LOM), and EdNA and KEM which are the applications of Dublin Core [2] exist as the metadata associated with education.

Since these metadata are designed for use in a specific field, they are not suited for use as attributes applicable to tourism. The tourism field should also construct metadata with an emphasis towards the unique features of tourist destinations.

There are VRA Core, CDWA and EAD in works of art [3]. These are the metadata concerning art of visual expression including the architecture. Even though the study of metadata for the tourist field is in its infancy, the architecture expressed by VRA Core can be a POI itself. Therefore, the title, subject, and location designated as attributes in this metadata were referred to in this paper.

2.2 Personalized Recommendation System

Even though each researcher has a somewhat different definition of personalized service, most would agree that the term refers to providing valuable information to the individual user. In general, personalized service is provided by recommended information.

There are various methods to filter recommended information. The typical methods include content-based filtering, demographic filtering, collaborative filtering, and rule-based filtering.

Since each filtering method has its strong and weak points, rather than using a single method, more than two filtering methods are usually mixed and used [4][5]. Data Blurring [6] solves the sparseness and cold-start problems which are drawbacks of collaborative filtering. This is done by applying collaborative filtering and content-based filtering methods, and computing unknown preference values which are known as feature values. Additionally, there are special techniques that apply associative words [7] of data-mining, entropy [8], artificial intelligence and agent [9] for increasing the accuracy of recommendation systems.

However, the realization of a recommendation system for personalized POI is a difficult task with only complex probability and professional methods. Understanding relations between data attributes is one possible way to incorporate the recommendation system through application of suitable algorithms and methods.

3 Personalized Tourist Information

This paper provides a brief outline of a study conducted regarding personalization, particularly, POI information which significantly impacts on tourist scheduling. Personalized POI recommendations require two basic elements. These basic elements consist of information regarding the POI and information concerning the user.

3.1 POI Attributes

The analyzed POI information is divided according to attributes as shown in Fig 1. These attributes are objective and subjective. Objective attributes are based on general tourist information such as POI location. Subjective attributes include information based on the visitors' opinion with regards to the particular POI. This information influences the consideration given in constructing a final recommendation.

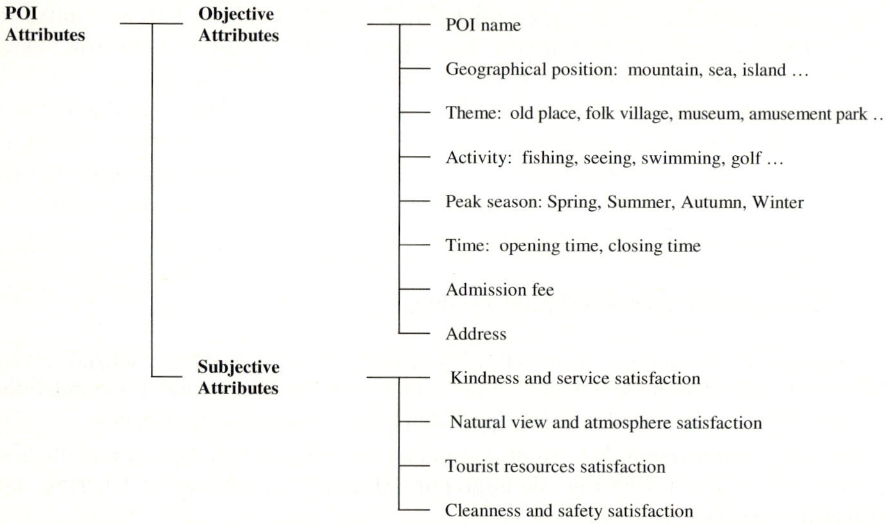

Fig. 1. POI Information

3.2 User Attributes

Even with the same tourist information, each user can choose different POIs. The attributes of the user which affect the selection of POI consist of personality, attitude, demographic, and social factors. Personality and attitude factors are defined as personal activities and preferences. Demographic factors include age, gender, and income. Also, social factors are conditions related to external motivations and companions [10]. User preference attributes about the POI information are contained by personality and attitude factors. They have a definite impact on POI recommendation in this method. Demographic factors and social factors are defined as general in nature and are distinguished from preference attributes.

It is desirable to minimize the input of the user when the characteristics of this system, as referred to in Section 1, are considered. As such, general attributes are classified into static attributes and dynamic attributes according to whether the user needs to set input data whenever he would like to travel (Fig 2). Static attributes only require input during initial registration because they don't change every time the user goes sightseeing.

However, dynamic attributes receive the user's input every time the user desires a recommendation because they often change according to conditions of the user at the time of tour.

After preference attributes are provided by the user initially, they are adjusted automatically with the tourist history of the user and the log in information. Users can modify preferences as they deem necessary. It had two steps to provide simple input for user.

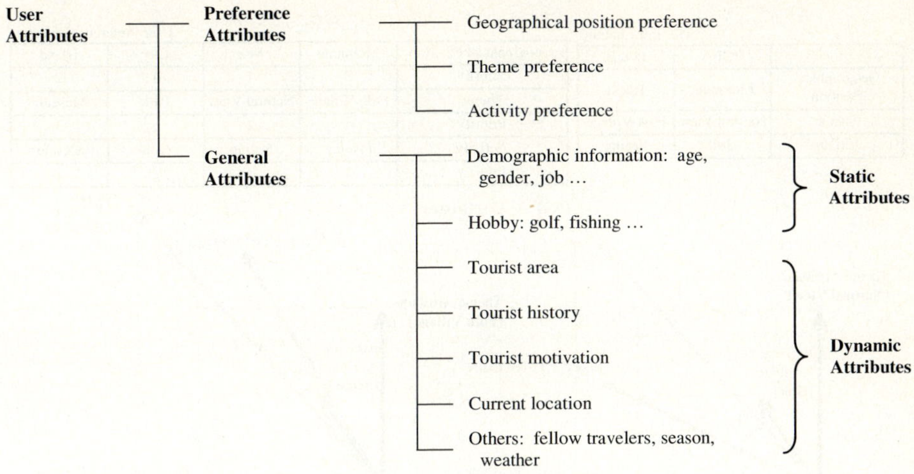

Fig. 2. User Information

- When user joins member in this system at first, they are required their necessary information on the Web.
- If they want to be recommended tourist POI, it requires only changing *optional values* about *parts of dynamic attributes* and *priorities of Preference attributes.*

4 Personalized POI Recommendation Method

4.1 Computing Algorithm of the Similarity

An example can be provided with attributes as outlined above. It is supposed that there are POI A, B, and individual preference attributes as shown in Fig 3. Then, attributes of POI are assumed to have the best preference value. And then, the user preference was compared with it.

The formula for computing similarity applies vector similarity that shows good performance on measurement of similarity [11]. Vector similarity (1) measures the similarity with cosine value between two vectors. In this case, it has a problem in that it computes the distributed pattern similarity between one vector set and the other vector set rather than the value similarity between individual vector elements. To resolve this problem, the similarity formula that applies length ratio of vectors, considering their size is shown in Formula (2).

$$VS(\overrightarrow{P_u}, \overrightarrow{O_t}) = \cos\theta = \frac{\overrightarrow{P_u} \bullet \overrightarrow{O_t}}{|\overrightarrow{P_u}| \times |\overrightarrow{O_t}|} \qquad (1)$$

$$VS(\overrightarrow{P_u}, \overrightarrow{O_t}) \times \frac{|\overrightarrow{P_u}|}{|\overrightarrow{O_t}|} = \frac{\overrightarrow{P_u} \bullet \overrightarrow{O_t}}{|\overrightarrow{O_t}| \times |\overrightarrow{O_t}|} \qquad (2)$$

$\overrightarrow{P_u}$: *Preference of user u,* $\overrightarrow{O_t}$: *Objective attribute of POI t*
$\cos\theta$: *Pattern similarity of user u to POI t*

POI Information

	POI A	POI B
Geographical Position	Mountain	Island
Theme	Natural View	Folk Village
Activity	Golf	Seeing

User Preference Attribute

Geographical Position	Mountain	Sea	City	Island
Priority	3	4	1	2
Theme	Folk Village	Natural View	Park	Museum
Priority	4	2	1	3
Activity	Fishing	Seeing	Golf	Casino
Priority	4	3	2	1

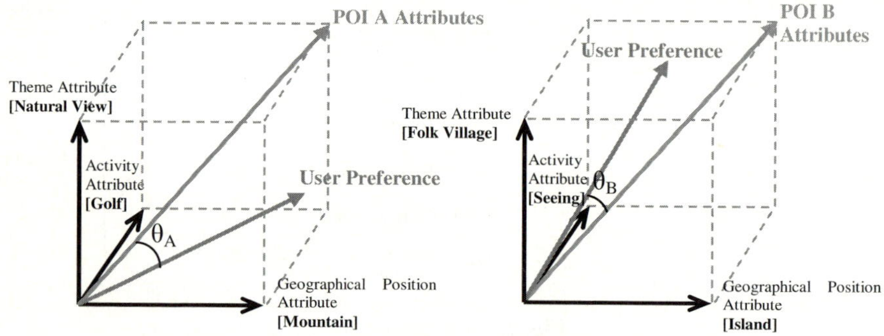

Fig. 3. Computing POI Similarity

4.2 The First POI Selection

- **Selection based on POI Objective and User Preference Attribute**

Recommendation should be made according to similarity. The similarity between the user preference attribute and the POI objective attribute was computed. Then, the first selection of POI was made. Results of the above formula show the closer to '1', the greater the similarity between the POI attribute and the user preference attribute. Through this, top N of the most similar POI was first selected.

- **Selection based on Similar User**

There are various methods to locate and apply similar users. In this system, similar users are divided into similar users of general attributes and preference attributes.

Users with similar general attributes are users with similar demographic and social factors. Users with similar preference attributes are users with similar preferences for objective attributes with regards to POI. A particular POI is selected by using data-mining of the tourist history of these similar users. Thus, the second selection is made through collaborative methods.

4.3 Final POI Recommendation

POIs as a result of the 1st and 2nd selections are sorted by weight to construct a list (Fig 4). Throughout the entire process, log in information is accumulated for future reference.

Fig 5 presents the user interface of the above recommendation system. If the user inputs his preference, as shown right of Fig 5, similarity is computed. The user has about 0.58 of similarity in POI A, and 0.75 in B. Consequently, POI B is recommended to the user.

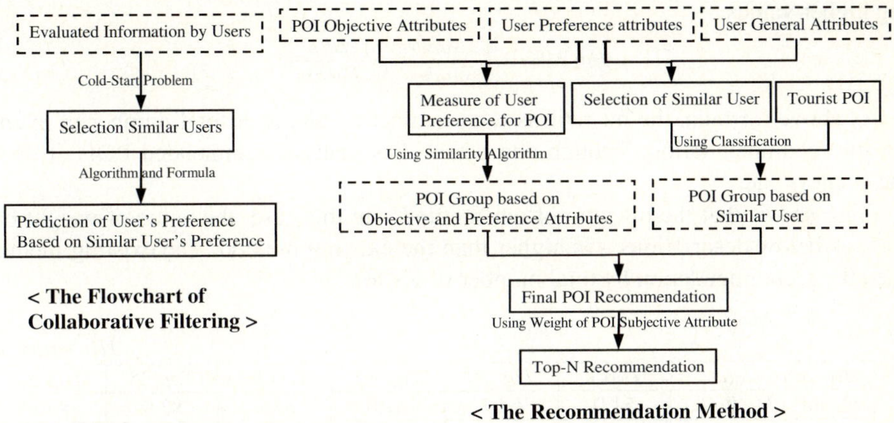

Fig. 4. The Flow of Recommendation Method

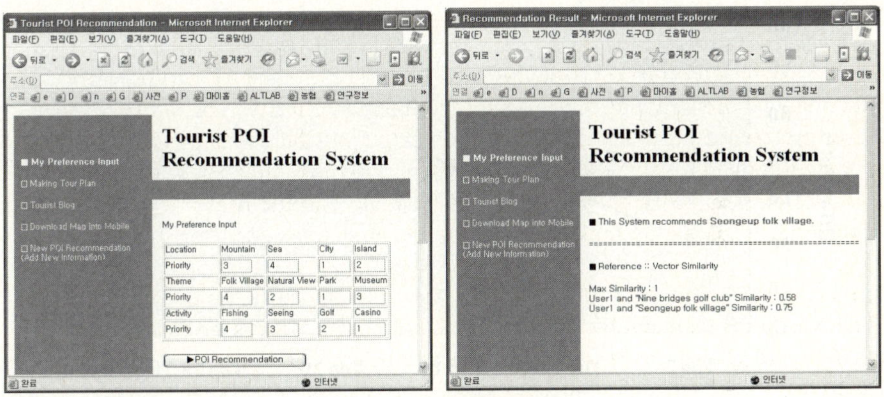

Fig. 5. The User Interface of POI Recommendation System

5 Results

In this section, we present results about the proposed tourist recommendation method. We used the tourist data in Jeju-do Tourist Association, Republic of Korea. And the user profiles are collected by the survey system on the web. With these data, we evaluate the hit-ratio and find out the optimal weight for our methods.

The evaluation proceeds as follows: First of all, we have tests for hit-ratio. We define the tourist features of POIs according to the metadata. Also, the user data are collected by using the survey system on the Web. And then we apply the Personalized

POI Recommendation Method to these data. With the results of Top-N recommendation, we assume hit-ratio between the recommended POIs and the selected POIs by the user. In this paper, the Top-N defines the number of recommended POIs by sorting higher values.

$$Hit\ ratio = \frac{Number\ of\ Hits}{Number\ of\ Events}$$

As above formula, the hit-ratio is total number of hits over total number of events for the recommendation. Though one person has many recommended POIs at once, the event is one.

Fig 6 shows that the hit-ratio of our method – in this case, the value of weight has '0.1' as below descriptions – is higher than the existing method. The existing method used the recommendation by total number of visitor.

Hit-ratio(%)

Hit-ratio	*Top 7*	*Top 10*	*Top 15*	*Top 20*	*Top 25*	*Top 30*
Existing Method	51.0	62.2	67.8	83.9	83.9	83.9
Our Method	53.3	68.9	77.8	84.4	91.1	97.8

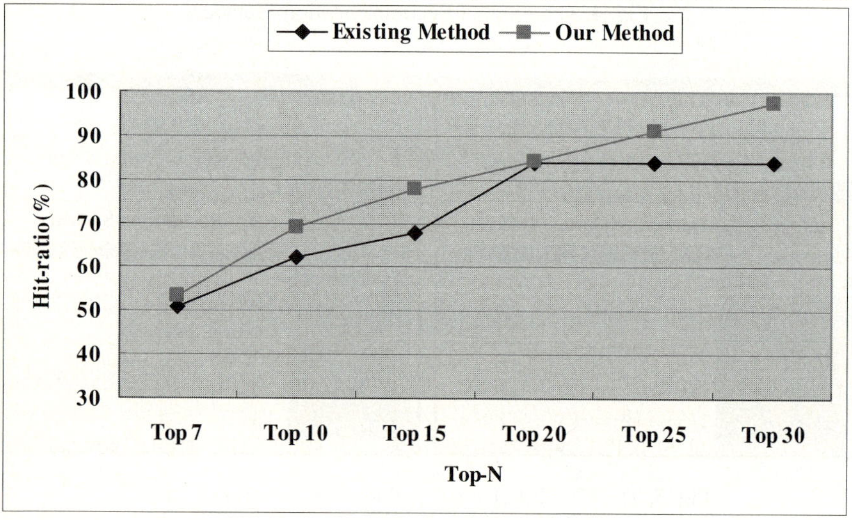

Fig. 6. Hit-ratio between the Our Method and The Existing Method according to Top-N

The second evaluation gets an optimal weight for tourist history of the user-profile. In the user point of view, we need to consider revisiting rate to tourist resorts. Our methods have the minus weight about visited POIs. And relatively, the new POIs will have advantages. The efficient weight brings higher hit-ratio but the bad weight has lower hit-ratio. Therefore, it is very important and difficult that the optimal weight decides. As changing the weight value, we obtain the optimal weight. In Fig 7, our methods have the highest hit-ratio average at '0.1' of the weights.

Hit-ratio(%)

Weight	*0.05*	*0.1*	*0.2*	*0.4*	*0.8*
Top 7	53.3	53.3	53.3	40.0	37.8
Top 10	60.0	68.9	71.1	55.6	51.1
Top 15	73.3	77.8	75.6	66.7	62.2
Top 20	86.7	84.4	82.3	75.6	71.1
Top 30	95.6	97.8	93.3	91.1	80.0
Average	73.8	76.4	75.2	65.8	60.4

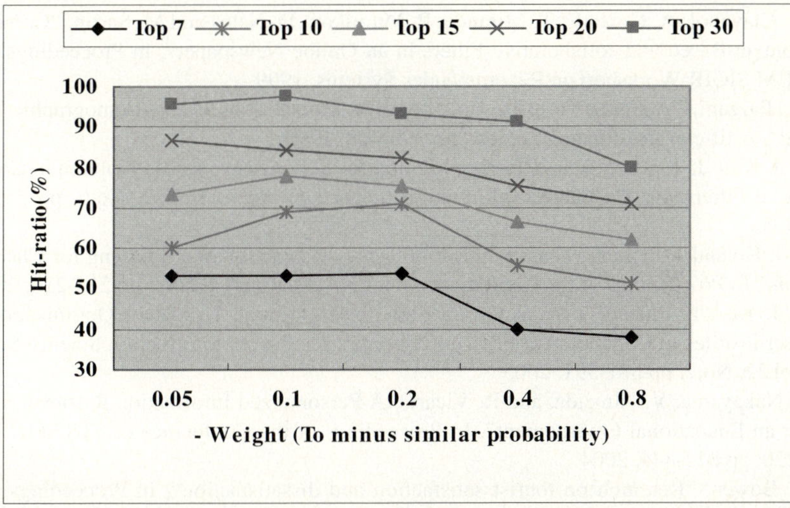

Fig. 7. Hit-ratio according to the Value of Weight

6 Conclusion

Currently, the system to introduce POI is frequently offered in the Web. However, the system is weak personalized service. Therefore, this paper has studied a method to provide a personalized POI recommendation. Through this system, individual users would enjoy an effective tour. The system also contributes to the development of related industries. Future research topics would be the realization of the real system to connect the Web services with this method. Performance evaluations of POI recommendation with real-data are necessary.

Acknowledgement

This research was supported by the MIC (Ministry of Information and Communication), Korea, under the ITRC (Information Technology Research Center) support program supervised by the IITA (Institute of Information Technology Assessment) (IITA-2005-C1090-0502-0009).

References

[1] L. A. Holm, "IFLA Model for Bibliographic Records: ELAG OO Edition 2", EALG`99-Managing Multimedia Collection-Bled, Slovenia, 1999.

[2] Y. S. Seo et al., "e-Learning Standardization Roadmap", KERIS, 2003.

[3] M. L. Zeng, "Metadata Element for object Description and Representation: A case Report from a Digitized Historical Fashion Project", Journal of the American society for Information Science, Vol.50, No.12, pp 1193-1208, 1999.

[4] M. Claypool, A. Gokhale, T. Miranda, P. Murnikov, D. Netes and M. Sartin, "Combining Content-Based and collaborative Filters in an Online Newspaper", In Proceedings of the ACM SIGIR Workshop on Recommender Systems, 1999

[5] M. Pazzani, "A Framework for Collaborative, Content-based and Demographic Filtering", Artificial Intelligence Review, pp 393-408, 1999.

[6] H. I. Kim, J. T. Kim, "Data Blurring Method for Solving Sparseness Problem in Collaborative Filtering", The Korea Information Science Society , Vol.32, No.6, pp 542-553, 2005.

[7] S. J. Ko and J. H. Lee, "Feature Selection using Association Word mining for Classification," In Proceedings of the Conference on DEXA2001, LNCS2113, pp 211-220, 2001.

[8] S. J. Ko, "Extracting Typical Group Preferences through User-Item Optimization and User Profiles in Collaborative Filtering System", The Korea Information Science Society, Vol.32, No.7, pp 581-591, 2005.

[9] L. Nakayama, V. Almeida, and R. Vicari, "A Personalized Information Retrieval Service for an Educational Environment", In Proceedings of the Conference on ITS2004, LNCS 3220, pp 842-844, 2004.

[10] D. Bowen, "Research on tourist satisfaction and dissatisfaction", In Proceedings of the Journal of Vacation Marketing, Vol.7, No.1, pp 31-40, 2001.

[11] J.Breese, D.Heckerman, and C. Kadie, "Empirical Analysis of Predictive Algorithms for Collaborative Filtering", In Proceedings of the Fourteenth Annual conference on Uncertainty in Artificial Intelligence, pp 43-52, 1998.

Architecture of RETE Network Hardware Accelerator for Real-Time Context-Aware System

Seung Wook Lee[1], Jong Tae Kim[1], Hongmoon Wang[1], Dae Jin Bae[1], Keon Myung Lee[2], Jee Hyung Lee[1], and Jae Wook Jeon[1]

[1] School of Information and Communication Engineering, Sungkyunkwan University, Korea
jtkim@skku.ac.kr
[2] School of Electrical and Computer Engineering, Chungbuk National University, Korea

Abstract. Context-aware systems, such as intelligent home-care systems or mobile communication devices that are aware of the channel environment, need reasoning ability with numerous rules to manage the current context. Reasoning techniques based on rule-based systems can be used for the efficient reasoning method of these numerous rules. The RETE algorithm has been used for the matching of reasoning rules in rule-based systems. However, the characteristics of the RETE algorithm cause it to have poor performance in Von Neumann computer systems. In this paper, we propose a novel architecture for the RETE network hardware accelerator, which provides efficient reasoning processing performance. Using the parallel RETE network hardware architecture, this accelerator can overcome the architectural constraints imposed by Von Neumann computer systems.

1 Introduction

Context-aware systems can be defined as systems which provide proper services through their grasp of the current context which is obtained from information acquired from externally sensed data[1]. These context-aware systems use the processes of combining and reasoning in conjunction with stored rules, in order to increase the amount of internal information and to select appropriate services with the context information. These reasoning functions need the same operations as rule-based systems. Namely, operations such as the updating of new information and the generation of reasoning results with the stored rules and acquired information are the same as those used in rule-based systems. The RETE algorithm proposed by Forgy is an efficient algorithm for reasoning processing in rule-based systems[2][3][4]. The purpose of this algorithm is to increase the computation performance by eliminating overlapped operations through the construction of a network of rules. However, when implementing software rule-based systems in Von Neumann computer systems, the sequential computational characteristics of the RETE algorithm cause it to incur inherent performance degradation[5][6]. To increase the computation speed of the RETE algorithm, a parallel implementation architecture based on a multiprocessor platform has been proposed[3]. However, using a multiprocessor platform for

B. Gabrys, R.J. Howlett, and L.C. Jain (Eds.): KES 2006, Part I, LNAI 4251, pp. 401 – 408, 2006.
© Springer-Verlag Berlin Heidelberg 2006

rule-based systems increases the cost required for their implementation. Namely, this implementation method is inefficient for battery powered portable device applications.

In [5][6], we discussed the need for real time context aware systems based on a hard-ware architecture and proposed the real-time system on chip architecture for rule-based context-aware computing. In this paper, we improved our conventional hardware architecture, in order to render it more efficient for real time context aware applications. We propose the architecture of the RETE network hardware accelerator (RETENHA), which can process the RETE algorithm in real-time. This RETENHA architecture can provide for the parallel processing of the RETE algorithm by interfacing with a general purpose processor such as an ARM processor. Also, RETENHA can improve the over-all system performance by reducing the computational load on the host processor. To accomplish this, the RETE network must be converted into a specific structure which can be interpreted in RETENHA. The RETENHA development system provides for this process of conversion with a software tool in the PC environment.

2 Architecture of RETE Network Hardware Accelerator

RETENHA is a hardware accelerator for processing typical rules, which is designed to keep the constraints associated with the scale and presentation structure of a given rule to a minimum. That is, by removing the constraints associated with the rule's presentation in the conventional hardware rule based system architecture, RETENHA can provide the same degree of flexibility as that provided by systems based on software platforms. Figure 1 shows the interfacing structure of RETENHA and the application system.

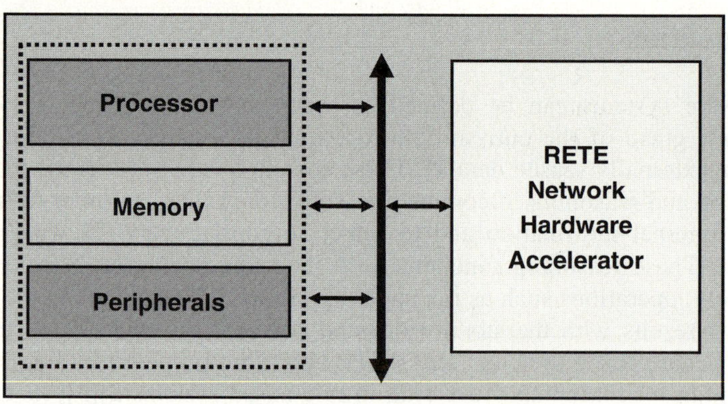

Fig. 1. Overview of Real-Time Rule-Based Systems Architecture

In figure 1, the application system provides RETENHA with the data which is used for the reasoning process, and then RETENHA transfers the reasoned results which were computed using the RETE algorithm to the application system. If this application system is an intelligent robot system which must produce the proper reaction with the current context information, the main control system in the robot converts the information acquired from the various external sensors into a specific data format which can be used in the reasoning process, and provides RETENHA with these data through the system bus. RETENHA sends the results generated by the real-time parallel reasoning process to the main control system in the robot and then the main control system in the robot converts the transferred data into an acceptable data format and initiates the appropriate reaction.

2.1 RETE Network

Figure 2 shows the typical structure of the RETE network.

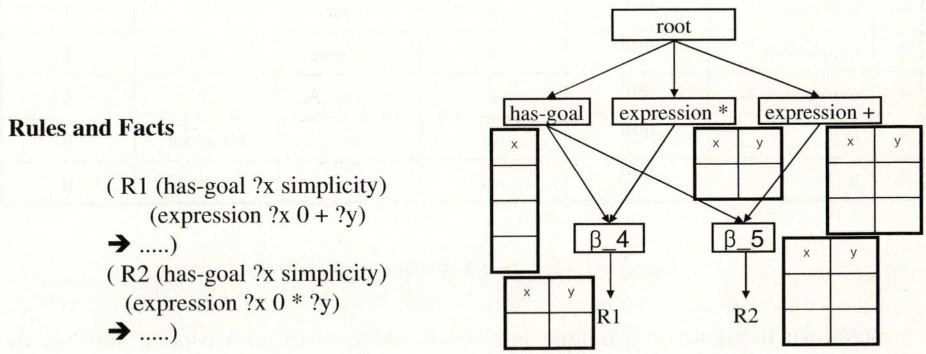

Fig. 2. Example of RETE network

In figure 2, R1 and R2 are rules to reason, which share common patterns of each rules and have both α-memory and β-memory. Here, by storing that part of the input data which has only new information in α-memory, the α-memory composes β-memory with other α-memory by means of a matching operation. This matching operation occupies 80%~90% of the overall computation load in the RETE algorithm[2]. That is, if this operation were performed in a Von Neumann computer system, it would be inefficient. When new data arrives at each α-memory, Von Neumann computer systems first confirm that this new data is not overlapped with the α-node's contents by sequentially comparing the computation with the existing data in the registers, and then compute the binding operation with the data in the remainder of the α-memory in sequential order. As described above, this process requires overlapped load/store operations between the memory and the registers, and this data transfer overhead determines the performance of the whole system. Increasing the number of rules in order to obtain more correctly reasoned results, gives rise to a rapid reduction in the performance of the whole system.

2.2 RETE Network Management Memory

In order to eliminate the overlapped data transfer operation in Von Neumann computer systems and the parallel RETE algorithm processing in RETENHA, the conventional RETE network should be converted. Figure 3 shows that the contents of the RETE network shown in figure 2 in section 2.1 are transformed in the RETE network management memory of RETENHA.

RETE network management memory						
Attribute		address of last contents	compared node	operation between nods	node to store results	number of variable to bind
1	has-goal	null	2	==	4	1
2	has-goal	null	3	==	5	1
3	expression *	null	1	==	4	1
4	expression +	null	1	==	5	1
5	β_4	null	no	==	terminal	0
6	β_5	null	no	==	terminal	0

Fig. 3. RETE network management memory

As shown in figure 3, a unique number is assigned to each α-node and β-node, which indicates the connecting relation between each node, the number of variables required for matching and the kinds of operations used in the matching process. Also, the address of the last content within each node is stored in both the α-node and β-node, in order to assign the locations of the contents in each node. By storing the address of the previous data and current data of each node in α-memory and β-memory, the system can determine the location of the previously stored contents from the address of the last content.

2.3 Parallel Matching Processing Module

RETENHA can perform the matching operation between the nodes in the RETE network by means of hardware operations. Figure 4 shows the structure of the parallel matching processing module (PPM) in RETENHA. The PPM, which is a parallel processing module, processes arithmetic computations and logic computations between the nodes in RETENHA. The matching processes can be performed in parallel and at high speed by using several PPMs. The data to match is input into the node and the new data in the node is transferred into the upper crossbar switch network, and then the switching controller configures the variables to be matched and the kinds of operations to perform in the parallel matching process by making use of the data stored in the RETE network management memory. For this configuration process, the switching operations of the crossbar switching network provide an active data transferring path. According to the results of the PPMs, each β-node determines whether the matching process was successful or not, and the lower crossbar switching network transfers the newly generated data to the RETE network management memory.

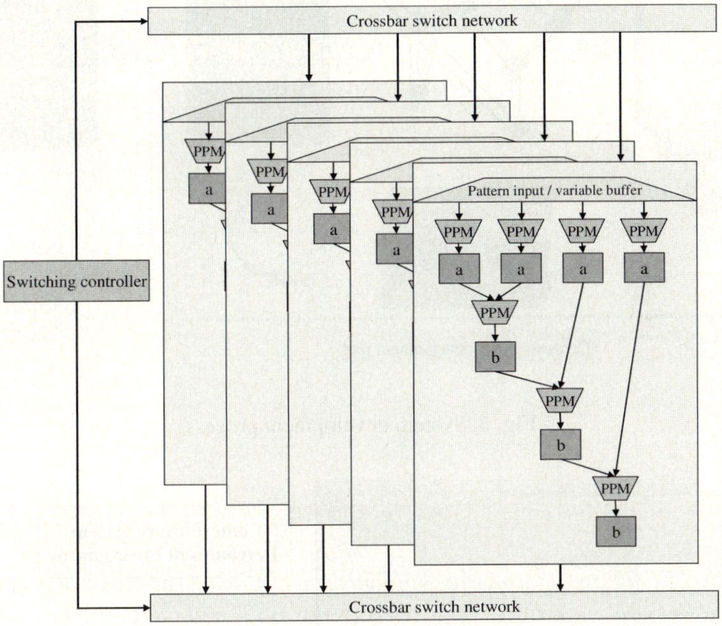

Fig. 4. Parallel matching operation module

3 System Development Process

A process of configuration is required before applying RETENHA to real application systems. Above all, rules have to be generated for the application systems. In general, the generated rules are in the form of *if~then* rules. The development tool constructs the RETE network from the generated rules and extracts the data for the RETE network management memory. Also, this development tool sets up RETENHA by converting the extracted data into a transferable format. Figure 6 shows the GUI interface of the development tool. As shown in figure 6, the development tool describes the rule's preprocessor and executable forms, converts it into a data format which is acceptable in RETENHA, and then transfers it through the external data link of the PC environment. A separate PC environment takes charge of these processing stages, as described above. After these initial processing stages, the real application system can operate as a rule-based system with RETENHA at high speed.

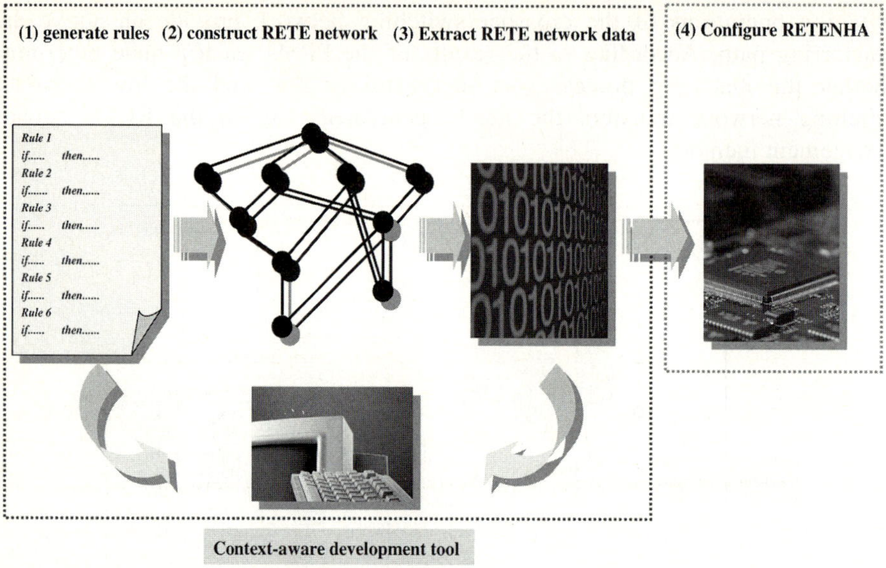

Fig. 5. System development process

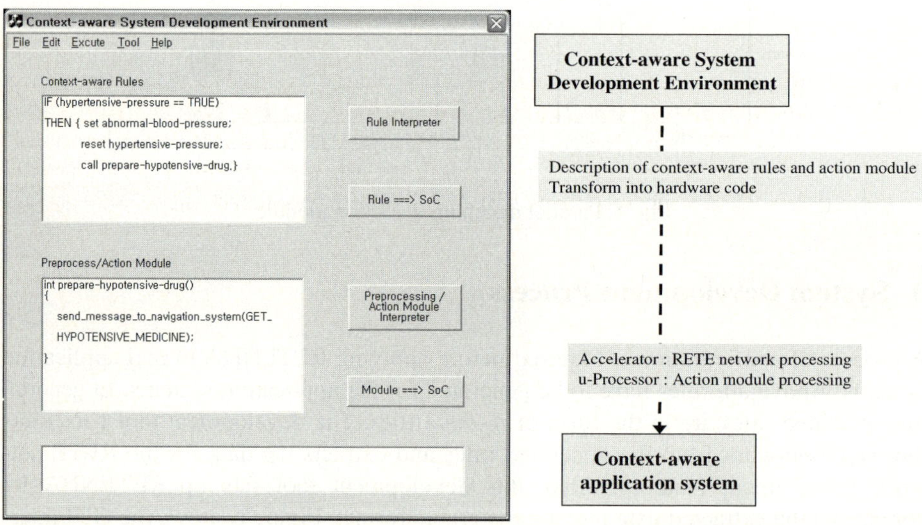

Fig. 6. GUI interface of automated compiler tool

4 Performance Evaluations

RETENHA was designed and evaluated with SystemC version 2.0, which is a hardware program language used to describe and verify systems. For the performance evaluation, we generated 500 rules which can become aware the current context of the external environment, and then compared their execution speed with that of JESS and optimal C language code. Table 1 shows the experimental results.

Table 1. Performance Evaluation Results

Estimation environment		Elapsed time
Software Rule-Based system environment	Pentium-4 2.4GHz (V-Tune)	Average 190 [msec] (OS+JESS)
	ARM Processor 400MHz (ADS)	Average 450 [msec] (OS + optimal C code)
RETE network hardware accelerator	ARM Processor 400MHz + RETENHA 10MHz (ADS)	Average 30 [msec] (OS + RETENHA)

The performance evaluation involved 500 rules described with JESS and C language that were executed on a Pentium-4 2.4GHz system, and the elapsed time was measured from the inputting of the context information to the deduction of the context aware result. The average elapsed time for the matching operation for a total of 7 rules was obtained for one deduced context aware result. We measured the elapsed time using V-Tune, which is a performance evaluation tool for Intel CPUs, for JESS and using an embedded code analysis tool within the ARM platform (ADS : ARM development system) for the optimal C code. Also, by using the verifying function of SystemC, the elapsed time for RETENHA was measured at a system frequency of 10MHz. As shown in table 1, RETENHA can operate 15 times faster than optimal C language code. Considering that this optimal C language code is operated at a system frequency of 400MHz, the performance of RETENHA operated at a system frequency of 10MHz, is impossible with the system implemented in software.

5 Conclusions

In this paper, we proposed a novel architecture for the RETE network hardware accelerator, which can be applied to real time context aware systems for reasoning computation. In RETENHA, the development tool converts the rules into the format required for the RETE network and configures the internal control path of RETENHA, which can be used for various application systems in common processes. Also, as demonstrated by the experimental results, the performance of RETENHA is superior to that of the Von Neumann computer system.

References

1. Kanter, T.G.: Attaching context-aware services to moving locations Internet Computing. IEEE , Volume: 7 , Issue: 2 , March-April 2003 Pages:43 – 51
2. C.L. Forgy: RETE : A Fast Algorithm for the Many Pattern/Many Object Pattern Match Problem. Artificial Intelligence (1982) vol. 19 17-37.

3. A. Gupta, C. Forgy, A. Newell: High Speed Implementations of rule-based systems. ACM Transactions on Computer Systems, Volume 7 Page 119-146, 1989
4. C.L. Forgy, "RETE : A Fast Algorithm for the Many Pattern/Many Object Pattern Match Problem. Artificial Intelligence (1982) vol. 19 17-37.
5. K. M. LEE, et al.: An SoC-based Context-aware System Architecture. Lecture Notes in Computer Science, Volume 3215 / 2004, KES 2004, Wellington, New Zealand, September 20-25, 2004, Proceedings, Part III
6. S. W. Lee, et al.: Real-Time System-on-a-Chip Architecture for Rule-Based Context-Aware Computing. Lecture Notes in Computer Science, Volume 3681 / 2005, KES 2005, Proceedings, Part I

A Combination of Simulated Annealing and Ant Colony System for the Capacitated Location-Routing Problem

Lyamine Bouhafs, Amir hajjam, and Abder Koukam

Equipe Informatique, Laboratoire Systèmes et Transport, Université de Technologie de Belfort-Montbéliard, 90010 Belfort Cedex France
{Lyamine.Bouhafs, Amir.Hajjam, Abder.Koukam}@utbm.fr

Abstract. Location-Routing Problem (LRP) can model several life situations. In this paper we study The Capacitated Location Routing Problem (CLRP) which is defined as a combination of two problems: the Facility Location Problem (FLP) and the Vehicle Routing problem (VRP). We propose a two-phase approach to solve the CLRP. The approach is based on Simulated Annealing algorithm (SA) and Ant Colony System (ACS). The experimental results show the efficiency of our approach.

Keywords: Location-Routing, Ant Colony Optimization, Simulated annealing.

1 Introduction

In this paper, we consider the Capacitated Location Routing Problem (CLRP) defined as a combination of two difficult problems: the Facility Location Problem (FLP) and the Vehicle Routing Problem (VRP). The CLRP can be divided in two levels: a set of potential capacitated distribution centers (DC) and a set of customers. The problem is also constrained with capacities on the vehicles. Moreover, there is a homogeneous fleet of vehicles, where each customer is visited just once. The objective is to minimize the routing and the location costs.

Many different location-routing problems have been described in the literature, and they tend to be very difficult to solve since they merge two NP-hard problems: facility location and vehicle routing. In [7], a review of early works on location routing problems presents and summarizes the different types of formulations, solution algorithms and computational results of work published prior to 1988. More recently, Min et al. [8] developed a hierarchical taxonomy and classification scheme used to review the existing location routing literature. In [10], three heuristics for the LRP are proposed; they explore the effects of several environmental factors on the algorithm performance. Tuzun and Burke [11] introduced a novel two-phase architecture that integrates the location and routing decisions of the LRP. The two-phase approach coordinates two tabu search mechanisms - one seeking for a good facility configuration, and the other a good routing that corresponds to that configuration – in

B. Gabrys, R.J. Howlett, and L.C. Jain (Eds.): KES 2006, Part I, LNAI 4251, pp. 409 – 416, 2006.
© Springer-Verlag Berlin Heidelberg 2006

a computationally efficient algorithm. In [2], a cluster analysis based on sequential heuristic that uses simple procedures was presented. Moreover, four grouping techniques (hierarchical and non hierarchical) and six proximity measures were used to obtain several versions of the heuristic.

In this paper we present a Simulated Annealing (SA) and Ant Colony System (ACS) approach to the CLRP. The approach coordinates a SA and ACS where the first (SA) seeking the good facility configuration, and the second (ACS) a good routing that corresponds to this configuration. The remainder of this paper is organized as follows. In the section 2, the CLRP is formulated. Section 3 describes the approach proposed. Section 4 describes the computational results. Section 5 concludes the work.

2 Problem Formulation

The CLRP can be represented by a graph $Gr = (S, A)$, where $S = G \cup H$. G denotes the set of nodes of potential distribution centers and H the set of nodes for the customers. For each $i \in G$, let f_i be the cost for opening the distribution center i. A facility located at site r has a capacity p_r. For each $j \in H$, q_j denotes the demand of client j. Associated to each edge $(i, j) \in A$ there is a routing cost d_{ij} which represent the distance between nodes i and j. An example of the CLRP model is shown in Fig.1.

The objective is to find a set of distribution centers to be opened and a set of routes to service the clients from the opened distribution centers in such way that the opening costs plus the routing costs are minimize. We assume that 1) the capacity of each potential depot $i \in G$ is known, 2) the fleet of vehicles is homogeneous, 3) the routes start and end at the same facility, 4) each customer is serviced by one and only one vehicle, 5) the total load of the routes assigned to a depot must fit the capacity of that depot. The objective function minimizes the total cost of routing an locating the depots.

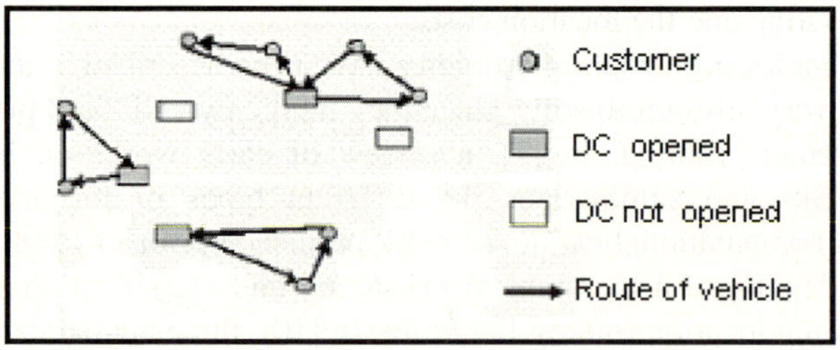

Fig. 1. Example of solution of CLRP (number of customers: 8, number of depots: 5)

3 A Simulated Annealing and Ant Colony System to the CLRP

In our approach, the CLRP is divided into two phases: facility location phase and routing phase. We propose a meta-heuristic approach based on the simulated annealing (SA) and ant colony system (ACS) to solve the combined location routing problem.

The SA is a search technique inspired by the physical process of annealing in metals, it is viewed as a probabilistic decision making process in which a control parameter called temperature is employed to evaluate the probability of accepting a move. The ACS is based on the behaviour of real ants searching for food. Real ants communicate with each other using an aromatic essence called pheromone, which they leave on the paths they traverse. If ants sense pheromone in their vicinity, they are likely to follow that pheromone, thus reinforcing this path. The pheromone trails reflect the 'memory' of the ant population. The quantity of the pheromone deposited on paths depends on both, the length of the paths as well as the quality of the food source found. The SA is used to find a good configuration of distribution centers (DC), and the ant colony system (ACS) is used to find the good routing corresponding to this configuration. The two phases will be tackled repeatedly until the total costs justify the algorithm termination. This two phase approach offers a simple and a natural representation of the CLRP. The two meta-heuristics are coordinated so that an efficient exploration of the solution space is performed.

3.1 Initialization

We represent a state of the facilities by a vector $D=\{D_1, D_2,..., D_m\}$ where $D_i \in D$, $i \in \{1,...,m\}$ if the depot D_i is opened. Initially, a randomly selected distribution center is opened, and all of the other candidate depots closed. In this case, all the clients are affected to the facility opened; the initial solution is represented by S_0 . An estimation of the location and routing cost, $Cost(S_0)$, is computed by the sum of the distances of the client to the depot plus the capacity of that depot. If the constraint capacity of the facility selected is verified (the selected facility can serve all the clients) we use the ACS algorithm to obtain the initial routing for the open facility denoted Y_0 , The total cost of the objective function, $TCost$, is the sum of the cost of the facility opened $FixedCost(S)$ and the routing cost $Cost(Y_0)$ obtained by the ACS. We denote the courant best solution by $BestSol$ and the according best cost by $BestCost$.

3.2 Location Phase

The location phase tries to find a good configuration of distribution centers that allow reducing the cost of the last configuration. For the neighborhood system, in this phase, we apply two different types of moves: swap moves and add moves. For the affectation of customers to the depots, we take a simplistic approach, and assume that each customer is assigned to the closest open facility without violating the capacity constraint of facility. A solution S represents the opened depots and the set of customers assigned to each depot without considering the routes of vehicles.

We select randomly a solution $S' \in Neighborhood(S)$, this solution is accepted if: $\Delta S (= Cost(S') - Cost(S)) \prec 0$ O $\exp(-\Delta S / T) \succ \mu$ where $\mu \in [0,1]$ a random number uniformly distributed. $Cost(S)$ is the estimation of the routing cost corresponding to the courant configuration of depots, and $\exp(-\Delta S / T)$ is Metropolis criterion. When a configuration S' is accepted, the ACS algorithm is applied to find a routing solution Y', and then a total cost $TCost'$ is calculated. If the total cost $TCost'$ is improved (i.e. less than the courant cost $TCost$) then the best solution is modified (*BestSol* ← Y', *BestCost* ← $TCost'$). The process is repeated with gradually lowering the temperature T for certain number of iteration until minimum is reached and stopping condition is verified. In the Following we detail the neighborhood system used in the location phase:

Swap moves. These types of moves open one facility currently closed and close one of the facilities simultaneously.

The swap moves explore the neighborhood to keep the number of open facilities in the solution constant, and search for a good configuration for a certain number of facilities, the location phase searches for the best swap move to perform. In order to select the best swap move, it is necessary to evaluate the cost of a swap move. The difference in the fixed cost when we open one facility and close another is straightforward (fixed cost of the facility to open - fixed cost of the facility to close). However, the difference in the routing cost is difficult to estimate. To do this, we take a simplistic approach, and assume that each customer is assigned to the closest open facility without violating the capacity constraint of facility. The difference in routing cost is then estimated using the difference in the direct distance between the customer and the facility according to the new and old assignments. The swap move evaluation is the sum of this routing cost estimate and the difference in the fixed cost. The swap move which yields the smallest evaluation is then performed, after the swap move is performed, the search resumes to the routing phase to update the routing according to the swap move.

Add moves. Having explored the configurations with the current number of facilities using the Swap moves, the search mechanism then increases the number of facilities by applying an add move.

An add move opens one of the currently closed facilities, and therefore increases the number of facilities by one. Here, the routing cost is again estimated using the difference in direct distances for the customer assignments before and after the add move. Since opening a facility can only improve the routing cost estimate, this cost is always negative. The fixed cost of the facility to be opened is then added to the routing estimate in order to calculate the overall cost estimate. As in the swap move, the search again returns to the routing phase in order to update the routing after the add move. After one add move, the search continues with a series of swap moves until the termination criterion is satisfied.

3.3 Routing Phase

After each swap or add move is performed, the routing phase is started from the best routing found for the previous facility configuration in order to modify the routing according to the current facility configuration. First, the customers are reassigned to the closest open facility with respecting the capacity constraint of the facilities. We obtain a certain number of clusters. A cluster is composed of one facility and a set of clients which are affected to this facility. We note that, the sum of customers' demand of each cluster doesn't exceed to the capacity of the facility. For each cluster, an ACS is executed to calculate the routing cost. The total cost is equal to the sum of routing cost of each cluster and the cost of opened facilities. If the total cost is smaller than the current best, we update the value of the best to the new value found and we go to the location phase. The algorithm is repeated for certain number of iterations.

In this section we describe the algorithm based on the Ant Colony System which is applied in each cluster to evaluate the routing cost:

Construction of vehicle routes. Initially, m ants are positioned on n customers randomly and initial pheromone trail levels are applied to arcs. In order to solve the capacitated vehicle routing problem (CVRP), artificial ants construct solutions by successively choosing a customer to visit, continuing until each customer has been visited. When constructing routes if all remaining choices would result in an infeasible solution due to vehicle capacity being exceeded then the depot is chosen and a new route is started. Ants choose the next city to visit using a combination of heuristic and pheromone information. During the construction of a route the ant modifies the amount of pheromone on the chosen arc by applying a local updating rule. Once all ants have constructed their tours then the amount of pheromone on arcs belonging to the best solution, as well as the global best solution, are updated according to the global updating rule.

The probabilistic rule used to construct routes is as follows. Ant k positioned on node i chooses the next customer j to visit with probability $p_k(i,j)$ given in Equation (1).

$$j = \begin{cases} \arg\max\{(\tau_{iu})^{\alpha}.(\eta_{iu})^{\beta}.(\gamma_{iu})^{\lambda}\} \text{ } if \text{ } q \le q_0 \\ J \text{ } if \text{ } q \succ q_0 \end{cases} \tag{1}$$

J is a random variable generated according to the probability distribution function given by:

$$p_{ij} = \begin{cases} \dfrac{(\tau_{ij})^{\alpha}.(\eta_{ij})^{\beta}.\ (\gamma_{ij})^{\lambda}}{\sum_{u \ne Fk} (\tau_{iu})^{\alpha}.(\eta_{iu})^{\beta}.(\gamma_{iu})^{\lambda}} & \text{if } u \in F_k \\ 0 & \text{otherwise} \end{cases} \tag{2}$$

where q is a random number uniformly distributed in $[0,1]$ and q_0 is a parameter $(0 \le q_0 \le 1)$. τ_{ij} is the pheromone associated with arc (i, j) , η_{ij} is the heuristic desirability, known as visibility, and is the local heuristic function which is the inverse of the distance between customer i and j. The selection probability is then further extended by problem specific information. There, the inclusion of savings leads to better results. The saving function [9] is represented by $\gamma_{ij} = d_{i0} + d_{0j} - g.d_{ij} + f.|d_{i0} - d_{0j}|$ where f and g are two parameters. F_k is the set of feasible customers that remain to be visited by ant k and α , β and λ are a parameters which determine the relative importance of the trails, distance and the savings heuristic, respectively. The parameter q_0 determines the relative importance of exploitation against exploration. Before an ant selects the next customer to visit the random number q is generated. If $q \le q_0$ then exploitation is encouraged, whereas $q \succ q_0$ encourages biased exploration.

Update pheromone trail. While an ant is building its solution, the pheromone level on each arc (i, j) that is visited is updated according to the local updating rule given in Equation (3).

$$\tau_{ij} = (1-\rho) + \rho.\Delta\tau_{ij} \tag{3}$$

Where ρ is a parameter $(0 < \rho < 1)$ and $\Delta\tau_{ij} = \tau_0$ the initial pheromone trail.

Once all ants have built their tours then the global updating rule is applied. In the ACS method only the globally best ant is allowed to deposit pheromone in an attempt to guide the search. The global updating rule is given in the Equation (4).

$$\tau_{ij}^{new} = (1-\rho)\tau_{ij}^{old} + \rho\Delta\tau_{ij} \tag{4}$$

Where $0 < \rho < 0$ is the pheromone decay parameter, If arc (i, j) is used by the *best* ant then the pheromone trail is increased on that arc by $\Delta\tau_{ij}$ which is equal to $1/L^*$, where L^* is the length of the tour produced by the *best* ant.

After the ant system is initialised, the steps detailed above are repeated for a given number of iterations. Once each ant has produced a set of routes the local optimizer 2-opt is applied to improve the solutions if possible.

4 Computational Results

To evaluate our approach, computational tests were carried out on 11 CLRP instances (see Table 1) obtained from the literature. Data relative to the used instances are available in http://sweet.ua.pt/~iscf143/_private/SergioBarretoHomePage.htm.

Table 1 gives the computational results for the test problems obtained by our approach compared to the best solution published, where the CLRP instance column contains information about the author, the publication year and the author and the number of customers and potential Distribution centers. The best published cost column contains the best result obtained running a set of versions of a clustering heuristic in Barreto et al. [2], and Simul-Ant cost column contains the best result obtained running our approach.

The cost is the sum of distribution costs (Euclidean distances) with the fixed costs of depots installed.

Table 1. Computational results and comparisons

CLRP Instance	Best published cost	Simul-Ant cost
Gaskell67-21x5	435.9	**430.36**
Gaskell67-22x5	591.5	**586.69**
Gaskell67-29x5	**512.1**	**512.1**
Gaskell67-32x5	571.7	**569.3**
Gaskell67-32x5_2	511.4	**506.13**
Gaskell67-36x5	470.7	**470.42**
Min92-27x5	**3062**	**3062**
Min92-134x8	6238	**6208.78**
Perl83-12x2	**204**	**204**
Perl83-85x7	1656.9	**1651.30**
Perl83Cli55x15	1136.2	**1118.43**

From the table we observe that the preliminary results obtained by Simul-Ant improve, in the most of cases, the best results published obtained by clustering heuristic. Hence, our approach is efficient and competitive.

For the setting parameters we use 10 artificial ants, $\alpha = 1, \beta = 2, \lambda = 1$, $\rho = 0.1$, $q_0 = 0.75$ and $f = g = 2$ (to the savings function) for the ACS and $T = 100$ for the temperature parameter in SA.

5 Conclusion

This paper has focused on the development of a new algorithm for the CLRP. The algorithm combines two meta-heuristics, a simulated annealing and Ant colony system which cooperating together to optimize the cost of location and routing. Preliminary results show that our algorithm outperforms other algorithms by producing the best results for the set of tested instances. The paper shows the successful application of the SA search and the ACS system to the capacitated location routing problem.

References

1. Aydin, M. E.: A simulated annealing algorithm for multi-agents systems: A job shop scheduling application, Journal of Intelligent Manufacturing, (2002)15 (6), 805 – 814.
2. Barreto, S. S., Ferreira, C. M., Paixão, J. M.: Using clustering analysis in a capacitated location-routing problem. Comunication presented at XIV Meeting of the European Working Group on Locational Analysis, (2003a) September 11-13, Corfu, Greece.
3. Bouhafs, L., Hajjam, A., Koukam, A.: A Hybrid Ant Colony System Approach for the Capacitated Vehicle Routing Problem. ANTS Workshop 2004, (2004) 414-415.
4. Clark, G., and Wright, J. W.: Scheduling of vehicles from a central depot to a number of delivery points. Operations Research 14, (1964) 568-581.
5. Dorigo, M., Gambardella, L.: Ant Colony System: A Cooperative Learning Approach to the Travelling Salesman Problem. IEEE Transactions on Evolutionary Computation, (1997) 1(1):53–66.
6. Kirkpatrick, S., Gelatt, Jr., Vecchi, M.P.: Optimization by simulated annealing, *Science*, 220, (1983) 671–680.
7. Laporte, G., Nobert, Y., Taillefer, S.: Solving a family of multi-depot vehicle routing and location-routing problems. Transportation Science 22 (3), (1988) 161–172.
8. Min, H., Jayaraman, V., Srivastava, R.: Combined location-routing problems: A synthesis and future research directions. European Journal of Operational Research 108 (1), (1998) 1–15.
9. Paessens, H.: The savings algorithm for the vehicle routing problem, Eur. J. Oper. Res. 34, (1988) 336-344.
10. Srivastava, L.: Alternate solution procedures for the location-routing problem. Omega International Journal of Management Science 21 (4), (1993) 497–506.
11. Tuzun, D., Burke, L.I.: A two-phase tabu search approach to the location routing problem. European Journal of Operational Research 116 (1), (1999) 87–99.

An Automatic Approach for Efficient Text Segmentation*

Keke Cai, Jiajun Bu, Chun Chen, and Peng Huang

College of Computer Science, Zhejiang University,
Hangzhou, China, 310027
{caikeke, bjj, chenc, oiabm6211}@zju.edu.cn

Abstract. This paper presents a domain-independent approach for partitioning text documents into a set of topic-coherent segment units, where the structure of segments reflects the patterns of sub-topics of the processed text document. The approach adopts similarity analyses, which is based on Shannon Information Theory, to determine topic distribution among text documents without incorporating thesaurus information and other auxiliary knowledge bases. It first observes the documents in terms of consistency of distribution from the viewpoint of individual word and then constructs a number of segmentation proposals accordingly. Furthermore, it employs the K-means clustering technique to get a consensus from these proposals and finally partition text into a set of topic coherent paragraphs. Through extensive experimental studies based on real and synthetic data sources, the performance analysis illustrates the effectiveness of the approach in text segmentation.

Keywords: Text segmentation, Shannon Information Theory, K-means.

1 Introduction

Observing the continuing growth of electronic text materials on Internet, end users are still demanded to make a lot of efforts to manually locate and extract useful data from web sites, even with various search engines available. Traditional search engines normally offer users a large number of retrieved documents and leave them to users to make selection. However, the increasing length and number of texts nowadays makes it a tedious process, which calls for more efficient solutions. In literature, many approaches have been proposed, such as passage-based retrieval, multi summarization and etc. One of the most fundamental prerequisite of such techniques is the recognition of the structure of text topics. The structure of texts can be characterized as a sequence of sub-topic discussions that occur in the context of a few main topic discussions [1]. In some documents, such as technical articles and academic dissertations, the sub-topic discussions can be differentiated through predefined boundaries labels. However, few explicit structural demarcations can be easily found in most

* This work is supported by the Key Science and Technology Plan of Zhejiang Province, China (Grant no. 2005C23047).

B. Gabrys, R.J. Howlett, and L.C. Jain (Eds.): KES 2006, Part I, LNAI 4251, pp. 417 – 424, 2006.
© Springer-Verlag Berlin Heidelberg 2006

general text documents. In such cases, an automatic recognition of text topical structure becomes much more important. One of the methods to address this issue is text segmentation, which partitions text into a set of relatively independent segments, each of which consists of a series of topic coherent sentences or paragraphs.

Several efficient text segmentation methods have been proposed, such as Texttiling [2], Dotplotting [3]. Although the details of the adopted strategies of above approaches are different, they all focus on locating the boundary points in text and dividing text into a series of consecutive segments. All of these methods are based on a strong assumption that sub-topics in text are presented orderly. Unfortunately, text documents are not always well structured due to many reasons, e.g., the complexity of thoughts expression. One sub-topic may be mentioned repetitiously in separated places where other sub-topics or auxiliary descriptions are intervened. These irregularities of text writing seriously weaken the effect of traditional text segmentation and consequently result in the appearance of several information-isolated islands.

In this paper, we investigate the diversity of text organization and explore the idea of text segmentation. By adopting Shannon information theory [4], a novel approach is developed for text segmentation. This approach is unsupervised and domain-independent, which requires no other auxiliary knowledge such as a thesaurus. Any keyword in text is considered as a voter and can propose its own segmentation proposal. The final segmentation schema is obtained according to the opinion majority, and each segment consists of a set of paragraphs involved in same sub-topic.

The rest of this paper is organized as follows. Section 2 reviews the previous text segmentation approaches in literature. Section 3 presents our proposed text segmentation method in details. The experimental results are presented in section 4 and section 5 concludes the paper.

2 Background

The main task of text segmentation is to break a document into topic-coherent multi-paragraph subparts. Several approaches have been proposed to solve this problem.

TextTiling [2] segments expository texts into multi-paragraph subtopics, called 'tiles', by evaluating the similarity of adjacent text blocks (normally 3 to 5 sentences). In 1994, based on lexical item repetition, Reynar proposed the Dotplotting methodology [3] to find discourse boundaries and argued that the items occurring more frequently within regions of a text describe the same topic. By combining Dotplots with divisive clustering, Choi [5] further proposed linear text segmentation by employing image-processing techniques to observe topic boundaries. Some previous works have demonstrated that lexical repetition is useful in determining the degree of cohesion between a pair of text blocks under some circumstance. However, [6] pointed out that those approaches depending on redundancy on word level will fail on the case of terse texts in which words are not repeated enough times. Thus, a topic-based text segmentation approach is proposed to make use of query expansion technique to discover common features of pairs of sentences. Lexical Cohesion Profile (LCP) [7] is another well-recognized method, which records mutual similarity of words in a sequence and further partitions the text to guarantee that words in a segment are linked together via lexical cohesion relations. Though this method is efficient in correlation identification, it needs the accessibility to existing linguistic resources.

All of the above methods assume that a certain topic always centers in a consecutive context, which consequently generate the linear text segments. In [8], Reem Khalil Al-Halimi introduced the concept of topic signals and pointed out that each topic can be randomly distributed through the whole text. However, this method is restricted to the knowledge of certain predefined topics and related descriptive information, which makes it inapplicable in many realistic conditions.

3 Text Segmentation Algorithm

The goal of our text segmentation process is to identify the same topical paragraphs throughout the text document. Firstly, a text is pre-processed by removing the words that have no specific meaning in the text [9], [10]. Secondly, the words are represented as vectors of their frequency appeared in paragraphs. The text documents are segmented based on these vectors respectively. Finally, by using the clustering technique the global segmentation of text gets consensus.

3.1 Shannon Information Theory

Shannon information theory proves that the equal probability of each information unit gives the largest possible value for the entropy. When used in the domain of text processing, it means that a word evenly distributed in a document collection implies its less importance. Let $D = \{d_1, d_2 \ldots d_n\}$ be a document set. The noise of a word k occurred in D is defined as formula 1, where fre_{ik} and $totalfreq_k$ respectively represent the frequency of k in text d_i and the total occurrence of k in D. Here, the noise of a word implies its importance to texts. The larger the noise is, the less importance it has. This definition also shows that the noise of a word is in inverse proportion to its centralization. If a word is evenly distributed in D, in other words, if the probability of its occurrence in each document is equal, its noise gets maximal. Shannon information theory gives us an inspiration that from the viewpoint of word text documents are similar if the word gets maximal noise with respect to these texts.

$$noise_k = \sum_{i=1}^{n} \frac{fre_{ik}}{totalfre_k} \log_2 \frac{totakfre_k}{fre_{uk}} . \qquad (1)$$

3.2 Segmentation Vote from Words of Text Document

In our approach each word is a voter, which is able to freely pronounce its suggestion for text segmentation and finally propose its own segmentation scheme. Figure 1 shows an example. The horizontal lines indicate the four different segmentation schemes corresponding to four words respectively. As shown in Figure 1, segmentations are not always consistent with each other completely. Each line segment is defined as a segment vote, which is generated based on the consistency of distribution identified by Shannon information theory. Formally, each word is represented as a vector, each item of which is the frequency of the word appeared in each paragraph. In the remainder of the paper, a text is denoted by T and P_1, P_2 … P_k denotes the paragraphs of T. A word w occurring in T is associated with the vector $PF_w = \{fre_{w1}, fre_{w2} \ldots fre_{wk}\}$, where fre_{wi} represents the frequency of w in paragraph i. PF_w describes the distribution of w throughout the text. According to discussion in the previous section, if a word has the maximal noise in a set of paragraphs, it is of least use for the partition of these paragraphs. In the context of PF_w, a local segmentation based on Shannon information theory is applied to P_1, P_2 … P_k and a set of segments with respect to w are formed. In each segment, w satisfies the maximal noise. Such local segmentation will be implemented to each word of the text and finally a segment vote set is constructed within which the consensus of segmentation is reached. Here the noise of term w for the paragraph set C is defined as formula 2.

$$AEntropy_w = \sum_{i=1}^{k} \frac{fre_{wi}}{totalfre_w} * \log \frac{totalfre_w}{fre_{wi}} . \tag{2}$$

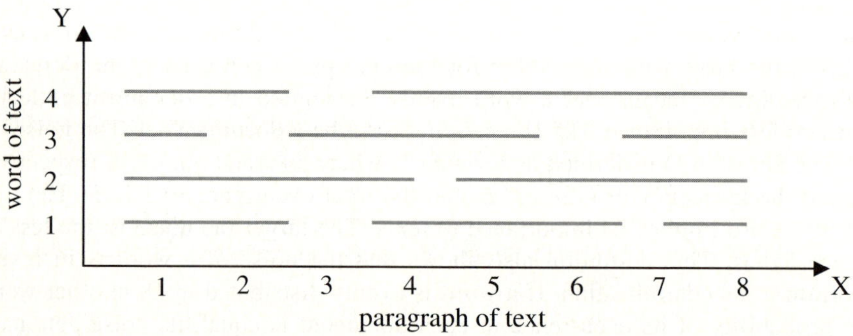

Fig. 1. Segmentation scheme from the words perspective. Horizontal coordinate indicates the paragraphs of a text and vertical coordinate shows the four distinct words in text.

Algorithm 1 describes pseudo code of the whole process of local segmentation. In this algorithm, the theoretical maximal noise $MaxEntropy_w(C)$ is defined as $log_2(k)$. If the difference between $MaxEntropy_w(C)$ and $AEntropy_w(C)$ is lower than the deviation variable σ, paragraphs of set C is seemed similar with respect to w.

Algorithm 1

```
Input:   term w, paragraphs P1, P2 … Pk of text T, vec-
         tor of w's frequency in each paragraph PFw and
         the predefined threshold
Output: a number of segment votes w related
```

```
Word_Based_Segmentation (P1, P2 … Pk, PFw, w,  ){
    sort items of PFw ascendingly;
    for each paragraph Pi{
        for each existing segment vote C{
            consider adding Pi to C
            if MaxEntropyw (Ci) - AEntropyw(Ci) <  {
                add Pi to Ci;
                break;
            }
        }
    }
    if Pi cannot be added to existing segment votes{
        construct a new segment vote G;
        add Pi to G;
    }
}
```

3.3 Paragraphs Clustering and Document Segmentation

In this paper, text segmentation is finally implemented through the technique of clustering. Here we use K-means [11], which uses a statistical test to disclose the appropriate number k of k-means. Let SV be the segment votes generated in pre-subsection. $m=|SV|$ is the number of segment votes. In the space of SV, each paragraph p is firstly characterized as an m-dimensional vector $pbv = <\mathrm{b}(p, sv_i)>$, where $i \in \{1 \ldots m\}$. If $p \in sv_i$, the value of $\mathrm{b}(p, sv_i)$ is equal to 1 otherwise 0. This vector indicates whether or not the occurrence of paragraph p is in the segment vote sv_i. To realize the similarity identification within paragraphs, the K-means clustering algorithm is applied on SV. Finally, paragraphs with the similar representation in the space of the segment votes will be grouped to the same cluster. Along with the completion of paragraphs clustering, text segmentation is simultaneously achieved and can be eventually defined as the clusters of the paragraphs.

4 Experimental Study

A simulation model of the proposed system is built to empirically evaluate the performance of the approach proposed in this paper. The test data is generated from two data sources. The first data source, as shown in Table 1, is generated according to the class definitions of Yahoo and contains documents extracted from six different classes. Each class is composed of five to ten sub-classes, from which one document is selected. The second data source includes six lengthy technical dissertations. Every one has very good topical structure characterized by a sequence of headings and sub-headings. Two test datasets are built from these two data sources respectively. The first dataset contains six lengthy documents, each of which is corresponding to one class of the first data source. The second dataset contains all documents in the second data source. Documents in the first dataset cannot be obtained directly and should be constructed as follows. A void document C_{di} with respect to each class is initially generated. Secondly, the document of certain subclass is randomly selected, from which one or more paragraphs are drawn. These paragraphs are then sequentially appended to C_{di}. Operation of the second step is iterated until the length of C_{di} reaches the maximum threshold or C_{di} has already contained all paragraphs of documents.

We use the average correctness of segments, which is defined as formula 3, to evaluate the segmentation generated by the algorithm. In this function, n is the number of texts concatenated for segmentation. Set $S = \{S_1, \ldots, S_m\}$ contains all segments after segmentation. A text T_j may be segmented into several segments of S, here N_j denotes the number of segments that T_j involves. This function considers the performance of segmentation from two aspects respectively. The first is the dispersion of each text and the second is the consistency of each segment. According to this function, the less number of the segments a text deals with and the less number of texts each segment relates, the higher the performance of the segmentation is. Here the average correctness of segmentation is calculated and the value is range from 0 to 1.

$$f(C) = \sum_{j=1}^{n} \sum_{S_i \cap T_j \neq \phi} \frac{|S_i \cap T_j|}{|S_i|} \Big/ N_j^{\,2} \Big/ n. \tag{3}$$

Table 1. A list of the 6 classes and characteristics of the six classes

ID	Six Datasets of Different Classes
1	Computer Science. Artificial Intelligence Computer Science. Database Computer Science. Distributed computing Computer Science. Graphics
2	Education. Bibliographies Education. Conference Education. Graduatio Education. Job Education. Journals Education. Teaching Education. History
3	Entertainment. Music Entertainment. Actors Entertainment. Humor Entertainment. Television Shows Entertainment. Movies and Film
4	Health. Fitness Health. Medicine Health. Nutrition Health. Disabilities Health. Diseases and Conditions Health. Emergency Services
5	Science. Artificial Life Science. Biology Science. Ecology Science. Energy Science. Chemistry Science. Agriculture Science. Geography
6	Culture. Advice Culture. Crime Culture. Fashing Culture. People Culture. Social Organization Culture. Employment and Work

Table 2 presents the segmentation results for texts in the first test dataset. From this table, it is found that the segmentation can effectively capture the distribution of each topic and exactly recognize the same-topic paragraphs. The difference of values *f(C)* in Table 2 is partially due to that the texts of some classes contain more refined subtopics and are always segmented according to the measure of the subtopics.

In the experiments, the segmentation on long text documents is also tested. The results are compared with the original pre-marked topic labels. Figure 2~6 demonstrates part of the results of the algorithm on the second test datasets. The numbers in this figure indicate the paragraphs position where one original break occurs. The difference between this algorithm segmentation and original segmentation is reflected in terms of insertions, deletions and moves. An insertion, which is reflected by the bold and underlined number, is a break that the algorithm produces but does not line up with the real break. A deletion refers to the real break that the algorithm does not recognize. In these figures, they are labeled as the bold strikethrough numbers.

Finally, a move that is represented as bold and italic number means a paragraph similar to other paragraph in different places. From these figures, it is observed that most of the correspondences were correct although there are many off-by-one-or-two paragraphs errors. Some inserted breaks are near to the true breaks. It is expected since in some dissertations some paragraphs exist as a connecting link between the preceding and the following content. These paragraphs may simultaneously involve multi-topics. Therefore, they can be assigned to different segments. If these connecting paragraphs are removed beforehand, the segmentations will be consistent. From these figures, it is also found that a number of move errors exist. It occurs because in these dissertations some topics are always mentioned multi-times although they are placed in different positions. From the topics perspective, they should be segmented together.

Table 2. Number of segments of the first test dataset and the correctness of these segments

Text Class	Number of segments	*f*(*C*)
Computer Science	6	.85
Education	12	.68
Entertainment	8	.75
Health	14	.54
Science	10	.71
Culture	9	.63

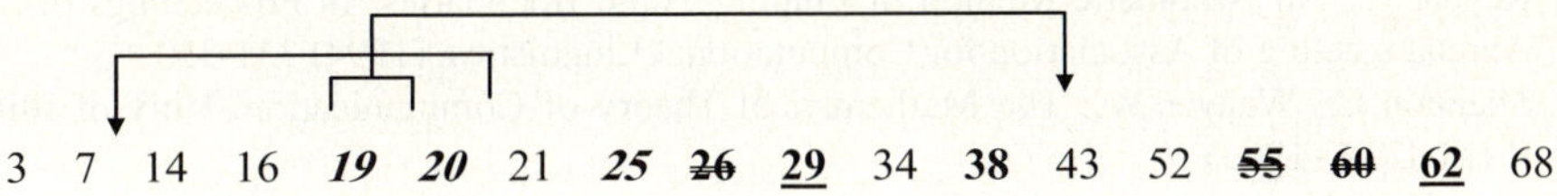

Fig. 2. Comparison of Segmentations on Dataset 1 of the second test dataset

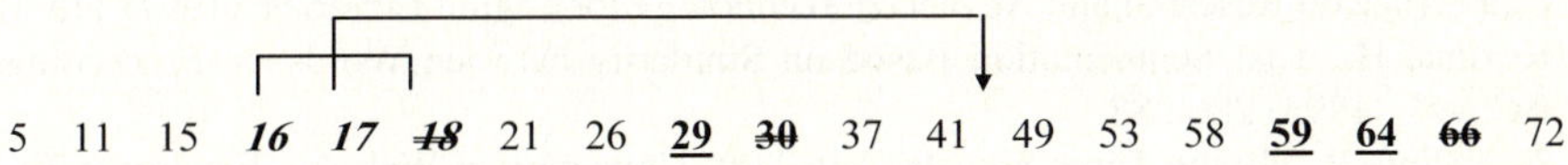

Fig. 3. Comparison of Segmentations on Dataset 2 of the second test dataset

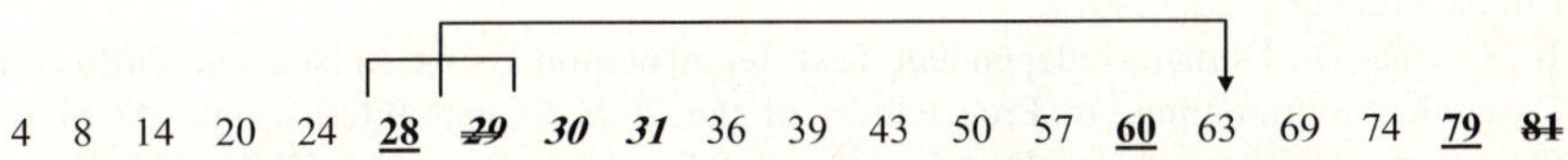

Fig. 4. Comparison of Segmentations on Dataset 3 of the second test dataset

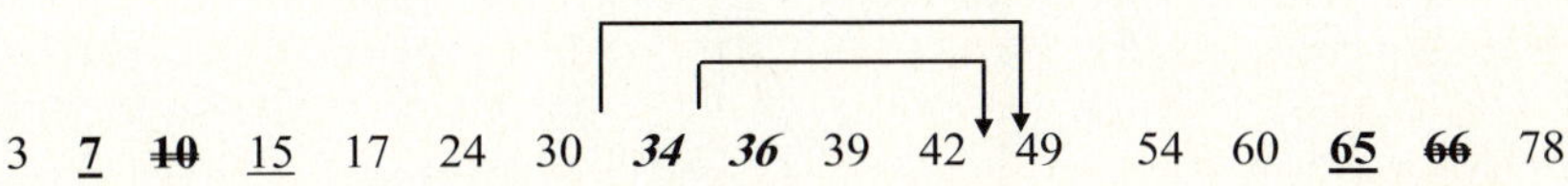

Fig. 5. Comparison of Segmentations on Dataset 4 of the second test dataset

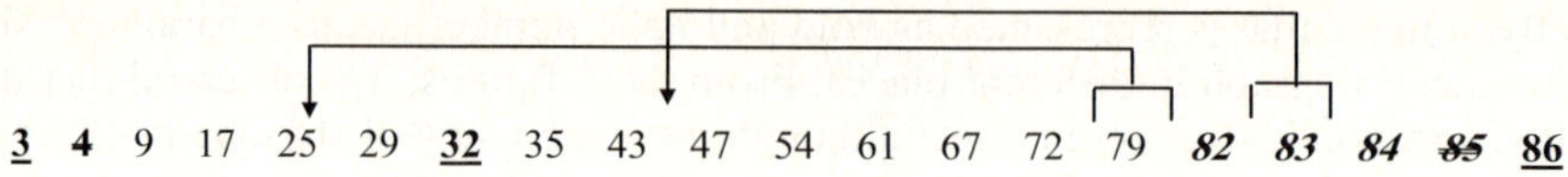

Fig. 6. Comparison of Segmentations on Dataset 5 of the second test dataset

5 Conclusions

In this paper, an efficient algorithm for the text document segmentation is presented, where the segments can reflect the topic structure of the original text document. We adopt the notion of Shannon Information Theory as well as the technique of K-means to recognize the information distribution throughout the processed text document. One advantage of this algorithm is that it can be fully implemented without using any other auxiliary knowledge. The experimental results over various texts have demonstrated the flexibility and applicability of our approach in text segmentation.

References

1. Hearst, M.: Multi-paragraph segmentation of expository texts. In Proceedings of 32nd Annual meeting of Association for Computational Linguistics, (1994) 9-16.
2. Hearst, M.: TextTiling: Segmenting text into multi--paragraph subtopic passages. Computational Linguistics Vol 23(1). (1997) 33-64.
3. Reynar, J.: An Automatic Method of Finding Topic Boundaries. In Proceedings of 32nd Annual meeting of Association for Computational Linguistics, (1994) 331-333.
4. Shannon C., Weaver W.: The Mathematical Theory of Communication. Univ of Illinois Press, USA (1963).
5. Choi F.: Advances in domain independent linear text segmentation. In Proceedings of the North American Chapter of the ACL, (2000) 26-33.
6. Ponte J., Croft W.: Text Segmentation by Topic. In Proceedings of the First European Conference on Research and Advanced Technology for Digital Libraries, (1997) 113-125.
7. Kozima, H.: Text Segmentation Based on Similarity between Words. In Proceedings of ACL-93, (1993) 286-288.
8. Al-Halimi, R. Mining Topic Signals from Text. University of Waterloo Electronic Theses, (2003).
9. Fox, C.: Lexical Analysis and stoplists. Information Retrieval: Data Structures and Algorithms, Prentice Hall, (1992).
10. Ji, X., Zha, H.: Domain-independent Text Segmentation Using Anisotropic Diffusion and Dynamic Programming. In Proceedings of the 26th Annual International ACM SIGIR Conference on Research and Development in Information Retrieval, (2003) 322-329.
11. Hamerly, G., Elkan, C.: Learning the k in k-means. In Thrun, S.(eds.): Advances of the Neural Information Processing Systems 16. Cambridge, MIT Press, (2004) 281-289.

Design and Configuration of a Machine Learning Component for User Profiling in a Declarative Design Environment

Georgios Bardis[1,2], Georgios Miaoulis[2], and Dimitri Plemenos[1]

[1] Université de Limoges, Laboratoire Méthodes et Structures Informatique
83, Rue d'Isle, 87060 Limoges cedex, France
{bardis, plemenos}@msi.unilim.fr

[2] Technological Education Institute of Athens, Department of Computer Science
Ag. Spyridonos, Egaleo, 122 10 Athens, Greece
{gbardis, gmiaoul}@teiath.gr

Abstract. MultiCAD is a design environment that generates geometric models of buildings based on abstract declarative descriptions. The increased number of solutions produced has called for an intelligent module selecting those closer to user's preferences. We have proposed and implemented such a module, featuring two components: a Decision Support Component, capturing user preferences based on attribute weight assignment techniques (used in SMART, AHP and via RR), and a Machine Learning Component, learning preferences by incrementally training a neural network committee based on user evaluated solutions. Alternative configurations must be compared before actual use of the ML Component takes place. Due to the practical limitation on the number of solutions that can be inspected and evaluated by human users, an automated mechanism plays the role of a group of virtual users. The best performing configuration, regarding virtual users' preferences, will be integrated to the system and evaluated against actual human evaluation results.

Keywords: Machine Learning, Decision Support, Declarative Modelling, Image Synthesis.

1 Introduction

Human behaviour can be confusing, contradicting or even irrational in several occasions and for diverse reasons. Although universal notions like the *survival instinct* or *love* and their variations are usually considered as the controlling forces of our behaviour, they often fail to explain the way we react to important as well as trivial matters. Either because *human entities* and their environment are too complicated to be simulated and measured respectively, or because *free will* cannot be accurately predicted (otherwise it should not really be considered *free*) it seems that, at least considering the present status of human knowledge and the corresponding technical capacity, a deterministic non-human system is not able to simulate human behaviour in its entire range. The latter becomes also evident by the fact that, to this

B. Gabrys, R.J. Howlett, and L.C. Jain (Eds.): KES 2006, Part I, LNAI 4251, pp. 425 – 434, 2006.
© Springer-Verlag Berlin Heidelberg 2006

day, Turing's Imitation Game has yet to be successfully passed by any artificial mechanism unlike its inventor's predictions [24].

It follows that, if we still need to simulate human behaviour for experimental or other reasons, we have to accept the idea of a compromise represented by *reduced model accuracy* and/or a set of *approximative assumptions*. The effort to capture and apply user morphological preferences in a design environment forms part of the effort to simulate and predict human behaviour. In the following, we present an intelligent user profile module with this aim, emphasising on its Machine Learning Component. We focus on the degree of *accuracy* we demand from our model and the *assumptions* we make while pursuing it, in order to deal with the problem of lack of adequate and reliable training data in the context of declarative modelling.

2 MultiCAD Design Framework

At the early stage of the *design* process, the designer is usually only aware of an abstract outline of the object he/she intents to create but not necessarily of its explicit geometric representation. It is usually the case that the object to be designed, depending on its functionality, cost, etc., has to comply with a minimum set of requirements, which, however, does not strictly suggest its form or the values of other attributes not affected by these requirements. In addition, the designer, at this stage, may practically be exploring alternative geometric representations of the same object idea, seeking an innovative or uncommon design, trying to conform, at the same time, to the predefined set of requirements.

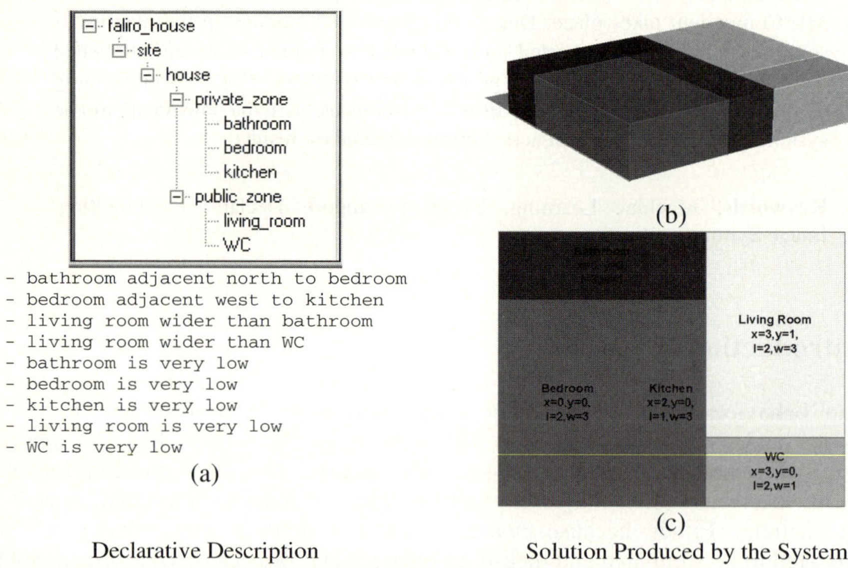

Fig. 1. Declarative description as a set of part-of (tree structure) and declarative (text) relations and properties (a) and the image of one of the corresponding solutions as produced by the system (b), schematically explained in (c)

MultiCAD [15] comprises a design framework that facilitates the transition from the abstract description of an object to its concrete visualised counterpart. At its current stage, MultiCAD is concentrated on the domain of Architecture by accepting descriptions of building *scenes*, subsequently generating and visualising alternative images representing valid geometric *solutions* for the submitted descriptions.

The scene descriptions that constitute the system's input are formulated according to the principles of Declarative Modelling by Hierarchical Decomposition (DMHD) [17]. DMHD in the current context requires the scene to be described as a tree of nodes, each representing a building object (e.g. bedroom, habitation etc.) that *encloses* its children in the form of a bounding box. This tree of *part-of* relations is enriched by declarative relations (e.g. *adjacent west, longer than*) among these objects, as well as with declarative object properties (e.g. *very high*). This information is practically the aforementioned set of requirements: every produced solution has to comply with them. An example is shown in Fig. 1 where a declarative description (a) comprised by a set of part-of (tree structure) and declarative relations and properties (text) has generated alternative geometric representations, the image of one of them appearing in (b). Fig. 1(c) provides an explanation of objects' dimensions to facilitate the cross-checking of the validity of the solution with respect to the declarative description. Notice, for example, that the bathroom is *adjacent-north* to the bedroom and the living room is *wider-than* WC (width spans along the *yy'* axis).

The solution generator mechanism of MultiCAD is an evolution of the CSP mechanism used in MultiFormes project [17], explored in detail in [4]. In particular, every declarative relation and property is translated to a set of restrictions regarding the participating objects' geometrical properties. Next, a CSP resolution mechanism is employed in order to produce all alternative solutions. The MultiCAD framework has been defined as an integrated information system in [15] and most of the modules described therein have already been or are currently being developed through a number of parallel projects. In particular, architectural styles have been modelled and incorporated to the system through an alternative solution generation mechanism, based on Genetic Algorithms [14]. Moreover, the reverse direction, i.e. the transition from the geometric representation of a *solution* to the declarative description of a *scene* that would produce this solution as output, is explored in [10]. The work in [7] presents a collaborative module for declarative description creation and manipulation. Modules for scene normalisation according to Architectural principles [13] as well as offering the ability to incorporate other domains of knowledge to the system through alternative ontology structures [20] have also been introduced. Machine Learning in the current context has previously been explored in [18] by proposing explicit modelling of each scene as a dedicated neural network.

3 Intelligent User Profile Module

User preferences have been largely explored in the domains of hypermedia and the WWW [5]. In the context of architecture, [12] presents an approach using a guided genetic algorithm and [16] a heuristic evaluation of building assemblies. [22] presents a Web system for house market assistance whereas [8] describes a module for preliminary structural building design, creating and improving a set of fuzzy rules through genetic techniques. [6] proposes an unsupervised solution classifier which, subsequently, allows the user to evaluate class representatives in order to capture his/her preferences, thus making the assumption that geometrically similar solutions share a similar degree of user preference. Our work is focused on the acquisition and maintenance of users' preferences with respect to the solutions produced by a declarative design environment, requiring minimal intervention to the existing MultiCAD system and transparency during normal system use. We allow the user maximum flexibility with respect to the approved/rejected solutions and their characteristics through a hybrid user profile module, combining Decision Support with Machine Learning methodologies in its corresponding components.

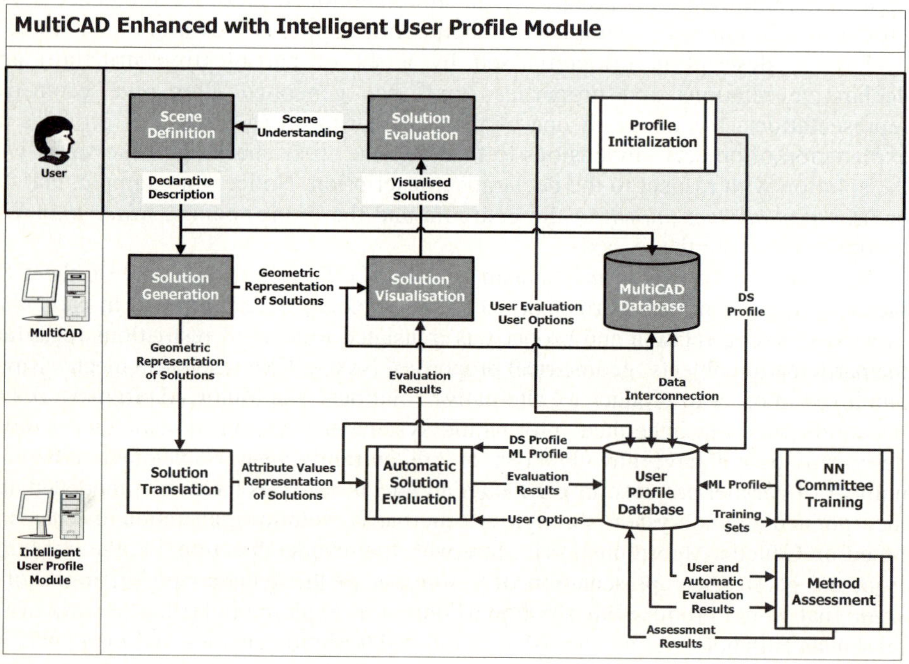

Fig. 2. Typical scenario of system use incorporating the Intelligent User Profile Module. Shaded tasks represent regular system performed by the existing system or the user.

In particular, each solution is mapped by the Intelligent User Profile module to a *vector of values* (Solution Translation task of Fig. 2) representing its performance against a set of observed morphological and geometric *attributes* [1],[2] chosen according to the corresponding building properties used in [9], [14] and [22]. The observed morphological attributes and their parameters vary for different kinds of buildings, i.e. different *project types.* Both components of the User Profile module operate on this *attribute encoded* version of the solutions.

3.1 Decision Support Component

The Decision Support Component relies on three alternative weight assignment techniques originating from the SMART [11] and AHP [23] multi-criteria methodologies as well as on the idea of standard weights assignment based exclusively on ranking, according to the Rank Reciprocal method [21]. The user initialises his/her Decision Support profile by completing a questionnaire regarding the relative importance of the observed attributes according to the aforementioned methodologies. He/she also suggests a value that represents adequate (middle) performance for each attribute. At the end of this process, three alternative *weight vectors* are available for the specific user, statically representing his/her preferences together with a *middle-values vector*. The performance of each solution is calculated according to each weight vector, as well as the middle values vector, yielding the corresponding *grade*. Each user's weight and middle values vectors constitute his/her Decision Support Profile which is available immediately after its initialisation, offering instant Automatic Solution Evaluation (Fig. 2) without prior training as well as serving as a reference threshold for the performance of the Machine Learning Component. Notice that the Decision Support Profile initialisation (appearing as a User task in Fig. 2) takes place only once for every user, at first system use.

3.2 Machine Learning Component

This is the second component of the Intelligent User Profile Module we have proposed and implemented for the MultiCAD environment. As the name states, it is responsible for adapting the system's response to the user by gradually learning the latter's preferences. It operates concurrently and transparently, during regular system use, automatically evaluating produced solutions while learning from actual user's evaluations. Due to the nature of the environment, we have chosen to apply methodologies of *learning by examples* to reflect user preferences. We have based our approach on the assumption that the average user is not – and probably not willing to be – fully aware of the exact nature of his/her preferences. Moreover, we have aimed, throughout this work, to an open system with the potential to operate and learn from examples originating from alternative sources with minimal modifications and without the need for prior user engagement.

During the solution visualisation stage of regular system use, the user suggests an overall grade for each solution ranging from 0 to 1. The Machine Learning Component creates and enhances a Machine Learning profile for each user based on the actual user's evaluation and the corresponding encoded solutions (NN Committee Training task of Fig. 2). In particular, the Machine Learning Component operation on the produced solutions can be distinguished as four distinct tasks:

1. Newly generated solutions are automatically evaluated by the Machine Learning Component according to the current state of the Machine Learning profile. At this point the Machine Learning Component is *unaware* of the actual user's evaluation for the *specific* set of solutions.
2. User evaluated solutions are used by the Machine Learning Component for further refinement of the Machine Learning profile. The Machine Learning Component

enhances the user's already existing Machine Learning profile based on the new set of evaluated solutions. The mechanism is described below.

3. The automatic solution evaluation results of Task 1. are compared to the actual user evaluation of Task 2. The current ability of the Machine Learning Component to generalise is assessed at this point.
4. The automatic solution evaluation results are compared to the Decision Support Component results. Assuming that the Decision Support Component offers a minimal consistent representation of the user's preferences, it is checked whether the Machine Learning profile outperforms its Decision Support counterpart.

Notice that training samples are submitted to the Machine Learning Component in *bursts,* a fact imposed by the MultiCAD functionality. As a result, the context itself has suggested that the use of machine learning techniques, where all training data are readily available, would not be appropriate and *incremental learning* [3] techniques are required. One of the most efficient mechanisms in this context is the family of *boosting algorithms* [25]. We have chosen to propose and implement a variation of the Learn++ algorithm presented in [19], modified to conform to the needs of the current context. The learning mechanism is a *committee of neural networks*, each representing a *weak-learner,* able to classify, with a relatively high error margin, the input samples. Every time a new set of samples arrives, the committee is extended with additional members focused on the newest set of samples. The overall evaluation of any sample is the result of *voting* of all members of the committee.

4 Machine Learning Component Configuration

At the next phase, we deal with the problem of lack of adequate and reliable training data for the Machine Learning Component. In particular, a number of parameters with respect to the internal architecture of each network separately as well as to the overall committee, may affect the performance in the long term, hence we need to compare and select the most efficient configuration by applying alternative adjustments before the network is exposed to real use. In order to achieve this, we create artificial training sets based on actual feedback from a few *test* users combined with the same users' Decision Support profiles. The methodology followed in order to create an artificial training set for the alternative network configurations is presented in Table 1. Notice that this series of steps refers to the *configuration* of the system and *not* to its actual use. This is the reason the generation of solutions is considered as a pre-processing step at this stage. In particular, the solutions originate from carefully created descriptions, tailored to cover a wide range of the attribute values and produce a variety of combinations. The advantage is that this set will be well-studied and pre-processed; therefore there is no need for real-time solution generation.

The partition of the artificially evaluated solutions to training subsets forms the core of this approach. This partition aims to simulate the way evaluated solutions will be produced during the actual use of the system. The advantage is that graded solutions of varying populations will be readily available since no user intervention is required. Each candidate Machine Learning Component configuration will be trained and subsequently tested based on all these subsets. The architecture eventually selected will be the one with the best overall performance. The key assumption of our method is that the Decision Support profile adequately – yet not perfectly – reflects the preferences of the average user. Better performance of a network committee with respect to a number of diverse users' profiles, created based on real user feedback, constitutes strong evidence that this configuration will also prove efficient in real use.

Table 1. Parameter Configuration for the Machine Learning Component. *Pre-processing* tasks take place only once at the beginning of the module's lifetime. *Parameter Configuration Processing* tasks take place during the initial stage of the module's lifetime before it is exposed to real use. *Post-processing* tasks represent exploitation of the results of the previous stages.

Pre-processing	
Step 0.	Generation and attribute encoding of solutions for pre-defined declarative descriptions (test scenes).
Step 1.	Decision Support Profiles (SMART, AHP, RR weights and middle values) initialisation by actual test users.
Step 2	Automatic Evaluation of corresponding solutions based on Decision Support Profiles.
Parameter Configuration Processing	
Step 3. For each alternative configuration of the Machine Learning Component	
For each (test scene, test user) combination	
Step 3.1.	Creation of the artificial training subset for the current description based on Decision Support Profiles evaluations.
Step 3.2.	Testing of current form of Network Committee against current training subset.
Step 3.3.	New sub-committee trained with current training set under alternative adjustments.
Step 3.4.	Network Committee enhanced with the new sub-committee tested against current training subset.
Post-processing	
Step 4.	Selection of the best performing configuration with respect to all test users and all test scenes.

5 Preliminary Results

Based on the proposed configuration methodology, we have performed a series of experiments using alternative adjustments for the various module parameters. These parameters include the number of members of each new sub-committee (which is created every time a new scene is processed by the system) and the configuration of the neural networks participating in the committee. Notice that the incremental training suggested by the current context is one of the toughest for learning mechanisms [19] in the sense that each session introduces representatives from only one category of examples. In other words, solutions corresponding to a specific scene are submitted only once to the mechanism, grouped together and separately from other scenes, in order to simulate regular system use. Table 2 provides a detailed view of one series of these experiments where the ML Component was set to add 5 new members to the committee for each new scene. Solutions from 5 scenes (one in each column of the table) were submitted in tandem for training and the shaded cells (on and over the diagonal) represent the error regarding *seen* solutions (training error).

After each training session, *unseen* solutions where evaluated; these results appear in the non-shaded cells (under the diagonal) representing the generalisation error.

The diagonal represents the error corresponding to the scene most recently used for training. We observe that although the error for the current scene is low, as soon as new scenes arrive, the performance for previously seen scenes degrades, which is, however, common for this style of learning. Moreover, the overall training error is more than satisfactory and the generalisation error seems promising when considering the incremental nature of the training [19] and under the assumption that the user remains consistent to his/her preferences during system use. Adding 3 new members for each scene instead of 5, while maintaining the exact same settings for the rest of the configuration has yielded 9.34%, 17.93%, 13.21% for the overall Average Training Error, Average Generalisation Error and Average Error respectively, representing a decrease in the ML Component's performance.

Table 2. Training session using evaluated solutions sets from 5 scenes, submitted as separate groups to the ML Component for training. Each solution set causes the addition of 5 members to the committee. Cells on and over the diagonal demonstrate the error with respect to *seen* solutions whereas cells under the diagonal demonstrate error with respect to unseen solutions.

Scene ID	Naxos House	Rentis House	Pireaus House	Kalamaki House	Patras House	
Naxos House	**1.94%**	11.04%	6.57%	9.55%	3.73%	
Rentis House	21.90%	**4.38%**	13.38%	9.49%	11.68%	
Pireaus House	28.99%	12.56%	**1.45%**	15.14%	14.49%	
Kalamaki House	41.89%	2.26%	10.38%	**0.94%**	2.08%	
Patras House	10.76%	17.42%	0.59%	17.42%	**3.33%**	
Average Training Error	1.94%	7.71%	7.13%	8.78%	7.06%	**6.53%**
Average Generalisation Error	25.88%	10.75%	5.48%	17.42%	-	**14.88%**
Average Error	21.09%	9.53%	6.47%	10.51%	7.06%	**10.93%**

6 Conclusions

The combinatorial explosion implied by the MultiCAD architecture has called for a module for intelligent solution evaluation able to capture and apply user preferences. Due to the nature of the MultiCAD environment, training data are obtained in discrete subsets corresponding to a single scene, at distinct and distant times. Depending on the declarative description submitted as input, the population of produced solutions may span from a few hundreds to hundreds of thousands making actual human evaluation unfeasible. In order to evaluate and select the most efficient configuration for a Machine Learning Component, we have employed the multi-criteria decision support methods used in the Decision Support Component of the User Profile Module. The Decision Support profile is built based on the actual user's feedback and offers instant approximation of the user's preferences. We operate under the assumption that this profile adequately – yet not perfectly – represents the average user's preferences for the needs of the experiments. Based on this assumption, we propose a methodology for the creation of an artificial training set. This set corresponds to solutions generated based on a wide range of carefully designed test scenes. These solutions become attribute-encoded, are evaluated by each user's Decision Support Profile and are submitted to the Machine Learning Component in subsets originating from the same description for testing and subsequent training. Alternative parameter configurations are tested in this manner in order to select the most appropriate for actual use.

Our method of parameter configuration sacrifices a certain degree of realism by accepting the Decision Support profile as representative of the average user for the needs of the experiments. In a sense, this represents an effort to concentrate on users who are *consistent* in their preferences throughout system use. In return, it offers some important advantages. In particular, it does not require real-time generation and encoding of solutions since they are based on pre-selected scenes. Moreover, the feedback required by the user – the initialisation of the Decision Support profile – is minimal, required only once and may subsequently be utilised during normal system use. Last but not least, it offers flexibility with respect to the number and the quality of the solutions available for the experimental training.

The preliminary experimental results seem promising, in the sense that the proposed Machine Learning Component, based on incremental learning from training subsets each originating from only one scene, exhibits low training error and adequate generalisation capability in the current context. The latter implies that the proposed Intelligent User Profile Module exhibits, at least according to these preliminary results based on the artificial training sets, the capability to capture and apply each user's preferences in the current context.

7 Future Work

Alternative parameters of the Machine Learning Component configuration as well as alternative *policies* applied to the incorporation of each training subset to the network's training comprise the main topics of subsequent experimentation. The selected architecture(s) will be exposed to real system use and actual users' data will be collected and compared with the experimental results to verify the Intelligent User Profile Module's ability to adequately model user preferences in the current context.

Acknowledgements. This study was co-funded by 75% from the European Union and 25% from the Greek Government under the framework of the Education and Initial Vocational Training Program – 'Archimedes'.

References

1. Bardis G., Miaoulis G., Plemenos D., An Intelligent User Profile Module for Solution Selection Support in the Context of the MultiCAD Project, 7e Infographie Interactive et Intelligence Artificielle (3IA) conference, pp. 25-37, Limoges, France, 2004.
2. Bardis G., Miaoulis G., Plemenos D., Intelligent Solution Evaluation Based on Alternative User Profiles, Conf.Area: Artificial Intelligence and Decision Support Systems, pp. 74-82, International Conference of Enterprise Information Systems (ICEIS), Miami, USA, 2005.
3. Elman J.L., Learning and Development in Neural Networks: The Importance of Starting Small, Cognition 48, pp. 71-99, 1993.

4. Bonnefoi P.-F., Plemenos D., Constraint Satisfaction Techniques for Declarative Scene Modeling by Hierarchical Decomposition, 3IA conference, Limoges, France, 2002.
5. Brusilovsky P., Adaptive Hypermedia, User Modelling And User-Adapted Interaction 11: pp. 87-110, 2001.
6. Champciaux L., Classification: a basis for understanding tools in declarative modelling, Computer Networks and ISDN Systems 30, pp. 1841–1852, 1998.
7. Dragonas J., Makris D., Lazaridis A., Miaoulis G., Plemenos D., Implementation of a Collaborative Environment in MultiCAD Declarative Modelling System, 3IA conference, Limoges, France, 2005.
8. Freischlad M., Schnellenbach-Held M., A Machine Learning Approach for the Support of Preliminary Structural Design, Advanced Engineering Informatics 19, p. 281–287, 2005.
9. Fribault P., Modelisation Declarative d'Espaces Habitable, (in French), Doctoral dissertation, University of Limoges, France, 2003.
10. Golfinopoulos V., Miaoulis G., Plemenos D., A Semantic Approach for Understanding and Manipulating Scenes, 3IA conference, Limoges, France, 2005.
11. Goodwin P., Wright G., Decision Analysis for Management Judgement, Second Edition, Wiley, 1998.
12. Joan-Arinyo R., Luzon M.V., Soto A., Genetic Algorithms for Root Multiselection in Constructive Geometric Constraint Solving. Computers and Graphics 27, pp. 51-60, 2003.
13. Makris D., Ravani I., Miaoulis G., Skourlas C., Fribault P., Plemenos D., Towards a domain-specific knowledge intelligent information system for Computer-Aided Architectural Design, 3IA conference, Limoges, France, 2003.
14. Makris D., Etude et réalisation d'un système déclaratif de modélisation et de génération de styles par algorithmes génétiques. Application à la création architecturale, Doctoral dissertation, University of Limoges, France, 2005.
15. Miaoulis G., Contribution à l'étude des Systèmes d'Information Multimédia et Intelligent dédiés à la Conception Déclarative Assistée par l'Ordinateur – Le projet MultiCAD (in French), Professorial dissertation, University of Limoges, France, 2002.
16. Nassar K., Thalet W., Beliveau Y., A Procedure for Multicriteria Selection of Building Assemblies, Automation in Construction 12, pp. 543-560, 2003.
17. Plemenos D., Declarative modelling by hierarchical decomposition. The actual state of the MultiFormes project, International Conference GraphiCon'95, St Petersburg, Russia, 1995.
18. Plemenos D., Miaoulis G., Vassilas N., Machine Learning for a General Purpose Declarative Scene Modeller. International Conference GraphiCon'2002, Nizhny Novgorod, Russia, 2002.
19. Polikar R., Udba L. Udba S., Honavar V., Learn++: An Incremental Learning Algorithm for Supervised Neural Networks, IEEE Transactions on Systems, Man, and Cybernetics—Part C: Applications And Reviews, Vol. 31, No. 4, pp. 497-508, November 2001.
20. Ravani J., Makris D., Miaoulis G., Plemenos D., 2002. Concept-Based Declarative Description Subsystem for CADD. 3IA conference, Limoges, France, 2004.
21. Roberts R., Goodwin P., Weight Approximations in Multi-attribute Decision Models, Journal of Multicriteria Decision Analysis 11, pp. 291-303, 2002.
22. S. Shearin and H. Lieberman: "Intelligent Profiling by Example," International Conference on Intelligent User Interfaces (IUI 2001), pp. 145-152, 2001.
23. Saaty, T.L., The Analytic Hierarchy Process, RWS Publications, Pittsburgh, USA, 1990.
24. Turing, A.M. 1950. "Computing Machinery and Intelligence." From Mind LIX, no. 2236 : 433-460, Oct. 1950.
25. Witten I.H., Frank E., Data Mining – Practical Machine Learning Tools and Techniques, 2nd Ed., Elsevier, 2005.

Emotional Intelligence: Constructing User Stereotypes for Affective Bi-modal Interaction*

Efthymios Alepis and Maria Virvou

Department of Informatics, University of Piraeus, 80 Karaoli & Dimitriou St.
18534 Piraeus, Greece
{mvirvou, talepis}@unipi.gr
http://www.unipi.gr/ faculty/mvirvou/index.htm

Abstract. The need of incorporating stereotypes concerning users' emotional characteristics in affective multi-modal user interfaces is important. To this end we present user stereotypes that we have constructed concerning the emotional behavior of users while they interact with computers. The construction of these user stereotypes have been based on an empirical study. The empirical study has been conducted among different kinds of computer users in terms of how they express emotions while they interact with a through a keyboard and a microphone in a bi-modal kind of interaction. The stereotypes have been incorporated in a user modeling component underlying an affective bi-modal user interface that uses a microphone and a keyboard. Actions of the users in terms of what they said and/or what they typed computer in the context of typical interaction situations are interpreted in terms of their feelings so that appropriate affective messages may be generated.

1 Introduction

Recently, recognition of emotions of users while they interact with software applications has been considered quite important, so that affective user interfaces may be constructed. At the same time, many researchers acknowledge that affect has been overlooked by the computer community in general [9]. However, how people feel may play an important role on their cognitive processes as well [2].

Picard [8], claims that one of the major challenges in affective computers is to try to improve the accuracy of recognising people's emotions. Picard also argues that people's expression of emotion is so idiosyncratic and variable, that there is little hope of accurately recognising an individual's emotional state from the available data. Ideally evidence from many modes of interaction should also be combined by a computer system so that it can generate as valid hypotheses as possible about users' emotions. This view has been supported by many researchers in the field of human computer interaction (e.g. [6], [7]).

* Support for this work was provided by the General Secretariat of Research and Technology, Greece, under the auspices of the PENED program.

B. Gabrys, R.J. Howlett, and L.C. Jain (Eds.): KES 2006, Part I, LNAI 4251, pp. 435–442, 2006.
© Springer-Verlag Berlin Heidelberg 2006

The need of incorporating stereotypes concerning users' characteristics in modern multi-modal application interfaces is important; individuals can more effectively understand universal emotions expressed by members of a cultural group to which they have greater exposure [1]. The need of incorporating user stereotypes is acknowledged by other researchers as well. For example, in [4],[5] it is suggested that the incorporation of stereotypes in emotion-recognition systems improves the systems' accuracy. Despite this importance, in 2001 there were only a few studies based on emotion stereotypes but the interest in such approaches was rapidly growing [4],[5].

In view of the above, in this paper we describe the user stereotypes that we have constructed for affective bi-modal interaction. The two modes of interaction are based on two input devices respectively: microphone and keyboard. The user stereotypes have been resulted from an empirical study that we conducted among 100 users. The empirical study aimed at finding out common user reactions in the context of typical interaction situations that express user feelings while they interact with computers. As a next step we associated the reactions with particular feelings within the context of certain user groups depending on their age, level of computer familiarity, educational level etc. The stereotypes have been incorporated in a user modeling mechanism that performs user affect perception. The aim of this paper is to improve the accuracy of an emotion-recognition system. Stereotypes provide additional evidence supporting that a guess about a user's feelings is correct or incorrect.

2 Empirical Study Concerning User Emotions

The empirical study involved 100 users (male and female) of varying educational backgrounds, ages and levels of computer experience. The users' characteristics were recorded in order to create the users' stereotypes. In our study users were given questionnaires to fill in concerning their emotional reactions to several typical situations of computer use in terms of their actions using the keyboard and the microphone. In some cases the participants had to imagine an interaction with a computer while using an educational application. Then they were asked to imagine certain situations and the resulting voice and keyboard actions that express emotions. Participants were also asked to determine possible actions in certain emotional states during their interaction. We found this part of the study very important for many reasons. Firstly, this kind of study provided an insight on personal user feelings as the users themselves perceive them and remember them throughout their experience in interacting with computers. Secondly, the data from this study were collected and then analyzed to construct a basic prototype stereotypic model used by the user modeling component. In most cases our hypotheses from the empirical study were confirmed by the users' real actions while using an application monitored by the user modeling component.

Our aim was to be able to recognize changes in the users' behavior and then to associate these changes with emotional states like anger, happiness, boredom, etc. Within a more general scope we tried to recognize positive and negative feelings from the bi-modal interaction.

2.1 Categories of Participants in Terms of Their Characteristics

People's behavior while doing something may be affected by several factors concerning their personality, age, experience, etc. For example, experienced computer users may be less frustrated than novice users or older people may have different approaches in interacting with computers, as compared with younger people, etc. Thus for the purpose of analyzing the results of our empirical study taking into account important characteristics of users we categorized them in several groups. These groups, presented below, comprise the basic stereotypes of our study that are analyzed in the subsequent sections. Analyzing the distribution of participants in the empirical study in terms of their age, there were 12,5 % of participants under the age of 18, approximately 20% of participants between the ages of 18 and 30. A considerable percentage of our participants were over the age of 40. Considering the participants' computer knowledge level stereotypic characteristic we have 20 % of inexperienced users, 30 % of novice users, 32 % of intermediate users and 18 % of expert computer users. Moreover, participants were also categorized in terms of the general educational background and level they had.

2.2 Results of the Empirical Study

The results of the empirical study have been collected and analyzed in terms of the categories of users as explained above and then they were analyzed in terms of each mode of interaction examined, namely speech and keyboard input as well as the combination of the two modes.

From the six major emotions (happiness, sadness, anger, fear, disgust and surprise), our study came up with considerable results for the emotions of happiness, anger and surprise. A sample of the results is illustrated in figure 1. Participants argued that fear and disgust would probably have the same evidence in their voice as surprise. The participants did not give considerable statistical data in terms of specific words or characteristic phrases for the emotion of sadness. They suggested that sadness could be detected mostly in changes of their voices' pitch and tone and by measuring changes in the speed they were talking.

It is interesting to notice that a very high percentage (85%) of young people (under 30 years old) who are also inexperienced with computers interacted with the oral mode very well. On the contrary participants belonging to the postgraduate level of education group and who were also computer experts didn't give us considerable data for the affect perception during the oral communication with their computer.

While using the keyboard most of the participants agreed that when they were nervous the possibility of making mistakes increased rapidly. This was also the case when they had negative feelings. Mistakes in typing are followed by many backspace-key keyboard strokes and concurrent changes in the emotional state of the user in a percentage of 82 %. Yet users under 20 years old and users who are over 20 years old but have low educational background seem to be more prone to making even more mistakes as a consequence of an initial mistake and lose their concentration while interacting with an application (67%). They also admit that when they are angry the rate of mistakes increases, the rate of their typing becomes slower 62 % (on the contrary, when they are happy they type faster 70 %) and the keystrokes on the keyboard become harder (65 %). Of course the pressure of the user's fingers on the keyboard is something that can not be measured without the appropriate hardware. Similar effects to the keyboard were reported when the emotion is boredom instead of anger.

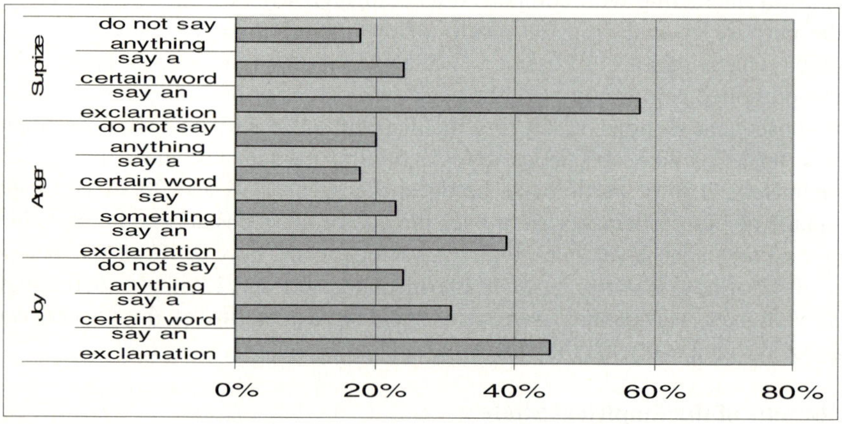

Fig. 1. Speech reactions in the feelings of joy, anger and surprise

A conclusion concerning the combination of the two modes of interaction(through the keyboard and the microphone) in terms of emotion recognition is that the two modes are complementary to each other to a high extent. In many cases the system can generate a hypothesis about the emotional state of the user with a higher degree of certainty if it takes into account evidence from the combination of the two modes rather than one mode. For example, when the rate of typing backspace of a user increases, this may mean that the user makes more mistakes due to a negative feeling. However this hypothesis can be reinforced by evidence from speech if the user says something bad that expresses negative feelings.

3 Stereotypes for Affect Perception

The stereotypes for user modeling were first introduced by Rich [10] in the context of a recommender system, called GRUNDY, that took into account different characteristics of users in order to recommend to them what book to buy. Stereotype-based reasoning takes an initial impression of the user and uses this to build a user model based on default assumptions [3]. Moreover, the stereotypes are largely based on the following basic components:

(a) A set of trigger conditions,
(b) A set of retraction conditions,
(c) A set of stereotypic inferences, and

Trigger conditions are Boolean expressions that activate a specific stereotype. The action of a stereotype is to make large numbers of inferences when a trigger becomes true. The retraction conditions are responsible for deactivating an active stereotype. Once the user is assigned to a stereotype, the stereotype inferences of this particular stereotype serve as default assumptions for the user. In our study, we have classified our users into stereotypes concerning their age, their educational level, their computer knowledge level and their gender. For each user there is a value that corresponds to a four-dimensional stereotypic vector of the form:

(User_Name, Stereotypic Characteristic1, Stereotypic Characteristic2, Stereotypic Characteristic3, Stereotypic Characteristic4).

Stereotypic Characteristic1 refers to the user's age and is an element of the following set concerning ages: [(10-18), (18-30), (30-40), (40-50), (over 50)]. Stereotypic Characteristic 2 refers to the user's computer knowledge level and is an element of the following set concerning the user's computer experience in months (using a personal computer): [(less than 1), (1-6), (6-12), (over 12)]. Similarly we have defined stereotypic characteristics 3 and 4 that refer to the user's educational level and to the user's gender.

Though it is very difficult to have high percentages of probability while trying to make predictions about the emotional state of a user interacting with a personal computer the aim of this paper is to improve the accuracy of an emotion-recognition system. Stereotypes help us have more evidence supporting that a guess is correct or incorrect. In many cases data from the vocal interaction or the interaction through the keyboard gives evidence of different emotions with quite similar degrees of certainty. For example the system may have evidence that a user is either angry while saying or typing something or stressed or even confused. The incorporation of stereotypes in the system provides a reasoning mechanism that may recognize an emotion among others or give evidence for distinguishing between emotions. Evidence for the character or the personality of a particular user may raise the degree of certainty for a particular emotion recognized. Moreover having evidence about basic characteristics of users may lead emotion recognition algorithms to focus on a certain mode of interaction. If the system knows that it is more likely to recognize the emotional state of a user by analyzing data from a specific mode (e.g. his/her voice through the microphone) then it should take that mode into more consideration than the other.

A simple example of reasoning based on stereotypes in the emotion-recognition system is the following. If a user is characterized as inexperienced in using computers and is also part of the "10-18 years old" group of stereotypes then we have strong evidence (deriving from the empirical study) that s/he may say an exclamation while being angry or happy and that the system should rely mostly on the oral mode of interaction (instead of the keyboard input) while trying recognize the user's feelings. On the contrary the system should try to make hypotheses for the emotional state of an experienced, over 40 years old user, by analyzing data based basically on the keyboard mode of interaction.

4 Trigger Conditions and Inferences

Considering the interaction of users with the system we have come up with 5 basic actions in the context of typical interaction situations that may generate emotions to a user during his/her interaction with an application. These are: a) a mistake (the user may receive an error message by the application or navigate wrongly) b) many consecutive mistakes c) absence of user action d) action unrelated to the main application e) correct interaction.

In order to specify the trigger conditions for user stereotypes we have also categorized user's input actions from the keyboard and the microphone. In particular considering the keyboard we have: a) user types normally b) user types quickly (speed higher than the usual speed of the particular user) c) user types slowly (speed lower than the usual speed of the particular user) d) user uses the backspace key often e) user hit unrelated keys on the keyboard f) user does not use the keyboard. Considering the users' basic input actions through the microphone we have 7 cases: a) user speaks words from a specific list of words showing an emotion b) user uses exclamations c) user does not say anything d) user speaks with a low voice volume (low than the average recorded level) e) user speaks in a normal voice volume f) user speaks with a high voice volume (higher than the average recorded level).

Each one of the triggering conditions described above is associated with stereotypic inferences. User's actions are categorized and for each action we have calculated the probabilistic value that corresponds to the likelihood of occurrence of each emotion (happiness, sadness, anger, fear, disgust and surprise). The likelihood of emotion occurrence is based on the results of the empirical study. The inferences are then incorporated in the stereotypic model of the user modeling component in order to provide a reasoning mechanism for affective bi-modal interaction. A part of the triple stereotype of users-emotions-input device is illustrated in table 1.

Table 1. Triple stereotype of users' category, emotions and device actions

Stereotype of users	**Emotion of Joy**			
Category of users	Keyboard		Microphone	
	Device actions	Likelihood of occurrence	Device actions	Likelihood of occurrence
Age(10-18), inexperienced	Type faster	0.36	Say a word from a specific list of words	0.31
	Type normally	0.52	Say an exclamation	0.45
	Not use the keyboard	0.12	Do not say anything	0.24

Particularly, the values that correspond to the device actions from the microphone are related to the percentages in figure 1. The numbers for the likelihood of occurrence of each device action give as the weights in order to calculate the likelihood of occurrence of an emotion. The same category of users will provide the system with corresponding values of likelihood of occurrence for each one of the five remaining emotions.

As an example using the stereotypic data from table 1 we demonstrate the way of calculation of the likelihood of occurrence for the emotion of joy and for the "age (10-18), inexperienced" category of users, using the above values: When a user types faster than her/his usual speed, this action will be weighted with value 0.36. Accordingly, if s/he says an exclamation while interacting with an application, this action will be recorded and weighted with value 0.45. The absence of an action zeroes the weight of that action. Finally, after having calculated all the weights for the six basic emotions, we can compare them and make a decision of which one of them is more likely to be occurring at this time of interaction.

5 User Modeling Component

The stereotypes have been used in a prototype user modeling system that monitors the users' actions from the keyboard and the microphone and interprets them in terms of emotions. The basic function of this application is to capture all the data inserted by the user either orally or by using the keyboard and the mouse of the computer and interpret them in terms of the users' feelings. The data is recorded to a database and then returned to the basic application the user interacts with. Figure 2 illustrates the monitoring function of the user modeling component that records the user's input and the exact time of each event.

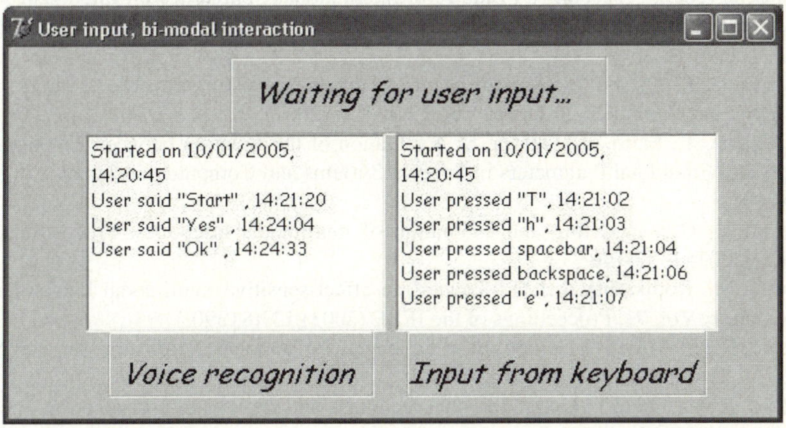

Fig. 2. Snapshot of operation of the user modeling component

The empirical study has supplied us with a significant number of words, phrases and exclamations, which are used in the creation of an active vocabulary for emotion recognition. That made the application more accurate in recognizing feelings thus the database that had been resulted from the empirical study could be made more accurate. In particular the results from the empirical study could be either confirmed by the recognition of the users' emotions in real-time (by using the user modeling component), or in some cases take specific results into reconsideration.

6 Conclusions

In this paper we have presented the user stereotypes that we have constructed concerning the emotional states of users while they interact with a computer. Our presentation was focused on basic emotions such us anger, happiness, surprise, fear, disgust. The user stereotypes have been based on an empirical study that we conducted among 100 users of different ages and backgrounds. As a result, the stereotypes were used in a user modeling mechanism that performs affect perception and uses the stereotypes to increase the accuracy of the hypotheses generated concerning users' emotions. This paper has a lot of empirical work that aims at the construction of a prototype system. The main evaluation and refinement of the system is in our future plans.

References

1. Elfenbein,H.A., Ambady, N.: When Familiarity Breeds Accuracy: Cultural Exposure and Facial Emotion Recognition, Journal of Personality and Social Psychology, Vol. 85, No. 2, (2003) 276–290
2. Goleman, D.: Emotional Intelligence, Bantam Books, New York (1995)
3. Kay, J.: "Stereotypes, student models and scrutability" In: G. Gauthier, C. Frasson and K. VanLehn (eds.): Proceedings of the Fifth International Conference on Intelligent Tutoring Systems, Lecture Notes in Computer Science, Springer-Verlag, Berlin, Heidelberg, Vol. 1839 (2000) 19-30
4. Moriyama, T., Ozawa, S.: Measurement of Human Vocal Emotion Using Fuzzy Control, Systems and Computers in Japan, Vol. 32, No. 4, (2001)
5. Moriyama, T., Saito, H., Ozawa, S.: Evaluation of the Relation between Emotional Concepts and Emotional Parameters in Speech, Systems and Computers in Japan, Vol. 32, No. 3, (2001)
6. Oviatt, S.: User-modeling and evaluation of multimodal interfaces. Proceedings of the IEEE (2003) 1457-1468
7. Pantic, M., Rothkrantz, L.J.M.: Toward an affect-sensitive multimodal human-computer interaction. Vol. 91, Proceedings of the IEEE (2003) 1370-1390
8. Picard, R.W.: Affective Computing: Challenges, Int. Journal of Human-Computer Studies, Vol. 59, Issues 1-2, (July 2003) 55-64
9. Picard, R.W., Klein, J.: Computers that recognise and respond to user emotion: theoretical and practical implications, Interacting with Computers 14, (2002) 141-169
10. Rich, E.: Users are individuals: individualizing user models. International Journal of Man-Machine Studies 18, (1983) 199-214

Garbage Collection in an Embedded Java Virtual Machine

Chang-Il Cha[1], Hyung-Jun Kim[1], Kyu-Jeong Hwang[1],
Sang-Wook Kim[1], Sang-Yun Lee[2], and Hee-Sun Won[2]

[1] Division of Information and Communications, Hanyang University,
Haengdang-dong Sungdong-gu, Seoul 133-791, Korea
cha8680@korea.com, kimhyungjun79@hanmail.net,
lingohkc@lycos.co.kr, wook@hanyang.ac.kr

[2] Embedded S/W Research Division, Electronics and Telecommunications Research Institute,161, Gajeong-dong, Yuseong-gu, Daejeon, 305-350, Korea
{syllee, hswon}@etri.re.kr

Abstract. Garbage collection in the Java virtual machine is a core component that relieves application programmers of difficulties related to memory man-agement. It should also take into account the characteristics of embedded envi-ronments. In this paper, we propose a garbage collector that meets the require-ments for embedded environments. The proposed garbage collector is primarily based on generational garbage collection where a heap is composed of young and old generations. A semi-space copying collector is employed for a young generation, and an incremental copying collector is employed for an old genera-tion. Owing to its contiguous allocations, the proposed method provides fast al-locations and the locality of references. By adjusting the size of a young genera-tion, the block size, and the number of blocks in a frame, it is able to control the delay time caused by garbage collection, and thus guarantee the real time con-straints. Also, it employs a novel write barrier mechanism for efficiently deter-mining reachable objects in a specific frame. With all these characteristics to-gether, the proposed garbage collector can reclaim all the garbage objects precisely as well as timely. To reveal its superiority, we perform a series of ex-periments.

1 Introduction

Java is one of the most widely-used programming languages owing to its support of object-oriented concepts, safety, and flexibility. The Java platform, which is for running applications written in Java, defines configurations for organizing environments appropriate for the target system's memory, CPU, power, and applications[4]. The configuration defines the Java virtual machine and core APIs, and is also classified into J2EE/J2SE/J2ME. Among these, J2ME(Java2 platform, micro edition) is to support embedded environments for network-enabled mobile and embedded devices[4]. The Java virtual machine plays a role of interfaces between hardware and the compiled Java bytecodes for program execution[10]. The garbage collector, an important component of the Java virtual machine, searches for memory objects that are no longer used by application programs and reclaims them. Consequently, the

B. Gabrys, R.J. Howlett, and L.C. Jain (Eds.): KES 2006, Part I, LNAI 4251, pp. 443 – 450, 2006.
© Springer-Verlag Berlin Heidelberg 2006

garbage collector significantly relieves the burden of memory management from applica-tion programmers and allows programs to run in a more stable manner. This paper addresses an efficient garbage collector for an embedded Java virtual machine.

The proposed garbage collector is based on generational garbage collection where the heap area is divided into two parts: young and old generations[6, 7]. A new object is placed in the young generation which is relatively smaller in size. Most objects die and thus are reclaimed in the young generation. A few long-lived objects are moved into the old generation. Since objects in two generations differ in characteristics, different garbage collectors are employed. We adopt a semi-space copying collector for the young generation because object allocations and deallocations are frequent[3]. On the other hand, for the old generation whose size is larger than that of the young generation, we adopt an incremental garbage collection which divides a whole area into smaller pieces and performs a garbage collection on one piece if needed in order to meet the real time requirements. Incremental garbage collection requires recognizing live objects efficiently in a target area. For this purpose, we propose a novel write barrier mechanism suitable for embedded environments[9]. Our garbage collector satisfies the following requirements due to the physical limitations of embedded devices: fast execution, real time support, precise garbage collection, elimination of fragmentation, and high locality.

2 Overview of Proposed Garbage Collector

Our garbage collector is primarily based on generational garbage collection that maintains a heap with two asymmetrical generations: young and old generations[6, 7]. In generational garbage collection, a large portion of garbage objects are reclaimed in a young generation[1, 2, 8] since most of objects live during a short time. Thus, garbage collection in a young generation occurs more frequently, and only a small number of live objects are moved into an old generation[5]. This makes programs run fast owing to small delay caused by garbage collections.

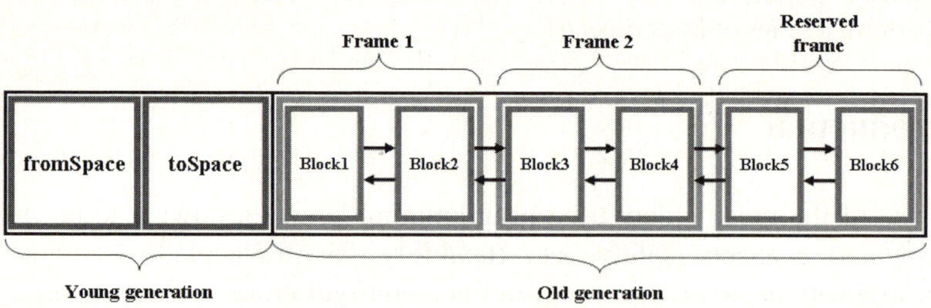

Fig. 1. Organization of a heap

For garbage collection in a young generation, we adopt the semi-space coping garbage collector that is easy to implement and has a small overhead for execution[3]. On the other hand, the heap for the old generation is quite large. So, garbage collection over the entire heap requires long time, and thus, is unable to meet the real time requirements of applications running in embedded devices. To resolve this problem, we propose a method that divides the old generation into multiple small pieces, and incrementally apply a copying collector to each piece. This method makes the real time requirements met and also enables the system to enjoy the nice properties of the copying collector. As depicted in Figure 1, the old generation consists of a set of blocks that are equal-sized and are connected to one another by a doubly linked list. The block is a unit for object allocations. A fixed number of blocks form a frame, which is a basic unit for garbage collection. When a heap is initialized, the sizes of a block and a frame are set by referencing the values stored in environment variables. These sizes affect the delay time caused by the execution of the garbage collector. Each block is assigned an age to keep its allocation order. Also, one frame is reserved for garbage collection, and thus is not used in object allocations.

Garbage collection in the old generation is initiated when there is not enough space for moving aged objects from the young generation. At this time, the target frame for garbage collection is the oldest one. This is because the oldest frame has a high probability of containing a large number of garbage objects[13,14,15]. The garbage collector copies a set of reachable objects[3] from the root set to the reserved empty frame mentioned earlier. The root set is a set of reference pointers to live objects, which are directly managed by programs in execution, and usually is stored in system stacks or global variables. After the copying, the oldest frame gets empty and is used for the next garbage collection. The age of the new frame is set to the youngest, and is used for objects moved from the young generation. When garbage collection for a frame completes, programs resume their execution.

In the proposed method, objects tend to be allocated contiguously in both the young and old generations. So, objects are allocated fast because there are no extra costs in computing allocation addresses. Since the objects allocated together have a high probability of being accessed together, this contiguous allocation increases the reference locality[2]. In a young generation, fragmentations do not occur. In an old generation, however, fragmentations may occur since a new block is allocated in case the object to be allocated is larger than the remaining area in an available block. However, the size of a fragmentation is not large. Let us consider the real time re-quirements, which are the most important in embedded environments. Since the young generation is not that large in size and most of objects die in the young genera-tion, its garbage collection satisfies the real time requirements. In the old generation whose size is relatively large, garbage collection performs over a small unit of a frame. Thus, the real time requirements can be met. As mentioned before, the sizes of a block and a frame are set by environment variables when a heap is initialized. Also, we note that garbage collection performs from the oldest frame. This means that a lot of objects are collected as garbage in a target frame and that the number of ob-jects moved between frames is small. Therefore, the speed of garbage collection is fast.

3 Write Barriers

3.1 Definition

The write barrier refers to a mechanism that detects write attempts to a certain memory area and executes specific operations related to memory management[9]. The reason why we need the write barrier is that our garbage collector does operate on just a part of a heap rather than a whole heap, that is, a young generation or a single frame in an old generation. In order to perform garbage collection on a part of a heap, we need a method to identify all the reachable objects in the target area efficiently. The write barrier is employed for this purpose.

3.2 Solution

In order to search for all the reachable objects quickly, whenever references among other generations or other frames occur, we record information related to such reference status. At the time of garbage collection, we only need to examine this information to determine all the reachable objects within the target area.

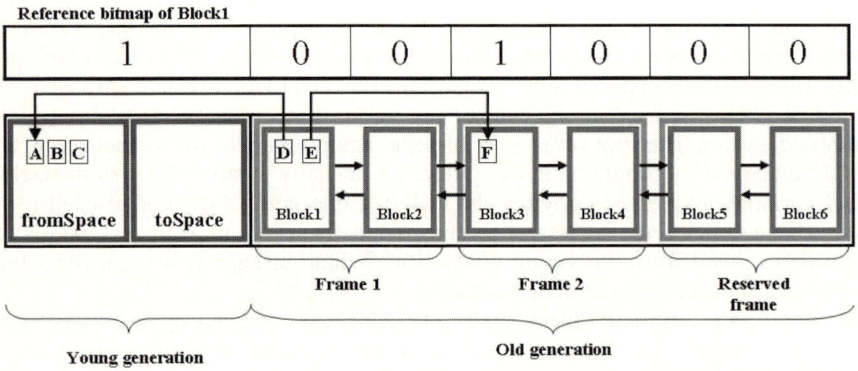

Fig. 2. A reference bitmap

To the end, we propose a two-step filtering method. The first filter recognizes the blocks to which the objects within a target block reference. For this first filtering, the header of each block contains a reference bitmap as in Figure 3. Bits in the reference bitmap are in one-to-one correspondences to blocks in a heap. Thus, each reference bitmap has bits as many as blocks in a heap. When an object within block i references another object in block j, the corresponding bit of the reference bitmap for block i is set to 1. The garbage collector can find out all the blocks referenced by the target block via investigating the reference bitmap of the target block. In Figure 2, it is shown that some objects in block 1 reference some objects in block 2 and also in a young generation. When a reference is established from block i to block j, the j-th bit of block i's reference bitmap is immediately set to 1. The reference bitmap of block i is initialized right after garbage collection for block i is finished.

Even though the first filter identifies the blocks that may contain reachable objects, it takes a substantial amount of time to track all the objects in such blocks since there are a lot of objects in each block. Thus, we adopt the second filter as in Figure 4 to distinguish only the objects in a block that reference objects in other blocks. For this, we partition a heap into a physical unit of pages and store information related to references for each page. As reference information, each page has a value of **CLEAN**, **DIRTY**, and **SUMMARIZE***. The value of **CLEAN** indicates that no objects within the page reference other objects outside the page, and there is no need to track the objects within the page in garbage collection. The value of **DIRTY** implies that many objects in the page reference other objects outside the page, and all those objects need to be tracked individually. The value of **SUMMARIZE** indicates that some objects reference others outside the page, but the number is smaller than a given threshold. In this case, we store the pointers to those referencing objects in a separate address table in order to directly access them in the page.

In summary, the garbage collector first finds out the blocks referenced by objects in the target area using the first filter, and then efficiently determines the reachable objects in those blocks by using the second filter.

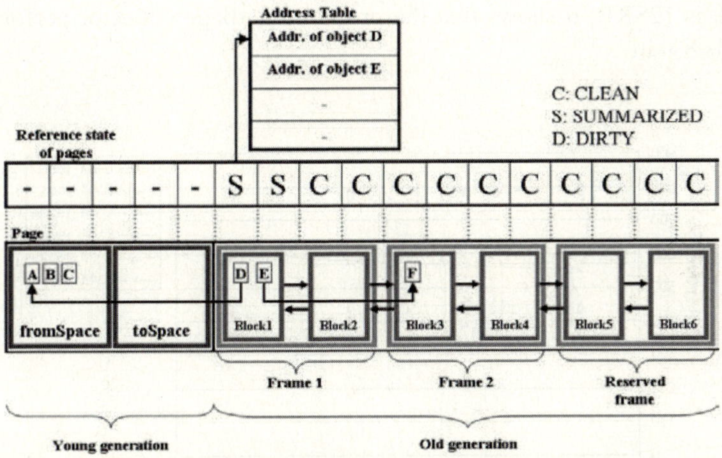

Fig. 3. Reference status in a page

4 Performance Evaluation

4.1 Experimental Environment

We used SUN Microsystems's CVM[4] that follows the CDC specification as a platform for performance evaluation. We installed the CVM on Linux fedora core version 4(kernel version 2.6.13) in this experiment and we used Intel Xeon 3.0GHz Dual CPU. The size of L1 and L2 Caches were 8KB and 1 MB respectively, and the size of main memory is DDR400 3GB. In this paper after installing our garbage collector to the existing CVM we measured and compared the performances of two gar-bage collectors.

We attached prefix CVM_ to represent experimental results of the original CVM's collector and the prefix EVM(Enhanced VM) for our garbage collector. In this performance evaluation, we used 201_ compress, _202_jess, _213_javac, and _228_jack from the SPECjvm98[16]. We set the size of whole heap to 32MB for CDC environment. Also, we fixed the heap size of fromspace and tospace of the young generation as 512KB and remaining 31MB for old generation. We set the minimum size of a block to 128KB and evaluated the performance by doubling the size up to 1,024KB.

4.2 Results

Figure 4 shows the garbage collection times taken when executing each program.

We can see that garbage collection time becomes smaller as the size of block gets smaller in our garbage collector. On the other hand, because the CVM's garbage col-lector performs garbage collection on the whole heap instead of dividing the heap by unit of blocks, the graph appears uniform regardless of the size of the block. Com-pared with the garbage collector of the CVM, when the size of a block is 128KB, it shows that the proposed garbage collector performs better as much as 8% in

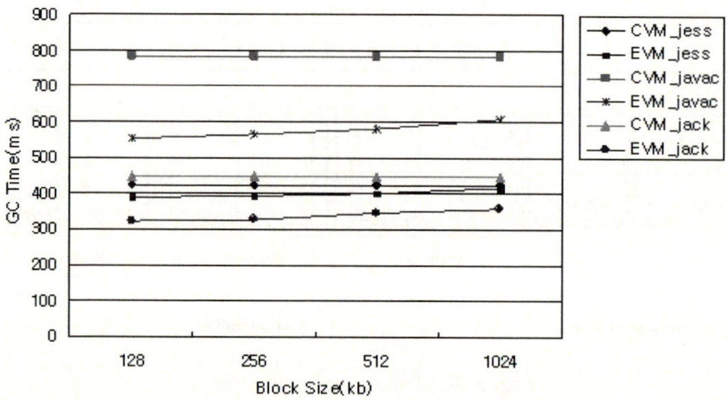

Fig. 9. Garbage Collection time

The improvement of performance can be explained as follows. Firstly, when the garbage collection fails in the young generation, the CVM's garbage collector for old generation starts immediately. On the other hand, in the proposed garbage collector, the garbage collection for old generation starts only when the old generation's heap is overcrowded in order to give objects enough time to die. Secondly, the reference bit-map which we adopted in the proposed garbage collector provides efficient filtering effects.

Figure 5 shows the pause time in the program execution due to garbage collection. For this, we used _202_compress the only program in the SpecJVM98 that invokes a garbage collector in old generation. We set the size of a whole heap to 32MB, and the size of the half of a young generation to 2MB As we can see from figure, the pause time of the proposed garbage collector is maintained within range of 0ms to 2ms. On the other hand, we can see that the pause time of the CVM's garbage collector rises abruptly when the garbage collector in the old generation is invoked.

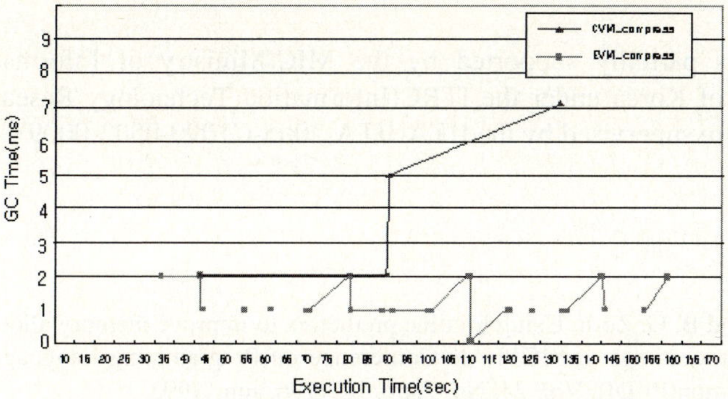

Fig. 5. Pause time due to garbage collection

5 Conclusions

This paper has proposed a novel garbage collector that meets the requirements in embedded environments. The proposed garbage collector is primarily based on generational garbage collection where a heap is managed with two separate parts: young and old generations. A semi-space copying collector is employed for a young generation, and an incremental copying collector is employed for an old generation. Owing to its contiguous allocations, the proposed method provides fast allocations and the locality of references. By adjusting the size of a young generation, the block size, and the number of blocks in a frame, we can control the pause time caused by garbage collection, and thus guarantee the real time constraints. Also, we have adopted a write barrier method for efficiently determining reachable objects in a frame. With all these methods together, we are able to reclaim all the garbage objects precisely.

To evaluate the performance of the proposed garbage collector after we installed our garbage collector to the existing CVM, we measured and compared the performances of two garbage collectors.

The proposed garbage collector and existing CVM's garbage collector show little difference in regard to the execution time of write barrier. However the proposed garbage collector outperforms the CVM's original one as much as 8% in _202_jess, 30% in _213_javac, and 18% in _228_jack in regard to the garbage collection time. The pause time of the proposed garbage collector is kept within range of 0ms to 2ms. On the other hand, the pause time of the CVM's garbage collector rises abruptly when the garbage collector is invoked in the old generation. So, we can infer from the experimental results that the proposed garbage collector works well in embedded environments. For future work, we plan to conduct more extensive experiments to evaluate performances in time and space perspectives, and we are going to port the proposed garbage collector in real embedded systems.

Acknowledgement

This research was partially supported by the MIC(Ministry of Information and Communications) of Korea under the ITRC(Information Technology Research Center) support program supervised by the IITA(IITA-2005-C1090-0502-0009).

References

1. D. A. Barrett and B. G. Zorn, Using lifetime predictors to improve memory allocation performance, In Proceedings of SIGPLAN Conference on Programming Languages Design and Implementation(PLDI), Vol. 24, No. 7, pp. 187-196, June 1993.
2. S. M. Blackburn, P. Cheng, and K. S. McKinley, Myths and reality: The performance impact of garbage collection, In Proceedings of International Conference on Measurement and Modeling of Computer Systems, pp. 25-36, June 2004.
3. C. J. Cheney, A non-recursive list compacting algorithm, Communications of the ACM, Vol. 13, No. 11, pp. 677-678, Nov. 1970.
4. Sun Microsystems, Java2 Platform, Micro Edition, Connected Device Configuration(CDC), http://java.sun.com/products/cdc/index.jsp 2005.
5. R. E. Jones and R. Lines, Garbage collection: Algorithms for automatic dynamic memory management, Wiley, Chichester, July 1996.
6. H. Lieberman and C. E. Hewitt, A real-time garbage collector based on the lifetimes of objects, Communications of the ACM, Vol. 26, No. 6, pp. 419-429, 1983.
7. D. M. Ungar, Generation scavenging: A non-disruptive high performance storage reclamation algorithm," ACM SIGPLAN Notices, Vol. 19, No. 5, pp. 157-167, Apr. 1984.
8. B. G. Zorn, Comparative performance evaluation of garbage collection algorithms, PhD thesis, University of California at Berkeley, Technical Report UCB/CSD 89/544, Mar. 1989.
9. B. Zorn, Barrier methods for garbage collection, University of Colorado, Technical Report CU-CS-494-90, Nov. 1990.
10. Tim Lindholm and Frank Yellin, The Java™ Virtual Machine Specification (second edition). Addison-Wesley. 1999.
11. G. Chen, R. Shetty, M.Kandemir, N. Vijaykrishnan, M.J. Irwin, M. Wolczko, Tuning Garbage Collection in an Embedded Java Environment, High-Performance Computer Architecture, 2002. Proceedings. Eighth International Symposium, pp. 92-103, February 2002.
12. CDC reference implementation 1.1 corresponding to JSR 218, http://java.sun.com/products/cdc/index.jsp
13. D. Stefanović, M. Hertz, S. M. Blackburn, K. S. McKinley, J. Eliot B. Moss , Older-first garbage collection in practice: evaluation in a Java Virtual Machine, ACM SIGPLAN Notices MSP '02, Vol. 38 No. 2, pp. 25-36, June. 2002.
14. D. Stefanović, K. S. McKinley, J. Eliot B. Moss, Age-based garbage collection, ACM SIGPLAN Notices , OOPSLA '99, Vol. 34 , No. 10, pp. 370-381, October 1999.
15. L. T. Hansen, W. D. Clinger, An experimental study of renewal-older-first garbage collection, ACM SIGPLAN Notices , ICFP '02, Vol. 37 No. 9, pp. 247-258, September 2002.
16. STANDARD PERFORMANCE EVALUATION CORPORATION. SPECjvm98 Documentation, release 1.03 ed., March 1999.
17. L. T. Hansen, W. D. Clinger, An experimental study of renewal-older-first garbage collection, ACM SIGPLAN Notices , ICFP '02, Vol. 37 No. 9, pp. 247-258, September 2002.

Design of Flexible Authorization System and Small Memory Management for XML Data Protection on the Server

Sun-Moon Jo[1] and Weon-Hee Yoo[2]

[1] Department of Information Technology Education, Paichai University
439-6 Doma-2Dong, Seo-Gu, Daejeon, Korea
sunmoonpink@hanmail.net

[2] Department of Computer Science and Information Engineering, Inha University
253 Yonghyun-Dong, Nam-Gu, Incheon, Korea
Whyoo@inha.ac.kr

Abstract. XML can supply the standard data type in information exchange format on a lot of data generated in running database or applied programs for a company by using the advantage that it can describe meaningful information directly. Therefore, as it becomes more and more necessary to manage and protect massive XML data in an efficient way, the development of safe XML access control techniques needs a new method. Access control techniques should be flexible enough to support all protection granularity levels. Since access control policies are very likely to be specified in relation to document types, it is necessary to properly manage a situation in which documents fail to be dealt with by the existing access control policies. In terms of XML documents, it is necessary to describe policies more flexibly beyond simple authorization models and to consider access control methods which can be selected. In this paper, we present design of flexible authorization system and small memory management for XML data protection on the server.

1 Introduction

After XML was presented as a standard for data exchange and representation on Internet, many new data have been made in the XML type and the existing data have been transformed into the XML type; consequently, the amount of XML data is increasing drastically [9]. XML can use its merit of describing meaningful information immediately to provide a standard data type in the form of exchanging information on a lot of data generated in the process of companies' database or applied program operation. It is therefore very appropriate for a component label and document management system that needs definition and description of detailed information and its meaning. As a large amount of XML-type information was provided on web environment, developers and users became more concerned about the issue of XML document security.

As for researches in XML document security and relevant products, control of access to information in web environment and transmission layer security protocols

B. Gabrys, R.J. Howlett, and L.C. Jain (Eds.): KES 2006, Part I, LNAI 4251, pp. 451 – 458, 2006.

© Springer-Verlag Berlin Heidelberg 2006

including electronic signature and coding is mostly related to HTML documents, and couldn't deal with access control according to the meaning of partial information, which is the main advantage of XML as file-based access control. So access control applying the advantage of XML became necessary [6]. The existing access control first parses XML documents to get DOM trees if a user demands XML documents. After the parsing, it sets the sign value that means permission (+) or denial (-) of access to nodes of DOM trees in reference to authorization of relevant database and XML documents. Nodes with the sign value set at - in DOM trees are removed and only those with the value set at + are shown to the user in the XML type [4], [5], [6].

However, it has problems that DOM trees should be loaded on memory, that a large amount of memory is used to search for trees repetitively to set access authorization on all nodes of DOM trees, and that the system becomes inefficient due to complicated authorization assessment.

In this paper, we suggest XML authorization system and small memory management for XML data protection on the server. It is therefore possible to make it easy to manage information on access authorization and users and remove unnecessary parsing and DOM tree searching, consequently obtaining better efficiency than the existing access control model.

The paper is organized as follows. Section 2 examines studies and problems about XML access control. Section 3 defines an Authorization and small memory management for authorization policy rule to describe document tree labeling algorithm. Section 4 evaluates the authorization policy models and section 5 draws a conclusion and describes the future course of studies.

2 Related Work

Our XML authorization system is related to many of the previously available access control [1], [2], [4], [5], [5], [8]. DOM is a platform-independent and language-neutral interface which can express contents and structure of Internet documents including HTML and XML ones as objects and handle them, that is, which can dynamically change contents, structure, and style of documents. As the definition of interface that can be used by operating and accessing objects (tags) and documents, DOM Level 1 can express contents of HTML or XML documents parsed without a loss, support HTML 4.0 and XML 1.0, generate a hierarchical structure of documents, and easily are expanded to use high-level API with ease. Level 2 defines the function of supporting an object model applying style sheet and operating information on the style of documents. DOM higher than Level 2 also includes description of user interface usable in Windows environment. By using this, a user can also define the function and security level of operating DTD of documents. That is, DOM higher than Level 2 serves to design API that enables a user to define, operate, change, and access all the things of documents, including style, events, structure, contents, and security level[3].

An XML DOM tree provides API (Application Program Interface) to access elements of XML documents [3]. The existing access control models [2], [4], [5] uses such a DOM tree to set access authorization to elements of DTD and XML documents and control users' access to XML data according to information on access authorization set.

According to the process of changes in documents in [6], there is a request for seeing XML documents. As for all XML documents and DTD concerned, information on access authorization is specified in documents called XAS (XML Authorization Sheet). XML documents are parsed to obtain DOM trees; then, a value of sign is set which means admission (+) or rejection (-) of access to nodes of DOM trees based on XAS of DTD and XML documents. It is called labeling to set authorization to nodes of DOM trees. The nodes with the value of sign set as - are removed from the labeled DOM trees and only those with the value set as + are restored to the user [2], [4], [5], [6]. Here, although XML documents with nodes removed from DOM trees can fail to be valid (its solution requires the loosening process, with all elements and attributes set as optional in DTD), they can maintain the existing DTD despite the removal of nodes from DOM trees.

To solve the problem that XML documents with nodes removed from DOM trees can fail to be valid for DTD, the loosening technique is suggested to maintain the existing DTD despite the removal of nodes from DOM trees [4]. However, this method causes a semantic conflict due to the loss of information on the structure. The tree labeling technique is used to maintain information on the structure of documents, which has a problem that it can violate secrecy by showing the existence of data and information on the structure with rejection (-) labeling [4], [5]. Although it applies a strong labeling technique to prohibit access to nodes with rejection (-) labeling [2], this technique has limitations in usability of data by prohibiting access to sub-nodes of prohibited nodes.

Above-mentioned studies have a problem of reducing the efficiency of system as the entire DOM trees should be loaded in memory and much memory is used due to repetitive tree retrieval to set access authorization to all nodes in DOM trees.

3 Access Authorization Model for XML Data Security

3.1 Security Access Authorizations Specification

In our model, a subject is a user. It is not the purpose of this paper to give detailed information on how these subjects are organized. Each user has an identifier which can be the login name. Each user is the member of one or several user groups. For the sake of simplicity we assume that groups are disjoint but they can be nested.
An object is Resource pointer of target object such as a directory path expression. Since we deal with XML documents in this paper, we use an XPath[10] tree to specify the target objects.

XML document access authorizations are composed of subject, object, action, action label type group, sign, and type:

- Subject: User name, IP address, computer name; (A subject who accesses XML documents and provides user group and pattern);

- Object: XPath 1.0 (An element of XML documents, which is expressed in Xpath);
- Action: Read, write, create, delete (An operation the subject can implement)
- Action Label Type Group: R(read operator group), DSLG(Document and Structure Label Group; operator group);
- Sign: {+, -} is the sign of the authorization, which can be positive (allow access) or negative (forbid access);
- Type: {L, R, LDH, RDH, LS, RS, LD, RD} means an authorization type (Type means an attribute value of authorization; L (Local) is applied to an element with authorization propagation described and its attributes; R (Recursive) is applied to all subelements with authorization propagation described and their attributes; and D is authorization to access DTD; Recursive DTD Hard; Local Soft; Recursive Soft; Local DTD).

The subject of authorization can be described as id or the access-requested position. The object means a resource to protect access. XPath language, which is a W3C standard of path expression, or expanded Uniform Resource Identifier (URI), is used to express the object [10]. Path expression is a list of pre-defined functions or element names differentiated by a divider (/) on the structure of document tree. Action refers to an operation the subject can implement; according to how much action label type group operators change XML, the operator group can be classified as follows:

- Read Operator Group: A set of operators that read but never change documents in XML (Read).
- Document Structure Label Group: A set of operators that change XML documents and structures (Insert, Delete, Rename).

3.2 Propagation Policy and Default Policy

A Propagation policy rule is a security policy to use for regulating authorization conflicts to set access authorization. As for authorization interpretation, the final sign (+ or -) is determined by reflecting propagation and overriding in each element. If there are both permission and denial for the same subject, only one access authorization is determined according to the conflict settlement principle. The following steps are rules to determine precedence of authorization in case of authorization conflicts.

Rule 1: Authorization on the most specific subject described according to partial order of subjects takes precedence.
Rule 2: Directly described authorization rather than that occurring by transmission takes precedence.
Rule 3: Authorization directly described on XML documents rather than that described on DTD takes precedence.
Rule 4: Authorization on nodes rather than that of its forefather takes precedence.

When there is no permission(grant or deny) for an access request or when the confliction resolution policy "nothing takes precedence" is enforced, we need to make a decision according to the specified default policy. This can be specified in the <default> element for each action the default policy is "deny" by default for every action [8].

3.3 XML Authorization System Strategy for Browsing Access Request

Algorithm is an algorithm to enforce a document tree labeling process. In general access control information, access authorization on an XML document is provided to A.xml and that for DTD is separated by A_dtd. If a security manager explicitly set no authorization on a specific element, a predefined basic access authorization is assigned. Given a requester rq and an XML document URI, the algorithm first initializes variable T to the tree representing document and r to the root of T. Therefore, the first step of Initial_label is composed of determination of the authorization set An applied to the requester rq and defined for the document URI at the instance and schema level. Default() function returns authorization assigned when there is no explicit authorization. Propagation_rule() function returns authorization with the highest precedence by predefined propagation rules if a conflict occurred in the same mode. Such a labeling process makes a preorder visit to trees. If the sum of sets for L1r and L2r is an empty set, that is, if there is no explicit access authorization information, the predefined basic access authorization value is set for Lr. Otherwise, authorization with the highest precedence of explicit ones is set. While root node doesn't have parent node, all kinds of node but root node has it. To preserve the structure of the document, the portion of the document visible to the requester will also include start and end tags of elements with a negative or undefined label that have a descendant with a positive level. Since the original tree includes even those portions a requester is not entitled to see, the view of the document is created which shows only the portions permissible for the requester through the tree pruning process according to signs of each node.

Input: A requester rq and an XML document URI
Output: The document to be returned to rq
(1) A.xml A = {a= <subject, object, action, action label type group, sign, type> | a ∈ authorization, rq ≤ AS subject, uri(object) = =URI OR uri(object) = = (URI)}
(2) Let r be the root of the tree T corresponding to the document URI, n is a node other than r, p is the parent node of n
(3) AM( ): returns the ALTG of a node specified in the authorization rule,
(4) Type(): returns the type specified in the authorization rule,
(5) Propagation_rule(): returns the ALTG determined by propagation rules
(6) For each c ∈ children(r) do label(c, r)
(7) For each c ∈ children(r) do prune(T, c)
(8) L1r = AM(r) in A.dtd , L2r = AM(r) in A.xml
(9) initial_label(r)
(10) For each c ∈ children(r) do label(n,p)

```
Procedure initial_label(n)
   if L1r ∪ L2r ={ }, Lr=default(r)
   else Lr = propagation_rule([L1r, L2r])
Procedure label(n,p)
   if type (p) in [L, R, LDH, RDH, LS, RS, LD, RD]
      /* the set of authorizations of each type on the node */
      if L1n & L2n = { }, Ln = Lp
```

```
      else Ln = propagation_rule([Lp, L1n, L2n])
   else
      if L1n & L2n = { }, Ln = default(n)
      else Ln = propagation_rule([L1n, L2n])
Procedure prune(T, n)
    /* Determines if n has to be removed from T */
   For each c ∈ children(n) do prune(T, c)
   if children(n) = { } and Ln ≠ '+'
   then remove the current node from T
```

Existing XML access control techniques determine whether to allow a query to access or not after labeling the DOM tree. Thus the system has to keep all information necessary for right tests to the end unnecessary right tests were repeated[6]. Such extra tasks slow down the speed of access control

4 XML Authorization and Memory Management Evaluation

Our processor takes as input a valid XML document requested by the user, together with its XAS listing the associated access authorizations at the instance level. The processor operation also involves the document's DTD and the associated XAS specifying schema-level authorizations. Processor output is valid XML documents including only information a user can access. To provide a uniform representation of XASs and other XML-based information, the syntax of XASs is given by the XML DTD depicted in Figure 1[8].

```
<!ELEMENT set of authorizations (authorization)+>
<!ELEMENT authorization (subject, object, ALTG, action, sign, type, priority)>
<!ELEMENT subject (#PCDATA)>
<!ELEMENT object (#PCDATA)>
<!ELEMENT ALGT empty>
<!ELEMENT action empty>
<!ELEMENT sign empty>
<!ELEMENT type empty>
<!ATTLIST set of authorizations about CDATA #REQUIRED>
<!ATTLIST ALTG(R, DSLG) #REQUIRED>
<!ATTLIST action value (read, write create, delete) #REQUIRED>
<!ATTLIST sign value (+ | – ) #REQUIRED>
<!ATTLIST type value (L|R|LDH|RDH|LS|RS|LD|RD) #REQUIRED>
<!ATTLIST priority value (hard | soft)  #IMPLTED>
```

Fig. 1. DTD Syntax of XML Authorization Sheet

In this study we designed an authorization policy management to access XML documents safely. Apache XML parser, XML, and Internet Explorer 6.0 in the Windows XP operating system were used as system implementation tools to design a prototype of the access control system with Java 5.0. Figure 2 shows the structure of the suggested access control system. The process output is a valid XML document including only the information the user is allowed to access. In this system documents and DTD are internally expressed according to DOM technology. As for access control, if a certified user requests desired data, the system will identify the user's authorization and determine if it will provide the requested data.

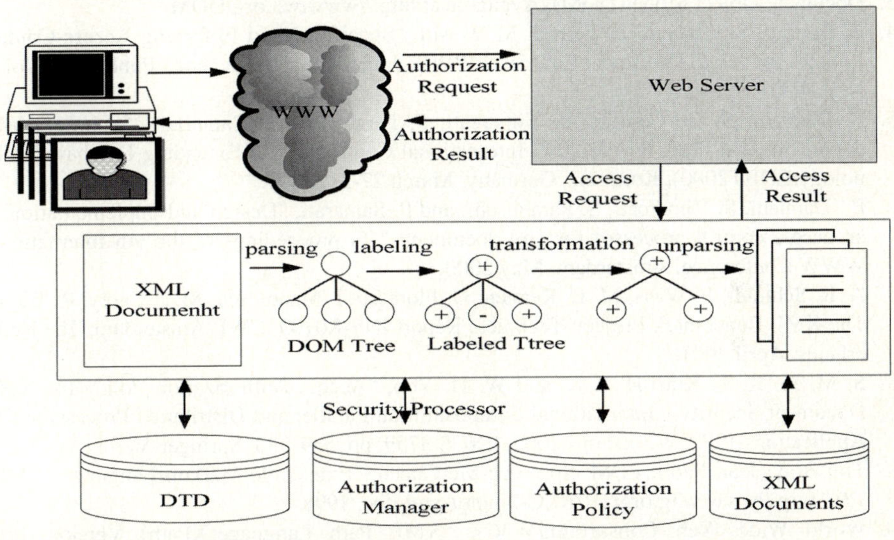

Fig. 2. Architecture of the Access Authorization and Memory Management

5 Conclusion

As for access control for an XML document, a certified user also asks the system for a desired XML document. A security processor checks the user's authorization from authorization database. If the user is authorized to access all or some portions of the requested XML document, they will go through an XML document conversion process and then be delivered in the view type suitable for the user.

In this study is primarily characterized by consideration of who would exercise what access privileges on a specific XML document and by good adjustment of organization-wide demands from a policy manager and a single document writer. It is therefore possible to reduce the costs for authorization evaluation which was complex and repetitive in the existing access control. Authorization rules are also applied when several subjects are in conflict with each other over the same object or when several authorizations for the same subject are overlapped as well as when common users or objects are authorized. By defining authorization, it was possible to provide rapid access control by removing unnecessary parsing and DOM tree retrieval. System performance lowering was reduced which had been caused by the labeling process to evaluate authorizations at each request by a user and repetitive visits to DOM trees as a defect of access control.

References

1. A. Anderson. A Comparison of EPAL and XACML. Sun Microsystems. Inc., 2004. Available at http://research.sun.com/projects/xacml/CompareEPALandXACML.html.
2. A. Gabillon and E. Bruno, "Regulating Access to XML Documents," In Proc. IFIP WG11.3 Working Conference on Database Security, 2001
3. Document Object Model(DOM), Available at http://www.w3.org/DOM/
4. E. Bertino, S. Castano. E. Ferrari, M. Mesiti, "Specifying and Enforcing Access Control Policies for XML Document Sources," WWW Journal, Baltzer Science Publishers, Vol.3, N.3, 2000.
5. E. Damiani, S. De Capitani di Vimercati, S. Paraboschi, P. Samarati, "Securing XML documents," in Proc. Of the 2000 International Conference on Extending Database Technology(EDBT2000), Konstanz, Germany, March 27-31, 2000.
6. E. Damiani, S. Vimercati, S. Paraboschi, and P. Samarati, "Design and implementation of an access control processor for xml documents," In proceedings of the 9th International WWW Conference, Amsterdam, May 2000.
7. A. R. Schmidt, F. Waas, M. L. Kersten, D. Florescu, I. Manolescu, M. J. Carey, R. Busse. The XML Benchmark Project. Technical Report INS-R0103, CWI, Amsterdam, The Netherlands, April 2001.
8. S. M. Jo, K. T. Kim, H. J. K, and W. H. Yoo, "Access Authorization Policy for XML Document Security," International Symposium on Parallel and Distributed Processing and Applications-ISPA Workshops 2005, LNCS 3759, pp. 589-598, Springer-Verlag
9. Tim Bray, Jean Paoli, C. M. Sperberg-Mc-Gueen, "Extensible Markup Language (XML) 1.0," http://www.w3g.org/TR/REC-xml#dt-xml-doc, 1998
10. World Wide Web Consortium(W3C), "XML Path Language(XPath) Version 1.0," http://www.w3.org/TR/PR-XPath 19991008, (October 1999).
11. Wang, Y. and Tan, K. A Scalable XML access control system. In Proceedings of the 10th International World Wide Web Conference(Hong Kong, 2001), 150-151

A Hybrid Heuristic Algorithm for HW-SW Partitioning Within Timed Automata*

Geguang Pu[1], Zhang Chong[2], Zongyan Qiu[2], Zuoquan Lin[2], and He Jifeng[1]

[1] Software Engineering Institute, East China Normal University, Shanghai, China, 200062
[2] LMAM, Department of Informatics, School of Math. Peking University, Beijing, China, 100871
{zc, lz}@is.pku.edu.cn, {ggpu, jifeng}@sei.ecnu.edu.cn, zyqiu@pku.edu.cn

Abstract. Hardware/Software (HW-SW) partitioning is a critical problem in co-design of embedded systems. This paper focuses on the synchronous system model, and formalizes the partitioning problem using timed automata (TA), which captures the key elements of the partitioning problem. Based on the TA model, we propose a hybrid heuristic algorithm to obtain near-optimal solutions effectively and efficiently. The experiments conducted show that our approach can deal with large applications with hundreds of nodes in task graph.

Keywords: hardware/software partitioning, timed automata, GRASP, tabu search, scheduling algorithm.

1 Introduction

Combined hardware/software implementations are common in embedded systems. In general, software is easy to modify, maintain, and upgrade, though it is slow and cheap compared to hardware . Thus, it is a critical state in co-design [12] to partition an application onto hardware (HW) and software (SW) execution units. This process is called hardware-software partitioning. One of the objectives of hardware/software partitioning is to search a reasonable composition of hardware and software components which not only satisfies the constraints such as timing, but also optimized desired quality metrics, such as communication cost, power consumption and so on.

Partitioning is a well-known NP-complete problem and during the last years some approaches based on algorithms have been developed, as described, for example, in [11,13,1]. In our previous work [9,10], we proposed a formal approach which modelled the partitioning process by means of timed automata tool [3], then obtained an optimal solution using the optimal reachachibility algorithm provided by model checker UPPAAL [7].

* Geguang Pu and Jifeng He are supported by 973 project No. 2005CB321904, and Zongyan Qiu is supported by NNSFC No. 60573081. Zhang Chong and Zuoquan Lin are supported by NNSFC No. 60373002, 60496322.

B. Gabrys, R.J. Howlett, and L.C. Jain (Eds.): KES 2006, Part I, LNAI 4251, pp. 459–466, 2006.
© Springer-Verlag Berlin Heidelberg 2006

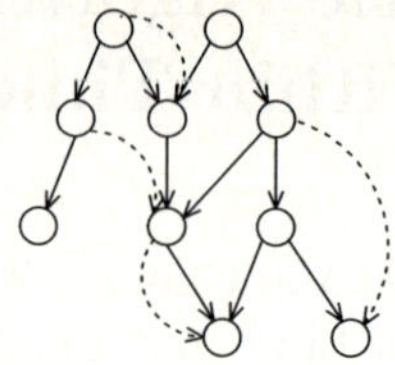

Fig. 1. Task Graph

However, due to the nature of NP-complete of the partitioning problem, the UPPAAL tool can only deal with the toy-size problem less than 40 nodes within reasonable time [9]. In this paper, we improve the approach proposed in [9] by integrating the TA model into the heuristic algorithm that is a hybridization of GRASP(Greedy Randomized Adaptive Search Procedures) [4] and tabu search [5]. This algorithm is used to search for near-optimal solutions in timed automata model. The experimental results in Section 4 show our algorithm can be applied to real applications effectively. Here, we consider the problem of minimizing execution time of an application for a system with hardware area constraints. The application can be formulated as a task graph shown in Figure 1, where the solid lines stand for the control relations between processes, while dotted lines denote data relations.

This paper is organized as follows. Section 2 describes a formal model of hardware/software partitioning using timed automata. In Section 3, we propose the hybrid heuristic algorithm to explore the design space of timed automata. Some partitioning experiments are conducted in Section 4. Finally, Section 5 is a short summary of the paper.

2 Formal Model Construction

In this section, we model the hardware/software partitioning formally by means of timed automata. The timed behavior of each process is modelled as a timed automaton, and the whole system is composed of a network of timed automata. After the model is constructed, a hybrid heuristic algorithm is applied to search for the allocation plan in the design space formed by timed automata network. Table 1 lists the variables used in our TA model together with their intended meaning.

2.1 Model Construction

The TA model of process P_i is depicted in Figure 2. Suppose process P_i is decided to run in software. Before P_i has occupied the software, it tests whether all its predecessors have been finished, i.e., $X_i = Token_i$ is established, then it sets St_i to 1 to denote it is put into software. Next, it has two possible branches selected: one represents that the communication is supposed to take place; the

Table 1. Notation

P_i	Process P_i
$Token_i$	The number of control predecessors of P_i
$Tcom_i$	The estimated communication time of P_i
$STexe_i$	The estimated execution time of P_i if implemented in software
$HTexe_i$	The estimated execution time of P_i if implemented in hardware
Rh_i	The estimated hardware resources needed for P_i
St_i	Variable indicating the state of P_i
X_i	Variable recording the number of the terminated control predecessors of P_i
SR	Variable indicating if the processor of software is occupied
HR	Variable indicating if the hardware is occupied
$Hres$	Constant storing the total available hardware resources
CS_i	Software clock for P_i
CH_i	Hardware clock for P_i
CC_i	Communication clock for P_i
T_i	The set of the processes from which P_i reads data from
D_i	The set of the indexes of the processes that P_i reads data from

other denotes no communication does. The details will be explained later. Before executing its computation task itself, it sets the global variable SR to 0 to prevent other processes from occupying the processor. It sets the software clock CS_i to 0 as well. The transition in the rightmost can only be taken place when the value of the clock CS_i equals $STexe_i$. As soon as the transition is taken place, variable X_j will be incremented if P_i is the predecessor of P_j. The situation is similar if P_i is implemented in hardware.

Let us consider the communication. Recall variable St_i is introduced to denote that process P_i is implemented in the hardware or software components. P_i then checks the processes that will transfer data to it (i.e. the processes in T_i) are in software or hardware. If at least one of them is in the hardware, the communication should be taken into account. In Figure 2, the condition $\Sigma_{j\in D_i} St_j < |T_i|$ is a guard to express at least one process that P_i reads data from is in hardware. The situation is the same when P_i is located in hardware and $\sum_{j\in D_i} St_j > 0$.

Next, when the communication really occurs between the software and hardware, it should occupy both the software and hardware components. That is to say, no other processes would be performed until the communication is finished. Accordingly, SR, HR are set to 0 simultaneously as long as the communication takes place. The clock CC_i is set to 0 when the communication begins.

3 Hybrid Heuristic Algorithm Within TA Model

In this section, a hybrid heuristic algorithm is introduced into the partitioning problem with TA model. This approach first generates heuristic indexes $hs_i \in [0, 1]$ on processes P_i, which indicates the possibility putting P_i into software or hardware. Following that, a GRASP style algorithm is used to serve as the

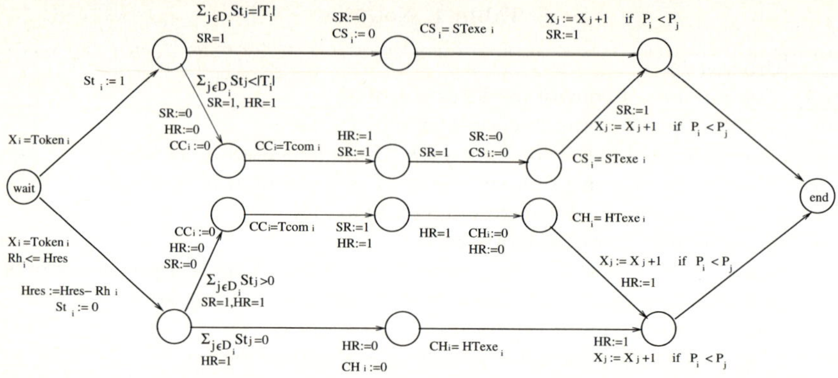

Fig. 2. The Timed Model of Process P_i

main algorithm frame. It generates initial solutions according to $\{hs_i\}$ and then a tabu search algorithm is used to refine those initial solutions and produce near-optimal solutions.

3.1 Heuristic Indexes for HW-SW Partitioning

Here we transform the partitioning problem presented with timed automata into a suit form which could be dealt with by our hybrid heuristic algorithm effectively.

Each TA model of process stands for the behaviors of P_i, and denotes P_i is allocated into hardware or software. All the TA models compose the network of timed automata. Vector Vec is introduced to represent the network of TA, and each location of vector Vec is assigned 1 or 0, where 1 denotes P_i is put in hardware, and 0 in software. The solution space with TA model can then be transformed into $\{0,1\}^{\|P\|}$. TA model provides a suitable solution space for this hybrid heuristic algorithm and the final solution found could be explained in TA model either. The suitable solution space is critical for the success of heuristic algorithm. Furthermore, the TA model provides the details of partitioning information, which will be used in the cost function of our algorithm.

The general approach of GRASP is composed of two steps. The first step constructs initial solutions, and the second step employs some refinement methods to improve the quality of the initial solutions further. We first introduce the way to obtain heuristic partitioning indexes hs_i on P_i used to generate good quality solutions.

The task graph can be represented by an acyclic directed graph G. We use C to denote the index set of processes in G and $\|C\| = N$. Let us consider the scenario where there is only one processing unit with limited resources named "hardware". The processing time for $\forall i \in C$ relies on the amount of "hardware" given to it. We denote U_i as the maximum amount of "hardware" required by

i, $e_i \in [0, U_i]$ as the amount of "hardware" given to P_i, and $r_i = \frac{e_i}{U_i}$ as the proportion P_i is satisfied. $STexe_i$ means the running time when given P_i no "hardware" and $HTexe_i$ means the running time when U_i amount of "hardware" is given to P_i. When there is no synchronized communication in G, equation (1) gives the processing time t_i of P_i.

$$t_i = STexe_i - (STexe_i - HTexe_i)r_i \qquad r_i \in \mathbb{R},\ r_i \in [0, 1] \tag{1}$$

Our goal is to minimize the total processing time under given limited amount of "hardware". This could be achieved according to

Theorem 1. *If the processes in G are satisfied as full as possible in the descendant order of* $\frac{STexe_i - HTexe_i}{U_i}$*, the total processing time will be minimized.*

Accordingly, $\frac{STexe_i - HTexe_i}{U_i}$ can serve as some heuristic weight and be used to construct heuristic method. It is easy to see these indexes are consistent with our intuition. It is supposed to put the processes into hardware whose acceleration uses less hardware resources.

When there is synchronized communication in G, the processing time of P_i is quite different from above. Let $P_i^j, j \in [1, m]$ represents neighbors which have synchronized communication with P_i in G and $q_i^j, j \in [1, m]$ represents the communication latency between P_i and P_i^j. Note that synchronized communication only happens when two processes located in different processing unit. We use the average communication latency to serve as the overall latency. We may regard P_i and P_i^j are all composed of two parts, one has "hardware" and one has none. Communication only happens between two different parts of P_i and P_i^j. The average latency between P_i^j and P_i is described by equation (2).

$$q_i^j \left(r_i(1 - r_i^j) + r_i^j(1 - r_i) \right) = q_i^j \left(r_i + r_i^j - 2r_i r_i^j \right) \tag{2}$$

The overall processing speed of P_i, therefore, is

$$t_i = STexe_i - (STexe_i - HTexe_i)r_i + \sum_{j=1}^{m} q_i^j \left(r_i + r_i^j - 2r_i r_i^j \right) \tag{3}$$

The total processing time of C is

$$\begin{aligned} t &= \sum_{i=1}^{N} t_i \\ &= \sum_{i=1}^{N} \left(STexe_i - (STexe_i - HTexe_i)r_i + \sum_{j=1}^{m} q_i^j \left(r_i + r_i^j - 2r_i r_i^j \right) \right) \\ &= -2R^T \cdot Q \cdot R + R^T(H - S + Q \cdot I + Q^T \cdot I) + I^T \cdot S \end{aligned} \tag{4}$$

where $R = (r_1, r_2, \cdots, r_n)^T$, $H = (h_1, h_2, \cdots, h_n)^T$, $S = (s_1, s_2, \cdots, s_n)^T$, $I = (1, 1, \cdots, 1)^T$. $Q = (q_{ij})_{N \times N}$ and q_{ij} represents synchronized communication amount from P_i to P_j. If we regard Equation (4) as the objective function and it is subject to $0 \le r_i \le 1$, it becomes a quadratic programming problem. We can solve it and use the result R as heuristic indexes for the hardware/software partitioning model.

3.2 Hybrid Heuristic Algorithm for TA Model

Once we obtain the heuristic indexes, we can use them to construct the hybrid heuristic algorithm for the partitioning problem.

Listing 1.1 shows the schema of our algorithm. The function "GetOneInitialSol" is executed according to the following steps. We first set $Vec[i] = 0$ if $R[i] = 0$ and $Vec[i] = 1$ if $R[i] = 1$. The rest items $Vec[k]$ are set to 1 or 0 with the possibility of $R[k]$. This process will not stop until no more items could be put into hardware. This process generates the initial solution which is a reasonable partitioning plan. We fix this initial result and use "Purify" function to obtain a near-optimal scheduling in the task graph. The result scheduling is used to evaluate the quality of this partition plan according to the minimum scheduling time.

```
p* = NULL;
while (!StopingCondition)
{
      p' = GetOneInitialSol(R);
      p̂ = Purify(p');
      if( D_p* > D_p̂ )
           p* = p̂;
}
```

Listing 1.1. Basic Scheme of Hybrid Heuristic Algorithm for H/S Partition

The basic purifying process is called "tabu search algorithm". It moves from the current solution to the next one according to some rules. First, it cannot move to those tabu solutions in the current one's neighborhood. The tabu solutions are determined according to the searching history information named tabu rules. Secondly, when a tabu solution is good enough, for example, better than the best solution found so far, it is possible to be accepted as the next step. This is known as aspiration rule.

To make the purifying method more effective, we employ a modified robust tabu search[5] to generate near optimal scheduling in G. One move defined in this tabu algorithm is inserting process P in hardware or software to another available place in hardware or software. The "available place" has two requirements. The first one is that P cannot be inserted in front of any process which is predecessor of it or insert at the back of any processes which is successor of it. The second one is the so-called "tabu rule". We call inserting P in front of P' as "forward surpass" and inserting P at the back of P' as "backward surpass". If P is inserted forward, all the processes it surpasses cannot be those that have been "back

Table 2. Experimental results of partitioning

Size	Cost (sec.)	Run time (sec.)	Num. of Process in HW
90	141.8	7.37	12
120	136.3	18.19	21
150	133.7	39.32	31
180	131.2	173.24	66

surpassed" in the recent $\mathcal{L}$ steps. If P is inserted backward, all the processes it surpasses cannot be those that have been "forward surpassed" in the recent $\mathcal{L}$ steps. $\mathcal{L}$ is the so-called *tabu length.* The evaluation function is the total running time of the current scheduling plan. We also adopt the aspiration criteria that a tabu solution is accepted despite the tabu state when its evaluation function value is better than the best solution ever found before.

4 Experimental Results

In this section, we use the technique in previous sections to find near-optimal solution for some hardware/software partitioning cases. "Lingo®" is used to solve the quadratic programming problem. All the test programs are coded in ANSI C and complied with gcc-3.2 on Ret Hat Linux 9.0. Optimization level "-O3" is used. The hardware platform is a single 2.4GHz Pentium4 machine with 512M Memory. The model checker UPPAAL is used as timed automata modelling tool.

The tabu length is randomly selected from $[0.9\sqrt{N}, 1.1\sqrt{N}]$ for every N steps. When the best solution ever found is not updated in $4 * N$ steps, the purify function is terminated. For each example, we test $0.1N$ initial solutions.

In order to validate our partitioning approach, we performed some experiments on a real-life model: the *Ethernet network coprocessor* [6]. The coprocessor transmits and receives data frames over a network under CSMA/CD protocol. Its purpose is to off-load CPU from managing communication activities. In terms of the different granularity of task graph, the number of nodes in task graph generated ranges from 90 to 180.

To estimate the required resources in hardware of each process, the hardware compiler [2] technique is used in our experiment as well. This technique can estimate the hardware resources such as *gate, register* very quickly. We also assume the HW execution time of a process is faster than the SW execution time by a number between 5 and 10 times.

The results are summarized in Table 2. The first column in the table is the number of the processes, the second column stands for the cost of the current partitioning found, which means the execution time of the whole system based on the allocation plan. The third column shows the execution time of our algorithm counted in seconds. The last column presents the number of processes allocated in hardware in terms of current partitioning plan.

5 Summary

This paper presents a novel approach to hardware/software partitioning supporting the abstract architecture in which the communication takes place synchronously. After the designer decides the process granularity of the initial specification, the partitioning process can be carried out automatically. We first model this partitioning problem with timed automata formally, and the behaviors of processes can be captured by TA model precisely. The whole TA network composes the design partitioning space. Due to the NP-complete nature of this problem, we propose an effective hybrid heuristic algorithm applied to the TA model. By integrating Tabu and GRASP algorithms, the proposed algorithm is very encouraging, and the experiments have clearly demonstrated it can explore larger design space up to 200 nodes in task graph. As a result, this approach can help designers to make quick and good decisions at design stage in co-design.

References

1. S. Banerjee and N. D. Dutt. Efficient search space exploration for HW-SW partitioning. *CODES+ISSS'04*, pp122-127, 2004.
2. J. Bowen and He Jifeng. An approach to the specification and verification of a hardware compilation scheme. *Journal of Supercomputing*, 19(1):pp23-29, 2001.
3. R. Alur and D. L. Dill. A Theory for Timed Automata. *Theoretical Computer Science*, Vol.125, pp183-235, 1994.
4. T. Feo and M. Resende. Greedy randomized adaptive search procedures. *Journal of Global Optimization*, Vol.6, pp109-133, 1995.
5. F. Glover and M. Laguna. Tabu Search. *Handbook of Combinatorial Optimization(Vol.3)*, pp621-757, 1998
6. Handle-C Language Overview. Celoxica Limited, 2002.
7. K. G. Larsen, P. Pettersson and Wang Yi. UPPAAL in a Nutshell. *Int.Journal of Software Tools for Technology Transfer 1*,1-2(Oct), pp134-152, 1997.
8. N. Mladenovic and P. Hansen. Variable neighborhood search. *Computers Oper. Res.*, 24:pp1097-1100, 1997.
9. Geguang Pu, Dang Van Hung, Jifeng He and Wang Yi. An optimal approach to hardware/software partitioning for synchronous model. *IFM'04*, LNCS 2999, pp363-381, 2004.
10. Geguang Pu, Zhao Xiangpeng, Wang Shuling, Qiu Zongyan, Jifeng He and Wang Yi. An approach to hardware/software partitioning for multiple hardware devices model. *IEEE SEFM'04*, pp376-385, 2004.
11. G. Quan, X. Hu and G. W. Greenwood. Preference-driven hierarchical hardware/software partitioning. In *Internatitional conference on Computer Design(IEEE)*, pp652-657, 1999.

Performance Analysis of WAP Packet Transmission Time and Optimal Packet Size in Wireless Network

Il-Young Moon

School of Internet Media Engineering,
Korea University of Technology and Education, Republic of Korea
iymoon@kut.ac.kr

Abstract. With the increase of data communication service by mobile devices, WAP is a protocol proposed to efficiently access the internet contents by user request through wireless condition that has a high error rate and mobility. In this paper, it has analyzed transmission time for wireless application protocol (WAP) in wireless CDMA network. In order for segmentation and reassembly (SAR) to improve the transfer capability, the transmission of messages have been simulated using a fragmentation that begins with the total package and incremental fragmentation for each layer using the wireless transaction protocol (WTP) to define the resultant packet size and the level of fragmentation for each proceeding layer. SAR decreases transmission time of radio link protocol (RLP) baseband packets by sending packets. From the results, it was able to obtain packet transmission time and optimal WTP packet size in wireless network.

1 Introduction

Recently, conventional telecommunications have been dependent on voice communications as the major application of technology. As data communication services have become more broadly available, there is growing interest to provide services that take advantage of data capabilities as well. WAP was shaped to create a standards-based structure in which value-added data services can be deployed, ensuring some degree of interoperability. In addition, WAP and HTML offers an interoperable presentation platform for end-user interfaces. Applications that will be increasingly popular in the future, such as man-machine and machine-machine interfaces, will drive WAP in wireless CDMA network. But a problem of WAP is limited transmission time by severe mobile environment. So it takes much time to transmit WAP packet and causes packet loss.

In this paper, it is analyzed that WAP packet transmission time to improve performance of WAP using SAR algorithm and variable RLP layer in wireless CDMA network [1],[2]. In order for SAR to progress the transfer capability, the whole messages are fragmented in WTP layer and segmented further as it passes through each layer towards the physical layer, where the actual packets are sent serially.

Also, to achieve an appropriate WTP packet size, it is considered the BER in a wireless channel.

B. Gabrys, R.J. Howlett, and L.C. Jain (Eds.): KES 2006, Part I, LNAI 4251, pp. 467 – 474, 2006.
© Springer-Verlag Berlin Heidelberg 2006

2 WAP Protocol

With the open system interconnection (OSI) in mind, the WAP stack basically is divided into five layers. They are a wireless application environment (WAE) as transaction layer, wireless session protocol (WSP) as session layer, WTP as transaction layer, wireless transport layer security (WTLS) as security layer and wireless datagram protocol (WDP) as transport layer [3],[4]. Figure 1 shows how it relates to the protocols on the internet. Each layer of the WAP protocol stack specifies a well-defined interface, meaning that a certain layer makes lower layers invisible in figure 1. The layered architecture allows other applications and services to utilize the features provided by the WAP stack as well. This makes it possible to use the WAP stack for service and applications that currently are not specified by WAP.

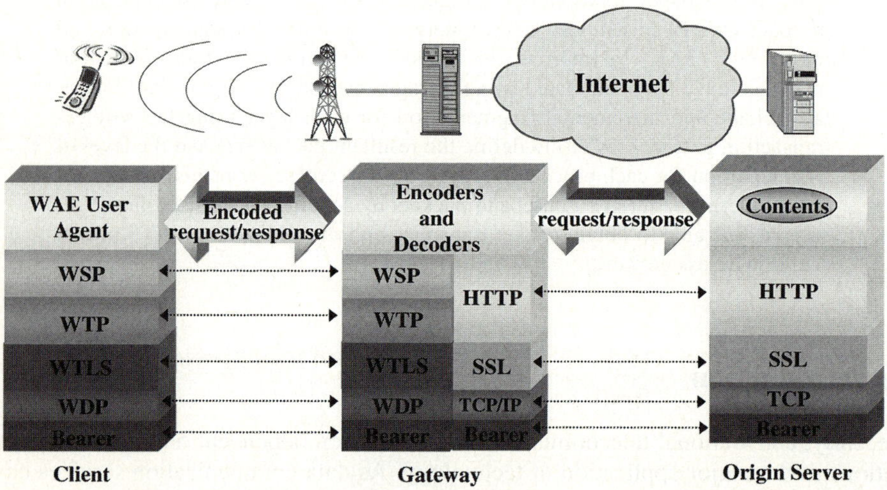

Fig. 1. The WAP Protocols on the Internet

3 Transmission Modeling for WAP Packet in Wireless Network

The SAR operation is used to increase efficiency by supporting a maximum transmission unit (MTU) larger than the maximum allotment of a single packet [5],[6]. This reduces overheads by spreading the packets used by higher layer protocols over several packets. In order for SAR to progress the transfer capability, the whole messages are fragmented in WTP layer and segmented further as it passes through each layer towards the physical layer, where the actual packets are sent sequentially [7]. Figure 2 shows a flow control of SAR using TPI (Transport Information).

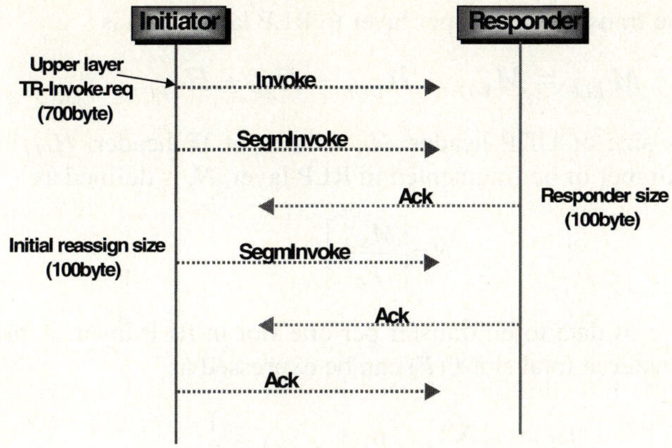

Fig. 2. Flow control of SAR

In this paper, to valuate WAP packet transmission time, it is simulated in Rayleigh fading channel [8]. Rayleigh probability density functions can be written as

$$p(R) = \frac{R}{\sigma_R^2} \exp\left(\frac{-R^2}{2\sigma_R^2} \right) \tag{1}$$

where, R is defined as envelope of a received signal and σ_R^2 means a distribution of a received signal .

4 Analysis of WAP Packet Transmission Time Using SAR Algorithm and RLP Layer

To be fragmented to WTP of WAP, the number of total message packet (K) follows by equations

$$K = \left\lceil \frac{M_{TOTAL}}{M_{SEG}} \right\rceil, \tag{2}$$

where, $\lceil x \rceil$ is integer less than $x+1$ and M_{TOTAL} is size of total message of upper layer WTP.

The total size of WTP packet, M_{WTP}, consist of the size of message to be fragment at upper layer, $M_{SEG,}$ and the size of WTP header, H_{WTP}, and can be expressed as

$$M_{WTP} = M_{SEG} + H_{WTP}, \tag{3}$$

Also, the size of last packet to be fragmented to WTP, L_{WTP} is obtained by

$$L_{WTP} = M_{TOTAL} - (K-1)M_{SEG} + H_{WTP}, \tag{4}$$

The size to be transfer from upper layer to RLP layer M_{RLP} is

$$M_{RLP} = M_{WTP} + H_{UDP} + H_{IP} + H_{PPP}, \tag{5}$$

where, H_{UDP} is size of UDP header, H_{IP} is size of IP header, H_{PPP} is size of PPP header, and a number of be fragmented in RLP layer, N, is defined as

$$N = \left\lceil \frac{M_{RLP}}{F_D} \right\rceil, \tag{6}$$

where, F_D is size of data to be transfer per one slot in RLP layer. A probability to be successfully transfer at total slot $E(F)$ can be expressed as

$$E(F) = \sum_{m=1}^{\infty} m \cdot P(F = m) = \frac{1}{p}, \tag{7}$$

where, $P(\cdot)$ is a probability of data frame to be successfully transfer and p is a probability to be successfully transfer. And, the number of average time slot to transfer WAP packet is as follows

$$E(P) = N \cdot E(F) = \frac{N}{p}, \tag{8}$$

Using the equation (7) and slot time S_{TIME}, the transmission time of WAP packet to be fragment is obtained as

$$T_{PKT}(N) = E(P) \cdot S_{TIME} = \frac{S_{TIME} N}{p} (ms) \tag{9}$$

Therefore, it is achieved the total message transmission time $T_{MSG}(ms)$, and is expressed as

$$\begin{aligned} T_{MSG} &= (K-1) T_{PKT}(q) + T_{PKT}(r) \\ &= (K-1) \frac{S_{TIME} \times q}{p} + \frac{S_{TIME} \times r}{p}, \end{aligned} \tag{10}$$

$$q = \left\lceil \frac{M_{WTP} + 36}{F_D} \right\rceil, \; r = \left\lceil \frac{L_{WTP} + 36}{F_D} \right\rceil, \tag{11}$$

where, q and r are the number of time slot to computed by F_D.

Also, it value for the BER of the payload part in the receiver part has been calculated. In order to simulate the BER performance, an independent, static Rayleigh fading and AWGN channel were assumed for every packet in CDMA/QPSK environment. A BER of CDMA/QPSK can be written as

$$p_e = \frac{1}{2} erfc(\sqrt{SNR}) \tag{12}$$

$$SNR = \frac{1}{\left(\frac{E_b}{N_0}\right)^{-1} + \frac{2(U-1)}{3PG}} \tag{13}$$

U : Number of multiple access user,
PG : Processing gain of system,
E_b/N_o : The ratio of signal energy per bit to noise power density.

Figure 3 depicts a process of segment transmission for WTP packet for the above equations.

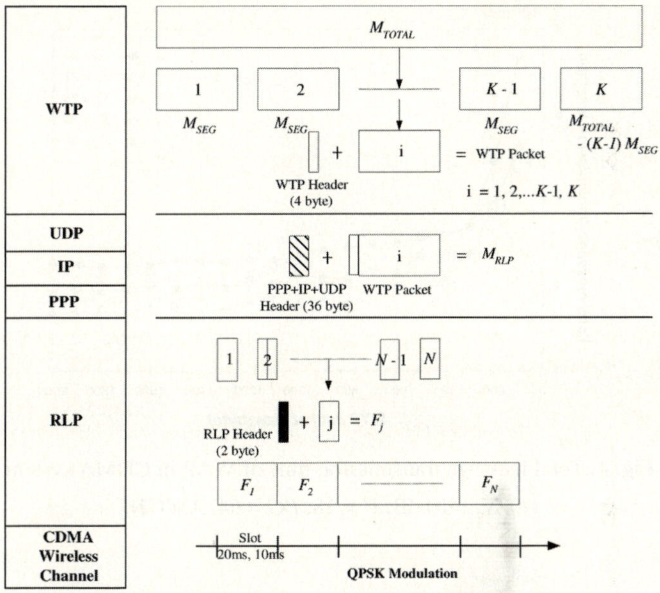

Fig. 3. Process of segment transmission for WTP packet

In the wireless channel, it gains transmission time of packets and analyze BER performance of the WAP in wireless CDMA network using SAR algorithm in AWGN and Rayleigh fading channel. It is considered the transmission time of WAP packet with variable RLP layer size using SAR algorithm on the CDMA wireless channel. So, it is analyzed RLP layer size from F_D = 26 to F_D = 30. M_{WTP} is analyzed fragmenting from 100 to 2000 bytes. In addition, in the case of CDMA and WCDMA network, S_{TIME} is set to 20ms, 10ms, respectively. For achieving transmission time of packets for WAP in wireless CDMA network, the total message transmission time is simulated at M_{TOTAL} = 5000 byte, E_b/N_o = 10 dB, U = 18, PG = 64 in AWGN channel and at M_{TOTAL} = 5000 byte, E_b/N_o = 10 dB, U = 18, PG = 64 in Rayleigh fading channel.

In figure 4 and 5, the parameter E_b/N_o is set at 10 dB using SAR algorithm in wireless CDMA and WCDMA AWGN channel, respectively. In figure 4, when the packet size increases from 100 to 2000 bytes, transmission time is less than the transmission time of a 100 bytes. Also, in the case of CDMA, transmission time is increased more about 2 times than WCDMA transmission time. Also, in figure 6 and 7, the result is approximately the same as figure 4 and 5, but one is AWGN channel and the other is Rayleigh fading channel.

From the results of figure 4,5,6 and 7, it finds that the WTP packet size should increase in order for the transmission time in wireless channel to decrease. To obtain an appropriate WTP packet size, it is also considered the BER in a wireless channel.

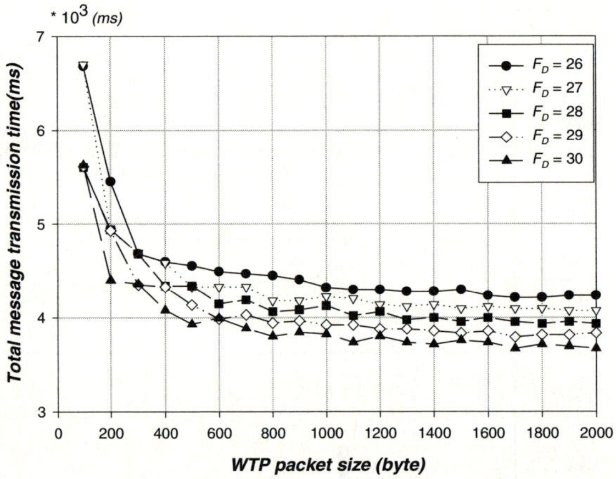

Fig. 4. Total message transmission time of WAP in CDMA system (E_b/N_o = 10 dB, U = 18, PG = 64, AWGN)

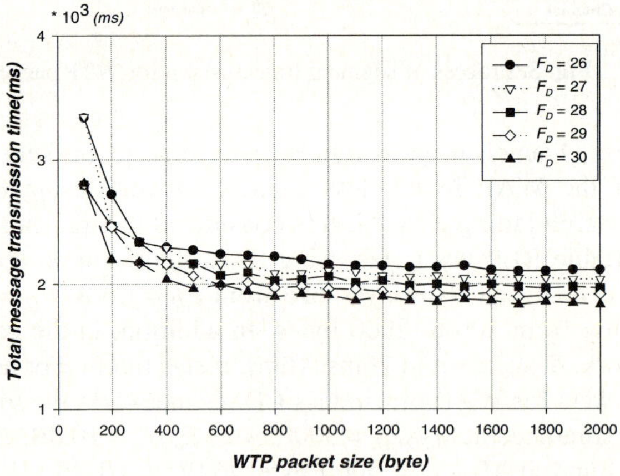

Fig. 5. Total message transmission time of WAP in WCDMA system

(E_b/N_o = 10 dB, U = 18, PG = 64, AWGN)

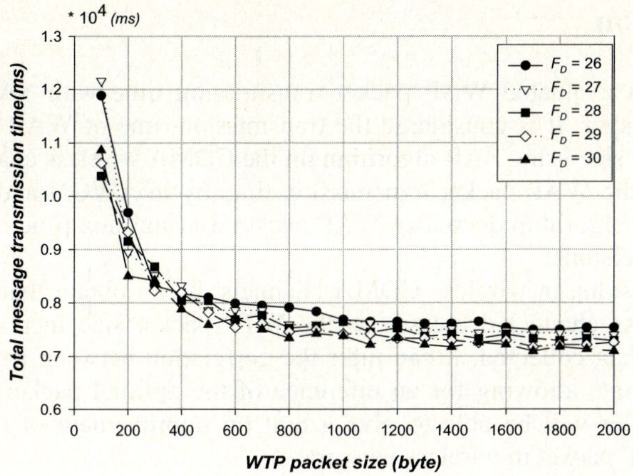

Fig. 6. Total message transmission time of WAP in CDMA system

(E_b/N_o = 10 dB, U = 18, PG = 64, Rayleigh fading)

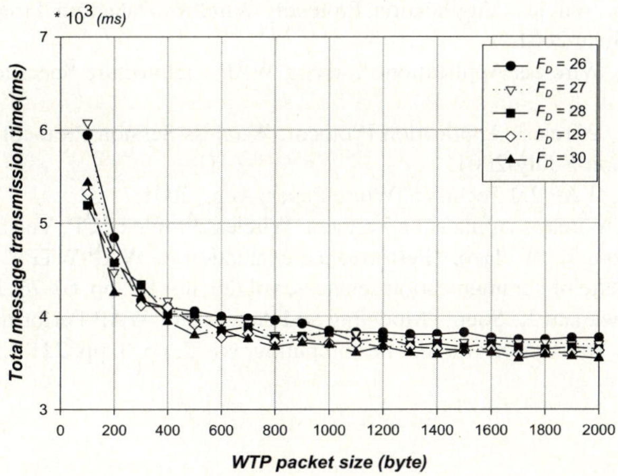

Fig. 7. Total message transmission time of WAP in WCDMA system

(E_b/N_o = 10 dB, U = 18, PG = 64, Rayleigh fading)

Furthermore, in the case of optimal WTP packet size (between about 500 byte and 600 byte) in AWGN and Rayleigh fading channel, the WAP packet transmission time (considering trade-off between total message transmission time and WTP packet size) is about 4000~4400 ms (CDMA), 2000~2400 ms (WCDMA) in AWGN and is about 7400~7700 ms (CDMA), 3700~4100 ms (WCDMA) in Rayleigh fading channel.

5 Conclusion

This paper has evaluated WAP packet transmission time with SAR algorithm and variable RLP layer. It is considered the transmission time of WAP packet with variable RLP layer size using SAR algorithm on the CDMA wireless channel. In addition, it is analyzed the WAP packet transmission time by in AWGN and Rayleigh fading channels. SAR algorithm decreases WAP packet transmission time of baseband packets in wireless channel.

From the results, in wireless CDMA channels, it can obtain that the transmission time in wireless channel decreases as the WTP packet size increases. As a result, based on the data collected, it can infer the correlation between packet size and the transmission time, allowing for an inference of the optimal packet size in the WTP layer. This paper will be able to give a hint on improvement of quality of service (QoS) for WAP packet in wireless network.

References

1. WAP forum, Wireless Application Protocol: Wireless Transaction Protocol Specification, Version 10-July-2001.
2. WAP forum, Wireless Application Protocol: Wireless Datagram Protocol Specification, Version 14-June-2001.
3. WAP forum, Wireless Application Protocol: WAP Architecture Specification, Version 12-July-2001.
4. WAP forum, Wireless Application Protocol: Wireless Session Protocol Specification, Approved Version 5-July-2001.
5. WAP forum, WAP 2.0 Technical White Paper, Aug., 2001.
6. WAP forum, Wireless Application Protocol: Wireless Profiled TCP, Version 31-Mar.-2001.
7. H. S. Park and K .W. Heo, "Performance evaluation of WAP-WTP," The journal of the Korean institute of communication sciences, vol. 26, no. 1A, pp. 67-76, Jan. 2001.
8. H. Rutagemwa and X. Shen, "Modeling and Analysis of WAP Performance over Wireless Link," IEEE Transactions on mobile computing, vol. 2, no. 3, pp. 221-232, Jul-Sep. 2003.

Using Artificial Intelligence in Wireless Sensor Routing Protocols

Julio Barbancho, Carlos León, Javier Molina, and Antonio Barbancho

Department of Electronic Technology, University of Seville
C/ Virgen de África, 7. Seville 41011, Spain
Tlfno.: (+034) 954 55 71 92, Fax: (+034) 954 55 28 33
{jbarbancho, cleon, fjmolina, ayboc}@us.es

Abstract. This paper represents a dissertation about how an artificial intelligence technique can be applied to wireless sensor networks. Due to the constraints on data processing and power consumption, the use of artificial intelligence has been historically discarded in these kind of networks. However, in some special scenarios the features of neural networks are appropriate to develop complex tasks such as path discovery. In this paper, we explore the performance of two very well known routing paradigms, *directed diffusion* and *Energy-Aware Routing*, and our routing algorithm, named **SIR**, which has the novelty of being based on the introduction of neural networks in every sensor node. Extensive simulations over our wireless sensor network simulator, OLIMPO, have been carried out to study the efficiency of the introduction of neural networks. A comparison of the results obtained with every routing protocol is analyzed.

Keywords: Wireless sensor networks (WSN); Ad hoc networks, Quality of service (QoS); Artificial neural networks (ANN); Routing; Self-Organizing Map (SOM), ubiquitous computing.

1 Introduction

Goals like efficient energy management, high reliability and availability, communication security, and robustness have become very important issues to be considered in wireless sensor networks (WSN). This is one of the many reasons why we can not neglect the study of the collision effects and the noise influence.

We present in this paper a new routing algorithm which introduces artificial intelligence (AI) techniques to measure the quality of service (QoS) supported by the network.

This paper is organized as follows. In section 2, we relate the main routing features we should consider in a network topology. A description of the defined network topology is given. Section 3 introduces the use of neural networks in sensors for determining the quality of neighborhood links, giving a QoS model for routing protocols. The performance of the use of this technique in existing routing protocols for sensor networks is evaluated by simulation in section 4. Concluding remarks and future works are given on section 5.

B. Gabrys, R.J. Howlett, and L.C. Jain (Eds.): KES 2006, Part I, LNAI 4251, pp. 475–482, 2006.
© Springer-Verlag Berlin Heidelberg 2006

2 Designing the Network Topology

The WSN architecture as a whole has to take into account different aspects, such as the protocol architecture; Quality-of-Service, dependability, redundancy and imprecision in sensor readings; addressing structures, scalability and energy requirements; geographic and data-centric addressing structures; aggregating data techniques; integration of WSNs into larger networks, bridging different communication protocols; etc.

Due to the desire to cover a large area, a communication strategy is needed. there are many studies that approach the problem of high connectivity in wireless ad hoc networks [1], [2]. In our research we consider a random distribution of sensors.

In general, routing in WSNs can be divided into *flat-based* routing, *hierarchical-base* routing, and *location-based* routing. In this paper we study networks where all nodes are supposed to be assigned equal roles or functionalities. In this sense, flat-based routing is best suited for this kind of networks.

Among all the existing flat routing protocols, we have chosen *directed diffusion* and *Energy-Aware Routing* (*EAR*) to evaluate the influence of the use of AI techniques.

In directed diffusion [3], sensors measure events and create gradients of information in their respective neighborhoods. The base station request data by broadcasting interests. Each sensor that receives the interest sets up a gradient toward the sensor nodes from which it has received the interest. This process continues until gradients are set up from the sources back to the base station.

EAR [4] is similar to directed diffusion. Nevertheless it differs in the sense that it maintains a set of paths instead of maintaining or enforcing one optimal path at higher rates. These paths are maintained and chosen by means of a certain probability. The value of this probability depends on how low the energy consumption that each path can achieve is. By having paths chosen at different times, the energy of any single path will not deplete quickly.

3 Introducing Neurons in Sensor Nodes

The necessity of connectivity among nodes introduces the routing problem. In a WSN we need a multi-hop scheme to travel from a source to a destiny. The paths the packets have to follow can be established based on a specific criterion. Possible criteria can be minimum number of hops, minimum latency, maximum data rate, minimum error rate, etc. For example, imagine that all the nodes desire to have a path to route data to the *base station*[1]. In this situation, the problem is solved by a technique called *network backbone formation.*

Our approach to enhance this solution is based on the introduction of artificial intelligence techniques in the WSNs: expert systems, artificial neural networks, fuzzy logic and genetic algorithms. Due to the processing constraints we have

[1] In WSN, we often consider two kind of nodes, base stations and sensor nodes. There is usually only one base station.

to consider in a sensor node, the best suited, among all these techniques, is the *self-organizing-map* (SOM). This is kind of artificial neural network based on the self organization concept.

SOM is an unsupervised neural network. The neurons are organized in an unidirectional two layers architecture. The first one is the input or sensorial layer, formed by m neurons, one per each input variable. These neurons work as buffers distributing the information sensed in the input space. The input is formed by stochastic samples $\boldsymbol{x}(t) \in \mathcal{R}^m$ from the sensorial space. The second layer is usually formed by a rectangular grid with $nxxn'y$ neurons. Each neuron (i, j) is represented by an m-dimensional weight or reference vector called *synapsis*, $\boldsymbol{w'}_{ij} = [w'_{ij1}, w'_{ij2}, \ldots, w'_{ijm}]$, where m is the dimension of the input vector $\boldsymbol{x}(t)$. The neurons in the output layer -also known as the competitive Kohonen layer- are fully connected to the neurons in the input layer, meaning that every neuron in the input layer is linked to every neuron in the Kohonen layer. In SOM we can distinguish two phases: the ***learning phase***, in which, neurons from the second layer compete for the privilege of learning among each other, while the correct answer(s) is (are) not known; and the ***execution phase***, in which every neuron (i, j) calculates the similarity between the input vector $\boldsymbol{x}(t)$, $\{x_k \mid 1 \leq k \leq m\}$ and its own synaptic-weight-vector $\boldsymbol{w'}_{ij}$.

3.1 Network Backbone Formation

This problem has been studied in mathematics as a particular discipline called *Graph Theory,* which studies the properties of graphs.

A *directed graph* G is an ordered pair $G := (V, A)$ with V, a set of vertices or nodes, v_i, and A, a set of ordered pairs of vertices, called *directed edges*, *arcs*, or *arrows*.

An edge $v_{xy} = (x, y)$ is considered to be directed from x to y; where y is called the head and x is called the tail of the edge.

In 1959, E. Dijkstra proposed an algorithm that solves the single-source shortest path problem for a directed graph with nonnegative edge weights.

We propose a modification on Dijkstra's algorithm to form the network backbone, with the minimum cost paths from the base station or *root*, r, to every node in the network. We have named this algorithm Sensor Intelligence Routing, **SIR** [5].

3.2 Quality of Service in Wireless Sensor Networks

Once the backbone formation algorithm is designed, a way of measuring the edge weight parameter, w_{ij}, must be defined. On a first approach we can assume that w_{ij} can be modelled with the number of hops. According to this assumption, $w_{ij} = 1 \; \forall \; i, j \in \mathcal{R}, i \neq j$. However, imagine that we have another scenario in which the node v_j is located in a noisy environment. The collisions over v_j can introduce link failures increasing power consumption and decreasing reliability in this area. In this case, the optimal path from node v_k to the root node can be p', instead of p. It is necessary to modify w_{ij} to solve this problem. The evaluation of the QoS in a specific area can be used to modify this parameter.

The traditional view of QoS in communication networks is concerned with end-to-end delay, packet loss, delay variation and throughput. Numerous authors have proposed architectures and integrated frameworks to achieve guaranteed levels of network performance [6]. However, other performance-related features, such as network reliability, availability, communication security and robustness are often neglected in QoS research. The definition of QoS requires some extensions if we want to use it as a criterion to support the goal of controlling the network. This way, sensors participate equally in the network, conserving energy and maintaining the required application performance.

We use a QoS definition based on three types of QoS parameters: timeliness, precision and accuracy. Due to the distributed feature of sensor networks, our approach measures the QoS level in a spread way, instead of an end-to-end paradigm. Each node tests every neighbor link quality with the transmissions of a specific packet named *ping*. With these transmissions every node obtains mean values of latency, error rate, duty cycle and throughput. These are the four metrics we have defined to measure the related QoS parameters.

Once a node has tested a neighbor link QoS, it calculates the distance to the root using the obtained QoS value. The expression 1 represents the way a node v_i calculates the distance to the root through node v_j, where qos is a variable whose value is obtained as an output of a neural network.

$$d(v_i) = d(v_j) \cdot qos \tag{1}$$

4 Performance Evaluation by Simulation

Due to the desire to evaluate the SIR performance, we have created two simulation experiments running on our wireless sensor network simulator OLIMPO [7]. Every node in OLIMPO implements a neural network (SOM) running the execution phase (online processing).

Noise influence over a node has been modelled as an Additive Gaussian White Noise, (AWGN), originating at the source resistance feeding the receiver. According to the radio communication parameters we can determine the signal-to-noise ratio at the detector input. This signal-to-noise ratio can be expressed as an associated BER (Bit Error Rate). An increase of the noise can degrade the BER. In another way, due to the relation between E_b/N_o and the transmission rate (R), $E_b/N_o = (S/R)/N_o$, an increase of R can also degrade the BER.

To evaluate the effect of noise we have defined a node state declared as *failure*. When the BER goes down below a required value (typically 10^{-3}) we assume this node has gone to a failure state. We measure this metric as a percentage of the total lifetime of a node.

Our SOM has a first layer formed by four input neurons, corresponding with every metric defined in section 3.2 (latency, throughput, error rate and duty cycle); and a second layer formed by twelve output neurons forming a 3x4 matrix.

Next, we detail our SOM implementation process.

4.1 Learning Phase

In order to organize the neurons in a two dimensional map, we need a set of input samples $\boldsymbol{x}(t)$=[latency(t), throughput(t), error-rate(t), duty-cycle(t)]. This samples should consider all the QoS environments in which a communication link between a pair of sensor nodes can work. In our research we create several WSNs over OLIMPO with 250 nodes and different levels of data traffic. The procedure to measure every QoS link between two neighbors is detailed as follows: every pair of nodes (eg. v_i and v_j) is exposed to a level of noise. This noise is introduced increasing the noise power density N_o in the radio channel in the proximity of a determined node. Hence, the signal-to-noise ratio at the detector input of this selected node decreases and consequently the BER related with its links with every neighbor gets worse.

In order to measure the QoS metrics related with every N_o, we run a ping application between a selected pair of nodes (eg. v_i and v_j). Node v_i sends periodically a ping message to node v_j. Because the ping requires acknowledgment (ACK), the way node v_i receives this ACK determines a specific QoS environment, expressed on the four metrics elected: latency (seconds), throughput (bits/sec), error rate (%) and duty cycle(%). This process is repeated 100 times with different N_o and d. This way, we obtain a set of samples which characterize every QoS scenario.

With this information, we construct a self-organizing map using a high performance neural network tool, such as MATLAB®, on a Personal Computer. This process is called *training*, and uses the learning algorithm. Because the training is not implemented by the wireless sensor network, we have called this process *offline processing.*

Once we have ordered the neurons on the Kohonen layer, we identify each one of the set of 100 input samples with an output layer neuron. According to this procedure, the set of 100 input samples is distributed over the SOM.

The following phase is considered as the most difficult one. The samples allocated in the SOM form groups, in such a way that all the samples in a group have similar characteristics (latency, throughput, error rate and duty cycle). This way, we obtain a map formed by clusters, where every cluster corresponds with a specific QoS and is assigned a neuron of the output layer. Furthermore, a synaptic-weight matrix $\boldsymbol{w'}_{ij} = [w'_{ij1}, w'_{ij2}, \ldots, w'_{ij4}]$ is formed, where every synapsis identifies a connection between input and output layer.

In order to quantify the QoS level, we study the features of every cluster and, according to the QoS obtained in the samples allocated in the cluster, we assign a value between 0 and 10. As a consequence, e define an output function $\Theta(i,j), i \in [1,3], j \in [1,4]$ with twelve values corresponding with every neuron $(i,j), i \in [1,3], j \in [1,4]$. The highest assignment (10) must correspond to that scenario in which the link measured has the worst QoS predicted. On the other hand, the lowest assignment (0) corresponds to that scenario in which the link measured has the best QoS predicted. The assignment is supervised by an engineer during the offline processing.

4.2 Execution Phase

As a consequence of the learning phase, we have declared an output function, that has to be run in every sensor node. This procedure is named the *wining neuron election algorithm.*

In the execution phase, we create a WSN with 250 nodes. Every sensor node measures the QoS periodically running a ping application with every neighbor, which determines an input sample. After a node has collected a set of input samples, it runs the wining neuron election algorithm. After the winning neuron is elected, the node uses the output function Θ to assign a QoS estimation, qos. Finally, this value is employed to modify the distance to the root (eq. 1). Because the execution phase is implemented by the wireless sensor network, we have called this process *online processing.*

Our SIR algorithm has been evaluated by the realization of three experiments detailed as follows:

Experiment #1: No node failure. The purpose of this experiment is to evaluate the introduction of AI techniques in a scenario were there is no node failure. This means that no node has gone to a failure state because of noise, collision or battery fail influence.

To simulate this scenario, a wireless sensor network with 250 nodes is created on our simulator OLIMPO. Node # 0 is declare as a sink and node # 22 is declared as a source. At a specific time, an event (eg. an alarm) is provoked in the source. Consequently, the problem now is how to route the event from the specified source to the declared sink.

As detailed in section 2 we solve this problem with three different routing paradigms: SIR, directed diffusion and EAR. We choose two metrics to analyze the performance of SIR and to compare it to others schemes. These metrics are: the average dissipated energy, which computes the average work done by a node a in delivering useful tracking information to the sinks (this metric also indicates the overall lifetime of sensor nodes); and the average delay, which measures the average one-way latency observed between transmitting an event and receiving it at each sink.

We study these metrics as a function of sensor network size. The results are shown in figures 1.a and 1.b.

Experiment #2: 20 % simultaneous node failures. The purpose of this experiment is to evaluate the introduction of AI techniques in a scenario where there is a 20 % of simultaneous node failures. This means that at any instant, 20 % of the nodes in the network are unusable because of noise, collision or battery failure influence.

To simulate these situations we create a WSN with 250 nodes. Amongst all of them, we select 20 % of the nodes (50) to introduce one of the following effects:

- S/N ratio degradation. Due to battery energy loss, the radio transmitter power decays. Consequently, the S/N ratio in its neighbors radio receivers

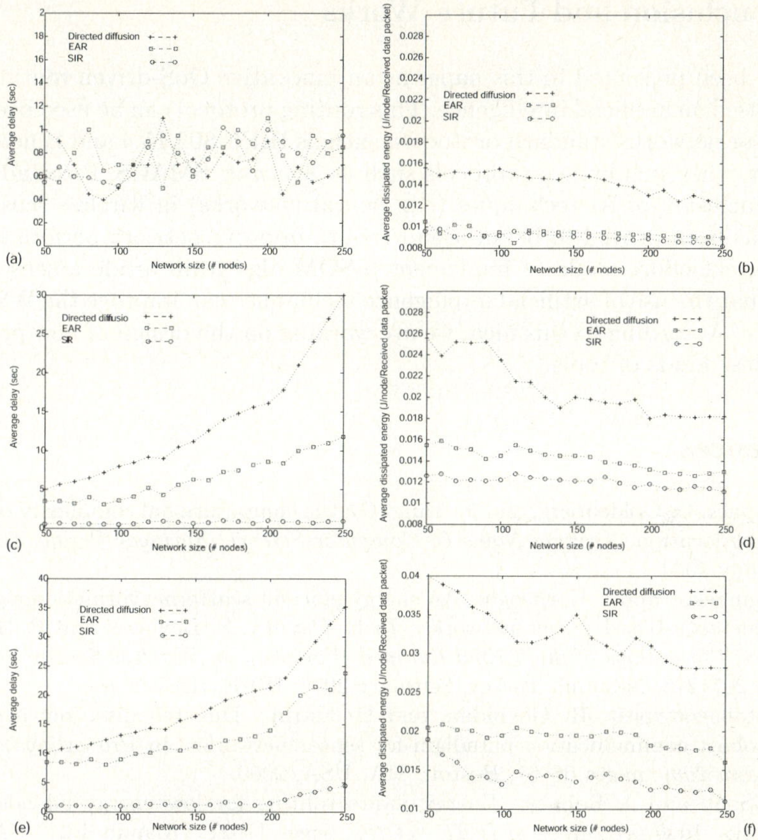

Fig. 1. Average latency and average dissipated energy in a scenario with no simultaneous node failure [(a) and (b)]; with 20 % simultaneous node failures [(c) and (d)]; and with 40 % simultaneous node failures [(e) and (f)]

is degraded, causing no detections with a certain probability, P. In this situation, we can assume that the node affected by the lack of energy is prone to failure with probability P.

- In many actual occasions, sensor nodes are exposed to high level of noise, caused by inductive motors. Furthermore, the radio frequency band is shared with other applications that can interfere with our WSN.

In these scenario we analyze the problem studied described in experiment #1 with the three paradigms related. The results are shown in figures 1.c and 1.d.

Experiment #3: 40 % simultaneous node failures. This experiment simulates a scenario with a 40 % of simultaneous node failures. The results are shown in figures 1.e and 1.f.

5 Conclusion and Future Works

SIR has been presented in this paper as an innovative QoS-driven routing algorithm based on artificial intelligence. This routing protocol can be used over wireless sensor networks standard protocols, such as IEEE 802.15.4 and Bluetooth®, and over other well known protocols such as *Arachne*, SMACS, PicoRadio, etc.

The inclusion of AI techniques (e.g. neural networks) in wireless sensor networks has been proved to be an useful tool to improve network performances.

The great effort made to implement a SOM algorithm inside a sensor node means that the use of artificial intelligence techniques can improve the WSN performance. According to this idea, we are working on the design of new protocols using these kinds of tools.

References

1. K. Aspnes, D. Goldenberg, and Y. Yang. On the computational complexity of sensor network location. *Lectune Notes In Computer Science, Springer Verlag*, 3121:235–246, July 2004.
2. S. Saginbekov and I. Korpeoglu. An energy efficient scatternet formation algorithm for bluetooth-based sensor networks. In E. Çayircy, Ş. Baydere, and P. Havinga, editors, *Proceedings of the Second Europen Workshop on Wireless Sensor Networks*, pages 207–216, Istanbul, Turkey, February 2005. IEEE, IEEE Press.
3. C. Intanagonwiwat, R. Govindan, and D. Estrin. Directed diffusion: a scalable and robust communication paradigm for sensor networks. In *Proceedings of ACM Mobicom 2000*, pages 56–67, Boston, MA, USA, 2000.
4. R.C. Shah and J. Rabaey. Energy aware routing for low energy ad hoc sensor networks. In *Proceeedings of IEEE WCNC*, pages 17–21, Orlando, FL, USA, 2002.
5. J. Barbancho, C. León, F.J. Molina, and A. Barbancho. SIR: A new wireless sensor network routing protocol based on artificial intelligence. *Lecture Notes in Computer Science, Springer Verlag*, 3842:271–275, January 2006.
6. B. Sabata, S. Chatterjee, M. Davis, J.J. Sydir, and T.F. Lawrence. Taxonomy for QoS specifications. In *Proceedings of the third International Workshop on Object-Oriented Real-Time Dependable Systems*, pages 100–107. IEEE, IEEE Press, 1997.
7. J. Barbancho, F.J. Molina, D. León, J. Ropero, and A. Barbancho. OLIMPO, an ad-hoc wireless sensor network simulator for public utilities applications. In E. Çayircy, Ş. Baydere, and P. Havinga, editors, *Proceedings of the Second Europen Workshop on Wireless Sensor Networks*, pages 419–424, Istanbul, Turkey, February 2005. IEEE, IEEE Press.

Design and Evaluation of a Rough Set-Based Anomaly Detection Scheme Considering Weighted Feature Values

Ihn-Han Bae, Hwa-Ju Lee, and Kyung-Sook Lee

School of Computer and Information Communication Eng.,
Catholic University of Daegu, GyeongSan 712-702, Korea

Abstract. The rapid proliferation of wireless networks and mobile computing applications has changed the landscape of network security. Anomaly detection is a pattern recognition task whose goal is to report the occurrence of abnormal or unknown behavior in a given system being monitored. This paper presents an efficient rough set based anomaly detection method that can effectively identify a group of especially harmful internal attackers – masqueraders in cellular mobile networks. Our scheme uses the trace data of wireless application layer by a user as feature value. Based on this, the use pattern of a mobile's user can be captured by rough sets, and the abnormal behavior of the mobile can be also detected effectively by applying a roughness membership function considering weighted feature values. The performance of our scheme is evaluated by a simulation.

1 Introduction

The nature of mobile computing environment makes it very vulnerable to an adversary's malicious attacks. The use of wireless links renders the network susceptible to attacks ranging from passive eavesdropping to active interfering. Unlike wired networks where an adversary must gain physical access to the network wires or pass through several lines of defenses at firewalls and gateways, attacks on a wireless network can come from all directions and target at any node. Damages can include leaking secret information, message contamination, and node impersonation [1].

In general, two complementary approaches exist to protect a system: prevention and detection. Prevention-based techniques, like authentication and encryption, can effectively reduce attacks by keeping illegitimate users from entering the system. They are usually based on some symmetric and asymmetric mechanisms to assure that users conform to predefined security polices. Nevertheless, in cellular wireless networks, the mobile devices are not physically secure: they can be lost or stolen. Since tamper-resistant hardware and software are still too costly for most users, such insecurity makes all secrets of the device open to malicious attackers. An attacker, once possesses the device as well as all secrets associated with the device, he becomes an internal user and is able to cause great damage to the whole network. At this time, intrusion detection approaches, utilizing different techniques to model the users' normal behavior and system vulnerabilities, come into place to help identify malicious activities [2].

B. Gabrys, R.J. Howlett, and L.C. Jain (Eds.): KES 2006, Part I, LNAI 4251, pp. 483 – 489, 2006.
© Springer-Verlag Berlin Heidelberg 2006

In this paper, we propose the intrusion detection technique that the normal profile of each mobile is constructed by using rough sets and anomaly activities are effectively detected by using a rough membership function. Our scheme uses the traveled cell ID, the duration of the service, and the service class of a user as the feature value. When an intrusion occurs, the attacker masquerading the legitimate user trends to have a different use pattern. Therefore, we can detect anomaly by comparing the use patterns.

The rest of this paper is organized as follows. Section 2 gives a brief description of related works for intrusion detection techniques in mobile networks. Section 3 describes rough set theory. Section 4 presents a rough set-based technique for anomaly detection in mobile networks. Section 5 presents the simulation study of our proposed detection approach. In Section 6, we conclude this paper and point out future work.

2 Related Works

The number of intrusions into computer system is growing because new automated intrusion tools are appearing every day. Although there are many authentication protocols in cellular mobile networks, security is still a very challenging task due to the open radio transmission environment and the physical vulnerability of mobile devices.

Generally, there are two intrusion detection techniques, misuse based detection and anomaly-based detection. A misuse-based technique encodes the known attack signatures and system vulnerabilities. An anomaly-based detection technique creates normal profiles of system states and user behaviors and compares then against current activities.

Y. Zhang proposed new model for Intrusion Detection System (IDS) and response in mobile, ad-hoc wireless networks. Each IDS agent runs independently and monitors local activities. It detects intrusion from local traces and initiates response. If anomaly is detected in the local data, or if the evidence is inconclusive and a broader search is warranted, neighboring IDS agents will cooperatively participate in global intrusion detection actions [1]. B. Sun proposed a mobility-based anomaly detection scheme in cellular mobile networks. The scheme uses cell IDs traversed by a user as the feature value [2]. O. Kachirski proposed multi-sensor intrusion detection system employing cooperative detection algorithm. By efficiently merging audit data from packet level, user level and system level sensors, the entire ad hoc wireless network for intrusion is analyzed [3]. J. Gomez proposed a technique to generate fuzzy classifiers using genetic algorithms that can detect anomalies and some specific intrusions [4]. In [5], based on fuzzy view of rough sets instead of exact rules, a soft computing approach that uses fuzzy rules for anomaly detection is proposed.

3 Rough Set Theory

The rough set is the approximation of a vague concept by a pair of precise concept, called lower and upper approximation, which are classification of domain of interest into disjoint categories. The rough set approach to processing of incomplete data is based on these approximations [6].

Let U be a finite set of objects called Universe, and $R \subseteq U \times U$ be an equivalence relation on U. The pair A=(U, R) is called approximation space, and equivalence classes of the relation R are called elementary sets in A.

For $x \in U$, let $[x]_R$ denote the equivalence class of R, containing x. For each $X \subseteq U$, X is characterized in A by a pair of sets – its lower and upper approximation in A, defined as:

$$\underline{A}X = \{x \in U \mid [x]_R \subseteq X\}$$

$$\overline{A}X = \{x \in U \mid [x]_R \cap X \neq \varnothing\}$$

The objects in $\underline{A}X$ can be with certainty classified as members of X on the basis of knowledge in R, while the objects in $\overline{A}X$ can be only classified as possible members of X on the basis of knowledge in R. The set $BN_A X = \overline{A}X - \underline{A}X$ is called the A-boundary region of X, and thus consists of those objects that we cannot decisively classify into X on the basis of knowledge in A.

Rough set can be also characterized numerically by the following coefficient called the accuracy of approximation, where *Card* denotes the cardinality.

$$\alpha_A(X) = \frac{Card\underline{A}X}{Card\overline{A}X}$$

Obviously $0 \le \alpha_A(X) \le 1$. If $\alpha_A(X)=1$, X is crisp with respect to A, and otherwise, if $\alpha_A(X)<1$, X is rough with respect to A.

Of course some other measures can also be defined in order to express the degree of exactness of the set X. It is possible to use a variety of $\alpha_A(X)$ defined as

$$\rho_A(X) = 1 - \alpha_A(X)$$

and referred to as a A-roughness of X. Roughness as opposed to accuracy represents the degree of incompleteness to knowledge A about the set X [7].

4 Rough Set-Based Anomaly Detection

In this section, we introduce the construction of a rough set-based anomaly detection method. For each mobile user, the features of the user activities from the wireless application layer are captured. We use the traveled cell ID, the duration of the service, and the service class of a user as the feature value. The feature values are stored in the user's feature profile database. In a cellular mobile network, this feature profile database is stored together with the mobile user's personal information, such as billing information, in the Home Location Register (HLR). We assume that HLR is secure and the feature information is accurate. Usually, because of its importance, HLR is protected with highly secure measure, and thus it is extremely hard to attack HLR.

The equivalence classes from the user's feature profile database are computed by using rough sets. For a mobile user, based on both the user activity information and the equivalence class information, a deviation number is computed by a roughness membership function considering weighted feature values, where the deviation number represents the degree that the user behavior is deviated from the normal behavior. When a user activity occurs, if the deviation number is greater than the deviation threshold that is a system parameter, an alert message occurs, otherwise the user activity identified as normal. The whole scheme is illustrated in Fig. 1.

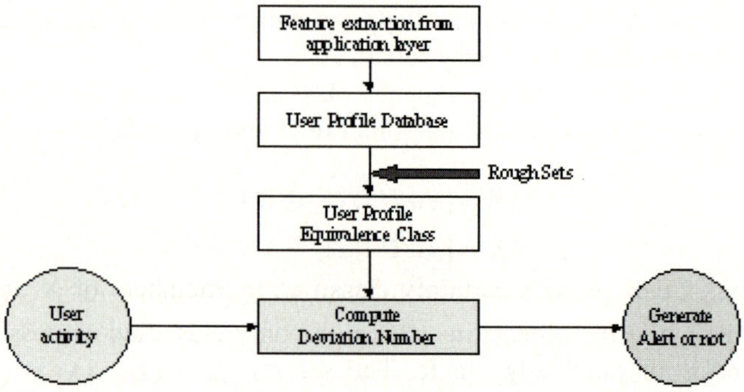

Fig. 1. The structure of rough set-based anomaly detection scheme

A relation/view instance is a snap shot of a relational database, which represents user's instant perception of entities or objects represented in the database. An information system is such an instance. We should note that the information system is an extension of relational databases without the entity integrity constraint.

The user feature profile database used in our scheme is shown in Table 1, where REQ#, CELL, DUR and CLASS represent the service request number, the traveled cell ID, the requested service duration and the requested service class, respectively.

Table 1. User feature profile information system

REQ#	CELL	DUR	CLASS
1	a	1	α
2	a	1	α
3	a	2	α
4	b	2	β
5	b	2	β
6	b	2	β
7	b	2	β
8	b	3	γ
9	c	3	γ
10	c	3	γ
11	a	1	α
12	a	1	α
13	a	2	α
14	b	2	β
15	b	3	β

In Table 1, let the attribute CLASS is decision attribute. Then we have the following three equivalence classes, called decision classes.

DE1={1, 2, 3, 11, 12, 13}={α}, DE2={4, 5, 6, 7, 14, 15}={β},
DE3={8, 9, 10}={γ}

For the conditional attributes (CELL, DUR), we have the following five equivalence classes, called condition classes.

CE1={1, 2, 11, 12}, CE2={3, 13}, CE3={4, 5, 6, 7, 14}, CE4={8, 15},
CE5={9, 10}

Comparing condition and decision classes, we get the inclusions.

$$CE1 \subseteq DE1,\ CE2 \subseteq DE1,\ CE3 \subseteq DE2,\ CE5 \subseteq DE3$$

The inclusion relation between CE4 and DE2 can be represented by a fuzzy inclusion. The fuzzy inclusions are represented by the inequalities of membership functions. Further, we will allow certain errors as long as they are within the radius of tolerance. Let $V_1 = \bar{R}X_1 \cup \bar{R}Y$, $V_2 = \bar{R}X_2 \cup \bar{R}Y$, $W_1 = \bar{R}X_1 \cap \bar{R}Y$ and $W_2 = \bar{R}X_2 \cap \bar{R}Y$, where X_i and Y represent the *i*-th condition attribute and decision attribute, respectively. The fuzzy inclusion is computed by Eq. (1), roughness membership function.

$$\rho = \min(1-\alpha(V_1,W_1), 1-\alpha(V_2,W_2)) \tag{1}$$

where $\alpha(V_1,W_1) = \dfrac{Card(W_1)\cdot(w_{x_1}+w_y)}{Card(V_1)\cdot(\max(w_x)+w_y)}$, $\alpha(V_2,W_2) = \dfrac{Card(W_2)\cdot(w_{x_2}+w_y)}{Card(V_2)\cdot(\max(w_x)+w_y)}$, and w_{x_i} and w_y represent the weighted values of the *i*-th condition attribute and the decision attribute of a user, respectively. The $max(w_x)$ gets the maximum weighted value among condition attributes.

For CE4 and DE2 those are not in inclusions, let X=CE4={b, 3}, Y=DE2={β}, w_{cell}=0.45, w_{dur}=0.25 and w_{class}=0.3, $\bar{R}X_1 = \{4,5,6,7,8,14,15\}$, $\bar{R}X_2 = \{8,9,10,15\}$ and $\bar{R}Y = \{4,5,6,7,14,15\}$, so V_1={4,5,6,7,8,14,15}, V_2={4,5,6,7,8,9,,10,14,15}, W_1={4,5,6,7,14,15} and W_2={15}. Therefore, the computed roughness, ρ=*min*(0.14, 0.92)=0.14 by Eq. (1), and so that CE4 and DE2 is in 0.14-fuzzy inclusion.

We assume that the deviation threshold (ε) is 0.4. In case that a user activity (c, 2, β) is occurred, we consider two cases those are the user weighted feature values, Table 2.

Table 2. The cases of weighted feature values

Case	w_{cell}	w_{dur}	w_{class}
I	0.45	0.25	0.3
II	0.25	0.45	0.3

Let X={c, 2} and Y={β}, $\bar{R}X_1 = \{9,10\}$, $R\bar{X}_2 = \{3,4,5,6,7,13,14\}$ and $\bar{R}Y = \{4,5,6,7,14,15\}$. If the weighted feature values of the user are setting by case I, the deviation number of the user, ρ=*min*(1.0, 0.54)=0.54. Accordingly, the user activity is evaluated as anomalous because that ρ>ε, and an alert message is generated. Otherwise, in case that the weighted feature values of the user is setting by case II, the deviation of the user, ρ=*min*(1.0, 0.375)=0.375. Accordingly, the user activity is identified as normal because that ρ≤ε.

5 Performance Evaluation

We use the following two metrics to evaluate the performance of our proposed anomaly detection scheme:

- Detection Rate: It is measured over abnormal itineraries. Suppose *m* abnormal itineraries are measured, and *n* of them are detected, detection rate in defined as *n/m*.
- False Alarm Rate: It is measured over normal itineraries. Suppose *m* normal itineraries are measured, and *n* of them are identified as abnormal, false alarm rate is defined as *n/m*.

Table 3. Simulation parameters

Parameters	Values	
The number of user activities	1,000	
The cases of weighted feature values, (w_{cell}, w_{dur}, w_{class})	A	(0.45, 0.25, 0.3)
	B	(0.25 0.45, 0.3)
	C	(0.3, 0.25, 0.45)
The traveled cell IDs	Random (1,4)	
The type of service durations	Random (1,4)	
The type of service classes	Random (1,3)	

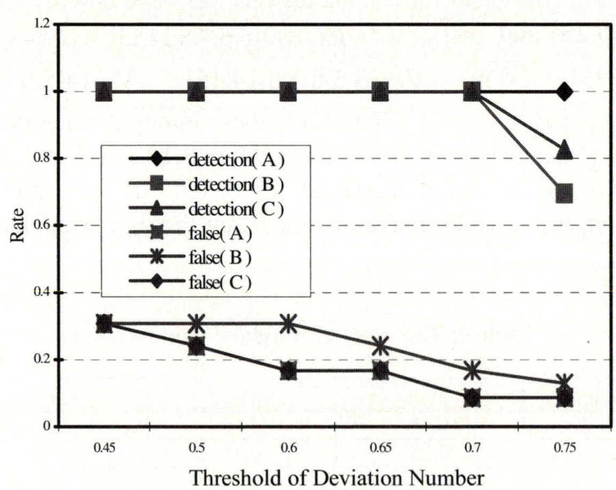

Fig. 2. Detection rate and false alarm rate at different deviation threshold

We present and analyze the simulation results at different deviation threshold. In the simulation, we assume that the user activity that has more than two feature data (records) those *cell* and *class* values match with the feature values of the user profile or more than three feature data those *dur* and *class* values match with the feature values of the user profile is normal, otherwise the user activity is anomalous. Table 3 shows the parameter values for the simulation.

Simulation results of the detection rate and the false alarm rate over deviation threshold are illustrated in Fig. 2. The performance of the case A that assigns large weighted value in the *cell* condition attribute is better than those of other cases regardless of threshold of deviation number. From the result, when deviation threshold is setting to 0.7, we can get the best performance for the proposed anomaly detection scheme.

6 Conclusions

This paper presents the rough set-based anomaly detection scheme for mobile networks. Our scheme uses the user activity information from the application layer as the feature value. Therefore, our scheme is used for all users because the feature values are selectively applicable to users. Simulation results demonstrate that anomalies are detected well by the method that assigns different weighted values to feature attributes depending on a factor of importance.

Future works include development of a new roughness membership function that reflects precise deviation number of user activities, applying aging to user feature profile information, and development of a rough set-based anomaly detection scheme for mobile ad-hoc networks.

References

1. Yongguang Z, Wenke L., Yi-An H.: Intrusion Detection Techniques for Mobile Wireless Networks, ACM Wireless Networks Journal, 9(5) (2003): 545-556
2. Bo S., Fei Y., Kui W., Vicot C-.M. L.: Mobility-Based Anomaly Detection in Cellular Mobile Networks, Workshop on Wireless Security, (2004): 61-69
3. Oleg K., Ratan G.: Effective Intrusion Detection Using Multiple Sensors in Wireless Ad Hoc Networks, HICSS, (2003):57.1
4. Jonathan G., Dipankar D.: Evolving Fuzzy Classifiers for Intrusion Detection, GECCO, (1) (2004): 1150-1161
5. Tsau Y. L.: Anamoly Detection - A Soft Computing Approach, Proceedings in the ACM SIGSAC New Security Paradigm Workshop, (1994): 44-53
6. Richard J., Qiang S.: Rough and Fuzzy Sets for Dimensionality Reduction, Proceedings of the 11th Int. Conf. on Fuzzy Systems, (2002):29-34
7. Zdzilaw P.: Rough Sets Theoretical Aspects of Reasoning about Data, Kluwer Academic Pub., (1991)
8. Ron K. and Brian F.: Useful Feature Subsets and Rough Set Reducts, In the International Workshop on Rough Sets and Soft Computing , (1994):310-317

ATM Based on SPMF

Dae-Young Choi

Dept.of MIS, Yuhan College, Koean-Dong, Sosa-Ku, Puchon, Kyoungki-Do, 422-749, Korea
dychoi@yuhan.ac.kr

Abstract. We propose an approximate transformation method (ATM) based on the standardized parametric membership functions (SPMF). The ATM provides fuzzy systems with the platform for the relatively fast computations by transforming the non-parametric membership functions (NPMF) into the parametric membership functions (PMF). In this paper, we apply the ATM to linguistic approximation.

1 Introduction

Based on their behavioral experiment, Zwick et al [8] recommended the five good distance measures, i.e., S_4, q_∞, q_*, Δ_∞, Δ_*, between fuzzy subset A and B of a universe of discourse U. We note that the five good distance measures concentrate their attention on a single value rather than performing some sort of averaging or integration. In the case of S_4, attention focuses on the particular x-value where the membership function of $A \cap B$ is largest; in q_∞ and Δ_∞, attention focuses on the α-level set where the x-distance is largest; in q_* and Δ_*, attention focuses on the x-distance at the highest membership grade. Considering the result of their behavioral experiment, we know that the reduction of complicated membership functions to a single 'slice' may be the intuitively natural way for human beings to combine and process fuzzy concepts. Moreover, we know that the distance measures between two fuzzy subsets can be efficiently represented by a limited number of features. From these ideas, we propose an ATM based on SPMF.

A key problem in the application of fuzzy set theory to real time control, expert systems, natural language understanding [6], etc., is devising relatively fast methods. From this engineering viewpoint, the parametric membership functions (PMF) play an important role in many applications related to the development of fuzzy systems. The reason is that it enables the fuzzy systems to be processed relatively fast compared to the fuzzy systems with non-parametric membership functions (NPMF). However, there has been little effort to transform the NPMF into the PMF in the development of fuzzy systems. In order to handle this problem we propose an ATM based on SPMF. In this paper, we suggest approximate transformation algorithms for fuzzy sets defined by n discrete elements and discuss its related properties. The ATM provides fuzzy systems with the platform for the relatively fast computations by transforming the NPMF into the PMF. It is derived from the idea that the fuzzy sets with a

B. Gabrys, R.J. Howlett, and L.C. Jain (Eds.): KES 2006, Part I, LNAI 4251, pp. 490–497, 2006.
© Springer-Verlag Berlin Heidelberg 2006

non-linear shaped NPMF can be transformed into the fuzzy sets with a piece-wise linear shaped PMF at the structural level by using the ATM. In this case, these parameters may be regarded as feature points for determining the structure of a fuzzy set. Thus, the difference between fuzzy sets is efficiently computed by using the parameters of the PMF.

2 Typical Shapes of SPMF

Every person has his/her own conception of the meaning of words. For instance, the membership function of '*tall men*' in the eastern nations is different from that in the western nations. Existing membership functions are generally developed with their own conception independently by using an ad-hoc method. Moreover, if several membership functions are necessary in order to describe the subjects of many persons, the question arise, how many membership functions are needed. If one wants to use a very small number of membership functions to implement the fuzzy systems, then these membership functions have to be adaptable to the specific context. This can be achieved by parameterization. In this connection, we introduce triangular-type, trapezoidal-type, Π-type membership functions as the SPMF. Let A be a fuzzy set for a linguistic term and be a subset of the universal set X, then, for $x \in X$, typical shapes of the SPMF can be defined in [7] :

2.1 Triangular-Type Membership Function

It can be represented by using 3 points $\mu_A(x_L, \boldsymbol{x_M}, x_H)$, where $x_L < x_M < x_H$, and if the result of this membership function is normalized to [0, 1] then $\mu_A(x_L, x_M, x_H) = 0$ for every $x \in [-\infty, x_L] \cup [x_H, \infty]$ and $\mu_A(x_L, x_M, x_H) = 1$ at x_M.

2.2 Trapezoidal-Type Membership Function

It can be represented by using 4 points $\mu_A(x_L, \boldsymbol{x_{I1}}, \boldsymbol{x_{I2}}, x_H)$, where $x_L < x_{I1} < x_{I2} < x_H$, and if the result of this membership function is normalized to [0, 1] then $\mu_A(x_L, x_{I1}, x_{I2}, x_H) = 0$ for every $x \in [-\infty, x_L] \cup [x_H, \infty]$ and $\mu_A(x_L, x_{I1}, x_{I2}, x_H) = 1$ at $[x_{I1}, x_{I2}]$.

2.3 Π-Type Membership Function

Slightly more complicated is the Π-type membership function (i.e., bell-shape) [7]. It can be represented by using 6 points $\mu_A(x_L^1, x_L^2, \boldsymbol{x_L^3}, \boldsymbol{x_R^3}, x_R^2, x_R^1)$, where $x_L^1 < x_L^2 < x_L^3 < x_R^3 < x_R^2 < x_R^1$, and if the result of this membership function is normalized to [0, 1] then $\mu_A = 0$ for $x < x_L^1$ or $x > x_R^1$ and $\mu_A = 1$ for $x_L^3 \le x \le x_R^3$ (see Fig. 2).

3 ATM Based on SPMF

We propose the ATM for fuzzy sets defined by n discrete elements. Let A be a fuzzy set and be a subset of the universal set X, then, for $x \in X$, a fuzzy set may be defined as follows :

$$A = \{(x_L^1, \mu_A(x_L^1)), (x_L^2, \mu_A(x_L^2)), \ldots, (x_L^{m-1}, \mu_A(x_L^{m-1})), (\boldsymbol{x}_L^m, \boldsymbol{\mu}_A(\boldsymbol{x}_L^m)), (\boldsymbol{x}_R^n, \boldsymbol{\mu}_A(\boldsymbol{x}_R^n)), (x_R^{n-1}, \mu_A(x_R^{n-1})), \ldots, (x_R^2, \mu_A(x_R^2)), (x_R^1, \mu_A(x_R^1))\} \quad (1)$$

In the ATM, we use Euclidean distance to find the positions for the parameters on the NPMF curve when transforming the NPMF into the PMF. For fuzzy sets defined by n discrete elements as in Eq. (1), we try to find the turning points that may be the parameters of the PMF according to the following condition :

$$\sqrt{(\mu_A(x^{i+1}) - \mu_A(x^i))^2} \geq \alpha,\ 0 < \alpha \leq 1, \text{ where } i = 1, 2, \ldots, m\text{-}1 \text{ for the left-side and } i = 1, 2, \ldots, n\text{-}1 \text{ for the right-side.} \quad (2)$$

In Eq. (2), for convenience sake, the subscript L and R in Eq. (1) are omitted. If Eq. (2) is satisfied, then $(x^{i+1}, \mu_A(x^{i+1}))$ is determined as a turning point. In the meantime, it is assumed that the leftmost and rightmost points, i.e., $(x_L^1, \mu_A(x_L^1))$ and $(x_R^1, \mu_A(x_R^1))$, and the kernel points, i.e., $(x_L^m, \mu_A(x_L^m))$ and $(x_R^n, \mu_A(x_R^n))$, are the default turning points. The ATM algorithm with the sampling interval 1 is as follows :

Algorithm 1: ATM algorithm with the sampling interval 1.
/* Finding the turning points with the sampling interval 1 regarding the left-side and right-side of the NPMF curve. */
/* Let L[1:m] and R[1:n] be arrays. These are used to indicate the state of flag regarding the Eq. (2). */
/* For the default turning points, let L[1] = 1, L[m] = 1, R[1] = 1, R[n] = 1*/
for i = 1 **to** m-1 **by** 1 **do** /* Finding turning points on the left-side */
if Eq. (2) is satisfied **then** L[i+1]=1 /* Position (i+1) is selected as a turning point */
else L[i+1]=0 **end**
for i = 1 **to** n-1 **by** 1 **do** /* Finding turning points on the right-side */
if Eq. (2) is satisfied **then** R[i+1]=1 /* Position (i+1) is selected as a turning point */
else R[i+1]=0 **end**

It should be noted that we can control the number of turning points by adjusting the value of α in Eq. (2). Thus, the value of α determines the level of approximation. That is, the higher the value of α, the higher is the level of approximation. For example,

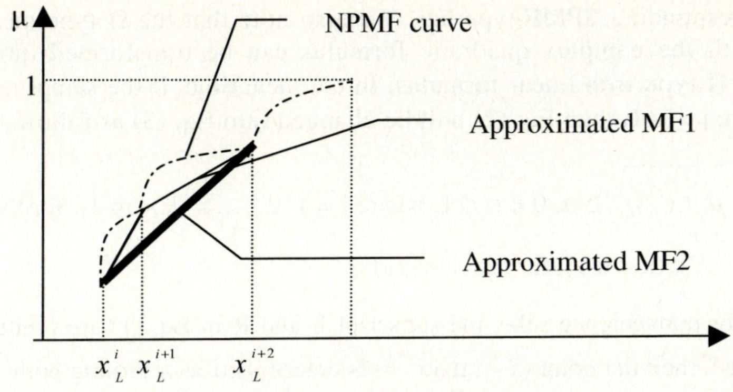

Fig. 1. An example using the ATM

In Fig. 1, if x_L^{i+1} and x_L^{i+2} are the turning points according to Eq. (2), then the NPMF curve will be approximated by the approximated MF1. However, if x_L^{i+1} is not the turning point by using the higher value of α, then the NPMF curve will be approximated by the approximated MF2. In this case, the approximated MF2 is more inaccurate than the approximated MF1 when compared to the NPMF curve. For another example, we assume that a NPMF curve is similar to the Π-type membership function curve and its corresponding SPMF is defined by using 6 points. In this case, if the turning points regarding the left-side and right-side are determined as shown in Fig. 2 by using the Eq. (2), then the following approximation will be obtained.

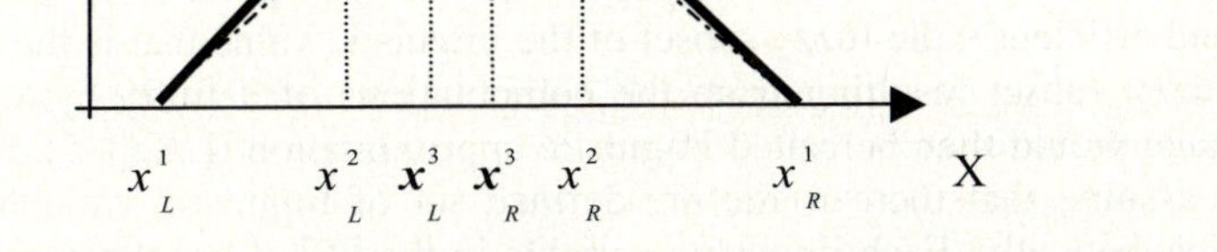

Fig. 2. An example using the ATM based on Π-type SPMF

Consequently, the fuzzy set with a non-linear shaped NPMF is transformed into the fuzzy set with a piece-wise linear shaped PMF at the structural level by using the ATM. In this case, we note that the approximated MF should have the same number of parameters regarding the left-side and right-side, respectively that its corresponding SPMF-type has. We also note that the Π-type membership function with the complex quadratic formulas can be transformed into the approximated Π-type with linear formulas. In the meantime, if the sampling interval is |j-i|, where j > i+1, then Eq. (2) will be changed into Eq. (3) as follows :

$$\sqrt{(\mu_A(x^j)-\mu_A(x^i))^2} \geq \alpha,\ 0 < \alpha \leq 1, \text{ where } i = 1, 2, \ldots, k\ (k < m-1), \text{ for the left-side and } i = 1, 2, \ldots, k\ (k < n-1), \text{ for the right-side.} \quad (3)$$

In Eq. (3), for convenience sake, the subscript L and R in Eq. (1) are omitted. If Eq. (3) is satisfied, then the point $(x^j, \mu_A(x^j))$ is determined as a turning point. It should be noted that we can control the number of turning points by adjusting the value of α or |j-i| in Eq. (3). Thus, the value of α or |j-i| determines the level of approximation when using Eq. (3). That is, the higher the value of α or |j-i|, the higher is the level of approximation. The ATM algorithm with the sampling interval |j-i| is as follows :

Algorithm 2: ATM algorithm with the sampling interval |j-i|.
/* Finding the turning points with the sampling interval |j-i| regarding the left-side and right-side of the NPMF curve. */
/* Let L[1:m] and R[1:n] be arrays. These are used to indicate the state of flag regarding the Eq. (3). */
/* For the default turning points, let L[1] = 1, L[m] = 1, R[1] = 1, R[n] = 1*/
for i = 1 **to** k **by** |j-i| **do** /* Finding turning points on the left-side where j > i+1 and k < m-1 */
if Eq. (3) is satisfied **then** L[j] = 1 /* Position (j) is selected as a turning point */
else L[j] = 0 **end**
for i = 1 **to** k **by** |j-i| **do** /* Finding turning points on the right-side where j > i+1 and k < n-1 */
if Eq. (3) is satisfied **then** R[j] = 1 /* Position (j) is selected as a turning point */
else R[j] = 0 **end**

4 Application to Linguistic Approximation

We try to find efficiently the fuzzy subset of the linguistic value that is the most similar to the fuzzy subset resulting from the computations of a fuzzy system. Such a linguistic value would then be called linguistic approximation (LA)[1-5]. For a fuzzy system, we assume that there is the pre-defined set of linguistic variables (PSLV) sorted by alphabetically. Each linguistic variable in the PSLV has the pointer to indicate its own table with the relevant linguistic values. The table regarding the linguistic variable '*age*' may be consisted of linguistic values such as '*young*', '*very young*', '*old*' represented by fuzzy subsets. In this case, we assume that these fuzzy subsets are defined by the SPMF. In order to avoid exploring the irrelevant linguistic values in the process of linguistic approximation, we suggest the partitioning concept that disjoint the linguistic variables used in a fuzzy system as in Fig. 3. The resulting fuzzy subset obtained from the computations of a fuzzy system can be easily associated with its linguistic variable. For example, given a rule '*If* X is a *then* Y is b', the linguistic value 'a' is associated with the linguistic variable 'X' whereas the linguistic value 'b' is associated with the linguistic variable 'Y'. Hence, we can find the related linguistic variable easily by referencing the firing rule in the rule-base. After the related linguistic variable (LV) is determined in the PSLV, its corresponding table (TBL) is obtained by using the pointer (P_i, i = 1, 2, …, n) associated with the linguistic variable. Based on the relevant fuzzy subsets represented by the SPMF in the table, the distances between all relevant fuzzy subsets in the table and the resulting fuzzy subset approximated by the ATM are computed by using Euclidean distance. Thus, the linguistic value with the minimum distance would be selected as the linguistic approximation. We note that each table (TBL_i, i = 1, 2, …, n) is consisted of the relevant linguistic values represented by the SPMF. For example, the triangular-type or the trapezoidal-type should be defined as (x_L, x_M, x_H) or $(x_L, x_{I1}, x_{I2}, x_H)$, respectively.

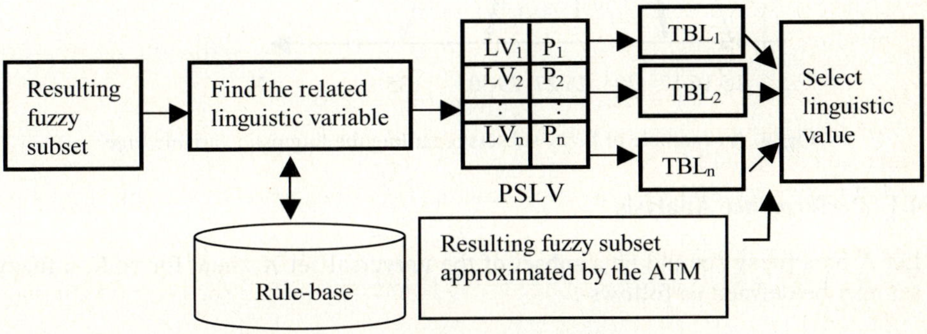

Fig. 3. The process of linguistic approximation

Example 1. We consider the linguistic variable '*age*'. For simplicity, we assume that the table regarding the linguistic variable '*age*' is consisted of 2 linguistic values such as '*young*' and '*very young*'. They are defined as $(x_L, x_{I1}, x_{I2}, x_H)_1 = (15, 20, 30, 35)_1 =$ T_1('*young*') and $(x_L, x_{I1}, x_{I2}, x_H)_2 = (17, 20, 27, 30)_2 = T_2$('*very young*') respectively, by using the trapezoidal-type. We also assume that the resulting fuzzy subset is parameterized to $(x_L', x_{I1}', x_{I2}', x_H') = (16, 20, 25, 29) = T'$ by applying the ATM algorithm. The relativity between these fuzzy subsets is represented as in Fig. 4. In Fig. 4, the fuzzy subset '*young*' is represented by the thin membership function, whereas the fuzzy subset '*very young*' is depicted by the thick membership function. Thus, linguistic approximation will be achieved by using Euclidean distance.

$$d_1(T_1, T') = \sqrt{(X_L - X_L')^2 + (X_{I1} - X_{I1}')^2 + (X_{I2} - X_{I2}')^2 + (X_H - X_H')^2}$$
$$= \sqrt{(15-16)^2 + (20-20)^2 + (30-25)^2 + (35-29)^2} = \sqrt{62}$$
$$d_2(T_2, T') = \sqrt{(17-16)^2 + (20-20)^2 + (27-25)^2 + (30-29)^2} = \sqrt{6}$$

In this case, for the resulting fuzzy subset T′, the linguistic value '*very young*' is selected as the linguistic approximation because $d_1 > d_2$. In a similar way, it may be extended to i, (i =1, 2, …, m), relevant linguistic values in the table by applying Euclidean distance repeatedly. Thus, the linguistic value with min (d_1, d_2, …, d_m) is selected as the linguistic approximation.

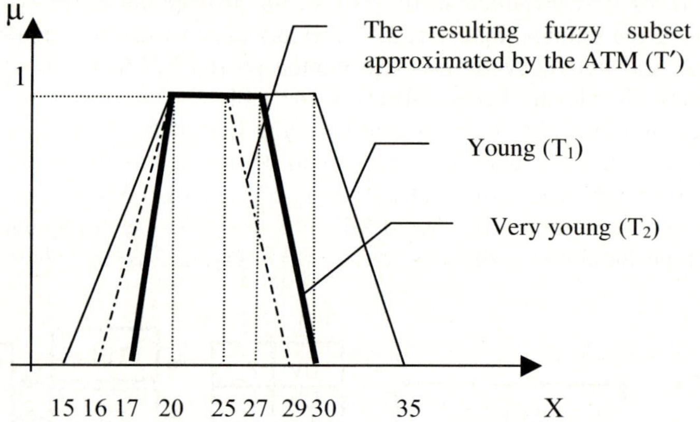

Fig. 4. An example of fuzzy subsets regarding the linguistic variable '*age*'

4.1 Performance Analysis

Let A be a fuzzy set and be a subset of the universal set X, then, for $x \in X$, a fuzzy set may be defined as follows :

$$A = \{(x_1, \mu_A(x_1)), (x_2, \mu_A(x_2)), \ldots, (x_n, \mu_A(x_n))\} \quad (4)$$

In the proposed LA, let m and n be the number of relevant linguistic values in the table and the number of discrete elements for representing a fuzzy subset as in Eq. (4), respectively. Then the efficiency of the proposed LA is obtained as follows :

Table 1. The number of comparisons for LA

Existing methods	The proposed method
m × n	m × The number of parameters in the SPMF

For example, if we use the trapezoidal-type membership function as the SPMF, the number of parameters in the SPMF is 4 as in Section 2. It should be noted that the larger the value of m and n, the higher the efficiency of the proposed LA.

5 Conclusions

We propose the ATM based on SPMF and discuss its related properties. The ATM provides fuzzy systems with the platform for the relatively fast computations by transforming the NPMF into the PMF. In the meantime, we present an efficient linguistic approximation method based on the ATM. This linguistic approximation is efficiently obtained by using the parameters of the SPMF. Moreover, in order to avoid exploring the irrelevant linguistic values in the process of linguistic approximation, we suggest the partitioning concept that disjoint the linguistic variables used in a fuzzy system. These features enable this linguistic approximation method to be processed relatively fast as compared with the previous works [1-5]. From the engineering viewpoint, it may be a valuable advantage.

Acknowledgement

The author wishes to thank Prof. L. A. Zadeh, University of California, Berkeley, for his inspirational address on the perceptual aspects of humans in the BISC (Berkeley Initiative in Soft Computing) seminars and group meetings.

References

1. Batyrshin, I., Wagenknecht, M.: Towards a Linguistic Description of Dependencies in Data, Int. J. Appl. Math. Compt. Sci., Vol. 12, No. 3 (2002) 391-401
2. Degani, R., Bortolan, G.: The Problem of Linguistic Approximation in Clinical Decision Making, Int. J. Approx. Reasoning 2(2) (1988)143-162
3. Eshragh. F., Mamdani, E. H.: A General Approach to Linguistic Approximation, International Journal of Man-Machine Studies, Vol. 11(1979) 501-519
4. Kowalczyk, R.: On Linguistic Approximation with Genetic Programming, Lecture Notes In Computer Science, Vol. 1415(1998) 200-209
5. Wenstop, F.: Deductive Verbal Models of Organization, International Journal of Man-Machine Studies, Vol. 8(1976) 293-311
6. Zadeh, L. A.: The Concept of a Linguistic Variable and Its Application to Approximate Reasoning -1, In : Yager, R. R., Ovchinnikov, S., Tong, R. M., Nguyen, H. T. (Eds.), Fuzzy Sets and Applications : Selected papers by Zadeh, L. A.(John Wiley & Sons, 1987) 219-269
7. Zimmermann, H.-J.: Fuzzy Set Theory and Its Applications (Kluwer-Nijhoff Pub., 1986)
8. Zwick, R., Carlstein, E., Budescu, D. V.: Measures of Similarity among Fuzzy Concepts : A Comparative Analysis, Int. J. of Approximate Reasoning, Vol. 1 (1987) 221-242

Using Rough Set to Induce More Abstract Rules from Rule Base

Feng Honghai[1,2], Liu Baoyan[3], He LiYun[3], Yang Bingru[2], Li Yueli[4], and Zhao Shuo[4]

[1] Urban & Rural Construction School, Hebei Agricultural University, 071001 Baoding, China
liuby@tcmcec.com
[2] University of Science and Technology Beijing, 100083 Beijing, China
[3] China Academy of Traditional Chinese Medicine, 100700 Beijing, China
[4] Hebei Agricultural University, 071001 Baoding, China

Abstract. In fault diagnosis and medical diagnosis fields, often there is more than one fault or disease that occur together. In order to obtain the factors that cause a single fault to change to multi-faults, the standard rough set based methods should be rebuilt. In this paper, we propose a decernibilty matrix based algorithm with which the cause of every single fault to change to multi-faults can be revealed. Additionally, we propose another rough set based algorithm to induce the common causes of all the single faults to change to their corresponding multi-faults, which is a process of knowledge discovery in rule base, i.e., not the usual database. Inducing more abstract rules in knowledge base is a very challenging problem that has not been resolved well.

1 Introduction

In fault diagnosis and medical diagnosis fields, often there is more than one fault or disease that occur together. With the multi-faults or multi-diseases, there may be more symptoms or the symptoms may be aggravated in comparison to that of the single fault or disease occurring, which is a kind of important knowledge for experts. But the standard rough set method only induces the rules for classification. The rules are simplified and generalized, namely there is no redundancy, and so the rules are not easier understood. Furthermore, with the standard rough set method, the cause of the formation of multi-faults cannot be induced. In order to obtain this kind of knowledge, the standard rough set based methods should be rebuilt.

Decernibility matrix is a valid method for attribute reduction and rule generation whose main idea is to compare the examples that are not in the same class. However, generally, the decernibility matrix is used to the whole data set [1], can it be used to partial data to induce the knowledge with which we can find the factors that cause multi-faults? The answer is affirmative. In this paper, we propose a decernibilty matrix based algorithm with which the cause of every single fault to change to multi-faults can be revealed.

B. Gabrys, R.J. Howlett, and L.C. Jain (Eds.): KES 2006, Part I, LNAI 4251, pp. 498 – 505, 2006.
© Springer-Verlag Berlin Heidelberg 2006

It is often the case that the knowledge discovered by a data mining algorithm needs to undergo some kind of post-processing. These post-processing techniques are such as pruning of rules, clustering of rules and generalization of rules. Pruning rules can eliminate the redundancy in the rule set [2, 3, 4]; clustering rules can help users to understand and process the rules more easily [5]; generalization of rules can generate more abstract rules that are new kinds of knowledge for the users [6].

Inducing more abstract rules from knowledge base is a very challenging problem that has not been resolved well. Yang Bingru has proposed an induction logic based schema for inducing the new knowledge [7]. Generally, as an inductive learning method, rough set theory can be used to discover knowledge in database, whereas has not been used to induce new knowledge in rule base. In this paper, we propose another rough set based algorithm to induce the common causes of all the single faults to change to their corresponding multi-faults, which is a process of knowledge discovery in knowledge base, i.e., not the usual database.

2 Basic Concepts of Rough Set

In a decision table DT=$\langle U, A, V, f \rangle$, to every subset of attributes $B \subseteq A$, a binary relation, denoted by $IND(B)$, called the B-indiscernibility relation, is associated and defined as follows:

$$IND(B) = \left\{ \begin{matrix} (x_i, x_j) \in U \times U : a \in B, f(x_i, a) \\ = f(x_j, a) \end{matrix} \right\} \tag{1}$$

The rough sets approach to data analysis hinges on two basic concepts, namely the lower and the upper approximations of a set, referring to:

-The elements that doubtlessly belong to the set, and
-The elements that possibly belong to the set.

Let X denotes the subset of elements of the universe U ($X \subset U$). The lower approximation of X in B ($B \subseteq A$), denoted as $IND(B)$, is defined as the union of all these elementary sets that are contained in X. More formally:

$$POS_B(X) = B\_(X) = \{x_i \in U \| [x_i]_{IND(B)} \subset X\}$$

The upper approximation of the set X, denoted as $B^-(X)$, is the union of these elementary sets, which have a non-empty intersection with X:

$$B^-(X) = \{x_i \in U \| [x_i]_{IND(B)} \cap X \neq 0\}$$

The difference:

$$BN(X) = B^{-}(X) - B_{\_}(X)$$

is called a boundary of X in U.

Discernibility matrix of DT, denoted $M_{DT}(m_{ij})$ a $n \times n$ matrix is defined as

$$m_{ij} = \begin{cases} \{a \in C : f(x_i, a) \neq f(x_j, a) \wedge (d \in D, f(x_j, d) \neq f(x_i, d))\} \\ 0, d \in D, f(x_i, d) = f(x_j, d) \\ \phi, f(x_i, a) = f(x_j, d) \wedge (d \in D, f(x_i, d) \neq f(x_j, d)) \end{cases} \quad (2)$$

Where $i, j = 1, 2 \cdots, n$

Thus entry m_{ij} is the set of all attributes that classify objects x_i and x_j into different decision classes in U. From formula (2), all of the distinguishing information for attributes is contained in above discernibility matrix.

3 Algorithm

(1) Algorithm for Deriving the Factors that Make the Single Fault to Multi-faults

- a) Select the examples with multi-faults and the examples with a single fault that will coexist with other faults.
- b) Use discernibility matrix method to compare the single fault examples and the coexistent faults examples and induce the rules for multi-faults.

(2) Algorithm for Inducing More Abstract Rules in Rule Base

- (a) Select the rules with which we want to get more abstract rules.
- (b) Transform the rules into data, and get a decision table.
- (c) Use the concept of positive region to generate more abstract rules.

4 Fault Diagnosis in Power Systems

4.1 Fault Diagnosis Decision System of Power Systems

Table 1 is a fault diagnosis decision system of power systems. The states of relays and circuits breakers are used as the attributes describing the related faults examples. In Table 1, there are 19 condition attributes and 16 fault classes. Table 1 consists of 26 faults cases, including 5 single faults and 10 double faults with no failure relay and breaker; 5 single faults with one failure breaker; 5 single faults with one failure breaker; 1 special case- “NO” represents no fault occurred.

Table 1. System fault cases and associated alarm patterns

U	CB_1	CB_2	CB_3	CB_4	Am	Bm	Cm	L_1Am	L_1Bm	L_2Bm	L_2Cm	L_1Ap	L_1Bp	L_2Bp	L_2Cp	L_1As	L_1Bs	L_2Bs	L_2Cs	Fault
1	1	0	0	0	1	0	0	0	0	0	0	0	0	0	0	0	0	0	0	A
2	0	1	1	0	0	1	0	0	0	0	0	0	0	0	0	0	0	0	0	B
3	0	0	0	1	0	0	1	0	0	0	0	0	0	0	0	0	0	0	0	C
4	1	1	0	0	0	0	0	1	1	0	0	0	0	0	0	0	0	0	0	L_1
5	0	0	1	1	0	0	0	0	0	1	1	0	0	0	0	0	0	0	0	L_2
6	1	1	0	0	0	0	0	0	1	0	0	1	0	0	0	0	0	0	0	L_1
7	1	1	0	0	0	0	0	1	0	0	0	0	1	0	0	0	0	0	0	L_1
8	0	0	1	1	0	0	0	0	0	1	0	0	0	0	1	0	0	0	0	L_2
9	0	0	1	1	0	0	0	0	0	0	1	0	0	1	0	0	0	0	0	L_2
10	1	0	1	0	0	1	0	0	0	0	0	0	0	0	0	1	0	0	0	B
11	1	0	0	1	0	0	0	0	0	1	1	0	0	1	0	1	0	0	0	L_2
12	0	1	0	0	1	0	0	0	0	0	0	0	0	0	0	0	1	0	0	A
13	0	1	0	0	0	0	0	0	0	0	0	0	0	0	0	0	1	0	0	A
14	0	0	1	0	0	0	0	0	0	0	0	0	0	0	0	0	0	1	0	C
15	0	1	0	1	0	1	0	0	0	0	0	0	0	0	0	0	0	0	1	B
16	1	0	0	1	0	0	0	1	1	0	0	0	0	0	0	0	0	0	1	L_1
17	1	1	1	0	1	1	0	0	0	0	0	0	0	0	0	0	0	0	0	AB
18	1	0	0	1	1	0	1	0	0	0	0	0	0	0	0	0	0	0	0	AC
19	1	1	0	0	1	0	0	1	1	0	0	0	0	0	0	0	0	0	0	AL_1
20	1	0	1	1	1	0	0	0	0	1	1	0	0	0	0	0	0	0	0	AL_2
21	0	1	1	1	0	1	1	0	0	0	0	0	0	0	0	0	0	0	0	BC
22	1	1	1	0	0	1	0	1	1	0	0	0	0	0	0	0	0	0	0	BL_1
23	0	1	1	1	0	1	0	0	0	1	1	0	0	0	0	0	0	0	0	BL_2
24	1	1	0	1	0	0	1	1	1	0	0	0	0	0	0	0	0	0	0	CL_1
25	0	0	1	1	0	0	1	0	0	1	1	0	0	0	0	0	0	0	0	CL_2
26	1	1	1	1	0	0	0	1	1	1	1	0	0	0	0	0	0	0	0	L_1L_2
27	0	0	0	0	0	0	0	0	0	0	0	0	0	0	0	0	0	0	0	NO

4.2 Induce the Factors That Cause Multi-faults

(1) For examples 1, 12 and 13, the fault is A, and for example 17, 18, 19 and 20, the faults include A. Which are the factors that cause the multi-faults including A? Now, with the partial decernibility matrix we can find the factors.

Firstly, the examples with faults A, AB, AC, AL_1 and AL_2 are selected. Secondly, generate the decernibility matrix with these examples. Thirdly, use the decernibility matrix to extract the factors that cause multi-faults including A.

Table 2 is the information table for inducing the causes of the formation of multi-faults that include A. Table 3 gives the results of the decernibility matrix.

The factors that cause fault A to change to multi-fault AB can be induced as follows (see example 17 in Table 2):

$$(CB_2 \vee CB_3 \vee Bm) \wedge (CB_1 \vee CB_3 \vee Bm \vee L_1Bs) \wedge (CB_1 \vee CB_3 \vee Am \vee Bm \vee L_1Bs) = CB_3 \vee Bm$$

The factors that cause fault A to change to multi-fault AB are $CB_3 \vee Bm$, in other words, CB_3=1 or Bm=1 are the facors that cause fault A to change to fault AB.

Table 2. The information table for inducing the causes of multi-faults that include A

U	CB_1	CB_2	CB_3	CB_4	Am	Bm	Cm	L_1Am	L_1Bm	L_2Bm	L_2Cm	L_1Ap	L_1Bp	L_2Bp	L_2Cp	L_1As	L_1Bs	L_2Bs	L_2Cs	Fault
1	1	0	0	0	1	0	0	0	0	0	0	0	0	0	0	0	0	0	0	A
12	0	1	0	0	1	0	0	0	0	0	0	0	0	0	0	0	1	0	0	A
13	1	1	1	0	1	1	0	0	0	0	0	0	0	0	0	0	0	0	0	AB
17	1	0	0	1	1	0	1	0	0	0	0	0	0	0	0	0	0	0	0	AC
18	1	1	0	0	1	0	0	1	1	0	0	0	0	0	0	0	0	0	0	AL_1
19	1	0	1	1	1	0	0	0	0	1	1	0	0	0	0	0	0	0	0	AL_2

Table 3. Decernibility matrix for the examples with faults A, AB, AC, AL_1 and AL_2

	1	12	13
17(AB)	CB_2 CB_3 Bm	CB_1 CB_3 Bm L_1Bs	CB_1 CB_3 Am Bm L_1Bs
18 (AC)	CB_4 Cm	CB_1 CB_2 CB_4 Cm L_1Bs	CB_1 CB_2 CB_4 Am Cm L_1Bs
19(AL_1)	CB_2 L_1Am L_1Bm	CB_1 L_1Am L_1Bm L_1Bs	CB_1 Am L_1Am L_1Bm L_1Bs
20(AL_2)	CB_3 CB_4 L_2Bm L_2Cm	CB_1 CB_2 CB_3 CB_4 L_2Bm L_2Cm L_1Bs	CB_1 CB_2 CB_3 CB_4 Am L_2Bm L_2Cm L_1Bs

The factors that cause fault A to change to multi-fault AC can be induced as follows (see example 18 in Table 2):

$(CB_4 \vee Cm) \wedge (CB_1 \vee CB_2 \vee CB_4 \vee Cm \vee L_1Bs) \wedge (CB_1 \vee CB_2 \vee CB_4 \vee Am \vee Cm \vee L_1Bs) = CB_4 \vee Cm$

So we can conclude that:

The factors that cause fault A change to multi-fault AC are CB4 $\vee$Cm.

Similarly, we can induce that L_1Am =1 or L_1Bm =1 are the factors that cause fault A to change to fault AL_1 (see example 19 in Table 2). And CB_3=1, CB_4=1, L_2Bm=1 or L_2Cm=1 are the factors that cause fault A to change to fault AL_2 (see example 20 in Table 2).

To sum up, the factors that cause faults A, B, C, L_1 and L_2 to change into corresponding multi-faults are as follows:

(1) $CB_3 \vee Bm \rightarrow (A \Rightarrow AB)$

(2) $CB_4 \vee Cm \rightarrow (A \Rightarrow AC)$

(3) $L_1Am \vee L_1 Bm \rightarrow (A \Rightarrow A L_1)$

(4) $CB_3 \vee CB_4 \vee L_2Bm \vee L_2 Cm \rightarrow (A \Rightarrow A L_2)$

(5) $CB_1 \vee Am \rightarrow (C \Rightarrow AC)$

(6) $CB_2 \vee CB_3 \vee Bm \rightarrow (C \Rightarrow BC)$

(7) $CB_1 \vee CB_2 \vee L_1Am \vee L_1 Bm \rightarrow (C \Rightarrow CL_1)$

(8) $CB_3 \vee L_2Bm \vee L_2 Cm \rightarrow (C \Rightarrow C L_2)$

(9) $Am \rightarrow (L_2 \Rightarrow A L_2)$

(10) $CB_2 \vee Bm \rightarrow (L_2 \Rightarrow B L_2)$

(11) $Cm \rightarrow (L_2 \Rightarrow C L_2)$

(12) $CB_2 \vee L_1Am \vee L_1 Bm \rightarrow (L_2 \Rightarrow L_1 L_2)$

(13) $Am \rightarrow (B \Rightarrow AB)$

(14) $Cm \rightarrow (B \Rightarrow BC)$

(15) $L_1Am \vee L_1 Bm \rightarrow (B \Rightarrow B L_1)$

(16) $L_2Bm \vee L_2 Cm \rightarrow (B \Rightarrow B L_2)$

(17) $Am \rightarrow (L_1 \Rightarrow A L_1)$

(18) $CB_3 \vee Bm \rightarrow (L_1 \Rightarrow B L_1)$

(19) $CB_4 \vee Cm \rightarrow (L_1 \Rightarrow C L_1)$

(20) $CB_3 \vee L_2Bm \vee L_2 Cm \rightarrow (L_1 \Rightarrow L_1 L_2)$

4.2 Induce More Abstract Rules from the Above Rules

From rule (1) mentioned above, we can conclude that: if CB_3=1 or Bm=1 the fault B is added, and from rule (11), if Cm=1 the fault C is added etc. So we can get the following table that can be regarded as the varietal form of information table of rough set.

Table 4. Using information table to represent the rules

U	CB_1	CB_2	CB_3	CB_4	Am	Bm	Cm	L_1Am	L_1Bm	L_2Bm	L_2Cm	A^+	B^+	C^+	L^+_1	L^+_2
1			1			1							1			
2				1			1							1		
3								1	1						1	
4			1	1						1	1					1
5	1				1							1				
6		1	1			1							1			
7	1	1						1	1						1	
8			1							1	1					1
9					1							1				
10		1				1							1			
11							1							1		
12		1						1	1						1	
13					1							1				
14							1							1		
15								1	1						1	
16										1	1					1
17					1							1				
18			1			1							1			
19				1			1							1		
20			1							1	1					1

In Table 4, U is the set of rules. CB_1, CB_2, CB_3, CB_4, Am, Bm, Cm, L_1Am, L_1Bm, L_2Bm and L_2Cm are the condition attributes whose values only take 1 which means that breakers open or relays operate. A^+, B^+, C^+, L^+_1 and L^+_2 are decision attributes whose values only take 1 that means that the fault has been added. The condition attributes values and decision attributes values are only take the value 1 is because that we only induce the causality between breakers openning and faults or between relays operating and faults.

With rough set theory, we can conclude that:

$$IND(Am) = IND(A) = POS_{Am}(A) = \{5,9,13,17\},$$

$$IND(Bm) = IND(B) = POS_{Bm}(B) = \{1,6,10,18\},$$

$$IND(Cm) = IND(C) = POS_{Cm}(C) = \{2,11,14,19\},$$

$$IND(L_1Am) = IND(L_1Bm) = IND(L_1) = POS_{L_1Am}(L_1) = POS_{Bm}(L_1) = \{3,7,12,15\},$$

$$IND(L_2Bm) = IND(L_2Cm) = IND(L_2) = POS_{L_2Bm}(L_2) = POS_{L_2Cm}(L_2) = \{4,8,16,20\}$$

Whereas $IND(CB_1)$, $IND(CB_2)$, $IND(CB_3)$ and $IND(CB_4)$ cannot be a positive region of anyone of A^+, B^+, C^+, L^+_1 and L^+_2.

So the following more abstract rules can be induced:

(21) Am is the common factor that causes fault B to change into BA, C to change into CA, L_1 to change into L_1A and L_2 to change into L_2A (see rules (13), (5), (17) and (9)).

(22) Bm is the common factor that causes fault A to change into AB, C to change into CB, L_1 to change into L_1B and L_2 to change into L_2B (see rules (1), (6), (18) and (10)).

(23) Cm is the common factor that causes faults A to change into AC, B to change into BC, L_1 to change into L_1C and L_2 to change into L_2C (see rules (2), (14), (19) and (11)).

(24) L_1Am or L_1Bm are the common factors that cause fault A to change into AL_1, B to change into BL_1, C to change into CL_1 and L_2 to change into L_1L_2 (see rules (3), (15), (7) and (12)).

(25) L_2Bm or L_2Cm are the common factors that cause faults A to change into AL_2, B to change into BL_2, C to change into CL_2 and L_1 to change into L_1L_2 (see rules (4), (16), (8) and (20)).

5 SARS Data Application

After generalizing the rules such as “the nth day of the patient having been ill”=58 -71 → “headache” = 0 with 29 records, we can get the following general rules.

(1) The patients' symptoms such as “headache”, “muscle ache”, “dry cough”, “nausea”, “head is heavy”, “desudation”, and “thirst” disappear during the period of “the nth day of the patient having been ill” being 58 -71 days.

(2) The patients' symptoms such as “breathe hard”, “breathe hurriedly”, “expectorate”, “heart-throb”, “inappetence”, and “symptom in psychosis” disappear during the period of “the nth day of the patient having been ill” being 63 -71 days.

6 Discussions

(1) The idea of decernibility matrix can be used to not only the whole data set but also the partial data to generate special knowledge.

(2) Obviously, rules (21)-(25) are more abstract than the rules (1)-(20), which mean that the rough set theory can be used to induce more abstract knowledge in knowledge base.

(3) For inducing easily understood knowledge, the rough set theory has an advantage over the black-box based machine learning methods such as ANN, SVM etc.

(4) These kinds of knowledge help experts understand the correlation of rules or of attribute values better and more clearly.

Acknowledgements

The paper is supported by the Special Foundation of SARS of the National High-Tech Research and Development Major Program of China (863)- "Clinical Research on the Treatment of Severe Acute Respiratory Syndrome with Integrated Traditional and Western (No.2003AA208101)".

References

1. Pawlak Z. Rough sets. Int. J. of Computer and Information Science. 11 (1982): 341-356
2. Raymond T. Ng Laks V.S. Lakshmanan Jiawei Han. Exploratory Mining and Pruning Optimizations of Constrained Associations Rules. SIGMOD '98 Seattle, WA, USA.
3. Agotnes, T. Filtering large propositional rule sets while retaining classifier performance. Master's thesis, Norwegian University of Science and Technology,February 1999.
4. H. Toivonen, M. Klemettinen, P. Ronkainen, and H. Mannila. Pruning and Grouping Discovered Association Rules. In MLnet Workshop on Statistics, Machine Learning and Discovery in Databases, 4 (1995), 47–52
5. Gray B., and Orlowska M.E. CCAIIA: Clustering Categorical Attributes into Interesting Association Rules, in Proceedings of the second Pacific-Asia Conference, PAKDD-98, Lecture Notes in Artificial Intelligence 1394 (1998), 132-143.
6. Srikant R., and Agrawal R. Mining Generalized Association Rules, Research Report IBM Research Division. (1995).
7. Yang Bingru, Shen Jiangtao and Chen Hongjie. Recearch on the Structure Model and mining Algorithm for Knowledge Discovery Based on Knowledge Base (KDK). Engineering Science (Chinese) 5 (2003): 49-53

Dynamic Calculation Method of Degree of Association Between Concepts

Noriyuki Okumura, Takayoshi Araki, Hirokazu Watabe, and Tsukasa Kawaoka

Dept. of Knowledge Engineering & Computer Sciences, Doshisha University
Kyo-Tanabe, Kyoto, 610-0394 Japan

Abstract. This paper proposes calculation method of a degree of association between words. Moreover, 'Two words Association Mechanism' is constructed and evaluated as the application. Using Concept-base constructed from electronic dictionary and newspaper mechanically, this paper proposes an achievement method for association mechanism of concepts to accomplish key role by conversational processing of computer. In the calculation method of degree of association which is the kernel of association mechanism, the attributes which is characterized concepts is dynamically changed; the common feature of the concept is emphasized, and evaluated. As a result, calculation method of degree of association close to human beings sense is obtained. Moreover, as a sample of association mechanism, using 'Two words Association Mechanism', the effectiveness of the proposal method was verified.

1 Introduction

Human beings have achieved a wide conversation and communications by various associations in daily life. This paper proposes the Construction Method of Association Mechanism to achieve human beings' association ability on computer.

Publicly, as Association Mechanism, it has two functions: one is selection some words to relate other words from various words; another one is associating similar words.

The function of the former is called Recollection Word Processing (*RWP*). A latter function is called Unknown Word Processing (*UWP*). This is the reason why it is often used to substitute an unknown word for an already-known word. *RWP* is the mechanism to present words with strong relation which Human beings common sensibly associates, for example, if Human input is 'Apple', then *RWP* answers 'Red, Sweet, Fruit, … '. In contract, *UWP* is the mechanism to substitute the word that is not defined Knowledge base [1] [2] (Unknown Word) for words which are defined Knowledge base (Already Known Word).

These mechanisms are constructed with Concept-base and Calculation Method of Degree of Association [3]. Calculation Method of Degree of Association is the method to evaluate relationship from Degree of Match between concepts. For example, not only similar words (Ex: Red, Purple), but also relative words (Ex: Red, Blood), Calculation Method of Degree of Association is able to quantify the relation. Using Association Mechanism, it is able to associate relative concepts from some words or some sentences, and to use knowledge flexibility.

B. Gabrys, R.J. Howlett, and L.C. Jain (Eds.): KES 2006, Part I, LNAI 4251, pp. 506 – 514, 2006.
© Springer-Verlag Berlin Heidelberg 2006

2 Concept-Base

Concept-base is constructed from Japanese Electronic Dictionary and Newspaper mechanically. It is composed, the headword is assumed to be a concept, the content words included in the explanation are assumed to be attributes. It is the database of important factor to Association Mechanism for Calculation Method of Degree of Association. Concept-base is set of concepts (words), and concept (*A*) is sets of pairs of some attributes (*i a*) characterizing concept (*A*) and weights (*w i*) of importance of each attributes. (Eq.1)

$$A = \{(a_1, w_1), (a_2, w_2), \cdots, (a_i, w_i), \cdots, (a_{znum}, w_{znum})\} \tag{1}$$

Attributes are also defined in Concept-base as concepts; so one concept is defined as *n* dimension attribute chain model. In this paper, Concept-base (thirty thousand concepts) made form electronic dictionary is extended to wide Concept-base (ninety thousand concepts) added from electronic newspaper. Calculation Method of Degree of Association is evaluated with extended Concept-base. In the case of deriving concepts and attributes from electronic newspaper mechanically, comparing with the case of deriving concepts and attributes from electronic dictionary, relationship between concepts is not good, so Concept-base's quality is lower. For this reason, it is wanted that Calculation Method of Degree of Association is high quality.

3 Calculation Method of Degree of Association

For measure method between concepts, Calculation Method of Degree of Association [3] is proposed. In Calculation Method of Degree of Association, concepts are defined as attributes sets with weight, and Degree of Association is calculated in consideration of Degree of Match. For the calculation of Degree of Match, pairs of attributes sets which are Match or Similar are defined between the concepts. After the process, for the pairs of attributes, it is calculated in consideration of weights of the attributes.

3.1 Degree of Match in Consideration of Weight Ratio (*DM/WR*)

For two concepts: *A* and *B*, the first-order attributes and the weights are defined as follows (Eq.2).

$$\begin{aligned} A &= \{(a_i, u_i) \mid i = 1 \sim L\} \\ B &= \{(b_j, v_j) \mid j = 1 \sim M\} \end{aligned} \tag{2}$$

Then, it is defined of Degree of Match in consideration of Weight's Ratio (*MatchWR(A, B)*) between concepts: *A* and *B*. In addition, each attributes' weights are normalized to the total of sum of weights equal 1.0.

$$\begin{aligned} MatchWR(A, B) &= \sum_{a_i = b_j} Min(u_i, v_j) \\ Min(u_i, v_j) &= \begin{cases} u_i (u_i \le v_j) \\ v_j (v_j < u_i) \end{cases} \end{aligned} \tag{3}$$

DM/WR is defined by Eq.3, because if common attributes are detected, then effective weight is smaller weight of both common attributes. In addition, *DM/WR* is from 0.0 to 1.0.

3.2 Static Calculation Method of Degree of Association (*SCM/DA*)

Refined Eq.2, lower number of attributes is A ($A(L \le M)$). Rank of first-order attributes of concept A draw in order.

$$A = \{(a_1, u_1), (a_2, u_2), \cdots, (a_L, u_L)\} \tag{4}$$

First, it searches attributes which become $a_i = b_j$ ($MatchWR(a_i, b_j) = 1.0$). The case is searched which becomes $a_i = b_j$ (Completely Match), if $u_i > v_j$, then $u_i' = u_i - v_j$, if $u_i > v_j$, then $v_i' = v_i - u_j$, then using u_i' or v_i', it is made to correspond to other attribute. The relation is more correctly appreciable.

Second, rank of first-order attributes of concept B is decided with concept B as constructed with concept A by Eq.2 considered with Completely Match. The number of Completely Match is α . In addition, $B_{L+\alpha+1}, B_{L+\alpha+2}, \cdots, B_M$ is not used.

$$\begin{aligned} B_x &= \{(b_{x_1}, v_{x_1}), (b_{x_2}, v_{x_2}), \cdots, (b_{x_{L+\alpha}}, v_{x_{L+\alpha}})\} \\ x_k &= \{\text{Number of attributes to which } a_i \text{ is decided correspondence}\} \end{aligned} \tag{5}$$

Third, *SCM/DA* (*ChainWR(A, B)*) is defined as follows, because $L+\alpha$ attributes correspond.

$$\begin{aligned} ChainWR(A,B) &= \sum_{i=1}^{L+\alpha} MatchWR(a_i, b_{x_i}) \times \frac{(u_i + v_{x_i})}{2} \times \frac{Min(u_i, v_{x_i})}{Max(u_i, v_{x_i})} \\ Min(u_i, v_{x_i}) &= \begin{cases} u_i (u_i \le v_{x_i}) \\ v_{x_i} (v_{x_i} < u_i) \end{cases} \qquad Max(u_i, v_{x_i}) = \begin{cases} u_i (u_i \ge v_{x_i}) \\ v_{x_i} (v_{x_i} > u_i) \end{cases} \end{aligned} \tag{6}$$

ChainWR(A, B) corrects Degree of Association by multiplying average ($(u_i + v_{x_i})/2$) of both attribute's weight and ratio ($Min(u_i, v_{x_i}) / Max(u_i, v_{x_i})$) of both attributes' weights to *MatchWR(A, B)*. Using *ChainWR(Desk, Chair)* as sample, *SCM/DA* is described. Table.1 shows first-order attributes of Desk and Chair, Table.2 shows second-order attributes of Desk and Chair.

Table 1. First-order attributes of desk and chair

Concept	First-order attributes		
Desk	(School,0.6)	(Study,0.3)	(Bookshelf,0.1)
Chair	(Study,0.5)	(Classroom,0.3)	(Tree,0.2)

Table 2. Second-order attributes

First-order attributes	Second-order attributes		
School	(College,0.4)	(Schoolhouse,0.4)	(Wooden,0.2)
Study	(Preparation,0.5)	(Examination,0.3)	(Book,0.2)
Bookshelf	(Library,0.6)	(Page,0.3)	(Book,0.1)
Classroom	(Teacher,0.4)	(Schoolhouse,0.4)	(Student,0.2)
Tree	(Timber,0.5)	(Wooden,0.4)	(Leaf,0.1)

MatchWR(Bookshelf, Study) by Table.1 is as follow using Eq.3 and Table.2.

$$MatchWR(Bookshelf, Study) = Min(0.2_{Book}, 0.1_{Book}) = 0.1$$

MatchWR(A, B) is calculated by sum of lower weight of Completely Matched attributes. In case of *MatchWR(Bookshelf, Study)*, (Book) is the only Completely Match of attribute. In consequence, *MatchWR(Bookshelf, Study)* is 0.1. Based on this case, all of combination of *DM/WR* is shown in Table.3.

Table 3. Matrix of *DM/WR*

	School	Study	Bookshelf
Study	0.0	**1.0**	**0.1**
Classroom	**0.4**	0.0	0.0
Tree	0.2	0.0	0.0

(Study) is Completely Matched of attribute ($MatchWR(Study, Study) = 1.0$) of Desk and Chair. Given this factor, difference of weight (Difference of Completely Match: *Dif/CM*) is used again. $ChainWR(Desk, Chair)$ is calculated using Eq.6 and Table.3 as follows.

$$ChainWR(Desk, Chair) = \underbrace{1.0 \times \frac{0.3+0.3}{2} \times \frac{0.3}{0.3}}_{\substack{\text{Completely Match} \\ \text{(Study, Study)}}} + \underbrace{0.4 \times \frac{0.3+0.6}{2} \times \frac{0.3}{0.6}}_{\substack{\text{Normally Match} \\ \text{(School, Classroom)}}} + \underbrace{0.1 \times \frac{0.2+0.1}{2} \times \frac{0.1}{0.2}}_{\substack{\text{DCM} \\ \text{(Study, Bookshelf)}}}$$

$$= 0.465$$

SCM/DA is from 0.0 to 1.0.

3.3 Dynamic Calculation Method of Degree of Association (*DCM/DA*)

SCM/DA is calculated intended concets with attributes of *s* pieces rank order of weights. It aims reducing calculation time, because max number of attributes is about four hundred (Section 4.3.).

DCM/DA does not use attributes based on rank order of weights, use attributes based on rank order of *DM/WR*. Especially, Completely Matched attributes are picked up in advance. In the case of odd attributes, *DCM/DA* uses attributes based on rank of weights. From here onwards, common attributes are made point between concept *A* and concept *B*. It is gotten that Degree of Association which is near human beings' sense.

Ultimately sample using *X-ABC evaluation data*, all of concept has seven attributes. In this case, three attributes are used based on rank of weights. *X-ABC evaluation data* is described in Section 4.1. But no common attributes are three attributes based on rank of weights. In consequence, Degree of Association is not calculated rightly.

Given this factor, the number of attributes is three, but attributes are used based on rank of *DM/WR* (Fig.1). For example, attribute: Automobile of concept *X* is made to corresponded attribute: Cab of concept *B*. As just described, this method uses attributes dynamically by the concept of becoming an object of comparison. Using attributes dynamically, Degree of Association is calculated rightly between concepts. In this sample, weights are not normalized (Fig.1).

Car(*X*)		Automobile(*A*)		Horse(*B*)		Egg(*C*)	
Automobile *(A: Automobile)* *(B: Cab)*	0.96	Car *(X: Car)*	0.8	Animal	0.9	Chicken	0.8
Car *(A: Car)*	0.75	Carriage *(X: Carriage)*	0.7	Mammals	0.77	Egg yolk	0.75
Riding	0.51	Auto truck	0.66	Blood horse	0.72	Food	0.7
Carriage *(A: Carriage)* *(B: Carriage)*	0.42	Automobile *(X: Automobile)*	0.6	Horse	0.6	Egg	0.64
Wheel	0.4	Engine	0.4	Cab *(X: Cab)** *(X: Automobile)***	0.55	Alimentation	0.52
Cab *(B: Cab)*	0.15	Run	0.15	Donkey	0.3	Albumen	0.31
Transit	0.05	Road	0.14	Carriage *(X: Carriage)*	0.25	Female	0.3
Cargo	0.05	Driving	0.11	Graminivorous	0.1	Bird	0.3

*Completely Match   ** Def of Completely Match

Fig. 1. Dynamic Calculation Method of Degree of Association

4 Evaluation

4.1 Data for Evaluation

As a Measure for evaluation of Degree of Association, *X-ABC evaluation data* is readied which is composed of foursome (Table.4). It is constructed with concept *A* which is most relative concept, with concept *B* that is relative concept, with concept *C* which is not relative, as contrasted with basis concept *X* that is one of any concepts defined on Concept-base. 1780 foursomes are readied.

Table 4. Data for Evaluation (*X-ABC evaluation data*)

X	*A*	*B*	*C*
Music	Musical	Sound	Train
Sea	Ocean	Salt	Car
Student	Pupil	Academics	Apple
…	…	…	…

4.2 The Method of Evaluation

Degree of Association between *X* and *A* is *ChainWR(X, A)*, Degree of Association between *X* and *B* is *ChainWR(X, B)*, Degree of Association between *X* and *C* is *ChainWR(X, C)*, shown in Table.4. When the following are filled, it is assumed which it answers correctly.

$$ChainWR(X,A) - ChainWR(X,B) > AveChainWR(X,C)$$
$$ChainWR(X,B) - ChainWR(X,C) > AveChainWR(X,C)$$

$$AveChainWR(X,C) = \frac{\sum_{l=1}^{1780} ChainWR(X_l,C_l)}{1780}$$

ChainWR(X, C) must be 0.0 ideally. But for the characteristic of Calculation of Degree of Association, the case of that only one attribute pair exists between *X* and *C*, no relative concept *C* as contrasted with *X* is not calculated to 0.0. Given this factor, Calculation of Degree of Association has some error. Consequently, when Degree of Association between *ChainWR(X, A)* and *ChainWR(X, B)*, between *ChainWR(X, B)* and *ChainWR(X, C)* has domination difference more than error margin, evaluation is correct. The correct ratio of this evaluation method is called C-Average Rank Rate (*CARR*).

4.3 Evaluation of Calculation Method of Degree of Association

SCM/DA and *DCM/DA* are evaluated by *CARR*. Each concept has some attributes, in the case of maximum about 400 attributes are given, another case of minimum only 1 attribute is given. Given this factor, *SCM/DA* is evaluated with limited 50 attributes of rank of weights, and *DCM/DA* is evaluated with limited 50 attributes of rank of *DM/WR*. It is shown as follows that *SCM/DA* and *DCM/DA* are evaluated by *CARR* (Fig.2, Fig.3).

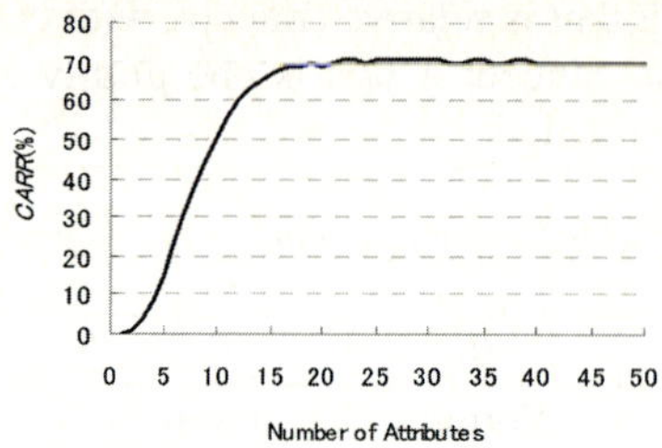

Fig. 2. *SCM/DA*

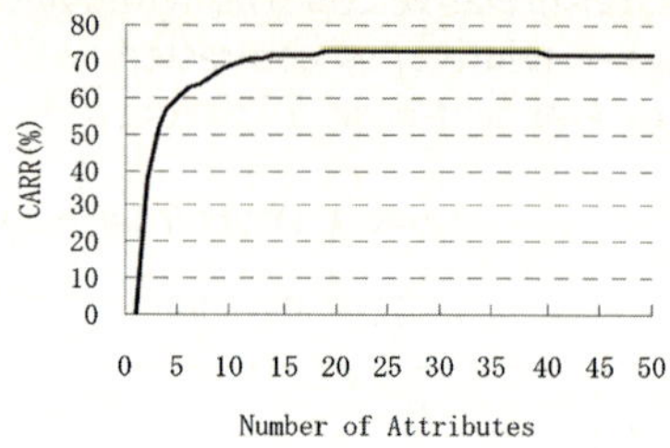

Fig. 3. *DCM/DA*

It is shown in Fig 2, Fig 3, using each Method of Degree of Association, number of attributes is from twenty to thirty, and *CARR* converges. In addition, the best number of attributes which are used by each method is thirty, because average of number of attributes defined Concept-base is about thirty. In case of that the number of attributes is thirty, CARR is shown as follow (Fig.4). In addition, it is shown that the average of *ChainWR(X, A)*, *ChainWR(X, B)*, *ChainWR(X, C)* by *SCM/DA* and *DCM/DA* (Table.5).

Table 5. Each average of Static and Dynamic Calculation Method

SCM/DA			*DCM/DA*		
ChainWR(X,A)	*ChainWR(X,B)*	*ChainWR(X,C)*	*ChainWR(X,A)*	*ChainWR(X,B)*	*ChainWR(X,C)*
0.242 ↔	0.048 ↔	0.002	0.302 ↔	0.065 ↔	0.002
	0.194	0.046		0.237	0.063

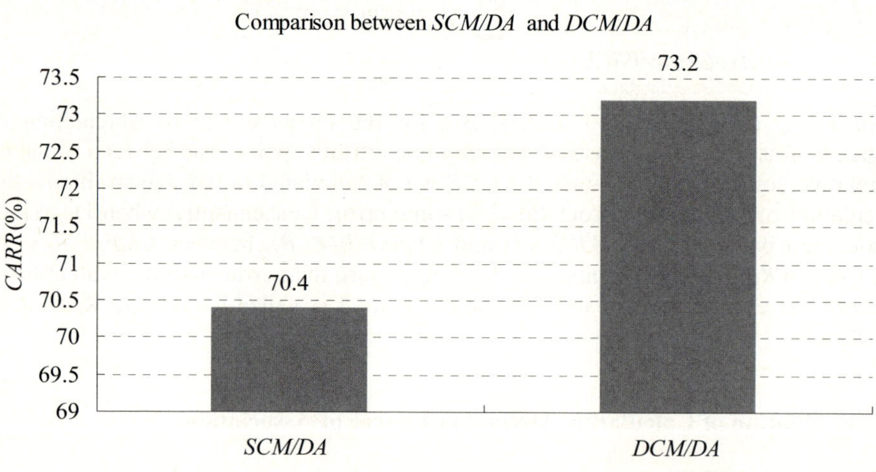

Fig. 4. Comparison between Static and Dynamic Method

DCM/DA has higher value (*ChainWR(X, A,) ChainWR(X, B)*) than *SCM/DA* by Table.5. Given this factor, most relative concept has higher value of Degree of Association than relative concept. For that reason, *DCM/DA* calculates righter Degree of Association than *SCM/DA*.

In next section, 'Two Words Association Mechanism' is reported for application.

5 Two Words Association Mechanism

5.1 Two Words Association

Using Static and Dynamic Method proposed this paper, 'Two Words Association Mechanism' is constructed that corresponds outputs (Ex: (Umbrella, Rain cape ...)) out of two words input (Ex: (Rain, Boots)).

In the 'Two Words Association Mechanism', first-order attributes and second order-attributes are gotten from Concept-base. In addition, first-order concepts and second-order concepts are gotten, too, because Concept-base is constructed as *s* dimension attributes model. Gotten words are selected by Static and Dynamic Method .

Degree of Association is calculated between some outputs and two inputs. By calculating average of all two Degree of Association, ten outputs are gotten by the upper rate of average.

5.2 Evaluation of 'Two Words Association Mechanism'

A hundred test sets (Noun, Noun) is readied for evaluation of 'Two Words Association Mechanism'. As the evaluation method, three persons measure the outputs. If all of three answered right, then the outputs are correct. Average of (Correct words)/ (number of outputs) with each test sets is degree of accuracy. In addition, it is common words gotten by questionnaire that the upper rate of each test sets' associative words. It is recall ratio that is (outputs comprised common answer)/ (a hundred sets times ten words) (Fig.5).

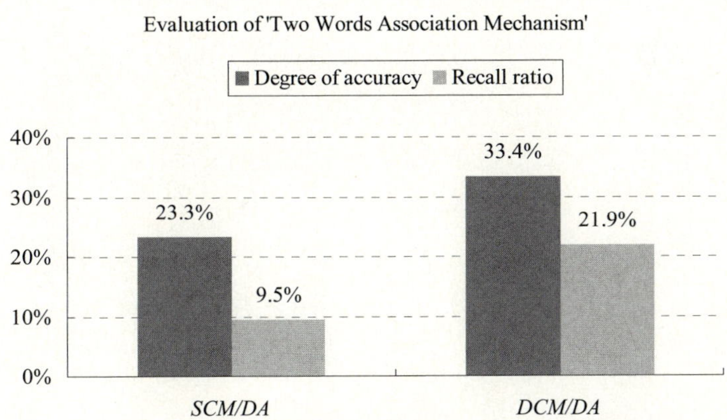

Fig. 5. Evaluation of 'Two Words Association Mechanism'

Dynamic Method is effective in 'Two Words Association Mechanism' by Fig.5. However, 'Two Words Association Mechanism' needs to refine, because Concept-base needs to refine much more.

6 Conclusions

In this paper, Static and Dynamic Calculation Method of Degree of Association based on ninety thousand Concept-base are proposed and evaluated. It is shown that Dynamic Method is more common human beings' sense than Static Method. In addition, Dynamic Method is more effective by the application of 'Two Words Association Mechanism' out of two words inputs.

Acknowledgements

This work was supported with the Aid of Doshisha University's Research Promotion Fund.

References

1. S. Tsuchiya, H. Watabe, and T. Kawaoka : A Time Judgement System Based of an Association Mechanism, *Proc. of KES2005 (Knowledge-Based Intelligent Information and Engineering Systems)*, Part III, pp.742-748 (2005).
2. A. Horiguchi, H. Watabe, and T. Kawaoka : Constructing a Sensuous Judgment System Based on Conceptual Processing, *Computational Linguistics and Intelligent Text Processing (Proc. of CICLing-2002)*, Vol. pp.86-95 (2002).
3. H. Watabe, and T. Kawaoka : The Degree of Association between Concepts using the Chain of Concepts, *Proc. of SMC2001 (IEEE International Conference on Systems Man & Cybernetics)*, pp.877-881 (2001).

Combining Fuzzy Cognitive Maps with Support Vector Machines for Bladder Tumor Grading

Elpiniki Papageorgiou, George Georgoulas, Chrysostomos Stylios[1], George Nikiforidis[2], and Peter Groumpos

Laboratory for Automation and Robotics, Department of Electrical and Computer Engineering, University of Patras, Rion 26500, GREECE
Tel.: +30 2610 997293; Fax: +30 2610 997309
epapageo@ee.upatras.gr, georgoul@ee.upatras.gr, groumpos@ee.upatras.gr

[1] Department of Communications, Informatics and Management, TEI of Epirus, 47100 Artas, Greece
Tel.: +30 6974638830
stylios@teleinfom.teiep.gr

[2] Computer Laboratory, School of Medicine, University of Patras, Rion 26500, Greece

Abstract. Fuzzy Cognitive Map (FCM) is an advanced modeling methodology that provides flexibility on the system's design, modeling, simulation and control. This research work combines the Fuzzy Cognitive Map model for tumor grading with Support Vector Machines (SVMs) to achieve better tumor malignancy classification. The classification is based on the histopathological characteristics, which are the concepts of the Fuzzy Cognitive Map model that was trained using an unsupervised learning algorithm, the Nonlinear Hebbian Algorithm. The classification accuracy of the proposed approach is 89.13% for High Grade tumor cases and 85.54%, for tumors of Low Grade. The results of the proposed hybrid approach were also compared with other conventional classifiers and are very promising.

Keywords: Fuzzy Cognitive Maps, Support Vector Machine, statistical learning, bladder tumor grading.

1 Introduction

For most tumor types, including bladder tumors, malignancy characterization and classification is expressed in pathological terms based on the morphology of tissues as viewed through a light microscope. Due to the subjective nature of this classification several approaches based on computer assisted methods have been used, trying to increase the diagnostic and/or prognostic value of tumors' characterization [1], [2]. Two main approaches have been reported in the literature: Data-driven models -such as statistical classifiers, Support Vector Machines (SVMs), artificial Neural Networks, Decision Trees, [3-9] and the more knowledge oriented models such as belief networks [10]. In this work, we explore the potential of a workable advanced system based on Fuzzy Cognitive Maps (FCMs) combined with SVMs to assist tumour characterization.

B. Gabrys, R.J. Howlett, and L.C. Jain (Eds.): KES 2006, Part I, LNAI 4251, pp. 515–523, 2006.
© Springer-Verlag Berlin Heidelberg 2006

The proposed system is based on FCMs, which is a soft computing methodology. FCMs are appropriate to explicit the knowledge and experience accumulated for years on the decision process of human experts [11], [12]. The flexibility of FCMs in system design and modeling, as well as their learning properties [13], [14], make their choice attractive for a variety of modeling and decision support tasks.

In a previous work, an FCM model has been proposed for the characterization of tumours' grade based on the analysis of conventional histopathological criteria [15], [16]. In this research work we are proposing the combination of FCMs with SVMs. SVMs have gained great attention and have been used extensively and, most important, successfully in the field of pattern recognition [17-21]. This hybrid approach has been proven quite successful and the results are very promising.

This paper is structured as follows; sections 2 and 3 briefly introduce the fundamentals of FCMs and SVMs respectively. Section 4 describes the developed FCM for the tumor grading model as well as the integration of SVMs in the whole procedure. Section 5 presents the implementation and the experimental results and in section 6 some conclusions and ideas for future work are discussed.

2 Fuzzy Cognitive Maps Background and Description

The FCM structure is similar to recurrent artificial neural networks, where concepts are represented by neurons and causal relationships by weighted links connecting the neurons. Concepts reflect attributes, characteristics, qualities and senses of the system. Interconnections among concepts of FCM signify the cause and effect relationship that a concept has on the others. These weighted interconnections represent the direction and degree with which concepts influence the value of the interconnected concepts. The directional influences are presented as all-or-none relationships, so the FCMs provide qualitative as well as quantitative information about these relationships. In [11], an analytical description of FCMs is presented.

Generally, the value of each concept is calculated, computing the influence of other concepts to the specific concept, by applying the following calculation rule:

$$A_i^{(k+1)} = f(\sum_{\substack{j \neq i \\ j=1}}^{N} A_j^{(k)} \cdot e_{ji}) \tag{1}$$

where $A_i^{(k+1)}$ is the value of concept C_i at time $k+1$, $A_j^{(k)}$ is the value of concept C_j at time k, e_{ji} is the weight of the interconnection between concept C_j and concept C_i and f is the logistic sigmoid threshold function.

The methodology for developing FCMs is based on a group of experts who are asked to define concepts, describe relationships among concepts, use IF-THEN rules to justify their cause and effect suggestions among concepts and infer a linguistic weight for each interconnection [11]. Every expert describes each interconnection with a fuzzy rule; the inference of the rule is a linguistic variable, which describes the relationship between the two concepts according to every expert and determines the grade of causality between the two concepts.

The inferred fuzzy weights are aggregated and through the defuzzification method of Center of Area (CoA) an overall linguistic weight is produced, which is transformed to a numerical weight e_{ji}, belonging to the interval [-1, 1] and representing the overall suggestion of experts [11].

3 Support Vector Machines Background

Support Vector Machines (SVMs) are learning systems that are trained using an algorithm based on optimization theory [17-21]. For real life problems, given l observations $D=\{(\mathbf{x}_i, y_i)\}_{i=1}^{l}$, the SVM solution finds the hyperplane in feature space that keeps both the empirical error small and maximizes the margin between the hyperplane and the instances closest to it. This can be done by minimizing:

$$\frac{1}{2}\|\mathbf{w}\|^2 + C\sum_{i=1}^{n}\xi_i \tag{2}$$

subject to

$$y_i((\mathbf{w}\cdot\phi(\mathbf{x}_i))+b) \geq 1-\xi_i\,,\ \xi_i \geq 0 \quad i=1,\ldots,n \tag{3}$$

where ξ_i are slack variables, which are introduced to allow the margin constraints to be violated, and $\boldsymbol{\varphi}(\cdot)$ is the nonlinear mapping from the input space to the feature space. Parameter C controls the trade off between maximizing the margin and minimizing the error and it is usually determined through a cross-validation scheme [19], [21-22].

The class prediction for an instance $\mathbf{x}$ is given by:

$$f(\mathbf{x}) = sign\left(\sum_{i=1}^{l} y_i\alpha_i \langle\boldsymbol{\varphi}(\mathbf{x}_i)\cdot\boldsymbol{\varphi}(\mathbf{x})\rangle + b\right) \tag{4}$$

where the coefficients a_i are calculated by maximizing the Lagrangian:

$$\sum_{i=1}^{l} a_i - \frac{1}{2}\sum_{i,j=1}^{l} a_i a_j y_i y_j \langle\boldsymbol{\varphi}(\mathbf{x}_i)\cdot\boldsymbol{\varphi}(\mathbf{x}_j)\rangle \tag{5}$$

subject to $\sum_{i=1}^{l} y_i\alpha_i = 0$ and $0 \leq \alpha_i \leq C \quad, i=1,2,\ldots,l$

The points for which a_i>0, are called Support Vectors and are the points lying closest to the hyperplane. If the nonlinear mapping function is chosen properly, the inner product in the feature space can be written in the following form:

$$\langle\boldsymbol{\varphi}(\mathbf{x}_i)\cdot\boldsymbol{\varphi}(\mathbf{x}_j)\rangle = K(\mathbf{x}_i,\mathbf{x}_j) \tag{6}$$

where K is called the inner-product kernel [17-21].

One of the most commonly used kernel is the Radial Basis Function (RBF) kernel

$$K(\mathbf{x},\mathbf{x}_i)=\exp\left(-\frac{1}{2\sigma^2}\|\mathbf{x}-\mathbf{x}_i\|^2\right) \tag{7}$$

where the width σ^2 is specified a priori by the user and is common for all kernels [22].

4 Fuzzy Cognitive Map Grading Model

Histopathologists with deep knowledge and great clinical experience were our experts whom we asked to develop and construct the FCM model for tumor grading using the methodology presented in section 2 and described analytically in [15]. Experts defined the main histopathological features (concepts) that determine the final grade characterization. More specifically, eight well documented in the bibliography histopathological criteria (features) essential for tumour grading (Table1) [23-24] were selected. In every day practice, each tissue section (patient slide) is evaluated retrospectively by histopathologists using these features. These considered features are the causative variables or factors of the tumour grading system that have been selected by experts to construct the FCM for tumour grading [15], [16].

The FCM tumor grading model was developed consisting of the following 9 concepts: Concept C_1 represents the cell distribution, C_2 represents the cell size, C_3 the cell number, C_4 the cytoplasm, C_5 the nuclei, C_6 the nucleoli, C_7 the necrosis, C_8 the mitoses and C_9 the degree of tumour grade.

Table 1. Main factors for grading

Histological feature	**Assessment Possible**
Cell distribution	Even, clustered
Cell size	Uniform, pleomorphic
Cell number	Numerous, variable
Cytoplasm	Homogeneous, variable
Nuclei	Uniform, irregular, very irregular, bizarre
Nucleoli	Inconspicuous, evident, prominent
Necrosis	Inconspicuous, frequent
Mitosis	Absent-rate, occasional, numerous

In order to use the FCM model histopathologists were asked to examine each tissue section retrospectively and estimate the value of the eight histopathological variables (Table 1); these values were transformed in the range [0, 1], and were assigned to the corresponding concepts.

Furthermore, histopathologists were asked to explain the cause-effect relationships among these concepts using IF-THEN rules that were described in section 2. Thus after defuzzification of the linguistic weights with GoA, the initial weights of the FCM tumor-grading model were determined and given in Figure 1.

The tumor grading procedure is based on the determination of the values of two concepts: concept "Nuclei" which represents the characteristics of nuclear appearance and may be considered as figuring out the nuclear grading concept, and concept "Grade" that characterizes the degree of tumor malignancy. These two concepts are considered the main attributes for the final degree of tumor characterization.

The unsupervised Nonlinear Hebbian Learning algorithm is used to modify the weights of the FCM grading model according to the initial values of concepts for each examined case of urinary bladder tumors. Also, according to the NHL algorithm, experts were asked to select the input and output concepts that determine the triggering process. The concepts C_5 and C_9 were defined as the Decision Output Concepts (DOCs), which determine the tumor grade. All the other concepts have been defined as input concepts.

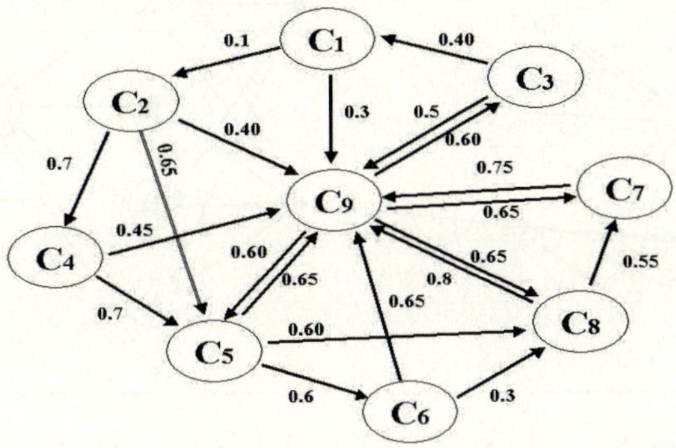

Fig. 1. The FCM tumor grading model consisting of 9 concepts and 21 weight relations

5 Description of the Tumour Grading Procedure Using SVMs

The FCM tumor-grading model can be used, after the development of the FCM model and the determination of the necessary specifications for the implementation of the NHL algorithm. For each case the training procedure is applied and then, through the use of an SVM with RBF kernels, the system determines the grade of the tumor (either "low" or "high").

The experimental data consisted of one hundred twenty-nine tissue sections (slides) each one from a different patient with superficial transitional cell carcinoma were retrieved from the archives of the Department of Pathology of University Hospital of Patras, Greece. Tissue sections were routinely stained with Haematoxylin-Eosin. Every case was reviewed independently by the doctors-experts to safeguard reproducibility. Histopathologists had classified the cases following the World Health Organization (WHO) grading system as follows: eighty-three as Low Grade, and forty-six as High Grade. Because of the restricted number of cases, we used the leave one out method [22] and the extreme case of multifold cross validation in order to evaluate the performance of the proposed methodology. Therefore, at each experiment we used 128 of the cases to build the model and then the model was tested on the example left out. However in order not to use the same data set for both building the model and estimating its performance [27], for each one of the 129 subsets we employed again the leave one out method within the 128 cases that consisted the training set in order to select the model's parameters. Once the parameters were optimized, the SVM was retrained for the whole training set (128 cases) and its performance was evaluated using the corresponding case that was originally left out. Figure 2 illustrates the integrated method of FCMs with SVMs for tumor grading.

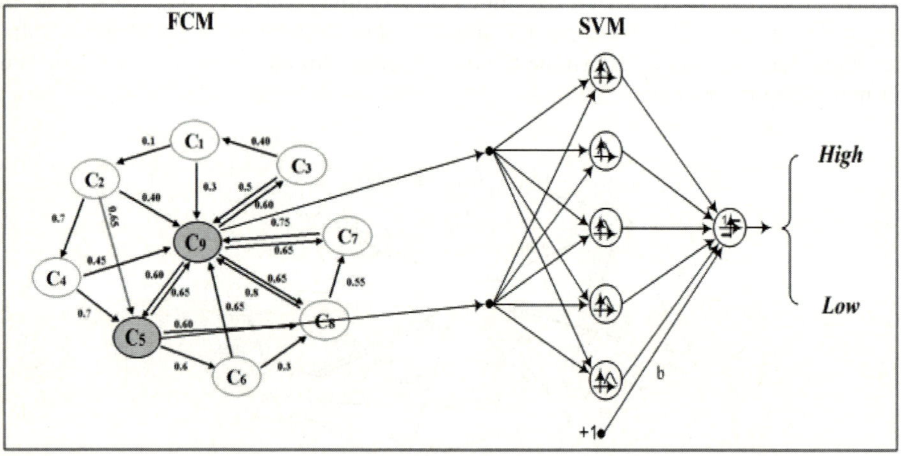

Fig. 2. The process of tumor grading combining FCM with SVM

It must be mentioned that due to the imbalanced nature of the available data, we introduced a modified formulation of the SVM algorithm. We used different error weights C^+ and C^-, so that to penalize more heavily the undesired type of error, and/or the errors related to the class with the smallest population [20], [26]. Therefore, the optimization problem is modified as follows:

$$\text{Minimize } \frac{1}{2}\|\mathbf{w}\|^2 + C^- \sum_{i:y_i=-1} \xi_i + C^+ \sum_{i:y_i=+1} \xi_i \qquad (9)$$

Selecting higher penalty value for the class with the smallest population, which in most cases is the class that needs to be correctly identified, we induce a boundary that is more distant from that class. In our experiments, we tested different values for the error weights C^+ and C^- keeping their ratio equal to 1/(46/83) (the inverse ratio of the corresponding cardinalities).

In previous works, our research group introduced the FCM-grading tool, working exclusively on the qualitative assessments of histopathological variables. The FCM grading tool had achieved a classification accuracy of 80%, and 87.5% for tumors of low and high grade respectively [16]. Here, this work further extends this research exploring means to improve classification accuracy by utilizing SVM classifier to assist medical diagnosis.

Table 2. Comparative results

Tumor category	FCM using SVMs	FCM-GT	quadratic classifier	linear classifier	9-nn
Low-Grade	85.54%	80 %	86.75%	85.56%	90.36%
High-Grade	89.13%	87.5%	78.26%	84.78%	78.26%
Overall-Accuracy	86.82%	84.5%	83.72%	85.27%	86.06%

The proposed hybrid grading tool, where the FCM-grading tool is combined with the SVM algorithm has achieved a classification accuracy of 85.54% for low grade tumors and 89.13% for high grade tumors. Thus, the introduction of the SVM stage improves the classification accuracy of the FCM-grading tool which had used initially the minimum distance method and only one target concept [16].

For comparison reasons, experiments were also conducted using three conventional classifiers: the k-nearest neighbor, the linear and the quadratic classifiers [27] and the results are summarized in Table 2.

6 Conclusions

In this research effort, a new extended structure is proposed where SVM classifier is added to the FCM-tumor grading model to improve the characterization of urinary bladder tumors. The hybridization of FCM model with SVMs exhibited high performance in correctly classifying tumors into two categories, low-grade and high-grade respectively, utilizing all the available diagnostic information. The proposed method could be considered as an efficient classification procedure able to make decisions with highly diagnostic accuracy.

Furthermore, the tumor characterization procedure based on the integration of the soft computing technique of FCMs and the statistical classification technique of SVMs has been compared with other conventional classifiers and the results are very promising. The current approach yield balanced performance for both classes and outperforms the other methods tested in terms of correctly classified high-grade instances. The hybrid tumor grading tool is fast, easily implemented in clinical practice and performs high accuracy in terms of specificity and sensitivity.

Acknowledgements

This work was supported by the "PYTHAGORAS II" research grant co funded by the European Social Fund and National Resources.

References

1. Parker, S.L., Tony, T, Bolden, S, Wingo, P.A: Cancer Statistics, Cancer Statistics. CA Cancer J Clinics, **47** 5 (1997) 5-27.
2. Ooms, E, Anderson, W, Alons C., Boon M, Veldhuizen R: "Analysis of the performance of pathologists in grading of bladder tumours", Human Pathology, **14** (1983) 140-143.
3. Gil, J, Wu, H, Wang, B.Y.: Image analysis and morphometry in the diagnosis of breast cancer, Microsc. Res. Techniq. **59** (2002) 109-118.
4. Catto J, Linkens D, Abbod M, Chen M, Burton J, Feeley K, Hamdy F: Artificial Intelligence in Predicting Bladder Cancer Outcome: A Comparison of Neuro-Fuzzy Modeling and Artificial Neural Networks. Artif Intell Med **9** (2003) 4172-4177.
5. Choi H, Vasko J, Bengtsson E, Jarkrans T, Malmstrom U, Wester K, Busch C.: Grading of transitional cell bladder carcinoma by texture analysis of histological sections. Anal Cell Pathol **6** (1994) 327-343.
6. Jarkrans T, Vasko J, Bengtsson E, Choi H, Malmstrom U, Wester K, Busch C. Grading of transitional cell bladder carcinoma by image analysis of histological sections. Anal Cell Pathol **18** (1995) 135-158.
7. Michie, D., Spiegelhalter, D. J., Taylor, C.C.: Machine Learning, Neural and Statistical Classification. Englewood Cliffs, N.J.: Prentice Hall (1994). Data available at http://www.ncc.up.pt/liacc/ML/statlog/datasets.html.
8. Podgorelec, V, Kokol, P, Stiglic, B, Rozman, I.: Decision Trees: An Overview and Their Use in Medicine, Journal of Medical Systems, **26** 5 (2002) 445-463.
9. Dreiseit, S, Ohno-Machado, L, Kittler, H, Vinterbo, S, Billhardt, H, Binder, M: A Comparison of Machine Learning Methods for the Diagnosis of Pigmented Skin Lesions, Journal of Biomedical Informatics **34** (2001) 28–36.
10. Antala, P., Fannesa, G., Timmermanb, D., Moreaua, Y, Moor B.D.: Bayesian applications of belief networks and multilayer perceptrons for ovarian tumor classification with rejection, Artificial Intelligence in Medicine **29** (2003) 39–60.
11. Stylios, C. D. and Groumpos, P.P.: Fuzzy Cognitive Maps in Modelling Supervisory Control Systems, J. of Intelligent & Fuzzy Systems, **8**, (2000) 83-98.
12. Papageorgiou, E., Stylios, C., Groumpos, P: An Integrated Two-Level Hierarchical Decision Making System based on Fuzzy Cognitive Maps (FCMs), IEEE Transactions on Biomedical Engineering, **50** 12 (2003) 1326-1339.
13. Papageorgiou, E.I., Stylios, C.D., Groumpos, P.P: Active Hebbian Learning to Train Fuzzy Cognitive Maps, Int. J. Approx. Reasoning, **37** (2004) 219-249.
14. Papageorgiou, E.I., Groumpos, P.P.: A weight adaptation method for fine-tuning Fuzzy Cognitive Map causal links, Soft Computing Journal, **9** (2005) 846-857.
15. Papageorgiou, E.I., Spyridonos, P., Ravazoula, P., Stylios, C.D., Groumpos, P.P., Nikiforidis, G: Advanced Soft Computing Diagnosis Method for Tumor Grading, Artif Intell Med, **36** (2006) 59-70.
16. Papageorgiou, E.I., Spyridonos, P. Stylios, C.D., Nikiforidis, G., Groumpos, P.P: The Challenge of Using Soft Computing Techniques for Tumor Characterization, *16th Intern. Conf. on Artificial Intelligence and Soft Computing (ICAISC)* Zakopane, Poland, 7-11 June, Lecture Notes in Computer Science 3070, (2004) 1031-1036.
17. Burges, C.J.C.: A Tutorial on Support Vector Machines for Pattern Recognition, Data Mining and Knowledge Discovery, **2** 2 (1998) 121-167.
18. Vapnik, V. N: The nature of statistical learning theory. New York: Springer-Verlag, 1995.
19. Muller, K. R., Mika, S., Ratsch, G., Tsuda, K., Scholkopf, B: An Introduction to Kernel-Based Learning Algorithms. IEEE Trans. Neural Networks, **2** 2 (2001) 181-201.

20. Veropoulos, K., Cambell, C. and Cristianini, N.: Controlling the sensitivity of support machines, Proc. Int. Joint Conf. on Artificial Intelligence (IJCAI99), Stockholm, 1999. pp. 55-60.
21. Cristianini, N. and Shawe-Taylor, J: An Introduction to Support Vector Machines. Cambridge University Press, (2000).
22. Haykin, S: *Neural Networks: A Comprehensive Foundation*. 2nd ed. Englewood Cliffs, NJ: Prentice Hall, (1999).
23. Spyridonos, P, Ravazoula, P, Cavouras, D, Berberidis, K, Nikiforidis, G.: Computer-based grading of haematoxylin-eosin stained tissue sections of urinary bladder carcinomas, Med Inform & Intern Med, **26** 3 (2001) 179-190.
24. Bostwick, G, Ramnani, D, Cheng, L.: Diagnosis and grading of bladder cancer and associated lesions, Urologic Clinics of North America, **26** (1999) 493-507.
25. Salzberg, S.L: On Comparing Classifiers: Pitfalls to avoid and a Recommended Approach, *Data Mining and Knowledge Discovery*, **1** (1997) 317-328.
26. Osuna, E.E., Freund, R., and Girosi, F: Support Vector Machines: Training and Applications, MIT, A.I. Memo No. 1602, March (1997).
27. R. O. Duda, P. E. Hart, and D. G. Stork, *Pattern Classification*. 2nd ed. John Wiley & Sons, (2001).
28. Murphy, WM.: Urothelial neoplasia, Pathology and pathobiology of the urinary bladder and prostate, Baltimore, Williams & Wilkins, (1992).

Image Classification by Fusion for High-Content Cell-Cycle Screening

Tuan D. Pham[1,2] and Dat T. Tran[3]

[1] Bioinformatics Applications Research Centre
[2] School of Information Technology
James Cook University
Townsville, QLD 4811, Australia
tuan.pham@jcu.edu.au
[3] School of Information Sciences and Engineering
University of Canberra,ACT 2601, Australia
dat.tran@canberra.edu.au

Abstract. We present a fuzzy fusion approach for combining cell-phase identification results obtained from multiple classifiers. This approach can improve the classification rates and allows the task of high-content cell-cycle screening more effective for biomedical research in the study of structures and functions of cells and molecules. Conventionally such study requires the processing and analysis of huge amounts of image data, and manual image analysis is very time consuming, thus costly, and also potentially inaccurate and poorly reproducible. The proposed method has been used to combine the results from three classifiers, and the combined result is superior to any of the results obtained from a single classifier.

1 Introduction

High-content cell-cycle screening by automated fluorescence microscopy is becoming one of the most widely used research tools to assist scientists in understanding the complex process of cell division or mitosis [1]-[5]. An essential task for high content screening is to measure cell cycle progression (interphase, prophase, metaphase, and telophase) in individual cells as a function of time. The progress of a cell cycle can be identified by measuring nuclear changes. Stages of an automated cellular imaging analysis consist of segmentation, feature extraction, classification, and tracking of individual cells in a dynamic cellular population; and the classfication of cell phases is considered the most difficult task of such analysis [6]-[8].

In time-lapse microcopy images are usually captured in a time interval of more than 10 minutes. During this period dividing nuclei may move far away from each other and daughter cell nuclei may not overlap with their parents. The consecutive image subframes from an image sequence show nuclear size and shape changes during cell mitosis. As an example, Figure 1 shows the nuclear migration during cell division in which there are two cases of cells splitting into

B. Gabrys, R.J. Howlett, and L.C. Jain (Eds.): KES 2006, Part I, LNAI 4251, pp. 524–531, 2006.
© Springer-Verlag Berlin Heidelberg 2006

two. Given the advanced fluorescent imaging technology, there still remain technical challenges in processing and analyzing large volumes of images generated by time-lapse microscopy. The quantity and complexity of image data from dynamic microscopy renders manual analysis unreasonably time-consuming. Therefore, automatic techniques for analyzing cell-cycle progress are of considerable interest in the drug discovery process.

Being motivated by the desire to study drug effects on HeLa cells, an ovarian cancer cell line, we have developed several computational techniques for identifying individual cell phase changes during a period of time, where different methods yield different results. The attempt addressed in this paper is to utilize the concept of fuzzy measures and fuzzy integral to fuse these results in order to improve the classification rate. In the following sections, we will first describe the vector-quantization (VQ) and Markov based methods that are used as the computational framework for the implementations of the k-nearest neighbor (k-NN) [9], LBG [10], fuzzy c-means (FCM) [11], and fuzzy entropy (FE) [12] algorithms for classifying cell phases. Secondly, we then describe how to implement the fuzzy fusion method in an adaptive procedure [13] to improve the classification rate. Finally, experiment and discussion on the proposed method will be addressed.

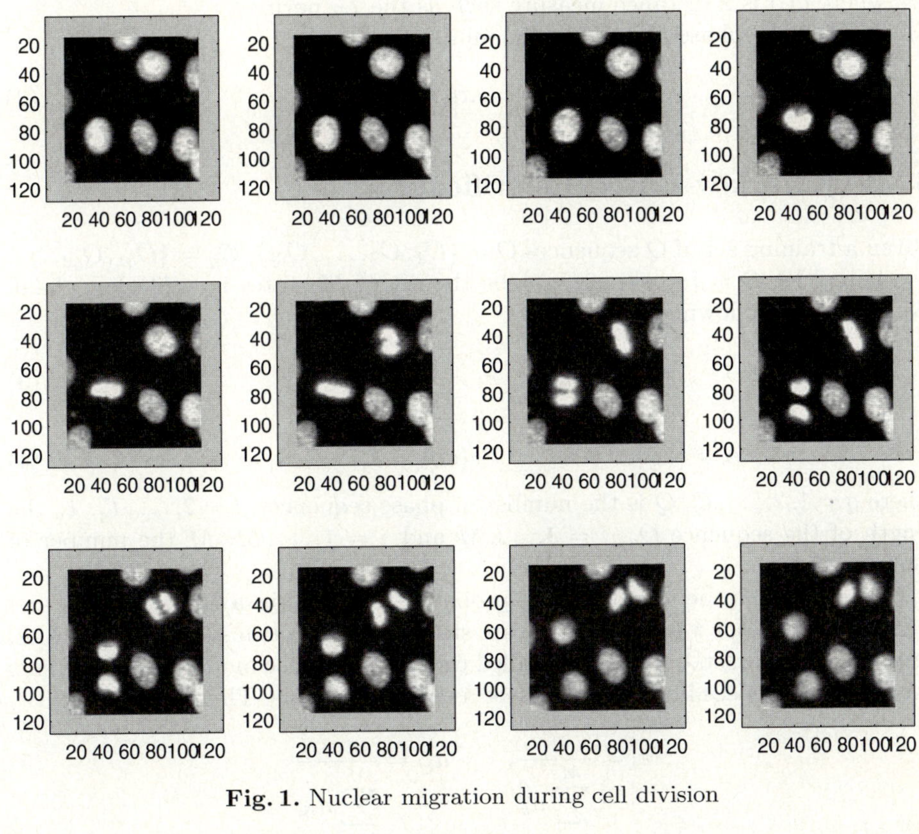

Fig. 1. Nuclear migration during cell division

2 VQ-Based Identification

Let p be a particular cell phase and N the size of the VQ codebook. Given a cell-phase training set $\mathbf{X}^p$, the VQ codebook can be designed for $\mathbf{X}^p$ given N. The identity of this template codebook is then used for identifying an unknown pattern. The training and classification procedures of this VQ-based algorithm can be designed as follows.

Training

1. Given a cell phase set $\mathbf{X}^p$ of phase p: $\mathbf{X}^p \in \mathcal{X}$, where $\mathcal{X}$ is the universe of cell phases.
2. Build a codebook $\mathbf{C}^p$ of N codewords $(\mathbf{c}_1^p, \mathbf{c}_2^p, \ldots, \mathbf{c}_N^p)$ for $\mathbf{X}^p \in \mathbf{Z}$.

Identification

1. Given a cell of an unknown phase $\mathbf{x}$.
2. Calculate the minimum distance between $\mathbf{x}$ and $\mathbf{C}^p, p = 1, \ldots, P$, where P is the number of phases:
$$d_p = \min_n d(\mathbf{x}, \mathbf{c}_n^p) \tag{1}$$
where $d(\cdot)$ is a distance measure such as the L_2 norm.
3. Assign $\mathbf{x}$ to phase p that has the minimum distance:
$$p^* = \arg\min_p (d_p) \tag{2}$$

3 VQ-Markov-Based Identification

Given a training set of Q sequences $\mathcal{O} = \{\mathcal{O}_1, \mathcal{O}_2, \ldots, \mathcal{O}_Q\}$, $\mathcal{O}_q = \{O_{q1}, O_{q2}, \ldots, O_{qT}\}$, and let $\mathcal{S} = \{s_1, s_2, \ldots, s_M\}$ be the set of M states in a Markov chain. We define the following parameters

$$\pi = [\pi_i], \qquad \pi_i = Pr(O_{q1} = s_i) \tag{3}$$

$$\mathbf{A} = [a_{ij}], \qquad a_{ij} = Pr(O_{qt} = s_j | O_{q(t-1)} = s_i) \tag{4}$$

where $q = 1, 2, \ldots, Q$, Q is the number of phase sequences, $t = 2, \ldots, T_q$, T_q the length of the sequence O_q, $i = 1, \ldots, M$ and $j = 1, \ldots, M$, M the number of phases.

Cell phases can be considered as probabilistic states in a Markov chain and values in the vector π and matrix $\mathbf{A}$ are state initial and transition probabilities, respectively. The set $\lambda = (\pi, \mathbf{A})$ is called the Markov phase model that represents the phase sequences in the training set as Markov chains. Thus, we have

$$\pi_i = \frac{n_i}{\sum_{k=1}^{M} n_k}, \qquad a_{ij} = \frac{n_{ij}}{\sum_{k=1}^{M} n_{ik}} \tag{5}$$

Thus, given a cell-phase training set X^p, a VQ codebook can be designed for X^p given N. The identity of this template codebook is then used for identifying an unknown pattern. The training and classification procedures of this VQ-Markov algorithm can be designed as follows.

Training

1. Given $\mathcal{X}$ as the universe of cell phases.
2. Train VQ-based phase models
   (a) Divide the set $\mathcal{X}$ into P distinct subsets $\mathbf{X}^1, \mathbf{X}^2, \ldots, \mathbf{X}^P$, where each $\mathbf{X}^p$ contains cells of phase p.
   (b) For each subset $\mathbf{X}^p$, train a VQ-based phase model using the training algorithm previously described.
3. Train Markov model for all phases
   (a) Align cells in the set $\mathcal{X}$ as sequences of cells
   (b) Extract Q phase sequences $\mathcal{O}_1, \mathcal{O}_2, \ldots, \mathcal{O}_Q$ from the set $\mathcal{X}$
   (c) Using Q phase sequences, calculate π and $\mathbf{A}$ according to (5)

Identification

1. Given an unknown sequence of cells $\mathbf{X} = \{\mathbf{x}_1, \mathbf{x}_2, \ldots, \mathbf{x}_T\}$.
2. Identify phase for the first cell $\mathbf{x}_1$ in the sequence as follows
   (a) Calculate the minimum distance between $\mathbf{x}_1$ and $\mathbf{C}^p, p = 1, \ldots, P$, where P is the number of phases:
   $$d_p = \min_n d(\mathbf{x}_1, \mathbf{c}_n^p) \tag{6}$$
   (b) Calculate the similarity score $S(\mathbf{x}_1, p)$
   $$S(\mathbf{x}_1, p) = \frac{\pi_p}{\sum_{k=1}^{P} (d_p/d_k)^{1/(m-1)}} \tag{7}$$
   where $m > 1$.
   (c) Assign $\mathbf{x}_1$ to the phase p^* that has the maximum score:
   $$p^* = \arg\max_p S(\mathbf{x}_1, p) \tag{8}$$
3. For each cell $\mathbf{x}_i$, $i = 2, \ldots, T$, identify it as follows
   (a) Calculate the minimum distance between $\mathbf{x}_i$ and $\mathbf{C}^p, p = 1, \ldots, P$, where P is the number of phases:
   $$d_p = \min_n d(\mathbf{x}_i, \mathbf{c}_n^p) \tag{9}$$
   (b) Calculate the similarity score $S(\mathbf{x}_i, p)$
   $$S(\mathbf{x}_i, p) = \frac{a_{p^*p}}{\sum_{k=1}^{P} (d_p/d_k)^{1/(m-1)}} \tag{10}$$
   where $m > 1$ and p^* is the identified phase of the previous cell.
   (c) Assign $\mathbf{x}_i$ to the phase p^* that has the maximum score:
   $$p^* = \arg\max_p S(\mathbf{x}_i, p). \tag{11}$$

4 Fuzzy Fusion

In the previous sections we have described how to implement VQ-based and VQ-Markov-based algorithms for classifying cell phases. We now address how to implement the mathematical concepts of fuzzy measure and fuzzy integral to combine results obtained from multiple classifiers.

Let $Y = \{S_1, \ldots, S_n\}$ be a set of the scores obtained from n classifiers over the same cell phase. A fuzzy measure g over a finite set Y satisfies the following conditions [14]:

1. $g(\emptyset) = 0$; and $g(Y) = 1$
2. If $A \subset B$, then $g(A) \leq g(B)$

The Sugeno measure or the g_λ fuzzy measure [14] satifies the following additional condition for some $\lambda > -1$:

$$g_\lambda(A \cup B) = g(A) + g(B) + \lambda g(A)g(B) \tag{12}$$

The value of λ can be calculated regarding to the condition g(Y)=1:

$$\lambda + 1 = \prod_{i=1}^{n}(1 + \lambda g^i) \tag{13}$$

where g^i is the fuzzy density of element $S_i \in Y$, and determined as the average classification rate of classifier i with respect to a particular cell phase p.

Thus, the fuzzy density of a classifier is adaptive in that it is variable depending on its overall classification rate for a particular cell phase.

Given the values of the fuzzy measures, the Sugeno (fuzzy) integral can perform as an aggregation operator. Let $\mathcal{C}$ be a set of classifiers, $h : \mathcal{C} \to [0, 1]$, and let $h(c_i)$ denote the overall classification rate of classifier c_i. The fuzzy integral of h over $A \subset \mathcal{C}$ with respect to the fuzzy measure g can be calculated as follows:

$$\int_A h(c_i) \circ g = \sup_{\alpha \in [0,1]} [\alpha \wedge g(A \cap H_\alpha] \tag{14}$$

where $H_\alpha = \{c_i | h(c_i) \geq \alpha\}$.

For a finite set of elements $\mathcal{C} = \{c_1, c_2, \ldots, c_n\}$ where the elements are sorted so that $h(c_i)$ is a descending function, the fuzzy integral, which represents the fused result, can be calculated as follows:

$$\int_A h(c_i) \circ g = \vee_{i=1}^{n} [h(c_i) \wedge g(H_i)] \tag{15}$$

where $H_i = \{c_1, \ldots, c_i\}$.

5 Experimental Results

Four nuclear sequences were provided by the Department of Cell Biology at the Harvard Medical School to test the efficiency of the proposed system. Each sequence consists of 96 frames over a duration of 24 hours.

Imaging was performed by time-lapse fluorescence microscopy with a time interval of 15 minutes. Two types of sequences were used denoting drug treated and untreated. Cell cycle progress was affected by drug and some or all of the cells in the treated sequences were arrested in metaphase. Cell cycle progress in the untreated sequences was not affected. Cells without drug treatment will usually undergo one division during this period of time.

After the nuclear segmentation has been performed, it is necessary to perform a morphological closing process on the resulting binary images in order to smooth the nuclear boundaries and fill holes insides the nuclei. These binary images are then used as a mask on applied the original image to arrive at the final segmentation. From this resulting image, features can be extracted. The ultimate goal for feature selection is to assign correct phase to cells via the training of some identification technique. In this work, a set of cell-nuclear features are extracted based on the experience of biologists. To identify the shape and intensity differences between different cell phases, a set of 7 features are extracted. These features include maximum intensity, mean, stand deviation, major axis, minor axis, perimeter, and compactness [6].

Because the feature values have different ranges, the scaling of features is therefore necessary by calculating the z-scores:

$$z_{tk} = \frac{x_{tk} - \bar{m}_k}{s_k} \tag{16}$$

where x_{tk} is the k-th feature of the t-th nucleus, $\bar{m}_k$ the mean value of all n cells for feature k, and s_k the mean absolute deviation, that is

$$s_k = \frac{1}{n} \sum_{t=1}^{n} |x_{tk} - \bar{m}_k| \tag{17}$$

The seven features extracted for each cell are considered as the feature vectors for training and testing several identification approaches including the k-NN algorithm; VQ approaches based on the k-means, fuzzy c-means (FCM), fuzzy entropy (FE), LBG algorithms, and the combinations of VQ and Markov models. The parameter k was set to be 6 for the k-NN algorithm. The degree of fuzziness m and degree of fuzzy entropy m_F were set to be 1.1 and 0.05, respectively.

There are 5 phases to be identified: interphase, prophase, metaphase, anaphase, and arrested metaphase. We divided the data set into 5 subsets for training 5 models and a subset for identification. Each of the 5 training sets for 5 phases contains 5000 cells, which were extracted from the cell sequences labeled from 590 to 892. These sequences were also used to calculate the Markov

Table 1. Cell-phase identification rate

Classifier	Identification rate (%)
k-NN	82.04
k-means	85.25
k-means and Markov	86.65
LBG-VQ	85.54
LBG-VQ and Markov	86.61
FE-VQ	86.13
FE-VQ and Markov	87.32
FCM-VQ	88.24
FCM-VQ and Markov	88.35
Fuzzy fusion	92.71

models. The identification set contains sequences labeled from 1 to 589. There are 249,547 cells in this identification set.

The identification results obtained from 9 individual classifiers and the fuzzy fusion model that combines the results from the three classifiers: LBG-VQ-Markov, FE-VQ-Markov, and FCM-VQ-Markov are presented in Table 1. The FCM-VQ-Markov model yields the better classifcation rate (88.35%) over other individual classifiers. The fuzzy fusion model gives the best result (92.71%).

The experimental results can be generally noted that the probabilistic modeling of the VQ code vectors by Markov chains always increases the identification rates in all VQ partitioning strategies being either k-means, fuzzy c-means, fuzzy entropy, or LBG algorithms. Moreover, the computational time for any of the VQ methods was significantly less than that for the k-NN method, particularly when the value for k of the k-NN rule increases.

The incorporation of probabilistic analysis using Markov chains into the template matching using vector quantization approach helps improve the identification rates with various clustering criterion (k-means, fuzzy c-means, and LBG algorithms). From the experimental results, it can be seen that the fuzzy vector quantization (either fuzzy entropy or fuzzy c-means) is superior to either the k-means or LBG based vector quantization methods. The FCM algorithm seeks to partition a data set into a specified number of fuzzy regions which are represented by the corresponding fuzzy prototypes. The degrees of each cellular-image feature vector that belong to different clusters are characterized by the corresponding fuzzy membership grades taking real values between 0 and 1. Thus, the use of the fuzzy c-means algorithm provides more effective analysis of the present problem where the image boundaries of different classes are vaguely defined.

Making use of the different strenghts of the three Markov-based classifiers in the sense that each classifier tends to make different mistakes, the fuzzy fusion model has been able to improve the classification rate by combining the scores assigned by the three classifiers in terms of the variable fuzzy densities and the reliability values of the three classifiers.

6 Conclusions

Several classification techniques for identifying cell phases using time-lapse fluoresence microscopic image sequences have been addressed. The proposed fuzzy fusion based on the concepts of fuzzy measure and fuzzy integral is able to increase the classification rate over other individual classifiers. The result can certainly be further improved by extracting more effective features of the cell images. Further applications and additional fusion of other machine learning methods such as support vector machines and neural networks will be investigated in our future research.

Acknowledgments. The image data set was provided by our collaborator, the HCNR-Center for Bioinformatics, Harvard Medical School. The first author acknowledges the financial support received from the ARC-DP grant DP0665598.

References

1. Fox, S.: Accommodating cells in HTS. *Drug Discovery World*, **5** (2003) 21-30.
2. Feng, Y.: Practicing cell morphology based screen. *European Pharmaceutical Review*, **7** (2002) 7-11.
3. Dunkle, R.: Role of image informatics in accelerating drug discovery and development. *Drug Discovery World*, **5** (2003) 75-82.
4. Yarrow, J.C., et al.: Phenotypic screening of small molecule libraries by high throughput cell imaging. *Comb Chem High Throughput Screen*, **6** (2003) 279-286.
5. Hiraoka, Y., and Haraguchi, T.: Fluoresence imaging of mammalian living cells. *Chromosome Res*, **4** (1996) 173-176.
6. Chen, X., Zhou, X., and Wong, S.T.C.: Automated segmentation, classification, and tracking cancer cell nuclei in time-lapse microscopy. *IEEE Trans. on Biomedical Engineering*, in press.
7. Pham, T.D., Tran, D., Zhou, X., Wong, S.T.C.: An automated procedure for cell-phase imaging identification. *Proc. AI-2005 Workshop on Learning Algorithms for Pattern Recognition*, pp. 52-29, 2005.
8. Pham, T.D., Tran, D.T., Zhou, X., and Wong, S.T.C.: Classification of cell phases in time-lapse images by vector quantization and Markov models. *Neural Stem Cell Research*, ed. Erik V. Greer, Nova Science, New York, 2006.
9. Cover, T.M. and Hart, P.E.: Nearest neighbor pattern classification. *IEEE Trans. Information Theory*, **13** (1967) 21-27.
10. Linde, Y., Buzo, A., and Gray, R.M.: An algorithm for vector quantization. *IEEE Trans. Communications*, **28** (1980) 84-95.
11. Bezdek, J.C.: *Pattern Recognition with Fuzzy Objective Function Algorithms.* Plenum Press, New York, 1981.
12. Tran, D., and Wagner, M.: Fuzzy entropy clustering. *Proc. Int. Conf. FUZZ-IEEE 2000*, vol. 1, pp. 152-157, 2000.
13. Pham, T.D.: Combination of multiple classifiers using adaptive fuzzy integral. *Proc. IEEE Int. Conf. Artificial Intelligence Systems* (ICAIS'02), pp. 50-55, 2002.
14. Sugeno, M.: The theory of fuzzy integrals and its applications. PhD thesis, Tokyo Institute of Technology, 1974.

Adaptive Context-Aware Filter Fusion for Face Recognition on Bad Illumination

Nam Mi Young, Md. Rezaul Bashar, and Phill Kyu Rhee*

Intelligent Technology Laboratory
Dept. of Computer Science & Engineering, Inha University,
Incheon 402-751, Korea
{rera, bashar}@im.inha.ac.kr, pkrhee@inha.ac.kr

Abstract. At present, the performance of face recognition system depends much on the variations in illumination. To solve this problem, this paper presents an adaptable face recognition approach that uses filter fusion representation. The key idea is to use context-aware filter fusion to get better image from a bad illumination one. Genetic algorithm is the tool for adaptation for individual context category. These can provide robust face recognition on illumination context-awareness under uneven environments. Gabor wavelet representation can also provide a robust feature for image enhancement. Using these approaches, we have developed a robust face recognition technique that can recognize with a notable success and it has been tested on Inha DB and FERET face images.

1 Introduction

Face recognition [1] technologies have been motivated from the application area of physical access, face image surveillance, people activity awareness, visual interaction for human computer interaction, and humanized vision. Much progress in face recognition has been made in the past decade, however, it remains some unsolvable problems: pose and illumination that creates a poor performance rate.

An image is expressed in terms of its illumination and reflectance components [2]. As a consequence, no image processing system is free from illumination problem. Most of the researcher developed their system: either ignoring or taking the illumination variation uniform. Unfortunately, illumination plays a vital rule for robust recognition, especially, shadow or light. For example, one face image is bright while other is dark as shown in Fig.1. As a result, for dark image, the edges of the face will be blurred and recognition rate will be dropped. However, none of the representation is sufficient by itself to overcome image variations caused by illuminations.

Now a day, many researchers have felt interest on solving illumination constraints. Among them, Hong Liu et. Al. [2] developed a multi-method integration (NMI) scheme by using grey, log, normal histogram and logabout techniques to compensate variations of illumination in face detection with maximum performance 91.16% and 69.09%. Face database2 and Yale Faces respectively. Jinho Lee et. al [3] have generated an illumination subspace for arbitrary 3D faces based on the statistics of

B. Gabrys, R.J. Howlett, and L.C. Jain (Eds.): KES 2006, Part I, LNAI 4251, pp. 532–541, 2006.
© Springer-Verlag Berlin Heidelberg 2006

(a) (b)

Fig. 1. Effect of illumination on image: (a) low illumination, and (b) high illumination

measured illuminations under variable lighting conditions. Marios Savvides et.al [4] presents illumination tolerant Face recognition system by using minimum average correlation energy (MACE) combining with PCA and got a tremendous success. Laiyun et. al.[5] modeled a face recognition system with the relighting process and applied to CMU-PIE database and has achieved very high recognition rate. Haitao Wang et. al.[6] proposed a face recognition system with varying illumination by using PCA and their own developed system. They have tested their system with Yale and FERET database and have got a notable success.

2 Our Approach

In our adaptive Face recognition system, the filter fusion guided by an evolutionary approach has been employed to adapt the system for variations in illumination. The proposed approach employs filter fusion, which is generated from the retinex, histogram equalization, and contrast stretching. Gabor wavelet provides the nice properties for face recognition; however, it cannot provide sufficiently reliable solution inchanging environments such as variations in illumination. Our system adopts the adaptive strategy by the illumination filter fusion that adapts itself by reorganizing its structure and parameters. An illuminated input face image is preprocessed through context –awareness and filter fusion to achieve a higher quality image, to that an adaptive filter technique is employed so that the system can create itself necessary

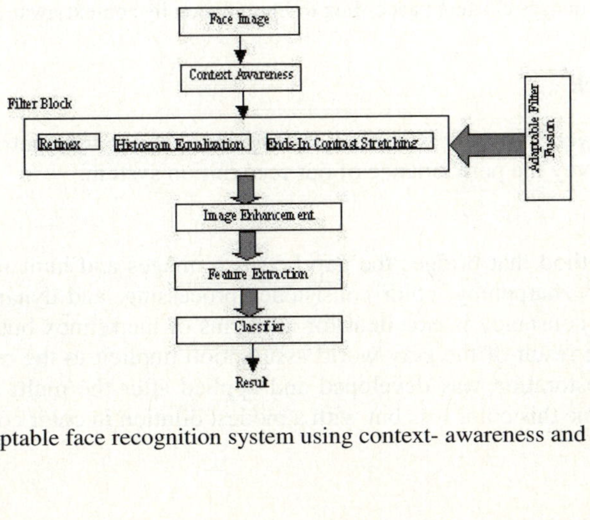

Fig. 2. Adaptable face recognition system using context- awareness and filter fusion

steps to classify an incoming object. After preprocessing the input face image, its features are extracted to assist the classifier. The general scheme of our system is depicted in Fig. 2.

2.1 Context-Awareness

To make robust face recognition, adaptive preprocessing and feature representation is taken into account. We have employed the method of context-awareness in order to provide the capability of adaptation in preprocessing, and feature representation stages and have also combined with adaptive Gabor feature space that can perform well under uneven environments. The system learns with changing environments in the context-awareness stage and adapts itself by restructuring its structure and/or parameters. Fuzzy ART is a synthesis of ART algorithm and Fuzzy operator that can only accept binary input pattern, but Fuzzy ART allows both binary and continuous input patterns. However, the clustering of illumination variations is very subjective and ambiguous. This is why; we have adopted Fuzzy ART to achieve an optimal illumination clustering architecture. In this paper, we have tried to improve the clustering performance by iterative learning. Fig. 3 shows the clustering result of face images by Fuzzy ART (FART).

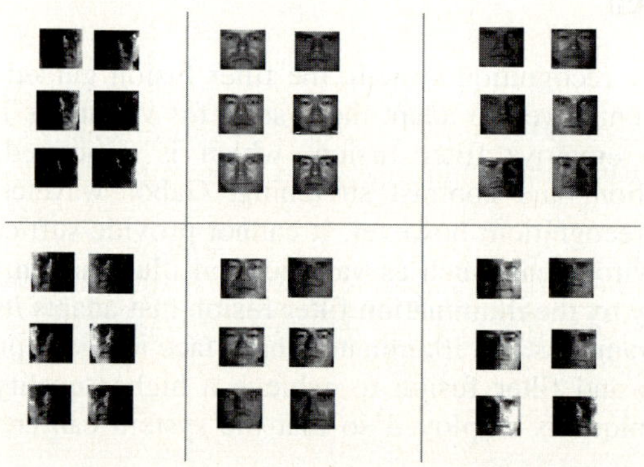

Fig. 3. Face images clustered according to illumination by context-awareness module

2.2 Filter Block

We have employed three filters: retinex, histogram analysis and contrast stretching to study and to survey the performance of our recognition system.

i) Retinex

Retinex is a method that bridges the gap between images and human observation of scenes that does sharpening, color consistence processing, and dynamic range compression. Color constancy is excellent for all forms of the retinex but color rendition was elusive as a result of the gray world assumption implicit to the retinex computation. A color restoration was developed and applied after the multi scale retinex in order to overcome this color loss but with a modest dilution in color constancy.

The single-scale retinex is given by [6,7,8,9].

$$R_i(x,y) = \log I\ (x,y) - \log[F(x,y) \times I_i(x,y)] \tag{1}$$

Where, $I_i(x,y)$: Image distribution,

i_{th} : Color band, and $F(x,y)$: The normalized surround function

ii) Histogram Equalization Filter

Histogram equalization plays a vital role to improve image contrast. If the distribution of gray level was biased to one direction or scaled value was not uniformly distributed, histogram equalization is a good solution for image enhancement.

iii) Ends-In Contrast stretching

The contrast stretching of the image is distribution of light and dark pixels and applied to an image to stretch a histogram to fill the full dynamic range of the images [9].

$$output(x) = \begin{cases} 0 & \text{for } x < low \\ 255 \times (x - low)/(high - low) & \text{for } low \le x \le high \\ 255 & \text{for } high < x \end{cases} \tag{2}$$

Ends-In Contrast stretching is possible intensities but have a pixel concentration in one part of the histogram. Results of different filtering approach with comparing original pictures are shown in Fig.4.

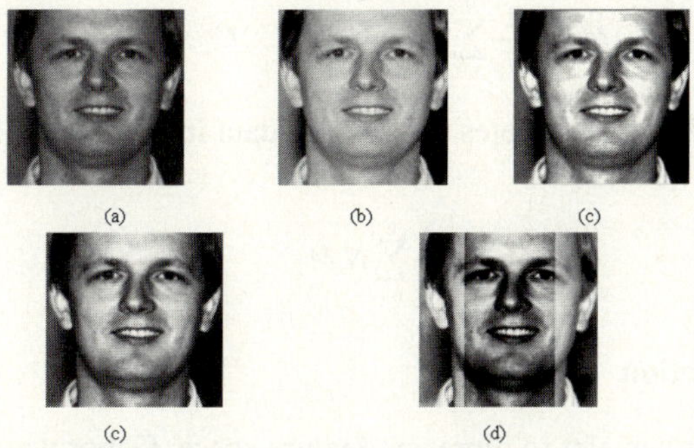

Fig. 4. (a) Original Image, (b) using retinex, (c) using ends-in-contrast stretching, (d) using histogram equalization, and (e) using local histogram equalization

2.3 Adaptive Filter Fusion

The GA is employed to search among the different combinations of feature representations (if any) and combining structure of classifiers. The optimality of the chromosome is defined by classification accuracy and generalization capability. Fig. 5 shows a possible encoding of chromosome description where FB_1 and FB_2 are histogram preprocessing filter flag and contrast stretching preprocessing filter respectively. This parameter is initialized random value. And population size is 1000, chromosome bits are 8-bit, mutation rate is 0.2, and crossover rate is 0.7.

FB_1	FB_2	…	FB_n	FB flag

Fig. 5. A possible chromosome description of the proposed scheme

As the GA searches the genospace, the GA makes its choices via genetic operators as a function of probability distribution driven by fitness function. The genetic operators used here are selection, crossover, and mutation [9]. The GA needs a salient fitness function to evaluate current population and chooses offspring for the next generation. The evolutionary module derives the classifier being balanced between successful recognition and generalization capability. The fitness function adopted here is defined as follows:

$$\eta(V) = \lambda_1 \eta_s(V) + \lambda_2 \eta_g(V) \tag{3}$$

where $\eta s^{(V)}$ is the term for the system correctness, i.e., successful recognition rate and $\eta_g^{(V)}$ is the term for class generalization. λ_1 and λ_2 are positive parameters that indicate the weight of each term, respectively.

Let $\omega_1, \omega_2, .., \omega_k$ be the classes and $N_1, N_2, \ldots, N_k$ be the number of images in each class, respectively. Let $M_1, M_2, .., M_k$ be the means of corresponding classes, and $M_{\omega k}$ be the total mean in the Gabor feature space. Then, M_i can be calculated,

$$M_i = \frac{1}{N_i} \sum_{j=1}^{N_i} S_j^{(i)} i = 1,2,3,\ldots\ldots, L \tag{4}$$

Where $S_j^{(i)} = j = 1,2, \ldots, N$, denotes the sample data in class w_i, n is images numberand

$$M_{avg} = \frac{1}{n} \sum_{i=1}^{k} N_i M_i \tag{5}$$

2.4 Feature Extraction

We have used Gabor wavelet to represent feature space. Gabor wavelet representation is a set of Gabor functions. Gabor wavelet can efficiently extract orientation selectivity, spatial frequency, and spatial localization. It is a simulation or approximation to the experimental filter response profiles in visual neurons [12]. The Gabor filter is known to be efficient in reducing redundancy and noise in images [13]. Gabor wavelet is biologically motivated convolution kernels in the shape of plane waves restricted by Gabor kernel. The convolution coefficients for kernels of different frequencies and orientations starting at a particular fiducial point are calculated. The Gabor kernels for a fiducial point are defined as follows:

$$\phi_{u,v}(\overrightarrow{x}) = \frac{\left\|\overrightarrow{(k)}_{\mu,\nu}\right\|^2}{\sigma^2} \exp^{-\frac{\left\|\overrightarrow{(k)}_{\mu,\nu}\right\|^2 \left\|\overrightarrow{(x)}\right\|^2}{2\sigma^2}} [\exp^{i\overrightarrow{(k)}_{\mu,\nu}\overrightarrow{(x)}} - \exp^{-\frac{a^2}{2}}] \tag{6}$$

Where μ and ν denote the orientation and dilation of the Gabor kernels, $\overrightarrow{(x)} = (x, y)$, $\| \bullet \|$ denotes the norm operator, and the wave vector $\overrightarrow{(k)}_{u,v}$ is defined as follows:

$$\overrightarrow{(k)}_{\mu\nu} = (k_\nu \cos\phi_\nu k_\nu \;\; k_\nu \sin\phi_\nu)^t \tag{7}$$

Where $k_\nu = 2^{-y+z}$ and $\phi_u = \pi\mu/8$.

The family of Gabor kernels is similar each other since they are generated from one mother wavelet by dilation and rotation using the wave vector $\overrightarrow{(k)}_{\mu,\nu}$. The first term in the brackets in Eq. (6) determines frequency part of the kernel and the second term compensates for the DC value, which makes the kernels DC-free. We can improve the performance by employing the genetic algorithm [19]. Gabor wavelet is usually used at eight different frequencies, ν = 0, . . . 7, and eight orientations, μ = 0, . . . , 7 [14] shown in Fig. 7. The kernels show desirable characteristics of spatial locality and orientation selectivity, a suitable choice for face image feature extraction for classification.

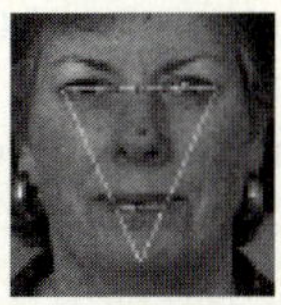

Fig. 6. Eleven (11) Facial landmarks used for Gabor Wavelet

Fig. 7. Gabor wavelet representation for 8 frequencies and 8 orientations

2.5 Algorithm for Our Developed System

1. Filter, and derive Gabor representation for each fiducial point, and normalize it.
2. Concatenate the Gabor representations for fiducial points to generate total Gabor vector.

$$V = (F^{(\ell)}(\overrightarrow{x_1}) F^{(\ell)}(\overrightarrow{x_2}) \cdots F^{(\ell)}(\overrightarrow{x_n})) \tag{8}$$

3. Start from the optimization of the preprocessing and feature representation and progress until a criterion is met i.e. the performance does not improve anymore or the predefined maximum trial limitation is encountered.
   a) Generate an initial filter block configuration, parameters, and feature space.
   b) Evaluate the performance evaluation function, $\eta(V) = \lambda_1 \eta_s(V) + \lambda_2 \eta_g(V)$ using the newly adapted filter block and feature space.
   c) Search for the new filter block configuration, parameters, and feature space that maximize the evaluation function.
   c) Applying GA's genetic operators to generate new population of filter block and Gabor feature space. Go to Step 3(a).
4. Perform recognition using the constructed new filter block and Gabor feature space from Step 3.

3 Face Recognition Experiments

The standard face image dataset: FERET with its illuminated portion fafb, and fafc in addition with our lab dataset INHA DB are used for our experiment. As shown in Fig.6, we have used 11 facial landmarks (feature points) for the performance evaluation. Therefore, size of the feature vector is 11×40. At first, we have tested the classier, that is, we have used SVM, Manhattan and Cosine distance measurement to understand which technique shows best recognition and by which we shall continue our next experiments. We have tested on 5 fiducial points face images of FERET c (fafc) dataset and have achieved the following results.

Table 1. Misclassified faces with different methods

# of subjects		Non filter	H.E.	Retinex
SVM	No. of faces	193	193	181
	%	99.48%	99.48%	93.30%
Manhattan	No. of faces	193	180	78
	%	99.48%	92.78%	40.21%
Cosine	No. of faces	193	175	73
	%	99.48%	90.21%	37.63%

Table 1 shows the misclassified performance of different classifiers on different filters and it is obviously seen that Cosine distance measurement is the best, which has been taken for our next consideration. There is no way to deny the role of kernel size for an efficient recognition system. To determine proper kernel size, we have tested on 33×33 kernel of 128×128 sized face images and 17×17 kernel size of 64×64 sized face images with histogram equalization and different normalized retinex filters with B-spline and super sampling techniques and have got the results shown in table 2, from where it is easy to understand that larger the kernel more the performance. For our remaining experiments, we have used e 33×33-sized kernel having 11 fiducial points of an image with 5 frequencies, 8 orientations and sigma with 2Π values.

Table 2. Misclassification rate with different kernel size and filtering

Kernel	Image Size	Non Filter		H.E.		Retinex (64)	
		# of Miss	% of Miss	# of Miss	% of Miss	# of Miss	% of Miss
33× 33	128 × 128	191	98.45	126	64.95	69	35.57
17 ×17	64 ×64	191	98.45	159	81.96	117	60.31

Kernel	Image Size	Retinex (16)		Retinex (32)		Retinex (64)		Retinex (128)		Sampling
		# of Miss	% of Miss	# of Miss	% of Miss	# of Miss	% of Miss	# of Miss	% of Miss	Remarks
33 ×33	128×128	69	35.57	69	35.57	69	35.57	69	35.57	B-Spline
17 × 17	64×64	117	60.31	117	60.31	117	60.31	117	60.31	B-Spline
17 × 17	64×64	103	53.09	103	53.09	103	53.09	103	53.09	Super

Table 3 shows face recognition using filter fusion each cluster by discrimination FART. In our experiment, test data are clustered into five context categories according to their illumination contexts by the context awareness module. Comparison of our proposed adaptive system with others non-adaptive system is shown in table 3.

Table 3. Comparison of face recognition rates of the proposed adaptive method to those of non-adaptive methods

Data Set	Proposed Method		H.E.		Retinex (32)	
	Wrong Recog. (WR)	Acceptance Rate (AR %)	WR	AR (%)	WR	AR (%)
fafc	32	83.51	157	19.07	42	78.35
fafb	110	90.79	246	79.41	297	75.15
Lab	20	98.41	35	97.22	154	87.77
Data Set	**Retinex (32)+HE**		**HE +Retinex (32)**		**CS (10%)**		**HE +Retinex (32) +CS(10)**	
	WR	AR (%)	WR	AR(%)	WR	AR(%)	WR	AR(%)
fafc	157	19.07	53	72.68	139	28.35	95	51.03
fafb	245	79.50	281	76.49	292	75.56	252	78.91
Lab	34	97.30	88	93.01	37	97.06	46	96.35

As shown in Table 4, the proposed method based on adaptive feature space outperforms the performance over the non-adaptive methods for the integrated data set. On the other hand, Histogram equalization based method has not shown any good performance under normal illumination, i.e. FERET fafb, or bad illumination, i.e. Table 4 shows the comparison of the proposed method to other previous approaches. It becomes apparent that tour developed system is better than any other previous developed system.

Table 4. Performance evaluation of the proposed system comparing with other approaches

Algorithm/Method	Integrated data set (FERET fafb + fafc)
Ef_hist_dev_anm	0.5055
ef_hist_dev_ml1	0.5625
ef_hist_dev_ml2	0.5405
mit_sep_96	0.634
umd_mar_97	0.775
usc_mar_97	0.885
Proposed method	0.907

4 Conclusion

We have shown a context-aware preprocessing and feature representation methods with adaptive filter fusion for robust face recognition technique under uneven illumination. Through out this experiment, retinex shows nice performance under bad illumination, while, it provides very poor performance under normal illumination, and however, histogram equalization method works very well on normal image comparison with bad illumination image. Consequently, we came up to conclude that the performance of individual filtering methods for image enhancement is highly depending upon application environments. Most of the cases, we have achieved higher success by adaptable system with the concept of Genetic algorithm. Our proposed system has shown the best result not only every features of images with different image data set but also it is the best comparing with existing systems have ever developed.

References

1. Seung Yonng Lee.: Illumination Direction and Scale Robust Face Recognition using Log-polar and Background Illumination Modeling. Thesis, Inha University, Korea, (2002).
2. H. Liu et. al.: Illumination Compensation and Feedback of Illumination Feature in Face Detection. Proc. International Conferences on Information-technology and Information-net, Beijing, vol. 3,(2001) pp.444-449.
3. Jinho Lee, etc.: A Bilinear Illumination Model for Robust Face Recognition. 10th IEEE International conference on Computer Vision (ICCV'05), 2005.
4. Marios Savvides et.al.: Corefaces- Robust Shift Invariant PCA based Correlation Filter for Illumination Tolerant Face Recognition. IEEE Computer Society Conference on Computer Vision and Pattern Recognition (CVPR'04), 2004.
5. Laiyun, et. al.: Face Relighting for Face Recognition Under Generic Illumination. IEEE International Conference on Acoustic, Speech and Signal Processing (ICASSP'04), 2004.
6. Haitao Wang, Stan Z. Li, et. al.: Illumination Modeling and Normalization for Face Image. IEEE International workshop on Analysis and Modeling of Faces and Gestures (AMFG'2003), 2003.
7. Rafael C. Gonzalez and Richard E. Woods.: Digital Image Processing. Addison-Wesley, (1992)

8. Andrew Moore.: Real-time Neural System for color Constancy. IEEE Transactions on Neural Networks, vol. 2, no.2, (1991)
9. Funt, B.V., Ciurea, F., and McCann, J.: Retinex in Matlab. In Proc. of IS&T/SID Eighth Color Imaging Conference, (2000), pp. 112– 121
10. Daniel J. Jobson, Zia-ur Rahman, Glenn A. Woodell.: The Spatial Aspect of Color and Scientific Implications of Retinex Image Processing.
11. Brian Funt, Kobus Barnard.: Luminance-Based Multi-Scale Retinex. rmalize proceedings AIC Colour 97 8th Congress of the International Colour Association, 1997.
12. Bossmaier, T.R.J.: Efficient image representation by Gabor functions - an information theory approach. J.J. Kulikowsji, C.M. Dicknson, and I.J. Murray(Eds.), Pergamon Press, Oxford, U.K. pp. 698-704
13. Marios Savvides , B.V.K. Vijaya Kumar, P.K. Khosla.: Robust, Shift-Invariance Biometric Identification from Partial Face Images. Proc of SPIE, Biometric Technologies for Human Identifications (OR51), Orlando, FL, (2004).
14. Marios Savvides, B.V.K. Vijaya Kumar. Quad Phase Minimum Average Correlation Energy Filters for Reduced Memory Illumination Tolerant Face Authentication. Lecture Notes in Computer Science, Springer-Verlag, vol. 2688/2003, (2003), pp. 19-26.
15. S. Georghiades, P. N. Belhumeur, and D. J. Kriegman.: From Few to Many: Illumination One Models for face recognition under Variable Lighting and Pose. IEEE Trans. on PAMI, vol.23, no.6, June(2001) , pp.643-660
16. P.Phillips.: The FERET database and evaluation procedure for face recognition algorithms. Image and Vision Computing, vol.16, no.5, (1999) , pp.295-306.
17. Ramuhalli, P., Polikar, R., Udpa L., Udpa S. "Fuzzy ARTMAP network with evolutionary learning.: Proc. of IEEE 25th Int. Conf. On Acoustics, Speech and Signal Processing (ICASSP 2000), Vol. 6. Istanbul, Turkey, (2000) pp.3466-3469.

A Study on Content Based Image Retrieval Using Template Matching of Wavelet Transform

Duck Won Seo[1], Kang Soo You[2], and Hoon Sung Kwak[1]

[1] Department of Computer Engineering, Chonbuk National University,
664-14, Dukjin-dong, Dukjin-gu, Jeonju, Jeonbuk, 561-756, Korea
{duck0419, hskwak}@chonbuk.ac.kr
[2] School of Liberal Arts, Jeonju University,
1200, 3-ga, Hyoja-dong, Wansan-gu, Jeonju, Jeonbuk, 560-759, Korea
gsyou@jj.ac.kr

Abstract. Recently, issues on the internet searching of image information through various multimedia databases have drawn a tremendous attention and several researches on image information retrieval methods are on progress. By incorporating wavelet transform and correlation metrics, we propose a novel and highly efficient feature vector extraction algorithm that has a capability of a robust similarity matching. Through a number of simulations using thousands of images as a database, our presented method has been tested and verified. The simulation results have yielded a faster and highly accurate candidate image retrieval performance in comparison to those of the conventional algorithms. Such an improved performance can be obtained because the used feature vectors were compressed to 256:1 while the correlation metrics are incorporated to provide a finer information for the better matching.

Keywords: Image Retrieval, Template Matching, Wavelet Transform.

1 Introduction

Recently, issues on the internet searching of image information through various multimedia databases have drawn a tremendous attention and several researches on image information retrieval methods are on progress. Up to now, there are two major directions in research. One is a technique that involves human-supplied text annotations describing image semantics. These text annotations are then used as the basis for searching. This method has an advantage that the retrieval accuracy is relatively high within limited numbers, but has some disadvantages that the manual annotation of text phrases for each image in a large database takes a lot of time and effort. That is because different people may input different textual annotations for the same image and it is sometimes difficult to describe the context of images by a keyword due to the complicated nature within an image.

The other method involves content-based image indexing and retrieval technique that perform the feature vectors[1], [2], [3]. This technique has an advantage that time and costs are reduced, because of extracting feature vectors from image itself. For that

B. Gabrys, R.J. Howlett, and L.C. Jain (Eds.): KES 2006, Part I, LNAI 4251, pp. 542 – 549, 2006.
© Springer-Verlag Berlin Heidelberg 2006

matter, researches for searching efficient feature vectors have been active. Things such as color, pattern, texture, contour, form and space-relationship are the representative feature vectors including image characteristics. The content based retrieval system development can be classified into two areas. In one area, investigations on feature of image data and transform method with a viewpoint of image processing have been undertaken and on the other area, studies on the efficient image data indexing method using feature vectors extracted from image with a viewpoint of database have been active.

Here, in our case, basically being a content-based image data retrieval technique, we are focused on investigating the characteristics of image data and transform method that lead to a robust retrieval of candidate image data stored in the database based on the feature vectors. By incorporating wavelet transform[4] and correlation metrics, we propose a novel and highly efficient capability of a robust similarity matching.

2 Wavelet Transform

Wavelet, developed in mathematics, quantum physics, and statistics are functions that decompose signals into different frequency components and analyze each component with a resolution matching its scale. Applications of wavelets to signal de-noising, image compression, image smoothing, fractal analysis and turbulence characterization are active research topics. They have advantages when the periodic signal contains many discontinuities or sharp changes. Wavelet transform decomposes original image into low frequency part. The lower frequency bands in the wavelet transform usually represent object configurations in the images and the higher frequency bands represent texture and local color variation. Fig. 1 shows an image decomposition by 3-level decomposed wavelet transform and Fig. 2 shows the decomposed images of Bird and Lenna by wavelet transform using Adelson's 5 tap QMF filter[5].

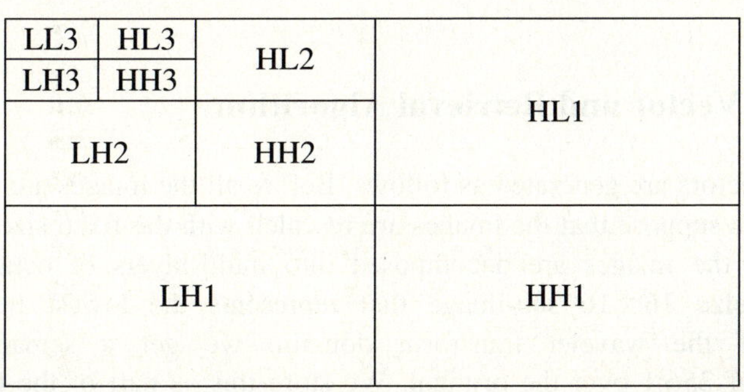

Fig. 1. Example of a 3-level wavelet decomposition of an image signal. The height and width of the decomposition are the same as that of the original image. LL3 is the low frequency approximation of the original image. HL3, LH3, and HH3 represent the three lowest level detail signals. HL2, LH2, and HH2 add additional high frequency content, or detail, and HL3, LH3, and HH3 add the highest frequency content in each of the three spatial orientations and make the transformation lossless.

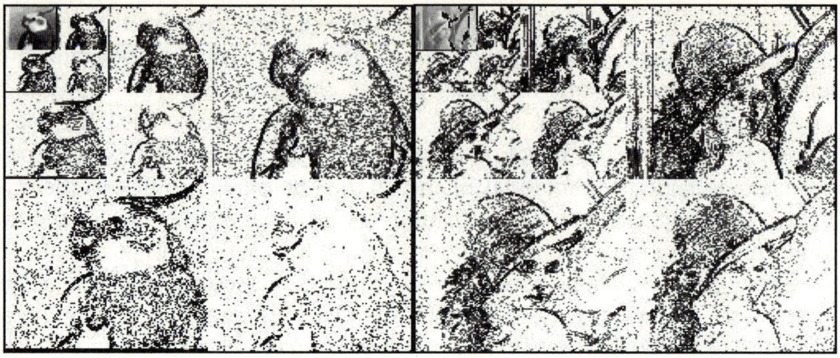

(a) Bird (b) Lenna

Fig. 2. The Decomposed images

3 Correlation of Images

In field of signal processing, correlation is used to measure the degree of similarity between two signals. And so, our goal in computing the correlation between the two images is to measure the degree to which the two images *f(x, y)* and *w(x, y)*. The crosscorrelation of *f(x, y)* and *w(x, y)* is defined as

$$C_{fw} = \sum_x \sum_y f(x,y)w(x,y) \tag{1}$$

In case of the autocorrelation, the correlation value is the highest when *f(x, y)* and *w(x, y)* images are binary image.

And the autocorrelation of *f(x, y)* is defined as

$$C_{ff} = \sum_x \sum_y f(x,y)f(x,y) \tag{2}$$

4 Feature Vector and Retrieval Algorithm

The feature vectors are generated as follows. Before all the images are inserted to the database, let us suppose that the images are rescaled with the fixed size of 256 x 256. Subsequently, the images are decomposed into multi-layers of octave bands. By extracting a size 16 x 16 sub-image that represents the lowest frequency band (LLLLLL) of the wavelet transform domain, we get a semantic-preserving compression of 256:1 over the original. We store this as part of the feature vector. And at the same time, we compute the energy value of this 16 x 16 sub-image and store the value as a part of the feature vector too. On the other hand, the other surrounding frequency bands(i. e. LLLLLH, LLLLHL and LLLLHH) are combined in such a way to obtain a close edge information. This edge information of a size

16 x 16 is also stored in the database for each one of the images and later used to compute correlation with respect to the query image[6].

When a user submits a query image for candidate matching, we must compute the feature vectors of the query image following the procedures described above and match it to the pre-computed feature vectors of the images in the database. The retrieval order consists of three steps. In the first step, we compare the energy value stored for the query image with the energy values stored for each image in the database. Consider the energy value computed for the querying image as E and the energy value stored in the database indexing for an image as E^*, and the selection criterion is as follows:

$$E - E(k_1/100) \le E^* \le E + E(k_1/100) \tag{3}$$

If the above condition holds, we select that image as a preliminary candidate.

In the second step, we compute the crosscorrelation of the edge information between the query image and the preliminarily selected candidate images. Consider the highest crosscorrelation value out of candidate images as C and the crosscorrelation values form the remaining candidate image as C^*, and the better filtering out of candidates obtained in the first step can be obtained as follows:

$$C - C(k_2/100) \le C^* \le C \tag{4}$$

Only the preliminary candidates that meet the above condition survive in this second step.

In the final step, we compute the Euclidean distance between the query image and surviving candidate images only using the 16 x 16 lowest sub-image computed and stored as a feature vector, previously. After sorting the Euclidean distance in an ascending order, the final best candidates are determined and displayed.

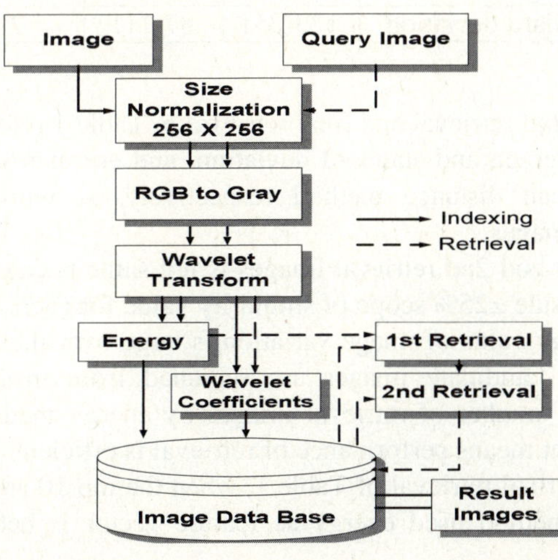

Fig. 3. The proposed image retrieval system

5 Experiment Result

In this paper, the proposed method has been tested and verified through a number of simulations with thousands of images. The images are classed as flowers, cars, people and sunset.

In order to evaluate objectivity the proposed method, we have used images supported by Web's Corel Draw Digitalriver. All the tested images and query images are 256 x 256 pixels in size with RGB.

The performance of the retrieval results is measured by Precision and AVRR(Average Rank of Relevant images).

Table 1. The retrieval result

Query image	Retrieval method	1st retrieval	2nd retrieval	Final retrieval	
				The top ten	The top fifteen
Flower	Energy	80(24)	63(20)	8	12
	Average	145(39)	100(28)	7	10
	Standard deviation	95(31)	66(27)	7	10
Car	Energy	110(26)	79(24)	6	7
	Average	176(38)	130(36)	6	6
	Standard deviation	126(33)	87(31)	5	7
People	Energy	75(23)	60(21)	6	11
	Average	115(27)	87(26)	7	8
	Standard deviation	117(7)	83(7)	2	3
Sunset	Energy	100(23)	69(7)	2	3
	Average	173(32	125(12)	2	3
	Standard deviation	123(25)	87(11)	2	3

The 1st retrieval, 2nd retrieval and final retrieval in Table 1 refer to retrieval results using energy, the average and standard deviation, and correlation from 1st retrieval results, and Euclidean distance method respectively. A numeral in parenthesis represents relevant images.

The number of 1st and 2nd retrieval images is not same because candidate images are searched from inside ±25% scope of similarity value for each. Generally, it is well known that the energy value of image variation is large over the average or standard deviation. So when candidate images are searched from inside ±25% scope of similarity value, the number of retrieval images by energy method is less of small number. After all, that means performance of retrieval is efficient.

As can be seen in final retrieval of Table 1, when the top 10 and the top 15 images are compared, the method used energy as feature vector is better retrieval results rather than others.

Table 2. The comparison Precision and AVRR about Table 1

Retrieval method	The top ten		The top fifteen	
	Precision	AVRR	Precision	AVRR
Energy	0.8	3.8	0.8	5.87
Mean value	0.7	3.5	0.67	4.8
Standard deviation	0.7	3.3	0.67	4.8

The resulting Precision and AVRR from Table 1 are shown in Table 2.

Table 3. The number comparison WBIIS and feature vector

	1st retrieval	2nd retrieval	3rd retrieval	Total
WBIIS	3	192	768	963
Our method	1	64	x	65

Table 4. The comparison WBIIS, Precision and AVRR

Retrieval method	The top ten		The top fifteen	
	Precision	AVRR	Precision	AVRR
WBIIS	0.70	3	0.67	4.67
Our method	0.80	3.2	0.80	4.90

Table 3 and Table 4 show the comparison of feature vector numbers, Precision and AVRR between James Ze Wang's WBIIS(Wavelet-Based Image Indexing and Searching)[3], [9] and the proposed method.

Fig. 4 shows the retrieval results of the people and cars. As can be seen in Fig. 4(a) and 4(b), the people images are well retrieved through the templet of wavelet transform, and similar experiment results are observed in Fig. 4(c). The white box in Fig. 4 indicates the objects(i.e. people, cars) in the retrieved image, it means the image marked the box is exactly retrieved.

With these results, we can see that our proposed method reduces the number of feature vector by 90% and is efficient and present a good performance in image retreival using template matching of wavelet transform, when compared to WBIIS.

6 Conclusion

In this paper, we have proposed content based image retrieval using template matching of wavelet transform which is widely used in image compression, digital signal analysis and preserving a spatial domain. Therefore we applied this theory to the proposed method. It is used that the feature vectors as lower frequency and correlation calculated from wavelet transform measure the similarities between query example and images in the database.

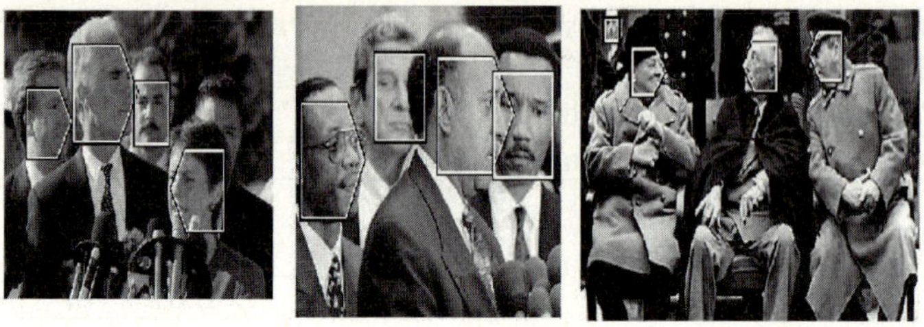

(a) The people

(b) The people

(c) The cars

Fig. 4. The retrieval results

From the simulations, it is verified that our proposed method using feature vector as lower frequency energy improved retrieval over conventional method using that as the average or standard deviation. In the measurement of Precision with the top 10 and 15, our proposed method shows improvement in performance more than 10% and 13% respectively. Also, in the AVRR, our proposed method achieves a better retrieval performance up to 20~30% over conventional method. Besides, the proposed method reduces the number of feature vector by 90% and is efficient a good performance in image retrieval, when compared to WBIIS.

In the future, the proposed method can be utilized and applied to retrieving of images and recognizing of objects using feature vectors as color and texture. And also it can be used video retrieval efficiently.

References

1. John R. Smith, Shih-Fu Chang : VisualSEEK - a fully automated content-based image query system, http://www.ctr.columbia.edu/VisualSEEK
2. R.W. Picard : A Society of Model for Video and Image Libraries, http://www.media.mit.edu/~picard/
3. James Ze Wang, Gio Wiederhold, Oscar Firschein, Sha Xin Wei : Applying Wavelets in Image Database Retrieval, http://www-leland.sanford.edu/~zwang
4. E.H. Adelson : Orthogonal Pyramid transform for image coding, In Proc. SPIE, Vol. 845 (1987) 50-58
5. Charles K. Chui : An Introduction to Wavelet, Academic Press, (1995)
6. Rafael C. Gonzalez and Richard E. Woods : Digital Image Processing, Addison-Wesley Publishing Company (1993)
7. Zw-Nian Li and Bing Yan : Recognition Kernel for Content-based Search, IEEE International Conf. on SMC, Vol. 1 (1996) 472 – 477
8. M. K. Mandal, T. Aboulnasr : Image Indexing Using Moments and Wavelets, IEEE Transactions on Consumer Electronics, Vol.42, No.3 (1996) 557-565
9. James Ze Wang and Sha Xin Wei : Wavelet-Based image indexing techniques with partial sketch retrieval capability, Proceeding of the 4th Forum on Research and Technology Advances in Digital Libraries (1997) 13-24

Semantic Decomposition of LandSat TM Image

Miguel Torres, Giovanni Guzmán, Rolando Quintero,
Marco Moreno, and Serguei Levachkine

Geoprocessing Laboratory – Centre for Computing Research – National Polytechnic Institute, Mexico City, Mexico
{mtorres, jguzmanl, quintero, marcomoreno, sergei}@cic.ipn.mx
http://geo.cic.ipn.mx, http://geopro.cic.ipn.mx

Abstract. In this paper, we propose a *semantic supervised clustering* approach to classify multispectral information in geo-images. We use the *Maximum Likelihood Method* to generate the clustering. In addition, we complement the analysis applying *spatial semantics* to determine the training sites and to improve the classification. The approach considers the *a priori* knowledge of the multispectral geo-image to define the classes related to the geographic environment. In this case the spatial semantics is defined by the spatial properties, functions and relations that involve the geo-image. By using these characteristics, it is possible to determine the training data sites with *a priori* knowledge. This method attempts to improve the supervised clustering, adding the intrinsic *semantics* of the geo-images to determine the classes that involve the analysis with more precision.

1 Introduction

The integration of remote sensing and geographic information systems (GIS) in environmental applications has become increasingly common in recent years. Remotely sensed images are an important data source for environmental GIS-applications, and conversely GIS capabilities are being used to improve image analysis procedures. In fact, when image processing and GIS facilities are combined in an integrated system vector data, they can be used to assist in image classification and raster image statistics. In addition, vectors are used as criteria for spatial queries and analysis [1].

Frequently the need arises to analyze mixed spatial data. These data sets can consist of satellite spectral, topographic and other point form data, which are registered geometrically [2].

In addition several works have been published such as [3], [4] and [5]. Theses works are oriented to make supervised classification in geo-images, with classical methods. However, this paper proposes a semantic supervised clustering algorithm applied to LandSat TM images to classify the land cover, according to the *a priori* knowledge. The semantics of geo-images is used to detect adequate *training sites* considering the geometrical and topological properties. Our approach attempts to overcome some of the limitations associated with computational issues from previous supervised clustering methods, such as number of classes, quantitative values processing, classes that may not be useful for users, treatment of post-refinement classes before they can be used in the classification and so on. Our approach is based on the

B. Gabrys, R.J. Howlett, and L.C. Jain (Eds.): KES 2006, Part I, LNAI 4251, pp. 550 – 558, 2006.
© Springer-Verlag Berlin Heidelberg 2006

spatial semantics[1], which is used to determine the behavior of a certain geographic environment by means of spatial properties, functions and relations [6].

The rest of the paper is organized as follows. In section 2 we describe the supervised clustering approach and the proposed algorithms to determine the training sites in a semantic way and the semantic supervised clustering. Section 3 shows the results obtained by applying the approach to LandSat TM images. Our conclusions are outlined in section 4.

2 Supervised Clustering Approach

2.1 Supervised Clustering Method

The supervised clustering is the procedure used for quantitative analysis of remote sensing image data [7]. In this work, we have implemented the *Maximum Likelihood Classification* method. In common cases, the supervised clustering is defined through training sites, which are determined by pixels according to *a priori* semi-automatic selection. When the clustering process has been finished, we obtain a set of *M* classes.

We have considered five basic steps to generate a supervised clustering: (I) Determine the number and type of classes to use for the analysis, (II) Choose training regions (sites) for the classes, according to the *spatial semantics* to identify the spectral characteristics for each specific class, (III) Use these training regions to determine the parameters of the supervised clustering, (IV) Classify all the pixels from the geo-image, assigning them to one of the defined classes by the training regions, and (V) Summarize the results of the supervised clustering [8]. An important assumption in supervised clustering usually adopted in remote sensing is that each spectral class can be described by a probability distribution in a multispectral space. A multidimensional normal distribution is described as a function of a vector location in multispectral space by Eqn 1.

$$p(x)=\frac{1}{(2\pi)^{N/2}|\Sigma|^{1/2}}\exp\left\{-\frac{1}{2}(x-\mu)^t\Sigma^{-1}(x-\mu)\right\}, \tag{1}$$

where:

N is the number of components of vector x.
x is a vector location in the N dimensional pixel space.
μ is the mean position of the spectral class.
Σ is the covariance matrix of the distribution.

The multidimensional normal distribution is completely specified by its *mean vector* and its *covariance matrix* (Σ). In fact, if the mean vectors and covariance matrices are known by each spectral class, it is possible to compute the set of probabilities that describe the relative *likelihoods* of a pattern, at a particular location belonging to each

[1] To define *spatial semantics* [2], we use the essential characteristics that involve the remote sensing imagery. Our definition is based on providing a set of rules that describe the geo-images. This set is composed of relations, properties and functions that define the behaviors of the raster image. This definition starts with a spatial description of the geographical objects that consists of topological, logical, geometrical and statistical properties, which define conceptually spatial semantic rules that involve the geo-image.

of these classes [9]. This can be considered as a member of the class, which indicates the highest probability. Therefore, μ and ∑ are known by every spectral class in an image, also every pixel can be examined and labeled according to the probabilities computed for the particular location of every pixel. Before the classification, μ and ∑ are estimated for each class from a representative set of pixels, commonly called *training sites*.

2.2 Detection of Semantic Training Sites Algorithm

The main issue of our research is to obtain *semantic training sites* (classes) to determine the areas that involve the classification. In addition, we attempt to define these areas by means of the spatial semantics[2] of the geo-images to improve the classification method. The semantic clustering algorithm is described as follows:

[Step 1]. Let O be the sampling set of objects. See Eqn. 2.

$$O = \{o_i \mid o_i \in Z^2\}, \tag{2}$$

Hence each vector o_i is a 3-tuple, which contains the RGB values[3] at the position (x, y) of the object randomly selected in the geo-image. According to the tests, the number of *seeds (i)* that provides good results is shown in Eqn. 3.

$$k_1(m \cdot n) \leq i \leq k_2(m \cdot n), \tag{3}$$

where:
m is the number of rows in the geo-image.
n is the number of columns in the geo-image.
$K_1 = 0.01$is the proposed value of seeds for every class (ω).
$K_2 = 0.03$ is the proposed value of seeds for every class (ω).

[Step 2]. Transform each vector o_i from spatial domain Z^2 to the semantic domain S by means of Eqn. 4.

$$S_O = \{s_i \mid s(o_i) \vdash o_i \in O\} \tag{4}$$

Therefore, $s(o_i)$ is the semantic characteristic vector, which is obtained by applying Eqn 5.

$$s = < s_1, s_2, s_3, s_4, s_5, s_6 >, \tag{5}$$

where: s_1, s_2, s_3 are the mean of the original seed and their 8-neighbors[4] in the spatial domain Z^2, according to the RGB spectral bands respectively. Additionally s_4, s_5, s_6 are the values of the standard deviation[5] of the original seed and their 8-neighbors in the same domain.

---

[2] In the semantic clustering algorithm, we consider the "essential" properties that describe the semantics of the geo-images. The essential features for this case study are the topological and geometrical properties, which are considered as *a priori* knowledge.

[3] In other types of images, we can use other spectral components.

[4] In this view, it is important to define the topological property of the geo-image, because this property is considered the semantics of the raster data.

[5] Standard deviation represents the variability of the reflectance in every selected spectral band.

[Step 3]. Obtain the dissimilitude coefficient (DSC) $d(s_i, s_j)$ for s_i, s_j that belongs to S_o, which is defined in Eqn. 6.

$$d(s_x, s_y) = \sqrt{\left(s_{x1}^2 - s_{y1}^2\right) + \left(s_{x2}^2 - s_{y2}^2\right) + \ldots + \left(s_{x6}^2 - s_{y6}^2\right)} \quad (6)$$

[Step 4]. Let $D = \{d(s_i, s_j) \mid s_i, s_j \in S_o \forall i \neq j\}$ be the set of DSCs, and $d_{max} = Max(D)$ and $d_{min} = Min(D)$. The threshold of maxim similarity U is obtained by applying Eqn. 7.

$$U = \lambda(d_{max} - d_{min}), \quad (7)$$

where:
λ is the discrimination coefficient[6].

[Step 5]. Let s_i, s_j and s_k be vectors that belong to S_o, $s_k = f(s_i, s_j)$ in which S_o is the merging process of s_i and s_j; if Eqn. 8 is accomplished.

$$s_k = \left\langle \frac{s_{i1} + s_{j1}}{2}, \frac{s_{i2} + s_{j2}}{2}, \ldots, \frac{s_{i6} + s_{j6}}{2} \right\rangle \quad (8)$$

Let $g: S_o \rightarrow S_o$ be the minimal dissimilitude function (MDS) in which $g(s_i)$ is the most similar vector to s_i. Then an evolution of the set S_o is defined by Eqn. 9.

$$\begin{aligned} S_o' &= \{s_i \mid g(g(s_i)) \neq s_i \forall s_i \in S_o\} \cup \\ &\{s_f = f(s_i, s_j) \mid g(s_i) = s_j, g(s_j) = s_i \forall s_i, s_j \in S_o\} \end{aligned} \quad (9)$$

If the dissimilitude distance of a vector s_i is greater than the threshold U regarding the rest of the vectors, then s_i goes to the next generation of vector. Otherwise, vector s_i must be merged with other vectors, whose dissimilitude distance is less than the threshold U.

[Step 6]. If $card(S'_o) < M$ (in which M is the number of desired classes), then repeat from step 4, with $\lambda = \lambda / 2$.

[Step 7]. Determine the proportional ratio of the classes in the semantic domain S, which is given by Eqn. 10.

$$P_r = \frac{d_{min}}{d_{max}}, \quad (10)$$

While the proportional ratio of the classes is closer to 1, the partition of the semantic domain will be more proportional; that means that the dissimilitude distances between vectors are closer.

[Step 8]. Since the process is iterative, it should be repeated with the new generation of vectors, which have been obtained; that is, $S_o = S_o'$

[Step 9]. Repeat the process from step 3, until $card(S_o) = M$.

2.3 Semantic Supervised Clustering (SSC) Algorithm

The result of the last stage (section 2.2) is the set of semantic characteristic vectors ω_i of the classification. From the Ω set, it is necessary to compute the mean and

[6] The initial value for λ is 0.5. This value provides good results according to the tests.

covariance matrix for every $\omega_i \in \Omega$. By attempting to determine the class or category for every pixel at a location x, it is necessary that each pixel contains a conditional probability, denoted by Eqn. 11:

$$p(\omega_i \mid x), i = 1,\ldots, M \tag{11}$$

In [7] we describe the supervised clustering method for more details. Therefore, Bayes theorem provides potential means of converting knowledge of predictive correlations. The constraint (Eqn. 12) is used in the classification algorithm, since the $p(x)$ are known by training data, we assume that it is conceivable that the $p(\omega_i)$ is the same for each ω_i, due to this, the comparison is $p(x \mid \omega_i) > p(x \mid \omega_j)$.

$$x \in \omega_i \text{ if } p(x \mid \omega_i)p(\omega_i) > p(x \mid \omega_i)p(\omega_j) \text{for all } j \neq i \tag{12}$$

In this analysis, we assume that the classes have multidimensional normal distributions and each pixel can assign a probability of being a member of each class. After computing the probabilities of a pixel being in each of the available classes, we assign the class with the highest probability. The algorithm consists of the following steps:

[Step 1]. Determine the number of classes ω_i by means of the semantic training sites algorithm (section 2.2).

[Step 2]. Compute the maximum likelihood distribution and the covariance matrix for each generated class ω_i.

[Step 3]. For each image pixel, determine its semantic vector, applying Eqn. 5.

[Step 4]. Compute the probability of vector s to know if it belongs to each class ω_i.

[Step 5]. Obtain the coordinates (x, y) of the pixel, if the constraint (see Eqn. 13) is accomplished, then the pixel belongs to the class ω_i.

$$p(x \mid \omega_i)p(\omega_i) > p(x \mid \omega_j)p(\omega_j) \text{for all } j \neq i \tag{13}$$

[Step 6]. Repeat from step 3, until all pixels in the geo-image can be classified.

3 Results

By using this approach, we can perform a semantic supervised clustering in LandSat TM images. The algorithm has been implemented in C++ Builder. The segment of Tamaulipas State LandSat TM image is shown in Fig. 1. In Fig. 1a, we appreciate the LandSat TM image, which is composed of three spectral bands (4 3 2). In addition, we have overlapped vector data to identify the urban and land areas. Fig. 1b depicts the results of the semantic supervised clustering.

We have considered that to obtain a good estimation of class statistics, it is necessary to choose several trainings fields for the one cover type, located in different regions of the image. The three band signatures for the classes are obtained from the training fields, which are given in Table 1.

In addition, we made other supervised clustering using PCI Geomatica software. We use the *Parallelepiped* method (PM) depicted in Fig 2a, and the *Minimum*

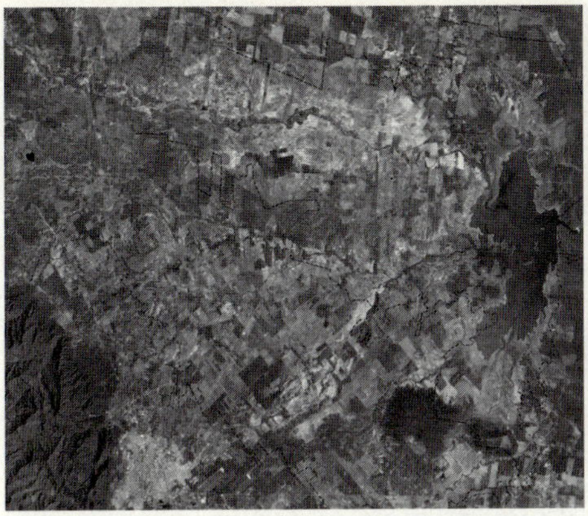

(a)

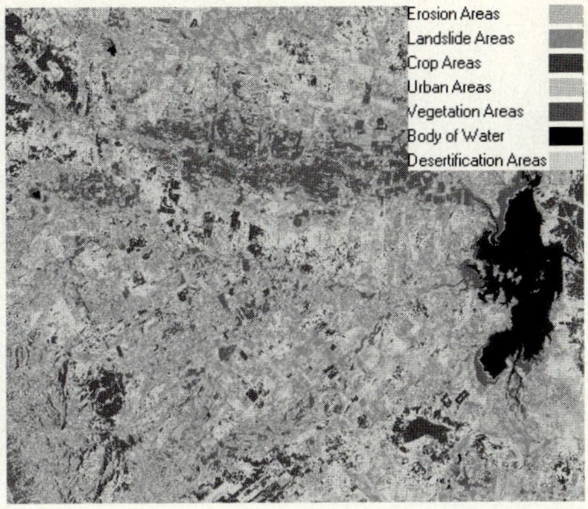

(b)

Fig. 1. (a) LandSat image of Tamaulipas basin[7]. (b) SSC of LandSat image.

Table 1. Class signatures generated from the training areas in Fig. 1b

Class	Mean vector	Standard deviation	Covariance matrix		
Body of Water (blue)	39.16	1.35	1.82	0.00	0.00
	34.30	1.74	1.83	3.03	0.00
	21.38	2.43	2.24	3.40	5.91
Urban areas (yellow)	52.27	5.11	26.15	0.00	0.00
	58.40	7.61	37.77	58.02	0.00
	58.50	7.31	25.61	37.84	53.56
Crop areas (brown)	46.89	1.01	1.02	0.00	0.00
	54.74	1.64	1.02	2.72	0.00
	62.52	2.55	0.94	1.28	6.50
Landslide areas (red)	39.68	0.93	0.88	0.00	0.00
	37.07	1.50	0.90	2.25	0.00
	63.65	5.14	-0.76	-0.64	26.43
Vegetation areas (green)	44.83	1.30	1.69	0.00	0.00
	49.26	2.32	2.42	5.38	0.00
	62.79	5.36	4.65	8.36	28.76
Desertification areas (pink)	44.08	1.27	1.63	0.00	0.00
	43.08	2.44	2.36	5.97	0.00
	83.87	5.89	-0.80	-5.69	34.75
Erosion areas (gray)	41.66	0.85	0.72	0.00	0.00
	42.81	1.28	0.50	1.66	0.00
	54.45	4.84	1.73	0.38	23.46

Distance method (MDM) shown in Fig 2b. In the three supervised methods, we use the same number of training sites. The PM method does not classify several quantity pixels of the geo-image. On the other hand, the MDM method classifies the pixels with some mistakes. For instance, the areas of vegetation and desertification classes are less than the areas generated for the SSC method. Our approach improves the classification process using the semantics to compute the training sites. The values for supervised clustering methods are described in Table 2.

---

[7] Basin is the entire geographical area drained by a river and its tributaries.

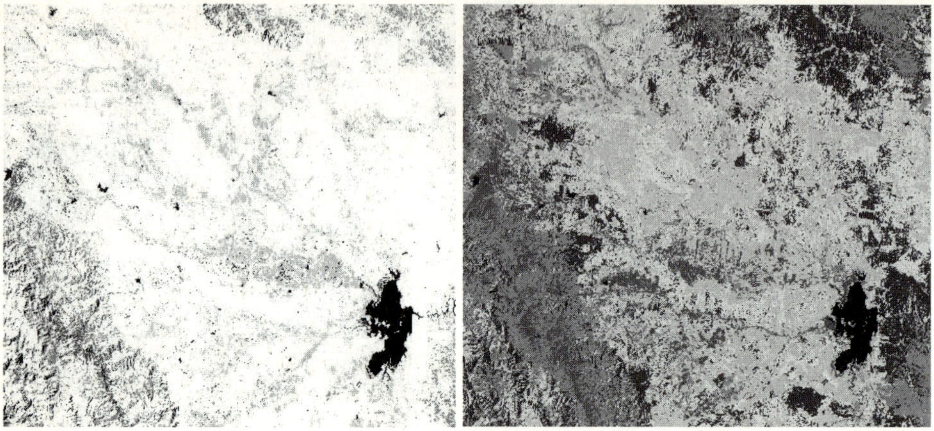

(a) (b)

Fig. 2. (a) Parallelepiped supervised clustering. (b) Minimum Distance supervised clustering.

Table 2. Comparative results[8] of three supervised clustering methods

Class	Mean vector	Standard deviation	Covariance matrix		
Body of Water (SSC)	39.16	1.35	1.82	0.00	0.00
	34.30	1.74	1.83	3.03	0.00
	21.38	2.43	2.24	3.40	5.91
Body of Water (PM)	23.53	1.23	2.81	0.00	0.00
	22.50	1.14	3.82	7.55	0.00
	17.36	1.05	4.99	1.36	18.66
Body of Water (MDM)	32.62	1.99	3.99	0.00	0.00
	30.27	3.17	5.91	10.05	0.00
	9.78	2.78	6.43	9.76	14.35

4 Conclusions

In the present work the *semantic supervised clustering* approach for Landsat TM (7 bands, resolution 25 meters per pixel) images is proposed. In addition, we use *spatial semantics* to improve the classification by means of *a priori* knowledge that consists of geometrical and topological properties. This knowledge is considered into the selection criteria of the training sites.

By applying this algorithm, it is essential to define the *a priori knowledge* involved in the training sites. We assume that the set of properties provides a starting point for the clustering process. The algorithm aims to preserve the intrinsic properties and relations between geospatial information and spectral classes. The method is used to make a quantitative analysis of sensing spatial data, although we consider qualitative properties to define the classes.

Moreover, the semantic clustering algorithm consists of transforming the problem from the spatial domain to the semantic domain. In this context, the most similar

[8] Due to space limitations, we only present the results for one class.

semantic classes will be merged according to their properties. This process is based on computing the dissimilitude coefficient (DSC) and merging pairs of classes. With this approach, it is possible to reduce the number of classes, according to the desired number.

The semantic supervised clustering approach is essential for the decision making process. Additionally, it is used to improve the spatial analysis in different geographical environments. The approach allows us to make more accurate the spatial analysis with data fusion (vector and raster). This algorithm can be incorporated to detect flooding and landslide areas.

On the other hand, this approach can be used to know other useful spatial and attributive properties, which define the *spatial semantics* of geo-images.

Our future works are related to compare this method with others supervised clustering to evaluate the performance and the results of the raster data classification. In addition, we are looking for defining an *automatic* methodology to classify geo-images by means of *semantic unsupervised clustering*. For this purpose, it is necessary to use the *spatial semantics* to know the properties and behavior of raster data.

Acknowledgments

The author of this paper wishes to thank the CIC, SIP, IPN and CONACYT for their support.

References

1. Unsalan, C., Boyer, K.L.: Classifying land development in high-resolution Satellite imagery using hybrid structural-multispectral features. *IEEE Transactions on GeoScience and Remote Sensing*, Vol. 42, No. 12, (2004), pp. 2840-2850.
2. Torres, M., Levachkine S.: *Generating spatial ontologies based on spatial semantics*, in: Levachkine S., Serra J., Egenhofer M. (Eds.), Research on Computing Science, Semantic Processing of Spatial Data, Vol. 4, (2003), pp. 169-178.
3. Nishii, R., Eguchi, S.: Supervised image classification by contextual AdaBoost based on posteriors in neighborhoods. *IEEE Transactions on GeoScience and Remote Sensing*, Vol. 43, No. 11, (2005), pp. 2547-2554.
4. Bandyopadhyay, S.: Satellite image classification using genetically guided fuzzy clustering with spatial information. *International Journal of Remote Sensing*, Taylor & Francis, Vol. 26, No. 3, (2005), pp. 579-593.
5. Jianwen, Ma., Bagan H.: Land-use classification using ASTER data and self-organized neutral networks. *International Journal of Applied Earth Observation and Geoinformation*, Elsevier, Vol. 7, No. 3, (2005), pp. 183-188.
6. Morgan, J.T., Ham, J., Crawford, M.M., Henneguelle, A., Ghosh, J.: Adaptative feature spaces for land cover classification with limited ground truth data. *International Journal of Pattern Recognition and Artificial Intelligence*, World Scientific Publishing Company, Vol. 18, No. 5, (2004), pp. 777-799.

7. Torres M., Moreno M., Quintero R., Guzmán G.: Applying Supervised Clustering to Landsat MSS Images into GIS-Application, *Advances in: Artificial Intelligence, Computing Science and Computer Engineering*, Research on Computing Science, Vol. 10, (2004), pp. 167-176.
8. Chung, K.F., Wang, S.T.: *Note on the relationship between probabilistic and fuzzy clustering*. Soft Computing, Lecture Notes in Computer Science, Springer-Verlag, Berlin Heidelberg, Vol. 8, No. 7, (2003), pp. 523-526.
9. Bandyopadhyay, S., Maulik, U., Pakhira, M.K.: Clustering using simulated annealing with probabilistic redistribution. *International Journal of Pattern Recognition and Artificial Intelligence*, World Scientific Publishing Company, Vol. 15, No. 2, (2001), 269-285.

Vision-Based Semantic-Map Building and Localization

Seungdo Jeong[1], Jounghoon Lim[2], Il Hong Suh[2], and Byung-Uk Choi[2]

[1] Department of Electrical and Computer Engineering, Hanyang University
17 Haengdang-dong, Sungdong-gu, Seoul, 133-791 Korea
sdjeong@mlab.hanyang.ac.kr
[2] School of Information and Communications, Hanyang University,
lightinday@hanafos.com, {ihsuh, buchoi}@hanyang.ac.kr

Abstract. A semantic-map building method is proposed to localize a robot in the semantic-map. Our semantic-map is organized by using SIFT feature-based object representation. In addition to semantic map, a vision-based relative localization is employed as a process model of extended Kalman filters, where optical flows and Levenberg-Marquardt least square minimization are incorporated to predict relative robot locations. Thus, robust SLAM performances can be obtained even under poor conditions in which localization cannot be achieved by classical odometry-based SLAM.

1 Introduction

To navigate and plan a path in an environment, the intelligent mobile robot has to be able to build a map of the environment and to recognize its location autonomously. SLAM(Simultaneous Localization And Mapping) is a popular technique to accurately localize the robot and simultaneously build a map of the environment. Numerous studies of SLAM have been performed, as this has been one of the important issues in the intelligent mobile robot community for a long time. Thus, a number of sensors and algorithms have been proposed to realize the SLAM technique. Since the 1990s, methods using extended Kalman filters have been focused on by a number of researchers interested in SLAM [3, 8]. Most algorithms using extended Kalman filters combine two methods. One is relative localization, which is the method used to compute the current position with respect to an initial location. The other is the method using a map composed of landmarks.

Odometry is often used for relative localization. However, it has many errors caused not only by systematic factors such as a difference in wheel diameter, inaccurate gauging of wheel size, and others, but also non-systematic factors such as slip, poor condition of the floor, etc. [4] proposed an algorithm to reduce such systematic errors. Their algorithm consists of three steps: odometry error modeling, error parameter estimation using the PC-method, and estimation of the covariance matrix. Even so, it is hard to overcome the error of estimation caused by non-systematic factors.

B. Gabrys, R.J. Howlett, and L.C. Jain (Eds.): KES 2006, Part I, LNAI 4251, pp. 559–568, 2006.
© Springer-Verlag Berlin Heidelberg 2006

The Harris three-dimensional vision system DROID is another algorithm for relative localization [5]. This system uses the visual motion of image corner features for 3D reconstruction. Kalman filters are used for tracking features, and from locations of the tracked image features, DROID determines both the camera motion and the 3D position of the features. Ego-motion determination by matching image features is generally very accurate in the short to medium term.

Building the map of the environment is another issue with SLAM. The map consists of landmarks which are used for robot localization. Davision uses three-dimensional corner points as landmarks, which are obtained by a Harris corner detector and stereo matching [5]. Se uses keypoints obtained from SIFT(Scale Invariant Feature Transform) [6, 9]. However, simple coordinates or features within the image are not enough to facilitate interaction with humans.

In this work, we use known objects as landmarks through object recognition and then build the semantic-map with these recognized object landmarks. This semantic-map is very helpful for interaction between a human and the robot. We also use vision-based relative localization process as the process model of our extended Kalman filter. This approach is robust enough for environments which cannot be supported by encoders that measure values such as the number of rotations of the robot's wheels.

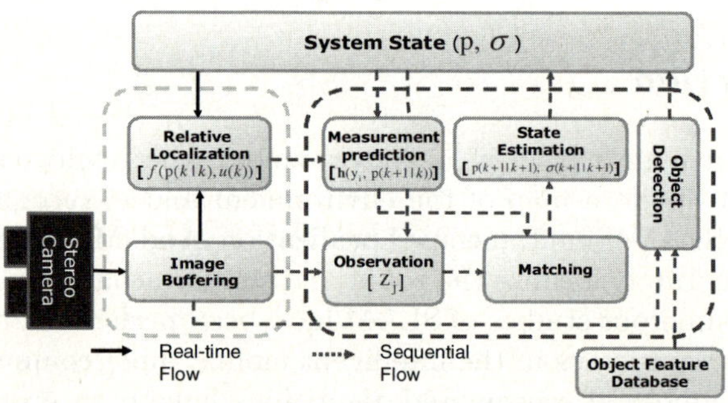

Fig. 1. Block diagram of the proposed method

2 OFM-Based Robot Localization and Semantic-Map Building

In this work, we propose a three-dimensional object feature model (OFM) with essential properties and propose a method to use the OFM for improving the performance of vision-based SLAM. Figure 1 shows the block diagram of our SLAM method using a three-dimensional OFM.

We use images taken by the stereo camera as the only sensor information in this work. Our system is composed of three parts: the real-time relative localization part which estimates robot location in real-time, the landmark recognition part which builds up and/or observes landmarks by using SIFT-based object recognition, and the data fusion part which combines two observation data using an extended Kalman filter.

2.1 Real-Time Relative Localization

Image-based relative localization is a method that applies the motion between corresponding points within consecutive image sequences to localize the robot. Thus, matching between feature points is a crucial factor for localization performance.

Tracking of Corner Features. To track the camera motion, it is necessary to obtain the feature points in the current frame corresponding to the feature points in the previous frame. Note that the corner points satisfy the local smoothness constraint. Lucas-Kanade optical flow performs well in tracking corner points with that property [7]. However, Lucas-Kanade optical flow is best suited to small motion tracking. It is not sufficient to follow the movement of the robot. Thus, pyramid Lucas-Kanade (PLK) optical flow using a Gaussian image pyramid is used for relative localization, where it is known to track a relatively wide area.

Localization. The feature points extracted from the initial frame are used as 3D point landmarks. Note that we use the stereo camera as the input sensor, and we obtain a stereo image pair. The 3D coordinates of the extracted feature point are calculated using the disparity of the stereo camera images. The stereo camera is previously calibrated. When we know the exact relationship between the 3D coordinates of the landmark and the corresponding 2D coordinates which are projected to the image, the projection matrix indicates the mapping relationship between points of 3D space and points of the image.

The projection matrix is composed of two parts. One is the intrinsic parameter matrix which includes internal information of the camera. This matrix represents the relationship between the camera coordinate system and the image coordinate system. The other part of the projection matrix is the extrinsic parameter matrix. This matrix represents the translational and rotational relationship between the 3D space coordinate system and the camera coordinate system.

Therefore, if the projection matrix is estimated with coordinates of corresponding points in 3D space and in the 2D image, we can obtain the motion of the camera and the relative location of the robot. It is a non-linear problem to estimate the projection matrix with numerous corresponding points. In this work, we estimate the rotation and translation components using the Levenberg-Marquardt least square minimization (LM LSM) method.

2.2 Extended Kalman Filter-Based SLAM Using Object Landmarks

In most previous research works on SLAM, landmarks were composed of lines of the environment. They were obtained using a range finder or feature points of camera images. These maps only involve coordinates of feature points. Thus, the use of these maps for high level purposes, such as delivery tasks or interactions with humans, requires an anchoring technique to link point-based landmarks with their associated semantics. Alternatively, in this work, objects are recognized using SIFT and then features of recognized objects are registered as landmarks. Therefore, we can create the semantic-map supporting the location of main objects in the environment without any additional anchoring process.

In most SLAM methods using Kalman filters, odometry-based robot kinematics or dynamics are used as the process model. However, those performances are very poor because of systematic factors such as the difference of wheel diameter and non-systematic factors such as slip. Moreover, in the case that a wheel encoder does not exist, such as for a humanoid robot, odometry-based methods are difficult to apply.

Therefore, to resolve such drawbacks, we propose the extended Kalman filter-based SLAM that integrates PLK optical flow and LM algorithm-based relative localization with object landmarks.

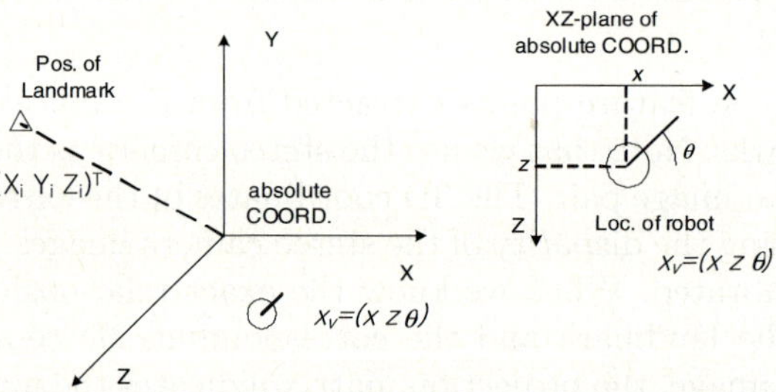

Fig. 2. System coordinate for the location of the robot x_v and the position of the landmark y_i

The State Vector and Covariance. We assume that the robot is located on a plane as in Fig. 2, so position and orientation of the robot is represented by $x_v = (x\ z\ \theta)^T$. The position of each landmark is denoted as $y_i = (\,X_i\ Y_i\ Z_i\,)^T$. Here, SIFT keypoints of objects recognized by the three dimensional OFM are represented as absolute coordinates.

To regulate uncertainty of landmarks and relationships among them we use the system state vector and covariance model proposed by [1]. Thus, we can represent the state vector p and covariance matrix Σ as

$$p = \begin{pmatrix} x_v \\ y_1 \\ y_2 \\ \vdots \end{pmatrix}, \Sigma = \begin{pmatrix} \sigma^2_{x_v x_v} & \sigma^2_{x_v y_1} & \sigma^2_{x_v y_2} & \cdots \\ \sigma^2_{y_1 x_v} & \sigma^2_{y_1 y_1} & \sigma^2_{y_1 y_2} & \cdots \\ \sigma^2_{y_2 x_v} & \sigma^2_{y_2 y_1} & \sigma^2_{y_2 y_2} & \cdots \\ \vdots & \vdots & \vdots & \ddots \end{pmatrix}. \tag{1}$$

Three Dimensional Object Feature Model. The three dimensional object feature model(3D OFM) is defined as the model which is composed of the three dimensional SIFT keypoints extracted from object images. To make the three dimensional object feature model, we rotate and object by 20 degrees with respect to the center of gravity and then take the images using the calibrated stereo camera. This process is repeated to get 18 views images. SIFT keypoints are extracted from each image and are given three dimensional coordinates calculated with the stereo vision technique. SIFT keypoints having three dimensional coordinate are called as the three dimensional SIFT keypoints. The three dimensional SIFT keypoints in objects recognized by using 3D OFM are used as landmarks.

Process Model. In EKF, we define the process model as the estimation of current location and its covariance for the robot and landmarks referring to the state vector and its covariance matrix for the previous system. We apply the displacement $\Delta x, \Delta z, \Delta\theta$ obtained by PLK optical flow and LM algorithm-based relative localization to the process model. Let the process model of our work be given as

$$x_v(k+1|k) = f(x_v(k|k), u(k)) = \begin{bmatrix} x \\ z \\ \theta \end{bmatrix} + \begin{bmatrix} \Delta x_r \\ \Delta z_r \\ \Delta\theta \end{bmatrix}. \quad (2)$$

In (2), Δx_r and Δz_r are given as

$$\begin{bmatrix} \Delta x_r \\ \Delta z_r \end{bmatrix} = \begin{bmatrix} cos(\theta+\Delta\theta) & sin(\theta+\Delta\theta) \\ -sin(\theta+\Delta\theta) & cos(\theta+\Delta\theta) \end{bmatrix} \begin{bmatrix} \Delta x \\ \Delta z \end{bmatrix}. \quad (3)$$

The current location for the robot $x_v(k+1|k)$ is estimated by function f as shown in (2). $x_v(k|k)$ and $u(k)$ represent displacement of the previous location and motion of the robot respectively. Δx_r and Δz_r are the displacement which are transformed to the reference coordinate system of the map from the estimated displacement Δx and Δz referring to the camera coordinate system of the previous location.

$$y_i(k+1|k) = y_i(k|k), \forall i. \quad (4)$$

In (4), $y_i(k+1|k)$ represents the estimated current position of the i-th landmark, and we assume that landmarks are fixed.

The covariance for the system $\sigma^2_{x_v x_v}, \sigma^2_{x_v y_i}$ and $\sigma^2_{y_i y_j}$ are obtained as

$$\begin{aligned} \sigma^2_{x_v x_v}(k+1|k) &= \nabla_{x_v} f \sigma^2_{x_v x_v}(k|k) \nabla_{x_v} f^T + \nabla_u f \sigma^2_u(k|k) \nabla_u f^T, \\ \sigma^2_{x_v y_i}(k+1|k) &= \nabla_{x_v} f \sigma^2_{x_v y_i}(k|k), \\ \sigma^2_{y_i y_j}(k+1|k) &= \sigma^2_{y_i y_j}(k|k), \end{aligned} \quad (5)$$

where $\nabla_{x_v} f$ and $\nabla_u f$ represent the Jacobian of the estimation function f for the state vector of the robot and the displacement respectively. These are given as

$$\begin{aligned}
&\nabla_{x_v} f = \left[\frac{\partial f}{\partial x} \; \frac{\partial f}{\partial z} \; \frac{\partial f}{\partial \theta} \right], \quad \nabla_u f = \left[\frac{\partial f}{\partial \Delta x} \; \frac{\partial f}{\partial \Delta z} \; \frac{\partial f}{\partial \Delta \theta} \right], \\
&where \\
&\partial f / \partial x = \left[1\, 0\, 0 \right]^T, \quad \partial f / \partial z = \left[0\, 1\, 0 \right]^T, \\
&\frac{\partial f}{\partial \theta} = \begin{bmatrix} -sin(\theta + \Delta\theta)\Delta x + cos(\theta + \Delta\theta)\Delta z \\ -cos(\theta + \Delta\theta)\Delta x - sin(\theta + \Delta\theta)\Delta z \\ 1 \end{bmatrix}, \\
&\partial f / \partial \Delta x = \left[cos(\theta + \Delta\theta) \; -sin(\theta + \Delta\theta) \; 0 \right]^T, \\
&\partial f / \partial \Delta z = \left[sin(\theta + \Delta\theta) \; cos(\theta + \Delta\theta) \; 0 \right]^T, \\
&\frac{\partial f}{\partial \Delta\theta} = \begin{bmatrix} -sin(\theta + \Delta\theta)\Delta x + cos(\theta + \Delta\theta)\Delta z \\ -cos(\theta + \Delta\theta)\Delta x - sin(\theta + \Delta\theta)\Delta z \\ 1 \end{bmatrix}.
\end{aligned} \tag{6}$$

Here, $(\Delta x \; \Delta z \; \Delta\theta)$ is the movement of the robot estimated by the LM algorithm.

In (5), σ_u^2 is the covariance matrix due to the noise in the process of camera motion estimation.

Measurement Model. Let the measurement model be given as $h_i(k+1) = h(y_i, x(k+1|k))$. Observation for landmark is the three dimensional coordinate of landmarks based on the calibrated stereo camera coordinate system. Measurement prediction model of h_i is given as

$$\begin{aligned}
&h_i(k+1) = [R][y_i - x'_v], \\
&where \quad x'_v = (x \; 0 \; z)^T, R = \begin{bmatrix} cos\theta & 0 & -sin\theta \\ 0 & 1 & 0 \\ sin\theta & 0 & cos\theta \end{bmatrix}.
\end{aligned} \tag{7}$$

Three dimensional absolute coordinate of landmark is denoted as y_i. And h_i is the function to transform absolute coordinate to camera coordinate system. x'_v represents planar coordinate of the robot based on the absolute coordinate system. Rotation between the absolute coordinate system to the camera coordinate system is denoted by R. θ in matrix R denotes the orientation of the robot.

Observation and Matching. Matching between landmarks and observed three dimensional coordinate of SIFT $z_j = [X_j \; Y_j \; Z_j]$ is accomplished through SIFT matching algorithm. To determine searching area for matching, we can use the innovation matrix and its covariance.

The innovation matrix $v_{ij}(k+1)$ between the predicted measurement $h_i(k+1|k)$ for landmark y_i and observation z_j is given as

$$\begin{aligned}
v_{ij}(k+1) &= [z_j(k+1) - h(y_i, p(k+1|k))], \\
\sigma^2_{IN,ij} &= \nabla_{x_v} h_i \cdot \sigma^2_{x_v x_v} \cdot \nabla_{x_v} h_i^T + \nabla_{x_v} h_i \cdot \sigma^2_{x_v y_i} \cdot \nabla_{y_i} h_i^T \\
&+ \nabla_{y_i} h_i \cdot \sigma^2_{y_i x_v} \cdot \nabla_{x_v} h_i^T + \nabla_{y_i} h_i \cdot \sigma^2_{y_i y_i} \cdot \nabla_{y_i} h_i^T + \sigma^2_{R,i},
\end{aligned} \tag{8}$$

where $\sigma^2_{R,i}$ represents the covariance of the measurement.

Searching area is determined with the Mahalanobis distance and threshold constant g^2 such as

$$v_{ij}^T(k+1) \cdot \sigma_{IN,ij}^{-2} \cdot v_{ij}(k+1) \leq g^2 \tag{9}$$

Using SIFT matching algorithm, matching is accomplished between the landmark y_i and the observation satisfying the criterion shown in (9).

Estimation of the System State Vector and Covariance. Kalman gain K can be calculated as

$$K = \sigma^2 \nabla_{x_v} h_i^T \cdot \sigma_{IN_i}^{-2} = \begin{pmatrix} \sigma_{x_v x_v}^2 \\ \sigma_{y_i x_v}^2 \\ \vdots \end{pmatrix} \frac{\partial h_i^T}{\partial x_v} \sigma_{IN_i}^{-2} + \begin{pmatrix} \sigma_{x_v y_i}^2 \\ \sigma_{y_i y_i}^2 \\ \vdots \end{pmatrix} \frac{\partial h_i^T}{\partial y_i} \sigma_{IN_i}^{-2}, \tag{10}$$

where $\sigma_{x_v x_v}^2$, $\sigma_{x_v y_i}^2$ and $\sigma_{y_i y_i}^2$ are 3×3 blocks of the current state covariance matrix Σ. $\sigma_{IN_i}^2$ is the scalar innovation variance of y_i.

The updated system state and its covariance can be computed as

$$\begin{aligned} p(k+1|k+1) &= p(k+1|k) + K \cdot v_i(k+1), \\ \sigma^2(k+1|k+1) &= \sigma^2(k+1|k) - K \cdot \sigma_{IN_i}^2(k+1) \cdot K^T. \end{aligned} \tag{11}$$

This update is carried out sequentially for each innovation of the measurement.

Registration and Deletion of Landmarks. When new three dimensional SIFT feature points are found on a recognized object using 3D OFM, absolute coordinates of those are computed using $y_n(x_v, h_n)$ of the measurement model. The computed absolute coordinates are added into the system state vector as in [1];

$$p_{new} = \left(x_v \; y_1 \; y_2 \cdots y_n\right)^T. \tag{12}$$

Updated covariance of the system state is given as

$$\Sigma_{new} = \begin{pmatrix} \sigma_{x_v x_v}^2 & \sigma_{x_v y_1}^2 & \cdots & \sigma_{x_v x_v}^2 \frac{\partial y_n}{\partial x_v}^T \\ \sigma_{y_1 x_v}^2 & \sigma_{y_1 y_1}^2 & \cdots & \sigma_{y_1 x_v}^2 \frac{\partial y_n}{\partial x_v}^T \\ \vdots & \vdots & \vdots & \vdots \\ \frac{\partial y_n}{\partial x_v} \sigma_{x_v x_v}^2 & \frac{\partial y_n}{\partial x_v} \sigma_{x_v y_1}^2 & \cdots & N \end{pmatrix}, \tag{13}$$

$$\textit{where} \quad N = \frac{\partial y_n}{\partial x_v} \sigma_{x_v x_v}^2 \frac{\partial y_n}{\partial x_v}^T + \frac{\partial y_n}{\partial h_n} \sigma_R^2 \frac{\partial y_n}{\partial h_n}^T.$$

We can delete landmarks by deleting all rows and columns related to target landmarks from the covariance matrix.

3 Experimental Results

3.1 Performance Evaluation for Vision-Based Relative Localization

To evaluate the performance of the vision-based relative localization proposed by this work, we compare the movement of the robot with the correct answer. For the localization experiment, the robot moves 900 millimeters straight forward. Calculation interval of the robot location is 50 millimeters. Table 1 shows the error between estimation using our algorithm and the real value.

Table 1. Error of the vision-based relative localization

	PLK optical flow and LM algorithm		
	Maximum	Minimum	Average
X(millimeter)	8.785	0.122	3.310
Z(millimeter)	5.153	0.880	3.254
θ(degree)	0.969	0.013	0.366

3.2 Performance Evaluation for Object Landmark-Based SLAM

Compensation Using the Extended Kalman Filter-Based Localization It is hard to register a landmark in real-time by the object recognition. Thus, we have to do localization and semantic-map building asynchronously.

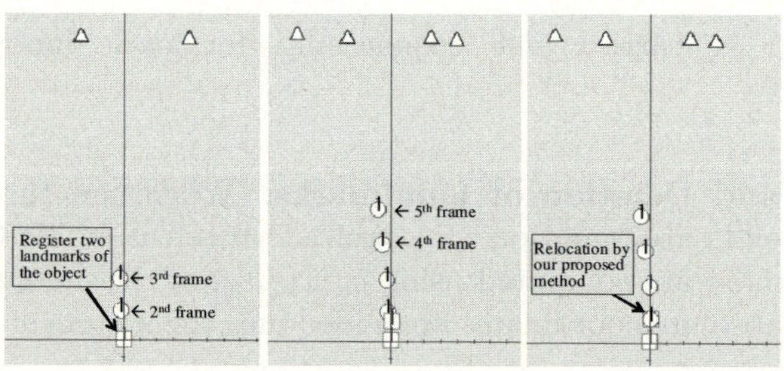

Fig. 3. The compensation process using extended Kalman filter

In Fig. 3, a circle and a rectangle indicate location of robot to be estimated by vision-based relative localization and our proposed EKF-based localization respectively. After the vision-based relative localization process, objects are recognized in a 3rd frame as shown in Fig. 3(left), and then SIFT features of the objects are registered as landmarks. On the other hand, robot localizations for 4th and 5th frames are estimated by the vision-based relative localization in real-time as shown in Fig. 3(center), while updating location of robot for 2nd frame by EKF-based localization. Fig. 3(right) represents the performances of relocation by our proposed EKF-based localization.

It is observed from Table 2 that performances of location by our proposed EKF is much better than those of vision-based relative localization.

Table 2. Error compensation result using EKF-based localization

	x (mm)	z (mm)	θ (degree)
True	0	900	0
Vision-based	-30.662	958.572	-2.383
EKF-based	3.173	911.351	-0.472

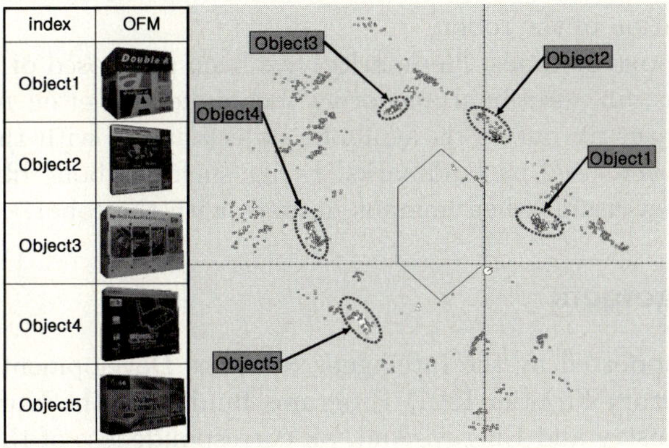

Fig. 4. Semantic-map built by robot

Semantic-Map Building. For our experiment, five objects are arranged as shown in Fig. 4. And, 3D SIFT OFMs for the five objects are also shown in Fig. 4. Keypoints of objects are recognized in the scene by using 3D SIFT OFM and are registered as landmarks with associated symbols to represent each object as shown in Fig. 4. Semantic-map indicates positions of objects as symbols associated with each object. In addition, estimated regions of each object is shown as ellipses in the semantic-map. Figure 4 shows the semantic-map formed by the robot while moving.

It is noted that object regions are informed with map-building for environment simultaneously. In traditional SLAM algorithm, however, additional process such as object recognitions and their anchorings must be required to complete this work.

Simultaneous semantic-map building with the robot localization can support new ability of the robot. First, the robot is able to simultaneously build up object-based semantic-map while carrying out delivery tasks in an unknown environment. Second, Semantic-map can be used to establish environmental ontology to be required in path planning or task planning.

4 Concluding Remarks

In this work, we have proposed autonomous semantic-map building combined with vision-based robot localization. We have used nonsymbolic SIFT-based object features with symbolic object names as landmarks and we have used vision-based relative localization as the process model of our EKF. Thus our method is able to be applied successfully to environments in which encoders are not available, such as humanoid robots.

For real-time localization, we have used the Harris corner detector and PLK optical flow. From the relationship between three dimensional Harris corner points computed by stereo vision and corner points tracked by the PLK optical flow, we have been able to obtain the motion of the camera as well as the relative localization of the robot.

In most previous methods, landmarks have been composed of points without semantics. Thus, an additional anchoring technique was often required for interaction. However, in our work, symbolic object names with their 3D feature location have been used as landmarks. Using such symbolic object names as landmarks is very useful when humans interact with the robot.

Acknowledgement

This work is supported by the Intelligent Robotics Development Program, one of the 21st Century Frontier R&D Programs funded by the Korea Ministry of Commerce, Industry and Energy. And, all correspondences of this work should be addressed to I.H. Suh.

References

1. A.J. Davison and W. Murray: Simulatoneous Localization and Map-Building Using Active Vision. IEEE Transaction on Pattern Analysis and Machine Intelligence. (2002) 865–880
2. A.J. Davision: Real-Time Simultaneous Lacalisation and Mapping with a Single Camera. Proceedings of Ninth IEEE International Conference on Computer Vision. (2003) 1403–1410
3. M.W.M. G. Dissanayake, P. Newman, S. Clark, and H. F. Durrant-Whyte: A Solution to the Simultaneous Localization and Map Building (SLAM) Problem. IEEE Transaction on Robotics and Automation. (2001) 229–241
4. N. Doh, H. Choset, and W.K. Chung: Accurate Relative Localization Using Odometry. Proceedings of IEEE International Conference on Robotics and Automation. (2003) 1606–1612
5. C. Harris and M.J. Stephens: A Combined Corner and Edge Detector. Alvey Vision Conference. (1988) 147–152
6. D.G. Lowe: Distinctive Image Features from Scale-Invariant Keypoints. International Journal of Computer Vision. (2004) 91–110
7. B. Lucas and T. Kanade: An Interative Image Registration Technique with an Application To Stereo Vision. Proceedings of DARPA IU Workshop. (1981) 121–130
8. P.S. Maybeck: The Kalman Filter:An Introduction to Concepts. Spinger-Verlag. (1990)
9. S. Se, D.G. Lowe, and J. Little: Mobile Robot Localization And Mapping with Uncertainty using Scale-Invariant Visual Landmarks. International Journal of Robotics Research. (2002) 735–758

Merging and Arbitration Strategy for Robust Eye Location

Eun Jin Koh and Phill Kyu Rhee

Department of computer science & Engineering Inha University
Biometric Engineering Research Center Young-Hyun Dong, Incheon, Korea
supaguri@im.inha.ac.kr, phrhee@inha.ac.kr

Abstract. The difficulties of eye location are mainly caused by the variations of intrinsic eye pattern characteristics from people to people, scale, pose, glasses frame, illumination, etc. To prevail from these problems, this paper addresses a novel and precise robust eye location method. It employs appearance based Bayesian framework to relive the effect of uneven illumination. The appearance of eye patterns is represented by 2D Haar wavelet. It also employs a sophisticated merging and arbitration strategy in order to manage the variations in geometrical characteristics of ambient eye regions due to glasses frames, eye brows, and so on. The located eye candidates are merged or eliminated according to the merging rule. If the merged regions are more than one, we apply the arbitration strategy. The arbitration strategy is based on a minimizing energy function by probabilistic forces and image forces that pull it toward eyes. The experimental results show that the proposed approach can achieve superior performance using various data sets to previously proposed methods.

1 Introduction

Realizing robust automatic face recognition has been a hot topic of computer vision research area for decades [9] [10]. It can be applied in many application areas such as intelligent human-computer interface, smart environment, ubiquitous computing, etc. Automated face recognition system usually consists of three stages: the detection of face, the alignment of facial feature points, and the identification of the face. The face detection stage determines whether or not there are any faces in the image and returns the location of the face (if any). The second stage, the alignment of salient facial feature points, is crucial for extracting of proper and beneficial face features in the subsequent stage of face identification. The third stage identifies or verifies persons using some traditional pattern classification method [2]. Most facial feature alignment methods require the location of the eyes as a crucial step, since the characteristics of the left and right eyes are relatively prominent and distinguishable features in face images. Some of them facilitate further localization of facial feature pointes in most face recognition systems, and allow focusing attention on salient facial configurations to achieve efficient face recognition.

The difficulties of object location are mainly caused by the variations of intrinsic object characteristics, viewpoint, viewing distance, illumination, etc. [11]. Locating the left and right eyes in a face image is relatively tractable owing to their prominent feature characteristics. However, the precise and robust location of eyes is a

B. Gabrys, R.J. Howlett, and L.C. Jain (Eds.): KES 2006, Part I, LNAI 4251, pp. 569 – 581, 2006.
© Springer-Verlag Berlin Heidelberg 2006

challenging mission as much as other problems of object detection such as face detection [12], car detection [13], etc. The appearance of eye patterns varies due to various factors such as size, rotation, tilt, the closure and opening of eyes, illumination conditions, the reflection of glasses, the occlusion by glass frames and/or hair, etc. Many researches for developing eye location algorithms has been carried out in recent years for solving these problems. They employed standard pattern classification methodology to locate the eyes patterns using extracted discrete local features. Hough transform, symmetry detector, projection analysis etc. have been employed based on priori knowledge of face structure to detect eyes [14] [15]. Sermaldi [16] has proposed sparse Gabor filter response method considering eye and eye neighborhood. Their method achieved relatively good result on M2VTS datasets. The appearance-based methods of detecting eyes are studied based on the intensity distribution of the objects [17]. They adopted statistical pattern analysis techniques, and usually require a huge training data set under various conditions. They tried to find the relevant models showing characteristics of positive (eyes) and negative (non-eye) samples. Real-time eye location has been proposed by combing its appearance, the effect of bright pupil, and motion characteristics in video stream. However, this technique is suffered from uneven lighting conditions and varying size of the pupils.

Most of eye location methods are based on the intensity distribution of eye patterns rather than frequency distribution. Thus, they show intrinsically brittleness in performance under uneven illumination. Many of them can only deal with part of variations of eye patterns or can be feasible under some constraints. They very often suffer from uneven illumination conditions, eyebrows and the frames of glasses. For example, eyebrows or thick spectacle frames sometimes are confused with eyes, and leading to make an incorrect decision. So, both the eye window, i.e. image regions including eyes, and the ambient region of eyes should be considered. This paper presents a novel approach for precise and robust locating eyes based on the wavelet transformation of eye patterns to relieve the effect of uneven illumination. It also employs the sophisticated merging and arbitration strategy for analyzing the ambient eye region to avoid incorrect decision due to variations of geometrical characteristics such as eyebrows and glasses frames.

The proposed merging and arbitration strategies relieve the problem of eyebrows and/or thick spectacular rims, which bother many eye detection and tracking algorithms. The experiments have been carried out using the BioID, FERET and Inha face data sets. We achieved very encouraging results, especially for the face images with glasses and uneven illumination. Detailed explanation of training images collection and methods are given in Section 2. The classifier architecture, merging and arbitration strategy are given in Section 3. In Section 4, the experimentation of the system is described. Conclusions and directions for future research are presented in Section 5.

2 The Outline of the Proposed Eye Location Method

In this session, the outline of the proposed eye location method is described. We assume that the located face images are upright frontal, however, it can be readily extended to other poses. It employs not only 2D Haar wavelet based Bayesian framework to relive the effect of uneven illumination but also sophisticated postprocessing. The proposed postprocessing, i.e. merging and arbitration strategies, solves partially the bothering problem of eyebrows and/or thick spectacular rims in previous approaches.

The proposed eye location method consists of the face location with multi-resolution (5 levels), the feature vector generations using 2D Haar wavelet transform, and the detection of eye candidate windows using the Bayesian classifier. 2D Haar wavelet transform allow system to be strong for illumination problem. There appear usually multiple eye candidate windows. The derived multiple eye candidate windows should be merged and arbitrated into a single eye window since there is only one eye in either left or right eye region. There are the effects of eyebrows and glasses frames, a simple merging and arbitration strategy nevertheless may gain its robustness. In this paper, the derived eye candidate windows are converted into eye candidate centroids. The final decision of the eye centroid among candidate centroids is by the stochastic arbitration strategy, which is based on the minimizing energy with image forces and probabilistic forces of eye candidate centroids. The block diagram of the proposed method is shown in Fig. 1.

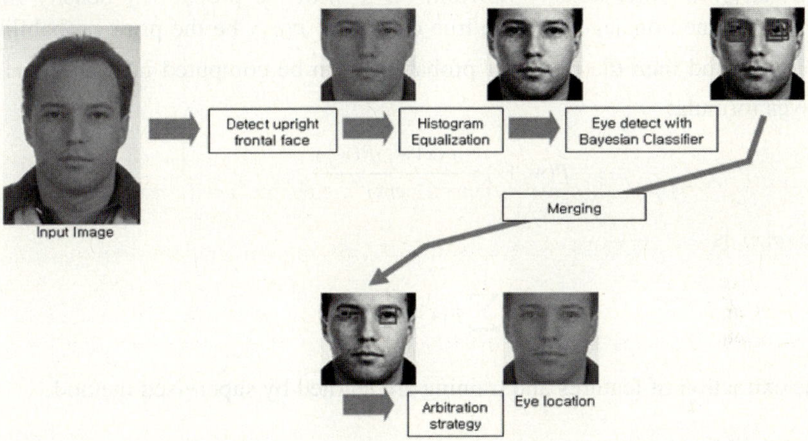

Fig. 1. A block diagram of the proposed eye location method

3 The Decision of Eye Candidate Window Using Haar Feature Based Bayesian Classifier

In this paper, we adopt 2D Haar wavelet representation for generating the feature vectors for eye windows [4]. Large number of eye images and huge number of non-eye images are needed to train the Bayesian classifier [1]. These samples are normalized to the same scale of 16×16. These samples are transformed to the feature vectors by 2D Haar wavelet transform, measured by Mahalanobis distance [8] using Bayesian discriminant method which PCA [7] is applied for reducing dimension. At least two class models [1] are required to classify eye and non-eye using Bayesian method: one is eye class model and the other is non-eye class model. We employ the two class models here.

3.1 Feature Generation Using 2D Haar Wavelet Transformation

We can also take advantage of the intelligence of frequency domain using 2D Haar wavelet transform. It is establish the eyes into bands, which are districted in orientation and frequency. In each sub-bands, each coefficient is spatially localized. We use a wavelet transform based on 4 level decomposition producing 13 sub-bands. In general, low frequency component has more discriminate power than higher. Also too high component has some noisy information.

3.2 Eye and Non-eye Class Modeling

In general, classification is the issue of predicting the class w of a sample x with probabilistic theories. The standard Bayesian method produces a posterior probability $P(w_j \mid x)$ i.e., the probability of the class w_j given that feature value x has been measured. Let w_1 and w_2 be the finite set of classes, and let X be the feature vector of eye. We adopt X as a d-dimension component vector and let $p(x \mid w_j)$ be the conditional probability density function for x with the probability density function for x conditioned on w_j being the true class. Let $P(w_j)$ be the prior probability that class is w_j, and then the posterior probability can be computed in terms $p(x \mid w_j)$ of by Bayes formula:

$$P(w_j \mid x) = \frac{p(x \mid w_j)P(w_j)}{p(x)} \tag{1}$$

where $p(x)$ is

$$p(x) = \sum_{j=1}^{c} p(x \mid w_j)P(w_j) \tag{2}$$

The extraction of features and training are learned by supervised method.

3.2.1 Eye Class Modeling

The posterior probability density of the eye class, w_e, is assumed (modeled) as a normal distribution [1]:

$$p(x \mid w_e) = \frac{1}{(2\pi)^{n/2} \mid \Sigma \mid^{1/2}} \exp\{-\frac{1}{2}(x - m_e)^t \Sigma_e^{-1} (x - m_e)\} \tag{3}$$

Where m_e and Σ_e are the mean and the covariance matrixes of eye class, w_e. The covariance matrix, Σ_e, can be factorized into the following by the principal component analysis (PCA).

Notion of distance is needed to construct righteous model. This paper refers to Mahalanobis distance instead of Euclidean distance. Mahalanobis distance is based on correlations between variables - different features - can be classified or tested. It is useful measure of determining similarity of an unknown test set to a known one. It is different from Euclidean distance in that it takes into compute the correlations of the data set. Formally, the Mahalanobis distance from a group of values with mean $m=(m_1,m_2,m_3,...,m_n)$ and covariance matrix Σ for a multivariate vector is defined as:

$$d(x)=\sqrt{(x-m)^t\,\Sigma^{-1}(x-m)} \tag{4}$$

It follows from (3) that the traces of points of steady density are hyper ellipsoids for which the form (4) is steady. The principal axes of these hyper ellipsoids are given by the eigenvectors of Σ ; the eigenvalues determine the length of these axes. Mahalanobis distance can be a good measure to Bayesian classifier, because it has a property of the mean and the covariance of data set as normal distribution.

3.2.2 Non-eye Class Modeling

Non-eye class modeling should be constructed during the training period. Practically any image can be offered as a non-eye sample because the domain of non-eye example is wider than the domain of eye example. However, selecting a "representative" set of non-eye is troublesome task. Non-eye class modeling starts with collecting random images those do not contain any human faces at all before the training is started, and non-eye images that similar to eyes are collected during the training in the following manner:

Step 1) Select an initial set of non-eye images by creating thousands of random images.

Step 2) Construct eye class modeling and train the system with eye images which are modeled, and get a threshold (θ) of true eyes.

Step 3) Test the selected non-eyes images in Step 1 by Bayesian classifier and collect images which return true output separately.

Step 4) Make a vector file made up of all theses incorrectly classified images. Get the maximum difference (τ) of distance of these images from θ . Now, we get two thresholds those are θ and τ .

3.3 Eye Candidate Window Using Haar Feature Based Bayesian Classifier

We employ Mahalanobis distance for each window. Let $d(x)_e$ be the Mahalanobis distance of the eye class, and $d(x)_n$ be that of for non-eye class. $d(x)_e$ and $d(x)_n$ can be calculated from the input pattern x , the eye class parameters (the mean eye, and the corresponding covariance matrix), and the non-eye class parameters (the mean non-eye, the corresponding covariance matrix) respectively. We use two thresholds, θ and τ as follows:

$$\begin{aligned}\theta &= \max(d(\alpha)_e)\\ \tau &= \max(d(\beta)_e - d(\beta)_n)\end{aligned} \tag{5}$$

where α is training samples of eye class and β is training samples of non-eye class. The two thresholds are constant values, which are calculated in the training time. The Bayesian classifier offers the classifying rule to the eye detection system as follows:

$$x\in\begin{cases} w_e & \textit{if } d(x)_e<\theta) \textit{ and } \left(d(x)_e-d(x)_n<\tau\right) \\ w_n & \textit{otherwise}\end{cases} \tag{6}$$

Note that there are some false locations; they will be eliminated by methods presented in the next section.

4 Merging and Arbitration Strategy for Deciding True Eye Centroid

The result from the Bayesian classifier usually contains some false and multiple detections. We employ merging and arbitration strategy to improve the credibility of the eye location by removing false detections. Extensive experiments discover that there is usually multiple candidate windows detected nearby eyes, while false detections usually arise with less frequency. This discovery gives us a reasonable decision rule that can eliminate false detections, and merge true detections into a single merged detection.

4.1 The Proposed Merging Strategy

For each eye centroid, the number of detected windows within the ambient of that position can be counted. If the number is greater than the specified number, the position is classified as the eye region. The result position is indicated by the centroid of the neighbor detections, therefore duplicated detections can be eliminated.

This method is good at not only improving accept rate but also at decreasing false detection. If a specific position is correctly classified as an eye, then all other detected positions, which overlap it, are regarded as errors. Therefore, these can be eliminated. The position with the higher number of detections is conserved, and the position with the lower detections is eliminated. This method is controlled by two variances: the density of overlapping and the size of spread out. We will only accept a detection if there are at least more than threshold detections within a region (spread out along x, y, and scale) in the detections phase. The size of the region is determined by a variable, which is the number of pixels from the center of the region to its edge. The proposed merging strategy is outlined as follows:

Step 1) The eye candidate windows are decided by the Haar based Bayesian classifier.

Step 2) The center of eye candidate window is calculated.

Step 3) Each center is spread out to size of virtual window.

Step 4) At each cluster of virtual windows, the density of overlapping virtual windows is counted, and the centroids of the virtual window in the cluster are collapsed into a single merged centroid.

Step 5) Only the merged centroid(s), with its density higher than merging threshold, survives.

Step 6) Each surviving merged centroid has expanded region. The region is called merged window. Size is defined by average size of eye candidate windows, which participate in merging. If there exists overlapping between merged windows, the merged window with the lower density will be eliminated. We call this step overlap elimination.

Each detected centroid at a particular position and scale is marked in image. Then, each position in image is replaced by the number of detections in a specified neighborhood of that position. This has the effect of "spreading out" the detections. Examples are shown in Fig. 2 and Fig. 3. Fig. 2 shows the case of simple merging strategy without overlap elimination process, and Fig. 3 is case of merging strategy with overlap elimination process.

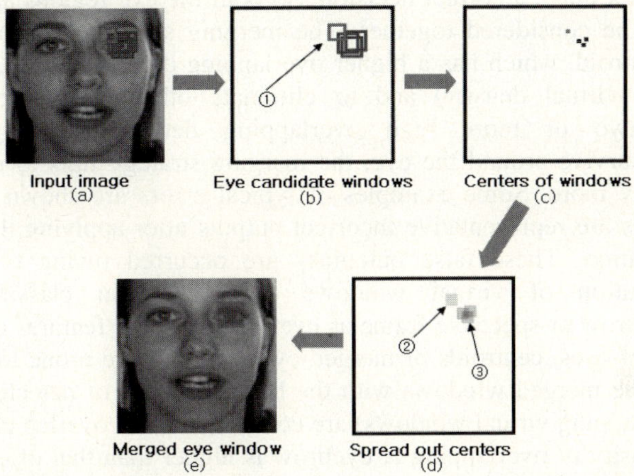

Fig. 2. The flow of merging strategy. (a) Input image. (b) Eye candidate windows by Bayesian classifier. ① is false detection. (c) The centers of eye candidate windows. (d) The centers of windows are "spread out" and a threshold is applied. If threshold is 2, ② is obliterated since density is lower than threshold.

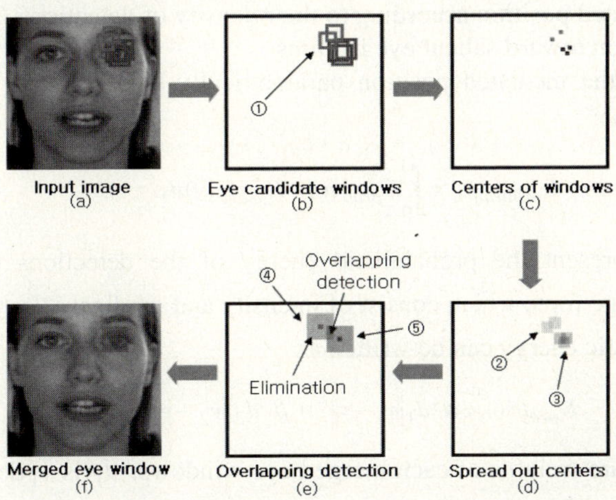

Fig. 3. The flow of merging and overlap elimination method. (a) Input image. (b) Eye candidate windows by Bayesian classifier. ① is false detections. (c) The center of eye candidate window. (d) The detections are "spread out" and a threshold is applied. If threshold is 2, ② is conserved because density is not lower than threshold. (e) Eliminating overlapping detections. ④ is eliminated since density of ⑤ is higher than density of ④.

4.2 The Proposed Arbitration Strategy

After the merging strategy is applied, sometimes two or more merged eye centroids appear. As a matter of fact, eyebrows or thick rims of spectacle often look so similar with closed eyes that the proposed Bayesian classifier with the merging strategy sometimes cannot make a correct decision. So both the eye regions and ambient eye regions should be considered together. The merging strategy enforces to remain a merged eye centroid, which has a higher overlapping detections (i.e., higher density of overlapping virtual density) and to eliminate other merged eye centroid(s). However, if two or more high overlapping detections (merged centroids) independently survive around the eye, the merging strategy fails to produce correct eye location any more. Some examples of typical errors are shown in Fig. 4. Red marks of images are representative incorrect outputs after applying the merging and overlap elimination. These false outcomes are occurred owing to their peculiar density distribution of virtual windows. The Bayesian classifier sometimes misclassify eyebrow or spectacle frame as eyes. Because the features of eyebrows are similar to that of eyes, centroids of merged eye windows are prone to appear around the eyebrow. The merged windows with the higher density of detections (i.e. higher density of overlapping virtual windows) are conserved at the overlap elimination step. Thus, if the density of overlapping at eyebrow is higher than that of eye, the merged eye windows around eyebrow are reserved but on the other hand detections around eye-center will be eliminated. The distribution of the multiple merged windows is analyzed to resolve the above problem.

Our arbitration strategy is a method, which finds a mediated position among false detections. The arbitration strategy is based on a minimizing energy function by probabilistic forces and image forces. The probabilistic forces serve to impose a reasonable mediated position according to their density of detections. The image force pushes the position toward salient eye features.

Representing the mediated position parametrically by $v = (x, y)$, we can use its energy function as:

$$E_{arbitration} = \int_0^1 E_{prob}(v) + E_{image}(v) ds \tag{7}$$

where E_{prob} represent the probabilistic energy of the detections due to density, E_{image} gives image force which consist of intensity and gradient.

The probabilistic energy can be written as:

$$E_{prob}(v) = \alpha \cdot d_1 \|v_1 - v\|^2 + \beta \cdot d_2 \|v_2 - v\|^2 \tag{8}$$

where α and β are weights of each merged eye windows, d_1 is v_1 density and v_1 is centroid position of first merged window, d_2 and v_2 are also those of second merged window. By adjusting the weights α and β, we can control the relative importance of the merged windows.

In order to make useful decision for eye location we need energy functions, which make fit position in center of true eyes. The image forces can be depicted as a weighted union of the two energy functions.

$$E_{image} = w_{int} E_{intensity} + w_{grad} E_{gradient} \tag{9}$$

By adjusting the weights, various eye positions can be created. The simplest useful image function is the intensity of itself. $E_{intensity}$ is defined as:

$$E_{intensity} = I(x, y) \tag{10}$$

Depending on the sign of w_{int}, the position will be attracted either to light region or dark region. w_{int} must be positive, because we want position to fit center of eyes.

In general, because gradient energy function is used to find edge, it has negative sign, but we use it as an opposite concept. Because we endeavor to find region, which has quite small gradients, $E_{gradient}$ is depicted as follows:

$$E_{gradient} = \|G(I(x, y))\|^2 \tag{11}$$

where $G(I(x, y))$ is gradient around position (x, y). The energy functions introduction should be scaled so that each component of function contains comparable values. This process is referred to as regularization. Each of the energy function is adjusted to the range [0, 1] as follows:

$$e'_{xy} = \frac{e_{xy} - e_{min}}{e_{max} - e_{min}} \tag{12}$$

where e_{min} and e_{max} are the minimum and maximum valued elements, respectively, in each energy function component.

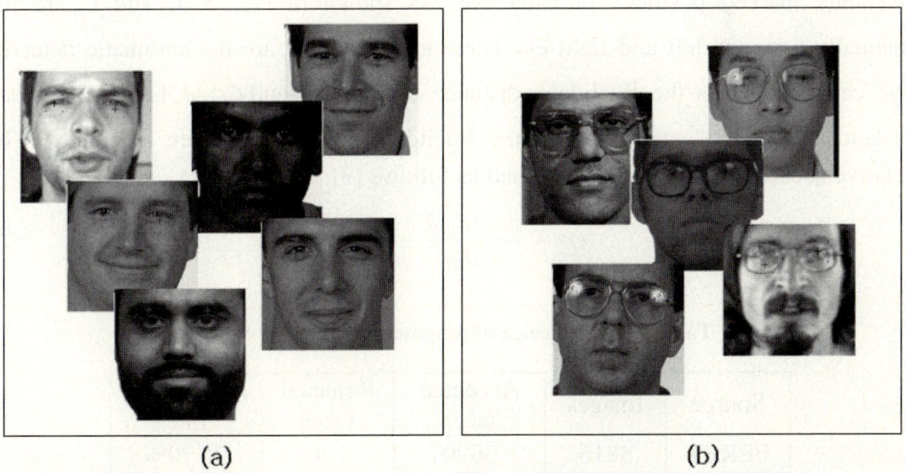

Fig. 4. The distribution of multiple centroid patterns: (a) type 1 false merged windows, which are influenced by eyebrows, (b) type 2 false merged windows, which are influenced by spectacle frames

5 Experiments

The training set is obtained from FERET database, and totally 1970 eyes of 985 faces are extracted and normalized for training. Experimental test set consists of BioID (1521 images), FERET (3816 images), and Inha (1200 images) data sets, and totally 6537 face images are used in our experiment. These three test data sets are obtained from diverse sources to cover diverse eye variations in view angles, sizes, illumination, and glasses.

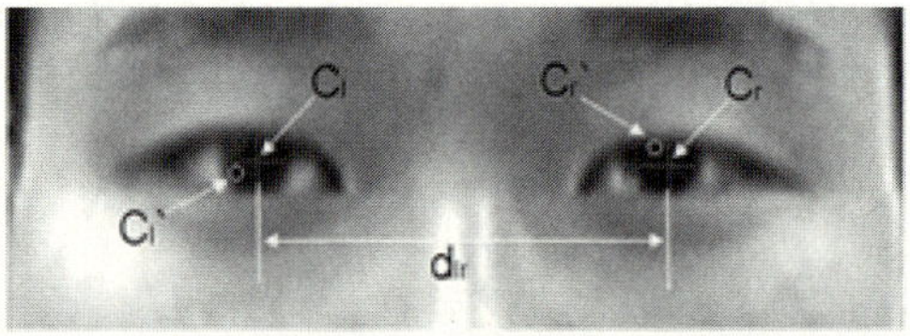

Fig. 5. An illustration of the centroids of left and right eyes

There are several methods to measure the accuracy of eye location used in previous research [6] [3]. In this paper, we adopt a scale independent localization measurement, called relative accuracy of eye location, to estimate the accuracy of eye location [3]. The ideal eye location is represented by the centroid of eye region. The centroid of eye region is defined as the center of the pupil of the located eye (see Fig. 5). The relative accuracy measurement of located eye compares the automatic located eye centroid, which produces results of the proposed system with the manually marked positions of each eye. As shown in Fig. 5, C_l and C_r are the manually assigned left and right eye centroids, $C_l^{'}$ and $C_r^{'}$ are the automatic detected eye centroids. d_l is the Euclidean distance between $C_l^{'}$ and C_l, d_r is the Euclidean distance between $C_r^{'}$ and C_r. d_{lr} is the Euclidean distance between d_l and d_r. The relative accuracy of detection is defined as follows [3]:

$$err = \frac{\max(d_l, d_r)}{d_{lr}} \tag{13}$$

Table 1. Performance of proposed method($err < 0.14$)

Source	Images	Accepted faces	Rejected faces	Acceptance rate
FERET	3816	3690	126	96.70%
BioID	1521	1458	63	95.86%
Inha	1200	1148	52	95.67%
Total	6537	6296	241	96.31%

The eye location acceptance rates of the system under *err* <0.14 is shown in Table 1. The average processing time per face of system on an AMD Barton 2500+ PC system is 58 ms without special code optimization. In Fig. 6, we offer error curves of three test databases. From the Figure, this system can show satisfactory performance in the vicinity of *err* <0.14.

Table 2. Comparative Eye Location Performance of the proposed method and others

Method	*err* < 0.2 (all dataset)	*err* < 0.25 (BioID)
Y. Ma [6]	99.1%	N/A
H. Zhou [5]	N/A	95.86%
The proposed method	99.53%	98.82%

To make a comparison with our detection rate for the same *err* rates reported by other systems, as in Y. Ma [6] and H. Zhou [5], we select these papers as targets of comparison because they use identical evaluation protocol (*err*) and datasets with us. We achieve better performance than the others as we adopt merging and arbitration strategy. Table 2 lists the detection rates of *err* rates for our system as well as for other published systems as reported in [5], [6]. It can be observed that our method compares favorably with the others, especially for low err rates. This shows that our approach separates eye and non-eye region in a robust and balanced way. In paper [6], the detection rate is 99.1% under *err* <0.20. On the other hand, our detection rate is 99.53% under *err* <0.20 at all test sets. In paper [5], the detection is considered to be correct if *err* <0.25.

Their detection rate on BioID dataset is 94.81%. We evaluate method 2 on BioID under the same evaluation protocol. The detection rate of our system is 95.86% if *err* <0.14, and the detection rate is 98.82 % if *err* <0.25. And their system achieve on 97.18% of JAFFE data set. But backgrounds and illumination conditions of JAFFE are not as complex and diverse as these of BioID.

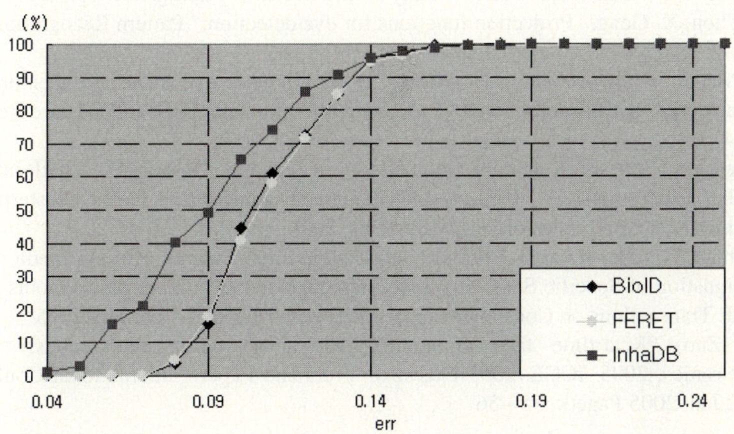

Fig. 6. Cumulative distribution of localization errors on three dataset

6 Conclusion

This paper describes a novel merging and arbitration strategy for robust eye location. The method, which is trained on images from only a portion of one database, yet works on test images from diverse sources, displays robust generalization performance. The novelty of this paper comes from the combination of the 2D Haar based Bayesian classifier, the statistical modeling of eye and non-eye classes, and the Arbitration strategy for growing performance. First, the system use 2D Haar wavelet representation, which is appropriate for 2D image. Second, statistical modeling evaluates the conditional probability density functions of the eye and non-eye classes. Finally, Arbitration strategy is applied to improve accuracy of system. The Arbitration strategy applied Bayesian classifier is trained with 1970 eye images and 3844 random natural (non-eye) images. Experimental results using 6537 images (containing a total of 13074 eyes) from various image sources. The novel Arbitration strategy applied Bayesian classifier achieves 96.31 percent eye detection accuracy under $err < 0.14$.

In addition, because the Arbitration strategy and the Bayesian discriminant method doesn't localize for case of eye, the proposed method in this paper can be applied for location of other face organs such as nose or mouth.

References

[1] C. Liu, "A Bayesian Discriminating Features Method for Face Detection" IEEE Trans. Pattern Analysis and Machine Intelligence, vol. 25, no. 6, pp. 725-740, 2003.

[2] C. Liu and H. Wechsler, "Gabor Feature Based Classification Using the Enhanced Fisher Linear Discriminant Model for Face Recognition," IEEE Trans. Image Processing, vol. 11, no. 4, pp. 467- 476, 2002.

[3] O. Jesorsky, K. Kirchberg, R. Frischholz, "Robust face detection using the Hausdorff distance," In: J. Bigun, F. Smeraldi Eds. Lecture Notes in Computer Science 2091, Berlin: Springer, 2001, pp.90-95.

[4] Henry A. Rowley, Shumeet Baluja, Takeo Kanade, "Neural Network-Based Face Detection," IEEE Trans. Pattern Analysis and Machine Intelligence, vol. 20, NO. 1, 1998

[5] H. Zhou, X. Geng, "Projection functions for eye detection," Pattern Recognition, 2004, in press.

[6] Y. Ma, X. Ding, Z. Wang, N. Wang, "Robust precise eye location under probabilistic framework," IEEE International Conference on Automatic Face and Gesture Recognition, 2004.

[7] Hidetomo Ichihashi, Katsuhiro Honda, Noboru Wakami, "Robust PCA with Intra-sample Outlier Process Based on Fuzzy Mahalanobis Distances and Noise Clustering," IEEE International Conference on Fuzzy Systems, 2005.

[8] Akihiko Watabe, Kazumi Komiya, Jun Usuki, Kayo Suzuki, Hiroaki Ikeda, "Effective Designation of Specific Shots on Video Service System Utilizing Mahalanobis Distance," IEEE Transactions on Consumer Electronics, Vol. 51, No. 1, February 2005

[9] Fei Zuo, "Real-time face recognition for smart home applications," Consumer Electronics, 2005. ICCE. 2005 Digest of Technical Papers. International Conference on 8-12 Jan. 2005 Page(s):35 - 36

[10] Yongsheng Gao, Leung, "Face recognition using line edge map," Pattern Analysis and Machine Intelligence, IEEE Transactions on Volume 24, Issue 6, June 2002 Page(s):764 - 779

[11] J. Huang, X.H. Shao, H. Wechsler. "Pose Discrimination and Eye Detection Using Support Vector Machines," Proceeding of NATO-ASI on Face Recognition: From Theory to Applications, 1998.

[12] P.Viola, M. Jones, "Rapid object detection using a Boosted cascade of simple features," Proc. of IEEE Conf. on CVPR, pp. 511 –518, 2001.

[13] H.Schneiderman, T.Kanade. "A statistical model for 3D object detection applied to faces and cars," Proc. of IEEE Conf. on CVPR, 2000.

[14] T.Kawaguchi, D.Hikada, and M.Rizon, "Detection of the eyes from human faces by hough transform and separability filter," Proc. of ICIP, pp.49-52, 2000.

[15] S.Baskan, M. M. B., V.Atalay, "Projection based method for segmentation of human face and its evaluation," Pattern Recognition Letters 23, 1623-1629, 2002.

[16] F.Smeraldi, J.Bigun, "Retinal vision applied to facial features detection and face authentication," Pattern Recognition Letters, Vol. 23, 2002: 463-475.

[17] S.Lucey, S.Sridharan, V.Chandran, "Improved facial feature detection for AVSP via unsupervised clustering and dicriminant analysis," EURASIP Journal on Applied Signal Processing, vol 3, pp.264-275, 2003.

Head Gesture Recognition Using Feature Interpolation

Yeon Gu Kang and Phill Kyu Rhee

Dept. of Computer Science & Engineering Inha University,
253 Yong-Hyun Dong, Nam-Gu, Incheon, South Korea
cky@inhatc.ac.kr, pkrhee@inha.ac.kr

Abstract. This paper addresses a technique of recognizing a head gesture. The proposed system is composed of eye tracking and head motion decision. The eye tracking step is divided into face detection, eye location and eye feature interpolation. Face detection obtains the face region using integrated feature space. Multiple Bayesian classifiers are employed for selection of face candidate windows on integrated feature space. Eye location extracts the location of eyes from the detected face region. Eye location is performed at the region close to a pair of eyes for real-time eye tracking. If a pair of eyes is not located, the system can estimate feature vector using mean velocity measure(MVM). After eye tracking, the coordinates of the detected eyes are transformed into the normalized vector of the x-coordinate and the y-coordinate. Head gesture recognition using HMMs. Head gesture can be recognized by HMMs those are adapted by a directional vector. The directional vector represents the direction of head movement. The HMMs vector can also be used to determine neutral as well as positive and negative gesture. The experimental results are reported. These techniques are implemented on a lot of images and a notable success is notified.

1 Introduction

Face and gesture are considered as a way of communication among people for a long time. For communication between person and computer, many researchers have studied face processing such as face tracking, face detection, recognizing face and facial expression, lip reading, eye blinking, etc. Therefore, many algorithms and techniques are invented, but it remains a difficult problem yet. The automatic face processing becomes a significant topic towards developing an effective visual HCI(Human-Computer Interface)[1].

In a system developed by Morimoto[2], movements in the facial plane are tracked by evaluating the temporal sequence of image rotations. These parameters are processed by a dynamic vector quantization scheme to form the abstract input symbols of a discrete HMM which can differentiate between four head gestures (*yes, no, maybe* and *hello*). Based on the IBM PupilCam technology, Davis[3] proposed a real-time approach for detecting user acknowledgments. Motion parameters are evaluated in a finite state machine which incorporates individual timing parameters. Using optical flow parameters as primary features. Cascia, Isidoro, and Sclaroff[4] proposed a novel

B. Gabrys, R.J. Howlett, and L.C. Jain (Eds.): KES 2006, Part I, LNAI 4251, pp. 582 – 589, 2006.
© Springer-Verlag Berlin Heidelberg 2006

method for 3D head tracking. Tang[5] proposed an algorithm for head gestures recognition. They formulate tracking in terms of image registration in the texture map of a 3D surface model. They got very good performance from images taken with low quality. Among head gestures, nodding and shaking can be used as a gesture to fulfill a semantic function, to communicate emotions and as conversational feedback. A system that could detect head nods and head shakes would be an important component in an interface that is expected to interact naturally with people[6][7].

In this paper, we discuss the techniques of eye tracking and head gesture recognition. We use the 54 sub-regions, Haar wavelet transform, and HMMs. Fig. 1. shows the description of the proposed system.

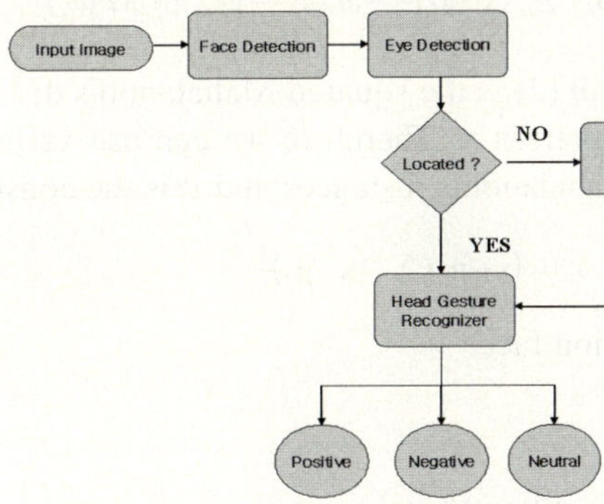

Fig. 1. The architecture of the proposed system

The organization of this paper as follows: section 2 presents eye tracking, section 3 discusses head gesture recognition, section 4 illustrates the experimental results. Finally conclusion is resided

2 Eye Tracking

Eye tracking is a repetitive operation of eye location. Eye location is performed in a detected face region. It is performed in the region close to a pair of eyes to save eye location time. If a pair of eyes is not located, face detection is performed again, and eye location is repeated. Preprocessing steps that enhance the quality of an image are required.

2.1 Face Detection

The printing Feature extraction method has much affect in face detection performance. The feature extraction method that was optimized by an experiment, as a result, raises good performance. Therefore, we can improve performance through efficient feature extraction and dimension decrease. We use two types of feature extraction method. One uses the 54 sub-regions. Other is the method using Haar wavelet transform. We use 3 level decomposition of Haar wavelet transform. [8]

We assume that distributions of face and nonface samples form the shape of Gaussian distribution. The likelihood of face and nonface can be modeled as multivariate normal distribution as (1).

$$p(\mathbf{x} \mid \omega_i) = \frac{1}{(2\pi)^{d/2} |\boldsymbol{\Sigma}_i|^{1/2}} \exp\left[-\frac{1}{2} (\mathbf{x} - \boldsymbol{\mu}_i)^t \boldsymbol{\Sigma}_i^{-1} (\mathbf{x} - \boldsymbol{\mu}_i) \right] \cdot \tag{1}$$

$\mathbf{x}$ is the d-dimensional feature, Σ_i is the covariance matrix of the samples, and $\boldsymbol{\mu}_i$ is the d-dimensional mean vector. Then two discriminant functions have the following form:

$$g_i(\mathbf{x}) = -\frac{1}{2} (\mathbf{x} - \boldsymbol{\mu}_i)^t \Sigma_i^{-1} (\mathbf{x} - \boldsymbol{\mu}_i) - \frac{d}{2} \ln 2\pi - \frac{1}{2} \ln |\Sigma_i| + \ln P(\omega_i) \cdot \tag{2}$$

First term of the formula in (2) is the squared Mahalanobis distance. And rest terms are not dependent of input pattern $\mathbf{x}$. Therefore we can use variable r and α to simplify the formula. r is the Mahalanobis distances and α is the constant.

$$r_i^2 = (\mathbf{x} - \boldsymbol{\mu}_i)^t \Sigma_i^{-1} (\mathbf{x} - \boldsymbol{\mu}_i) \cdot \tag{3}$$

If r > threshold recognition face.

2.2 Eye Location

When an image contains an object having homogeneous intensity and a background with a different intensity level, the image can be segmented in two regions by using image segmentation by thresholding[9]. The following equation is the definition of segmentation adopted in this paper.

$$g(x, y) = \begin{cases} 1 & \text{if } e(x, y) > T \, . \\ 0 & \text{other wise} \end{cases} \tag{4}$$

where T is threshold whose choice can be based on the image histogram.

The gray level of facial components such as hair, eyebrow, pupil, nostril, lip, etc. is darker than the gray level of the skin. The threshold T can be obtained using this property. The threshold is used to discriminate pixels belonging to eye regions.

The labeling algorithm examines a state of connection between a pixel and its 8-neighborhood pixels, and labels the objects in the image. The labeling algorithm is used to obtain connected and isolated regions[10]. The labeled facial components are recorded as eye candidates.

The information obtained by the labeling process as well as the information of face region is used to extract a pair of eyes. A pair of eyes is detected by utilizing the heuristic rules based on eye's and face's geometrical characteristics. The width of eye candidate has relations with the width of face candidate. The distance between the left eye and the right eye is used as important geometrical characteristic.

Heuristic rules based on eye's geometrical information are defined as follows. k_x is heuristically predefined constants. The eye candidate best matching with these rules is determined on a pair of eyes. The evaluation performance of eye location is described in the section of experimental results.

Rules for determining a pair of eyes are follows:

1) The area of eye candidate $\times k_1 \leq$ the number of pixels in the face region $\leq$ the area of eye candidate $\times k_2$

2) The width of face $\times k_3 \leq$ the width of eye candidate $\leq$ the width of face $\times k_4$

3) The width of eye candidate $\times k_5 \leq$ the height of eye candidate $\leq$ the width of eye candidate $\times k_6$

4) The distance between the left eye and the right eye $\geq k_7$

2.3 Eye Feature Interpolation

To obtain accuracy of feature vector sequence variation when the feature vector generation module failed in tracking appropriate feature, we referenced mean velocity of feature vector. The system can estimate feature vector $\mathbf{p}_t$ at time t using mean velocity measure(MVM). The MVM from time t to $t+k$ has window scale factor k and velocity $\mathbf{d}$ and is shown in equation.

$$a_{(t,t+k)} = \frac{1}{t+k}\sum_{i=t}^{k}\mathbf{d}_i \cdot \tag{5}$$

The velocity $\mathbf{d}$ is defined as:

$$\mathbf{d}_t = \mathbf{p}_t - \mathbf{p}_{t-1} \cdot \tag{6}$$

We can determine the feature vector $\mathbf{p}_t$ which could not be obtained from eye location module using the equation.

$$\mathbf{p}_t = \mathbf{p}_{t-1} + a_{(0,t)} - \mathbf{p}_0 \cdot \tag{7}$$

The feature vector variation of head gesture has sinusoid-like-pattern because the head moves up/down or left/right in positive, negative respectively. If the MVM applied in this pattern we cannot have accurate $\mathbf{p}_t$. To solve this problem, the peak of feature vector sequence $\mathbf{p}_{\max}$ is needed because a must be recalculated to model direction of acceleration. We calculate the feature vector $\mathbf{p}_{\max}$ as equation.

$$\mathbf{p}_{\max} = \begin{cases} \text{if } a_{(t,t+k)} < 0 \text{ then } \mathbf{p}_{\max} \\ \text{if } a_{(t,t+k)} \geq 0 \text{ then continue} \end{cases} \cdot \tag{8}$$

Since the average failure of eye location was $k < 3$ in experiment, the proposed model could be applied with satisfaction.

3 Head Motion Decision

After eye tracking is performed, the coordinates of eye are transformed into a vector suitable for head gesture recognition. Head gesture is recognized by direct observation of the variation of the vector. HMMs are used to absorb a few errors which can be generated in eye tracking process, for more accurate recognition.

3.1 Head Gesture Patterns

There exists much difference among people's head motions such as the start-point, the end-point, the distance of head movement, etc. If a system requires a user response, the system acquires consecutive images from video camera for 1~2 seconds. In the system, three head gestures, i.e. positive, negative, and neutral gesture, are recognized. The positive response is to nod the head, and the negative to shake the head. Other head gestures are neutral. After performing eye location, the system calculates a vector representing the user's head gesture. Owing to the difference of people's head motions, the vector must be normalized. The normalized vector can be denoted in the charts with the x-axis, the y-axis, and the time-axis as follows. When a user responds positive gesture, the normalized vector of the x-coordinate is distributed in the 0~63 range value.

Positive gesture, nodding head, has a little variation of the x-coordinate, but much variation of the y-coordinate. Negative gesture, shaking head, has much variation of the x-coordinate, but a little variation of the y-coordinate. If there exists much variation of the x-coordinate and the y-coordinate at the same time, it can be considered as a neutral gesture.

3.2 Head Gesture Recognition

The average of the normalized vector of the x-coordinate or the y-coordinate is used to roughly determine positive or negative head gesture. As comparing the average of the x-coordinate with that of the y-coordinate, we can know that the average of which coordinate is higher. A vector with higher average has sinusoid pattern. Simple recognition of head gesture is possible by determining which one has sinusoid pattern.

Two left-right models of HMMs with 64 symbols and 4 states are used to produce more accurate recognition result. The input to HMMs is a normalized vector obtained. These are trained by the Baum-Weiss procedure[11]. One recognizes sinusoid pattern of the vector; the other recognizes irregular linear-like pattern of the vector. Two HMMs are trained respectively with appropriate pattern, i.e sinusoid or irregular linear-like pattern. If the vector at the x-axis has sinusoid pattern and the vector at the y-axis has irregular linear-like pattern, the system recognizes a user acting negative head gesture.

One of HMMs will output a result of recognition of sinusoid pattern or irregular linear-like pattern of vector. If HMMs recognize the sinusoid pattern, we can believe the fact that the selected axis by the first method has activity of gesture. Suppose that the x-axis is selected, we now have strong confidence that the head gesture of a user must be an intention of negative. Max selector selects HMM-x or HMM-y. If the output value of HMM-x or HMM-y is less than the predefined threshold, the system outputs neutral gesture. Otherwise, according to the result of the comparison of HMM-x value with HMM-y value, the system reports positive or negative gesture.

It is difficult for the previous methods to recognize neutral gesture, because neutral gesture is recognized based on the threshold only. This paper proposes another method using HMMs to clearly recognize neutral gesture. These are designed using the directional vector of the center of two eyes as input. The HMMs generate the appropriate result of head gesture.

The movement of the face is converted into 8-directional vector codes. The number stands for the code representing the direction of face movement. The directional codes from 1 to 8 indicate that a face moves to each direction, the directional code 0 indicates that a face does not move at all. 8 directional codes obtained from the result of video image sequences are converted into a vector code. The vector code is inputted into the HMMs to decide the negative, positive, or neutral gesture of the user. Fig. 2. shows the block diagram of the head gesture recognition.

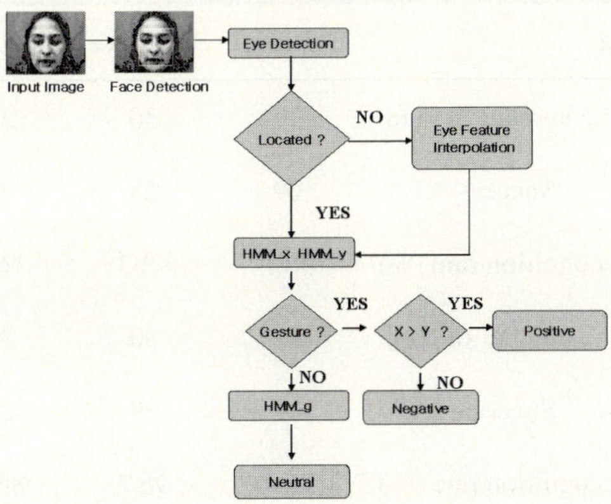

Fig. 2. The block diagram of head gesture recognition

4 Experimental Results

Experiments have been performed with sequential images which are captured from video camera. The proposed algorithm is implemented using MFC(Microsoft foundation class) at Pentium Ⅳ 2.4GHz processor having Windows XP as operating system and compiler was Visual C++ 6.0. Experimental images were 320×240 size and 256 gray levels.

Table 1. shows the result of eye location. For 35 people, 400 images are captured and considered for the experiment. In this experiment, the factor of spectacles was considered and experimental images were classified by the factor of spectacles. Therefore, the experimental data for head gesture recognition is captured under good illumination.

Table 1. The result of eye location

Spectacles	The number of image	Without MVM		With MVM	
		Success	Successful rate	Success	Successful rate
Not Wearing	247	232	93.92%	243	98.38%
Wearing	153	139	90.84%	143	93.46%
Total Sum	400	371	92.75%	386	96.50%

We use 6 training data for HMM learning for positive, negative, and neutral gestures. Each data contains 30 frames. We use 90 data for testing our system. Table 2 and Fig. 3 show the result of head gesture recognition methods.

Table 2. The result of head gesture recognition methods

Data		Positive	Negative	Neutral	Total
Without MVM	The number of data	30	30	30	90
	Success	29	28	23	80
	Recognition rate (%)	96.7	93.3	76.7	88.9
With MVM	The number of data	30	30	30	90
	Success	29	29	26	84
	Recognition rate (%)	96.7	96.7	86.7	93.3

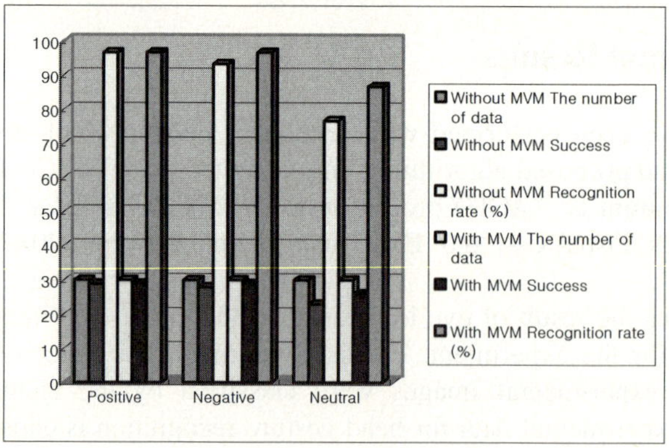

Fig. 3. The result of head gesture recognitions

5 Concluding Remarks

This paper addresses head gesture recognition system consisting of face detection, eye location, and HMMs. Due to the 54 sub-regions and Haar wavelet transform for detecting a face, the proposed system runs fast. Since HMMs absorb some errors of eye location, very encouraging experimental result is obtained. Thus, this recognition system will be able to effectively applied to visual user interface. One weak point of our system is that when a user moves his face right and left, recognition result is negative. Likewise, when moving his face up and down, recognition result is positive. Other features of face and face rotation are observed to overcome this fault. Another weak point is that a pair of eyes of an extremely tilted face cannot be located in many cases. It is necessary to focus positive area rather negative.

References

1. P.Ekman, T.Huang, T.Sejnowski, and J. Hager, Final Report to NSF of the planning Workshop on Facial Expression Understanding, Technical report, National Science Foundation, Human Interaction Lab., UCSF, CA 94143, (1993)
2. Morimoto, C., et al.: Recognition of head gestures using hidden markov models. Proc. of IEEE Int. Conf. on Pattern Recognition (1996)
3. Davis, J., et al.: A perceptual user interface for recognizing head gesture acknowledgements. In WS on Perceptive User Interfaces (PUI 01) (2001)
4. Marco La Cascia, John Isidoro, and Stan Sclaroff.: Head Tracking via Robust Registration in Texture Map Images, IEEE Conputer Sociey Conference on Computer Vision and Pattern Recognition, Santa Barbara, California, pp.508-514, June, (1998)
5. Jinshan Tang and Ryohei Nakatsu.: A Head Gesture Recognition Algorithm, ICMI 2000, LNCS 1948, pp.72-80, (2000)
6. Ashish Kapoor, Rosalind W. Picard.: A Real-Time Head Nod and Shake Detector, Proceedings from the Workshop on Perceptive User Interfaces, November (2001)
7. Wenzhao Tan, Gang Rong: A real-time head nod and shake detector using HMMs, Expert Systems with Applications 25, pp.461-466 (2003)
8. Mi Young Nam and Phill Kyu Rhee.: An Efficient Face Location using Integrated Feature Space, KES 2005, LNAI pp.327-335, (2005)
9. I. Pitas.: Digital image processing algorithms, Prentice Hall, (1993)
10. R. L. Lumina, G. Shapiro, and O. Zuniga.: A New Connected Components Algorithm for Virtual Memory Computers, Computer Vision, Graphics, and Image Processing, Vol. 22, pp. 287-300, (1983)
11. L. Baum, T. Petrie, S. G, and N. Weiss.: A maximization techniques occurring in the statistical analysis of probabilistic functions of Markov chains, Ann. Math. Stat., 41(1): 164-171, (1970)

A Stereovision Based Advanced Airbag System

Seok-Joo Lee[1,3], Yong-Guk Kim[2], Min-Soo Jang[1],
Hyun-Gu Lee[1], and Gwi-Tae Park[1,*]

[1] Dept. of Electrical Engineering, Korea University, Seoul, Korea
*gtpark@korea.ac.kr
[2] School of Computer Engineering, Sejong University, Seoul, Korea
ykim@sejong.ac.kr
[3] Hyundai Autonet Co. Ltd., Korea
gidung@haco.co.kr

Abstract. Occupant classification in the car is an essential issue for an advanced airbag system. The present paper describes a stereovision based occupant classification system (OCS) within an embedded system, by which triggering of the airbag deployment can be intelligently controlled. The embedded system consists of dual Digital Signal Processors; one is for stereo matching algorithm and the other is for calculating an SVM algorithm for the OCS. Performance was evaluated using our stereo image database. Results suggest that the system is satisfactory as an embedded OCS system.

1 Introduction

Introduction of airbag into automobiles has significantly improved the safety of the occupants. Unfortunately, report indicates that airbag itself can also cause fatal injuries if the occupant is a child smaller than a 6 year old.

In response to this, the National Highway Transportation and Safety Administration (NHTSA) has mandated starting in the 2006 model year all automobile be equipped with an automatic suppression system to detect the presence of child or infant and suppress the airbag. For this, in May 2001 the U.S NHTSA defined the Federal Motor Vehicle Safety Standard (FMVSS) No. 208 that mandates automatic airbag suppression when an occupant smaller than a 6 year old child is in the passenger seat [1].

For protection of child sitting on the passenger seat, an automaker may employ either of two options: (a) Dynamic automatic suppression (DAS) feature or (b) Low risk deployment (LRD) feature that does not completely disable the airbag but ensures that the airbag deploys in a way that minimizes the risk of airbag-inflicted injuries. But current airbag technologies are not sufficiently advanced to satisfy all requirements for low-risk deployment. So automakers adopt automatic suppression featured airbag system. It means that the system must have an Occupant Classification System (OCS) that enables it to determine the class of a passenger. Therefore the airbag suppression problem can be defined as a 4-class problem as table 1.

B. Gabrys, R.J. Howlett, and L.C. Jain (Eds.): KES 2006, Part I, LNAI 4251, pp. 590 – 597, 2006.
© Springer-Verlag Berlin Heidelberg 2006

There exist several OCS technologies for satisfying occupant crash protection regulations. For example, using the sensor mats installed in the seat cushion which measure the two dimensional pressure profiles in the seat area. The electric filed sensing system detects the relative mass and position of occupant. Another example is using strain gauge installed under the seat which measures the weight of occupant. But these systems have limitations for accurate occupant classification.

Table 1. Classes of Passenger

Class	Name	Description
0	RFIS	Rear Facing Infant Seat
1	Child	<25.6kg in weight & <124.5cm tall
2	Adult	> 46.7 kg in weight & >139.7cm tall
3	Empty	Empty seat

Many automotive electronic supplier and engineers predict that the camera will be the next generation sensor for active (e.g. pre-crash detection, lane departure recognition, night vision, drowsy warning, and etc.) and passive (e.g. OCS, occupant pose recognition, pedestrian detection, and etc.) car safety system. Some powerful advantages of vision sensor, such as diverse information about target object, high accuracy, and low restriction of installation are that reasons. Especially, for passive safety system, they spend their effort to develop classification algorithm and embedded system.

Early research, we present a new system that consists of stereovision and occupant classification system using a SVM (Support Vector Machine) classifier [3, 4]. For the stereovision system, the fast SAD algorithm is adopted to calculate the disparity map of the stereo images. As the disparity map image contains 3-D information of the occupant, it is possible to classify the class of the occupant by analyzing the disparity map image.

In this paper, we describe stereovision based OCS and intelligent algorithm with embedded system. The embedded system consists of dual DSP, one is for stereo matching algorithm and the other is for SVM algorithm calculation. The paper consists of several sections as follow. The upgraded OC algorithms are discussed in section 2. The embedded system is described in section 3. Result of experiments is reported in section 4. Finally, in section 5 summarize the result and the performance of the whole system is discussed.

2 Occupant Classification

Figure 1(a) shows an overall sequence of the occupant classification. First, we generate a disparity map image by using a fast sum of the absolute difference (SAD) algorithm. There are much unnecessary information, such as dashboard, windows and door, in the disparity map image. To eliminate the unnecessary information, we define a cubic model, whose origin is the right bottom point of the passenger seat and apply it to the disparity map image, as shown in figure 1 (b). Finally, the SVM classifier takes the disparity map image as an input and classifies. Since the disparity map of an adult and a child are similar compared to other cases, we compose the SVM classifier with two stages: in the first stage, it classifiers the occupant into two classes, such as the adult or the child as the first class, the RFIS or the empty seat as the second class; in the second stage, the first class case is classified as either the adult or the child, and the second class case is classified as either the RFIS or the empty seat.

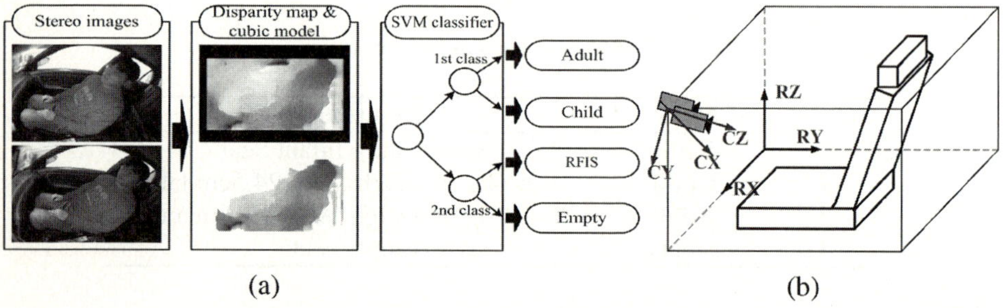

Fig. 1. Operation sequence of occupant classification (a) and the coordinate system transformation:[cx, cy, cz]→[rx, ry, rz] (b)

2.1 Stereo Matching Algorithm

We use the SAD algorithm but other similarity measure algorithms (e.g. Sum of Squared Difference (SSD) or Normalized Cross Correlation (NCC)) could be used, because the SAD algorithm is comparatively simple to other algorithms. The similarity measure of the SAD algorithm is as follows:

$$SAD(x, y, d) = \sum_{i,j=-n}^{n} \left|L(x+i, y+j) - R(x+d+i, y+j)\right| \tag{1}$$

The most expensive task performed by the SAD algorithm is the computation of SAD scores, which are needed to carry out the direct matching phase. To avoid the redundant calculation, we adopt a fast SAD algorithm [6].

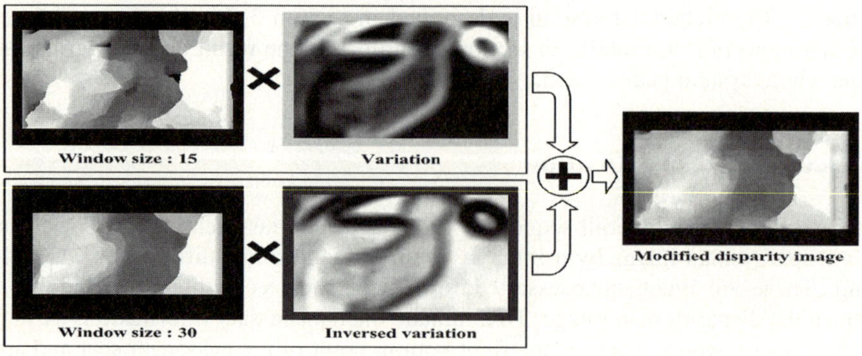

Fig. 2. Dual window stereo matching

The size of the matching window affects the quality of the disparity map image. The larger the size of the matching window, the disparity map image is more blurred. In the opposite case, the disparity map shows detail information but poor quality in non-patterned area. To improve the matching window problem, a dual window stereo matching technique is adopted. As shown in figure 2, we accomplish the stereo matching algorithm by using two different sizes of the matching window (15, 30) and adopt the disparity value by the variation of the image. The equation is as follow:

$$d = d_s \frac{\sigma}{255} + d_l \frac{255-\sigma}{255} \tag{2}$$

where σ is variation of the input image, d_s is disparity value by small size window, d_l is disparity value by large size window, and d is final disparity value. For calculating the SAD score for large window, we use the SAD score for small window again, as shown in equation 3.

$$\begin{aligned} SAD_L(x, y) &= SAD_S(x-n, y-n) + SAD_S(x+n, y-n) \\ &+ SAD_S(x-n, y+n) + SAD_S(x+n, y+n) \end{aligned} \tag{3}$$

2.2 Support Vector Machine as a Classifier

The fundamental idea of SVM is to construct a hyperplane as the decision line, which separates the positive (+1) classes from the negative (-1) ones with the largest margin [2, 7]. In a binary classification problem, let us consider the training sample $\{(\boldsymbol{x}_i, d_i)\}_{i=1}^{N}$, where $\boldsymbol{x}_i$ is the input pattern for the i-th sample and d_i is the corresponding desired response (target output) with subset $d \in \{-1,+1\}$. The equation of a hyperplane that does the separation is

$$\boldsymbol{w}^T \boldsymbol{x} + b = 0 \tag{4}$$

where $\boldsymbol{x}$ is an input vector, $\boldsymbol{w}$ is an adjustable weight vector, and b is a bias.

The margin, distance of the nearest point to the typical hyperplane, equals to $1/\|\boldsymbol{w}\|$. So, the problem turns into a quadratic programming:

$$\begin{aligned} &\text{minimize} \quad \frac{1}{2}\|\boldsymbol{w}\|^2 \\ &s.t. \quad y_i(\boldsymbol{w} \cdot x_i + b) \geq 1 \quad , i = 1,2,3,\cdots,N \end{aligned} \tag{5}$$

Introducing Lagrange multipliers $\alpha_i \geq 0, i = 1,\ldots,n$, one for each of the constraints in (3), we get the following Lagrangian:

$$L(\boldsymbol{w}, b, \boldsymbol{\alpha}) = \frac{1}{2}\|\boldsymbol{w}\|^2 - \sum_{i=1}^{n} \alpha_i \left(y_i (\boldsymbol{w} \cdot x_i + b) - 1 \right) \tag{6}$$

We get the dual quadratic optimization problem:

$$\begin{aligned} &\max_{\alpha} \quad \sum_{i=1}^{n} \alpha_i - \frac{1}{2} \sum_{i,j=1}^{n} \alpha_i \alpha_j y_i y_j x_i^T x_j \\ &\text{subject to } \alpha_i \geq 0, i = 1, \ldots, n, \quad \sum_{i=1}^{n} \alpha_i y_i = 0 \end{aligned} \tag{7}$$

Thus, by solving the dual optimization problem, one obtains the coefficients α_i . This leads to the following decision function.

$$f(x) = \operatorname{sgn}\left(\sum_{i=1}^{n} y_i \alpha_i (x \cdot x_i) + b \right) \tag{8}$$

The other important property of SVM is that it can use diverse kernels in dealing with the input vectors. The standard kernels are polynomial, Gaussian and sigmoid kernels.

2.3 Exceptional Conditions

We assume that there are two exceptional conditions, such as camera saturation and occlusion problem. The camera saturation caused by sunlight or its reflection is one of major problem of the vision system in outdoor, as depicts in figure 3. In case of figure 3 (a) or (b), we can use it as an input of the system, but in case of figure 3 (c), it isn't suitable. We filter out the unsuitable images by using the average and variance of the image. In a car, many situations that can bring about the occlusion problem can occur, such as, if the occupant covers a sight of camera by hands, burden, or newspaper, or if the passenger in the back seat throws some objects to the passenger seat like balls. To complement the problem, we set the region of interest in the disparity map image and compute the average distance from the camera in the ROI. If the distance is smaller than a predefined threshold, the system ignores the disparity map image and acquires new image, otherwise the system runs to next stage.

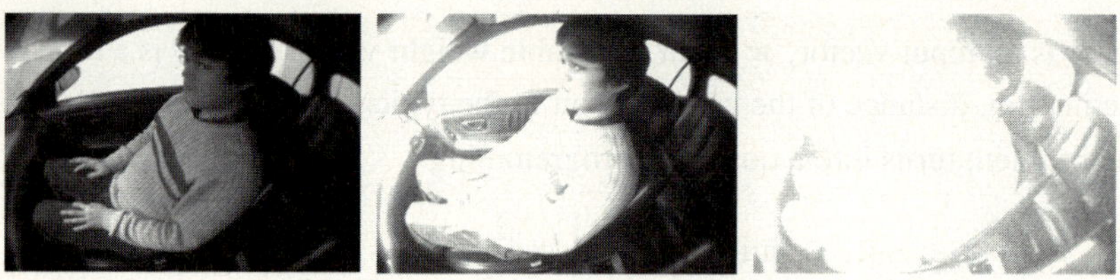

Fig. 3. Camera saturation caused by sunlight or its reflection

3 Embedded System

Early research, we have developed PC based VOC system, and accomplish algorithm test about stereo matching, camera calibration, and SVM based OC. In this section, we briefly describe hardware and software architecture for embedded system.

Figure 4 shows hardware block diagram and software architecture. Hardware consists of two different DSP, TMS320DM642 and TMS320C6713. TMS320DM642, fixed point DSP(4000MIPS, 500MHz) take the responsibility of stereo matching, camera calibration and basic image preprocessing algorithms. The C64x DSP generation features TI's VelociTI.2 VLIW architecture extensions that include support for packed data processing and special purpose instructions to accelerate broadband infrastructure and imaging applications. The C64x generation is shipping DSPs with clock speeds available up to 1 GHz and can incorporate multiple memory, peripheral and voltage combinations to address a wide range of high performance applications.

A SVM based OC algorithm plant in TMS320C6713, floating point DSP (1600MIPS/1200MFLOPS, 200MHz). The 8 functional units of the C6000 DSP core, including 2 multipliers and 6 arithmetic units, are highly orthogonal, providing the compiler and assembly optimizer with many execution resources.

The results of each DSP shared by using McBSP (Multi-channel buffered Serial Ports) and final result transfer to PC through 10/100M Ethernet MAC. Also device driver, BIOS and NDK are optimized and planted in embedded DSP board.

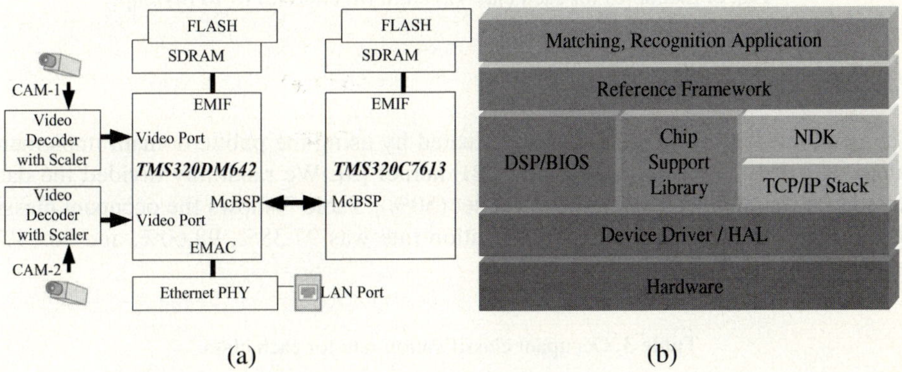

Fig. 4. VOC hardware block diagram (a) and software architecture (b)

4 Experiments

4.1 Stereo Image Database

The performance of the system was evaluated with our stereo image database, which is captured within the experimental car. The total of the stereo images are 1746 and the resolution of each image was 320x240. Table 2 shows composition of the database, and figure 5 depicts several examples for each class, such as adult (or 5th%tile female, a), child (b), RFIS (c), and empty seat (d), respectively. We used the disparity map image as an input of the SVM classifier, it was down-sampled (40x30) to reduce the dimension.

Table 2. Composition of the database

Class	Occupant	# of images	
0	Adult	1448	5 people, 165~180cm
	5th%tile female	30	a dummy, 140cm
1	Child	104	a dummy, 120cm
2	RFIS	118	15 products
3	Empty seat	46	not occupied

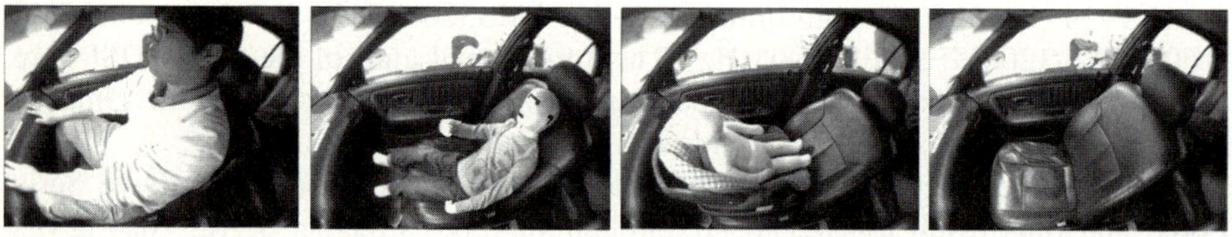

(a) (b) (c) (d)

Fig. 5. Examples for each class (a) adult (b) child (c) RFIS (d) empty

4.2 Performance of the System

The performance of the system was evaluated by using the public domain implementation of SVM, called LibSVM with RBF kernel [5]. We randomly divided the data sets into a training set (50%) and a test set (50%). Table 3 shows the occupant classification rate on each stage, the classification rate was 97.38%, 98.60%, and 96.04%, respectively.

Table 3. Occupant classification rate for each class

Class	Classification rate
adult, RFIS vs. child, empty	97.38%
adult vs. RFIS	98.60%
child vs. empty	96.04%

Table 4 shows a comparison result of the processing time of the PC and the embedded system. The PC runs on 3.0GHz(Pentium-4) and has 1 GB of memory. Stereo matching phase of the table 4 includes the stereo matching algorithm and preprocessing algorithms, such as warping for camera calibration and histogram equalization, and the classification phase include the cubic model and SVM classification. The total processing time of the PC and the embedded system was about 285ms and 958ms, respectively.

Table 4. Comparison of the processing time of the PC and the embedded system

	Processing time	
	PC	Embedded system
Stereo matching	282 ms	780.99 ms
Classification	2.6 ms	177.33 ms

5 Conclusions and Discussion

The study presents a stereovision based OCS with an embedded system. The OCS algorithm consists of the stereo matching with the fast SAD algorithm and the SVM classifier for classifying the occupant into four classes, such as adult, child, RFIS, and empty. The embedded system consists of dual DSP, one is for the stereo matching algorithm and another is for the classification algorithm. The system shows about 97% of classification rate and about 960ms of the processing time. Experimental results suggest that the system is feasible as an embedded OCS system.

Acknowledgments

This work was supported by Hyundai Autonet Co.

References

1. National Highway Traffic Safety Administration. Air Bag Fatal and Serious Injury Summary Reports. http://www-nrd.nhtsa.dot.gov/departments/nrd-30/ncsa/TextVer/SCI.html
2. V. N. Vapnik, The Nature of Statistical Learning Theory, Springer-Verlag, NY, USA, pp.45-98, 1995
3. H.-G. Lee et al., Occupant Classification for Smart Airbag Using Stereovision and Support Vector Machines, Springer Lecture Notes in Computer Science, Vol. 3173, pp.642-647, 2004
4. M.-S. Jang et al., Vision Based Automatic Occupant Classification and Pose Recognition for Smart Airbag Deployment, Springer Lecture Notes in Computer Science, Vol. 3643, pp.410-415, 2005
5. http://www.csie.ntu.edu.tw/~cjlin/libsvm/index.html
6. L. Di Stefano and S. Mattoccia, Fast Stereo Matching for the VIDET System Using a General Purpose Processor with Multimedia Extensions, IEEE International workshop on Computer Architecture for Machine Perception, pp.356-362, 2000
7. S. Haykin, "Neural Network", Prentice Hall, 1999

Context-Based Approach for Human Gesture Analysis*

Chil-Woo Lee[1], Jae-Yong Oh[1], and Yang-Weon Lee[2]

[1] Department of Computer Engineering, Chonnam National University,
300 Yongbong-Dong, Buk-Gu, Gwangju, 500-757, Korea
leecw@chonnam.ac.kr, ojyong@image.chonnama.ac.kr
http://image.chonnam.ac.kr

[2] Department of Information and Communication Engineering, Honam University,
59-1 Seobong-Dong, Gwangsan-Gu, Gwangju, 506-714, Korea
ywlee@honam.ac.kr

Abstract. In this paper, we propose a state transition model using context based approach for gesture analysis. This method defines the analysis situation as five different states; NULL, OBJECT, POSTURE, LOCAL, and GLOBAL. We first infer the situation of the system by estimating the transition of the state model, and then apply different analysis algorithms according to the system state. Gestures are analyzed with the queue-based matching method which is newly proposed in the paper instead of general gesture spotting algorithms. In the algorithm, movement of feature points; face and both hands, is compared directly, so gesture can be recognized quickly without applying any constraints for real world application. The transition of the states can be interpreted as the context of motion and that is estimated with conditional probability between the states in the algorithm.

1 Introduction

Recently human gesture analysis has received much interest in computer vision field for the applications such as human friendly interfaces, intelligent surveillance, virtual reality, and so on. Especially, natural communication method between human and computer is becoming one of the key issues in the various fields of industry because computer is getting more complicated contrary to user's ability.

We can assume that human instinctively understands behaviors by watching the temporal sequence of body movement and then estimating some meanings of the temporal movement by using the previously accumulated knowledge and motion data. However, it is very difficult to analyze temporal changes by computer, namely historical meaning of bodily motion automatically because a human body is a three-dimensional object with very complicated structure and flexibility.

In the early works, many researchers tried to measure the configuration of the human body with sensors attached to the joints of limbs, and to recognize gesture by analyzing the variation of joint angles [1][2]. But, this requires the user to put on

* This research has been supported by research funding of "Center for High-Quality Electric Components and Systems", Chonnam National University, Korea, and "Development of humanoid technology based on network", KIST, Korea.

B. Gabrys, R.J. Howlett, and L.C. Jain (Eds.): KES 2006, Part I, LNAI 4251, pp. 598 – 605, 2006.
© Springer-Verlag Berlin Heidelberg 2006

irritating devices, and it usually needs long cable connections from the devices to the computers. To overcome these problems, video-based recognition techniques are proposed and it uses a set of video cameras and computer vision techniques to interpret gestures.

One of the famous algorithms using global motion feature information is *Motion History Image (MHI)* method. In the method, sequentially accumulated image data of object silhouettes is used for analyzing the human motion [3]. Many other algorithms which use feature points or silhouette of body are also proposed [4][5], but all the methods could not solve the generalization problems for gesture recognition. That means that many systems can understand the pre-trained gestures, but if input image sequence includes any gestures which are not defined or trained previously, the systems can not recognize or understand the gestures. Moreover, because the human gestures can not be trained for all cases, it is very difficult to develop the situation independent system. To solve these problems, we propose a new gesture recognition algorithm using attentive features; probability of state, global and local motion and positional information. In this algorithm, the system classifies the situation of behaviors before recognizing the individual meaning of gestures. So, proposed algorithm can analyze the human gestures more precisely and it has a feature that can expand the application of recognition system.

The paper is structured as follows: Section 2 and 3 present a probabilistic state transition model and gesture interface system using the state transition model respectively. And in Section 4, we suggest the example of gesture recognition system. Finally, the paper is closed with mentioning the conclusion and further works in Section 5.

2 Overview of the Gesture Analysis Algorithm

Before taking up the main subject, we need to define the terminology "*context*" clearly. In the Cambridge dictionary, "*context*" is defined as *the situation within which something exists or happens,* and it also means that *variation of circumstance or locational information of robot in engineering fields.*

In this paper, we define the "*context*" as *intention of behaviors between human and computer.* And in the algorithm, we use only cameras for communication, and do not include the situation using sound or any other sensors. We define the human gestures as five different states based on subject movement. And then, we can represent the human gestures by continuous sequence of states as shown in Figure 1, also we can understand intention of actor with the state information. The state transition model adopting probabilistic change obtained through the training process is depictured in Figure 2. Also, the figure shows the overall process of our algorithm. In the first step, the system is trained with the image sequences. Through the training process, we can get the probability of state transition and weights of feature vectors. And then, the system decides the state transition with this information. Finally, we can apply the reasonable recognition algorithm to the system according to the context information.

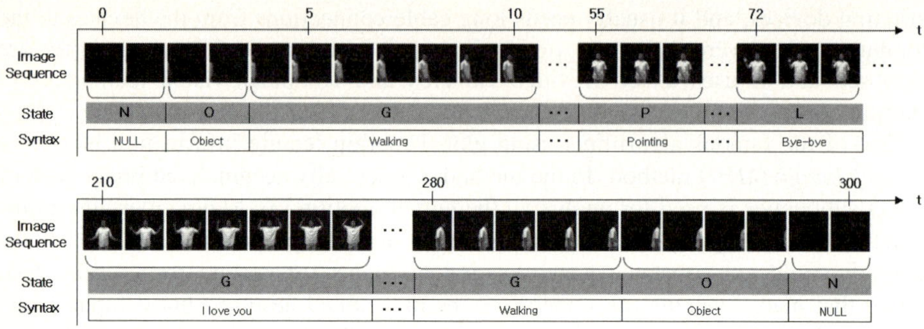

Fig. 1. Definition of states with image sequences (N: Null, O: Object, P: Posture, G: Global Motion, L: Local Motion)

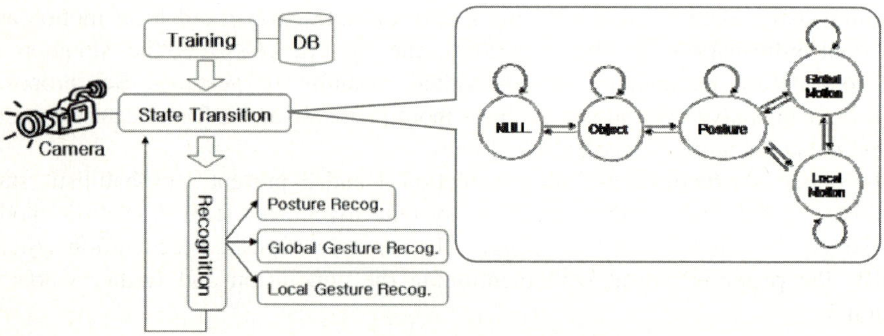

Fig. 2. Flowchart of the recognition system and the state transition model

3 Probabilistic State Transition Model

As previously described, the model consists of five states, and these states are represented as probabilistic model by training the process with image sequences. In this section, we describe the state transition model using the probabilistic method.

3.1 Feature Extraction and Training

In this paper, we use the three kinds of image features to decide a transition of state. The features are motion information in image, distance from subject, and three dimensional trajectories of head and hands. The system can extract and track the head and hand regions of subject every frame from stereo image. And then, we can get the image features using this information.

The training image sequence includes the state information, and system can be established the probabilistic model using these sequences. From the training process, we can get the probabilities of each image features and its weights.

Let the three image features f_i , ($i = 1,2,3$), feature vector set F and its probabilities $P(F)$ can be denoted as Equation (1). Also, if the set of states is S_k , ($k = 1,2,3,4,5$), the probabilities of feature vector set on each state can be represented as $P(F \mid S_k)$.

We can get not only probabilities of states transition, but also weight information through the training process. The weight information means importance rate of image features when state transition is occurred, and can reduce the negative effects from meaningless image features.

$$F = \{f_1, f_2, f_3, f_4, f_5\} \tag{1}$$

$$P(F) = \prod_i P(f_i)$$

3.2 State Transition

We introduce *Bayes' Theorem* to decide transition between each state. First, the probabilities of set F on the each state (a prior probability) are calculated from training process. And then, we can estimate the probabilities of transition to the each state (a posterior probability). Finally, the system is updated to the state information which has the maximum probabilities. The probabilities of transition with feature vector set F are denoted as Equation (2).

$$P(S_k \mid F) = \frac{P(S_k)P(F \mid S_k)}{\sum_i P(S_i)P(F \mid S_i)} \tag{2}$$

Also, the system adjusts the weights of features (w_i , $i = 0,1,2,3,4$) by using the weight information as shown in Equation (3).

If the feature vector set is extracted on the state S_j^t at time t, the probabilities of transition to each state is calculated with adjusted feature vector set F' and prior probabilities. And then, the system is updated to the state S_k^{t+1} which has the maximum probabilities as shown in Equation (4).

$$P(F') = \prod_i w_i P(f_i) \tag{3}$$

$$k = \arg\max_j \; P(S_j^{t+1} \mid F') \tag{4}$$

4 Recognition System

The proposed analysis algorithm is marked by recognizing the gestures according to the situational information. It is easy to develop the restricted and system dependant algorithms, because all of human gestures can not be classified using only a small number of features. In this section, we propose the recognition system that can reduce limitations using context awareness.

4.1 Posture Recognition

"Pose" can be defined as a particular shape or attitude of body. Also, it contains little movement of the body, and can describe using the geometric relations of subject's head and hands. Therefore, we can understand the poses by interpreting positional information of head and both hands. In this paper, we use the 3D position of head and hands for recognizing the poses, and these data can be extracted from the pre-processing procedure. If the 3D positions of head and hands are $H(x_h, y_h, z_h)$, $L(x_l, y_l, z_l)$ and $R(x_r, y_r, z_r)$ respectively, then the feature vector is denoted by Equation (5).

$$F_{pose} = \{D_l, D_r, N, A\} \tag{5}$$

where

$$D_l = H - L,\ D_r = H - R,\ N = D_l \times D_r$$

$$A = a\cos\left(\frac{|D_l| \cdot |D_r|}{D_l \bullet D_r}\right)$$

This feature vector set is compared with the set of model poses M as shown in Equation (6). Finally, the pose P_k $(P_k \in M)$ that has the maximum similarity R_i is selected as the result of the pose recognition.

$$M_i = \{D_{lM_i}, D_{rM_i}, N_{M_i}, A_{M_i}\} \ (M_i \in M) \tag{6}$$

$$R_i = (D_l \bullet D_{lM_i}) + \left\| A_i - A_{M_i} \right\| + (N_i \bullet N_{M_i}) \tag{7}$$

$$k = \arg\max_i (R_i)$$

4.2 Recognition of Local and Global Gestures

The local gesture includes user's specific intention with local motions in view. For example, in the case of [waving hands], local motion is more important than whole trajectories of hands. Therefore, local gesture can be classified with the shape, local motion and size of hand itself. In this paper, we use cyclic information of hand shapes for recognition. However, this method is not suitable for all cases of local gestures and there are various methods for local gestures such as hand sign method for hearing disabled person. It can be applied to the local gesture recognition too.

The global gesture contains user's intention in motions of whole body. Especially, trajectories of hands include much information. So, we adopt the different recognition method which uses trajectories of both hands as features. Many researchers have tried to develop the matching algorithm for the trajectories in a number of ways. Generally, the methods are used for recognition of handwritten character. But, it is not effective to apply into the gesture recognition because it is difficult to decide the start and end point of meaningful gestures. Therefore, many researchers are trying to solve the problem; Gesture Spotting [6].

In this paper, we propose the simple queue matching method instead of general gesture spotting algorithm. And this method has the advantage of fast processing and easy implementation.

The basic concept of this algorithm is as follows. Assume that the model set M has N models. Also, direction vectors represent the trajectories of hands, and these vectors are stored continuously with each gesture models G_j as shown in Equation (10).

$$Q_{\text{model}} = \{N_1, N_2, N_3, \ldots\ldots, N_k\} \tag{8}$$

$$M = \{G_1, G_2, \ldots\ldots, G_j, \ldots G_N\} \tag{9}$$

$$G_j = \{D_1^j, D_2^j, D_3^j, \ldots\ldots, D_k^j\} \tag{10}$$

$$D_i^j = N_{i+1}^j - N_i^j \;\; (N_{i+1}^j, N_i^j \in Q_{\text{model}})$$

We can get directional vectors from each frame. And, input queue I with the length of l is a set of these vectors as shown in Equation (12). ($l > k$)

$$Q_{input} = \{N_1, N_2, N_3, \ldots\ldots, N_l\} \tag{11}$$

$$I^j = N_{i+1}^j - N_i^j \;\; (N_{i+1}^j, N_i^j \in Q_{input})$$

$$I = \{I^1, I^2, I^3, \ldots\ldots, I^l\} \tag{12}$$

If the meaningful gesture of subject is in the input queue, it can be assumed that this queue includes the subject's intention. And then, input queue is compared with each model gesture G_j. Finally, we can decide the gesture G_k , $G_k \in M$, as a recognition result with the Equation (13). Figure 3 shows the appearance of the implemented system. And Figure 4 depicts a result of global gesture recognition using the queue matching algorithm.

$$R_j = \max_m \left(\sum_{i=1}^{k} (I^{m+i} \bullet D_i^j) \right) \tag{13}$$

$$k = \arg\max_j \, (R_j)$$

Fig. 3. The Example of recognition system (Similarities between gestures are on the left side, and yellow line means trajectories of hands)

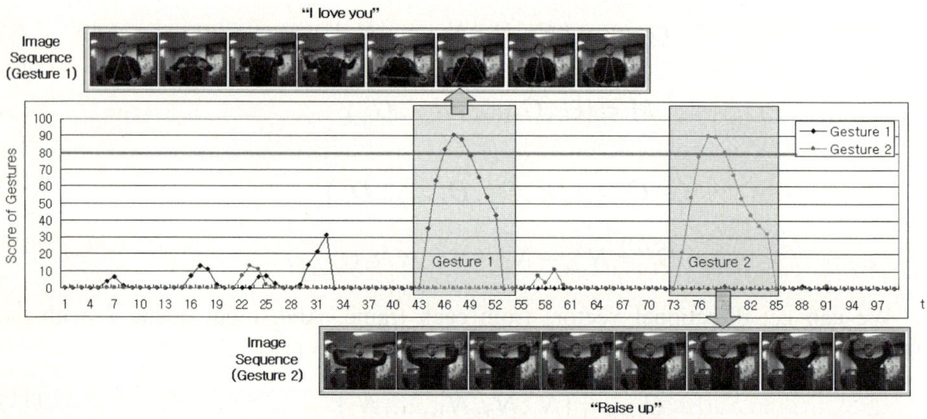

Fig. 4. The result of recognition using queue matching algorithm (Each curve means similarities between input gesture and model gestures, gesture 1(blue) and gesture 2(red))

5 Conclusion

In this paper, we described the gesture analysis algorithm using context awareness. We define five different states of behavior, *NULL*, *OBJECT*, *POSTURE*, *LOCAL* and *GLOBAL*. The transition of the states can be interpreted as the context of motion and that is estimated with conditional probability between the states in the algorithm. In the training process, system can be established the probabilistic model from the predefined image sequences. And we can recognize the context of gestures and catch the intention of user by analyzing the state transition model. Also, we can apply the different recognition algorithms to the system according to the context information. And, for the recognition of global gestures, we proposed the fast and simple queue matching method instead of general gesture spotting algorithm.

However, there are some problems in the case of analyzing multiple persons in image because the system can not track user's head and hand region due to occlusion.

If we can use efficient tracking algorithm to solve the problems, and also develop a human focusing or selection algorithm for deciding one specific communication subject from multiple person, proposed algorithm can be used for constructing intelligent interface system.

References

1. Baihua Li and Holstein, H., "Articulated point pattern matching in optical motion capture systems", Qinggang Meng; Control, Automation, Robotics and Vision, 2002. ICARCV 2002. 7th International Conference on , Volume: 1 , 2-5 Dec. 2002
2. Yu Su, Allen, C.R., Geng, D., Burn, D., Brechany, U., Bell, G.D. and Rowland, R., "3-D motion system (data-gloves): application for Parkinson's disease", Instrumentation and Measurement, IEEE Transactions on , Volume: 52 , Issue: 3 , June 2003
3. James Davis, "Recognizing Movement using Motion Histograms", MIT Media Lab. Technical Report No. 487, March 1999
4. Ross Cutler and Matthew Turk, “View-based Interpretation of Real-time Optical Flow for Gesture Recognition”, Third IEEE International Conf. on Automatic Face and Gesture Recognition, 1998.
5. Chil-Woo Lee, Hyun-Ju Lee, Sung H. Yoon, and Jung H. Kim, “Gesture Recognition in Video Image with Combination of Partial and Global Information”, in Proc. of VCIP 2003, July, 2003
6. Ho-Sub Yoon, Byung-Woo Min, Jung Soh, Young-Iae Bae, and Hyun Seung Yang, "Human computer interface for gesture-based editing system", Image Analysis and Processing, pp. 969 – 974, 1999

Adaptive Implicit-Camera Calibration in Photogrammetry Using Anfis

Erkan Beșdok[1] and Pınar Çivicioğlu[2]

[1] Erciyes University, Enginering Faculty, Geodesy and Photogrammetry Engineering Dept., Kayseri, Turkey
ebesdok@erciyes.edu.tr

[2] Erciyes University, Civil Aviation School, Aircraft Electrics and Electronics Dept., Kayseri, Turkey
civici@erciyes.edu.tr

Abstract. Camera Calibration (CC) is required in many Photogrammetry and Computer-Vision applications, where 3D information is extracted from images and CC is also employed for pose determination of imaging sensors. In this paper, a novel implicit-CC model (**ICC**) based on Adaptive Neuro Fuzzy Inference System has been introduced. The **ICC** is particularly useful for back-projection in the applications that do not require internal and external camera calibration parameters in addition to the expert knowledge. The **ICC** supports multi-view back-projection in intelligent-photogrammetry. In this paper, the back-projection performance of the **ICC** has been compared with the Modified Direct-Linear-Transformation (MDLT) on real-images in order to evaluate the success of the proposed **ICC**. Extensive simulation results show that the **ICC** achieves a better performance than the MDLT in the 3D reconstruction of scene.

1 Introduction

Camera Calibration (CC) is defined as estimating the parameters of imaging systems, in order to determine the exact mapping between 3D world to 2D image plane. Therefore, CC explains the transformation of 3D world coordinates into 2D image coordinates [1, 2, 3, 4, 5, 6, 7, 8]. In the traditional methods, this transformation is made using several camera calibration parameters that include rotation angles (ω, ϕ, κ), translations (X_0,Y_0, Z_0), the coordinates of principal points (u_0, v_0), scale factors (β_u, β_v) and the skewness (λ) between image axes. Several methods have been implemented in the literature that use various camera calibration parameters such as radial lens distortion coefficients (k_1, k_2), affine image parameters (A, B) and decentering lens parameters (p_1, p_2). Most of the traditional approaches for camera calibration use various non-linear optimization algorithms to obtain camera calibration parameters [1, 2, 3, 4, 5, 6, 7, 8].

The **ICC** provides transformation of 2D world coordinates into 3D image coordinates using an Adaptive Neuro-Fuzzy Inference System (Anfis) structure [9, 10, 11] without restricting imaging models. Hence, in this paper, the camera calibration parameters have been described as the Anfis parameters such as weights, transfer functions and membership values.

B. Gabrys, R.J. Howlett, and L.C. Jain (Eds.): KES 2006, Part I, LNAI 4251, pp. 606–613, 2006.
© Springer-Verlag Berlin Heidelberg 2006

The Intelligent Photogrammetry (Photogrammetron) [2, 3] consists of ideas, methods and applications from digital photogrammetry, intelligent agents and active vision. Full automation of the Photogrammetry which is referred as *Intelligent Photogrammetry* can only be possible with an autonomous and intelligent agent system such as Artificial Neural Networks, Fuzzy Systems, Genetic Algorithms, and other intelligent computing techniques. Photogrammetron is basically a Photogrammetric system that has a full functionality of Photogrammetry in addition to an intelligent agent system and a physical structure of active vision. It may have different forms as coherent stereo photogrammetron, separated stereo photogrammetron and multi-camera network photogrammetron. Photogrammetron can be used in various applications including photogrammetry-enabled robotic guidance, intelligent close-range photogrammetry, intelligent 3D multimedia-video indexing and real-time digital Videogrammetry [12].

A number of CC methods that implement an intelligent agent have been introduced in the literature [2, 3, 4, 6, 7, 12]. Most of these methods implement the structure of an intelligent agent either to learn the mapping from 2D world to 3D image coordinates, or to improve the performance of other existing methods. Knowing that the CC parameters are important in various computer vision applications such as stereo-reconstruction, the **ICC** goes beyond the existing ones by providing 3D reconstruction from multi-view images besides stereo-images [2, 3, 4, 6, 7, 12]. In this paper, the camera calibration problem is addressed within an Anfis. The **ICC** adaptively models the imaging-sensor for each back-projection points by using fuzzy methods in order to achieve 3D to 2D mapping more accurately. The **ICC** can be used with automated active lenses and does not require a good initial guess of classical CC parameters.

The rest of the paper is organized as follows: *An Anfis-Based Novel Approach for Camera Calibration* is explained in Section 2. *Experiments and Statistical Analysis* are given in Section 3 and finally, *Conclusions* are given in Section 4.

2 An Anfis-Based Novel Approach for Camera Calibration

A number of calibration methods implementing intelligent agents have been introduced recently [2, 3, 4, 6, 7, 12]. The intelligent agents based camera calibration methods introduced in the literature generally require a set of image points with their 3D world coordinates of the control points and the corresponding 2D image coordinates for the learning stage. The **ICC** can merge multisource camera images, for the reconstruction of a 3D scene, due to its high flexible properties of adaptivity and fuzzyness. In this paper, the preparation steps of the learning and training data used in the training Anfis structure of the ICC (Fig. 1) to obtain the 3D world coordinates (X,Y,Z) of a position (T) are given below:

1. Determine the control points on the Planes-1 and 2 seen in Figure 2-a.
2. Start moving Plane-3 (seen in Fig. 2-b) towards the camera throughout the volume comprised by the object which will be measured and compute the 3D coordinates of the points on the Plane-3.

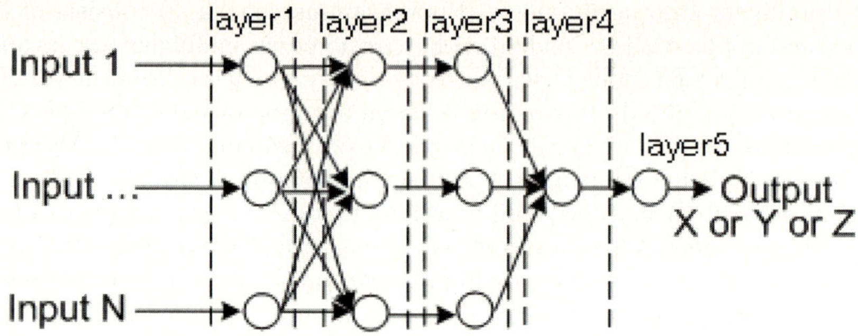

Fig. 1. The Anfis structure of the **ICC**

3. Repeat Step 2 until the end of the volume comprised by the object which will be measured is reached.
4. Use the 2D image coordinates and their corresponding 3D world-coordinates of the Plane-3 obtained as a result of the steps 2-3 as the training data.
5. Train Anfis-x, Anfis-y and Anfis-z structures using the training data of Step 4. The input data of the Anfis structures are the image coordinates of the Plane-3 and the output data are the corresponding X world-coordinates of the input data for Anfis-x , Y world-coordinates of the input data for Anfis-y, and Z world-coordinates of the input data for Anfis-z.
6. Apply the image coordinates to the Anfis structures (Anfis-x for X, Anfi-y for Y, and Anfis-z for Z) in order to compute the 3D (X,Y,Z) coordinates of an image point T.

As shown in Fig. 1, the **ICC** is used to translate the image coordinates $(u,v)_T$ into the world coordinates $(X,Y,Z)_T$. All the Anfis structures possess 2 triangular membership functions and they have been trained with 2000 epochs.

2.1 Adaptive Neuro-fuzzy Inference System: Anfis

Neuro-fuzzy systems [9, 10, 11] are hybrids of fuzzy systems and neural networks. The goal of neuro-fuzzy systems is to combine the learning capability of a neural network with the intuitive representation of knowledge found in a fuzzy system. This may be accomplished by designing a network architecture to mimic a fuzzy system, by incorporating linguistic terms into the computations performed by the network, by means of an explanation mechanism for the network, and so forth. A good introduction to neuro-fuzzy systems may be found in [9, 10, 11]. Anfis is the well-known neuro-fuzzy system, which mimics the operation of a TSK fuzzy system, whose rules are at the form of,

$$y = a_0 + \sum_{i=1}^{n} a_i x_i \tag{1}$$

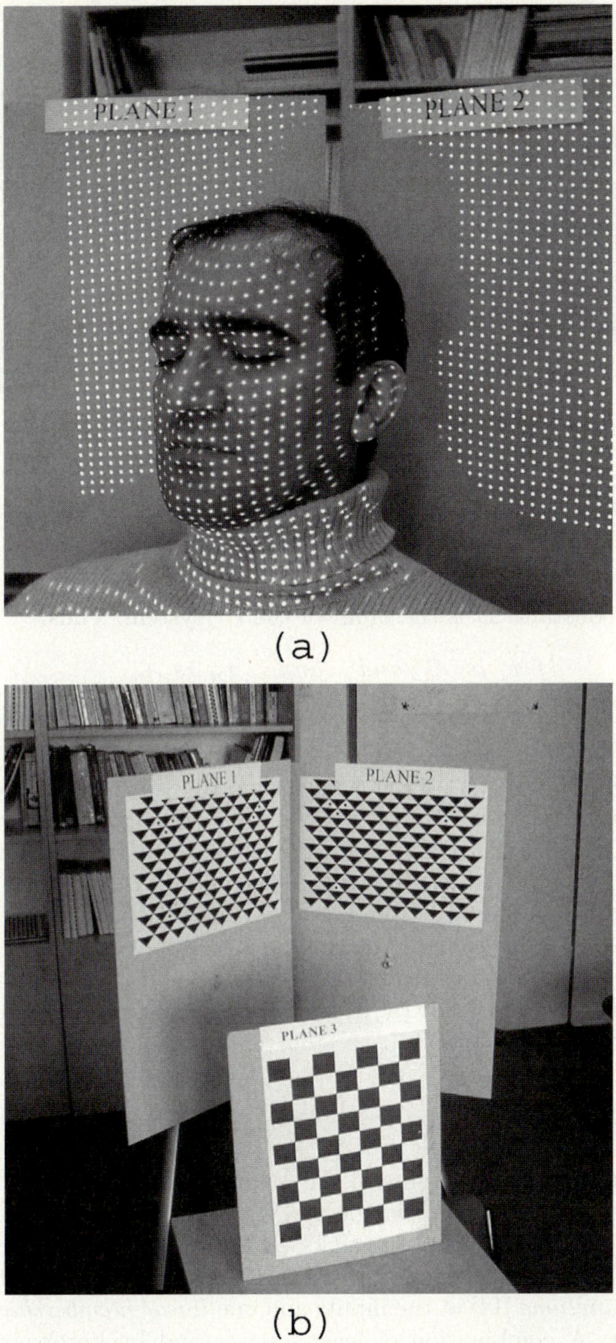

Fig. 2. (a) The author's face used in the experimental setup of **ICC** (b) Experimental setup for the training data of Anfis used in **ICC**

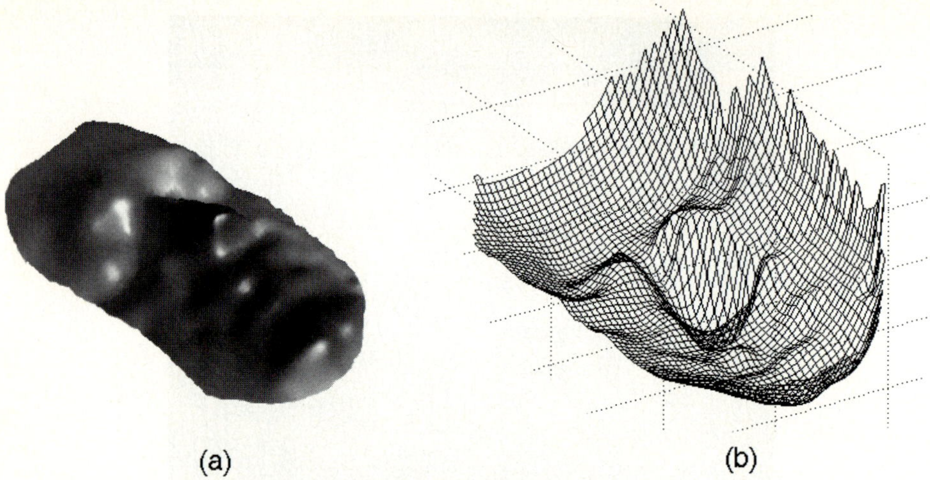

Fig. 3. (a) The 3D-solid model obtained by using radial basis functions [13]. (b) The 3D-Mesh model of the author's face.

where a_i is a constant, x_i is the input of the ith system. Thus,

$$\begin{array}{l} if\ x_1\ is\ A_1\ \ and\ ...\ and\ ...\ x_n\ \ is\ A_n \\ \qquad\qquad then\ \ y = a_0 + a_1 x_1 + ... + a_n x_n \end{array} \tag{2}$$

where A_i is a fuzzy set [9, 10, 11].

The number of inputs (N) of the Anfis structures used in this paper is twice the number of the images (see Fig.1) [14]. The Anfis network is a 5-layered network, in which the layers are not fully connected. The transfer function of a neuron is determined according to the layer that the neuron is in. The Anfis structure used in this study is illustrated in Fig.1 and implemented in Matlab, which is a highly approved technical computing language [14]. The fuzzy structures used in the proposed method are trained by using a hybrid learning algorithm [14] to identify the parameters of the single-output of the triangular membership function, which is defined as [14],

$$f\left(q;a,b,c\right) = \max\left(\min\left(\frac{q-a}{b-a},\frac{c-q}{c-b}\right),0\right) \tag{3}$$

where q denotes the set of inputs of the membership function. The parameters a and b locate the feet of the triangle and the parameter c locates the peak.

The parameters have been obtained for each *back-projected T point* by training one Anfis structure with 2000 epochs. All the fuzzy structures have two *triangular membership functions* [14] at the inputs and one *linear membership function* [14] at the output. A combination of least-squares and backpropagation gradient descent methods [14] is used in training of the fuzzy structures, in order to compute the parameters of the triangular membership functions (a, b, and c) which are used to model given set of inputs and single-output data f where

f denotes one of the spatial parameters (X,Y,Z) of the *back-projected T point.* Only one distinct fuzzy structure has been used for each of the spatial parameters because Anfis supplies only one output [14].

Table 1. Mean (μ) and standard deviation (σ) of the differences between the results obtained using the **ICC** and MDLT

	ΔX =ICC$_X$-MDLT$_X$	ΔY =ICC$_Y$-MDLT$_Y$	ΔZ =ICC$_Z$-MDLT$_Z$
μ	0.014	0.018	0.063
σ	0.013	0.015	0.068

Table 2. Mean Square Error (MSE) and Pearson correlation coefficient (Corr) values between the results obtained using the **ICC** and MDLT

	MSE	Corr
ICC$_X$; MDLT$_X$	0.000	0.987
ICC$_Y$; MDLT$_Y$	0.008	0.982
ICC$_Z$; MDLT$_Z$	0.011	0.941

3 Experiments and Statistical Analysis

A set of real images have been employed in the analysis of the **ICC**. The obtained results are then compared with those of the MDLT method [5, 15]. A Casio QV 3000EX/ır camera has been used in the study. For the analysis of the **ICC**, the images obtained from the camera have been employed directly. The image coordinates of the control points have been extracted by employing the well-known least square matching algorithm and no deformation corrections have been applied to the images. Before using the MDLT method, the required corrections have been applied to the coordinates obtained as a result of the image matching. On the other hand, no corrections have been applied to the corresponding coordinates in the **ICC** method. The results obtained using the **ICC** and MDLT methods are then compared statistically with each other. As a conclusion, 96.05% of the Δx , 93.07% of the Δy and 83.48% of the Δz values that are obtained with the **ICC** are found between ($2\mu \pm \sigma$) where Δx, Δy and Δz denote the differences between the corresponding **ICC** and MDLT coordinates. The 3D-mesh model and radial basis functions based 3D models of the **ICC** are illustrated in Fig.3-a and Fig.3-b, respectively. Extensive simulations show that **ICC** supplies statistically acceptable and more accurate results as seen in Tables 1-2.

4 Conclusions

An Anfis based camera calibration method for 3D information recovery from images is proposed in this paper. The obtained results have been compared with the traditional MDLT. The main advantages of the **ICC** are that it does not require the knowledge of complex mathematical models and an initial estimation of camera calibration, it can be used with various cameras by producing correct outputs, and it can be used in dynamical systems to recognize the position of the camera after training the ANN structure. Therefore, the **ICC** is more flexible and straightforward than the methods introduced in the literature.

The advantages of the **ICC** may be summarized as follows:

- Does not require expert knowledge.
- Suitable for multi-view geometry.
- Offers high accuracy.
- Simple to apply.
- Suitable for computer vision and small scale desktop photogrammetry.
- Does not use traditional calibration methods and parameters.

Acknowledgments

This study has been supported by the Research Foundation of Erciyes University under the project codes of FBT- 05-31, FBT-06-25 and FBA-05-25.

References

[1] G., Klir, Z., Wang, D., Harmanec, Geometric Camera Calibration Using Circular Control Points, IEEE Transactions on Pattern Analysis and Machine Intelligence, 22, 1066-1076, 2000.

[2] H.P., Pan, A Basic Theory of Photogrammetron, International Archives of Photogrammetry and Remote Sensing XXXIV (3), 2002.

[3] H.P., Pan, C.S. Zhang, System Calibration of Intelligent Photogrammetron, International Archives of Photogrammetry and Remote Sensing XXXIV (2), 2002.

[4] M.B., Lynch, C.H., Dagli, M., Vallenki, The Use of Fedforward Neural Networks for Machine Vision Calibration, Int. Journal of Production Economics 60-61, 479-489, 1999.

[5] H., Hatze, High-Precision Three-Dimensional Photogrammetric Calibration and Object Space Reconstruction Using a Modified Dlt-Approach., J. of Biomechanics, 21, 533-538, 1988.

[6] M.T., Ahmed, E., Hemayed, A., Farag, Neurocalibration: A Neural Network that can Tell Camera Calibration Parameters, Proc. of the International Conference on Computer Vision, Korfu, Greece, 1, 463-468, 1999.

[7] J., Jun, K., Choongwon, Robust Camera Calibration Using Neural Network, IEEE Region 10 Conference TENCON 99, 1, 1694-697, 1999.

[8] L. Lucchese, Geometric Calibration of Digital Cameras through Multi-View Rectification, Image and Computing, 23, 517-539, 2005.

[9] J.S.R., Jang, Anfis: Adaptive-Network-Based Fuzzy Inference System, IEEE Transactions on Systems, Man and Cybernetics, 23 (3), 665-685, 1993.

[10] E., Beşdok, P., Çivicioğlu, M., Alçı, Using an Adaptive Neuro-Fuzzy Inference System-Based Interpolant for Impulsive Noise Suppression from Highly Distorted Images, Fuzzy Sets and Systems, 150, 525-543, 2005.

[11] E., Beşdok, P., Çivicioğlu, M., Alçı, Using Anfis with Circular Polygons for Impulsive Noise Suppression From Highly Distorted Images, AEU-International Journal of Electronics and Communications, 59 (4), 213-221, 2005.

[12] P., Heping, Z., Chusen, System Structure and Calibration Models of Intelligent Photogrammetron, Wuhan University Journal 2 (48), 2003.

[13] FarField Technology, FastRBF Toolbox, MATLAB Interface Version 1.4, http://www.farfieldtechnology.com/products/toolbox/ 2004.

[14] Mathworks Inc., Matlab Neural Networks Toolbox and Fuzzy Toolbox, Mathworks, 2005.

[15] Y.I., Abdel-Aziz, H.M., Karara, Direct Linear Transformation from Comparator Coordinates into Object Space Coordinates in Close-Range Photogrammetry, Proceedings of the Symposium on Close-Range Photogrammetry, 1-18, Falls Church, VA: American Society of Photogrammetry, 1971.

Applying A-Priori Knowledge for Compressing Digital Elevation Models

Giovanni Guzmán, Rolando Quintero, Miguel Torres, and Marco Moreno

Centre for Computer Research, National Polytechnical Institute, Mexico City, Mexico
{jguzmanl, quintero, mtorres, marcomoreno}@cic.ipn.mx
http://geo.cic.ipn.mx, http://www.cic.ipn.mx

Abstract. Up-to-date, some algorithms related to compress digital elevation models (DEMs) or high-resolution DEMs, use wavelet and JPEG-LS encoding approaches to generate compressed DEM files with good compression factor. However, to access the original data (elevation values), it is necessary to decompress whole model. In this paper, we propose an algorithm oriented to compress a digital elevation model, which is based on a sequence of binary images encoded using RLE compression technique, according to a specific height (contour lines). The main goal of our algorithm is to obtain specific parameters of the DEM (altitudes and contours lines) without using a decompression stage, because the information is directly read from the compressed DEM.

1 Introduction

The term digital elevation model or DEM is frequently used for referring to any digital representation of a topographic surface. However, the most often it is used to refer specifically to a raster or regular grid of spot heights. Digital terrain model or DTM may actually be a more generic term for any digital representation of a topographic surface, but it is not so widely used [1]. DEM has gained popularity in applications to determine terrain's attributes, such as elevation at any point, slope and aspect, finding features of the terrain, and so on. The most important features are the following: drainage basins and watersheds, drainage networks and channels, peaks and pits, other landforms, modeling hydrologic functions, energy flux and forest fire [2].

On the other hand, the main purpose of a compression algorithm is to reduce the amount of information needed to describe the original data. To retrieve the original data, we apply a process commonly named *decompression*. If the original information is fully-retrieved without any modification or alteration, then we can talk about a lossless compression algorithm. Also, it is possible to have compression algorithms with loss of information [9]. Many compression approaches first proceed to inspect the input image and attempt to detect different types of redundancy such as: statistical, psycho-visual or by correlation.

A digital elevation model requires a huge amount of data, up to 100 or more megabytes of storage space. Nowadays, compression techniques are used to compress Digital Elevations Models with a high compression coefficient, in [4], [5] and [6] the authors propose to use wavelets and JPEG-LS encoding approaches. However, if the user requires to access the file or to obtain certain information about the image, it is necessary to decompress all DEM information.

B. Gabrys, R.J. Howlett, and L.C. Jain (Eds.): KES 2006, Part I, LNAI 4251, pp. 614 – 622, 2006.
© Springer-Verlag Berlin Heidelberg 2006

In this paper, we propose an algorithm oriented to compress a digital geo-image of a digital elevation model, which is based on a sequence of images with a specific height. The sequence is compressed by applying a binary compressor. The main goal of our algorithm is to obtain the specific parameters of the DEM (altitudes and contours lines) without using a decompression stage, because the information is directly read from the compressed DEM.

The rest of this paper is organized as follows: Section 2 describes the proposed compression algorithm. Section 3 sketches the manipulation of compressed DEM file to obtain both elevations and contours lines, without applying a decompression process. Section 4 presents the results obtained by applying our approach, and Section 5 outlines the conclusions and future works.

2 Description of the Compression Algorithm

Nowadays, there are some types of DEMs such as: 7.5-minute, 15-minute, 2-arc-second, and 1-degree units [7]. For implementation purposes, we have proposed to choose the 1-degree DEM variant.

An important characteristic is that the frequency of the DEM is not high, because the pixels do not present changes in their structure. The value of each pixel is correlated to their neighbor values. Due to this, it is possible to apply this kind of compression. Moreover, this approach is not useful for common images.

Basically, a DEM file is integrated by three types of records, usually called A, B and C. The structure of these records is as follows [1]:

- *Record A* contains information, which defines the general characteristics of DEM, it includes descriptive header information related to the DEM's name, boundaries, units of measurement, minimum and maximum data values, number of type B records and projection parameters. There is only one type A record for each DEM file, and it appears as the first record in the data file.
- *Record B* contains elevation data and associated header information. All type B records of the DEM files are made up data from one-dimensional bands, called points. Therefore, the number of complete points covering the DEM area is the same as the number of type B records in the DEM.
- *Record C* contains statistics on the accuracy of the data in the file.

In a Digital Elevation Model, the altitude describes elevations greater or equal to 0 in a specific area. In consequence, the minimum altitude is 0 meters (sea level) and the maximum is 8,850 meters (Everest Mountain). In the most of cases, the DEM is displayed by using a 3-D render approach or applying a transformation function to obtain a gray level image. To detail the proposed algorithm, it is necessary to give some definitions that are important to describe our method.

The altitude or height associated to one element that belongs to Record B in the DEM, is defined by Eqn. 1.

$$h(x,y)=\alpha \tag{1}$$

To associate all points with the same height α_k, in which the obtained set is defined by Eqn. 2.

$$S_{\alpha_k} = \{(x, y) \neg h(x, y) = k\} \tag{2}$$

The number of elements in each S_{α_k} will be denoted by $card(S_{\alpha_k})$. Considering these definitions, the DEM altitudes are described by using Eqn. 3.

$$S = \{S_{\alpha_{min}} \cup S_{\alpha_2} \cup \ldots \cup S_{\alpha_{max}}\}, \tag{3}$$

where:

α_{MAX} denotes the maximum height in the original DEM.

Additionally, it is possible to define the join operation, according to Eqn. 4.

$$\bigcup_i^j S_\alpha = \{S_{\alpha_i} \cup S_{\alpha_{i+1}} \cup \ldots \cup S_{\alpha_{j-1}} \cup S_{\alpha_j}; i \le m \le j\} \tag{4}$$

To obtain the difference between two sets, we can use the minus operation defined in Eqn. 5.

$$S_{\alpha_a} - S_{\alpha_b} = \{(x, y) \neg (x, y) \in S_{\alpha_a} \text{ and } (x, y) \notin S_{\alpha_b}\} \tag{5}$$

With these definitions, we can formally state our proposed algorithm.

2.1 Compression Approach

In this section, we provide a detailed explanation of the required steps to compress a 1-degree DEM with altitudes expressed as integer values.

Step 1. Records A and C are stored without modification in the output.

Step 2. Find minimum and maximum altitudes $(\alpha_{min}, \alpha_{max})$ presented in the DEM and generate the header described in Table 1 in the output.

Table 1. Header description of the compressed DEM

Header of Compressed DEM	Type of variable to store information
Image width (columns)	integer (2 bytes)
Image height (rows)	integer (2 bytes)
α_{min}	integer (2 bytes)
α_{max}	integer (2 bytes)

The process starts with $k = \alpha_{min+1}$ instead of α_{min} due to all heights in the image are greater or equal to α_{min}, it is necessary to apply an iterative process from step 3 to step 5.

Step 3. Determine the digital image associated to join operation, it is necessary to use the following two operations:

1. Compute the join operation composed of all altitudes lower than α_k, that is, inside the range $[0, k-1] \rightarrow i = 0, j = k-1$.
2. Generate a binary image applying the transformation function of the Eqn. 6.

$$g_{\alpha_k}(x,y) = \begin{cases} 1 \text{ if } (x,y) \notin \bigcup_{i=\alpha_{\min}}^{k-1} S_{\alpha_i} \\ 0 \text{ otherwise} \end{cases} \tag{6}$$

3. If the new binary image is equal to previous altitude, it is necessary to skip the step 4, and use the value -1 in the encoded DEM file and continue with step 5.

Step 4. According to the binary image obtained in step 3, we proceed to use the run length encoding (RLE) method [8] [10]. Basically, RLE is a straightforward way of encoding data so that it can take less space. It relies on the string being encoded containing runs of the same character. In the RLE approach, it is important to indicate both, the frequency and the intensity encoded, but considering that the resulting image contains only two values (0 or 1), the value or intensity is not required in the final sequence. To determine the encoded image the following steps are applied.

1. By convention, the first frequency denotes the number of continuous zeros in the image.
2. By applying a top-down inspection, it is indispensable to count the number of pixels with same intensity and send this value to the encoded array, according to Eqn. 7.

$$value = \begin{Bmatrix} byte(value) \mid 80_{HEX} \text{ if } value < 128 \\ integer(value) \text{ if } value \geq 128 \end{Bmatrix}, \tag{7}$$

where:
byte (value) returns the byte representation of *value*. Additionally, we establish the most significative byte (MSB) to 1 using the Boolean OR operation.
integer (value) computes the integer representation of *value*.

By using this adjustment, it is possible to increase the compression factor because in some cases the total number of continuous pixels with same intensity should not be more than 128. To indicate this condition, the MSB of *value* is established to one; otherwise, an integer representation of *value* is stored in the new array.

3. Repeat the previous step until the end of the image is reached.
4. At the beginning of encoded array, we append the total number of encoded characters, which is the number of transitions between 0 and 1 in *g(x, y)*.

Step 5. Make a unitary increment to *k* variable, and repeat it from Step 1. The stop condition is used when $k > \alpha_{\max}$.

2.2 Decompression Method

The compressed DEM files contain three headers: Records A and C, DEM header, and the total of $\alpha_{max} - \alpha_{min} - 1$ RLE arrays. Retrieving original binary values from any RLE array do not provide the specific altitude at some point. If a point *p(x, y)* of source DEM has an altitude equals to *k*, this point has an intensity of 1 in total of

$(k-\alpha_{\min})\,g(x,y)$ images. When we perform the join operation in the range [$\alpha_{\min+1}$, $\alpha_{\max}$] a pixel *p(x, y)* has a 0 intensity. If Eqn. 6 is not accomplished $([k+1,\alpha_{\max}])$, then the gray-level in *g(x, y)* will be 1 $([\alpha_{\min+1},k])$. In conclusion the elevation of one point *p(x, y)* is determined by using Eqn. 8.

$$h(x,y)=\alpha_{\min}+\sum_{i=\alpha_{\min+1}}^{\alpha_{\max}} g_{\alpha_i}(x,y) \tag{8}$$

In addition, the required steps of decompression stage are the following:

Step 1. Both Records A and C are retrieved directly from the compressed file.

Step 2. Process the compressed DEM header, which contains the *width* and *height* of DEM image, as well as the minimum and maximum elevations. To generate an integer *width x height* matrix (referred as DEM image) and establish all the matrix values to minimum elevation. It is necessary to apply an iterative process from step 3 to step 5 for each RLE encoded array.

Step. 3. Obtain the number of transitions in current encoded array (*n*), and remember that the value of *n* has been written before each RLE encoded array. If the value of *n* is equal to -1, then we apply the next steps with the first preceded valid RLE array.

Step 4. Both the absolute pixel position (*position* variable) and the current intensity encoded *(intensity)* are initialized to zero.

Step 5. Read the next byte (b_1) from the compressed DEM file. If the most significative byte of this data is -1, then read next byte (b_2) and apply Eqn. 9 to obtain the value (n_0) of continuous encoded intensities. In the compression algorithm, the first n_0 refers to zero intensity. Additionally,

$$n_0=\begin{cases} b_2+(b_1\times(2\wedge 8)) & \text{if } MSB(b_1)=1 \\ b_1\,\&\,7F_{HEX} & \text{otherwise} \end{cases} \tag{9}$$

If the current intensity is equal to 0, we only increment with the value of n_0, the absolute position *(position=position + n_0)*. In other case, we need to make an n_0-unitary increment both the value of the DEM image at the position *p(x, y)*, and to the absolute position, as described in Eqn. 10.

$$\begin{aligned} x &= (position - width\times\lfloor position/width\rfloor) \\ y &= \lfloor position/width\rfloor \\ p(x,y) &= p(x,y)+1 \\ position &= position+1 \end{aligned} \tag{10}$$

It is necessary to obtain the logical complement of the intensity value. Repeat step 5 until process all RLE values.

3 Compressed DEM Management

In section 1, we cited the most important characteristics of this compression algorithm. We assume that it is not indispensable to apply a decompression method to obtain information about the elevation from the DEM (read data). Additionally, we can generate the contour layer, which has several map applications related to Geoprocessing area. These two operations are described in the next section.

3.1 Accessing Data Elevations

In some cases a DEM file requires up to 100 MB of storage space. Moreover, some person can require an application that works with 10, 20, or more DEMs. It is possible that any user does not have storage limitations, but in conditions that the use of lower memory, or the use of the hard disk space is crucial, the compressed DEM file could be useful. To retrieve the elevation at some point *p(x, y)* we cited Eqn. 8, however it requires to process all RLE encoded array. To reduce processing time, we apply the Eqn. 11.

$$p(x,y)=\alpha_h \neg g_{\alpha_h}(x,y)\neq 0,\left(g_{\alpha_{h+1}}(x,y)=0 \vee \alpha_h=\alpha_{\max}\right) \tag{11}$$

This equation basically consists of a lineal-search in all RLE encoded arrays, in the cases when the desired elevation is near α_{min+1}, the elevation is rapidly obtained, otherwise all arrays (closer to α_{max}) will be processed. In consequence, we have $\Theta(n/2)$ algorithm, where *n* denotes the total number of bytes to process. To improve this process the *bin-search* can be used. When *bin-search* starts, the total RLE arrays will be equal to *n* (see Eqn. 12).

$$A=\{a_0,a_1,\ldots,a_n\} \tag{12}$$

Formally, the binary-search function (BS) is described in Eqn. 13.

$$BS\left(A,k_1,k_2,m,g_{\alpha_m}(x,y)\right)=\left\{\begin{array}{l}\alpha_{\min}+k_1+1,\\ \text{if } g_{\alpha_{k_1}}(x,y)=0, g_{\alpha_m}(x,y)=1, k_2=m+1\\ \alpha_{\max}\ \text{if} g_{\alpha_{k_2}}(x,y)=1, k_2=n\\ BS\left(\{a_{k_1}\ldots a_{m'}\},k_1,m'-1,m',g_{\alpha_{m'}}(x,y)\right),\\ \text{if } g_{\alpha_{m'}}(x,y)=0\\ BS\left(\{a_{m'+1}\ldots a_{k_2}\},m'+1,k_2,m',g_{\alpha_{m'}}(x,y)\right),\\ \text{if } g_{\alpha_{m'}}(x,y)=1\end{array}\right\}, \tag{13}$$

where:

A denotes the RLE arrays to search.
k_1 and k_2 are the number of RLE array $k_1, k_2 \in \{0, 1, \ldots, n\}$.
m is the middle value of previous binary-search.
$g_{\alpha_m}(x,y)$ is the altitude at middle position in previous search.
m' is the new middle value, computed as $m'=\lfloor (k_2-k_1)/2 \rfloor + k_1$.

In the first execution of the search function $k_1=0,\ k_2=n,$ $m=0, g_{\alpha_m}(x,y)=0$. A special condition is reached when the *k*-RLE array is empty, because of $g_{\alpha_k}(x,y) = g_{\alpha_{k-1}}(x,y)$. To solve this ambiguity, the method searches the first previous RLE array, which is not empty.

3.2 Generating Contours Without Compression

Contours are lines drawn on a map that connect points of equal elevation. Contour lines are useful, because they allow us to show the shape of the land surface (topography) on a map [3].

The methodology to obtain contour lines form in a Digital Elevation Model is not complicated; some commercial tools to handle cartographic data make this task. However, when we work with DEMs, these systems require all elevation data, but our algorithm generates the contour lines directly form compressed DEM file. In this section we point out this process.

The single required parameter is the altitude interval to sample DEM (Δ), all additional information is available in the compressed data.

Step 1. Generate an *N x M* matrix and establish all values to 0, this matrix will represent the contour lines image *f(x, y)*. The size of the matrix is obtained from DEM compressed header.

Step 2. Take the first RLE encoded array (i.e. encoded data of $g_{\alpha_{\min+1}}$). Obtain original binary image by using Eqn. 8. If the array is empty, then apply the same criteria as we defined it in previous sections. With the original binary image, it is necessary to obtain its negated version.

Step 3. Compute the 8-connected contour of binary image. Strictly, an object pixel *p(x, y)* is part of contour, if the number of 8-neighbors with intensity equal to zero (background pixels) is at least equal to 1.

Step 4. Remove redundant contour pixels. All pixels that accomplish one of the masks defined in Fig. 1 are removed.

X	1	X
1	1	X
X	X	0

X	1	X
X	1	1
0	X	X

0	X	X
X	1	1
X	1	X

X	X	0
1	1	X
X	1	X

Fig. 1. Set of redundant pixels in detection masks

Step 5. Establish value of one at coordinates *(x, y)* in contour lines image, if the pixel *p(x, y)* is a member of non-redundant contour set obtained in the step 4.

Step 6. Process each remembered encoded array applying from step 4.

4 Tests and Results

In this section, we show in a detailed way, the obtained results with our approach, and compare it with commercial WinZip program (it uses LZW coding method). In this case the standard-compression option has been applied. The DEMs used for the tests are displayed in Fig. 2, and all DEM sources have the same spatial resolution (1,200 x 1,200 pixels) and their sizes are 10MB approximately. In Table 2 the numerical information about the obtained results appears.

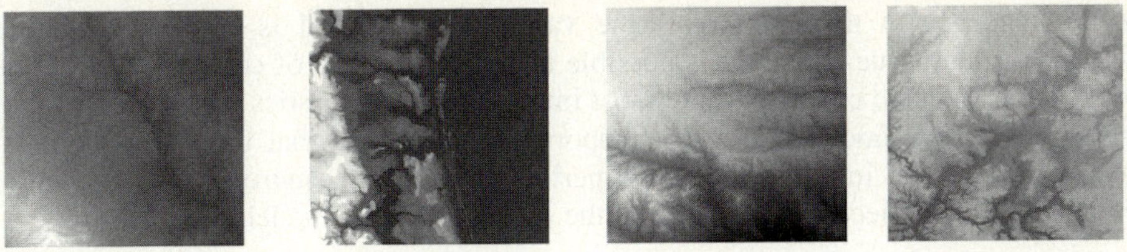

(a) Eagle (b) Jacksonville (c) Pratt (d) Cincinatti

Fig. 2. DEMs samples used in compression tests

Table 2. DEM compression results

Name	WinZip	Compression	Our Technique	Compression
Eagle	917,504	90.67%	1,645,939	83.27%
Jacksonville	262,144	97.33%	236,126	90.07 %
Pratt	868,352	91.17%	1,306,272	86.73 %
Cincinatti	819,200	91.67%	1,642,829	83.31 %

Regarding the contour line computation, in Fig. 3 (b) we show the obtained result with the DEM source that appears in Fig. 3 (a). The altitude interval is 25 meters.

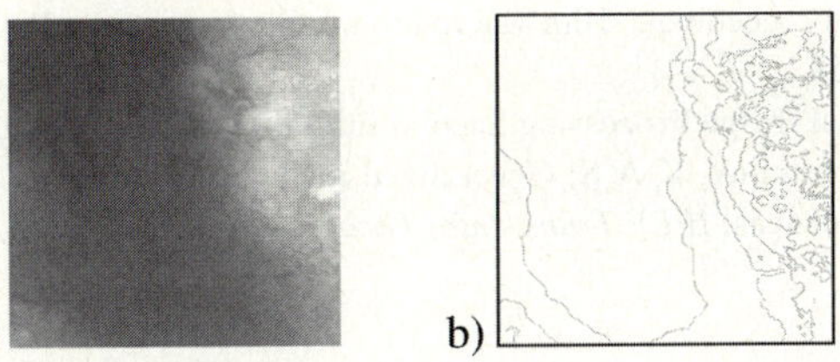

a) b)

Fig. 3. Contour lines result: (a) Original geo-image (b) Contour lines obtained from the compressed DEM file

5 Conclusions

In the present work, we develop a loss-less compression algorithm using RLE encoding approach. Although the compression factor is not the same, we can obtain this factor applying other techniques, it is important to mention that encoded elevations can be directly read from the compressed DEM file, and the computation of contours lines is easy and fast. The most important part of the test is the compression factor, which has been 80% high. To make some enhancements, this algorithm should be adapted as a new format to store and describe digital elevations models. Future works are oriented to study other encoding formulations that allow us to increase the compression coefficient and attempt to implement developed compression algorithms to compare them with our method. An important characteristic is that the frequency of the DEM is not high, because the pixels do not present changes in their structure; the value of each pixel is correlated to their neighbor values. Due to this, it is possible to apply this kind of compression. On the other hand, we have considered the most important characteristics of digital elevation models in our method, because it is important to point out that DEMs are complex geo-images, which involve several properties of a certain environment. In addition these properties reflect the *semantics* of the digital elevation models.

References

1. Maune D.F., Digital Elevation Model Technologies and Applications: The DEM Users Manual, Asprs Pubns, USA, (2001).
2. Application of digital elevation models to delineate drainage areas and compute hydrologic characteristics for sites in the James River Basin, North Dakota, USGS, USA, (1990).
3. Gousie M. B, Randolph F., Converting Elevation Contours to a Grid, Proceedings of the Eighth International Symposium on Spatial Data Handling, (1998).
4. Randolph F., Amir S., Lossy Compression of Elevation Data, USA, (1995).
5. Shantanu D. R., Sapiro G., Evaluation of JPEG-LS, the New Lossless and Controlled-Lossy Still Image Compression Standard, for Compression of High-Resolution Elevation Data, *IEEE Transactions On Geoscience And Remote Sensing*, Vol. 39, No. 10, (2001), pp. 2298-2306.
6. Creusere C. D., Compression Of Digital Elevation Maps Using Nonlinear Wavelets, IEEE, (2001), pp. 824-827.
7. Standards for Digital Elevation Models – Part 1, U.S. Geological Survey (USGS), USA, (2002).
8. Hinds, S.C, et al, A document skew detection method using run-length encoding and the Hough transform, IEEE, Proceedings, 10th International Conference on Pattern Recognition, Vol. 1, (1992), pp. 464-468.
9. Gonzalez R, Digital Image Processing Second Edition, Prentice Hall, USA, (2001).
10. Beenker, G F M, Immink, K A S, Generalized method for encoding and decoding run-length-limited binary sequences, *IEEE Trans. Info. Theory*, Vol. IT-29, No. 5, (1983), pp. 751-753.

Evaluation of the Perceptual Performance of Fuzzy Image Quality Measures

Ewout Vansteenkiste[1], Dietrich Van der Weken[2], Wilfried Philips[1], and Etienne Kerre[2]

[1] Image Processing & Interpretation Group, Ghent University
Sint-Pietersnieuwstraat 41, 9000 Ghent, Belgium
ervsteen@telin.ugent.be
[2] Fuzziness & Uncertainty Modelling, Ghent University
Krijgslaan 281 (building S9), 9000 Ghent, Belgium
dietrich.vanderweken@ugent.be

Abstract. In this paper we present a comparison of fuzzy instrumental image quality measures versus experimental psycho-visual data. A psycho-visual experiment we recently performed at our departments was used to collect data on human visual perception. The Multi-Dimensional Scaling (MDS) framework was applied in order to test which of our fuzzy image similarity measures correlates best to this human visual perception. Based on Spearman's Rank Order Correlation coefficient we will show that the M_6^p and M_{i3}^h measures outperform their peers as well as the commonly used MSE and PSNR measures, in the case where image distortions are less trivial to distinguish with the bare eye.

1 Introduction

In this paper we focus on instrumental image quality measures that are based on similarity measures initially introduced to express the degree of equality between two fuzzy sets. Fuzzy similarity measures can be applied in different ways to digital images. First of all, we used the different similarity measures to construct neighborhood-based similarity measures which also incorporate homogeneity [13]. Secondly, we have illustrated how fuzzy similarity measures can be applied to image histograms [12]. In this case similarity measures turned out to be useful for the comparison of two different kinds of histograms.

In a next step we now want to confirm these results based on experimental psycho-visual data. In former research, a Multi-Dimensional Scaling (MDS) approach towards analyzing and modeling image quality variations has been shown successful [3]. This approach is based on the fact that image quality is often determined by several underlying attributes (such as noise and blur e.g.) and uses a multidimensional geometric model to describe the mutual relationships between different perceptual attributes, as well as the relationships between these attributes and the overall image quality. The dimensionality of this model is determined by the number of independently varying perceptual attributes.

B. Gabrys, R.J. Howlett, and L.C. Jain (Eds.): KES 2006, Part I, LNAI 4251, pp. 623–630, 2006.
© Springer-Verlag Berlin Heidelberg 2006

The output or geometrical stimulus configuration can be used to create a perceptual distance matrix for the different stimuli, which can then be compared to the distance matrices created by our own fuzzy similarity measures. Herefore, Spearman's Rank Order coefficient is used. We will first explain our psycho-visual experiment in Section 2. Then we will introduce the fuzzy similarity measures in more detail in Section 3. We present our results in Section 4 and end with a concluding Section 5.

(a) Barbara (b) Face (c) Hill

Fig. 1. Test images used in our new psycho-visual experiment

2 Experimental Setup

Three scenes (Barbara, Face and Hill), Fig. 1, were used in an experiment where we wanted to check the visual quality of 8 state of the art noise-reduction filters. Therefore, the original image, a noisy one where artificial gaussian noise was added with a standard deviation $\sigma = 15$, together with 8 images denoised by 8 state of the art filters [2, 4, 5, 7, 8, 9, 15] were shown to 35 subjects, on the same calibrated display. An example of the test images can be seen in Fig. 2 for Barbara. First, 40 images were shown separately one after the other, of which the first 10 are used to train the subject. The subjects were asked to score the attributes noisiness, bluriness and overall image quality on a discrete scale from 0 to 5. Then 140 image couples were shown where image dissimilarity (how different one thinks the images are), on a scale from 0 to 5, and preference scores (which of the two images, left or right, one prefers in image quality), on a scale from -3 to 3, were asked. Finally, the data resulting from this experiment were processed using the MDS framework.

3 Fuzzy Similarity Measures

Many different similarity measures found in literature, originally introduced to express the degree of comparison between two fuzzy sets, were subject to an extensive theoretical study. We have investigated the similarity measures w.r.t.

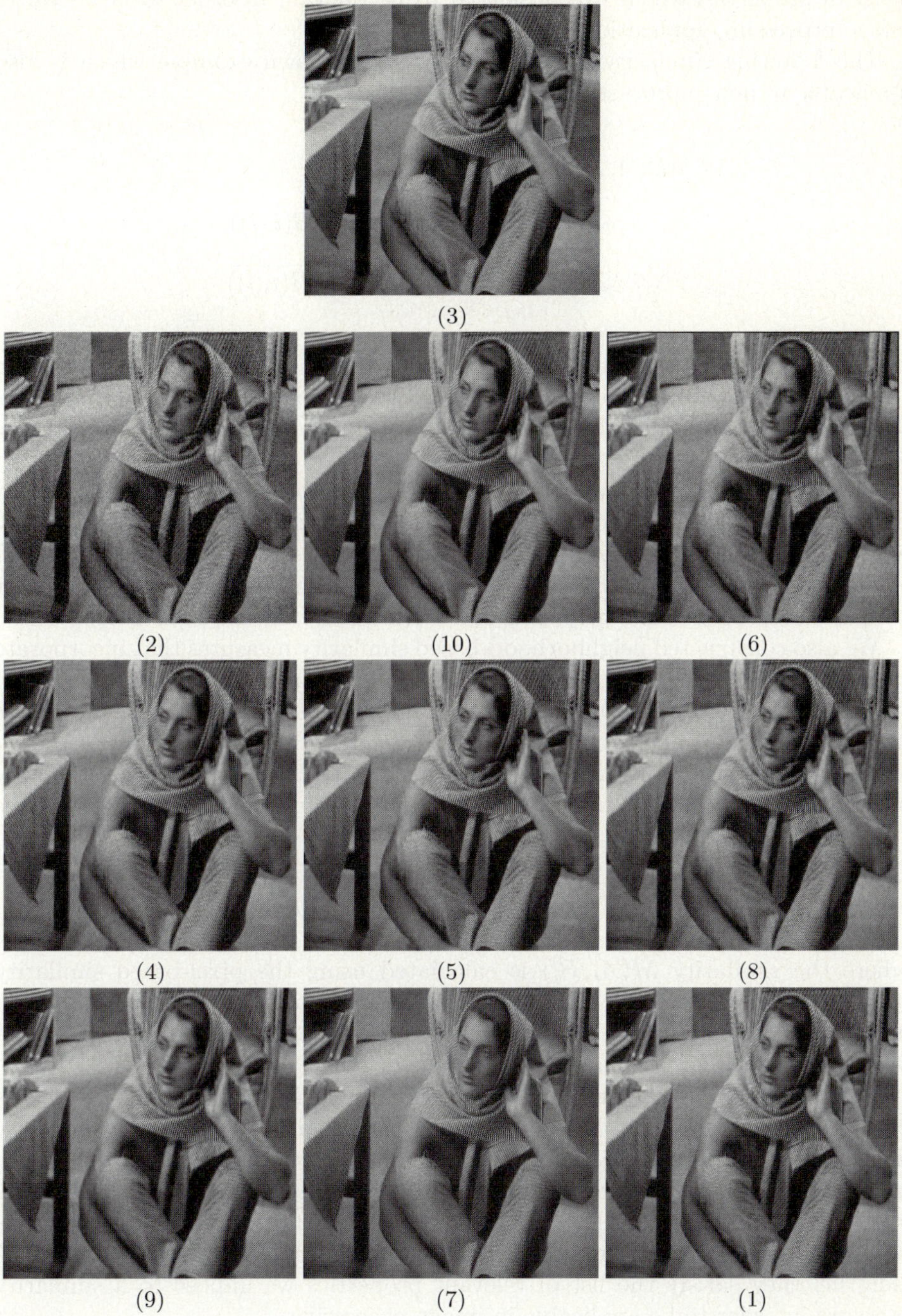

Fig. 2. The test images for Barbara as presented in our psycho-visual experiment

a list of properties which we considered to be relevant in order to be useful in image processing applications [10, 11].

The following similarity measure [1] is a well-known example which is also applicable in image processing:

$$M_6(A,B) = \frac{|A \cap B|}{|A \cup B|} = \frac{\sum\limits_{(i,j)\in X} \min(A(i,j), B(i,j))}{\sum\limits_{(i,j)\in X} \max(A(i,j), B(i,j))}$$

with A and B two digital images over the universe X of pixels. Note that we use the minimum to model the intersection between two fuzzy sets, the maximum to model the union and the sigma count to model the cardinality of a fuzzy set.

Unfortunately, similarity measures which are applied directly to normalized digital images did not show a convincing perceptual behaviour. That's why we need more advanced image comparison methods. First of all, we applied the pixel-based similarity measures to partitioned images in order to construct neighborhood-based similarity measures with a more robust behaviour.

We also constructed neighborhood-based similarity measures that incorporate characteristics of the human visual system [13]. Suppose the image is divided into N image parts of size 8×8 pixels, and the similarity between the image part A_i of image A and the image part B_i of image B is denoted by $M(A_i, B_i)$, then the similarity between the two images A and B is given by the weighted average of the similarities in the corresponding disjoint image parts. So, we have that

$$M^h(A,B) = \frac{1}{N}\sum_{i=1}^{N} w_i \cdot M(A_i, B_i),$$

where the similarity $M(A_i, B_i)$ is calculated using the pixel-based similarity measures restricted to the image parts A_i and B_i and the weight w_i is defined as the similarity between the homogeneity h_{A_i} of image part i in image A and the homogeneity h_{B_i} of image part i in image B.

Besides applying the similarity measures directly to the pixels of the considered images or neighborhoods of pixels, we investigated whether the similarity measures, having their origin in fuzzy set theory, could be applied in a meaningful way to the histograms of the considered images. A profound experimental study [12] of the applicability of similarity measures to normalized histograms resulted in 15 similarity measures which are appropriate for histogram comparison, i.e. they satisfy the list of relevant properties we impose to a similarity measure in order to be applicable in image processing.

Similarity measures can also be applied in a second way to associated histograms of digital images. The values of a histogram can be ordered in such a way the least occurring gray value is placed in the first position of the histogram and the remaining frequencies are ordered in increasing order. Then we can apply the different similarity measures to these ordered and normalized histograms. If the similarity measures are applied to ordered histograms we obtain 22 similarity measures which satisfy the list of relevant properties.

In this experiment we selected the 7 most relevant measures of all those described above. Therefore, we considered the similarity measure M_6 applied in different ways to a pair of images:

- applied neighborhood-based (M_6^p);
- applied neighborhood-based and incorporating homogeneity (M_6^h);
- applied to normalized histograms (H_6);
- applied to ordered normalized histograms (OH_6);

Furthermore, we considered the neighborhood-based similarity measures M_1^h, M_{18c}^h and M_{i3}^h, that also incorporate homogeneity. Again we refer to [10, 11, 12, 13, 14] for the exact details.

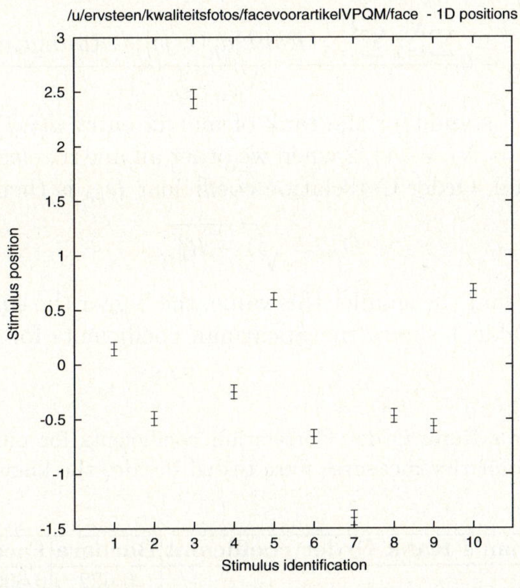

Fig. 3. This figure shows the 1D-geometrical output of the MDS framework for Face. On the X-axis the different filters as numbered in Fig. 2, on the Y-axis the perceived quality.

4 Results

Fig. 3 shows the 1D geometrical output configuration as optimized by the MDS framework from the combined results of the 35 subjects in our psycho-visual experiment. Each point corresponds to one of the filters shown in Fig. 2. The standard deviations on the positions are also plotted. All stimulus positions were found statistically significant and similar configurations were obtained for all three scenes. The Y-axis shows the perceived overall image quality and gives us a relative ranking. In that way, one can easily see that the original image (3) comes out best, followed by the 3Dwdft (10) en SADCT-AVGW (5) filter. The GOA filter (6) and BiShrink 3 filter (7) are ranked worst.

What we now are interested in is to know which of our fuzzy similarity measures corresponds best to the obtained psycho-visual configuration. We proceed as follows. From the stimulus positions, the euclidean distance matrix can be computed. Otherwise, using the similarity measures, we can also compute the different distance matrices directly from the input images. By comparing these matrices to the experimental euclidean distance matrix we obtain the similarity measure that fits the experiments best, or in other words that is closest related to human visual quality assessment.

The equivalence between two matrices is computed by Spearman's Rank Order Correlation coefficient. Let D_p be the psycho-visual distance matrix and D_s be a $N \times N$ similarity measure matrix, N being the number of input images used, let

$$R_{ps} = 1 - 6\frac{\sum_{i=2}^{N}\sum_{j=1}^{i-1}(Rank[d_p(i,j)] - Rank[d_s(i,j)])^2}{N_D(N_D^2 - 1)}$$

where $Rank[d(i,j)]$ stands for the rank of matrix entry $d(i,j)$ and is a number between 1 and $N_D = N(N-1)/2$ when we order all matrix elements ascendingly. The Spearman Rank Order Correlation coefficient D_{ps} is then given by

$$D_{ps} = \sqrt{1 - R_{ps}^2}.$$

One can easily see that the smaller this value, the bigger the equivalence between the matrices is. Table 1 shows the Spearman coefficients for our psycho-visual experiment.

Table 1. Spearman's Rank Order Correlation coefficients for our psycho-visual experiment. 7 fuzzy similarity measures were tested besides the known MSE and PSNR measures.

Spearman's Rank Order coefficient	Barbara	Face	Hill
M_1^h	0.622	0.509	0.622
M_{18c}^h	0.627	0.505	0.538
M_6^h	0.607	0.506	0.514
M_{i3}^h	0.589	0.497	0.516
H_6	0.738	0.584	0.68
M_6^p	0.478	0.531	0.516
OH_6	0.588	0.59	0.69
PSNR	0.586	0.642	0.634
MSE	0.586	0.642	0.634

5 Conclusions

From the results in Table 1 we can see that for the Barbara image the measure M_6^p performs best, that the measures M_{18c}^h, M_6^h and M_{i3}^h perform best for the Face image and the measures M_6^p, M_6^h and M_{i3}^h perform best for the Hill image. So these measures outperform the classical measures PSNR and MSE.

This can be explained by the nature of the different distortions. In a former experiment described in [6] where the distortions are rather large, the classical measures PSNR and MSE can grasp all differences best. In our experiment the difference in visual quality between the different images is much more subtle and therefore more sophisticated measures are necessary to detect the differences in visual quality.

It is of course not surprising that the neighborhood-based similarity measures perform better than the histogram similarity measures, since the histogram similarity measures do not incorporate the spatial properties of the different gray values, but only consider the frequency of occurrence. In an overall conclusion M_6^p is the best alternative to the classical similarity measures when it comes to human visual perception. In future we want to continue testing this by comparing it to a full reference model known to have good performance such as NTIA-VQM or VQEG.

Acknowledgements

We would like to thank Prof. Dr. ir. Jean-Bernard Martens for the assistance on the MDS package and ir. Stefaan Lippens for the creation of the web-interface on which the psycho-visual experiment was performed.

References

1. Chen S.M., Yeh M.S. & Hsiao P.Y., "A comparison of similarity measures of fuzzy values", *in: Fuzzy Sets and Systems*, vol. 72, 1995, pp. 79-89.
2. Dabov K., Foi A., Katkovnik V., & Egiazarian K., "Image denoising with block-matching and 3D filtering", *To appear in Image Processing: Algorithms and Systems V, 6064A-30, IST/SPIE Electronic Imaging*, 2006, San Jose, CA., 2006.
3. Escalante-Ramirez B., Martens J.B. & de Ridder H., "Multidimensional characterization of the perceptual quality of noise-reduced computed tomography images", *J. Visual Comm. Image Representation 6*, December 1995, pp. 317-334.
4. Foi A., Katkovnik V. & Egiazarian K., "Pointwise Shape-Adaptive DCT as an overcomplete denoising tool", *Proc. Int. TICSP Workshop Spectral Meth. Multirate Signal Process.*, SMMSP 2005.
5. Guerrero-Colon J.A. & Portilla J, "Two-Level Adaptive Denoising Using Gaussian Scale Mixtures in Overcomplete Oriented Pyramids", *Proceedings of IEEE ICIP conference*, Genova, Italy, Sept 2005, pp 105-108.
6. Kayagaddem V. & Martens J.B., "Perceptual characterization of images degraded by blur and noise: experiments", *Journal of Opt. Soc. Amer. A 13*, June 1996, pp. 1178-1188.

7. Pizurica A., Philips W., Lemahieu I. & Acheroy M., "A Joint Inter- and Intrascale Statistical Model for Bayesian Wavelet Based Image Denoising", *IEEE Transactions on Image Processing*, vol. 11, no. 5, 2002, pp. 545–557.
8. Sendur L. & Selesnick I.W., "Bivariate Shrinkage With Local Variance Estimation", *IEEE Signal Processing Letters*, vol. 9, no. 12, 2002, pp. 438-441.
9. Sendur L. & Selesnick I.W., "Bivariate Shrinkage Functions for Wavelet-Based Denoising Exploiting Interscal Dependency", *IEEE Trans. on Signal Processing*, vol 50, no. 11, 2002, pp. 2744-2756.
10. Van der Weken D., Nachtegael M. & Kerre E.E., "The applicability of similarity measures in image processing", *Intellectual Systems*, vol. 6 (1-4), 2001, pp. 231-248 (in Russian).
11. Van der Weken D., Nachtegael M. & Kerre E.E., "An overview of similarity measures for images", *Proceedings of ICASSP'2002 (IEEE International Conference on Acoustics, Speech and Signal Processing)*, Orlando, United States, 2002, pp. 3317-3320.
12. Van der Weken D., Nachtegael M. & Kerre E.E., "Using Similarity Measures for Histogram Comparison", *Lecture Notes in Artificial Intelligence*, vol. 2715, 2003, pp. 396-403.
13. Van der Weken D., Nachtegael M. & Kerre E.E., "Using Similarity Measures and Homogeneity for the Comparison of Images", *Image and Vision Computing*, vol. 22 (9), 2004, pp. 695-702.
14. Van der Weken D. "The use and the construction of similarity measures in image processing", *PhD thesis, Ghent University*, 2004 (in Dutch).
15. Van De Ville D., Nachtegael M., Van der Weken D., Kerre E.E., Philips W., Lemahieu I., "Noise Reduction by Fuzzy Image Filtering", *IEEE Transactions on Fuzzy Systems*, vol. 11, no. 4, 2003, pp. 429-436.

A New Color Image Enhancement Algorithm for Camera-Equipped Mobile Telephone*

Zhengyou Wang[1,2], Quan Xue[3], Guobin Chen[3], Weiming Zeng[1], Zhijun Fang[1], and Shiqian Wu[1]

[1] School of Information Technology, Jiangxi University of Finance & Economics, Nanchang, 330013, China
[2] Key Laboratory of High-Performance Computing Technology, Jiangxi Province, Nanchang330022, China
wangzhengyou@jxufe.edu.cn, zengweiming2004@163.com, fangzhijun@21cn.com, shiqian.wu@gmail.com
[3] Department of Information Science and Electronic Engineering, Zhejiang University, Hangzhou 310027, China
jackyxue2002@163.com, chenguobin@zju.edu.cn

Abstract. We present in this paper a color image enhancement algorithm with high speed, high quality, and low memory usage for the camera-equipped mobile telephone. To achieve this, we first introduce a propagation scheme that estimates and computes the statistical characteristics of an image, then we apply a piecewise nonlinear transformation in the R/G/B channels separately. If the value of a pixel is below minimum or above maximum, it can be ignored by using range stretching, then a narrow intensity range between minimum and maximum may be expanded to a much broader range defined by an exponent function, which consists of mean and variance of the pixel values, and its power is chosen by the trials. The optimal strategies for fast processing in the code style are also discussed. We demonstrate our color image enhancement system and compare it with other feasible techniques on the camera-equipped mobile telephone. The proposed algorithm can be carried out in about 2 seconds using the ARM7 processor and generally performs much better than traditional contrast enhancement methods with higher image quality.

1 Introduction

Nowadays camera-equipped mobile telephones are widely used because of their convenience, but the images from them usually suffer from poor illumination and color distortion, so contrast enhancement of the images is necessary. However, the computational cost of traditional effective approaches is generally quite expensive and

* This work was partially supported by Innovation Fund of Jiangxi University of Finance and Economics; Fund of Key Laboratory of High-Performance Computing Technology (JXHC-2005-004), Jiangxi Province; and the Science and Technique Program of Educational Department of Jiangxi Province (No: [2006]232).

B. Gabrys, R.J. Howlett, and L.C. Jain (Eds.): KES 2006, Part I, LNAI 4251, pp. 631 – 638, 2006.
© Springer-Verlag Berlin Heidelberg 2006

requires high speed processors and vast memory space. With the continuous development of hardware, PCs can easily achieve such high speed processing and provide abundant memory space. In a mobile telephone however, the situation is not so good for the platform is a handheld device and it has very limited computational power and memory space compared with PCs. Therefore, the color image enhancement algorithm to be applied for the mobile telephone must be sped up and shrunk down in terms of computational complexity, program size and memory usage.

There are a number of techniques for color image enhancement including histogram equalization [1], histogram modification [2], Curvelet transformation [3], Land's Retinex theory [4], nonlinear filtering technique [5], clustering method [6], Wavelet transformation [7], fuzzy logic [8], and decimation method [9]. Although some algorithms have proved most useful and effective, when we consider the limitations discussed above, these methods cannot be applied directly and successfully for lack of effective platform support. This fact must be taken into account when we want to find a feasible solution on the mobile telephone.

Generally, spreading out the range of scene illumination can increase the contrast in the gray and color image, and this kind of technique is called contrast stretch or named transform stretch. As a simple method, contrast stretch and manipulation [10] is based on pixel-by-pixel stretching and can be chosen as one of the possible approaches. It is an early method and has been proven not the best way for color image enhancement, but this method can avoid the computationally intensive process because of its low-complexity [11]. With the proper modification and optimization, a real-time color image contrast enhancement system may possibly be implemented on the mobile telephone.

The proposed algorithm in our paper has two processes. Statistical parameters are firstly retrieved from the picture through a fast approximate computation from R/G/B channels separately, that is, the minimum pixel value, maximum pixel value, average pixel value and variance pixel value of the whole image. After analyzing the characteristics of a picture, every pixel is then treated by the contrast stretch which, unlike the conventional techniques, is a piecewise nonlinear transformation based on the statistical information from the first step. In this way, color contrast and fine details can be optimally enhanced, while dynamic range expansion can be controlled with high image quality.

The paper is structured as follows: in Section 2 the statistical characteristics of original image are analyzed, then based on these parameters a piecewise nonlinear transformation is given in Section 3, and in Section 4 software optimization strategies including trade off between storage space and computational power are discussed. Performance shows a nearly real-time and low complexity color image enhancement system is realized on the mobile telephone in Section 5. The last section summarizes our conclusion.

2 Statistical Characteristics Analysis

The motivation of our proposed algorithm comes from the observation of many images from camera-equipped mobile telephones. Since most cameras in mobile telephones have low-resolution and no zoom lens, the pictures are gloomy and blurry, especially when the light is not enough, e.g. night time.

The signal intensity in this kind of image is usually distributed in the mid-range of pixel intensity, and only a few pixels are out this range, so the transformation stretch is effective to expand the original local value to the whole value.

Here we defined twelve parameters to judge the picture characteristic and act as the threshold criteria for the transfer function as follow. The minimum, maximum, mean, and variance of pixel values with R/G/B channels separately, described as min/max/avg/var. In fact, the whole computational process in the mobile platform takes a lot of time to search for these parameters in the whole picture. So in the following paper we use a propagation scheme to get the min/max/avg, similar to the method used in [12].

The propagation rule from a neighbor pixel $(m-1,n)$ to the pixel (m,n) is defined as follows:

$$lavg_{m,n} = (1-C)\times lavg_{m,n} + C\times lavg_{m-1,n} \tag{1}$$

where C is called conductivity factor, ranging from 0 to 1. The matrix $lavg$ stands for the average map, which is initialized with the image intensity values. The above propagation rule is sequentially applied in row and column directions in the picture..

In order to compute the min/max maps, we introduce in the following a conditional propagation scheme. Assume that $l\min$ and $l\max$ stand for the min/max maps, respectively, and are initialized with the image intensity values. Let conditional propagation scheme from $(m-1,n)$ to (m,n) be defined as follows:

$$\begin{cases} if\,(l\min_{m-1,n} < l\min_{m,n}) \\ l\min_{m,n} = (1-C)\times l\min_{m,n} + C\times l\min_{m-1,n} \\ if\,(l\max_{m-1,n} > l\max_{m,n}) \\ l\max_{m,n} = (1-C)\times l\max_{m,n} + C\times l\max_{m-1,n} \end{cases} \tag{2}$$

Through this propagation model, we can conveniently get the values of min/max/avg. using the average value as the middle result, the variance representation of the red, green and blue channels may be computed directly by their computational formula.

3 Determining Stretching Function

In the following section, we will describe the key transform stretch function in our proposed contrast enhancement algorithm. A new intensity is assigned to each pixel according to a piecewise transfer function that is designed on the basis of the statistical parameters from the picture.

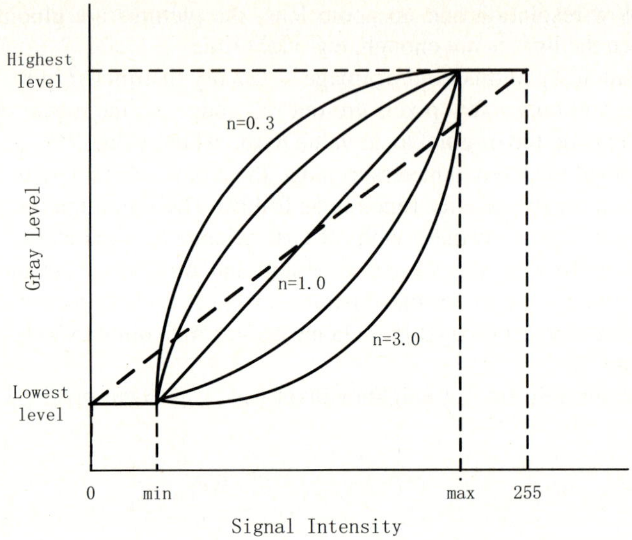

Fig. 1. Piecewise nonlinear transformation function from the signal intensity to the color level

Ordinarily, one can not obtain a good picture by a simple linear transform of signal intensity into a gray level as shown by the dashed line in Fig.1, which transforms the strongest intensity into highest gray level and the weakest signal into the lowest gray level directly. As a result, if objects in the dark background have lower contrasts, they would be omitted and blurred.

In such cases, we define a piecewise nonlinear transformation with three segments identified as the solid line in Fig. 1 to improve the contrast.

Base on the twelve statistical parameters for every pixel component in the image, a stretching function from pixel to pixel is designed as followed:

$$y_i = \begin{cases} \min \\ (\max - \min) \times \left(\dfrac{x_i - avg + 2\,\mathrm{var}}{4\,\mathrm{var}} \right)^n + \min \\ \max \end{cases} \qquad (3)$$

The essential idea behind this stretch function is to take advantage of range stretching. The narrow intensity range seen in the original image is often expanded to a much broader range.

The value of the signals below the threshold min and above the threshold max can be ignored. The coefficients avg and var are calculated with the min and max in the first step, and the exponent n in the definition (3) is determined by the trials so that significant information in the image is preserved. The trials should be repeated many times in our experiments until most of the outputs are acceptable and enhanced substantially.

4 Software Optimization

Our color image contrast enhancement can be developed further from software optimization in ANSI C in the Linux operation system. The optimal implementation makes no assumption about any specific hardware platform, and it can work well in any operating system. File operation and memory management is accomplished with standard C library functions too. The instruction extensions and processor features are unable to be used here, although it is very effective for acceleration.

Our strategies include the use of efficient and simple data structures with minimum pointers; the reduction and improvement of input and output operations; using short byte data type and integer type; using look-up tables to avoid repetition and so on. Especially, a trade-off between the amounts of storage space and direct computation is made for higher speed computation. The Richard book [13] gives detailed information about software optimization.

5 Performance Results

The proposed method for color image enhancement was implemented on the DoEasy E898 mobile telephone produced by the Chinabird Company in China, which has an ARM-7 embedded processor with 26 M core frequency using the Linux operating system. Experiments indicate that our proposed color image enhancement system can be carried out in about 2 seconds, nearly real-time application.

Subjective comparison in image quality was tested here. In fact, we tested our method on various types of scenes and illumination. In this section, however, we have to only show several typical pictures as examples to demonstrate the enhanced results by our proposed program. To verify the effects caused by different enhancement techniques, the comparison experiments were also conducted with the global contrast enhancement and histogram equalization. Here we only select the feasible methods used in the mobile telephone. The algorithms that are too slow are not included, for they lack significance in the actual application.

(a) (b) (c) (d)

Fig. 2. Applying our proposed algorithm on the nature scene in sun light (a) Original image. (b) Image enhanced by our algorithm. (c) Image enhanced by global contrast enhancement. (d) Image enhanced by histogram equalization.

Figure 2 (a) is an original nature scene with low contrast in the sun, and the enhanced image after applying piecewise nonlinear transformation is shown in figure 2 (b). Contrasting the original picture, we can find the green color is compensated obviously, so the whole scene is more beautiful and realistic. This image is clearly better than the ones in fig 2 (c) and (d), which used the traditional methods. Figure 3 (a) shows an example of an office image illuminated by the artificial light at day time. In contrast to the original picture, we find that the image enhanced by our method is clearer and brighter. The image enhanced by the traditional methods also get good results in the figure 3 (c) and (d). Possibly, this kind of image can be regarded as an easy picture for the enhancement.

(a) (b) (c) (d)

Fig. 3. Applying our proposed algorithm on an office scene with artificial light. (a) Original image. (b) Image enhanced by our algorithm. (c) Image enhanced by global contrast enhancement. (d) Image enhanced by histogram equalization.

Finally we discuss the robust property of our method in figure 4 and figure 5. The original image, figure 4 (a) was taken in a bus at night. Figure 4 (b) illustrates that our method reduces the enhanced image details, and a higher image quality than the other approaches in (c) and (d). On the contrary, in figure 5(a) and (b), sometimes we can notice that our proposed algorithm does not reduce an enhanced result under very poor conditions without necessary light, figure 5 (c) and (d) show that the simple methods can not reduce a good result either. The picture in figure 5 (b) is not a successful application. It proves that our proposed algorithm is not very effective in images with very low contrast.

(a) (b) (c) (d)

Fig. 4. Applying our proposed algorithm on the bus scene at night with artificial light. (a) Original image. (b) Image enhanced by our algorithm. (c) Image enhanced by global contrast enhancement. (d) Image enhanced by histogram equalization.

Note that there is a murky-gray effect in our processed night time images. We think there are two reasons. One is due to some mean offset of pixels gray levels, and the other is that contrast enhancement is done on the red, green and blue channels separately, so the hue of the image is not preserved effectively. We should do further research on this phenomenon for an additionally improved image.

(a) (b) (c) (d)

Fig. 5. Applying our proposed algorithm in a night scene with very low contrast (a) Original image. (b) Image enhanced by our algorithm. (c) Image enhanced by global contrast enhancement. (d) Image enhanced by histogram equalization.

6 Conclusion

In this paper we present a fast approach for color image contrast enhancement based on a piecewise nonlinear transformation. Our approach is not only fast to implement, but also it has the adaptive enhancement property by analyzing the statistical information in the picture. We demonstrated the performance of our method on four types of images within different scenes and illuminations. Our proposed algorithm has good performance to deal with color images compared with traditional methods, and a nearly real time implementation is realized on the mobile telephone.

At the same time, we notice that our proposed algorithm is not very effective in all trials, that is, the enhanced results depend on the illumination conditions. In the very dark scenes which lack necessary illumination, the piecewise nonlinear transformation function needs to be changed or adjusted adaptively. That means a new feasible transfer function or different parameters should be adopted as a supplement in the very low contrast image.

References

1. Stark, J.A.: Adaptive Image Contrast Enhancement Using Generalizations of Histogram Equalization. IEEE Trans. on Image Processing. Vol. 9, No. 5. (2000) 889-896
2. Caselles, V., Lisani, J.L., Morel, J.M.: Shape Preserving Local Histogram Modification. IEEE Trans. on Image Processing, Vol.8, No.2, (1999) 220-230
3. Starck, J.J., Murtagh, F., Candes, E.J.: Gray and Color Image Contrast Enhancement by the Curvelet Transform. IEEE Trans. on Image Processing. Vol.12, No. 6. (2003) 706-717
4. Jobson, D.J., Rahman, Z., Woodell, G.A.: Properties and Performance of a Center/Surround Retinex. IEEE Trans. Image Processing, vol. 6, no. 3, (1997) 451-462
5. Kao, W.C., Chen, Y.J.: Multistage bilateral Noise Filtering and Edge Detection for Color Image Enhancement. IEEE Trans. On Consumer Electronics, Vol.51, No.4, (2005) 1346-1351
6. Hanmandlu, M., Jha, D., Sharma, R.: Localized contrast enhancement of color images using clustering. IEEE International Conference on Information Technology: Coding and Computting, Las Vegas, NV, USA (2001)
7. Chambolle, A., DeVore, R.A., Lee, N.: Nonlinear wavelet image processing: variational problems, compression, and noise removal through wavelet shrinkage, IEEE Trans. Image Process., (1998) 319-335

8. Hanmandlu, M., Jha, D., Sharma, R.: Color image enhancement by fuzzy intensi-fication. International Conference on Pattern Recognition, Vol. 3. Barcelona (2000)
9. Cho, H.H., Kim, S.H., Cho, T.K.: Efficient Image Enhancement Technique by Decimation Method. IEEE Trans. On Consumer Electronics, Vol.51, No.2, (2005) 654-659
10. Russ, J.C.: The Image Processing Handbook. 4th edn. CRC Press (2002).
11. Jain, A.K.: Fundamentals of Digital Image Processing. Englewood Cliffs, NJ. Prentice-Hall. (1989)
12. Yun, Z.Y., Bajaj, C.: A Fast and Adaptive Method for Image Contrast Enhancement. IEEE International Conference on Image Processing, Vol. 1. Singapore (2004)
13. Richard, G.: The software optimization Cookbook: High-performance Recipes for the Intel Architecture. Intel Press, USA (2002)

Human Arm-Motion Classification Using Qualitative Normalised Templates

Chee Seng Chan, Honghai Liu, and David J. Brown

Institute of Industrial Research, The University of Portsmouth
Portsmouth PO1 3HE, England, UK
{cheeseng.chan, honghai.liu, david.j.brown}@port.ac.uk

Abstract. This paper proposes an approach to classify human arm motion using qualitative normalized templates. The proposed method consists of construction of human arm model, qualitative representation of prior knowledge of human arm motion and a search algorithm. First, convention robotic model is employed to build up a generic vision model for a human arm; Secondly, qualitative robotic model in [1] is used to construct qualitative normalised templates; Finally a search algorithm is provided to match the vision model with the templates in image frames. Experimental evaluation demonstrates that the proposed method is effective for the classification of human-arm motion. Future work will focus on extending the proposed method to the classification of a full human-body motion.

1 Introduction

Classification of human motion accurately is still difficult despite the recent work in Section 2. Besides, human body is richly articulated, for an example, a simple stick model describing the pose of arms, legs, torso and head requires more than 20 degrees of freedom. In general, the more complex a human body model, the more accurate the tracking results, but the more expensive the computation. In theory, joint angles are sufficient for recognition of people by their motion. However, accurately recovering joint angles from a human-motion video is still an unsolved or not well-solved problem. In addition to that, computational cost of the model-based approaches is quite high. Furthermore, humans typically wear clothing which may be loose or textured, and with significant variations in the appearance of the human throughout the sequence, this makes the task even difficult to identity limb boundaries and segment the main parts of the body. Nevertheless, illumination, viewing conditions, self-occlusions, relative position and orientation, all contribute to make the task more challenging.

In this work, we focus on the classification of human-arm motion. The proposed method consists of construction of human arm model, qualitative representation of prior knowledge of human arm motion and a search algorithm. First, conventional robotic model is employed to build up a generic vision model for a human arm, that is to say, we use robotic kinematics to construct a stick model; Secondly, qualitative robotic model in [1] is used to construct qualitative

B. Gabrys, R.J. Howlett, and L.C. Jain (Eds.): KES 2006, Part I, LNAI 4251, pp. 639–646, 2006.
© Springer-Verlag Berlin Heidelberg 2006

normalised templates; Finally a search algorithm is provided to match the vision model with the templates in image frames. Our task aims to classify human complex motions using the extension of the proposed approach.

The rest of the paper is organized as follows: Section 2 presents related work in this research line. Section 3 introduces the qualitative normalised templates for human-arm motion. Section 4 adapts conventional robotic kinematics to the model of human arms. Section 5 evaluates the proposed method. Finally, concluding remarks is provided.

2 Related Work

Generally speaking, human motion classification can be divided into three phases, detection, model of prior knowledge and tracking. Given the complicated nature of human shape and motion, it could be very difficult to perform robust human tracking without the use of a prior model. Modeling the human body as rigid parts linked in a kinematic structure is a simple yet accurate model for tracking purpose where different approaches have been employed to provide dense information and obtain robust estimation of motion parameters. Approaches and their applications cover stick models [2], ellipsoids models [3], 2D models [4] and 3D [5], and even deformable models [6]. However, not only do most of these approaches require manual initialisation in the first frame, but also need the fusion of spatial and temporal information which usually leads to expensive computation. In Bregler and Malik [7] work, arbitrary 3D human motion under scaled orthographical projection is encoded in a 6 dimensional twist vector. Estimation of the motion parameters are similar to [3] but with an additional constraint to reduce the search space. The advantages of such solution to parameterised motion model are where Euler angles suffer from singularities and as opposed to quaternion, such that it is not necessary to constraint a number of parameters larger than the degrees of freedom to a set of admissible values. Bissacco [8] performed tracking by minimizing a cost function using gradient descent scheme with respect to unknown position and shape of body parts. The difference between this approach to standard approach [7] is where a visibility function that model self-occlusion was proposed. Although in both [7, 8] works are able to track under self occlusion and estimate scale changes but the need of initialisation of first frame is impractical to real time tracking.

Using a prior human model for the classification of human motion is a effective strategy. First, it makes classification algorithms intrinsically robust; Secondly human body's structure, constraints and other prior knowledge could be fused for its tracking task; Finally it will make easier the combination of 2D and 3D models. However, two major disadvantages of methods using a prior model are model construction and high computational cost. For instance, automatic initialisation using particle filter scheme [9] leads to expensive computation. On the other hand, approaches without prior models, e.g., [10, 11] usually rely on heuristic procedures to find correspondences of body parts between frames of video sequences.

3 Qualitative Normalised Templates for Arm Motion

Models for the description of the geometric structure of human body are stick figures, 2-D contour, volumetric models and hierarchical models. We proposed a method of representing human body using qualitative normalised stick figure. The proposed model serves as prior knowledge to predict motion parameters, to interpret and recognize human behaviours. Recent research shows that tracking and localizing the human body accurately is still difficult despite the recent work on structure-based methods. In theory, joint angles are sufficient for recognition of people by their motion. However, accurately recovering joint angles from an arm-movement video is still an unsolved or not well-solved problem. In addition, the computational cost of the model-based approaches is quite high. The qualitative normalised stick model is come up with negotiating the tradeoffs between accurately recovering trajectories and computational cost.

We adapt the qualitative model of kinematic robots proposed by Liu et al in [1] into the stick figure model. For instance, a n-link planar robotic manipulator, the qualitative representation of the end-effector can be described as,

$$\begin{cases} qp_l = \overset{n}{\underset{i=1}{\oplus}} qp_l^i \,|qp_l^i \in UC_{ql} \\ qp_\theta = \overset{n}{\underset{i=1}{\oplus}} qp_\theta^i \,|qp_\theta^i \in UC_{q\theta_i} \\ C_{dof} = UC_{dof} \end{cases} \tag{1}$$

where

$$\begin{aligned} UC_{qp_l} &= \left[\frac{qp_1^i}{\sum_{i=1}^{n} l_i}, \frac{qp_2^i}{\sum_{i=1}^{n} l_i}, \cdots, \frac{qp_{(s_i-1)}^i}{\sum_{i=1}^{n} l_i}, \frac{l_i}{\sum_{i=1}^{n} l_i} \right] \\ UC_{qp_\theta^i} &= \left[\frac{q\theta_1^i}{2\pi}, \frac{q\theta_2^i}{2\pi}, \cdots, \frac{q\theta_{(r_i-1)}^i}{2\pi}, 1 \right] \\ UC_{dof} &= [qp_\theta^i = i],\ i \in (0,\ 1, \cdots,\ n) \end{aligned} .$$

Here $UC_{qp_l^i}$ stands for the translation component of the ith link segment in a unit circle, $UC_{qp_\theta^i}$ for the orientation component. The constraint function C_{dof} is employed to define the degrees of freedom constraints between components. $qp_\theta^0 = 0$ stands for the base of the robot when i is equal to zero. The qualitative representation of the end effector is a qualitative addition of translation and orientation components of each link segment based on its constraints on degrees of freedom. The role of qualitative addition can be played by a variety of qualitative techniques. For example, fuzzy arithmetic can be employed given the components are described by fuzzy numbers. In this case fuzzy qualitative trigonometry [12] is employed to calculate qualitative arithmetic where fuzzy qualitative quantity space Q_X is introduced to represent qualitative states of a Cartesian translation Q_X^d and orientation Q_X^a (see Figure 1). It means that distance and angle measurement in fuzzy qualitative coordinate are depends on the numbers and fuzzy characteristic of the elements of a fuzzy qualitative quantity space. For example, a fuzzy qualitative version of a Cartesian position given for an orientation range $[0\ \Theta]$ and a translation range [0 L],

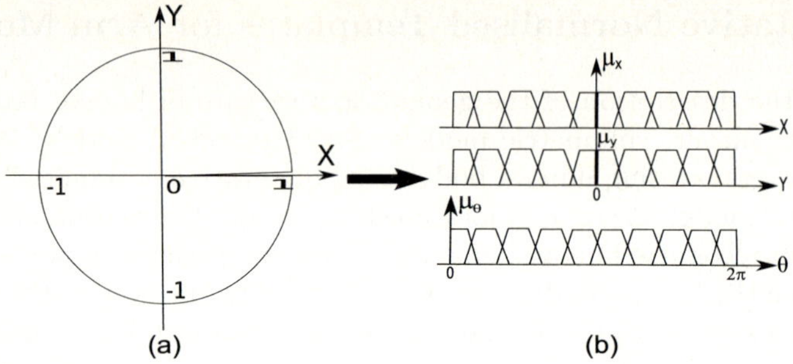

Fig. 1. (a) Conventional Unit Circle and (b) Fuzzy Qualitative Circle

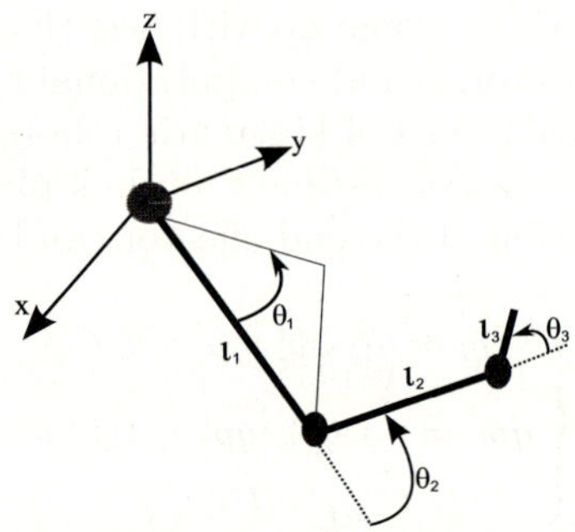

Fig. 2. A simplified model for a human arm

$$Q_X = \{Q_X^a, Q_X^d\} \tag{2}$$

where

$$Q_X^a = [QS_a(\theta_1), ..., QS_a(\theta_i), ...QS_a(\theta_m)]$$
$$Q_X^d = [QS_d(l_1), ..., QS_d(l_i), ...QS_d(l_n)]$$

$QS_a(\theta_i)$ denotes the state of an angle θ_i, $QS_d(l_j)$ denotes the state of a distance l_j; m and n are the number of elements of the two quantity spaces. It is noted that θ_1, l_1 and θ_m and l_n are equal to 0 and Θ, L respectively to ensure that the description is closed. The measurement of qualitative position P can be denoted by $P(QS_a(\theta_i), QS_d(l_j))$. Therefore, each human motion pose can be defined by using a qualitative normalised template.

For a simplified model of a human arm shown in Figure 2, let us consider it as a three-link planar robot whose links from the hand to the shoulder are l_1, l_2 and l_3, its joint angles are θ_1, θ_2 and θ_3. Equation 1 is used to map this parameters to a normalized state, fuzzy qualitative trigonometry can be used to do the calculation among this parameters. Hence, we can normalise links qp_l^i and joint angles qp_θ^i in fuzzy qualitative terms to construct a stick model for the arm. The tradeoffs of accurate models and computational cost is negotiated by the setting of the orientation and translation components of a fuzzy qualitative unit circle. The actual values of the links p^i and joint angles θ^i in their normalised terms are,

$$p_l^i = \frac{l_i}{l}, \quad p_\theta^i = \frac{\theta_i}{2\pi} \tag{3}$$

where $l = \sum_{i=1}^{3} l_i$ and i=1, 2 and 3.

4 Human-Arm Motion Model

Kinematics is a science of movement without focusing on force, which affects the movement. The construction of kinematic model for robot manipulators is a well establish field in robotics perspective [13]. The unknown kinematic parameters are most commonly identified from end effector pose or robot joint position reading. Unfortunately, such information is not available when constructing a kinematic model of a human. But, this limitation doesn't constraint researchers from computer vision as identification of kinematic models for human has been heavily studied since then as well as biomechanics. Bregler and Malik [7] were the first to exchange affine motion model, with kinematic model for motion estimation. With this in mind, we extend the framework into exploiting kinematic model as human motions representation. In this paper we focus on the problem of finding a kinematic model of human motion in a video sequence. The ultimate goal is to build a system that, if properly initialised, can reliably model human motions from a sequence of monocular images. We do not consider the issue of model initialisation, instead we assume that the configuration of the object in the first frame is given, for example, by manual initialisation.

Let us denote a n-link spatial robot with its home position as $QS_d(\mathbf{p_O})$, its end-effector position $QS_d(\mathbf{p})$ can be obtained by the transformations of its link components $\mathbf{H}_i$ or the transformations ${}^{i-1}_{i}\mathbf{H}_{DH}$ in D-H parameter terms, respectively,

$$\mathbf{p} = \prod_{i=1}^{n} \mathbf{H}_i \cdot \mathbf{p_O} = \prod_{i=1}^{n} {}^{i-1}_{i}\mathbf{H_{DH}} \cdot \mathbf{p_O} \tag{4}$$

where ${}^{i-1}_{i}\mathbf{H}_{DH}$ is Denavit-Hartenberg kinematic structure whose general form is given in the following,

$${}^{i-1}_{i}\mathbf{H}_{DH} = \begin{bmatrix} C\theta_i & -S\theta_i & 0 & a_{i-1} \\ S\theta_i C\alpha_{i-1} & C\theta_i C\alpha_{i-1} & -S\alpha_{i-1} & -d_i S\alpha_{i-1} \\ S\theta_i S\alpha_{i-1} & C\theta_i C\alpha_{i-1} & C\alpha_{i-1} & d_i C\alpha_{i-1} \\ 0 & 0 & 0 & 1 \end{bmatrix} \tag{5}$$

where C stands for cos; S stands for sin, where θ_i is rotation angle from X_{i-1} to X_i measured about Z_i, joint variable from revolute joints, α_{i-1} is twist angle from Z_{i-1} to Z_i measured about X_{i-1}, d_i is distance from X_{i-1} to X_i measured along Z_i, joint variable for prismatic joints, a_{i-1} is link offset, distance from Z_{i-1} to Z_i measured along X_{i-1}. X_i, Y_i, Z_i is the coordinate axises of ith coordinates system mounted on ith link segment. For a graphical representation please see Craig's [14]. Further, robot kinematics in equation 5 can be obtained as,

$$\mathbf{p}_e = \prod_{i=1}^{n} \mathbf{H}_{DH} \cdot \mathbf{p}_0 = \begin{bmatrix} \mathbf{R}_{3\times3} & \mathbf{P}_{3\times1} \\ 0_{1\times3} & 1 \end{bmatrix} \cdot \mathbf{p}_0 \tag{6}$$

where $\mathbf{R}_{3\times3}$ represents the rotation about the axis and $\mathbf{P}_{3\times1}$ represents translation along the axis. This is a de facto standard in robotic industries to model robot manipulators [13]. As a fundamental, this kinematic model shown in equation 4 is employed to represent human motion in planar direction. We model the human skeleton as a kinematic chain of rigid bodies. Each body segment is represented by a rigid link, and the links are connected together by joints. To restrict the set of admissible motions, we assume that each joint allows for one single degree of freedom, a rotation around its axis. This constraint is justified by the fact that typically the motion of the limbs can be approximated as planar around an axis perpendicular to the direction of motion as shown in Figure 2, the arm's DH structure is given in Table 1.

Table 1. DH-parameters for a Simplified Human-Arm

i	α_{i-1}	a_{i-1}	d_i	θ_i
1	0	0	l_1	θ_1
2	0	0	l_2	θ_2
3	0	0	l_3	θ_3

5 Experimental Evaluation

In this section, we describe preliminary experiments in employing robotic solution to the human-arm motion representation. We collected a sequence of human arm planar motion as shown in Figure 3. The sequences have length of 11 frames. In this sequence, we modelled the planar movement of the human arm using a kinematic model with 3 links, pictured in Figure 2.

Manual initialisation in the first frame of the sequence by specifying the geometry of the links and by finding their position in the image. Once this initialisation is completed, the system performed the human motion analysis. In this solution, we employed a standard de facto robotic solution in modeling 2 links robotic arm motion in planar to 3 links human arms with assumption that shoulder position is fix all the time. The idea here is assumed that, given a human is successfully segmented from the scene, the kinematic model will be able to model and classify the motion performed (eg. moving upward, moving downward) in each time frames or selected time frames with respect to the qualitative normalised templates.

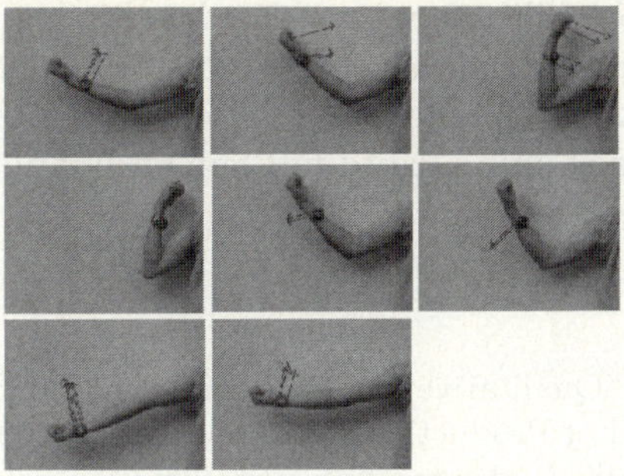

Fig. 3. Planar Arm Motion

Here, quatitative model (Table 2) found in experiment is derived in qualitative model where the orientation and translation properties of the arm and their links segments are derived in terms of qualitative position and orientation as Equation 1. The reason is that accurate model building is a difficult and time consuming process whereas is also a difficult machine vision challenge for complex domains such as human body structure. So, the choice of qualitative model is employed. The tradeoff between tracking precision and computational cost are negotiated by setting of the orientation and translation components of a fuzzy qualitative unit circle (see [12] for detailed).

Table 2. DH-parameters for a Human-Arm

$Frame$	i	α_{i-1}	a_{i-1}	$d_i(cm)$	$\theta_i(degree)$
1	1	0	0	25	4
	2	0	0	25	0
	3	0	0	10	0
2	1	0	0	25	1
	2	0	0	25	10
	3	0	0	10	0
..	1	..	..	..	..
	2	..	..	..	..
	3	..	..	..	..
11	1	0	0	24	1
	2	0	0	25	18
	3	0	0	10	0

6 Concluding Remarks

We have proposed a novel approach for modeling human motion in video sequences. Firstly, we have described human motion using qualitative normalized templates, which is introduced based on fuzzy qualitative trigonometry. Then we have employed conventional robotic DH parameters to model the motion of a human arm. Finally we developed an algorithm to match joint trajectories from images frames to qualitative normalised motion templates. This method has been inspired by fuzzy qualitative trigonometry to choose the tradeoffs between tracking precision and computational cost. Our future work targets the extension of the proposed approach to full human-body motion.

References

1. H. Liu, G.C., Xu, H.: Qualitative modelling of kinematic robots for fault diagnosis. International Journal of Production Research **43**(11) (2005) 2277–2290
2. Guo, Y., G.Xu, S.Tsuji: Understanding human motion patterns. In: Proceedings of the 12th IAPR International Conference on Pattern Recognition. Volume 2. (1994) 325–329
3. Yamamoto, M., Koshikawa, K.: Human motion analysis based on a robot arm model. In: CVPR'91:Proceedings of IEEE Computer Society Conference on Computer Vision and Pattern Recognition. (1991) 664–665
4. Ju, S.X., Black, M.J., Yacoob, Y.: Cardboard people: A parameterized model of articulated image motion. In: FG '96: Proceedings of the 2nd International Conference on Automatic Face and Gesture Recognition (FG '96), Washington, DC, USA, IEEE Computer Society (1996) 38
5. Kakadiaris, I., Metaxas, D.: Model-based estimation of 3d human motion. IEEE Transactions on Pattern Analysis and Machine Intelligence **22**(12) (2000) 1453–1459
6. P.Tissainayagam, D.Suter: Contour tracking with automatic motion model switching. Pattern Recognition Society **36** (2003) 2411–2427
7. Bregler, C., Malik, J.: Tracking people with twists and exponential maps. In: CVPR '98: Proceedings of the IEEE Computer Society Conference on Computer Vision and Pattern Recognition, Washington, DC, USA, IEEE Computer Society (1998) 8
8. Bissacco, A.: Visual tracking of human body with deforming motion and shape average. Technical report, UCLA (2002)
9. Lee, M.W., Cohen, I., Jung, S.K.: Particle filter with analytical inference for human body tracking. In: MOTION '02: Proceedings of the Workshop on Motion and Video Computing, Washington, DC, USA, IEEE Computer Society (2002) 159
10. Yonemoto, S., Arita, D., Taniguchi, R.: Real-time human motion analysis and ik-based human figure control. In: HUMO '00: Proceedings of the Workshop on Human Motion (HUMO'00), Washington, DC, USA, IEEE Computer Society (2000) 149
11. Sundaresan, A., Chellappa, R., RoyChowdhury, A.: Multiple view tracking of humans modelled by kinematic chains. In: International Conference on Image Processing, 2004. (2004)
12. Liu, H., Coghill, G.: Fuzzy qualitative trigonometry. IEEE International Conference on Systems, Man and Cybernetics (2005)
13. Murray, R.M., Shastry, S., Li, Z., Sastry, S.S.: A Mathematical Introduction to Robotic Manipulation. CRC Press (1994)
14. Craig, J.: Introduction to robotics: Mechanics and control. 2nd edition,Addison-Wesley (1989)

A Fuzzy Extension of Description Logic ALC with Comparison Expressions*

Dazhou Kang, Baowen Xu, Jianjiang Lu, and Yanhui Li

Dep.of Computer Science and Engineering, Southeast University,
Nanjing 210096, China
Jiangsu Institute of Software Quality, Nanjing, 210096, China
swws@seu.edu.cn

Abstract. The fuzzy description logics extend the classical description logics to support expressive representation and reasoning for fuzzy knowledge. But the current fuzzy description logics are not be accomplished in expressing comparisons between fuzzy memberships. The paper proposes $\mathcal{ALC}_{FC}$, a fuzzy extension of description logic $\mathcal{ALC}$ with comparison expressions. $\mathcal{ALC}_{FC}$ is able to express the comparisons between fuzzy assertions in ABoxes and define the comparison cut concepts for more complex comparison expressions. It also gives reasoning algorithms for satisfiability of $\mathcal{ALC}_{FC}$ concepts w.r.t. empty TBoxes.

1 Introduction

Description logics (DLs) are a class of knowledge representation languages with well-defined semantics and determinable reasoning methods, which are widely used in a variety of logic-based applications on the semantic web [1]. Many web ontology languages use DLs as its logic fundament. But it often needs to represent vague knowledge in web applications, especially when dealing with text, multimedia or uncertain data. Classical DLs are insufficient to deal with such knowledge. Fuzzy extensions of DLs import the fuzzy theory to enable the capability of dealing with fuzzy knowledge [2].

Straccia provided a reasoning algorithm for fuzzy $\mathcal{ALC}$ based on tableau rules [3], transformed fuzzy $\mathcal{ALC}$ into classical $\mathcal{ALC}$ [4], and recently presented fuzzy concrete domain [5] as well as the semantics of fuzzy $\mathcal{SHOIN}(D^+)$ [6]. Höldobler introduced the membership manipulator constructor to present $\mathcal{ALC}_{FH}$ [7]. Sanchez generalized the quantification in fuzzy DLs [8]. Stoilos provided pure ABoxes reasoning algorithms for the fuzzy extensions of $\mathcal{SHIN}$ [9] and $\mathcal{SHOIN}$ [10]. Li presented a family of extended fuzzy description logics ($\mathcal{EF}$DLs) [11], and provided reasoning algorithms for $\mathcal{EFALCN}$ [12].

* This work was supported in part by the NSFC (60373066, 60425206 and 90412003), National Grand Fundamental Research 973 Program of China (2002CB312000), Innovation Plan for Jiangsu High School Graduate Student, Advanced Armament Research Project (51406020105JB8103) and Advanced Research Fund of Southeast University (XJ0609233), High Technology Research Project of Jiangsu Province (BG2005032).

B. Gabrys, R.J. Howlett, and L.C. Jain (Eds.): KES 2006, Part I, LNAI 4251, pp. 647–654, 2006.
© Springer-Verlag Berlin Heidelberg 2006

The current fuzzy description logics are not be accomplished in expressing comparisons between fuzzy memberships. For example, they can only describe assertions like "Tom is very tall and Mike is not very tall", but cannot support a comparison expression like "Tom is taller than Mike", or even more complex comparison expressions like "no close friend of Tom is taller or stronger than him". This paper presents $\mathcal{ALC}_{FC}$, a fuzzy extension of description logic $\mathcal{ALC}$ with comparison expressions. Section 2 introduces the syntax and semantics of $\mathcal{ALC}_{FC}$ and defines its reasoning problems. Section 3 gives reasoning algorithms for satisfiability of $\mathcal{ALC}_{FC}$ concepts w.r.t. empty TBoxes.

2 The Fuzzy Description Logic $\mathcal{ALC}_{FC}$

2.1 A Quick Look to Fuzzy $\mathcal{ALC}$

Fuzzy $\mathcal{ALC}$ [3] is a fuzzy extension of the basic description logic $\mathcal{ALC}$.

Definition 1. *Let N_I, N_C and N_R be three disjoint sets: N_I is a set of individual names, N_C is a set of fuzzy concept names and N_R is a set of fuzzy role names. Fuzzy $\mathcal{ALC}$-concepts are defined as*

$$C, D ::= \top | \bot | A | \neg C | C \sqcup D | C \sqcap D | \exists R.C | \forall R.C;$$

where $A \in N_C$, $R \in N_R$. Fuzzy $\mathcal{ALC}$ semantics is defined by a fuzzy interpretation $\mathcal{I} = \langle \Delta^{\mathcal{I}}, \cdot^{\mathcal{I}} \rangle$, where $\Delta^{\mathcal{I}}$ is a nonempty set and $\cdot^{\mathcal{I}}$ is an function which maps every $a \in N_I$ to an element $a^{\mathcal{I}} \in \Delta^{\mathcal{I}}$, maps every $B \in N_C$ into a function $B^{\mathcal{I}} : \Delta^{\mathcal{I}} \to [0,1]$, and maps every $R \in N_R$ into a function $R^{\mathcal{I}} : \Delta^{\mathcal{I}} \times \Delta^{\mathcal{I}} \to [0,1]$. Furthermore, for any fuzzy $\mathcal{ALC}$-concepts C and D, $R \in N_R$ and $x \in \Delta^{\mathcal{I}}$, $\cdot^{\mathcal{I}}$ satisfies the following equations:

$$\begin{aligned}
\top^{\mathcal{I}}(x) &= 1; \bot^{\mathcal{I}}(x) = 0;\\
(\neg C)^{\mathcal{I}}(x) &= 1 - C^{\mathcal{I}}(x);\\
(C \sqcap D)^{\mathcal{I}}(x) &= \max(C^{\mathcal{I}}(x), D^{\mathcal{I}}(x));\\
(C \sqcup D)^{\mathcal{I}}(x) &= \min(C^{\mathcal{I}}(x), D^{\mathcal{I}}(x));\\
(\exists R.C)^{\mathcal{I}}(x) &= \sup_{y \in \Delta^{\mathcal{I}}}\{\min(R^{\mathcal{I}}(x,y), C^{\mathcal{I}}(y))\};\\
(\forall R.C)^{\mathcal{I}}(x) &= \inf_{y \in \Delta^{\mathcal{I}}}\{\max(1 - R^{\mathcal{I}}(x,y), C^{\mathcal{I}}(y))\};
\end{aligned}$$

Definition 2. *A fuzzy $\mathcal{ALC}$ knowledge base is composed of a TBox and an ABox:*

- *A TBox is a finite set of terminology axioms of the form $C \sqsubseteq D$. Any interpretation $\mathcal{I}$ satisfies $C \sqsubseteq D$ iff for any $x \in \Delta^{\mathcal{I}}$, $C^{\mathcal{I}}(x) \leq D^{\mathcal{I}}(x)$. $\mathcal{I}$ is a model of TBox $\mathcal{T}$ iff $\mathcal{I}$ satisfies all axioms in $\mathcal{T}$.*
- *An ABox is a finite set of assertions of the form $\langle \alpha \bowtie n \rangle$, where $\bowtie \in \{>, \geq, <, \leq\}$, $n \in [0,1]$, α (called a fuzzy assertion) is either of the form $a : C$ or $(a,b) : R$, $a, b \in N_I$. Any interpretation $\mathcal{I}$ satisfies $\langle a : C \bowtie n \rangle$ iff $C^{\mathcal{I}}(a^{\mathcal{I}}) \bowtie n$, satisfies $\langle (a,b) : R \bowtie n \rangle$ iff $R^{\mathcal{I}}(a^{\mathcal{I}}, b^{\mathcal{I}}) \bowtie n$. $\mathcal{I}$ is a model of ABox $\mathcal{A}$ iff $\mathcal{I}$ satisfies all assertions in $\mathcal{A}$.*

Reasoning algorithms for fuzzy $\mathcal{ALC}$ and their proofs can be found in [3].

2.2 Comparison Expressions

In fuzzy $\mathcal{ALC}$ ABoxes, assertions can only be of the form $\langle \alpha \bowtie n \rangle$. For example, we can express "Tom is very tall and Mike is not very tall" by $\langle Tom : Tall \geq 0.8 \rangle$ and $\langle Mike : Tall < 0.8 \rangle$; but cannot express either "Tom is taller than Mike", or "no friend of Tom is taller or stronger than him".

In order to enable the representation of comparisons between fuzzy assertions and more complex comparison expressions, we extend fuzzy $\mathcal{ALC}$ in the following ways to define a fuzzy description logic with comparison expressions, $\mathcal{ALC}_{FC}$:

- A new assertion form is added in ABoxes to enable assertions of the form $\langle \alpha \bowtie \beta \rangle$, where both α, β are fuzzy assertions. Then one can use the assertion $\langle Tom : Tall > Mike : Tall \rangle$ to tell that "Tom is taller than Mike."
- The comparison cut concepts (cuts for short) and two new concept constructors $\exists R.P$ and $\forall R.P$, where P is a cut, are defined.
- There are three kinds of basic cuts $(C, n)_{\bowtie}, (C, D)_{\bowtie}$ and $(C, D^{\uparrow})_{\bowtie}$ ($C_{\bowtie}$ is short for $(C, C^{\uparrow})_{\bowtie}$), for example
  - $(C, n)_{\bowtie}$ represents any individual x such that $\langle x : C \bowtie n \rangle$. For example, $\langle Tom : \exists friend.(Tall, 0.8)_{<} \geq 0.9 \rangle$ means that "there is a close friend of Tom not very tall", i.e. $\exists x, friend^{\mathcal{I}}(Tom, x) \geq 0.9$ and $Tall^{\mathcal{I}}(x) < 0.8$.
  - $(C, D)_{\bowtie}$ represents any individual x such that $\langle x : C \bowtie x : D \rangle$. The assertion $\langle Tom : \forall friend.(Absolutist, Liberalist)_{<} > 0.1 \rangle$ means that "all close friends of Tom prefer liberalism to absolutism", i.e.$\forall x, friend^{\mathcal{I}}(Tom, x) \geq 0.9 \rightarrow Absolutist^{\mathcal{I}}(x) < Liberalist^{\mathcal{I}}(x)$.
  - $(C, D^{\uparrow})_{\bowtie}$ provides the ability of describe the comparisons between two individuals. $\langle Tom : \exists friend.Strong_{\leq} \geq 0.9 \rangle$ means that "there is a close friend of Tom not stronger than Tom", i.e. $\exists x, friend^{\mathcal{I}}(Tom, x) \geq 0.9$ and $Strong^{\mathcal{I}}(x) < Strong^{\mathcal{I}}(Tom)$.
- Two cuts P, Q can be combined in a single cut by $P \sqcap Q$ or $P \sqcup Q$. For example, $\langle Tom : \exists friend.((Tall, 0.8)_{<} \sqcup Strong_{\leq}) \geq 0.9 \rangle$ means "there is a close friend of Tom not very tall or not stronger than Tom".
- Another new assertion form $a : P_b$ is added in ABoxes, where P is a cut. The assertion $\langle Tom : (Tall_{>} \sqcup Strong_{>})_{Mike} \rangle$ means "Tom is either taller or stronger than Mike".

2.3 The Fuzzy Description Logic $\mathcal{ALC}_{FC}$

$\mathcal{ALC}_{FC}$ extends fuzzy $\mathcal{ALC}$ with two new concept constructors, the comparison cut concepts and new assertion forms in ABoxes.

Definition 3. *$\mathcal{ALC}_{FC}$-concepts (concepts for short) are defined as*

$$C, D ::= \top | \bot | A | \neg C | C \sqcup D | C \sqcap D | \exists R.C | \forall R.C | \exists R.P | \forall R.P;$$

where $A \in N_C$, $R \in N_R$, and P is a comparison cut concept. The semantics of $\exists R.P$ and $\forall R.P$ is showed in the following equations:

$$(\exists R.P)^{\mathcal{I}}(x) = \sup_{y \in P^{\mathcal{I}}(x)} R^{\mathcal{I}}(x, y);$$
$$(\forall R.P)^{\mathcal{I}}(x) = \inf_{y \in (\neg P)^{\mathcal{I}}(x)} (1 - R^{\mathcal{I}}(x, y)).$$

The comparison cut concepts (cuts for short) are defined as

$$P, Q ::= (C,n)_{\bowtie}|(C,D)_{\bowtie}|(C,D^{\uparrow})_{\bowtie}|\neg P|P \sqcap Q|P \sqcup Q;$$

where C, D *are concepts,* $n \in [0,1]$ *and* $\bowtie \in \{>, \geq, <, \leq\}$. $\cdot^{\mathcal{I}}$ *maps a cut* P *into a function* $P^{\mathcal{I}} : \Delta^{\mathcal{I}} \to 2^{\Delta^{\mathcal{I}}}$, *which satisfies that*

$$(C,n)^{\mathcal{I}}_{\bowtie}(x) = \{y|C^{\mathcal{I}}(y) \bowtie n\};$$
$$(C,D)^{\mathcal{I}}_{\bowtie}(x) = \{y|C^{\mathcal{I}}(y) \bowtie D^{\mathcal{I}}(y)\};$$
$$(C,D^{\uparrow})^{\mathcal{I}}_{\bowtie}(x) = \{y|C^{\mathcal{I}}(y) \bowtie D^{\mathcal{I}}(x)\};$$
$$(\neg P)^{\mathcal{I}}(x) = \Delta^{\mathcal{I}} \backslash P^{\mathcal{I}}(x);$$
$$(P \sqcap Q)^{\mathcal{I}}(x) = P^{\mathcal{I}}(x) \cap Q^{\mathcal{I}}(x);$$
$$(P \sqcup Q)^{\mathcal{I}}(x) = P^{\mathcal{I}}(x) \cup Q^{\mathcal{I}}(x).$$

It is clear that only $(C,D^{\uparrow})^{\mathcal{I}}_{\bowtie}(x)$ lies on x, while $(C,n)^{\mathcal{I}}_{\bowtie}(x)$ and $(C,D)^{\mathcal{I}}_{\bowtie}(x)$ do not. But we always interpret a cut w.r.t. an individual to make it uniform.

Any concept or cut can be transformed into the Negation Normal Form (NNF), where $\neg$ can only occur in front of a concept name. The transformation can be done in polynomial time by the following rules: $\neg W_{\bowtie} \to W_{\bowtie^{\ominus}}$; $\neg(X \sqcup Y) \to \neg X \sqcap \neg Y$; $\neg(X \sqcap Y) \to \neg X \sqcup \neg Y$; $\neg(\exists R.X) \to \forall R.\neg X$; $\neg(\forall R.X) \to \exists R.\neg X$; where $W \in \{(C,n),(C,D),(C,D^{\uparrow})\}$, X, Y are concepts or cuts, $>^{\ominus} = \leq$; $<^{\ominus} = \geq$; $\geq^{\ominus} = <$; $\leq^{\ominus} = >$. From now on, we assume that all concepts are in NNF.

Definition 4. *An* $\mathcal{ALC}_{FC}$ *knowledge base is composed of a TBox and an ABox:*

- *A TBox is a finite set of terminology axioms of the form* $\langle C \sqsubseteq D \rangle$. *Any interpretation* $\mathcal{I}$ *satisfies* $\langle C \sqsubseteq D \rangle$ *iff for any* $x \in \Delta^{\mathcal{I}}$, $C^{\mathcal{I}}(x) \leq D^{\mathcal{I}}(x)$. $\mathcal{I}$ *is a model of TBox* $\mathcal{T}$ *iff* $\mathcal{I}$ *satisfies all axioms in* $\mathcal{T}$.
- *An ABox is a finite set of assertions of the form* $\langle \alpha \bowtie n \rangle$, $\langle \alpha \bowtie \beta \rangle$ *or* $\langle a : P_b \rangle$, *where* $\bowtie \in \{>, \geq, <, \leq\}$, $n \in [0,1]$, α, β *(called fuzzy assertions) are either of the form* $a : C$ *or* $(a,b) : R$, $a, b \in N_I$. *Let* $\alpha^{\mathcal{I}} = C^{\mathcal{I}}(a^{\mathcal{I}})$ *if* $\alpha = a : C$; $\alpha^{\mathcal{I}} = R^{\mathcal{I}}(a^{\mathcal{I}}, b^{\mathcal{I}})$ *if* $\alpha = (a,b) : R$. *Any interpretation* $\mathcal{I}$ *satisfies* $\langle \alpha \bowtie n \rangle$ *iff* $\alpha^{\mathcal{I}} \bowtie n$, *satisfies* $\langle \alpha \bowtie \beta \rangle$ *iff* $\alpha^{\mathcal{I}} \bowtie \beta^{\mathcal{I}}$, *satisfies* $\langle a : P_b \rangle$ *iff* $a^{\mathcal{I}} \in P^{\mathcal{I}}(b)$. $\mathcal{I}$ *is a model of ABox* $\mathcal{A}$ *iff* $\mathcal{I}$ *satisfies all assertions in* $\mathcal{A}$.

Definition 5. *The basic reasoning problems of* $\mathcal{ALC}_{FC}$ *include*

- satisfiability of concepts*: a concept* C *is satisfiable w.r.t. TBox* $\mathcal{T}$ *to a given degree* n, *iff there exists an interpretation* $\mathcal{I}$ *such that* $\mathcal{I}$ *is a model of* $\mathcal{T}$, *and* $\exists x \in \Delta^{\mathcal{I}}, C^{\mathcal{I}}(x) \geq n$.
- consistency of ABoxes*: an ABox* $\mathcal{A}$ *is consistent w.r.t. TBox* $\mathcal{T}$, *iff there exists an interpretation* $\mathcal{I}$ *such that* $\mathcal{I}$ *is a model of both* $\mathcal{T}$ *and* $\mathcal{A}$.
- entailment*: a knowledge base* $\mathcal{K} = \langle \mathcal{A}, \mathcal{T} \rangle$ *entails an assertion* ϕ *or an axiom* φ, *iff any model of* $\mathcal{K}$ *satisfies* ϕ *or* φ, *written* $\mathcal{K} \models \phi$ *or* $\mathcal{K} \models \varphi$.
- bounds of assertions*: the greatest lower bound of an fuzzy assertion* α *w.r.t. knowledge base* $\mathcal{K}$ *is defined as* $\mathrm{glb}(\mathcal{K}, \alpha) = \sup\{n|\mathcal{K} \models \langle \alpha \geq n \rangle\}$; *the least upper bound of an fuzzy assertion* α *w.r.t.* $\mathcal{K}$ *is defined as* $\mathrm{lub}(\mathcal{K}, \alpha) = \inf\{n|\mathcal{K} \models \langle \alpha \leq n \rangle\}$.

$\bowtie_t$ for any variables m, n, and a variable or a fuzzy assertion l
if $X \in \mathcal{A}$, and $t(X) \notin \mathcal{A}$
then $\mathcal{A} \to \mathcal{A} \cup \{t(X)\}$

$$\text{where } t(X) = \begin{cases} \langle n \bowtie^- m \rangle & \text{if } X = \langle m \bowtie n \rangle \\ \langle m^- \bowtie^- n^- \rangle & \text{if } X = \langle m \bowtie n \rangle \\ \langle m \leq n \rangle & \text{if } X = \langle l \geq m \rangle, \langle l \leq n \rangle \\ \langle m < n \rangle & \text{if } X = \langle l > m \rangle, \langle l \lhd n \rangle \\ \langle m < n \rangle & \text{if } X = \langle l \rhd m \rangle, \langle l < n \rangle \end{cases}$$

$\bowtie_n$ if $\langle \alpha \bowtie \beta \rangle \in \mathcal{A}$, and
there is no n with $\{\langle \alpha \leq n \rangle, \langle \alpha \geq n \rangle, \langle \beta \bowtie^- n \rangle\} \subseteq \mathcal{A}$
then $\mathcal{A} \to \mathcal{A} \cup \{\langle \beta \bowtie^- n \rangle\}$, if there is n with $\{\langle \alpha \leq n \rangle, \langle \alpha \geq n \rangle\} \subseteq \mathcal{A}$;
$\mathcal{A} \to \mathcal{A} \cup \{\langle \alpha \leq n \rangle, \langle \alpha \geq n \rangle, \langle \beta \bowtie^- n \rangle\}$ with a new n, otherwise

C_1 if $\langle x : (C, n)_{\bowtie z} \rangle \in \mathcal{A}$, and $\langle x : C \bowtie n \rangle \notin \mathcal{A}$
then $\mathcal{A} \to \mathcal{A} \cup \{\langle x : C \bowtie n \rangle\}$

C_2 if $\langle x : (C, D)_{\bowtie z} \rangle \in \mathcal{A}$, and $\langle x : C \bowtie x : D \rangle \notin \mathcal{A}$
then $\mathcal{A} \to \mathcal{A} \cup \{\langle x : C \bowtie x : D \rangle\}$

C_3 if $\langle x : (C, D^{\uparrow})_{\bowtie z} \rangle \in \mathcal{A}$, and $\langle x : C \bowtie z : D \rangle \notin \mathcal{A}$
then $\mathcal{A} \to \mathcal{A} \cup \{\langle z : D \bowtie^- x : C \rangle\}$

$C_\sqcap$ if $\langle x : (P \sqcap Q)_z \rangle \in \mathcal{A}$, and $\{\langle y : P_z \rangle, \langle y : Q_z \rangle\} \nsubseteq \mathcal{A}$
then $\mathcal{A} \to \mathcal{A} \cup \{\langle x : P_z \rangle, \langle x : Q_z \rangle\}$

$C_\sqcup$ if $\langle x : (P \sqcup Q)_z \rangle \in \mathcal{A}$, and $\{\langle x : P_z \rangle, \langle x : Q_z \rangle\} \cap \mathcal{A} = \emptyset$
then $\mathcal{A} \to \mathcal{A} \cup \{\langle x : P_z \rangle\}$, or $\mathcal{A} \to \mathcal{A} \cup \{\langle x : Q_z \rangle\}$

Fig. 1. Completion rules for $\mathcal{ALC}_{FC}$, part I

3 Reasoning for Satisfiability of $\mathcal{ALC}_{FC}$ Concepts w.r.t. Empty TBoxes

In this section, we will design a algorithm for satisfiability of $\mathcal{ALC}_{FC}$ concepts w.r.t empty TBoxes. The algorithm is developed in the style of tableau algorithms [1]. For satisfiability of $\mathcal{ALC}_{FC}$ concept C_0 w.r.t. an empty TBox to a given degree n_0, our algorithm starts with an ABox $\mathcal{A} = \{\langle x_0 : C_0 \geq n_0 \rangle\}$. The completion rules for $\mathcal{ALC}_{FC}$ are showed in Fig.1 and Fig.2. In the figures, $\bowtie \in \{>, \geq, <, \leq\}$, $\rhd \in \{>, \geq\}$, $\lhd \in \{<, \leq\}$, $>^- = <$, $<^- = >$, $\geq^- = \leq$, $\leq^- = \geq$. C, D be concepts, R be a role, P, Q be cuts, and x, y be individuals.

The completion rules of fuzzy $\mathcal{ALC}$ [3] can only deal with the assertions of the form $\langle \alpha \bowtie n \rangle$, while cannot support the assertions of the form $\langle \alpha \bowtie \beta \rangle$, which are necessary in $\mathcal{ALC}_{FC}$ ABoxes. Our algorithm for $\mathcal{ALC}_{FC}$ transforms $\langle \alpha \bowtie \beta \rangle$ into $\langle \alpha \leq n \rangle, \langle \alpha \geq n \rangle, \langle \beta \bowtie^- n \rangle$ (see $\bowtie_n$-rule in Fig.1), where n is a variable.

In order to deal with variables in the algorithm, we allow using variables in assertions and extend a new assertion form $\langle n \bowtie m \rangle$. Let $U = \{0, 1, 0.5\} \cup \{n | n \in \mathbb{R}$ and n or $1 - n$ occurs in $C_0\}$. For any $m, n \in U$, if $m < n$, then let $\mathcal{A} \to \mathcal{A} \cup \{\langle m < n \rangle\}$. For any $n \in X$, let $n^- = 1 - n$. From now on, all real numbers can be seemed as variables. For any variable n, let $n^- = 1 - n$ be a variable, and $(1-n)^- = n$. Their comparisons are performed by $\bowtie_t$-rule in Fig.1.

$\neg_{\bowtie}$ if $\langle x : \neg A \bowtie n\rangle \in \mathcal{A}$, and $\langle x : A \bowtie^{-} n^{-}\rangle \notin \mathcal{A}$
then $\mathcal{A} \to \mathcal{A} \cup \{\langle x : A \bowtie^{-} n^{-}\rangle\}$

$\sqcap_{\rhd}$ if $\langle x : C \sqcap D \rhd n\rangle \in \mathcal{A}$, and $\{\langle x : C \rhd n\rangle, \langle x : D \rhd n\rangle\} \nsubseteq \mathcal{A}$
then $\mathcal{A} \to \mathcal{A} \cup \{\langle x : C \rhd n\rangle, \langle x : D \rhd n\rangle\}$

$\sqcap_{\lhd}$ if $\langle x : C \sqcap D \lhd n\rangle \in \mathcal{A}$, and $\{\langle x : C \rhd n\rangle, \langle x : D \lhd n\rangle\} \cap \mathcal{A} = \emptyset$
then $\mathcal{A} \to \mathcal{A} \cup \{\langle x : C \lhd n\rangle\}$, or $\mathcal{A} \to \mathcal{A} \cup \{\langle x : D \lhd n\rangle\}$

$\sqcup_{\rhd}$ if $\langle x : C \sqcup D \rhd n\rangle \in \mathcal{A}$, and $\{\langle x : C \rhd n\rangle, \langle x : D \rhd n\rangle\} \cap \mathcal{A} = \emptyset$
then $\mathcal{A} \to \mathcal{A} \cup \{x : C \rhd n\}$, or $\mathcal{A} \to \mathcal{A} \cup \{x : D \rhd n\}$

$\sqcup_{\lhd}$ if $\langle x : C \sqcup D \lhd n\rangle \in \mathcal{A}$, and $\{\langle x : C \rhd n\rangle, \langle x : D \lhd n\rangle\} \nsubseteq \mathcal{A}$
then $\mathcal{A} \to \mathcal{A} \cup \{\langle x : C \rhd n\rangle, \langle x : D \lhd n\rangle\}$

$\exists_{\rhd}$ if $\langle x : \exists R.X \rhd n\rangle \in \mathcal{A}$, and
there is no y with $\{\langle (x,y) : R \rhd n\rangle, Y\} \subseteq \mathcal{A}$
then $\mathcal{A} \to \mathcal{A} \cup \{\langle (x,y) : R \rhd n\rangle, Y\}$ with a new y

$\exists_{\lhd}$ if $\langle x : \exists R.X \lhd n\rangle \in \mathcal{A}$, and there is y with $\langle (x,y) : R \rhd m\rangle \in \mathcal{A}$, and $\{\langle (x,y) : R \lhd n\rangle, Y\} \cap \mathcal{A} = \emptyset$
then $\mathcal{A} \to \mathcal{A} \cup \{\langle (x,y) : R \lhd n\rangle\}$, or $\mathcal{A} \to \mathcal{A} \cup \{Y\}$

$\forall_{\rhd}$ if $\langle x : \forall R.X \rhd n\rangle \in \mathcal{A}$, and there is y with $\langle (x,y) : R \rhd m\rangle \in \mathcal{A}$, and $\{\langle (x,y) : R \rhd^{-} n^{-}\rangle, Y\} \cap \mathcal{A} = \emptyset$
then $\mathcal{A} \to \mathcal{A} \cup \{\langle (x,y) : R \rhd^{-} n^{-}\rangle\}$, or $\mathcal{A} \to \mathcal{A} \cup \{Y\}$

$\forall_{\lhd}$ if $\langle x : \forall R.X \lhd n\rangle \in \mathcal{A}$, and
there is no y with $\{\langle (x,y) : R \lhd^{-} n^{-}\rangle, Y\} \subseteq \mathcal{A}$
then $\mathcal{A} \to \mathcal{A} \cup \{\langle (x,y) : R \lhd^{-} n^{-}\rangle, Y\}$ with a new y

$$\text{where } Y = \begin{cases} \langle y : C \rhd n\rangle & \text{if } X = C \text{ in } \exists_{\rhd} \text{ or } \forall_{\rhd} \text{ rule} \\ \langle y : C \lhd n\rangle & \text{if } X = C \text{ in } \exists_{\lhd} \text{ or } \forall_{\lhd} \text{ rule} \\ \langle y : P_x\rangle & \text{if } X = P \end{cases}$$

Fig. 2. Completion rules for $\mathcal{ALC}_{FC}$, part II

The other rules in Fig.1 deal with the cuts. From the semantics of cuts, it is easy to see the correctness of the rules. The rules in Fig.2 are similar to the completion rules for fuzzy $\mathcal{ALC}$ [3], except $\exists_{\lhd}$-rule and $\forall_{\rhd}$-rule. Since n is a variable, we may not know whether two assertions are conjugated with or not.

For any variable n, variable or fuzzy assertion l, if $\langle n < n\rangle \in \mathcal{A}$ or $\langle l < 0\rangle \in \mathcal{A}$ or $\langle l > 1\rangle \in \mathcal{A}$, then $\mathcal{A}$ contains a *clash*; otherwise $\mathcal{A}$ is *clash-free*. If none of the completion rules can be applied to $\mathcal{A}$, then $\mathcal{A}$ is *complete*.

While $\mathcal{A}$ is clash-free and not complete, apply completion rules to any individual x in $\mathcal{A}$ exhaustively. It can yield a complete and clash-free $\mathcal{A}$ iff C_0 is satisfiable to a degree n_0 w.r.t. an empty TBox.

Theorem 1. *The algorithm is a decision procedure for satisfiability of $\mathcal{ALC}_{FC}$ concepts w.r.t. empty TBoxes.*

Proof. For any concept or cut X, the size of X (written size(X)) is the number of symbols used to write down X, where every name in N_I, N_C, N_R, every constructor ($\neg, \sqcap, \sqcup, \exists, \forall, \bowtie$) and every variable n is written in one symbol. The set sub(X) is defined below. Hence $|\,\text{sub}(X)| \le \text{size}(X)$, and for any $Y \in \text{sub}(X)$, $\text{size}(Y) \le \text{size}(X)$. Let $N = \text{size}(C_0)$ be the input size of the algorithm.

$$\text{sub}(X) =_{def} \{X\} \cup \begin{cases} \text{sub}(Y) & \text{if } X = \neg Y, \exists R.Y \text{ or } \forall R.Y \\ \text{sub}(Y) \cup \text{sub}(Z) & \text{if } X = Y \sqcap Z \text{ or } Y \sqcup Z \\ \text{sub}(C) & \text{if } X = (C, n)_{\bowtie} \\ \text{sub}(C) \cup \text{sub}(D) & \text{if } X = (C, D)_{\bowtie} \text{ or } (C, D^{\uparrow})_{\bowtie} \end{cases}$$

Termination is unusual to the fuzzy $\mathcal{ALC}$ case, because it may create new variables. If a rule applied to individual x creates a new individual y, then y is a successor of x. For any $\mathcal{A}$, its individuals can form a completion tree $T_{\mathcal{A}}$, which root node is x_0 called a level 0 node. For any level i node x in $T_{\mathcal{A}}$, its child-nodes are the successor of x, called level $i+1$ nodes. For any individual x, let $L(x) = \{C | \langle x : C \bowtie n\rangle \in \mathcal{A}\} \cup \{P | \langle x : P_z\rangle \in \mathcal{A}\}$. From the completion rules, $L(x) \subseteq \text{sub}(C_0)$; $l(x) =_{def} \max\{\text{size}(X) | X \in L(x)\} \le \text{size}(C_0) \le N$; and for any x and its successor y, $l(y) < l(x)$. So the depth of $T_{\mathcal{A}}$ is no more than $N+1$. Let $\mathcal{L}(x)$ be the set of assertions containing x, V_x be the set of variables occuring in $\mathcal{L}(x)$. From $\bowtie_n$-rule which is the only rule to create a new variable, for any individuals, no more than N new variables can be created. So $|V_{x_0}| \le 2N$, For any level i node x, $|V_x| \le (i+2) \times N$ variables; then $|\mathcal{L}(x)| \le |\,\text{sub}(C_0)| \times 4 \times (i+2) \times N$. Since the deepst level is N, for any x, $|\mathcal{L}(x)| \le |\,\text{sub}(C_0)| \times 4 \times (N+2) \times N = O(n)$. A $\exists_{\lhd}$- or $\forall_{\rhd}$-rule creates an new individual. So x has no more than $|\mathcal{L}(x)|$ successors. Both the depth and the breadth of $T_{\mathcal{A}}$ is finite. There should be finite assertions in $\mathcal{A}$. Every completion rule add at least one assertion in $\mathcal{A}$. Therefore, the algorithm terminates.

The reminder of the proof is very similar to the fuzzy $\mathcal{ALC}$ case. The algorithm can yield a complete and clash-free $\mathcal{A}$ iff C_0 is satisfiable to a degree n_0 w.r.t. an empty TBox. For the "if" direction, we can build a model of $\mathcal{A}$ from a complete and clash-free ABox $\mathcal{A}$ with $\langle x_0 : C_0 \ge n_0\rangle$; hence C_0 is satisfiable to a degree n_0. For the "only if" direction, if C_0 is satisfiable to a degree n_0, then there is a model of $\mathcal{A}_0 = \{\langle x_0 : C_0 \ge n_0\rangle\}$; we can indeed construct a complete and clash-free ABox $\mathcal{A}$ from $\mathcal{A}_0$. □

A PSpace algorithm can be yielded by adopting a depth-first manner and removing redundant variables. It is shown in Depth($x, \mathcal{A}$):

1. For any $\alpha = x : C$ such that $C \in \text{sub}(X), X \in L(x)$, if there is no n with $\{\langle \alpha \le n\rangle, \langle \alpha \ge n\rangle\} \subseteq \mathcal{A}$, then $\mathcal{A} \to \mathcal{A} \cup \{\langle \alpha \le n\rangle, \langle \alpha \ge n\rangle\}$ with a new n.
2. While $\mathcal{A}$ is clash-free, apply completion rules to x in $\mathcal{A}$ exhaustively. Record V_t, the set of all variables in $\mathcal{A}$.
3. For any successor y of x, execute $\mathcal{A} \to \text{Depth}(y, \mathcal{A})$.
4. Removes from $\mathcal{A}$ any assertion containing variables not in V_t or descendants of x, if the assertion is not a clash. Return $\mathcal{A}$.

And since satisfiability of fuzzy $\mathcal{ALC}$ concepts w.r.t. empty TBoxes is already PSpace-complete [3], it implies

Theorem 2. *Satisfiability of $\mathcal{ALC}_{FC}$ concepts w.r.t. empty TBoxes is PSpace-complete.*

4 Conclusions

This paper presents $\mathcal{ALC}_{FC}$, a fuzzy extension of description logic $\mathcal{ALC}$ with comparison expressions, and a reasoning algorithms for satisfiability of $\mathcal{ALC}_{FC}$ concepts w.r.t. empty TBoxes. The future work is to consider general TBoxes, extend comparison expressions in more expressive fuzzy description logics and design their reasoning algorithms.

References

1. Baader, F., Calvanese, D., McGuinness, D., Nardi, D., Patel-Schneider, P.F., eds.: The Description Logic Handbook: Theory, Implementation, and Applications. Cambridge University Press (2003)
2. Ye, J.: Generalizing term subsumption languages to fuzzy logic. In: Proceedings of the 12th International Joint Conference on Artificial Intelligence, Sidney, Australia (1991) 472–477
3. Straccia, U.: Reasoning within fuzzy description logics. Journal of Artificial Intelligence Research **14** (2001) 137–166
4. Straccia, U.: Transforming fuzzy description logics into classical description logics. In: Proceedings of the 9th European Conference on Logics in Artificial Intelligence, Lisbon (2004) 385–399
5. Straccia, U.: Description logics with fuzzy concrete domains. In: Proceedings of the 21st Conference on Uncertainty in Artificial Intelligence, Edinburgh (2005) 385–399
6. Straccia, U.: Towards a fuzzy description logic for the semantic web (preliminary report). In: Proceedings of the 2nd European Semantic Web Conference. Volume 3532 of Lecture Notes in Computer Science. Heraklion, Greece, Crete, Springer Verlag (2005) 167–181
7. Hölldobler, S., Störr, H.P., Tran, D.K.: The fuzzy description logic alcfh with hedge algebras as concept modifiers. Journal of Advanced Computational Intelligence and Intelligent Informatics **7** (2003) 294–305
8. Sánchez, D., Tettamanzi, A.G.: Generalizing quantification in fuzzy description logic. In Reusch, B., ed.: Proceedings of the 8th Fuzzy Days in Dortmund: Computational Intelligence, Theory and Applications. Advances in Soft Computing Series, Berlin, Springer (2004)
9. Stoilos, G., Stamou, G., Tzouvaras, V., Pan, J.Z., Horrocks, I.: The fuzzy description logic f-shin. In: Proceedings of the International Workshop on Uncertainty Reasoning for the Semantic Web, Galway, Ireland (2005)
10. Stoilos, G., Stamou, G., Tzouvaras, V., Pan, J.Z., Horrocks, I.: Fuzzy owl: Uncertainty and the semantic web. In: Proceedings of the International workshop on OWL: Experience and Directions, Galway, Ireland (2005)
11. Li, Y., Xu, B., Lu, J., Kang, D., Wang, P.: A family of extended fuzzy description logics. In: Proceedings of the IEEE 29th Annual International Computer Software and Applications Conference, Edinburgh, Scotland (2005) 221–226
12. Li, Y., Xu, B., Lu, J., Kang, D., Wang, P.: Extended fuzzy alcn and its tableau algorithm. In Wang, L., Jin, Y., eds.: Proceedings of Fuzzy Systems and Knowledge Discovery, Second International Conference, Part I. Volume 3613 of Lecture Notes in Computer Science. Changsha, China (2005) 232–242

A Distributed and Fuzzy Extension of Description Logics*

Yanhui Li[1,2], Baowen Xu[1,2], Jianjiang Lu[3], and Dazhou Kang[1,2]

[1] School of Computer Science and Engineering, Southeast University, Nanjing 210096, P.R. China
[2] Jiangsu Institute of Software Quality, Nanjing 210096, P.R. China
[3] Institute of Command Automation, PLA University of Science and Technology, Nanjing 210007, P.R. China
bwxu@seu.edu.cn

Abstract. By development of the Semantic Web, demands for increasing expressive power of ontology languages has triggered a mass of extending description logics in fuzzy and distributed cases. However, current solutions are proposed respectively on one of these two aspects. By integrating E-connection into extended fuzzy description logics (EFDLs), this paper proposes a novel logical approach distributed EFDLs (DEFDLs) to couple both fuzzy and distributed features within description logics. In DEFDLs, cut links are introduced to describe fuzzy relations among various fuzzy areas, hence connecting distributed EFDLs knowledge bases to construct a global DEFDLs knowledge base. We make a primitive discussion about reasoning issues within DEFDLs, where some current fuzzy reasoning technique can be reused in restricted DEFDLs knowledge bases.

1 Introduction

The term "Semantic Web" stands for the idea of a future Web, which aims to increase machine support for interpretation and integration of Web data in the semantic level [1]. To achieve a machine-understandable representation way, knowledge is usually structured in the form of ontologies in the Semantic Web [2]. Description logics are often named as main ontology languages that can support ontologies and help to make ontologies with formal semantics [3]. The proposed web ontology language recommendation OWL Lite and OWL DL are basically very expressive description logics SHIF(D) and SHOIN(D) with RDF syntax [4].

By the Web's evolving into the Semantic Web, more and more knowledge that covers all aspects and smoothes all links in daily life has been going into the Semantic Web. Increasing demands for expressive power of ontology languages has triggered a

* This work was supported in part by the NSFC (60373066, 60425206, 90412003), National Grand Fundamental Research 973 Program of China (2002CB312000), Excellent Ph.D, Thesis Fund of Southeast University, Advanced Armament Research Project (51406020105JB8103), High Technology Research Project of Jiangsu Province (BG2005032) and Advanced Armament Research Project(51406020105JB8103).

B. Gabrys, R.J. Howlett, and L.C. Jain (Eds.): KES 2006, Part I, LNAI 4251, pp. 655 – 662, 2006.
© Springer-Verlag Berlin Heidelberg 2006

mass of extensions of description logics that makes them convenient to express knowledge in some nonstandard cases, such as fuzzy and distributed cases.

There are some vague concepts, such as "Tall" and "Hot", in application area of the Semantic Web. Straccia adopted a fuzzy interpretation framework to construct a representative fuzzy extension of description logics, and designed a constraint propagation calculus for this extension [5]. Stoilos et al introduced Straccia's fuzzy framework into the Semantic Web, hence getting a fuzzy ontology language Fuzzy OWL [6]. They also gave a restricted reasoning technique (not suitable for general fuzzy concept inclusions [7]) for Fuzzy OWL reasoning tasks. However, to some degree, Straccia's fuzzy framework only brings limited fuzzy expressive power in some complex fuzzy cases. To overcome its insufficiency, we pointed out a new family of extended fuzzy description logics (EFDLs), in which cut sets of fuzzy concepts and fuzzy roles are considered as the atomic concepts and atomic roles [8]. Some complexity results and reasoning techniques in EFDLs were proposed in [9-10], and in these papers, expressive advantages of our framework were detailed discussed.

For the number of independently developed ontologies increases rapidly in current Semantic Web, working with multiple ontologies brings a growing body of work in distributed research of description logic. Borgida et al proposed distributed description logics (DDLs), in which multiple description logic knowledge bases are combined by inter-ontology bridge rules [11]. A primary attempt for distributed ontology languages resulted in C-OWL, a straight extension of OWL that follows the DDLs formalism [12], but no reasoning supports are available for it. E-connection is another popular technique for coupling different ontologies [13], which is based on a wide logic framework Abstract Description System [14]. Cuenca Grau et al integrated the E-connections formalism into OWL in a compact and natural way by defining "links" that stand for the inter-ontology relations [15]. They also pointed out that their extension is strictly more expressive than C-OWL. Parsia et al made some more expressive extension of "links" and gave tableau calculus for reasoning within such links [16].

Although the extensions of description logics in fuzzy and distributed aspects respectively have done a lot, to our best knowledge, no published work has considered on the both of these two extensions. In this paper, we combine EFDLs and E-connections, thus getting distributed EFDLs (DEFDLs). The remainder of this paper is organized as follows: a brief introduction of EFDLs is given in section 2; by introducing cut links that couple different EFDL knowledge bases, we propose a general definition of syntax and semantics of DEFDLs in section 3; section 4 considers on reasoning issue within DEFDLs, where some reasoning problems and techniques are detailed discussed; finally section 5 concludes this paper and talks about further work.

2 Extended Fuzzy Description Logics

Compared with other fuzzy extensions of description logics, the main feature of EFDLs is that they introduce cut sets of fuzzy concepts and fuzzy roles as atomic concepts and atomic roles, called atomic cut concept and atomic cut role respectively. In fuzzy set theory [17], a fuzzy set S w.r.t a universe U is defined as a membership function f_s: $U \rightarrow [0,1]$, and the n-cut set $S_{[n]}$ of S is defined as $S_{[n]}=\{d \in U | f_s(d) \geq n\}$, where $0<n\leq 1$. In EFDLs, fuzzy concepts and fuzzy roles are interpreted as fuzzy sets and their cut sets are interpreted as cut sets of corresponding fuzzy sets. Note that cut sets of fuzzy sets are indeed crisp set. Therefore, it facilitates a crisp description logic theory for simulating the fuzzy theory. Based on atomic cut concepts and atomic cut roles, EFDLs inherit concept and role constructors from crisp description logics to create a new fuzzy knowledge representation system.

2.1 Atomic Cut Concepts and Atomic Cut Roles

Let N_C and N_R be two disjoint and countably infinite sets of atomic fuzzy concepts (denoted B) and of atomic fuzzy roles (denoted R, S). For any $B \in N_C$, $R \in N_R$ and $1 \geq n > 0$, we call $B_{[n]}$ an atomic cut concept and $R_{[n]}$ an atomic cut role, where B and R are the prefixes of n, and n is the suffix of B and R. The semantics of atomic fuzzy concepts and roles, and their cut sets are defined in terms of a fuzzy interpretation $\mathcal{I} = <\Delta^{\mathcal{I}}, \cdot^{\mathcal{I}}>$. The domain $\Delta^{\mathcal{I}}$ is a nonempty set and the interpretation function $\cdot^{\mathcal{I}}$ maps every fuzzy atomic concept B to a membership function $B^{\mathcal{I}} : \Delta^{\mathcal{I}} \rightarrow [0,1]$, and every fuzzy atomic role R to a membership function $R^{\mathcal{I}} : \Delta^{\mathcal{I}} \times \Delta^{\mathcal{I}} \rightarrow [0,1]$. Additionally, $\cdot^{\mathcal{I}}$ maps $B_{[n]}$ and $R_{[n]}$ to n-cuts of $B^{\mathcal{I}}$ and $R^{\mathcal{I}}$, which are subsets of $\Delta^{\mathcal{I}}$ and $\Delta^{\mathcal{I}} \times \Delta^{\mathcal{I}}$:

$$\begin{aligned} (B_{[n]})^{\mathcal{I}} &= \{d \mid d \in \Delta^{\mathcal{I}} \wedge B^{\mathcal{I}}(d) \geq n\} \\ (R_{[n]})^{\mathcal{I}} &= \{(d,d') \mid d,d' \in \Delta^{\mathcal{I}} \wedge R^{\mathcal{I}}(d,d') \geq n\} \end{aligned} \tag{1}$$

Obviously $B^{\mathcal{I}}$ and $R^{\mathcal{I}}$ are fuzzy sets w.r.t $\Delta^{\mathcal{I}}$ and $\Delta^{\mathcal{I}} \times \Delta^{\mathcal{I}}$, while their cuts $(B_{[n]})^{\mathcal{I}}$ and $(R_{[n]})^{\mathcal{I}}$ are actually crisp sets. A collection of $(B_{[n_1]})^{\mathcal{I}}$, $(B_{[n_2]})^{\mathcal{I}}$,..., $(B_{[n_k]})^{\mathcal{I}}$ ($(R_{[n_1]})^{\mathcal{I}}$, $(R_{[n_2]})^{\mathcal{I}}, \cdots, (R_{[n_k]})^{\mathcal{I}}$) are enough to describe $B^{\mathcal{I}}$ ($R^{\mathcal{I}}$) completely or at an acceptable degree.

2.2 Concept and Role Constructors

Starting with atomic cut concepts $B_{[n]}$ and atomic cut roles $R_{[n]}$, so-called cut concept descriptions and cut role descriptions (cut concepts and cut roles, for short) can be inductively defined by applying concept and role constructors in classical description logics. The syntax and semantics of these constructors are summarized in Table 1.

Here $C_{[n_1,\ldots,n_h]}$ is an abbreviation expression of cut concepts by collecting all the suffixes in a suffix vector. For example, $\exists \text{friend}_{[0.7]}.(\text{Tall}_{[0.8]} \sqcap \text{Strong}_{[0.9]})$ can be rewritten as: $\exists \text{friend}.(\text{Tall} \sqcap \text{Strong})_{[0.7,0.8,0.9]}$, where $\exists \text{friend}.(\text{Tall} \sqcap \text{Strong})$ is called the prototype of the cut concept and [0.7, 0.8, 0.9] is called the suffix vector. Throughout this paper, we will use abbreviation expressions to express cut concepts. And for a set M, $\#M$ denotes its cardinality in Table 1.

Table 1. Concept and role constructors

Name	Syntax	Semantics
Top concept	$\top$	$\Delta^{\mathcal{I}}$
Bottom concept	$\bot$	$\varnothing$
Conjunction	$C_{[n_1,\dots,n_h]} \sqcap D_{[n_{h+1},\dots,n_k]}$	$(C_{[n_1,\dots,n_h]})^{\mathcal{I}} \cap (D_{[n_{h+1},\dots,n_k]})^{\mathcal{I}}$
Disjunction	$C_{[n_1,\dots,n_h]} \sqcup D_{[n_{h+1},\dots,n_k]}$	$(C_{[n_1,\dots,n_h]})^{\mathcal{I}} \cup (D_{[n_{h+1},\dots,n_k]})^{\mathcal{I}}$
Complement	$\neg C_{[n_1,\dots,n_k]}$	$\Delta^{\mathcal{I}} \setminus (C_{[n_1,\dots,n_k]})^{\mathcal{I}}$
Value restriction	$\forall R_{[n_1]}.C_{[n_2,\dots,n_k]}$	$\{d \in \Delta^{\mathcal{I}} \mid \forall d' \in \Delta^{\mathcal{I}}, (d,d') \in (R_{[n_1]})^{\mathcal{I}} \to d' \in (C_{[n_2,\dots,n_k]})^{\mathcal{I}}\}$
Existential restriction	$\exists R_{[n_1]}.C_{[n_2,\dots,n_k]}$	$\{d \in \Delta^{\mathcal{I}} \mid \exists d' \in \Delta^{\mathcal{I}}, (d,d') \in (R_{[n_1]})^{\mathcal{I}} \wedge d' \in (C_{[n_2,\dots,n_k]})^{\mathcal{I}}\}$
Unqualified number restriction	$\geq NR_{[n_1]}$	$\{d \in \Delta^{\mathcal{I}} \mid \#\{d' \mid d' \in \Delta^{\mathcal{I}}, (d,d') \in (R_{[n_1]})^{\mathcal{I}}\} \geq N\}$
	$\leq NR_{[n_1]}$	$\{d \in \Delta^{\mathcal{I}} \mid \#\{d' \mid d' \in \Delta^{\mathcal{I}}, (d,d') \in (R_{[n_1]})^{\mathcal{I}}\} \leq N\}$
Qualifying number restrictions	$\geq NR_{[n_1]}.C_{[n_2,\dots,n_k]}$	$\{d \in \Delta^{\mathcal{I}} \mid \#\{d' \mid d' \in \Delta^{\mathcal{I}}, (d,d') \in (R_{[n_1]})^{\mathcal{I}} \wedge d' \in (C_{[n_2,\dots,n_k]})^{\mathcal{I}}\} \geq N\}$
	$\leq NR_{[n_1]}.C_{[n_2,\dots,n_k]}$	$\{d \in \Delta^{\mathcal{I}} \mid \#\{d' \mid d' \in \Delta^{\mathcal{I}}, (d,d') \in (R_{[n_1]})^{\mathcal{I}} \wedge d' \in (C_{[n_2,\dots,n_k]})^{\mathcal{I}}\} \leq N\}$
Inverse role	$R^{-}_{[n_1]}$	$\{(d,d') \in \Delta^{\mathcal{I}} \times \Delta^{\mathcal{I}} \mid (d',d) \in (R_{[n_1]})^{\mathcal{I}}\}$
Transitive role	Trans($R_{[n_1]}$)	$(R_{[n_1]})^{\mathcal{I}}$ is a transitive closure

2.3 Extended Fuzzy Description Logic Knowledge Bases

A EFDL knowledge base *K(T, R, A)* contains terminological axioms about cut concepts and roles in TBox *T* and RBox *R*, and assertions about individuals in ABox *A*.

TBox: A TBox *T* consists of cut concept axioms, which have the following form:

$$C_{[n_1,\dots,n_k]} \sqsubseteq D_{[f_1(n_1,\dots,n_k),\dots,f_m(n_1,\dots,n_k)]} \quad [n_1,\dots,n_k] \in X \tag{2}$$

where X is a continuous subset of $[0,1]^k$, $f_i(n_1,\dots,n_k)$ is a linear function from X to $[0,1]$, $i=1,\dots,m$, $C_{[n_1,\dots,n_k]}$ and $D_{[f_1(n_1,\dots,n_k),\dots,f_m(n_1,\dots,n_k)]}$ are two cut concepts. For example, $B_{1[n]} \sqsubseteq B_{2[1-n]}$ $n \in (0,0.5]$ is a simple cut concept axiom. For any interpretation $\mathcal{I}$, $\mathcal{I}$ satisfies a cut concept axiom $C_{[n_1,\dots,n_k]} \sqsubseteq D_{[f_1(n_1,\dots,n_k),\dots,f_m(n_1,\dots,n_k)]}$ $[n_1,\dots,n_k] \in X$, iff for any $[n_1,\dots,n_k] \in X$, the inclusion $(C_{[n_1,\dots,n_k]})^{\mathcal{I}} \subseteq (D_{[f_1(n_1,\dots,n_k),\dots,f_m(n_1,\dots,n_k)]})^{\mathcal{I}}$ holds. $\mathcal{I}$ satisfies a TBox *T*, iff $\mathcal{I}$ satisfies all cut concept axioms in *T*, such $\mathcal{I}$ is called a model of *T*.

RBox: An RBox *R* is a finite set of cut role axioms, whose expressions are:

$$R_{[n]} \sqsubseteq S_{[f(n)]} \quad n \in X \tag{3}$$

where X is a continuous subset of $[0,1]$, $f(n)$ is a linear function from X to $[0,1]$, $R_{[n]}$ and $S_{[f(n)]}$ are cut roles. An interpretation $\mathcal{I}$ satisfies the above cut role axiom, iff for any $n \in X$, $(R_{[n]})^{\mathcal{I}} \subseteq (S_{[f(n)]})^{\mathcal{I}}$ holds. $\mathcal{I}$ satisfies an RBox R, iff $\mathcal{I}$ satisfies all cut role axioms in R; such $\mathcal{I}$ is called a model of R.

ABox: An ABox A consists of cut assertions, which have the following two forms:

$$a : C_{[n_1,\ldots,n_k]} \text{ and } (a,b) : R_{[n]} \tag{4}$$

where a and b are individuals and their interpretation $a^{\mathcal{I}}$ and $b^{\mathcal{I}}$ are elements in $\Delta^{\mathcal{I}}$. $\mathcal{I}$ satisfies $a : C_{[n_1,\ldots,n_k]}$ ($(a,b) : R_{[n]}$), iff $a^{\mathcal{I}} \in (C_{[n_1,\ldots,n_k]})^{\mathcal{I}}$ ($(a^{\mathcal{I}},b^{\mathcal{I}}) \in (R_{[n]})^{\mathcal{I}}$). $\mathcal{I}$ satisfies an ABox A, iff $\mathcal{I}$ satisfies all cut assertions in A; such $\mathcal{I}$ is called a model of A.

$\mathcal{I}$ satisfies a knowledge base $K(T, R, A)$ (written $\mathcal{I} \models K$) iff $\mathcal{I}$ is a model of T, R and A; such $\mathcal{I}$ is also called a modal of K.

3 Distributed Extended Fuzzy Description Logics

In this section, we will integrate E-connection into our EFDLs. We define "cut links" to denote fuzzy relations between two EFDL knowledge bases. By using cut links, we couple various distributed knowledge bases, hence getting a combinational DEFDL knowledge base.

3.1 Cut Links Between Two Knowledge Bases

Let K^1 and K^2 be two EFDL knowledge bases, $\mathcal{I}_1 = <\Delta^{\mathcal{I}_1}, \cdot^{\mathcal{I}_1}>$ and $\mathcal{I}_2 = <\Delta^{\mathcal{I}_2}, \cdot^{\mathcal{I}_2}>$ are their interpretation. N_L^{12} is a set of fuzzy links (denoted E) that connect K^1 and K^2. Let $\mathcal{I}_{12} = <\Delta^{\mathcal{I}_{12}}, \cdot^{\mathcal{I}_{12}}>$ be the interpretation of E, where $\Delta^{\mathcal{I}_{12}} = \Delta^{\mathcal{I}_1} \times \Delta^{\mathcal{I}_2}$. For any E, $E^{\mathcal{I}_{12}}$: $\Delta^{\mathcal{I}_1} \times \Delta^{\mathcal{I}_2} \rightarrow [0,1]$, for its cut set $E_{[n]}$ (called cut link), $E_{[n]}{}^{\mathcal{I}_{12}} = \{ (d,d') \mid E^{\mathcal{I}_{12}}(d,d') \geq n\}$. To describe constraints among cut links, we propose cut link axioms:

$$E_{[n]} \sqsubseteq E'_{[f(n)]} \quad n \in X \tag{5}$$

where X is a continuous subset of $[0,1]$, $f(n)$ is a linear function from X to $[0,1]$, and $E_{[n]}$ and $E'_{[f(n)]}$ are cut links. An interpretation $\mathcal{I}_{12}$ satisfies the above cut link axiom, iff for any $n \in X$, $(E_{[n]})^{\mathcal{I}_{12}} \subseteq (E'_{[f(n)]})^{\mathcal{I}_{12}}$. An LBox L^{12} is a finite set of cut link axioms, $\mathcal{I}_{12}$ satisfies L^{12}, iff it satisfies every axiom in L^{12}.

These two EFDL knowledge bases K^1 and K^2 and their LBox L^{12} construct a simple DEFDL knowledge base Σ=($\{K^1, K^2\}$, $\{L^{12}\}$), where the formal definition of DEFDL knowledge bases will be given in next subsection. By introducing cut link, we allow two new cut concepts in K^1: $\exists E_{[n_1]}{}^{12}.C_{[n_2,\ldots,n_k]}{}^2$ and $\forall E_{[n_1]}{}^{12}.C_{[n_2,\ldots,n_k]}{}^2$, where $E_{[n_1]}{}^{12}$ is a cut link in L^{12} and $C_{[n_2,\ldots,n_k]}{}^2$ is a cut concept in K^2. These two cut concepts are considered as normal cut concepts in K^1, hence they can appear in TBox and ABox of K^1. For example, let K^1 and K^2 be two knowledge bases about animal and person respectively. Dog^1 and Person^2 are fuzzy concepts in K^1 and K^2, and lovewith^{12} is a fuzzy link in

$N_L{}^{12}$. Now we need define a new fuzzy concept Friendly-dog^1. By using the new cut concept $\exists E_{[n_1]}{}^{12}.C_{[n_2,\ldots,n_k]}{}^2$, we can define cut sets of Friendly-dog^1 in K^1's TBox:

$$\text{Friendly-dog}_{[n]}{}^1 \equiv^* \exists(\text{lovewith}_{[n]})^{12}.\text{Person}_{[1]}{}^2 \sqcap \text{Dog}_{[1]}{}^1 \qquad n\in(0,1] \tag{6}$$

Note that, Person^2 and Dog^1 are indeed crisp concepts, so we only considered on their 1-cut sets in above definition. This definition guarantees that for any dog *a* (obviously, if *a* is not a Dog, it is totally not a Friendly-dog) in K^1, the membership of *a* belonging to Friendly-dog depends on its "lovewith" relation with Persons.

3.2 Distributed Extended Fuzzy Description Logic Knowledge Bases

In above subsection, we discuss the fuzzy links between two EFDL knowledge bases and give a simple example of DEFDL knowledge bases. Now we will give a general definition of it.

DEFDL knowledge base: A DEFDL knowledge base is a pair $\Sigma=(K_S, L_S)$, where K_S is a set of EFDL knowledge bases: $K_S=\{K^1,\ldots,K^m\}$, and L_S is a set of LBoxes that connect two knowledge bases in K_S: $L_S=\{L^{ij}|1\le i, j\le m \text{ and } i\ne j\}$. For any cut concept $C_{[n_2,\ldots,n_k]}{}^j$ in K^j and any cut link $E_{[n]}{}^{ij}$ in L^{ij}, the following expressions $\exists E_{[n]}{}^{ij}.C_{[n_2,\ldots,n_k]}{}^j$ and $\forall E_{[n]}{}^{ij}.C_{[n_2,\ldots,n_k]}{}^j$ are also considered as cut concepts in K^i.

An interpretation of a DEFDL knowledge base is a pair $\mathcal{I}=(\{\mathcal{I}_i\},\{\mathcal{I}_{ij}\})$, where $\mathcal{I}_i$ is an interpretation of K^i and correspondingly $\mathcal{I}_{ij}$ is an interpretation of L^{ij}. For any cut concept $C_{[n_2,\ldots,n_k]}{}^j$ (cut role $R_{[n]}{}^j$) in K^j, $(C_{[n_2,\ldots,n_k]}{}^j)^{\mathcal{I}}=(C_{[n_2,\ldots,n_k]}{}^j)^{\mathcal{I}_j}$ ($(R_{[n]}{}^j)^{\mathcal{I}}=(R_{[n]}{}^j)^{\mathcal{I}_j}$), any cut link $E_{[n]}{}^{ij}$ in L^{ij}, $(E_{[n]}{}^{ij})^{\mathcal{I}}=(E_{[n]}{}^{ij})^{\mathcal{I}_{ij}}$, any individual a^j in K^j, $(a^j)^{\mathcal{I}}=(a^j)^{\mathcal{I}_j}$. And for $\exists E_{[n]}{}^{ij}.C_{[n_2,\ldots,n_k]}{}^j$ and $\forall E_{[n]}{}^{ij}.C_{[n_2,\ldots,n_k]}{}^j$, their interpretation are inductively defined as:

$$\begin{aligned}(\exists E_{[n_1]}{}^{ij}.C_{[n_2,\ldots,n_k]}{}^j)^{\mathcal{I}} &= \{d\in\Delta^{\mathcal{I}_i} \mid \exists d'\in\Delta^{\mathcal{I}_j},(d,d')\in(E_{[n_1]}{}^{ij})^{\mathcal{I}} \wedge d'\in(C_{[n_2,\ldots,n_k]}{}^j)^{\mathcal{I}}\}\\ (\forall E_{[n_1]}{}^{ij}.C_{[n_2,\ldots,n_k]}{}^j)^{\mathcal{I}} &= \{d\in\Delta^{\mathcal{I}_i} \mid \forall d'\in\Delta^{\mathcal{I}_j},(d,d')\in(E_{[n_1]}{}^{ij})^{\mathcal{I}} \rightarrow d'\in(C_{[n_2,\ldots,n_k]}{}^j)^{\mathcal{I}}\}\end{aligned} \tag{7}$$

An interpretation $\mathcal{I}$ is a model of $\Sigma=(K_S, L_S)$, iff $\mathcal{I}$ satisfies every K^i in K_S and every L^{ij} in L_S.

4 Reasoning Issues

A DEFDL knowledge base $\Sigma=(K_S, L_S)$ merge a collection of distributed EFDL knowledge bases K^1, …, K^m into a global knowledge system. Besides enabling expressions of fuzzy and distributed knowledge, $\Sigma=(K_S, L_S)$ also supports several reasoning problems:

DEFDL knowledge base satisfiability: $\Sigma=(K_S, L_S)$ is satisfiable iff it has a model.

Cut concept satisfiability: a cut concept $C_{[n_1,\ldots,n_k]}{}^j$ is satisfiable w.r.t $\Sigma=(K_S, L_S)$, iff there is a model $\mathcal{I}$ of $\Sigma=(K_S, L_S)$ and $(C_{[n_1,\ldots,n_k]}{}^j)^{\mathcal{I}}\ne\varnothing$.

---

* In this axioms, "≡" is an abbreviation of both "⊑" and "⊒", where ⊒ is the inversion of ⊑.

Cut concept subsumption: a cut axiom ${C_{[n_1,...,n_k]}}^j \sqsubseteq {D_{[f_1(n_1,...,n_k),...,f_m(n_1,...,n_k)]}}^j$ $[n_1,...,n_k] \in X$ holds w.r.t $\Sigma=(K_S, L_S)$, iff for any model $\mathcal{I}$ of $\Sigma=(K_S, L_S)$, $\mathcal{I}$ satisfies this axiom. And cut role subsumption is defined analogously.

Individual checking: an individual a belongs to ${C_{[n_1,...,n_k]}}^j$ w.r.t $\Sigma=(K_S, L_S)$, iff for any model $\mathcal{I}$ of $\Sigma=(K_S, L_S)$, $a^{\mathcal{I}} \in ({C_{[n_1,...,n_k]}}^j)^{\mathcal{I}}$.

Unfortunately, within an unrestricted DEFDL knowledge base, no practical reasoning technique is available for these reasoning problems. How to deal with fuzzy general concept inclusion axioms still remains an open problem in fuzzy concept languages [6]. And our cut concept axioms are more complex than fuzzy general concept inclusion axioms [8]. However, if we make some constraints of DEFDL knowledge bases (mainly on TBox form), we will get some reasoning support from current reasoning calculus. Under acyclic TBox, which is a constraint form of our TBox defined in section 2.3, a PSPACE bounded calculus in Extended Fuzzy ALCH [9] can be considered as a recommendation solution, where a novel approach to compute role-successor and to construct a "simple TBox" [18] can be directly applied in our DEFDL knowledge bases. Furthermore, Straccia adopted "knowledge base expansion" [20] to eliminate acyclic TBox [19], such elimination technique can also be reused in our distributed framework. Another noteworthy method is PTIME bounded translation from fuzzy description logics into crisp ones [21, 7], but no believable evidences guarantee that these translation techniques can be straightforwardly used in distributed cases.

5 Conclusion and Future Work

By development of the Semantic Web, fuzzy and distributed extensions of ontology languages attract more and more attention. Current solutions are proposed respectively on one of these two aspects. This paper proposes a novel logical approach to couple fuzzy and distributed features. By integrating E-connection into our EFDLs, we point out DEFDLs. In DEFDLs, we introduce cut link to describe fuzzy connections among distributed fuzzy areas, and make use of cut link to connect fuzzy knowledge bases to construct a global DEFDL knowledge base. We also make a primitive discussion about reasoning issue within DEFDLs, where some current reasoning technique can be reused in restricted DEFDLs knowledge bases. Our future work will consider on two points: extending cut link with more expressive form and developing new reasoning method in unrestricted DEFDLs knowledge bases.

References

[1] Berners-Lee, T., Hendler, J., Lassila, O.: The Semantic Web. Scientific American. vol. 284, no.5 (2001) 34-43

[2] Lassila, O., R. Swick, R.: Resource Description Framework (RDF) Model and Syntax Specification–W3C Recommendation 22 February 1999. Technical report, World Wide Web Consortium

[3] Baader, F., Calvanese, D., McGuinness, D.L., Nardi, D., Patel-Schneider, P.F.(Eds.): The Description Logic Handbook: Theory, Implementation, and Applications. Cambridge University Press (2003)

[4] Horrocks, I., Patel-Schneider, P.F.: Reducing OWL entailment to description logic satisfiability. In: Proceedings of the 2003 Description Logic Workshop, (2003) 1-8

[5] Straccia, U.: A Fuzzy Description Logic. In: Proceedings of AAAI-98, 15th National Conference on Artificial Intelligence, Madison, Wisconsin, (1998) 594-599

[6] Stoilos, G., Stamou, G., Tzouvaras, V., Pan, J., Horrocks, I.: Fuzzy OWL: Uncertainty and the Semantic Web. In: International Workshop of OWL: Experiences and Directions, Galway, (2005)

[7] Straccia, U.: Transforming fuzzy description logics into classical description logics. In: Proceedings of the 9th European Conference on Logics in Artificial Intelligence, Lisbon (2004) 385-399

[8] Li, Y. H., Xu, B. W., Lu, J. J., Kang, D. Z., Wang, P.: A family of extended fuzzy description logics. In: Proceedings of the 29th Annual International Computer Software and Applications Conference, Edinburgh, (2005) 221-226

[9] Li, Y. H., Lu, J. J., Xu, B. W., Kang, D. Z., Jiang J. X.: A Fuzzy Extension of Description Logic ALCH. Lecture Notes in Artificial Intelligence, 2005, Volume 3789: 152-161.

[10] Lu, J. J., Xu, B. W., Li, Y. H., Kang, D. Z., Wang, P.: Extended fuzzy ALCN and its tableau algorithm. Lecture Notes in Artificial Intelligence, 2005, Volume 3613: 232-242.

[11] Borgida, A., Serafini, L.: Distributed description logics: Assimilating information from peer sources. Journal of Data Semantics. (2003) 1:153-184.

[12] Bouquet, P., Giunchiglia, F., van Harmelen, F., Serafini, L., Stuckenschmidt H.: C-owl: Contextualizing ontologies. In: Proceedings of the 2003 International Semantic Web Conference (ISWC 2003) (2003)

[13] Kutz, O., Lutz, C., Wolter, F., Zakharyaschev, M.: E-connections of abstract description systems. Artificial Intelligence. 156(1) (2004) 1-73

[14] Baader, F., Lutz, C., Sturm, H., Wolter, F.: Fusions of description logics and abstract description systems. Journal of Artificial Intelligence Research, vol.16 (2003) 1-58

[15] Cuenca Grau, B., Parsia, B., Sirin, E.: Working With Multiple Ontologies on the Semantic Web. In: Proceedings of the 3thrd International Semantic Web Conference. Volume 3298 Lecture Notes in Computer Science (2004).

[16] Parsia, B., Cuenca Grau, B.: Generalized Link Properties for Expressive E-Connections of Description Logics. In Proceedings of the Twentieth National Conference on Artificial Intelligence (AAAI-2005), Pittsburg, P.A (2005).

[17] Zadeh, L. A.: Fuzzy sets. Information and Control. vol.8 no.3 (1965) 338-353

[18] Calvanese, D. Reasoning with inclusion axioms in description logics: Algorithms and complexity. In: Proceedings of the Twelfth European Conference on Artificial Intelligence (ECAI-96), Budapest (1996) 303-307

[19] Straccia, U. Reasoning within fuzzy description logics. Journal of Artificial Intelligence Research. vol.14 (2001) 137-166

[20] Nebel, B. Computational complexity of terminological reasoning in BACK. Artificial Intelligence. vol.34 (1988) 371-383

[21] Li, Y. H., Xu, B. W., Lu, J. J., Kang, D. Z., Xu, J.: Reasoning Technique for Extended Fuzzy Description Logics. In: Proceedings of the 17th IEEE International Conference on Tools with Artificial Intelligence, Hong Kong, China (2005)

Similarity Metrics for Set of Experience Knowledge Structure

Cesar Sanin and Edward Szczerbicki

Faculty of Engineering and Built Environment, University of Newcastle
University Drive, Callaghan, NSW 2308, Australia
{Cesar.Maldonadosanin, Edward.Szczerbicki}@Newcastle.edu.au

Abstract. When referring to knowledge forms, collecting *formal decision events* in a knowledge-explicit way becomes an important development. Set of experience knowledge structure can assist in accomplishing this purpose. However, to make set of experience knowledge structure useful, it must be classifiable and comparable. The purpose of this paper is to show similarity metrics for set of experience knowledge structure, and within, similarity metrics for its components: variables, functions, constraints, and rules. A comparable and classifiable set of experience would make explicit knowledge of formal decision events useful elements in multiple systems and technologies.

1 Introduction

Knowledge is a priceless asset of incalculable worth and has been considered as the only true source of competitive advantage of a company [2]. Hence, managers have turned to knowledge administration and companies want technologies that make possible all forms of controlling knowledge. Knowledge management can be considered as the key for the success or failure of a company. One of the most complicated issues about knowledge is its representation, because it determines how knowledge is acquired and how knowledge is transformed from tacit knowledge to explicit knowledge, that is, knowledge must be obtained and represented in an understandable and shareable form for the agents that experienced it.

One theory proposes that experienced decision-makers base most of their decisions on situation assessments [10]. In other words, decision-makers principally use experience for their decisions. They extract the most significant characteristics from the current circumstances, and relate them to comparable situations that have worked well in the past. In consequence, tools for representing and store *formal decision events* in a knowledge-explicit way are evidently necessary, understanding that a *formal decision event* is a decision occurrence that was made following procedures that make it structured and formal [12].

Set of experience knowledge structure has been developed to facilitate representation of formal decision events. It is a structure developed as part of a platform for transforming information into knowledge named Knowledge Supply Chain System (KSCS) [11]. In brief, the KSCS takes information from different technologies that

B. Gabrys, R.J. Howlett, and L.C. Jain (Eds.): KES 2006, Part I, LNAI 4251, pp. 663 – 670, 2006.
© Springer-Verlag Berlin Heidelberg 2006

make formal decision events, integrates them and transforms them into knowledge making use of sets of experience. Because different technologies can be both source and target of sets of experience, it must be possible to compare and classify them. Therefore, similarity techniques for set of experience must be explored.

In knowledge discovery, similarity between represented objects is one of the fundamental concepts, and its estimation must be computed based on the data structure that represents the object. Comparisons among these data structures provide, usually, a scalar measure for the similarity or proximity between each pair of objects, and most of the well-known similarity measures are related to geometrical functions such as Minkowskian or Euclidean distance metrics. Commonly, data objects are represented by a vector containing qualitative or quantitative values. Thus, one way to generate a similarity measure is by applying a comparative measure to the vectors that describe the pair of objects.

Because set of experience has different elements that comprise it, and each one has its own characteristics, similarity must be analyzed separately for each of them. Afterward, a unified unique similarity measure can be defined.

The purpose of this paper is to show an effective similarity metric for sets of experience knowledge structure, and more specifically, similarity among variables, functions and constraints as part of a unique element. This developed technique can be applied in different systems and technologies, and more specifically in the KSCS. Therefore, set of experience would have the potential to improve the way knowledge is managed as an asset in current decision-making environments.

2 Background

2.1 Set of Experience Knowledge Structure

In this text a concise idea of set of experience and its components is offered, for additional information Sanin and Szczerbicki [12] should be examine.

Arnold and Bowie [1] claim that "the mind's mechanism for storing and retrieving knowledge is transparent to us... somehow, during this process, all the essential qualities of an object are stored. Later, when someone mentions it, our senses are activated from within, and we see, smell, touch, and taste the object all over again". Computers, unfortunately, are not yet capable of forming representations of the world in this way, and even simpler, representations of just formal decision events. This presents a problem of how to adequately, efficiently, and effectively represent information and knowledge inside a computer.

Set of experience has been developed to store formal decision events in an explicit way [12]. It is a model based upon existing and available knowledge, which must adjust to the decision event is built from. Four basic components surround decision-making events: *variables, functions, constraints,* and *rules.* They are stored in a combined dynamic structure that comprises set of experience (see Figure 1).

Variables usually involve representing knowledge using an attribute-value language (that is, by a vector of variables and values) [7]. This is a traditional approach

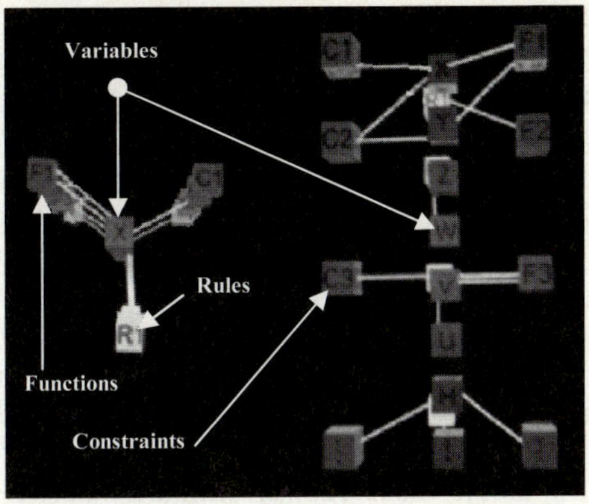

Fig. 1. Structure of the Set of Experience

from the origin of knowledge representation, and is the starting point for set of experience. Variables that intervene in the process of decision-making are the first component of the set of experience. These variables are the centre root of the structure, because they are the origin of the other components.

Based on the idea of Malhotra [8] who maintains that "to grasp the meaning of a thing, an event, or a situation is to see it in its relations to other things", variables are related among them in the shape of functions. Functions, the second component, describe associations between a dependent variable and a set of input variables; moreover, functions can be applied for reasoning optimal states, because they come out from the goals of the decision event. Therefore, set of experience uses functions, and establishes links among the variables constructing multiobjective goals.

According to Theory of Constraints (TOC), Goldratt [5] affirms that any system has at least one constraint; otherwise, its performance would be infinite. Thus, constraints are another way of relationships among the variables; in fact, they are functions as well. A constraint, as the third component of set of experience, is a restriction of the feasible solutions in a decision problem, and a factor that limits the performance of a system with respect to its goals.

Finally, rules are suitable for associating actions with conditions under which the actions should be performed. Rules, the fourth component of set of experience, are another form of expressing relationships among variables. They are conditional relationships that operate in the universe of variables. Rules are relationships between a condition and a consequence connected by the statements IF-THEN-ELSE.

In conclusion, the set of experience consists of variables, functions, constraint and rules, which are uniquely combined to represent a formal decision event. Set of experience can be used in platforms to support decision-making, and new decisions can be made based on sets of experience.

2.2 Similarity

Nowadays, improving strategic and operational efficiency and effectiveness on the companies requires technologies to develop tools that allow using information and knowledge in an effective manner. On developing these tools, computers are used to recognize patterns automatically in large multivariate data sets for supporting decision-making [3]. Analyzing such data is costly and tedious, thus we need efficient methods to classify, compare, and understand data. Data mining (DM) and knowledge discovery (KDD) are research areas in computer science that study these issues.

In searching for such patterns, it is normally not enough to consider only equality or inequality of data. Instead, it needs to consider how similar or different two data objects are, that is, it must be able to calculate how much these two objects differ from each other. This is why similarity between data objects is a central concept in DM and KDD. Similarity between objects can help, for example, in predictions, hypothesis testing, and rule discovery [14]; additionally, it can offer ranking of results for choosing best matching objects. During the last years, there has been significant interest in defining computable measures for similarity between objects in different application areas.

Evaluating similarity between data objects depends essentially on the type of the data. Nevertheless, it is possible to have several types of similarity measures on a single set of data. Different similarity measures can reveal different aspects of the data, and as a result, two data objects can be classified to be similar by one measure and different by another measure. This means that it is significant to choose the one that best suits your purposes. Following, this section tries to establish some general ideas for defining a similarity metric.

A typical data object is a relation that consists of several attributes. Approaches generally used for defining similarity between two data objects consider their attributes in pairs and compare them as isolated elements. Hence, similarity metrics can be calculated based on a composed addition of many different comparisons. A common approach for similarity between objects is defined in terms of a notion of distance (i.e. a geometric approach). A data object is represented by its coordinates in a *similarity space*. Thus, similarity, in that case, is given by distance between objects in the space. The relation between distance and similarity deals with mapping associations among data objects into a chosen distance measure, such that similar data objects are plotted closer to one another in the multi-dimensional attribute space than less similar ones [4]. This technique assumes that data objects can be represented by values on continuous dimensions, and that similarity can be represented by distance in a coordinate space. A problem with this model is that the method seems inappropriate for data objects that have a number of qualitative attributes. For this reason, some researchers have assumed that an object, which is represented by such set of qualitative attributes, may be prearranged as binary variables, i.e. existence or not.

Other techniques to estimate a similarity metric are in the group of event sequences. Similarity between event sequences is based on the idea that a similarity notion should reflect how much work is needed to transform an event sequence into another. Similarity between types of events occurring in sequences explains that two event types are similar if they occur in similar contexts [9]. Additionally, more specialized techniques have been developed. Additive trees, for instance, represents data objects as terminal nodes in a tree, thus similarity between two objects is calculated by the length of the path between them; and additive clustering, which defines a number of clusters with associated weights, thus similarity between two objects is estimate by the sum of the weights of their common clusters.

In conclusion, similarity measures establish element of analysis among data objects making them comparable and classifiable. Consequently, data objects or in our case, sets of experience containing knowledge about formal decision events, can be used as comparable elements for helping in the decision-making process.

Following, similarity metrics applied to the set of experience knowledge structure are exposed.

3 Similarity Metrics for Sets of Experience Knowledge Structure

Initially, let us define some necessary elements for the estimation of similarity among sets of experience.

Definition 3.1. Let U be the universe of sets of experience, and d a measure of a distance between objects in the universe U. The similarity measure d is called a metric, if it satisfies the following conditions for all sets of experience s_i, s_j and s_k in the universe U:

1. $d(s_i,s_j)\geq 0$ (non-negativity)
2. $d(s_i,s_j)=0,\Leftrightarrow s_i=s_j$ (identity)
3. $d(s_i,s_j)=d(s_j,s_i)$ (symmetry)
4. $d(s_i,s_k)\leq d(s_i,s_j)+d(s_j,s_k)$ (triangle inequality)

Our object structure set of experience, is identifiable by the whole set of elements that comprises it: variables, functions, constraints and rules. Besides, there are additional fields added to set of experience that comprise the head of it such as date, time, areas or subareas, aim function involved in the decision event among others (see [13] for more information), and they help to contextualize and reduce the universe of comparisons.

Definition 3.2. Let C be a subset of the universe U of sets of experience (i.e. $C\subseteq U$) named context set. A boolean expression θ containing one or many restrictions on elements of the head of set of experience is called a selection condition on sets of experience, e.g. θ = (area = "Human Resources" AND subarea = "Salary Office" AND aim function = "Payment Level"). The context set is a subset of U that satisfies θ. It is defined as:

$$C(\theta)=\{(s_1,s_2,...,s_k)\mid\forall s_i\in(s_1,s_2,...,s_k)satisfies\,\theta\wedge s_i\in U\}\ . \quad (1)$$

Once the universe has been reduced by using the definition 3.2, similarity metrics are estimated by comparing our query with the elements of C(θ). Because each set of experience has different elements that comprise it and each element has its own characteristics, similarity is examined separately for each of the elements; afterwards, a unique similarity measure is offered by combining their separate results, i.e. a Heterogeneous Euclidean Metric. Following, similarity metrics are considered and applied for each of the components of set of experience knowledge structure.

3.1 Variables

Variables are the most common element of set of experience. Because variables can be quantitative(continuous or discrete) and/or qualitative, prior to the estimation of the similarity measure, a prearranging process must be done over the qualitative variables such as assigning numerical values to each category.

Given a pair of sets of experience s_i, s_j Є U, it is possible to generate a similarity metric of the variables called S_V Є [0,1] by calculating the distance measure between each of the pairwise attributes k Є s_i and s_j. The Euclidean distance measure has been selected based upon its simplicity and extended use. Besides, a normalization form was included following the notion of range of comparison, that is, the maximum function. The similarity metric takes the following equation:

$$S_V = \sum_{k=1}^{n} w_k \left[\frac{\left| s_{ik}^2 - s_{jk}^2 \right|}{\max(|s_{ik}|,|s_{jk}|)^2} \right]^{1/2} \quad \forall k \in s_i \wedge s_j \,. \tag{2}$$

s_{ik}, s_{jk} are the kth attribute of the sets s_i and s_j, w_k is the weight given to the kth attribute, in this case, variable; and n is the number of variables on s_i.

3.2 Functions

Functions are more complicated elements of comparison than variables. Functions are composed by elements called factors, and according to the number of factors it contains, the function can be atomic or complex, i.e. one factor makes an atomic function, and two or more factors make a complex function. Therefore, similarity of function is estimated by comparing their factors. Moreover, functions can be qualitative or quantitative; thus, a prearranging process is necessary for the estimation of a similarity metric S_F Є [0,1]. Firstly, qualitative functions are compared as they are. They follow the next function of similarity.

$$S_{FQl} = \sum_{k=1}^{n} w_{ik} \cdot \left\{ \frac{1 \Leftrightarrow F_k \in s_i \wedge F_k \notin s_j}{0 \Leftrightarrow F_k \in s_i \wedge F_k \in s_j} \right\} \,. \tag{3}$$

F_k is the function being analyzed. w_k is the weight associated to the kth function in the set of experience i.

Secondly, each quantitative function is converted into a vector of factors, thus comparison can be performed easily among all functions that belong to the sets of experience. Therefore, similarity metric of a factor is evaluated according to its components; they are variable-potency joint in one, and coefficient. The calculation of the similarity value is as follows:

$$S_{FQt} = \sum_{k=1}^{n} \sum_{l=1}^{m} \sum_{h=1}^{p} \sum_{g=1}^{q} \frac{w_{ik}}{p} \left[\frac{0.5 \left[\frac{\left| c_{ikh}^2 - c_{j\lg}^2 \right|}{\max(|c_{ikh}|,|c_{j\lg}|)^2} \right]^{1/2} \Leftrightarrow (v_{ikh} \in F_{ik} \in s_i) \wedge (v_{j\lg} \in F_{jl} \in s_j)}{1 \Leftrightarrow (v_{ikh} \in F_{jk} \in s_i) \wedge (v_{j\lg} \notin F_{jl} \in s_j)} \right] . \tag{4}$$

Letters k and l correspond to the functions belonging to s_i and s_j respectively. Letters h and g correspond to the factor within the functions F_{ik} and F_{jl} respectively. Elements c_{ikh}, c_{jlg}, v_{ikh}, and v_{jlg} correspond to the hth and g^{th} coefficient and variable-potency values in the kth and l^{th} function of s_i and s_j respectively. w_{ik} is the weight associated to the function k of s_i, while w_{ikh} is the assigned weighted value of the variable-potency h, associated to the function k in s_i.

S_F can be finally calculated as the sum of the qualitative and quantitative similarities exposed previously: $S_F = S_{FQl} + S_{FQt}$.

3.3 Constraints

Constraints are functions as well, and they are expressed alike except for the use of comparative relationships. Similarity of constraints S_C is calculated in the same form as in functions, but having into account a comparison between the relationships (that is, a different relationship produces S_C=1). S_C can be calculated as the sum of the qualitative and quantitative similarities exposed in equations (3) and (4): $S_C = S_{CQl} + S_{CQt}$.

3.4 Rules

Rules are elements that regulate the decision event, but they are not elements of comparison due to its conditional characteristic. Rules are part of the decision event that influence the variables according to certain conditions, but because not all of them are executed, they are not part of the similarity estimation.

Finally, defining a similarity metric for set of experience knowledge structure is as simple as adding up all the estimations for each of previous elements: variables, functions, and constraints, and assuming that all of them are assigned with the same weight due to their equal importance on influencing the formal decision event.

$$S(i, j) = (S_V + S_F + S_C)/3 \ . \tag{5}$$

Having exposed the similarity metrics for set of experience knowledge structure, which make it comparable and classifiable, the possibilities are open for being extendedly used to and within many different systems and technologies.

4 Conclusion

A comparable and classifiable set of experience knowledge structure able to store formal decision events would advance the notion of administering knowledge in the current decision making environment. Similarity measures such as the ones defined for comparing variables, functions, and constraints within set of experience knowledge structure enables us to extend it to different applications, and in that form, companies which are expanding the knowledge management concept externally, can explore new ways to put explicit classifiable knowledge in the hands of employees, customers, suppliers, and partners. Even though, preliminary experiments have been done, a further research on comparing different methods should be completed.

References

1. Arnold, W. and Bowie, J.: Artificial Intelligence: A Personal Commonsense Journey. Prentice Hall, Inc., Englewood Cliffs (1985) 46.
2. Drucker, P.: The Post-Capitalist Executive: Managing in a Time of Great Change. Penguin, New York (1995).
3. Duda, R.O., Hart, P.E., and Stork, D.G.: Pattern Classification. Wiley, New York, N.Y. (2001).
4. Fabrikant, S.I. and Buttenfield, B.P.: Formalizing Semantic Spaces for Information Access. Annals of the Association of American Geographers, vol. 91, no. 2 (2001) 263-280.
5. Goldratt, E.M. and Cox, J.: The Goal. Grover, Aldershot, Hants (1986).
6. Lin, B., Lin, C. et al.: A Knowledge Management Architecture in Collaborative Supply Chain. Journal of Computer Information Systems, Vol. 42 (5). (2002) 83-94.
7. Lloyd, J.W.: Logic for Learning: Learning Comprehensible Theories from Structure Data. Springer, Berlin (2003).
8. Malhotra, Y.: From Information Management to Knowledge Management: Beyond the 'Hi-Tech Hidebound' Systems. In Srikantaiah, K. and Koening, M. E. D (Eds.): Knowledge Management for the Information Professional, Information Today, Inc., (2000) 37-61.
9. Moen, P.: Attribute, Event Sequence, and Event TypeSimilarity Notions for Data Mining. Department of Computer Science Series of Publications A Report A-2000-1, Helsinki University Printing House, Helsinki (2000).
10. Noble, D.: Distributed Situation Assessment. In proceedings: FUSION '98 International Conference, (1998).
11. Sanin, C. and Szczerbicki, E.: Knowledge Supply Chain System: A Conceptual Model. Knowledge Management: Selected Issues. Szuwarzynski, A. (Ed), Gdansk University Press, Gdansk, (2004) 79-97.
12. Sanin, C. and Szczerbicki, E.: Set of Experience: A Knowledge Structure for Formal Decision Events. Foundations of Control and Management Sciences, Vol. 3 (2005) 95-113.
13. Sanin, C. and Szczerbicki, E.: Using XML for Implementing Set of Experience Knowledge Structure In Proceedings on Int. Conf. on Knowledge-base and Intelligent Information and Engineering Systems, KES, Part I LNCS 3681, Springer (2005) 946-952.
14. White, D.A. and Jain, R.: Algorithms and strategies for similarity retrieval. Technical Report VCL-96-101, Visual Computing Laboratory, University of California, San Diego, CA, USA, July (1996).

Knowledge Management Based on Dynamic and Self-adjusting Fuzzy Models

Cezary Orłowski and Zdzisław Kowalczuk

[1] Faculty of Management and Economics
Gdansk University of Technology, 80-952 Gdansk, Poland
cor@zie.pg.gda.pl
[2] Faculty of Electronics, Telecommunication and Computer Science
Gdansk University of Technology, 80-952 Gdansk, Poland
kova@pg.gda.pl

Abstract. The objective of this paper is to present a knowledge management based on dynamic and self-adjusting fuzzy models. Specifically, the proposed approach is founded on a knowledge-based methodology and theories of dynamic systems and fuzzy sets. A preliminary design of our fuzzy model takes advantage of the experience from managing of *IT* projects. Relevant experimental data resulting from these projects are utilized for adjusting the membership functions of the fuzzy model applied in the knowledge-based decision-support system.

1 Introduction

The Software Project Management (*SPM)* as a knowledge management problem will be analyzed by using various modeling techniques in order to construct appropriate formal models. A white-box approach will be used to model the *SPM* task, and a hybrid (gray) method (including a black-box part) will be applied with reference to the project team SPM_T issue, which is the primary object of our deliberations. An establishing point for such modeling procedures (based on knowledge management) [2] is an experimental setting (or a measurement basis) allowing us to evaluate, verify and/or tune the developed models representing the hypothetical 'real *SPM* system' under consideration [8].

Consequently, first, a suitable hierarchical consortium model will be shown and, next, several structural models of the project team will be developed. Within the project team model, a useful analytical integrated dynamic model will be synthesized, which next will be shown with the use of a black-box model and a simple dynamic model. To build the black-box model of practical use, elementary findings of the fuzzy control theory [5] and the rules of Mamdani type [7] will be utilized, resulting in a rule-fuzzy subsystem. Moreover, dynamic state variables will be introduced to describe the states of management in terms of team knowledge, infrastructure, design processes, and supporting technologies. This approach will yield a suitable dynamic state-space subsystem. Thus the idea preliminary proposed in [1] will be developed here to yield an effective integrated vector-matrix dynamic model of SPM_T with rule-fuzzy and state-space subsystems incorporated. Certainly, we shall have to fix the

B. Gabrys, R.J. Howlett, and L.C. Jain (Eds.): KES 2006, Part I, LNAI 4251, pp. 671 – 678, 2006.
© Springer-Verlag Berlin Heidelberg 2006

values of all model parameters to obtain a fully applicable one. This will be done by using a procedure of parameter-tuning based on real project data, gathered in the form of a knowledge base, resulting from the project-team managers' knowledge. Clearly, the completeness and consistency of the knowledge base applied have to be verified beforehand. By repeating this procedure for a continuously growing knowledge base (suitably aggregated), we implement a learning characteristic of the knowledge-based decision support system (*KB-DSS*).

It is clear that *IT* projects of this type can be implemented in many ways and with a varying quality of management. Consequently, the management data obtained from a variety of systems, when used for tuning the model, would lead to an overly averaged system of little use. Therefore, a pre-selection of projects, whose rules are admitted to the knowledge base, should be done in accordance with different management quality levels, attributed to each of the advising *SPM* experts. In this way we can aggregate the data into "successful" and "unsuccessful" variants of the projects, for instance. Taking into account the above precautions, we propose constructing a collection of the *Software Project Management Rule Fuzzy Model* (SPM_T*–RFM*), which:

(1) are different, because they refer to diverse variants of the projects,
(2) allow each project manager to select one model, which matches his/her knowledge of the quality of management of the project being analyzed, and, in particular,
(3) support the project managers in their decision-making.

It we can also interpret the developed utilitarian models as the result of a search for new solutions to the problem of building fuzzy models supporting complex knowledge-based systems, which are classified as embedded [3] within the conception of the *Cost Construction Model* (*COCOMO*).

2 Models of Software Project Management Based on Team Mangers' Knowledge

2.1 Hierarchical Consortium Model

The proposed consortium model presents a hierarchy in terms of the structure of management applied within the whole project enterprise, including project teams. A general structure of the consortium model is shown in Figure 1.

There are two principal management levels assigned to the respective functions of project management (with the assumed emphasis on team management):

1. The project coordinator derives his decisions (Levels II and III in Fig. 1) by analyzing the results of the comparison of the scheduled tasks with the budgeted tasks (performed in Level I shown in Fig.1).

2a. The project team manager (Level IV) makes plans on the basis of the evaluation of the *IT* methods and tools used in the team part of the project implementation.

2b. The project team manager can initiate technological innovations, i.e. changes in the *IT* methods and tools (and in accordance with the results of Levels II and III).

2c. The project team manager introduces proposed solutions, follows up their execution, and evaluates the results of their implementation.

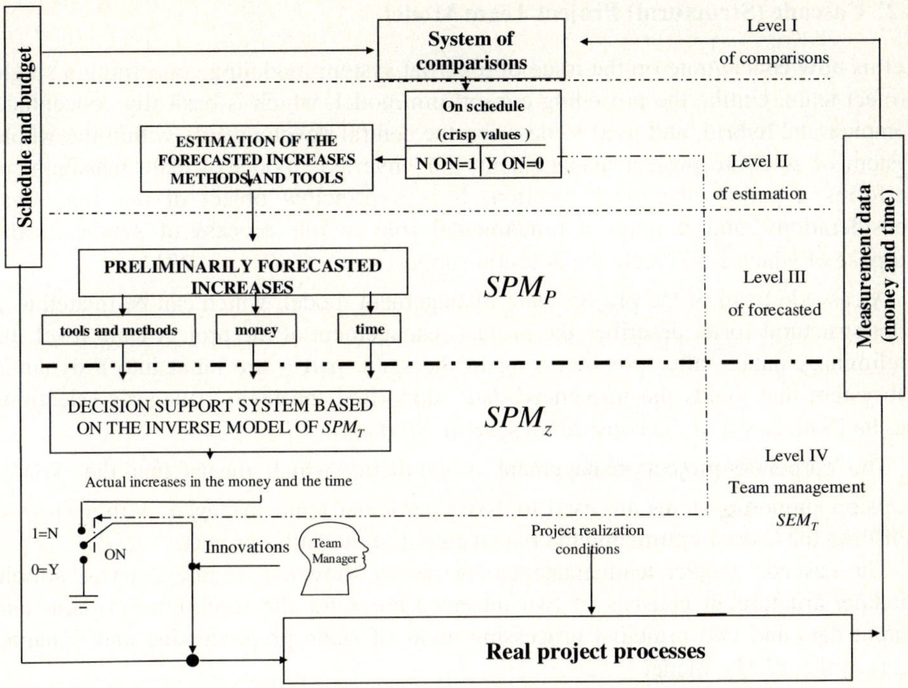

Fig. 1. Hierarchical model of a project consortium (SPM_P) with emphasis on the team management and its decision support system SPM_T

Preliminary forecasts are made at the third Level to support the decision-making system at the Level of individual project teams (IV). These forecasts involve *IT* methods and tools, money, and time in terms of their preliminary increases when projects are behind schedule. They should be processed by the respective team manager or should be fed to the $SPM_T - RFM$ system, initiating his decision-making. By taking into consideration the closed-loop conditions, depicted in Figure 1, it is clear that the designed $SPM_T - RFM$ system substitutes the principal functions of the team manager by emulating an 'inverse' to the project team process in terms of the effects of management (schedule and budget targets). In particular, the purpose of the decision-making support system is to generate *'inferenced*' changes in the resources (money and time), which can be used as command data for the team manager, who can either accept or innovate them. As a result, the team manager is informed about the inferenced changes in the time and money allocated for the project tasks and the increases in the *IT* methods and tools necessary for completing them. 'Natural' processing of this information by a real working project team (in actual project conditions) yields measurement data (see Figure 1), which can be sampled according to the management needs (bi-monthly, for instance).

2.2 Cascade (Structural) Project Team Model

Let us now concentrate on the issue of a partial system modeling concerning a single project team. Unlike the preceding consortium model, which is basically conceptual, complex and hybrid, and used to describe the central developments within the whole system of software project management, the 'inverted' model of team management performs a purely analytical function. It is a principal object of our functional considerations, and it plays a fundamental role in our process of synthesis, the purpose of which is to create the decision support system $SPM_T - RFM$.

A cascade form of the project team management model, which can be treated as a pilot structural form, describes the project management at the project team level. Its preliminary data – after preprocessing in the static part – are input into a dynamic subsystem that yields the inferenced data supporting the team-manager's selections (at the IV-th Level of the consortium system SPM_P).

The corporate project management is synthetic, which means that the SPM_T decision support systems are used to assist particular team managers with a view to fulfilling the task of optimizing the management of the entire project.

The cascade project team management model shown in Figure 2 has a simple module structure. It consists of two interface areas for the (preliminary) input and output data and two principal processing areas of static preprocessing and dynamic parts of the SPM_T model.

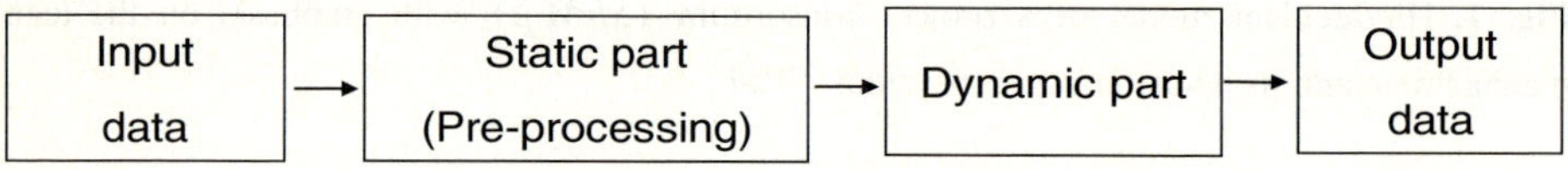

Fig. 2. The cascade model of $SPM_T - RFM$

2.3 Integrated (Structural) Project Team Model

As we built the integrated model, appropriate simplifying procedures were used to eliminate some descriptive variables (especially those at the project enterprise/consortium management level) and to group some elements of the $SPM_T - RFM$ model.

As shown in Figure 3 (parallel to Figure 2), the discrete-time integrated version of the team model processes the input variables, describing the preliminarily forecasted increases in money and time (which can be called the resource innovations) as well as in *IT* methods and tools (referred to as the technological innovations).

The team-management model is composed of two formal (static and dynamic) submodels founded on state variables (knowledge, infrastructure, supporting technologies, and project processes). The model output variables refer to the inferenced increases in the resources (project money and time) that are suggested to the team manager.

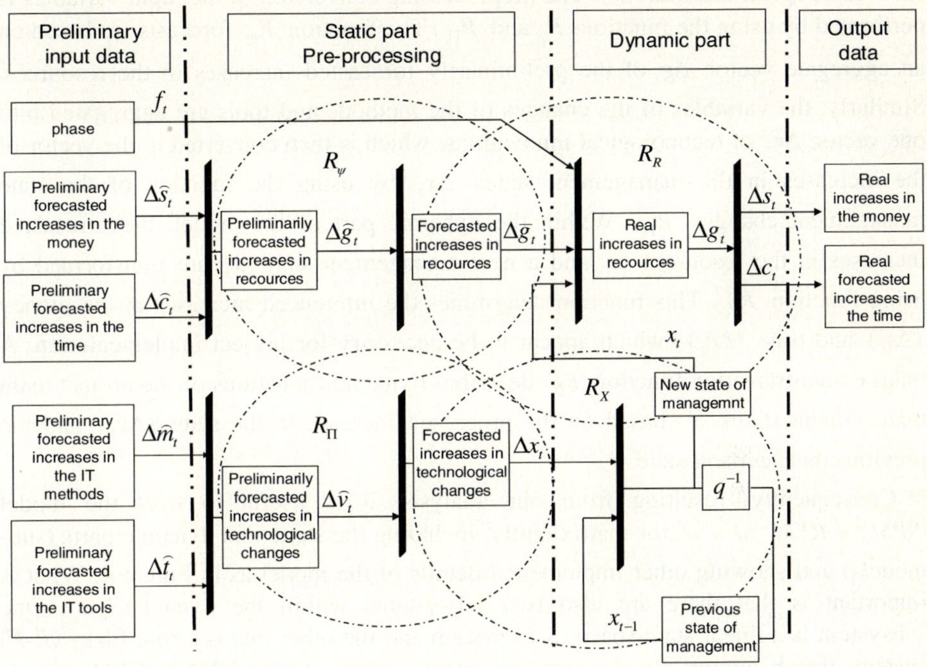

Fig. 3. Integrated structural model of $SPM_T - RFM$ describing team management with the use of selected state and input variables

By introducing formal symbols of the data (variables) and operators (discrete functions), the integrated model of $SPM_T - RFM$ (the decision support for the managers of the fourth Level of the consortium model depicted in Figure 1) can be described as follows. The (preliminary) input data include the technological innovations (the preliminarily forecasted increases in the *IT* methods $\Delta\hat{m}_t$ and the tools $\Delta\hat{t}_t$) and the resource innovations (the preliminarily forecasted increases in the money $\Delta\hat{s}_t$ and the time $\Delta\hat{c}_t$,where the subscript $t \geq 0$ denotes a discrete-time index). Please note that the resource innovations originate in the area of technological innovations (Figure 1).

As presented in Figure 3 resulting from the modeling decomposition applied, the integrated model of the project management at the team level makes use of static and dynamic (state) variables, as well as static functions (R_ψ, R_Π, R_R) and dynamic (R_X) ones.

A project phase is represented by a decision variable f_t which reprograms the *KB-DSS* accordingly when it is used by the project managers (see Figures 1 and 3). Static preprocessing performs an analysis of the preliminarily forecasted increases in the resources (necessary to implement or modify the *IT* methods and tools as declared by the technological innovations). The preprocessing conversion of the input variables is performed by using the functions R_ψ and R_Π. The function R_ψ forecasts $\hat{g}_t$ based on an aggregate vector $\Delta\hat{g}_t$ of the preliminarily forecasted increases in the resources. Similarly, the variables of the changes of the methods and tools are aggregated into one vector $\Delta\hat{v}_t$ of technological innovations, which is then converted to the vector of the increases in the management states Δx_t by using the function of the state management changes R_Π. Within the dynamic part of the model, the forecasted increases in the resources $\hat{g}_t$ and a new management state x_t are transformed by using function R_R. This function determines the inferenced increases in the money (Δs_t) and time (Δc_t), which appear to be necessary for project implementation. A matrix state-transition function R_X describes dynamic transitions of the project team management states x_t based on the processed increase in the state (Δx_t) and the previous management state x_{t-1}.

Consequently, resulting from our analysis, it is useful to treat the model $SPM_T - RFM$ as a vector-matrix entity, including the static and dynamic parts (sub-models) and showing other implemental details of the model as in Figure 4. What is important is that there are also two subsystems within the dynamic part. One subsystem is a linear state-space (*S-S*) system and the other one is a rule-fuzzy (*R-F*) system, which outputs the resource inferences (Δg) of the $SPM_T - RFM$ system, supporting the work of a team manager.

According to the cascade (Figure 2) and integrated (Figure 3) structural models, the implemental (matrix-vector) structure of the $SPM_T - RFM$ model describes both the static (preprocessing) and dynamic parts. The black line in Figure 4 separates these parts, while the dotted line divides the dynamic part into the state-space *(S-S)* subsystem and rule-fuzzy (*R-F*) subsystem.

3 Conclusions

In this paper we propose the example of knowledge management based on rule-fuzzy mechanisms (*RFM*) in an integrated $SPM_T - RFM$ model of software project management (*SPM*) concerning particular *IT* project teams. Our presentation includes a complete description of the concept, formalization and a method of modeling and designing a decision support system, incorporating a fuzzy subsystem.

Within the cascade (pilot structural) model $SPM_T - RFM$, a distinction was made between the input and output data and the preliminary processing of the resource and technological innovations. Technological innovations, concerning the *IT* methods and tools, underwent a preprocessing, aimed at a suitable transition of the states of management (expressed in terms of the knowledge, supporting technology, design processes and infrastructure utilized in the project team processes).

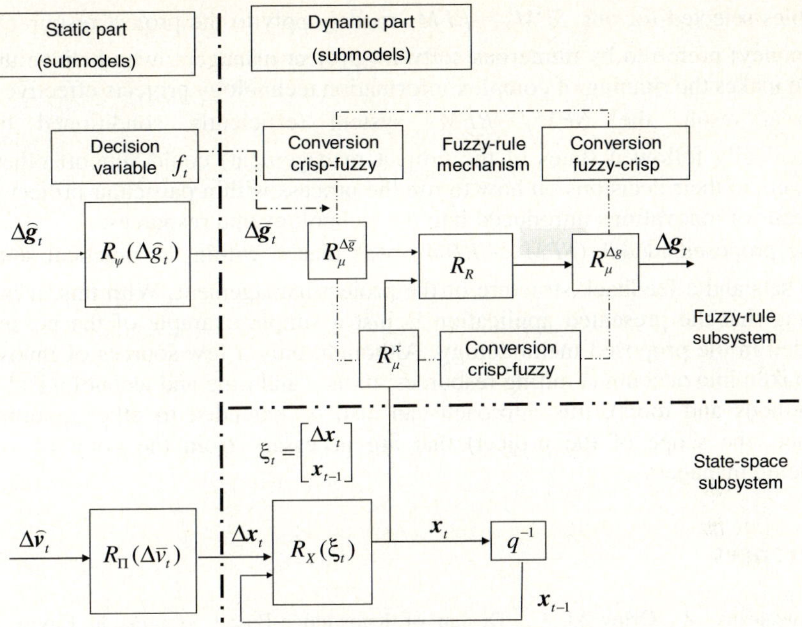

Fig. 4. Implemental structure of the $SPM_T - RFM$ model

Thus, within the structural model, we assumed that the management states are interpreted as state variables of a discrete-time dynamic model. They represent momentary values of management levels in the aforementioned aspects. By adopting the concept of state variables, we can model the changes in management and take into consideration the *IT* methods and tools applied in the project. This solution reflects the idea that the project manager controls the selection of the methods and tools, whereas the resources are derivatives of their changes, considered technological innovations. The use of the state variables is well justified because the values of the distinguished variables depend on their previous state. The consequences of this assumption for the SPM_T are that the management levels of the knowledge, infrastructure, design process, and supporting technology go up or down depending on how the project advances.

The output data represent inferenced changes in the project resources versus the scheduled ones. Building the hierarchical project consortium model and the cascade team model was based on different descriptive variables. They can also represent various model-management quality indices, the number of team members, as well as other exceptional conditions of project management. With the introduction of new input, output, and state variables, however, one has to be aware of the known effect of the "curse of dimensionality". In the case in question, the resulting solution would be a fuzzy model with a greater number of rules, which makes its application and adaptation more difficult. A similar experience with the fuzzy-controller designs [6] suggests that the model subjected to an effective tuning should contain no more than four to six input variables and one or two output variables. The input and output variables selected for our $SPM_T - RFM$ model apply to the project resources (time and money) preferred by numerous software project managers, who believe that this choice makes the running of complex information technology projects effective [4]

As a result, the $SPM_T - RFM$ system (efficiently conditioned by the automatically followed states of the project management) could support other team managers in their decisions on how to run the process within particular project phases by means of innovations introduced into the technology and resources.

The proposed model $(SPM_T - RFM)$ was placed within the practical setting of fuzzy sets and a feedback structure of the project management. With this, it is worth noticing that the presented application is just a simple example of the possibilities included in the proposed methodology. Although, only a few sources of innovations were taken into account (limiting resources: money and time and technological means *IT* methods and tools), this approach can also be extended to other qualities (for instance, the scope of the project) that are necessary from the point of view of practicing managers.

References

1. Kowalczuk, Z., Orłowski, C.: Design of Knowledge-Based Systems in Environmental Engineering. Proceedings of the Computer Science Convention Conference, Gdańsk (CD-ROM) (2003).
2. Orłowski C., Szczerbicki E.: Evaluation of Information Technology Projects. Cybernetics and Systems, 2002 vol. 33 nr 6, pp. 659-673.
3. Paulk, M.C.: Software Capability Maturity Model, Version 2., Draft Technical Report. Software Engineering Institute, Carnegie Mellon University Pittsburgh (1997)
4. Szyjewski, Z..: Software Project Management. Methodology of IT-system Creation. Agencja Placet, Warsaw (in Polish) (2001)
5. Tong, R.M.: The Construction and Evaluation of Fuzzy Models. Advances in Fuzzy Set Theory and Applications, North Holland Amsterdam (1979) pp. 559-576
6. Yager, R. and Filew, D.: Essentials of Fuzzy Modeling and Control. John Wiley and Sons New York. (1994)
7. Zadeh, L.: A. Fuzzy Sets as a Basis for Theory of Possibility. Fuzzy Sets and Systems vol. 1, (1978) pp. 1762-1784
8. Ziegler, B.: Theory of Modeling and Simulation, Springer Verlag Berlin, London (1984)

Knowledge Based Tools to Support the Structural Design Process

Carlos Toro[1], Jorge Posada[1], Maite Termenón[1], Joaquín Oyarzun[2], and Juanjo Falcón[3]

[1] VICOMTech Research Centre, Mikeletegi Pasealekua 57
20009 Donostia, Spain
{ctoro, jposada, mtermenon} @vicomtech.es
http://www.vicomtech.es
[2] LANIK S.A, Paseo Mundaiz, 8, 20012 Donostia, Spain
[3] SOME S.L. Avda. Navarra s/n (oficina 10), 20500 Mondragón, Spain

Abstract. In this paper we present an architecture and a system implementation for the exploitation of semantic aspects in the computer assisted design of Steel Detailing Structures (Structural Design). We support our approach in domain specific standardization efforts (CIS/2) by modeling the knowledge of the product structure and the design process as OWL ontologies in order to provide the designer with additional tools for his everyday work. As test case we present a collection of applications based in the proposed schema and developed in the frame of an R&D project together with an actual structural engineering company.

1 Introduction

The consideration of semantic aspects in traditional computer applications brings new possibilities as it increases on the one side accurate knowledge sharing, and on the other side the reliability and performance of such systems. The main advantages of using embedded semantics are: (*i*) the improved information management, (*ii*) searching and sharing enhancements, and (*iii*) the empowerment of the intrinsic knowledge embedded in the elements being described.

In this paper we introduce an architecture intended for the exploitation of semantic aspects embedded in a CAD (Computer Aided Design) model in the steel detailing domain. Our implementation handles its storage model in an OWL (Web Ontology Language) compliant ontology where the concepts and relations are mapped from an international standardization initiative in the domain called CIS/2 [2].

As test case, we present a set of tools developed in the frame of a research project with an actual structural design company, who has identified some critical stages in the design flow that can be clearly benefited by an implementation based on semantics. The tools implemented allow the designer to improve the design workflow by exploiting the semantics both in the product model and the design process.

This paper is organized as follows: In chapter 2, a review of related works and some background information is presented. In chapter 3 we introduce our architecture to implement knowledge based tools for the Steel Detailing process. In chapter 4 we

B. Gabrys, R.J. Howlett, and L.C. Jain (Eds.): KES 2006, Part I, LNAI 4251, pp. 679–686, 2006.
© Springer-Verlag Berlin Heidelberg 2006

show some results as tools developed following our architecture, explaining briefly some highlights, and in chapter 5, we present our conclusions and lines for future work.

2 Conceptual Basis and Background

In recent times, different approaches to extend an existing system capability using semantics in the engineering domain have been presented by the scientific community. Marcos [10] discusses a methodology to perform multimedia retrieval of engineering data by explicitly modeling and queering the semantics of the specific domain. Malkewitz [9] presents an alternative for the interactive design of 3D objects in specific domains (e.g. furniture or toys) where semantics are exploited to model explicitly the concepts and relations between the related objects in a context design approach. In a similar initiative, Sevilmis [14], presents a framework for the use of semantic web based techniques in the workflow of a community of engineers and designers in the car design industry. In the work by Posada [12], semantics are used as a basis for the consideration of user profile, intention and available resources in the Large Model Visualization of massive CAD models representing industrial facilities.

2.1 Standardization Focused on Steel Detailing Design: CIS/2 LPM/6

CIS/2 (CIMsteel Integration Standards, Version 2) [2] is an effort led by Leeds University (UK) and SCI (Steel Construction Institute). This approach is based on the ISO-STEP (10303) protocol, extending the standard to cover conditions encountered in structural steel design. The CIS/2 data model is called the Logical Product Model (LPM) and it is currently on its 6th revision (LPM/6). The main purpose is to support the engineering of low, medium and high rise constructions for domestic, commercial and industrial contexts. Although CIS/2 has been developed primarily to enable the engineering of building frames, it can also be applied to other types of steel frames, such as process plant installations, transmission towers, and (to some degree) offshore structures. To refer some examples of use of CIS/2, we can mention the work by Danso-Amoako [4] who presents a case study for the support of steel supply chain processes, showing that the standard can be used not only for information exchange in design but also for information management during long term stages of the supply chain. In contrast to Danso-Amoako, we focus on the geometry modeling aspect of the design process and not in the supplier-customer relationship. Lipman [8] presents an approach to use CIS/2 as a kind of storage model for an immersive virtual reality system inspiring one of our test tools; we follow a similar approach and introduce some enhancements provided by the use of semantics, like the adaptive coloring of elements according to the design stage. Alda [1] uses the CIS/2 framework for the supporting of collaborative design in a peer to peer basis, showing the importance of a common base in a Virtual Reality (VR) design scenario. Our work is not focused on collaborative design, but could be extended in a near future to support such scenario.

2.2 The Analytical, Design and Manufacturing Views in CIS/2

The structural steelwork lifecycle supported by CIS/2 is shown in Figure 1. This is a high level overview of the process model. CIS/2 describes three major stages needed to design a structural steel frame for a building: *(i)* Analysis, *(ii)* Design and *(iii)* Fabrication, each phase has its own data model that serves for information exchange between each other and possible external applications. The three different models of a steel structure are called "Views" [3]. These views served as the initial source for our process ontology implementation where they are used as a simplified model of the structural design process (this process ontology will be extended in a future work to reflect a more explicit representation).

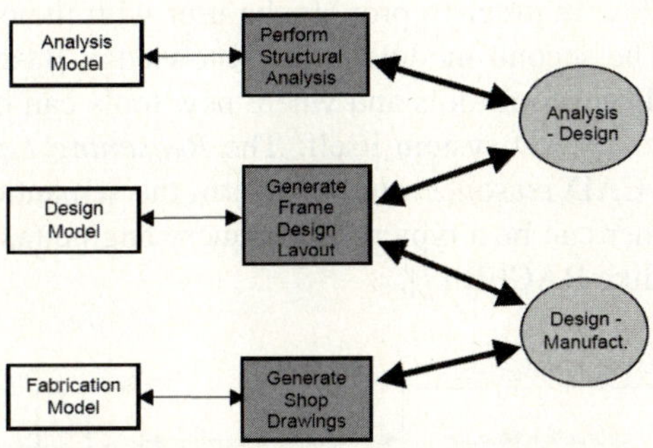

Fig. 1. Simplified process model identifying the various CIS/2 views and major exchanges, supported by high-level conformance classes [2]

2.3 Knowledge Representation Using Ontologies – The OWL Approach

An ontology is the explicit *specification* of a conceptualization [5]. In simple terms, it is a description of the concepts and relations in a domain for the purpose of enabling knowledge sharing and reuse. A body of formally represented knowledge is based on a conceptualization: the objects, concepts and other entities that are assumed to exist in some area of interest and the relations that hold among them [5]. The existence of tools for ontology definition, querying and reasoning gives interesting possibilities to several application domains, including the engineering domain. A widely used ontology language is OWL, which is a semantic mark-up language for publishing and sharing ontologies on the World Wide Web. OWL is developed as a vocabulary extension of RDF (the Resource Description Framework) and it is derived from the DAML+OIL Web Ontology Language allowing more expressiveness than RDF depending on the OWL flavour used [7]. There exist several OWL API's to handle computationally an ontology, between the most known ones, we can mention the Protégé OWL API [7], KAON2 [6] and the WonderWeb OWL API [11].

3 An Architecture for the Implementation of Semantically Based Tools Aimed for the Steel Detailing Design Process

We present in this chapter our proposed architecture for the semantic enhancement of the structural design process.

3.1 Modules Description

The architecture can be divided in 5 layers (figure 2). The *User Layer* is in charge of the bidirectional interaction between the user and different application modules of the system, which belong to the *Application Layer*. The latter has three modules: first one is the Process Stage Identifier Module, whose work is to determine the state of the process in the work flow in order to provide the user with the collection of available tools for that stage. The second module is the Knowledge Based Tools Module that actually stores the collection of tools and where new tools can be added a *posteriori*. The third module is the CAD system itself. The *Reasoning Layer* contains both an OWL reasoner and a CAD reasoner which perform the semantic processing over the ontologies. The reasoner can be a typical OWL query engine (as in our case study) or a more powerful tool like RACER [7].

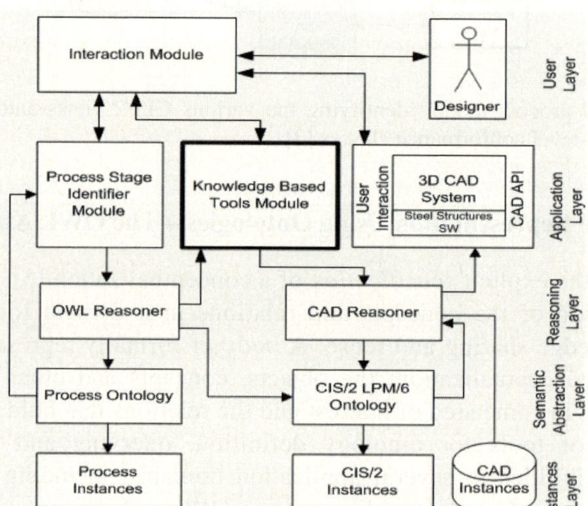

Fig. 2. Our Semantic Based Proposed Architecture

Next layer is called *Semantic Abstraction Layer* and it contains the Process Ontology and the CIS/2 Ontologies that actually represent the knowledge contained in the domain. Last layer is called *Instances Layer* and its main purpose is to contain the Process Instances and the CIS/2 Instances, as well as the concrete CAD model.

3.2 Interfacing a Steel Detailing CAD Applications and OWL

Most of today's CAD programs offer an API (Application Programming Interface) in C++ to extend the system capabilities. On the other hand ontology programming is mostly done thru a Java interface. Interfacing C++ with Java is a technical problem that can be partially solved using JNI (Java Native Interface) allowing Java code to be called from a C++ based application. The problem arises when third party applications, like databases and specialized plug-ins with their own API's are used. In order to communicate from a specialized CAD for Structural Design and an OWL API, we developed an interface that allows the ontology instantiation of elements from the CAD program and the queering of the model from both environments. The interface uses socket technology in order to send and receive tokens that represent function calls and returns. Using this technique we are able to handle OWL API manipulations programmatically right from the CAD coding environment.

4 Case Study – Semantic Based Tools

We present in this chapter three tools based on the architecture introduced in the last chapter. These cases were chosen specifically to solve particular problems reported by an actual Structural Design Company.

4.1 Tool 1 – Parts Generation and Verification from FEA (Finite Element Analysis) Results

In the design of a steel structure, usually different engineering software packages are used. Some of them are used to calculate stresses, allowed forces and other material static and dynamic issues. Those calculations are commonly performed by FEA packages who give as output the suggested dimensions of the needed elements. A common design process starts with a 3D wire frame sketch drawn by the designer; this sketch passes through the FEA software and produces the mentioned outputs. The information obtained is not always a 3D CAD model, and even if that is the case, these elements are not 100% compatible with other stages of the design without the use of importers and exporters. The semantic based tool we developed is applied to this stage of the design to automatically generate the CAD elements that FEA software recommends in their proper places. The generated elements are CIS/2 compliant, allowing the data to be backward and forward compatible with other software tools. The OWL Reasoner Module is used for query purposes over the CIS/2 instances using description logics. The initial sketch drawn by the designer and the post- FEA model are compared by the Reasoner (see Fig. 3).

If the designer committed a mistake drawing the wire frame, the system can be used to check element intersections (collision detection) in order to avoid potential problems during the assembly phase of the structure. This checking is done not based on the geometry but via a query to the ontology thru the OWL API.

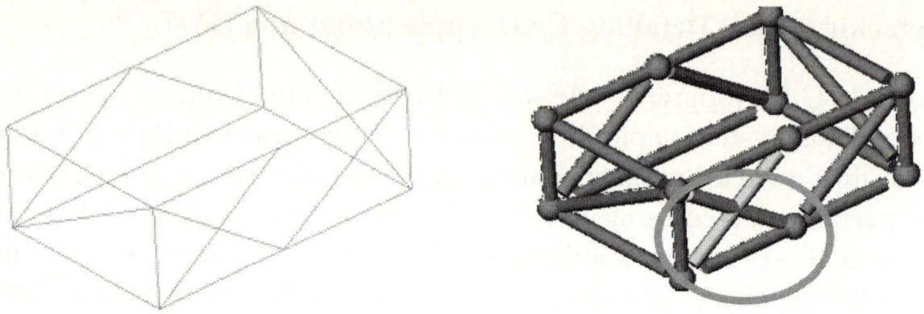

Fig. 3. Parts generation tool. Left: Frame drawn by designer. Right: CIS/2 element generation with automatic collision detection between created elements.

4.2 Tool 2 – Semantic Visualization of the CAD Model in Virtual Reality

The model can be visualized in a VR environment using the geometries and relations between elements. User intention is taken into account as the system presents this tool in the Design stage of the structure (see Fig 4). A fast visualization helps the user to locate certain parts in the drawing to modify or add elements from a renderized view that gives a better impression that the common wire frame used in traditional approaches. Although it is true that modern systems allow faster visualizations, our implementation deals directly with the ontology and its instantiation of CIS/2 elements, given an extra aid as every entity is handled semantically. This tool is similar to the approach presented by Lipman [8] and Posada [13]. The novel aspect in this case is the exploitation of semantic aspects to generate simplified versions of the model according to certain stages in the Design Process, enhancing specific parts (e.g. coloring) and adapting the camera position (see Fig. 4).

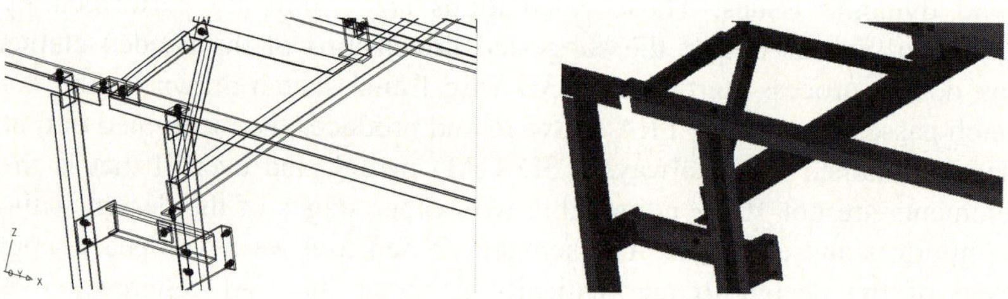

Fig. 4. Visualization tool, left wire frame in CAD, right VR output (CIS/2 compliant), this screenshot was taken in the design stage. Beam's color is automatically changed to help the differentiation of elements.

The interaction module allows the simultaneous interaction with the user interface inside the 3D CAD system and with the VR environment. The possibility to change an entity's color upon the stage of the design is taken by the OWL Reasoner. The values can be returned to the CAD in a bidirectional interaction (e.g. camera position, etc). This parameter exchange of non geometric data between the knowledge model and CAD is one of the clear benefits of our proposed schema.

4.3 Tool 3 – Special Beams Dimensioning for Workshop Manufacturing

Besides Steel Structures, some companies in the steel detailing field manufacture wood based frames for small facilities. These wood-based structures are designed with both a functional and an aesthetic intention. Usually, wood beams manufacture is still carried out in large spaces by artisans, without the support of NC (numerically controlled) machines. These structures should be dimensioned considering that the actual workers, who cut the pieces, have difficulties in the interpretation of traditional CAD dimensioning. In many cases, they only have a metric tape to measure distances in a frame where very long beams are being shaped (in some cases reaching 20 meters long). The semantic system is activated in this case when a free-form beam sketch is drawn in the CAD program. When the beam doesn't match with a steel shape modeled in the CIS/2 LPM/6 Ontology, the semantic system activates a non steel frame and the user is presented with new dimensioning tools intended for the manufacturing workshop described above. Factors like the lamination direction of the wooden beams and special considerations that must be fulfilled in critical zones like corners and attachments are immediately taken into account (see Fig. 5).

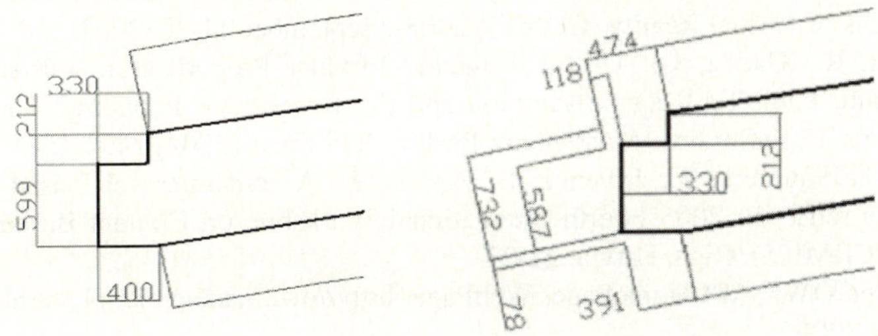

Fig. 5. Dimensioning tool. Left traditional CAD dimensioning. Right using our tool.

5 Conclusions and Acknowledgements

We presented in this paper an architecture and a system implementation for the exploitation of semantic aspects in the computer assisted design of Steel Detailing Structures (Structural Design). The architecture is based on two ontologies (process and domain model) whose elements and relations are instances based on a standardization effort called CIS/2. As test cases, we developed three tools intended to aid in particular stages of the structure's design process that were presented by a real Steel Detailing company in the frame of an industrial research project. As future work, we will extend the Process Ontology in order to identify stages where new tools can be developed from a deeper process oriented philosophy. We will also explore the collaborative design possibilities of our architecture. We want to thank the Basque Government for the partial financing of the MiroView Project (INTEK-4140/2004). A special mention is given to our Steel Detailer partner LANIK S.A, who served as test users, providing key points in their design workflow to apply the results presented in this paper. We also thank SOME for their participation in this project.

References

1. Alda, S., Bilek, J., Hartmann, D.: Support of Collaborative Structural Design Processes through the Integration of Peer-to-Peer and Multiagent Architectures, Proceedings of ICCCBE 2004, Weimar, Germany.
2. CIS/2 – CIMSteel integration Standard, WebPage: http://www.cis2.org/ Last access 26 January 2006.
3. Georgia Tech CIS/2 HomePage, WebPage: http://www.coa.gatech.edu/~aisc/ Last access 26 January 2006.
4. Danso-Amoako, M., O´Brien, W., Issa, R.: A Case Study of IFC and CIS/2 Support for Steel Supply Chain Processes. Proceedings of the of the 10th International Conference on Computing in Civil and Building Engineering ,Weimar, Germany.
5. Gruber, T.R.: Toward principles for the design of ontologies used for knowledge sharing. International Journal of Human-Computer Studies. 1995. Volume 43 , Issue 5-6 Nov./Dec. 1995. Pages: 907 - 928. ISSN 1071-5819.
6. KAON2, WebPage: http://kaon2.semanticweb.org/ Last access 26 January 2006.
7. Knublauch, H., Musen, M., Noy, N.: Tutorial: Creating Semantic Web (OWL) Ontologies with Protégé. International Semantic Web Conference (ISWC2003).
8. Lippman, R.: Immersive Virtual Reality for Steel Structures. Conference on Construction Applications of Virtual Reality (CONVR 2003). September 24-26, 2003.
9. Malkewitz, R., Gadda, G.: Useful Semantic Product Properties in Virtual Worlds.In: Cunningham, Paul (Ed.) u.a.: eAdoption and the Knowledge Economy, Part 2 : Issues, Applications, Case Studies. Amsterdam; Berlin : IOS Press (2004) 1494-1501
10. Marcos, G., Smithers, T., Jiménez, I., Posada, J.: A semantic web based approach to multimedia retrieval. 2005 Fourth International Workshop on Content-Based Multimedia Indexing (CBMI05). Riga, Latvia, 21-23.
11. WonderWeb OWL API HomePage, WebPage: http://owl.man.ac.uk/api.shtml/ Last access 26 January 2006.
12. Posada, J., Toro, C., Wundrak, S., Stork, A..: Ontology Supported Semantic simplification of Large Data Sets of industrial plant CAD models for design review visualization. In LNCS - Lecture Notes in Computer Science / Artificial Intelligence (Selected papers from KES05). Springer Verlag GmbH. 184-194.
13. Posada, J., Toro, C., Wundrak, S, "Using Ontologies and STEP Standards for the Semantic Simplification of CAD Models in Different Engineering Domains". Proceedings of FOMI (Formal Ontologies Meet Industry) (2005)
14. Sevilmis, N., Stork, A., Smithers, T., Posada, J., Pianciamore, M., Castro R., Jimenez, I., Marcos, G. et al: Knowledge Sharing by Information Retrieval in the Semantic Web. In Lecture Notes of Computer Science (Vol. 32/2005, p. 471). (Selected papers from ESWC 2005). ISBN: 3-540-26124-9. Springer Verlag, Germany.

Project Situations Aggregation to Identify Cooperative Problem Solving Strategies

Chaker Djaiz and Nada Matta

Charles Delaunay Institute, Tech-Cico, University of technology of Troyes,
12 rue Marie Curie- BP2060-10010 Troyes Cedex- France
{chaker.djaiz, nada.matta}@utt.fr
http://www.utt.fr/labos/TECH-CICO/index.php

Abstract. The knowledge engineering offers a rational framework allowing a representation of knowledge obtained through the experiences. This technique found a great application in knowledge management and especially to capitalize knowledge. In fact, the rational representation of knowledge allows their exploitation and their re-use. It is a necessary condition to allow a re-use and a knowledge appropriation. The knowledge management must take into account this dimension, since its first concern is to make knowledge persistent, ready to be re-used. In this paper, we study the traces classifications of the design project achievements in order to have a knowledge aggregation and to thus provide a representation of handled knowledge: directives and competences organization as well as negotiation strategies and cooperative problems solving.

1 Introduction

The knowledge engineering offers a rational framework allowing a representation of knowledge obtained through the experiments. This technique found a great application in knowledge management and especially to capitalize knowledge. In fact, the rational representation of knowledge allows their exploitation and their re-use. It is a necessary condition to allow a re-use and a knowledge appropriation. Behaviour laws provide strong semantics to observe as well as an argumentation of this behaviour, ready to be reproduced to solve new problems [23]. The knowledge management must take into account this dimension of knowledge, since its first concern is to keep a persistent knowledge, ready to be re-used and adapted. We present, in this paper, a form of knowledge management, keeping track and capitalization of project knowledge. This memorizing follows two essential steps: project traceability [3] and a knowledge capitalization.

The traceability makes it possible to keep track of the episodically memory in which space-time associations of events are described [14]. Project Memory is a form of this memory where cooperative problem solving and their context are represented. These events are a source for epistemic constructions, intended to build interpretations [23] which can be represented in the semantic memory.

Classifications of these tracks can be made. These classifications can be guided, either by a structure of representations and / or by typologies and generic knowledge.

We try to reproduce this step to represent the "collective" knowledge of organization. Indeed, the situations of problems solving arise through the projects.

B. Gabrys, R.J. Howlett, and L.C. Jain (Eds.): KES 2006, Part I, LNAI 4251, pp. 687 – 697, 2006.
© Springer-Verlag Berlin Heidelberg 2006

Traceability and a structuring of these projects "context and design rationale" provide a knowledge asset structured by situations problems. These various situations must then be analyzed to identify cooperation strategies as well as the knowledge handled by the organization. We will thus obtain a representation of the body "knowledge handled", directing actions "competences of the organization" and behaviour laws "negotiation strategies and cooperative problem solving". Classifications can be based on similarities of events and hierarchy aggregation of concepts (type of problems, conflicts …). Here, we present this latter type of classifications.

2 Project Knowledge

The realization of a project in a company implies several actors, if not also other groups and companies. For example, in concurrent engineering, several teams of several companies and in several disciplines collaborate to carry out a project of design. The several teams are regarded as Co-partners who share the decision-making during the realization of the project. This type of organization is in general dissolved at the end of the project [16]. In this type of organization, the knowledge produced during the realization of the project has a collective dimension which is in general volatile. The documents produced in a project are not sufficient to keep track of this knowledge which even the head of project cannot explain. This dynamic character of knowledge is due to the cooperative problem solving where various ideas are confronted and with a cooperative definition of the produced solution.

A project memory describes "the history of a project and the experience gained during the realization of a project" [16]. It must consider mainly [3]:

- The project organization: different participants, their competences, their organization in sub-teams, the tasks which are assigned to each participant, etc.
- The reference frames (rules, methods, laws ...) used to carry out the various stages of the project.
- The realization of the project: the potential problem solving, the evaluation of the solutions as well as the management of the incidents met.
- Main goal of the project: the negotiation strategy which guides the making of the decisions as well as the results of the decisions.

Often, there are interdependence relations among the various elements of a project memory. Through the analysis of these relations, it is possible to make explicit and relevance of the knowledge used in the realization of the project. The traceability of this type of memory can be guided by design rationale studies [14]. In the same way, work in knowledge management study techniques of traceability and definition of project memory [3], [17]. In this paper, we propose to study classification of project track as a mean to capitalize the organization knowledge.

3 Classification

Classification consists in gathering various objects (individuals) in subsets of objects (classes). It can be supervised where the classes are known, they have in general an associated semantics [9], or not supervised where the classes are created via the structure of the objects, semantics associated with the classes is then more difficult to determine [19]. In both cases, we need to define the concept of distance between two classes, which will be made by the means of criteria; these criteria of aggregation will be explained thereafter. Classification methods form part of the whole of the multi-dimensional descriptive methods, and their purpose is to clarify the structure of a whole (in our case, tracks of design projects).

The main objective of the methods of classification is to distribute the elements of a whole into groups, i.e. to establish a partition of this unit [12]. Several constraints in the form of assumptions are imposed and validated progressively, each group having to be the most homogeneous possible, and groups having to be the most different possible between them. The methods of classification clarify the structure of an important data entity. It is then possible to classify a whole group of individuals who make it up, characterized by similarities criteria [13]. There are many methods of classification, we present some of these methods, the most adequate for our work, and one can distinguish 5 types of methods of classification suggested in the literature [4].

1. Hierarchical classification: Where hierarchy is obtained either by an agglomerative (upward) method, maybe by a divisive (or downward) method.
2. Classification by partition: Where the number of classes is fixed at the beginning, the aim is to gather N individuals in K classes. The individuals must similar and the classes must be separate. It makes it possible to treat sets of relatively high manpower by optimizing relevant criteria.
3. Method of the densities: This seeks in the workspaces the high density zones if they exist.
4. Fuzzy classification: Method providing by the means of algorithms of the not disjoined classes (encroaching).
5. Analyze factorial Q-SORT: Based on the spectral decomposition of matrix XX' where X, of dimension (N, p), is the table of figures containing N individuals to be classified.

We chose the hierarchical method which is defined as a search for a valued hierarchy, or also a hierarchical classification (hierarchical clustering) by methods listed, it was offered the most possibility of Accessibility and collection of information in project [5]. Such a research is based on a concept of distances between individuals or objects (elements of classification) which induces a measurement of the heterogeneity of a part based on the distances among individuals objects which are in a class and a measurement of dissimilarity between two parts based on the distance between an individual of a class (i) and another individual of another class (j). Moreover, one is not satisfied with a partition, but one seeks a hierarchy of parts, that constitute a tree called the dendrogramme [4]. In our case, we would like to establish causal relations (explicitly or implicitly) between the concepts of the preliminary project memories (the suggestions, discussed arguments, the competences, the nature of the task, initial knowledge on the problem, their kinds, the number of participants and their roles) and aggregations criteria which one will call dependent variables (the sociological criteria, psycho-cognitive, and relating to co-operative work), in order to define aggregation strategies, that can be summarized as follows:

Criteria link Concepts to obtain (Strategy R).

4 Process of Aggregation

Our aggregation process (Fig 1) follows a number of steps and functionalities in order to allow firstly an enumeration of all the concepts, and secondly the potential use of regrouping mechanism of the project memory concepts. This allows facilitating the re-use, the interpretation and reading of knowledge intervening within the framework of the design.

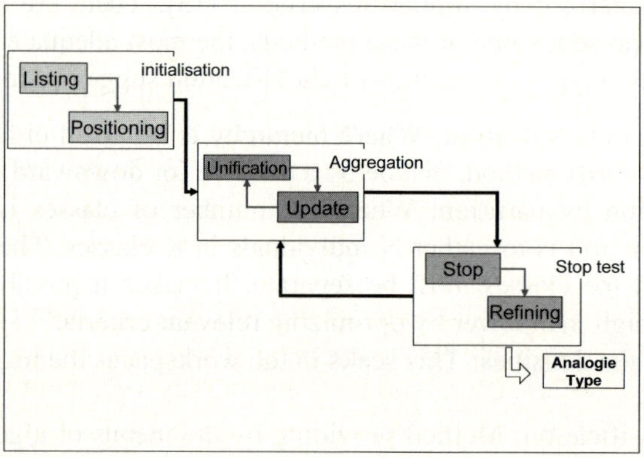

Fig. 1. Architecture for knowledge concept aggregation

The initialization phase: it relates to two sub-phases which make it possible to support the concepts emerging during the meetings of projects, it consists, amongst other things, to give a micro sight on the meeting of project by peeling all these elements (concepts) and by making the first separation, to be able to distinguish the aggregation thereafter.

- Listing: listing all the concepts of the project of design, all participants' arguments suggestions, as well as the decisions, to try thereafter to see possible separations for example: distinct problems, tasks not implying the same resources, participants not carrying out the same tasks, arguments…
- Positioning: identifying the provisions of the concepts, those which can be gathered, for example, arguments with the suggestions corresponding to a participant... then to build a starting grid, what we will call thereafter matrix of aggregation.

The phase aggregation: it will contains to two sub-phases which will be much more functional, it will act as a phase of gathering and a second phase of update the matrix built and that the concepts bound by a similarity of selection criteria are together.

- Unification: this process operates in the following way, if the criterion characterizes the same concepts suggested, then they must be gathered, while taking as a starting point the hierarchical methodology of classification. We defined strategies of aggregation based on criteria extracted from our study of team work. We will present (paragraphs 4.1) these criteria.
- Update: this sub-phase of aggregation, will allow bringing up to date the matrix, to saying that the gathered segments will be unified under the same criterion.

The stop phase: We proceed in an iterative way until obtaining classes:

- Stop test: if there is not any more concept to be gathered, or if the criteria are not valid any more (criteria available and non-active to carry out gathering) then continue to the following step, and stop the process.
- Refining: to name the classes and to verify redundancy.

4.1 Criteria of Aggregation

It should be noted in this paragraph where we will present the criteria of aggregations of our process of classification, that several criteria can be released according to various aspects of the study of the cooperation and the negotiation in a project: sociological, psycho-cognitive and cooperative work.

4.1.1 Criteria from the Sociological Point of View

"Among the various types of communicative interactions which can be produced in situations of group, it is noted that the argumentative and explanatory interactions are particularly favourable with the Co-development"[2], in addition to "the multiple relations which exist between the explanation and the argumentation" [21], "the first dimension (the explanation): corresponds to the degree of subdivision of the responsibilities for the realization for the tasks for the problem solving in the organizations" [10]. In other words, it acts as the spontaneous distribution of the work of cooperative problem solving. These tasks link on the one hand the problem to be solved, and on the other hand, the activity to cooperate with itself [2], in and by the dialogue. When the responsibilities for the realization for these tasks are assumed spontaneously by the participants in a relatively stable way through the interaction, one will describes sociological criteria consequently depending on the task of the problem to solve parts played by the participants and related to enunciated arguments [7]. In these studies, we can propose, concept of organization of the tasks and argumentative and explanatory interactions.

The argumentative criteria allow interpretation, vision and evolution of the suggestive arguments; one currently can distinguish 5 types [2], [21], [25], [10]:

1. Dialectical criteria types: to strongly bind to the argumentative play of intervention such as attacks, defence, counteract.
2. Evolutionary criteria types: evolutions of attitude noted at the time of discussion such as for, against, neutral.
3. Epistemological criteria types: the nature of knowledge concerned in the dialogue, origins and the perceptual plans (how knowledge and perceived by the various participants)

4. Conceptual Criteria types: how universe of reference, evolution of the concepts concerned.
5. Interactive criteria types: approach the cognitive-linguistic, transformations (reformulation).

On another plan, the distribution of the tasks and the roles can be symmetric or asymmetric. The distribution is symmetric in case, the role will assume completely andglobally the task, and conversely for the asymmetry role / task. Should the oppositeoccur this distinction reflects a clear vision for the later decision-making. It is to note that in this first approach we notice that the organizational approach (Basing itself on the actors, the roles, the visions and the group) favours the exchange by means of interaction and explanation in organizations, and it for the argumentation and the suggestion of the tasks for the cooperative problem solving.

The criteria of task can be distinguished on an organisational level throughout the evolution of the projects of design, knowing that a project of design forwards by three phases, we present the criteria below to bind to these phases: Preparatory phase: imply two criteria types:

1. Criteria of specification: criteria to bind to the problems dealt with the implied participants.
2. Criteria of planning: strongly bound to time and the fixed stakes, for the problem solving.

Realization phase: it also puts two standard criteria:

1. Criteria of execution: executive criteria which use average the techniques and human (taking part, their roles) and estimated tasks
2. Criteria of piloting: criteria of piloting of project putting the Masters of project(directing and development).

Phase of finalization: allows exposing criteria of feedbac:

1. Design criteria: criteria implying average technique, taking part, stake of time …
2. Criteria of debriefing: feedback and analysis.
3. Criteria of filing: average technique, method of traceability, memory of project, data bases etc.

4.1.2 Criteria from the Psycho-cognitive Point of View

It was noted that research on distributed cognition, carried out in cognitive psychology these last decades, stressed the study of the acquisition of the procedures [1], instead of the development of "comprehension" on the conceptual level. In the case of certain types of problems, the performance can be dissociated from comprehension: the participants can solve successfully without thorough control of the concepts concerned. However, in the case of the production of certain types of exchanges ("epistemic"), like the explanation and the argumentation, it cannot have such dissociation: competence and performance as regards explanations coincident [20]. The psychology-cognitive sight supports the individual aspect, which relates to participant and competence. We can distinguish from this study the criteria of the competence. We have two types of competences functional competences related to the direct functions of the design project, organisational competences always related to the management of the project:

1. Criterion of execution.
2. Criterion of manufacture or creation.
3. Criterion of planning.
4. Criterion of management.
5. Criterion of piloting.

4.1.3 Criteria from Cooperative Work Point of View

The study of the processes of collaboration must lie within a broader scope, that of a model of cooperative problem solving in and by the dialogue. A starting point impossible to circumvent is of course the "traditional" model of the individual problem solving [8], who will consists on the six following phases: the search for a problem, development of a representation of the problem to be solved, the planning of the solving strategies, the generation of possible solutions, the checking of the solutions, the feed back in order to integrate knowledge (reorganization of knowledge). In the case of a team, these stages integrate other criteria such as the roles and their evolution (change and modification) [11], collaboration, interactions argumentative, conflicts, alliances of teams, negotiations.

It should be noted that it is very probable that the cooperative problem solving in the projects of design, at a working group follows a relational diagram where all concepts are connected.

A number of participative role one distinguished by PLETY [22], relations between these roles will come to be added to support a number of creation from knowledge based on the negotiation such as relations of the complementarities type/alliance/ conflict. The roles can be the objects of the implicit or explicit negotiations. [22]. PLETY identified the four roles: Questioner, Verificator, independent and animator. The role of the "Questioner" operates on the compromise plans and interdiscursifs (to ask the problem of the direction of the solution). The "Verificator" on the compromise level, against-argue the statements to push one second time the questioner to reargue his suggestion. The "Animator" on the international plan manages the exchange. "The independent" corresponds to a role more fluid than the others, or with the absence of a determined role, in a sociological point of view; it is passive, it does not interact for two reasons: it does not have competence necessary for a given task, or it is not concerned with the task. PLÉTY noticed that "the animator" and the "verificator" have a relation of alliance or competition: "they are not unaware of...". The true cooperation occurs if the roles operating for problem solving and interaction. For example, to criticize a solution suggested by its partner, he checks the problem (the checking) and the interaction (to express its agreement or dissension compared to what it is proposed by its interlocutor). Criteria chosen from this point of view can be: Role, Interaction argumentative and taking part. Criteria of role [11] as:

1. To assign: a role to a participant.
2. To take: catch of a role directly by a participant.
3. To offer: a role to a participant by a hierarchy.
4. To change: change of a role.
5. To define new: creation and definition of a new role.
6. Conflict of role: competition, opposition of participant on the same role.

We gathered these criteria to exploit them in the definition of the aggregation strategies.

4.1.4 Tree of Criteria

In this section, we will present a summary of the criteria enumerated above in the three selected fields (Sociology, psychology-cognitive, Co-operative work). We have to gather these criteria in the form of tree while putting especially ahead: argumentative, task, role, participant and competence (Fig 2).

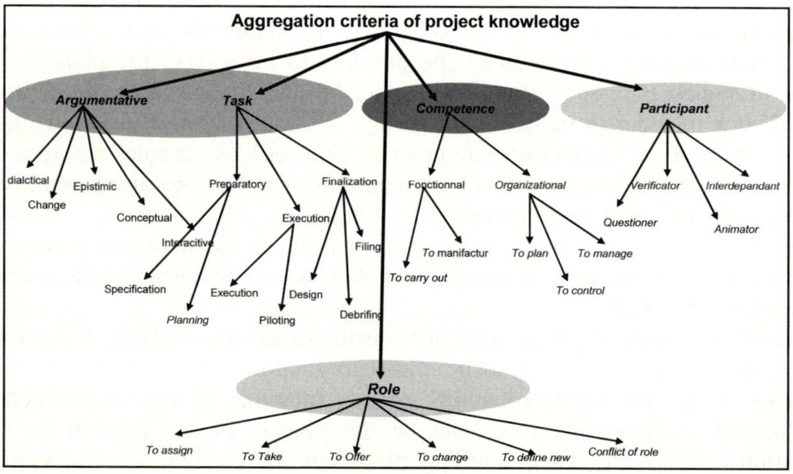

Fig. 2. Scheme of tree criteria

4.2 Aggregation Strategies

Traceability allows keeping track of situations of cooperative problem solving and emphasizing characteristics of these situations. The capitalization of knowledge which we recommend is based on the similarity of these characteristics to make explicit the strategies learned by the organization through these projects.

To define aggregation strategies allowing this classification, we proceed by stages according to the process of aggregation presented paragraph 3.

While analyzing, bonds between the sociological points of view, psycho-cognitive and co-operative work, we defined aggregation strategies on these bonds.

A first type of bonds appears between the concepts argument, resources, competences and product. It makes it possible to propose the realization of the tasks and the evolution of the design. These bonds request criteria like argumentative of change, dialectical and epistemic like competences of manufacture and task of project piloting.

One can notice that a first unification is made with criteria; this first unification is shown by dotted lines on (Fig 3).

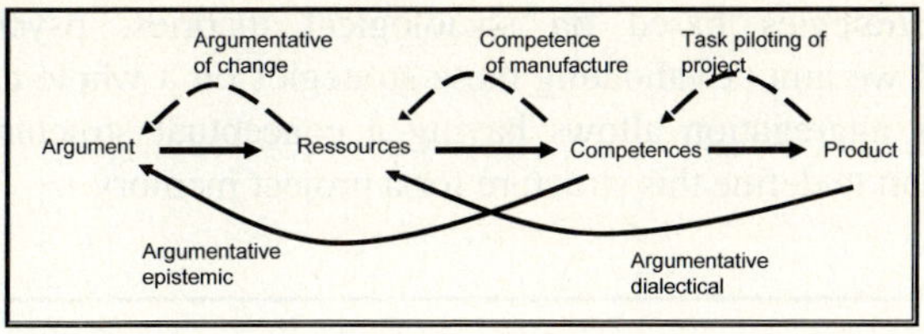

Fig. 3. The aggregation strategy based on the carry out task

A second aggregation strategy related to the negotiation can be identified considering the dialectical exchanges and based on the negotiation which is for a number of researcher an interaction where the speakers seek to conclude an agreement starting from an initial situation where such an agreement misses (if there is a conflict or not). This definition is closer to that adopted in linguistics studies [18], [15], [6], [24]. The aggregation strategy based on the negotiation can emerge starting from the relation between participant, role, suggestion and argument. Adding to, dialectical criteria argumentative, we distinguished the criteria from questioner and verificator defining participating in a negotiation. In the same way, role criteria (to take, to offer and change...) can reveal bonds of negotiation. In addition the argumentative criterion of change must be considered to illustrate the changes advanced in a negotiation (Fig 4).

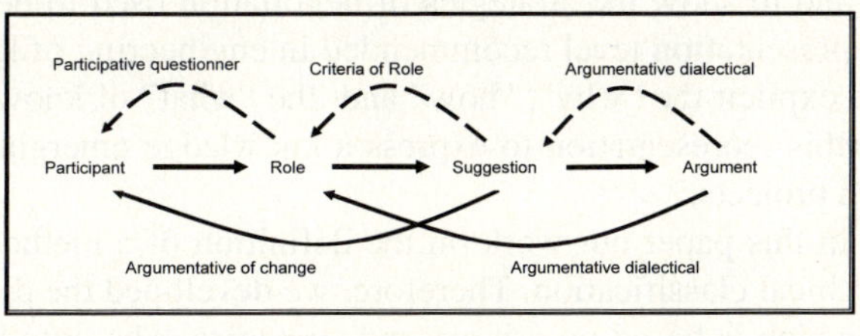

Fig. 4. The aggregation strategy based on the negotiation

From a knowledge management point of view, one will say that the relations indirectly to create between (Argument←→Role) and (Suggestion←→Participant) allow a tacit reading of the meeting. Conceptually we will say that the classification criteria build interdependent relations between the concepts.

We noticed also a strong bond connecting task suggestion and role, this concerns organization thanks to the two criteria (Fig 5):

- Criterion of specification: a task criterion which is related to much more the phases of the project management, and attributions, assignments.
- Criterion of role: (assignment, to offer) strong and direct functionality allowing to assign a role with a task.

In this paper, we show our first studies and outlines of definition of aggregation strategies for the cooperative problem solving on design projects. We will try thereafter to build other strategies based on sociological theories, psycho-cognitive and organisational and we aim at validating these strategies on a whole design application. The result of this aggregation allows having a conceptual structure of the projects knowledge. We plan to define this structure for a project memory.

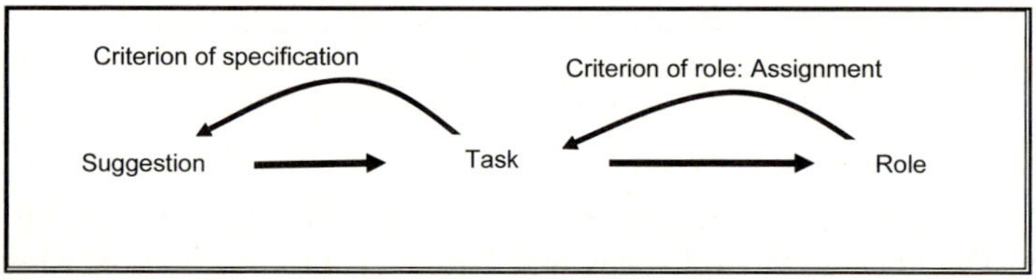

Fig. 5. The aggregation strategy based on organization of project

5 Conclusion

The traceability of project knowledge is only one first stage for the representation of the cooperative problem solving. Indeed, tracks including/understanding of the concepts like participants, suggestions, competences, constraints, tasks, arguments, decisions are only representation of situations of cooperative problem solving. Aggregation using classifications of this information must be carried out in order to emphasize deep knowledge and to show the strategies of negotiation used to deal with environments. We use representation level recommended in engineering of knowledge which proposes to make explicit the "why", "how" and the "what" of knowledge. We study the projection of this representation to express a knowledge emerging from of a collective like design projects.

We presented in this paper our work on the definition of a method of aggregation based on a hierarchical classification. Therefore, we developed the different phases of this aggregation which is based on criteria and strategies. The criteria and the strategies were identified after a study of the project knowledge under three points of view: sociological, psycho-cognitive and cooperative work. This study made it possible to propose the principal concepts as well as the bonds between these concepts to be considered to represent a project memory. We plan to look further into this study and to determine other aggregation strategies. We apply strategies on an example extracted from a project track of design of a wind mill. We plan to validate the process of classification we defined, on the whole of these tracks and on other project tracks, for instance design of vehicle prototypes.

References

1. Anderson, J., the Architecture of Cognition, Harvard University Press, (1983).
2. Baker, M.J., Recherche sur l'élaboration de connaissances dans le dialogue, HDR, Univ Nancy2, (2004).

3. Bekhti, S., Matta, N., Project memory: An approach of modelling and reusing the context and the design rationale, Proceedings of IJCAI'03 (International joint of conferences of Artificial Intelligence) Workshop on knowledge management and organisational memory, Acapulco, (2003).
4. Benzecri, J.P., Data analyzes. T1: La taxinomie. Dunod. (1973).
5. Chavent, M., Guinot, C., Lechevallier, Y., “divisives Methode of classification”. Inria., Rocquencourt, (1999).
6. Clark, H.H., Schaefer, E.F., Contributing to Discourse. Cognitive Science 13 , 259 -294, (1989).
7. Conein, B., “Ethologie and sociologie: la contribution de l'éthologie à la théorie de l'interaction sociale”, French Revue of Sociologie, Vol. XXXIII, n°1, Janvier-Mars (1992), pp. 87104.
8. Darses, F., et Falzon, P., collective design : une approche de l’ergonomie cognitive, Coopération et conception, sous la direction de G. De Terssac et E. Friedberg aux éditions Octares, (1996).
9. Diday, E., Une représentation visuelle des classes empiétantes. Rapport INRIA, n- 291. Recquoncourt 78150, France, (1884).
10. Ducrot, O., Les échelles argumentatives, Paris, Minuit, (1982).
11. Th. Herrmann, I. Jahnke, The Role Concept as a Basis for Designing Community Systems. (2004).
12. Jain, A.K., R. Dubes, Algorithms for clustering data. Prentice-Hall, Englewood Cliffs, (1988).
13. Johnson, S.C., Hierarchical clustering schemes, Psychometrika, 32, 241-254, (1967).
14. Karsenty, L., An Empirical Evaluation of Design Rationale Documents, Proceedings of CHI, R. Bilger, S. Guest, and M. J. Tauber (Eds), (1996).
15. Kerbrat-Orechioni, C., Le discours en interactions, Paris : A. Colin, (2005).
16. Matta, N., M. Ribiere, O. Corby, M. Lewkowicz, M. Zacklad, “Project Memory in Design”, Industrial Knowledge Management - A Micro Level Approach, Rajkumar Roy (Eds), Springer-Verlag, (2000).
17. Matta, N., B. Eynard, L.Roucoules, M. Lemercier, Continuous capitalization of design knowledge, Proceedings of ECAI'02 (European conferences of Artificial Intelligence) Workshop on knowledge management and organisational memory, Lyon, (2002).
18. Moeschler, J., Modélisation du Dialogue. Hermès, Paris, (1989).
19. Nakache, J.P., J. Confias, “Méthodes de Classification”, CISIA. CERESTA edition Mars, p 5-54. (2000).
20. Ohlsson, S., Learning to do and learning to understand: A lesson and a challenge for cognitive modeling, dans P. Reiman et H. Spade (dirs.), Learning in Humans and Machines: Towards an interdisciplinary learning science, Oxford, Elsevier Science, p. 37-62. (1996).
21. Plantin, C., L’argumentation. Paris : Seuil, (1996).
22. Pléty, R., cooperative learning. Lyon : Presses Universitaires de Lyon, (1996).
23. Richard, J.F., Les activités mentales, Comprendre, raisonner, trouver des solutions, Armand Colin, Paris, (1990).
24. Roulet, E., On the Structure of Conversation as Negotiation. In H. Parret & J. Verschueren (Eds.) (On) Searle on Conversation, pp. 91-100. Amsterdam: John Benjamins, (1992).
25. Toulmin, S., The uses of argument, Cambridge, Cambridge, University Press, (1958).

Incorporating Semantics into GIS Applications

Karina Verastegui, Miguel Martinez, Marco Moreno,
Sergei Levachkine, and Miguel Torres

Geoprocessing Laboratory-CIC- National Polytechnic Institute, Mexico City, Mexico
{kverastegui, miguelrosales}@sagitario.cic.ipn.mx,
{marcomoreno, sergei, mtorres}@cic.ipn.mx
http://geo.cic.ipn.mx

Abstract. An approach focused on incorporating semantic content into Geographic Information Systems is proposed. Our methodology is based on a conceptualization of a geospatial domain. The concepts are extracted automatically by analyzing the properties of geospatial objects (in this paper we are primarily focusing on geometrical and topological properties). The concepts are stored in the spatial database to support subsequent processing. The conceptualization of geometric properties is based on measures of geospatial objects. The measurement results obtained by the measurements are classified to obtain representative clusters of values in order to describe these properties. The values are used to define which concept better represents the properties of each object. The conceptualization of topological relations (special case of topological properties) is used to analyze and represent the topology in two levels: intrinsic and extrinsic. The use of semantics is fruitful in problems traditionally treated by numerical approaches.

1 Introduction

The use of semantics to solve problems related to the conceptualization of heavily mixed geographical databases has recently gained increased attention among researchers in geographic information sciences. In this paper, we present a conceptualization oriented to generate ontological information and thus structure or homogenize the databases. We are looking for a format of data representation, which does not depend on scale, format, references, etc. According to [2] a body of formally represented knowledge is based on a *conceptualization*: the objects, concepts, and other entities, which exist in some area of interest and the relationships that hold among them. A conceptualization is an abstract, simplified view of the world that we wish to represent for some purpose [3]. Ontologies encode semantic relations between concepts and hence facilitate the detection of associations between related terms. In modern information systems based on ontologies, one seeks canonical descriptions of knowledge domains and associated classificatory theories. In such fields as artificial intelligence and computer science, ontology typically refers to a vocabulary or classification system that describes the concepts operating in a given domain through definitions, which are sufficiently detailed to capture the semantics of that domain [5].

B. Gabrys, R.J. Howlett, and L.C. Jain (Eds.): KES 2006, Part I, LNAI 4251, pp. 698 – 705, 2006.
© Springer-Verlag Berlin Heidelberg 2006

Traditionally, geographic data models explicitly represent a set of basic objects, their geometry and their properties in numeric format. But much of the geographic world's semantics appears in the relations linking objects [6]. Nevertheless, most of these relations are not explicitly represented in data models describing geographic databases. Usually, these relations only implicitly appear when one is displaying of a geographic database [7]. Geographic Information Systems (GIS) handle geo-spatial data at different levels of details. Due to page limit our literature overviews stopped at this point.

In this work, we generate a conceptualization of geometrical and topological properties of spatial objects and explicitly represent them into spatial databases. The paper is organized as follows: Section 2 describes the measures to quantify the geometric characteristics and describes how the conceptualizations of relations are defined. Additionally, Section 3 describes the conceptualization method. Some preliminary results are shown in Section 4, and Section 5 sketches out our conclusion and future work.

2 Conceptualization of Geometrical Characteristics and Topological Relations

This research has as main purpose to design a Geographic Information System (GIS) oriented to explicitly represent the geometry and topology of the spatial data. The explicit representation is based on the conceptualization of the geometrical and topological properties (See Fig. 1). The conceptualization of geometrical properties is based on measures of geospatial objects. The values obtained by the measurements are classified to obtain representative clusters of values in order to describe these properties. The values are used to define which concept better represents the properties of each object. The conceptualization of topological relationships (special case of topological properties) is oriented to analyze and to represent the topology in two levels: intrinsic and extrinsic.

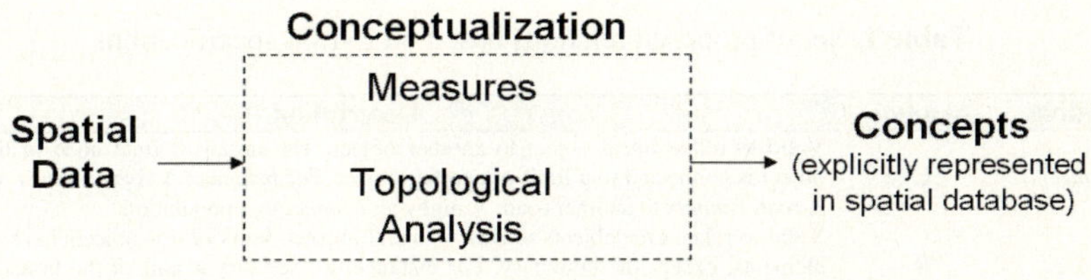

Fig. 1. Process to generate concepts explicitly represented

2.1 Measures

The measurements are numeric values assigned to an observation that reflects a magnitude, quantity or a characteristic. For this reason, we can make measurements according to their geometric form. It is used to analyze characteristics as size, sinuosity / complexity, elongation / eccentricity, compactness as well as other important aspects.

Fig. 2 shows a classification of measures. Shape describes the geometric representation of spatial objects according to [8].

Shape Geometry	*Size*	Length Area Perimeter
	Sinuosity/ Complexity	Measures on Angularity Max Bend height Sinuosity Number of Bends
	Elongation/ Eccentricity	Brown-Eccentricity Elongation Spreadness
	Compactness	Convex Deficiencies Compactness-Measures Squareness
	Important Aspects	Bend Shape Bend Description Number of Points Characterization using a Distance

Fig. 2. Classification of Measures

2.2 Conceptualization of Topological Relationships

The conceptualization is performed by means of the relationships defined in 9-intersection model [1] and INEGI[1]specification [4]. Each relationship is denominated by a specific name (it does not depends on the language i.e. English or Spanish). INEGI specification only defines the relationships *connect* and *share*. But the *share* relationship is ambiguous. Due to it, we propose other relationships to be more explicit.

According to the interpretation, each relationship represents a subset of the 9-intersection model. Table 1 depicts a set of six relationships defined to describe the topology of topographic maps. This set is composed of other relationships, which represent a specific meaning according to the properties in the data dictionary.

Table 1. Set of proposed relations based on INEGI specifications

Relations	Symbol	Description
Connect	C	Valid to relate lineal object to another object. The initial or final node of the lineal object is connected to a limit of another object. For instance: A river *connect* to a prey; A road *connect* to another road; A highway *connect* to a population.
Share	S	Valid to relate area objects to area or lineal objects. Pairs of this objects have common elements, except the boundary. For instance, a river that is part of the boundary of a country.
Share limit	Sl	Valid to relate area objects to another area objects. The only common element is the boundary. For instance, the boundary between two states.
Cross	X	Valid to relate lineal objects to areal or lineal objects. (1) part of lineal object is inside of an area object; (2) two lineal objects are intersected, but the flow is not shared. For instance, a highway *cross* a railroad.
Intersect	Y	Valid to relate pairs of lineal object. This relations describe an intersection and their flow is shared. For instance, an street *intersect* with another street.
Inside	I	Valid to relate any kind of objects, if they are inside of an area object. For instance, a populations *inside* an state.

[1] National Institute of Geography, Statistics and Informatics, National Mapping Agency.

3 Methodology

The explicit representation is based on the conceptualization of the geometrical and topological properties. This method do not depends on the scale as the traditional GIS approaches. In the following subsections, we outline our methodology.

3.1 Methodology of Geometrical Conceptualization

In our model, the geometry of geospatial objects will be described by means of concepts that represent geometrical properties, such as sinuosity, elongation, eccentricity, size, etc. We consider three types of geospatial data; point-like objects (i.e. well or tree), linear objects (i.e. a road or river) and the area objects (i.e. state boundary or lake).

Fig. 3 depicts a workflow diagram of the conceptualization process. It consists of four steps: (1) to separate objects in layers that contain the same primitive of representation (point-like, linear and areal layers), (2) to obtain the measurements, we use algorithms that evaluate geometric characteristics, (3) the values obtained by the measurements are classified to obtain representative clusters of values in order to describe the geometric characteristics in a qualitative way (4) each cluster will be used to assign a concept according to the classification of intervals of the measures.

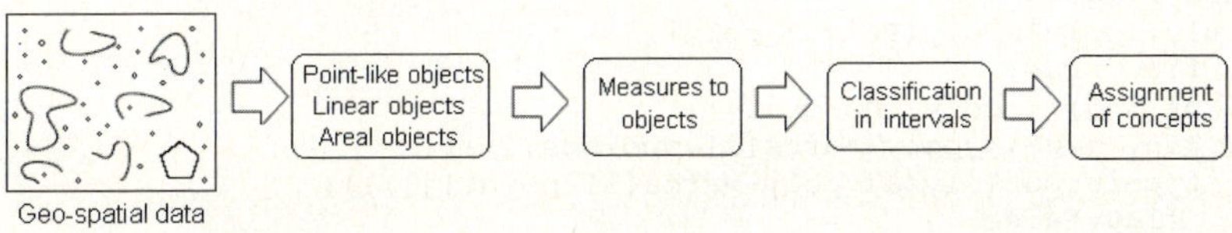

Fig. 3. Workflow diagram to obtain the conceptualization

In the case of *sinuosity measure* (*S*), it is obtained by the equation 1, which is computed by dividing curve length with Euclidean distance from the initial node d_0 to the terminal node *N* ($P(d_0,N)$) (See Fig. 4a). For example, *S* for different lines (L_0, L_1) is $S(L_0) = 1.0034$ and $S(L_1) = 3.3905$. It means that this property will be represented by the concepts *straight* and *very sinuosity* respectively (See Fig. 4b).

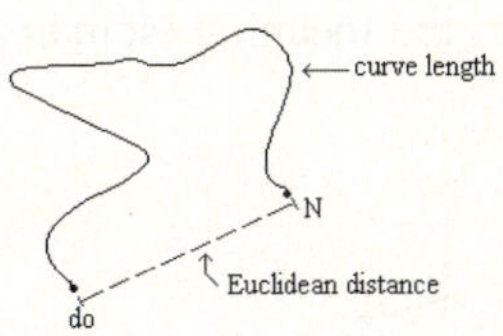

$$Sinuosity = \frac{Curve\ length}{P(d_0, N)}$$

a)

Classification of S	Concept
1.0 – 1.4	Straight
1.5 – 2.0	little sinuous
2.1 – 2.4	half sinuous
2.5 – 3.5	very sinuous

b)

Fig. 4. Compute of Sinuosity Measure

3.2 Topological Description

The conceptual schema to generate a topological description is shown in Fig. 5. It is necessary to perform an analysis of topological relationships. This analysis is composed of two levels: (1) analysis of intrinsic relationships and (2) analysis of extrinsic relationships. The intrinsic relationships; are those that exist between objects that compose the same thematic, e.g., relations that exist in a thematic of hydrology.

On the other hand, the extrinsic relations are those that exist between different thematics, e.g., relations between objects that belong to hydrology and roads.

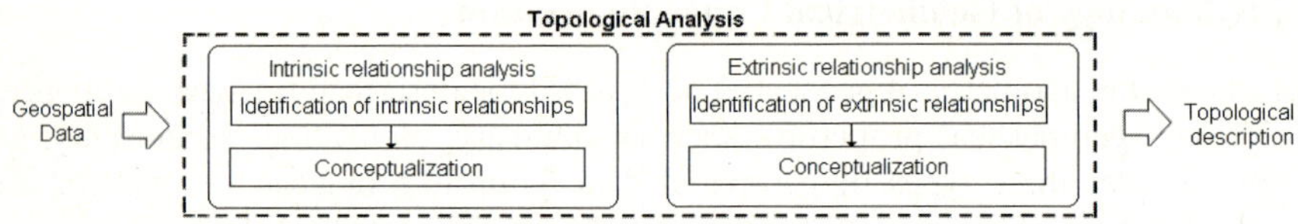

Fig. 5. Conceptual schema to generate a topological description

The functions in C++ to identify the *Inside relation* between polygons are:

```
bool Relations::Inside(Poly *_p1,Poly *_p2){
  bool flag=false;
  if(_p2->polyInPoly(_p1))
    flag=true;
  return flag;}
bool Poly::polyInPoly(Poly *_poly){
  bool flag=true;
  for(int i=0;i<_poly->narcs;i++)
    for(int j=0;j<_poly->arcs[i].npoints;j++){
      if(pointOutPoly(&(_poly->arcs[i].point[j]))){
        flag=false;
        break;      }
    } return flag; }
```

3.2.1 Intrinsic and Extrinsic Relationships Analysis

To carry out this analysis, we are developed tables to represent the relationships between different thematics. Fig. 6 shows an example, where a River *connect* Body of water, both objects belong to the thematic of hydrology. Fig. 7 shows an example, where Road *share* Border. Border belongs to communications road thematic and Road belongs to hydrology thematic. It represents that the river and the highway "share" the same geographical space. The relation *share* defines the meaning (semantics) of this situation.

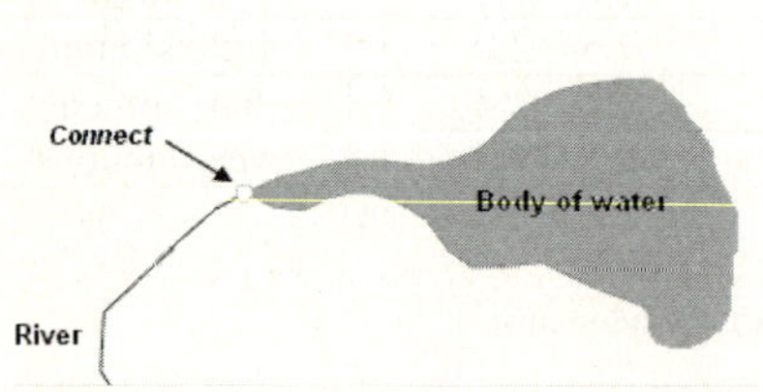

Fig. 6. River *connect* Body of water, both objects belong to the thematic of hydrology

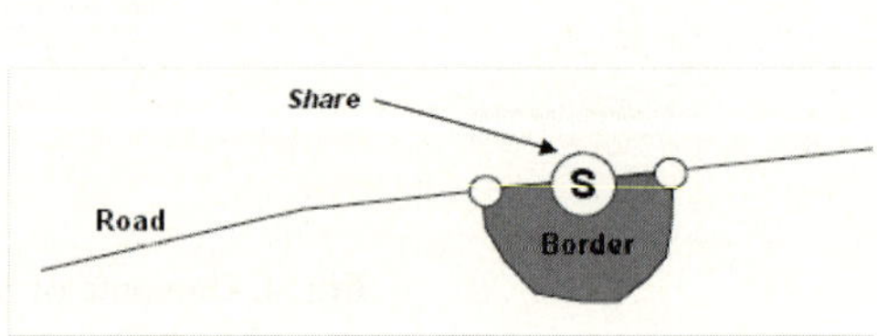

Fig. 7. Road *share* Border

4 Experimental Results

To obtain the measures and to make the topological analysis was developed a library of classes that store and manage the spatial object in a proprietary format. These classes are implemented in Borland C++ Builder. The implementation of the classes is focused on working on vector data.

The classes implement methods to obtain the basic topology (connectivity and adjacency), the topological relationships between pairs of spatial objects provide methods for computing geometrical measures, For instance, sinuosity measure of an arc.

The methods to obtain the topological relationships and to compute the sinuosity measure are described in Table 2.

Table 2. Methods to obtain the topological relationships and geometry measures

Method	Description
Connect()	Analyze *Connect* relationships between two objects, according to its definition.
Share()	Analyze *Share* relationships between two objects, according to its definition.
ShareLimit()	Analyze *Share limit* relationships between two objects, according to its definition.
Inside()	Analyze *Inside* relationships between two objects, according to its definition.
Cross()	Analyze *Cross* relationships between two objects, according to its definition.
Intersect()	Analyze *Intersect* relationships between two objects, according to its definition.
Sinuosity()	Compute the sinuosity measure of any arc object.

Fig. 8 depicts an example in which we show *connect* relationships between two spatial objects; "River" and "Body of water" (explicitly represented in the database). In addition, Fig. 9 depicts the computation of *sinuosity measure* for the River.

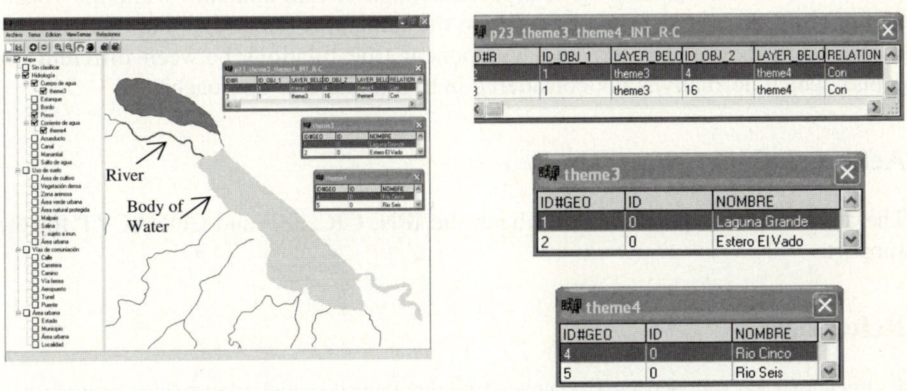

a) b)

Fig. 8. a) River *connect* Body of water, b) tables that store the conceptualization

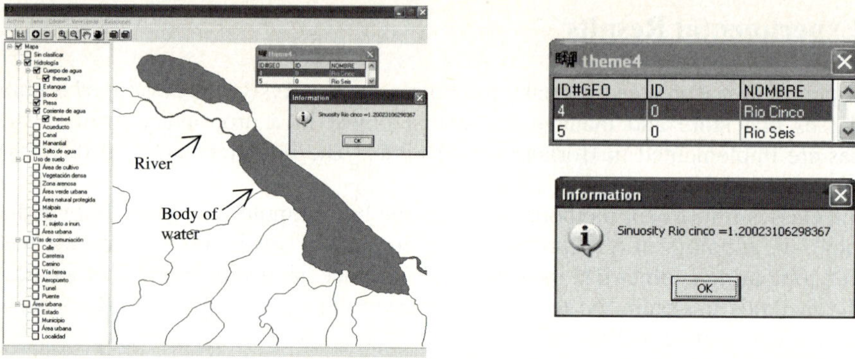

Fig. 9. Sinuosity Measure

5 Conclusions and Future Work

In this work an approach to incorporate *semantic content* into Geographic Information Systems has been presented. The semantic data content is expressed by concepts. These concepts are generated analyzing the datasets in a geographic domain, and represented by concepts. The concepts represent the interpretation of the spatial data, and the meaning of River Body of Water River Body of water geospatial objects. Our method is based on explicit representation of topological and geometrical properties of geographic data by means of concepts. By using this approach, we attempt to catch or dimension *semantic contents*, which implicitly contain the spatial data and do not depend on the scale. In addition, it is possible to hierarchize the objects according to geometric characteristics.

In the future work we will measure the semantic similarity among different geospatial datasets. In our humble opinion, a domain conceptualization is useful to build ontologies, which represent (globally) the context of that domain, while the vocabulary of concepts and its relations describe the semantics (locally). Ontologies are very useful since they add a semantic component (the relations between different concepts), which normally is not considered in traditional GIS approaches.

Acknowledgements

The authors of this paper wish to thank the IPN, CIC, SIP and CONACYT for their support.

References

1. Egenhofer M. and Herring J.: Categorizing topological spatial relationships between point, line and area objects, The 9-intersections: formalism and its use for natural language spatial predicates, *Technical report* 94-1, National and Analysis, Santa Barbara, (1994).
2. Genesereth, M. R. and Nilsson, N.: Logical Foundation of Artificial Intelligence. Morgan Kaufmann, Los Altos, California, (1987).

3. Gruber T.: Toward principles for the design of ontologies used for knowledge sharing, International Journal of Human-Computer Studies archive, Volume 43, Issue 5-6 Nov./Dec. 1995 table of contents, Special issue: the role of formal ontology in the information technology, Pages: 907 - 928, Year of Publication, ISSN:1071-5819, (1995).
4. INEGI, Base de datos geográficos. Diccionario de datos topográficos, Instituto Nacional de Estadística Geografía e Informática (INEGI), (1995).
5. Mark D., Egenhofer M., Hirtle S., and Smith B.: Ontological Foundations for Geographic Information Science. University Consortium for Geographic Information Science. Emerging Themes in GIScience Research 2000, White Paper. (http://www.ucgis.org/emerging/), (2001).
6. Molenaar, M., An Introduction to the Theory of Spatial Object Modelling for GIS, Department of Geo-Informatics, International Institute for Aerospace Survey and Earth Science, Enschede, The Netherlands, 1998.
7. Papadias D. and Theodoris Y.: Spatial relations, minimum bounding rectangles, and spatial data structures, "International Journal of Geographical Information Science", vol. 11, n.2, pp.111-138 (1997).
8. Ubeda T. and Egenhofer M., Topological Errors Correcting in GIS, Advances in Spatial Databases, Fifth International Symposium on Large Spatial Databases, SSD '97, Lecture Notes in Computer Science, Vol. 1262, Springer-Verlag, pp. 283-297, (1997).

Intelligent Reduction of Tire Noise

Matthias Becker and Helena Szczerbicka

University Hannover
Welfengarten 1
30167 Hannover, Germany
xmb@sim.uni-hannover.de

Abstract. In this paper we report about deployment of intelligent optimisation algorithms for noice reduction in tire manufacturing. Since the complexity of the problem grows exponentially (the search space is typically of the order of a 65-dimensional vector space), a complete search for the optimal tread profile is not possible even with today's computers. Thus heuristic optimization algorithms such as Genetic Algorithms and Simulated Annealing are an appropriate means to find (near) optimal tread profiles. We discuss approaches of speeding up the generation and analysis of tread profiles, and results using various optimization algorithms.

1 Introduction

Modern tires have to fulfill a lot of basic requirements concerning adhesion, abrasion, stability, etc. If the basic design of a tire is finished, there are still some fine points to be taken into account. One area of research is the noise characteristics of a tire.

This paper is concerned with application of intelligent optimization algorithms in order to find designs of tread profiles for tires, that will produce a low, unobtrusive noise inside a car.

Though the noise design comes after determination of the basic functional properties, the noise characteristics are an important marketing issue. The subjective impression of the quality of a car is highly influenced by a nice low sound inside the car. This lets the car appear comfortable and of high quality, thus car manufacturers are interested to choose a tire for their cars that makes them appear more valuable. It is therefor important for tire manufacturers to design low noise tires, in order to get a share of the large market of supplying car manufacturers.

2 Problem Statement

A tread profile usually consists out of a number of given basic building blocks (called 'pitches', normally an elevated part followed by a 'hole'). There is a set of different pitch types (two to approx. ten different types). Pitches of different types normally have the same structure, but vary in their length. The length is

B. Gabrys, R.J. Howlett, and L.C. Jain (Eds.): KES 2006, Part I, LNAI 4251, pp. 706–713, 2006.
© Springer-Verlag Berlin Heidelberg 2006

usually given as a length ratio, the first pitch type is assigned length 1, the second e.g. length 1.1 etc. Typically a tread profile consists out of 50 to 70 pitches. This is called a 'pitch sequence' (of length n). A tread profile (or the corresponding

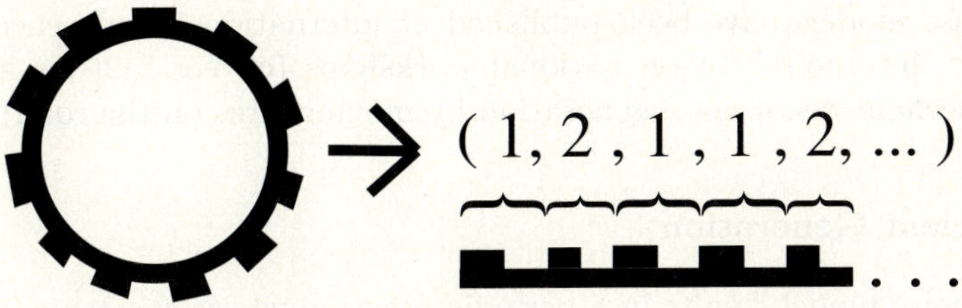

Fig. 1. Mathematical representation of a tread profile

pitch sequence of length n) can be represented by an n-dimensional vector (see fig. 1), thus the problem can be mathematically formulated as a search in an n-dimensional vector space. Let k be the number of different pitch types, then the number of different possible pitch sequences is equal to k^n. Such search space is too large for an exhaustive search in a reasonable amount of time, even with todays computers. The pure generation (without evaluation) of sequences reaches rates of millions of pitch sequences per second, however even at this rate the complete search would last thousands of years.

The rest of the paper is organized as follows. In the next section we will show how to make the generation and analysis of pitch sequences more efficient. This is to be used for enumerative search. In section 4, a more directed search using heuristic optimization algorithms is presented.

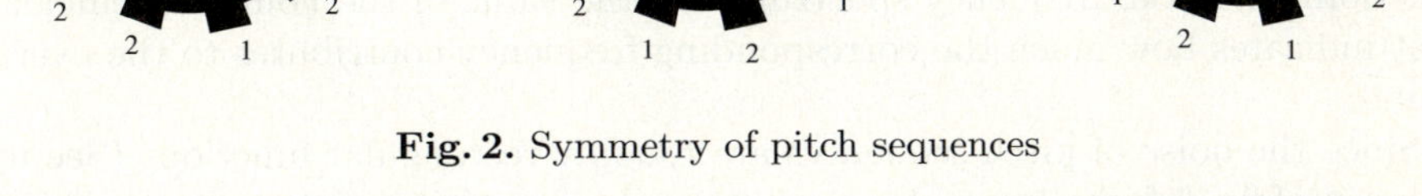

Fig. 2. Symmetry of pitch sequences

3 Efficient Analysis and Generation of Pitch Sequences

The prediction of the noise characteristics of a pitch sequence is done by spectral analysis using the Fourier transform (cf. e.g. [3] or any mathematical standard literature). Roughly the noise model assumes that each pitch issues a noise impulse when hitting the ground. The details of the noise model vary among the tire manufacturers and are confidential or covered by patents (e.g. tire designs where the individual pitch lengths do not have a common divisor, or are prime numbers, EP 0118059/13.02.84, EP 0246996/14.05.87). Some works about details of noise models have been published at international conferences such as the regular 'Internoise', or on national workshops (cf. e.g. [2] for a study of physical mechanisms of noise generation by moving tires on the road).

3.1 Efficient Generation

Since the noise model results in a periodic function whose spectrum is then analyzed, two rotational symmetric pitch sequences result in the same spectrum. Furthermore the noise spectrum remains the same, if the tire is moved forward or backward, or in other words, two pitch sequences that are mirror symmetric result in the same noise spectrum (see fig. 2). Thus for efficiency, it is enough to generate only one representative pitch sequence for each equivalence class of rotational and mirror symmetry. Algorithms for that can be found in mathematical books of permutation theory, under the keyword 'bracelet' and 'necklace', e.g. in [1]. Other keywords are 'Lyndon word' (= an aperiodic necklace representative). It is important to use an algorithm that produces exactly one (and only one) representative of each equivalence class, instead of e.g. generating all sequences and checking for symmetry afterwards, and throwing equivalent sequences away then.

Furthermore there are a number of constraints to be taken into account. Common constraints on pitch sequences are a maximal number of pitches of the same time in series, or how many pitches of each type should occur in a pitch sequence maximally/minimally, etc.

3.2 Efficient Analysis

The assessment of the quality of a given pitch sequence (analysis) is based on Fourier analysis, i.e. the noise spectrum of the tire is estimated.

The noise model for a pitch sequence is an irregular rectangular function, that results in an impulse function, if discretized via sampling. Fourier analysis of that function yields the frequency spectrum, i.e. the value of the Fourier coefficients that indicates how much the corresponding frequency contributes to the overall sound.

Since the noise of pitch sequences are roughly rectangular functions (See upper part of fig. 3 for an example of a rectangular function with amplitude $A1$ over time t.), the noise spectrum of a sequence has the peak around the Fourier coefficient equal to the number of pitches, and sub-peaks at multiples of it (harmonics) (See lower part of fig. 3 where the amplitude $A2$ for the corresponding fourier coefficients FC is shown.). Ideally there should not be one peak frequency but the energy of the noise should be distributed rather equally over all frequencies.

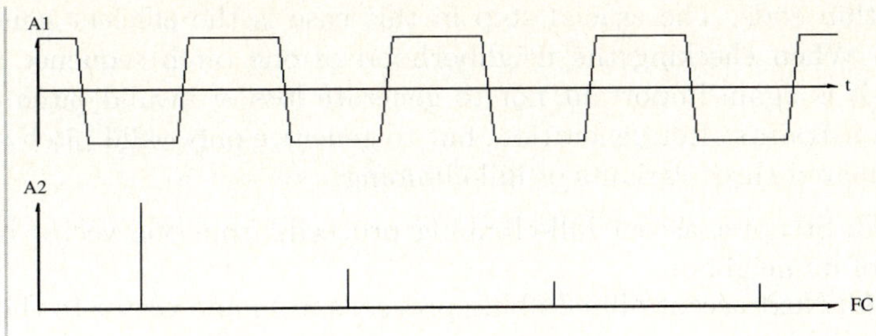

Fig. 3. Square function and Fourier spectrum

Evaluation Order of Fourier Coefficients. Since we are interested in a pitch sequence that has the lowest maximal Fourier coefficient (that mostly is located around the basic period equal to the pitch sequence length) one obvious improvement of efficiency is not to calculate the Fourier coefficients from 1 to lets say 200, but to start with the nth Fourier coefficient, then evaluate $n + / - 1$ etc. (given a pitch sequence length of n).

Sine Table. The calculation of each Fourier coefficient is a sum of sines. We achieve a significant speed up when using a table with pre-calculated values for sines, instead of calculating each value again and again with much too high precision.

Re-use of Partial Sums. If pitch sequences are enumerated, so that the ith pitch sequence differs only in one or two digits from the $i - 1$th, then we can re-use most of the calculations/of the sum evaluated for the ith sequence. This measure also increase efficiency a lot. Note that this is not possible if pitch sequences are generated randomly.

4 Heuristic Optimization of Pitch Sequences

Often, there exists a pitch sequence that is in use already, and tire manufacturers would like to improve the existing pitch sequence. This can be done by local heuristic optimization algorithms, such as hill-climbing. Global heuristic optimization algorithms can also be used as alternative to random search or enumeration of pitch sequences. For more information on heuristic optimization algorithms see standard books on optimization algorithms, e.g. [4].

4.1 Local Optimization

In order to increase the quality of existing pitch sequences, we implemented local optimization algorithms, namely hill-climbing variants. The basic mechanism of hill-climbing is to check all neighbors of the momentarily best solution, and advance to the next better neighbor. If no better neighbor is found, then the algorithm ends. The crucial step in this case is the efficient generation of neighbors. When checking the neighborhood of one pitch sequence for better solutions, it is again important not to generate lots of invalid pitch sequences and test constraints after generation, but to generate only valid pitch sequences.

We evaluated three variants of hill-climbing:

- **SAHC:** Steepest ascent hill-climbing proceeds from one vector to the *best of all* of its neighbors.
- **NAHC:** Next ascent hill-climbing proceeds from one vector to the *next best found* neighbor.
- **RMHC:** Random mutation hill-climbing modifies the current vector randomly and proceeds to the next best found vector which, different to NAHC, does not have to be a neighbor of the current vector. RMHC is not a strictly local search anymore, since by the random mutations also not direct neighbors might be chosen as next vector.

4.2 Global and Combined Optimization

Local heuristic algorithms get stuck easily in a local optimum. Thus global heuristic algorithms are an alternative, if not a local optimization is wanted, but a global search (more intelligent than random search) should be accomplished. We implemented Simulated Annealing (SA), Genetic Algorithms (GA), and also combined algorithms, that first do a global search via random search/SA/GA, and afterwards do a local optimization using hill-climbing. These combined algorithms are known to achieve good results in terms of quality of solution, and execution speed quite often (see. e.g. [5]). We studied hill-climbing, SA, GA and two hybrid variants of both GA and RMHC. In the first variant a hill-climber is used for fine-tuning after the global search stops, and in the second a hill-climber is used to generate fine-tune new points during one step of the global search. All parameters used in the algorithms cannot be given. However we point out some import ones for GA: gene size of three, gene drift and crossover has been used, population size was 500.

In detail, the algorithms used in the following test cases were:

1 RMHC
2 SAHC
3 NAHC
4 SA
5 GA
6 RMHC with SAHC for optimizing each new neighbor
7 RMHC with SAHC after RMHC has stopped
8 GA with SAHC for optimizing new members of the population
9 GA with SAHC after GA has stopped

4.3 Test Cases

We tested the optimization algorithms on two problems, one smaller problem (pitch vector size of 21) where the optimal solutions is known, and on a large problem (pitch vector size of 66).

Small Vector Size. This is a small test problem which is not very realistic, but which is small enough to search the complete search space for the global optimum, giving a known reference point for the performance of the individual algorithms. The vector size is 21, and there are three different types of pitches. Constraints were that each element should occur between one and seven times, each pitch type may occur maximal six times in a row. The global optimum is 112112222333331213312 with a value of 6.44. Each heuristic optimization algorithm has been run 50 times, with different random seeds.

Results for Small Problem. Fig. 4 shows the results. It shows that often HC based algorithms have difficulties to find valid neighbors with regard to the constraints, therefore they get stuck and stop with a bad result (this explains the large standard deviation in the right graph for some of the HC based results.). Best results showed the GA based algorithms, in 18% the global optimum has been reached, the median is only 1.1% away from the optimum. Pure GA was slightly better than hybrid GA. SA performed well, the found vectors were about 7.7% away from the optimum. NAHC performed worst, being around 16.6% away from the optimum, and getting stuck very often because of the constraints.

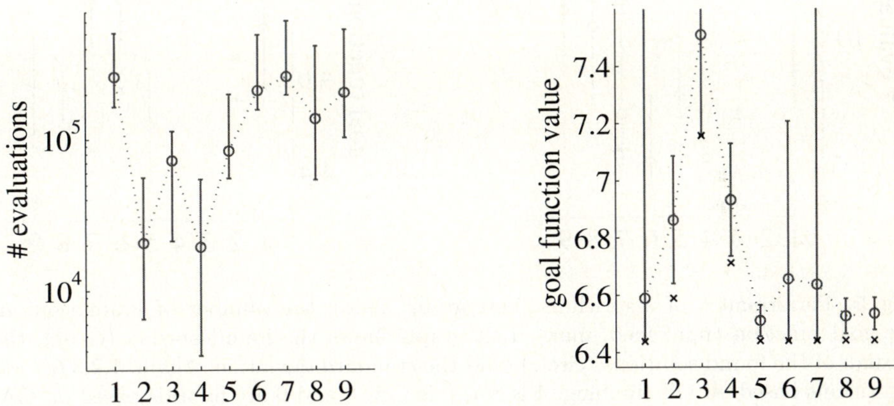

Fig. 4. Performance of algorithms. Left graph shows the number of evaluations of the goal function (minimum, median, maximum). Right graph shows the found optima (cross), median of found solution (circle), and standard deviation. 1,2,3,6,7 are algorithms based on hill-climbing, 4 is SA, 5 is pure GA, 8 and 9 are hybrid algorithms based on GA.

Large Vector Size. Here the vector size is 66 with three different types of pitches. There are several constraints. Each vector should contain exactly 26 elements of type one, 22 of type two and 18 of type three. For each pitch type, there should not be more than six in a row. Pitches of type one and three may not be neighbors.

Results for Large Problem. The global optimum is not known, we tried to find the best vector by a very long search prior to the 25 (shorter) test for each algorithm. The best found vector is *123233211 1232122111 1112211111 2333332212 3212111233 3333233212 2211112* with a value of 4.32. The outcome is shown in fig. 5. Best results showed again GA, finding a vector only 0.9% away from the optimum (mean distance to optimum is 4.1%). Second was RMHC, the best solution was 5.9% away (mean distance to optimum 5.9%). Both algorithms yielded good results with a small standard deviation in short time. Hybrid algorithms of GA and RMHC let the standard deviation become larger, but also increase the quality of the best found solution. SA got stuck in the same local optimum, despite including randomness in the search. Obviously this optimization problem was relatively hard, regarding the structure of the search space. Some very attractive suboptimal solutions exist and make it harder to find the best solution.

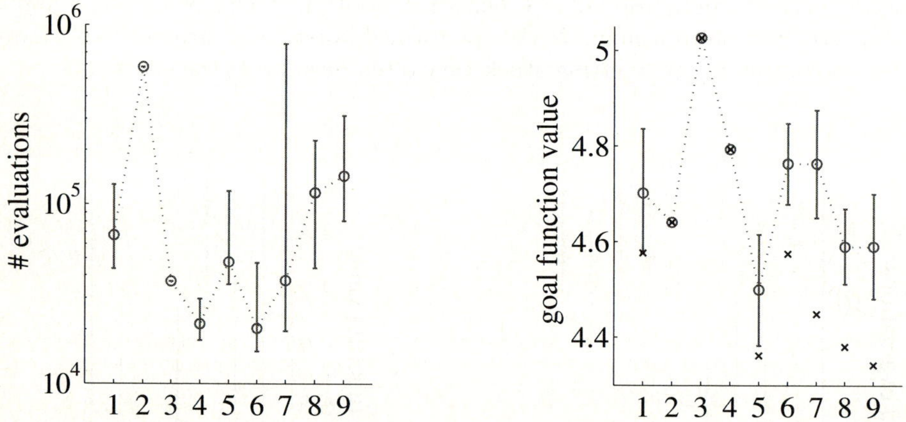

Fig. 5. Performance of algorithms. Left graph shows the number of evaluations of the goal function (min, med, max). Left graph shows the found optima (cross), the median of the found solutions (circle), and the standard deviation. Again, 1,2,3,6,7 are algorithms based on hill-climbing, 4 is SA, 5 is GA, 8 and 9 are hybrids based on GA.

5 Conclusion

In this work, we studied the applicability of heuristic optimization algorithms in order to design low noise tire profiles. It showed that this particular problem should be best approached with genetic algorithms. We furthermore give hints on how to improve efficiency of generation and analysis of pitch sequences. We implemented a prototype software from scratch with regard to efficiency (see fig. 6), and we achieved a speed that is several magnitudes faster (in terms of generated and analyzed pitch sequences per second) than a grown bundle of existing software.

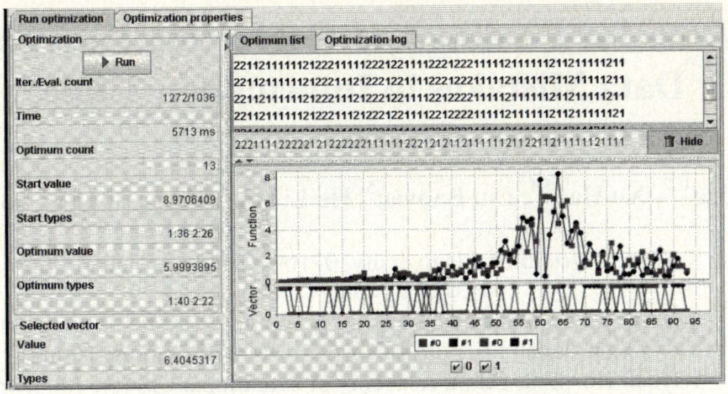

Fig. 6. Screen shot of prototype

In the future, we will try to exploit the mathematical properties of the Fourier analysis to possibly find good pitch sequences by construction.

References

1. Combinatorical object server. *http://www.theory.csc.uvic.ca/˜cos/inf/neck/NecklaceInfo.html*
2. Beckenbauer, T.: Reifen-Fahrbahn-Geräusche – Minderungspotentiale der Strassenoberfläche. In: *Proceedings of the 'Deutsche Arbeitsgemeinschaft für Akustik' (DAGA 03)*, Aachen, Germany, 2003
3. Briggs, W., Henson, V.E.: *The DFT.* SIAM, Philadelphia, 1995
4. Michalewicz, Z., Fogel, D.: *How to solve it: Modern Heuristics.* Springer, London., 1999
5. Szczerbicka, H., Syrjakow, M., Becker, M.: Genetic algorithms, a tool for modelling, simulation and optimization of complex systems. *Cybernetics and Systems: An International Journal, Special Issue: Intelligent modelling and simulation for complex systems*, **II**(7): 639–660, 1998

Algorithms for Finding and Correcting Four Kinds of Data Mistakes in Information Table

Feng Honghai[1,2], Xu Hao[3,4], Liu Baoyan[5], He LiYun[5], Yang Bingru[2], and Li Yueli[1]

[1] Hebei Agricultural University, 071001 Baoding, Hebei, China
liuby@tcmcec.com
[2] University of Science and Technology Beijing, 100083 Beijing, China
[3] China_Japan Friendship Hospital, 100029, Beijing, China
[4] Xiyuan Hospital, China Academy of Traditional Chinese Medicine
[5] China Academy of Traditional Chinese Medicine, 100700 Beijing, China

Abstract. In a real world data set there are usually four kinds of mistaken values, the first one is the mistake in unit; the second one is the mistake of putting the radix points in wrong place, the third one is a scribal error, and the fourth one is a computational mistake. In this paper, we propose two algorithms for finding these four kinds of mistaken data. SARS and coronary heart disease data sets experimental results show that the two algorithms are available, that is, using the two algorithms we find some mistakes in the SARS and coronary heart disease data sets, and the results correspond to that found manually by medical experts.

1 Introduction

Due to the "garbage in, garbage out" principle, data quality problems can be very expensive - "losing" customers, "misplacing" billions of dollars worth of equipment, misallocated resources due to glitched forecasts, and so on. Solving data quality problems typically requires a very large investment of time and energy - often 80% to 90% of a data analysis project is spent in making the data reliable enough so that the results can be trustworthy. So finding and correcting mistakes in the data set is a necessary and important step during the course of handling information systems [1].

Now several software vendors are addressing the data quality problem with new data quality software products that can analyze database extracts and point to errors and inconsistencies in data using statistical analysis and knowledge-based systems technology [2-5]. The statistics based methods are according to the Chebyshe theorem and the mean value and variance of attribute values are used to take the confidence interval into account and identify the exceptional records. The clustering based methods use Euclidean or other distance to identify the exceptional records. Pattern based methods can be used to detect the records that cannot accord with the existing pattern [6]. The association rules based methods induce the rules between attributes and can detect the exceptional records [7].

When inducing rules from the SARS data set we find some mistaken values that would influence the validation of the induced knowledge.

B. Gabrys, R.J. Howlett, and L.C. Jain (Eds.): KES 2006, Part I, LNAI 4251, pp. 714–719, 2006.
© Springer-Verlag Berlin Heidelberg 2006

2 Different Kinds of Data Mistakes

In a real world data set, there are various kinds of mistaken values. These mistakes can be divided into the following types.

(1) Scribal errors. These kinds of mistakes are scattered in the data table, namely, they do not cluster in a place in the information table. These kinds of mistakes can be found by rules between attributes one of which the mistaken values belong to, that is, we can use rough set theory to induce dependencies between any two attributes and using the dependencies to find the mistaken values, if most of the attribute values accord with the dependency rules, whereas only a few attribute values do not, we can assume that these few values are mistakes.

(2) Calculation mistakes. Most of these kinds of mistakes are scattered too in the data table and can be found in the same way as the scribal errors.

(3) Mistakes in unit, that is, some values are in a particular unit, and some values are in another unit. Usually, this kind of mistaken values cluster together and the differences between correct values and mistaken values are much more than the ones between the correct values or between the mistaken values. So we can calculate the differences between attribute values. If the difference is very large, we can assume that there is a mistaken value between the two values.

(4) Mistakes where the radix points have been put in wrong place are scattered too in the data table and can be found in the way the mistakes in unit are found, because if a value's radix points have been put in wrong place, the difference between the wrong value and its neighbor value is much more than the difference between two correct values. So we can calculate the differences between two neighbor values to identify this kind of mistakes.

3 Rough Set Concepts

Rough set is a new knowledge discovery theory. Rough set can generate easily understood knowledge without additional information outside of the data set.

In rough set theory, knowledge about objects is represented in the form of an information table, the rows of the table are labeled by objects, columns are labeled by attributes and entries of the table are attribute-values, called descriptors. Formally, by an information table we understand a 4-tuple $S = \langle U, Q, V, f \rangle$, where U is a finite set of objects, Q is a finite set of attributes, $V = \cup_{q \in Q} V_q$, where V_q is the domain of attribute q, and $f : U \times Q \rightarrow V$ is a total function such that $f(x,q) \in V_q$ for every $(x,q) \in U \times Q$, called an information function .The set Q is, in general, divided into the set C of condition attributes and the set D of decision attributes.

Suppose we are given a set of objects U called the universe and an indiscernibility relation (attribute) $R \subseteq U \times U$, representing our lack of knowledge about elements of U. The equivalence class of R determined by element x will be denoted by $R(x)$ that are the set of objects whose values of attribute R are the same as the value of x. Let X be a subset of U. We want to characterize the set X with respect to R.

Rough sets can be defined employing rough membership function

$$\mu_X^R : U \to \langle 0,1 \rangle$$

where

$$\mu_X^R(x) = \frac{|X \cap R(x)|}{|R(x)|}$$

and $|X|$ denotes the cardinality of X.

The rough membership function expresses conditional probability that x belongs to X given R and can be interpreted as a degree that x belongs to X in view of information about x expressed by R.

μ_X^R denotes the dependency of X on $R(x)$, so we can use this measure to induce the dependencies between two equivalence classes of two attributes and to identify the mistakes mentioned above.

4 Algorithms

(1) Data preprocessing

If there are more than two attributes whose values are temporal ones, calculate the differences of two neighbor values and take these differences as input values to build a new decision table.

(2) Calculate i (x)// i (x) is the equivalence class of i determined by element x.

```
  For (i=0; i<n; i++)// i denotes an attribute and n denotes the amount of attributes
    For (j=0; j<m-1; j++)// j denotes an example and m denotes the amount of
                                  //examples
      For (k=j+1; k<m; k++)// k denotes another example.
        If i (j) = i (k)
          Put i (j) and i (k) into a set.
```

(3) Algorithm for finding scribal errors and calculation mistakes

Input: A decision table $T = \langle U, C \cup D \rangle$, where U is the finite universe of objects; C, the finite set of condition attributes, and D, the finite set of decision attributes; threshold of $\mu(x)$.

Output: offer the values that may be mistaken ones to experts to identify whether they are mistaken values

```
For (i=0; i<n; i++)// i denotes an attribute and n denotes the amount of attributes.
  For (j=i+1; j<n; j++)// j denotes another attribute
```

Calculate $\mu_i^j(x) = \frac{|i(x) \cap j(x)|}{|i(x)|}$ and $\mu_j^i(x) = \frac{|i(x) \cap j(x)|}{|j(x)|}$

// if the values of x is numeric, it should be discretized first.

If $\mu_i^j(x)$ or $\mu_j^i(x)$>=95%// 95% is a threshold that can be chosen by //user.

Then x with $\mu_i^j(x)$ or $\mu_j^i(x)$ <5% are offered to experts to identify whether they are mistaken values.

(4) Algorithm for finding mistakes in unit and mistakes where the radix points have been put in wrong place

Input: A decision table $T = \langle U, C \cup D \rangle$, where U is the finite universe of objects, C the finite set of condition attributes, and D, the finite set of decision attributes; the threshold of the difference and the quotient.

Output: offer the values that may be mistake ones to experts to identify whether they are mistaken values.

For (i=0; i<n; i++)// i denotes an attribute and n denotes the amount of attributes.

Sort the values of attribute i by ascending order.

Calculate the differences and the quotients of the neighbor values

If a difference or a quotient >= threshold

Then x is offered to experts to identify whether they are mistaken values.

5 Clinical SARS and Coronary Heart Disease Data Experiments

In the SARS data set, "the nth day of the patient being in hospital" and "the nth day of the patient being sick" are two attributes that have a strong correlation. For example, for a patient, if "the nth day of the patient being in hospital" is 3, and the corresponding "the nth day of the patient being sick" is 7, then when "the nth day of the patient being in hospital" is 4, the corresponding "the nth day of the patient being sick" should be 8. If the attribute value is not 8, we can conclude that the value is a mistake. Obviously, the two attributes are temporal ones, so we calculate the differences of two neighbor values and take these differences as input values to build a new decision table with the two attributes. As a result, we find 211 mistakes among 7665 cases that are such as the above mistakes.

Another example in SARS data set is the attribute "the suffered disease in the past" and "healthy or not in the past". If the attribute "the suffered disease in the past"takes on a certain value meaning that the patient have suffered this disease, the attribute "healthy or not in the past" should take on the value "1" (1 means not healthy), else the value is mistake. In the SARS data set we find 4 mistakes among 66 cases that are such as the above mistake.

The third case in SARS data set is the attribute "coprostasis" and "diarrhea". The value of $\mu_{\text{coprostasis}}^{\text{diarrhea}}(x)$ = 2/ 220 =0.009% and $\mu_{\text{diarrhea}}^{\text{coprostasis}}(x)$= 2/ 158 =0.012%that are far less than 5%, so the two values that in the intersection are mistakes, i.e., if a patient suffers from coprostasis, he cannot suffer from diarrhea at the same time. Actually, coprostasis and diarrhea are mutually exclusive.

The fourth case in SARS data set is the attribute "the damage of heart" and "MODS" (multiple organ dysfunction syndrome). The value of $\mu_{\text{the damage of heart}}^{\text{MODS}}(x)$ = 1/ 612 =0.016% $\mu_{\text{MODS}}^{\text{the damage of heart}}(x)$ = 1/ 36 =0.027%that are far less than 5%, so we assume that the value in the intersection is a mistake.

The fifth case in SARS data set is the attributes “dilatation pressure” and “diastolic pressure”. We set the threshold of difference and quotient are 5 and 1.5 respectively and after sorting the values of attribute “dilatation pressure” in ascending order we find that the difference and the quotient of certain two neighbor values that are 25.33 and 81 is very large. Similarly, for attribute “diastolic pressure”, the difference of two neighbor values 20 and 40 is much more than the differences between any other two neighbor values. After offering these values to experts, they are all identified as the mistakes caused by using different unit, that is, the attribute values of “dilatation pressure” that are less than or equal to 25.33 take the unit of “mmHg”, whereas the values that are larger than or equal to 81 take the unit “kpa” . The case for attribute “diastolic pressure” is the same as the attribute “dilatation pressure”. As a result, there are 555 mistaken values of “diastolic pressure” and “dilatation pressure” among the total 4019 values respectively.

The other cases in SARS data set are for the attributes “lymph”, “AST”, ”TBIL” and ”LDH”. The neighbor values whose differences and quotients are larger than the thresholds are 21 and 165, 1133 and 3760, 184 and 809, 3 and 20, but after being identified by experts they are all the correct values.

In coronary heart disease data set, for “dilatation pressure” there are four values 1750, four values 1440, four values 1020, and three values 974. After checking the original source data set, we find that 1750, 1440, and 1020 should be 175, 144, and 102 respectively, and 974 is 174 originally. The other mistakes are mistakes in unit. That is, after ascending sort the amount of neutral granular cell rise from 8 to 46 abruptly; red blood cell count (RBC) rise from 6 to 352 abruptly; white blood cell count (WBC) rise from 26 to 4500 abruptly.

6 Conclusions and Discussions

(1) Before data mining in database we should find and correct the mistakes to avoid the problem of “garbage in, garbage out”.

(2) Usually scribal errors, computation mistakes, and the mistakes of putting the radix points in wrong place are scattered, so the amount of the mistaken values are much less than the correct values. But the amount of the mistaken values in wrong unit may be very large. If we cannot find them, they must influence the results of data mining.

(3) The algorithms should be combined with the experts and the mistakes should be identified and corrected by experts finally.

Acknowledgements

The paper is supported by (1) the Special Foundation of SARS of the National High-Tech Research and Development Major Program of China (863)- “Clinical Research on the Treatment of Severe Acute Respiratory Syndrome with Integrated Traditional and Western (No.2003AA208101)”. (2) Beijing Science and Technology Program-Study on the Diagnosis and Treatment Rules Discovery and Treatment Regimens in the Coronary Heart Disease (No. H020920010490).

References

1. Johnson, Theodore, Dasu, Tamrapami, Data Quality and Data Cleaning: An Overview Proceedings of the ACM SIGMOD International Conference on Management of Data, (2003)
2. Deng Zhong-guo, Zhou Yi-xin, Research of the Data Cleaning Technique. Journal of Shandong University of Science and Technology (Natural Science) (Chinese). Vol.2(2004) 55-57
3. Simoudis. E., Livezey. B. And Kerber. R., Using Recon for Data Cleaning. Proceedings of KDD 1995, (1995) 282-287
4. Levitin, A. and Redman, T.: A Model of the Data (life) Cycles with Application to Quality. Information and Soft Ware Technology, vol. 4(1995) 217-223
5. Jonathan, I., Maletic Andrian Marcus. Data Cleaning: Beyond Integrity Analysis. Division of Computer Science, vol.2(2000)
6. Guyon, I., Matic, N and Vapnik V: Discovering Information Patterns and Data Cleaning. In Advances in Knowledge Discovery in Data Ming. MIT Press/ AAAI Press, (1996)
7. Andrian Marcus, Jonathan. I., and Maletic: Utilizing Association Rules for the Identification of Errors in Data. Technical Report. CS-00-04

Relevance Ranking of Intensive Care Nursing Narratives

Hanna Suominen[1,2], Tapio Pahikkala[1,2], Marketta Hiissa[1,4], Tuija Lehtikunnas[3,5], Barbro Back[1,4], Helena Karsten[1,2], Sanna Salanterä[3], and Tapio Salakoski[1,2]

[1] Turku Centre for Computer Science, Lemminkäisenkatu 14 A, FIN-20520 Turku, Finland

[2] University of Turku, Department of Information Technology, Lemminkäisenkatu 14–18 A, FIN-20520 Turku, Finland
firstname.lastname@utu.fi

[3] University of Turku, Department of Nursing Science, FIN-20014 University of Turku, Finland
firstname.lastname@utu.fi

[4] Åbo Akademi University, Department of Information Technologies, Lemminkäisenkatu 14, FIN-20520 Turku, Finland
firstname.lastname@abo.fi

[5] Turku University Hospital, PL 52, FIN-20521 Turku, Finland
firstname.lastname@tyks.fi

Abstract. Current computer-based patient records provide many capabilities to assist nurses' work in intensive care units, but the possibilities to utilize existing free-text documentation are limited without the appropriate tools. To ease this limitation, we present an adaptation of the Regularized Least-Squares (RLS) algorithm for ranking pieces of nursing notes with respect to their relevance to breathing, blood circulation, and pain. We assessed the ranking results by using Kendall's τ_b as a measure of association between the output of the RLS algorithm and the desired ranking. The values of τ_b were 0.62, 0.69, and 0.44 for breathing, blood circulation, and pain, respectively. These values indicate that a machine learning approach can successfully be used to rank nursing notes, and encourage further research on the use of ranking techniques when developing intelligent tools for the utilization of nursing narratives.

1 Introduction

Current computer-based patient records offer many possibilities to support nurses' work in intensive care units (ICUs). They have eased the way in which data are recorded, and they provide also new ways to utilize the recorded data. Numerical data are recorded directly from, e.g., monitoring and respiration devices, and the data are used to automatically generate different kinds of curves and diagrams. Much data are recorded also as free-text notes, but without appropriate tools, the utilization of these narratives is limited.

A possible solution to enhance using the information saved as narratives is to develop tools that process them automatically and provide nurses with targeted

B. Gabrys, R.J. Howlett, and L.C. Jain (Eds.): KES 2006, Part I, LNAI 4251, pp. 720–727, 2006.
© Springer-Verlag Berlin Heidelberg 2006

and relevant information, e.g., when they need to build an overview about issues related to the patient's blood circulation. To serve as a basis for these kinds of tools, we need the classification of the narratives according to their content. Furthermore, if the classification results are ranked so that each text piece is assigned a relevance score reflecting the strength of the relation to the given class, the weaker statements could be used, e.g., to focus the nurse's attention to some particular issue in order to examine it more carefully, and the stronger statements could be used, e.g., to summarize the most essential information in the record.

In health care, classification has recently been applied, e.g., to categorize texts such as chief complaint notes, diagnostic statements, and injury narratives into different kinds of syndromic, illness and cause-of-injury categories [1, 2, 3, 4, 5]. In most of these studies, the main goal was to serve the needs of different kinds of surveillance tasks. Our approach differs from these in two major ways. First, the aim is to develop tools that assist nurses in the direct caring process, and second, instead of building a binary classifier, we apply relevance ranking with respect to the given class, i.e., the more strongly the statement relates to the given class, the higher rank it gets.

In this paper, we present an adaptation of the Regularized Least-Squares (RLS) algorithm to ranking pieces of nursing notes according to their relevance to breathing, blood circulation, and pain. We chose to use the RLS algorithm, also known as the Least-Squares Support Vector Machine [6], because the Support Vector Machine approach has recently shown encouraging results, e.g., in improving the retrieval performance of WWW search engines [7], and parse ranking on a set of biomedical sentences [8].

2 Materials and Methods

2.1 Materials

As data we used anonymized Finnish nursing narratives gathered with proper permissions from 16 ICUs in the spring of 2001. In total, we had 43 copies of patient records with nursing notes for one day and night. The documents contained both very long sentences describing different issues, and very short ones with the focus only on one issue. In order to standardize their style, we divided the long sentences into smaller pieces consisting of one matter or thought. This was done manually by one of the authors with nursing experience, and resulted in 1363 pieces, with an average length of 3.7 words.

In ranking the text pieces, we considered the classes of *Breathing*, *Blood Circulation*, and *Pain*. Breathing and Blood Circulation were chosen because the emphasis in intensive care nursing is on monitoring, assessment and maintenance of vital functions [9]. Pain was selected for two reasons. First, we think that pain assessment in critically ill patients could be supported by automatically extracting pieces of related text. This is because nurses must often perform this task relying only on implicit behavioral and physiological indicators due to the patient's inability to communicate [10], and because many potential pain indicators are documented in ICU nursing notes [11]. Second, due to the nature of pain assessment generally, and especially in critically ill patients, we assume that pain is documented differently from breathing and blood circulation.

In order to achieve the ranking that can be used in the training of the RLS algorithm and in the evaluation of the results, the text pieces were labelled independently by three nurses according to the classes of Breathing, Blood Circulation, and Pain. All three nurses were specialists in nursing documentation; two had a long clinical experience from intensive care units (N_1 and N_2), and one was an academic nursing science researcher (N_3). The classes were considered separately, and the nurses were advised to label a phrase if they considered it to include relevant information about the given class. The agreement between the nurses was measured by using *Cohen's* κ [12], which is a chance-corrected agreement measure, and appropriate especially in situations like ours when there are no criteria for the correctness of judgments. The values of κ showed some disagreements between the nurses, in particular in the classes of Breathing and Pain (Table 1)[13].

Table 1. The values of Cohen's κ and respective 95 % confidence intervals (CI) for comparisons between nurses N_1, N_2, and N_3 [13]. These statistical analyses were performed with SPSS 11.0 for Windows.

	Breathing		Blood Circulation		Pain	
	κ	95 % CI	κ	95 % CI	κ	95 % CI
$N_1 - N_2$	0.73	(0.68 – 0.78)	0.89	(0.85 – 0.92)	0.88	(0.82 – 0.94)
$N_1 - N_3$	0.67	(0.62 – 0.72)	0.81	(0.77 – 0.86)	0.79	(0.73 – 0.86)
$N_2 - N_3$	0.85	(0.82 – 0.89)	0.87	(0.83 – 0.90)	0.76	(0.69 – 0.83)

We considered the disagreements between the nurses to reflect various interpretation possibilities as well as different experiences, knowledge and expertise areas. Thus, we regarded them as a valuable resource, upon which the ranking system was based. For each phrase, we counted the number of nurses who had labelled it belonging to the given class, and as a result, we got ranked data, where each phrase had a rank from zero to three corresponding to the number of nurses who had assigned the phrase into the class. This ranking was considered to reflect the relevance level of the expressions according to the given class. The ranked data were divided into *training set* and *test set* so that 708 pieces of text belonged to the training set, and the remaining 655 pieces to the test set. No patient record was divided between the two sets.

2.2 Methods

The automated classification was performed by using a *Regularized Least-Squares (RLS) algorithm.* Let $\{(x_1, y_1), \ldots, (x_m, y_m)\}$, where X is a space of bag-of-words vectors, $x_i \in X$, and $y_i \in \{0, 1, 2, 3\}$, be the set of training examples. The target output value y_i is the rank of the corresponding example, set according to the labellings of the three nurses. We considered the RLS algorithm as a special case of the following regularization problem known as the Tikhonov regularization (for a more comprehensive introduction, see, e.g., [6], or [14]):

$$\min_{f} \sum_{i=1}^{m} l\left(f(x_i), y_i\right) + \lambda \|f\|_k^2, \tag{1}$$

where l is the loss function used by the learning machine, $f : X \rightarrow \mathbb{R}$ is a function, $\lambda \in \mathbb{R}_+$ is a regularization parameter, and $\|\cdot\|_k$ is a norm in a reproducing kernel Hilbert space defined by a positive definite kernel function k. The second term in (1) is called a *regularizer*. The loss function used with RLS is called *least-squares loss* and is defined as

$$l\left(f(x_i), y_i\right) = \left(y_i - f(x_i)\right)^2 . \tag{2}$$

By the Representer Theorem, the minimizer of equation (1) with the least-squares loss function has the form

$$f(x) = \sum_{i=1}^{m} \alpha_i k(x, x_i), \tag{3}$$

where $\alpha_i \in \mathbb{R}$ and k is the kernel function associated with the reproducing kernel Hilbert space mentioned above. The kernel function which we used is the cosine of the word feature vectors, i.e.,

$$k(x, x_i) = \frac{\langle x, x_i \rangle}{\sqrt{\langle x, x \rangle \langle x_i, x_i \rangle}}. \tag{4}$$

The RLS algorithm was trained for each of the classes separately. To unify the data, we separated punctuation marks and special characters from words with spaces, converted all letters to lower case, and used the *Snowball stemming algorithm* [15] for Finnish text. In order to optimize the ranking of the training examples, we divided the output values into four separate intervals corresponding to the four labelling levels. The threshold values that determined the output intervals and the regularization parameter were selected by using a leave-one-out cross-validation on the training set with a rank correlation coefficient *Kendall's* τ_b [16] as the optimization criterion. We chose this measure of the association because, according to [16], it is appropriate in situations like ours when there is a considerable amount of tied items in the data.

3 Results and Discussion

We measured the association between the rankings produced by nurses and the RLS algorithm by using Kendall's τ_b. According to these values (Table 2), the strongest associations between the desired ranking and the one produced by the RLS algorithm were reached in the classes of Blood Circulation (τ_b=0.69) and Breathing (τ_b=0.62). In the class Pain, the value of τ_b was 0.44.

Table 2. The values of Kendall's τ_b and respective 95 % confidence intervals (CI), calculated with SPSS 11.0 for Windows, for rankings in the classes of Breathing, Blood Circulation, and Pain

	τ_b	95% CI
Breathing	0.62	(0.56 – 0.68)
Blood Circulation	0.69	(0.61 – 0.76)
Pain	0.44	(0.30 – 0.59)

Next, a possibility to utilize the ranking is illustrated in Fig. 1 by considering the case in which there is a need to quickly build an overview about issues related to blood circulation in our test set. In this case, by choosing the sensitivity level from the output of the RLS algorithm, e.g., the fifty or hundred highest ranked phrases of nursing notes could be returned. In our first experiment, when fifty phrases were returned, a great majority of them (45 phrases) belonged to the blood circulation rank three in the desired ranking. Two phrases (*Situation stable.* and *Lift in infusion.*) were considered as irrelevant in the desired ranking, and the number of returned phrases with ranks one and two in the desired ranking were one and two, respectively. In our second experiment with hundred phrases, 55 out of 65 phrases with the rank three in the desired ranking were returned. Here, however, the number of phrases that were considered as irrelevant in the desired ranking increased to 39, and the number of returned phrases with ranks one and two in the desired ranking were one and five, respectively. In summary, this example demonstrates how ranking tools could enhance the utilization of existing narratives.

Let us now consider the manual ranking upon which the RLS algorithm was based. In this ranking, the classes of Breathing, Blood Circulation, and Pain had similar general characteristics of the ranks. For example, in the class Breathing, the phrases with the rank three were related to issues such as oxygenation and the use of different kinds of respiration devices, whereas the phrases with lower ranks described mucus in patients' lungs, coughing, and issues related to pleural tubes. In the class Blood Circulation, the phrases with the highest rank were mostly about the progression of various measurement values, such as pulse or blood pressure. The phrases with the blood circulation ranks two or one described issues such as patients' body temperature and bluish skin shade. In the class Pain, the phrases with the rank three commonly contained keywords such as *pain* or *ache*, whereas the phrases with rank one or zero included neither the word *pain* nor its derivatives or synonyms. The phrases with the pain rank one or two covered implicit pain indicators such as the quality of sleep and reactions to nursing interventions. These examples illustrate that the phrases which all the nurses considered to contain relevant information about the particular class were the most evident notes about the class. The connection to the topic became more implicit when the relevance rank decreased. Hence, it seems that the manual ranking logically demonstrates the connection of the phrases and the given class.

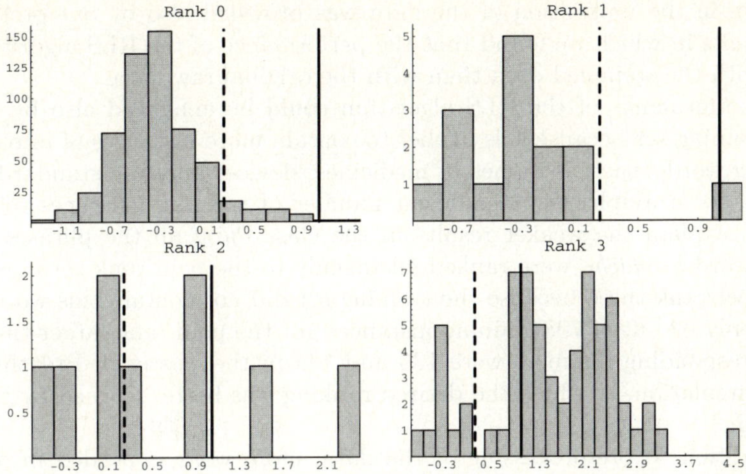

Fig. 1. Histograms illustrating the association between the desired ranking and the output of the RLS algorithm for each of the four human ranks separately in the class Blood Circulation. The horizontal axis corresponds the output of the RLS algorithm in real values, and the vertical axis represents the number of phrases in the corresponding range. Vertical lines at 1.05 (solid) and at 0.24 (dashed) separate the output of the RLS algorithm so that 50 or 100, respectively, of the highest ranked phrases are on the right-hand side of the line.

Although a reliable reference standard can be composed on the basis of several experts' aggregated opinions [17], more work is required in order to obtain a manual ranking that reliably reflects the real relevance of the phrases with respect to the desired classes. Based on the method of how the manual ranking was compiled, the correct interpretation about a relevant phrase could have been outvoted if only one of the three nurses labelled the phrase. However, the aim of the study was not to evaluate the correctness of nurses' classifications but to assess the possibility of automatically ranking nursing notes according to their relevance to a given topic. Encouraged by the results of this study, we will focus in the future on the reliability of the manual ranking by considering, e.g., the number of experts, the required level of expertise and alternatives to compose manual ranking on the basis of reference material.

Let us now discuss the results of the RLS algorithm. In all three classes, the differences between the manual and automated rankings seemed to be mostly due to the sparse data. Thus, the performance of the RLS algorithm might be enhanced by reducing the sparseness of the data through the use of pre-processing algorithms that find the base form of the word instead of its stem. For example, this kind of an algorithm may enhance the performance of the RLS algorithm by recognizing the key word *kipu* (pain) from its derivatives *kivulias* and *kipuja*, and from rare compounds, such as *kipulääke* (painkiller) and *kipukynnys* (pain threshold). Evidence of the importance of the pre-processing algorithm in the unification of the data was provided also by our preliminary experiments in which we found that the performance of the RLS algorithm was better with the stemmed data than with the original raw data.

The performance of the RLS algorithm could be enhanced also by using a larger training set because it is likelier to contain more instances of infrequently occurring words, such as names of medicines, devices and non-standard abbreviations. For example, the insufficient number of training instances related to pain can explain the weaker results in the class Pain; all the phrases including the word *headache* were ranked mistakenly to the pain rank zero instead of the correct rank three because the training set did not contain this word. Altogether, only 62 out of 708 training instances got the pain rank larger than zero. The corresponding numbers were 136 and 144 in the classes of Breathing and Blood Circulation, in which the desired ranking was better learned by the RLS algorithm.

In addition to pre-processing of the data, more work is needed in order to divide long sentences into smaller pieces automatically. Even though the division we used was based on the point of view of one nursing science researcher, it can be considered quite objective since only few text pieces belonged to two classes and none of the pieces belonged to three classes in any of the manual classifications.

4 Conclusions

The association between the output of the RLS algorithm and the desired ranking was somewhat weaker in the class Pain than in the classes of Breathing and Blood Circulation. We consider the differences in the strength of the associations to be due to the different nature of the documentation related to the classes, and anticipate this to affect the details of the ranking application.

Our results indicate that a machine learning approach can successfully be used to rank nursing notes, and encourage further research on the use of ranking techniques when developing intelligent tools for the utilization of recorded nursing narratives. In the future, tools of this kind may enable, e.g., automated highlighting of nursing narratives according to their relevance to the particular topic. In this way, the utilization of recorded narratives can be supported without losing the original context of text pieces. Moreover, the ranking result reflecting the relevance of the text pieces with respect to the desired classes is more informative than the classification output which only separates relevant text pieces from irrelevant parts of the documentation.

Acknowledgements

We gratefully acknowledge the financial support of Tekes (grant no. 40435/05) and the Academy of Finland (grant no. 104639). We thank Filip Ginter and Heljä Lundgrén-Laine for their contributions.

References

1. Chapman, W.W., Dowling, J.N., Wagner M.M.: Fever detection from free-text clinical records for biosurveillance. J Biomed Inform **37** (2004) 120–127
2. Pakhomov, S.V., Buntrock, J.D., Chute, C.G.: Using compound codes for automatic classification of clinical diagnoses. Medinfo **11** (2004) 411–415
3. Wellman, H.M., Lehto, M.R., Sorock, G.S., Smith, G.S.: Computerized coding of injury narrative data from the National Health Interview Survey. Accid Anal Prev **36** (2004) 165-171
4. Chapman, W.W., Christensen, L.M., Wagner, M.M., Haug, P.J., Ivanov, O., Dowling, J.N., Olszewski, R.T.: Classifying free-text triage chief complaints into syndromic categories with natural language processing. Artif Intell Med **33** (2005) 31–40
5. Pakhomov, S.V., Buntrock, J., Chute, C.G.: Prospective recruitment of patients with congestive heart failure using an ad-hoc binary classifier. J Biomed Inform **38** (2005) 145–153
6. Rifkin, R.M.: Everything old is new again: A fresh look at historical approaches in machine learning. PhD Thesis, Massachusetts Institute of Technology (2002)
7. Joachims, T.: Optimizing search engines using clickthrough data. In: Proceedings of the Eighth ACM SIGKDD International Conference on Knowledge Discovery and Data Mining, New York, NY, USA, ACM Press (2002) 133–142
8. Tsivtsivadze, E., Pahikkala, T., Pyysalo, S., Boberg, J., Mylläri, A., Salakoski, T.: Regularized Least Squares for parse ranking. In A.F. Famili et al. (Eds.): IDA 2005, LNCS 3646. Springer-Verlag Berlin Heidelberg (2005) 464–474
9. Ward, N.S., Snyder, J.E., Ross, S., Haze, D., Levy, M.M.: Comparison of a commercially available clinical information system with other methods of measuring critical care outcomes data. J Crit Care **19** (2004) 10–15
10. Payen, J.-F., Bru, O., Bosson, J.-L., Lagrasta, A., Novel, E., Deschaux, I., Lavagne, P., Jacquot, C.: Assessing pain in critically ill sedated patients by using a behavioral pain scale. Crit Care Med **29** (2001) 2258–2263
11. Gélinas, C., Fortier, M., Viens, C., Fillion, L., Puntillo, K.: Pain assessment and management in critically ill intubated patients: a retrospective study. Am J Crit Care **13** (2004) 126–135
12. Cohen, J.: A coefficient of agreement for nominal scales. Educ Psychol Meas **20** (1960) 37–46
13. Hiissa, M., Pahikkala, T., Suominen, H., Lehtikunnas, T., Back, B., Karsten, H., Salanterä, S., Salakoski, T.: Towards automated classification of intensive care nursing narratives. To be presented at: The 20th International Congress of the European Federation for Medical Informatics (MIE 2006), Maastricht, the Netherlands, 27–30 August (2006)
14. Poggio, T., Smale, S. The mathematics of learning: Dealing with data. Amer. Math. Soc. Notice **50** (2003) 537–544
15. Snowball, the Finnish stemming algorithm. Martin Porter; c2001 and (for the Java developments) Richard Boulton; c2002 [updated 2005 May 23; cited 2006 Jan 10]. http://snowball.tartarus.org/algorithms/finnish/stemmer.html
16. Kendall, M.G.: Rank correlation methods. 4th edition, Griffin, London (1970)
17. Hripcsak, G., Wilcox, A.: Reference standards, judges, and comparison subjects: roles for experts in evaluating system performance. J Am Med Inform Assoc **9** (2002) 1–15

Unification of Protein Data and Knowledge Sources

Amandeep S. Sidhu[1], Tharam S. Dillon[1], and Elizabeth Chang[2]

[1] Faculty of Information Technology, University of Technology, Sydney, Australia
{asidhu, tharam}@it.uts.edu.au
[2] School of Information Systems, Curtin University of Technical University, Perth, Australia
Elizabeth.Chang@cbs.curtin.edu.au

Abstract. Advances in technology and the growth of life sciences are generating ever increasing amounts of data. High-throughput techniques are regularly used to capture thousands of data points in an experiment. The results of these experiments normally end up in scientific databases and publications. Although there have been concerted efforts to capture more scientific data in specialist databases, it is generally acknowledged that only 20 per cent of biological knowledge and data is available in a structured format. The remaining 80 per cent of biological information is hidden in the unstructured scientific results and texts. Protein Ontology (PO) discussed in this paper provides a common structured vocabulary for this structured and unstructured information and provides researchers a medium to share knowledge in proteomics domain. It consists of concepts, which are data descriptors for proteomics data and the relations among these concepts. Protein Ontology provides description for protein domains that can be used to describe proteins in any organism.

1 Introduction

In accelerating quest for disease biomarkers, the use of high-throughput technologies, such as DNA microarrays and proteomics experiments, has produced vast datasets identifying thousands of genes whose expression patterns differ in diseased versus normal samples. Although many of these differences may reach statistical significance, they are not biologically meaningful. For example, reports of mRNA or protein changes of as little as two-fold are not uncommon, and although some changes of this magnitude turn out to be important, most are attributes to disease-independent differences between the samples. Evidence gleaned from other studies linking genes to disease is helpful, but with such large datasets, a manual literature review is often not practical. The power of these emerging technologies – the ability to quickly generate large sets of data – has challenged current means of evaluating and validating these data. Thus, one important example of a data rich but knowledge poor area is biological sequence mining. In this area, there exist massive quantities of data generated by the data acquisition technologies. The bioinformatics solutions addressing these data are a major current challenge. However, domain specific ontologies such as Gene Ontology [1], MeSH [2] and Protein Ontology (PO) [3, 4, 5, and 6] exist to provide context to this complex real world data.

B. Gabrys, R.J. Howlett, and L.C. Jain (Eds.): KES 2006, Part I, LNAI 4251, pp. 728 – 737, 2006.
© Springer-Verlag Berlin Heidelberg 2006

2 Protein Ontology Conceptual Framework

Advances in technology and the growth of life sciences are generating ever increasing amounts of data. High-throughput techniques are regularly used to capture thousands of data points in an experiment. The results of these experiments normally end up in scientific databases and publications. Although there have been concerted efforts to capture more scientific data in specialist databases, it is generally acknowledged that only 20 per cent of biological knowledge and data is available in a structured format. The remaining 80 per cent of biological information is hidden in the unstructured scientific results and texts. Protein Ontology (PO) [3, 4, 5, and 6] provides a common structured vocabulary for this structured and unstructured information and provides researchers a medium to share knowledge in proteomics domain. It consists of concepts, which are data descriptors for proteomics data and the relations among these concepts. Protein Ontology has (1) a hierarchical classification of concepts represented as classes, from general to specific; (2) a list of attributes related to each concept, for each class; and (3) a set of relations between classes to link concepts in ontology in more complicated ways then implied by the hierarchy, to promote reuse of concepts in the ontology. Protein Ontology provides description for protein domains that can be used to describe proteins in any organism. Protein Ontology Framework describes: (1) Protein Sequence and Structure Information, (2) Protein Folding Process, (3) Cellular Functions of Proteins, (4) Molecular Bindings internal and external to Proteins and (5) Constraints affecting the Final Protein Conformation. Protein Ontology uses all relevant protein data sources of information. The structure of PO provides the concepts necessary to describe individual proteins, but does not contain individual protein themselves. A database based on PO acts as instance store for the PO. PO uses data sources include new proteome information resources like PDB, SCOP, and RESID as well as classical sources of information where information is maintained in a knowledge base of scientific text files like OMIM and from various published scientific literature in various journals. PO Database is represented using XML. PO Database at the moment contains data instances of following protein families: (1) Prion Proteins, (2) B.Subtilis, (3) CLIC and (4) PTEN. More protein data instances will be added as PO is more developed. The Complete Class Hierarchy of Protein Ontology (PO) is shown in **Figure 1**. More details about PO is available at the website: **http://www.proteinontology.info/**

Semantics in protein data is normally not interpreted by annotating systems, since they are not aware of the specific structural, chemical and cellular interactions of protein complexes. Protein Ontology Framework provides specific set of rules to cover these application specific semantics. The rules use only the relationships whose semantics are predefined to establish correspondence among terms in PO. The set of relationships with predefined semantics is: {SubClassOf, PartOf, AttributeOf, InstanceOf, and ValueOf}.

The PO conceptual modeling encourages the use of strictly typed relations with precisely defined semantics. Some of these relationships (like SubClassOf, InstanceOf) are somewhat similar to those in RDF Schema but the set of relationships

- *ProteinOntology*
    - *AtomicBind*
    - *Atoms*
    - *Bind*
    - *Chains*
    - *Family*
    - *ProteinComplex*
        - *ChemicalBonds*
            - *CISPeptide*
            - *DisulphideBond*
            - *HydrogenBond*
            - *ResidueLink*
            - *SaltBridge*
        - *Constraints*
            - *GeneticDefects*
            - *Hydrophobicity*
            - *ModifiedResidue*
        - *Entry*
            - *Description*
            - *Molecule*
            - *Reference*
        - *FunctionalDomains*
            - *ActiveBindingSites*
            - *BiologicalFunction*
                - *PathologicalFunctions*
                - *PhysiologicalFunctions*
            - *SourceCell*
        - *StructuralDomains*
            - *Helices*
                - *Helix*
                    - *HelixStructure*
            - *OtherFolds*
                - *Turn*
                    - *TurnStructure*
            - *Sheets*
                - *Sheet*
                    - *Strands*
        - *Structure*
            - *ATOMSequence*
            - *UnitCell*
    - *Residues*
    - *SiteGroup*

Fig. 1. Class Hierarchy of Protein Ontology

that have defined semantics in our conceptual PO model is small so as to maintain simplicity of the system. The following is a description of the set of pre-defined semantic relationships in our common PO conceptual model.

SubClassOf: The relationship is used to indicate that one concept is a subclass of another concept, for instance: SourceCell SubClassOf FunctionalDomains. That is any instance of SouceCell class is also instance of FunctionalDomains class. All attributes of FunctionalDomains class (_FuncDomain_Family, _FuncDomain_SuperFamily) are also the attributes of SourceCell class. The relationship SubClassOf is transitive.

AttrributeOf: This relationship indicates that a concept is an attribute of another concept, for instance: _FuncDomain_Family AttributeOf Family. This relationship also referred as PropertyOf, has same semantics as in object-relational databases.

PartOf: This relationship indicates that a concept is a part of another concept, for instance: Chain PartOf ATOMSequence indicates that Chain describing various residue sequences in a protein is a part of definition of ATOMSequence for that protein.

InstanceOf: This relationship indicates that an object is an instance of the class, for instance: ATOMSequenceInstance_10 InstanceOf ATOMSequence indicates that ATOMSequenceInstance_10 is an instance of class ATOMSequence.

ValueOf: This relationship is used to indicate the value of an attribute of an object, for instance: "Homo Sapiens" ValueOf OrganismScientific. The second concept, in turn has an edge, OrganismScientific AttributeOf Molecule, from the object it describes.

3 Comparing GO and PO

Gene Ontology (GO) [1] defines a structured controlled vocabulary in the domain of biological functionality. GO initially consisted of a few thousand terms describing the genetic workings of three organisms and was constructed for the express purpose of database interoperability; it has since grown to a terminology of nearly 16,000 terms and is becoming a de facto standard for describing functional aspects of biological entities in all types of organisms. Furthermore, in addition to (and because of) its wide use as a terminological source for database-entry annotation, GO has been used in a wide variety of biomedical research, including analyses of experimental data [1] and predictions of experimental results [7]. Characteristics of GO that we believe are most responsible for its success: community involvement; clear goals; limited scope; simple, intuitive structure; continuous evolution; active curation; and early use.

It is clear that organisms across the spectrum of life, to varying degrees, possess large numbers of gene products with similar sequences and roles. Knowledge about a given gene product (i.e., a biologically active molecule that is the deciphered end product of the code stored in a gene) can often be determined experimentally or inferred from its similarity to gene products in other organisms. Research into different biological systems uses different organisms that are chosen because they are amenable to advancing these investigations. For example, the rat is a good model for the study of human heart disease, and the fly is a good model to study cellular differentiation. For each of these model systems, there is a database employing curators who collect and store the body of biological knowledge for that organism. This enormous amount of data can potentially add insight to related molecules found in other organisms. A reliable wet-lab biological experiment performed in one organism can be used to deduce attributes of an analogous (or related) gene product in another organism, thereby reducing the need to reproduce experiments in each individual organism (which would be expensive, time-consuming, and, in many organisms, technically impossible). Mining of Scientific Text and Literature is done to generate list of keywords that is used as GO terms. However, querying heterogeneous, independent databases in order to draw these inferences is difficult: The different database projects may use different terms to refer to the same concept and the same terms to refer to different concepts. Furthermore, these terms are typically not formally linked with each other in any way. GO seeks to reveal these underlying biological functionalities by providing a structured controlled vocabulary that can be used to describe gene products, and shared between biological databases. This facilitates querying for gene products that share biologically meaningful attributes, whether from separate databases or within the same database.

Challenges faced while developing GO from unstructured and structured data sources are addressed while developing PO. Protein Ontology is a conceptual model that aim to support consistent and unambiguous knowledge sharing and that provide a framework for protein data and knowledge integration. PO links concepts to their interpretation, i.e. specifications of their meanings including concept definitions and relationships to other concepts. Apart from semantic relationships defined in Section 2, PO also model relationships like Sequences. By itself semantic relationships described in Section 2, does not impose order among the children of the node. In applications using Protein Sequences, the ability of expressing the order is paramount. Generally Protein Sequences are a collection of chains of sequence of residues, and that is the format Protein Sequences have been represented unit now using various data representations and data mining techniques for bioinformatics. When we are defining sequences for semantic heterogeneity of protein data sources using PO we are not only considering traditional representation of protein sequences but also link Protein Sequences to Protein Structure, by linking chains of residue sequences to atoms defining three-dimensional structure. In this section we will describe how we used a special semantic relationship like *Sequence(s)* in Protein Ontology to describe complex concepts defining Structure, Structural Folds and Domains and Chemical Bonds describing Protein Complexes. PO defines these complex concepts as *Sequences* of simpler generic concepts defined in PO. These simple concepts are *Sequences* of object and data type properties defining them. A typical example of *Sequence* is as follows. PO defines a complex concept of *ATOMSequence* describing three dimensional structure of protein complex as a combination of simple concepts of *Chains*, *Residues*, and *Atoms* as: *ATOMSequence Sequence (Chains Sequence (Residues Sequence (Atoms)))*. Simple concepts defining ATOMSequence are defined as: *Chains Sequence (ChainID, ChainName, ChainProperty)*; *Residues Sequence (ResidueID, ResidueName, ResidueProperty)*; and *Atoms Sequence (AtomID, Atom, ATOMResSeqNum, X, Y, Z, Occupancy, TempratureFactor, Element)*. Semantic Interoperability Framework used in PO is depicted **Figure 2**.

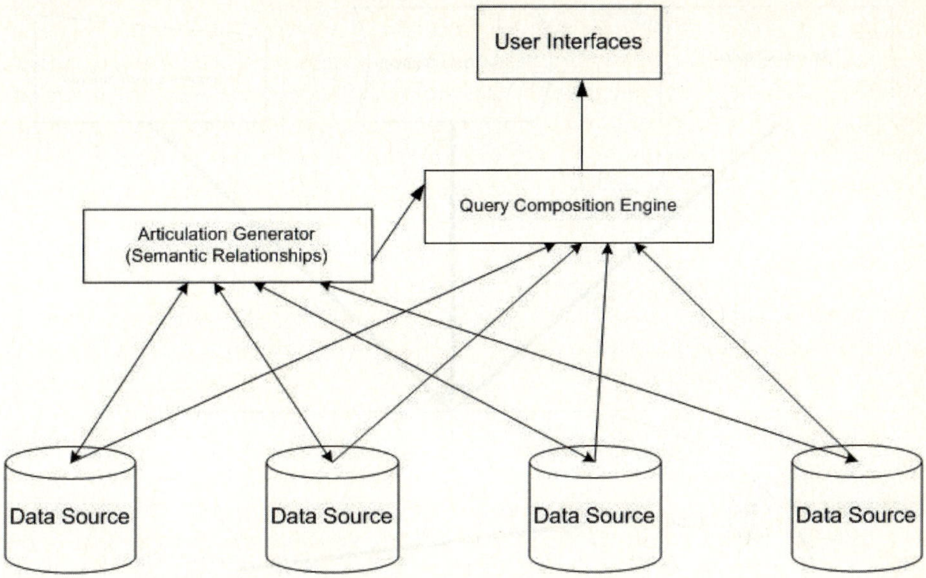

Fig. 2. Semantic Interoperability Framework for PO

Therefore, PO reflects the structure and relationships of Protein Data Sources. PO removes the constraints of potential interpretations of terms in various data sources and provides a structured vocabulary that unifies and integrates all data and knowledge sources for proteomics domain (**Figure 3**). There are seven subclasses of Protein Ontology (PO), called Generic Classes that are used to define complex concepts in other PO Classes: Residues, Chains, Atoms, Family, AtomicBind, Bind, and SiteGroup. Concepts from these generic classes are reused in various other PO Classes for definition of Class Specific Concepts. Details and Properties of Residues in a Protein Sequence are defined by instances of Residues Class. Instances of Chains of Residues are defined in Chains Class. All the Three Dimensional Structure Data of Protein Atoms is represented as instances of Atoms Class. Defining Chains, Residues and Atoms as individual classes has the benefit that any special properties or changes affecting a particular chain, residue and atom can be easily added. Protein Family class represents Protein Super Family and Family Details of Proteins. Data about binding atoms in Chemical Bonds like Hydrogen Bond, Residue Links, and Salt Bridges is entered into ontology as an instance of AtomicBind Class. Similarly the data about binding residues in Chemical Bonds like Disulphide Bonds and CIS Peptides is entered into ontology as an instance of Bind Class. All data related to site groups of the active binding sites of Proteins is defined as instances of SiteGroup Class. In PO the notions classification, reasoning, and consistency are applied by defining new concepts or classes from defined generic concepts or classes. The concepts derived from generic concepts are placed precisely into class hierarchy of Protein Ontology to completely represent information defining a protein complex.

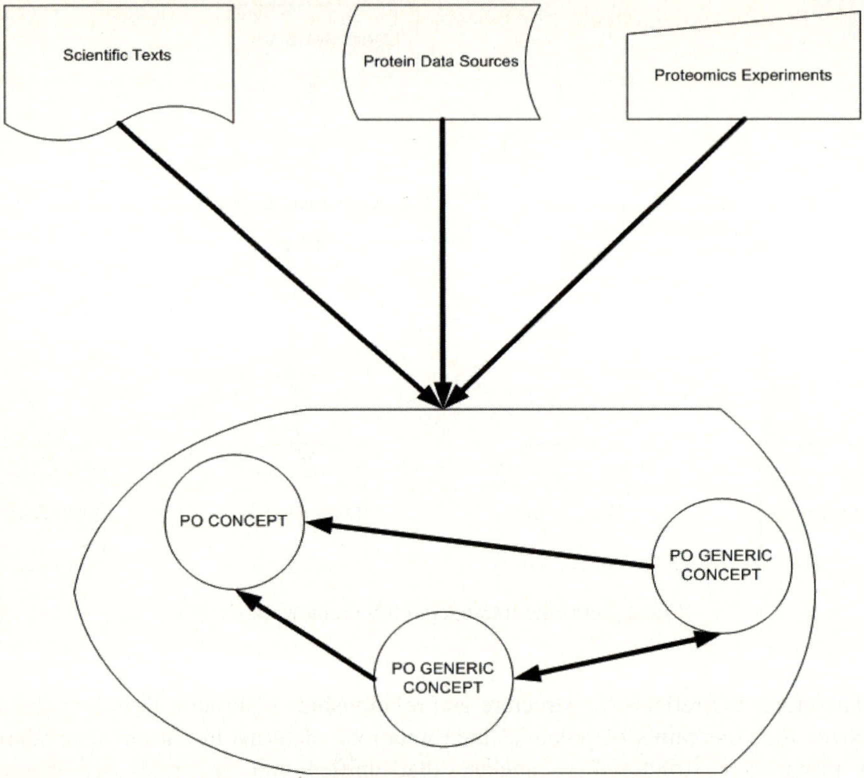

Fig. 3. Unification of Protein Data and Knowledge

As such PO can be used to support automatic semantic interpretation of data and knowledge sources, thus providing a basis for sophisticated mining of information.

4 Mining Facilitated by Protein Ontology

The Protein Ontology Database is created as an instance store for various protein data using the PO format. PO provides technical and scientific infrastructure to allow evidence based description and analysis of relationships between proteins. PO uses data sources like PDB, SCOP, OMIM and various published scientific literature to gather protein data. PO Database is represented using OWL. PO Database contains data instances of following protein families: Prion Proteins, B.Subtilis, and Chlorides. More protein data instances will be added as PO is more developed. The PO instance store at moment covers various species of proteins from bacterial and plant proteins to human proteins. Such a generic representation using PO shows the strength of PO format representation.

We used some standard hierarchical and tree mining algorithms [8] on the PO Database. We compared MB3-Miner (MB3), X3-Miner (X3), VTreeMiner (VTM) and PatternMatcher (PM) for mining embedded subtrees and IMB3-Miner (IMB3),

FREQT (FT) for mining induced subtrees of PO Data. In these experiments we are mining Prion Proteins dataset described using Protein Ontology Framework, represented in XML. For this dataset we map the XML tags to integer indexes. The

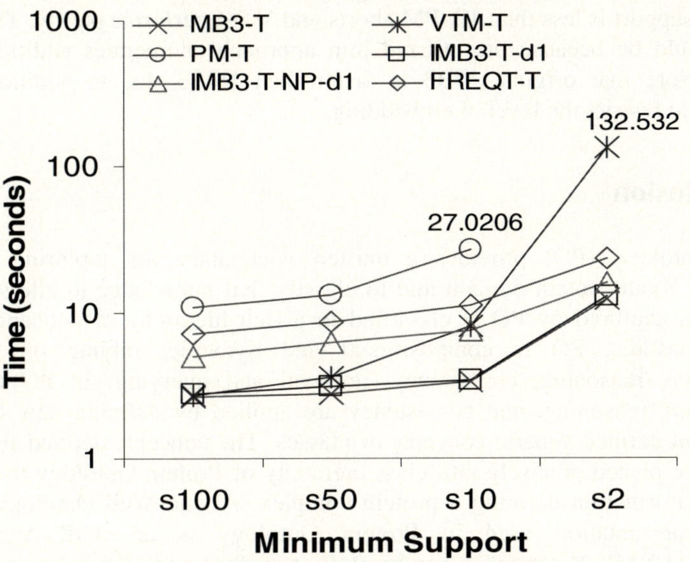

Fig. 4. Time Performance for Prion dataset of PO Data

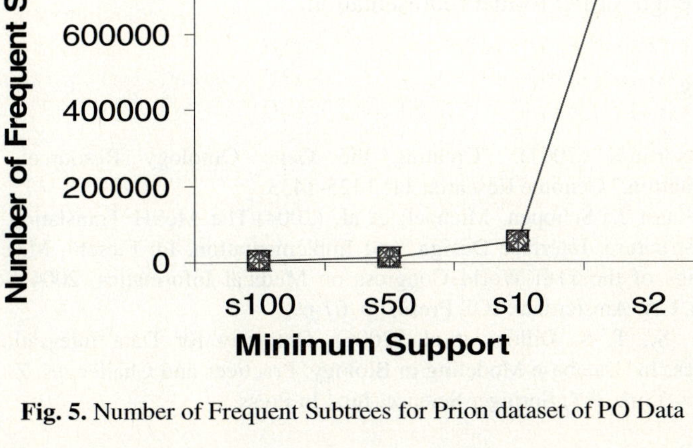

Fig. 5. Number of Frequent Subtrees for Prion dataset of PO Data

maximum height is 1. In this case all candidate subtrees generated by all algorithms would be induced subtrees. **Figure 4** shows the time performance of different algorithms. Our original MB3 has the best time performance for this data.

Quite interestingly, with Prion dataset of PO the number of frequent candidate subtrees generated is identical for all algorithms (**Figure 5**). Another observation is that when support is less than 10, PM aborts and VTM performs poorly. The rationale for this could be because the utilized join approach enumerates additional invalid subtrees. Note that original MB3 is faster than IMB3 due to additional checks performed to restrict the level of embedding.

5 Conclusion

Protein Ontology (PO) provides a unified vocabulary for capturing declarative knowledge about protein domain and to classify that knowledge to allow reasoning. Information captured by PO is classified in a rich hierarchy of concepts and their inter-relationships. PO is compositional and dynamic, relying on notions of classification, reasoning, consistency, retrieval and querying. In PO the notions classification, reasoning, and consistency are applied by defining new concepts or classes from defined generic concepts or classes. The concepts derived from generic concepts are placed precisely into class hierarchy of Protein Ontology to completely represent information defining a protein complex. As the Web Ontology Language (OWL) representation used in Protein Ontology is an XML-Abbrev based (Abbreviated XML Notation), it can be easily transformed to the corresponding RDF and XML formats without much effort using the available converters. Our Protein Ontology (PO) is the first ever work to integrate protein data based on data semantics describing various phases of protein structure. PO helps to understand structure, cellular function and the constraints that affect protein in a cellular environment. The attribute values in the PO are not defined as text strings or as set of keywords. Most of the Values are entered as instances of Concepts defined in Generic Classes. PO Database contains data instances of following protein families: Prion Proteins, B.Subtilis, and Chlorides. More protein data instances will be added as PO is more developed. The PO instance store at moment covers various species of proteins from bacterial and plant proteins to human proteins. Such a generic representation using PO shows the strength of PO format representation.

References

[1] GO Consortium (2001). "Creating the Gene Ontology Resource: Design and Implementation." Genome Research 11: 1425-1433.

[2] Nelson, Stuart J.; Schopen, Michael; et al. (2004).The MeSH Translation Maintenance System: Structure, Interface Design, and Implementation. In: Fieschi, M. et al., editors. Proceedings of the 11th World Congress on Medical Informatics; 2004 Sep 7-11; San Francisco, CA. Amsterdam: IOS Press; pp. 67-69.

[3] Sidhu, A. S., T. S. Dillon, et al. (2006). Ontology for Data Integration in Protein Informatics. In: Database Modeling in Biology: Practices and Challenges. Z. Ma and J. Y. Chen. New York, NY, Springer Science, Inc.: In Press.

[4] Sidhu, A. S., T. S. Dillon, et al. (2006). Protein Ontology Project: 2006 Updates (Invited Paper). Data Mining and Information Engineering 2006. A. Zanasi, C. A. Brebbia and N. F. F. Ebecken. Prague, Czech Republic, WIT Press.

[5] Sidhu, A. S., T. S. Dillon, et al. (2005). Ontological Foundation for Protein Data Models. First IFIP WG 2.12 & WG 12.4 International Workshop on Web Semantics (SWWS 2005). In conjunction with On The Move Federated Conferences (OTM 2005). Agia Napa, Cyprus, Springer-Verlag. Lecture Notes in Computer Science (LNCS).

[6] Sidhu, A. S., T. S. Dillon, et al. (2005). Protein Ontology: Vocabulary for Protein Data. 3rd IEEE International Conference on Information Technology and Applications (IEEE ICITA 2005). Sydney, IEEE CS Press. Volume 1: 465-469.

[7] GO Consortium and S. E. Lewis (2004). "Gene Ontology: looking backwards and forwards." Genome Biology 6(1): 103.1-103.4.

[8] Tan, H., T.S. Dillon, et. al. (2006). IMB3-Miner: Mining Induced/Embedded Subtrees by Constraining the Level of Embedding. Accepted for Proceedings of PAKDD 2006.

Ontology Fusion with Complex Mapping Patterns

Jinguang Gu and Yi Zhou

College of Computer Science and Technology,
Wuhan University of Science and Technology, Wuhan 430081, China
simongu@ieee.org, simongu@acm.org

Abstract. Ontology mapping plays a key role in information integration systems. This paper discusses the patterns of complex ontology mapping. It proposes the definition and properties of ontology mappings, and discusses the procedure and algorithm of ontology fusion with complex ontology mapping.

1 Introduction

The development of information technologies such as Internet technology, Web based commercial business technology and their applications, demands for complete access to available information, which is often heterogeneous and distributed. Problems that might arise due to heterogeneity of the data are already well known within the distributed database systems community: *structural heterogeneity* and *semantic heterogeneity*. Ontology based semantic integration is a possible approach to overcome the problem of semantic heterogeneity, it uses ontologies to describe the semantics of the information sources and to make the content explicit. The major bottleneck in semantic integration is the mapping discovery between different ontologies which represent different local concepts in the distributed and heterogeneous information environment. Heuristics-based, machine learning based[1] or Bayesian network based[2] methods were discussed in recent years, a survey of this kind of research is discussed in paper [3]. However, most of current research are focused on one to one ontology mapping, the patterns of complex ontology mapping is discussed insufficiently. This paper discuss a fusion mechanism with complex ontology mapping patterns.

2 The Patterns of Ontology Mapping

2.1 The Definition of Complex Ontology Mapping

The mapping patterns can be categorized into four expressions: direct mapping, subsumption mapping, composition mapping and decomposition mapping. A direct mapping relates the local ontology concept to global ontology concept directly, and the cardinality of direct mapping could be one-to-one. A subsumption mapping is used to denote concept inclusion relation especially in the multiple IS-A inclusion hierarchy. The composition mapping is used to map one concept to

B. Gabrys, R.J. Howlett, and L.C. Jain (Eds.): KES 2006, Part I, LNAI 4251, pp. 738–745, 2006.
© Springer-Verlag Berlin Heidelberg 2006

combined concepts. For example, the mapping *address=contact(country, state, city, street, postcode)* is a composition mapping, in which the concept *address* is mapped to combined concept "*contact, country, state, street, and postcode*" of local schema elements. The decomposition mapping is used to map a combined concept to one local concept. These four mapping patterns can be described in the figure 1.

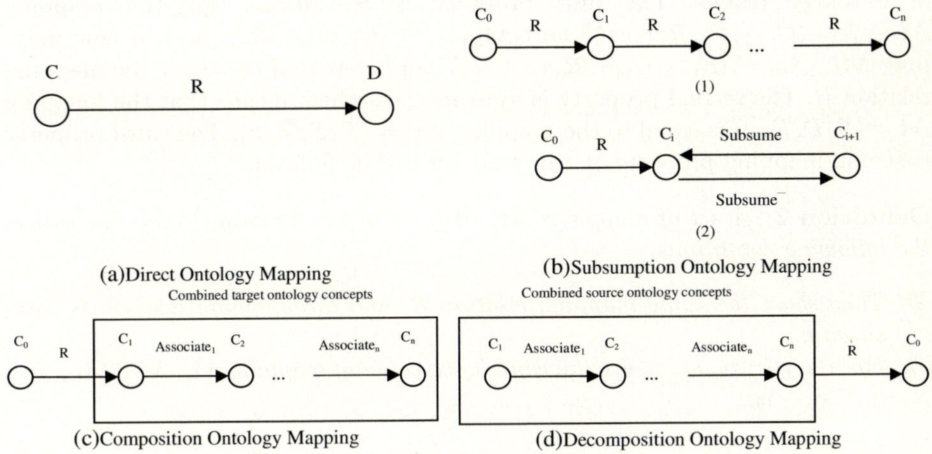

Fig. 1. The Patterns of Ontology Mapping

This paper extends the definitions of ontology mapping proposed in paper [4] with semantic similarity enhanced, and discussed the properties of ontology mappings. The patterns of the ontology mapping can be defined as:

Definition 1. *A* ***mapping*** *is a structure* $\mathcal{M} = (\mathcal{S}, \mathcal{D}, \mathcal{R}, v)$, *where* $\mathcal{S}$ *denotes the concepts of source ontology,* $\mathcal{D}$ *denotes the concepts of target ontology,* $\mathcal{R}$ *denotes the relation of the mapping and* v *denotes the confidence value of the mapping,* $0 \leq v \leq 1$.

A **subsumption ontology concept mapping** is a 6-tuple $\mathcal{S}_\mathcal{M} = (\mathcal{D}_m, \mathcal{R}_m, \mathcal{B}_m, \preceq_m, \mathcal{I}_m, v)$, where $\mathcal{D}_m$ is a direct mapping expression; $\mathcal{R}_m$ is the first target concept, which is the most specialized ontology concept. The mapping between the source ontology and $\mathcal{R}_m$ is denoted as **Root ontology concept mapping**; $\mathcal{B}_m$ is the last target concept, which is the most generalized ontology concept. The mapping between the source ontology and $\mathcal{B}_m$ is denoted as **Bottom ontology concept mapping**; $\preceq_m$ is inclusion relation between target ontology concepts; $\mathcal{I}_m$ is the inverse mapping. A **composition ontology concept mapping** is a 4-tuple $\mathcal{C}_\mathcal{M} = (\mathcal{F}_m, \mathcal{A}_m, \mathcal{B}_m, v)$, where $\mathcal{F}_m$ is a direct mapping expression; $\mathcal{A}_m$ is chaining of role(s) between target ontology concepts; $\mathcal{B}_m$ is the last target symbol, which is the node of chaining target role(s); A **decomposition ontology concept mapping** is a 4-tuple $\mathcal{C}_\mathcal{M} = (\mathcal{A}_m, \mathcal{B}_m, \mathcal{L}_m, v)$, where

$\mathcal{A}_m$ is chaining of role(s) between source ontology concepts; $\mathcal{B}_m$ is the last target symbol, which is the node of chaining source role(s); $\mathcal{L}_m$ is a direct mapping expression.

2.2 The Property of Ontology Mapping

It defines some properties of semantic mapping which are useful in the task of ontology fusion. The first property is *transitivity*, for the mapping $\mathcal{M}_{i-1,i} = (C_{i-1}, C_i, \mathcal{R}, v_{i-1,i})$ and $\mathcal{M}_{i,i+1} = (C_i, C_{i+1}, \mathcal{R}, v_{i,i+1})$, a new mapping $\mathcal{M}_{i-1,i+1} = (C_{i-1}, C_{i+1}, \mathcal{R}, v_{i-1,i+1})$ can be created to satisfy the mapping relation R. The second property is **symmetric**, which means that the mapping $\mathcal{M} = (\mathcal{S}, \mathcal{D}, \mathcal{R}, v)$ is equal to the mapping $\mathcal{M}' = (\mathcal{D}, \mathcal{S}, \mathcal{R}, v)$. The third property is strong mapping property, it can be described as follows.

Definition 2. *A set of mappings $\mathcal{M}_i$ $(0 \leq i \leq n)$are **strong** if they can satisfy the following conditions:*

i). They share the same mapping relation R, and the mapping relation is transitivity;
ii). For $\forall(i, j, k)$,v_i, v_j, v_k are the confidence value of mapping $\mathcal{M}_i, \mathcal{M}_j, \mathcal{M}_k$, then $v_i \leq v_j + v_k$.

3 The Ontology Fusion Mechanism

Some concepts are introduced before discussing the ontology fusion mechanism, the first one is **Fusion Ontology**, it can be expressed as[5]:

Definition 3. *Suppose O_i $(1 \leq i \leq n)$are n different ontologies and suppose **IC** is a finite set of interoperation constraints, an ontology O is said to be an **fusion** of O_i $(1 \leq i \leq n)$ iff there are n injective mapping $\mathcal{M}_i(1 \leq i \leq n)$from O_i $(1 \leq i \leq n)$ respectively to O such that:*

i. $(\forall i \in \{1, 2, \ldots, n\})(O_i : x \preceq O_i : y \rightarrow \mathcal{M}_i(O_i : x) \preceq \mathcal{M}_i(O_i : y))$;
ii. $(\forall x, y)((x \in O_i \wedge y \in O_j) \wedge (x \;\mathbf{op}\; y) \in \mathbf{IC} \rightarrow (\mathcal{M}_i(x) \;\mathbf{op}\; \mathcal{M}_j(y)) \in \mathbf{IC})$.

Axiom *i*) above says that the fusion ontology must preserve the ordering associated with each of the input ontology; Axiom *ii*) above says that they must preserve the interoperation constraints.

Definition 4. ***Fusion Connection*** *is a structure $\mathcal{F}_c(O_1 : C_1, O_2 : C_2, \mathcal{M})$, where C_1 denotes a concept or concept set of ontology O_1, C_2 denotes a concept or concept set of Ontology O_2, $\mathcal{M}$ denotes the direct mapping relationship between C_1 and C_2.*

The function of Ontology fusion is to add connection tag between the concepts which has direction mapping relationship. In direction mapping $\mathcal{M} = (\mathcal{S}, \mathcal{D}, \mathcal{R}, v)$, the fusion connection adds connection tag between the elements of

$\mathcal{S}$ and $\mathcal{D}$; in subsumption mapping $\mathcal{S_M} = (\mathcal{D}_m, \mathcal{R}_m, \mathcal{B}_m, \preceq_m, \mathcal{I}_m, v)$, the fusion connection adds connection tag between the concepts which have mapping relation $\mathcal{D}_m$, and in composition mapping $\mathcal{C_M} = (\mathcal{F}_m, \mathcal{A}_m, \mathcal{B}_m, v)$ or decomposition mapping $\mathcal{C_M} = (\mathcal{A}_m, \mathcal{B}_m, \mathcal{L}_m, v)$, the fusion connection adds connection tag between the concepts which have relations $\mathcal{F}_m$ or $\mathcal{L}_m$ respectively. $\mathcal{F}_{cd}$, $\mathcal{F}_{cs}$ and $\mathcal{F}_{cc}$ are used to denote fusion connection of direct mapping, subsumption mapping and composition or decomposition mapping respectively. The *fusion connection list* is a list and its elements are the fusion connections, denotes it as FL.

1. The first step of ontology fusion is **Ontology Fusion for Direct Mapping**, which creates fusion list FL from the mapping list of different local ontologies, it can be described by algorithm 1.

Algorithm 1. Direct_Fusion(S_M)

Input: S_M is the set of mapping
Output: FL is the fusion connection list

$FL \leftarrow \emptyset$;
foreach M *in* S_M **do**
  switch M **do**
    case M *belongs to* $\mathcal{M}$
      // if the mapping is a direct mapping.
      $FL \leftarrow FL + \mathcal{F}_{cd}(O_1 : C_1, O_2 : C_2, M)$, $O_1 : C_1 \in \mathcal{S} \wedge O_2 : C_2 \in \mathcal{D}$;
    case M *belongs to* $\mathcal{S_M}$
      // if the mapping is a subsumption mapping,
      // and $\mathcal{D}_m$ is the direct mapping expression of $\mathcal{S_M}$
      let $\mathcal{D}_m = (\mathcal{S}, \mathcal{D}, \mathcal{R}, v)$;
      $FL \leftarrow FL + \mathcal{F}_{cs}(O_1 : C_1, O_2 : C_2, \mathcal{D}_m)$, $O_1 : C_1 \in \mathcal{S} \wedge O_2 : C_2 \in \mathcal{D}$;
    case M *belongs to* $\mathcal{C_M}$
      // if the mapping is a composition mapping,
      // and $\mathcal{F}_m$ is the direct mapping expression of $\mathcal{C_M}$
      let $\mathcal{F}_m = (\mathcal{S}, \mathcal{D}, \mathcal{R}, v)$;
      $FL \leftarrow FL + \mathcal{F}_{cc}(O_1 : C_1, O_2 : C_2, \mathcal{F}_m)$, $O_1 : C_1 \in \mathcal{S} \wedge O_2 : C_2 \in \mathcal{D}$;
    case M *belongs to* $\mathcal{D_M}$
      // if the mapping is a decomposition mapping,
      // and $\mathcal{L}_m$ is the direct mapping expression of $\mathcal{D_M}$
      let $\mathcal{L}_m = (\mathcal{S}, \mathcal{D}, \mathcal{R}, v)$;
      $FL \leftarrow FL + \mathcal{F}_{cc}(O_2 : C_2, O_1 : C_1, \mathcal{L}_m)$, $O_1 : C_1 \in \mathcal{S} \wedge O_2 : C_2 \in \mathcal{D}$;
    otherwise
      Errors Handler;
    end
  end
end
return FL

2. The second step of ontology fusion is **Ontology Fusion for Complex Semantic Mapping**, which is used to find the mappings of the concept that are

not in the mapping list, the basic idea of this step is to find the semantic similarity of the mapping relations, and create new mappings between this relations. This step can be discussed in two situations, one is to find the mapping between two subsumption mapping relations, and the other is to find the mapping between two composition mapping relations or decomposition mapping relations.

Subsumption mapping situation is discussed at first, suppose $\mathcal{M}_1$ and $\mathcal{M}_2$ are two subsumption mappings, and R_1 and R_2 are the mapping relations of $\mathcal{M}_1$ and $\mathcal{M}_2$ respectively, C_{10} and C_{20} are the concepts of source ontology, C_{1i} $(1 \leq i \leq n)$ and C_{2j} $(1 \leq j \leq m)$ are the concepts of target ontology. If there exists $JS(R_1, R_2) \geq 1 - \epsilon$, which means that the two relations of the mappings can match each other, so creates a new mapping named M' and creates the fusion connection list FL with algorithm 2.

Algorithm 2. Subsumption_Fusion(M_1, M_2, ϵ)

Input: M_1, M_2 are two subsumption mappings, ϵ is the threshold of semantic similarity
Output: FL is the fusion connection list

$i = 1, j = 1, FL \leftarrow \emptyset$;
foreach C_{2j} *in* M_2 **do**
  switch *the semantic similarity between* C_{1i} *and* C_{2j} **do**
    case $JS(C_{1i}, C_{2j}) \geq 1 - \epsilon$
      // create the direct fusion connection.
      $FL \leftarrow FL + \mathcal{F}_{cd}(C_{1i}, C_{2j}, M')$, $C_{1i} \in M_1$;
      $i = i + 1$;
    case $MSP(C_{1i}, C_{2j}) \geq 1 - \epsilon$
      // create the subsumption fusion connection.
      $FL \leftarrow FL + \mathcal{F}_{cs}(C_{1i}, C_{2j}, M')$, $C_{1i} \in M_1$;
    case $MGC(C_{1i}, C_{2j}) \geq 1 - \epsilon$
      // create the subsumption fusion connection.
      $FL \leftarrow FL + \mathcal{F}_{cs}(C_{2j}, C_{1i}, M')$, $C_{1i} \in M_1$;
    otherwise
      $i = i + 1$;
    end
  end
end
return FL

Then the composition and decomposition mapping situation are discussed, this kind of mapping is very complex because the concatenate between two concepts of target ontology is different with each other, the concatenate of target concepts are divided into two kinds. One kind of concatenate does not have meanings, just expresses that there exist concatenate between these concepts (maybe most of the concatenate can be expressed as this kind), for example the mapping: *address*=*contact*(*country*, *state*, *city*, *street*, *postcode*), the concatenate among *contact*, *country*, *state*, *city*, *street* and *postcode* are this kinds of concatenate, this kinds of concatenate is named as **None-Meanings Concatenate**;

The other kind of concatenate does have meanings between the concepts, for example the mapping: *payment* = *total*(*commodity_amount* * (*unit_price* + *tax*)) is this kind of concatenate, named it as **Full-Meanings Concatenate**. For the first kind concatenate, it only need to identify whether the related concepts satisfy the definition of semantic similarity or not, if it satisfies the definition, adds a new fusion connection to the fusion connection list. And it need to introduce the concept of Fusion Equivalence and Fusion Subsumption.

For composition (or decomposition) mappings $\mathcal{M}_1$ and $\mathcal{M}_2$, $\mathcal{A}_{m1}$ and $\mathcal{A}_{m2}$ represents the concatenate relationship of $\mathcal{M}_1$ and $\mathcal{M}_2$ respectively, it uses $|\mathcal{A}_{m1}|$ and $|\mathcal{A}_{m2}|$ to denote the amount of the concepts in $\mathcal{A}_{m1}$ and $\mathcal{A}_{m2}$, $\mathcal{A}_{m1}$ and $\mathcal{A}_{m2}$ are **Fusion Equivalence** iff,

i). $\mathcal{A}_{m1}$ and $\mathcal{A}_{m2}$ are Full-Meaning Concatenates, and $|\mathcal{A}_{m1}|=|\mathcal{A}_{m2}|$;
ii). For $\forall i(C_{1i} \sqsubseteq \forall Concatenate_{1,i+1}.C_{1,i+1}, C_{2i} \sqsubseteq \forall Concatenate_{2,i+1}.C_{2,i+1})$ exists $(JS(C_{1i}, C_{2i}) \geq 1-\epsilon) \wedge (JS(C_{1,i+1}, C_{2,i+1}) \geq 1-\epsilon) \wedge (JS(Concatenate_{1,i+1}, Concatenate_{2,i+1}) \geq 1-\epsilon)$.

It uses $\mathcal{A}_{m1} \Leftrightarrow \mathcal{A}_{m2}$ to denote the fusion equivalence of $\mathcal{A}_{m1}$ and $\mathcal{A}_{m2}$, if $|\mathcal{A}_{m1}| > |\mathcal{A}_{m2}|$, suppose $|\mathcal{A}_{m1}| = m$ and $|\mathcal{A}_{m2}| = n$, then $m > n$. If $\mathcal{A}_{m1,n}$ is the first n concepts of $\mathcal{A}_{m1}$ and $\mathcal{A}_{m1,n} \Leftrightarrow \mathcal{A}_{m2}$, then we say $\mathcal{A}_{m1}$ and $\mathcal{A}_{m2}$ are **fusion subsumption**, denotes it as $\mathcal{A}_{m1} \Rightarrow \mathcal{A}_{m2}$, or $\mathcal{A}_{m2} \Rightarrow \mathcal{A}_{m1}$ if $n > m$.

It uses AU_1 and AU_2 to denote the concepts of $\mathcal{A}_{m1}$ and $\mathcal{A}_{m2}$ respectively, and whose concatenate is None-Meanings Concatenate, use AN_1 and AN_2 to denote the concepts whose concatenate is Full-Meanings concatenate, use AE_1 and AE_2 to denote the concepts of AN_1 and AN_2, and which keep fusion equivalence or fusion subsumption relationship with each other. the fusion connection can be expressed with algorithm 3.

3. The last step of ontology fusion is **Canonical Fusion**, which merges the concepts of the fusion connection into one concept if the fusion connection type is $\mathcal{F}_{cd}$ or $\mathcal{F}_{cc}$, and add a real relation connection to the concepts if the fusion connection type is $\mathcal{F}_{cs}$. For example the fusion connection $\mathcal{F}_{cd} = (C_1, C_2, M)$, C_1 and C_2 are concepts of different ontologies, merge it to a concept (C_1, C_2) and all the hierarchy of the concepts will be kept. But not all the concepts with the same mapping relation can be merge into one concepts, only the concepts which have strong mapping relation can be merged. For example, if the mappings $\mathcal{M}(C_1, C_2, \mathcal{R}, v_1)$,$\mathcal{M}(C_2, C_3, \mathcal{R}, v_2)$ and $\mathcal{M}(C_1, C_3, \mathcal{R}, v_3)$ satisfy the strong mapping property, we can merge the concept C_1, C_2, C_3 into one concept (C_1, C_2, C_3), otherwise, we have to merge them into two concepts (C_1, C_2) and (C_2, C_3). The result of canonical fusion is the global ontology of mediator.

4 Comparison with Related Works

It compares this mechanism with other three approaches, the comparison is based on four different categories, named *mapping method*, *complex mapping support*, *pattern support*, and *semantic similarity support*. As shown in table 1, different approach uses different method to map and integrate ontology, IF-Map[6]

Algorithm 3. Composition_Fusion(M_1, M_2, ϵ)

Input: M_1, M_2 are two composition(decomposition) mappings, ϵ is the threshold of semantic similarity
Output: FL is the fusion connection list

$i = 1, j = 1, FL \leftarrow \emptyset$;
// Process the Full-Meanings concatenate
foreach $C_1 \in AE_1, C_2 \in AE_2$ **do**
    if $C_1 \Leftrightarrow C_2$ **then**
        // C_1 and C_2 are fusion equivalence.
        $FL \leftarrow FL + \mathcal{F}_{cc}(C_1, C_2, M')$;
    else
        if $|C_1| > |C_2|$ **then**
            let $C_1' \subset C_1$ and $C_1' \Leftrightarrow C_1$, $FL \leftarrow FL + \mathcal{F}_{cc}(C_1', C_2, M')$;
        else
            let $C_2' \subset C_2$ and $C_1 \Leftrightarrow C_2'$, $FL \leftarrow FL + \mathcal{F}_{cc}(C_1, C_2', M')$;
        end
    end
end
// Process the None-Meanings concatenate
foreach $C_{2j} \in AU_2$ **do**
    foreach $C_{1i} \in AU_1$ **do**
        if $JS(C_{1i}, C_{2j}) \geq 1 - \epsilon$ **then** $FL \leftarrow FL + \mathcal{F}_{cc}(C_{1i}, C_{2j}, M')$
    end
end
return FL

and FCA-Merge[7] are the most mature methods which have been accepted in knowledge management community widely, OBSA focuses on a mediator-wrapper based distributed environment, just like FCA-Merge approach, it uses the bottom-up mapping method to integrate different local ontologies. Most of the approaches do not define the ontology mapping patterns, and they can not define the relationship between the local ontologies and the integrated global ontologies with formal method, and this is the reason why they can not apply ontology integration to support semantic level query rewriting, MBL[4] approach defines the mapping patterns, but it uses logic mapping method.

Table 1. Comparison of proposed approach with related works

	mapping method	complex mapping	pattern	semantic similarity
IF-MAP	Information Flow	NA	NA	Support
FCA-Merge	Bottom-Up	NA	NA	Support
MBL	Logic	Support	Support	Not Support
OBSA	Bottom-Up	Support	Support	Support

5 Discussion and Conclusion

It discusses the patterns of complex ontology mapping based on semantic similarity, and elaborate on the algorithm of ontology fusion. The advantage of the fusion ontology with these patterns can be expressed as: (1)It can match the semantic similar concepts more exactly, especially for the concepts which is a part of concept hierarchy; (2)It can reduce the semantic inconsistent by solving problem semantic absent. (3)It can reduce the redundancy of the global ontology by finding more semantic matching in subsumption and composition (decomposition) mappings. A mediator-based implementation of the mechanism is introduced in the paper [8].

References

1. Doan, A., Madhavan, J., Domingos, P., Halevy, A.: Learning to map between ontologies on the semantic web. In D.Lassner, D.D.Roure, A., ed.: Proceedings of the eleventh international conference on World Wide Web, ACM Press (2002) 662–673
2. Pan, R., Ding, Z., Yu, Y., Peng, Y.: A Bayesian Network Approach to Ontology Mapping. In: Proceedings of the 4th International Semantic Web Conference. (2005) 1–15
3. Kalfoglou, Y., Schorlemmer, M.: Ontology Mapping: The State of the Art. The Knowledge Engineering Review **18** (2003) 1–31
4. KWON, J., JEONG, D., LEE, L.S., BAIK, D.K.: Intelligent semantic concept mapping for semantic query rewriting/optimization in ontology-based information integration system. International Journal of Software Engineering and Knowledge Engineering **14** (2004) 519–542
5. Hung, E., Deng, Y., V.S.Subrahmanian: TOSS: An Extension of TAX with Ontologies and Simarity Queries. In G.Weikum, ed.: Proceedings of the 2004 ACM SIGMOD international conference on Management of data, Paris, France, ACM Press (2004) 719–730
6. Kalfoglou, Y., Schorlemmer, M.: IF-Map: An Ontology-Mapping Method Based on Information-Flow Theory. Journal on Data Semantics, LNCS **2800** (2003) 98–127
7. Stumme, G., Maedche, A.: FCA-MERGE: Bottom-up merging of ontologies. In: Seventeenth International Joint Conference on Artificial Intelligence, Seattle, WA (2001) 225–234
8. Gu, J., Chen, H., Yang, L., Zhang, L.: OBSA:Ontology-based Semantic Information Processing Architecture. In Liu, J., Cercone, N., eds.: Proceedings of IEEE/WIC/ACM International Conference on Web Intelligence 2004, Beijing, IEEE Computer Society Press (2004) 607–610

Adaptive QoS for Using Multimedia in e-Learning

Hee-Seop Han and Hyeoncheol Kim

Korea University
Department Of Computer Science Education, College of Education
Anam-dong, Seoul, 136-701, Korea
anemon@korea.com, hkim@comedu.korea.ac.kr

Abstract. Web-based instruction has a lot of potential power in constructivism learning theory, and multimedia-based learning is able to provide natural learning environments. However, existing e-learning platforms have a limitation to provide multimedia learning objects on the web because of network traffic problems, which make it difficult to use a large amount of multimedia learning objects or a synchronous video streaming in e-learning. In this paper, we propose an adaptive QoS architecture to use the large amount of multimedia and realtime multimedia pedagogically for e-learning. First, we introduce the pedagogical issues and adaptive issues. And then we describe the adaptive strategies and architecture to support adaptive QoS. The purpose of this paper is to contribute towards understanding how to support adaptive QoS for service-oriented learning environments of e-learning.

1 Introduction

e-Learning has a lot of potential power in education, and constructivism learning is expected to be realized in computer-based learning environments. There have been reported many successful stories of e-Learning in enterprize education as well as K-12 education. There have been also many studies and standardizations about learning objects and meta-data in e-Learning(i.e. SCORM, RLO, NLO)[2].

The studies of Najjar *et al.* show that computer-based multimedia helps people to learn more information in less time than traditional classroom lectures, and he presented several useful methods to use multimedia resources [9][10][11]. According to these studies, we can find important facts that multimedia can be more effective when it is used to learners with low prior knowledge or aptitude in the domain being learned. And multimedia helps learners to build cognitive models of abstract knowledge. To use the multimedia effectively, we should consider cognitive models as well as prior knowledge of learners for the learning subjects. From these facts we think that the use of multimedia should be considered with pedagogical goals.

In this paper, we propose an adaptive QoS for multimedia-based e-learning that uses multimedia learning objects with lots of bandwidth. We focus on two issues as follows:

B. Gabrys, R.J. Howlett, and L.C. Jain (Eds.): KES 2006, Part I, LNAI 4251, pp. 746–752, 2006.
© Springer-Verlag Berlin Heidelberg 2006

– Pedagogical issue to provide learning objects according to learner's prior knowledge.
– Adaptive QoS issue that includes learning environments such as network type, bandwidth and leaner's computer capability.

We also describe strategies and architecture for adaptive QoS that will improve quality of multimedia-based e-Learning.

2 Main Issues and Service Framework

2.1 Pedagogical Issues

Many studies have shown that multimedia-based learning improves effectiveness of learning. However, what is more important for effective multimedia-based learning is about how to use multimedia objects pedagogically, not just simple use of multimedia itself [9]. Our pedagogical issues in this paper address which learning contents should be provided and how natural learning environments should be built up for learners. These pedagogical issues are associated with the next adaptive issues closely. In the proposed architecture, we consider the following pedagogical issues;

– **Prior knowledge Level.** Najjar *et al.* showed that multimedia information appears to be more effective for learners with low prior knowledge or aptitude in the domain being learned [9]. When the learners try to construct new knowledge, visual materials are very effective to make learners to understand and memorize. In order to present pedagogically appropriate learning objects, e-Learning system such as LMS(learning management system) must take the learner's level of prior knowledge into consideration.
– **Natural Learning.** Natural learning is supported by constructivism theory, and it is known that collaboration and visualization are good for natural learning [6][7][8][12][14]. Multimedia is useful to consist of natural learning environments because it can support effective visualization for learners and A/V-based collaboration. In e-Learning environment, it is important that learners are able to construct new knowledge for themselves through natural learning.

When multimedia is utilized for learning, this media can support dual coding of information, in which information is processed through one of two generally independent channels such as verbal and non-verbal information [5]. According to the Najjar *et al.*, poor readers performed better with multimedia illustration while good readers were not affected by the illustrations [11]. Therefore, it is important to consider pedagogically which multimedia objects should be provided to learners according to their individual profiles.

2.2 Adaptive QoS Issues

The learning materials that we consider in this paper are multimedia learning objects and streaming data for on-line conference. The video-audio conferencing is synchronous learning objects while most learning objects in e-Learning are generally asynchronous. Both cases suffer bandwidth problems in order to support these multimedia data on good service. Allison *et al.* showed that there is network-bandwidth differences in multimedia sources (in Table 1) and estimated possibility of adaptive QoS through using the agents and traffic repository[1]. Our adaptive QoS issues are the duration and ubiquity for good service regardless of the network bandwidth problems.

- **Duration.** The goal of our QoS is to support the stable learning environments and the suitable learning objects for learners in the given network environments. The proposed architecture is preserved for service duration adaptively in spite of the changes of network environments and the difference of learner's client system such as PC, notebook, mobile phone, PDA, etc. [1][4].
- **Ubiquity.** More flexible access to educational resources should be provided, often referred to as 'anytime/anywhere' learning. And multiple different types of devices, interfaces, and network connection types should be supported where possible [1].

Table 1. Comparison of sample multimedia related to network-bandwidth[1]

Resource	Bandwidth	Rate	Delay	Jitter	Loss Tolerant
Video	High	15fps	<250ms	Yes	Yes
Audio	Low-Medium	8000hz	<250ms	Yes	Yes

2.3 Service Framework

Our service framework is designed to take the pedagogical and adaptive issues into consideration.

When using the multimedia synchronously such as conferencing, traffic information of network bandwidth is updated immediately. We propose the solution that collects and utilizes the traffic information through learner's agents in the network environment based on RTP Session. *Traffic data repository* stores the traffic information that includes location information of traffic multimedia data to support the adaptive service statically. Allison *et al.* showed that RTCP sender and receiver were able to report location information to be obtained congestion information from realtime traffic[1]. RTCP sender and receiver were the agents that locate learner's computer and management server, respectively. This system helped good collaborative environments to be kept continuously using the traffic information.

Secondly, it is the asynchronous case when multimedia learning objects in educational digital library or LMS are used. Learning objects include meta-data concerning subjects, difficulty, file size, etc. There are several meta-data standardization such as SCO, RLO, NLO, etc. Furthermore, inner meta-data of

H.Shi *et al.* was designed to include the adaptive elements. The adaptation meta-data can be included in the meta-data sub-layer[15]. If the required bandwidth information of learning objects were stored in the adaptation meta-data, we can serve suitable learning objects according to leaner's knowledge level as well as multimedia based learning environments can be maintained in spite of change of network traffic circumstance.

3 Strategies for Adaptive QoS

The adaptive QoS addresses how to select and provide the suitable learning objects based on the learner's knowledge level and network status. In this paper we propose the strategies that connect learning objects to learners dynamically based on network environments. To realize this strategies, we construct three different models such as *learner model*, *learning object model* and *network model.*

- **Learner Model** consider the prior knowledge level of learner on given domain. In other words, *learner model* is constructed by meta-data which is stored at learner's profile as learning history. This model includes learner's knowledge level in a given domain. If multimedia learning objects should be supported according to the knowledge level of *learner model*, a suitable *learning object model* is linked by *network model* dynamically. By the Najjar *et al.* learner's prior knowledge level become a basis for decision whether learner is supported multimedia learning objects or not [11].
- **Network Model** is numerically calculated with traffic information of network bandwidth and capability of client computer. These information is collected through RTCP and stored into the *traffic data repository* by learner's agents as time-stamped data. This model is constructed by these information such as learner's location, previous bandwidths and learner's system capability at time-stamped history(stored at Traffic Data Repository in figure 1). Allison *et al.* showed that *Network model* will provide users with the most suitable multimedia learning objects based on several bandwidths[1].
- **Learning Object Model** includes bandwidth information of learning objects. In the same subject, there are several learning objects according to the required bandwidth. Adaptation is to select *Learning object model* according to *learner model* and *network model.*

Adaptation Strategy is summarized briefly as follows:

- **Strategy to provide the most suitable learning objects** is to connect the most suitable *learning object model* to *learner model* adaptively. From learner's profile, we can estimate the learner's prior knowledge level in a given learning domain. When connecting *learning object model* to *learner model*, we take the learner's prior knowledge level into consideration pedagogically.
- **Strategy to provide the most stable learning environments** guarantees that user can learn continuously regardless of network bandwidth status. In the *traffic data repository*, the adaptive QoS manager(see figure 1) finds

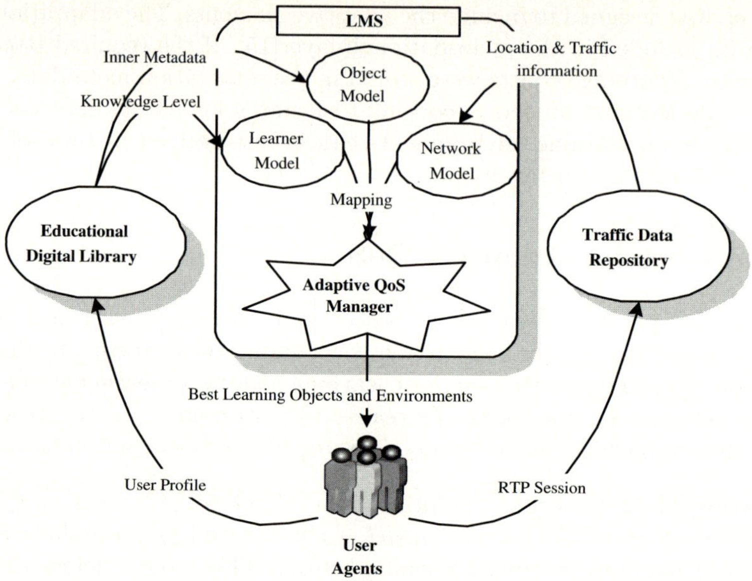

Fig. 1. The proposed adaptive QoS manager

the stable bandwidth for learning duration compared with *network model* and *learning object model.* It is difficult to adapt the change of bandwidth dramatically, but it is possible to find the stable bandwidth using *traffic data repository* according to Allison *et al.*[1].

In conclusion, the adaptive strategies are to provide suitable learning objects for learner pedagogically as well as to guarantee the learning duration in stable environments.

4 Architecture

A proposed system architecture is based on distributed environments of learners and learning objects. There are several studies to propose the architecture of e-Learning Grid platform and to provide adaptive QoS for multimedia data [1][3][4][13]. This architecture is proposed for a successful combination of adaptive grid-based e-Learning platform and adaptive QoS for multimedia-based learning.

Figure 2 describes the proposed architecture for adaptive QoS in e-Learning. This architecture is based on the grid environment which facilitates solutions for a wide range of application and the agent systems which manages an adaptive QoS for multimedia-based learning.

To guarantee the stable bandwidth in multimedia-based learning as well as web-based realtime conferencing, learner's agent stores the feasible bandwidth at *traffic data repository* continuously. *Adaptive QoS manager* finds out the stable

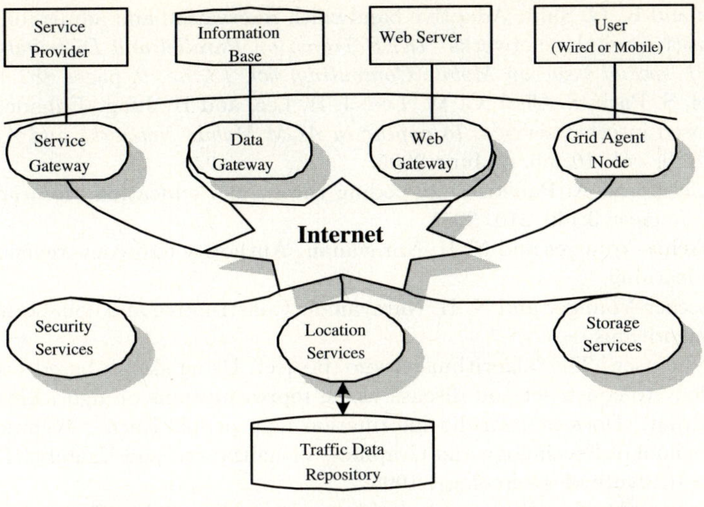

Fig. 2. The architecture for adaptive QoS in e-Learning

bandwidth to be calculated using the gathered bandwidth information statistically. Of course when multimedia learning objects are supported to learner, the learner's prior knowledge level must be considered as an educational point of view by *Adaptive QoS manager* .

5 Conclusion

This paper presents grid computing-based adaptive QoS for multimedia-based e-learning using A/V-based realtime media or multimedia learning objects with lots of bandwidth in pedagogical approach. In order to solve problems, we introduced the pedagogical issues and adaptive issues. We considered two factors for the issues. The first one is to provide the suitable multimedia learning objects pedagogically because it is important to consider the learner's prior knowledge level for multimedia-based learning. Another is to guarantee the learning duration in spite of network traffic because it is important to learn just in time.

We proposed the adaptive QoS architecture for multimedia-based e-Learning in grid computing environments. Since it is to provide multimedia contents adaptively according to the *learner model*, *learning object model* and *network model*, we can take the pedagogical issues and adaptive issues into account.

References

1. C. Allison, M. Bateman, R. Nicoll, and A. Ruddle. Adaptive qos for collaborative service-oriented learning environments. In *Proceeding of CCGrid2004*, 2004.
2. M. D. Bas, B. Giesbers, J. Janssen, J. Daniels, and R. Koper. Transforming existing content into reusable learning objects.

3. S. Choi and K. G. Shin. Adaptive bandwidth reservation and admission control in qos-sensitive cellular networks. *IEEE Trans. on Parallel and Distributed Systems (TPDS), special issue on Mobile Computing, vol. 13, no. 9*, pages 882–897, 2002.
4. Y. Choi, S. Park, S. Choi, G. W. Lee, J. H. Lee, and H. Jung. Enhancement of a wlan-based internet service. *to appear in ACM Mobile Networks and Applications (MONET), vol. 10, no. 3*, June 2005.
5. J. M. Clark and A. Paivio. Dual coding theory and education. *Educational Psychology Review*, 3:149–210, 2001.
6. T. Hubscher-Younger and N. H. Narayanan. Authority and convergence in collaborative learning.
7. T. Hubscher-Younger and N. H. Narayanan. Constructive and collaborative learning of algorithms.
8. C. Hundhausen. The 'algorithms studio' project: Using sketch-based visualization technology to construct and discuss visual representations of algorithms.
9. L. J. Najjar. Does multimedia information help people learn? Technical Report 95-28, School of Psychology and Graphics, Visualization, and Usability Laboratory Georgia Institute of Technology, 1995.
10. L. J. Najjar. Dual coding as a possible explanation for the effects of multimedia on learning. Technical Report 95-29, School of Psychology and Graphics, Visualization, and Usability Laboratory Georgia Institute of Technology, 1995.
11. L. J. Najjar. The effects of multimedia and elaborative encoding on learning. Technical Report 96-05, School of Psychology and Graphics, Visualization, and Usability Laboratory Georgia Institute of Technology, 1996.
12. T. Naps, G. RoBling, J. Anderson, S. Cooper, W. Dann, R. Fleischer, B. Koldehofe, A. Korhonen, M. Kuittinen, C. Leska, L. Malmi, M. McNally, J. Rantakokko, and R. J. Ross. Evaluating the educational impact of visualization, 2003.
13. N.Bogonikolos, M. Chrysostalis, K. Giotopoulos, S. Likothanassis, and K.Votis. Adaptive e-learning grid platform. In *LEGE-WG International Workshop on Educational Models for GRID Based Services*, pages 1–6. eWiC, 2002.
14. R. Schank. Teaching architectures. Technical report, The Institute for the Learning Sciences, Northwestern University, Evanston, IL, 1990.
15. H. Shi, O. Rodriguez, S.-S. Chen, and Y. Shang. Open learning objects as an intelligent way of organizing educational material. *International Journal on E-Learning*, pages 51–63, 2004.

User-Centric Multimedia Information Visualization for Mobile Devices in the Ubiquitous Environment

Yoo-Joo Choi[1,2], Seong Joon Yoo[3], Soo-Mi Choi[3],
Carsten Waldeck[4,5], and Dirk Balfanz[4]

[1] Institute for Graphic Interfaces, Ewha-SK telecom building
11-1 Daehyun-Dong, Seodaemun-Ku, Seoul, Korea
choirina@ewhain.net
[2] Seoul Univ. of Venture and Information, Seoul, Korea
[3] School of Computer Engineering, Sejong Univ., Seoul, Korea
{sjyoo, smchoi}@sejong.ac.kr
[4] ZGDV Computer Graphics Center, Dept. Mobile Information Visualization,
[5] Infoverse.org, Darmstadt, Germany
{carsten.waldeck, dirk.balfanz}@zgdv.de

Abstract. This paper suggests a user-centric browsing approach for large amounts of multimedia information in small mobile devices. The proposed approach allows users to be able to easily access the target information by interactively rearranging large amounts of information according to two user-centric sorting criteria, while overcoming the problems of the display size and the information overlapping. In this paper, a service migration concept is also proposed to support the task mobility which allows for the transfer of a started visualization task between different devices. In order to show the usefulness of this browsing approach, we apply the proposed browser to visualization of the query results in the heterogeneous multimedia retrieval system for the ubiquitous environment.

1 Introduction

Over many years, research on mobile information systems concentrated heavily on technical issues like device capabilities, computing power and communication networks. As a result, mobile networks and computing power of mobile devices are very powerful today. On the other hand, the display size of the mobile device has been minimized to increase the device mobility, while the information has been diversified and become huge.

Users usually want to effectively find the target information from the distributed heterogeneous resources using mobile devices. Various retrieval systems have been introduced to support effective resource retrieval in the distributed heterogeneous resources environment. Most of information retrieval systems have shown the query results in the table-style view. However, the results of the query itself have often included a huge dataset. Therefore, an effective browsing approach is necessary to easily find the target data by effectively rearranging large amount of data according to criteria that each user wants.

B. Gabrys, R.J. Howlett, and L.C. Jain (Eds.): KES 2006, Part I, LNAI 4251, pp. 753 – 762, 2006.

© Springer-Verlag Berlin Heidelberg 2006

In this paper, we introduce a user-centric browsing approach for large amounts of multimedia information in small mobile devices that have rapidly spread in the ubiquitous environment. The proposed approach basically uses a 2D scatter space of information bubbles in order to overcome the limitations of table-style data browsing in small mobile devices. This approach also overcomes the information overlapping problem that has been a drawback of previous data browsing approaches based on the 2D scatter plot space. This allows users to be able to easily access the wanted information by interactively rearranging large amounts of information according to two user-centric sorting criteria. Since the proposed browser analyzes and reads the information in the XML format without any specific limitations, any types of the multimedia information can be browsed in the proposed 2D scatter space. Furthermore, a service migration concept is proposed to support the task mobility which allows for the transfer of a started visualization task between different devices, e.g. a desktop PC and a PDA. In order to show the usefulness of this browsing approach, we apply the proposed browser to visualization of the query results in the heterogeneous multimedia retrieval system for the ubiquitous environment.

The rest of the paper is organized in the following manner. We briefly summarize the features of the graphical information visualization in Section 2 and address the multi-dimensional information browsing concept in section 3. Section 4 presents the liquid effect to overcome the drawback of previous browsing approaches and service migration to support the seamless service mobility among various types of computing devices. In section 5, we show a heterogeneous multimedia retrieval system using the proposed browser as a viewer and reorganizer of query results. Finally, we conclude the paper in Section 6.

2 Related Works

Information visualization[1-5] uses the powerful human visual abilities to extract meaning from graphical information. Color, size, shape position or orientation is mapped to data attributes. Experimental studies usually have been able to show significant task completion time reduction and recall rate improvements when using graphical displays instead of tabular text displays [3]. Graphical displays mean that composite data attributes are represented by color-coded or size-coded dots in the multi-dimensional space.

Sarkar and Brown [1] firstly presented the Fisheye view which represent the information as a fixed size plot on the 2D scatter plot space with zoom lens to allow the partial enlargement of the information. Ahlberg and Shneiderman [2] connected the dynamic query filters concept and the starfield displays concept in order to allow query parameters to be rapidly adjust with sliders, buttons and maps, and also allow the query results to be displayed on 2D scatter plot space. Dunlop et al.[5] presented a city guide application and a conference guide application using starfield displays on palmtop computers. However, they didn't propose the solution for the overlapping problem of information plots in the plot space. Some information plots could not be selected due to the overlapping problem. Nevertheless, the information visualization approach based on 2D scatter space has been widely used in many application domains. For example, Craig et al.[4] applied the scatter-plot views to monitoring the activity of the target genes in parallel across multiple stages of a biological process.

Most of applications have used the fixed criteria axes for the specific analysis form and ignored the information overlapping problem because the overall distribution of the objects has been a main analysis target. In order to apply the data or information browser based on the 2D scatter space to more general application domain, not only general methods in getting the input information and in changing the scatter space but also the reliable method for the object selection should be provided.

3 Multi-dimensional Information Browsing

The multi-dimensional information browsing technique can effectively and dynamically rearrange the information according to user's criteria. The multi-dimensional information browsing technique is especially suited to the small mobile devices which have spread quickly in the ubiquitous environment.

A table view is very popular to browse information records, but the common table view on a mobile device such as PDA allows users to only look at 10-20 records at a time without scrolling while allowing to sort and to compare according to one single criterion. A multi-dimensional information browsing approach allows much larger amounts of information to be visualized and is simultaneously sortable based on two criteria. The proposed approach represents an information record by a bubble in an interactive 2D scatter plot space. The size factor of each bubble can represent a scalar-typed information attribute. Fig. 1 compares the table view and multi-dimensional view in sorted display of large amounts of information.

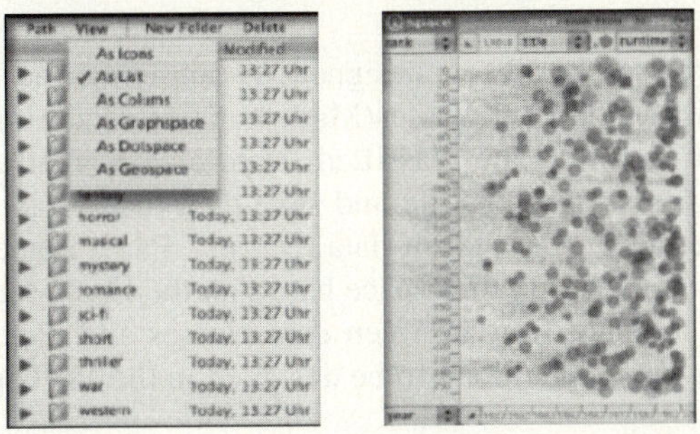

Fig. 1. Comparison of Table view(left) and multi-dimensional view (right). 15 vs. 250 information objects are displayed on table and multi-dimensional views respectively.

In the multi-dimensional view approach, positions of information bubbles in a 2D scatter space are interactively changed according to user's selection to two axes criteria, i.e, x-axis and y-axis criteria. For example, if we use the proposed browser to browse a movie database, movie bubbles can be dynamically rearranged in the increasing(decreasing) order of the rank and the showing year values, or in alphabetic order of the genre and in increasing order of the runtime by changing the values of the list boxes for x- and y-axes. Users can flexibly change not only x- and y-axes criteria, but also the criteria of the title and bubble size using attribute list boxes. Therefore, one display shot at a time can represent four attributes of each information bubble, i.e, x-axis, y-axis, title and bubble size, simultaneously. The proposed browser includes the XML analyzer in order to more general method to import the information to the browser. If the information can be represented in XML format, any kind of information can be imported to the proposed browser. Fig. 2 shows 250 or more image bubbles to be spread in the 2D scatter space that is defined by the title and the preference fields.

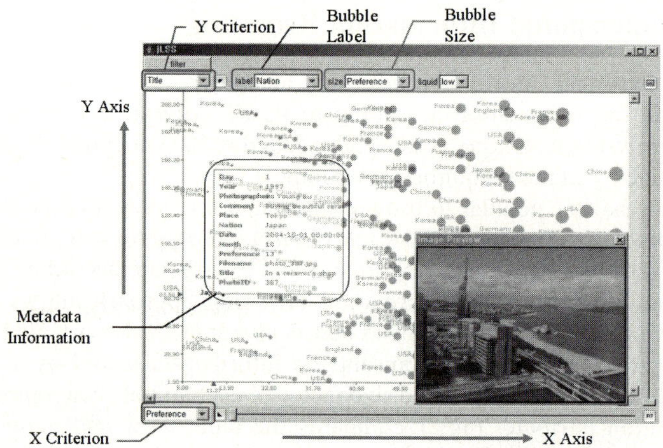

Fig. 2. The photo data bubbles to be displayed by the multi-dimensional browser

The proposed browser has been designed by applying the MVC(Model-View-Controller) pattern in which data classes(Model), panel rendering classes(View) and application classes(Controller) are classified. Data classes read an XML(eXtensible Markup Language) file as an input file and save data records in the HashMap structure in order to effectively treat a lot of data records. Panel rendering classes display information bubbles in the 2D scatter space based on the HashMap structure. Application classes control the data flow between data classes and panel rendering classes. The relations among principal classes to be designed in the MVC pattern are as shown in Fig. 3.

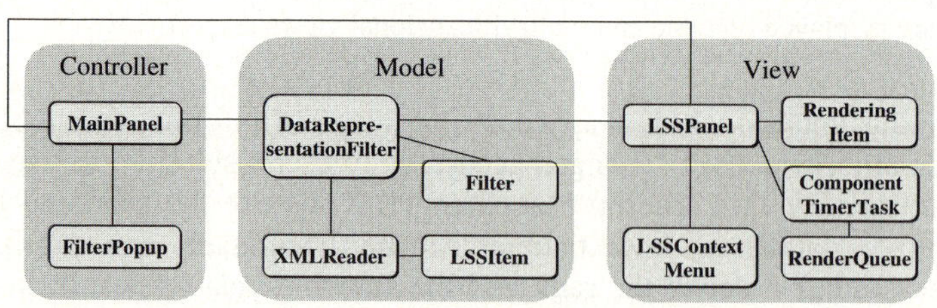

Fig. 3. Principal classes to be designed in the MVC pattern

4 Liquid Browsing and Service Migration

One disadvantage of a 2D scatter plot approach however, is the possibility of the information objects to overlap to some extent hindering readability and perceptibility of the information or, worst case, even making it impossible. To solve this problem, zoom and lens functions[1] have been implemented, which allow the (partial) enlargement of the information to counter this problem. Still, browsing high information densities in this way is a very challenging task, because a simple enlargement based fisheye lens does not reduce the overlapping and very often has to deal with heavy distortion issues.

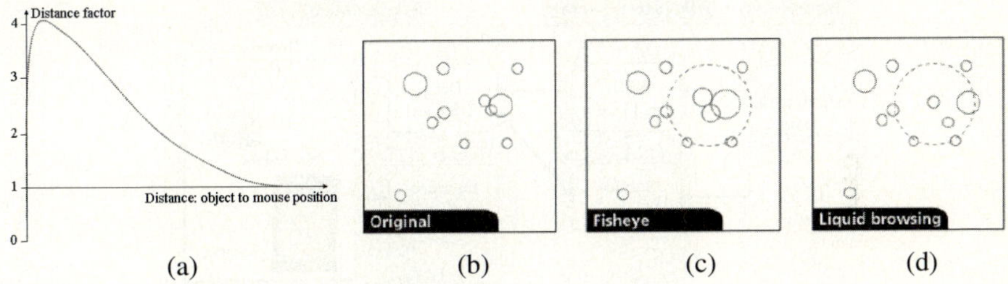

Fig. 4. Comparison of fisheye approach and liquid browsing. (a) Adaptive distance factor according to the distance to mouse position for liquid browsing, (b) overlapping information bubbles, (c) the fisheye approach, (d) the liquid browsing approach.

To address this problem, we apply a distance manipulation based expansion lens[6]. The overlapping of information bubbles can be solved much more effectively by no enlarging the information objects themselves, but rather the spaces between them. The space between information bubbles is decided by applying the Gaussian distance filter in the circle lens area which is defined by the current stylus pens or mouse position. The Gaussian distance filtering controls the distance factor of information bubbles according to the distance of objects from the current mouse position. That is, near distant objects from the mouse position move far to the boundary of a circle, while far distant objects move in the relatively less distance. The distance filtering shows oil- or liquid-like moving effect of information bubbles according to the interactive mouse movement. Fig. 4(a) shows the distance factor to be controlled by the distance between the information object and the mouse position. Fig. 4 (b),(c) and (d) show the results of fisheye approach and liquid browsing approach to solve the overlapping of information bubbles. The overlapping bubbles can not be separated from each other even though bubbles are magnified by the enlargement lens in fisheye approach, while the distance based Gaussian lens makes the overlapping bubbles to be separated.

As various types of computing devices have widely spread, users want to immerse in a multi-platform environment with the possibility of interacting with an application while freely moving from one device to another. That is, when user moves from one device to another device in use of an application, he or she wants to resume the application from the stop point without any manual interface initialization. We call this function the service migration. The service usage history for each user is managed by the service manager to support the service migration. That is, the service information such as the interface parameters and the current selected data is notified to the service manager whenever user changes the application status. Therefore, the service server can trace which kind of device is used by a user or from which point should the service be resumed. Fig. 5 depicts the system architecture for the service migration. Whenever the liquid browser is launched, the resume point and the last interface setup status are transferred from the service manager in the server to the interface adapter in the client. The interface adapter adjusts the interface parameters and the active data bubble to be suitable for the current device based on the information from the server.

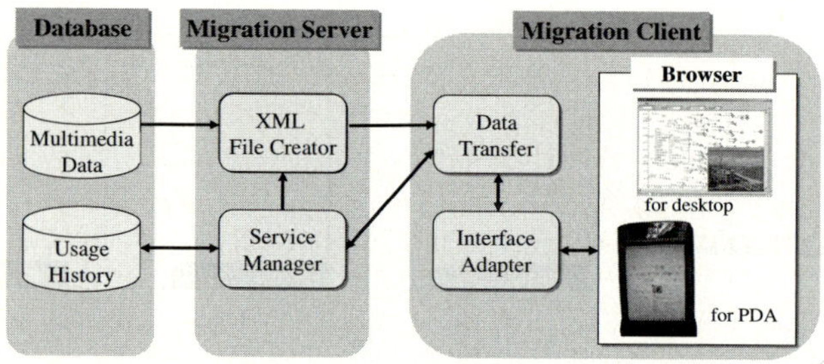

Fig. 5. System architecture for the service migration

5 Implementation and Experimental Results

5.1 Integrated Retrieval of Heterogeneous Multimedia Resources

This section describes the web service based multimedia integration framework. It gives users more flexibility than previous multimedia integration systems, since the prototype system is implemented using web services technology. The integration framework basically uses universal media access (UMA) metadata[7] that have been defined for ensuring compatibility with the current international metadata standards such as MPEG-7 Multimedia Description Scheme[8] and TV Anytime metadata[9]. The framework is composed of three layers- application layer, mediation layer and resource layer. The web service APIs provided by the mediation layer and the resource layer are used for data transfer between layers.

***Application Layer*:** The application layer transfers a query and receives the results. For example, a user can generate a general query concerning to name, genre, ID, or keyword through an application query window in order to search the target data that she or he wants to watch. The application layer calls the web service API provided by the mediation layer to transfer the query to the mediation layer in XML format using SOAP protocol. The web service API of the mediation layer is also called to get the results of the query from the mediation in XML format.

***Mediation Layer*:** The mediation layer includes a core engine to integrate heterogeneous multimedia data in the ubiquitous environment. The query from the application layer can be defined based on various multimedia metadata such as MPEG-7 MDS or TV-AnyTime. In the mediation layer, the query is translated to various types of metadata queries by using the UMA metadata mapping table and broadcasted to resource servers in the local query form to be translated for each resource server. The results from resource layer of each resource server are integrated and transferred to the application layer. The mediation layer is composed of query processor, rule manager, and global schema manager.

The query processor receives the query from the application layer and validates the query. If the query is valid, the query processor analyzes the composed elements of the query and translates to the local query forms for MPEG-7 MDS or TV-AnyTime using the mapping table in the global schema manager. The global schema manager includes the core mapping table based on the UMA schema. The mapping table defines the matching relations among MPEG-7 MDS, TV-AnyTime and UMA. The global schema manger also manages the locations of the local resource DBs. The web service API of the mediation layer allows the wrapper of the resource layer to register the wrapper information of each resource server. The rule manager collects, validates and integrates the results of the query. Finally the integrated query results are transferred to the application layer.

***Resource Layer*:** The resource layer is connected to a local DB through wrappers. Each resource server includes a local DB which stores the multimedia data to be represented in the form of MPEG-7 MDS or TV-AnyTime metadata et al. This layer retrieves relevant data from the local DB on which the XML-type metadata are saved. The resource layer is composed of a wrapper manager and multiple wrappers. The wrapper manager searches multimedia data by choosing and communicating with the proper wrapper among MPEG-7 wrapper, TV-AnyTime wrapper and UMA wrapper. Since the major functions are provided with web services APIs, users can build their own wrapper with ease.

5.2 Multimedia Information Browser

The prototype multimedia information browser was implemented using the Java platform for the device platform independence. That is, this framework was implemented using Java 2 Platform Standard Edition(J2SE) for the devices of Windows XP, Linux and MAC, and using Java 2 Platform Micro Edition(J2ME) for PocketPC devices such as PDA.

A user is sending the initial query to collect the wanted multimedia data. The collected multimedia dataset is transferred to the proposed liquid browser in the XML format including the MPEG-7 MDS metadata and user-defined metadata as shown in Fig. 6. The multi-dimensional liquid browser analyzes the XML file that includes a lot of multimedia information. Each multimedia object to be read from the XML file, such as an image or a movie, is displayed as a bubble in the interactive 2D scatter space. The image bubbles can be interactively rearranged if the user changes criteria for the X-axis, Y-axis, the bubble size or the bubble title. A user can also select a subset of bubbles by using filtering query. The Fig. 7 depicts the filtering query windows in which a user can input the filtering queries for a user defined metadata items. Users can access the wanted multimedia objects more easily and rapidly by reselecting them using the filtering query.

```
<?xml version="1.0" encoding="iso-8859-1"?>
<Mpeg7 xmlns="urn:mpeg:mpeg7:schema:2001"
    xmlns:xsi="http://www.w3.org/2001/XMLSchema-instance"
    xmlns:mpeg7="urn:mpeg:mpeg7:schema:2001"
    xmlns:xml="http://www.w3.org/XML/1998/namespace"
    xsi:schemaLocation="urn:mpeg:mpeg7:schema:2001 Mpeg7-2001.xsd">
        <Description xsi:type="ContentEntityType">
            <MultimediaContent xsi:type="ImageType">
                <image id =1>
               ........user defined metadata for each image
                </image> .....
            </MultimediaContent></Description></Mpeg7>
```

Fig. 6. MPEG-7 XML file format for the image dataset

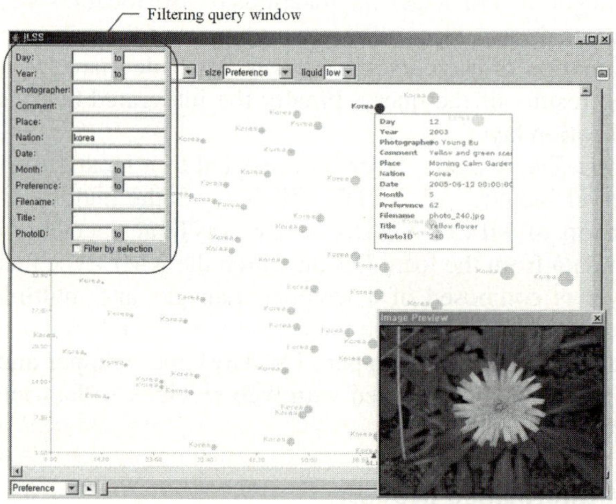

Fig. 7. Input window for the filtering query

In order to prove the efficiency of the proposed browser, we performed the usability test to ten users. In this test, we compared the average access time of the same photo data set when users used the table-typed browser with that when users used the proposed browser. Table 1 shows the comparison result of the access time. When the proposed browser was used, most of users searched the wanted photo more efficiently.

The service server for the service migration was also implemented based on J2SE for facilitation of platform independence. The prototype migration was tested on three different H/W platforms, i.e, laptop(1280 x 1024), PDA(640 x 480), and tablet PC(800 x 600). Fig. 8 shows results of the service and interface migration when a user moves from a laptop to PDA and tablet PC. Users can resume the multimedia data retrieval task without nuisance interface resetting affairs after moving to different computing device.

Table 1. Comparison result of the average photo access time

	Usr1	Usr2	Usr3	Usr4	Usr5	Usr6	Usr7	Usr8	Usr9	Usr10
The Proposed browser	5.6	6.8	7.6	3.6	13.6	6.6	10.8	4.8	6.8	4.6
Table-typed browser	5	12.8	9.6	4.8	19	10.8	13.8	8.6	10.8	11

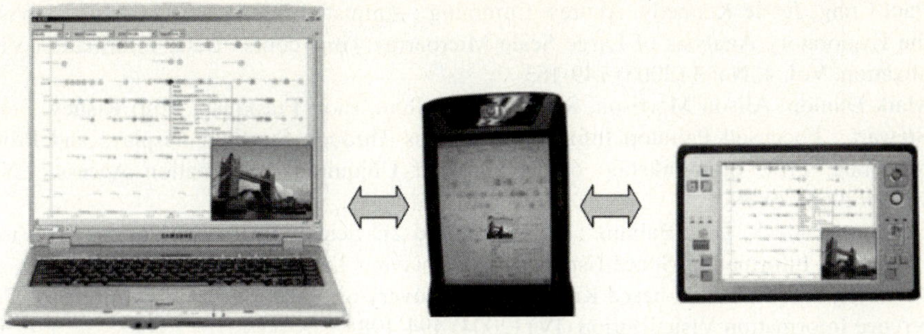

Fig. 8. The results of the service migration among a laptop, PDA and tablet PC

6 Conclusions

This paper suggested a user-centric browsing approach for large amounts of multimedia information on small mobile devices and introduced the application of the proposed browser for interactively retrieving large amounts of multimedia objects in a lot of different screen situations. The proposed browsing method applies XML-based information porting approach, and thus the browser was ported to the application framework with ease. In the multimedia retrieval framework, a user can easily access the target data from the query results by dynamically rearranging the retrieved information according to user centric criteria. Moreover, since all filtering and interface settings are stored on a central server, they can accompany the user to any supported device in the ubiquitous environment

Acknowledgements

I would like to thank So-Young Bu and Hyun-Joo Yun for giving useful help and comments. This work was supported in part by the ubiquitous autonomic computing and network project under the Ministry of Information and Communication (MIC) 21st Century Frontier R&D Program in Korea. This research was also supported in part by the Seoul R&BD (Research and Business Development) program.

References

1. Manojit Sarkar, Marc Brown : Graphical Fisheye Views of Graphs. Proc. Of CHI 1992, ACM, New York (1992) 317-324
2. Christopher Ahlberg, Ben Shneiderman : Visual Information Seeking: Tight Coupling of Dynamic Query Filters with Starfield Displays. Prof. of CHI (1994) 313-317
3. D. Lindwarm, A. Rose, C. Plaisant, K. Norman : Viewing personal history records: A comparison of tabular format and graphical presentation using LifeLines , Behaviour & Information Technology (1998)

4. Paul Craig, Jessie Kennedy, Andrew Cumming : Animated Interval Scatter-plot Views for the Exploratory Analysis of Large Scale Microarray Time-course Data. Information Visualization, Vol. 4, No. 3 (2005) 149-163
5. Mark Dunlop, Alison Morrison, Stephen McCallum, Piotr Ptaskinski, Chri Rishey, Fraser Stewart : Foccused Palmtop Information Access Through Starfield Displays and Profile Matching. Prof. of Workshop on Mobile and Ubiquitous Information Access, LNCS v2954(2004) 79-89
6. Carsten Waldeck, Dirk Balfanz : Mobile Liquid 2D Scatter Space (ML2DSS) : A Visual Interactive Information Space (ispace) for Displaying Large Amounts of Information and Allowing Simple Vision-based Knowledge Discovery on Small Screen Sizes. Proc. of Conference Information Visualization (IV) (2004) 494-498
7. C. W. Song, S. J. Yoo : UMA Metadata for Integrating Hetefogeneous Multimedia Resources. Asia Information Retrieval Symposium (AIRS), Oct. 13-15 (2005)
8. J. M. Martinez : Overview of the MPEG-7 Standard(version 5.0), ISO/IECJTC1/SC29/WG11 N4031, Singapore, March(2001):
9. TV-Anytime Forum : http://www.tv-anytime.org/

DAWN – A System for Context-Based Link Recommendation in Web Navigation

Sebastian Stober and Andreas Nürnberger

Institute for Knowledge and Language Engineering,
School of Computer Science,
Otto-von-Guericke-University Magdeburg, D-39106 Magdeburg, Germany
{stober, nuernb}@iws.cs.uni-magdeburg.de

Abstract. In this paper, we present the system "DAWN" (direction anticipation in web navigation) that learns navigational patterns to help users navigating through the world wide web. We motivate the purpose of such a system and the approach taken and point out relations to other approaches. Further, we briefly outline the architecture of the system and focus on the prediction model and the algorithm for link recommendation. Evaluation on real-world data gave promising results.

1 Introduction

Users that browse the world wide web often need to choose between multiple options on how to continue navigation. The dynamic nature of the steadily growing and constantly changing web makes this a very hard task. Depending on how well hyperlinks are chosen, it will take users less or more time to finally get to the information they desire. Whilst some users might find this task quite easy, others may get lost, especially if they are facing unfamiliar web content. Although there is most likely no general rule on how to most efficiently navigate to the desired information, there may be certain navigational patterns that can be used as heuristics. Users may unconsciously develop such patterns. A system that watches users navigating the web could learn their navigational patterns and use them to provide navigational support by suggesting the best matching hyperlinks.

This paper introduces a prototype of such a system, named DAWN (direction anticipation in web navigation). Differences to related work are pointed out in Sect. 2. Sect. 3 gives an overview on the system architecture and addresses model induction and hyperlink recommendation as the two crucial points of the system in detail. In Sect. 4 we present some promising results of a first evaluation of the system with real-world data. Finally, we summarize our work and give an outlook on future developments.

2 Related Work

The general idea to support users browsing the world wide web is not new. Systems that accomplish this task by suggesting hyperlinks or web pages are e.g.

B. Gabrys, R.J. Howlett, and L.C. Jain (Eds.): KES 2006, Part I, LNAI 4251, pp. 763–770, 2006.
© Springer-Verlag Berlin Heidelberg 2006

discussed in [1, 2, 3, 4, 5]. The system presented here has much in common with these systems: Like Webmate [3], Broadway [4] or Personal Webwatcher [5] it uses an HTTP-proxy to log user actions and to manipulate the requested web pages. Documents are represented according to common practice as term vectors with TF/iDF-weights. Letizia [1] was one of the first systems that used background look-ahead crawling. However, Letizia was designed as a plug-in for the Netscape Navigator 3 and heavily relied on its API whereas the proxy-architecture chosen for DAWN allows the users to use their browsers of choice. Moreover, bandwidth and computational expensive tasks can be performed on the server which reduces the burden on the client machine. In contrast to DAWN, which uses a combination of navigational patterns and document similarities, Webmate, Personal Webwatcher and Letizia are purely content-based systems that rely on document-similarities to decide which web pages are interesting. Broadway relies on case-based reasoning to recommended web pages. Webwatcher [2] uses a collaboration-based approach by asking a user about the desired information and then suggesting links that users with similar information need have followed previously. Furthermore, Webwatcher is a server-side application that is restricted to one website, whereas DAWN works on a client-side proxy and therefore has no such local restriction.

DAWN stores the navigational patterns in a Markov Model. Such models have been successfully used for prediction of HTTP-requests to optimize web-caches [6]. Recently, they have also been applied to modeling of user navigational behavior in the context of adaptive websites [7, 8, 9] and web-usage mining [10]. However, these are solely server-side applications that are inherently locally confined, i.e. they are restricted to one website. To our knowledge, there have not been any client-side systems that use Markov Models so far. The algorithm for model induction has been derived from the one presented by Borges and Levene in [10] which has been designed to represent a collection of user navigation sessions of a website. Sect. 3.1 briefly summarizes the algorithm's concept and subsequently discusses our modifications and extensions.

3 System Architecture

An overview of the system architecture is shown in Fig. 1. The system has been integrated into CARSA, an architecture for the development of context adaptive retrieval systems [11]. All HTTP-request made by the user through the system's HTTP-proxy are recorded and stored in a database (DB). The requested web pages are analyzed and modified prior to forwarding them to the user. This allows tracking of browsing sessions including events such as clicks, form submits and closing of browser windows or the browser itself. From the information stored in the database a model-learner generates a model using the algorithm described in Sect. 3.1. It is possible to learn a separate model for each user as well as a global one. Based on a model, candidate pages collected by a web crawler that performs a background look-ahead crawl can be assessed given a user's current history of the most recently accessed web pages. The recommenda-

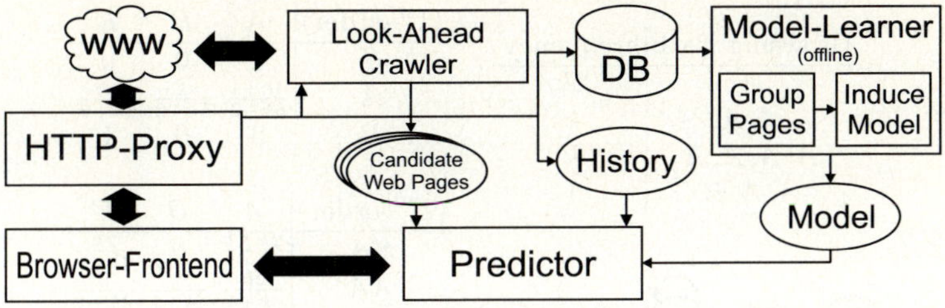

Fig. 1. DAWN: System Overview

tion algorithm is described in Sect. 3.2. Eventually, information about the most probable candidates is displayed in the browser-frontend. The display combines ranking, content and visual information to make it easier for the user to assess the candidates' relevance. Ranking information is provided by sorting the candidates according to the predicted visit probability. Displaying title, URL and keywords gives information about the content and automatically generated thumbnails give a general visual impression of the page layout.

3.1 Model Induction

The algorithm for model induction has been derived from the one proposed by Borges and Levene in [10] which has been designed to represent a collection of user navigation sessions of a website. This algorithm iteratively constructs an nth-order Markov Model by making use of the state cloning concept where a state is only duplicated if the nth-order probabilities diverge significantly from the corresponding 1st-order probabilities. Once a state needs cloning, the nth-order probability distributions induced by the different in-paths of the state are clustered. In-paths referring to probability distributions assigned to the same cluster can then be redirected to the same clone of the original state. This reduces the number of clones needed and reduces the model size.

The algorithm by Borges and Levene is intended for server-side application. To be able to use it in a client-side setting as addressed by DAWN, the algorithm had to be extended. Most importantly, we introduced an additional clustering step prior to model induction as an extra abstraction level. In this additional step similar pages are grouped together thus drastically reducing the size of the model's state space and combining different browsing paths into an abstract navigational pattern. As a side effect, cycles of length one may occur when consecutive pages in a browsing path are assigned to the same cluster. These cycles are something the original algorithm cannot deal with. As a result, an nth-order model (with $n \geq 3$) may contain paths of length n that do not exist in the data and thus induce undefined transition probabilities as shown in Fig. 2. State A is not cloned because for large k all transition probability distributions (table rows) dependent on k converge to $(\frac{1}{2}, 0, \frac{1}{2})$.

We propose the following extension of the algorithm to solve this problem: Prior to induction of an nth-order model from an $(n-1)^{\text{th}}$-order model, the

Browsing Path	Frequency
A	$2k$
A, A	k
A, A, A, A	k
A, A, B	1

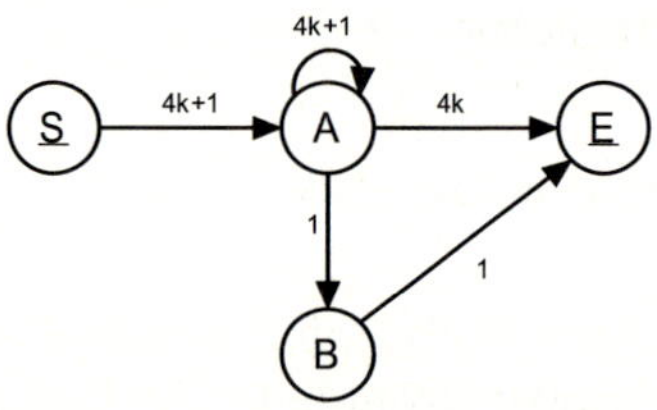

(1st-order)	A	B	E
S	1	0	0
A	$\frac{4k+1}{8k+2}$	$\frac{1}{8k+2}$	$\frac{4k}{8k+2}$
B	0	0	1

(2nd-order)	A	B	E
SA	$\frac{2k+1}{4k+1}$	0	$\frac{2k}{4k+1}$
AA	$\frac{2k}{4k+1}$	$\frac{1}{4k+1}$	$\frac{2k}{4k+1}$
AB	0	0	1

(3rd-order)	A	B	E
SAA	$\frac{k}{2k+1}$	$\frac{1}{2k+1}$	$\frac{k}{2k+1}$
AAA	$\frac{k}{2k}$	0	$\frac{k}{2k}$
AAB	0	0	1

Fig. 2. An incorrect 3rd-order model induced by the algorithm presented in [10]. The path S, A, B exists in the model but not in the data. (upper left: data, lower left: graph representation of the induced model, right: 1st, 2nd and 3rd-order transition probabilities)

$(n-1)^{th}$-order model is checked for paths of length n that do not exist in the data. Let $\boldsymbol{l} = (l_{i_1}, \ldots, l_{i_n})$ be such a path and $l_{i_{n-1}}$ the state preceding the last one in the path. Further, let M be the set of all paths in the model that comprise $(n-1)$ states and have $l_{i_{n-1}}$ as the last state. All paths in M must be contained in the data because they were checked in the previous iteration of the algorithm. For the path $\boldsymbol{l}' \in M$ with $\boldsymbol{l}' = (l_{i_1}, \ldots, l_{i_{n-1}})$ the $(n-1)^{th}$-order modeled transition probability $P(l_{i_n}|\boldsymbol{l}')$ should be 0 because $\boldsymbol{l}$ does not exist in the data. Obviously, this is not the case. Otherwise $\boldsymbol{l}$ would not exist in the $(n-1)^{th}$-order model. Hence, there must be at least one path $\boldsymbol{m} \in M$ with $P(l_{i_n}|\boldsymbol{m}) > 0$. Thus, M can be partitioned into two non-empty subsets, $M^{(0)} = \{\boldsymbol{m} \in M | P(l_{i_n}|\boldsymbol{m}) = 0\}$ and $M^{(\bar{0})} = M \setminus M^{(0)}$. To fix the problem, the paths in $M^{(0)}$ need to be separated from those in $M^{(\bar{0})}$ by using the method for in-path-separation of the original algorithm. Fig. 3 shows the modified 2nd-order model and the correct 3rd-order model for the example in Fig. 2.

As another modification of the original algorithm, we replaced the originally proposed K-Means-algorithm for clustering of the transition-probability-vectors by an agglomerative one which always finds the optimal clustering in a single run. Apart from that, several smaller modification were made that mainly address performance issues and are beyond the scope of this paper. They are discussed in detail in [12].

3.2 Hyperlink Recommendation

In the preceding section we have presented how a higher-order Markov Model can be induced. Having learned an nth-order Markov Model that represents the navigational patterns from training data, candidate pages can be assessed in the

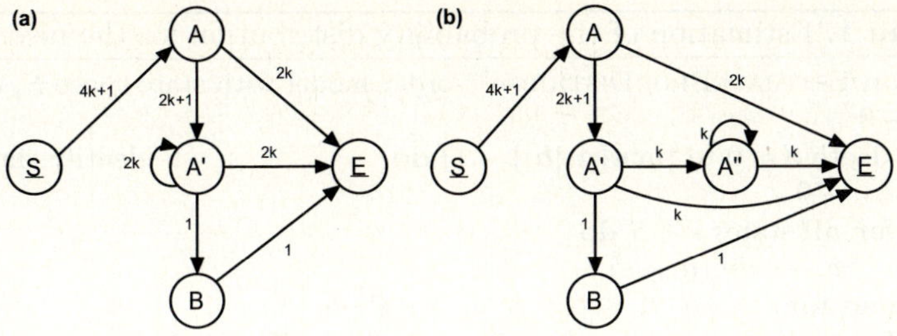

Fig. 3. Modified 2nd-order model (a) and the correct 3rd-order model (b) for the example in Fig. 2

context of a user's current history of the last n accessed web pages. For this, a probability distribution for the next state has to be estimated by mapping the history onto the model as shown in Alg. 1. In the first part of the algorithm (lines 1–21), all navigational paths that are similar to the history are identified, weighted and stored in L. To compute the similarity of a web page and a state, which is a cluster of web pages from the training data that can be represented by its center, the cosine similarity measure is used. The number of states to be regarded as similar is controlled by the similarity threshold λ and bounded by the parameter k. The weight $w_{\boldsymbol{l}}$ of a path $\boldsymbol{l}$ then corresponds to the similarity of the path with the history, which is the minimum of all one-on-one state similarities. Obviously, a navigational path can only be similar to the history, if all of its sub-paths are similar to the corresponding parts of the history. Thus, the set L_d containing all similar paths with $d+1$ states can be constructed by extending all paths contained in L_{d-1}. (Considering that during a browsing session the history is only extended by one web page at a time, the computation of L can be sped up by storing the sets $L_0, \ldots, L_{n-2}$ from the previous call.)

In the second part of the algorithm (lines 22–30), the probability distribution for the next state is computed as the weighted average of the probability distributions induced by all weighted paths in L. For each path $\boldsymbol{l} \in L_d$ each set L_i with $i < d$ contains a suffix of $\boldsymbol{l}$. All these suffixes are taken into account for the computation of the probability distribution, too. At the same time the weight of a suffix cannot be smaller than the weight of the whole path because of the minimum operation that is used to compute the path weight in Alg. 1, line 14. This leads to an intentional higher weighting of the most recently accessed web pages in the history that is motivated by the assumption that these web pages have a greater impact on the choice of the next hyperlink [13].

Once the probability distribution for the next state has been computed, the candidate web pages collected by the crawler can be assessed. For this, each candidate is mapped onto the model's states analogously to the mapping of the elements of the history identifying a set of similar states. The probability of the candidate is the maximum probability of all these states according to the previously estimated probability distribution for the next state. This probability can then be used to rank the candidate pages.

Algorithm 1. Estimation of the probability distribution for the next state.

function EstimateProbDistrib(n^{th}-order model with state space S, n, $\boldsymbol{h}$, λ, k)
  $L \leftarrow \emptyset$
  for depth $d = 0$ to $(\min\{n, |\boldsymbol{h}|\} - 1)$ **do** ▷ identify similar paths
    $L_d \leftarrow \emptyset$
    for all states $s \in S$ **do**
      $w_s \leftarrow sim(\boldsymbol{h}_{[d]}, s)$
    end for
    for the maximal k most similar states s with $w_s \geq \lambda$ **do**
      if $d = 0$ **then**
        $L_0 \leftarrow L_0 \cup (s, w_s)$
      else
        for all weighted paths $(\boldsymbol{l}, w_{\boldsymbol{l}}) \in L_{d-1}$ **do**
          if s predecessor of $\boldsymbol{l}$ **then**
            $\boldsymbol{l}' \leftarrow s, \boldsymbol{l}$
            $w_{\boldsymbol{l}'} \leftarrow \min\{w_s, w_{\boldsymbol{l}}\}$
            $L_d \leftarrow L_d \cup (\boldsymbol{l}', w_{\boldsymbol{l}'})$
          end if
        end for
      end if
    end for
    $L \leftarrow L \cup L_d$
  end for
  for all states $s \in S$ **do** ▷ compute probability distribution for the next state
    $p_s \leftarrow 0$
  end for
  for all weighted path $(\boldsymbol{l}, w_{\boldsymbol{l}}) \in L$ **do**
    for all successors s of $\boldsymbol{l}$ **do**
      $p_s \leftarrow p_s + w_{\boldsymbol{l}} \cdot P(s|\boldsymbol{l})$
    end for
  end for
  $\boldsymbol{p} \leftarrow \left\{ \left(s, \frac{p_s}{\sum_{s \in S} p_s} \right) \middle| \, s \in S \right\}$
end function

The crucial point is, that the model can be used for prediction, even if, as generally the case, a web page has not been visited during the training period. By mapping the candidate web pages and the elements of the history to multiple similar states of the model instead of using a sharp one-to-one mapping, the system can interpolate between several states in case newly encountered web pages differ significantly from those in the training data.

4 Evaluation

To get a first impression of the system's performance, we have measured the prediction accuracy on server log files. The prediction accuracy is however only an indicator for the systems usefulness. In some cases a user might have followed a link to a resource that was not useful at all. Correct prediction of such links results in high prediction accuracy but a useful system should not make such predictions. Predictions that do not match the actual link chosen may, on the other hand, still lead to useful resources.

During the clustering step that groups similar pages for model induction and for the computation of the similarities in the prediction process (Alg. 1, line 5) the web page content, at least in bag-of-words representation, is needed. Thus, simple log files are not sufficient. Either the content of the accessed web pages has to be included in the evaluation data or at least the accessed URLs must not be outdated, i.e. the URLs must still be valid and accessible and the content should not have changed since the access was logged. To our knowledge, there is currently no data set publicly available that meets these requirements. Therefore, we used anonymized log files that contained requests on web pages hosted by the University of Magdeburg recorded during six consecutive days for evalutaion.

The log files were filtered as follows: Any requests with a response status code other than 200 (OK) and 304 (not modified) were removed as well as requests of URLs with non-HTML content. Additionally, several possible denial of service attacks on specific URLs were detected and removed. Afterwards, sessions were identified as, e.g., described in [14]. The prediction model was learned from the data of the first five days. For the evaluation of the model the data of the 6th day was used. Table 1 shows the number of sessions, requests and unique URLs in the data.

Table 1. Number of sessions, requests and unique URLs in the data used for training and evaluation

dataset	sessions	requests	URLs
train	58314	237098	25771
test	13437	60461	5657
total	71751	297559	27877

The requested web pages in the training data were clustered into 250 clusters resulting in an average cluster size of about 100. From the 250 clusters and the sessions extracted from the training data, a 2nd-order Markov Model was induced. This model was used to estimate the probability of all outgoing links for each web page in the test sessions. Table 2 shows the results.

Table 2. Predicted ranks of the actually followed links. The number of candidates (number of different outbound links in a web page) ranged from 1 to 607 with a mean of 10.77

rank	1st	2nd	3rd	1st–3rd	in 1st third
frequency	20%	5%	5%	31%	45%

In about 31% of all test cases, the candidate that had actually been chosen by the user was amongst the three highest ranked candidates. In these situations the recommendations could have helped the user. However, as mentioned above, this does not necessarily mean that the recommendations would have been useless in the other cases. The real usefulness of the system's recommendations remains to be assessed in a user study.

5 Conclusions and Future Work

In this paper, we have presented a prototype system that provides navigation support for users browsing the world wide web. A first evaluation with real-world data has shown that the system is able to learn navigational patterns to predict a user's browsing path. As future work we plan an evaluation of the system in a user study.

References

1. Lieberman, H.: Letizia: An Agent That Assists Web Browsing. IJCAI, 1995.
2. Joachims, T., Freitag, D., Mitchell, T.: WebWatcher: A Tour Guide for the World Wide Web. In: IJCAI, 1997.
3. Chen, L., Sycara, K.: WebMate: A Personal Agent for Browsing and Searching. In: Proc. 2nd Intl. Conf. on Auton. Agents and Multi Agent Sys., AGENTS '98, 1998.
4. Jaczynski, M., Trousse, B.: Broadway, A Case-Based Browsing Advisor for the Web. In: ECDL '98: Proc. of the 2nd Europ. Conf. on Research and Adv. Technology for Digital Libraries, 1998.
5. Mladenic, D.: Machine learning used by Personal WebWatcher. In: Proc. of ACAI-99 Workshop on Machine Learning and Intelligent Agents, 1999.
6. Sarukkai, R.: Link Prediction and Path Analysis using Markov Chains. Computer Networks, Vol. 33, pp. 337–386, 2000.
7. Anderson, C., Domingos, P. Weld, D.: Relational Markov Models and their Application to adaptive Web Navigation. In: Proc. 8th ACM SIGKDD Intl. Conf. on Knowl. Discovery and Data Mining, 2004.
8. Zhu, J., Hong, J., Hughes, J.: Using Markov Chains for Link Prediction in Adaptive Web Sites. In: Soft-Ware 2002: Comp. in an Imperfect World: 1st Intl. Conf., 2002.
9. Cadez, I., Heckerman, D., Meek, C., Smyth D., White, S.: Model-Based Clustering and Visualization of Navigation Patterns on a Web Site. Data Min. Knowl. Discov., Vol. 7, pp. 399–424, 2003.
10. Borges, J., Levene, M.: Generating Dynamic Higher-Order Markov Models in Web Usage Mining. In: Proc. of the 9th Europ. Conf. on Principles and Practice of Knowl. Discovery in Databases (PKDD 05), 2005.
11. Bade, K., De Luca, E., Nürnberger, A., Stober, S.: CARSA - An Architecture for the Development of Context Adaptive Retrieval Systems. In: Adaptive Multimedia Retrieval: User, Context, and Feedback: 3rd Intl. Workshop, 2005.
12. Stober, S.: Kontextbasierte Web-Navigationsunterstützung mit Markov-Modellen. Diplomarbeit, Otto-von-Guericke-University Magdeburg, Dec. 2005.
13. Schechter, S., Krishnan, M., Smith, M.: Using path profiles to predict http requests. In: Proc. of the 7th intl. conf. on World Wide Web, 1998.
14. Cooley, R., Mobasher, B., Srivastava, J.: Data Preparation for Mining World Wide Web Browsing Patterns. Knowl. and Information Sys., Vol. 1, No. 1, pp. 5–32, 1999.

Modeling Collaborators from Learner's Viewpoint Reflecting Common Collaborative Learning Experience

Akira Komedani, Tomoko Kojiri, and Toyohide Watanabe

Department of Systems and Social Informatics,
Graduate School of Information Science, Nagoya University
Furo-cho, Chikusa-ku, Nagoya, 464-8603, Japan
Phone: +81-52-789-2735; Fax: +81-52-789-3808
{k-akira, kojiri, watanabe}@watanabe.ss.is.nagoya-u.ac.jp

Abstract. In collaborative learning through the Internet, learners sometimes cannot communicate with other learners smoothly because they cannot focus on particular learners. Since a learner tends to focus on others whom he thinks understands exercise well, we have already proposed a mechanism for inferring other learners' understanding levels from their utterances using a solution network which represents understandability of knowledge contained in the exercise. However, the understandability of knowledge may differ among learners. In this paper, understanding levels for general knowledge are introduced as a permanent learner model. The permanent learner model is updated based on understanding levels acquired after each collaborative learning. Then, the understandability in the solution network is generated by using the permanent model.

1 Introduction

In the last decade, research into CSCL (Computer Supported Collaborative Learning) has become a hot topic. Collaborative learning is a learning style in which various learners exchange their opinions and solve common exercises through a network [1]. Gridcole [2] is a new collaborative learning system which can be easily tailored by educators to support their own CSCL scenarios. Furthermore, Raitman, et al. [3] proposed a powerful new environment whereby learners can discover, create, query and share their knowledge and resources.

In collaborative learning, a learner acquires knowledge about exercises through interactions with other learners. In particular, the learner often focuses on other learners who are expected to understand these exercises, and acquires knowledge from the learners positively. However, when interacting through a network it is difficult for the learner to collaborate with other learners effectively because the means of communication. In response to this problem, our objective is to construct a system which automatically provides user with learning information on other learners who are expected to understand exercises. In this paper, we

B. Gabrys, R.J. Howlett, and L.C. Jain (Eds.): KES 2006, Part I, LNAI 4251, pp. 771–779, 2006.
© Springer-Verlag Berlin Heidelberg 2006

focus on one learner and regard him as a user. Other learners who are expected to understand exercises by the user are called *target learners* throughout this paper. In addition, other learners who participate in collaborative learning are defined as *collaborators*.

In order to support the interaction between user and target learners, it is necessary to be able to identify the target learners of the user automatically. So, the following three mechanisms are needed:

- a mechanism which models collaborators' levels of understanding from user's viewpoint,
- a mechanism which models user's level of understanding, and
- a mechanism which identifies the target learners by comparing the understanding of the collaborators with that of the user.

We have previously proposed a method for realizing the first mechanism; that is, to model collaborators' levels of understanding from user's viewpoint during the collaborative learning [4],[5]. The collaborators' levels of understanding are modeled as an *temporal learner models*. In the temporal learner model, collaborator's level of understanding is represented as *understanding level*. In addition, a *trust level* which indicates the degree of trustworthiness of understanding level is also introduced. In this approach, the understandabilities of collaborators for knowledge which constitutes an exercise are expressed on the basis of Bayesian network, called a *solution network*. The solution network represents sequential relations of knowledge which constitute an exercise. Moreover, a degree of association between the knowledge is defined as CPT(Conditional Probability Table) [6] which indicates understandability for a target node when a linked node is derived. When the utterance has been occurred, the understanding level of its target knowledge is determined and those of related knowledge are estimated on the basis of the solution network.

Currently, CPTs are the same for all collaborators. It means that user thinks all collaborators have the same understandability for exercises. However, naturally each collaborator has a different understanding level for knowledge and user recognizes such differences when he learns with the collaborators repeatedly. Therefore, in this paper, we propose a method for changing CPTs for collaborators according to learning experiences. The collaborators' understanding levels for generalized knowledge change during the learning. So, *permanent learner model* which represents learner's understanding levels for generalized knowledge is introduced. The permanent learner model is updated after collaborative learning has been finished. By extracting collaborators' appropriate permanent learner models, personalized CPT for a solution network of exercise for each collaborative learning can be generated.

This paper is organized as follows. Chapter 2 presents the framework and modeling method of the temporal learner model for collaborative learning in mathematics. Chapter 3 then describes the method for constructing the permanent learner model from temporal collaborative learning. Chapter 4 gives the experimental result and a brief discussion of them, and Chapter 5 concludes this paper.

2 Approach

2.1 Temporal Learner Model

The temporal learner model represents learner's level of understanding for knowledge which constitutes exercise and is created during collaborative learning. Several studies on constructing learner models have been carried out [7],[8]. The target of these studies is the learner who directly uses the system. However, in this paper, our objective is to model not user but collaborators.

This research focuses on collaborative learning of mathematical exercises in high school. There are several answering steps which correspond to the specific answering methods, and those are associated with each other in the context of solving exercise. We define an answering method specialized to an exercise as *exercise knowledge* in contrast to generalized knowledge called *domain knowledge.*

In the collaborative learning, collaborators do not always utter all the knowledge that they understand. A user estimates the collaborators' understanding levels over exercise knowledge on the basis of utterances and related solution steps. The solution network is introduced in order to represent the exercise knowledge of the exercise associated with the solution steps. Fig.1 shows an example of the solution network. Nodes correspond to exercise knowledge. Each node has keywords, which characterize the exercise knowledge, and CPT ,which shows degrees of association with previous exercise knowledge. There are two types of degrees of association. One is a conditional probability which represents the degree that this exercise knowledge is understood when previous exercise knowledge was understood. The other is a conditional probability which indicates the degree that this exercise knowledge is understood when previous exercise knowledge was not understood. Each link corresponds to the sequence relations between exercise knowledge.

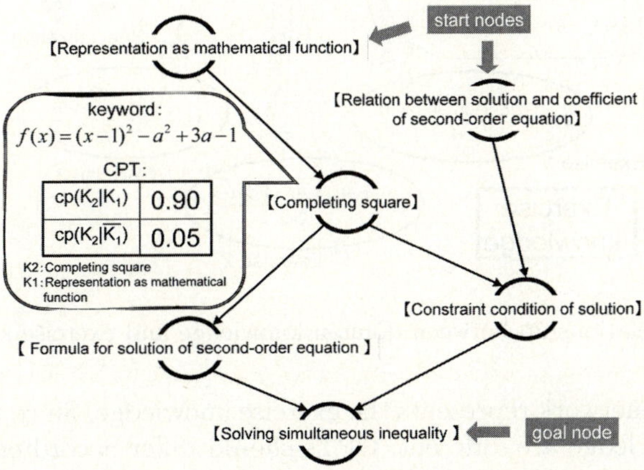

Fig. 1. Example of solution network

The temporal learner model consists of understanding level and trust level for individual exercise knowledge. The understanding level and the trust level take values from 0 to 1. The higher the value of the understanding level is, the better the corresponding exercise knowledge is understood. Also, the higher the value of the trust level is, the more trustworthy the corresponding understanding level is.

When an utterance has been occurred, the target exercise knowledge in the solution network is determined on the basis of the keywords. The understanding level of the target exercise knowledge is estimated according to type of utterance and its trust level is set to 1. Then, the understanding levels of the related exercise knowledge are calculated by understanding level of the target exercise knowledge and the CPT. Their trust levels are also set on the basis of the distance between the target exercise knowledge and related exercise knowledge.

2.2 Domain Knowledge and Exercise Knowledge

Domain knowledge is general knowledge of the mathematical area which corresponds to the solution methods or formulas, while the exercise knowledge is the knowledge which is specialized to mathematical exercises. Fig.2 shows an example of domain knowledge and exercise knowledge. The domain knowledge corresponds to solution methods, formulas and so on. The exercise knowledge is generated from the domain knowledge by assigning appropriate values which fit to an exercise.

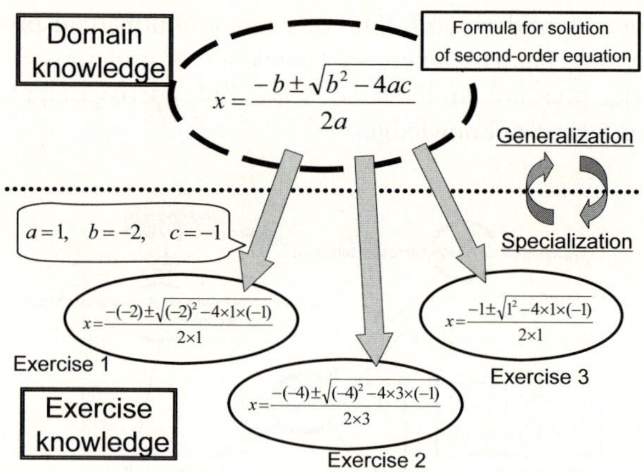

Fig. 2. Relationship between domain knowledge and exercise knowledge

The solution network represents the exercise knowledge. Since understandings of domain knowledge are different, CPTs should differ according to the collaborators. CPT can be estimated based on collaborators' understandings between two corresponding domain knowledge. The understandings of the domain knowledge can be grasped by collaborators' understanding levels for each exercise.

Therefore, in this work permanent learner models of collaborators which show the association between the domain knowledge are created after collaborative learning has been finished. Learners who understand the domain knowledge are considered to understand the exercise knowledge. With the permanent learner models of the collaborators, the solution network can be personalized to each collaborator. Whereby, the temporal learner models of collaborators can be created to be more suitable for the collaborators.

2.3 Collaborators' Understandings from User's Viewpoint

Impressions of the collaborators which the user has depend on their behaviors in the latest collaborative learning. So, the user recognizes current collaborators' levels of understandings as that was recognized in the latest collaborative learning with the collaborators. Fig.3 shows an example of collaborators' impressions along the time sequence. In this example, the learners A, B and C participated in the collaborative learning 1. Then, the learners A, B and D participated in the collaborative learning 2. After these learnings, the learners B and D have impression of the learner A as those they have gotten in the collaborative learning 2. On the other hand, the learner C has impression of the learner A as that he has gotten in the collaborative learning 1. That is, the impression of the collaborators corresponds to the permanent learner model of the collaborator which was updated after the last collaborative learning has been finished.

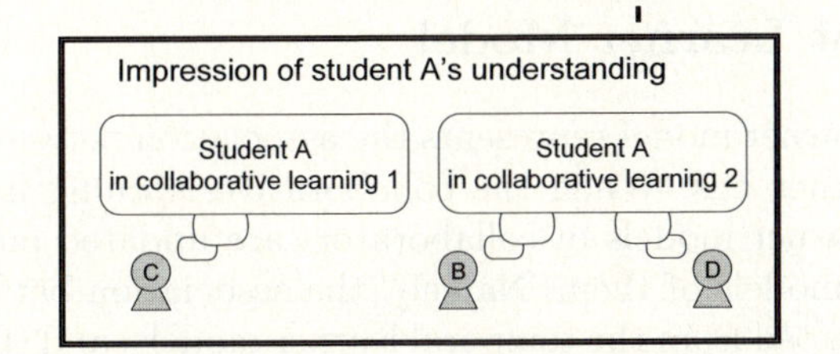

Fig. 3. Example of collaborators' impressions

2.4 Overview of Proposed System

Fig.4 shows an overview of proposed system. The server has each learner's permanent model along with the time sequence. The permanent learner models show the association between the domain knowledge. After the collaborative

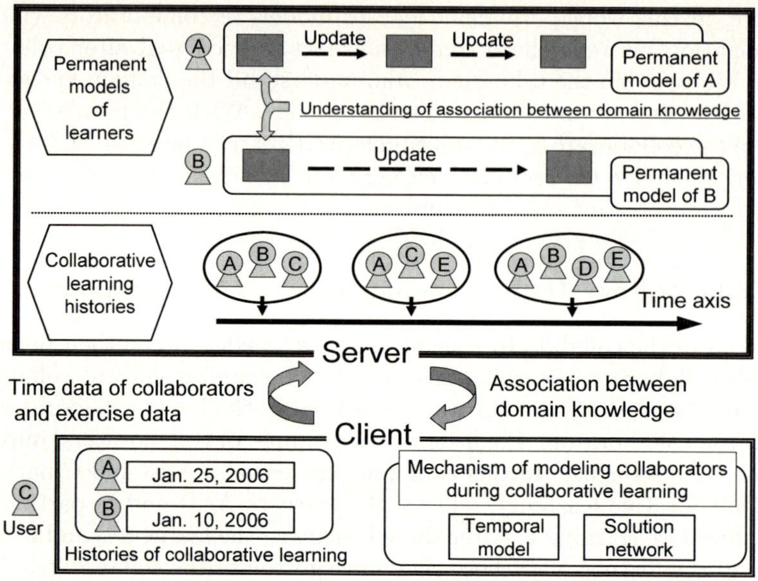

Fig. 4. Conceptual image of system

learning has been finished, these permanent models are updated on the basis of the temporal learner models.

On the other hand, the client has histories of collaborative learning in which when and who the user learned with are described. In the beginning of the collaborative learning, the client sends latest time data which the user learned with the corresponding collaborator and the domain knowledge about the exercise which is used in this collaborative learning. On the basis of the association between the domain knowledge which is sent from the server, the client constructs the personalized solution network.

3 Permanent Learner Model

The permanent learner model represents the association between domain knowledge which a learner has. When the collaborative learning has been finished, the permanent learner models of collaborators are updated on the basis of the temporal learner models of them. Namely, the association between the exercise knowledge is extracted from the temporal learner models, and the association between the corresponding domain knowledge within the permanent learner model is updated.

The following equation describes an updating method of the association between the domain knowledge a and b. The domain knowledge a is assumed to be solved before the domain knowledge b. $AD_{new}(a,b)$ is the new association between the domain knowledge a and b after the update, while $AD_{old}(a,b)$ shows the old association. $\Delta AD(a,b)$ represents the degree of the change of the association between the domain knowledge a and b during the collaborative learning.

$$AD_{new}(a,b) = AD_{old}(a,b) + \Delta AD(a,b) \tag{1}$$

$\Delta AD(a,b)$ reflects the temporal learner model which represents the understanding between the exercise knowledge. That is, if collaborator understands the exercise knowledge, the collaborator also understands its domain knowledge. The exercise knowledge which corresponds to the domain knowledge a and b are represented as a' and b', respectively. The association between the exercise knowledge a' and b' is represented as $AE(a',b')$. $AE(a',b')$ is indicated as the conditional probability $cp(b'|a')$.

$$AE(a',b') = cp(b'|a') = \frac{U(a' \wedge b')}{U(a')} \tag{2}$$

$U(a')$ is the understanding level of the exercise knowledge a'. Since the exercise knowledge a' should be derived before the exercise knowledge b', a' has a strong association with b'. So, the following equation is derived.

$$U(a' \wedge b') \simeq U(b') \tag{3}$$

From the equations 2 and 3, $AE(a',b')$ is represented as the equation 4.

$$AE(a',b') \simeq \frac{U(b')}{U(a')} \tag{4}$$

The user cannot be aware of the collaborators' understandabilities if they have not performed extremely good/bad. So, the extended sigmoid function in which the threshold value is 0.15 and the gain value is 20 is applied. Finally, with $T(a')$ and $T(b')$ which are the trust level of $U(a')$ and $U(b')$, $\Delta AD(a,b')$ is defined as follow.

$$\Delta AD(a,b) = \frac{T(a') + T(b')}{2} \times \Big(dif_{ab} \times sigmoid(20\,(|dif_{ab}| - 0.15))\Big) \tag{5}$$

$$\Big(dif_{ab} = AE(a',b') - AD_{old}(a,b)\Big)$$

The equation 5 reduces the update of the association between not-uttered exercise knowledge because the trust level for the not-uttered exercise knowledge is small.

4 Experiment

The proposed mechanism is embedded into the our system in which other learners' understanding levels are modeled from user's viewpoint [4],[5]. This system provides users with communication tool so as to exchange their opinions efficiently. In collaborative learning, the proposed mechanism creates the temporal learner models of collaborators on the basis of utterances. When the learning has been finished, this mechanism creates and updates the permanent learner models of collaborators on the basis of their temporal learner models.

We experimented in order to evaluate the effectiveness of permanent learner models. In this experiment, four students in our laboratory were asked to form groups randomly and to solve mathematical exercises of high school collaboratively four times. After these collaborative learnings, the four students were asked to study collaboratively. The exercise used in this learning is based on the solution network shown in Fig.1. Students mainly solved the exercise by using the exercise knowledge "representation as mathematical function", "completing square", "constraint condition of solution" and "solving simultaneous inequality" in Fig. 1. After the learning, we showed utterance of each student that are prepared experimentally and the students were asked to estimate the understanding level of other students who utter such utterances. In this experiment, two types of solution network were created. CPTs of one solution network were set on the basis of permanent learner models. CPTs of the other were prepared statically beforehand. We compared two temporal learner models which are modeled on each solution network.

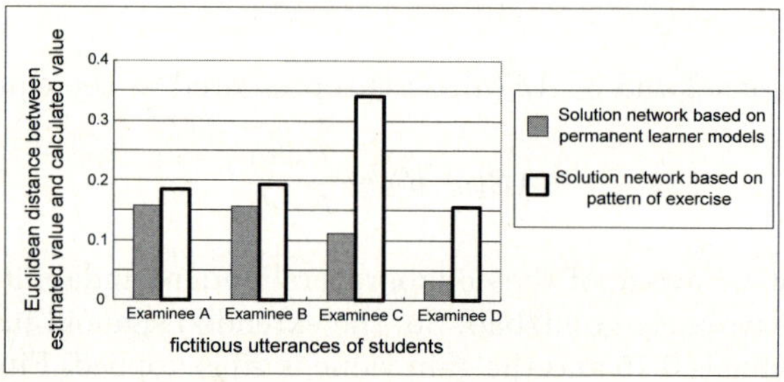

Fig. 5. Estimated understanding level for each exercise knowledge

Fig.5 shows experimental results with the Euclidean distance for the averages of values estimated by students and those calculated by the system. The experimental results show that the solution network based on the permanent learner models represents examinees' viewpoints for collaborators more precisely in all the cases. Especially, since the examinees C and D uttered the exercise knowledge "solving simultaneous inequality" in the past collaborative learnings, the other examinees can estimated that the students C and D understand the exercise knowledge "solving simultaneous inequality" more correctly. The experimental results also show that proposed system can estimate understanding levels for the exercise knowledge which was uttered in past collaborative learnings.

5 Conclusion

In this paper, we improved the mechanism which models collaborators in order to personalize solution networks which reflect common collaborative learning experiences. The permanent learner models which are based on the domain knowledge were introduced for that purpose. By updating the permanent learner models in the end of a collaborative learning and extracting the association between the domain knowledge when generating solution network, the understandabilities of collaborators can be extracted more precisely. In our future work, we should perform a long-term experiment in which many learners form different groups and study collaboratively.

Acknowledgement

The authors would like to thank the 21st Century COE(Center of Excellence) Program for 2002, a project titled Intelligent Media (Speech and Images) Integration for Social Information Infrastructure, proposed by Nagoya University.

References

[1] P. Dillenbourg (eds): Collaborative Learning - Cognitive and Computational Approaches, Elsevies Science Ltd. (1999).

[2] M. L. Bote-Lorenzo, L. M. Vaquero-Gonzalez, G. Vega-Gorgojo, Y. A. Dimitriadis, J. I. Asensio-Perez, E. Gomez-Sanchez, and D. Hernandez-Leo: A Tailorable Collaborative Learning System That Combines OGSA Grid Services and IMS-LD Scripting, Proc. of CRIWG 2004, pp.305-321 (2004).

[3] R. S. Raitman, W. Zhou, and P. Nicholson: Exploring the Foundations of Practicing Online Collaboration, Proc. of ICWL 2003, pp.532-541 (2003).

[4] A. Komedani, T. Kojiri, and T. Watanabe: Modeling Understanding Level of Learner in Collaborative Learning Using Bayesian Network, Proc. of KES 2005, pp.665-672 (2005).

[5] A. Komedani, T. Kojiri, and T. Watanabe: Modeling Learner in Collaborative Learning Based on Bayesian Network, Proc. of INAP 2005, pp.87-96 (2005).

[6] S. Russell and P. Norvig: Artificial Intelligence -A Modern Approach, Prentice Hall (2002).

[7] J. Reye: A Belief Net Backbone for Student Modeling, Proc. of ITS 1996, pp.596-604 (1996).

[8] A. S. Gertner and K. Vanlehn: Andes: A Coached Problem Solving Environment for Physics, Proc. of ITS 2000, pp.133-142 (2000).

A Multi-Criteria Programming Model for Intelligent Tutoring Planning

Ruihong Shi[1,2] and Peng Lu[3]

[1] Key Laboratory of Intelligent Information Processing, Institute of Computing Technology, Chinese Academy of Sciences
[2] Graduate School of the Chinese Academy of Sciences
Ruihong.shi@gmail.com
[3] College of Computer Science and Technology, Beijing Institute of Technology
mym_chn@hotmail.com

Abstract. This paper proposed a practical approach to personalized tutoring planning by exploiting existing tutoring resources (e.g., a book, a courseware). More exactly, it does not build an instructional course from scratch - from the domain curriculum, as most Intelligent Tutoring Systems (ITS) do. Instead, information in the curriculum model is used as metadata, together with other metadata, to describe tutoring resources. Given a learning requirement (learning objectives and/or constraints), it finds out the most appropriate tutoring resource(s) and proposes a proper learning sequence of them. Based on the proposed fuzzy instructional model and learner model, the tutoring planning problem is defined as a multi-criteria programming (MCP) model.

1 Introduction

Intelligent Tutoring Systems (ITS) are a generation of Computer Aided Instruction (CAI) systems that provide individualized tutoring or instruction. Based on the expectations of ITS, researches of ITS mainly focus on the diagnosis of student's knowledge structures, skills, and learning styles, course generation to meet the requirements and preferences of individual users, adaptation of the instruction according to the practical learning progress, and effective and efficient editorial and delivery techniques.

ITS provide a cost-effective way of personalized learning. Conceptualization of the instructional components (e.g., knowledge objects, pedagogical strategies, cognitive model), however, is time-consuming and expertise-intensive. And the maintenance cost is high. Consequently, ITS fail to prove their usefulness in wider academic or business environments. While some researchers endeavor to address issues like reusability, shareability and interoperability of ITS components (for example [1], [2], [3]), we propose a method, called Macro Tutoring Planning (MTP), to deal with the practicability problem. Instead of building an instructional course directly from the curriculum - a structured set of learning objects (as did in most ITS), MTP finds out one or a bundle of most appropriate tutoring resources and arranges a proper learning sequence of them according to individual's personalized learning requirements (i.e., learning objectives and/or constraints) and characteristics (using a user model). Here,

B. Gabrys, R.J. Howlett, and L.C. Jain (Eds.): KES 2006, Part I, LNAI 4251, pp. 780 – 787, 2006.
© Springer-Verlag Berlin Heidelberg 2006

a tutoring resource can be either traditional (such as a book, a face-to-face classroom training) or intelligent (such as an animated web page or a virtual courseware). At this very first stage, the proposed approach is not designed to be able to integrate different parts of multiple tutoring resources into one singular tutoring resource, for the heterogeneous nature of various tutoring resources makes this task even more difficult than the problems faced by present ITS. Rather, our basic motivation is to make Intelligent Tutoring Planning (ITP) more practical, in a sense that ITP can guide a learner to the right resource, among many others, at the right time.

This idea is supported by a recent report [4] which states that the activity in ITS has transitioned, over the last decade, from a focus on complete systems and their evaluation to exploring ways to improve specific functional components and their interfaces (e.g., natural language input). This is, in our opinion, more reasonable. For diverse system architectures may be derived when looked from different viewpoints, and there is no omnipotent tutoring strategy.

The metadata for describing tutoring resources and the fuzzy instructional model (FIM) are described in Section 2 and Section 3, respectively. Section 4 gives the multi-criteria programming (MCP) model. And an application example is depicted in Section 5. Conclusion is drawn in Section 6.

2 Metadata

We generalize the definition of *tutoring resource* to any available thing that provides knowledge about certain learning object, either formally (i.e., ITS) or informally (e.g., human teachers) planned. And we adopt an application-oriented view in defining the metadata, that is, only data useful for choosing and ordering tutoring resources are considered. Tutoring Resource Metadata (TRM) consists of three disjoint groups. The first group (e.g., Language, Cost, Version) is used to describe general information about tutoring resources.

The second group comprises properties describing the contents of tutoring resources. In our model, a tutoring resource's content refers to a set of learning objects involved in it. There are many types of *learning objects* (LOs), such as entities (objects in the world), concepts (abstract artifacts), skills (abilities to solve a problem), and so on. Note that a tutoring resource may not teach a topic from the most basic learning objects. So, with regard to a tutoring resource, we consider:

- Contributions: a set of learning objects that can be learned and corresponding mastery levels that can be reached after learning, and
- Backgrounds: a set of learning objects (together with related mastery levels) that should be known before learning

The third group includes some pedagogical properties, such as Format (e.g., text, image, video, simulation), TeachingStyle (e.g., narrative, teach-learn, partner), and LearningTime (the typical learning time length needed to learn a tutoring resource).

And each tutoring resource has a unique identifier and a description. The metadata are used for both content planning (i.e., choosing tutoring resources) and delivery planning (i.e., sequencing tutoring resources).

3 Fuzzy Instructional Model

ITP is based on the library of learning objects (LOs) and dependences (i.e., prerequisite relation) between them. Besides, several learning objects can aggregate into more general objects, called *learning topics* (LTs). LTs provide a succinct manner to refer to a set of LOs from certain aspect. With the aspect being general (containing more LOs) or specific (containing less LOs), we have different granular learning topics. That is, the learning topics are hierarchically organized. A topic in higher level of the hierarchy covers several topics in the lower level (called *subtopics*). In addition, a super topic (x is called *supertopic* of y, if y is subtopic of x) may include LOs outside the union of its sub topics – LOs that are not restricted to either sub topic.

It is worth mentioning that LTs are fuzzy concepts. That is, different LOs are not equally important to one learning topic. We model a LT as a fuzzy set, and the membership function represents how important a LO is with respect to it. And we assume that the more important a LO is relative to a LT, the more possible the LO belongs to the LT.

The instructional domain is modeled as $D = (T, O, M, R, L, f, \alpha, \beta, \gamma)$ where:

T is a set of learning topics;
O is a set of learning objects;
M is a set of tutoring resources;
$R \subset T \times T$ is a partial order relation (called subtopic relation) on T, $(t', t) \in R$ means that t' is a sub topic of t (t is a super topic of t');
L is a totally ordered set, elements of which are *mastery levels*;
$f : O \times L \to 2^{O \times L}$ is the prerequisite relation. $(o', l') \in f(o, l)$ means that one must master o' to at least level l' before he can master o to level l. The multiple prerequisites (if any) of a learning objects are "and" interrelated;
$\alpha : O \times T \to [0, 1]$ is the membership function;
$\beta : M \to 2^{O \times L}$ is the background function. $\beta(m)$ is the required knowledge that makes one capable of learning m well;
$\gamma : M \to 2^{O \times L}$ is the contribution function. By $(o, l) \in \gamma(m)$ we mean that m is designed to make learner master learning object o to level l;

The *subtopic* relation R is defined using the inclusion relation on fussy sets. Given two learning topic $t_1, t_2 \in T$, $t_1 = \{(o, \mu_1(o))\}$, $t_2 = \{(o, \mu_2(o))\}$, t_1 is called subtopic of t_2, denoted as $t_1 R t_2$, if $\forall o \in O$, $\mu_1(o) \le \mu_2(o)$, where $\mu_i(o) = \alpha(o, t_i)$ (i =1, 2).

Before going on to the problem definition, we first introduce some notations. In the fuzzy instructional model, for succinctness, f consists of only direct prerequisite relations. And indirect prerequisites can be computed using the composition operation with f. $\Gamma_f(o,l)$, which denotes all the prerequisites (direct and indirect) of learning object o of level l, is calculated as follows:

$$f^i = f \circ f^{i-1} . \quad (1)$$

$$\Gamma_f = \bigcup_i f^i . \quad (2)$$

We use symbol " $\prec$ " to represent the *not-worse-than* relation between mastery levels. Note that L is a totally ordered set, the anterior element in the order is considered better than the posterior element. We distinguish five grades of mastery levels, L = {perfect, good, average, poor, not-at-all}. So, we have "perfect" $\prec$ "perfect", "good" $\prec$ "average", and so forth. If $x \prec y$ does not hold (x is worse than y), we denote as $x \not\prec y$ (e.g., "good" $\not\prec$ "perfect"). Then we can identify all the tutoring resources that can teach a learning object o to at least mastery level l:

$$\Omega(o,l) = \{m \in M \mid \gamma(m,o) \prec l\} \ . \tag{3}$$

$\pi : L \times L \to L$ is defined to get the better (in the sense of $\prec$) element of a pair of mastery levels. For example, π(good, average) = good.

4 Multi-Criteria Programming Model

Personalized tutoring planning generates a learning plan with respect to: (1) learning requirements (learning objectives and preferences), (2) background knowledge (learning objects has already known and how well they are mastered), (3) learning preferences (which kind of tutoring resources the user is shown to be more receptive, and (4) constraints (number of spare hours and monetary budget).

Learning requirements are discussed in 4.1. User's background knowledge and learning preferences are considered in the user model (4.2). 4.3 describes the multi-criteria programming model, taking into consideration the constraints.

4.1 Learning Requirements

A learning requirement is a triple $A = (X, W, C)$ where $X \subset T \cup O$ is a set of learning objectives, $W : X \to [0, 1]$ represents user preferences on X, C is a set of constraints. We consider two types of constraints: total learning time E and cost budget B.

Before inputted to the tutoring planning module, learning objectives are firstly weighted. Weight of explicitly specified learning object ($o \in O$) equals its preference value, whereas weight of implicitly mentioned learning object ($t \in T$, $\alpha(o, t) > 0$) is the product of user preference value and its importance coefficient. For each $o \in O$,

$$W'(o) = \begin{cases} W(o) & o \in X \cap O \\ \max\limits_{t \in X \cap T} (W(t) \cdot \mu_t(o)) & else \end{cases} \ . \tag{4}$$

Then weighted learning objectives are filtered: $A' = (X', W', C)$, where $X' = \{o \in O \mid W'(o) > \varepsilon\}$. Threshold ε is set empirically or dynamically, e.g., to make $|X'| \leq k$.

4.2 User Model

Two classes of personal information are considered in the user model: learning preferences and cognitive states. Learning preferences manifest of which kind of tutoring resources an individual is more receptive. The two metadata, Format and TeachingStyle, are used for this purpose. However, for the time being, we do not include these factors in our formal model.

Cognitive states refer to learning ability and background knowledge. Learning ability influences real learning effect of a tutoring resource (may differ from the contributions of the resource). We indirectly model this influence by introducing the concept of *time effort* (actual time length elapsed to learn a resource). Given a tutoring resource *m*, the time effort one invests for *m* is estimated as follows:

$$coe = \frac{1}{1+(\frac{\theta_1}{d-\theta_2})^2} . \tag{5}$$

$$te(m) = \frac{LearningTime(m)}{coe} . \tag{6}$$

In (5), d (a number) is the learning ability, and θ_1, θ_2 are two empirically set constants. As can be seen from Figure 1., the higher the learning ability, the more approximate the time effort is to the typical learning time length. On the contrary, if one is slow in studying things, he will spend more time. Then the total time one uses to learn a set of tutoring resources $\{m_1, \ldots, m_k\}$ is $\Sigma te(m_i)$ (i = 1, .., k), which is compared with the constraint of total learning time E in the MCP model (section 4.3).

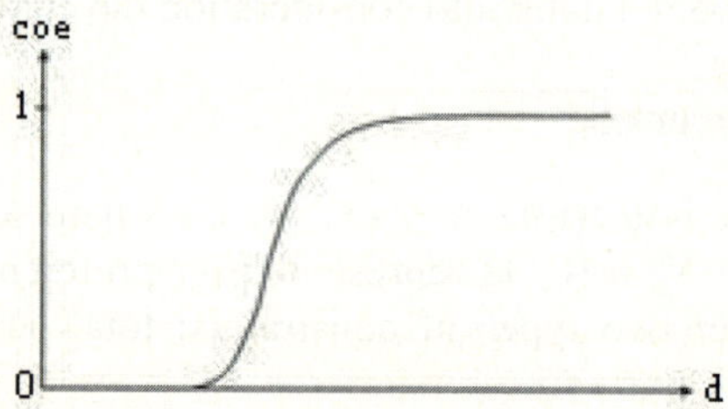

Fig. 1. Coefficient for estimating the time effort for learning a tutoring resource

Learner's background knowledge is a labeled set of learning objects, with the label being the corresponding mastery level. And it evolves as the learning process proceeds. Suppose that user knowledge at time index i is K^i, after learning a tutoring resource, say m, user knowledge is updated according to (7) or (8):

$$K^{i+1}(K^i, m) = \{(o,l) \mid o \in O, l = \pi(\Lambda(K^i, o), \Lambda(\gamma(m), o))\} . \tag{7}$$

$$K^{i+1}(K^i, m) = K^i . \tag{8}$$

where

$$\Lambda(X, o) = \begin{cases} l & (o,l) \in X \\ not-at-all & else \end{cases} . \tag{9}$$

(7) describes the "good" situation, that is, user knowledge satisfies the prerequisite conditions of m ($\forall(o, l) \in \beta(m), \Lambda(K^i, o) \prec l$). And (8) describes the "bad" situation - user knowledge does not satisfy at least one of the prerequisite conditions of m ($\exists(o, l) \in \beta(m), \Lambda(K^i, o) \nprec l$).

There are three ways to obtain user's initial knowledge K^0: self-estimation, pre-test, or retrieval from the user database (in case the user ever used the system).

4.3 Intelligent Tutoring Planning

Given a learning requirement, the ITP problem is to find a sequence of tutoring resources $\{m_1, m_2, \ldots, m_p\}$ such that, after learning them one by one, learner's final knowledge K^p is optimal.

Definition 1: An *intelligent tutoring planning domain* is a triple $S = (D, A, U)$, where $D = (T, O, M, R, L, f, \alpha, \beta, \gamma)$ is the instructional domain, $A = (X, W, C)$ is a learning requirement, and $U = (d, K^0)$ is the user model.

Definition 2: Given an intelligent tutoring planning domain $S = (D, A, U)$, a sequence of tutoring resources $M^p = \{m_1, m_2, \ldots, m_p\}$ is called a *learning plan* of S.

Definition 3: Given an intelligent tutoring planning domain $S = (D, A, U)$ and a learning plan $M^p = \{m_1, m_2, \ldots, m_p\}$ of S, $K^p(((K^0, m_1)\ldots), m_p)$ is called the *learning effect* of M^p.

Given a learning effect K^p, the knowledge that a learner mastered (ΣK^p) and the knowledge level (ΠK^p) are defined as follows, respectively:

$$\Sigma K^p = \{o \mid \Lambda(K^p, o) \prec \text{"average"}\} \ . \tag{10}$$

$$\Pi K^p = \sum_{o \in \Sigma K^p, (o,l) \in K^p} \tau(l) \ . \tag{11}$$

$\tau(l)$ is the function that allows to switch from qualitative labels to quantitative values.

We propose three different strategies to compare learning effects. Suppose there is a pair of learning effects K^p, K^q. By $K^p \prec K^q$, we mean that "K^p is better than K^q".

- *Maximal knowledge strategy* (MKS): maximize the knowledge amount

$$K^p \prec K^q \quad \mathit{iff} \quad card(\Sigma K^p) \geq card(\Sigma K^q) \ . \tag{12}$$

- *Maximal knowledge level strategy* (MKLS): maximize the knowledge level

$$K^p \prec K^q \quad \mathit{iff} \quad \Pi K^p \geq \Pi K^q \ . \tag{13}$$

- *Hybrid strategy* (HS): maximize knowledge amount as well as knowledge level

$$K^p \prec K^q \quad \mathit{iff} \quad card(\Sigma K^p) \geq card(\Sigma K^q) \wedge \Pi K^p \geq \Pi K^q \ . \tag{14}$$

Definition 4: Given an intelligent tutoring planning domain $S = (D, A, U)$ and a set Z of learning plans, $M^* \in Z$ is called the *optimal learning plan*, if there exists no other learning plan $M^r \in Z$ such that $K^r \prec K^*$.

Definition 5: Given an intelligent tutoring planning domain $S = (D, A, U)$, the *intelligent tutoring planning problem* is to find the optimal learning plan M^*.

The above defined optimal learning plan considers only the learning objectives. When taking into consideration total learning time E and cost budget B, the tutoring planner must strikes a "delicate balance" among these competing objectives. A constraint can be dealt with as either hard (a tutoring resource sequence that violates the constraint is not considered as a learning plan, e.g., exceeds the time limit E) or soft (to maximize the utility along the constraint, e.g., as less payment as possible). Therefore, we get a multi-criteria programming (MCP) model. And the overall optimal learning plan can be obtained in two ways. One method is to give each competing objective a weight, and reformulate the qualitative comparison into quantitative ones (e.g., relative/absolute distance between the actual value and the goal value). Another approach is to apply the pareto-optimal strategy, i.e., a learning plan is better (than the other) if it is better for at least one objective and, at the same time, is not worse for the other objectives. Additionally, note that learner's preferences (W of A) can be easily added into the model.

5 Application Example

Almost all ITS build their applications in education and training environments. However, we would like to discuss a potential application in the tourism domain. In particular, we consider a trip as a learning process. On the one hand, we want to verify that the proposed model is domain-independent (though simplified). On the other hand, we attempt to introduce the achievements in ITS to new application fields.

The viewpoint that trip can be seen as learning is more and more accepted, since tourists are no longer satisfied by just going somewhere. Instead, they aspire to obtain something from their trips, e.g., escape from a perceived mundane environment, recovery from physical stress. We focus on a type of trips motivated by the need of widening one's horizons, or more briefly, cultural traveling. In other words, the tourists want to know something new to their knowledge. In addition, tourists have limited idle time and expense budget for a trip.

At an early stage of our work, we aim to automate pre-trip itinerary planning. Tourism attractions (things that attract people to go to a destination) are learning topics (e.g., culture) or learning objects (e.g., a building). And the tutoring resources correspond to tourist spots, places developed for tourists to visit. A learning plan is an itinerary (a sequence of tourist spots). The prerequisite relationships include causal relationships and time sequence relationships between tourism attractions. Various tourists prefer different visiting modes. One prefers to see a tourism attraction itself, while others prefer to listen about explanation for it or participate in activities revealing it. In addition, there are suggested visiting hours for each tourist spot.

Some domain-specific metadata are added, organized in a tourism ontology. In fact, we also defined an upper ontology to describe the representation framework of the tourism ontology, including some semantic restriction axioms and inference rules.

While introducing the defined MCP model into tourism domain, some hypotheses are no longer proper. For example, in the model, we force that learner's background knowledge must meet the prerequisite conditions of a tutoring resource in order to acquire the claimed learning effect. However, in tourism, the prerequisite relation only makes understanding of a learning topic more smoothly. Another main issue, not stressed in the MCP model, is the spatial relation between tourist spots which has a great impact on itinerary planning (through time and expense restrictions).

6 Conclusion

In this paper, we proposed an intelligent tutoring planner based on multi-criteria programming model, which designs a learning plan with respect to personalized learning objectives, user data and constraints. Content planning (choosing resources) and delivery planning (ordering resources) are accomplished in an integrated manner.

Many other researches study intelligent course composition. Knolmayer [5] did not employ the user model. Nkambou [6] also mentioned the role of resources in achieving instructional objectives, but he builds courses under ideal environments (without constraints). Xu et al. [7] also used labeled prerequisite relations (a label is a mastery level) between topics to dynamically design a learning plan. Its plan, however, must fully satisfy the learning goal (also under ideal conditions). The main contribution of our work is that it suggests an alternative perspective for developing ITS, i.e., building courses with existing tutoring resources instead of from scratch. Thus our approach can be considered as secondary-level instruction planning, in a sense that it leaves the specifics of tutoring resource development to other ITS applications. In this way, our approach is apt to provide more flexible configuration and general applicability in real-world learning situations, by taking advantage of merits of existing conventional as well as computer-aided tutoring resources.

References

1. Mühlenbrock, M., Tewissen, F., Hoppe, H.U.: A Framework System for Intelligent Support in Open Distributed Learning Environments. Int. J. Artificial Intelligence in Education, Vol. 9. The International Society for Artificial Intelligence in Education (1998) 256-274
2. Rodrigues, M., Novais, P., Santos, M.F.: Future Challenges in Intelligent Tutoring Systems - A Framework. Recent Research Developments in Learning Technologies: Proc. 3rd Int. Conf. Multimedia and Information & Communication Technologies in Education (2005) 129-134
3. Rosić, M., Glavinić, V., Stankov, S.: Distance Learning System based on Distributed Semantic Networks. EUROCON 2003, Computer as a Tool, Region 8, Vol. 2. IEEE (2003) 26-29
4. MYMIC LLC: Outstanding research issues in Intelligent tutoring systems. scientific and technical report (2004-4-21)
5. Knolmayer, G.F.: Decision support models for composing and navigating through e-learning objects. Proc. 36th Hawaii Int. Conf. System Sciences. IEEE (2003)
6. Nkambou, R.: Using fuzzy logic in ITS-course generation. Proc. 9th Int. Conf. Tools with Artificial Intelligence. IEEE (1997) 190-193
7. Xu, D.M., Wang, H.Q., Su, K.L.: Intelligent student profiling with fuzzy models. Proc. 35th Hawaii Int. Conf. System Sciences. IEEE (2002).

An Implementation of KSSL Recognizer for HCI Based on Post Wearable PC and Wireless Networks

Jung-Hyun Kim and Kwang-Seok Hong

School of Information and Communication Engineering, Sungkyunkwan University, 300, Chunchun-dong, Jangan-gu, Suwon, KyungKi-do, 440-746, Korea
kjh0328@skku.edu, kshong@skku.ac.kr
http://hci.skku.ac.kr

Abstract. A traditional study on recognition and representation of sign language based on the desktop PC have general restrictions conditionality in space and limitation of motion according to the technology of wire communication, and then the vision technology-based traditional sign langage recognition systems, recognition performance has the possibility of change and diversity according to the performance of the camera, change of background (colors) and illumination condition. To overcome these restrictions and problems, we implement embedded-sentential the Korean Standard Sign Language (hereinafter, KSSL) recognizer based on the post wearable PC for mobile and ubiquitous computing. The KSSL is a dialog system and interactive elements in the South Korea deaf communities. The advantages of our approach are as follows: 1) it improves efficiency of the KSSL input module according to the technology of wireless communication and user need not be constrained by the limitations of a particular interaction mode at any given moment, 2) it recognizes and represents continuous the KSSL of users with flexibility in real time and can offer to user a wider range of personalized and differentiated information more effectively, and 3) it is possible more effective and free interchange of ideas and information between deaf person and hearing person (the public). Experimental result shows an average recognition rate of 93.3% for the KSSL sentences and 96.9% for the KSSL words with significant, dynamic and continuous the KSSL.

1 Introduction

The sign language is defined any form of communication using gestures to represent words and ideas, especially an official system of hand gestures used by an aphasiac or a deaf people. That is, a sign language is a language which uses manual communication instead of sound to convey meaning - simultaneously combining hand shapes, orientation and movement of the hands, arms or body, and facial expressions to fluidly express a speaker's thoughts [1]. The related studies about recognition and representation technology of sign language is progressing actively in many different countries such as the Americas, Japan, Europe, South Korea etc. to a hand-signal recognition for crane operation of the construction field, the control system of mobile robot and avatar using hand action recognition, the communication with the hearing people (the public) through sign language recognition, the hand action recognition in

B. Gabrys, R.J. Howlett, and L.C. Jain (Eds.): KES 2006, Part I, LNAI 4251, pp. 788 – 797, 2006.
© Springer-Verlag Berlin Heidelberg 2006

virtual reality and so on. Especially, "standard sign language translation system" developed jointly by the professors of the KIST (The Korea Institute of Science and Technology) and the Samsung Electronics Co., Ltd. in the South Korea is sign language translation system that recognize and represents the sign language of word or morpheme level, and it is possible recognition about 20 basic sign language and 31 finger-spelling. Also, the Hitachi Ltd. in the Japan announced "Japanese - sign language translation technology" that a sign language animation is created automatically after input Japanese sentence [2].

But, the general purposes and features of these studies are as follows: 1) it has general purposes to the control of optional hand signal and sign language recognition based on the desktop computer and the technology of wire communication, 2)) because traditional sign language recognition system used image capture system or video processing system for an acquisition of sign language, it is impossible correct measuring with gesture data and needs complex computation algorithm, and 3) it is pursuing free communication with hearing people (the public) through sign language recognition and representation based on word and morpheme.

In this paper, to implement user interface to guarantee mobility of portable terminal and efficient dialog system between a deaf person and hearing person, we proposes and implements embedded-sentential the KSSL recognizer based on the ubiquitous environment using the post wearable PC platform, blue-tooth module and wireless haptic devices. The advantages of our approach are as follows: 1) it improves efficiency of the KSSL input module according to the technology of wireless communication and user need not be constrained by the limitations of a particular interaction mode at any given moment, 2) it contributes to user's the convenience and recognizes and represents continuous the KSSL of users with flexibility and presented in real time aurally and visually to maximize comprehension, and 3) Because the ability of communication and representation with sentential the KSSL recognizer are very superior more than recognizer based on word and morpheme, it is possible more effective and free interchange of ideas and information between a deaf person (and aphasiac) and hearing person (the public).

The composition of this paper are 1) regulation of components the KSSL in Section 2, 2) input module of the KSSL using wireless haptic devices in Section 3, 3) training and recognition models using RDBMS in Section 4, 4) fuzzy max-min module for the KSSL recognition in Section 5, and 5) experiments and results in Section 6. Finally, this study is summarized in Section 7 together with an outline of challenges and future directions.

2 Regulation and Components of the KSSL

Sign language differs from oral language in its relation to writing. The phonemic systems of oral languages are primarily sequential: that is, the majority of phonemes are produced in a sequence one after another, although many languages also have non-sequential aspects such as tone. As a consequence, traditional phonemic writing systems are also sequential, with at best diacritics for non-sequential aspects such as stress and tone. The signs are formed by 7 sign formation principles: indication; imitation; metonymy; metonymy-indication; metonymy-movement; blending; home-sign according to structure method of sign and gestures. Also, we prescribed basic elements (meaningless units) that correspond to phoneme of spoken language by "sign language morpheme" in the Korean sign language. That is, there are 5 classes of cheremes (a motor analogue to a phoneme) in the KSSL: configuration of the hand-DEZ (designator); position of the hand-TAB (tabulator); movement of the hand-SIG (signation); direction of the hand; and non-manual markers (or facial expression) [3]. To implement significant the KSSL recognition interface in real time, in this study, we selected 25 basic KSSL through analysis of "Korean Standard Sign Language Tutor (hereafter, KSSLT) [4]" and according to a classification standard of 'sign language morpheme'. And 23 hand gestures necessities of the KSSL are classified as hand shapes, pitch and roll degree. Consequently, we developed 117 significant the KSSL recognition models (with 45 the KSSL sentence and 72 the KSSL word) according to associability and presentation of hand gestures and basic KSSL. The example of associability of hand gestures and basic KSSL about "the date (numbers-day-month)" in 117 significant the KSSL recognition models are shown in Table 1.

Table 1. The example of the KSSL about "the date (finger numbers-day-month)"

Object Language	Description of the Korean Sign Language
Finger-Number	1~30(or 31) : Signing of numbers
Month	Represent signing of numbers that correspond to "1(one)" with a left hand (indication of month), and draw crescent with thumb, index finger of a right hand
Day	Spread out thumb, index fingers of a both hands and rises on in front of chest.
Hand Gesture	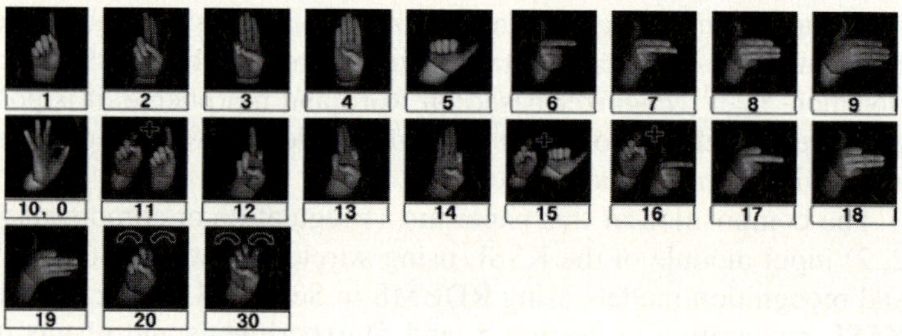

3 Input Module of the KSSL

In vision technology for image processing and recognition, recognition performance has the possibility of change and diversity according to the performance of the camera, change of background (colors) and illumination condition. In order to overcome these problems, in this paper, we use 5DT company's 5th data glove system (wireless) and fastrak® which is one of the most popular input devices in the haptic application field.•5th data glove system is basic gesture recognition equipment that can capture the degree of finger stooping using fiber-optic flex sensor and acquires data through this. Also, because it has pitch and roll sensor inside, the pitch and roll of wrist can be also recognized without other equipment possible. For motion information of 5th data glove system's each finger, user's dynamic gesture data is saved by f1=thumb, f2=index, f3=middle, f4=ring and f5=little in regular sequence. Each flexure value has a decimal range of 0 to 255, with a low value indicating an inflexed finger, and a high value indicating a flexed finger. Also, motion tracker is the accurate electromagnetic motion tracking system available, and it is the solution for accurately computing position and orientation through space. With real time, six degrees of freedom - position (X, Y, and Z Cartesian coordinates) and orientation (azimuth, elevation, and roll) - tracking and virtually no latency, this system is ideal for head, hand, and instrument tracking [5].

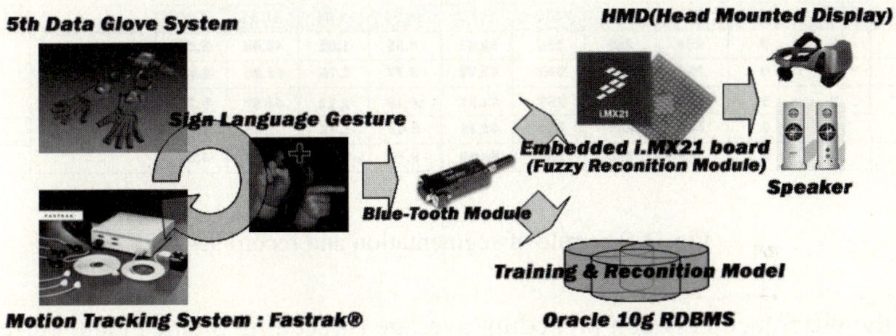

Fig. 1. The architecture of sign language input module

The captured dynamic the KSSL data of various users is transmitted to embedded i.MX21 board which is the post wearable PC platform for mobile and ubiquitous computing and database server for RDBMS module using blue-tooth module. The KSSL data that is transmitted to server is used to as a training data for the KSSL recognition model by analytic function of RDBMS SQL. And, the KSSL data that is transmitted to i.MX21 board is used as the input to fuzzy recognition module for significant the KSSL recognition. The architecture of the KSSL input module is shown in Fig. 1.

4 Training and Recognition Models Using RDBMS

The RDBMS is the main stream database management system that maintains data records and indices in tables and their relationships may be created and maintained across and among the data and tables. Also, often used for transaction processing and data warehouses and has the capability to recombine the data items from different tables, providing powerful tools for data usage [6], [7]. The RDBMS is used to classify saved the KSSL document data from the sign language input module (data gloves and motion tracker) into valid gesture record set and invalid record set (that is, status transition record set) and to efficiently analyze valid record set. The analytic function features, which are recently introduced in the SQL language, perfectly provide the analysis power for gesture validity analysis. According to the logic in source code, even though a record set is, based on the above process, decided as valid, the record set is regarded as a status transition gesture record set. If the valid record set contains less than 5 valid records in sequence. The continuous the KSSL from data gloves and motion tracker are segmented into the significant record and transition record in 98% believability by the analytic function of SQL. A rule to segment valid gesture record set and invalid record-set (changing gesture set) is shown in Fig. 2.

1. Difference Between Preceding Average(preceding 3 and 1) and Current Row Value
2. Decide validity : Preceding Average - Current Value = 52-17 = 35

Average Between 3 Preceding and 1 Preceding from Current Row : 52

1. Difference between Preceding Average and Current Row Value > 5
2. 'X'(Invalidity Record) check

SerialNO	F1THUMB	F1INDEX	F1MIDDLE	F1RING	F1LITTLE	F1X-POSI.	F1Y-POSI.	F1Z-POSI.	F1_Px	F1_Py	GESTURE DATA...	VALIDITY
22	73	86	255	255	255	50.90	-14.53	-4.74	55.42	-5.71		X
23	51	48	255	255	255	52.05	-11.97	-2.13	53.29	-4.95		X
24	32	15	255	255	255	53.99	-9.50	0.01	50.02	-4.75		X
25	17	0	255	255	255	51.03	-10.50	-1.05	51.11	-4.83	Omission...	X
26	7	0	255	255	255	52.01	-9.85	-1.02	49.85	-5.23		
27	0	0	255	254	255	52.78	-8.77	-1.75	51.85	-6.02		
28	0	0	253	254	255	49.87	-10.18	-2.12	48.97	-5.27		
29	0	0	255	255	254	48.71	-9.43	-1.92	52.85	-4.89		
30	0	0	255	255	255	51.09	-8.75	-1.75	47.93	-4.94		

Fig. 2. The rule of segmentation and record set

- If the difference between preceding average (preceding 3 and 1) and current row value is over 5, the current value is regarded as transition sign language record.
- If one of data gloves and motion tracker data value are over 5, current value data is also regarded as changing sign language record.

5 Fuzzy Max-Min Module for the KSSL Recognition

The fuzzy logic is a powerful problem-solving methodology with a myriad of applications in embedded control and information processing, and is a paradigm for an alternative design methodology which can be applied in developing both linear and non-linear systems for embedded control and has been found to be very suitable for embedded control applications [8] [9]. The fuzzy logic system consists of the fuzzy set, fuzzy rule base, fuzzy reasoning engine, fuzzification, and defuzzification. Also, the design of a FLS includes the design of a rule base, input scale factors, output scale factors and the membership functions.

The training and recognition model by the RDBMS is used with input variable of fuzzy algorithm (fuzzy max-min composition), and recognizes user's dynamic the KSSL through efficient and rational fuzzy reasoning process. Therefore, we decide to take the characteristics of gesture input data as an average value over repeated experiment result values, where the repetition number is controlled by several parameters. Input scale factors transform the real inputs into normalized values, and output scale factors transform the normalized outputs into real values. Also, we decide to give a weight to each parameter and do fuzzy max-min reasoning through comparison with recognition model constructed using RDBMS module. The proposed fuzzy max-min CRI for fuzzy relations in this paper is defined in Fig. 3 and membership function of fuzzy set is defined as in the Equation (1).

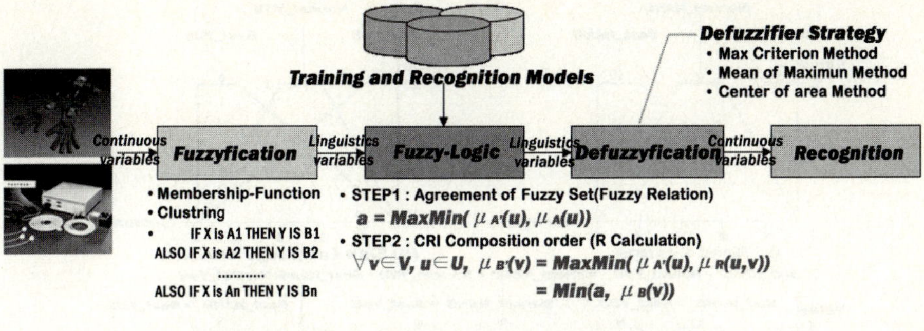

Fig. 3. Fuzzy Max-Min CRI (Direct Method)

$$\mu_{tz} = \begin{bmatrix} \frac{1}{(s-p)}(x-s)+1, & p<x\le s \ ; \textit{Slopes-up} \\ 1 \qquad , & s<x\le t \ ; \textit{Horizontality} \\ -\frac{1}{(q-t)}(x-t)+1, & t<x\le q \ ; \textit{Slopes-down} \end{bmatrix} \tag{1}$$

Also, Fig. 4 describes the Equation (1) pictorially; because fuzzy input variables are very various, we represent membership functions partially: "YOU" in KSSLT.

6 Experiments and Results

To implement embedded-sentential the KSSL recognizer in wireless-ubiquitous environment, we composed experiment environment using embedded LINUX-based i.MX21 board, blue-tooth module and wireless haptic devices. In the proposed embedded the KSSL Recognizer, data gloves transmit 14 kinds (10 fingers gesture data, 4 pitch & roll data) and motion tracker transmits 12 kinds of the KSSL data (position X, Y, and Z Cartesian coordinates and orientation-azimuth, elevation, and roll) in both hands to embedded i.MX21 board via blue-tooth module.

The proposed file size of embedded the KSSL recognizer is 521 Kbytes (including image and speech for representation of recognition results) and it can process and calculate 200 samples per seconds on i.MX21 board. The overall process of embedded the KSSL Recognizer consists of three major steps. In the first step, while the user inputs prescribed the KSSL into i.MX21 board using data gloves, motion tracker and blue-tooth module, the KSSL input module captures user's sign language data. And, in the second step, the KSSL recognizer changes characteristics of data (e.g. the degree of stooping and direction of hands, hand's position and gesture in spatial dimensions, etc.) by parameters for fuzzy recognition module. In the last step, it calculates and produces fuzzy value about user's dynamic the KSSL through a fuzzy

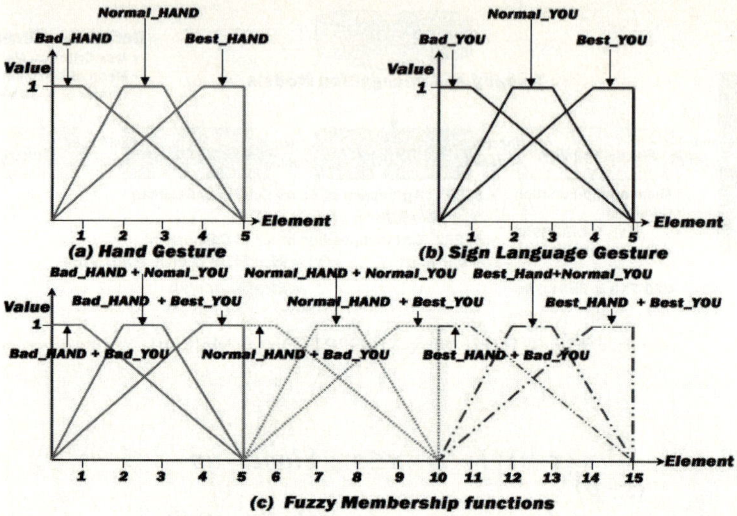

Fig. 4. The fuzzy membership functions. "YOU" sign language word in KSSLT consist of hand gesture that correspond to "paper" in "the game of paper, stone and scissors" and hand motion that correspond to "point at a person with one's finger". We prescribe gesture behavior such as the shape of one's hand (fingers) and the direction of back of the hand by 'Hand Gesture', and classified by three types of "Bad_HAND, Normal_HAND and Best_HAND" according to accuracy in "(a) Hand Gesture". Also, we prescribe hand's position and gesture in spatial dimensions by "Sign Language Gesture", and classified by three types of "Bad_YOU, Normal_YOU and Best_YOU" according to accuracy in "(b) Sign Language Gesture" in spatial dimensions.

reasoning and composition process. The recognition model by the RDBMS is used as independent variable of fuzzy reasoning, and we decide to give a weight to each parameter. The produced fuzzy value is used as a judgment of user action and the KSSL recognizer decides user's dynamic the KSSL according to degree of fuzzy value. Flow-chart of embedded the KSSL recognizer is shown in Fig. 5.

Experimental set-up is as follows. The distance between the KSSL input module and embedded i.MX21 board for processing of sign language recognition is about radius 10M's ellipse form. As the KSSL, we move the data gloves and receivers of the motion tracker to prescribed position. For every 18 reagents, we repeat this action 12 times about the KSSL sentence and words.

Experimental result, the GUI for visual representation of the KSSL recognition results is shown in Fig. 6 and Fig. 7 shows an average recognition rate of 93.3% for the KSSL sentences and 96.9% for the KSSL words with dynamic, significant and continuous 117 the KSSL recognition models. The root causes of errors between the KSSL recognition are various: that is, imprecision of prescribed actions, user's action inexperience, and the changes in experiment environment of physical transformation at data gloves fiber-optic flex sensor.

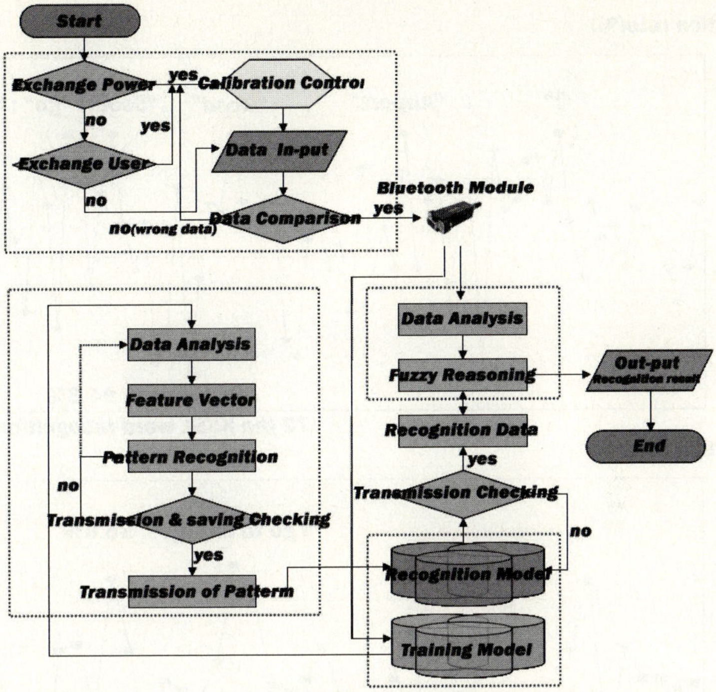

Fig. 5. The flow-chart of embedded-sentential the KSSL recognizer

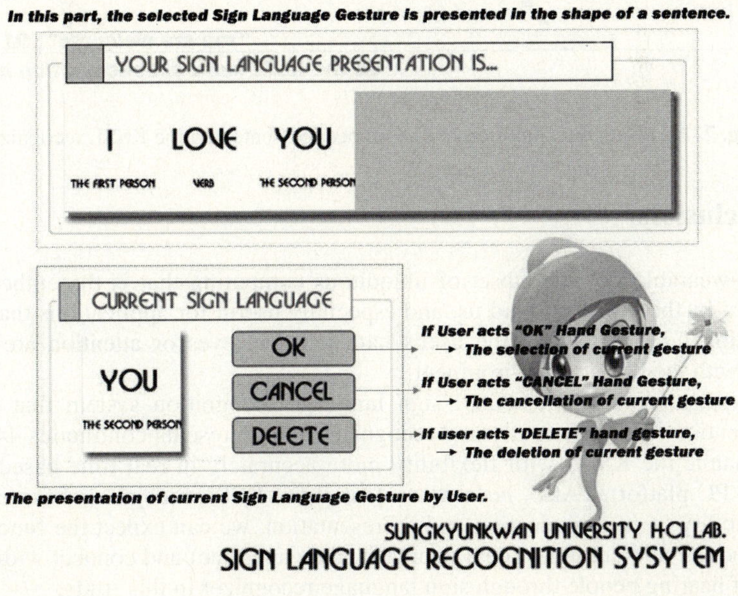

Fig. 6. The user interface for visual representation of the KSSL recognition

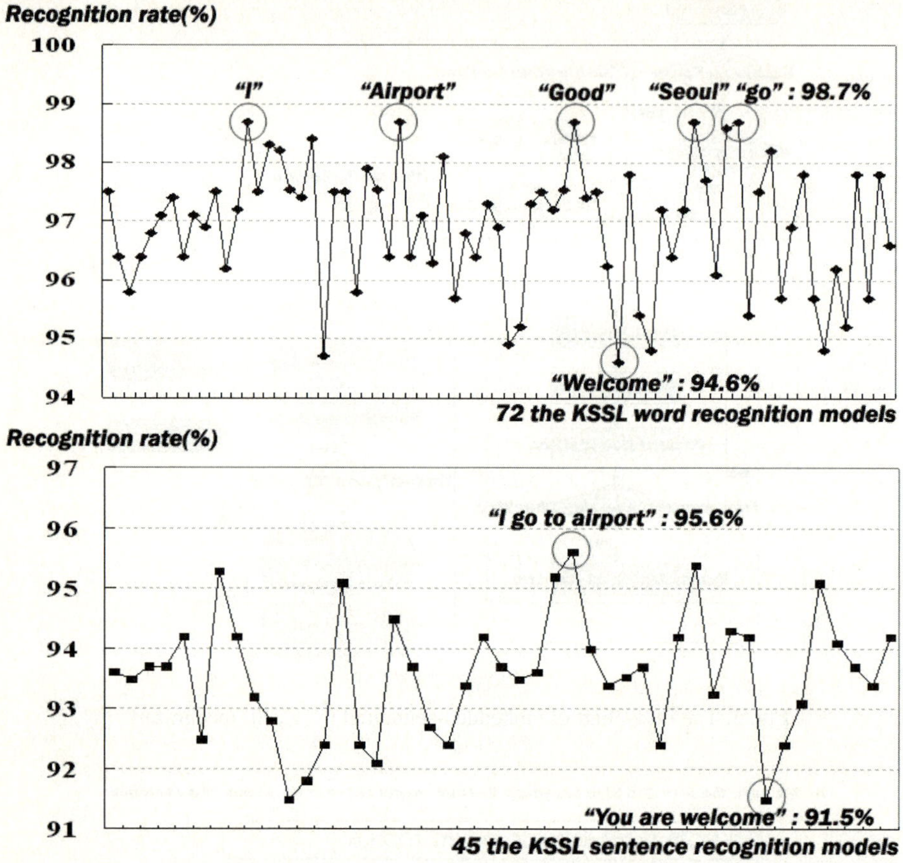

Fig. 7. An average recognition rate of embedded-sentential the KSSL recognizer

7 Conclusions

The post-wearable PC are subset of ubiquitous computing that is the embedding of computers in the world around us, and especially useful for applications that require computational support while the user's hands, voice, eyes or attention are actively engaged with the physical environment.

In this paper, we implemented sign language recognition system that analyzes user's intention more efficiently and recognizes and represents continuous 44 significant, dynamic the KSSL with flexibility more accurately in real time based on post wearable PC platform. Also, because sentential the KSSL recognizer has very superior the ability of communication and representation, we can expect the function that understand life and culture of deaf people (and an aphasiac) and connect with modern society of hearing people through sign language recognizer in this study.

The result of the experiment for the KSSL recognition was successful and satisfactory, and in the future, by integrating the KSSL recognition system and other human's five senses such as smell, taste, hearing and sight, we would like to develop advanced multi-modal HCI technology.

Acknowledgement

This research was supported by the MIC (Ministry of Information and Communication), Korea, under the ITRC (Information Technology Research Center) support program supervised by the IITA (Institute of Information Technology Assessment)" (IITA-2005- C1090-0501-0019) and this work was supported by the Brain Korea 21 Project in 2006.

References

1. Use of Signs in Hearing Communities.: http://en.wikipedia.org/wiki/Sign_language
2. H.-Y.Jang., D.-J.Kim., J.-B.Kim., and Z.-N.Bien.: A Study on Hand-Signal Recognition System in 3-Dimensional Space. Journal of IEEK, Vol. 2004-41CI-3-11. IEEK (2004)
3. S.-G.Kim.: Standardization of Signed Korean, Journal of KSSE, Vol. 9. KSSE (1992)
4. S.-G.Kim.: Korean Standard Sign Language Tutor. 1st edn. Osung Publishing Company, Seoul (2000)
5. 5DT Data Glove 5 Manual and FASTRAK® Data Sheet.: http://www.5dt.com
6. Relational DataBase Management System.: http://www.auditmypc.com/acronym/RDBMS.asp
7. Oracle 10g DW Guide.: SQL for Analysis and Reporting, http://www.oracle.com
8. J.-H.Kim., D.-G.Kim., J.-H.Shin, S.-W.Lee., and K-S.Hong.: Hand Gesture Recognition System using Fuzzy Algorithm and RDBMS for Post PC. FSKD2005. Lecture Notes in Artificial Intelligence, Vol. 3614. Springer-Verlag, Berlin Heidelberg New York (2005)
9. C.H.Chen.: Fuzzy Logic and Neural Network Handbook. 1st edn. McGraw-Hill, New York (1992)

Intelligent Multi-Modal Recognition Interface Using Voice-XML and Embedded KSSL Recognizer

Jung-Hyun Kim and Kwang-Seok Hong

School of Information and Communication Engineering, Sungkyunkwan University,
300, Chunchun-dong, Jangan-gu, Suwon, KyungKi-do, 440-746, Korea
kjh0328@skku.edu, kshong@skku.ac.kr
http://hci.skku.ac.kr

Abstract. A desktop PC and wire communications net-based traditional studies on pattern recognition and multimodal interaction have some restrictions (e.g. limitation of motion, conditionality in space and so on) and general problems according to using of the vision technologies for recognition and representation of the haptic-gesture information. In this paper, we propose and implement Multi-Modal Recognition Interface (hereinafter, MMRI) integrating speech using Voice-XML and gesture based on wireless networks, it have purposes that recognizes and represents the Korean Standard Sign Language (hereinafter, KSSL) which is a dialog system and interactive elements in the Korean deaf communities, and the need to dialogue with deaf person in their own language, sign language, is well recognized and is widely accepted as being a positive influence on communication. The advantages of our approach are as follows: 1) it improves efficiency of the MMRI input module according to the technology of wireless communication, 2) it shows higher recognition performance than unimodal recognition system 3) it recognizes and represents continuous sign language of users with flexibility in real time and offer to user a wider range of personalized and differentiated information using the MMRI more effectively. Experimental results, the MMRI deduces an average recognition rate of 96.23% for significant, dynamic and continuous the KSSL and speech of various users.

1 Introduction

While the PC has a relatively standard and fixed set of user-interaction devices (e.g. mouse, keyboard, speakers, and display), smart devices such as wireless data-glove for gestures recognition, controllable motion tracker and mobile phones are becoming more prevalent. Furthermore, while today's devices tend to be standalone, enabling only a single capability, faster networks and wireless technologies such as blue-tooth will allow these smart devices to be connected and behave in a coordinated fashion. That is, multimodal interaction is especially relevant in the context of mobile devices where traditional user interface peripherals may not be available or appropriate [1] [2]. The related studies on multimodal interaction and the sign language recognition are progressing actively in many countries such as the Americas, Asia, Europe, South Korea etc. Especially, going beyond today's telephony-based spoken language

B. Gabrys, R.J. Howlett, and L.C. Jain (Eds.): KES 2006, Part I, LNAI 4251, pp. 798 – 807, 2006.
© Springer-Verlag Berlin Heidelberg 2006

systems, researchers at AT&T Labs are addressing this challenge by developing technologies to support next-generation multimodal dialog systems, and "standard sign language translation system" developed jointly by the professors of the KIST (The Korea Institute of Science and Technology) and the Samsung Electronics Co., Ltd. in the South Korea is sign language translation system that recognize and represent small vocabulary-based sign language [3].

But, the general restrictions and features of these studies are as follows: 1) they have general purposes that recognize optional hand signal or sign language and integrate spoken and graphical elements simply based on the desktop computer and the technology of wire communication. Although, users can issue requests using speech, gesture, or dynamic combinations of the two based on embedded-ubiquitous environment, 2) because traditional sign language recognition system used image capture system or video processing system for an acquisition of sign language, it is impossible correct measuring with gesture data and needs complex computation algorithm, and 3) it is pursuing free communication with hearing people through restrictive and word-based sign language recognition and representation.

Therefore, we propose and implement improved the MMRI integrating of speech and the sign language (gesture) based on network, and the MMRI consists of the VXML for speech recognition and synthesis and embedded-sentential the KSSL recognizer based on the embedded-ubiquitous environment. The advantages of our approach are as follows: 1) it improves efficiency of the sign language input module according to the wireless communication and user need not be constrained by the limitations of a particular interaction mode at any given moment, 2) it contributes to user's the convenience and shows higher recognition performance than uni-modal recognition system(using gesture or speech), 3) it recognizes and represents continuous sign language of users with flexibility and presented in real time aurally and visually to maximize comprehension, and 4) because the ability of communication and representation of sentential the sign language recognizer are very superior more than sign language recognizer based on word and morpheme, it can offer to user a wider range of personalized and differentiated information using MMRI more effectively.

The composition of this paper are 1) implementation of the KSSL recognizer and a components in Section 2, 2) VXML module for speech recognition and synthesis in Section 3, 3) integration architecture of speech and the KSSL (gesture) for MMRI in Section 4, and 4) experiments and results in Section 5. Finally, this study is summarized in Section 6 together with an outline of challenges and future directions.

2 Embedded-Sentential KSSL Recognizer

2.1 Regulation of the KSSL

The sign language is a language which uses manual communication instead of sound to convey meaning - simultaneously combining hand shapes, orientation and movement of the hands, arms or body, and facial expressions to fluidly express a speaker's thoughts [4]. The KSSL consist of basic elements (meaningless units) that correspond to phoneme of spoken language by "sign language morpheme". That is, There are 5 classes of cheremes (a motor analogue to a phoneme) in the Korean sign language: configuration of the hand-DEZ (designator); position of the hand-TAB (tabulator); movement of the hand-SIG (signation); direction of the hand; and non-manual markers (or facial expression) [5].

In this paper, in order to implement significant embedded KSSL Recognizer in real time, we selected 25 basic cheremes according to a classification system of 'sign language morpheme' and analysis of "Korean Standard Sign Language Tutor (hereafter, KSSLT) [6]". And 23 hand gestures necessities of the KSSL are classified as hand shapes, pitch and roll degree. Consequently, we developed 32 significant the KSSL recognition models according to associability and presentation of hand gestures and basic cheremes. The examples of cheremes and hand gesture about "the date (numbers-day-month)" in the KSSL recognition models are shown in Table 1.

Table 1. The example of sign language about "the date (finger numbers-day-month)"

Object Language	Description of the Korean Sign Language
Finger-Number	1~30(or 31) : Signing of numbers
Month	Represent signing of numbers that correspond to "1(one)" with a left hand (indication of month), and draw crescent with thumb, index finger of a right hand
Day	Spread out thumb, index fingers of a both hands and rises on in front of chest.
Hand Gesture	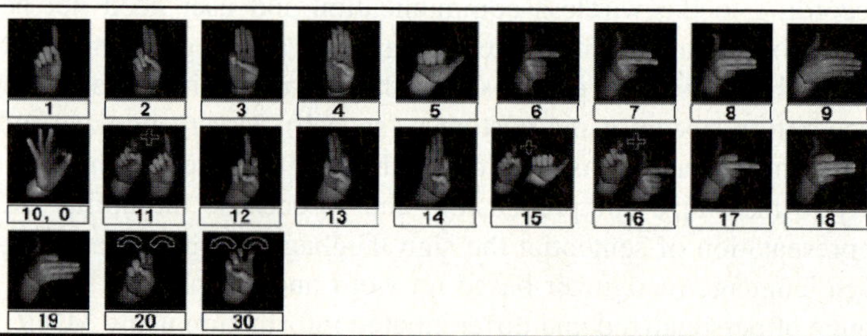

2.2 In-put Module of the KSSL and Recognition Models Using RDBMS

In vision technology for image processing and recognition, recognition performance has the possibility of change and diversity according to the performance of the camera, change of background (colors) and illumination condition. In order to overcome these problems, in this paper, we use 5DT company's 5th data glove system (wireless) and fastrak® which is one of the most popular input devices in the haptic application field. 5th data glove system is basic gesture recognition equipment that can capture the degree of finger stooping using fiber-optic flex sensor and acquires data through this. Also, because it has pitch and roll sensor inside, the pitch and roll of wrist can be also recognized without other equipment possible. Also, fastrak® is the accurate electromagnetic motion tracking system available, and it is the solution for accurately computing position and orientation through space. With real time, six degrees of freedom - position (X, Y, and Z Cartesian coordinates) and orientation (azimuth, elevation, and roll) - tracking and virtually no latency, this system is ideal for head, hand, and instrument tracking [7].

The captured dynamic the KSSL data of various users is transmitted to embedded i.MX21 board and database server through blue-tooth module. The KSSL data which is transmitted to server is used to as a training data for the KSSL recognition model by analytic function of RDBMS SQL. Also, it that is transmitted to i.MX21 board is used as the input-variables of fuzzy recognition module for significant the KSSL recognition. The architecture of sign language input module is shown in Fig. 1.

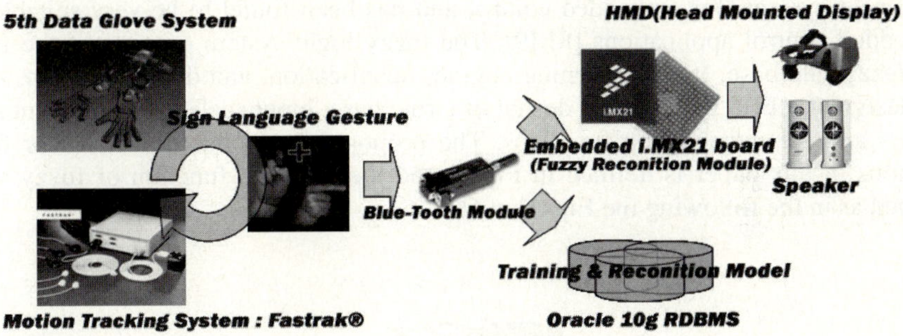

Fig. 1. The architecture of the KSSL input module

The RDBMS is used to classify sign language data that is inputted valid and invalid gesture record set (that is, invalid gesture records are records of transition state) from the KSSL input module, and to efficiently analyze valid record set. According to the logic in source code, even though a record set is, based on the above process, decided as valid, the record set is regarded as a status transition gesture record set. A rule to segment valid and invalid gesture record-set is shown in Fig. 2.

1. Difference Between Preceding Average(preceding 3 and 1) and Current Row Value
2. Decide validity : Preceding Average - Current Value = 52-17 = 35

Average Between 3 Preceding and 1 Preceding from Current Row : 52

1. Difference between Preceding Average and Current Row Value > 5
2. 'X'(Invalidity Record) check

SerialNO	F1THUMB	F1INDEX	F1MIDDLE	F1RING	F1LITTLE	F1X-POSI.	F1Y-POSI.	F1Z-POSI.	F1_Px	F1_Py	GESTURE DATA...	VALIDITY
22	73	86	255	255	255	50.90	-14.53	-4.74	55.42	-5.71		X
23	51	48	255	255	255	52.05	-11.97	-2.13	53.29	-4.95		X
24	32	15	255	255	255	53.99	- 9.50	0.01	50.02	-4.75		X
25	17	0	255	255	255	51.03	- 10.50	-1.05	51.11	-4.83	Omission...	X
26	7	0	255	255	255	52.01	- 9.85	-1.02	49.85	-5.23		
27	0	0	255	254	255	52.78	- 8.77	-1.75	51.85	-6.02		
28	0	0	253	254	255	49.87	- 10.18	-2.12	48.97	-5.27		
29	0	0	255	255	254	48.71	- 9.43	-1.92	52.85	-4.89		
30	0	0	255	255	255	51.09	- 8.75	-1.75	47.93	-4.94		

Fig. 2. The rule of segmentation and the KSSL record set

- If the difference between preceding average (preceding 3 and 1) and current row value is over 5, the current value is regarded as transition sign language record.
- If one of data gloves and motion tracker data value are over 5, current value data is also regarded as changing sign language record.

2.3 The KSSL Recognizer Using Fuzzy Max-Min Composition

The fuzzy logic is a powerful problem-solving methodology with a myriad of applications in embedded control and information processing, and is a paradigm for an alternative design methodology which can be applied in developing both linear and non-linear systems for embedded control and has been found to be very suitable for embedded control applications [8] [9]. The fuzzy logic system consists of the fuzzy set, fuzzy rule base, fuzzy reasoning engine, fuzzification, and defuzzification. Also, the design of a FLS includes the design of a rule base, input scale factors, output scale factors and the membership functions. The proposed fuzzy max-min CRI for fuzzy relations in this paper is defined in Fig. 3 and membership function of fuzzy set is defined as in the following the Equation (1).

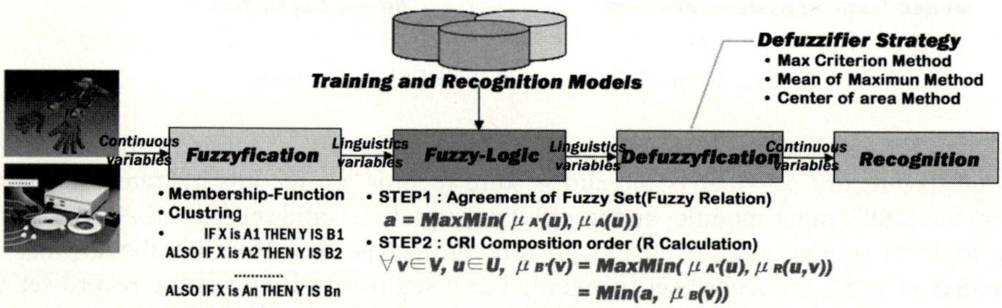

Fig. 3. Fuzzy Max-Min CRI (Direct Method)

$$\mu_{tz} = \begin{bmatrix} \frac{1}{(s-p)}(x-s)+1, & p<x\leq s \ ; \ Slopes\text{-}up \\ 1 \quad , & s<x\leq t \ ; \ Horizontality \\ -\frac{1}{(q-t)}(x-t)+1, & t<x\leq q \ ; \ Slopes\text{-}down \end{bmatrix} \tag{1}$$

Fig. 4 describes the Equation (1) pictorially (because fuzzy input variables are very various, we represent membership functions partially: "YOU" in KSSLT). The sign language "YOU" in KSSLT consist of hand gesture that correspond to "paper" in “the game of paper, stone and scissors” and hand motion that correspond to "point at a person with one’s finger". We prescribe gesture behavior such as the shape of one's hand (fingers) and the direction of back of the hand by ‘Hand Gesture', and classified by three types of “Bad_HAND, Normal_HAND and Best_HAND” according to accuracy in "(a) Hand Gesture". Also, we prescribe hand's position and gesture in spatial dimensions by "Sign Language Gesture", and classified by three types of "Bad_YOU, Normal_YOU and Best_YOU" according to accuracy in "(b) Sign Language Gesture" in spatial dimensions.

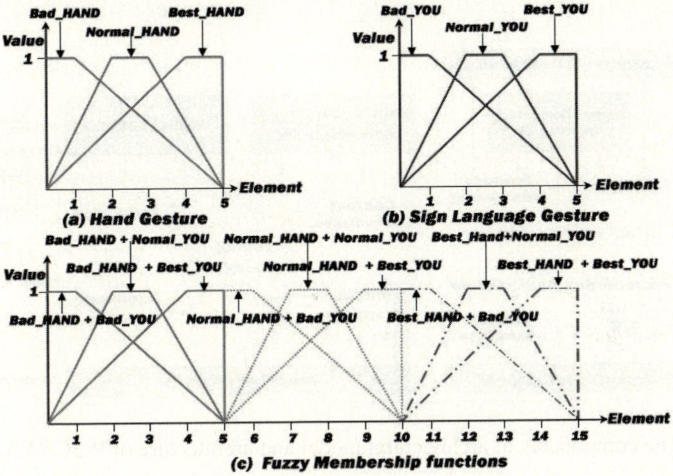

Fig. 4. The fuzzy membership functions

The KSSL recognition processes using fuzzy max-min composition consist of three major steps. In the first step, while the user inputs prescribed the KSSL into i.MX21 board using data gloves, motion tracker and blue-tooth module, the KSSL input module captures user's sign language data. And, in the second step, the KSSL recognizer changes characteristics of data (e.g. the degree of stooping and direction of hands, hand's position and gesture in spatial dimensions, etc.) by parameters for fuzzy recognition module. In the last step, it calculates and produces fuzzy value about user's dynamic the KSSL through a fuzzy reasoning and composition process. The recognition model by the RDBMS is used as independent variable of fuzzy reasoning, and we decide to give a weight to each parameter. The produced fuzzy value is used as a judgment of user action and the KSSL recognizer decides user's dynamic gesture according to degree of fuzzy value. The flow-chart of KSSL recognizer is shown in Fig. 7 (in Section 4) together with an outline and flow-chart of VMXL for MMRI.

3 Voice-XML for Speech Recognition and Synthesis

VXML is the W3C's standard XML format for specifying interactive voice dialogues between a human and a computer [10]. A document server (e.g. a Web server) processes requests from a client application, the VXML Interpreter, through the VXML interpreter context. The server produces VXML documents in reply, which are processed by the VXML interpreter. The VXML interpreter context may monitor user inputs in parallel with the VXML interpreter. For example, one VXML interpreter context may always listen for a special escape phrase that takes the user to a high-level personal assistant, and another may listen for escape phrases that alter user preferences like volume or text-to-speech characteristics. The implementation platform is controlled by the VXML interpreter context and by the VXML interpreter. The components of architectural model and architecture of VXML 2.0 by W3C are shown in Fig. 5.

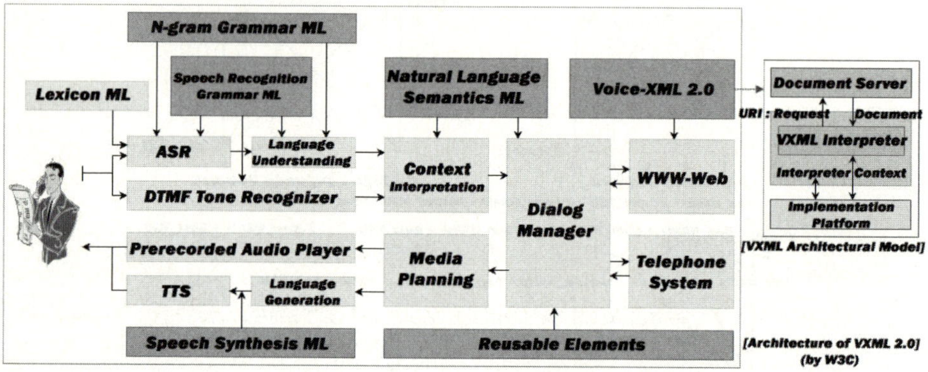

Fig. 5. The components of architectural model and architecture of W3C's VXML 2.0

4 Integration Architecture of Speech and Gesture for MMRI

In this paper, the proposed components and architecture of MMRI using speech and sign language (gestures) are shown in Fig. 6, and the flow-chart of the MMRI integrating VXML and KSSL Recognizer is shown in Fig. 7.

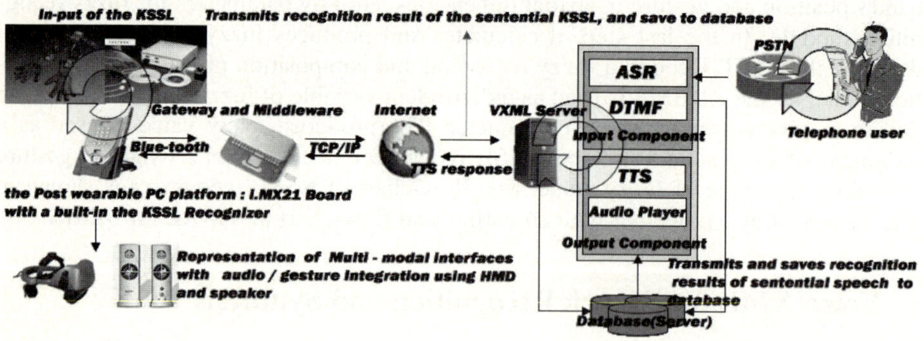

Fig. 6. The components and architecture of MMRI using speech and sign language (gestures)

The user connects to VXML server through internet and PSTN using i.MX21 board based on wireless networks (middleware) and telephone terminal, and input prescribed speech and KSSL. User's sentential speech data which is inputted into telephone terminal transmits to ASR-engine (we used 'HUVOIS-TTS' that is speech recognition / synthesis program for visually-impaired people and developed by KT Corp. in Korea),, and saves sentential ASR results to MMI database (in VXML server). Also, User's KSSL data which is inputted into embedded i.MX21 board is recognized by sentential KSSL recognizer, transmits and saves sentential recognition results to VXML server using TCP/IP protocol based on middleware and wireless sensor networks. Sentential ASR and KSSL recognition results execute comparison arithmetic by internal SQL logic, and transmit arithmetic results to MMRI. These arithmetic results are definite intention that user presents. Finally, user's intention is provided to user through speech (TTS) and visualization. The KSSL recognition processes of MMRI synchronize with the speech recognition and synthesis using VXML.

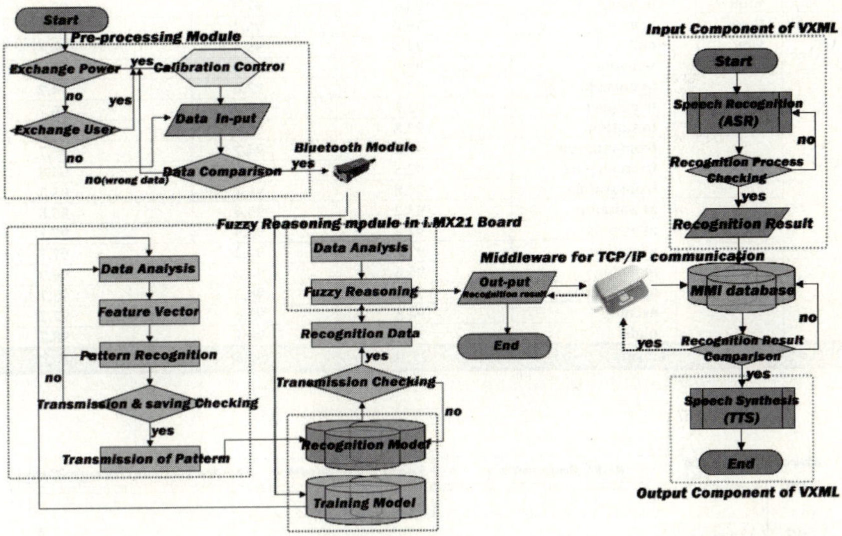

Fig. 7. The flow-chart of the MMRI integrating VXML and KSSL Recognizer

5 Experiments and Results

Experimental set-up is as follows. The distance between the KSSL input module and embedded i.MX21 board for processing of the KSSL recognition composed in about radius 10M's ellipse form. When user inputs the KSSL and speech, we move the 5th data glove system and receivers of the motion tracker to prescribed position. For every 20 reagents, we repeat this action 15 times. While user inputs the KSSL using data gloves and motion tracker, speak using blue-tooth headset of telephone terminal. Experimental results, the uni-modal and MMRI recognition rate about 32 sentential recognition models in this paper are shown in Table. 2. Also, the comparison charts are given in Fig.8 respectively.

Table 2. An uni-modal and MMRI recognition rate about 32 sentential recognition models

Sentential Recognition Model			Uni-modal recognition rate		MMRI recognition rate
			The KSSL (%)	Speech (%)	The KSSL + Speech (%)
I	go	to museum	92.3	93.7	94.1
		to airport	92.7	93.2	93.7
		to station	91.9	96.3	96.5
	come	from museum	94.4	96.2	96.7
		from airport	90.5	95.4	95.7
		from station	91.0	94.7	94.1
	arrive	at museum	93.7	95.1	95.8
		at airport	93.5	97.3	97.8
		at station	93.9	97.5	97.5
	love	you	94.7	96.2	96.7
	lost	my passport	95.1	94.5	95.4
	am	sorry	94.3	95.4	95.6
		fine	92.1	95.3	96.2
		a Korean	92.4	94.8	95.6
		from Korea	92.6	96.1	96.8
	want	to Seoul	93.4	97.3	97.6
-	thank	you	95.0	96.2	97.8
Are	you	ok?	92.5	95.4	95.7
You	are	welcome	93.5	93.9	93.6
We	go	to museum	92.3	92.6	94.2
		to airport	92.1	95.4	96.7
		to station	92.8	96.6	97.2
	come	from museum	90.5	94.7	95.7
		from airport	92.5	94.0	94.8
		from station	90.8	95.1	95.7
	arrive	at museum	93.2	95.4	95.8
		at airport	92.9	96.9	97.1
		at station	92.5	97.2	97.7
-	good	morning	95.6	94.8	96.9
		afternoon	95.1	98.3	98.7
		evening	94.8	95.6	97.2
		night	94.5	98.1	98.7
An average recognition rate			93.10	95.60	96.23

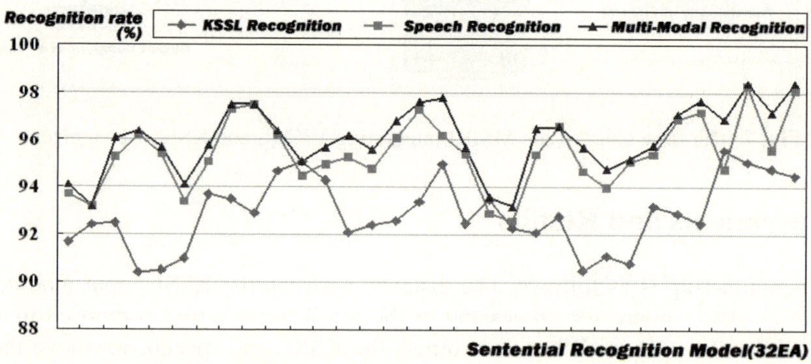

Fig. 8. An average recognition rate of the uni-modal and MMRI

6 Conclusions

The post wearable PC are especially useful for applications that require computational support while the user's hands, voice, eyes or attention are actively engaged with the physical environment. In this paper, we implemented MMRI that 1) guarantees mobility of portable terminal based on the embedded-ubiquitous environment with wireless sensor networks, 2) analyzes and communicates user's intention more efficiently, 3) and can recognizes and represents continuous 32 significant, dynamic the KSSL and speech of various users with flexibility in real time. In experiment results, MMRI is more efficient and powerful than uni-modal recognition system that uses one in the KSSL (gesture) or speech. Especially, while the average recognition rate of uni-modal recognition system that use KSSL (gesture) only is 93.10%, MMRI deduced an average recognition rate of 96.23% and showed difference of an average recognition rate as much as about 3.13%. And, because the MMRI has very superior the ability of communication and representation, we can expect the function that understand life and culture of deaf people (and an aphasiac) and connect with modern society of hearing people. The experiment results and implementation of the MMRI was successful and satisfactory, and in the future, by integrating sign language recognition system with other sensory channels such as smell, taste and sight, we would like to develop advanced multi-modal HCI technology.

Acknowledgement

This research was supported by the MIC (Ministry of Information and Communication), Korea, under the ITRC (Information Technology Research Center) support program supervised by the IITA (Institute of Information Technology Assessment) (IITA-2005- C1090-0501-0019) and this work was supported by the Brain Korea 21 Project in 2006.

References

1. Stéphane H. Maes et al.: Multimodal Interaction Requirements. W3C Note, http://www.w3.org (2003)
2. Jim Barnett et al.: Multimodal Interaction Activity-Multimodal Architecture and Interfaces. W3C Working Draft, http://www.w3.org (2005)
3. H.-Y.Jang., D.-J.Kim., J.-B.Kim., and Z.-N.Bien.: A Study on Hand-Signal Recognition System in 3-Dimensional Space. Journal of IEEK, Vol. 2004-41CI-3-11. IEEK (2004)
4. Use of Signs in Hearing Communities.: http://en.wikipedia.org/wiki/Sign_language
5. S.-G.Kim.: Standardization of Signed Korean, Journal of KSSE, Vol. 9. KSSE (1992)
6. S.-G.Kim.: Korean Standard Sign Language Tutor. 1st edn. Osung Publishing Company, Seoul (2000)
7. 5DT Data Glove 5 Manual and FASTRAK® Data Sheet.: http://www.5dt.com
8. J.-H.Kim., D.-G.Kim., J.-H.Shin, S.-W.Lee., and K-S.Hong.: Hand Gesture Recognition System using Fuzzy Algorithm and RDBMS for Post PC. FSKD2005. Lecture Notes in Artificial Intelligence, Vol. 3614. Springer-Verlag, Berlin Heidelberg New York (2005)
9. C.H.Chen.: Fuzzy Logic and Neural Network Handbook. 1st edn. McGraw-Hill, New York (1992)
10. Scott McGlashan et al.: Voice Extensible Markup Language (VoiceXML) Version 2.0. W3C Recommendation, http://www.w3.org (1992)

Subband Adaptive Filters with Pre-whitening Scheme for Acoustic Echo Cancellation

Hun Choi, Jae Won Suh, and Hyeon-Deok Bae

Dep. of Electronic Engineering,
Chungbuk National University, Korea
{eliga, sjwon, hdbae}@chungbuk.ac.kr

Abstract. The signal-decorrelating properties of the AP algorithm and the subband filtering improve the convergence speed of the conventional AP adaptive filter. In this paper, a new subband adaptive filtering with pre-whitening scheme for acoustic echo cancellation is presented. The proposed algorithm provides fast convergence and reduced computational complexity by combining merits of the affine projection (AP) algorithm and the subband filtering. The projection order of AP adaptive filter can be decreased by subband partitioning with the polyphase decomposition and the noble identity. The decreased projection order reduces the computational complexity in the proposed algorithm. Computer simulations illustrate the convergence rate improvements and the computational efficiency of the proposed algorithm and show the validity of the theoretical results.

1 Introduction

Adaptive filtering is essential for applications such as acoustic echo cancellation (AEC). However, the AEC problem is not as easily solved with conventional fullband adaptive filtering systems due to the requirements for faster convergence and for a much longer adaptive filter impulse response. In fullband adaptive filtering, there are two well-known problems. First, the convergence rate and tracking ability of an adaptive filter can be deteriorated if the input correlation matrix is ill-conditioned or equivalently if the input signal has wide spectral dynamic range such as that found in speech. Second, very high-order adaptive filters are computationally expensive [1]. To overcome these problems of the fullband adaptive filtering, the various affine projection(AP) algorithms and subband adaptive filtering(SAF) algorithms have been proposed [2,3,4,5,6,7]. For improving the convergence speed, these algorithms commonly use the signal-decorrelating technique when they update the weights of adaptive filter. In the AP algorithm, highly correlated input signals are pre-whitened by projection operation on a data-related affine subspace. However, for this pre-whitening, the AP algorithm requires more computational complexity that resulted from a matrix inversion. In [3,4], fast versions of the AP algorithm(FAP) have suggested. The FAP variants use various iterative methods to avoid the matrix inversion in weight-updating. However, they have a problem of performance deterioration

B. Gabrys, R.J. Howlett, and L.C. Jain (Eds.): KES 2006, Part I, LNAI 4251, pp. 808–816, 2006.
© Springer-Verlag Berlin Heidelberg 2006

by iterative approximation and also, they are still complex for implementing the systems that require long impulse responses like as AEC. Alternatively, the SAF may be good choice to solve the performance limitations in fullband adaptive filtering [5,6,7]. In SAF, each subband signal is decorrelated by the analysis filters and its eigenvalue disparity decreases. And also, the length of the adaptive sub-filters, which are polyphase components, is smaller than that of fullband adaptive filter. Therefore, each adaptive sub-filter is updated with pre-whitened input signals and rapidly converges to the steady-state. Pradhan [7] has suggested a new subband structure that uses polyphase decomposition, the noble identity, and maximal decimation. Pradhan's subband structure achieves a more rapid convergence rate without the aliasing problems and additional computation. In [8], convergence analysis of [7] was performed in the frequency domain. This paper showed that Pradhan's subband adaptive filter is always stable, and that its excess mean square error (EMSE) is reduced. Recently, the subband adaptive algorithms based on FAP have been proposed [9,10].

In this paper, we present a new subband adaptive filtering with pre-whitening scheme for AEC. The pre-whitening technique is accomplished by using subband structured adaptive filter and applying the AP algorithm to the polyphase decomposed adaptive filter of each subband. In subband adaptive sub-filter, its impulse response length is smaller than that of the fullband case, the projection order of the AP algorithm can be reduced. It provides reduced computational complexity. Therefore, we obtained improved convergence speed and reduced computational complexity by combining the advantageous characteristics of SAF and AP. We will call the proposed algorithm a subband affine projection (SAP) algorithm. The weight-updating formula of the proposed method is derived as a simple form as compared with the normalized least mean square (NLMS) algorithm. To evaluate the performance of the proposed method, computer simulations are performed for system identification of echo cancellation problem.

2 Decorrelating Properties of AP and SAF

Fig. 1 shows a AEC system and its adaptive system identification (ASID) model in fullband structure. In Fig. 1b, the impulse response of echo path is represented by $\mathbf{h}^*$ and the adaptive filter of the canceller is presented by the vector $\mathbf{h}(n)$. The input signal $x(n)$ is the sampled far-end signal, the desired signal $s(n)$ is the combination of the echo and the near-end talker's voice, and $f(n)$ is presented by the measurement noise. And it is the near-end signal which consists of speaker's voice and back ground noise. The response of the model $r(n)$ is subtracted from the desired signal $s(n)$. For suppressing the echo, the optimization criterion of the AP algorithm for designing the adaptive filter is given by [1]

$$\textit{minimize} \quad \|\mathbf{h}(n+1) - \mathbf{h}(n)\|^2 \quad \textit{subject to} \quad \mathbf{s}(n) = \mathbf{X}^T(n)\mathbf{h}(n+1). \qquad (1)$$

where $\mathbf{X}(n) = [\mathbf{x}(n)\ \mathbf{x}(n-1)\ \ldots\ \mathbf{x}(n-P)]$, $\mathbf{x}(n-l) = [x(n-l)\ x(n-l-1)\ \ldots\ x(n-l-N+1)]^T$. P is the projection order of the AP algorithm and N

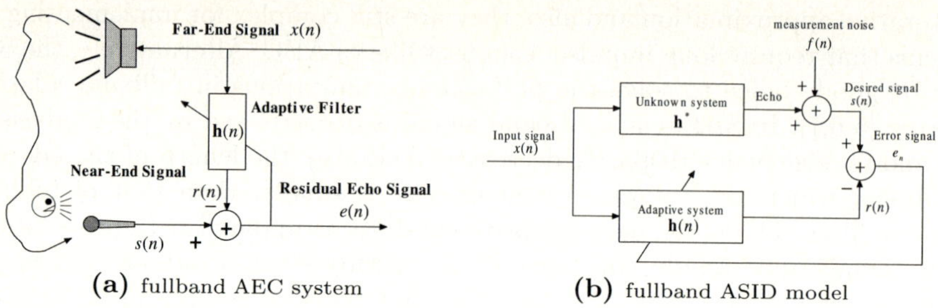

Fig. 1. Fullband system identification for adaptive AEC

is the length of adaptive filter. By using the *Lagrangian theory* [1], the weights of the adaptive filter are updated by the AP algorithm as

$$\mathbf{h}(n+1) = \mathbf{h}(n) + \mu\mathbf{X}(n)[\mathbf{X}(n)^T\mathbf{X}(n)]^{-1}\mathbf{e}(n), \tag{2}$$

$$\mathbf{e}(n) = \mathbf{X}^T(n)\mathbf{h}^* - \mathbf{X}^T(n)\mathbf{h}(n) + \mathbf{f}(n) = [e(n)\ e(n-1)\ \ldots\ e(n-P)]^T \tag{3}$$

where μ is the step size parameter. If the input signal is the auto-regressive (AR) process of order K, we can rewrite the AR signal as $x(n) = \sum_{l=1}^{K} a_l x(n-l) + z(n)$, where $z(n)$ is a wide sense stationary (WSS) white process with variance σ_z^2. With the weight-updating direction vector demonstrated in [11], the AP algorithm of (2) can be represented as

$$\mathbf{h}(n+1) = \mathbf{h}(n) + \frac{\boldsymbol{\Phi}(n)}{\boldsymbol{\Phi}^T(n)\boldsymbol{\Phi}(n)}\mathbf{e}(n), \tag{4}$$

$\boldsymbol{\Phi}(n)$ is the weight-updating direction vector of the adaptive filter [11] and it is given by

$$\boldsymbol{\Phi}(n) = [\mathbf{I} - \mathbf{P}(n)]\mathbf{z}(n) = \mathbf{P}_\perp(n)\mathbf{z}(n) = \mathbf{z}_\perp(n). \tag{5}$$

In (5), $\mathbf{I}$ is the $N \times N$ identity matrix and $\mathbf{z}(n) = \mathbf{z}_\mathbf{P}(n) + \mathbf{z}_{\mathbf{P}_\perp}(n)$, where $\mathbf{z}_\mathbf{P}(n) = \mathbf{P}(n)\mathbf{z}(n)$ and $\mathbf{z}_{\mathbf{P}_\perp}(n) = \mathbf{P}_\perp(n)\mathbf{z}(n)$. $\mathbf{P}(n) = \mu\mathbf{X}(n)[\mathbf{X}^T(n)\mathbf{X}(n)]^{-1}\mathbf{X}^T(n)$ is the relaxed projection operator onto the affine subspace spanned by the columns of $\mathbf{X}(n)$ and $\mathbf{P}_\perp(n) = \mathbf{I} - \mathbf{P}(n)$ is complementary. If we assume that $\mu = 1$, the AP algorithm performs the orthogonal projections onto the subspace spanned by columns of $\mathbf{X}(n)$ before it updates adaptive filter weights. $\boldsymbol{\Phi}(n)$ is a vector which consists of white signals. From these results, we can easily find that the AR(P) input signal is decorrelated by the projection operation and the weights of adaptive filter are updated with the pre-whitened input signals.

In the subband filtering, the colored input signal is decomposed by the orthogonal analysis filters like as the cosine modulated filter banks (CMFB) [6] and its power spectrum is approximately flat in each passband of the analysis filter bank. Therefore, in the decomposed subband input signal $x_i(n)$, the eigenvalues of its correlation matrix, $\mathbf{R}_i = E[\mathbf{X}_i^T(n)\mathbf{X}_i(n)]$, $i = 0, 1, \cdots, M-1$, will all be approximately equal to 1. The adaptive algorithm updates the weights of adaptive sub-filter with the subband input signal that pre-whitened by the

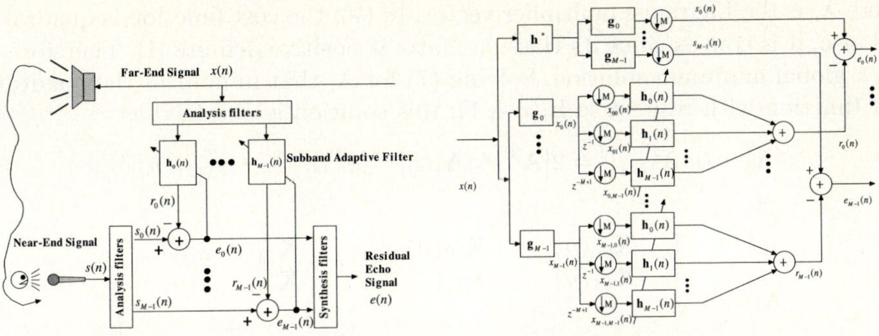

(a) subband AEC system **(b)** subband ASID model

Fig. 2. Subband system identification for the M-subband AEC

analysis filters. For a number of subbands not enough to decorrelate the input signal, there would be a degradation in the convergence speed of the adaptive algorithm. However, the algorithm convergence speed would still be better than that of the fullband adaptive algorithm, since the eigenvalue ratio $\lambda_{max}/\lambda_{min}$ of $\mathbf{R}_i$ would be smaller than the eigenvalue ratio of the fullband correlation matrix.

3 Subband Affine Projection Algorithm

The equivalent subband AEC system of Fig. 1 is given in Fig. 2. This structure is obtained by applying the polyphase decomposition, the noble identity, and the maximal decimation to the fullband structure [7]. Note that $x_i(n)$ and $s_i(n)$ are the subband components of the input and the desired signals that partitioned by the analysis filter $\mathbf{g}_i$. The $N_s \times 1$ vector, $\mathbf{h}_i(n)$, is the polyphase component of an adaptive filter, where N_s is its length, $\mathbf{x}_{ij}(n)$ is an input signal vector for $\mathbf{h}_i(n)$. The notation $(\downarrow M)$ means decimation with the rate M. Based on the principle of minimum disturbance [1], we formulate the criterion for the M-subband AP filters as one of optimization subject to multiple constraints, as follows:

$$\begin{aligned} minimize \quad & \textstyle\sum_{i=0}^{M-1} f[\mathbf{h}_i(n)] = \|\mathbf{h}_0(n+1) - \mathbf{h}_0(n)\|^2 \\ & \qquad + \cdots + \|\mathbf{h}_{M-1}(n+1) - \mathbf{h}_{M-1}(n)\|^2 \\ subject\ to \quad & \mathbf{s}_i(n) = \textstyle\sum_{j=0}^{M-1} \mathbf{X}_{ij}^T(n)\mathbf{h}_j(n+1) \quad for\ i = 0, 1, \cdots, M-1 \end{aligned} \tag{6}$$

where $\mathbf{s}_i(n) = [s_i(n)\ \ s_i(n-1)\ \ \ldots\ \ s_i(n-P_s+1)]^T$, $\mathbf{X}_{ij}(n) = [\mathbf{x}_{ij}(n)\ \mathbf{x}_{ij}(n-1)\ \cdots\ \mathbf{x}_{ij}(n-P_s+1)]$, and $\mathbf{x}_{ij}(n) = [x_{ij}(n)\ \ x_{ij}(n-1)\ \ \ldots\ \ x_{ij}(n-N_s+1)]^T$. The parameter N_s and P_s are the length of the adaptive sub-filter and the projection order in each subband, respectively. Applying the method of Lagrange multipliers with multiple constraints to (6), we define the following cost function for the AP algorithm in the M-subband structure of Fig. 2b as

$$J(n) = \sum_{i=0}^{M-1}\left(f[\mathbf{h}_i(n)] + [\mathbf{s}_i(n) - \sum_{j=0}^{M-1} \mathbf{X}_{ij}^T(n)\mathbf{h}_j(n+1)]^T \boldsymbol{\lambda}_i \right) \tag{7}$$

where $\boldsymbol{\lambda}_i$ is the Lagrange multiplier vector. In (7), the cost function is quadratic, and also, it is convex since its Hassian matix is positive definite [1]. Therefore, it has a global minimum solution. Solving (7) for $\boldsymbol{\lambda}_i$ that minimizes the quadratic cost function with respect to $\mathbf{h}_i(n+1)$, this solution is obtained as

$$[\boldsymbol{\lambda}_0^T \ \cdots \ \boldsymbol{\lambda}_{M-1}^T]^T = 2[\mathbf{A}^T(n)\mathbf{A}(n)]^{-1}[\mathbf{e}_0^T(n) \ \cdots \ \mathbf{e}_{M-1}^T(n)]^T \tag{8}$$

$$\mathbf{A}(n) = \begin{bmatrix} \mathbf{X}_{00}(n) & \mathbf{X}_{10}(n) & \cdots & \mathbf{X}_{(M-1)0}(n) \\ \mathbf{X}_{01}(n) & \mathbf{X}_{11}(n) & \cdots & \mathbf{X}_{(M-1)1}(n) \\ \vdots & \vdots & \ddots & \vdots \\ \mathbf{X}_{0(M-1)}(n) & \mathbf{X}_{1(M-1)}(n) & \cdots & \mathbf{X}_{(M-1)(M-1)}(n) \end{bmatrix} \tag{9}$$

$$\mathbf{e}_i(n) = \mathbf{s}_i(n) - \sum_{j=0}^{M-1} \mathbf{X}_{ij}^T(n)\mathbf{h}_j(n+1) \tag{10}$$

$\mathbf{A}(n)$ is $MN_s \times MP_s$ matrix. Let $\boldsymbol{\Phi}(n)$ be $\mathbf{A}^T(n)\mathbf{A}(n)$. It can be represented as

$$\boldsymbol{\Phi}(n) = \begin{bmatrix} \boldsymbol{\Xi}_0(n) & & \mathbf{C}(n) \\ & \ddots & \\ \mathbf{C}^T(n) & & \boldsymbol{\Xi}_{(M-1)}(n) \end{bmatrix} \tag{11}$$

In (11), $\boldsymbol{\Phi}(n)$ is $MP_s \times MP_s$ matrix, and the elements of off-diagonal component $\mathbf{C}(n)$ consist of sample cross-correlations between the signals for the ith and jth subband ($i \neq j$), whereas, the elements of diagonal component $\boldsymbol{\Xi}_i(n)$ consist of sample auto-correlations. Assuming that the input signal is WSS and ergodic, the cross-correlation at zero lag, $\gamma_{\mathbf{x}_{00}\mathbf{x}_{10}+\mathbf{x}_{01}\mathbf{x}_{11}}(k,l)$, can be expressed as

$$\gamma_{\mathbf{x}_{00}\mathbf{x}_{10}+\mathbf{x}_{01}\mathbf{x}_{11}}(0) = \left[\mathbf{x}_{00}^T(k)\mathbf{x}_{10}(k) + \mathbf{x}_{01}^T(k)\mathbf{x}_{11}(k)\right]/N_s. \tag{12}$$

For analytical simplicity, we further assume that the input signal is white giving us a flat spectrum. From these assumptions, $E\{\mathbf{x}_{00}^T\mathbf{x}_{00} + \mathbf{x}_{01}^T\mathbf{x}_{01}\} = \sigma_{\mathbf{x}_0}^2$ ($\sigma_{\mathbf{x}_0}^2$ is the variance of subband signal, $\mathbf{g}_0^T\mathbf{x}$) and $E\{\mathbf{x}_{00}^T\mathbf{x}_{10} + \mathbf{x}_{01}^T\mathbf{x}_{11}\} = 0$. For colored inputs, $E\{\mathbf{x}_{00}^T\mathbf{x}_{10} + \mathbf{x}_{01}^T\mathbf{x}_{11}\} \neq 0$. However, if the frequency responses of the analysis filters do not overlap significantly, it is always true that $E\{\mathbf{x}_{00}^T\mathbf{x}_{10} + \mathbf{x}_{01}^T\mathbf{x}_{11}\} \ll E\{\mathbf{x}_{00}^T\mathbf{x}_{00} + \mathbf{x}_{01}^T\mathbf{x}_{01}\}$. That is, the elements of off-diagonal component $\mathbf{C}(n)$ are remarkably small compared with the elements of diagonal component $\boldsymbol{\Xi}_i(n)$. Therefore, we can consider $\mathbf{C}(n) \approx \mathbf{0}$. From this, we can easily get $\boldsymbol{\Phi}^{-1}(n)$ of (11) following as

$$\boldsymbol{\Phi}^{-1}(n) = \begin{bmatrix} \boldsymbol{\Xi}_0^{-1}(n) & \mathbf{0} & \cdots & \mathbf{0} \\ \mathbf{0} & \boldsymbol{\Xi}_1^{-1}(n) & & \vdots \\ \vdots & & \ddots & \mathbf{0} \\ \mathbf{0} & \cdots & \mathbf{0} & \boldsymbol{\Xi}_{(M-1)}^{-1}(n) \end{bmatrix} \tag{13}$$

With the above approximations, the recursive relation for updating the coefficients of the subband adaptive filters can be obtained as

$$\mathbf{H}(n+1) = \mathbf{H}(n) + \mu\mathbf{A}(n)\boldsymbol{\Phi}^{-1}(n)\mathbf{E}(n), \tag{14}$$

Table 1. Comparison of the computational complexities

Algorithms	multiplications/iter.	multiplications/iter. for $L=64, N=512$ $M=4, P=4, D=2$
SNLMS[7]	$3N+2M(L+2)$	2,064
Fullband AP[2]	$P^3/2+3NP^2+NP+N$	27,168
Subband LC-GSFAP[10]	$M\left(P^2+P+N+(2P+N)/D+1\right)+2ML$	3,160
The proposed SAP	$P^3/(2M^3)+NP^2(M+1)/M^3+NP(P+M+1)/M^2+2ML$	$\approx 2,305$
The SSAP	$3N+2P(L+2)$	2,064

L : Length of the filter banks.
D : The size of data frame for the LC-GSFAP.

where $\mathbf{H}(n) = \left[\mathbf{h}_0^T(n)\ \mathbf{h}_1^T(n)\ \cdots\ \mathbf{h}_{M-1}^T(n)\right]^T$ and $\mathbf{E}(n) = \left[\mathbf{e}_0^T(n)\ \mathbf{e}_1^T(n)\ \cdots\ \mathbf{e}_{M-1}^T(n)\right]^T$. The proposed SAP algorithm shown in (14) has faster convergence speed than that of the fullband AP by combining signal pre-whitening properties of the AP and the SAF. And we note that when the size of the data matrix in the fullband is $N \times P$, in the subband, it becomes $N_s \times P_s = (N/M) \times (P/M)$. Therefore, the computational complexity for weight-updating is reduced in the proposed SAP. For practical implementation, by partitioning the P-order fullband AP into P-subbands, we obtain the simplified SAP (SSAP) with $N/P\times 1$ data vectors for weight-updating. Consequently, the weight-updating formula for each subband adaptive filter is similar to that of the NLMS adaptive filter.

Computational Complexity of the Proposed Methods

The computational complexities per iteration in terms of the number of multiplications for the proposed SAP, the SSAP, the fullband AP [2], the subband NLMS (SNLMS) [7], and the subband LC-GSFAP [10] are shown in Table 1. In the fullband AP and the SAP, matrix inversions were assumed to be performed with standard LUD (standard Lower and Upper triangular matrix Decomposition: $O^3/2$ multiplications)[9], where O is the rank of a square matrix, and it is equal to the projection order in AP ($O = P$ or P_s). From Table 1, it can be seen that the proposed algorithms are much more efficient than the other algorithms.

4 Simulations

We carry out computer simulations to evaluate the performance of the proposed algorithms in AEC scenario. The unknown system is an actual impulse response of the echo path in a room, sampled at 8kHz and truncated to 512 samples ($N = 512$). For signal decomposing, we use the CMFB for analysis and synthesis filters. For the input signal, we use the AR(4) process with its coefficients are $[1\ 0.999\ 0.99\ 0.995\ 0.9]^T$ and the real man's speech. The measurement noise is added to desire signal $s(n)$ such that $SNR = 30$dB. In all experiments, we assume that the double talk condition is not active.

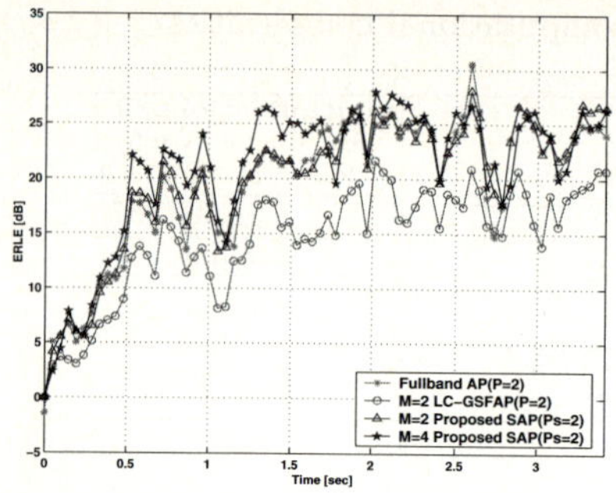

(a) Comparison of the ERLE curves for speech

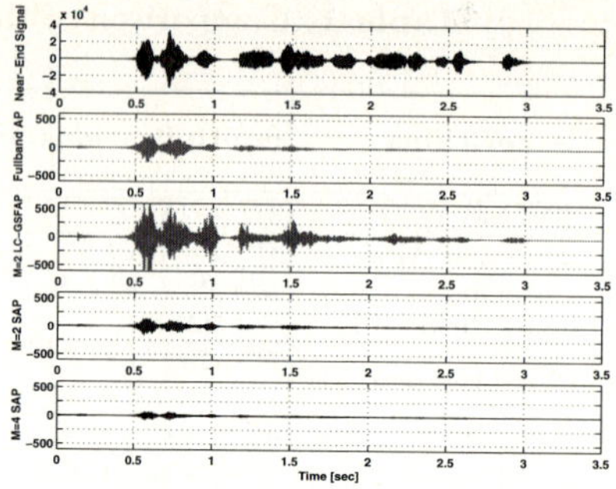

(b) Comparison of the residual error signals for speech

Fig. 3. ERLE curves of the fullband AP with $P = 2$, $M = 2$ LC-GSFAP with $P = 2$ and $D = 2$, $M = 2$ SAP with $P_s = 2$, and $M = 4$ SAP with $P_s = 2$ for speech ($N = 512,\ \mu = 1,\ SNR = 30\ dB$)

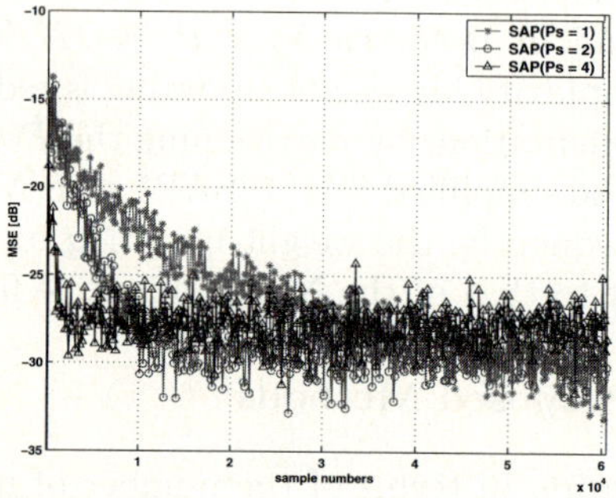

(a) MSE curves of SAP with $M = 2$ for different P_s

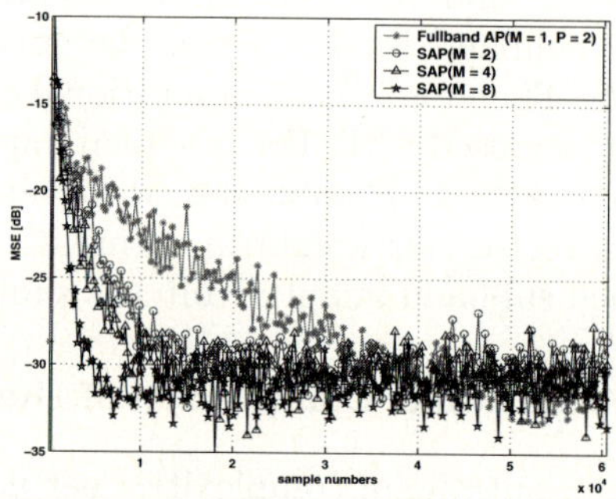

(b) MSE curves of SAP with $P_s = 2$ for different M

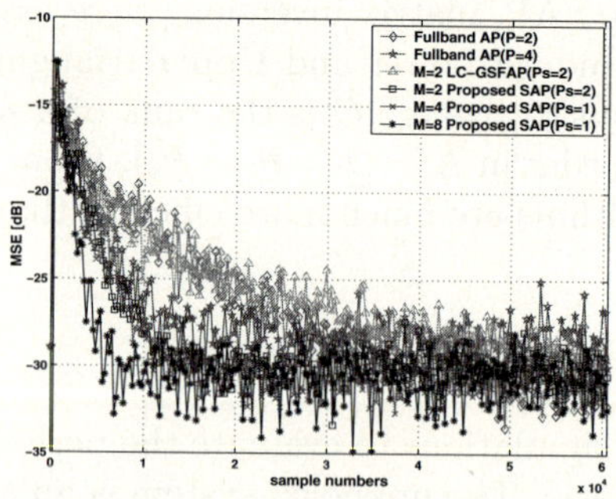

(c) MSE curves of SSAP for different M

Fig. 4. MSE curves of the proposed algorithms (SAP and SSAP) and the fullband AP with different projection orders for AR(4) inputs ($N = 512,\ \mu = 1,\ SNR = 30\ dB$)

Fig. 3 shows the echo return loss enhancement (ERLE) curves and the residual error signals (RES) for a man's voice sampled at 8kHz. In Fig. 3a, the proposed SAP is superior to other algorithms and its performance of ERLE is improved as the number of subband, M, increases. Fig. 4 shows the MSE curves of the proposed algorithms (the SAP and SSAP) and the fullband AP for different values of M and P. Fig. 4a shows the MSE curves of the SAP for different projection orders in two subband case. In the SAP, increasing the projection order improves the convergence rate. In Fig. 4b, the projection orders for the fullband AP and the SAP are the same ($P = 2$ and $P_s = 2$). In the proposed SAP, the convergence rate increases with the number of subbands. The results of Fig. 4a and Fig. 4b show that the convergence rate of the SAP depends on the number of subbands as well as the projection order. However, from these results, we readily expect that the increase in M is more efficient than that in P for improving the convergence of the AP. Moreover, in the proposed SAP, the increase in M allows the projection order P to be reduced. Fig. 4c shows the MSE performance of SSAP for different values of M. When $MP_s = P$, we can observe that the SSAP is similar in the view of convergence speed to the fullband AP. As described earlier, however, the SSAP is less computationally complex than the fullband AP. Consequently, we can conclude that the effect of the plenty subband partitioning is more effective than that of the high projection order for improving the convergence rate of the AP.

5 Conclusions

In this paper, we propose a new subband adaptive filtering with pre-whitening scheme for acoustic echo cancellation. By combining the pre-whitening properties of the subband structure and affine projection algorithm, the derived method gives a rapid convergence rate and reduced computational complexity. In addition, we show that the proposed algorithm can be derived to a simplified form, such as the NLMS by partitioning it into the same number of subbands as the projection order. This simplified algorithm, SSAP, is a good approach for implementing the proposed algorithm. Several simulation results were included to verify the theoretical results and to show the improved performance of the proposed method.

References

1. Hakin, S.: Adaptive filter theory. 4th edn., NJ:Prentice-Hall (2002)
2. Ozeki, K., Umeda, T.: An adaptive filtering algorithm using an orthogonal projection to an affine subspace and its properties. Electron, Comm. Jap. 67-A(5)(1984)19–27
3. Guy, S.L., Benesty, J..: Acoustic Signal Processing for Telecommunication. Kluwer Academic Press. (2000)
4. Albu, F., Kwan, H.K.: Fast block exact Gauss-Seidel pseudo affine projection algorithm. Electronics Lett. 40(22)(2004)1451–1453

5. Gilloire, A. Vetterli, M.: Adaptive filtering in subbands with critical sampling: analysis, experiments, and application to acoustic echo cancellation. IEEE Trans. on Sig. Proc.40(8)(1992)1862–1875
6. Vaidyanathan, P.P.: Multirate System and Filter Banks. Englewood Cliffs, NJ: Prentice-Hall(1993)
7. Pradhan, S.S., Reddy, V.U.: A new approach to subband adaptive filtering. IEEE Trans. Signal Proc.45(3)(1999)655–664
8. Miyagi, S., Sakai, H.: Convergence analysis of alias-free subband adaptive filters based on a frequency domain technique. IEEE Trans. Signal Proc.52(1)(2004)79–89
9. Bouchard, M.: Multichannel affine and fast affine projection algorithms for active noise control and acoustic equalization systems. IEEE Trans. Speech and Audio Proc. 11(1)(2003)54–60
10. Chau, E., Sheikhzadeh, H., Berennan, R.L.: Complexity reduction and regularization of a fast affine projection algorithm for oversampled subband adaptive filters. IEEE Proc. ICASSP(2004)V109–V112
11. Almeida, S.J.M., Bermudez, J.C.M., Bershad, N.J., Costa, M.H.: A statistical analysis of the affine projection algorithm for unity step size and autoregressive inputs. IEEE Trans. Circuits and Systems-I.52(7)(2005)1394-1405

A Noise Cancelling Technique Using Non-uniform Filter Banks and Its Implementation on FPGA

Sang-Wook Sohn[1], Hun Choi[2], Al-Chan Yun[1], Jae-Won Suh[2], and Hyeon-Deok Bae[1]

[1] Dep. of Electrical Engineering,
Chungbuk National University, Korea
{sohn6523, alchan, hdbae}@chungbuk.ac.kr
[2] Dep. of Electronic Engineering,
Chungbuk National University, Korea
{sjwon, eliga}@chungbuk.ac.kr

Abstract. The problem of estimating a signal that is corrupted by additive noise has been a interesting theme of digital signal processing field for several decades. Due to advantages over the linear methods, nonlinear methods based on wavelet transform have become increasingly popular. It has been shown that wavelet-thresholding algorithm generates near-optimal properties in the minimax sense. However, the wavelet-thresholding algorithm is very complex and difficult to implement it on hardware such as Field Programmable Gate Array(FPGA). Therefore, we need alternative simple approach for noise cancelling. In this paper, we propose a new noise cancelling algorithm with the binary tree structured filter banks and implement it on FPGA. To cancel the noise, we use the signal power ratio of each subband. For simple implementation, the filter banks are designed by Hadamard transform coefficients. From the results of simulations and hardware implementation, we show that the proposed algorithm produces a good results.

1 Introduction

In many application fields of signal processing, such as image processing, image coding, artifact free signal processing and biomedical signal processing, etc., noise removing is very important problem. The most popular solution for denoising is wavelet based algorithm [1, 3, 4, 5, 6, 7, 8, 9]. This solution decomposes the noisy signal into orthogonal wavelet basis, manipulates the wavelet coefficients using hard or soft thresholding, and inverses wavelet transform the modified signal back to the original domain [3, 4, 5, 6, 7, 8, 9]. In addition, uniform/non-uniform filter banks are well-known techniques in noise removing. The energy compaction property of subbands can be exploited for denoising [2, 6]. In this paper, we develop a noise cancelling technique by modifying each subband output signal of filter bank. To modify each subband output signal of filter bank,

B. Gabrys, R.J. Howlett, and L.C. Jain (Eds.): KES 2006, Part I, LNAI 4251, pp. 817–824, 2006.
© Springer-Verlag Berlin Heidelberg 2006

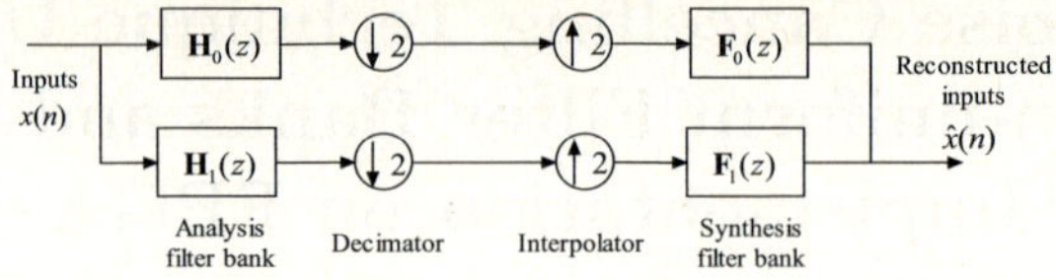

Fig. 1. Two-chancel QMF filter bank

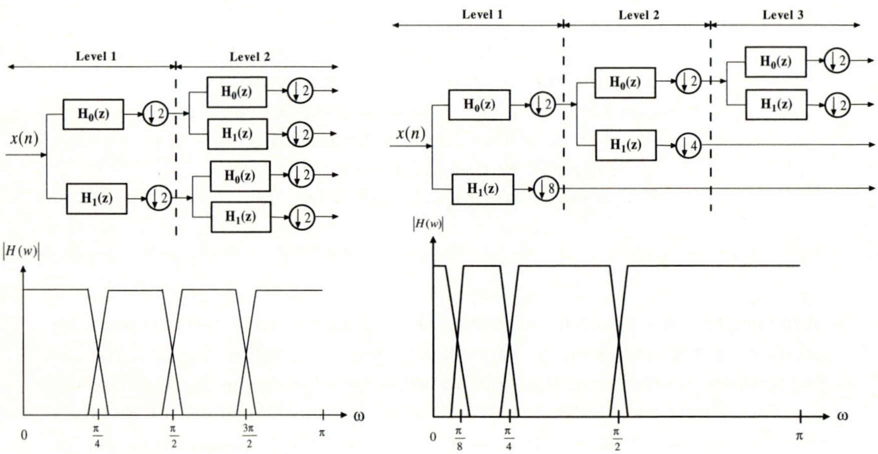

(a) Uniform (b) Non-uniform

Fig. 2. Uniform/Non-uniform binary-tree structured analysis filter banks

we use the ratio of subband signal power to input signal power. This method uses local characteristic of the signal in filter banks, that is, the signal power will often be concentrated in a few subbands, while the noise power is spread in all subbands. To reduce hardware complexity in implementing the proposed technique on Field Programmable Gate Array(FPGA), we employ non-uniform binary tree structured filter banks [2], and the Hadamard transform [8], which is adopted to design analysis and synthesis filters. The proposed method is designed by Very High Speed Integrated Circuit Hardware Descriptive Language(VHDL) and implemented on FPGA. The simulation and experimental results show that the efficiency of the proposed method.

2 Noise Cancellation Using Non-uniform Binary Tree Filter Bank

2.1 Binary-Tree Structured Filter Banks

As shown in Fig. 1, the filter bank is the primary tool used to perform the subband decomposition of signal. This is a well known Quadrature Mirror Filter(QMF) bank [2]. When the filterbank bandlimiting is non-ideal and the sub-sampling is critical, the analysis bank causes aliasing of the subband signals. This aliasing can be cancelled in the synthesis bank when certain conditions are met by the synthesis filters. Fig. 2 shows the uniform/non-uniform binary tree filter bank structures. These two configurations do the same number of filtering operations to cancel noise signal. The uniform binary tree structured filter bank produces the same delay in each subband filtering with the same bandwidth, While non-uniform binary tree filter bank can generate more detail resolution in low frequency.

2.2 Noise Cancelling Technique

The filter bank theory can be considered as analogy of the discrete wavelet [2]. A similar method, like as soft-thresholding of wavelet based algorithms, is also applied to filter banks based denoising technique [6]. In filter bank structures, the non-uniform binary tree structure is similar to wavelet decomposition tree [2]. Fig. 3 illustrates this filter bank tree, where H_0 and H_1 represent the low pass filter (LPF) and the high pass filter (HPF) of the analysis filter banks, and F_0 and F_1 are also LPF and HPF in the synthesis filter banks. In Fig. 3, $x_i(n)$ denotes a subband signal, and the additional down-sampler($\downarrow 2$) is added to produce the subband signals $x_{Di}(n)$ following as

$$x_{Di}(n) = \begin{cases} x_i(2^{M-(i+1)}n), & for\ i = 1, 2, ..., (M-2) \\ x_i(n), & for\ i = M-1, M \end{cases} \tag{1}$$

where M is the number of subband. The additional down samples match the down sampling rate of $x_i(n)$ and $x_{Di}(n)$. The sampling rate matching between subband signals is very important in filter bank implementation. The input signal $x(n)$ with noise is decomposed to each subband non-uniformly. The signal power may often be concentrated in a few subbands. Therefore, the subband signals which contain more information may have higher power than other subband signals. For denoising purpose, these subband signals are maintained and others are reduced. The noise reduction technique of this paper uses local characteristics which is the power ratio of subband to the input signal. In general, the stochastic properties of the input signal are not known fully, so we estimate the subband signal power as follows iteratively,

$$P_i(n) = (1-\alpha)P_i(n-1) + \alpha x_{Di}^2(n) \tag{2}$$

where α is positive constant. The total signal power is estimated as

$$P_{total}(n) = \sum_{i=1}^{M} P_i(n) \tag{3}$$

The power ratio in $n$the subband is defined by

$$\gamma_i(n) = \frac{P_i(n)}{P_{total}(n)} \tag{4}$$

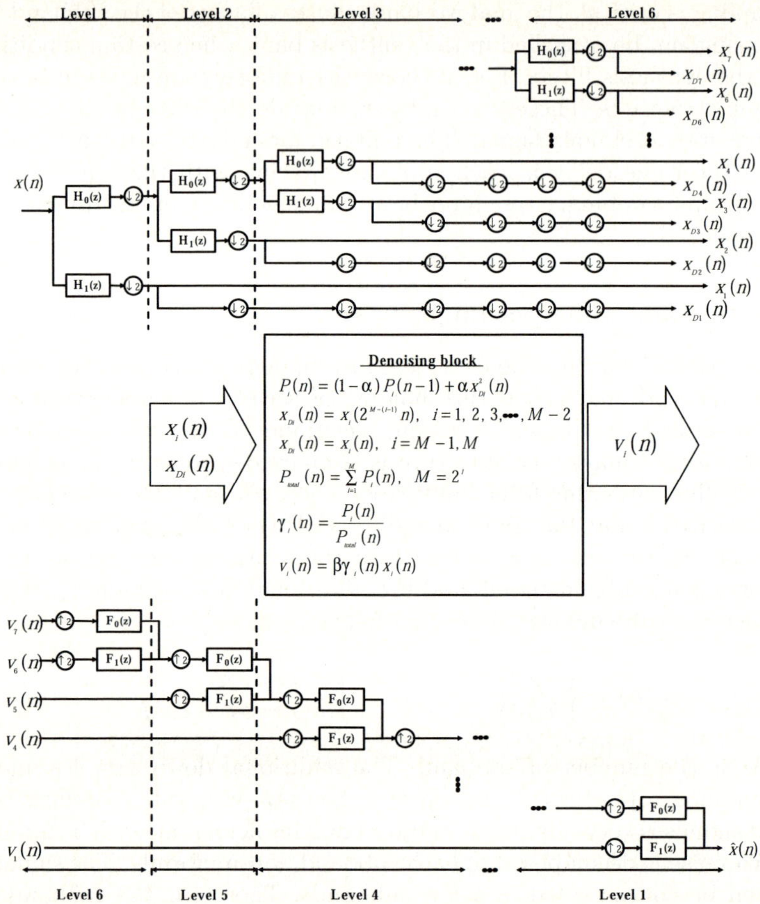

Fig. 3. Non-uniform binary tree structured filter banks ($M = 7$)

The power ratio, $\gamma_i(n)$, has following property when the signal power is higher then the noise.

$$\gamma_i \;=\; \begin{cases} > \frac{1}{M}, & \textit{in signal concentracted subbands} \\ \ll \frac{1}{M}, & \textit{otherwise} \end{cases} \tag{5}$$

The noise reduction is accomplished by modifying the subband signals as,

$$\nu_i(n) = \beta \gamma_i x_{Di}(n) \tag{6}$$

where β is the positive constant. The modified signals $\nu_i(n)$ is applied to synthesis filter bank, and therefore noise cancelled signal is reconstructed. The noise reduction technique of the proposed algorithm is shown in (1) to (6). And its properties can be considered similar to the soft thresholding method using wavelets. However, the proposed method is considerably simple in hardware implementing compared with soft thresholding technique.

Table 1. The SNR comparison of proposed and wavelet method (input signal SNR = 10dB)

algorithm	Doppler signal	block signal
wavelet method	46dB	34dB
proposed method	45.3dB	32.7dB

3 Simulations and Implementation

This section consider simulations, hardware implementations, and their results. We design the subband noise canceller with the non-uniform binary tree filter bank as shown in Fig. 3. The coefficients of analysis and synthesis filters are those of Hadamard transform. In computer simulations to evaluate the performance of the proposed noise cancellation algorithm, we use an white Gaussian noisy Doppler and block signal as excitation with $SNR = 10$dB. Finally, we implement the proposed algorithm on FPGA of Altera Stratix 1S25F672C6[11].

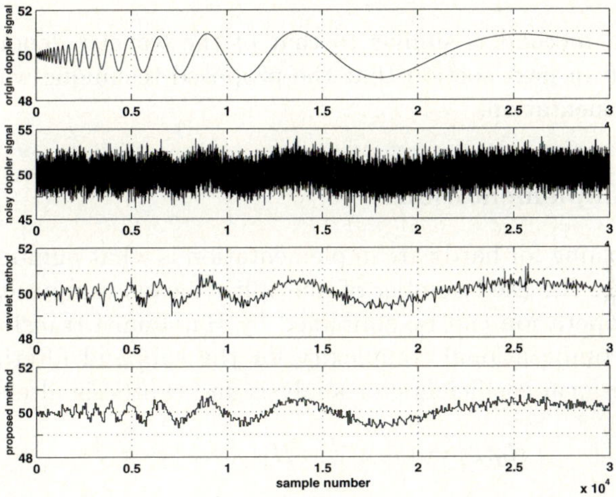

Fig. 4. Performance comparison of the proposed method and wavelet method with noisy Doppler signal as excitation($SNR = 10$dB, $M = 7$)

3.1 Simulations

We evaluate the denoising performances of the proposed technique as comparing with wavelet soft thresholding methods. Fig. 4 and Fig. 5 show that the simulate results for Doppler input signal and block signal input, respectively. In Fig. 4, we compare the result of the minmax algorithm [12] and the proposed algorithm. In Fig. 5, we compare the result of the Donoho algorithm [1] and the proposed algorithm. Table 1 shows SNR comparison result of proposed method and wavelet method. From these results, we can conclude that the performance

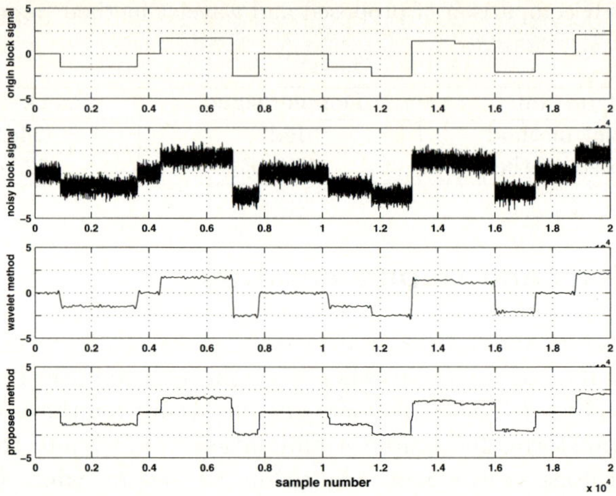

Fig. 5. Performance comparison of the proposed method and wavelet method with noisy Block signal as excitation($SNR = 10$dB, $M = 7$)

of the proposed method is similar to that of the wavelet denoising. However, it will be shown in next section that the proposed technique was advantage in hardware implementation.

3.2 FPGA Implementations

The important thing for hardware implementation is what number of multipliers are used, because the used number of multiplier means computational complexity. Multiplier operation can be simplified by Hadamard transform. Therefore, to reduce the computational complexity for the subband filtering, we use the Hadamard transform coefficients for analysis and synthesis filters as follows

$$H_0(z) = 1 + z^{-1}, \quad H_1(z) = 1 - z^{-1} \tag{7}$$

$$F_0(z) = 1 + z^{-1}, \quad F_1(z) = -1 + z^{-1} \tag{8}$$

Fig. 6 shows the block diagram of the procedure of subband noise canceller. Input signal is sampled at 500kHz and is converted to 8-bits digital signal by ADC(Analog to Digital Converter; MAX153). In noise cancelling system, signal are processed with 12-bits fixed point. It consist of 8-bits for the sampled signal and additional 4-bits for preventing the loss of information in the FIR filtering. The output signals reconstructed by the synthesis filters are represented by 18-bits digital signal, where the increased 6-bits are used for anti-overflows. In reconstructed signal, we take the most significant bit(MSB) that consists of 12-bits and it converted to analog signal using DAC(Digital to Analog Converter; AD7541).

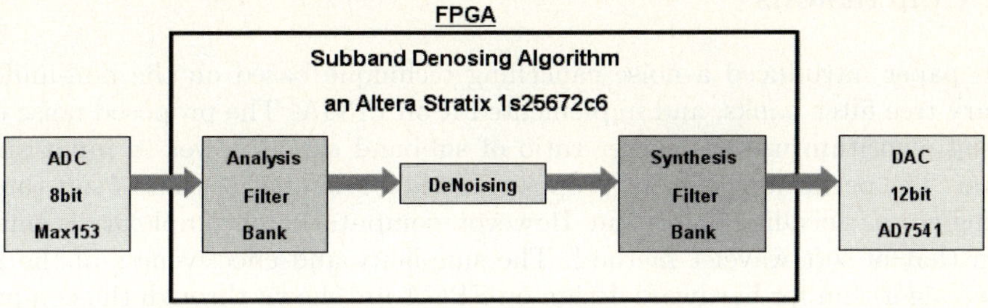

Fig. 6. The block diagram of subband noise canceller

Table 2. The subband noise canceller performance on FPGA

Logic Element(EA)	12, 700/25, 660
Memory(byte)	15, 336/1, 944, 576
DSP Block(EA)	18/85
Clock Speed(Mhz)	76.73

We summarize the contents of the FPGA report file at Table 2. It shows that the logic element is used 49.5%, memory is used 59.7%, and DSP block is used 18 element. The maximal clock speed of subband noise canceller has been measured 76.73Mhz. Therefore, this system don't have any limitations in real time processing. To prove the performance of the canceller implementation on the FPGA, we performed experiment to noisy Doppler signal. Experimental result by oscilloscope measured signal is shown in Fig. 7. This result proves that computer simulations using MATLAB is valid. From the computer simulations and experimental results, the proposed algorithm is efficient for noise cancellation.

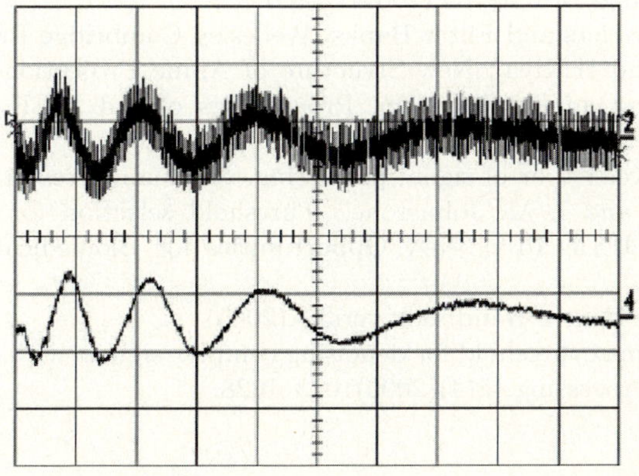

Fig. 7. Experimental noise cancelling result of the proposed method(SNR = 10dB, $M = 7$)

4 Conclusions

This paper introduced a noise cancelling technique based on the non-uniform binary tree filter banks, and implemented it on FPGA. The proposed noise cancelling algorithm use the power ratio of subband signal power to input signal power. The performance of the proposed method is similar to that of soft thresholding noise cancelling algorithm. However, computational complexity is smaller than that of soft wavelet method. The simplicity and effectiveness of the proposed algorithm for hardware design on FPGA are shown through the computer simulations and experiments.

Acknowledgement

The present study was supported by Regional Industrial Technology Development Project, Ministry of Commerce, Industry and Energy, Korea(Grant No. 10017508).

References

1. D.L.Donoho.: Denoising by soft-Thresholding. IEEE Trans. Information Theory.(1995)613–627
2. Vaidyanathan, P.P.: Multirate System and Filter Banks. Englewood Cliffs, NJ: Prentice-Hall(1993)
3. W.S.Lu.: Wavelet Approaches to still Image Denosing. Proceedings of 31st Asiloman Conference.(1997)1705–1709
4. Q. Pan, L. Zhang, D. Guanzhong, and Z. Hongai.: Two Denoising Method by wavelet Transform. IEEE Trans. Signal Processing. 47(12)(1999)3401–3406
5. W.Shenggian et al.: Load Characteristic based Wavelet Shrinkage Denoising Algorithm. IEEE Electronics Lett. 38(9)(2002)411-412
6. T.Tanaka and L. Duval.: Noise Reduction of Image with Multiple Subband Transforms. Proceedings of ICIP'04.(2004)1209–1212
7. G. Strang.: Wavelets and Filter Banks. Wellesley-Cambridge Press(1997)
8. K.Nishikawa and H.Kiya.: New Structure of Affine Projections Algorithm using a novel subband adaptive system. Proceedings of 3rd IEEE Signal Processing Workshop.(2001)364–367
9. S. Mallat.: Wavelet tour of signal processing. Academic Press(1999)
10. D. L. Donoho and I. M. Johnstone.: Threshold selection for wavelet shrinkage of noisy data. IEEE Proc. New Opportunities for Biomedical Engineers.(1994) A24– A25
11. Altera.: Stratix Device Handbook ver.3.3.(2005)
12. S. Sardy.: Minimax threshold for denoising complex signals with Waveshrink. IEEE Trans. Signal Processing.48(4)(2000)1023–1028

Spiking Neural Network Based Classification of Task-Evoked EEG Signals

Piyush Goel, Honghai Liu, David J. Brown, and Avijit Datta*

Institute of Industrial Research, The University of Portsmouth, Portsmouth, PO1 3HE, England, UK *Institute of Biomedical and Biomolecular Sciences, The University Of Portsmouth, Portsmouth, PO1 2DT,England, UK
{piyush.goel, honghai.liu, david.j.brown, avijit.datta}@port.ac.uk

Abstract. This paper presents an improved technique to detect evoked potentials in continuous EEG recordings using a spiking neural network. Human EEG signals recorded during spell checking, downloaded from the BCI Competition website, were pre-processed using a Wavelet Transform to remove the noise and to extract the low frequency content of the signal. Analysis of the signals was performed on the ensemble EEG and the task of the neural network was to identify positive and negative peaks of different shapes. The network has a time-warp invariance property, which means that an input linearly compressed or elongated in time is still recognisable by the network. This enabled the network to train on one peak shape and generalize it to recognise similarly shaped peaks. The neural network presented was trained on one epoch of filtered EEG and was tested on the remaining samples. A post hoc examination of the averaged evoked EEG signal pre-designated as target and non-target show a nadir in the non-target, but not in the target signals. A new supplementary template containing a nadir was therefore created and the effectiveness of this was tested on the ability of the network to correctly identify evoked EEG. After final testing 94.7% of the signals assigned as containing P300 by the paradigm used for the data on the website were correctly classified as P300s, and 83.7% of the non-P300s were also classified as non-P300s. The sensitivity of the technique, utilising the data from this paradigm was 94.7%, specificity 83.68%, and positive predictive value was 53.71%.

1 Introduction

The electroencephalogram (EEG) was first recorded by Hans Berger in 1924 [1]. Electrical potentials are produced within the human brain as a result of neural firing. The collective firing of many neurons in the brain leads to the formation of small currents which diffuse around the head and can be measured as EEG at the scalp using electrodes. They have been proved to be highly beneficial for detecting brain conditions like epilepsy and strokes [2, 3, 4, 5, 6], and for constructing Brain-Computer Interfaces (BCIs) [7, 8, 9, 10, 11, 12]. Most of the brain activity of interest falls in the range of 0.1 Hz to 45 Hz with an amplitude of the order of a few microvolts (μV). This frequency range is further divided into

B. Gabrys, R.J. Howlett, and L.C. Jain (Eds.): KES 2006, Part I, LNAI 4251, pp. 825–832, 2006.
© Springer-Verlag Berlin Heidelberg 2006

frequency bands of EEG activity - delta (0.1-4 Hz), theta (4-8 Hz), beta (8-13 Hz), alpha (13-25 Hz) and gamma (25-45 Hz).

A slow positive potential, known as the P300 [13], appearing typically between 250-500 ms from the time of presentation of a stimulus has been of great interest to EEG research, for error detection and prediction [14] and for building Brain-Computer Interfaces (BCI) [8, 9, 11]. Its shape can be represented as a Gaussian function with a standard deviation of 150ms and has amplitude of around 10 μV.

The present study aims to classify evoked potentials depending on whether the signal contained the P300 component or not. It also analysed the importance of a negative dip before the P300-like peak in the non-target signals. This nadir was much evident in the non-target signals as compared to the target signals, where it was not present in most cases. Following the discovery, an additional module was added to identify the dip and improve the classification accuracy.

To perform the recognition task, a spiking neuron model inspired by the mus silicium built by J.J. Hopfield and C. Brody [15, 16], was employed to detect the presence of P300 component in the EEG signal. The spiking neuron network comprises of three feed-forward (no backward connections) layers, namely, Layer-A, Layer-W and γ-Layer. Before the network starts its computation, the raw input signal is pre-processed (for details please refer [15, 16]) before it is fed to the first layer (Layer-A). The pre-processing method looks for the times of certain features in the raw signals. Let these times of occurrence of features be referred as events.Occurrence of each event triggers a fixed set of twenty decaying currents with different pre-set decaying times. Each of these decaying currents is represented by a neuron in Layer-A which drives a pair of Layer-W neurons. Layer-W consists of leaky-integrate-and-fire (LIF) neurons [17]. An LIF neuron sums (integrates) all the input currents, over a period of time, discharges (leaks) exponentially continuously, and sends out an action potential (fires) whenever its membrane potential reaches a particular threshold, thus giving it the name.As the currents flow into the neuron, the charge induced is stored in a capacitor, thus increasing the membrane potential. When this potential reaches a threshold the neuron fires and the membrane potential is reset back to a reset value. The final layer in the network, γ-Layer, consists of ten LIF neurons, one for each of the words to be classified. Being LIF neurons, they principally work in the same way as Layer-W neurons but with a difference that they have a shorter time constant. Each γ-neuron takes input from many neurons in Layer-W, however, neither does it have any lateral connections to other neurons in the γ-Layer nor any external current inputs. Due to its short time constant, the γ-layer neuron accumulates charges only over a very short period of time. Hence, for the neuron to fire, it is essential that a lot of simultaneous W-neuron firings take place. When numerous neurons in layer-W synchronize, they provide enough current for the γ-neuron to fire. Hence, the firing of a γ-Layer neuron just after the end of the word indicates the presence of the word associated with that neuron in the input to the network. If the trained network is presented with a word different from the training word, then the temporal relationship between the features will not match and there will be no synchrony in the W-Layer neurons, rendering the γ-neurons unfired.

Mus silicium was proposed primarily to recognise mono syllables as done by a biological brain. It uses the notion of transient synchrony [18] as a means of recognising the input. Testing it on speech data showed that the network was robust in the presence of noise. Moreover, it did not recognise a reversed sound for the same word, indicating that merely the presence of features was not sufficient, the temporal relationships between the features was critical for recognition. This network presents a time-warp invariant technique for feature detection, i.e., it would be able to recognise the input even it were stretched or compressed in time. These properties of the network provide a robust method for developing a template matching algorithm to detect features like the P300 in the on going EEG.

2 Data Pre-processing

The Wadsworth BCI Dataset (IIb) which is published on the BCI Competition 2003 Website [19] was used for this study. The data collected for one subject were examined. This dataset was collected using the BCI2000 platform (BCI2000 is a BCI research and development platform for research in brain signals, signal processing and user applications. For research purposes, it is freely available from the [20]). The following subsections describe the data in detail along with the collection method.

The data were collected using the P3 Speller paradigm mentioned by [8]. In this paradigm, the subject was presented with a 6-by-6 grid of characters on a computer screen. He/she was then instructed to concentrate on letters of a particular word, one by one. The rows and columns of the grid were lit randomly while the subject was focusing on a letter in the grid. For each letter of the word, there were 12 illuminations (6 for the rows and 6 for the columns), out of which only 2 (1 row and 1 column) contained the letter which had the subjects focus. There were 15 repetitions of this set of 12 illuminations for each letter. The idea behind the paradigm is that, whenever the row or column containing the relevant letter is lit, it stimulates a P300 response in the brain of the subject and can be detected in the recorded EEG signal.

To record the signals, 64 electrodes were placed on the scalp according to the standard 10-20 scheme. The signal was digitised at 240 Hz and recorded in three sessions, each consisting of a number of runs. The number of runs was equal to the number of words in the session and each run contained one word. The recording proceeded in the following manner: A blank matrix was displayed for 2.5s after which a row or column was illuminated for 100ms followed by 75ms of blank grid. This resulted in 12 such combinations for all rows and columns. This set of 12 intensifications was repeated 15 times for each letter of the word in a run making it a total of 180 intensifications for each letter. At the end of these 180 illuminations, the grid was blank for 2.5s indicating the end of intensifications for that letter. The letter to be concentrated on was shown in parentheses. The focus then shifted to the next letter of the word and the same process followed until the end of the word was reached.

A single channel Cz from over the vertex, was used in the analysis of the signals. Due to the low signal-to-noise ratio, features of interest are not immediately apparent in single trial EEGs and generally an average of multiple trials is required to see them clearly. This makes the features common to all trials more visible and cancels out the random noise present in the signal [21]. In the present study, averages of fifteen trials were performed on the EEG data. From the stimulus onset time 600 ms worth of data was extracted from EEG for each trial, for each letter. This was subjected to wavelets analysis (for a review on wavelets, please refer to [22]) using Daubechies-6 wavelet as the mother wavelet. Approximation level 4 was regenerated using the wavelet coefficients. The resulting data represented the lower frequencies (delta band) of the EEG. Figure 1 shows an EEG signal filtered using wavelets.

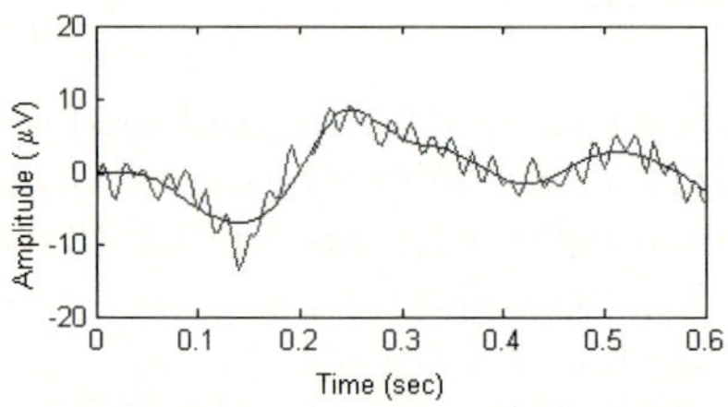

Fig. 1. The result of filtering with wavelets. The wavy line shows the original signal and the smooth line shows the signal after filtering it with Daubechies-6 wavelet.

3 Feature Extraction/Selection

Once the signals are filtered in the pre-processing stage, they are suitable to be analysed. The first step of the analysis of signals involved training the spiking neural network with a sample of an EEG signal containing the P300 component. An EEG signal containing an ostensible P300 potential was manually chosen from the training data which lay completely within the prescribed P300 latency period (250-450ms). The training algorithm then searched for the maximum value (peak) of the signal in the window of 250-450ms from the time of the stimulus. Then seven points were marked vertically downward from the peak at equal steps. The values of the signal to the left and right of these points were searched for, yielding two points on the signal for each vertical step. The times of all these fourteen points, together with the time of the peak formed the input feature vector to train the network. Figure 2 shows the method pictorially.

The above process trained the network to recognise peaks in the latency period of the P300 component. The trained network stored the temporal relationship between these features in the form of corresponding decaying rates for the different times of the points.

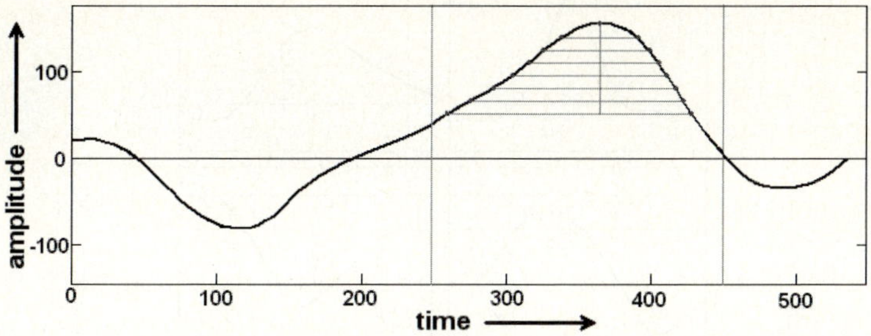

Fig. 2. The training algorithm determines 15 points times of which form the input feature vector for the spiking neural network. The green vertical lines mark the bounds of the P300 latency period, the blue horizontal line is the zero value line. The red dots on the signal are the points marked by the algorithm with the peak being at the top.

During the testing phase, the network searches for the highest value of the signal in the 250-450ms latency period, this is defined as the peak. If the peak is below a certain threshold, then the signal is classified as a non-target signal. If the peak happens to be above the threshold then the algorithm searches for the fourteen points as discussed above within the designated P300 time period. If the peak is not completely contained within the bounds of the designated period, then the points lying outside the bounds are ignored and are not included in the feature vector. This prevents the peaks that are near the P300 latency period but not within it, being classified as P300 responses.

A peak which is narrower will have these point times compressed together and dilated for a broader peak. Due to the time warp invariant nature of the network, both of these types of peaks will be recognised by the network as P300s. Figure 3 shows signals that are classified as P300s with a *tick* and those that are not, with a *cross*.

Figure 4 shows the averages of the filtered target and non-target signal responses. As can be seen from the figure, there is an evident negative dip in the P300 period for the non-target signals (dotted line in the figure). In many, non-target signals, the negative dip was followed by a positive peak. This positive peak is falsely classified as a P300 peak due to its shape and amplitude. In contrast, the positive peak in the target signals is not preceded by such a negative dip. Following this observation, an algorithm was devised to check for the nadir in a similar way in which the positive peak was detected for target signals as discussed above. The signals which had the positive peak in the P300 latency period but also had the negative preceding it, were classified as non-target signals.

4 Results and Discussion

The above mentioned method was applied to the BCI Competition dataset. Table 1 shows the results that were obtained without the application of nadir

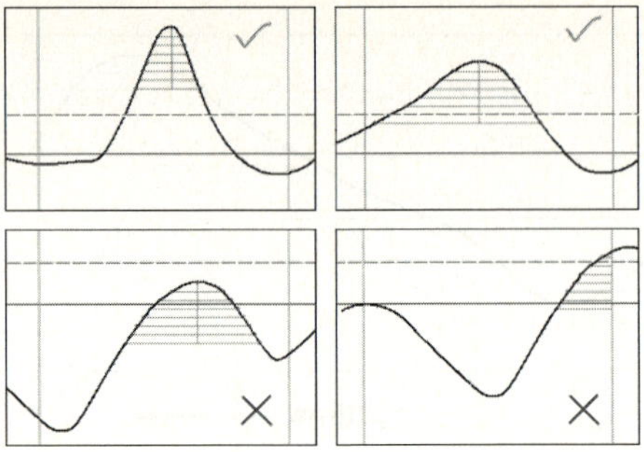

Fig. 3. The algorithm tested on four different types of EEG. The pattern of diagrams above follow that of Figure 2 and show only the region for the P300 latency period for each. The dashed line represents the threshold level (discussed in text). The peaks in the two signals on top are correctly recognised as P300s (marked with a tick) and the ones on the bottom are not (marked with a cross). The bottom left signal is below the threshold however the bottom right signal is above the threshold but out of bounds of the P300 region.

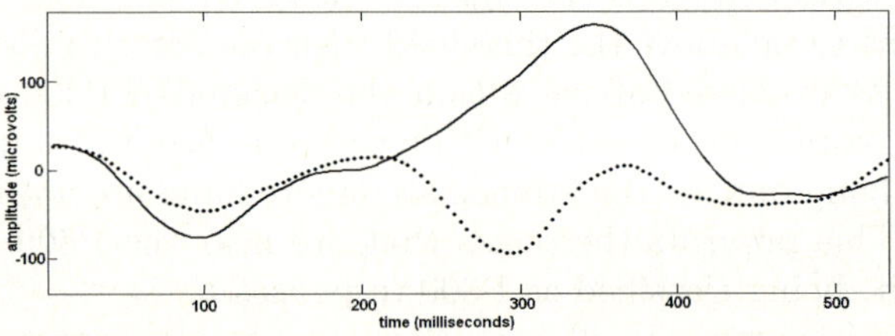

Fig. 4. Signals showing all the signals after filtering them with wavelets. The solid line shows the average of all the target responses and the dotted line shows the average of all the non-target responses.

detection [23] and Table 2 depicts the numbers as a result of negative dip detection. As can be seen from the table, there was no reduction to the rate of target signals being correctly identified but the number of falsely classified non-target signals was reduced significantly.

The new results above correspond to a sensitivity of 94.7%, specificity of 83.68%, positive predictive value of 53.71%, and negative predictive value of 98.75%. We plan to use this technique to investigate sleepiness, human error detection due to low alertness and hence reduction of the same by making the subject aware of it through feedback.

The present method shows promising results on the classification of EEG signals containing the stimulus response from the ones that do not appear to do so. The performance is close to the performance of the state of the art in the field [19] but has potential to improve as spiking neural network allows us to combine biological stimulus response into the networks and we are working on schemes to improve it.

Table 1. BCI Competition Dataset Results before nadir detection

	P300	Non-P300
Identified Correctly	36 (94.7%)	132 (69.5%)
Identified Incorrectly	2 (5.3%)	58 (30.5%)

Table 2. BCI Competition Dataset Results after nadir detection

	P300	Non-P300
Identified Correctly	36 (94.7%)	159 (83.68%)
Identified Incorrectly	2 (5.3%)	31 (16.32%)

The method we discuss here uses a single epoch of a stimulus signal containing the P300 component and is able to generalise it for peaks with similar but slightly different shapes. This can be seen as a method of template matching where one template can be used to identify many signals containing a similar peak. The performance of the process can be improved by involving many samples in the training stage to give multiple templates to be matched. These templates would run as parallel networks to identify the signal. The signal which is then recognised by the majority of this committee of networks would be deemed as a stimulus signal. This would reduce the number of non-target signals being recognised as target signals. Another direction for future work would be to use multiple channels to classify the signal.

References

1. H. Berger. On the electroencephalogram of man. *Archiv fur Psychiatrie und Nervenkrankheiten*, 87:527–570, 1929.
2. M. D'Alessandro, R. Esteller, G. Vachtsevanos, J. Hinson, A. andEchauz, and B. Litt. Epileptic seizure prediction using hybrid feature selection over multipleintracranial eeg electrode contacts: a report of four patients. *IEEE Transactions on Biomedical Engineering*, 50(5):603–615, May 2003.
3. M. Seeck, F. Lazeyras, C.M. Michel, O. Blanke, C.A. Gericke, J. Ives, J. Delavelle, X. Golay, N. Haenggeli, C.A. andde Tribolet, and T. Landis. Non-invasive epileptic focus localization using eeg-triggered functionalmri and electromagnetic tomography. *Electroencephalography and Clinical Neurophysiology*, 106(6):508–512, June 1998.
4. M.L.V. Quyen, J. Martinerie, C. Adam, and F.J. Varela. Nonlinear analyses of interictal eeg map the brain interdependences in humanfocal epilepsy. *Physica D*, 127(3):250–266, March 1999.
5. Z. Huang, W. Dong, Y. Yan, Q. Xiao, and Y. Man. Effects of intravenous mannitol on eeg recordings in stroke patients. *Clinical Neurophysiology*, 113(3):446–453, March 2002.
6. M. Molnar, G. Gacs, G. Ujvari, J.E. Skinner, and G. Karmos. Dimensional complexity of the eeg in subcortical stroke - a case study. *International Journal of Psychophysiology*, 25(3):193–199(7), April 1997.

7. J.R. Wolpaw, D.J. McFarland, G.W. Neat, and C.A. Forneris. An eeg-based brain-computer interface for cursor control. *Electroencephalography and Clinical Neurophysiology*, 78(3):252–259, March 1991.
8. E. Donchin, K.M. Spencer, and R. Wijensinghe. The mental prosthesis: Assessing the speed of a p300-based brain-computerinterface. *IEEE Trans. Rehab. Eng.*, (8):174–179, 2000.
9. N. Birbaumer, N. Ghanayim, T. Hinterberger, B. Iversen, I. andKotchoubey, A. Kübler, J. Perelmouter, E. Taub, and H. Flor. The thought translation device (ttd) for completely paralyzed patients. *IEEE Transactions on Rehabilitation Engineering*, 8(2):190–193, June 2000.
10. G.E. Birch and S.G. Mason. Brain-computer interface research at the neil squire foundation. *IEEE Transactions on Rehab. Eng.*, 8(2):193–195, June 2000.
11. J. D. Bayliss. The use of the p3 evoked potential component for control in a virtual apartment. *IEEE Transactions on Neural Systems and Rehabilitation Engineering*, 11(2), June 2003.
12. J.D.R. Millán, F. Renkens, J. Mouri no, and W. Gerstner. Brain-actuated interaction. *Artificial Intelligence*, 159:241–259, 2004.
13. T.W. Piction. The p300 wave of the human event-related potential. *J. Clin. Neurophysiol.*, 9:456–479, 1992.
14. A.K. Datta, A.K. Hawkins, J. Heutink, T. Manly, R. Cusack, C. Rorden, B. Davison, and I. Robertson. An electrophysiological predictor of human error. *J. Physiology*, 555P(PC107), 2004.
15. J.J. Hopfield and C.D. Brody. What is a moment? "cortical" sensory integration over a brief interval. In *Proceedings of the National Academy of Sciences of the USA*, pages 97(25): 13919–13924, 2000.
16. J.J. Hopfield and C.D. Brody. What is moment? transient synchrony as a collective mechanism for spatiotemporalintegration. In *Proceedings of the National Academy of Sciences of the USA*, pages 98: 1282–1287, 2001.
17. W. Gerstner and W.M. Kistler. *Spiking Neuron Models.* Cambridge University Press, first edition, August 2002.
18. A.K. Datta and J.A. Stephens. Short term synchronization of motor unit activity during voluntary contractionin man. *Physiol.*, (422):397–420, 1992.
19. B. Blankertz. Bci competition 2003, url: http://ida.first.fraunhofer.de/projects/bci /competition/index.html, 2003.
20. G. Schalk. Bci2000, url: http://www.bci2000.org, 2005.
21. T.W. Piction, O.G. Lins, and M. Scherg. The recording and analysis of event-related potentials. *Handbook of Neuropsychology*, (10):3–73, 1995.
22. S. Gilbert and T. Nguyen. *Wavelets and filter banks.* Wellesley-Cambridge Press, second edition, 1998.
23. P. Goel, D. Brown, H. Liu, C. James, and A. Datta. Analysis of evoked potentials using a spiking neural network. In *IEE Advances in Medical, Signal and Information Processing (Accepted)*, 2006.

Learning Control Via Neuro-Tabu-Fuzzy Controller

Sarawut Sujitjorn and Sudarat Khwan-on

School of Electrical Engineering, Suranaree University of Technology
Nakhon Ratchasima, 30000, Thailand
sarawut@sut.ac.th

Abstract. The paper presents a novel search algorithm named adaptive tabu search (ATS). The algorithm has been applied to enhance the learning process of neural networks. The paper presents a new neuro-tabu-fuzzy (NTF) control structure to demonstrate the capability of the ATS algorithm. The algorithm is general and can be applied to various problems including machine learning, optimization, etc. Our proposed algorithm and controller nicely stabilize the single- and double-inverted pendulum systems.

1 Introduction

Today's control technology offers a great deal of flexibility to industrial production lines. In most cases, dynamical systems and factory processes are complex, highly nonlinear, and sometimes unstable. Controlling them can be achieved by complicated conventional control or an alternative of intelligent control. Intelligent control has become a worldwide interest because of its simplicity. It does not require an explicit model of the system to be controlled. But an accurate model, if exists, is helpful to obtain very high quality and stable performance. Vast amount of intelligent control applications can be found in transportations, robotics, industry automation, etc.

Among those intelligent control techniques, Yi and Yubazaki proposed in 2000 [1] the single input rule module (SIRM) fuzzy controller. Its advantage is to have simple and less control rules. They successfully applied this control structure to stabilize single, and double inverted pendulum systems [1, 2]. While the control parameter namely the dynamic importance degree (DID) was arbitrarily tuned in the previous works, the authors of this article proposes an enhanced structure that helps to automatically tune the DID parameter with learning capability. The proposed structure contains neural network, and adaptive tabu search (ATS) [3]. The structure is somewhat more complicated than the previous one in the exchange of much better stabilized performance.

The article contains five sections including the introduction. Sections 2 and 3 give the reviews of the SIRM fuzzy controller, the neural network, and the ATS. The new proposed control structure so called neuro-tabu fuzzy (NTF) control is explained in section 3. Section 4 gives some simulation results and discussions of the stabilizing single and double inverted pendulum systems by using the SIRM fuzzy controller and the NTF controller. Conclusion follows in section 5.

B. Gabrys, R.J. Howlett, and L.C. Jain (Eds.): KES 2006, Part I, LNAI 4251, pp. 833–840, 2006.
© Springer-Verlag Berlin Heidelberg 2006

2 SIRM Fuzzy Controller

Since fuzzy sets and fuzzy logic have been well-known, their reviews are omitted. This section presents a brief explanation of the single input rule module (SIRM) fuzzy controller [1] as the basis of our proposed new control structure. Figure 1 depicts the SIRM fuzzy control structure that contains the following main components: norm, DIDs, SIRMs, and OSF. The norm block normalizes and fuzzifies the error signals and produces the signals $x_1, x_2, \ldots, x_i$ within the given universe of discourse. The modules SIRMs process these signals x_i according to the rule

$$\text{SIRM-}i\ :\ \left\{R_i^j : \text{if}\ \ x_i = A_i^j\ \ \text{then}\ \ u_i = C_i^j\right\}_{j=1}^{m_i} \tag{1}$$

where SIRM-*i* is the fuzzy rule module of the ith input, R_i^j is the jth rule of the SIRM-*i*, x_i is the ith input, u_i is the ith output of the SIRM-*i*, A_i^j is the membership function of the input x_i, C_i^j is the membership function of the output u_i, $i = 1, 2, \ldots, n$ is the counter of inputs, and $j = 1, 2, \ldots, m_i$ is the counter of rules of the SIRM-*i*.

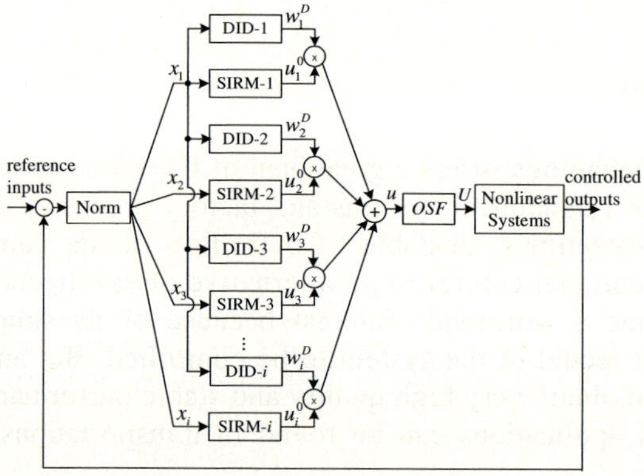

Fig. 1. SIRM fuzzy control

The output u_i^0 is fuzzy singleton and can be computed via the simplified fuzzy reasoning [4] as follows

$$u_i^0 = \frac{\sum_{j=1}^{m_i} A_i^j(x_i) C_i^j}{\sum_{j=1}^{m_i} A_i^j(x_i)} \tag{2}$$

The dynamic importance degree (DID) indicates the weight of each input x_i. The relation (3) is used to calculate the DID-values, w_i^D.

$$w_i^D = w_i + B_i \cdot \Delta w_i^0 \tag{3}$$

where w_i is the base value (constant), B_i is the breadth value (constant), and Δw_i^0 is the output of fuzzy rules according to some dynamic variables Δw_i . B_i and w_i are control parameters that can be obtained by trial-and-error or random optimization search according to Yi et. al. [1, 2]. The actual control U issued to the system under control can be obtained from (4) and (5), respectively.

$$u = \sum_{i=1}^{n} w_i^D . u_i^0 \tag{4}$$

$$U = u \,.OSF \tag{5}$$

OSF stands for output scaling factor.

3 Enhanced Fuzzy Control

The enhanced fuzzy control as our proposal has been named neuro-tabu-fuzzy (NTF) control. This section thus presents some reviews of the neural network, and the adaptive tabu search, respectively. Then it presents the structure of the NTF control.

3.1 Neural Network

It has been known for many years that learning capability can be incorporated into control via artificial neural network or neural network (NN) in short. The diagram in figure 2 represents the neural network model in which

$$net_j = \sum_{i=1}^{n} w_{ij} x_i \tag{6}$$

$$y_j = f(net_j + \theta_j) \tag{7}$$

where x_i is the input, y_i is the output, w_{ij} is the weight from the ith input to the jth cell, θ_j is the bias at the jth cell, and $f(\cdot)$ is the activation function.

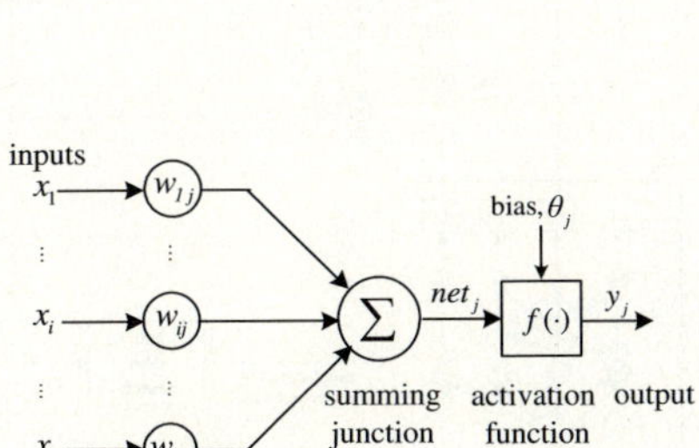

Fig. 2. Neural network model

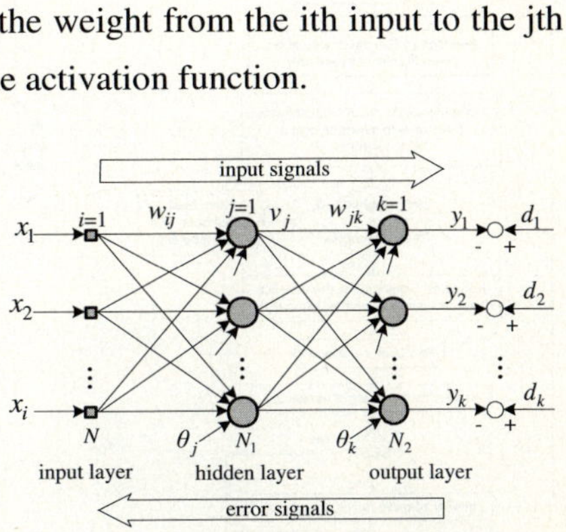

Fig. 3. Backpropagation of multilayer NN

Several forms of activation function are available such as step, sigmoid, linear, etc. NN learns through an iterative process of weight, and bias adjustment. Learning can be classified as supervised, and unsupervised learning. In this work, we employ a specific supervised learning method called backpropagation learning algorithm. The diagram in figure 3 represents the backpropagation learning of multilayer NN. Referring to figure 3, x, y, w, and θ carry the meanings of input, output, weight, and bias as previously described. i, j, and k represent the NN cells in the input, hidden, and output layers, respectively. d stands for the required output. The error signal e_k in the backpropagation process is the difference between d_k and y_k, i.e. $e_k = d_k - y_k$. The adjustment of weight and bias values can be achieved via the gradient descent method, as one possibility. Good explanations on the subject of neural networks can be found in [5].

3.2 Adaptive Tabu Search

The adaptive tabu search (ATS) has been one of the efficient AI search methods. It had been developed on the practical concept of simple tabu search. The ATS has two important mechanisms namely back-tracking and adaptive search radius that help to accelerate the solution convergence, and to escape from local solution lock. It also possesses the tabu list as in the conventional case. The tabu list has first-in-last-out property and keeps the history of solution movement. Figure 4 summarizes the ATS algorithms. When the number of solution-cycling meets the maximum allowance, the ATS activates the back-tracking mechanism. This means that a local solution lock occurs. To escape from the lock, an old solution stored in the tabu list is selected to restart a new search path that could lead to a new local solution. When a current solution is

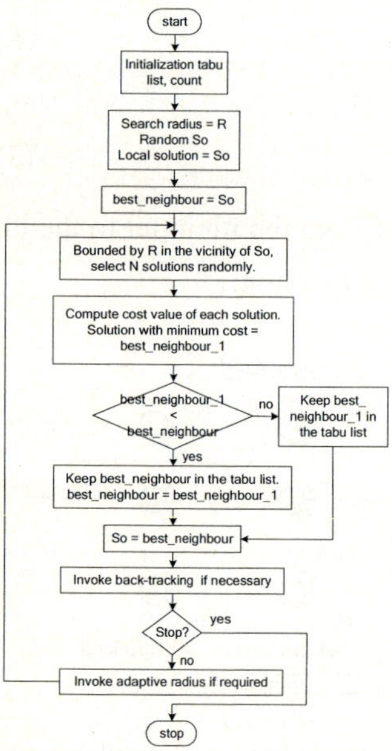

Fig. 4. ATS algorithms

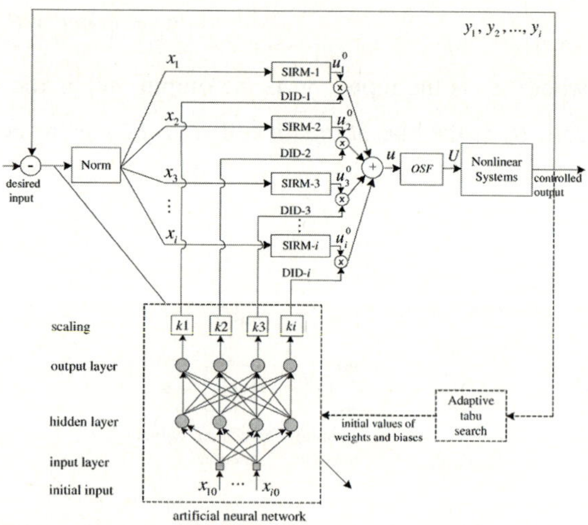

Fig. 5. Neuro-tabu-fuzzy (NTF) control structure

relatively close to a local solution, the ATS activates the adaptive search radius. This mechanism decreases the search radius according to the best cost function obtained. The lower the cost function, the shorter the radius. This helps the search to focus the minimum solution. Detailed explanation, and performance evaluation of the ATS can be found in [3].

3.3 Neuro-Tabu-Fuzzy Control

The neural network (NN) and the ATS have been incorporated into the SIRM fuzzy controller. The new structure is named neuro-tabu-fuzzy (NTF) control. Figure 5 illustrates the NTF control structure. With DIDs obtained from the NN learning capability, better system control could be achieved. The ATS searches for the optimum initial parameters of the NN. This leads to better performance of the NN. The structure of NN in the proposed controller used feedforward multilayer type.

4 Results and Discussions

The performance comparison between the NTF and the SIRM fuzzy controllers is to stabilize an inverted pendulum to an upright position. The considered pendulums are single and double rods situated on a moving cart.

For the SIRM case, the DID values can be obtained from the equation (3).The base and breadth values can be obtained from trial-and-error [1]. For the NTF case, the DID values are obtained from the NN-learning that starts with initial weights and biases obtained from the ATS. These weights and biases are optimum under some specific situations. During the operation of the pendulum system, the DID values are interactively adjusted through the NN-learning process. Therefore, the DIDs can be adjusted correspondingly to situation changes.

Simulations have been conducted extensively for the stabilization of single inverted pendulum, figure 6, of which models can be formed in [1]. Some of the simulation results are illustrated in figures 8, and 9, respectively. Referring to the figures, P(0.1,0.5,1.0) indicates the values of the pendulum mass, pendulum rod half length, and cart mass, respectively, while S(30.0,0.0) indicates the initial pendulum angle, and cart position, respectively. As a result, the NTF controller stabilizes the pendulum within 3.58 sec which is less than half of the stabilization time used by the SIRM fuzzy controller

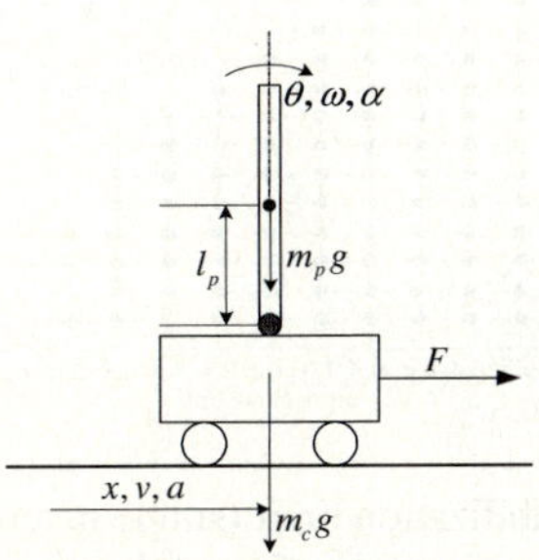

Fig. 6. Single inverted pendulum system

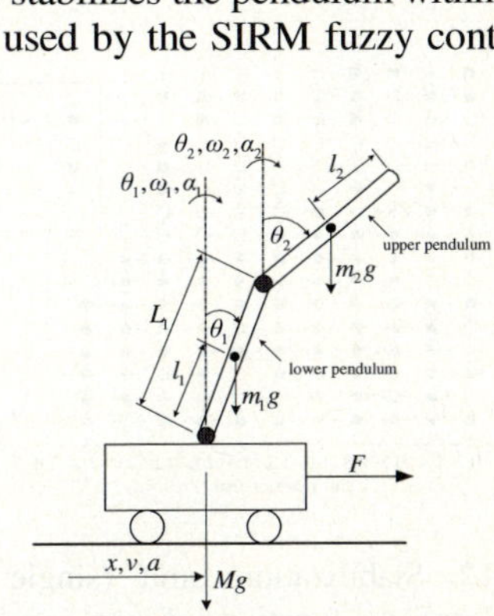

Fig. 7. Double inverted pendulum system

(8.24 sec). Figures 10 and 11 show the stabilization limits in terms of stabilization time of various simulation situations corresponding to different initial pendulum angles, and pendulum lengths. The symbols are located to indicate the complete stabilization time. It is noticed that the NTF controller provides shorter stabilization time, and wider useful ranges of operation. In most cases, the NTF controller consumes the stabilization time of less than 5 sec.

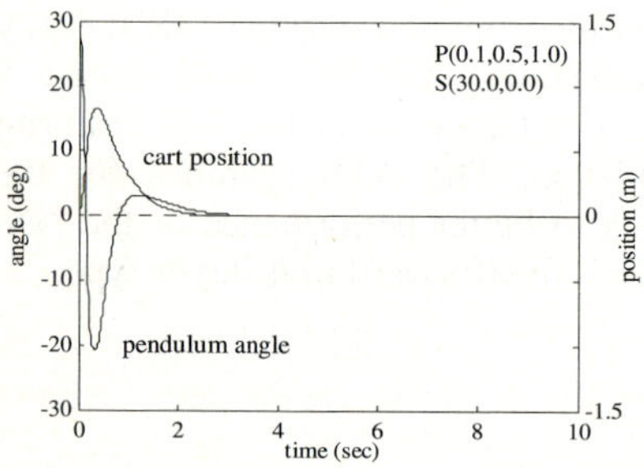

Fig. 8. Stabilization of single inverted pendulum: NTF controller

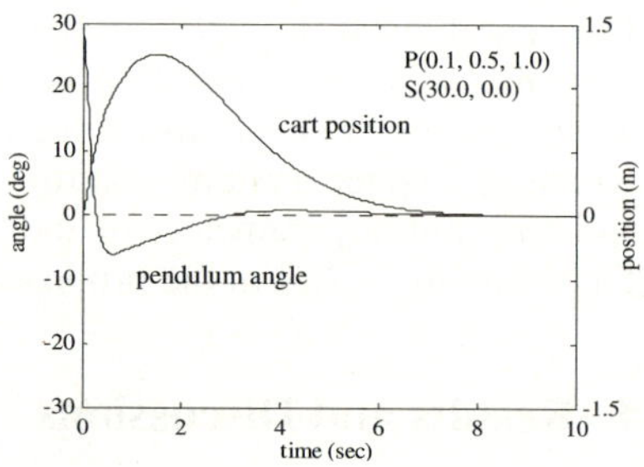

Fig. 9. Stabilization of single inverted pendulum: SIRM fuzzy controller

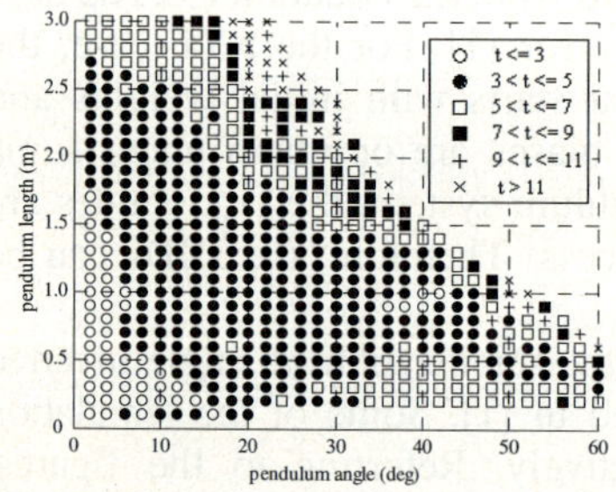

Fig. 10. Stabilization limit (single inverted pendulum): NTF controller

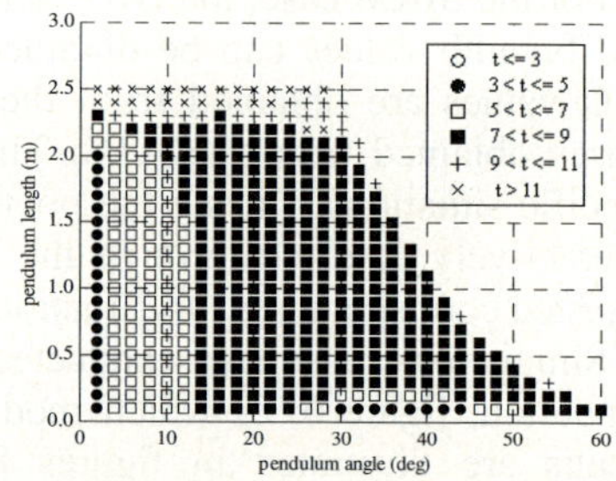

Fig. 11. Stabilization limit (single inverted pendulum): SIRM fuzzy controller

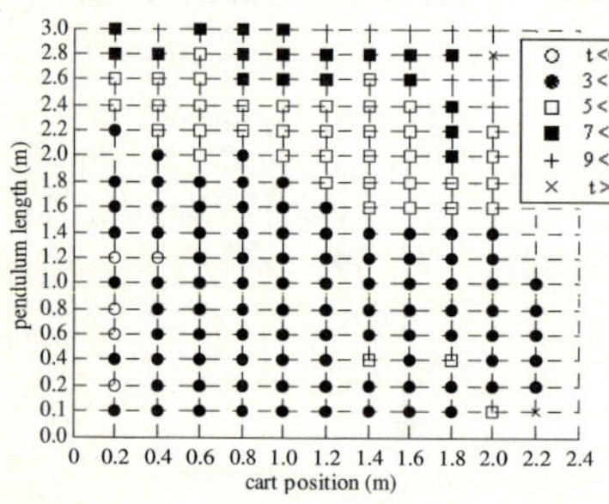

Fig. 12. Stabilization limit (single inverted pendulum) as function of cart position and pendulum length: NTF

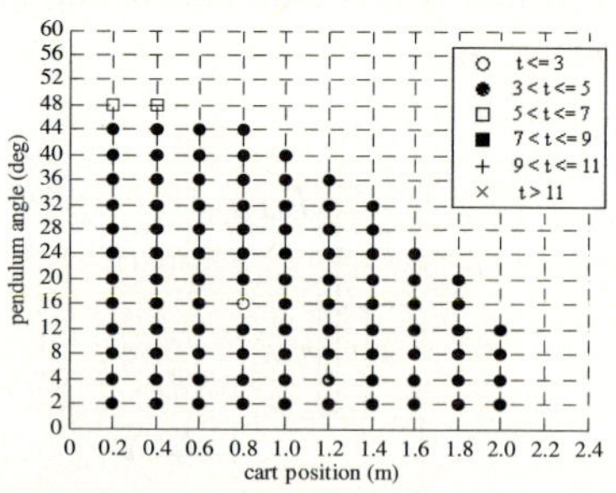

Fig. 13. Stabilization limit (single inverted pendulum) as function of pendulum angle and cart position: NTF

To assess the ability of the NTF controller, the controller must operate under various situations. Figure 12 discloses the stabilization limit as the function of pendulum length and the initial cart position where the initial pendulum angle is 0.0 deg. Figure 13 shows the stabilization limit as the function of the initial angle of the pendulum with the initial position of the cart when the half length of pendulum rod is 0.5 m. These results show that the proposed NTF controller has ability to rapidly stabilize the single inverted pendulum system of wide ranges. Table 1 summarizes the results according to the NTF and the SIRM fuzzy controllers operations to stabilize the single inverted pendulum.

Table 1. Stabilization time (single inverted pendulum)

Situations	Stabilization time (sec)		Shorter stabilization time consumed by NTF (in %)
	SIRM	NTF	
1 P(0.1,0.5,1.0) S(30.0,0.0)	8.24	3.58	56.55
2 P(0.1,0.1,1.0) S(30.0,0.0)	6.65	4.48	32.63
3 P(0.1,2.0,1.0) S(30.0,0.0)	8.95	6.31	29.50
4 P(0.1,0.5,1.0) S(0.0,2.0)	7.16	3.30	53.91
5 P(0.1,0.5,1.0) S(30.0,1.5)	unstable	3.36	100.0

Table 2. Stabilization time (double inverted pendulum)

Situations	Stabilization time (sec)		Shorter stabilization time consumed by NTF (in %)
	SIRM	NTF	
1 P(0.2,0.4,0.1,0.2) S(15.0,10.0,0.0)	7.19	4.47	37.83
2 P(0.2,0.4,0.1,0.2) S(15.0,15.0,0.0)	8.46	6.00	29.08
3 P(0.2,0.4,0.1,0.2) S(15.0,20.0,0.0)	9.27	5.36	42.18
4 P(0.1,0.5,0.1,0.5) S(15.0,15.0,0.0)	10.13	8.41	16.98
5 P(0.2,0.4,0.1,0.2) S(25.0,25.0,2.5)	unstable	9.20	100.0

Extensive simulations have been conducted for the stabilization of double inverted pendulum, figure 7, of which models can be formed in [2]. The learning of neural network results in the DIDs. Some of the results illustrated in the figures 14 and 15 are for the cases of that the initial angles of the lower, and the upper rods are 15, and 10 deg, respectively. The initial cart position is at 0.0 m. It can be noticed that the NTF controller consumes the stabilization time of about 4.5 sec while the SIRM fuzzy controller consumes about 7.2 sec. Figures 16 and 17 indicate that the stabilization limits of both controllers are somewhat the same, but the NTF controller consumes shorter stabilization times. Table 2 summarizes the results and confirms the advantage of our proposed NTF controller.

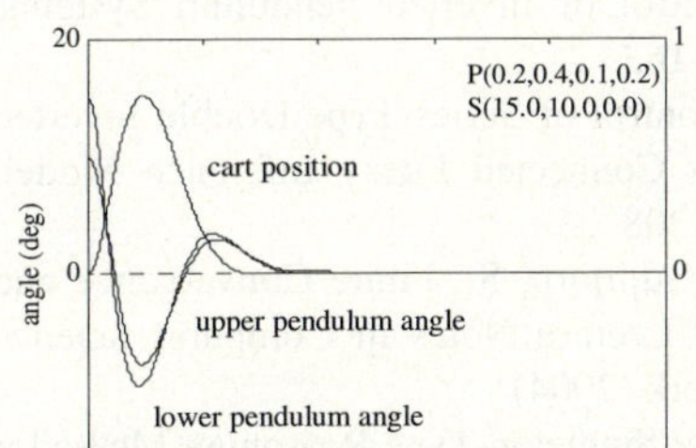

Fig. 14. Stabilization of double inverted pendulum: NTF controller

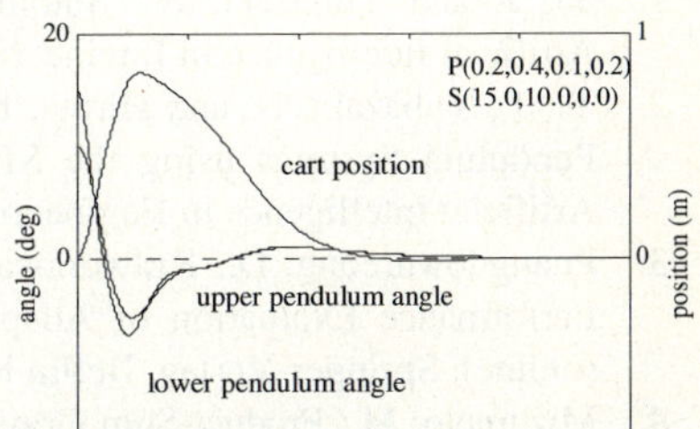

Fig. 15. Stabilization of double inverted pendulum: SIRM fuzzy controller

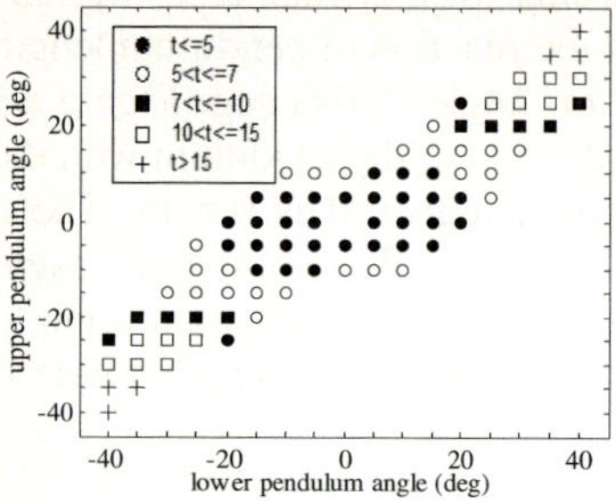

Fig. 16. Stabilization limit (double inverted pendulum): NTF controller

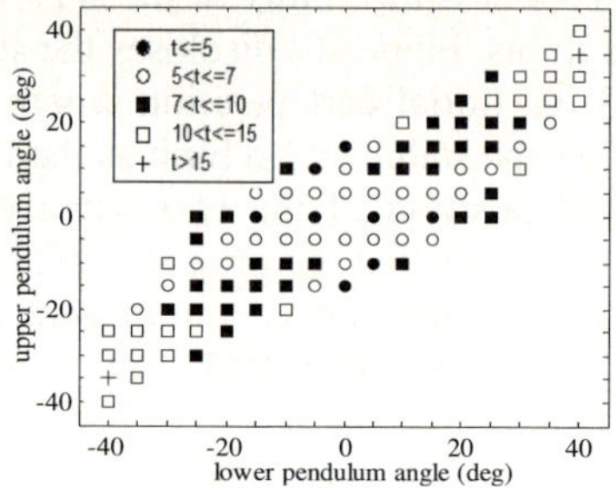

Fig. 17. Stabilization limit (double inverted pendulum): SIRM fuzzy controller

5 Conclusions

We have described the adaptive tabu search (ATS) algorithm that is general and possibly applied to machine learning, optimization, and so on. In this article, the ATS has been incorporated into neural network and a fuzzy controller. This results in a novel neuro-tabu-fuzzy (NTF) controller. The ATS enhances the learning capability of the controller. Extensive simulations on stabilization of single and double inverted pendulums have been conducted to confirm the performances and advantages of the NTF controller. The results obtained are summarized in the Tables 1 and 2.

Acknowledgments

The authors' thanks are due to Suranaree University of Technology for its financial support.

References

1. Yi, J. and Yubazaki, N.: Stabilization Fuzzy Control of Inverted Pendulum Systems, Artificial Intelligence in Engineering, 14 (2000) 153-163
2. Yi, J., Yubazaki, N. and Hirota, K.: Stabilization Control of Series-Type Double Inverted Pendulum Systems using the SIRMs Dynamically Connected Fuzzy Inference Model, Artificial Intelligence in Engineering, 15 (2001) 297-308
3. Puangdownreong, D., Kulworawanichpong, T. and Sujitjorn, S.: Finite Convergence and Performance Evaluation of Adaptive Tabu Search, Lecture Notes in Computer Science (online), Springer-Verlag, Berlin Heidelberg New York (2004)
4. Mizumoto, M.: Product-Sum-Gravity Method = Fuzzy Singleton-Type Reasoning Method = Simplified Fuzzy Reasoning Method, Proc. IEEE Int. Conf. on Fuzzy Systems, 3 (1996) 2098-2102
5. Fausett, L.: Fundamentals of Neural Networks Architectures, Algorithms, and Application, Prentice-Hall, New Jersey (1994)

DSP Based Implementation of Current Mode Fuzzy Gain Scheduling of PI Controller for Single Phase UPS Inverter*

Emine Doğru Bolat**

Kocaeli University, Faculty of Technical Education,
Department of Electronics & Computer, Kocaeli, Turkey
ebolat@ieee.org

Abstract. This paper presents Control of UPS inverter Using Fuzzy Gain Scheduling of PI Controller. This control scheme has Double Loop Current Mode Control Scheme in core. This scheme includes two control loops as inner and outer control loops. Inner control loop uses inductor current of inverter filter as feedback while outer control loop uses inverter output voltage as feedback. Fuzzy Logic Controller, FLC is used to adjust the voltage loop proportional-integral, PI controller parameters. The voltage error, *e* and its derivative, *de* are used as input variables of the FLC. In this study, Fuzzy Gain Scheduling of PI Controller is simulated digitally using PSIM simulation package and C++ and implemented using Texas Instruments, TI's TMS320LF2407 DSP (Digital Signal Processor) under linear, rectifier type nonlinear and fluorescent loads. And the results show that Fuzzy Gain Scheduling of PI Controller can provide low Total Harmonic Distortion, THD and better regulation quality especially under rectifier type nonlinear and fluorescent loads.

1 Introduction

Uninterruptible Power Supplies, UPS are used to produce clean and uninterrupted power to the critical loads such as medical/life support systems, computer systems, communication systems, etc. A typical UPS system consists of a battery (dc source), a dc-ac inverter, and an LC filter. The role of the inverter for UPS is to regulate the output voltage waveform with constant voltage and constant frequency although the variations of the line source or loads. The quality of UPS can be evaluated by THD value of output voltage and characteristic of transient response.

UPS power quality depends on the choice of pulse width modulation, PWM inverter control methods. Traditional analogue control methods are generally used in PWM inverter design. However there are a number of disadvantages in an analogue system, for example, temperature drifts and aging effect of the components, more component numbers for the system, necessity for making adjustment to many

* This study with the number is supported by TUBITAK (Turkey Scientific and Technical Research Association).

** Associate Member, IEEE.

B. Gabrys, R.J. Howlett, and L.C. Jain (Eds.): KES 2006, Part I, LNAI 4251, pp. 841 – 849, 2006.
© Springer-Verlag Berlin Heidelberg 2006

physical parts, and sensibility to Electro Magnetic Interference, EMI. When an analogue circuit is affected by temperature drift or EMI noise, it could cause a number of problems such as dc offset in output voltage, change of output switching frequency, increase of output voltage harmonics and so on [1]. Therefore, to be able to avoid all these disadvantages the digital double loop control scheme is used in this study.

Basic inverter circuit with an LC filter and load, R_L is given in Fig..1. In this Figure, the full bridge inverter, LC filter, and load are considered as the plant to be controlled. r_C is the equivalent series resistor, ESR of the capacitor, while r_L is the ESR of the inductor. Single phase PWM inverter modulates a DC bus voltage V_{dc} into a cycle by cycle average output voltage V_a. The amplitude of V_a is directly proportional to the commanded duty cycle of the inverter and the amplitude of the dc bus voltage V_{dc}. Therefore, the range of V_a changes between $+V_{dc}$ and $-V_{dc}$ [2].

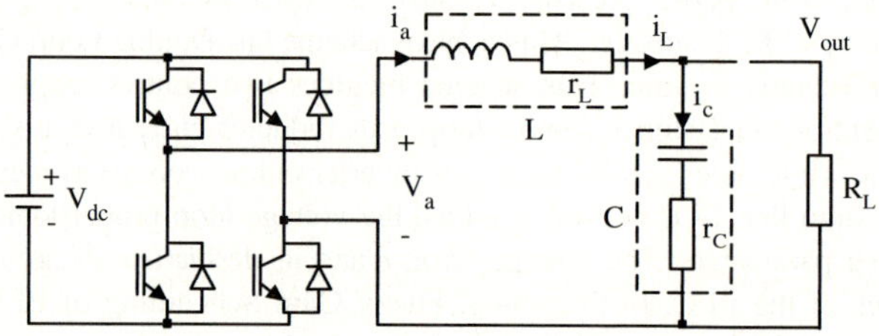

Fig. 1. Basic inverter circuit with an LC filter and R_L

In this study, An FLC is added to the double loop current mode control scheme to adjust the voltage loop PI controller parameters. This control scheme is simulated digitally using PSIM and C++. The input variables of the FLC are voltage error, *e* and its derivative, *de*. The output of the FLC is the voltage loop PI controller parameters.

2 Current Mode Fuzzy Gain Scheduling of PI Controller

Most industrial process control continues to rely upon 'classical', or 'conventional' proportional-integral-derivative, PID control. Gain scheduling is the most common PID advancement used in industry to overcome nonlinear process characteristics through the tailoring of controller gains over local operating bands [3]. When the controlled process is nonlinear, a fixed gain PID controller cannot usually give satisfactory control performance at some operating points, since the controller parameters must be adjusted following a change in operating conditions. Therefore, one way to improve the control performance of a PID controller on highly nonlinear process is to vary the controller parameters according to the process operating conditions. This is known as gain scheduling control [4].

Deadbeat-controlled PWM inverter has very fast response for load disturbances and nonlinear loads. But in deadbeat control approach, the control signal depends on a precise PWM inverter load model and the performance of the system is sensitive to parameter and load variations. Repetitive control generates high-quality sinusoidal output voltage whereas its dynamic response is very slow [5].

Current Mode Fuzzy Gain Scheduling of PI Controller presented in this paper is shown in Fig. 2. It has Double Loop Current Mode Control Scheme in core and includes two control loops as inner current and outer voltage loops. The filter inductor current and output voltage are sensed as feedback. In this scheme, note that the output inductor is hidden within the inner current control loop. This simplifies the design of the outer voltage control loop and improves UPS performance in many ways, including better dynamics and a feedforward characteristic which could be used to compensate DC bus ripple and dead time effect, etc. The objective of this inner loop is to control the state-space averaged inductor current. The current mode control requires a circuit for measurement of inductor current i_L, however, in practice such a circuit is also required in voltage mode control systems, for protection of the IGBT against excessive currents during transients and fault conditions [1]. In this control scheme, PI parameters of the voltage control loop are adjusted by FLC. In Fig. 2, K_1 and K_2 are taken as unit gains.

In this study, the advantages of FLC and Double Loop Current Mode Control are combined in the same control scheme. The FLC can handle nonlinearity and does not need accurate mathematical model. It is represented by if-then rules and thus can provide an understandable knowledge representation. FLC converts linguistic control strategy to an automatic control strategy. Linguistic control strategy is based on expert knowledge and experience [5].

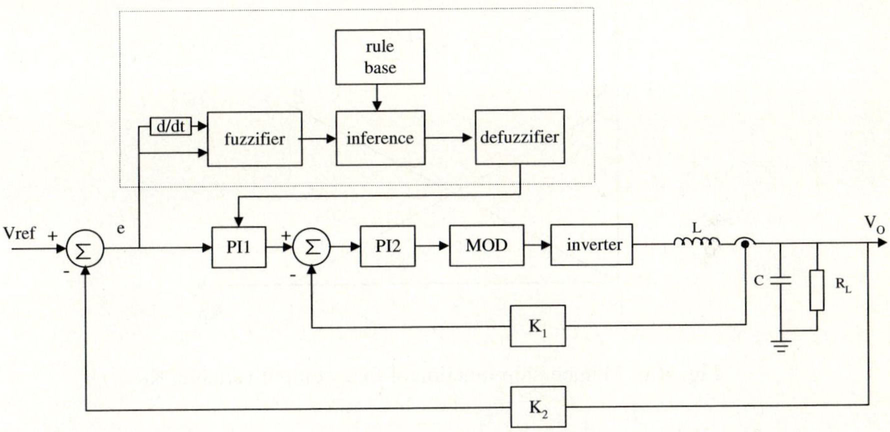

Fig. 2. Block diagram of Current Mode Fuzzy Gain Scheduling of PI Controller

Inverter output voltage error and change of this error are input variables of the FLC. The input variables, voltage error, *e*(k) and its derivative, *de*(k) are calculated using the equations (1) and (2) respectively. Two rule tables are composed for the gain K_p and K_i. These tables are based on expert knowledge and experience. FLC is used to adjust the PI parameters of voltage control loop.

$$e(k) = V_{ref}(k)-V_{out}(k) \tag{1}$$

$$de(k) = [e(k)-e(k-1)]/T \tag{2}$$

In equation (1) and (2),

$V_{ref}(k)$: reference sine wave,

$V_{out}(k)$: inverter output voltage,

T : sample time (100μs).

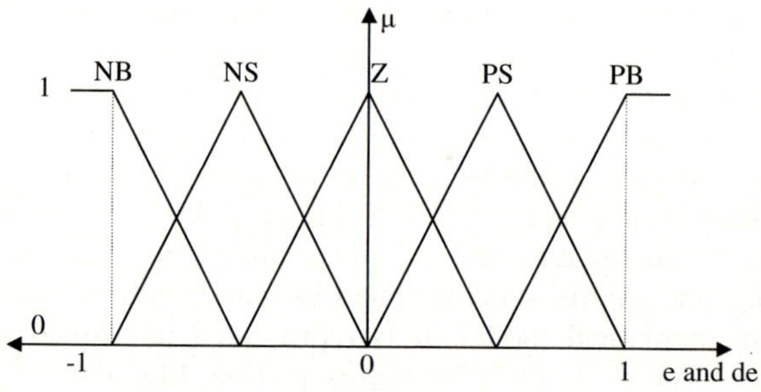

Fig. 3. Membership function of fuzzy input variables, *e* and *de*

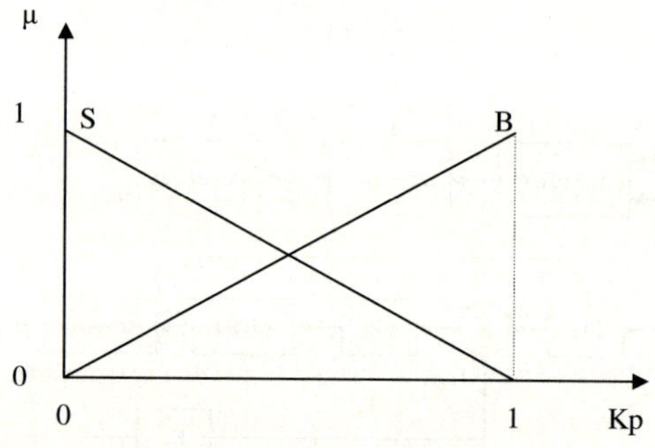

Fig. 4(a). Membership function of fuzzy output variable, K_P

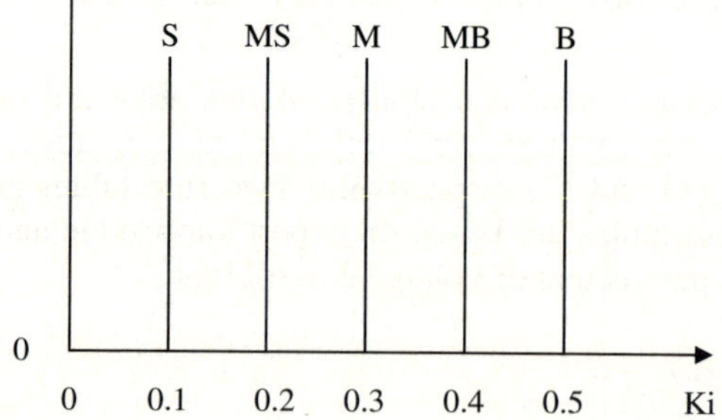

Fig. 4(b). Membership function of fuzzy output variable, K_i

A typical fuzzy process can be divided into three steps: the fuzzification, the inference and the defuzzification. The task of the fuzzification is to to transform the nonfuzzy measurements of the input variables into the fuzzy linguistic range [6]. In this study, universe of discourse of two inputs (*e* and *de*) and one of the outputs (K_i) are divided into five fuzzy subsets while universe of discourse of the output K_p is divided into two fuzzy subsets (Small, S and Big, B). The fuzzy subsets of *e* and *de* are Negative Big, NB, Negative Small, NS, Zero, Z, Positive Small, PS and Positive Big, PB. And, fuzzy subsets of the output, K_i are Small, S, Medium Small, MS, Medium, M, Medium Big and Big, B. The fuzzy subsets and the shape of membership function of FLC input and output variables are seen in Fig. 3, Fig. 4(a) and (b) respectively. The membership functions of input variables are trapezoidal and classical triangular shapes with 50% overlap. The value of each input is normalized in [-1,1] while the value of the output, K_p is normalized in [0,1] and the other output, K_i is normalized in [0,0.5] by using suitable scaling factors.

The next step of fuzzy process is the inference mechanism. As fuzzy control rely on knowledge or experience, process behavior is analyzed to establish a set of rules with IF…AND…THEN statements. A typical rule example is:

IF the voltage error (*e*) is **N**egative **B**ig (NB)
AND change of voltage error (*de*) is **P**ositive **S**mall (PS)
THEN the voltage loop proportional (K_P) is **B**ig (B)

These rules define a fuzzy relationship between the observed values of the voltage error, change of voltage error and the outputs, K_p and K_i. Fuzzy control rules are obtained from the behavior of the system and operator's expertise. The rule tables generated are shown in Table 1(a) and (b). MAX-MIN method is used as the inference method. The output membership function of each rule is given by MIN operator while the combined fuzzy output is given by MAX operator.

The last step is defuzzification process. In defuzzification process, the input for the defuzzification is the fuzzy set (the aggregate output fuzzy set) and the output is a crisp number. Centroidal defuzzification method (center of gravity method (COG)) is used for defuzzification in this paper. Output denormalization converts the normalized value of the control output variable into physical domain. The equation for the COG method is given in equation (3).

Table 1(a). Rule table for K_p

voltage error (e) \ change of voltage error (de)	NB	NS	Z	PS	PB
NB	B	B	B	B	B
NS	B	S	B	S	S
Z	S	S	B	S	S
PS	S	S	B	B	B
PB	B	B	B	B	B

Table 1(b). Rule table for K_i

change of voltage error (de)

voltage error (e)	NB	NS	Z	PS	PB
NB	B	B	B	B	B
NS	S	S	MS	B	MS
Z	M	M	M	M	M
PS	MS	B	MB	MB	B
PB	B	B	B	B	B

$$U_o = \frac{\sum_{j=1}^{n} U(U_j)U_j}{\sum_{j=1}^{n} U(U_j)} \quad (3)$$

In equation (3), U_o is determined by means of a gravity center of the area under the membership function curve of the fuzzy output and $U(U_j)$ is a membership grade of Uj [7].

3 Simulation and Experimental Results

Current Mode Fuzzy Gain Scheduling of PI Controller explained above is simulated using PSIM and C++ and implemented using TI's TMS320LF2407 DSP. Simulations and implementations are realized under resistive, rectifier type nonlinear and fluorescent loads. System parameters used in simulations and implementations are listed in Table 2. Rectifier type nonlinear load is used as nonlinear load and this load is shown in Fig. 5. In this figure, L is taken as 770μH.

Table 2. System parameters used in simulations

V_O (output voltage)	50Hz, 220 V_{RMS}
V_{DC} (DC bus voltage)	360 V
f_S (sampling frequency)	20 kHz
L (filter inductor)	700 μH
C (filter capacitor)	30 μF
r_L (ESR of the inductor)	0.05 Ω
r_C (ESR of the cap.)	0.02 Ω
Resistive load	10 Ω, (5 kVA, 100%)
Nonlinear load (rectifier type)	5 kVA (100%)
Fluorescent load	2.6 kVA (50%)

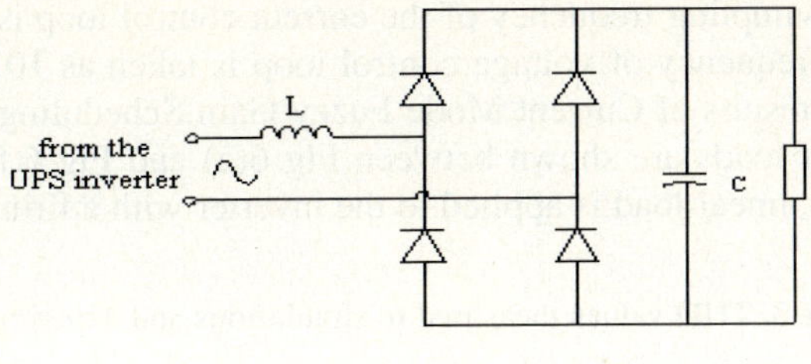

Fig. 5. Rectifier load with R_{load}-C

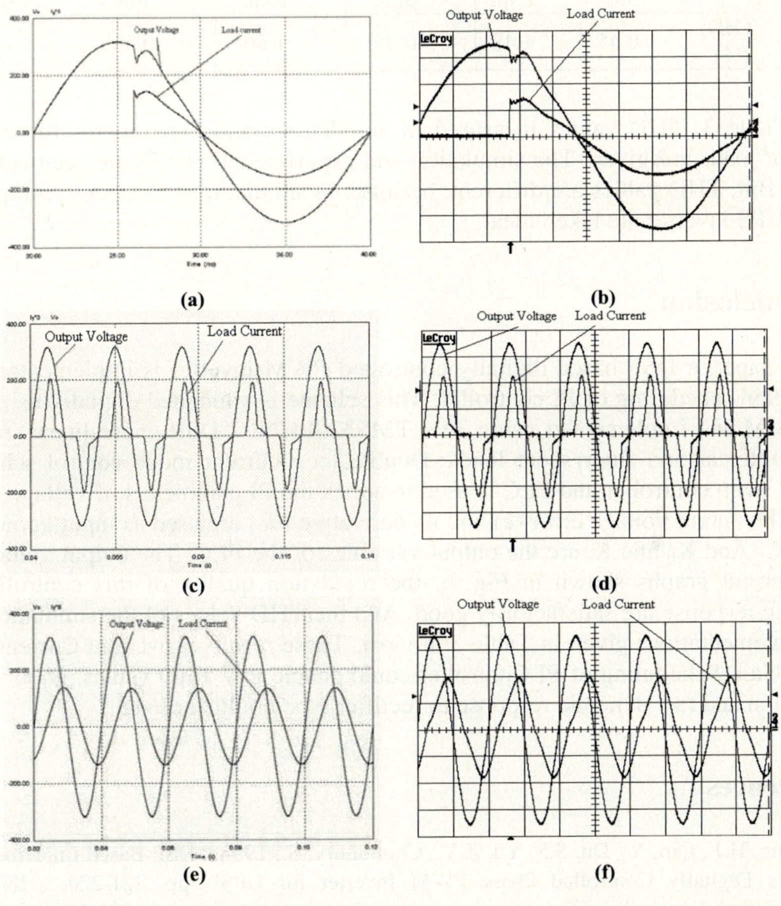

Fig. 6. Simulation and experimental results of output voltage and load current.**(a).** Simulation result for linear load (THD =0.45%) **(b).** Experimental result for linear load (THD =1.48%). **(c).** Simulation result for rectifier type nonlinear load (THD =0.95 %). **(d).** Experimental result for rectifier type nonlinear load (THD =1.66%). **(e).** Simulation result for fluorescent load (THD = 0.38%). **(f).** Experimental result for fluorescent load (THD =1.48%).

In this study, the sampling frequency of the current control loop is taken as 20 kHz while the sampling frequency of voltage control loop is taken as 10 kHz. Simulation and implementation results of Current Mode Fuzzy Gain Scheduling of PI Controller for different kinds of loads are shown between Fig 6(a) and Fig 6(f). In simulations and implementations, linear load is applied to the inverter with a firing angle of 108°.

Table 3. THD values measured in simulations and experiments

	Linear Load (100%)		Nonlinear Load (100%)		Fluorescent Load (50%)	
	Sim.	Exp.	Sim.	Exp.	Sim.	Exp.
THD (%)	0.45	1.48	0.95	1.66	0.38	1.48

In Table 3, THD values measured in simulations and experiments for different kinds of loads are given. The simulation and experimental results are seen similar in Fig. 6. But, THD values are different, because, in simulations, electronic components of the UPS inverter are taken ideal.

4 Conclusion

In this paper, a DSP based digitally controlled PWM inverter is implemented using fuzzy gain scheduling of PI controller. This scheme is simulated digitally using C++ and PSIM and implemented using TI's TMS320LF2407 DSP under linear, rectifier type nonlinear and fluorescent loads. Double loop current mode control scheme is used as main controller and FLC is used to adjust the PI parameters of voltage control loop. The output voltage error (*e*) and its derivative (*de*) are used as input variables of the FLC. And K_p and K_i are the output variables of the FLC. The output voltage and load current graphs shown in Fig. 6, the regulation quality of this controller and dynamic response are satisfactorily good. And the THD values of the simulations and the implementations given in Table 3 are low. These results prove that Current Mode Fuzzy Gain Scheduling of PI Controller could enable low THD values, good voltage regulation and fast dynamic response to rectifier type nonlinear load.

References

1. Jiang, H.J., Qin, Y., Du, S.S. Yu, Z.Y., Choudhury, S.: 1998, "DSP Based Implementation of a Digitally Controlled Phase PWM Inverter for UPS", pp. 221-224, , INTELEC Twentieth International Telecommunications Energy Conference, INTELEC.
2. Choudhury S.:1999, "Implementing Triple Conversion Single-Phase On-line UPS using TMS320C240", Texas Instruments Application Report.
3. Blanchett T.P., Kember G.C. and Dubay R., PID Gain Scheduling Using Fuzzy Logic, ISA Transactions 39 (2000), 2000, 317-325.
4. Santos M., and Dexter A.L., Control of a Cryogenic Process Using a Fuzzy PID Scheduler, Control Engineering Practice 10 (2002), 2002, 1147-1152.

5. Jian L., Yong K., Jian C.:2000, “Fuzzy-Tuning PID Control of an Inverter with Rectifier-Type Nonlinear Loads”, pp. 381-384 vol.1, PIEMC 2000, Power Electronics and Motion Control Conference.
6. Osterholz H., Simple Fuzzy Control of a PWM Inverter for a UPS System, 17th International Telecommunications Energy Conference, INTELEC '95, 1995, 565-570.M. Santos, and A. L. Dexter, Control of a Cryogenic Process Using a Fuzzy PID Scheduler, Control Engineering Practice 10 (2002), 2002, 1147-1152.
7. Kelkar N.D., A Fuzzy Controller for Three Dimensional Line Following of an Unmanned Autonomus Mobile Robot, Thesis of master of science, Department of Mechanical, Industrial, and Nuclear Engineering, University of Cincinnati, Ohio, 1997.
8. Manning C D., Control of UPS Inverters, IEE Colloquium, The Institution of Electrical Engineers, printed and published by the IEE, Savoy Place, London WC2R OBL,UK, 1994, pp. 3/1-3/5.
9. Bolat E.D., Ertunc H.M., Implementation of Current Mode Fuzzy-Tuning PI Control of Single Phase UPS Inverter Using DSP, LNAI 3682 (2005) 600-607.
10. Bolat E.D., Bolat Y., Ertunc H.M., Double Loop Current Mode Control Scheme for Single Phase UPS Inverter, Mechatronics 2004, The 9th Mechatronics Forum International Conference, Ankara, Turkey, 2004, 861-872.
11. Bolat E.D., Erkan K., Postalcioglu S., Using Current Mode Fuzzy Gain Scheduling of PI Controller for UPS Inverter, EUROCON 2005-The International Conference on “Computer as a Tool”, November 21-24 2005, Sava Center, Belgrade, Serbia & Montenegro.

Designing a Self-adaptive Union-Based Rule-Antecedent Fuzzy Controller Based on Two Step Optimization

Chang-Wook Han and Jung-Il Park

School of Electrical Engineering and Computer Science, Yeungnam University,
214-1 Dae-dong, Gyongsan, Gyongbuk, 712-749 South Korea
{cwhan, jipark}@yumail.ac.kr

Abstract. A self-adaptive union-based rule-antecedent fuzzy controller (SURFCon), which can guarantee a parsimonious knowledge base with reduced number of rules, is proposed. The SURFCon allows union operation of input fuzzy sets in the antecedents to cover bigger input domain compared with the complete structure rule which consists of AND combination of all input variables in its premise. To construct the SURFCon, we consider the union-based logic processor (ULP) which consists of OR and AND fuzzy neurons. The fuzzy neurons exhibit learning abilities as they come with a collection of adjustable connection weights. In the development stage, genetic algorithm (GA) constructs a Boolean skeleton of SURFCon, while stochastic reinforcement learning refines the binary connections of GA-optimized SURFCon for further improvement of the performance index. A cart-pole system is considered to verify the effectiveness of the proposed method.

1 Introduction

Fuzzy logic, proposed by Zadeh in 1965, is a logic with fuzzy truth, fuzzy connectives and fuzzy rules of inference rather than the conventional two-valued or even multi-valued logic [1]. It combines multi-valued logic, probability theory and a knowledge base to mimic human thinking by incorporating the uncertainty inherent in all physical systems. Relying on the human nature of fuzzy logic, an increasing number of successful applications have been developed, like automatic process control [2], pattern recognition systems [3] and so forth.

A common practice in traditional approaches to building fuzzy rule bases is to use all AND-combinations of input fuzzy sets as rule antecedents [4][5]. In this case, the number of such combinations increases exponentially with the input number [6]. To reduce the size of the rule bases, a self-adaptive union-based rule-antecedent fuzzy controller (SURFCon) is proposed in this paper. The SURFCon allows union operation of input fuzzy sets in the antecedents to cover bigger input domain compared with the complete structure which consists of AND combinations of fuzzy sets of all input variables in its premise [7]. Basically, the SURFCon is constructed with the aid of union-based logic processor (ULP) which consists of OR and AND fuzzy neurons. The fuzzy neurons exhibit learning abilities as they come with a collection of adjustable connection weights [6][8]. This paper aims at constructing a binary structure of

B. Gabrys, R.J. Howlett, and L.C. Jain (Eds.): KES 2006, Part I, LNAI 4251, pp. 850 – 857, 2006.
© Springer-Verlag Berlin Heidelberg 2006

SURFCon by genetic algorithm (GA) [9], and subsequently further refining the binary connections by random signal-based learning employing simulated annealing (SARSL) introduced in [10][11]. The effectiveness of the SURFCon is demonstrated by stabilizing an inverted pendulum system.

2 Structure of the SURFCon

Before discussing the architecture of the SURFCon, we will briefly remind AND and OR fuzzy neurons and then move on to the ULP which is the basic logic processing unit of the SURFCon.

2.1 OR and AND Fuzzy Neurons

As originally introduced in [6][8], fuzzy neurons emerge as result of a vivid synergy between fuzzy set constructs and neural networks. In essence, these neurons are functional units that retain logic aspects of processing and learning capabilities characteristic for artificial neurons and neural networks. Two generic types of fuzzy neurons are considered:

AND neuron is a nonlinear logic processing element with n-inputs x $[0, 1]^n$ producing an output y governed by the expression

$$y = \text{AND}(x; w) \tag{1}$$

where w denotes an n-dimensional vector of adjustable connections (weights). The composition of x and w is realized by an t-s composition operator based on t- and s-norms, that is

$$y = \mathop{\mathrm{T}}_{i=1}^{n} (w_i \, \mathrm{s} \, x_i) \tag{2}$$

with “s” denoting some s-norm and “t” standing for a t-norm. As t- norms (s-norms) carry a transparent logic interpretation, we can look at as a two-phase aggregation process: first individual inputs (coordinates of x) are combined or-wise with the corresponding weights and these results produced at the level of the individual aggregation are aggregated and-wise with the aid of the t-norm.

By reverting the order of the t- and s-norms in the aggregation of the inputs, we end up with a category of OR neurons,

$$y = \text{OR}(x; w) \tag{3}$$

that is

$$y = \mathop{\mathrm{S}}_{i=1}^{n} (w_i \, \mathrm{t} \, x_i) \tag{4}$$

We note that this neuron carries out some and-wise aggregation of the inputs followed by the global or-wise combination of these partial results.

Some obvious observations hold:

(i) For binary inputs and connections, the neurons transform to standard OR and AND gates.
(ii) The higher the values of the connections in the OR neuron, the more essential the corresponding inputs. This observation helps eliminate irrelevant inputs; the inputs associated with the connections whose values are below a certain threshold are eliminated. An opposite relationship holds for the AND neuron; here the connections close to zero identify the relevant inputs.
(iii)The change in the values of the connections of the neuron is essential to the development of the learning capabilities of a network formed by such neurons; this parametric flexibility is an important feature to be exploited in the design of the networks.

These two types of fuzzy neurons are fundamental building blocks used in the design of logic expressions supporting the development of logic-driven models.

2.2 SURFCon and Its Two-Step Optimizations

To construct the SURFCon, we first elaborate on the ULP which consists of OR and AND fuzzy neurons, as shown in Fig. 1, where, $\mu_N(x_i)$, $\mu_Z(x_i)$ and $\mu_P(x_i)$ are the membership grades of the fuzzy sets N (negative), Z (zero) and P (positive) for the input variable x_i, i=1,2,3,4, respectively. The OR and AND fuzzy neurons realize pure logic operations on the membership values and exhibit learning abilities as being introduced in [6][8]. In this paper, we consider these triangular norms and co-norms to be a product operation and probabilistic sum, respectively.

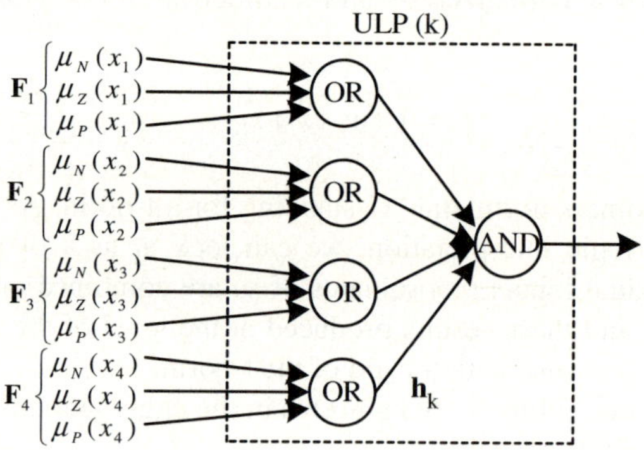

Fig. 1. Structure of an ULP

An important characteristic of ULP is that union operation of input fuzzy sets is allowed to appear in their antecedents, i.e., incomplete structure. For fuzzy system of complex processes with high input dimension, the ULP is preferable because it achieves bigger coverage of input domain compared with the complete structure. For example, consider a system with x_1, x_2 as its inputs and y as its output characterized by three linguistic terms, N, Z and P, respectively. The incomplete structure rule 'If x_1=N then y=N' covers the following three complete structure rules:

(i) If (x_1=N) and (x_2=N) then y=N
(ii) 'If (x_1=N) and (x_2=Z) then y=N
(iii) 'If (x_1=N) and (x_2=P) then y=N

Similarly, the rule 'If (x_1=N or Z) and (x_2=N or Z) then y=N' covers the following four complete structure rules:

(i) If (x_1=N) and (x_2=N) then y=N
(ii) If (x_1=N) and (x_2=Z) then y=N
(iii) If (x_1=Z) and (x_2=N) then y=N
(iv) If (x_1=Z) and (x_2=Z) then y=N

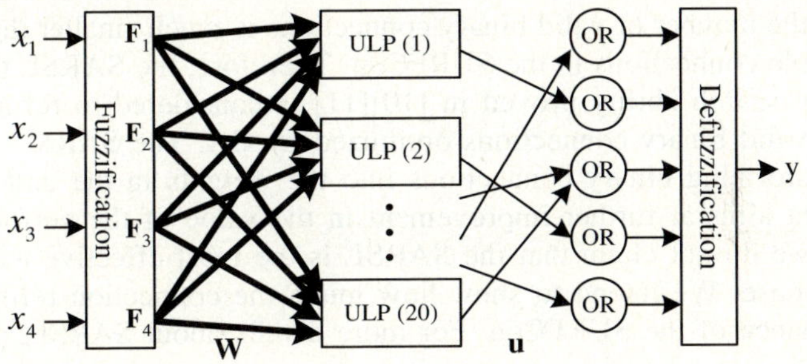

Fig. 2. Structure of a SURFCon with 4 input and 1 output variables characterized by 3 and 5 fuzzy sets, respectively (NU=20)

Fig. 2 describes the SURFCon constructed with the aid of ULPs. The OR neurons in the output layer are placed to aggregate the outputs of ULPs for each corresponding consequences. In Fig. 2, the connections to the ULPs are described as bold lines which contain a set of connection lines, refer to Fig. 1. The only parameter that has to be controlled in the SURFCon is the number of ULP (NU), which will be set large enough in the experiment. A conflict occurs in the rule bases if there exist two rules which have overlapping AND combinations but different linguistic consequences. Let us consider the following two rules to count the number of conflict (NC):

(i) If (x_1=N or Z) and (x_2=N or Z) then y=N
(ii) If (x_1=Z or P) and (x_2=N or Z) then y=Z

In this case, NC=2 because the antecedents '(x_1=Z) and (x_2=N)' and '(x_1=Z) and (x_2=Z)' have different consequences. Therefore, NC should be checked for all possible pairs of different consequences of the rule bases. It will be included in evaluating the performance index to remove the conflict from the rule bases later on.

For the development of the SURFCon, GA attempts to construct a Boolean structure of SURFCon by selecting the most essential binary connections that shape up the architecture. GA optimizes the binary connections **W** and **u**, as shown in Fig. 2. All the connections to AND neurons, $\mathbf{h}_k$, in the ULPs are initialized as zero (valid connection). It is worth noting that if the connection weights from $\mathbf{F}_i$ to an OR neuron in the ULP are all-one, that is, x_i is 'don't care' in this ULP, the corresponding connection to AND neuron should be modified as one (invalid connection). It is obvious that the following two cases lead to an invalid rule antecedent:

(i) All-one connection weights to an ULP, meaning that all input variables are ignored in the antecedent.
(ii) All-zero connection weights to an OR neuron in the ULP, i.e., empty fuzzy set for the corresponding input variable.

Therefore, these ULPs will be removed from the rule bases and finally the compact rule bases will be established. To avoid the rule confliction, the output of an ULP should be connected to only one of the OR neurons in the output layer. Since the GA is a very popular optimization algorithm considered in many application areas, we do not elaborate on this. For more details about the GA, please refer to [9].

Once a Boolean skeleton has been constructed by GA, we concentrate on the detailed optimization of the valid binary connections with the aid of SARSL. It is apparent that the number of valid binary connections is much smaller than the number of all possible connections in the SURFCon. Therefore, the SARSL that has an excellent local search ability proved in [10][11] is considered to refine the reduced number of valid binary connections optimized by GA. The SARSL refinement involves transforming binary connections into the weights in the unit interval. This enhancement aims at further improvement in the value of the performance index. Obviously we do not claim that the SARSL is the most effective learning method for this purpose. We intend to show how much the connection refinement affects the performance of the SURFCon. For more details about SARSL, please refer to [10][11].

3 Experimental Results

To show the effectiveness of the proposed SURFCon, a cart-pole system, a well-known nonlinear system, was considered. The objective is to bring the cart to center with a vertical pole. The system has four state variables which are θ (angle of the pole with the vertical), $\dot{\theta}$ (angular velocity of the pole), x (position of the cart) and $\dot{x}$ (linear velocity of the cart). The nonlinear differential equations for a cart-pole system are described as follows [7][12]:

$$\ddot{\theta} = \frac{(M+m)g\sin\theta - \cos\theta[f + mL\dot{\theta}^2\sin\theta - \mu_c\,\mathrm{sgn}(\dot{x})] - \dfrac{\mu_p(M+m)\dot{\theta}}{mL}}{\dfrac{4}{3}(M+m)L - mL\cos^2\theta} \tag{5}$$

$$\ddot{x} = \frac{f + mL(\dot{\theta}^2\sin\theta - \ddot{\theta}\cos\theta)}{M+m} - \mu_c\,\mathrm{sgn}(\dot{x})$$

The parameter values used in this simulation were set as the same as [11] except the failure conditions, where $|\theta|>0.2$rad or $|x|>0.5$m. The following optimization parameters were considered: (i) GA parameters are population size=100; generation number=200; crossover rate=0.9; mutation rate=0.03; (ii) SARSL parameters are learning rate=0.01; initial temperature=0.1; cooling rate=0.99; iteration number=500; (iii) others are time step=0.01s; simulating number of time steps for each initial condition q=500; NU=20. For the GA, standard version including two-point crossover based on the boundary between binary (connections **W** for antecedents) and integer (connections **u** for consequences) codes was used. GA optimized 240 (12x20) binary parameters and 20 integer parameters that have the integer values of {1,2,…,5}. The parameters of SARSL were set to perform fine learning of the binary connections. The following normalized performance index (fitness function) that has to be maximized was used [7]:

$$Q = \frac{1}{8}\sum_{cond1}^{cond8}\left\{\frac{q_s}{q}\left[1-\frac{1}{2q}\sum_{j=1}^{q}\left(\frac{|\theta_j|}{\theta_{fail}}+\frac{|x_j|}{x_{fail}}\right)\right]\right\}\left\{1-\frac{\mathrm{NC}}{\mathrm{NC}_{\max}}\right\} \tag{6}$$

where θ_{fail} is 0.2rad, x_{fail} is 0.5m, q_s is the survival time steps for each trial, and $\mathrm{NC}_{\max}$ is the maximum NC which is set as 10. If NC exceeds $\mathrm{NC}_{\max}$, NC= $\mathrm{NC}_{\max}$. For the evaluation of the performance index, eight different sets of initial conditions (cond1~cond8) were considered to cover wide range of input spaces. Because our focal point is SURFCon and its two-step optimization, we assume that fuzzy sets of the input/output variables are given in advance as 3/5-uniformly distributed triangular membership functions with an overlap of 0.5 and left unchanged. For the defuzzification, center of area method was used. Ten independent simulations for the optimization of SURFCon have been performed, and the results are shown in Table 1. This table describes the average best performance index after GA and SARSL over ten independent simulations as well as the maximum, minimum and average number of valid ULP (incomplete structure rules) for ten optimized SURFCons. As can be seen, the optimized SURFCon has at most 15 incomplete structure rules covering most of the essential input domain without linguistic conflict, and besides, the SARSL further refines the valid binary connections optimized by GA.

Table 1. Results of the optimized SURFCons

After GA	After SARSL	Max rule	Min rule	Average rule
0.912	0.947	15	13	14.1

Fig. 3(a) illustrates the results of testing simulations for the optimized SURFCon with a set of initial condition $(\theta, \dot{\theta}, x, \dot{x})$=(0, 0, -0.25, 0) which is independent of the sets for the optimization. To compare the results, the conventional fuzzy controller (CFC) which has all possible AND combinations (81 complete structure rules) as rule antecedents was considered. To prove the SURFCon to be practical, we have also performed experiment on the real cart-pole system manufactured by Realgain Co., Ltd. (Fig. 4). The experimental results are shown in Fig. 3(b).

Obviously the performance of CFC is better than that of SURFCon, but the control results indicate that the reduced rule bases of SURFCon are enough to center the cart with a vertical pole, and moreover, the performance of SURFCon outperforms that of [7] and [11]. Both the simulation and experimental results reveal that the SURFCon is really feasible.

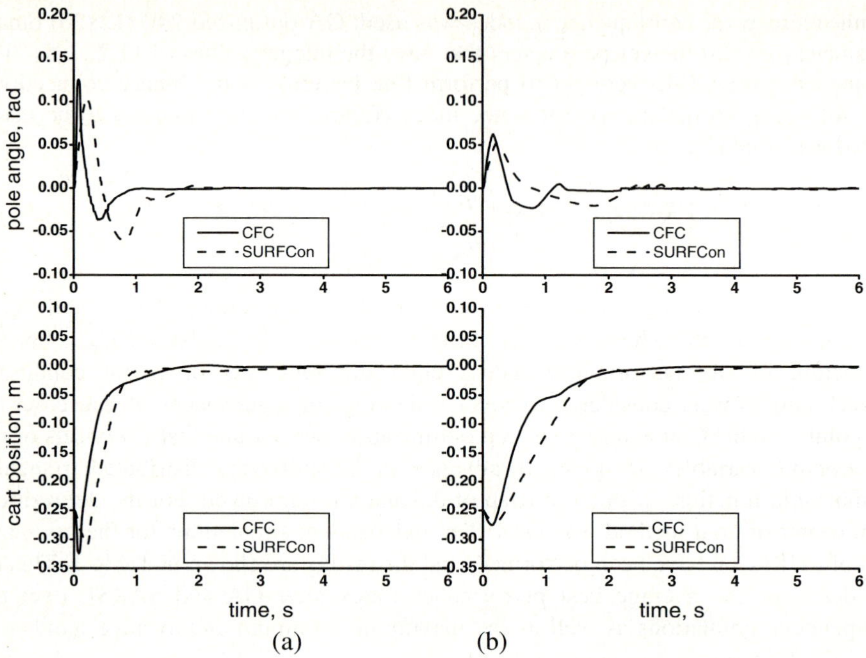

Fig. 3. Control results using a testing initial condition: (a) Simulation results, (b) Experimental results

Fig. 4. Real inverted pendulum system

4 Conclusions

The SURFCon that allows union operation to appear in their antecedents for the parsimonious rule bases has been demonstrated. It has been constructed with the aid of fuzzy neurons and two-step optimizations, where GA develops a Boolean skeleton and subsequently SARSL further refines the binary connections. As can be seen from the simulation and experimental results, the incomplete structure of rule allows the SURFCon to utilize only fewer rules without performance degrading, and also proved the SURFCon to be practical.

References

1. Zadeh, L. A.: Fuzzy sets. Inform. Control, Vol. 8 (1965) 338-353
2. King, P. J., Mamdani, E. H.: The application of fuzzy control systems to industrial processes. Automatica, Vol. 13, No. 3 (1977) 235-242
3. Pal, S. K., King, R. A., Hashim, A. A.: Image description and primitive extraction using fuzzy set. IEEE Trans. Syst. Man and Cybern., Vol. SMC-13 (1983) 94-100
4. Karr, C. L.: Design of an adaptive fuzzy logic controller using a genetic algorithm. Proc. Int. Conf. on Genetic Algorithms (1991) 450-457
5. Homaifar, A., McCormick, E.: Simultaneous design of membership functions and rule sets for fuzzy controllers using genetic algorithms. IEEE Trans. Fuzzy Systems, Vol. 3, No. 2 (1995) 129-139
6. Pedrycz, W., Reformat, M., Han, C. W.: Cascade architectures of fuzzy neural networks. Fuzzy Optimization and Decision Making, Vol. 3, No. 1 (2004) 5-37
7. Xiong, N., Litz, L.: Reduction of fuzzy control rules by means of premise learning-method and case study. Fuzzy Sets and Systems, Vol. 132, No. 2 (2002) 217-231
8. Pedrycz, W.: Fuzzy Neural Networks and Neurocomputations. Fuzzy Sets and Systems, Vol. 56 (1993) 1-28
9. Goldberg, D.E.: Genetic Algorithms in Search, Optimization and Machine Learning. Addison-Wesley, Reading, MA (1989)
10. Han, C. W., Park, J. I.: Design of a fuzzy controller using random signal-based learning employing simulated annealing. Proc. IEEE Conf. on Decision and Control (2000) 396-397
11. Han, C. W., Park, J. I.: A study on hybrid random signal-based learning and its applications. Int. Jour. of Systems Science, Vol. 35, No. 4 (2004) 243-253
12. Liu, B. D., Chen, C. Y., Tsao, J. Y.: Design of adaptive fuzzy logic controller based on linguistic-hedge concepts and genetic algorithms. IEEE Trans. Syst. Man and Cybern.-B, Vol. 31, No. 1 (2001) 32-53

Image-Based Robust Control of Robot Manipulators in the Presence of Uncertainty

Min Seok Jie and Kang Woong Lee

School of Electronics, Telecommunication and Computer Engineering, Hankuk Aviation University, 200-1, Hwajeon-dong, Deokyang-gu, Koyang-city, Kyonggi-do, 412-791, Korea
Tel.: +82 2 3158 0166, Fax: +82 2 3159 9257
tomsey@korea.com, kwlee@mail.hangkong.ac.kr

Abstract. In this paper, we propose a robust controller with visual feedback for tracking control of robot manipulators in the presence of bounded parametric uncertainties. The proposed control input consists of a nonlinear term to achieve robustness to the parametric uncertainty and feedback terms including the image feature and the joint velocity to guarantee the stability of the closed-loop system. The globally uniformly ultimate boundedness of the closed-loop system is shown by Lyapunov method. The effectiveness of the proposed method is shown by simulation and experiment results on the 5-link robot manipulators with two degree of freedom.

Keywords: robust control, image-based control, image Jacobian, robot manipulator, visual feedback.

1 Introduction

Recently, robot applications have increased in unstructured environments. When the robot is working in these environments, a vision system gives the position information for an object to the controller. The use of visual feedback in these applications can be an attractive solution for the position and motion control of robot manipulators [1]-[3]. A visual servo control scheme can be classified in two configurations: fixed-camera, where the visual servoing camera is fixed with respect to the world coordinate frame and camera-in-hand, where the camera is mounted on the end-effector of the robot manipulator. In the camera-in-hand configuration, the camera supplies visual information of the object in the environment to the controller. The objective of this visual feedback control scheme is to move the robot manipulator in such a way that the projection of an object be at a desired position in the image plane obtained by the camera [4].

The manipulator vision system whose dynamics do not interact with the visual feedback loop can not achieve high performance for high speed tasks. In order to overcome this drawback, the controller must take into account the dynamics of robot manipulators [5]. The robot dynamics, however, includes parametric uncertainties due to load variations and disturbances. A visual servo controller taking into account robot

B. Gabrys, R.J. Howlett, and L.C. Jain (Eds.): KES 2006, Part I, LNAI 4251, pp. 858 – 865, 2006.
© Springer-Verlag Berlin Heidelberg 2006

dynamics must be robust to the parametric uncertainties and disturbances. In [6], a robust tracking controller has been designed to compensate the uncertainty in the camera orientation and to ensure globally uniformly ultimate boundedness. Kelly [7] proposed an image-based direct visual servo controller for camera-in-hand robot manipulators, which is of a simple structure based on a transpose Jacobian term plus gravity compensation.

In this paper, we design a robust visual servo controller to compensate parametric uncertainties due to load variations or disturbances. Our main theoretical contribution is the extension of Kelly's results and Zergeroglu's results [6]. The closed-loop stability including the whole robot dynamics is shown by the Lyapunov method. The proposed robust visual servo controller is applied on two degree of freedom 5-link robot manipulator to show the performance of the closed-loop system. The experimental results show the convergence behavior of the image feature points.

This paper hereafter is organized as follows. In Sections 2 present the robotic system model and problem statement. The proposed robust control system is analyzed in Section 3. In Section 4 we introduce experiment results on two degrees of freedom robot manipulators. The paper is finally summarized in Section 5.

2 Robot Model and Problem Statement

In the absence of friction, the dynamic equation of an n-link rigid robot manipulators can be expressed as [8]

$$M(q)\ddot{q} + C(q,\dot{q})\dot{q} + G(q) = \tau \tag{1}$$

where $q \in R^n$ is the vector of joint displacements, $\tau \in R^n$ is the vector of torques applied to the joints, $M(q) \in R^{n\times n}$ is the symmetric positive definite inertia matrix, $C(q,\dot{q})\dot{q} \in R^n$ is the vector of centripetal and Coriolis torques, and $G(q) \in R^n$ is the vector of the gravitational torques. The above robot dynamics model has the following properties [8].

Property 1: The matrix $\dot{M} - 2C$ is skew symmetric.

Property 2: The matrix M and C and the vector G depend linearly on a $p\times 1$ vector θ of unknown constant parameters. This implies that we may write the dynamic equation (1) as

$$M(q)\ddot{q} + C(q,\dot{q})\dot{q} + G(q) = Y(q,\dot{q},\ddot{q})\theta = \tau \tag{2}$$

where $Y(q,\dot{q},\ddot{q})$ is an $n\times p$ matrix of known functions.

The motion dynamics of the image feature point described by the robot joint velocity as

$$\dot{\xi} = J(q,\xi,Z)\dot{q} \tag{3}$$

where

$$J(q,\xi,Z) = J_{img}(q,\xi,Z)\begin{bmatrix} {}^cR_w & 0 \\ 0 & {}^cR_w \end{bmatrix}\begin{bmatrix} I & 0 \\ 0 & T(q) \end{bmatrix} J_A(q) \tag{4}$$

$J_{img}(q,\xi,Z)$ is the so-called image Jacobian, $J_A(q)$ is the analytic Jacobian and $T(q)$ is the rotational matrix of end-effector.

In robot control with visual feedback, the control problem is to design a controller to move the end-effector in such a way that the actual image features reach the desired ones specified in the image plane. Let us denote with ξ_d the desired image feature vector which is assumed to be constant. We define the image feature error as

$$\tilde{\xi} = \xi_d - \xi \tag{5}$$

Using (3), the time derivative of (5) can be expressed by

$$\dot{\tilde{\xi}} = -J(q,\xi,Z)\dot{q} = -J(q,\xi,Z)(\dot{q}_d - \dot{\tilde{q}}) \tag{6}$$

where $\dot{q}_d$ is the desired joint velocity vector and $\dot{\tilde{q}} = \dot{q}_d - \dot{q}$.

We take the desired joint velocity $\dot{q}_d$ as

$$\dot{q}_d = J^+(q,\xi,Z)K\tilde{\xi} \tag{7}$$

where K is a positive definite gain matrix and $J^+(q,\xi,Z)$ is the pseudo-inverse matrix defined by $J^+(q,\xi,Z) = \left[J^T(q,\xi,Z)J(q,\xi,Z)\right]^{-1} J^T(q,\xi,Z)$.

Substituting (7) into (6) leads to

$$\dot{\tilde{\xi}} = -K\tilde{\xi} + J(q,\xi,Z)\dot{\tilde{q}} \tag{8}$$

3 Robust Control

In this section the visual feedback control problem of the rigid robot manipulators with bounded parametric uncertainties is considered. It is assumed that the uncertainty of the parameter vector θ is bounded as

$$\left\|\tilde{\theta}\right\| = \left\|\theta - \theta_0\right\| \le \rho, \ \rho > 0 \tag{9}$$

where θ_o the nominal value of θ and ρ is a known value.

The control objective is to design a robust visual feedback controller that performs tracking control of robot manipulator (1) in the presence of bounded parameter uncertainties. The proposed controller is given by

$$\tau = Y(q,\dot{q}_d,\ddot{q}_d)(\theta_0 + u) + K_v\dot{\tilde{q}} + J(q,\xi,Z)^T\tilde{\xi} \tag{10}$$

where K_v is the symmetric positive definite matrix, u is an additional control input that will be designed to achieve robustness to the parametric uncertainty and $Y(q,\dot{q}_d,\ddot{q}_d)\theta_0$ is represented by nominal values of $M(q)$, $C(q,\dot{q})$ and $G(q)$ such that

$$Y(q,\dot{q}_d\ddot{q}_d)\theta_0 = M_o(q)\ddot{q}_d + C_o(q,\dot{q})\dot{q}_d + G_o(q) \tag{11}$$

Substituting (10) into (1) and defining the state vector as $\left[\tilde{\xi} \quad \dot{\tilde{q}}\right]^T$ the closed-loop system equation is obtained

$$\begin{bmatrix} \dot{\tilde{\xi}} \\ \ddot{\tilde{q}} \end{bmatrix} = \begin{bmatrix} -K\tilde{\xi} + J(q,\xi,Z)\dot{\tilde{q}} \\ M^{-1}(q)\left[-C(q,\dot{q})\dot{\tilde{q}} - K_v\dot{\tilde{q}} - J(q,\xi,Z)^T\tilde{\xi} + Y(q,\dot{q}_d,\ddot{q}_d)(\tilde{\theta}+u)\right] \end{bmatrix} \tag{12}$$

We propose the additional control vector u as

$$u = \begin{cases} -\rho \dfrac{Y^T\dot{\tilde{q}}}{\left\|Y^T\dot{\tilde{q}}\right\|}, & \left\|Y^T\dot{\tilde{q}}\right\| > \varepsilon \\ -\rho \dfrac{Y^T\dot{\tilde{q}}}{\varepsilon}, & \left\|Y^T\dot{\tilde{q}}\right\| \le \varepsilon \end{cases} \tag{13}$$

where $\varepsilon > 0$ is a design parameter.

We now show the robust stability of the proposed control system despite parametric uncertainty exists. Consider the Lyapunov function candidate

$$V = \frac{1}{2}\dot{\tilde{q}}^T M(q)\dot{\tilde{q}} + \frac{1}{2}\tilde{\xi}^T\tilde{\xi} \tag{14}$$

Differentiation of V along the trajectory of (12) yields

$$\begin{aligned} \dot{V} &= \dot{\tilde{q}}^T\left[-C(q,\dot{q})\dot{\tilde{q}} - K_v\dot{\tilde{q}} - J(q,\xi,Z)^T\tilde{\xi} + Y(q,\dot{q}_d,\ddot{q}_d)(\tilde{\theta}+u)\right] \\ &\quad + \frac{1}{2}\dot{\tilde{q}}^T\dot{M}(q)\dot{\tilde{q}} + \tilde{\xi}^T\left[-K\tilde{\xi} + J(q,\xi,Z)^T\dot{\tilde{q}}\right] \\ &= -\dot{\tilde{q}}^T K_v\dot{\tilde{q}} - \tilde{\xi}^T K\tilde{\xi} + \dot{\tilde{q}}^T Y(q,\dot{q}_d,\ddot{q}_d)(\tilde{\theta}+u) \end{aligned} \tag{15}$$

If $\left\|Y^T\dot{\tilde{q}}\right\| > \varepsilon$, we have

$$\begin{aligned} \dot{V} &= -\dot{\tilde{q}}^T K_v\dot{\tilde{q}} - \tilde{\xi}^T K\tilde{\xi} + \dot{\tilde{q}}^T Y\Big(\tilde{\theta} - \rho\frac{Y^T\dot{\tilde{q}}}{\left\|Y^T\dot{\tilde{q}}\right\|}\Big) \\ &\le -x^T W x \end{aligned} \tag{16}$$

where $x = \begin{bmatrix}\tilde{\xi} & \dot{\tilde{q}}\end{bmatrix}^T$ and $W = \begin{bmatrix} K & 0 \\ 0 & K_v \end{bmatrix}$.

If $\left\|Y^T\dot{\tilde{q}}\right\| \le \varepsilon$, we have

$$\begin{aligned} \dot{V} &\le -\dot{\tilde{q}}^T K_v\dot{\tilde{q}} - \tilde{\xi}^T K\tilde{\xi} + \rho\left\|Y^T\dot{\tilde{q}}\right\| - \frac{\rho}{\varepsilon}\left\|Y^T\dot{\tilde{q}}\right\|^2 \\ &\le -\lambda_{\min}(W)\|x\|^2 + \frac{\rho\varepsilon}{4} \end{aligned} \tag{17}$$

where $\lambda_{\min}(W)$ denotes the minimum eigenvalue of W.

For $\| x \| \geq \frac{1}{2}\sqrt{\frac{\rho\varepsilon}{\lambda_{\min}(W)}}$, we have $\dot{V} \leq 0$. Hence the uniform ultimate boundedness of the closed-loop system is achieved[9]. Since we can choose sufficiently small ε, our conclusions are summarized in the following theorem.

Theorem 1: Given the robot system (1) with the bounded parametric uncertainty and for arbitrary small $\varepsilon > 0$, the control law (10) and (13) guarantees the uniformly ultimate boundedness of the tracking error.

4 Experimental Results

The proposed control method was implemented on 5-link robot manipulator manufactured by Samsung as shown in Fig. 1. The dynamic equation of the manipulators is given by

$$\begin{bmatrix} M_{11}(q) & M_{12}(q) \\ M_{12}(q) & M_{22}(q) \end{bmatrix}\begin{bmatrix} \ddot{q}_1 \\ \ddot{q}_2 \end{bmatrix} + \begin{bmatrix} 0 & C_{12}(q,\dot{q}) \\ C_{21}(q,\dot{q}) & 0 \end{bmatrix}\begin{bmatrix} \dot{q}_1 \\ \dot{q}_2 \end{bmatrix} + \begin{bmatrix} G_1(q) \\ G_2(q) \end{bmatrix} = \begin{bmatrix} \tau_1 \\ \tau_2 \end{bmatrix}$$

The entries $M_{ij}(q)(i, j = 1,2)$ of the inertia matrix $M(q)$ are

$$M_{11}(q) = m_1 l_{c2}^2 + m_3 l_{c3}^2 + m_4 l_1^2 + I_1 + I_3 = \theta_1$$
$$M_{12}(q) = (m_3 l_2 l_{c3} - m_4 l_1 l_{c4})\cos(q_2 - q_1) = \theta_2 \cos(q_2 - q_1)$$
$$M_{21}(q) = M_{12}(q) \quad M_{22}(q) = l_2 l_{c2}^2 + m_3 l_2^2 + m_4 l_{c4}^2 + I_2 + I_4 = \theta_3$$

The elements $C_{ij}(q,\dot{q})(i, j = 1,2)$ of the centripetal and Coriolis matrix $C(q,\dot{q})$ are

$$C_{11}(q,\dot{q}) = 0 \quad C_{22}(q,\dot{q}) = 0$$
$$C_{11}(q,\dot{q}) = -(m_3 l_2 l_{c3} - m_4 l_1 l_{c4})\sin(q_2 - q_1)\dot{q}_1 = \theta_2 \sin(q_2 - q_1)\dot{q}_1$$
$$C_{21}(q,\dot{q}) = (m_3 l_2 l_{c3} - m_4 l_1 l_{c4})\sin(q_2 - q_1)\dot{q}_1 = -\theta_2 \sin(q_2 - q_1)\dot{q}_1$$

The gravitational torque vector $G(q)$ is given by

$$G_1 = g(l_1 l_{c1} + m_3 l_{c3} + m_4 l_1)\cos q_1 = g \cdot \theta_4 \cdot \cos q_1$$
$$G_2 = g(l_2 l_{c2} + m_3 l_2 - m_4 l_{c4})\cos q_2 = g \cdot \theta_5 \cdot \cos q_2$$

The Y_{ij} $(i = 1,2\ j = 1,2,3,4,5)$ components of regressor $Y(q,\dot{q},\ddot{q})$ are described by

$$Y_{11} = \ddot{q}_1 \quad Y_{12} = \ddot{q}_2 \cdot \cos(q_2 - q_1) - \sin(q_2 - q_1) \cdot \dot{q}_2^2 \quad Y_{13} = 0 \quad Y_{14} = g \cdot \cos q_1$$
$$Y_{15} = 0 \quad Y_{21} = 0 \quad Y_{21} = \ddot{q}_1 \cdot \cos(q_2 - q_1) + \sin(q_2 - q_1) \cdot \dot{q}_2 \cdot \dot{q}_1 \quad Y_{23} = \ddot{q}_2$$
$$Y_{24} = 0 \quad Y_{25} = g \cdot \cos q_2$$

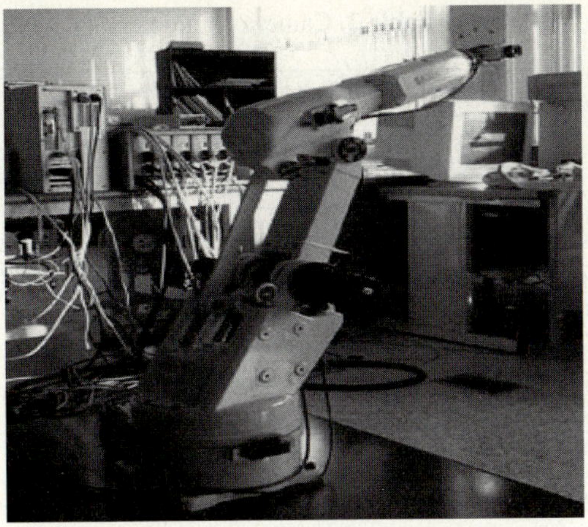

Fig. 1. Samsung robot manipulators

The physical parameters of the unloaded robot dynamics are given by Table 1.

Setup for experiments is shown in Fig. 2. A motion control board(MMC) mounted in a main computer is used to execute the control algorithm. The feature signal is acquired by an image processing board(Matrox Meteor II) mounted on a main computer which processes the image obtained from a CCD camera and extracts the image feature. The CCD camera is mounted on the end-effector.

Table 1. Parameters of the robot manipulators

m_1	5.5 Kg	l_2	0.350 m	I_1	0.040 Kgm^2
m_2	0.3 Kg	l_{c1}	0.175 m	I_2	0.003 Kgm^2
m_3	0.2 Kg	l_{c2}	0.070 m	I_3	0.001 Kgm^2
m_4	4.5 Kg	l_{c3}	0.175 m	I_4	0.060 Kgm^2
l_1	0.350 m	l_{c4}	0.250 m		

The variation interval of the parameter due to an unknown load is considered

$$0 \le \Delta m_4 \le 4.5kg$$

We choose a vector θ_o of nominal parameters

$$\theta_o = [1.0424 \ -0.5857 \ 0.4903 \ 3.3600 \ -1.6525]^T$$

With choosing nominal parameter vector θ_o and uncertainty range, the uncertainty bound ρ is defined as 1.0349. Since the image obtained by the image processor has a 640×480 pixels resolution, scale factors are slightly different. The numerical values of the camera model are listed in Table 2.

Table 2. Camera parameters

α	$6.941\times10^{-6}(m/\,pixel)$
β	$9.425\times10^{-6}(m/\,pixel)$
f	$0.016(m)$

In the experimental test, we have considered one object feature point on a whiteboard. The white board was located at a distance $Z=1$ m in front of the camera and parallel to the plane where the manipulator moves. The gain matrices of controller K and K_v are chosen as $K=50I$ and $K_v=1100I$, respectively. The initial positions of each joint are $q_1=\pi/2$ [rad], $q_2=\pi$ [rad]. It was assumed that the initial feature point of the object was $\xi=[100\ 100]^T$ pixels. The desired feature point coordinate was $\xi_d=[0\ \ 0]^T$ pixels.

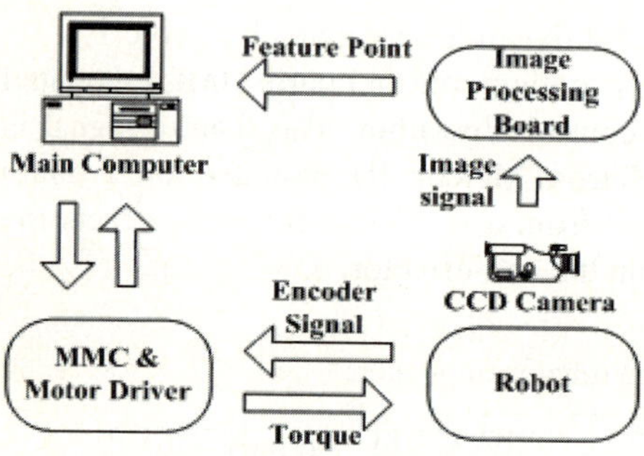

Fig. 2. Block diagram of vision systems

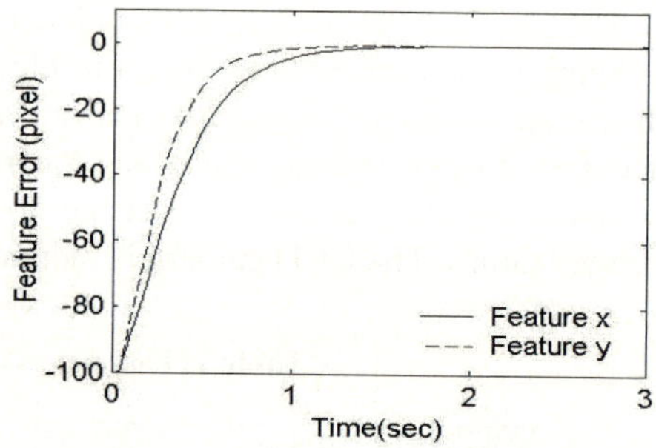

Fig. 3. Feature tracking errors

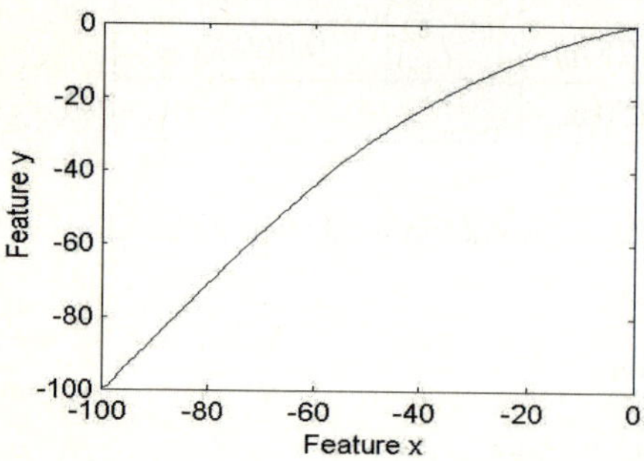

Fig. 4. Trajectory of feature point

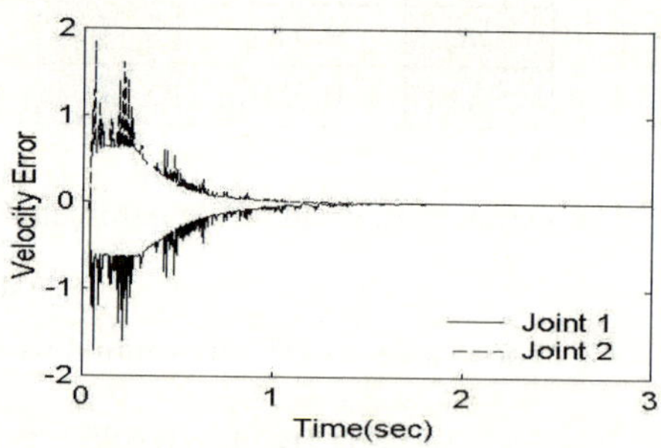

Fig. 5. Velocity errors of joint 1, 2

The experimental results are shown in Figs. 3, 4 and 5. Fig. 3. shows the time evolution of the components x and y of the feature error vector $\tilde{\xi}$. Each component of the feature error vector converges to a small neighborhood of zero after about 1.5

second. Fig. 4. illustrates the trajectory of feature point on the image plane which shows the convergence to the desired feature point. Fig. 5. represents the velocity errors of joint 1 and 2. The velocity error of each joint is also decreasing to a sufficiently small value.

5 Conclusions

In this paper, a robust controller with visual feedback for robot manipulators was proposed. The controller is of structure based on the image feature errors and the joint velocities fed back by the CCD camera and the encoder, respectively. This controller also includes the nonlinear control term to compensate the parametric uncertainty. The desired joint velocity is generated from the image feature error. The ultimate uniform stability of the overall closed-loop system is proved by using the Lyapunov method. Experiment results on a two degree of freedom manipulator have shown that the proposed control method has effectiveness to control robot manipulators with uncertainty.

References

1. Hutchinson, S., Hager, G. D., and P. I. Corke, P. I.:A tutorial on visual servo control. IEEE Trans. Robot. Automat. Vol. 12 (1996) 651-670
2. Espiau, E., Chaumette, F., Rives, P.: A new approach to visual servoing in robotics. IEEE Trans. Robotics and Automation, Vol. 8, No.3 (1992) 313-326
3. Hashimoto, K., Kimoto, T., Ebine, T., Kimura, H.: Manipulator control with image-based visual servo. IEEE International Conference on Robotics and Automation (1991) 2267-2272
4. Kelly, R,: Robust asymptotically stable visual servoing of planar robots. IEEE Trans. Robotics and Automation, Vol. 12, No.5 (1996) 759-766
5. Shen, Y., Xiang, G., Liu, Y. H.: Uncalibrated visual servoing of planar robots, IEEE International Conference on Robotics and Automation (2002) 580-585
6. Zergerglu, E., Dawson, D. M., de Queiroz, Nagarkatti, S.: Robust visual-servo control of robot manipulators in the presence of uncertainty, IEEE Conference on Decision and Control (1999) 4137-4142
7. Kelly, R., Carelli, R., Nasisi, O., Kuchen, B., Reyes, F.: Stable visual servoing of camera-in-hand robotic systems. IEEE/ASME Trans. Mechatronics, Vol. 5, No.1, (2000) 39-43
8. Spong, M. W., Vidyasagar, M.: Robot Dynamics and Control. Wiley, NewYork (1989)
9. Corless, M., Leitmann, G.: Continuous state feedback guaranteeing uniform ultimate boundedness for uncertain dynamics systems. IEEE Trans. Automatic Control, Vol.26 (1981) 1139-1144

Real Time Obstacle Avoidance for Mobile Robot Using Limit-Cycle and Vector Field Method

Min Seok Jie[1], Joong Hwan Baek[1], Yeh Sun Hong[2], and Kang Woong Lee[1]

[1] School of Electronics, Telecommunication and Computer Engineering, Hankuk Aviation University, 200-1, Hwajeon-dong, Deokyang-gu, Koyang-city, Kyonggi-do, 412-791, Korea
tomsey@korea.com, jhbaek@hau.ac.kr, kwlee@hau.ac.kr
[2] School of Aerospace and mechanical engineering, Hankuk Aviation University, 200-1, Hwajeon-dong, Deokyang-gu, Koyang-city, Kyonggi-do, 412-791, Korea
yshong@hau.ac.kr

Abstract. In this paper, we propose a novel navigation method combined limit-cycle method and the vector field method for navigating a mobile robot with avoiding unexpected obstacles in the dynamic environment. The limit-cycle method is applied to avoiding obstacles placed in front of the robot and the vector field method is applied to avoiding obstacles placed on the side of the robot. The proposed method is tested using Pioneer 2-DX mobile robot. Through both simulation and experiment, the effectiveness of proposed methods for navigating a mobile robot in the dynamic environment is demonstrated.

Keywords: limit-cycle, vector field method, mobile robot, real time avoidance.

1 Introduction

In the field of robot navigation research, there are many methods have been developed. Among them, there are two different approaches to the mobile robot navigation planning for application to a real environment, the deliberative approach and the reactive approach. In the deliberative approach, a mobile robot selects its own geometry model using sensors attached on its body and creates paths to navigate [1]. It allows a robot to display a desired smooth goal-directed behavior. However in practical application, such an approach lacks robustness in the face of the environmental uncertainty. To solve such disadvantage of the deliberative approach, the reactive approach was introduced [2][3][4]. This research generally requires relatively little computation and commits the robot to a particular action for only a short span of time. Thus the reactive system handles the uncertainty by continually updating its action. However, due to limited reasoning ability, purely reactive systems are not recommendable.

The new approach, using limit cycle method with 2nd order non-linear function adopts advantages of each approach. The robot soccer systems is a typical example where the limit cycle method is applied [5]. To prevent the robot from colliding with any uncertain obstacles, a route where the robot navigates should be defined before hand. The limit cycle navigation method proposed enables a robot to maneuver

B. Gabrys, R.J. Howlett, and L.C. Jain (Eds.): KES 2006, Part I, LNAI 4251, pp. 866 – 873, 2006.
© Springer-Verlag Berlin Heidelberg 2006

smoothly towards any desired destination by detecting the uncertain obstacles through sectionalizing each length of obstacles.

Another new approach, the vector field method was proposed where the limit cycle method cannot be assured of its safety of navigation, such as where the obstacles are placed on side of the robot. To effectively avoid the obstacles on side of the robot, the optimum vector is determined based on the measured positional data such as angle and minimum distance between the robot and obstacles.

In this paper we propose a novel navigation algorithm combined with limit-cycle method and the vector field method. The proposed method is applied to avoiding unexpected obstacles in a dynamic environment. When the robot starts navigating with limited data obtained from the initial stage of path planning, it may face an unexpected obstacle. Depending on the location of obstacles, customized path finding methods should be considered. For example, if an obstacle is located in front of the robot, then the limit-cycle method can be applied to create a new route. If an obstacle is placed on the side of the robot, the vector field method is considered. Using vector field method, the position of robot can be detected by measuring the angle and minimum length between the robot and the obstacle. Based on such position data, the optimum vector for navigation will be determined.

The robot's obstacle avoidance ability is proven with computer simulation and experiment.

2 Limit-Cycle Method for Navigation Scheme

2.1 Limit-Cycle

Consider the following 2nd order nonlinear system [5].

$$\begin{aligned}\dot{x}_1 &= x_2 + x_1(r^2 - x_1^2 - x_2^2)\\ \dot{x}_2 &= -x_1 + x_2(r^2 - x_1^2 - x_2^2)\end{aligned} \tag{1}$$

And the Lyapunov function candidate,

$$V(x) = x_1^2 + x_2^2$$

The derivative of V(x) along the trajectories of the system is given by

$$\begin{aligned}\dot{V}(x) &= 2x_1\dot{x}_1 + 2x_2\dot{x}_2\\ &= 2x_1x_2 + 2x_1^2(r^2 - x_1^2 - x_2^2) - 2x_1x_2 + 2x_1x_2(r^2 - x_1^2 - x_2^2)\\ &= 2V(x)(r^2 - V(x))\end{aligned} \tag{2}$$

The derivative of $V(x) > r^2$ for $\dot{V} < 0$, or $V(x) < r^2$ for $\dot{V} > 0$. Hence, on $V(x) = c_1$ with $0 < c_1 < r^2$, all the trajectories will be moving outward, while on $V(x) = c_2$ with $c_2 > r^2$, all the trajectories will be moving inward. In the equation (1), if r is 1 in $0 < c_1 < 1$, all the points in the $V(x) = c_1$ go toward to the outside, however, in $c_2 > 1$, all the points in the $V(x) = c_2$ go toward to the inside. It means this domain is a positive invariant, closed and bounded domain. The origin in this domain is the only equilibrium point. Therefore, in accordance with the Poincare-Bendixon theory[6], the periodic path, M can be existed.

$M = \{x \in R^2 \mid c_1 \le V(x) \le c_2\}$.

2.2 Setting Occupancy Box

To measure both a sectional length and a central point of the obstacle to the mobile robot, we place occupancy boxes, each box is 50mm wide and 1500mm long, on the left and the right side of the plane as shown in Fig. 1. We choose a value close to the robot's Y coordinate as a distance between the robot and the obstacle in front. If the obstacle is detected in the occupancy box, we measure a length and a central point to the obstacle up to 3 different cases, and by using the obtained data, the limit-cycle path is generated.

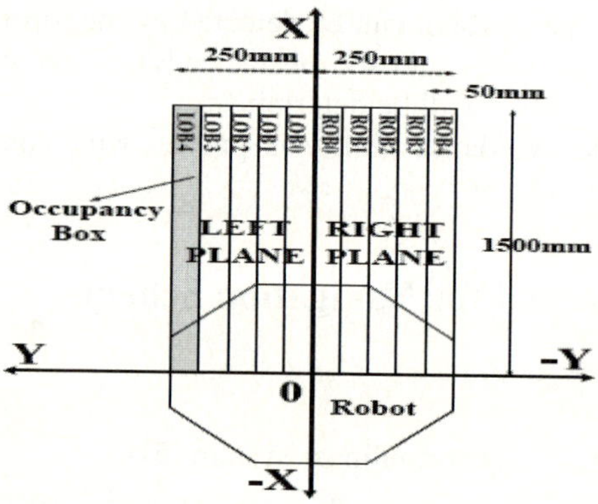

Fig. 1. Occupancy boxes placed on the coordination of the robot

In Fig. 1, if the obstacle is located in the occupancy box of the left plane, two ultrasonic values, the least and the most coordinate value in the occupancy box from the X coordinate are obtained. We can calculate the central coordinate of the straight line using the following equation (3) which is lined between two values, where the length of the straight line becomes the sectional length of the obstacle.

$$C_{x,y} = \frac{LOB_{\min}(x,y) + LOB_{\max}(x,y)}{2} \tag{3}$$

In Fig. 1, if the obstacle is located in the occupancy boxe of the right plane, two ultrasonic values, the least and the most coordinate value in the occupancy box from the X coordinate are obtained. We can calculate the central coordinate of the straight line using the following equation (4) which is lined between two values, where the length of the straight line becomes the sectional length of the obstacle.

$$C_{x,y} = \frac{ROB_{\min}(x,y) + ROB_{\max}(x,y)}{2} \tag{4}$$

In Fig. 1, if the obstacles are occupied in occupancy boxes of both the left & right plane, the least and the most coordinate value in the occupancy box from the X coordinate are obtained. We can calculate the central coordinate of the straight line using the following equation (5) which is lined between two values, where the length of the straight line becomes the sectional length of the obstacle.

$$C_x = \frac{LOB_{\max}(x, y) + ROB_{\max}(x, y)}{2}$$

$$C_y = LOB_{\max}(y) - \frac{|LOB_{\max}(y)| + |ROB_{\max}(y)|}{2} \tag{5}$$

In Fig. 1, LOB0 ~ LOB4 and ROB0 ~ ROB4 represent the number of occupancy boxes in the left and the right plane respectively.

2.3 The Limit-Cycle Navigation Method Using Occupancy Boxes

In Fig. 2, the sectional length is obtained after detecting the edge of an obstacle in the range of 500mm occupancy box. In Fig. 3, the central point of the obstacle, (C_x, C_y) is given with the equations (3), (4) and (5). The central point is used as an origin to the limit-cycle of the equation (1), and r is a radius derived by adding a half length of the obstacle to a half size of actual robot. Therefore, through the process shown in Fig. 2 and Fig. 3, the limit-cycle path where the robot navigates with avoiding obstacles is generated.

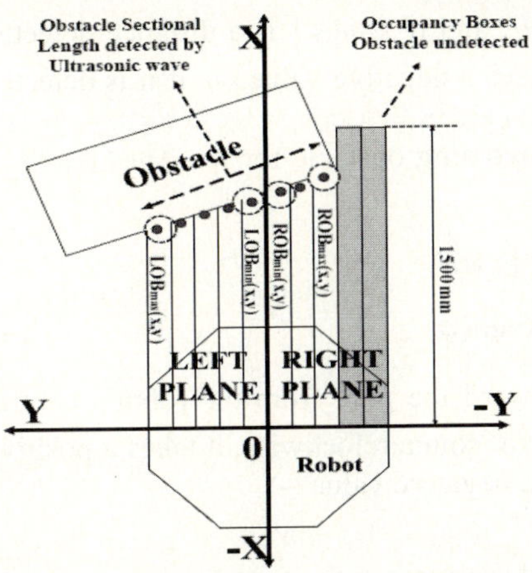

Fig. 2. Sectional length detection of obstacle in front of the robot

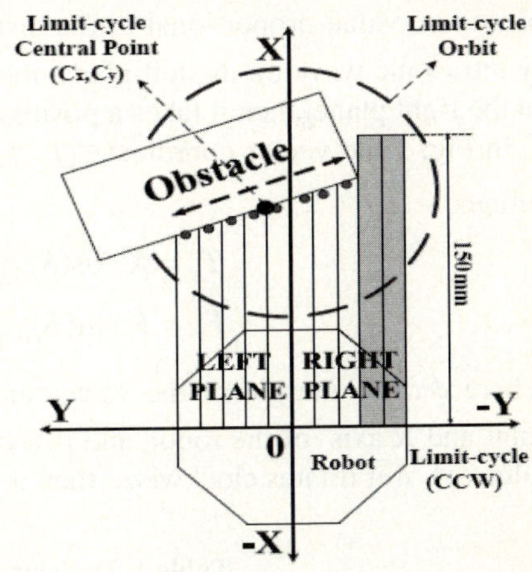

Fig. 3. Limit-cycle path planning for obstacle

For the direction of robot rotating, the robot rotates clockwise(CW) for $\sum_{i=0}^{3} S_i > \sum_{i=3}^{7} S_i$, or the robot rotates counterclockwise(CCW) for ,

$\sum_{i=0}^{3} S_i < \sum_{i=3}^{7} S_i$. In Fig. 4, where S_i denotes the distance of each of ultrasonic sensors. In this paper, the maximum distance is set to 3m. If there is no obstacle detected, the distance of ultrasonic sensor is set to 3m not to 0.

3 Vector Field Method for Navigation Scheme

To generate a limit-cycle path for obstacles placed on side of the robot, the vector field method is proposed. As shown in Table 1, after selecting the least value among the distance values detected by 8 groups of ultrasonic sensor, we pre-set the angle of avoidance. In Fig. 4 and Fig. 5, robot's avoidance vector is measured. For $Min\{S_i\} \le l_i$ use equation (6), and for $l_1 < Min\{S_i\} \le l_2$, use equation (7), where $Min\{S_i\}, i = 0,1,\cdots,6,7$ is the least value among the values detected by ultrasonic waves. The line l_1 and l_2 are the absolute distance between the robot and the obstacle, which represents the distance where the obstacle influences to robot's navigation, and the value is selected through experiment.

In Fig. 4, the vector coordinate (T_x, T_y) of avoiding obstacles for $Min\{S_i\} \le l_1$, defines

$$T_x = K\cos\left(S_{i\deg} \pm \alpha\right)$$
$$T_y = K\sin\left(S_{i\deg} \pm \alpha\right) \tag{6}$$

Where K is a constant to decide the magnitude of a vector of avoidance. In Table 1 $S_{i\deg}$ is an angle of avoidance of $Min\{S_i\}$ and α denotes $[(l_2 - S_1)(90°/l_1)]$, which is an inverse value proportional to the distance l_1, and if $Min\{S_i\}$ is a distance detected by ultrasonic wave on the left plane, then it takes a negative value (-), if it is detected on the right plane, then it takes a positive value (+).

In Fig. 5 the vector coordinate (T_x, T_y) of avoiding obstacle for $l_1 < Min\{S_i\} \le l_2$, defines,

$$T_x = K\cos(S_{i\deg}) + K\cos(\phi)$$
$$T_y = K\sin(S_{i\deg}) + K\sin(\phi) \tag{7}$$

Where ϕ is a margin of the vector angle toward the goal from the robot's central point and X axis of the robot, and if it rotates to counterclockwise, it takes a positive value (+), if it rotates clockwise, then it takes a negative value (-).

Table 1. Decision table of avoidance angle

Ultrasonic sensor number	0	1	2	3	4	5	6	7
$S_{i\deg}$ (deg)	0	-30	-50	-80	80	50	30	0

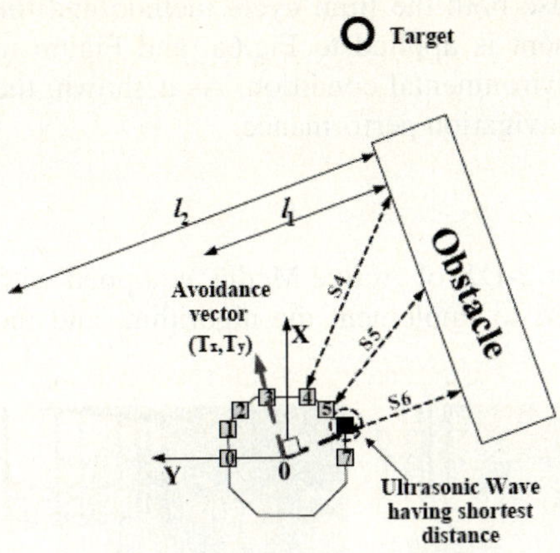

Fig. 4. Case for $Min\{S_i\} \leq l_1$

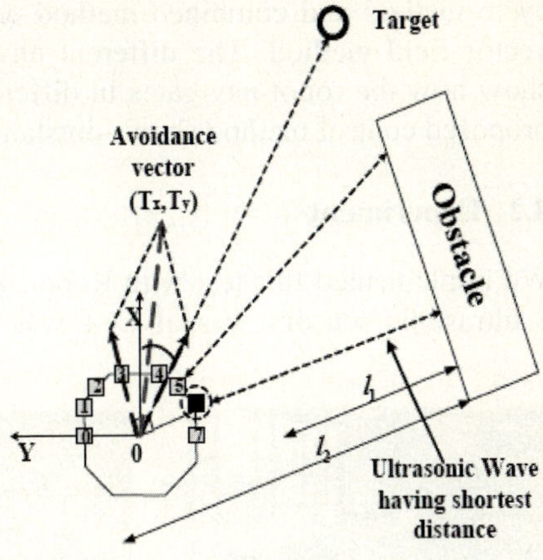

Fig. 5. Case for $l_1 < Min\{S_i\} \leq l_2$

4 Simulation and Experiment

4.1 Simulation

The proposed method is tested through simulation, which shows that the robot drives to the target and avoids unexpected irregular size obstacles. The parameters are set to $K = 500, l_1 = 60cm, l_2 = 100cm$, and the average speed of robot running is set to 0.5(m/sec). Fig. 6a and Fig. 6b show obstacle avoidance simulation by using limit

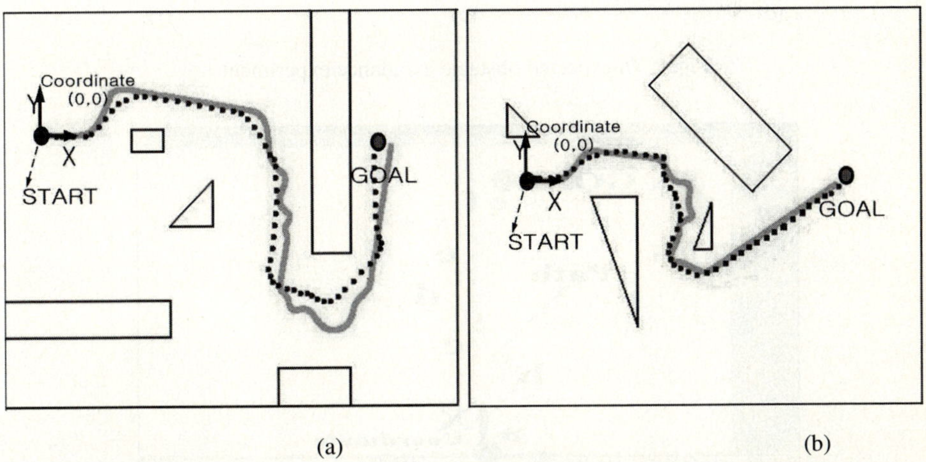

Fig. 6. Unexpected obstacle avoidance simulation: limit cycle method (solid) and Combined method(dashed)

cycle method and combined method which use both the limit cycle method and the vector field method. The different environment is applied to Fig.6a. and Fig.6b to show how the robot navigates in different environmental condition. As it shown, the proposed control method shows outstanding navigation performance.

4.2 Experiment

We implemented this test with Robot, Pioneer 2-DX of Active Media, equipped with 8 ultrasonic sensors. Visual C++ was utilized to implement the algorithm, and the

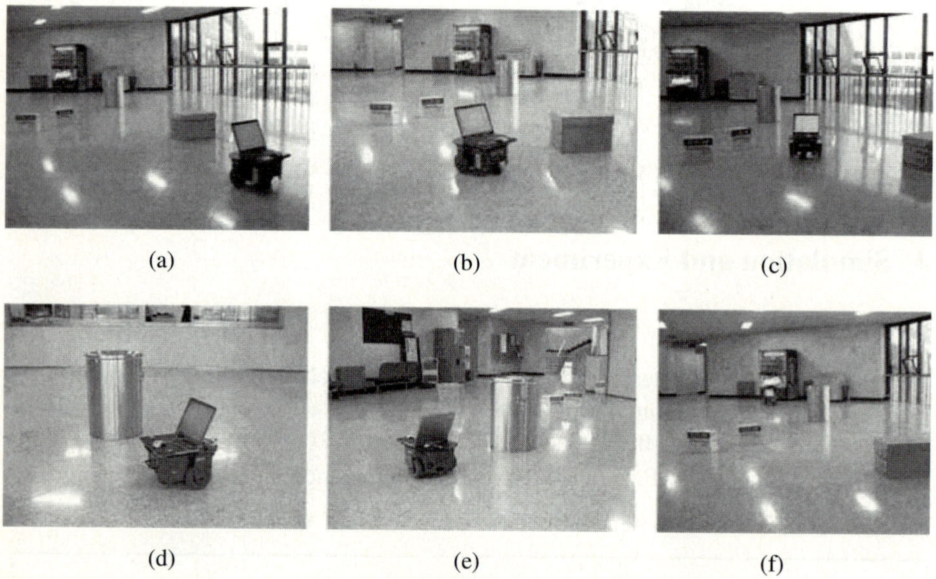

Fig. 7. Unexpected obstacle avoidance experiment

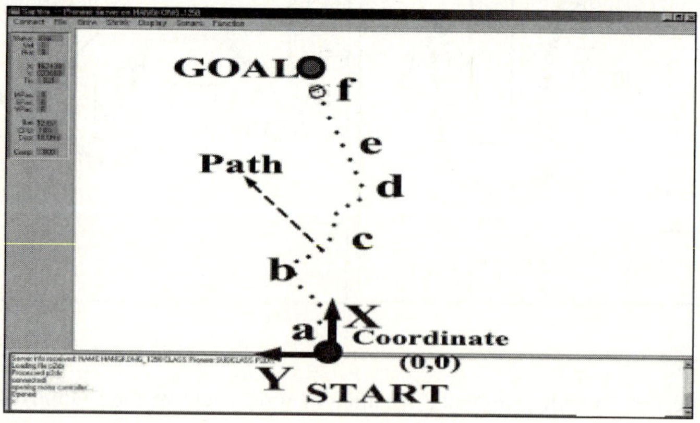

Fig. 8. Mobile robot moving path to the Fig. 7

650 MHz Pentium III laptop was used to transport data to the mobile robot through RS-232 serial port. Each parameter value set to $K = 500, l_1 = 50cm, l_2 = 100cm$, and the average speed of robot running is set to 0.45(m/sec).

Fig.7 is a snap shot captured when the robot drives to the goal located 10m away from the starting point through avoiding obstacles by detecting the unexpected irregular size obstacles which were placed on the way where the robot is suppose to pass. The Fig. 8 is an actual running path where the robot is passed.

5 Conclusion

In this paper, we propose a novel navigation method combined with the limit-cycle method and the vector field method for a mobile robot to avoid unexpected obstacles in a dynamic environment. The limit-cycle method is used to effectively avoid the obstacles in front, the vector field method is used for the obstacle on the side. We adopt advantages of each method and apply them to robot navigation with swiftness, accuracy, and safety. It is tested through both simulation and experiments. However since this method is not enough to create the optimum path, the study for the optimum path should be considered, as a next research subject.

References

1. Lefebvre, D., Saridis, G.: A Computer Architecture for Intelligent machines. Proceeding of the IEEE international Conference on Robotics and Automation, France (1992) 2245-2250
2. Brook, R.: A Robust Layered Control System for a Mobile Robot. IEEE Journal of Robotics and Automation, vol. RA-2, no. 1 (1986) 14-23
3. Arkin, R.C.: Behavior Based Robotics. The MIT Press (1988)
4. Borenstein, J., Koren, Y.: Real-time Obstacle Avoidance for Fast Mobile Robotics. IEEE Transactions on System, Man and Cybernetics, Vol.19, No. 5 (1989) 1179-1187
5. Kim, D.H., Kim, J.H.: A real-time limit-cycle navigation method for fast mobile robots and its application to robot soccer. In the Robotics and Autonomous System,1 (2003)
6. Khalil, H. K.: Nonlinear Systems, 2nd ed. Prentice-Hall, Chap 7.(1996) 289-312

Developing Knowledge-Based Security-Sense of Networked Intelligent Robots

M. Omar Faruque Sarker[1], ChangHwan Kim[1], Bum-Jae You[1], and Mohammed Golam Sadi[2]

[1] Intelligent Robotics Research Center
Korea Institute of Science and Technology(KIST)
P.O. Box 131, Cheongryang, Seoul, 130-650, Korea
writefaruq@gmail.com, {ckim, ybj}@kist.re.kr
[2] Computer Engineering Department
Hankuk Aviation University, Korea
sadi@hau.ac.kr

Abstract. Distributed processing capabilities of robotic-applications using state-of-the-art component-based middleware and mobile-agent software technologies are continuously helping to develop more powerful intelligent robots. These application development frameworks provide several attractive conceptual solutions for network-based operations, however, these benefits cannot be realized unless the appropriate security mechanisms are in place. In this paper, this important issue is addressed in a systematic way: firstly, by analyzing the key limitations of traditional security models, the new security requirements for agent-based networked robots are pointed out and secondly, an effective defense mechanism is devised by constructing a knowledge-based security-aware robot control model. The main contributions of this work are, to address the security problem in a knowledge-based approach and to show the roles and mechanisms of the proposed security architecture. The feasibility of our approach is examined through a case study on development of robot security-sense using Linux-based reliable security-tools.

1 Introduction

Network-based intelligent robots provide a new outlook to design and implement cost-effective and powerful solutions for a great number of users. Recent developments in Component Based Software Engineering (CBSE) and in state-of-the-art Mobile-Agent Software (MAS) [2] technologies are continuously creating new opportunities to build more powerful and cost-effective intelligent robots. These application development frameworks provide several attractive conceptual solution for application development: real-time capabilities, dynamic deployment, QoS etc., but they hardly satisfy the dynamic security requirement which is highly essential for implementing a mission critical networked robot. Generally security and privacy technologies are divided into three categories: prevention, avoidance and detection [3]. In this paper, we have covered mainly the security issues where privacy is implicitly considered in our security architecture.

B. Gabrys, R.J. Howlett, and L.C. Jain (Eds.): KES 2006, Part I, LNAI 4251, pp. 874–881, 2006.
© Springer-Verlag Berlin Heidelberg 2006

Rest of this paper: Section 2 describes the traditional security models along with their key pros and cons. New security requirements and preview of our knowledge-based approach have been presented at Section 3. Section 4 discusses our proposed model. Section 5 provides the details of our implementation. Section 6 describes our case study of Linux-based Netfilter package to develop security sense of our robot. Finally, Section 7 concludes the paper with summary and outlook.

2 Related Works

Recent security researches on distributed computing, in general, propose architectures that offer security in a traditional way like role-based access control (RBAC) or in context-aware form. Some modified versions have also been proposed like Usage Control Model (UCON), Ubiquitous Context-based Security Middleware (UbiCOSM) [3]. Though RBAC model simplifies management and review of access control by defining a set of roles, it assumes a centrally administered trusted computing environment. It is therefore difficult to specify security policy for a dynamic environment by this model. In UCON model, core components comprise subjects, objects and rights with additional components like authorizations, conditions and obligations. Here subjects hold rights on objects satisfying conditions. This model is very promising in ubiquitous environment as it covers both security and privacy. But the delegation of rights and administration issues makes it difficult to apply in an autonomous environment. In case of the component based agents, most are usually based on Java and are often kept simple in security design. As a result, they hardly meet the dynamic security demands caused by remote software agent interacting with other agent/platform [1] [2].

3 Knowledge-Based Approach for Robot Security

Security researchers have classified security threats for a networked system into two broad categories: general threats (e.g., unauthorized access, disclosure and corruption of information, denial of service etc.) and software agent based threats (e.g., masquerading, eavesdropping, alteration etc.) [1], [3]. Here, we have considered the later case where agents forms a complicated security-threat matrix among agent to platform, agent to agent and agent to others. These can cause serious damages to every element of surrounding sensor networks including human users and any other living objects. According to Section 2, it is seen that traditional models are not well suited for preventing new kinds of threats in a decentralized, autonomous and dynamic environment. Therefore, the knowledge-based approach has been considered as an alternate. For example, a fuzzy-logic based intrusion detection system has been successfully implemented by Botha and Solms [6]. Along with fuzzy-logic, Susan and Rayford have also used neural network and genetic algorithm for anomaly and misuse detection [7].

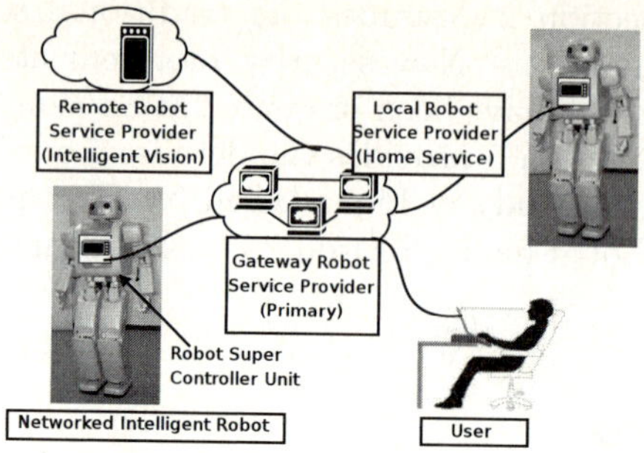

Fig. 1. Main elements of security-aware networked robot control model

However, those works are limited in network intrusion detection and trend-analysis after the actual attack has been performed. Since, they have no run-time protection mechanism, it is not possible to apply them in dynamic decision-making and protection of robot against agent based threats. So we have proposed an active knowledge-based system which is not only capable of detecting run-time security threats but also can infer a decision from its past experiences in an unknown situation.

In this paper, we have mainly demonstrated how our system can sense and react against the agent based threats. For example, suppose we have deployed a third-party vision agent which is responsible for providing intelligent motion-hints to the robot by observing an unknown environment through robot cameras. And we also have a specific network security agreement with the agent operating company. If the agent gets camera image as input (usually UDP streams) and provides motion-hints (TCP packets) as output, the network traffic pattern, CPU use, connection time etc. can easily be defined. If we use these settings as our sample security-policies, we can definitely build our knowledge-based system to sense and react on the violation of these policies. Furthermore, if the system can remember past histories of policy violation/reaction sequences, it can also infer new system reaction in an unknown violation case.

Our knowledge-based approach is highly adaptable on dynamic environments as it can decide autonomously using its security-senses. Consequently, there is no huge administrative overhead if the system is properly trained and tuned. Moreover, the system can also prove its robustness and efficient decision-making capabilities using its past experiences from knowledge-base.

4 Security-Aware Robot Control Model

In this section, we have described our knowledge-based security model, its architecture and mechanisms for policy and knowledge base generation and reuse.

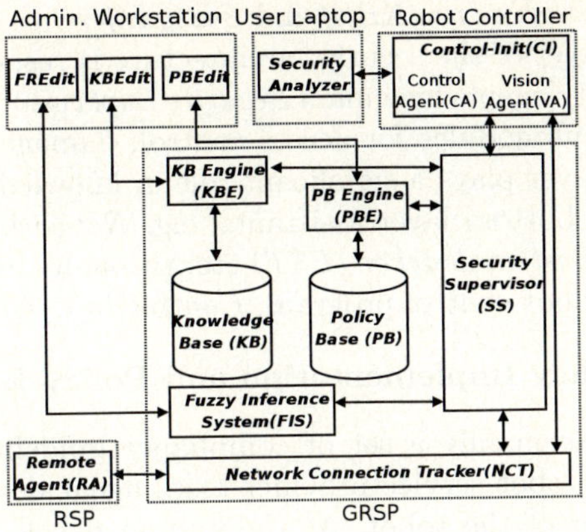

Fig. 2. Knowledge-based security architecture for a networked robot

4.1 Model Overview

The main role-players of our model, as shown in Fig. 1, are the *Networked Intelligent Robot, User* (a privileged user is termed as *superuser*), *Gateway to Robot Service Provider (GRSP), Robot Service Provider (RSP).* Here, GRSP have a central role for providing access to the services offered by the robot. RSPs are the network servers which provide various services to the users and robots via the GRSP. This model has two major distinct features.

1) Gateway Protected Security: Our model clearly introduces a new robot control architecture based on a mandatory gateway-layer for ensuring the maximum security and reliability. We choose this approach for two main reasons: firstly, a gateway-based security management is a well-known approach used by almost every secure network and secondly, the system resources (e.g. battery power) of robot are too expensive to utilize it in security handling..

2) Knowledge-Based Architecture: In our model, another important enhancement is the provision for knowledge development and dynamic security management using the acquired knowledge. The following subsections describe our knowledge-based architecture, policy and knowledge-development processes with necessary examples.

4.2 Knowledge-Based Security Architecture

The security system architecture of our model, as shown in Fig. 2, consists of three implicit layers: planning and decision, supervisory control and routine job management. The members of the planning and decision-making layer is a strategic layer and is only concerned about the global security policy and knowledge of robot and they help to maintain system security by a high-level application interface to user. For example, Policy-Base (PB) has a front-end server PB Engine (PBE) which can be accessed by *Policy-Base Editor (PBEdit)*. Similarly, *Knowledge-Base Editor (KBEdit)* and*Fuzzy Rule Editor (FREdit)* can access *Knowledge-Base Engine (KBE)* and *Fuzzy Inference System (FIS)* respectively. The middle layer of the security architecture is responsible for managing the lower level security agents by a main *Security Supervisor (SS)* program which may have different submodules for access control, damage detection, intrusion detection etc. This layer plays a significant role in knowledge development process by utilizing the bottom layer assistants, e.g. *Network Connection Tracker (NCT)*, *Emergency System Tracker (EST)* etc. through a close interaction with the chief executive robot control program, *Control-Init (CI)*.

4.3 Security Policy Implementation and Policy-Base Development

The security policy is mainly a set of compliance guidelines provided by the robot manufacturer, robot service provider (e.g., agent developer and/or operator) and user/owner of the robot. At the system development stage, several domain-experts gather these guidelines, e.g., the regular working patterns of a remote software agent like usage of network bandwidth, port, protocol etc. (by a network engineer). A knowledge engineer will take several interviews with these domain-experts to translate these high level system policies into the appropriate sentences of the Policy-Base (PB). These sentences will be fed to a deductive database engine, PBE, which is responsible for storing/retrieving the policy sentences to/from PB and answering queries from SS. PB is a relational database where raw data is stored. A knowledge-engineer or superuser will interact with the PBE via PBEdit (which is actually a client to the deductive database system).

From the sketch of our running system, as shown in Fig. 2, the PB development process is outlined as follows:

- When NCT initializes (at GRSP), it asks SS for a fresh copy of robot's current network security policies which include the remote agent interaction (port and protocol (TCP/UDP) usage, bandwidth characteristics, CPU use, time of interaction etc).
- SS immediately asks the ConrolInit(CI) for a set of current relevant settings of the robot, e.g., operational mode (6 operational modes are defined: Emergency Stop (ES), Graceful Stop (GS), Maintenance/Safe (MS), Limited Human Interaction (LI), Full Human Interaction (FI)), current motion state (standing/walking/running), human command under execution (if any) and so on.
- SS feeds these settings to the PBE as policy generating constraint (e.g., at MS mode, no remote agent connection can be alive).
- Consequently, PBE tries to find a set of policy which can satisfy these settings. If any prior policy data does not exist, it uses the inference rule-sets for generating a fresh copy of policy data.
- While serving the generated policy data to SS, PBE also saves a local copy to the PB for future use.

Thus PB can be filled with necessary policies for different run-time settings of the robot and they can also be reviewed by the superuser through PBEdit.

4.4 Knowledge-Base Development and Knowledge Reuse

Making a safe decision in an unknown situation is a big challenge for the robot. If any security policy is violated by a remote agent the robot needs to find out the anomalous behavior patterns. Here the Fuzzy Inference System (FIS) becomes handy to decide on a remote agent's suspicious behavior. If FIS reports that the agent's behavior is highly anomalous, SS can notify CI with an appropriate caution alarm (Unsafe/Critical/Emergency). Consequently, CI can take emergency decision to switch to a different mode where remote agent interaction will be limited or ceased. In such a case, SS again needs to trigger an update to NCT about the current security policy. These sequences contribute to create dynamic knowledge which is briefly mentioned below.

- Suppose a Remote Agent (RA) is maintaining a connectivity with the motion control agent CA which is transparent to NCT. NCT periodically collects the connection data and checks against the security policy (specific to that RA). In course of time, if NCT notices policy violation of RA it immediately reports to SS.
- SS then asks CI about its current settings and feeds the violation data to FIS and gets a decision on RA's behavior (degree of anomalousness).
- SS submits all these information to KBE to match a similar past experience (e.g. what was done at past in this kind of situation). If no past experience is found KBE generates a new experience.
- While serving the new experience to SS, KBE also saves a local copy to KB for future use.

Thus KB can be filled with necessary experiences for different run-time settings of the robot and they can also be modified by the superuser through KBEdit.

5 Implementation

The feasibility of our model is primarily verified by a prototype implementation as demonstrated in Fig. 2. For the sake of simplicity, a remote-agent (RA) from a RSP system has connected with two local agents: control-agent (CA) and vision-agent (VA) through the security-controller, GRSP. Both CA and VA have been implemented as the core modules of CI. All connections from remote-agents are transparently passed through (and intercepted by) NCT. SS is the main intelligent agent which actively collaborates with CI, as described in Section 4.3 and 4.4, for overall security implementation. It is also interfaced with the PB and KB through PBE and KBE. PB and KB are implemented by MySQL 4.1 relational database [5] and PBE and KBE are CORAL 1.5.2 deductive database system which supports a rich declarative language, and an interface to C++ [4]. Here, the CORAL database engine has two workspaces: relational database workspace and default workspace. Default workspace serves the CORAL query from SS (and also from PBEdit/KBEdit) and RDB workspace maps CORAL query into SQL query for MySQL database. A superuser can manipulate the PB and KB through the PBEdit and KBEdit respectively. All security related reports are periodically sent to SecurityAnalyzer for superuser's review.

Table 1. A KB-term and *iptables* match gives robot a security-sense

Fuzzy Term	Sets	Possible Feelings (Sense)	*iptables* Match
State	LOW	Someone tries to talk	NEW
	MEDIUM	Talk already started	ESTABLISHED
	HIGH	Talk has expired	INVALID
Defense	LOW	Let them talk	ACCEPT
	MEDIUM	Force to stop talking	REJECT
	HIGH	Force to stop talking	DROP
Logging	HIGH	Keep their details	LOG
	MEDIUM	Mark them	MARK
	LOW	Keep them awaiting	QUEUE

In our work, most of the software modules are developed using C++ and Linux. We have also used Matlab Fuzzy Logic Toolbox for initial development and verification of FIS. A user interface, FREdit, is used to tune the FIS by an expert knowledge engineer. Linux-based open-source security framework Netfilter's [9] packet filtering package *iptables* (version 1.2) module is integrated with our NCT. We have used our humanoid platform *Network Based Humanoid (NBH-1)* [10] with Xenomai real-time OS [11](former fusion branch of Real Time Application Interface RTAI/fusion) interfaced with our robot control program, CI.

6 A Case Study on Netfilter/Iptables

In this section, we have presented a simple case study to demonstrate the reactions of our system under some malicious activity of a remote agent. For this purpose, we have exploited the security-features offered by Netfilter's *iptables* package. From Section 4.2 we have already known that our NCT will report SS if it finds any security policy violation caused by remote agent RA, as outlined in Fig 2. For operating our FIS, we have used the similar techniques discussed in [6]. Three fuzzy sets: low, medium and high are defined to represent the degree of anomalousness of RA's behavior. So when the SS gets the anomalous behavior data of RA from NCT, it feeds them to FIS and gets a decision, i.e., the degree of anomalousness of the RA's behavior. Now we may examine the iptables options and combine them with fuzzy rules to get sample system reaction. Let us consider a simple case where our system has sensed that agent RA tries to talk and it has some anomalousness in its behavior. So KBE will read its rule base like this:

```
IF someone tries to talk AND his behavior is anomalous,
   THEN defense is high.
```

Here, *someone tries to talk* is, in fact, the sense obtained from the iptables *state* match options `NEW` (iptables has its internal mechanism to identify a new packet by looking SYN types in TCP packet header).This is mapped to KB term `state` as LOW (some possible robot senses, iptables options, fuzzy terms and sets are listed in Table 1). Similarly in output space, ***defense is high*** corresponds to the iptables target jump option DROP. A detail description on various iptables matches and jumps can be found at [8]. Thus, if we correctly map the iptables option to robot sense then it is also possible to define the desired system reaction on various anomalous behaviors of any remote agent.

7 Summary and Outlook

The networked intelligent robots are expected to detect quickly any violation of security policy of the environment as well as to defend the potential security threats according to its preprogrammed knowledge or previously developed security-sense without waiting for any user-interaction. The security defense approach described in this paper is dynamic and more robust comparing with the traditional security models. This behavior is very interesting and highly desired in a decentralized environment like sensor networks.

Although the merits of this knowledge development of robot may not be instantly visible to user, but after a long time, the developed intelligence will play a significant role to save human and all other elements of the environment. The improved behavior patterns of the robots will greatly increase the feelings of reliability and confidence of the users and thus it can create many new possibilities for implementing the networked intelligent robots in a challenging environment.

References

1. Brazier, F.M.T., Jonker, C. M.,Treur, J.: Principles of ←Component-Based Design of Intelligent Agents. http://www.few.vu.nl/ wai/Papers/DKE02.princ.pdf
2. Jansen, W., Karygiannis, T,: NIST Special Publication 800-19 Mobile Agent Security. http://www.mirrors.wiretapped.net/security/info/reference/nist/special-publications/sp-800-19.pdf
3. Yamada, S., Kamioka, E.: Access Control for Security and Privacy in Ubiquitous Computing Environments. IEICE Trans. Commun., Vol.E88-B, No.3. March 2005.
4. Ramakrishnan, R., Srivastava, D., Sudarshan, S., and Seshadri, P.: The CORAL deductive system. The VLDB Journal. 3 (1994) 161-21.
5. Widenjus, M., Axmark, D.: MySQL Reference Manual. O'Reilly & Associates, Inc. June 2002.
6. Botha, M. and von Solms, R.: Utilizing Fuzzy Logic and Trend Analysis for Effective Intrusion Detection. Computers and Security. Vol 22, No 5, (2003) 423-434.
7. Bridges, S.M. and Vaughn, R.B.: Fuzzy Data Mining and Genetic Algorithms Applied to Intrusion Detection. Proc. 23rd National Information Systems Security Conference (NISSC), October 2000, Baltimore, MD.
8. Andreasson Oskar: Iptables Tutorial 1.2.0. Available Online http://iptables-tutorial.frozentux.net/iptables-tutorial.html.
9. http://www.netfilter.org/
10. http://humanoid.kist.re.kr/
11. http://www.xenomai.org/

Autonomous Action Generation of Humanoid Robot from Natural Language

Hirokazu Watabe and Tsukasa Kawaoka

Dept. of Knowledge Engineering & Computer Sciences, Doshisha University
Kyo-Tanabe, Kyoto, 610-0394, Japan

Abstract. Communication between human and robot is necessary for an intelligent robot to be active in daily life. Conversation is the basis of communication. Most of the instructions given to a robot are related to some actions. This paper reports the method to generate the action of the humanoid robot by conversation. For this purpose, comprehension of action instruction written by natural language is necessary. The system to understand natural language for generating action is proposed. This system consists of a semantic understanding method to arrange input information, the knowledge base of vocabulary related to actions, the knowledge base of the parameter for robots to act, and association mechanism to handle a word, which is not known. The system is capable of understanding many input sentences by the association mechanism using Concept-Base and a degree of association.

1 Introduction

Intelligent humanoid robots, which can become a partner to us humans, need ability to act naturally. In actions of the robot, it is very important to understand natural language from humans.

Conventionally, the knowledge of action for the robot was mostly constructed by humans for the fixed number of input sentences. Therefore, the robot cannot act anymore if the given sentence is not acceptable one. In this paper, the system to understand natural language for action of the humanoid robot is proposed. This system consists of a semantic understanding method to arrange input sentences, the knowledge base of vocabulary related to particular actions, the knowledge base of the parameter for the robot to act, and the word association system to handle unknown words. The system is capable of understanding many input sentences by the word association system using Concept-Base and the degree of association among word concepts.

2 Structure of the Humanoid Robot

In this paper, it is used the humanoid robot called "Robovie"[1] shown in fig.1. The robot that has a human-like appearance has various sensors, such as vision, sense of touch and so on. It has two arms (4 degrees of freedom x 2), a head (3 degrees of freedom), and a mobile platform (2 driving wheels).

B. Gabrys, R.J. Howlett, and L.C. Jain (Eds.): KES 2006, Part I, LNAI 4251, pp. 882 – 889, 2006.
© Springer-Verlag Berlin Heidelberg 2006

Fig. 1. Robovie

The one pose is decided by 11 parameters of two arms and the head. The one action is decided by consecutive poses. The problem treated in this paper is to generate the action from natural language sentence input.

3 Structure of the Proposed System

Fig. 2 shows the structure of the proposed system. The following is the algorithm to generate action of the robot from input sentence.

(1) Given input sentence (Japanese typed text) is stored into semantic frame as separated words by semantic meaning understanding system.

(2) Each word in the semantic frame is compared with Word Knowledge-Base (WKB) for action. When there are no matched words in WKB, search semantic similar word using Word Concept Association System (WCAS). WCAS consists of Concept-Base (CB) and the calculation module of the degree of association between concepts [2][3].

(3) Action Generator generates 11 joint parameters of the robot and action data using Basic Action KB and Complex Action KB.

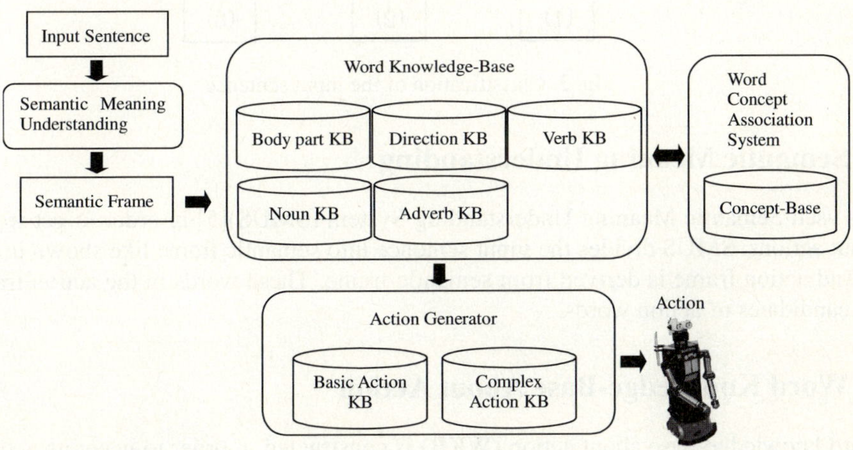

Fig. 2. System structure

4 Input Sentence

The proposed system treats a simple sentence as an input sentence. It has one subject and one predicate (verb). The input sentence is classified into the following three categories as the action generation (fig. 3).

(1) The subject is one of the body parts and the direction or the position is mentioned. (Ex: Put your right hand on your head.)

(2) The subject is one of the body parts, but the direction or the position is not mentioned. (Ex: Be your right hand up.)

(3) The subject is not one of the body parts. (Ex: Let's go. Hello.)

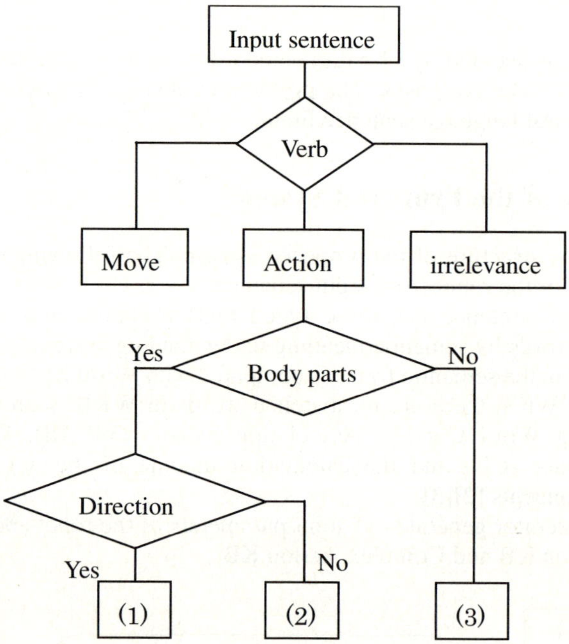

Fig. 3. Classification of the input sentence

5 Semantic Meaning Understanding

It is used Semantic Meaning Understanding System (SMUS)[5] in order to get words about actions. SMUS divides the input sentence into semantic frame like shown in fig. 4. And action frame is derived from semantic frame. These words in the action frame are candidates of action words.

6 Word Knowledge-Base About Action

Word knowledge-base about action (WKB) is constructed in order to generate actions from action words. WKB consists of five KB as shown in fig. 2. Table 1 shows the part of WKB. Body part KB has nouns representing human body parts with verb about possible actions. There are 128 nouns of human body parts and 263 verbs in body part KB. Verb KB has 154 verbs. Each verb in verb KB represents some action by itself, such as ‘stand up’, ‘nod’ and so on. Noun KB has 167 nouns. Each noun in noun KB represents some action by itself, such as ‘straightening one’s back’, ‘deep breathing’ and so on. Direction KB has 106 words representing directions or positions including body parts, such as ‘up’, ‘down’, ‘eye’, ‘head’ and so on. Adverb KB has 54 adverbs representing the degree of actions, such as ‘a little’, ‘pretty’, ‘well enough’ and so on.

Input sentence: 私は右手を腹に当てる. (I put my right hand on my belly.)

Semantic frame

Who	What	when	Where	why	How	Whom	Pred.
私	右手		腹				当てる
(I)	(right hand)		(belly)				(put)

Action frame

Part	Where	Pred.
右手	腹	当てる
(right hand)	(belly)	(put)

Fig. 4. Semantic frame (Action frame)

Table 1. Word knowledge base about action

Body part KB

Part	Verb	ID
頭	曲げる	52
頭	振る	10
手	上げる	53

Verb KB

Verb	kana reading	ID
起き上がる	おきあがる	22
立つ	たつ	-27
うなずく	うなずく	87

Noun KB

Noun	kana reading	ID
背伸び	せのび	22
握手	あくしゅ	-3
拍手	はくしゅ	9

7 Unknown Word Processing

Since the input sentence is arbitrary, there are many words, which are not stored in the WKB. These words are called unknown words. If there are unknown words in the input sentence, unknown word processing is executed and these unknown words are substituted with the words in WKB.

Fig. 5 shows the flow of the unknown word processing. If each word on action frame is not the word in the WKB, the degree of association [3] between the unknown word and words in WKB are calculated, and the word, which has the highest degree of association that is greater than threshold (0.03), is selected. The degree of association derives the semantic distance between two words.

Concept-Base

A Concept is defined as the following equation.

$$A = \{(a_1, w_1), (a_2, w_2), \cdots, (a_m, w_m)\} \tag{1}$$

Where, A is named the concept label (a word), a_i is called the attribute, and w_i is called the weight. The Concept-Base is a set of concepts.

$$\begin{aligned} A_1 &= \{(a_{11}, w_{11}), (a_{12}, w_{12}), \cdots, (a_{1m_1}, w_{1m_1})\} \\ A_2 &= \{(a_{21}, w_{21}), (a_{22}, w_{22}), \cdots, (a_{2m_2}, w_{2m_2})\} \\ &\vdots \\ A_n &= \{(a_{n1}, w_{n1}), (a_{n2}, w_{n2}), \cdots, (a_{nm_n}, w_{nm_n})\} \end{aligned} \tag{2}$$

In the concept-base, each attribute a_{ij} must be one of the concept labels A_k. Therefore each concept is defined by infinite chain of attributes. Each attribute expressed by the equation 1 is called the first order attribute, and each attribute of an attribute is called the second order attribute, and so on.

In this paper, the concept-base which is basically generated by the following ideas is used.

(1) Prepare several electronic Japanese dictionaries and newspapers.

(2) Take out the headword as a concept label, and take out each word in the sentences explaining the headword as an attribute.

(3) Calculate the weight of each attribute based on the frequency.

(4) Remove noises (improper attributes) by statistical procedures.

Fig. 5 shows an example of the concept -base.

Degree of Association

Each concept is defined as a set of attributes and each attribute is also a concept as described above. Each concept is defined by an infinite chain of attributes (concepts). In this paper, a method to derive the degree of association between concepts using up to second order attributes is used [3].

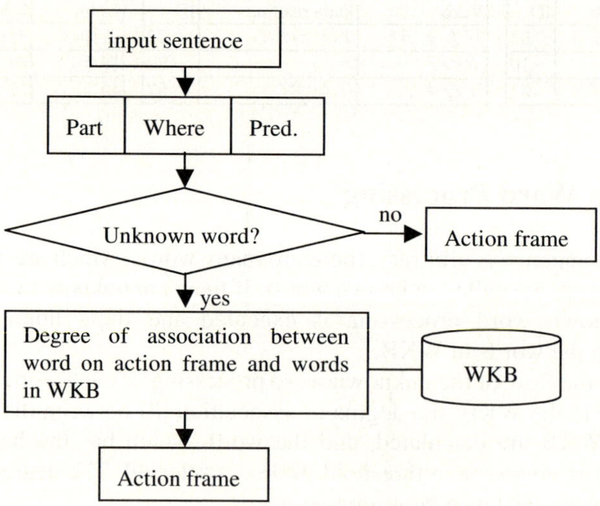

Fig. 5. Unknown word processing

bird {(egg 2) (feather 1) (wing 3) (hunt 1)}

egg {(female 3) (animal 1) (oval 2) (bird 1)}

oval {(egg 2) (shape 3) (ellipse 1)}

.

.

.

Fig. 6. Example of the concept-base

8 Action Generator

After action frame (Part, Where, Predicate) is decided, joint parameters for indicated action are made by using Basic Action KB or Complex Action KB. These knowledge bases are constructed by human. In the Basic Action KB, 11 joint parameters are defined for each two (Part, Where) or three (Part, Where, Pred.) words of action frame (Table 2). These 11 joint parameters generate a single action. In the Complex Action KB, combinations of basic actions or complex actions are defined for two words of action frame (Part, Pred.). These actions generate some consecutive basic actions. Furthermore, in the Complex Action KB, another types of actions are defined such as ID 25 in the Table 3. In this case (Other = "I am happy"), although there are no Part word nor Predicate word, it is possible that the action is defined using combination of the other actions.

Table 2. Basic Actions KB

ID	Part	Where	Pred.	J1	J2	...	J11
1	右手	上		5	-5	...	0
2	右手	下		0	-5	...	0
18	頭	右	傾ける	0	0	...	5

Ji: Joint parameter

Table 3. Complex Actions KB

ID	Part	Pred.	Other	Combination of Basic Action
2	右手	振る		3,4,repeat
10	頭	振る		22,23,repeat
17	両手	上げる		1,5
25			うれしい	3,21,7

9 Dynamic Generation of Joint Parameters

Basically, the one pose of the robot is decided by 11 joint parameters, which are defined by hand-coded. However, there are the cases that the position and orientation of the tip of hand are unknown in advance. In these cases, it must derive 4 joint parameters of the arm after the position and orientation of the tip of a hand is fixed. In this paper, 4 joint parameters of each arm corresponding to the position and orientation of the tip of a hand are derived by genetic algorithms (GA) [4].

Fig. 7 shows an example of static and dynamic generation of 4 joint parameters of right arm. The given pose is 'put right hand on the right eye'. When the right eye is in the original position, there is no problem. However, when the right eye (the head) is moved from the original position, since the objective position of the tip of a hand is different, 4 joint parameters of the right arm must be changed. By dynamic generation of joint parameters using GA is worked well.

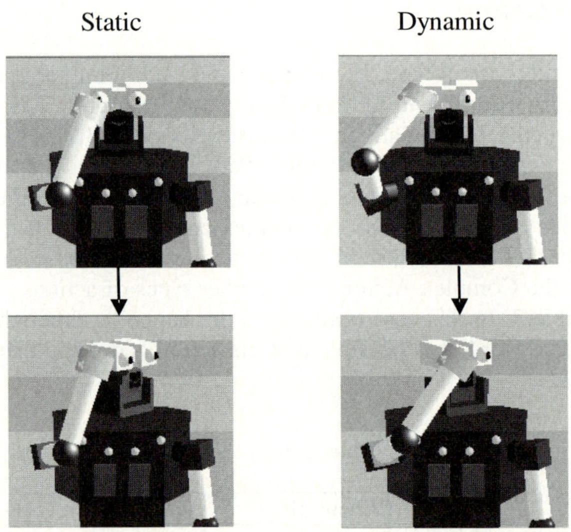

Fig. 7. Static and Dynamic generation of joint parameters

10 Evaluation of the System

The proposed system is evaluated by using 112 input sentences, which correspond some actions. The robot acts using action data that the system generates, and humans judge that the action is correct, wrong, or hard to say which. As a result, the rate of correct actions is 61%; the rate of wrong actions is 32%.

11 Conclusions

In this paper, the method to generate the action of the humanoid robot by natural language was proposed. Proposed system consists of a semantic meaning understanding module that arrange input sentence, the knowledge base of word related to actions, the knowledge base of the parameter for robots to act, and association mechanism to handle a word which is not known. By the evaluation, from 61% of input sentences, the robot made correct actions.

Acknowledgements

This work was supported with the Aid of Doshisha University's Research Promotion Fund.

References

1. H. Ishiguro, T. Ono, M. Imai, and T. Kanda: Development of an Interactive Humanoid Robot "Robovie" – An interdisciplinary approach, Robotics Research, STAR 6, pp.179-191, Springer-Verlag, 2003.
2. K. Kojima, H. Watabe and T. Kawaoka: Concept-base Refining with Thesaurus and Logical Relations for a Word Association-system, *Proc. of KES2001*, Part 2, pp.1590-1594, 2001.
3. H. Watabe and T. Kawaoka: The Degree of Association between Concepts using the Chain of Concepts, *Proc. of SMC2001*, pp.877-881, 2001.
4. D. E. Goldberg: *Genetic Algorithms in Search, Optimization and Machine Learning*, Addison-Wesley, 1989.
5. Y. Shinohara, H. Watabe and T. Kawaoka: A Conversation Semantic Understanding Method using the Commonsense Judgment System, *IPSJ SIG Notes*, NL-153, pp.89-96, 2003.

Robust Positioning a Mobile Robot with Active Beacon Sensors

JaeMu Yun, SungBu Kim, and JangMyung Lee

Department of Electronics Engineering, Pusan National University
San 30 Jangjeon-dong Kumjeong-ku, Busan, 609-735, Korea
{yunjaemu, zato, jmlee}@pusan.ac.kr
http://robotics.ee.pusan.ac.kr

Abstract. With the development of service robots and with the emerging concept of the ubiquitous world, localization of a mobile robot has become a popular issue. Even though several localization schemes have been previously, none of them is free from the economical constraints placed on commercial robots. To implement a real time commercial localization system, a new absolute position estimation method for a mobile robot in indoor environment is proposed in this paper. Design and implementation of the localization system comes from the usage of active beacon systems each of which is composed of an RFID receiver and an ultra-sonic transmitter. The RFID receiver gets the synchronization signal from the mobile robot and the ultra-sonic transmitter sends out a short-term signal to be used for measuring the distance from the beacon to the mobile robot. The position of a mobile robot in a three dimensional space can be calculated basically from the distance information from three beacons which have their own absolute position information. One of the interesting problems comes from the static localization scheme. That is, the collision avoidance of surrounding objects can be planned and controlled efficiently if the velocity of the moving object can be estimated. The main contribution of this paper is developing a dynamic localization scheme to be used for the moving objects, which is a new application in the field of active beacon sensors. In the dynamic localization scheme, sensor fusion techniques and extended Kalman filter have been used to improve the estimation accuracy. Experimental results demonstrate the effectiveness and feasibility of this scheme.

1 Introduction

Robot positioning, or the process of a robot finding its position in an unknown environment, is a major concern in mobile robot navigation. This process is crucial for a home or service robot because most robot operations require precise control of position. Mobile robots can be applied in various fields from home and office to the industry because they can conduct various missions with the two degrees of freedom motion. Particularly, mobile robots are inevitable for the tasks in dangerous and hazardous environments such as space and atomic plant. In many tasks assigned to mobile robots, the environment is partially known or inherently uncertain. This

B. Gabrys, R.J. Howlett, and L.C. Jain (Eds.): KES 2006, Part I, LNAI 4251, pp. 890 – 897, 2006.
© Springer-Verlag Berlin Heidelberg 2006

uncertainty also arises in modeling and planning of the mobile robot itself as well as the environment.

Accurate position estimation is essential for a mobile robot, especially in an unknown environment. The conventional dead-reckoning method has the problem of accumulating wheel slippage error. Therefore this method is not suitable for object carrying operations where a precise absolute position is required. Therefore, to prevent the error from being accumulated, many researchers have felt the necessity of the absolute position estimation. Robot positioning can be classified into two fundamental methods: absolute and relative positioning schemes. Relative positioning is usually based on dead reckoning (*i.e.*, measuring the wheel rotation angle to compute the offset from a known starting position). Dead reckoning is simple, inexpensive, and easy to implement in real time. The disadvantage of dead reckoning is its unbounded accumulation of errors. Absolute positioning schemes usually rely on navigation beacons, active or passive landmarks, map matching, or satellite based navigation signals [1].

When a robot does not know its initial position and orientation, the cost to find autonomously the position and orientation is very high. GPS (Global Positioning System) is one of the economical solutions for this problem [2]. But GPS is inaccurate for navigating mobile robots and is not accessible in a building. An indoor GPS system has been introduced, which consists of pseudo satellites and receivers to sense the relative position and attitude of a vehicle in a laboratory environment [3]. However, the system is too complex and expensive to be used for robotic applications.

Instead of pseudo satellites, beacon systems provide an economical solution for the indoor environment [4,5]. In this paper, as a kind of active beacon, it is proposed to use Active Beacon System (ABS) that consists of a radio frequency (RF) receiver and an ultrasonic transmitter. A mobile robot can select a specific beacon which has its own ID and position information during the navigation by sending a desired beacon code in RF. When a beacon receives its own ID from the robot, it sends back ultrasonic wave to measure the distance from the beacon to the robot based on the time of freight. Using the relative position information from the known beacons, the robot can estimate its own position. As there are so many error resources in measuring the position of a moving mobile robot, the Kalman filter is needed to obtain a more accurate estimated position.

The composition of this paper is as follows: First, section 2 explains the operation of ABS system. In section 3, the Kalman filter algorithm for position estimation of a mobile robot system is described. In section 4, computer simulations and experimental results to verify efficiency of system are shown. Finally, section 5 concludes this paper.

2 Dynamic Localization

It is possible to use different operating frequencies of ultrasonic generators in order to distinguish each signal for ultrasonic receiver modules. However in general it is not feasible since the total system cost becomes very high when each of the ultrasonic transmitter/receiver sensor has a different operating frequency. In this paper, a simple RF receiver module is added to each ultrasonic generator in order to solve the problem. That is, a simple RF receiver module is added to distinguish each ultrasonic generator. An RF transmitter module is installed on a mobile robot to invoke one of the ultrasonic generators as desired.

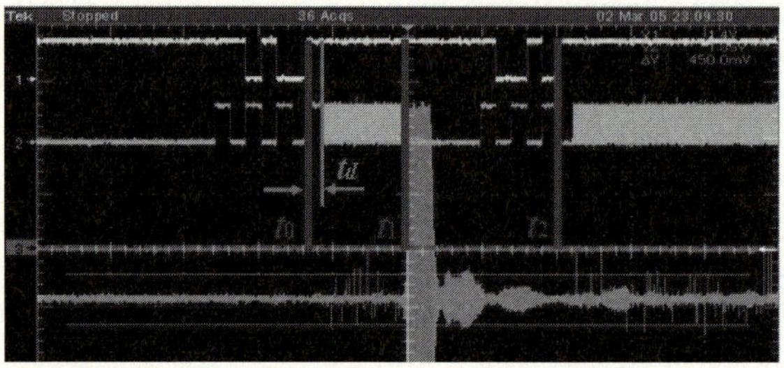

Fig. 1. Ultrasonic sensor and RFID transmitter timing

Figure 1 illustrates the localization procedure for the mobile robot. To initiate the localization process, a mobile robot transmits an RF signal to beacon #1 at time t0. When the beacon #1 receives the RF signal corresponding to its own ID, it sends out an ultrasonic signal to the mobile robot, which reaches the robot at time t1. Using the known traveling speed of the ultrasonic signal and the traveling time, t1-t0, the distance from the beacon #1 to the mobile robot can be calculated, which can be denoted as d1. While the mobile robot is traveling in the room, the distances to four beacons, d1, d2, d3, and d4, are measured continuously.

The distance from a beacon to the mobile robot can be defined in terms of mobile robot and sensor parameters as follows:

$$s = t_1 - t_0 - t_d \tag{1}$$

where t_d is the delay time in the beacon.

$$v = 331.5 + 0.6 \times T_a \tag{2}$$

where T_a is the room temperature in Celsius. Then the distance, d , between the beacon to the mobile robot can be calculated as

$$d[m] = v[m/s] \times s[\text{sec}] \tag{3}$$

To simplify and improve the filtering process, out-of-range data are pre-filtered out based on the following heuristics:

(1) When $d_i > d_{\max}$ or $d_i < d_{\min}$, discard the d_i , where $d_{\max}$ and $d_{\min}$ is the possible maximum and minimum distances, respectively, which can be estimated for a given room.

(2) When the distance between two consecutive ultrasonic signals is less than $d_{\min}$, discard the later one that can be considered as a reflected one.

When the mobile robot is stationary and there is no obstacle blocking the ultrasonic signal between a beacon and the mobile robot, the localization process can be done properly and accurately by utilizing three or four beacons. However, in real situations, there are several obstacles in the room, which prevents the mobile robot from receiving the beacon signal. Also, when the mobile robot is moving fast, the sequential operation of receiving the ultrasonic signals from the beacons one by one is not fast enough to localize the mobile robot.

As a main contribution of this paper, a dynamic localization scheme is proposed.

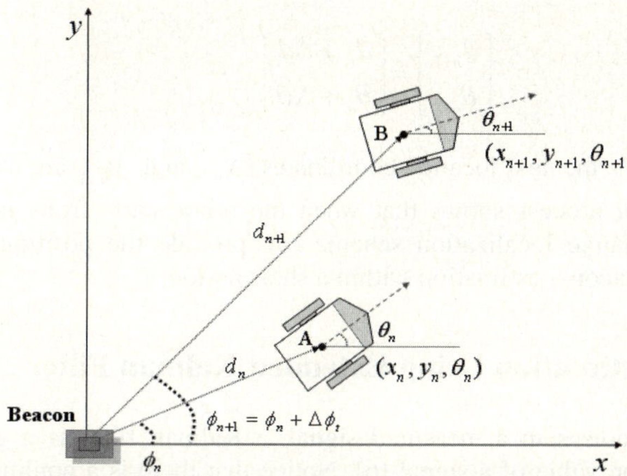

Fig. 2. Linear incremental algorithm

In Fig. 2, the initial location of the mobile robot is represented as, $P=[x_n, y_n, \theta_n]^t$, and the command velocity of the mobile robot, v, is defined as $v=[\dot{x}_n, \dot{y}_n]$. Note that in determining the location, it is assumed that the distance from the mobile robot to the beacon, d_n, is known *a priori,* and the distance information from a beacon, d_{n+1}, is measured by the mobile robot. Therefore, the position of the mobile robot can be represented as

$$\begin{bmatrix} x \\ y_n \end{bmatrix} = \begin{bmatrix} d_n \cos\theta_n \\ d_n \sin\theta_n \end{bmatrix}. \tag{4}$$

Differentiating both sides *w.r.t* time, the velocity can be represented as

$$\begin{bmatrix} \dot{x}_n \\ \dot{y}_n \end{bmatrix} = \begin{bmatrix} \cos\theta_n & -d_n \sin\theta_n \\ \sin\theta_n & d_n \cos\theta_n \end{bmatrix} \begin{bmatrix} \dot{d} \\ \dot{\theta} \end{bmatrix}. \tag{5}$$

This equation can be transformed into a discrete system equation by replacing $\dot{d}$ as $\Delta d / T$.

$$\frac{1}{T}\begin{bmatrix}\Delta d\\ \Delta\theta\end{bmatrix}=\begin{bmatrix}\cos\theta_n & \sin\theta_n\\ -\frac{1}{d_n}\sin\theta_n & \frac{1}{d_n}\cos\theta_n\end{bmatrix}\begin{bmatrix}\dot{x}_n\\ \dot{y}_n\end{bmatrix} \tag{6}$$

where $\Delta d = d_{n+1} - d_n$ can be obtained by the distance measurement to the beacon and T represents the sampling period for the distance measurement. Plugging the known d_n and θ_n into Eq. (6), $\Delta\theta$ can be obtained. Therefore, the new location can be estimated as

$$\begin{bmatrix}d_{n+1}\\ \theta_{n+1}\end{bmatrix}=\begin{bmatrix}d_n+\Delta d\\ \theta_n+\Delta\theta\end{bmatrix}. \tag{7}$$

Using equation (4), the next location coordinates, x_{n+1} and y_{n+1} are obtained.

This derivation process shows that when the robot starts from an initial known position, the dynamic localization scheme can provide the position —localization using only one beacon—estimation within a short period.

3 Position Estimation Using Extended Kalman Filter

When there are noises in a measured signal, a Kalman filter is a general tool for estimating the true value of a signal [6]. Notice that there is a nonlinear relationship between the measured distance and global coordinates of a mobile robot in this ABS system. Therefore, an extended Kalman filter for the nonlinear system is applied for this navigation to improve the estimation accuracy [7].

A mobile robot state vector, r, consists of the 2 dimensional position, (x, y), and a heading angle, θ. The state vector of a well-known point model can be defined as follows:

$$r_{k+1}=\begin{bmatrix}x_k+Tv_k\cos\theta+w_{1,k}\\ y_k+Tv_k\sin\theta+w_{2,k}\\ \theta_k+T\omega_k+w_{3,k}\end{bmatrix} \tag{8}$$

where T is the sampling period, $r=[x, y, \theta]^T$ is a state vector, $w_k=[w_{1,k}, w_{2,k}, w_{3,k}]^T$ is Gaussian noise with the 0 mean and Q variance, that is, $w_k \sim N(0,Q)$, and the subscript k means time-step. With this state definition, a control input, the commanded motion of the mobile robot, is defined as, $u_k=[v_k,\omega_k]^T$, where v_k is the linear and ω_k is the angular velocity. Note that the floor for the mobile robot is assumed to be flat.

Now, the ultrasonic observation equation is defined as

$$z_k = h(r_k, n_k) \tag{9}$$

where the measurement noise, n_k, is modeled as Gaussian, that is $n_k \sim N(0, G)$ and the function $h(r_k, n_k)$ is described as follows:

$$h(r_k, n_k) = r_k + n_k. \tag{10}$$

Now, we can describe an extended Kalman filter algorithm to estimate the state vector of a mobile robot as follows:

$$\hat{r}_{k+1}^{-} = f(\hat{r}_k, u_k, w_k) \tag{11a}$$

$$P_{k+1}^{-} = A_k P_k A_k^t + Q_k \tag{11b}$$

$$K_k = P_k^{-} H_k^{\,t} \left(H_k P_k^{-} H_k^{\,t} + G_k\right)^{-1} \tag{11c}$$

$$\hat{r}_k = \hat{r}_k^{-} + K_k \left(z_k - h(\hat{r}_k^{-}, n_k)\right) \tag{11d}$$

$$P_k = (I - K_k H_k) P_k^{-} \tag{11e}$$

where $\hat{r}_k^-$ and $\hat{r}_k$ represent state values of a mobile robot before and after correction, respectively, and K_k is a Kalman filter gain. Also P_k^- and P_k are error covariance matrices defined as follows:

$$P_k^- = E[(r_k - \hat{r}_k^-)(r_k - \hat{r}_k^-)^t] \tag{12a}$$

$$P_k = E[(r_k - \hat{r}_k)(r_k - \hat{r}_k)^t]. \tag{12b}$$

Matrices A_k and H_k are derived as Jacobian matrices about the state vector r of each nonlinear functions $f(\cdot)$ and $h(\cdot)$:

$$A_k = \frac{\partial f}{\partial r}(\hat{r}_k, u_k, 0) = \begin{bmatrix} 1 & 0 & -v_k \sin\theta_k \\ 0 & 1 & v_k \cos\theta_k \\ 0 & 0 & 1 \end{bmatrix} \tag{13}$$

$$H_k = \frac{\partial h}{\partial r}(\hat{r}_k, 0) = I_k. \tag{14}$$

4 Experiments

From real experiments, it is intended to show the possibility of dynamic localization. Assuming that only one beacon datum is reliable, the dynamic localization scheme is derived. This actually represents the dynamic situation where multiple distance data can not be used for localization since each datum is obtained at a different location.

For the experiments, a mobile robot is developed using DSP2406 as shown in Fig. 3.

Fig. 3. An experimental mobile robot

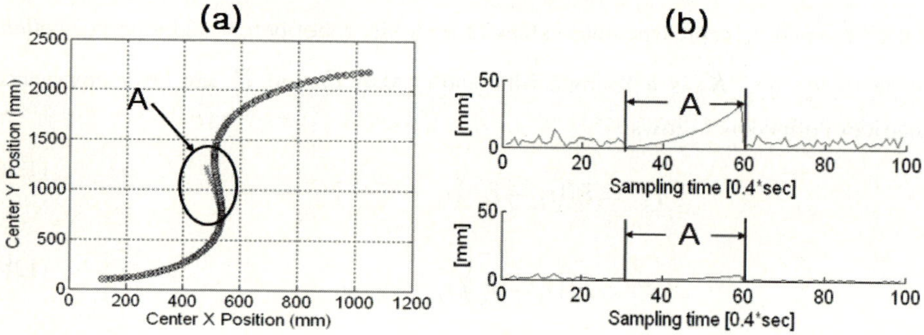

Fig. 4 (a). Trajectory of a mobile robot. **(b).** Error between real value and estimate value of a mobile robot.

Figure 4(a) illustrates a trajectory of the mobile robot. In the middle of the trajectory, region A, the mobile robot is moving fast and only one beacon is effective for the localization. Upper one of Fig. 4(b) shows the position estimation error without Kalman filtering. It shows that the estimation error increases rapidly with the time in the region A. However as it is shown in the lower one of Fig. 4(b), by applying the Kalman filtering, the estimation error can be kept small –less than 5 mm—for more than 25 sampling periods.

5 Conclusion

A new localization scheme for a mobile robot that is navigating indoor environments has been developed, which utilizes an ultrasonic signal for distance measurement and utilizes RF signals for synchronization. The general structure of the equipment and the algorithm for calculating a position have been described and experimentally verified. Characteristics of the error sources have been closely analyzed to achieve a cheap and accurate localization system. The motion of a mobile robot, which is one of the main error sources for the localization, has been estimated and utilized to overcome the sources of error. This enables dynamic localization of the mobile robot, which is definitely required for practical mobile robots where speed becomes a higher and higher requirement. A future direction will be to investigate communication capability among sensors [8], which is not limited by the environment.

Acknowledgement. This work was supported by "Research Center for Logistics Information Technology (LIT)" hosted by the Ministry of Education & Human Resources Development in Korea.

References

1. J. Borenstein and L. Feng, “UMBmark – A Method for Measuring, Comparing, and Correcting Dead-reckoning Errors in Mobile Robots,” The University of Michigan, 1994, Technical Report UM-MEAM-94-22.
2. Iowa State University GPS page. Web site at http://www.cnde.iastate.edu/gps.html.
3. Zimmerman, et al. “Experimental Development of an Indoor GPS Based sensing system for Robotic Applications,” Navigation, 1997, vol 43, No. 4, pp. 375-395.
4. L. Kleeman, “Optimal estimation of position and heading for mobile robots using ultrasonic beacons and dead reckoning,” IEEE International conference on Robotics and Automation, May 1992, pp. 2582-2587.
5. Jae. H. Kim and Hyung S. Cho, “Real time determination of a mobile robot`s position by linear scanning of a landmark,” Robotica, 1992, vol 10, pp. 309-319, Cambridge university press.
6. Kyoung C. Koh, Jae S. Kim and Hyung S. Cho, “A position estimation system for mobile robots using a monocular image of a 3-D landmark,” Robotica, 1994, vol 12, pp. 431-441, Cambridge university press.
7. T. Arai and E. Nakano, “Development of measuring equipment for location and direction (MELODI) using ultrasonic waves,” Trans. ASME, Journal of dynamic systems, Measurement and control, vol 105, pp. 152-156, 1983.
8. C.M. Clark, S.M. Rock, and J.C. Latombe, “Motion Planning for Multiple Mobile Robot Systems Using Dynamic Networks,” IEEE Int. Conference on Robotics and Automation, Taipei, Taiwan, 2003.

The Implementation of an Improved Fingerprint Payment Verification System

Woong-Sik Kim [1,2] and Weon-Hee Yoo[2]

[1] Department of Electronics & Information Engineering Konyang University #119 Nae-dong, Nonsan, Chungnam, South Korea
wskim@konyang.ac.kr
[2] Department of Computer Science & Information Engineering Inha University #253 YonhHyun Dong, Nam Ku, Incheon, South Korea
whyoo@inha.ac.kr

Abstract. In the world security market, the size of the biometric industry is gradually increasing, of which the fingerprint verification is recognized of its superiority and marketability in terms of relatively low cost, light weight and easiness to input bio-information, compared to other biometric methods[1,2].

This paper proposes a method to utilize the strengths of the fingerprint verification for credit card dealing. The example of the utilization of the credit card payment is the model business of fingerprint verification credit card payment applied for three months in large domestic marts in Korea, to discuss the possibility to settle the non-medium credit trading by the biometric technology and the applicability to the financial services.

1 Introduction

Non-medium payment is a new payment method eliminating credit cards or cash cards in the dealing process by the bio-information to get rid of the inconvenience to carry a medium and the risk from loss[3,4,5,6].

For the first use of the service in a credit card company, a bank or a store, the registration by the user of his or her fingerprint information together with personal and credit card information enables the payment only with the bio-information such as fingerprint and the residential number or password without any medium in the next credit transaction.

The introduction of such non-medium payment began in the stores where the risk of loss of the medium is high, the dealing time is required to be reduced and they have high frequency of transaction. The model service was executed in the credit card company which secures the highest number of clients in Korea.

The model test of the non-medium payment is defined to be used only by the client who has registered in the discount store in question. It is firstly used because the clients of the discount store have low probability to use another discount store, an secondly because the database can be integrated when it will be expanded to the nationwide network after the model service.

B. Gabrys, R.J. Howlett, and L.C. Jain (Eds.): KES 2006, Part I, LNAI 4251, pp. 898 – 905, 2006.
© Springer-Verlag Berlin Heidelberg 2006

1.1 System Configuration

The credit dealing system in the past has used POS system (POS: Point-of-sale) to make the calculation of the sales in a member store and to make payment with cash or credit card. For the payment by a credit card, they use the card reader of POS system to read the credit card, to request the authorization of the card company or the bank, and as authorized, to receive the authorization results.

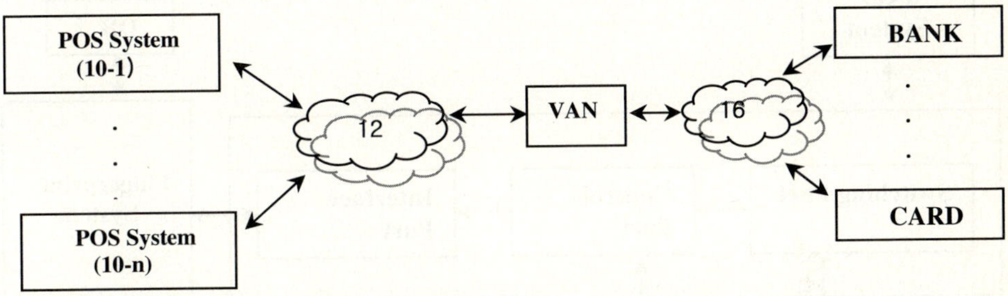

Fig. 1. Block Diagram of Rough Configuration of Credit Dealing System

<Fig.1> is a block diagram of rough configuration of the credit dealing system in the past, which is composed of POS systems (10-1 ~ 10-n), networks (12, 16), VAN (Value-Added Network) companies, credit card companies and banks.

POS systems are installed in the member stores, by which the sellers acquire the related data to the sales of products and the payment is made through cash or credit card. For the payment by credit card, they use the card reader of POS system to read the card, to transfer through the network and to receive the authorization information through the network . A VAN company transfers to the network the dealing authorization request (user info, dealing details and member store info etc.) input by the network, and then transfers to the network the authorization information input by the network. A credit car company or a bank compares the dealing authorization request input by the network with its own data to transfer the authorization information through the network. The network used among VAN companies, credit card companies and banks is a private network, and the network used between POS systems and VAN companies can be wired or wireless Internet or a private network.

1.2 Interface Device Between POS and Fingerprint Identification System

This chapter explains the interface device and its method between POS system and fingerprint identification system and the credit dealing system made by it. This device is composed of the switching part which transfers to the POS system the input signal from the keyboard of POS system, or ignores the input signal of the keyboard and transfers to the POS system the input signal from the fingerprint identification system, the control part which prints out the input signal from the keyboard to the fingerprint identification system, or when a signal is input from the fingerprint identification system, controls the switching part and transfers to the switching part the input signal from the fingerprint identification system, and the interface part for the signal transfer between the control part and the fingerprint identification system. Therefore, it is possible to integrate the fingerprint identification system to the existing POS system to enable the data transfer between them with only one keyboard. By using this device, it is possible to graft the biometric system without changing the existing POS system.

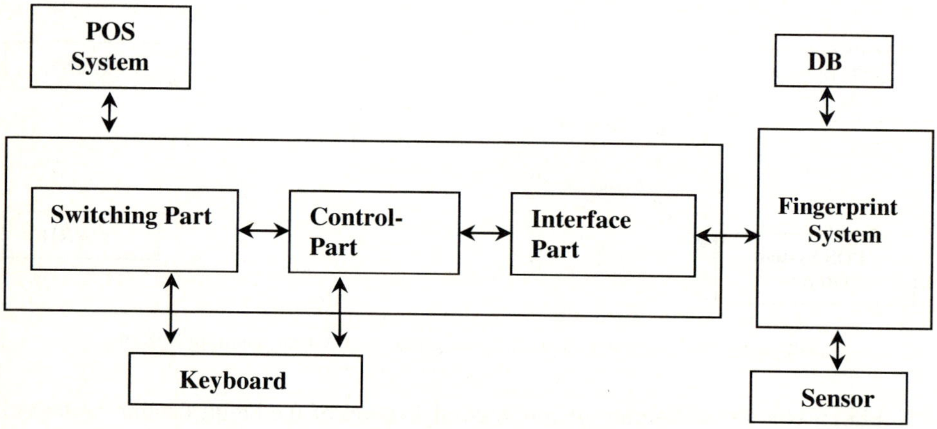

Fig. 2. System Configuration

By using this interface device, it is configured not to change the existing POS program and its operating system, with grafting the fingerprint identification system to the existing system without possibility of an error in financial dealings, so as to permit the consumers to more easily utilize the non-medium payment.

The followings are the operating method. First, assuming that a fingerprint has already been registered, if the user inputs his or her fingerprint without showing the credit card, the interface device sends this to the local verification device, and then if the authorization is made, it transfers to the POS system the customer's card information saved in the database in the same manner to read the card. Since it transfers the already-registered card number, a clerk who is not familiar to the credit card can more easily make billing and also the consumer is not required to show up the card from the wallet, resulting higher security and convenience of the card.

And the strong point of this system is the possibility to register also the point card of a discount store, so as to enable the point accumulation only with one input of the fingerprint, meaning that it will be very properly and usefully utilized in the discount stores which require shorter dealing time.

2 Fingerprint Verification Payment System

2.1 Fingerprint Registration and Matching Procedure

The user should pass the registration process when he or she needs the non-medium dealing in a discount store. The registration process is as follows;

(1) Registration of Credit Card Information (credit card, bonus card and ID card required)
- Swipe registration and dissemination of the credit card info using the card reader
- Registration and dissemination of the bonus card info using barcode reader
- Grant of an arbitrary group number from the server after registration
(2) Fingerprint Registration
- Continuous inputs of three times on the fingerprint sensor to register temporarily and check double application
- Verification test of the fingerprint to be registered and then transferring to the central server for registration
(3) Modification and Deletion
- Verification of the customer and requesting the server for modification or deletion of the customer fingerprint and info

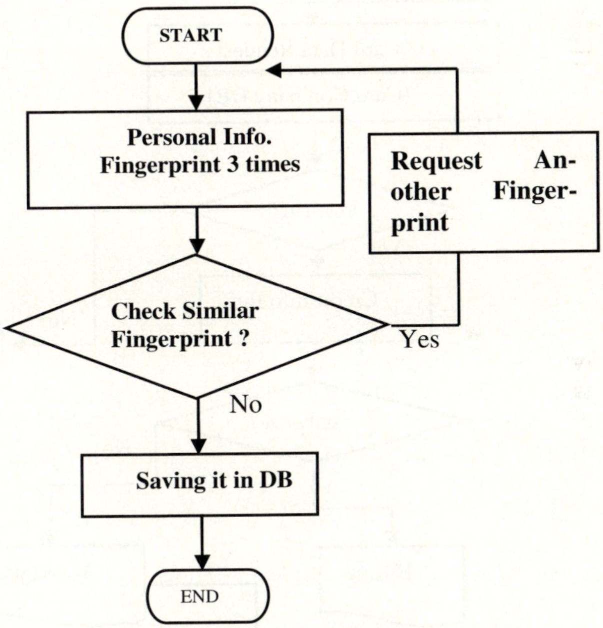

Fig. 3. Procedure of Fingerprint Registration

The fingerprint matching procedure is as follows;

(1) POS communication module installed to maintain the process of the existing POS system : Processing the card scanning input as a fingerprint verification input
(2) Fingerprint Verification Method : Inputting the customer's fingerprint group number at the credit dealing input screen of POS
- Inputting a specific key in POS to request fingerprint verification
- Inputting once on the fingerprint sensor to confirm the verification
- Receiving the credit info of the customer from the server and transferring to POS

- Using the existing authorization system of POS to complete the dealing
- Saving in a log the completed authorization information.

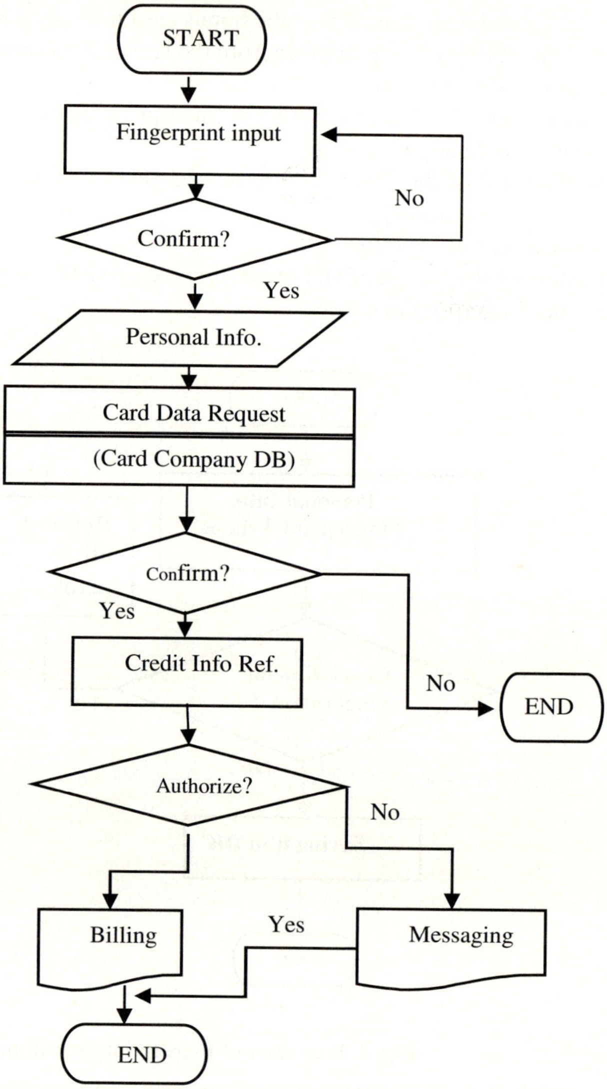

Fig. 4. Fingerprint Matching Procedure

2.2 Algorithm of Credit Dealing System

Three templates of fingerprint data are saved as one unit in order to reduce FNMR rate which was heightened as the FMR was adjusted as 0% under the characteristics of the credit dealing authorization system. By this, having reached 3.86%, the FNMR rate is reduced to 0.1% (FMR=0). The fingerprint matching is classified as 1:1 matching and 1:N matching, and this system applies 1:N matching[7,8].

Because compared to 1:1 matching, the 1:N matching requires shorter time to match the input templates with all the saved fingerprint templates, the Classification & Search Method which is developed on the basis of the actual user data affects the swiftness and accuracy of the matching results[5,6,11].

3 Experiments and Results

3.1 Fingerprint Verification System

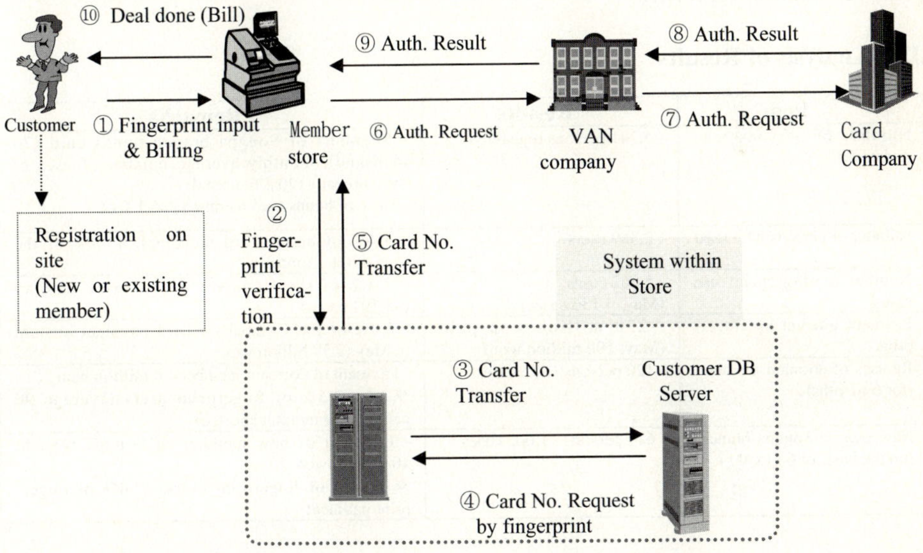

Fig. 5. Describes the Configuration of Fingerprint Verification System

3.2 Operating Results

Item	Results	Remarks
Number of persons who registered fingerprint	- Totally 5,341 persons - Average 137 persons per day	- This number decreases as time goes by - Now average 80 persons per day
Number of payment cases by fingerprint	- Totally 5,062 cases - Average 130 cases per day	- This number increases as time goes by - Now average 170 cases per day
Payment amount by fingerprint	- Totally 230 million won paid - Average 5.9 million won per day	- This number increases as time goes by - Now average 7.7 million won per day
Number of users & payment per customer	- Totally 2,389 users - Average 45 thousand won per customer	- 45% of the fingerprint-registered persons - The payment per customer hardly changes as time goes by

3.3 System Configuration

- Fingerprint Verification Server Installation within Store
• Info to be saved: group number, fingerprint template, residential number, name, credit card info and bonus card information.
• Use of the network within store, permitting cross verification of many stores (One server can combine many stores)
- 1:1 relation between respective POS within store and Fingerprint Verification Memory Pack
• Info saved in Memory Pack: group number, fingerprint template, credit card info and bonus card information.
• Designed to allow fingerprint payment despite troubled communication with fingerprint verification server[5].

3.4 Analysis of Results

Item	Results	Remarks
Number of persons who registered	- 5,341 persons registered	- Members of Songpa branch bonus card (70 thousand), monthly average number of uses of the Company (20.5 thousand) - 7.6% of bonus card members, 26.1% of
Number of persons who used	- 2,389 users	- 3.4% of bonus card members, 11.7% of the uses of the Company
Number of fingerprint use cases	- 5,062 cases (May: 4,136 cases)	- 7.5% of the Company's card uses in May (55,191 cases)
Payment amount by fingerprint	- 230 million won paid (May: 190 million won)	- 6.8% of the Company's card payment amount in May (2.77 billion won)
Re-uses of dormant member (for 6 months)	- 120 persons / 215 cases	- Payment of dormant members: 8 million won ※ about 5% of fingerprint users (average 90 persons per month expected)
New uses in Songpa branch (on the basis of 6 months)	- 611 persons / 1,002 cases	- Payment of new member of Songpa branch: 40 million won ※ 26.6% of fingerprint users, 17.4% of fingerprint payment

4 Conclusion

FAR(False Accept Rate) and FRR(False Reject Rate) are in inverse proportion, while the minimization of false acceptance is more important for maintaining the high performance. The probability of false acceptance in fingerprint verification is about one millionth, but it is necessary to make preventive solutions and ex post facto measures.
< Precautions >
- Granting the group numbers in accordance with the fingerprint categories.
- To minimize misconception rate by grouping 5,000 registered fingerprint clients in one unit and by matching them within such unit.
- Securing exact information while registering the fingerprints.
< Post-factum countermeasures >
- Checking the bill after settling accounts and then inducing to register the bonus card
- To confirm the transaction processing in real-time by sending SMS(Short Message Service by Mobile Phone).

References

1. Dario Mario, and David Maltoni, "Direct Gray-Scale Minutiae Detection In Fingerprints", IEEE Trans. Pattern Analysis and Machine Intelligence, vol. 19, no. 1, , pp. 27-39, Jan., 1997.
2. Lin Hong, Yifei Wan, and Anil Jain, "Fingerprint Image Enhancement: Algorithm and Performance Evaluation", IEEE Trans. Pattern Analysis and Machine Intelligence, vol. 20,no. 8,pp. 777-789,Aug., 1998.
3. A.K.Jain, L. Hong, R. Bolle, "On-Line Fingerprint Verification", IEEE Trans., Pattern Analysis and Machine Intelligence, vol. 19,no. 4,pp. 302-313, April., 1997.
4. A.K. Jain, S.Prabhakar, L. Hong, and S. Pankanti, "Filterbank-based fingerprint matching", IEEE Trans. On Image Processing. 9(5):846-859, May 2000.
5. Nalini K, and A.K. Jain, "A Real-Time Matching System for Large Fingerprint Database", IEEE Trans., Pattern Analysis and Machine Intelligence, vol. 18,no. 18,pp. 799-813, Aug., 1996.
6. Xudong Jiang and Wei-Yun Yau, "Fingerprint Minutiae Matching Based on the Local and Global Structures", International conference on pattern Recognition(ICPR'00)-Vol 2.,no. 18,pp. 1038-1042, Sep., 2000.
7. T. MAEDA, M. Matsushida and K. Sasakawa, "Identification Algorithm Using a Matching Score Matrix", IEICE Trans. INF. & SYST., Vol.E-84-D,no. 7, pp. 819-824, July 2001.
8. Xudong Luo, Jie Tian and Yan Wu, "A Minutia Matching algorithm in Fingerprint Verification", International conference on pattern Recognition (ICPR'00)-Vol 4., no. 18,pp. 4833-4841, Sep., 2000.
9. J. D. Stoszand L. A. Alyea, "Automated Systems for Fingerprint Authentication Using Pores and Ridge Structure," Proceedings of SPIE, Automatic Systems for the Identification and Inspection of Humans (SPIE Vol. 227 7), San Diego, 1994, p. 210- 223.
10. W.-S.Kim, W.-H.Yoo, "A Layered Fingerprint Method", LNCS3546, pp.702-709, 2005.
11. W.-S.Kim, W.-H.Yoo, "A Method For Acquiring Fingerprint by Linear Sensor", LNCS 3684, pp. 410-416, 2005.

Capacity Planning for Scalable Fingerprint Authentication

Byungkwan Park[1,3], Daesung Moon[2,3], Yongwha Chung[3,*], and Jin-Won Park[4]

[1] Department of Computer and Information Science, Sunmoon U., Korea
bkpark@sunmoon.ac.kr

[2] Biometrics Technology Research Team, ETRI, Korea
daesung@etri.re.kr

[3] Department of Computer and Information Science, Korea U., Korea
ychungy@korea.ac.kr

[4] School of Games, Hongik U., Korea
jinon@hongik.ac.kr

Abstract. As the fingerprint authentication technique becomes widely used at close range such as for the door lock, it will be applied for the remote user authentication as the next step. For large-scale remote user authentication services such as board control using bio-passport, the real-time as well as the security/privacy requirements should be satisfied. However, nothing is known about the collective performance of the fingerprint verification and the authentication protocol on the client-server model. In this paper, we analyzed the collective performance of the task assignment of the fingerprint authentication on the client-server model. To obtain the characteristics of the workload of both the fingerprint verification and the authentication protocol, we first consider three types of primitive operations on the client and the server; fingerprint verification module, cryptography module and communication module. Then, based on these primitive operations, the workload of each scenario of the task assignment was applied to the M/D/1 queueing model representing the client-server model, and the collective performance of each scenario was analyzed quantitatively. The modeling results showed that the server could handle the lowest level of workload when the server does decryption, verification and matching. On the contrary, the server can handle 9.57 times heavier workload when the clients do decryption, verification and matching. The modeling results also showed that the workload manageable by the server depends largely on the communication setup time.

Keywords: Performance Evaluation, Fingerprint Verification, Authentication Protocol.

1 Introduction

Traditionally, verified users have gained access to secure information systems, buildings, or equipment via multiple PINs, passwords, smart cards, and so on. However,

* Corresponding author.

B. Gabrys, R.J. Howlett, and L.C. Jain (Eds.): KES 2006, Part I, LNAI 4251, pp. 906–915, 2006.
© Springer-Verlag Berlin Heidelberg 2006

these security methods have important weakness that can be lost, stolen, or forgotten. In recent years, there is an increasing trend of using biometrics, which refers to the personal biological or behavioral characteristics used for verification[1].

The **fingerprint** is chosen as the biometrics for verification in this paper. It is more mature in terms of the algorithm availability and feasibility[2]. At present time, the personal authentication system is mainly used at close range, such as for in-house room entry control, access to safes, and systems operation. In the future, it will be widely applied and diversified, particularly for a variety of approvals and settlements over networks, information access control, e-commerce via Internet, and remote personal identification. Especially, we consider the traveler identification using the fingerprint in the board control systems. Furthermore, we consider a **client-server model**[3] for remote user authentication. That is, clients extract traveler's fingerprint information, perform the authentication by themselves, or send the data to the server and receive the authentication result from the server.

In this model, however, **security issues** ensure that the opponents will neither be able to access the individual information/measurements nor be able to pose as other individuals by electronically interjecting stale and fraudulently obtained biometric measurements into the system[2-6]. To guarantee the integrity/confidentiality of the fingerprint information transmitted between the client and the server, we employ The ANSI X.9.84 that is the security standard for biometric system.

The **ANSI X.9.84 Biometric Information Management and Security standard**[7] covers requirements for managing and securing biometric information – fingerprint, iris, voiceprint, etc. – for use in customer identification and employee verification, mainly for use in the financial industry. In addition, this standard identifies the digital signature and encryption to provide both integrity and privacy of biometric data. A comprehensive set of control objectives to validate a biometric system is also included. The standard was developed with other interested organizations including the BioAPI Consortium, the National Institute of Standards and Technology(NIST) Information Technology Laboratory's Common Biometric Exchange File Format Initiative, and the International Biometric Industry Association.

In this paper, to provide the board control service in real-time, we use the **three-way verification** method among the Machine Readable Travel Document(MRTD) biometric data, the live biometric data and the biometric data stored in the central DB which was recommended by ICAO Doc 9303 Part 3[8].

Under this environment where the X.9.84 is satisfied in the client-server model for remote user authentication using fingerprint by the three-way verification, the goal of this research is to examine what kind of work is most sensitive to guarantee the security/privacy as well as the real-time execution requirements. To derive an optimal task assignment for a given number of clients on the client-server model, we consider typical scenarios of the task assignment of the fingerprint verification and its corresponding authentication protocol to guarantee both the integrity and the confidentiality of the fingerprint information transmitted. To obtain the characteristics of the workload of both the fingerprint verification and the authentication protocol, we first measure the performance of primitive operations on the client and the server, respectively. Also, we examine the impact of the communication setup time on the total authentication time with different authentication methods. Then, based on these measured performance, the workload of each scenario of the task assignment is applied to the M/D/1 queueing model representing the client-server model because the request for user authentication can be assumed to happen in a random fashion, while the authentication process may be performed in deterministic fashion in a computer system, and the collective performance of each scenario is analyzed quantitatively.

The rest of the paper is structured as follows. Section 2 describes the authentication protocols based on the task assignment of the fingerprint verification for bio-passport service. The modeling methodology employed and results of performance evaluation are described in Section 3. Finally, conclusions are given in Section 4.

2 Fingerprint Authentication Scenarios for the Client-Server Model

In general, there are three steps involved in the verification phase[2]: Image Pre-Processing, Minutiae Extraction, and Minutiae Matching. Note that Image Pre-Processing and Minutiae Extraction steps require a lot of integer computations, and the computational workload of both steps occupies 96% of the total workload of the fingerprint verification[9].

As we explained in Section 1, we consider the client-server model for remote user authentication using fingerprint. Especially, we consider the board control system using bio-passport. The client-server model for remote user authentication using fingerprint must guarantee the security/privacy as well as the real-time execution requirements. To satisfy those requirements, we first consider possible scenarios for remote fingerprint authentication in terms of assigning the tasks of the fingerprint authentication to each entity(*i.e.*, client and server). Then, we evaluate the performance of each scenario.

2.1 Authentication Scenarios

To simplify the analysis, we make following assumptions;

1) Between the client and the server, the same master key is shared when the system is installed.
2) The entity authentication is completed using proper methods such as trusted Certificate Authority(CA).
3) The user authentication service is assumed to be requested by the client at which the board control investigator is working.
4) The execution times to perform some cryptography mechanisms to protect the fingerprint features stored in the bio-passport and in the server's database are not considered because this research focuses only on the protection of the fingerprint data transmitted.

In fact, these assumptions are reasonable, since the master key sharing operation(described in assumption 1) needs to be executed only once, and the time for protecting the stored fingerprint data(described in assumption 4) is negligible.

As we explained in Section 1, to analyze the workload of the board control system using the bio-passport, we use the three-way verification method among the biometric data stored in the bio-passport, the live biometric data and the biometric data stored in the central DB which was recommended by ICAO Doc 9303. For the purpose of explanation, we define first the following notations:

N : a nonce generated randomly in the client and used as a "challenge"
K_m : a master key shared by both the sensor and the client
$f(N)$: a simple function to generate a "response" for N
K_s : a shared session key generated for each transmission
P_Fe : a fingerprint feature stored in the bio-passport
L_Fe : a fingerprint feature extracted from the live fingerprint image
S_Fe : a fingerprint feature stored in the DB
L_Fi : a live fingerprint image
Mat_PL : a matching result between *P_Fe* and *L_Fe*
Mat_PS : a matching result between *P_Fe* and *S_Fe*

Scenario 1

In Scenario 1 as shown in Fig. 1, the sensor attached to the client captures a live fingerprint image, and the client extracts some features from the image. Then, the client sends *L_Fe* and *P_Fe* to the server after applying the encryption and digital signature with the same key received from the server.

After verifying the signature for *L_Fe* and *P_Fe* and decrypting these fingerprint features, the server performs two comparisons with *P_Fe* – *S_Fe* and *P_Fe* – *L_Fe*. After checking the two matching results, the server returns a final result to the client. Note that this is a typical scenario of assigning the fingerprint verification tasks to the client-server model and requires five times of the communications for the data transmission. This scenario can improve the security level of the fingerprint authentication system because the server can be more secure than the client. That is, compared to the server possibly protected by the security experts, the client maintained by an individual user is more vulnerable to several attacks such as Trojan horse[2]. On the other hand, the computational workload of the server in this scenario increases as the number of clients increases.

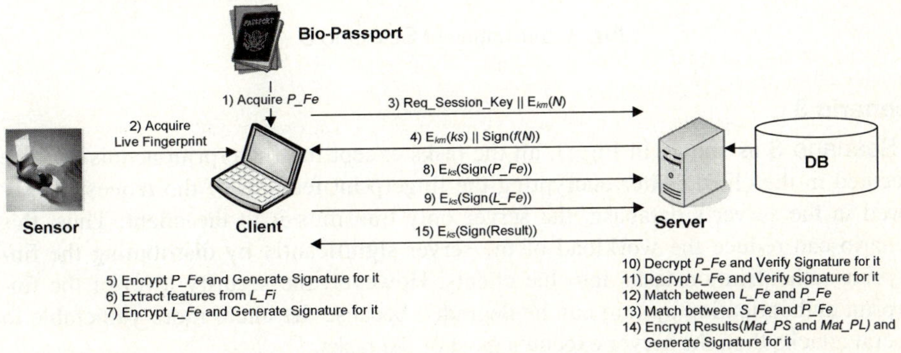

Fig. 1. Illustration of Scenario 1

Scenario 2

Unlike Scenario 1, in Scenario 2 as shown in Fig. 2, the comparison with *P_Fe – L_Fe* and *P_Fe – S_Fe* is executed in the client and the server, respectively. In this scenario, the client sends only *P_Fe* to the server and calculates the matching score between *Mat_PS* received from the server and *Mat_PL* resulted in the client.

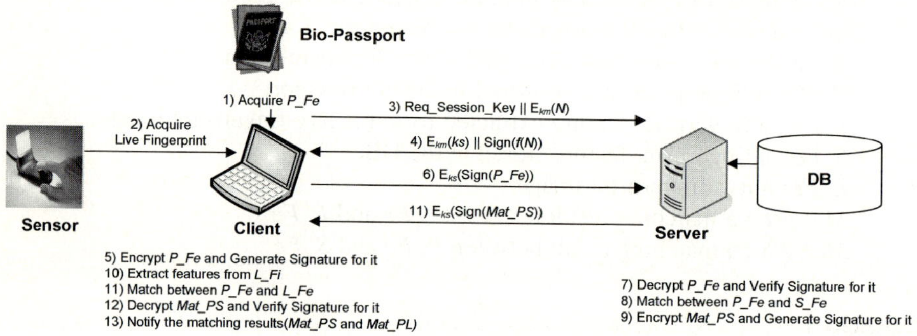

Fig. 2. Illustration of Scenario 2

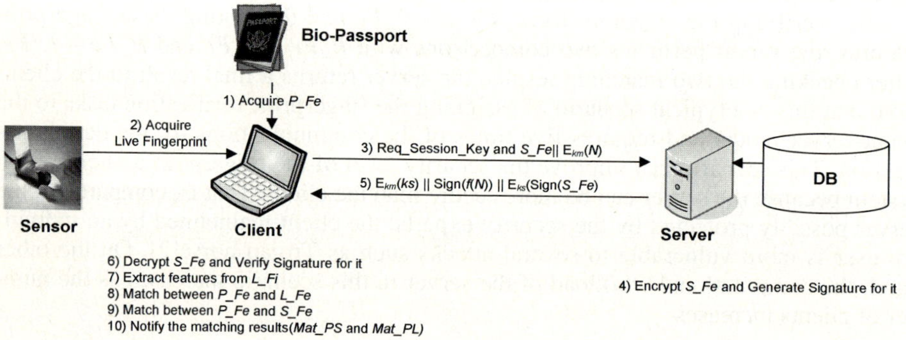

Fig. 3. Illustration of Scenario 3

Scenario 3

In Scenario 3 as shown in Fig. 3, all the tasks except the fingerprint acquisition are executed in the client. After encrypting the fingerprint features of the requested user stored in the server's database, the server only transmits it to the client. Thus, this scenario can reduce the workload of the server significantly by distributing the fingerprint authentication tasks into the clients. However, the security level of the fingerprint authentication system can be degraded because the client more vulnerable to several attacks than the server executes most of the tasks.

2.2 Configuration for Performance Evaluation

To evaluate the performance of each scenario, we assume the following configuration. First, the client and the server have `Pentium 4`(2 *GHz*) processor and `Xeon`(3 *GHz*) processor, respectively. We assume that the transmission speed of the Internet is 10 *Mbps*. To guarantee both the integrity and the confidentiality of the transmitted information, 128-bit `AES`, 1024-bit `ECC`, and `SHA-1`[10] are used as our symmetric encryption algorithm, digital signature algorithm, and hash algorithm, respectively. Also, we examined each scenario on the fingerprint images captured with an optical scanner manufactured by NitGen. The size of the fingerprint image and the fingerprint feature is about 1 *MByte* and 1 *KByte*, respectively. Finally, the disk access time of the fingerprint data stored in the server is assumed to be 50 *miliseconds*.

Table 1 shows the number of instructions of each task in the fingerprint verification measured on an instruction simulator `SimpleScalar`[11]. Based on Table 1, we can compute the estimated execution time of each task on each processor. Finally, the time to acquire a fingerprint image through the fingerprint sensor is assumed about 1 *second*.

Table 2 shows the measured execution times of several cryptography algorithms used in our evaluation. These data are measured by arithmetic mean of 1000 executions on each platform.

Table 1. Number of Instructions Required for Fingerprint Verification[9]

	Number of Instructions
Pre-processing	366,883,162
Feature Extraction	84,856,088
Feature Matching	21,164,578

Table 2. Evaluation Data of Cryptography Algorithms(*Byte*, *milisecond*)

		Generate Signature	Verify Signature	Size of Signature	AES Encrypt	AES Decrypt
Pentium 4	1KByte	6.359	33.656	48	3.0	4.0
	1MByte	74.703	98.922	47	234.0	328.0
Xeon	1KByte	3.656	19.797	48	1.0	2.0
	1MByte	42.844	59.046	47	125.0	218.0

3 Performance Evaluation of Three Scenarios

In the fingerprint authentication scenarios described in the previous section, the response time being less than 5 *seconds* has to be satisfied for real-time execution. As we might easily expect, the server processes most of the time-consuming tasks in Scenario 1, whereas the clients have the heavy workload in Scenario 2. The extreme case is Scenario 3 where the clients do almost everything. However, there is a security weakness in Scenario 3 as explained previously.

In this section, we will evaluate the three scenarios in terms of the response time versus workloads imposed on the server. In other words, we will investigate the maximum workload that the server can handle less than 5 *seconds* response time, or system time in queueing theory, in each scenario. We adopt M/D/1 queueing results assuming that the clients request services to the server in a random fashion, but the server processes the jobs in deterministic fashion.

In an M/D/1 system, the response time is given by

$$w = \frac{1}{\mu} + \frac{\lambda}{2\mu(\mu - \lambda)} \tag{1}$$

where W is the response time(or the systems time), μ is the service rate, and λ is the arrival rate. Here, λ is the job request rate to the server by the clients, and the value of λ is assumed to increase as the number of clients increases.

3.1 Scenario 1, The Server Does Everything

In Scenario 1, fingerprint acquiring(1,000 *ms*) is done by the sensor, whereas feature extraction(225 *ms*), generation of signature for feature(6.359 *ms*), and encryption(3.0 *ms*) tasks are done by the client. Additionally, we consider the communication setup time for sending and receiving data from the server as *st* and the data transmission time(1.6 *ms*, 2 × 0.8 *ms*, 2 times of 1 KB by 10 Mbps Internet transmission). Thus, the total sums to the total of 1,235.959 + 5*st* *ms*. We assume that *st* be 1, 5 or 50 *ms* depending on the communication environments. On the server side, decryption for live feature(2.0 *ms*), decryption for passport feature(2.0 *ms*), verifying the signs for live feature and passport feature(2 × 19.797 *ms*), matching twice(2 × 6.5 *ms*) and data transmission time of 1.6 *ms* and the communication setup time(5*st*) sum to 58.194 + 5*st*, which plays the service time($1/\mu$). Thus, we may build the response time(W) in the M/D/1 systems as eq. (2)

$$w = 1235.959 + 5st + \frac{1}{\mu} + \frac{\lambda}{2\mu(\mu - \lambda)} < 5000 \tag{2}$$

Fixing *st* be 1 *ms*, we may solve eq. (2), we have λ being less than 0.01569 in order to meet the total response time being less than 5 seconds. Here, the workload can be interpreted as the number of job requests by the clients to the server in unit time(*millisecond*).

3.2 Scenario 2, Extraction by the Client and Matching by the Server

Similar approach can be applied to Scenario 2, fingerprint acquiring(1,000 *ms*) is done by the sensor, whereas live feature extraction(225 *ms*), decryption for passport feature(4.0 *ms*), verifying sign for passport feature(33.656 *ms*) and matching passport versus live feature(10 *ms*) tasks are done by the client. Additionally, we consider the communication setup time for sending and receiving data from the server as *st* and the data transmission time(0.8 *ms*, 1 time of 1 KB by 10 Mbps Internet transmission). Thus, the total sums to 1,273.456 + 4*st* *ms*. We assume that *st* be 1, 5 or 50 *ms* depending on the communication environments. On the server side, decryption for passport feature(2.0 *ms*), verifying the sign for passport feature(19.797 *ms*), matching once(6.5 *ms*) and data transmission time of 0.8 *ms* and the communication setup time

($4st$) sum to 29.097 + $4st$, which plays the service time($1/\mu$). Thus, we may build the response time(W) in the M/D/1 systems as eq. (3)

$$w = 1273.456 + 4st + \frac{1}{\mu} + \frac{\lambda}{2\mu(\mu - \lambda)} < 5000 \quad (3)$$

Fixing st be 1ms, we solve eq. (3), we have λ being less than 0.03008 in order to meet the total response time being less than 5 *seconds*.

3.3 Scenario 3, The Client Does Everything

In Scenario 3, the server is doing practically nothing except encrypting the stored template and sending the encrypted data to the clients. Fingerprint acquiring(1,000 *ms*) is done by the sensor, whereas decryption for DB feature(4.0 *ms*), verifying signature for DB feature(33.656 *ms*), decryption for passport feature(4.0 *ms*), verifying sign for passport feature(33.656 *ms*), live feature extraction(225 *ms*) and 2 times of matching(2 × 10 *ms*) tasks are done by the client. Additionally, we consider the communication setup time for sending and receiving signals from the server as st. Thus, the total sums to the total of 1,320.312 + $2st$ *ms*. We assume that st be 1, 5 or 50 *ms* depending on the communication environments. On the server side, encryption for DB feature(1.0 *ms*), generating signature for DB feature(3.656 *ms*) and the communication setup time($2st$) sum to 4.656 + $2st$, which plays the service time($1/\mu$). Thus, we may build the response time(W) in the M/D/1 systems as eq. (4)

$$w = 1320.312 + 2st + \frac{1}{\mu} + \frac{\lambda}{2\mu(\mu - \lambda)} < 5000 \quad (4)$$

When we solve eq. (4) with fixing st be 1 *ms*, we have λ being less than 0.1501 in order to meet the total response time being less than 5 *seconds*.

3.4 Summary of Performance Evaluation

Summarizing the results obtained in this section is depicted in Table 3~5. As we see in Table 3, the server can handle the lowest level of workload with Scenario 1, whereas the server can handle 9.57 times heavier workload with Scenario 3. In most reasonable case of Scenario 2, the server can handle 1.92 times heavier workload compared to the case of Scenario 1.

Table 3. Summary of Performance Evaluation 1(st = 1 *ms*)

Scenario	1	2	3
Maximum Workload	0.01569	0.03008	0.15010
Relative Workload	1	1.92	9.57

Table 4. Summary of Performance Evaluation 2(st = 5 *ms*)

Scenario	1	2	3
Maximum Workload	0.01188	0.02023	0.06810
Relative Workload	1	1.70	5.73

Table 5. Summary of Performance Evaluation 3(*st* = 50 *ms*)

Scenario	1	2	3
Maximum Workload	0.00310	0.00422	0.00942
Relative Workload	1	1.36	3.04

Another interesting result comes when we vary the communication setup time from 1 *ms* to 5 *ms* and 50 *ms*, as shown in Tables 4, and 5. As we notice in Tables 4 and 5, the workload that the server can handle within the specified time constraint of 5 *seconds* changes dramatically as the communication setup time increases. When the communication setup time becomes 5 *ms*, the server can handle only 5.73 times heavier workload with Scenario 3 compared to those with Scenario 1. The things are becoming worse with the communication setup time being 50 *ms*, where the server can handle only 3.04 times heavier workload with Scenario 3 compared to those with Scenario 1 even though almost everything is performed in the clients with Scenario 3.

4 Conclusions

Biometrics is expected to be increasing widely used in conjunction with other techniques such as the cryptography on the network. For large-scale, remote user authentication services, the real-time issue as well as the security/privacy issue should be managed. In this paper, the performance of the fingerprint authentication service was examined as the number of clients increased.

To protect the fingerprint information transmitted, we considered typical scenarios of the task assignment of the fingerprint verification and its corresponding authentication protocol to guarantee both the integrity and the confidentiality of the fingerprint information. To obtain the characteristics of the workload of both the fingerprint verification and the authentication protocol, we first measured the performance of the primitive operations on the client and the server, respectively. Then, based on these measured performance, the workload of each scenario of the task assignment was applied to the M/D/1 queueing model representing the client-server model, and the collective performance of each scenario was analyzed quantitatively.

The modeling results showed that the server could handle the lowest level of workload when the server does decryption, verification and matching. On the contrary, the server can handle 9.57 times heavier workload when the clients do decryption, verification and matching. The modeling results also showed that the workload that the server can handle depends largely on the communication setup time.

Acknowledgement

This research was supported by the MIC(Ministry of Information and Communication), Korea, under the ITRC(Information Technology Research Center) support program supervised by the IITA(Institute of Information Technology Assessment).

References

[1] A. Jain, R. Bole, and S. Panakanti, *Biometrics: Personal Identification in Networked Society*, Kluwer Academic Publishers, 1999.

[2] D. Maltoni, et al., *Handbook of Fingerprint Recognition*, Springer, 2003.

[3] R. Bolle, J. Connell, and N. Ratha, "Biometric Perils and Patches", *Pattern Recognition*, Vol. 35, pp. 2727-2738, 2002.

[4] B. Schneier, "The Uses and Abuses of Biometrics", *Communications of the ACM*, Vol. 42, No, 8, pp. 136, 1999.

[5] S. Prabhakar, S. Pankanti, and A. Jain, "Biometric Recognition: Security and Privacy Concerns", *IEEE Security and Privacy*, pp. 33-42, 2003.

[6] S. Pan, et al., "A Memory-Efficient Fingerprint Verification Algorithm using A Multi-Resolution Accumulator Array for Match-on-Card", *ETRI Journal*, Vol. 25, No. 3, pp. 179-186, 2003.

[7] X.9.84, www.x9.org.

[8] Machine Readable Travel Documents – Doc. 9303, Part 3, www.icao.org.

[9] D. Moon, et al., "Performance Analysis of the Match-on-Card System for the Fingerprint Authentication", *Proc. of International Workshop on Information Security Applications*, pp. 449-459, 2001.

[10] W. Stallings, *Cryptography and Network Security: Principles and Practice*, Prentice Hall, 2003.

[11] D. Burger and T. Austin, "The SimpleScalar Tool Set, Version 2.0.", *Technical Report, University of Wisconsin*, 1997.

Security Analysis of Secure Password Authentication for Keystroke Dynamics*

Hyunsoo Song and Taekyoung Kwon

Information Security Lab.,
Sejong University, Seoul 143-747, Korea
hssong@sju.ac.kr, tkwon@sejong.ac.kr

Abstract. Password-based authentication and key distribution are important in today's computing environment. Since passwords are easy to remember for human users, the password-based system is used widely. However, due to the fact that the passwords are chosen from small space, the password-based schemes are more susceptible to various attacks including password guessing attacks. Recently, Choe and Kim proposed a new password authentication scheme for keystroke dynamics. However, in this paper, we cryptanalyze the Choe-Kim scheme and show it is vulnerable to various types of attacks such as server-deception attacks, server-impersonation attacks and password guessing attacks. We also comment on the scheme that more care must be taken when designing password-based schemes and briefly show how the standard like IEEE P1363.2 can be used for strengthening those schemes.

1 Introduction

Password-based authentication and key distribution are important in today's computing environment. The password-based schemes are more susceptible to various attacks because passwords are chosen from small space [1].

Verifier-based protocols useful since they resist server compromise attacks in an asymmetric model where a password is kept by the user and a verifier by the server. These protocols generate a common session key and perform key-verification steps about the session key. They include A-EKE [2], B-SPEKE [7,8], SRP [12], PAK [3,10,11] and AMP [9].

In secure password authentication for keystroke dynamics [4], Choe and Kim proposed efficient password authentication protocol. They also extended a two-party protocol to a three-party protocol, where each user only shares a password with a trusted server.

We present security analysis of the Choe-Kim scheme. This scheme is vulnerable to server-impersonation attacks and server-deception attacks. Also, it is vulnerable to password guessing attacks. In other words, an attacker can guess a valid password with repeated off-line queries [1,6]. We also comment on the

* This study was supported in part by a grant of the Korea Health 21 R&D Project, Ministry of Health & Welfare, Republic of Korea. (0412-MI01-0416-0002).

B. Gabrys, R.J. Howlett, and L.C. Jain (Eds.): KES 2006, Part I, LNAI 4251, pp. 916–923, 2006.
© Springer-Verlag Berlin Heidelberg 2006

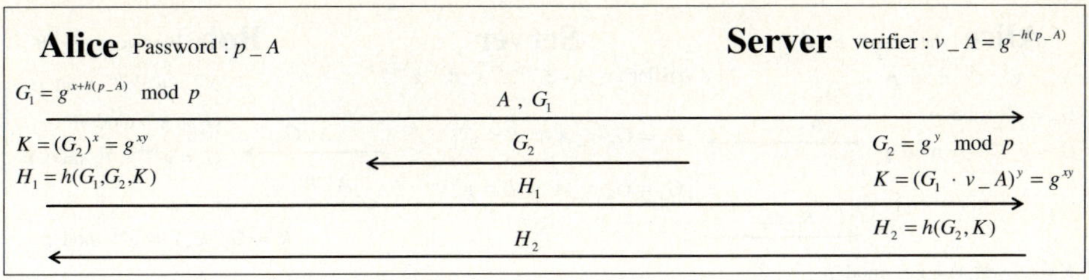

Fig. 1. 2-Party Protocol

scheme that more care must be taken when designing password-based schemes and briefly show how the standard like IEEE P1363.2 can be used for strengthening those schemes.

This paper is organized as follows. Section 2 introduces the password-based authentication scheme proposed by Choe and Kim. In Section 3, we cryptanalyze and show new attacks on them the security of the Choe-Kim scheme. In Section 4, we examine the possibility of password guessing attacks by experiments. Finally, Section 5 concludes this paper along with specific comments.

2 Review of Choe-Kim Schemes

This section briefly reviews the Choe-Kim scheme [4]. It is based on the discrete logarithm problem and Diffie-Hellman problem [5].

Let p, q be sufficiently large prime numbers such that $p = 2q + 1$ and this scheme use the following notations.

$\boldsymbol{A, B}$	User's identifier
$\boldsymbol{p\_A, p\_B}$	User's password
$\boldsymbol{v\_A, v\_B}$	User's verifier derived from the password
$\boldsymbol{Z_p^*}$	multiplicative group
$\boldsymbol{g}$	primitive element of Z_p^*
$\boldsymbol{Z_q}$	subgroup of Z_p^* (order q)
$\boldsymbol{x, y}$	random variables
$\boldsymbol{K}$	session key
$\boldsymbol{h()}$	one-way hash function

Alice or Bob selects a password p_A or p_B respectively, and the server stores verifier v_A or v_B, which is derive from the password.

2.1 2-Party Choe-Kim Scheme

The 2-party Choe-Kim scheme is illustrated in Figure 1 and works as follows: When Alice wants to communicate with the server,

1. Alice chooses a random value $x \in_R Z_q$, computes $G_1 = g^{x+h(p\_A)} \bmod p$ and sends G_1 and A to the server.

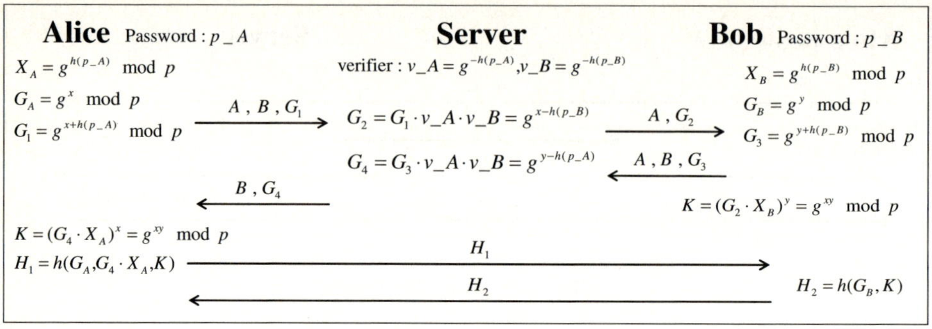

Fig. 2. Extended 3-Party Protocol

2. After receiving Alice's request, the server chooses a random value $y \in_R Z_q$, computes $G_2 = g^y \bmod p$ and sends G_2 to Alice.
3. Alice and the server compute a common session key $K = g^{xy} \bmod p$ using the values x and y.
4. Alice computes $H_1 = h(G_1, G_2, K)$ and sends H_1 to the server.
5. The server computes $h(G_1, G_2, K)$ and verifies whether $H_1 \stackrel{?}{=} h(G_1, G_2, K)$.
6. The server computes $H_2 = h(G_2, K)$, and then sends H_2 to Alice.
7. Alice computes $h(G_2, K)$ and verifies whether $H_2 \stackrel{?}{=} h(G_2, K)$. If being successful, Alice accepts a common session key.

2.2 3-Party Choe-Kim Scheme

In the Choe-Kim scheme, a 2-party protocol can be extended to a 3-party protocol which is achieved by letting the trusted server act as a relay. The 3-party Choe-Kim scheme is illustrated in Figure 2 and works as follows:

When Alice wants to communicate with Bob,

1. Alice computes $X_A = g^{h(p\_A)} \bmod p$ using password p_A, chooses a random value x and computes $G_1 = G_A \cdot X_A = g^{x+h(p\_A)}$ where $G_A = g^x \bmod p$. Alice then sends A, B and G_1 to the server.
2. After receiving Alice's request and the server computes $G_2 = G_1 \cdot v\_A \cdot v\_B = g^{x-h(p\_B)} \bmod p$, and sends A, G_2 to Bob.
3. Bob computes $X_B = g^{h(p\_B)} \bmod p$ using password p_B, chooses a random value y, and computes $G_3 = G_B \cdot X_B = g^{y+h(p\_B)}$ where $G_B = g^y \bmod p$. After receiving server's request, Bob sends A, B and G_3 to the server.
4. Bob computes $K = (G_2 \cdot X_B)^y = g^{xy} \bmod p$.
5. The server computes $G_4 = G_3 \cdot v\_A \cdot v\_B = g^{y-h(p\_A)} \bmod p$, and sends B and G_4 to Alice.
6. Alice computes $K = (G_4 \cdot X_A)^x = g^{xy} \bmod p$ and $H_1 = h(G_A, G_4 \cdot X_A, K)$ and sends H_1 to Bob.

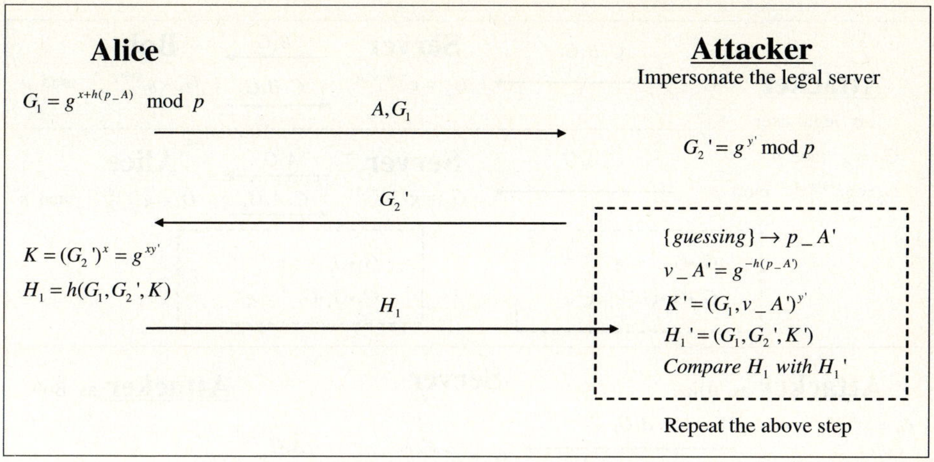

Fig. 3. Server-Impersonation Attack

7. Bob computes $h(G_A, G_B, K)$ and verifies whether $H_1 \stackrel{?}{=} h(G_A, G_B, K)$.
8. Bob computes $H_2 = h(G_B, K)$ and sends H_2 to Alice.
9. Alice computes $h(G_4 \cdot X_A, K)$, and verifies whether $H_2 \stackrel{?}{=} h(G_4 \cdot X_A, K)$. If being successful, Alice and Bob accepts a common session key between Bob.

3 Security Analysis of Choe-Kim Schemes

In this section, we present new attacks on the 2-party scheme and 3-party scheme, respectively.

3.1 Server-Impersonation Attack on 2-Party Scheme

This section shows that the Choe-Kim scheme is vulnerable to server-impersonation attacks. An attacker can perform the following as depicted in Figure 3:

1. The attacker impersonates the legal server after the attacker makes DoS attack to the server.
2. The attacker gets a message G_1 and A.
3. The attacker chooses a random value $y' \in_R Z_q$, computes $G_2' = g^{y'} \bmod p$ and sends G_2' to Alice. The attacker sends a forged value to Alice. However, Alice doesn't know whether the server or attacker sends the value.
4. Alice computes a common session key $K = g^{xy'} \bmod p$ using the value x. Alice then computes $H_1 = h(G_1, G_2, K)$ and sends H_1 to the attacker.
5. The attacker finds out a Alice's valid password using *Password Guessing Attack*. The detailed attack is given in the rest of this step.

   - Let $p\_ A'$ be a password-guessing value.

$$\{guessing\} \rightarrow p\_ A'.$$

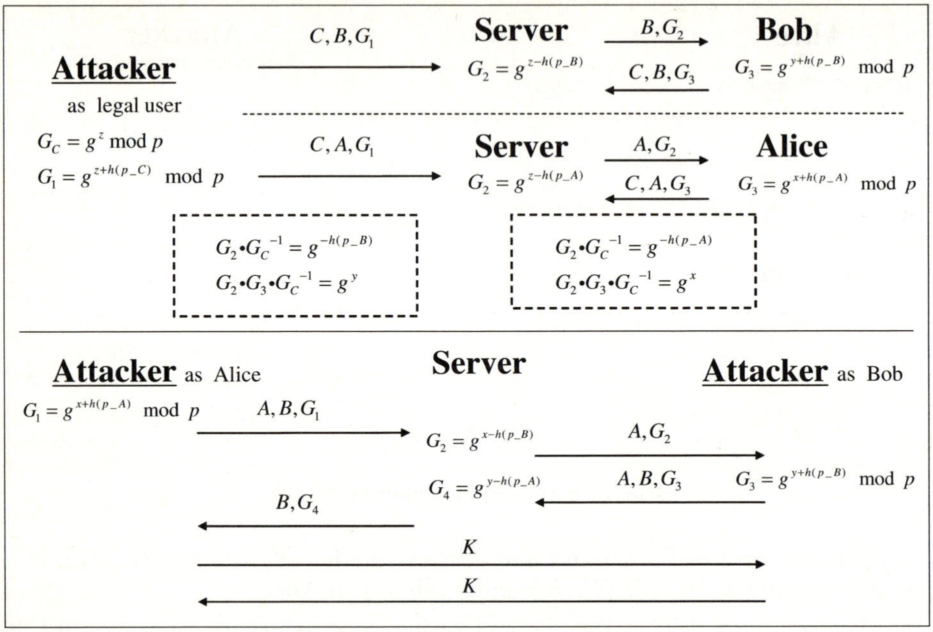

Fig. 4. Server-Deception Attack

- Compute a verifier $v\_A' = g^{-h(p\_A')}$ using the chosen password p_A'.
- Generate $K' = (G_1, v\_A')^{y'}$ by computing v_A'.
- Compare H_1 with $H_1' = (G_1, G_2', K')$.
- Repeat the above step 5 until the attacker finds Alice's valid password.

3.2 Server-Deception Attack on 3-Party Scheme

In this section, we show the vulnerability of the 3-Party scheme. As illustrated in Figure 4, the 3-party scheme is vulnerable to server-deception attacks. An attacker can perform the following:

1. The attacker in the 3-party scheme communicates with Bob as legal user.
2. The attacker computes X_C using password p_C, chooses a random value z and computes $G_C = g^z \bmod p$. Then an attacker sends C, B and G_1 to the server.
3. The attacker gets a message G_2 and G_3 by eavesdropping.
4. The attacker gets Bob's G_B and verifier v_B by computing

$$G_2 \cdot G_C^{-1} = g^{z-h(p\_B)} \cdot g^{-z} = g^{-h(p\_B)} \quad \textit{and}$$

$$G_2 \cdot G_3 \cdot G_C^{-1} = g^{z-h(p\_B)} \cdot g^{y+h(p\_B)} \cdot g^{-z} = g^y.$$

5. Like as the above, the attacker communicates with Alice as legal user. Then, the attacker gets Alice's G_A and verifier v_A.

6. The attacker practice deception on the server, using A, B, G_A, G_B, v_A, v_B

As a result, the server is deceived by the attacker (See Figure 4).

3.3 Password Guessing Attack on 3-Party Scheme

An attacker is able to find out user's (Alice or Bob) valid password by exploiting the values obtained from the server-deception attack. In other words, the attacker can further launch off-line password guessing attacks after the deception attack. The detailed attack is given in the rest of this section.

- Let p_A' be a password-guessing value.

$$\{guessing\} \rightarrow p\_A'.$$

- Compute a verifier $v\_A' = g^{-h(p\_A')}$ using the chosen password p_A'.
- Compare v_A with v_A'.
- Repeat the above steps until the attacker finds Alice and Bob's valid passwords.

4 Experimental Study

We implement the Choe-Kim schemes and examine the possibility of password guessing attacks by experiments.

4.1 Implementation Environment

The environment for our implementation is as follows.

1. Hardware Platform
   - 3 portable PCs : CPU-Intel PentiumM 1.0GHz, RAM-768MB
   - Communication channel : 100BaseT Ethernet

2. Software Platform
   - System: Microsoft Windows XP
   - Language: C++ (Visual Studio 6.0)
   - Crypto lib : OpenSSL

4.2 Implementation Scenarios

In practice, we assume the password space is limited by 2^{10}. This is only for our simple experiments. In real world, the password space is recognized as 2^{40}. Note that our attack is viable to 2^{40} space as well, and our experiments can be extended to the case of 2^{40} simply by setting the space variable as 2^{40}. For simplicity, we set a target password as a random integer chosen from the given space.

1. 2-party attack scenario with password guessing

Attack password guessing attack on 2-party protocol

Target password p_A: an integer $0 \leq p\_A \leq 1023$
Protocol run: Refer to Figure 3
1. Alice sends A, G_1 to the attacker.
2. The attacker sends G'_2 to Alice.
3. Alice sends H_1 to the attacker.

Off-line guessing:
4. For i from 0 to 1023 do the following:
   4.1 Compute $v\_A' = g^{-h(i)}$
   4.2 Compute $K' = (G_1, v\_A')^{y'}$.
   4.3 Compute $H'_1 = (G_1, G'_2, K')$.
   4.4 If $H'_1 = H_1$ then $i = p\_A$
5. The attacker obtains p_A.

For searching 2^{10} space, it takes about 16.047*sec*. In the case above, it took only about 3.343*sec* on average for finding the target password.

2. 3-party attack scenario with password guessing

Attack password guessing attack on 3-party protocol

Target password p_B: an integer $0 \leq p\_B \leq 1023$
Protocol run: Refer to Figure 4
1. The attacker sends C, B, G_1 to the server.
2. The server sends C, G_2 to Bob.
3. The server sends G_2 to the attacker.
4. Bob sends C, B, G_3 to the server.
5. The server sends B, G_4 to the attacker.
6. The attacker sends H_1 to Bob.
7. Bob sends H_2 to the attacker.

Off-line guessing:
8. Compute $g^{-h(p\_B)} = G_2 \cdot G_c^{-1}$
9. For i from 0 to 1023 do the following:
   9.1 Compute $v\_B' = g^{-h(i)}$
   9.2 If $v\_B' = g^{-h(p\_B)}$ then $i = p\_B$
10. The attacker obtains p_B.

For searching 2^{10} space, it takes about 3.172*sec*. In the case above, it took only about 1.398*sec* on average for finding the target password.

5 Conclusion

In this paper, we cryptanalyze the Choe-Kim scheme and show it is vulnerable to various types of attacks such as server-deception attacks, server-impersonation attacks and password guessing attacks.

We comment on the scheme that more care must be taken when designing password-based schemes and the standard like IEEE P1363.2 can be used for strengthening those schemes. This standard is against common attacks including a password-guessing attack and provides very efficient password authentication protocol blocks.

For example, the Choe-Kim protocol can be improved in the way of setting $G_1 = g^{x+h(p\_A)} \cdot f(p\_A) \bmod p$ where $f(p\_A)$ resides in a proper subgroup. This improvement follows IEEE P1363.2 so as to resist our attacks on the Choe-Kim scheme.

References

1. S. Bellovin and M. Merritt, "Encrypted key exchange: password-based protocols secure against dictionary attacks," In IEEE Symposium on Research in Security and Privacy, pp. 77-84, 1992.
2. S. Bellovin and M. Merritt, "Augmented encrypted key exchange: a password-based protocols secure against dictionary attacks and password-file compromise," In ACM Conference on Computer and Communications Security, pp. 244-250, 1993.
3. V. Boyko, P. MacKenzie and S. Patel, "Provably secure password authenticated key exchange using Diffie-Hellman," In Eurocrypt .00, pp.156-171, 2000.
4. Y. Choe and S. Kim, "Secure Password Authentication for Keystroke Dynamics," 9th International Conference on Knowledge-Based & Intelligent Information & Engineering Systems (KES 2005), LNAI 3683, pp. 317-324, 2005.
5. W. Diffie and M. Hellman, "New directions in cryptograpy," In IEEE Transactions on Information Theory, Vol. 22, no. 6, pp. 644-654, 1976.
6. L. Gong, M. Lomas, R. Needham and J. Saltzer, "Protecting Poorly Chosen Secrets from Guessing Attacks," IEEE Jpurnal on Selected Areas in Commuinications, vol.11. no. 5, pp. 648-656, 1993.
7. D. Jablon, "Strong password-only authenticated key exchange," ACM Computer Communications Review, Vol. 26, no. 5, pp. 5-26, 1996.
8. D. Jablon, "Extended password key exchange protocols immune to dictionary attacks," In WETICE.97 Workshop on Enterprise Security, pp. 248-255, 1997.
9. T. Kwon, "Authentication and Key agreement via Memorable Passwords," In Network and Distributed System Security Symposium Conference Proceedings, 2001.
10. P. MacKenzie, "More Efficient Password-Authenticated Key Exchange," In CT-RSA 2001, pp. 361-377, 2001.
11. P. MacKenzie, "The PAK suites: Protocols for Password-Authenticated Key Exchange," 2002, available from http://grouper.ieee.org/groups/1363/passwdPK/contributions.html#Ma c02.
12. T. Wu, "Secure remote password protocol," In Network and Distributed System Security Symposium Conference Proceedings, 1998.
13. IEEE P1363.2, http://grouper.ieee.org/groups/1363/passwdPK/index.html.

Security Techniques Based on EPC Gen2 Tag for Secure Mobile RFID Network Services

Namje Park[1,3], Jin Kwak[2], Seungjoo Kim[3], Dongho Won[3,*], and Howon Kim[1]

[1] Information Security Research Division, ETRI, 161 Gajeong-dong, Yuseong-gu, Daejeon, 305-350, Korea
{namjepark, khw}@etri.re.kr

[2] Faculty of Information Science and Electrical Engineering, Kyushu University, 6-10-1 Hakozaki, Higashi-ku, Fukuoka, 812-0053, Japan
jkwak@itslab.csce.kyushu-u.ac.jp

[3] Information Security Group, Sungkyunkwan University, 300 Cheoncheon-dong, Jangan-gu, Suwon, Gyeonggi-do, 440-746, Korea
{njpark, skim, dhwon}@security.re.kr
http://www.security.re.kr

Abstract. RFID (Radio Frequency Identification) technology will be a ubiquitous reality in daily life in the near future. The R&D groups in global now have paid attention to integrate RFID with mobile phone devices as well as to associate with the existing mobile telecommunication network. Such a converged technology and services would lead to make new markets and research challenges. However, privacy threats on RFID tags embedded into products has become stumbling block. Therefore, this paper propose a new security technique for mobile RFID service, which can be support to secure RFID service and uses the privacy protection in mobile RFID network services.

1 Introduction

Due to the rapid development of information technology, handheld terminal is evolving into a low-power, ultra-light, integrated, and intellectual terminal to support various information service and ubiquitous environment, and it will develop to a more advanced form current services. The wireless internet infrastructure integrated with the mobile communication system and RFID gave birth to mobile RFID to provide new services to users, and the standardization of mobile RFID information protection technology (such as the privacy protection, verification of personal information, authorization, and key management) and its technological development are being progressed along the way.

---

* Dongho Won is the corresponding author for this paper. The fourth author of the research was supported by the University IT Research Center Project funded by the Korean Ministry of Information and Communication.

B. Gabrys, R.J. Howlett, and L.C. Jain (Eds.): KES 2006, Part I, LNAI 4251, pp. 924–934, 2006.
© Springer-Verlag Berlin Heidelberg 2006

RFID reader has been mainly used as RFID tag recognizable unattended information production terminal, and now it is expanding into the mobile RFID service providing useful information to users by reading various RFID tag information through RFID tag chip and RFID reader chip installed to mobile devices (such as cellular phone). Mobile RFID service is defined as to provide personalized secure services such as searching the products information, purchasing, verifying, and paying for the products while on the move through the wireless internet network by building the RFID reader chip into the mobile terminal [1,2,4]. The service infrastructure required for providing such RFID based mobile service is composed of RFID reader, handset, communication network, network protocol, information protection, application server, RFID code interpretation, and contents development.

In this paper, we describe a way to incorporate its new technology to work with cell phones in particular as an external security reading device (replacing 900MHz) and same time as an added security service to manage all RFID mobile device mediums. This is new technology to RFID will provide a solution to protecting absolute confidentiality from basic tags to user's privacy information. The proposed expanded mobile service platform, the ETRI (Electronics and Telecommunications Research Institute) mobile RFID security middleware platform, is composed of AAL (Application Adaptation Layer) and RFID HAL (Handset Adaptation Layer) [5,6]. The proposed AAL is the core component of the security middleware platform. The security platform is a building block for the extended security API (Application Programming Interface) for secure mobile RFID and it has to be integrated with WIPI (Wireless Internet Platform for Interoperability) and mobile RFID security mechanisms for phone-based RFID services to be more secure mobile business applications. It enables business to provide new services to mobile customers by securing services and transactions from the end-user to a company's existing e-commerce systems and IT technologies.

2 Overview of Mobile RFID Technology

2.1 Mobile RFID Technology

RFID is expected to be the base technology for ubiquitous network or computing, and to be associated with other technology such as MEMS, telematics, and sensors. Meanwhile Korea is widely known that it has established one of the most robust mobile telecommunication networks. In particular, about 78% of the population uses mobile phones and more than 95% among their phones have Internet-enabled function. Currently, Korea has recognized the potential of RFID technology and has tried to converge with mobile phone. Mobile phone integrated with RFID can be expected that it create new markets, provide new services to end-users, and will be considered as an exemplary technology fusion. Furthermore, it may evolve its functions as end-user terminal device, or 'u-device (ubiquitous device)', in the world of ubiquitous information technology.

Actually, mobile RFID phone may represent two types of mobile phone devices; one is RFID reader equipped mobile phone, and the other is RFID tag attached mobile phone. Each type of mobile phone has different application domains, for example, the RFID tag attached one can be used as a device for payment, entry control, and identity authentication, and the feature of this application is that RFID readers existed in the fixed position and they recognize each phone to give user specific services like door opening. In the other hand, the RFID reader equipped mobile phone, which Korea is paying much attention now, can be utilized for providing end-users detailed information about the tagged object through accessing mobile wireless network.

Right now, there are several projects in regards to 'mobile RFID', as arranged in the table below [3,13,15]. In particular, Korea's mobile RFID technology is focusing on the UHF range (860~960MHz), since UHF range may enable longer reading range and moderate data rates as well as relatively small tag size and cost. Then, as a kind of handheld RFID reader, in the selected service domain the UHF RFID phone device can be used for providing object information directly to end-user using the same UHF RFID tags which have spread widely.

Table 1. Class of Mobile RFID Technology

	Nokia's Mobile RFID	KDDI's Mobile RFID		NFC	Korea's Mobile RFID
Frequency	13.56MHz	2.45GHz (Passive)	315MHz (Active)	13.56MHz	908.5~913.9MHz
Read Range	2~3cm	~5cm	~10m	~10cm	~100cm
Standards	ISO/IEC 14443 A	-	-	ISO/IEC 18092	ISO/IEC 18000-6 B/C

2.2 Privacy on Mobile RFID Services

Mobile RFID service structure is defined to support ISO/IEC 18000-6 A/B/C through the wireless access communication between the tag and the reader, however there is no RFID reader chip supporting all three wireless connection access specifications yet that the communication specification for the mobile phone will be determined by the mobile communication companies. It will be also possible to mount the RF wireless communication function to the Reader Chip using SDR technology and develop ISO/IEC 18000-6 A/B/C communication protocol in software to choose from protocols when needed.

Mobile RFID network function is concerned with the communication protocols such as the ODS (Object Directory Service) communication for code interpretation, the message transportation for the transmission and reception of contents between the mobile phone terminal and the application server, contents negotiation that supports mobile RFID service environment and ensures the optimum contents transfer between the mobile phone terminal and the application server, and session management that enables the application to create and manage required status information while transmitting the message and the WIPI extended specification which supports these communication services [3,15].

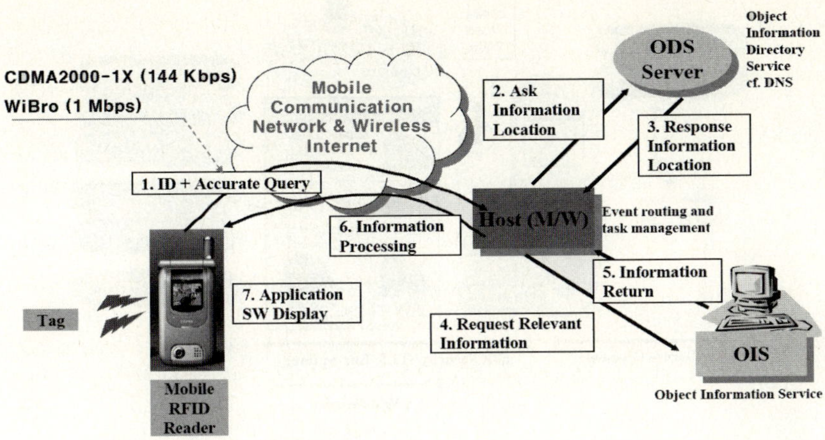

Fig. 1. Conceptual Network Model for Mobile RFID Service

The service model, as shown in figure 1, consists of tag, reader, middleware and information server. In the point of view of information protection, the serious problem for the RFID service is a threat of privacy [3,13,15]. Here, the damage of privacy is of exposing the information stored in tag and the leakage of information includes all data of the personal possessing the tag, tagged products and location. The privacy protection on RFID system can be considered in two points of view. One is the privacy protection between the tag and the reader, which takes advantage of ID encryption, prevention of location tracking and the countermeasure of tag being forged. The other is of the exposure of what the information server contains along with tagged items[7,8,9]. First of all, we will have a look about the exposure of information caused between tag and reader, and then discuss about the solution proposing on this paper.

3 Techniques for Secure Mobile RFID Service

3.1 Service Architecture Model for Secure Mobile RFID Network

The mobile RFID is a technology for developing a RFID reader embedded in a mobile terminal and providing various application services over wireless networks. Various security issues - Interdomain security, privacy, authentication, E2E (End-to-End) security, and untraceability etc. - need to be addressed before the widespread use of mobile RFID. Model of mobile RFID service as shown in figure 2 defines additional three entities and two relationships compared to that defined in RFID tag, RFID access network, RFID reader, relation between RFID tag and RFID reader, relation between RFID reader and application server.

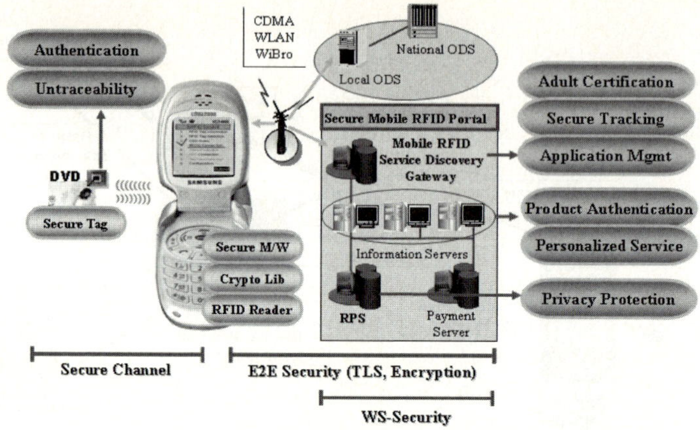

Fig. 2. Platform Architecture of Secure Mobile RFID Service

Generally, in mobile RFID application such as smart poster, Application Service Provider (ASP) has the ownership of RFID tags. Thus, mobile RFID users have to subscribe to both the ASP for these kinds of RFID services and mobile network operator for mobile communication service. Namely, there exist three potentially distrusted parties: user owned RFID reader, mobile network operator, and ASP. Accordingly, trust relationship among three parties must be established to realize secure mobile RFID service. Especially, when a RFID reader tries to read or change RFID service data stored in tag, the reader needs to get a tag access rights. Additionally, it is important that new tag access rights whenever some readers access a same tag must be different from the already accessed old one.

MRF-Sec 631 strategy represents 6 standard security functions at mobile RFID middleware, 3 major security service mechanisms using 6 security functions, and 1 secure mobile RFID application portal service in order to realize the above 3 security service mechanisms. What is the MRF-Sec 631 strategy? 6-standards security functions are mobile RFID Data encryption API function, mobile RFID secure communication API function, mobile RFID password management API function, EPC C1G2 security command API function, adult certification API function, and privacy protection API function. 3-security service mechanisms are authentication service mechanism, privacy protection service mechanism, and secure location tracking service mechanism. 1-secure application service is secure mobile RFID application portal service.

3.2 Phone-Based Mobile RFID Middleware System

In the WIPI specification, the core functions are the functions of handset hardware, native system software, handset adaptation module, run time engine, APIs, and application programs are the areas of the core functional specifications of WIPI. Actually, in the WIPI specifications, only the handset adaptation and APIs are included and the other parts of functions of the wireless Internet platform are considered as the requirements to the handset vendors whether they accept it or not. For example, the run time engine part is required as the mode of download of binary code for its maximum performance.

The core functions of the WIPI are the handset adaptation and APIs which are called 'Handset Adaptation Layer (HAL)' and 'Application Adaptation Layer (AAL)', respectively. The HAL defines an abstract specification layer to support hardware platform independence when porting applications. The AAL defines the specifications for API of the wireless Internet platform. The AAL support the C/C++ and Java programming languages.

Figure 3 depicts the mobile RFID middleware platform based on WIPI layers. The mobile RFID middleware platform concerns three layers of the common wireless Internet layers, mobile RFID API layer, RFID engine, and handset adaptation layer. The RFID engine is defined as a requirement to the implementations. The native system software defines the operating system of handset device.

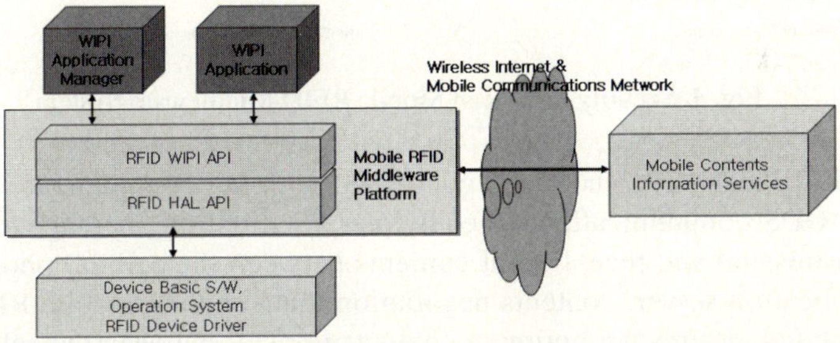

Fig. 3. Proposed Architecture for a WIPI-based Mobile RFID Middleware

The RFID device handler provides the definitions for functions of starting the platform and transferring the events from the upper layer of HAL to the RFID H/W Reader. The categories of RFID device handler API cover call, RFID device, network, serial communication, short message service, sound, time, code conversion, file system, input method, font, frame buffer, and virtual key. The AAL provides the definitions for functions of adaptive functions for RFID engine, C/Java API, crypto libraries, and RFID security components.

3.3 Security Enhanced Mobile RFID Middleware System

In this subsection, we design a security enhanced RFID middleware to support trust and secure m-business based on RFID [10,14]. Mobile RFID terminal's function is concerned with the recognition distance to the RFID reader chip built into the cellular phone, transmission power, frequency, interface, technological standard, PIN specification, UART communication interface, WIPI API, and WIPI-HAL API extended specification to control reader chip. RFID reader chip middleware functions are provided to the application program in the form of WIPI API as in figure 4. Here, 'Mobile RFID Device Driver' is the device driver software provided by the reader chip manufacturer.

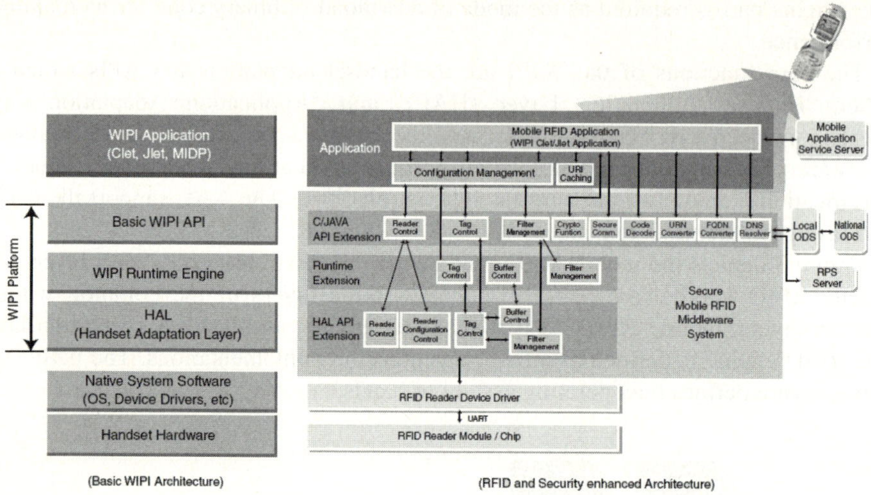

Fig. 4. Security Enhanced Mobile RFID Middleware System

Mobile RFID network function is concerned with the communication protocols such as the ODS communication for code interpretation, the message transmission for the transmission and reception of contents between the cellular phone terminal and the application server, contents negotiation that supports mobile RFID service environment and ensures the optimum contents transfer between the cellular phone terminal and the application server, and session management that enables the application to create and mange required status information while transmitting the message and the WIPI extended specification which supports these communication services.

In scenario, a mobile RFID client performs an update for its own middleware in the security server, where applying security modules to implement business processes satisfies security requirements. In this scenario, a RFID service provider already been authenticated can do 1-to-N businesses (from one provider to multiple providers) as well as 1-to-1 business (from one provider to one provider), because he can search and access to various RFID middleware registered in the RFID security server. To offer more flexible access to multiple business providers, distributed registries need to be integrated, but it causes the problems of user authentication and security vulnerability. By applying single sign-on scheme, we can simplify user authentication and overcome the problems. The detailed message communication processes are as follows.

1) The user clicks the application's button or starts the RFID reader application from the cellular phone's application menu.
2) Application calls RFID reader control API.
3) RFID reader calls HAL API.
4) Transmits the command to RFID reader's device driver.
5) Transmits the command to RFID reader to start reader
6) Obtains the result from the reader

7) Device driver returns RFID reader results to HAL
8) HAL returns the RFID reader results to reader control API
9) Reader control API returns the RFID reader results to application
10) Application calls RFID code-support API to interpret the RFID tag data code.
11) Code-support API calls HAL API for common memory access of the handheld terminal to interpret the code.
12) HAL returns the common memory access data to code-support API.
13) Code-support API interprets the tag data code read based on the code in the common memory contents and transmits the result to the application.
14) If the tag represents adult level, calls the security API for service authorization.
15) The result of authorization for adult service to the application.
16) Creates ODS question message by extracting the required codes through the code interpretation from [13] and calls network API.
17) Requests local ODS for the URI (Uniform Resource Identifier) related to the code extracted from 14 above through network API. (This type of network API is referred to as ODS resolver.)
18) If local ODS does not have the OIS (Object Information Service) information or contents server information related to the transmitted code, it sends the transmitted code to national ODS to request such information.
19) National ODS transmits the URI related to the transmitted code to local ODS.
20) Local ODS returns the transmitted URI to network API.
21) Network API returns this to application.
22) The application runs WAP/ME browser using the URI acquired.
23) WAP/ME browser accesses the OIS or Contents Server using the URI acquired to request for tag related information.
24) OIS or contents server transmits the product or service profile related to the OID of the transmitted tag so that WAP/ME browser can display it on the screen.
25) Let's the user determine the privacy level by displaying the policy configuration screen for the required privacy in the middle of the communication with OIS or contents server when privacy protection is required.

3.4 Mobile RFID Privacy Protection Service

Widespread deployment of RFID technology may create new threats to privacy due to the automated tracking capability. Especially, in the mobile RFID environment, privacy problem is more serious since RFID reader is contained in handheld device and many application services are based on Business-to-Costumer model. The RPS (RFID user Privacy management Service) provides mobile RFID users with information privacy protection service for personalized tag under mobile RFID environment [14,16]. When a mobile RFID user possesses an RFID tagged product, RPS enables the owner to control his backend information connected with the tag such as product information, distribution information, owner's personal information and so on.

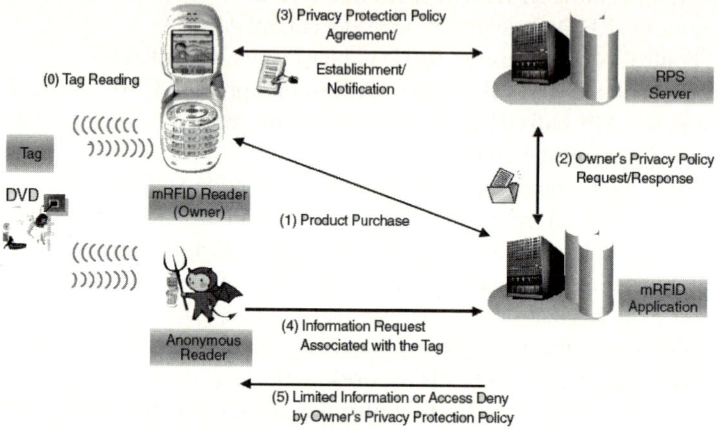

Fig. 5. Service Scenario of RPS Service

Main features of this service mechanism are owner's privacy protection policy establishment and management, access control for information associated with personalized tag by owner's privacy policy, obligation result notification service, and privacy audit service by audit log management.

3.5 Secure Mobile RFID Information Service System

The secure IS (Information Service) is a security-enhanced EPCglobal's IS system. It is able to strengthen security for managing events and providing tag information and provides security mechanism such as message security, authentication authorization, etc. It can be also used as a mobile RFID's IS server in the mobile RFID application environment.

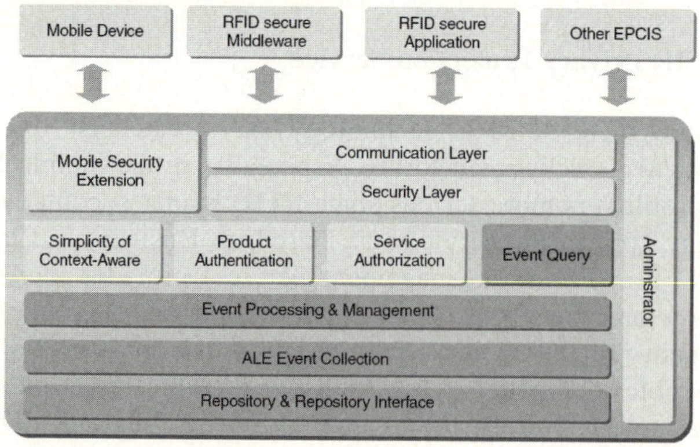

Fig. 6. Architecture of Mobile RFID Information Service System

3.6 Portal Service Platform for Secure Mobile RFID Application

Secure mobile RFID application portal is a secure service portal for various mobile RFID application services. The service provider using SMAP (Secure Mobile Application Portal) can easily deploy several mobile RFID applications guaranteed with security and privacy protection.

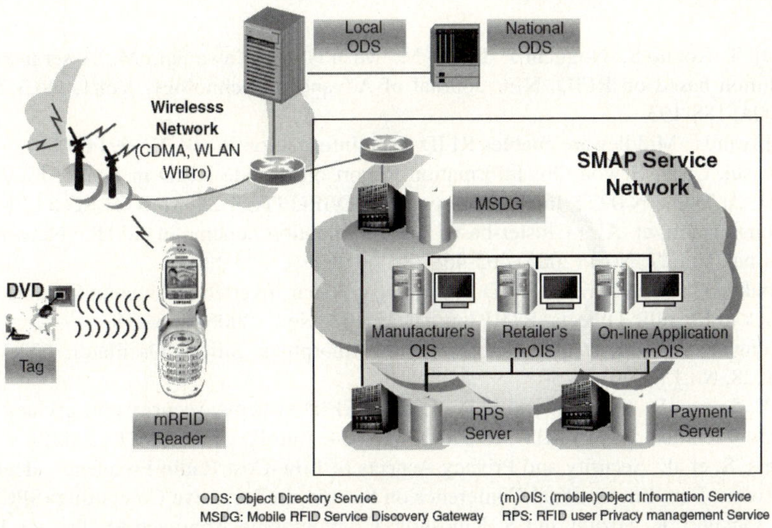

Fig. 7. Architecture of Secure Mobile RFID Application Portal

Main features of secure mobile RFID application portal service platform are Mobile RFID service discovery, Secure object traceability service, Application information service, mobile OIS generation & management service, mobile RFID privacy protection service, mobile RFID payment service, and mobile RFID security mechanisms - Authentication/Privacy/Untraceability.

4 Conclusion

In this paper, we describe a way to incorporate its new technology to work with cell phones in particular as an external security reading device (replacing 900MHz) and same time as an added security service to manage all RFID mobile device mediums. With this purpose, the application areas of this service platform are also briefly presented. By doing so, the customized security and privacy protection can be achieved. In this regard, the suggested mechanism is an effective solution for security and privacy protection in a mobile RFID system.

Acknowledgement

The authors are grateful to Haedong Lee and Dr. Dooho Choi for helpful discussions on the subject of this paper. We would also like to thank the anonymous referees for valuable comments.

References

1. Tsuji T. Kouno S. Noguchi J. Iguchi M. Misu N. and Kawamura M.: Asset management solution based on RFID. NEC Journal of Advanced Technology. Vol.1, No.3, Summer. (2004) 188-193
2. Sullivan L.: Middleware enables RFID tests. Information Week, No.991 (2004)
3. Jongsuk Chae, Sewon Oh: Information Report on Mobile RFID in Korea. ISO/IEC JTC 1/SC 31/WG 4 N 0922, Information paper, ISO/IEC JTC 1 SC 31 WG4 SG 5 (2005)
4. Seunghun Jin, et. Al.: Cluster-based Trust Evaluation Scheme in Ad Hoc Network. ETRI Journal, Vol.27, No.4 (2005) 465-468
5. Wonkyu Choi, et. Al.: An RFID Tag Using a Planar Inverted-F Antenna Capable of Being Stuck to Metallic Objects. ETRI Journal, Vol.28, No.2 (2006) 216-218
6. Byung Seong Bae, et. Al.: Stability of an Amorphous Silicon Oscillator. ETRI Journal, Vol.28, No.1 (2006) 45-50
7. S. E. Sarma, S. A. Weis, and D.W. Engels: RFID systems, security and privacy implications. Technical Report MIT-AUTOID-WH-014, AutoID Center, MIT (2002)
8. Weis, S. et al.: Security and Privacy Aspects of Low-Cost Radio Frequency identification Systems. First International Conference on Security in Pervasive Computing (SPC) 2003
9. M. Ohkubo, K. Suzuki and S. Kinoshita: Cryptographic Approach to "Privacy-Friendly" Tags. RFID Privacy Workshop (2003)
10. Namje Park, Howon Kim, Seungjoo Kim, and Dongho Won: Open Location-based Service using Secure Middleware Infrastructure in Web Services,. Lecture Notes in Computer Science, Vol. 3481. Springer-Verlag (2005) 1146–1155
11. Y. Yutaka, K. Nakao: A Study of Privacy information Handling on Sensor Information etwork. Technical report of IEICE (2002)
12. Nokia.: RFID Phones – Nokia Mobile RFID Kit. http://europe.nokia.com/nokia
13. Sangkeun Yoo: Mobile RFID Activities in Korea. The APT Standardization Program (2006)
14. Namje Park, Jin Kwak, Seungjoo Kim, Dongho Won, and Howon Kim: WIPI Mobile Platform with Secure Service for Mobile RFID Network Environment. Lecture Notes in Computer Science, Vol. 3842. Springer-Verlag (2006) 741–748
15. Byungho Chug, et. Al.: Proposal for the study on a security framework for mobile RFID applications as a new work item on mobile security. ITU-T, COM 17 – D 116 – E, Geneva (2005)
16. Jin Kwak, Keunwoo Rhee, Soohyun Oh, Seungjoo Kim, and Dongho Won: RFID System with Fairness within the Framework of Security and Privacy. Lecture Notes in Computer Science, Vol. 3813. Springer-Verlag (2005) 142-152

A Hardware Implementation of Lightweight Block Cipher for Ubiquitous Computing Security

Jong Sou Park, Sung-Hwan Kim, and Dong Seong Kim

Network Security Lab., Computer Engineering Department, Hankuk Aviation University
{jspark, elfeye, dskim}@hau.ac.kr

Abstract. Ubiquitous Computing applications have resource constraints and conventional block ciphers are infeasible to meet their requirements. This paper presents a hardware design and implementation of lightweight block cipher based on Reversible Cellular Automata (RCA), which is able to generate high pseudo random patterns with low computational overheads. The prototype implementation shows that the total number of gate is less than 3k and operates for 400 clock cycles in 82.156 MHz and it outperforms AES and NTRU. In addition, the cryptanalysis including Strict Avalanche Criterion (SAC), Differential Cryptanalysis (DC) and Linear Cryptanalysis (LC) to our implementation is satisfied.

1 Introduction

The ubiquitous computing applications have some resource constraints. In case of Radio Frequency Identification (RFID), the total number of gates in an implementation must be less than from 5K to 10K [1]. Note that a large number of data block encryption is not necessary or even impossible in most tiny ubiquitous computing devices. Therefore, our design will be an obvious choice for the design of a new block cipher aiming at extremely resource-constrained applications. This paper represents Reversible Cellular Automata (RCA) based lightweight block cipher. RCA is an extended form of Cellular Automata (CA), and CA generates high pseudo random patterns and has been utilized in various applications [2-6]. We adopt RCA, which has not only the previous merits of CA in terms of the computation and generation of high pseudo random patterns but also more efficient architecture to design a reversible block cipher than general CA [9]. We utilize superior rules proposed by F. Seredynski *et al.* [7]; especially, 5 rules including rule 86, 90, 101, 153 and 165, which generate higher pseudo random patterns than other rules. The design rationale of RCA based block cipher is based on the overall architecture of the block cipher module in [8]. We compare our approach with existing approaches; the comparison results show that the proposed approach outperforms the existing block ciphers such as AES, NTRU and so on, in terms of the number of gates and the processing time. The proposed approach also satisfies cryptanalysis such as Differential Cryptanalysis (DC), Linear Cryptanalysis (LC) and Strict Avalanche Criterion (SAC).

B. Gabrys, R.J. Howlett, and L.C. Jain (Eds.): KES 2006, Part I, LNAI 4251, pp. 935 – 942, 2006.
© Springer-Verlag Berlin Heidelberg 2006

2 Proposed Approach

2.1 Reversible Cellular Automata

Before presenting our proposed approach, we remind of reversible cellular automata (RCA) in brief. When analyzing elementary CA, it turns out that only a small number of rules have the property of being reversible. For instance, among all 256 radius 1 CA rules, only six rules are reversible. Different Reversible Cellular Automata (RCA) classes have been presented in [11]. Toffoli and Margolus [12] have proposed a simple explanation of the general idea of this method. In this paper, we use RCA class presented by Wolfram [13]. The algorithm of design for RCA with rule 153 can be summarized in the following way:

1. *If the step-back at t-1 is 0 (or 1).*
2. *The applied function is a rule 153 (or rule 102).*

The previous algorithm which provides random number patterns or symmetric keys can be applied to RCA8, RCA8LR and RCA8sng which operate as the S-box of AES. The overall architecture of proposed approach is presented in next subsection.

2.2 Overall Architecture for 16 Rounds

The overall architecture of encryption for 16 rounds is depicted in figure 1. Total size of key is 24 bits for RCA8LR, RCA8 and RCA8sng with the size of 8 bits. The architecture can encrypt or decrypt input data of 128 bits. *Initial data*, *Plaintext*, *Initial data1*, *Initial data2*, *LE* and *Clock* are inputted to the 1 round encryption or decryption. RCA algorithm needs both *Initial data* and *Plaintext* as inputted to encrypt or decrypt. *Initial data1* and *Initial data2* are necessary for subcomponent for one round.

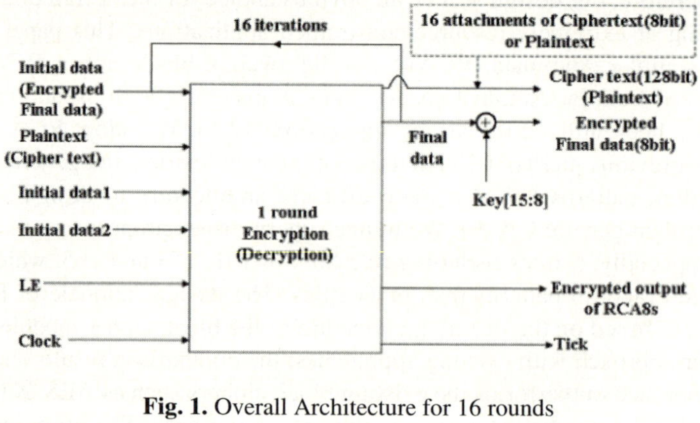

Fig. 1. Overall Architecture for 16 rounds

LE enables the inputs to be loaded to 1 round encryption. *Clock* brings synchronization to all devices of the block cipher. 16 iterations represent that 8-bit *Plaintext* or *Ciphertext* is respectively encrypted or decrypted over again for 16 counts. After 16 iterations, 128-bit *Cipher text* is outputted. Each *Final data* at every round is inputted over again as an *Initial data* for 16 rounds. The last *Final data* in 16th round is encrypted by XOR with key data from 9th bit to 16th bit. The 16th *Final data* is outputted as an output. *Tick* is connected to *LE* for every round. *Tick* instructs next encryption (or decryption) when previous one is finished.

2.3 The Subcomponents for One Round

Block diagram of encryption for single round is depicted in figure 2. It has five pages including Round inputs, RCA8 iteration, Substitution and Shift operation, RCA8LR iteration, and Next round initial data and cipher text. An algorithm is uniform when one rule is only applied to the algorithm. An algorithm with many rules is non-uniform. The data size of each block is 8 bits. RCA8, RCA8sng and RCA8LR are iterated for 11 steps. The data labeled c0 and f0 are input data for next process and respectively divided into to two blocks with 8-bit size. These blocks are shifted to the right or the left and substituted. After RCA8LR, cipher text and final data are outputted and initiated for the next round.

Decryption for single round is described in figure 3. It is reversed operation of encryption. Shift operations of encryption are replaced with each other in decryption and the final output of RCA8sng in encryption is decrypted by XOR with the key. Decryption does not have RCA8sng but needs the final output to shift in reverse order.

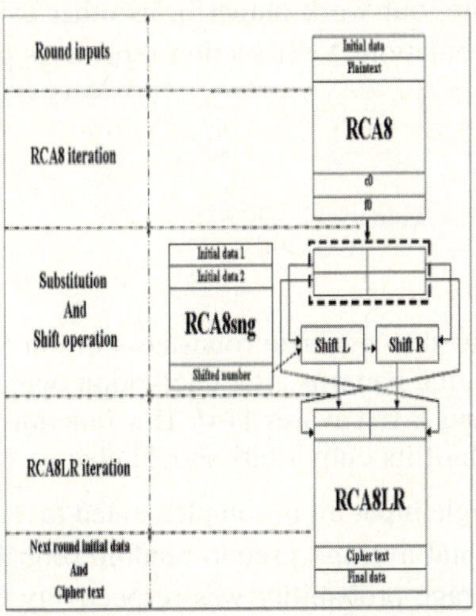

Fig. 2. Block diagram of encryption for one round

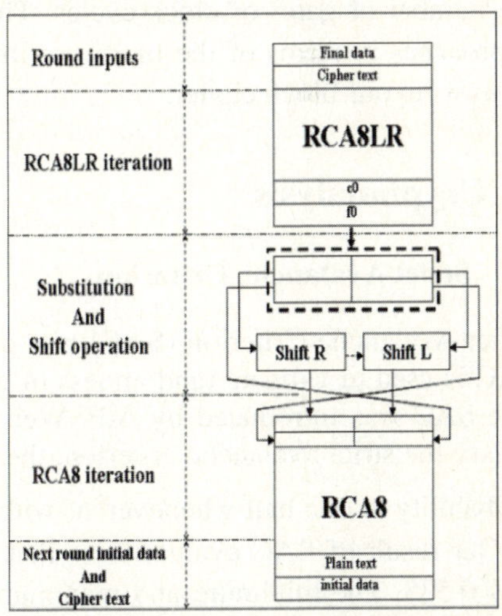

Fig. 3. Block diagram of decryption for one round

3 Comparison Results

The implementation of proposed work operates at max frequency 82.156 MHz and needs 400 clock cycles for encryption or decryption of each 128-bit data block. The required number of gates is estimated to be 2,898 gates for the encryption and to be 1,958 gates for the decryption. All presented results come from VHDL and FPGA Xilinx.

Table 1. A comparison of the implementation between proposed approach and previous approaches

	Gate	Clock cycle
Proposed approach	2,898	400
128-bit AES [10]	3,628	1,016
600-bit NTRU [11]	2,975	722,401
32-bit Mangard [12]	10,799	64
128-bit Verbauwhede [13]	173,000	10

Table 1 presents a comparison of our approach with existing implementations. This represents when comparing our work with existing ciphers such as AES, the number of gates was reduced by 20 % and the number of clock cycles was reduced by about 59%. When comparing with the other ciphers, there was a noticeable gap in terms of the number of gates or clock cycles. Therefore, our work outperforms other existing approaches in terms of the hardware implementation. Next section represents cryptanalysis to our block cipher.

4 Cryptanalysis

4.1 Strict Avalanche Criterion

Strict Avalanche Criterion (SAC) was performed to evaluate robustness of our work. SAC is used to validate randomness of generating patterns after encryption operation. The SAC was introduced by A.F. Webster and S.E Tavares [14]. If a function is to satisfy the strict avalanche criterion, then each of its output bits should change with a probability of one half whenever x with a single input bit is complemented to $\bar{x}$.

Our result of SAC evaluation is that the total average pseudo random probability was 0.535. The minimum and maximum average probability was respectively 0.453 and 0.609. At least, the random patterns by 45.3% were generated. These results satisfy SAC and it is not yet enough to satisfy cryptanalysis for a block cipher. Next subsections present the results of differential cryptanalysis and linear cryptanalysis.

4.2 Differential Cryptanalysis

Differential Cryptanalysis (DC) is an algorithm for cryptanalysis of block cryptography. DC is an attack which examines changes in the output of the block cipher in response to controlled changes in the input. When Boolean function is S: $\{0, 1\}^n \rightarrow \{0, 1\}^m$ (n and m represents respectively data length of input and output), each of equations of DC on S-box is summarized as following.

- XOR distribution of S-box, $\delta s(a,b)$:

$$\delta s(a,b) := \#\{x \mid S(x) \oplus S(x \oplus a) = b\} \quad (1)$$

- XOR difference probability, $DP^s(a \rightarrow b)$:

$$DP^s(a \rightarrow b) := \delta s(a,b) / 2^n \quad (2)$$

- Maximum difference probability, P_d or DP^s max :

$$P_d \text{ or } DP^s \max := \max_{a \neq 0, b} DP^s(a \rightarrow b) \quad (3)$$

In equation (1), a and b represent respectively difference of input and output. Each of x and S(x) represents input and output. P_d is also presented to DP^s max by (3). The algorithm presents that counts the total number when the sub-equation $S(x) \oplus S(x \oplus a) = b$ is true. After $\delta s(a,b)$ is researched by (1), each of $DP^s(a \rightarrow b)$s is calculated and among the values of $DP^s(a \rightarrow b)$, maximum difference probability called P_d or DP^s max is investigated where $\alpha \neq 0$.

Table 2. A comparison of robustness between RCA and SPN of H. M. Heys against DC

Comparison of Researches	RCA based block cipher	SPN of H. M. Heys [16]
P_D	$(\frac{3}{64})^{48}$	$(\frac{1}{2})^{52}$

If it is assumed that RCA-boxes (RCA8, RCA8LR, RCA8s) are similar to S-box of DES or AES, equation (1) is able to be performed for each of RCA-boxes. We evaluated XOR distribution with 256 initial data, inputs and outputs (for total 66536 times). After calculation of XOR probability distribution, by equation (3), P_d or $DP^s \max = \frac{3}{64}$.

Where the total number of S-box is n in a block cipher, $P_D < (P_d)^n$ in [16]. Each of P_D and $(P_d)^n$ represents total round differential probability and its boundary probability. The smaller P_D, i.e. $(P_d)^n$, the more known plaintexts required for the cryptanalysis. Namely, the larger n of $(P_d)^n$, a block cipher against DC is more robust. Since the number of RCA-boxes is 3 for single round, the total number of RCA-boxes for 16 rounds is 48 and $P_D < (\frac{3}{64})^{48}$. Table 2 represents a comparison our results with data of H. M. Heys *et al.* [16]. If P_D is getting smaller, proposed block cipher against DC is more robust. The comparison is that $(\frac{3}{64})^{48} < (\frac{1}{2})^{52}$.

4.3 Linear Cryptanalysis

In [15], Matsui has presented an effective linear cryptanalysis method for DES. The attack uses a known plaintext technique to extract key information by finding a linear equation consisting of plaintext, cipher text, and key terms which is statistically likely to be satisfied. When Boolean function is S: $\{0, 1\}^n \rightarrow \{0, 1\}^m$ (n and m represents respectively data length of input and output), LC equation on S-box is summarized as following:

- Linear approximation distribution of S-box, $\lambda_s(a,b)$:

$$\lambda_s(a,b) := \#\{x \mid a \bullet x = b \bullet S(x)\} - 2^{n-1} \tag{4}$$

- Linear approximation probability of S-box, P_ε:

$$P_\varepsilon := (\lambda_s(a,b) + 2^{n-1}) / 2^n \tag{5}$$

- Maximum linear approximation of S-box, Λ_s:

$$\Lambda_s := \max_{a,b \neq 0} |\lambda_s(a,b)| \tag{6}$$

Equation (6) represents if Λ_s is getting smaller, a block cipher against LC is getting more robust. We evaluated all linear approximation distributions with 256 initial data, inputs and outputs for total 66536 times. The maximum value of Λ_s was 40 and P_ε was $\frac{40+2^{47}}{2^{48}}$, i.e. $\frac{21}{32}$.

From Lemma 3 of [15], equation (7) and (8) are defined as follows:

$$|P_L - \frac{1}{2}| \leq 2^{n-1} |P_\varepsilon - \frac{1}{2}|^n \tag{7}$$

$$N_L = |P_L - \frac{1}{2}|^{-2} \tag{8}$$

If $|P_L - \frac{1}{2}|$ or $2^{n-1} |P_\varepsilon - \frac{1}{2}|^n$ are getting smaller and the *n* are getting larger, the block cipher against LC is more robust. In [16], the minimum $N_L = 2^{50}$. By (8), $|P_L - \frac{1}{2}| = \frac{1}{2^{25}}$. Where $P_\varepsilon = \frac{21}{32}$, by equation (7), $2^{n-1} |P_\varepsilon - \frac{1}{2}|^n = 2^{47} \times (\frac{5}{32})^{48}$.

Table 3. A comparison of robustness between RCA and SPN H. M. Heys against LC

Comparison of Researches	RCA based block cipher	SPN of H. M. Heys [16]
$\mid P_L - \frac{1}{2} \mid$	$2^{47} \times (\frac{5}{32})^{48}$	$\frac{1}{2^{25}}$

Table 3 represents a comparison of robustness between RCA with it of SPN against LC. Since $2^{47} \times (\frac{5}{32})^{48} < \frac{1}{2^{25}}$, the proposed work has the super robustness than SPN.

5 Conclusions

The existing CA-based ciphers are not feasible for resource constrained ubiquitous computing applications. This paper has represented a design and implementation of RCA based lightweight block cipher. The implementation has 2,898 gates for encryption and 1,958 gates for decryption and has operated for 400 clock cycles at max frequency of 82.156 MHz. When comparing this work with existing approach such as AES, the total number of gates was respectively reduced by 20 % and the number of clock cycles was reduced by about 59 %. When comparing our approach with the other ciphers, there has been a noticeable gap in terms of the number of gates or clock cycles. In addition, the proposed block algorithm has satisfied cryptanalysis such as SAC, DC, and LC. In conclusions, our implementation is applicable for resource constrained ubiquitous computing applications.

Acknowledgements

This research was supported by the MIC (Ministry of Information and Communication), Korea, under the ITRC (Information Technology Research Center) support program supervised by the IITA (Institute of Information Technology Assessment).

References

1. Dorothy S. E. Sarma, S. A. Weis and D. W. Engels, Radio-Frequency Identification, "Security Risks and Challenges", RSA Laboratories Cryptobytes, 6(1), pp.2-9, Spring 2003.
2. P. P. Chaudhuri, A. R. Chowdhury, S. Nandi, S. Chattopadhyay, "Additive Cellular Automata, Theory and Applications", IEEE Computer Society Press, Volume 1, 1997.
3. S. Bhattacharjee, S. Sinha, C. Chattohyay, P. P. Chaudhuri, "Cellular Automata based scheme for solution of Boolean equations", IEEE Proceedings, Computer and Digital Techniques, vol. 143, no. 3, 1996.
4. M. Mihaljevic, H. Imai, "A Family of Fast Keystream Generations based on Programmable Linear Cellular Automata over GF(q) and Time-Varient Table", IEICE Transactions on Fundamentals, vol. E82-A, no.1, pp.32-39, 1999.

5. M. Mihaljevic, Y. Zhang, H. Imai, "A Fast and Secure Stream Cipher based on Cellular Automata over GF(q)", IEEE Global Telecommunications Conference, GLOBECOM '98, vol.6, pp.3250-3255, 1998.
6. B. Srisuchinwong, T. A. York, Ph. Taslides, "A Symmetric Cipher using autonomous and non-autonomous cellular automata", IEEE Global Telecommunications Conference, GLOBECOM '95, pp.1172-1177, 1995.
7. F. Seredynski, P. Bouvry, Albert Y. Zomaya, "Cellular Automata computations and secret key cryptography", Parallel Computing 30 , pp. 753-766, 2004.
8. M. Seredynski, K. Pienkosz and P. Bouvry, "Reversible Cellular Automata Based Encryption", Network and Parallel Computing, IFIP International Conference, NPC 2004.
9. T. Toffoli, N. Margolus, "Invertible cellular automata", Physica D 45, pp. 229-253, 1997.
10. M. Feldhofer, S. Dominikus, and J. Wolkerstorfer, "Strong Authentication for RFID Systems Using the AES Algorithm", CHES 2004, LNCS 3156, pp. 357-370, 2004.
11. C. M. O'Rourke, "Efficient NTRU Implementations", M. S. Thesis, Electrical & Computer Engineering, Worcester Polytechnic Institute, 2002.
12. S. Mangard, M. Aigner, and S. Dominikus, "A Highly Regular and Scalable AES Hardware Architecture", IEEE Transactions on Computers, 52(4):483–491, April 2003.
13. I. Verbauwhede, P. Schaumont, and H. Kuo, "Design and Performance Testing of a 2.29 Gb/s Rijndael Processor", IEEE Journal of Solid-State Circuits, pages 569–572, March 2003.
14. A.F.Webster and S.E. Tavares, "On the Design of S-Boxes, Advances in Cryptology", Crypto'85 proceedings, Springer, 1986.
15. M. Matsui, "Linear cryptanalysis method for DES cipher", Advances in Cryptology, Proceedings of EUROCRYPT'93, Springer-Verlag, Berlin, pages 386-397, 1994.
16. H. M. Heys and S. E. Tavares, "Substitution-Permutation Network Resistant to Differential and Linear Cryptanalysis", Journal of Cryptology, page 148-155, 1996.

Intelligent Environment for Training of Power Systems Operators

G. Arroyo-Figueroa[1], Yasmin Hernandez[1], and Enrique Sucar[2]

[1] Gerencia de Sistemas Informaticos, Instituto de Investigaciones Electricas
Reforma # 113, Col. Palmira,
69042 Cuernavaca, Morelos, Mexico
{garroyo, myhp}@iie.org.mx
http://www.iie.org.mx/
[2] Instituto Nacional de Astrofísica Optica y Electrónica
Luis Enrique Erro #1,
72840 Sta. María Tonantzintla, Puebla, Mexico
esucar@inaoep.mx

Abstract. Training of operators has become an important problem to be faced by power systems: updating knowledge and skills. An operator must comprehend the physical operation of the process and must be skilled in handling a number of normal and abnormal operating problems and emergencies. We are developing an intelligent environment for training of power system operators. This paper presents the architecture of the intelligent environment composed by computer based training components labeled as reusable learning objects; a learning object repository; concept structure map of the power plant domain, a tutor module based on course planner, operator model based on cognitive and affective components and operator interface. The general aim of our work is to provide operators of complex industrial environments with a suitable training from a pedagogical and affective viewpoint to certify operators in knowledge, skills, expertise, abilities and attitudes for operation of power systems.

1 Introduction

Training of operators is an important problem faced by thermal power plants: updating knowledge and skills. The process of learning how to control, maintain and diagnose in a complex industrial environment takes years of practice and training. An operator must comprehend the physical operation of the process and must be skilled in handling a number of abnormal operating problems and emergencies. The problem increases when the complexity of the system obstructs the efficiency, reliability and safe operation. So the novice and experienced operators need continuous training in order to deal reliably with uncommon situations.

Distributed control system (DCS) and management information systems (MIS) have been playing an important role to monitor the plant status. These systems are generally employed to deal with information during normal operation. However, in abnormal operations such as equipment failures and extreme operation, human operators have to rely on their own experience. During disturbances, the operator must

B. Gabrys, R.J. Howlett, and L.C. Jain (Eds.): KES 2006, Part I, LNAI 4251, pp. 943 – 950, 2006.
© Springer-Verlag Berlin Heidelberg 2006

determine the best recovery action according to the type and sequence of the signals received. In a major upset, the operator may be confronted with a large number of signals and alarms, but very limited help from the system, concerning the underlying plant condition. Faced with a vast amount of raw process data, human operators find it hard to contribute a timely and effective solutions.

For safety efficiency reasons it is not recommend to train operators on real equipment. Several ways to cope with operators training has been proposed. Most proposals for operator training are based on Computer Based Training Systems (CBT) and simulators. CBT is a computer system that groups a set of programs and training methods that uses the computer capabilities to present the operator with the instructional material and to provide an environment for practical training. The main advantage of CBTs are their unlimited availability: they can be used whenever the trainee wants.

Due to necessity of training of operation skills in power plants, simulators have been a common alternative. A power simulator is composed of a set of quantitative and qualitative models which simulate the physical behavior of the plant. The aim of training with a simulator is to get trained operators not only in performing the common procedure, but in analyzing abnormal situations and performing emergency procedures as well. The next generation in training systems are computer-based training systems that provides instruction just like a human trainer does. The training requirements ask for powerful interfaces, a more efficient and better adaptive training, by means of incorporating artificial intelligent (AI) techniques, adaptive interfaces, simulations tools, learning objects based on multimedia and virtual reality components.

This paper describes the architecture and development of an intelligent environment (IE) for the training of power systems operators. In contrast with a traditional training system, the main goal of the intelligent environment is to certify operators in knowledge, skills, expertise, abilities and attitudes for operation of power systems.

2 Architecture of the Intelligent Environment

The architecture of the intelligent environment is based on dynamic course generating systems proposed by Brusilovsky [1]. The architecture of the intelligent environment, shown in Figure 1, is composed by five main modules: Learning Content Module, Domain Knowledge Module, Tutor, Operator Model, and Operator Interface.

2.1 Learning Content Module

This module contains tools for edition of teaching and testing materials based on learning objects. The teaching and testing materials consider both theoretical and practical concepts contained in: electronic books, simulation tools, multimedia and virtual reality tools, to present to the operator pedagogical actions such as explanations, exercises, tests, and so on. Each material is labeled as Reusable Learning Objects (RLO) or Shared Content Objects (SCO) according with the SCORM (Sharable

Content Object Reference Model) terminology [2]. Learning objects are self-contained learning components that are stored and accessed independently. RLO are any digital resource that can be reused to support Web-based learning and learning management system (LMS). The authors or learning developers can create, store, reuse, manage and deliver digital learning content. The same object may be used as many times and for as many purposes as is appropriate. XML servers to this function by separating content from programming logic and code [3].

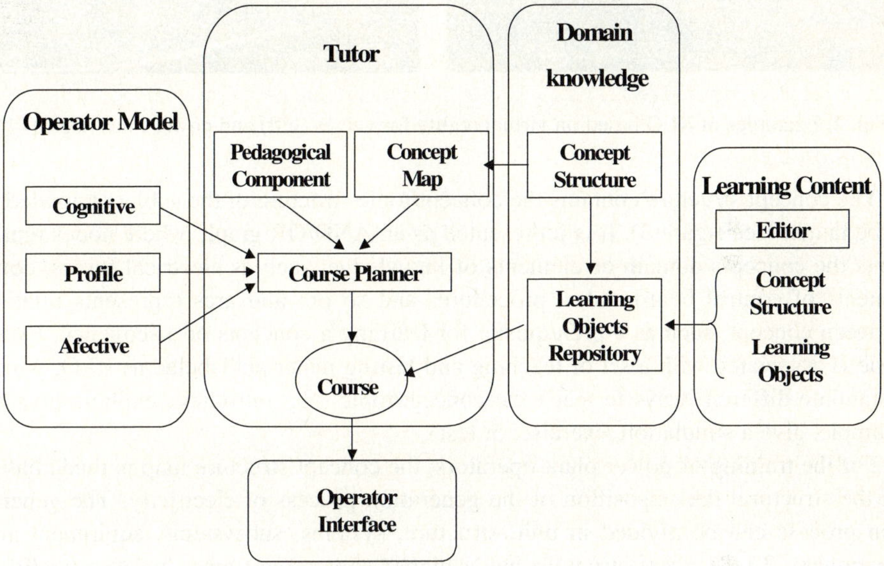

Fig. 1. Architecture of the intelligent environment

The learning objects are either dispensed to users individually or used as components to assemble larger learning modules of full courses, depending on individual learning needs. The instructional output many be delivered via the Web, CD-ROM or printed materials. We have developed reusable learning objects for two areas: valve maintenance and maintenance of energized power-lines, based on virtual reality [4]. Some examples are depicted in figure 2.

2.2 Domain Knowledge Module

The domain knowledge module has two main components: the concept structure map and the learning objects repository. The domain knowledge contains the expert's operation abilities in procedure and malfunction operations and its representation considers both theoretical and practical concepts. It is possible to organize the domain concepts/topics into a set of smaller, possibly interrelated AND/OR graphs, representing relatively independent sub-areas of the knowledge, different views, or different levels of granularity.

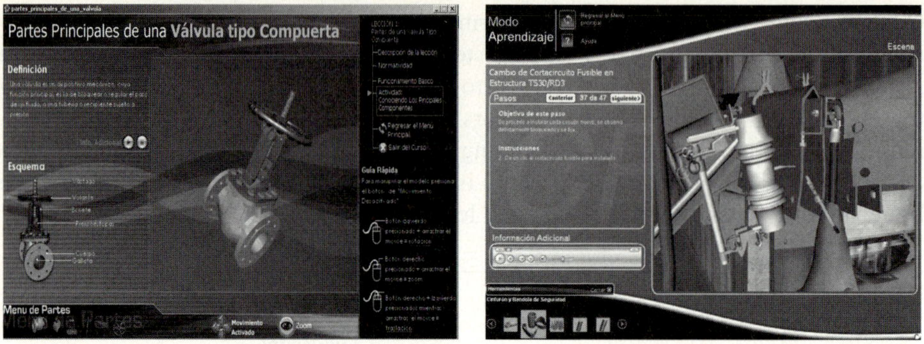

Fig. 2. Examples of RLO based on virtual reality for valves (left) and power lines (right) [4]

The concept structure contains the concept/topic structure of the subject knowledge to be taught (see figure 3). It is represented as an AND/OR graph, where nodes represents the concepts domain or elements of knowledge, such as electrical topics, components of control board, rules, procedures and so on; and arcs represents relation between concept, such as a prerequisite for learning a concepts or a sequence. Every node is associated with a set of teaching and testing materials labeled as RLO, which instantiate different ways to teach the concept/topic (e.g. introduce, explain, give an example, give a simulation, exercise, or test).

For the training of power plant operators, the concept structure map is made based on the structural decomposition of the generation process of electricity. The generation process can be divided in unit, structure, systems, subsystems, equipment and component. The concept structure has a higher expressive power because it allows representing not only prerequisites, but many different types of relations between concepts; and it enables the use of AI planning techniques for the generation of alternatives courses. Therefore, it guarantees a wide variety of different teaching goals and several courses for achieving these goals.

The learning content module has a Learning Object Repository (LOR). The LOR is a central database in which learning content (RLO) is stored and managed.

2.3 Tutor

The tutor is the component that generates the sequence of learning objects (teaching and testing materials) to be presented to operator in a form of course. Taking as a basis the AND/OR graph representing the concept map, the pedagogical component and the operator model, the training course planner generates a specific course. The concept plan is a sub-graph of the graph of knowledge domain (concept structure). The pedagogical component provides a set of teaching rules for the selection of content and presentation of the concept plans according with the cognitive style, affective state or learning preferences of the operator. The pedagogical component contains the learning goals and several pedagogical strategies to teach operators and select the pedagogical strategy to be used, based on operator model. This strategy represents the specific necessities of the operator from both, affective and pedagogical point of view.

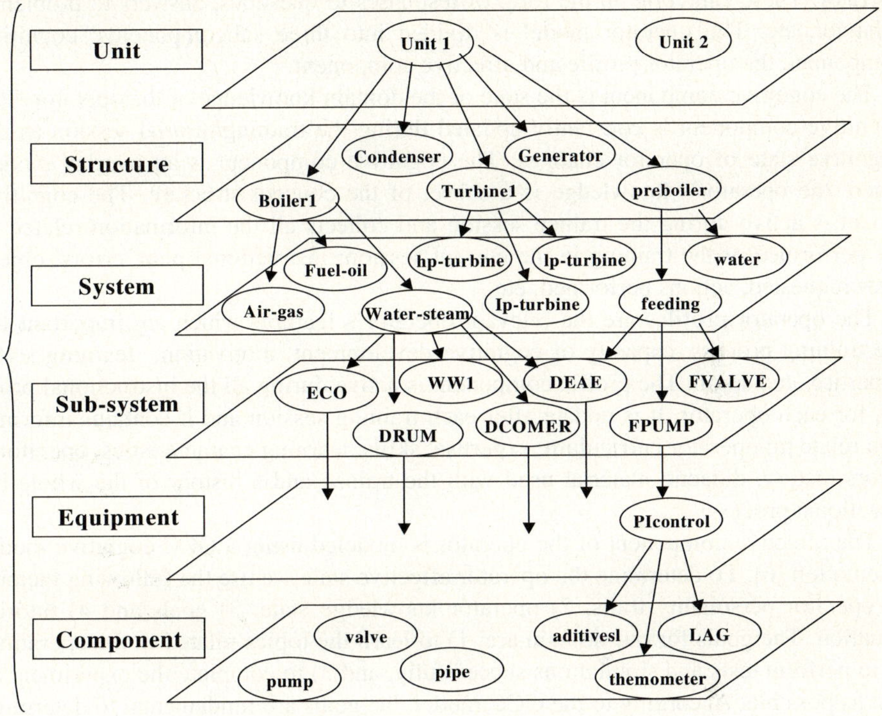

Fig. 3. Concept structure map for training of power plant

The training course planner (TCP) is a compromise between an Intelligent Tutorial (ITS) system and a traditional CBT system. A course generated by TCP looks like a traditional structured course. However, this course is generated individually for every operator to achieve a certain training goal (a concept or topic that has to be learned). The course planner is an AI for planning system based on the well known algorithm for graph planning, AO [5]. Each course is generated in a dynamic way based on particular characteristics of the operator, and it can be changed also dynamically to reflect changes in operator model. During the presentation of the course to the operator, if the operator answers the test items or simulator task correctly, i.e. demonstrates that she/he has acquired the concepts or skills, no changes to the course are necessary. However, if the operator fails to demonstrate knowledge of a concept, a re-planning of the course follows. The TCP generates a new sequence of concepts and teaching materials leading to the goal concept, starting from the current state of student knowledge as recorded in the operator model.

2.4 Operator Model

The operator model is used to adapt the intelligent environment (IE) to each specific operator. The operator model is built from observations that the IE makes about the operator. These can come in the form of responses to questions, answers to problems, behavior, etc. The operator model is divided into three subcomponents: cognitive component, the operator profile and affective component.

The cognitive component is the state of the domain knowledge of the operator. The cognitive component is constantly updated during the training/tutorial session as the cognitive state of operator changes. The cognitive component is an overlay model, where the operator's knowledge is a subset of the concept structure. The cognitive model is active during the trainee session and collects all the information related to the performed of the training in the current session: instructional plan, errors, objectives requested, actions performed, etc.

The operator profiles are the relevant operator's features which are important for the training process: capacity of cognitive development, motivation, learning style, expertise, and so on. The profile component is active during all the instructional process for each operator. It is update after each training session and it contains information relate to: operator curriculum, expertise, skills, learning characteristics, operator's errors history, didactic material used with the trainee and a history of the whole instructional process.

The affective component of the operator is modeled using a OCC cognitive model of emotion [6]. To determine the operator affective state we use the following factors: 1) operator personality traits, 2) operator knowledge state, 3) goals and 4) tutorial situation. The goals for our domain are: 1) to learn the topics related to the operation, 2) to perform tests and simulations successfully, and 3) to complete the experiment as fast as possible. According to the OCC model, the goals are fundamental to determine the affective state; we infer them by means of personality traits (profile).

The OCC model establishes emotional state as a cognitive appraisal between goals and situation. We represent this comparison with nodes *goals* and *training situation*, propagating evidence to nodes *goals satisfied*, and with these last nodes propagating evidence to the nodes *affective state*. According to the OCC model, the goals are fundamental to determine the affective state; we infer goals by means of personality traits and the knowledge state.

We based personality traits on the five factor model [7], which considers five dimensions for personality: *openness*, *conscientiousness*, *extraversion*, *agreeableness*, and *neuroticism*. Currently, we use only two of them (*conscientiousness*, *neuroticism*) to establish goals because a relationship exists between these two dimensions and learning [8].

In order to know if a goal has been satisfied, we include the nodes *goals satisfaction* (one for each goal), and they are influenced by the nodes *goals* and *tutorial situation*. The nodes *goals satisfied* represent the comparison between goals and situation as established by the OCC model. The nodes *tutorial situation* have the outcomes of the user actions, and their values are obtained from the performance of the student on the intelligent environment.

From the emotion set proposed by the OCC model [6], we consider four possible emotional states: *joy, distress, pride, shame*. We use these emotions because they emerge as a consequence of the actions of the students. A representation of the affective operator model based on a probabilistic network [10] is presented in Figure 4.

The affective state is used in combination with the knowledge state (in the cognitive component) to select the "best" tutorial action based on the current course.

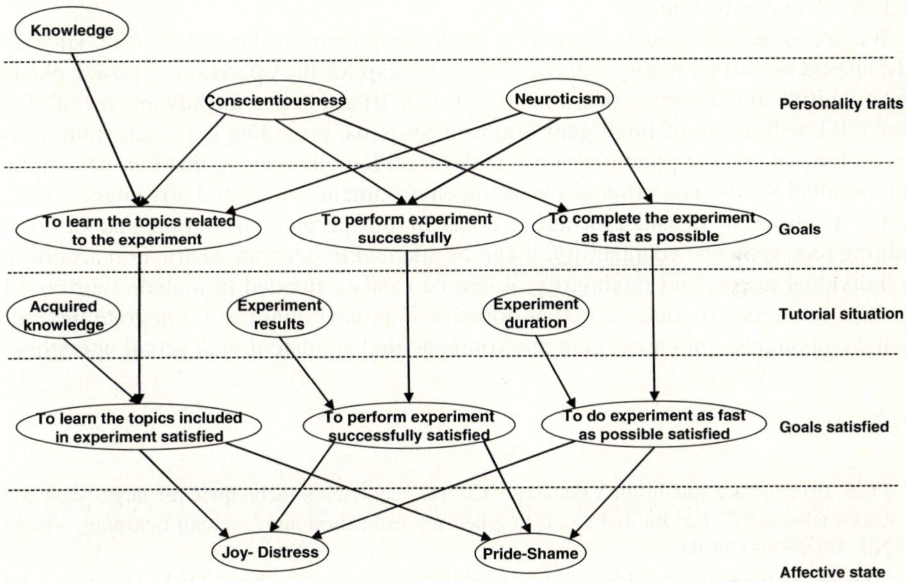

Fig. 4. Affective Operator Model

2.5 Operator Interface

The operator interface controls the interactions with the operator, including the dialogue and the screen layout The main purpose of this module is to present to the operator the teaching materials in the most effective way. One of major challenge of this module is to take advantage of new technologies such as pedagogical agents, virtual reality, natural language processing, affective wearables, among others. The user interface must provide the motivation for the operator to continue. The application of new technologies is promising, for example the use of animated pedagogical agents give students the sensation of someone behind the program and it can be used transmit enthusiasm and motivate them. The use of virtual reality provides to operator a mock up of the real work environment. This last one enables to train operator how to perform a task, guide him in the work and show him, for instance, where the work tools are.

3 Conclusions

In this paper we have presented the architecture of an Intelligent Environment for training of Power Systems Operators. The aim is build an advance training system which includes: computer based training components labeled as Reusable Learning Objects; a learning object repository; concept structure map of the power plant domain; a tutor module based on course planner and operator model based on cognitive and affective components.

We presented preliminary results for each component of the architecture; such as-RLO based on virtual reality, concept structure map for the operation of power plants, and cognitive and affective operator model. Our IE combines the advantages of classical CBT with those of Intelligent Tutoring Systems, providing a general framework for training of power plant operators with reusable components, that can be adapted to individual needs. The proposed learning environment has several advantages: flexibility, it allows to consider different areas of power plant operation and different training requirements; adaptability, it can be adapted to generate instruction according to individual needs; and modularity, it can be easily extended to include more trainees, and more experiments and other domains. The next phase is to integrate the individual components into a working environment, and validate it with actual operators.

References

1. Peter Brusilovsky and Julita Vassileva: "Course sequencing techniques for large-scale web based education", Int. Journal Cont. Engineering Education and Lifelong Learning, Vol 13, No. 1/2, (2003) 75-94.
2. ADL. Sharable Content Object Reference Model Version 1.2:The SCORM Overview. Advanced Distributed Learning, (2001) URL http://www.adlnet.org.
3. W3C. Extensible Markup Language (XML) 1.0. World Wide Web Consortium 2 ed. W3C Recommendation, (2000). URL http://www.w3c.org/TR/2000/REC-xml-2001006.
4. M. Perez, Virtual Reality: introduction, prototypes, applications and tendencies (In Spanish): Boletin IIE, Vol. 28, No. 2, (2004), Abril-Junio.
5. Darwyn R. Peachey and Gordon I. McCalla: "Using planning techniques in intelligent tutoring systems", Int. Journal Man-Machines Studies, Vol. 24, (1986) 77-98.
6. Ortony, A., Clore, G.L., Collins, A.: The Cognitive Structure of Emotions. Cambridge University Press (1988).
7. Costa, P.T., McCrae, R.R.: Four Ways Five Factors are Basic. Personality and Individual Differences, 13(1) (1992) pp. 653-665.
8. Y. Hernández, J. Noguez, Enrique Sucar and G. Arroyo-Figueroa, "A probabilistic model of affective behavior for Intelligent Tutoring Systems", A. Gelbukh, A. de Albornoz, and H. Terashima (Eds): MICAI 2005, LNAI 3789, (2005) 1175-1184.
9. J. Pearl: "Causality: Models, Reasoning and Inference",Cambridge University Press, (2000).

Building a Better Air Defence System Using Genetic Algorithms

Millie Pant[1] and Kusum Deep[2]

[1] Lecturer, BITS-Pilani, Goa Campus, India
[2] Associate Professor, IIT-Roorkee, India
millidma@gmail.com, kusumfma@iitr.ernet.in

Abstract. It is the aim of every country to have a good and strong defence system for the protection of its people and its assets. In this paper we have shown the application of Genetic Algorithms (GA'S) for optimizing the expected survival value of an asset subjected to air attacks. We have developed a mathematical model of the problem subjected to relevant constraints. We have solved this problem with the help of Binary Coded Genetic Algorithm or Simple Genetic Algorithm (SGA) and Real Coded Genetic Algorithm (RCGA). For RCGA we have developed a new crossover operator called the Quadratic Crossover Operator (QCX), which is multi parental in nature. This operator makes use of three parents to produce an offspring, which lies at the point of extrema of the quadratic curve passing through the three selected parents. The working of the operator is shown with help of a simple, steady state Genetic Algorithm having conditional elitism. After testing the validity of this algorithm on several test problems we applied it to the mathematical model of the air defence problem. The comparison of results show that although both the techniques are well suited for solving the above said problem, RCGA with QCX operator gives slightly better results then the SGA.

Keywords: Optimization, Genetic Algorithms, Elitism, Recombination Operator, Air defence models.

1 Introduction

GA's are perhaps the most popular form of Evolutionary Algorithms being used for solving complex optimization problems (Schwefel [1], Goldberg [2], Fogel [3], Bach et al. [4]). Their working mechanism is based on the principle of evolution in nature. In essence GA's consists of a population of solutions, which are transformed in each generation with the help of genetic operators. Traditionally the GA's consisted of the population of bit string. However during the past few years a lot of work has been done in the field of Real Coded Genetic Algorithms (RCGA's) where the individuals in the population are continuous floating-point numbers. Various RCGA's have been proposed so far (Davis [5], Radcliffe [6], Wright [7], Janikow [8], Michalewicz [9], Eshelman [10], Voigt et al [11], Ono and Kobayashi [12], Kita [13]).

As in Binary Coded Genetic Algorithms, in Real Coded Genetic Algorithms (RCGA) also, the types of reproduction may be divided into three categories viz. unisexual, bisexual and multisexual. Whereas unisexual reproduction is nothing but

B. Gabrys, R.J. Howlett, and L.C. Jain (Eds.): KES 2006, Part I, LNAI 4251, pp. 951 – 959, 2006.
© Springer-Verlag Berlin Heidelberg 2006

mutation, in which a gene goes on mutating itself to produce new offsprings, in bisexual reproduction two parents participate in the reproduction process to produce an offspring or offsprings. Similarly in a multisexual reproduction, an offspring is produced after the recombination of three or more parents.

In this paper we have suggested the use of GA's for solving the air defence problem, where the objective is to maximize the expected survival value of the asset. We have solved this problem using SGA having binary encoding and RCGA using a new crossover operator developed by us called the quadratic crossover operator (QCX) operator. This operator makes use of three parents to produce an offspring. In order to test the efficiency and robustness of RCGA-QCX operator we used it to solve the standard test problems available in literature and compared it with the original results as well as with the results obtained by SGA using binary encoding. The test problems varied from unimodal functions having unique global optima to multimodal problems having several global optima. The test problems and the numerical results obtained by the two algorithms are shown in Appendix I and Appendix II respectively.

In section 2, we have described the genetic algorithm using quadratic crossover operator (RCGA-QCX). In section 3, we have developed the mathematical model of the air defence problem and in section 4; we have given the numerical results and conclusions drawn, from the present work.

2 Genetic Algorithms Using QCX or the Quadratic Crossover Operator

In the recent past, many recombination/ crossover operators have been proposed for the RCGA's for solving optimization problems. In this paper a new crossover operator called the QCX is proposed. It is a multiparent crossover operator in which three parents participate in the reproduction process to produce an offspring, which lies on the point of extrema of the quadratic curve passing though the three selected parents. The three parents, for reproduction, are selected as:

$\vec{\xi}_L$ – Parent vector giving the best (minimum) function value $f\left(\vec{\xi}_L\right)$.

$\vec{\xi}_i$ and $\vec{\xi}_j$ – chosen randomly from the population (for $i \neq j$).

The three selected parents produce offspring

$$\vec{\varsigma}_K = .5\left(\frac{(\vec{\xi}_i^2 - \vec{\xi}_j^2) f(\vec{\xi}_L) + \left(\vec{\xi}_j^2 - \vec{\xi}_L^2\right) f\left(\vec{\xi}_i\right) + \left(\vec{\xi}_L^2 - \vec{\xi}_i^2\right) f\left(\vec{\xi}_j\right)}{(\vec{\xi}_i - \vec{\xi}_j) f(\vec{\xi}_L) + \left(\vec{\xi}_j - \vec{\xi}_L\right) f\left(\vec{\xi}_i\right) + \left(\vec{\xi}_L - \vec{\xi}_i\right) f\left(\vec{\xi}_j\right)}\right)$$

Most of the crossover operators available in the literature are parent centric in nature, i.e. the offspring produced lies in the vicinity of the parents. This behavior often leads to the premature convergence of the algorithm as the offspring's are not able to explore the existing search space. In the QCX operator the offspring lies at the point of extrema of the quadratic curve passing through the three selected parents; it sometimes lies far away from the selected parents. This gives a better opportunity to the offspring to explore the search space, which in turns prevents premature convergence.

The details of the algorithm are as follows

Initialization: an initial population (*parent population*) of feasible solution vectors is generated and the objective function value is calculated for each solution vector. These solution vectors along with their function values are stored in an array Ω (*initial gene pool*) in ascending order of function values. Size of Ω is taken as 10(n+1) for all test problems, where n is the number of decision variables.

Selection of individuals (parents): Select $\vec{\xi}_L$, the solution vector with the least function value $f\left(\vec{\xi}_L\right)$. Since $\vec{\xi}_L$ gave the best function value it is considered to be the most potent parent for producing healthy offsprings. Two other parents (solution vectors) say $\vec{\xi}_i$ and $\vec{\xi}_j$ $(i \neq j)$ are selected randomly fromΩ.

Mating of selected individuals using QCX operator: mating of individuals is done in order to produce offsprings for the next generation. $\vec{\xi}_L$, the most potent parent is mated with two randomly chosen parents $\vec{\xi}_i$and $\vec{\xi}_j$ $(i \neq j)$ with the help of QCX operator. All the individuals of the population participate in the reproduction process for different values of *i* and *j*.

New offsprings: if some of the offsprings produced after reproduction have illegal genes (i.e. the offspring lie outside the feasible domain), then instead discarding the offspring, repair mechanism is used in the following manner: offspring $\zeta_i = b_i$, if $\zeta_i > b_i$ and $\zeta_i = a_i$ if $\zeta_i < a_i$ where a_i and b_i are the lower and upper bounds of the feasible domain.

Elitism: elitism is the process in which the elite (best) solution of the previous generation is taken to the next generation also. Preserving of elite solutions from the previous generation assures that the best solution obtained so far will remain in the population and will have more opportunity to produce good offsprings. In our algorithm we have used the concept of conditional elitism i.e. the best solution of the previous generation is allowed to enter the next generation only if it is better then the worst solution of the next generation.

Termination criteria: we have taken into account two termination criteria and if either is satisfied the algorithm is terminated and the solution obtained is considered to be the best solution. The two criteria are:

(i) If $\left|\dfrac{f\left(\vec{\xi}_m\right)-f\left(\vec{\xi}_L\right)}{f\left(\vec{\xi}_m\right)}\right| < \varepsilon$ where $f\left(\vec{\xi}_m\right)$ and $f\left(\vec{\xi}_L\right)$ are the greatest and least function values corresponding to for solution vectors $\vec{\xi}_m$ and $\vec{\xi}_L$ respectively. ε is taken as 1×10^{-4} for all test problems. Or

(ii) When the number of function evaluations increases a certain predefined value (taken as 25,000 for all test problems in our algorithm) the algorithm is terminated.

The structure of the algorithm is given as:

Step 0: initialization (creation of initial gene poolΩ);

Step 0(a): gen_next = 0;

Step 1: selection of individuals for mating

Step 2: mating of individuals using QCX operator

Step 3: new offspring

Step 4: insertion of offspring into the population

Step 5: preserve the elite

Step 6:check the stopping criteria

Step 7(a): gen_next=gen_next+1

Step 8:finish

The working of GA-QCX operator is shown with the help of a simple two dimension Rastrigin's function [17], given as

Min $f(x_1, x_2, \ldots x_n) = 10n + \sum_{i=1}^{n}\left(x_i^2 - 10\cos(2\pi x_i)\right)$, for i = 1,2, …n. subject to: $-5.12 \le x_i \le 5.12$, I = 1,2, …, n.

This test problem is highly multimodal, however the locations of global minima are regularly distributed. Its global minima is at (0,0,…,0) with $f_{min} = 0$. To show the working of the algorithm graphically, we have considered it as a two dimension problem taking n = 2. Fig 1, 2, 3, 4 shows the working of GA-QCX algorithm after every 10 iterations. The other test problems tested with GA-QCX operator are shown in Appendix I and the numerical results obtained by SGA and GA using QCX operator are shown in Appendix II.

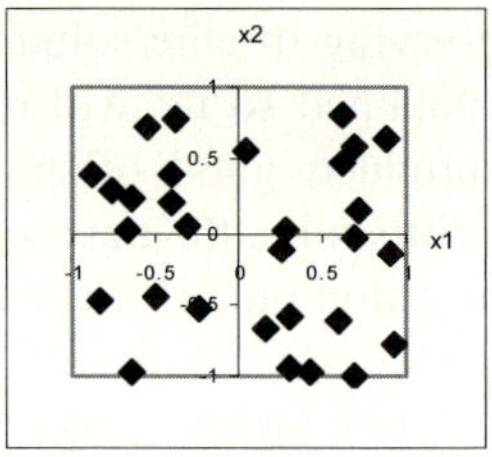

(1)

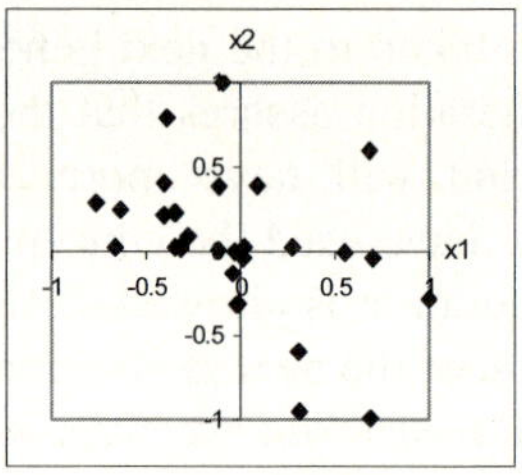

(2)

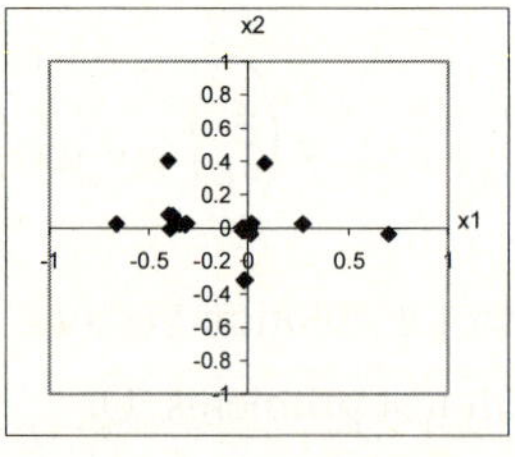

(3)

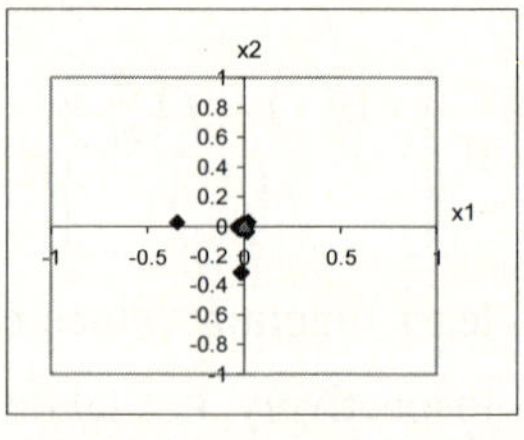

(4)

3 Mathematical Model of the Air Defence Problem

Mathematical models of Air Defence problems are usually of two types viz. maximizing the expected target damage or maximizing the expected survival value of an asset. In this paper we have considered the second case in which the objective is to maximize the expected survival value of an asset against air attacks. For various mathematical models for air defence please refer [22], [23], [24], [25], [26].

A country has a number of locations/ assets, which have to be protected against the attacks of the enemy. Every asset has an economic value, which is both related to their size and strategic importance. In this paper, we have taken into consideration, three types of weapon systems that can be deployed for protection of these assets namely, air craft in close air patrol and air defence, surface to air missiles of medium (around 25 – 28 km) and short (around 4 – 6 km) firing ranges and anti aircraft guns. The air craft primarily indulge in air to air combat around the site. The missiles are used for hitting and damaging the air craft leading to its kill. The air defence guns are used for creating a barrier around the target area, thereby restricting entry of enemy aircraft into its territory or around the asset area though the barrier. Enemy can launch an air attack from across the border near assets in order to damage them. We have developed a mathematical model of this problem, where the objective function is to maximize the expected surviving value of the asset, subjected to constraints like man power and weapon availability.

It has been assumed that the number and types of weapons of each type are known. For the sake of simplicity we have used certain notations given below:

d = type of defending weapon (d = 1, 2, 3,…D)
s = asset to be defended (s = 1, 2, 3,…S)
a = type of attacking weapon (a = 1, 2, 3…A)
Kd = kill probability of one defending weapon of type d.
W_{ds} = No. of defending weapons of type d deployed on asset s.
N_{sa} = No. of attacking weapons of type a aimed at asset a.
L_{sa} = probability that a single attacking weapon of type a completely destroys the asset s when it penetrates the defending weapons.
M_d = manpower required per defending weapon of type d.
B_d = number of defending weapons of type d
M_{maxs} = Maximum available manpower at asset s.

Depending on the criticality (how essential is the asset?), Vulnerability (how susceptible is the asset to the surveillance or attack?), Reconstructability (how hard it will be to recover from the inflicted damage?) and threat (how probable is an attack on this asset?) we have chosen 3 assets namely oil refineries, divisional headquarters and air base.

The expected probabilities for hitting the target and the weapon/ manpower required per weapon as well the as the other data has been obtained from Defence Research Development Organization (DRDO), New Delhi, India.

Table 1. Relevant information related to the assets

S. No.	Assets	Size: L×B	Asset Contents	Priority Value (Vs)	No. of attacking aircraft	Max. Man-power available
1	Oil refinery	500×1000	Oil wells, Tanks etc.	800	37	50
2	Div. Hq.	400×800	Bldgs, amm dump	600	32	55
3	Air base	500×800	Air strip, sheds	300	34	45

Table 2. Number of defensive weapons available for distribution over different assets, their probability of hitting the target and the manpower required per weapon

Weapon number	Nos. available	Probability of hitting (in %)	Man power required per weapon
ADG – 1	20	2.0	3
ADG – 2	15	1.8	4
ADMM – 1	30	70.0	3
ADMS – 1	40	85.0	2
ADMS-2	20	90.0	3
A/C – 1	10	65.0	0
A/C – 2	10	70.0	0

Table 3. The attacking plan N_{sa}

$N_{11} = 15$	$N_{12} = 12$	$N_{13} = 10$
$N_{21} = 10$	$N_{22} = 12$	$N_{23} = 10$
$N_{31} = 12$	$N_{32} = 12$	$N_{33} = 10$

Table 4. Probability that a single attacking weapon of type a completely destroys the asset s when it penetrates the defending weapons (L_{sa})

$L_{11} = 0.2302$	$L_{12} = 0.3504$	$L_{13} = 0.2987$
$L_{21} = 0.2900$	$L_{22} = 0.2059$	$L_{23} = 0.3333$
$L_{31} = 0.2814$	$L_{32} = 0.2667$	$L_{33} = 0.3108$

For constructing the mathematical model of the problem, we have considered the objective of maximizing the total expected surviving value of the asset. The constraints associated with problem are the ones restricting the man power and the constraint depending on the availability of the weapon. The mathematical model is given as:

Max $M_{tot} = \sum_s V_s H(s)$ (Objective is to maximize the total expected surviving value of the assets)

Where $H(s) = \prod_a \left[1 - \left\{ \prod_d (1 - Kd)^{(Wds / Nsa)} \right\} Lsa \right]^{Nsa}$

Subject to $\sum_{d} M_d W_{ds} \leq M_{\max s} \quad \forall \quad s$ (man power constraint)

$\sum_{s} W_{ds} \leq B_d \quad \forall \quad d$ (weapon availability constraint)

4 Numerical Results and Conclusions

In this section we have discussed the numerical results obtained for the air defence problem using RCGA-QCX and SGA. Since Genetic Algorithms are probabilistic in nature, both SGA and RCGA-QCX were executed 100 times out of which 10 best results are presented Table 5.

It can be observed that although both SGA and RCGA-QCX gave good results, the best value of the objective function obtained by RCGA-QCX is slightly better then the best value obtained by SGA. Also it can be seen that RCGA-QCX gave better results in terms of function evaluations.

Table 5. Comparison of numerical results obtained by RCGA using QCX operator and SGA

No. of Runs	RCGA- QCX		SGA using binary encoding	
	Best value of objective function	**No. of function evaluation**	**Best value of objective function**	**No. of function evaluation**
1	1581.92	1553	1567.73	2018
2	1573.56	1532	1566.55	1997
3	1557.99	1520	1575.37	1885
4	1557.99	1534	**1580.87**	**2101**
5	**1589.74**	**1782**	1575.34	1856
6	1561.89	1567	1565.33	2001
7	1580.73	1676	1559.08	1992
8	1581.92	1656	1575.37	2000
9	1589.74	1886	1562.57	1898
10	1589.56	1923	1575.87	2100

However this algorithm using QCX crossover operator is still in the stage of infancy and many improvements can be incorporated in it. A theoretical analysis of the fast convergence of the algorithm using QCX operator will be quite interesting. Also we can see what will be the effect on the algorithm when some suitable mutation operator is taken into account along with the QCX operator. Whether or not its hybridized version with techniques like Tabu Search, Simulated Annealing or Hill Climbing Method will be of some practical use is still under consideration. It should be tested for more complex problems and its performance should be compared with some standard crossover operator for real parameter like Arithmetic Crossover or UNDX crossover. The mathematical model of the air defence model can also be made more interesting and challenging by considering the attacking plan of the opponent's weapons and manpower as well.

Acknowledgements. The authors would like to thanks the unknown referees who gave valuable comments for the improvement of the paper.

References

1. Schwefel H.P.: Evolution and Optimum Seeking, John Wiley and Sons Inc (1995).
2. Goldberg D.E.: Genetic Algorithms in Search Optimization and Machine Learning, Addison Wesley (1989).
3. Fogel D.B.: Evolutionary Computation: Towards a New Philosophy of Machine Intelligence IEEE Press, Piscataway, NJ (1995).
4. Bach, T.F., Hoffmeister and H.P. Schwefel: A Survey Of Evolutionary Strategies. In R.K. Belew and Booker (Eds), Proceedings of the 4th International Conference on Genetic Algorithms, Morgan Kaufman (1991) 2-9.
5. Davis, L.: Adapting Operator Probabilities In Genetic Algorithms", in J.D. Schaffer (Ed.) Proceedings of the 3rd International Conference on Genetic Algorithms, Morgan Kaufman (1989) 61 – 69.
6. Radcliffe N.J.: Forma Analysis and Random Respectful Recombination, Proc. 7th ICGA, (1997) 246-253.
7. Wright A.: Genetic Algorithms for Real Parameter Optimization, FOGA, 31 – 36 (1991) 31 – 36.
8. Janikow C.Z., Michalewicz, Z.: An Experimental Comparison of Binary and Floating Point Representations in Genetic Algorithms, Proc., 4th ICGA, (1991) 31 – 36.
9. Michalewicz Z.: Genetic algorithms + Data Structures = Evolution Programs, 3rd edition Springer Verlag, New York (1996).
10. Eshelman, L., Schaffer J.D.: Real coded Genetic algorithms and Interval Schemata, In LD Whitley (Ed.), Foundations of Genetic Algorithms 2, Los Altos, CA, Morgan Kaufman (1993) 187-202.
11. Voigt, H. M., Muhlenbein, H., Gvetkovic, D: Fuzzy Recombination for Breeder Genetic Algorithm", Proc. 6th ICGA, (1995) 104-111.
12. Ono, I., Kobayashi: A Real Coded Genetic Algorithm for Function Optimization using Unimodal Distribution Crossover, Proceedings of 7th International Conference on Genetic Algorithms, (1997) 246-253.
13. Kita, H. Ono I, Kobayashi S. : Multiparental Extension of the Unimodal Normal Distribution Crossover for Real Coded Genetic Algorithms, Proceedings of 1999 congress on evolutionary computation (1999) 1581-1587.
14. Deb, K.: Multiobjective Optimization Using Evolutionary Algorithms, Chichester: Wiley (2001).
15. Herrera, F., Lozano, M., Verdegay, J.L.: Tackling Real Coded Genetic Algorithms. Operators and Tools for Behavioral Analysis, Artificial intelligence review 12(**4**), (1998) 265-319.
16. De Jong, K.: An Analysis Of The Behavior of a Class Of Genetic Adaptive Systems, Doctoral dissertation, University of Michigan, Dissertation Abstracts International, **36** (10), 5140B, University Microfilms No. 76-9381 (1975).
17. Storn R, Price K.: Differential Evolution a Simple and Efficient Heuristic for Global Optimization Over Continuous Spaces, J. of Global Optimization **11**, (1997) 341-359.
18. Levy, A.V., Montalvo, A., Gomez, S., Caderon, A.: Topics in Global Optimization", Lecture Notes in Mathematics, Numerical Analysis, 909, and (Eds.) A. Dold and B. Eckman, Proceedings, Cocyoo, Mexico, Springer Verlag, (1981) 18-33.

19. Bramin, F.H.: Widely Convergent Method for Finding Multiple Solutions of Simultaneous Non Linear Equations, I.B.M.J. R and D, New York (1972).
20. More J.J., Garbow B.S., Hillstorm K.E.: Testing Unconstrained Optimization Software, ACM Transactions on Mathematical Software 7(**1**) (1981) 17-41.
21. Himmelblau, D. M.: Applied Nonlinear Programming, McGraw-Hill, New York (1972).
22. Wood, K.:Deterministic Network Interdiction," *Mathematical and Computer Modelling*, **17**, (1993) 1-18.
23. Garcia, M.: *The Design and Evaluation of Physical Protection Systems*, Butterworth-Heinemann, Boston (2001).
24. GAMS-CPLEX.http://www.gams.com/solvers/solvers.htm#CPL EX. (2004).
25. Brown, G., M. Carlyle, D. Diehl, J. Kline, K. Wood,::How to Optimize Theater Ballistic Missile Defense," *Operations Research,* **53**(**5**) (2005).
26. Brown, G., Carlyle, M., Harney, R., Skroch, E., Wood, K.,: Interdicting a Nuclear Weapons Project, in review (2005).

Appendix I: Test Functions

F1 DeJong[16] $\sum_{i=1}^{n} x_i^2$

F2 Zakharov [17] $\sum_{i=1}^{n} x_i^2 + \left(\sum_{i=1}^{n} 0.5ix_i\right)^4$

F3 Rastrigin [17] $10n + \sum_{i=1}^{n}\left(x_i^2 - 10\cos(2\pi x_i)\right)$

F4 Rosenbrock [20]

$$\sum_{i=1}^{n}\left(100\left(x_{i+1} - x_i^2\right)\right)^2 + \left(1 - x_i^2\right)$$

F5 Griewank [17]

$$\sum_{i=1}^{n}\left(x_i^2 / 4000\right) - \Pi\left(\cos(x_i)/\sqrt{i}\right) + 1$$

F6 Ackley [17] 20

$$+ - 20e^{-0.2}\sqrt{1/n\sum_{i=1}^{n} x_i^2} - e^{1/n}\sum_{i=1}^{n}\cos(2\pi x_i)$$

F7 Himmelblau [21]

$$\sum_{i=1}^{2}\left[\log^2(x_i - 2) + \log^2(10 - x_i)\right] - (\Pi x_i)^{0.2}$$

F8 Six-hump camel back function [18]

$$\left(4 - 2.1x_1^2 + x_1^{4/3}\right)x_1^2 + x_1x_2\left(-4 + 4x_2^2\right)x_2^2$$

F9 Bramin's r cos function [19]

$$\left(x_2 - \frac{5.1}{4\pi^2}x_1^2 + \frac{5}{\pi}x_1 - 6\right)^2 + 10\left(1 - \frac{1}{8\pi}\right)\cos x_1 + 10$$

F10 Schubert[18]

$$\left[\sum_{i=1}^{5} i\cos\{(i+1)x_1 + i\}\right]\left[\sum_{i=1}^{5} i\cos\{(i+1)x_2 + i\}\right] + \beta\left[(x_1 + 1.42513)^2 + (x_2 + .80032)^2\right]$$

Appendix II: Numerical Results of Test Problems Using SGA and GA Using QCX Operator

Function	No. of variables	Genetic Algorithm using QCX operator: %age of success	Genetic Algorithm using QCX operator: Avg. function evaluations	Simple Genetic Algorithm: %age of success	Simple Genetic Algorithm: Avg. function evaluations
F1	2	100	80	100	876
	4	100	150	100	950
	6	100	226	100	924
	8	100	386	100	1287
	10	100	596	98	1296
F2	2	100	102	100	976
	4	100	286	100	1072
	6	100	390	100	1128
	8	100	484	98	1879
	10	100	698	98	4226
F3	2	100	98	100	884
	4	100	250	100	979
	6	98	376	98	1198
	8	94	480	92	1879
	10	92	586	90	3986
F4	2	100	120	100	996
	4	100	290	100	1038
	6	100	508	100	1186
	8	100	598	98	1972
	10	100	662	98	1996
F5	2	100	122	100	1097
	4	100	298	100	1165
	6	94	406	92	1294
	8	90	532	90	1997
	10	88	699	86	3099
F6	2	100	108	100	870
	4	100	290	100	958
	6	98	376	94	1478
	8	96	497	92	1886
	10	96	525	92	3972
F7	2	96	132	94	884
F8	2	96	98	96	1887
F9	2	94	96	94	898
F10	2	92	182	90	1176

Artificial Immune System-Based Customer Data Clustering in an e-Shopping Application

D.N. Sotiropoulos, G.A. Tsihrintzis, A. Savvopoulos, and M. Virvou

University of Piraeus
Department of Computer Science
Piraeus 185 34
Greece
{dsotirop, geoatsi, asavvop, mvirvou}@unipi.gr

Abstract. We address the problem of adaptivity of the interaction between an e-shopping application and its users. Our approach is novel, as it is based on the construction of an Artificial Immune Network (AIN) in which a mutation process is incorporated and applied to the customer profile feature vectors. We find that the AIN-based algorithm yields clustering results of users' interests that are qualitatively and quantitatively better than the results achieved by using other more conventional clustering algorithms. This is demonstrated on user data that we collected using *Vision.Com*, an electronic video store application that we have developed as a test-bed for research purposes.

1 Introduction

The large quantity of information that exists in an electronic shop, as well as the lack of human salesmen to assist customers, impose a need for adaptivity in e-shops. Adaptivity provides individualised assistance to users, which is dynamically generated. Adaptive responses to users are usually created using the technology of adaptive hypermedia [1]. To create such adaptive behaviour of the e-shop to individual users, the system needs information about them, so that it can generate hypotheses about what they need and how they can be assisted in the most appropriate way. This means that an adaptive e-shop needs user modelling components, which monitor the user's actions and generate hypotheses about his/her preferences based on his/her behaviour. According to Rich [12] a major technique people use to build models of other people very quickly is the evocation of stereotypes, or clusters of characteristics. Given the fact grouping people can provide quick default assumptions about their preferences and needs [12], the clustering of users' interests has drawn a lot of research energy for purposes of personalization of user interfaces.

One solution to the problem of grouping users' behaviour can be provided by clustering algorithms that may group users dynamically based on their behaviour while they use a system on-line. The main advantage of such an approach is that the categorization of user behaviour can be conducted automatically. Clustering algorithms help adaptive systems to categorize users according to their

B. Gabrys, R.J. Howlett, and L.C. Jain (Eds.): KES 2006, Part I, LNAI 4251, pp. 960–967, 2006.
© Springer-Verlag Berlin Heidelberg 2006

behaviour. More specifically adaptive e-commerce systems need not only to acquire information about users' interests in products but also to have the ability to group users with similar interests. By grouping users together systems can understand more clearly the user's intention and generate more efficient hypotheses about what a user might need. In this part, clustering algorithms undertake the role of grouping users in an efficient way, thus creating the bone structure of the user model. In this paper, we present an advanced, immune network-based clustering approach that has been used to provide adaptivity to a test-bed e-shop application constituting an e-video shop.

There are a lot of web-based recommendation applications that have used clustering algorithms (e.g. [2], [3], [4], [5], [6], [7]). All of these applications are recommendation systems, but are primarily concerned with the acquisition of user behaviour-related data. In contrast, our work has focused on incorporating an evolutionary clustering algorithm based on the construction of an Artificial Immune System (AIS), in order to personalize the developed e-shop system through grouping similar users' behaviour. This means that we have used a very different algorithm that has not been used by these systems. Similarly to our approach, [8] and [9] have also utilized an AIS in order to tackle the task of film and web site recommendation respectively by identifying a neighborhood of user preferences. Although, their systems have used a similar algorithm to the one presented in this paper, the algorithm that we have used is significantly more sophisticated. In particular, we have based our research on the construction of an Artificial Immune Network (AIN) which incorporates a mutation process that applies to the customer profile feature vectors. More specifically, our work has focused on how a significant amount of redundancy within the customer profile dataset can be revealed and reduced, how many clusters intrinsic to the customer dataset can be identified and what the spatial structure of the data within the identified clusters is.

Specifically, we develop an AIN for the problem of clustering a set of unlabelled multidimensional customer profile feature vectors generated in an electronic video store application in which customer preferences are quantified and groups of customer profiles are maintained. We have examined thoroughly the effectiveness of this algorithm by comparing it with three other clustering algorithms, namely agglomerative hierarchical clustering, fuzzy c-means clustering and spectral clustering. For this purpose, we have collected data through a video e-shop application and fed them into each of the four clustering algorithms and compare the corresponding results.

2 Operation of e-Shop System and Customer Data Description

Vision.Com (Fig 1) is an adaptive e-commerce video store that learns from customers preferences. Its aim is to provide help to customers, choosing the best movie for them. The web-based system ran on a local network with IIS playing the role of the web server. We used this technique in order to avoid network problems during peak hours. For every user the system creates a different record at the database. In Vision.Com, every customer may browse a large number of movies by navigating through four movie categories, namely social, action, thriller and comedy movies. Every customer has a personal shopping cart. A customer intending to buy a movie must simply move the movie into her/his cart by pressing the specific button. S/He may also remove one or more movies from his/her cart by choosing to delete them. After deciding on which movies to buy, a customer can easily purchase them by pressing the button buy. All navigational moves of a customer are recorded by the system in the statistics database. In this way Vision.Com maintains statistics regarding visits to various movie categories and individual movies. Same type statistics are maintained for every customer and every movie that was moved to a shopping cart. Moreover, the same procedure is followed for those movies that are eventually bought by every customer. All of these statistical results are scaled to the unit interval [0,1].

More specifically, Vision.Com interprets users actions in a way that results in estimates/predictions of users interests in individual movies and movie categories. Each users action contributes to the individual user profile by indicating relevant degrees of interest into a movie category or individual movie. If a user browses into a movie, this indicates interest of this user for the particular movie and its category. If the user puts this movie into the shopping cart, this implies *increased* interest in the particular movie and its category. Finally, if the user decides to buy the movie, then this shows highest interest of this user in the specific movie and its category and is recorded as an increase in an *interest counter*. On the other hand, if the user removes the movie from the shopping cart without buying it, the interest counter is unchanged. Apart from movie categories that are already presented, other movie features taken into consideration are *price range*, *leading actor* and *director*. The price of every movie belongs to one of the five price ranges in euro: 20 to 25, 26 to 30, 31 to 35, 36 to 40 and over 41. As a consequence, every customers interest in one of the above features is recorded as a percentage of his/her visits to movie pages. For example, interest of the

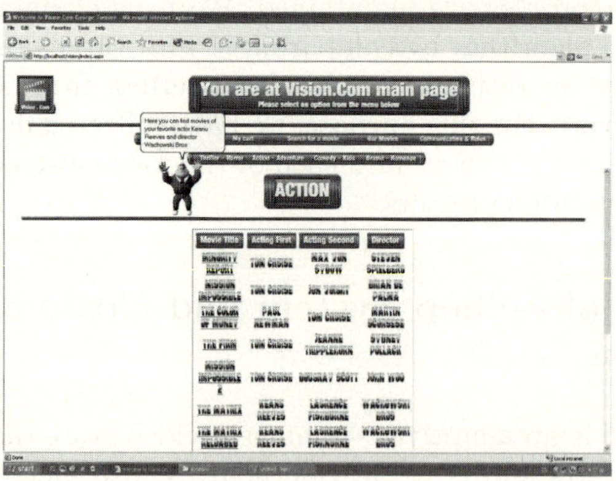

Fig. 1. Screenshot of Vision.Com User Interface

customer at a particular movie category is calculated as in Eq.(1). Every result can also be considered as a probability of a customer's intention to buy a movie.

$$InterestInMovieCategory = \frac{VisitsInSpecificCategory}{VisitsInAllCategories} \quad (1)$$

Vision.Com was used by 150 users to buy movies. The system collected data about the user's behaviour. The data collected consisted of three similar parts. The first part contains statistical data of the visits that every user made to specific movies. The second part contains data of the cart moves (i.e. which movies the user moved into his/her cart). The last part consists of statistical data concerning the preferences on the movies bought by every user. Every record in every part is a vector of the same 80 features that were extracted from movie features and represents the preferences of a user. The 80 features of these vectors are computed as the movie features we described above. Every 80 featured vector consists of the four movie categories, the five price ranges, all the leading actors and all the directors. The value of each feature is the percentage of interest of every individual customer in this particular feature (equation (1)).

In the context of Vision.Com, antibodies correspond to neighbouring user profiles that are represented by the feature vectors. Specifically, the immune network system clustered users interests as well as movies and represented each resulting cluster with corresponding antibodies.

3 AIS-Based Customer Data Clustering Algorithms

AIS-based clustering relies on a computational imitation of the biological process of *self/non-self discrimination*, that is the capability of the adaptive biological immune system to classify a cell as "self" or "non-self" cell. Any cell or even individual molecule recognized and classified by the self/non-self discrimination process is called an *antigen*. A non-self antigen is called a *pathogen* and, when identified, an immune response (specific to that kind of antigen) is elicited by the adaptive immune system in the form of *antibody* secretion. The essence of the antigen recognition process is the *affinity* (molecular complementarity level) between the antigen and antibody molecules. The strength of the antigen-antibody interaction is measured by the complementarity of their match and, thus, pathogens are not fully recognized, which makes the adaptive immune system *tolerant to molecular noise*. In the present e-shop application, the antigenic population consists of the initial set of customer profile feature vectors. The produced set of memory antibodies provide an alternative, more compact way of representing the original dataset while conserving their original spatial distribution.

Learning in the immune system is established by the *clonal selection principle* [11] which suggests that only those antibodies exhibiting the highest level of affinity with a given antigen be selected to proliferate and grow in concentration. Moreover, the selected antibodies also suffer a *somatic hypermutation* process [11], that is a genetic modification of their molecular receptors which allows them learn to recognize then given antigen more efficiently. This hypermutation process is termed *affinity maturation* [11]. To develop an AIN for data clustering, we generate a minimal set of representative points that capture the properties of the original dataset and can be interpreted as the centers of the initial feature vector distribution. In AIS terminology, this representative point set in a multidimensional feature space constitute a set of memory *antibodies* that recognize, in the sense of Euclidean distance proximity, the antigenic population that corresponds to the original dataset.

The AIN that we developed for clustering the customer profile feature vectors in the present application was an edge-weighted graph composed of a set of *memory antibodies* (i.e., *nodes*) and *edges* (i.e., sets of node pairs with an assigned weight or connection strength to reflect the affinity of their match). To quantify immune recognition, we consider all immune events as taking place in a shape-space S, constituting a multi-dimensional metric space in which each axis stands for a physico-chemical measure characterizing molecular shape [11].

Specifically, we utilized a real-valued shape-space in which each element of the AIN is represented by a real-valued vector of 80 elements, thus, $S = \Re^{80}$. The affinity/complementarity level of the interaction between two elements of the AIN was computed on the basis of the Euclidean distance between the corresponding vectors in $\Re^{80}$. The antigenic pattern set to be recognized by the AIN is composed of the set of 150 80-dimensional feature vectors. The memory antibodies produced by the AIN can be considered as an alternative compact representation of the original customer profile feature vectors set. The AIN learning algorithm [11] was as follows:

1. *Initialization*: Create a random initial population of network antibodies.
2. *Antigenic presentation*: For each antigenic pattern do:
   (a) *Clonal selection and expansion*: For each network element, determine its affinity with the antigen presented. Select a number of high affinity elements and reproduce (clone) them proportionally to their affinity.
   (b) *Affinity maturation*: Mutate each clone inversely proportionally to affinity. Reselect a number of highest affinity clones and place them into a clonal memory set.
   (c) *Metadynamics*: Eliminate all memory clones whose affinity with the current antigen is lower than a predefined threshold.
   (d) *Clonal interactions*: Determine the network interactions (affinity) among all the elements of the clonal memory set.
   (e) *Clonal suppression*: Eliminate those memory clones whose mutual affinity is lower than a predefined threshold.
   (f) *Network construction*: Incorporate the remaining clones of the clonal memory in all network antibodies.
3. *Network interactions*: Determine the similarities among all network antibody pairs.
4. *Network suppression*: Eliminate all network antibodies with affinity below a prespecified threshold.
5. *Diversity*: Introduce a number of new randomly generated antibodies into the network.
6. *Cycle*: Repeat steps 2 to 5 for a prespecified number of iterations.

4 Comparison of Customer Data Clustering Algorithms and Conclusions

We tested and compared three widely used clustering techniques [10], namely a) *agglomerative hierarchical clustering*, b) *fuzzy c-means clustering* and c) *spectral clustering*, against the less well-known AIS-based clustering. Specifically, we applied these four clustering methodologies on the 150 customer profile feature vectors collected as described in Section 3. In Fig. 2, we show and compare the dendrograms corresponding to the feature vectors produced by hierarchical (top left), spectral (top right), and AIN-based (center left) clustering, respectively. We observe, that spectral clustering does not provide a clearer revelation of the intrinsic similarities in the dataset over hierarchical clustering. On the other hand, the leaves in the AIN-based dendrogram are significantly fewer than the leaves in either the hierarchical or the spectral dendrograms in Fig. 1, which stems from the fact that the former corresponds to clustering only 22 *representative* points in the 80-dimensional feature space, while the latter two correspond

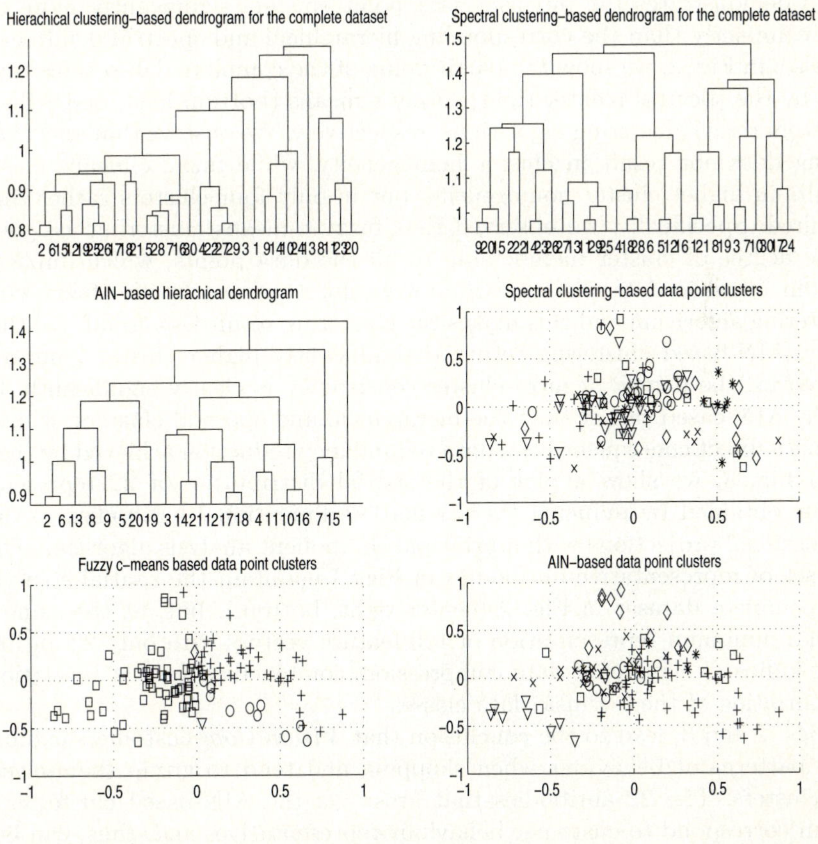

Fig. 2. Dendrograms and clusters of customer profile feature vectors returned by four clustering methodologies

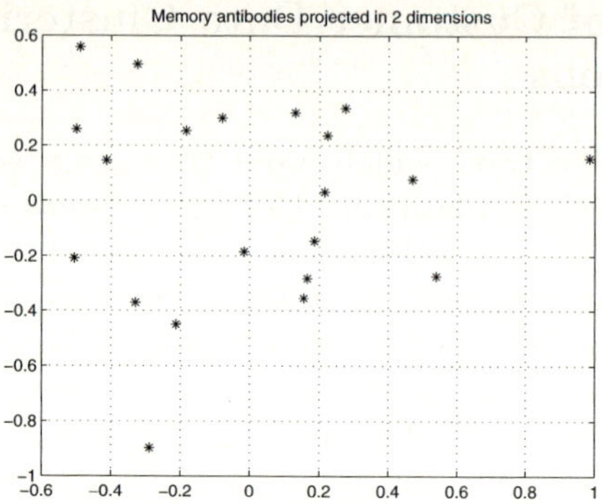

Fig. 3. Plot of the spatial distribution of the projection of the 22 AIN-produced memory antibodies onto two dimensions

to clustering the *complete* set of 150 data points. Thus, the AIN-based dendrogram demonstrates the intrinsic data point clusters significantly more clearly and compactly than the corresponding hierarchical and spectral dendrograms.

Also, in Fig. 2, we show the partitioning of the complete dataset into six clusters by the spectral (center right), fuzzy c-means (bottom left), and AIN-based (bottom right) clustering algorithms, respectively. We observe that spectral clustering does not result in cluster homogeneity, while fuzzy c-means clustering results in higher cluster homogeneity, but in only four clusters rather than six required. Specifically, we observed that fuzzy c-means clustering assigned the same degree of cluster membership to all the data points, which implies that certain intrinsic data *dissimilarities* were not captured by the fuzzy c-means clustering algorithm and this makes the clustering result less useful. On the contrary, AIN-based clustering returned significantly higher cluster homogeneity. Moreover, the degree of intra-cluster consistency is clearly significantly higher in the AIN-based rather than the hierarchical and spectral clusters, which is of course a direct consequence of a data redundancy reduction achieved by the AIN.

In Fig. 3, we show a plot of the spatial distribution of 22 representative points obtained by reducing the original 80-dimensional antibodies to their 2-dimensional projections with a principal component analysis algorithm. Clearly, the set of representative antibodies in Fig. 3 maintain the spatial structure of the complete dataset in Fig. 2 (center right, bottom), but, at the same time, form a minimum representation of 150 feature vectors with only 22 antibodies. This indicates significant data compression, combined with clear revelation and visualization of the intrinsic data classes.

Figs. 2 and 3, lead to the conclusion that *Vision.Com* customers exhibit certain patterns of behaviour when shopping and tend to group themselves into six clusters. The 22 antibodies that arose via the AIN-based clustering algorithm correspond to customer behaviour representatives and, thus, can be seen as important customer profiles, which eventually correspond to stereotypes in user models. This process is promising because it can provide recommendations based on the users' interests and the characteristics of a movie irrespective of whether this movie has ever been selected by a user before. Thus, the recommendation system can recommend new movies or movies newly acquired by the e-shop as efficiently as previously stored movies. This approach forms the basis of work which is currently in progress and will be reported on a future occasion.

References

1. Brusilovsky, P.: Adaptive Hypermedia, User Modeling and User-Adapted Interaction **11**. Kluwer Academic Publishers (2001) 87–110
2. Jin, X., Zhou, Y., Mobasher, B.: Unified Approach to Personalization Based on Probabilistic Latent Semantic Models of Web Usage and Content. Proceedings of the AAAI 2004 Workshop on Semantic Web Personalization (SWP'04). San Jose (2004)
3. Adil, C.S., Banaei, F.F, Faruque, K.J.: INSITE: A Tool for Real-Time Knowledge Discovery from Users Web Navigation. Proceedings of the 26th International Conference on Very Large Databases. Cairo Egypt (2000)
4. Menczer, F., Monge, A.E, Street, W.N.: Adaptive Assistants for Customized E-Shopping. Journal of IEEE Intelligent Systems (2002) 12–19
5. Jin, X., Zhou, Y., Mobasher, B.: Web usage mining based on probabilistic latent semantic analysis. Proceedings of ACM SIGKDD (KDD'04) (2004) 197–205
6. Ajith, A.: 6. Business Intelligence from Web Usage Mining. Journal of Information & Knowledge Management, Vol. 2, No. 4. iKMS & World Scientific Publishing Co (2004) 375–390.
7. Wang, Q., Makaroff, D.J., Edwards, H.K.: Characterizing Customer Groups for an Ecommerce Website. Proceedings of ACM Conference on Electronic Commerce (EC'04), New York USA (2004)
8. Cayzer, S., Aickelin, U.: A Recommender System based on the Immune Network. Proceedings of the 2002 Congress on Evolutionary Computation (2002)
9. Morrison, T., Aickelin, U.: An Artificial Immune System as a Recommender for Web Sites. Proceedings of the 1st Conference on ARtificial Immune Systems (ICARIS-2002) Canterbury UK (2002) 161–169
10. Theodoridis, S, Koutroumbas, K.: Pattern Recognition.3rd edn. Academic Press, San Diego (2006)
11. De Castro, L.,N., Timmis, J.: Artificial Immune Systems: A New Computational Inteligence Approach.1st edn. Springer-Verlag, London Berlin Heidelberg (2002)
12. Rich, H.: User Modeling via Stereotypes, Cognitive Science **3**. Morgan Kaufmann Publishers Inc. (1979) 329–354

A GA Driven Intelligent System for Medical Diagnosis

Grigorios Beligiannis[1], Ioannis Hatzilygeroudis[1], Constantinos Koutsojannis[1], and Jim Prentzas[1, 2]

[1] Department of Computer Engineering & Informatics, University of Patras
26500 Patras, Greece
[2] Dept of Informatics & Computer Technology, TEI of Lamia
35100 Lamia, Greece

Abstract. Manipulation of Male sexual dysfunction (or Impotence) that concerns 10% of the male population requires expertise and great experience. Different diagnostic approaches according to medical as well as to psychosocial and cultural characteristics of patients are usually followed. In this paper, a GA (genetic algorithm) driven intelligent system (GADIS) for diagnosis of male impotence is presented. The rule-base of GADIS has been constructed by using a genetic algorithm for rule extraction from a patients database. Experimental results show a very good diagnostic performance in terms of accuracy, sensitivity and specificity of the intelligent system. The rule-base can be refined each time the patient database is updated over a limit.

Keywords: GA driven intelligent system, intelligent medical diagnosis system, hybrid intelligent system.

1 Introduction

Human sexual dysfunction (or impotence) is characterized by disturbances in sexual desire and in psychophysiological changes associated with the sexual response cycle in men and women. An estimated 10% of the male population experience chronic Erectile Dysfunction (ED), a type of sexual dysfunction, however as few as 5% seek treatment. ED may affect 50% of men between the ages of 40 and 70 [1]. Most men experience this inability at some stage in their lives, usually by the age of 40, but are not psychologically affected by it. It has many causes, most of which are treatable. It is not an inevitable consequence of aging. Due to experts on this field, more men have been seeking help and returning to normal sexual activity because of improved, successful treatments for ED. Manipulation of the dysfunction requires expertise and great experience. Doctors, even urologists, cannot provide a typical evaluation and treatment strategy. A number of parameters and their possible impacts on the diagnosis and treatment are still under consideration and vogue. So, the creation of an intelligent system to assist non-expert doctors in making an initial diagnosis is quite desirable.

Rule-based intelligent systems are very popular in representing medical diagnosis processes. However, a problem with rule-based systems in general is the difficulty to acquire knowledge from experts, due to a variety of reasons [2]. Therefore, techniques for extracting rules from empirical data have been used. This can be done either

B. Gabrys, R.J. Howlett, and L.C. Jain (Eds.): KES 2006, Part I, LNAI 4251, pp. 968–975, 2006.
© Springer-Verlag Berlin Heidelberg 2006

indirectly (by constructing a neural network and then extracting rules from that) [3] or directly, by applying, for example, machine learning techniques (like the ID3, C4,5 etc algorithms) [4]. Recently, genetic algorithms (GAs) have been used for generating rules from empirical data. Most of existing efforts refer to generation of fuzzy rules from empirical data [5, 6]. A few of them have been used on medical data and have produced working systems [7]. However, generation of fuzzy rules is more complicated than that of simple symbolic rules, which are sometimes adequate for solving the classification problem at hand.

In this paper, we present a genetic algorithm driven intelligent system (GADIS) to make diagnoses and suggest appropriate treatments for male impotence. To acquire diagnosis rules from experts for such an application is not an easy task, given that not a consensus has been reached at yet [8]. GAs can help in such a task. GADIS uses a simple GA to evolve simple rules from a medical database with quite satisfactory results. The paper is organized as follows. In Section 2, we present the system architecture. In Section 3, the rule evolution process is described. Section 4 deals with the GA. Section 5 gives some implementation details, whereas Section 6 presents evaluation results. Finally, Section 7 concludes the paper.

2 System Architecture

The architecture of GADIS is depicted in Figure 1. It consists of the following units: user interface (UI), genetic algorithm unit (GA), rule extracting and updating unit (REU), medical database (MDB), rule base (RB) and inference engine (IE). Actually, RB and IE constitute the run-time system. GA, RE and MDB are used off-line.

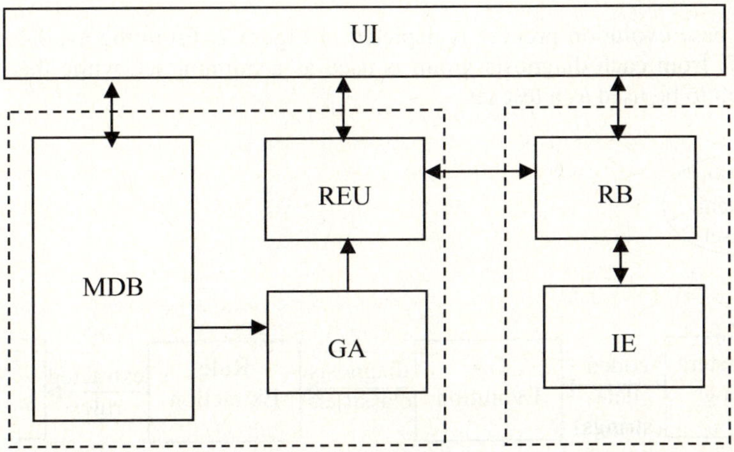

Fig. 1. GADIS Architecture

MDB contains patient data organized in groups. Each group refers to one of the possible diagnoses that the system can make. In the off-line mode, GA is successively applied to the data from each group in the MDB and produces a diagnosis classifier for each data group. The classifiers are then used by REU to produce the evolved rules, which are then suitably transformed and modify the RB. UI is the means for user communication and interference with the system, where and when it is necessary.

3 Rule Evolution Process

3.1 Medical Knowledge

GADIS concerns diagnosis and treatment of male impotence. There are four possible diagnoses, i.e. reasons, for male impotence: (a) psychogenic, (b) arteriopathy, (c) venooclusive insufficiency and (d) specific neuropathy. To get to a diagnosis, information about a number of parameters is needed. To extract those parameters we used a statistical analysis method (Pearson analysis) to evaluate which of the parameters recorded in the patient records are significant for the intermediate and final diagnosis. We analyzed 75 patient records from the patient database of the 'Andrology Laboratory' of the Department of Urology of the University Hospital of Patras [8].

We finally came up with twenty two (22) parameters that may play a role in the diagnosis of male impotence, which are the following: onset, non-coital erection, onanism, diabetes mellitus, coronary artery, prostate, neuropathies, chronology, age, depression, smoking, alcohol, blood pressure, overweight, hormonal evaluation, cholesterol, NPT, PIP, Doppler, DICC, neurophysiological (the last five refer to medical tests). Some of them, like overweight, are yes-no parameters, whereas others can take a number of values (e.g. age can take one of nine values that represent age intervals).

3.2 The Evolution Process

The rule base evolution process is depicted in Figure 2. From the available medical data, 40% from each diagnosis group is used as a training set, while the remaining 60% is left to be used as a test set.

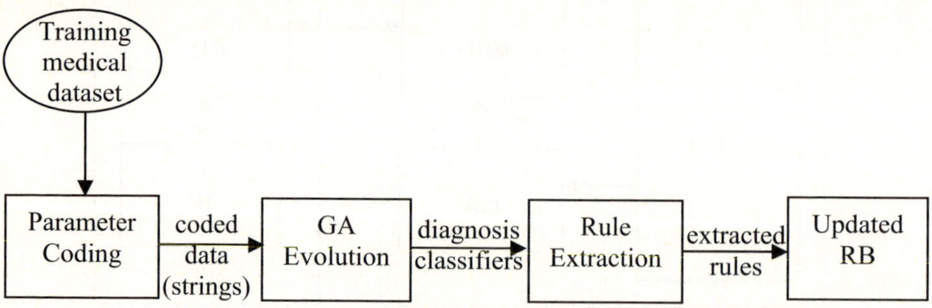

Fig. 2. GA Based Rule Evolution Process

The training dataset is given as input to the parameter coding unit (considered as part of the GA unit in the architecture of Figure 1). So, the system codes the training data in order to create the initial population of the GA and triggers the GA evolution process. After a few generations, the GA provides four different binary strings each corresponding to one of the four different possible diagnoses (psychogenic, arteriopathy, venooclusive insufficiency and specific neuropathy). Each binary string constitutes a diagnosis classifier pattern for the corresponding diagnosis category.

Each such diagnosis classifier comprises a binary string that matches most of the provided training data of a specific diagnosis category. In this way, each diagnosis classifier contains the necessary information required by a rule that would correctly classify most of the provided training data. After that, the rule extraction procedure is taking over. Based on the characteristics of the four different diagnosis classifiers, this procedure results in a set of four different production rules each corresponding to a different diagnosis category. After the extraction of the rules is completed, the rule base is updated. We use the term 'updated' and not the term 'created', because this off-line process can be periodically repeated, given that adequate new patient data has been added to MDB.

3.3 Rule Extraction

The rule extraction process is as follows: After the GA evolution process is completed and the classifier for each different diagnosis category is produced, the system tries to identify, for each different classifier, the most important parameters of each diagnosis category. This is done as follows: For each classifier, it tries to find the parameters for which the specific classifier has different values compared to the other three classifiers. For example, the classifier for the psychogenetic case has different values compared to the other three classifiers for the parameters in the places 14 (NON-COITAL), 15 (ONSET), 17 (CHOLESTEROL), 18 (NPT) and 21 (DICC). These five parameters are used in order to construct the final form of the classifier. So, the rule extraction process is able to extract only the important parameters, which are significant for a specific diagnosis category, and omit all the others. After that, the process evolves the set of rules corresponding to different combinations of the significant parameters and finally results in the one that better classifies the cases of the training set.

4 The Genetic Algorithm

GAs, in general, operate on binary string structures, analogous to biological creatures (genomes). These structures are evolving in time according to the rule of survival of the fittest, by using a randomized, yet structured, information exchange scheme. Thus, in every generation, a new set of binary strings is created, using parts of the fittest members of the old set [9]. GAs process a binary coding of the parameter space and work on it. This coding (which is an essential part of the GA design procedure) results in formation of binary strings.

In our case, the coding scheme comprises the 22 parameters, specified in the previous section, that represent the data on which diagnosis of male impotence is based on. Each parameter is coded into a binary string. This binary string is in fact the genome of the GA that will be evolved. Each parameter is coded using a 31-bits binary string. For example, parameter 'age', which can take nine different values (1 to 9), is coded into the binary strings $\underbrace{000...000}_{27}0001$ to $\underbrace{000...000}_{27}1001$ respectively (actually only the four last digits are required, the rest are used because of the implementation tool). On the other hand, parameter 'prostate', which is a yes-no parameter, is coded into binary strings $\underbrace{000...000}_{27}0001$ and $\underbrace{000...000}_{27}0010$ respectively. Each binary string constitutes a gene of the genome of the GA. As a result the whole genome length equals 682 (= 22 x 31) bits (Fig. 3). A set of such genomes constitutes the GA's population.

$$\underbrace{\underbrace{0000...0011}_{31} \quad ... \quad \underbrace{0000...1100}_{31}}_{22}$$

Fig. 3. The structure of the genome used in the evolution procedure

The steps of the GA are the classical ones. The initial population is generated using a set of random binary strings created by the parameter coding unit. Selection, crossover and mutation are used for the generation of new population, in each iteration. Evaluation of the objective function is made by applying it to the training set of the available medical data.

```
for i = 1 to m do
    fitness = 0;
    for j = 1 to 22 do
      for k =1 to n do
        if g[i, j] = d[k, j]
          then matches = matches+1;
        endif
      endfor
      if matches >= limit
        then fitness =fitness + 1;
      endif
    endfor
endfor
```

Fig. 4. Algorithm for fitness computation

The process is repeated for each different diagnosis category. The termination criterion is the number of generations, that is the evolution process is terminated when a specific number of generations (in our case 5) are completed. While the termination criterion is not met, the whole evolution process is taking place for the current population, i.e. selection, crossover and mutation are applied. When the termination criterion is met, the GA evolution process is terminated and results to one diagnosis classifier for each different diagnosis category.

The objective function estimates the goodness (fitness) of each individual (binary string) and is presented in Figure 4, where *g* is a two-dimension (*m*x22) array representing the *m* genomes of the current population (with 22 genes each representing 22 parameter values), *d* is also a two-dimension (*n*x22) array representing the *n* examples/patterns of the training medical dataset (with 22 parameter values each), *matches* is a counter that counts how many data patterns a gene of the current genome matches, *fitness* is a variable representing the fitness value of the current genome and *limit* is a constant representing the minimum number of matches that a gene must have with the corresponding genes of the patterns of the training dataset to add '1' to its genome's fitness value.

5 Implementation Details

As stated, the type of GA used is the classic simple GA. The representation used for the genomes of the genetic population is the classic binary string. As far as the reproduction operator is concerned, the classic biased roulette wheel selection is used. The crossover operator used is uniform crossover (with crossover probability equal to 0.9), while the mutation operator is the flip mutator (with mutation probability equal to 0.001. Except that, the size of the population is set to 50 while the GA uses linear scaling and elitism [10].

The GA was implemented using the C++ Library of Genetic Algorithms GAlib [11] and especially the GASimpleGA class for the implementation of the GA (non-overlapping populations) and the GABin2DecGenome class for the binary string genomes (an implementation of the traditional method for converting binary strings to decimal values). All the experiments were carried out on an Intel Pentium IV 2.7GHz PC with 256 MB RAM.

6 System Evaluation

We run GADIS using the test dataset (each case in the test dataset is accompanied by the correct diagnosis, which was really verified). The patient cases included in the test dataset were also presented to three urology residents to make their own diagnoses (based on the given data for each case). The results of GADIS were compared to the diagnoses of the three residents via three metrics, commonly used for this purpose: accuracy, sensitivity and specificity (abbreviated as Acc, Sen and Spec respectively), defined as follows:

$$Acc=(a + d)/(a + b + c + d),$$
$$Sen = a/(a + b),$$
$$Spec = d/(c + d)$$

where, *a* is the number of positive cases correctly classified, *b* is the number of positive cases that are misclassified, *d* is the number of negative cases correctly classified and *c* is the number of negative cases that are misclassified. By 'positive' we mean that a case belongs to the group of the corresponding diagnosis and by negative that it doesn't.

Table 1a. Evaluation results for Psychogenic diagnoses

Metrics	3 Residents (mean score)	GADIS
ACCURACY	61	88
SENSITIVITY	57	94
SPECIFICITY	65	78

Table 1b. Evaluation results for Arteriopathy diagnoses

Metrics	3 Residents (mean score)	GADIS
ACCURACY	63	89
SENSITIVITY	57	73
SPECIFICITY	72	96

Table 1c. Evaluation results for Venooclusive diagnoses

Metrics	3 Residents (mean score)	GADIS
ACCURACY	62	89
SENSITIVITY	57	73
SPECIFICITY	67	96

The comparison results are presented in Tables 1a, b and c and show a clear overall superiority of the performance of GADIS. GADIS is much better as far as all three metrics are concerned. However, the three residents have a better balance between sensitivity and specificity than GADIS for each type of diagnosis individually. Also, this balance is uniform in the three residents, but not in GADIS (Sen is larger than Spec in Psychogenic diagnoses -Table 1a, but the reverse is true in the other two types of diagnoses -Tables 1b and c). So, the three residents tend to better classify negative cases in all types of diagnosis, whereas GADIS does it for two of them, but the reverse for the other type of diagnosis (Psychogenic).

7 Conclusions

In this paper, a genetic algorithm driven intelligent system (GADIS) for diagnosis of male impotence is presented. The rule-base of GADIS has been constructed by using a genetic algorithm for rule extraction from a medical database. Experimental results show a very good diagnostic performance in terms of accuracy, sensitivity and specificity of the intelligent system. The rule-base can be refined each time the patient database is adequately updated. A further research step could be, to find a way to generate a a fuzzy rule-base and examine whether it would give better results.

Acknowledgements

We thank the European Social Fund (ESF), Operational Program for Educational and Vocational Training II (EPEAEK II), and particularly the Program PYTHAGORAS I, for partially funding the above work.

References

1. A. Jardin, G. Wagner, S. Khoury, F. Giuliano, H. Padman Nathan and R. Rosen, Erectile Dysfunction, ISSIR,Pr S. Khoury, (Eds)(1999), pp 115-138.
2. Gonzalez, A. and Dankel, D. (1993), The Engineering of Knowledge-Based Systems: Theory and Practice, Prentice-Hall, Inc.
3. Andrews, R., Diederich, J. and Tickle, A. (1995), "A Survey and Critique of Techniques for Extracting Rules From Trained Artificial Neural Networks", Knowledge-Based Systems, vol. 8(6), pp. 373-389.
4. R. S. Michalski, J. D. Carbonell and T. M. Mitchell, "Machine Learning, an AI approach", Tioga Publishing Company, 1983.
5. O. Gordon, F. Herrera and P. Villar (2001), "Generating the knowledge base of a fuzzy rule-based system by the genetic learning of the data base", IEEE Transactions on Fuzzy Systems, 9(4), 667-674.
6. R. A. K. Mehdi, H. Khali and A. Araar (2006), "Generating fuzzy rules for classification problems using genetic algorithms", Proceedings of the IASTED Conference on Artificial Intelligence and Applications (AIA-06), Innsbruck, Austria, Feb. 13-16.
7. A. Tsakonas, G. Dounias, J. Jantzen, H. Axer, B. Bjerregaard and D. G. von Keyserlingk (2004) "Evolving rule-based systems in two medical domains using genetic programming", AI in Medicine, 32, 195-216.
8. Perimenis P, Gyftopoulos K, Giannitsas K, Markou SA, Tsota I, Chrysanthopoulou A, Athanasopoulos A, Barbalias G. A comparative, crossover study of the efficacy and safety of sildenafil and apomorphine in men with evidence of arteriogenic erectile dysfunction. Int J Impot Res. 2004 Jan;16(1):2-7.
9. Z. Michalewicz (1999), Genetic Algorithms + Data Structures = Evolution Programs, Springer-Verlag, Berlin.
10. D. E. Goldberg (1989), Genetic Algorithms in Search, Optimization and Machine Learning, Addison-Wesley, Reading, Mass.
11. GAlib - A C++ Library of Genetic Algorithm Components, Matthew Wall, Massachusetts Institute of Technology (MIT) (http://lancet.mit.edu/ga/).

Combined Gene Selection Methods for Microarray Data Analysis

Hong Hu[1], Jiuyong Li[1], Hua Wang[1], and Grant Daggard[2]

[1] Department of Mathematics and Computing, University of Southern Queensland
QLD 4350, Australia
{huhong, jiuyong, hua}@usq.edu.au
[2] Department of Biological and Physical Sciences, University of Southern Queensland
QLD 4350, Australia
grant@usq.edu.au

Abstract. In recent years, the rapid development of DNA Microarray technology has made it possible for scientists to monitor the expression level of thousands of genes in a single experiment. As a new technology, Microarray data presents some fresh challenges to scientists since Microarray data contains a large number of genes (around tens thousands) with a small number of samples (around hundreds). Both filter and wrapper gene selection methods aim to select the most informative genes among the massive data in order to reduce the size of the expression database. Gene selection methods are used in both data preprocessing and classification stages. We have conducted some experiments on different existing gene selection methods to preprocess Microarray data for classification by benchmark algorithms SVMs and C4.5. The study suggests that the combination of filter and wrapper methods in general improve the accuracy performance of gene expression Microarray data classification. The study also indicates that not all filter gene selection methods help improve the performance of classification. The experimental results show that among tested gene selection methods, Correlation Coefficient is the best gene selection method for improving the classification accuracy on both SVMs and C4.5 classification algorithms.

1 Introduction

In the past decade, bioinformatics has been a fast growing research field due to the advent of DNA Microarrays technology. Amongst many active DNA Microarray researches, gene expression Microarray classification has been a hot topic in recent years and attracted the attention of many researchers from different research fields such as data mining, machine learning, and statistics.

The primary purpose of gene expression Microarray classification is to build a classifier from the categorized historical gene expression Microarray data, and then use the classifier to categorize future incoming data or predict the future trend of data. These methods encompass SVMs [2], bagging [21] and boosting [6], decision tree or rule based methods [20], etc. In practise, the gene expression Microarray classification has been extensively used in cancer research for classifying

B. Gabrys, R.J. Howlett, and L.C. Jain (Eds.): KES 2006, Part I, LNAI 4251, pp. 976–983, 2006.
© Springer-Verlag Berlin Heidelberg 2006

and predicting clinical cancer outcomes [22,23]. It is also applied to cancer diagnosis and prognosis [25,24]. In addition, classification can help researchers to discover the drug response for particular patients in order to use appropriate treatment for individuals [16].

Gene expression Microarray data is usually of very high dimensions and a small number of samples. This makes it very difficult for many existing classification algorithms to analyze this type of data. In addition, Gene expression Microarray data contain a high level of noise, irrelevant and redundant data. All these attribute to unreliable and low accuracy analysis results.

Since many classification methods are not scalable to the high dimensions, they are inapplicable to analyzing raw gene expression Microarray data. Consequently, reducing the number of genes is crucial for applying classification algorithms to analyzing gene expression data.

This paper is organized in six sections. In this introductory section, we give a brief introduction to some problems in gene expression Microarray data classification. In section 2, we explain the importance of data preprocessing for gene expression Microarray data. In section 3, we review the gene selection methods. In section 4, we present the design of methods for comparing the accuracy of SVMs and C4.5 using different gene selection methods. In section 5, we test the four different gene selection methods with three datasets, and present a discussion of the results. In section 6, we conclude the paper.

2 Gene Expression Microarray Data Preprocessing

An objective of gene expression Microarray data preprocessing is to select a small set of genes which can be used to improve the accuracy and efficiency of classification from a high dimensional gene expression dataset.

For example, normally a gene expression Microarray dataset contains less than 100 examples, but has tens of thousand of genes(attributes). High dimensionality may cause significant problems in Microarray data analysis.

1. Irrelevant and noise genes decrease the quality of classification. Gene expression Microarray is a newly developed technology, and the data stored in Microarray data often contain a great deal of noise which is caused by human errors, malfunctions and missing values. In addition, not all genes in a dataset are informative for classification. Using irrelevant and noise genes for classification only causes a greater risk of decreasing the accuracy of classification. The noise and irrelevant genes should be removed from Microarray data before classification takes place.
2. The huge number of genes causes great computational complexity in building classifiers. A Microarray dataset contains large number of genes. This high dimensionality makes many classification algorithms inapplicable or inefficient.

Microarray data preprocessing is a very important stage for classification. One crucial step of Microarray data preprocessing is gene selection. A good gene selection method can not only increase the accuracy of classification by eliminating the irrelevant genes from Microarray data, but also speed up the classification process by reducing the Microarray data size.

3 Gene Selection Methods

Gene selection is a process of selecting the most informative genes which are most predictive to its related class for classification. According to the dependency with classification algorithms, gene selection methods can be divided into wrapper and filter methods [12].

A filter method performs gene selection independently to preprocess the Microarray dataset before the dataset is used for classification analysis. In recent years, filter methods have become increasingly popular as these methods can reduce the dataset size before classification. For example, one of the most popular filter gene selection methods is called *ranking* and it has been applied to cancer classification. Within the ranking, Golub *et al* [22] used a Signal-to-Noise ratio method for gene selection in a leukemia dataset, while a correlation coefficient method was applied to a breast cancer dataset by Van't Veer *et al* [23].

However, using a one-gene-at-a-time ranking method does not take the relationships between genes into account. Some genes among the selected genes have similar expression levels among classes, and they are redundant since no additional information is gained for classification algorithms by keeping them all in the dataset. This redundancy problem affects the performance of classifiers. To eliminate this problem, Koller and Sahami [13] developed an optimal gene selection method called Markov blanket filtering which can remove redundant genes. Based on this method, Yu and Liu [26] proposed the Redundancy Based Filter(RBF) method to deal with redundant problems and the results are very promising.

In contrast, a wrapper method embeds a gene selection method within a classification algorithm. The wrapper methods are not as efficient as the filter methods due to the fact that an algorithm runs on the original high dimensional Microarray dataset. However, Kohavi and John [12] have discovered that wrapper methods could significantly improve the accuracy of classification algorithms over filter methods. This discovery indicates that the performance of a classification algorithm is dependent on the chosen gene selection method. Nevertheless, no single gene selection method can universally improve the performance of classification algorithms in terms of efficiency and accuracy. An example of a wrapper method is SVMs [9], which uses a recursive feature elimination(RFE) approach to eliminate the features iteratively in a greedy fashion until the largest margin of separation is reached.

In summary, in order to deal with gene expression Microarray data more effectively and efficiently, classification algorithms need to consider applying a combination of filter gene selection and wrapper methods for Microarray data classification.

4 Experimental Design and Methodology

In this paper, we examine if combined gene selection methods can enhance the performance of a classification algorithm. We conduct some experiments on different existing gene selection methods to preprocess Microarray data for classification by benchmark algorithms SVMs and C4.5.

First of all, we choose SVMs-lights classification system [10] and C4.5 [17] for experimental study. This choice is based on the following considerations.

Consideration of benchmark systems: SVMs and C4.5 have been regarded as bench mark classification algorithms. SVMs was proposed by Cottes and Vapnik [4] in 1995. It has been one of the most influential classification algorithms. SVMs has been applied to many domains, for example, text categorization [11], image classification [18], cancer classification [7,2]. SVMs can easily deal with the high dimensional datasets with a wrapper gene selection method. SVMs also can achieve a higher performance compared to most existing classification algorithms. C4.5 [20,19] was proposed by Quinlan in 1993. It is a benchmark decision tree classification algorithms. It has been widely used in ensemble decision tree methods for gene expression microarray classification and the results are very promising [21,5,6,15].

Considering of wrapper methods: SVMs and C4.5 are not only benchmark classification systems, but each of them contains a wrapper gene selection method. SVMs uses a recursive feature elimination(RFE) approach to eliminate the features iteratively in a greedy fashion until the largest margin of separation is reached. Decision tree method can also be treated as a gene selection method. It selects a gene with the highest information gain at each step and all selected genes appear in the decision tree.

In this study, we choose and implement four popular ranking methods collected by Cho and Won [3], namely Signal-to-Noise ratio (SN), correlation coefficient (CC), Euclidean (EU) and Cosine (CO) ranking methods. A ranking method identifies one gene at a time with differentially expressed levels among predefined classes and puts all genes in decreasing order. After a specified significance expressed level or number of genes is selected, the genes lower than the significance level or given number of genes are filtered out. The advantages of these methods are intuitive, simple and easy to implement.

To evaluate the performance of different gene selection methods, three datasets from Kent Ridge Biological Data Set Repository [14] are selected. These datasets were collected from some influential journal papers, namely breast cancer [23], lung cancer [8] and B-cell lymphoma [1]. Table 1 shows the summary of the three datasets.

During the gene expression Microarray data preprocessing stage, we define the number of selected genes as 20, 50 and 100 for all filter gene selection methods. In our experiments, a tenfold cross-validation method is also carried out for each method to test its accuracy.

Table 1. Experimental dataset details

Dataset	Genes	Class	Record
ALL-AML Leukemia	7129	2	72
Breast Cancer	24481	2	97
Lung Cancer	12533	2	181

5 Experimental Results and Discussions

Table 2 and Table 3 show the detailed results for SVMs and C4.5 tested on three different datasets preprocessed by four different filter gene selection methods.

From these experimental results, we make the following observations.

1. When datasets are preprocessed, SVMs improves its prediction accuracy by up to 15%. Among the four gene selection methods, Correlation coefficient and Signal-to-Noise methods are the most effective preprocessing method with 91% accuracy on average, followed by Cosine 84%, and Euclidean 73%. All tested gene selection methods except Euclidean have improved the prediction accuracy of the classifier; while the performance of C4.5 improves its prediction accuracy by up to 12%. Among the four gene selection methods, Correlation coefficient and Euclidean methods are the most effective preprocessing methods with 85% accuracy on average, followed by Cosine 82%, and Signal-to-Noise 60%. All tested gene selection methods except Signal-to-Noise have improved the prediction accuracy of the classifier.

   These results indicate that the gene selection methods in general improve the prediction accuracy of classification. As we mentioned before, Microarray data contains irrelevant and noise genes. Those genes do not help classification but reduce the prediction accuracy . Microarray data preprocessing is able to reduce the number of irrelevant genes in Microarray data classification and therefore can generally help to improve the classification accuracy.
2. The experimental results show that with preprocessing, the number of genes selected affects the classification accuracy. In Table 2, the highest accurate results for both lung cancer and lymphoma are base on 100 genes while the highest accurate results for Breast cancer is based on 20 genes. The overall performance is better when datasets contain 50 or 100 genes. In Table 3, the overall performance is better when datasets contain 50 or 100 genes.

   The results indicate that a smallest dataset does not necessarily guarantee the highest prediction accuracy. As a preprocessing method, the number of selected genes can not be too small. At this stage, the objective of gene selection is to eliminate irrelevant and noise genes. However, less informative genes can sometimes enhance the power of classification if they are co-related with the most informative genes. If the number of genes has been eliminated too harshly, it can also decrease the performance of the classification. So during the preprocessing, we need to make sure that a reasonable number of genes are left for classification.

Table 2. The accuracy results for SVMs

Dataset	Original data	100 gene				50 gene				20 gene			
		CC	SN	EU	CO	CC	SN	EU	CO	CC	SN	EU	CO
Breast cancer	.62	.75	.72	.46	.54	.74	.74	.48	.57	.77	.77	.49	.53
Lung cancer	.96	.98	1.0	.92	.98	1.0	1.0	.91	.99	.99	.99	.92	.99
Lymphoma	.92	.99	1.0	.81	.95	.99	.99	.78	.95	.97	.98	.76	.98
Average	.80	.91	.91	.73	.82	.91	.91	.72	.84	.91	.91	.72	.83

Table 3. The accuracy results for C4.5

Dataset	original data	100 gene				50 gene				20 gene			
		CC	SN	EU	CO	CC	SN	EU	CO	CC	SN	EU	CO
Breast cancer	.61	.73	.46	.72	.57	.64	.46	.68	.57	.69	.46	.72	.59
Lung cancer	.93	.96	.82	.96	.96	.96	.82	.96	.96	.96	.83	.95	.96
Lymphoma	.80	.87	.5	.82	.87	.85	.5	.9	.87	.87	.5	.62	.9
Average	.78	.85	.59	.83	.8	.82	.59	.85	.8	.84	.6	.76	.82

3. The experimental results also show that not all gene selection methods improve the performance of classification. Based on Euclidean gene selection method, the classification accuracy of SVMs decreases by up to 8% on average, while the classification accuracy of C4.5 improves by up to 5%. Signal-to-Noise is able to improve the performance of SVMs by 11%, but this method decrease the performance of C4.5 by up to 19%.

   Those results remind us that when selecting the gene select method for data preprocessing, we must consider which classification method that the gene selection is for. For example, if we select SVMs as a classification algorithm, then Correlation coefficient or Signal-to-Noise gene selection method is better for data preprocessing. An inappropriate choice can only harm the power of prediction.

6 Conclusions

In this paper, we have conducted some experiments on different existing gene selection methods for preprocessing gene expression Microarray data for classification by SVMs and C4.5, which themselves contain a wrapped method. We observed that in general the performance of SVMs and C4.5 are improved by using the preprocessed datasets rather than original datasets. The results indicated that not all gene selection methods improve the performance of classification. Among the different gene selection methods, correlation coefficient is the best gene selection method and improves the performance of SVMs and C4.5 significantly. Our results also implied that a small dataset usually does not produce high prediction accuracy. So during the preprocessing, we need to make sure a reasonable number of genes are chosen for classification. In future, we will evaluate more gene selection methods and classification algorithms. We will investigate how Microarray data size are impact the performance of Microarray data analysis.

References

1. A. Alizadeh, M. Eishen, E. Davis, and C. M. et. al. Distinct types of diffuse large b-cell lymphoma identified by gene expression profiling. *Nature*, 403:503–511, 2000.
2. M. Brown, W. Grundy, D. Lin, N. Cristianini, C. Sugnet, T. Furey, M. Jr, and D. Haussler. Knowledge-based analysis of microarray gene expression data by using suport vector machines. In *Proc. Natl. Acad. Sci.*, volume 97, pages 262–267, 2000.
3. S.-B. Cho and H.-H. Won. Machine learning in dna microarray analysis for cancer classification. In *CRPITS '19: Proceedings of the First Asia-Pacific bioinformatics conference on Bioinformatics 2003*, pages 189–198, Darlinghurst, Australia, Australia, 2003. Australian Computer Society, Inc.
4. C. Cortes and V. Vapnik. Support-vector networks. *Machine Learning*, 20(3):273–297, 1995.
5. M. Dettling. Bagboosting for tumor classification with gene expression data. *Bioinformatics*, 20(18):3583–3593, 2004.
6. T. G. Dietterich. An experimental comparison of three methods for constructing ensembles of decision trees: Bagging, boosting, and randomization. *Machine learning*, 40:139–157.
7. T. S. Furey, N. Christianini, N. Duffy, D. W. Bednarski, M. Schummer, and D. Hauessler. Support vector machine classification and validation of cancer tissue samples using microarray expression data. *Bioinformatics*, 16(10):906–914, 2000.
8. G. Gordon, R. Jensen, L.-L. Hsiao, and S. G. et. al. Translation of microarray data into clinically relevant cancer diagnostic tests using gege expression ratios in lung cancer and mesothelioma. *Cancer Research*, 62:4963–4967, 2002.
9. I. Guyon, J. Weston, S. Barnhill, and V. Vapnik. Gene selection for cancer classification using support vector machines. *Machine Learning*, 46(1-3):389–422, 2002.
10. T. Joachims. Making large-scale support vector machine learning practical. In *Advances in Kernel Methods: Support Vector Machines.*
11. T. Joachims. Text categorization with support vector machines: learning with many relevant features. In C. Nédellec and C. Rouveirol, editors, *Proceedings of ECML-98, 10th European Conference on Machine Learning*, number 1398, pages 137–142. Springer Verlag, Heidelberg, DE, 1998.
12. R. Kohavi and G. H. John. Wrappers for feature subset selection. *Artificial Intelligence*, 97(1-2):273–324, 1997.
13. D. Koller and M. Sahami. Toward optimal feature selection. In *International Conference on Machine Learning*, pages 284–292, 1996.
14. J. Li and H. Liu. Kent ridge bio-medical data set repository. http://sdmc.lit.org.sg/gedatasets/datasets.html, 2002.
15. J. Li, H. Liu, S.-K. Ng, and L. Wong. Discovery of significant rules for classifying cancer diagnosis data. In *ECCB*, pages 93–102, 2003.
16. O. Maron and T. Lozano-Pérez. A framework for multiple-instance learning. In M. I. Jordan, M. J. Kearns, and S. A. Solla, editors, *Advances in Neural Information Processing Systems*, volume 10, pages 570–576. The MIT Press, 1998.
17. T. M. Mitchell. *Machine Learning.* McGraw-Hill, 1997.
18. E. Osuna, R. Freund, and F. Girosi. Training support vector machines:an application to face detection. In *In Proceedings of the IEEE Computer Society Conference on Computer Vision and Pattern Recognition*, 1997.

19. J. Quinlan. Improved use of continuous attributes in C4.5. *Artificial Intelligence Research*, 4:77–90, 1996.
20. J. R. Quinlan. *C4.5: Programs for Machine Learning.* Morgan Kaufmann, San Mateo, California, 1993.
21. A. C. Tan and D. Gibert. Ensemble machine learning on gene expression data for cancer classification. *Applied Bioinformatics*, 2(3):s75–s83, 2003.
22. T.R.Golub, D.K.Slonim, P. Tamayo, and et.al. Molecular classification of cancer: Class discovery and class prediction by gene expression monitoring. *Science*, 286:531–537, 1999.
23. L. V. Veer, H. Dai, M. V. de Vijver, and et.al. Gene expression profiling predicts clinical outcome of breast cancer. *Nature*, 415:530–536, 2002.
24. M. West and C. B. et. al. Predicting the clinical status of human breast cancer by using gene expression profiles. *Proc Natl Acad Sci U S A*, 98:11462–11467, 2001.
25. C. Yeang, S. Ramaswamy, P. Tamayo, and et.al. Molecular classification of multiple tumor types. *Bioinformatics*, 17(Suppl 1):316–322, 2001.
26. L. Yu and H. Liu. Redundancy based feature selection for microarray data. In *Proceedings of the Tenth ACM SIGKDD International Conference on Knowledge Discovery and Data Mining, Seattle, Washington, USA,*, pages 737–742, 2004.

Toward an Intelligent Local Web Search in Telematics

Dae-Young Choi

Dept.of MIS, Yuhan College, Koean-Dong, Sosa-Ku, Puchon, Kyoungki-Do, 422-749, Korea
dychoi@yuhan.ac.kr

Abstract. A new approach for an intelligent local Web search in telematics is proposed. It is based on voice Web search, driver's physical location, search distance, search directory and driver's search intentions. In order to handle fuzzy query, perception index (PI) is also proposed. Kinking these issues is technologically demanding and potentially useful in telematics.

1 Introduction

GPS, which stands for Global Positioning System, is able to show a user his/her exact position on the earth anytime, anywhere. GPS satellites, 24 in all, orbit at 11,000 nautical miles above the earth. The satellites transmit signals that can be detected by anyone with a GPS receiver. Using the receiver, a user can determine his/her location with great precision [1, 8]. Radio signals are sent from orbiting satellites to earth. GPS receivers on the ground can collect and convert the radio signals into position, velocity, and time information.

Telematics originally comes from the combination of telecommunication and informatics. In many countries and major IT companies, it now represents the convergence of the IT and automotive industries. For example, IBM pervasive computing software provides key components that help enable telematics services such as vehicle diagnostics, location-based safety and security systems, navigation aids, concierge services, Internet and e-mail access, and entertainment. The great market potential of telematics is broadly recognized not only for the many benefits it offers the consumer, but also for its return. Beyond the manufacture and sale of original equipment, telematics services will create subscription-based, aftermarket revenue streams for auto manufacturers, their supplier partners, and content providers. Voice-enabled telematics solutions deserve special emphasis because the safety benefits of hands-off operation are obvious. A hands-free voice interface is the safest and most natural interface of all. IBM embedded ViaVoice® provides both voice recognition and speech synthesis [10]. In this paper, a new approach for an intelligent local Web search in telematics is proposed. It is based on voice Web search, driver's physical location, search distance, search directory, and driver's search intentions. For example, if a traveler is visiting San Francisco and try to find the nearest '*famous* sushi' restaurant from driver's physical location on the Web, he/she tells a fuzzy query

B. Gabrys, R.J. Howlett, and L.C. Jain (Eds.): KES 2006, Part I, LNAI 4251, pp. 984 – 991, 2006.
© Springer-Verlag Berlin Heidelberg 2006

'*famous* sushi' to the voice Web searching component in vehicle. The voice Web searching component could give the nearest '*famous* sushi' restaurant relevant to the current location while the vehicle is in motion, and show navigation to go there with the aid of GPS, Web technologies and GIS server (like MapQuest).

2 A Voice Web Searching Component for IVIS

There is a rising public demand for voice-enabled telematics applications, because the safety benefits of hands-off operation are obvious. We suggest a voice Web searching component for in-vehicle information systems (IVIS). It has a voice communication support module between drivers (users) and Web search engines for a voice recognition and speech synthesis, and checks the query until 'yes' by the driver as in Fig. 1. In order to start a voice Web searching in vehicles, drivers press 'start' button. Now, drivers can use their natural voice to dictate Web searching. For example, drivers select search directory such as 'restaurants', 'travel', 'health', 'recreation', etc., and give search distance for Web searching (i.e., distance from driver's physical location), and then tell keyword to the voice recognition and speech synthesis. Thus, locally targeted search results will be read back to them aloud.

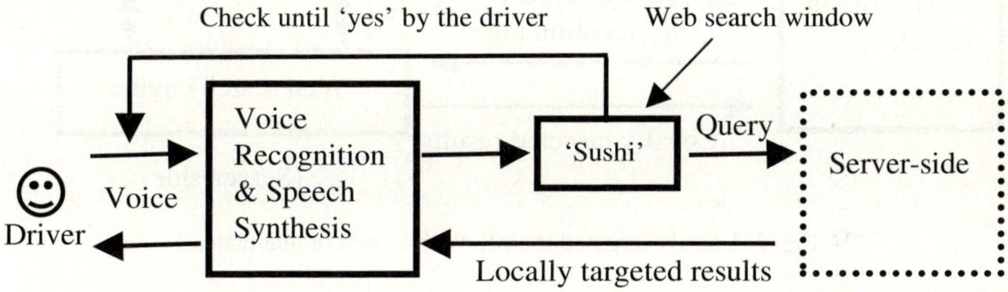

Fig. 1. A voice Web searching component (Client-side in vehicle)

3 Locally Targeted Mobile Web Search in Telematics

Web search technology companies including FAST Search & Transfer and Google are working with wireless phone companies to power mobile search that can benefit from locally targeted results. In addition, wireless Internet devices are more often equipped with GPS technology that can track the user's physical whereabouts. Both parties want to localize Web search to make it more relevant and draw more regional advertising dollars. These new technologies have attractive market potential for the wireless and hand-held sectors, specifically for any company providing location-based services. Geographical data in conjunction with the relevant keywords will give the mobile user much better search results. Any business owner who relies upon a physical presence for their business such as banks, gas stations, restaurants, even coffee shops, will benefit from this service.

We assume that each vehicle has the voice Web searching component and a GPS receiver. In this case, a driver in vehicle can search the Web with location tag (i.e., latitude, longitude, etc.). For example, if a traveler is visiting San Francisco and try to find the nearest 'sushi' restaurant from driver's physical location on the Web, he/she tells a query 'sushi' to the voice Web searching component in vehicle. The voice Web searching component could give the nearest 'sushi' restaurant relevant to driver's physical location by using the keyword 'sushi' and the location tag from GPS receiver in vehicle, and show navigation to go there with the aid of GPS, Web technologies and GIS server (like MapQuest). Consequently, driver's physical location in conjunction with the relevant keywords provide drivers with a local angle to make searching for local businesses on the Web.

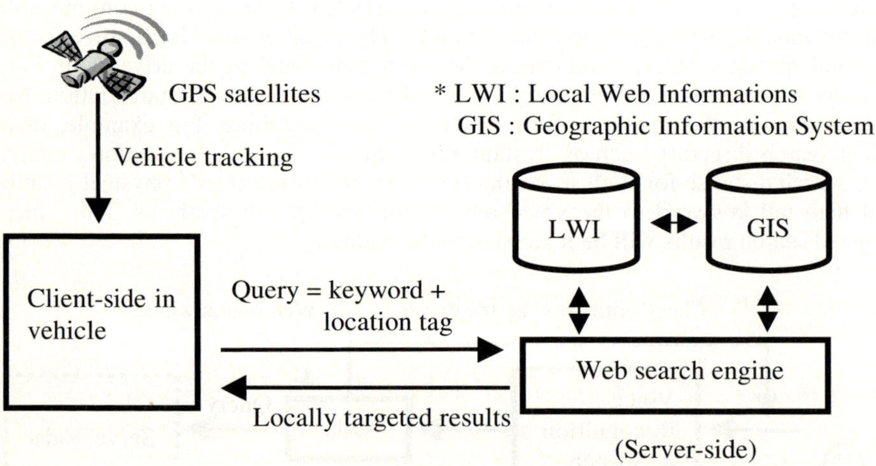

Fig. 2. Locally targeted mobile Web search in telematics

4 Personalized Local Web Search in Telematics

Although location tag in conjunction with the relevant keywords provide drivers with local angle to make searching for local businesses on the Web, it is still within the keyword-based local search as in SuperPages [9], Google 'local', etc. Now, SuperPages provides wireless phone service and it brings online yellow pages on the road, but not in vehicle service (telematics). Moreover, it has been defined which manage information only in a crisp way as in existing commercial Web search engines. Thus, their query languages do not allow the expression of preferences or vagueness which could be desirable for the following reasons : (i) to control the size of the results, (ii) to express soft retrieval conditions, (iii) to produce a discriminated answer [4].

Although commercial Web search engines such as Yahoo, Google help Internet users get to rich information, they generally return too many Web pages irrelevant to user's query. In addition, they do not properly handle fuzzy query. For example, if a traveler is visiting San Francisco and try to find '*famous* sushi' restaurants on the Web, he/she types a fuzzy query '*famous* sushi'. In this case, '*famous*' and 'sushi' are generally processed as keywords in existing commercial Web search engines. As a result, search engines return many Web pages (or URLs) irrelevant to user's query. It should be noted that fuzzy term '*famous*' is a constraint on the focal keyword 'sushi' rather than an independent keyword. In other words, the fuzzy term '*famous*' plays the role of a constraint on the fuzzy query. Thus, using fuzzy term(s), Internet users can narrow thousands of hits to the few that users really want. In this respect, the fuzzy terms in a query provide helpful hints for targeting queries that users really want. However, existing commercial Web search engines tend to ignore the importance of fuzzy terms in a query. Large-scale Web search engines such as Yahoo, Google effectively retrieve entire documents, but they are imprecise, and do not exploit and retrieve the semantic Web document content. We can not automatically extract such content from general documents yet [5].

A search engine, on receiving a query, would match the query against its document index (DI). All of the pages that match the user query will be selected into an answer set and be ranked according to how relevant the pages are for a given query. The DI is generally consisted of keywords that appear in the title of a page or in the text body. Based on the DI, commercial Web search engines help users to retrieve information in the Internet. Although location tags in conjunction with the relevant keywords provide users with a local angle to make searching for local businesses on the Web, it is still within the keyword-based local search as in SuperPages [9], Google 'local', Yahoo 'local'. In other words, they do not properly process fuzzy queries. For example, they do not properly handle a fuzzy query that finds '*famous* sushi' restaurants from driver's physical location within a specified area. A specified area is a search distance for Web searching. In this fuzzy query, we note that user want to find '*famous* sushi' restaurants, not just 'sushi' restaurants. Thus, commercial Web search engines do not properly reflect user's search intentions. Moreover, they have problems as follows : (i) large answer set, (ii) low precision, (iii) ineffective for general-concept queries [3]. In order to handle the fuzzy query, we introduce a perception index (PI). Perceptions are mainly manipulated based on fuzzy concepts. For processing a fuzzy query, the PI is consisted of attributes associated with a keyword restricted by fuzzy term(s) in a fuzzy query. In this respect, the restricted keyword is named as a focal keyword, whereas attribute(s) associated with the focal keyword is named as focal attribute(s). The PI can be mainly derived from the contents in the text body of a Web page or from the other sources of information with respect to a Web page. For example, the PI may be consisted of distance, size, weight, color, popularity, etc., on a keyword in the text body of a Web page. Using the PI, search engines can process fuzzy concepts (terms). Thus, if we integrate the DI used in existing commercial Web search engines with the proposed PI as in Table 1, search engines can handle fuzzy queries.

Table 1. An example of integrated index (DI + PI)

Document index (DI)	IPs	Perception index (PI)	FPs (Results)
Keywords	URLs	Size, No. of visitors, …	Targeted URLs

(IPs : Intermediate pointers, FPs : Final pointers, URLs : Uniform resource locators)

For example, consider a fuzzy query that finds '*famous* sushi'. In this case, the fuzzy term '*famous*' is processed by using the PI, whereas keyword 'sushi' is processed by using the DI.

In order to enhance the power of voice Web search-GPS-enabled telematics by means of fuzzy query, we introduce a search mechanism for integrating voice Web search, driver's physical location, search distance, search directory, and driver's search intentions (i.e., keywords, preferences, etc.) as shown in Fig. 3.

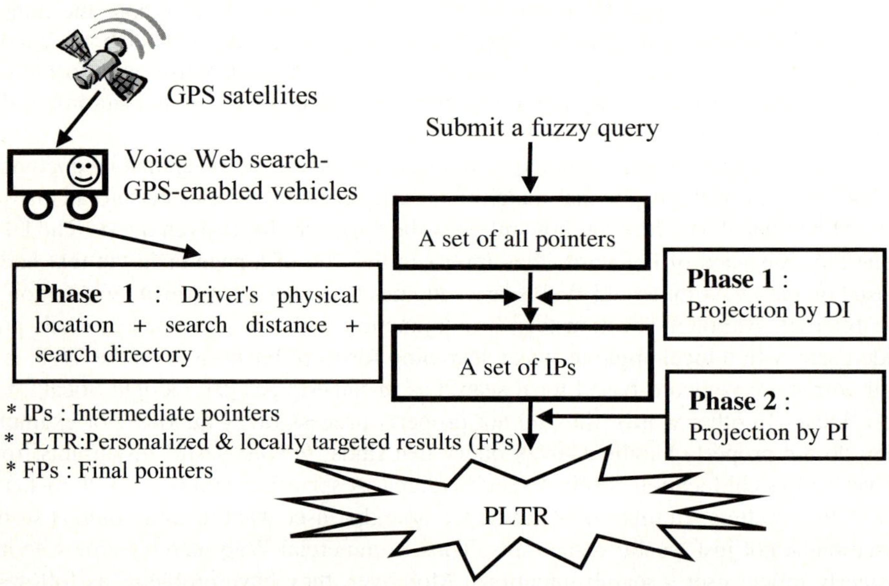

Fig. 3. Personalized local Web search in telematics

This search mechanism can be explained by SQL-like language : SELECT * FROM {a set of intermediate pointers that satisfies keyword(s) in the DI, driver's physical location, search distance and search directory} [WHERE the value(s) of focal attribute(s) in the PI are satisfied by the user]. We note that existing commercial Web search engines tend to ignore the importance of [WHERE] part. In this approach, driver's physical location, search distance, search directory, and the PI may be regarded as a constraint on the DI. Consequently, the proposed mechanism gives drivers more personalized and locally targeted search results (i.e., personalized local Internet in vehicles), and contributes to a significant convenience in location-based mobile Web searching.

4.1 Examples of Query

Let the set of 'sushi' restaurants from driver's physical location within a specified area be A = {A_1, A_2, …, A_{99}, A_{100}} and each A_i has its own page title or URL.

Example 1. Consider a crisp query that finds 'sushi' restaurants from driver's physical location within a specified area (Q_1). In this case, the PI in Fig. 3 is not used. Thus, the integrated index is made as in Table 2.

Table 2. A snapshot of integrated index (DI + PI) after processing Q_1

DI	IPs	PI	FPs (Results)
Sushi	A_1	No. of visitors, …	A_1
	A_2	No. of visitors, …	A_2
	…	…	…
	A_{100}+ Irrelevant URLs	No. of visitors, …	A_{100} + Irrelevant URLs

We note that IPs and FPs are equal. Thus, drivers receive locally targeted search results based on the focal keyword 'sushi' in the DI, driver's physical location, search distance and search directory 'restaurants'.

Example 2. Consider a fuzzy query that finds '*famous* sushi' restaurants from driver's physical location within a specified area (Q_2). In this case, the DI and PI in Fig. 3 are used. The DI and PI may be as follows : DI = {*sushi*, …}, PI = {*no. of visitors*, …}. We note that a focal keyword 'sushi' in the DI is restricted by a fuzzy term '*famous*'. In this case, the fuzzy term '*famous*' may be manipulated by the number of visitors (i.e., a focal attribute in the PI) per day, and represented by a membership function as in Fig. 4. For example, the proposed search mechanism finds '*famous* sushi' restaurants (i.e., 'no. of visitors' per day $\geq$ 100) from driver's physical location within 2 miles. The value of focal attribute 'no. of visitor' in the PI (i.e., 'no. of visitors' per day $\geq$ 100) is given by the driver. We assume that Q_2 is A_F, $A_F \in \{A_1, A_2, \ldots, A_{99}, A_{100}\}$, by using α-cut in Fig. 4. In this case, the integrated index is made as in Table 3. Consequently, drivers receive more personalized and locally targeted search results (A_F) based on the focal keyword 'sushi' in the DI, the value of focal attribute 'no. of visitor' in the PI, driver's physical location, search distance and search directory 'restaurants'.

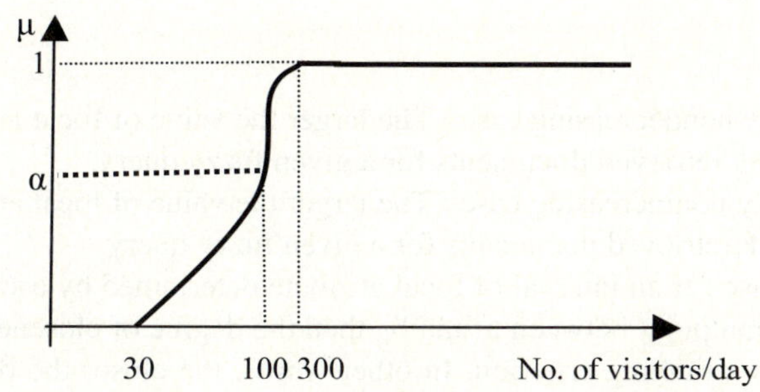

Fig. 4. A membership function of '*famous*'

Table 3. A snapshot of integrated index (DI + PI) after processing Q_2

Document index (DI)	IPs	Perception index (PI)	FPs (Results)
Sushi	$\{A_1, \ldots, A_{100}\}$ + Irrelevant URLs	*No. of visitors, …*	URLs w.r.t $\{A_F\}$

(Focal keyword : Sushi, Focal attribute : *No. of visitors*)

4.2 Personalized Ranking Algorithm Based on the PI

Ranking algorithm plays an important role in Web search engines. In existing Web search engines, however, detailed information regarding ranking algorithms used by major search engines is not publicly available. A simple means to measure the quality of a Web page, proposed by Carriere and Kazman [2], is to count the number of pages with pointers to the page. Google is a representative Web search engine that uses link information. Its rankings are based, in part, on the number of other pages with pointers to the page. In November 1999, Northern Light introduced a new ranking system, which is also based, in part, on link data. In other words, Google and Northern Light rank search results, in part, by popularity. In the meantime, HotLinks ranks search results based on the bookmarks of its registered users. In theory, more popular links indicate more relevant content, but if a user differs from the crowd, simply popularity-based ranking approaches dive deeply into other possibilities on the Web. Thus, they often give users many Web pages irrelevant to user's query. Moreover, they often tend to return unranked random samples in response to user's query. However, a ranking algorithm based on the PI makes the targeted search results displayed from the highest rank to the lowest rank by using the values of focal attributes in the PI. Yager [6, 7] identified three classes of linguistic quantifiers that cover most of these used in natural language : (i) A quantifier Q is said to be monotonically nondecreasing if $r_1>r_2$ then $Q(r_1)\geq Q(r_2)$. (ii) A quantifier Q is said to be monotonically nonincreasing if $r_1>r_2$ then $Q(r_1)\leq Q(r_2)$. (iii) A quantifier Q is said to be unimodal if there exists two values $a \leq b$ both contained in the unit interval such that for $r < a$, Q is monotonically nondecreasing, for $r > b$, Q is monotonically nonincreasing, and for $r \in [a, b]$, $Q(r) = 1$. Similarly, we can identify three classes of fuzzy terms that cover most of these used in natural language. In this respect, we design a new ranking algorithm based on the PI.

Algorithm 1
(i) Monotonically nondecreasing case : The larger the value of focal attribute in the PI, the higher the rank retrieved documents for a given fuzzy query.
(ii) Monotonically nonincreasing case : The larger the value of focal attribute in the PI, the lower the rank retrieved documents for a given fuzzy query.
(iii) Unimodal case : If an interval of focal attribute determined by α-cut is $[a_i, b_i]$, and let β denote the midpoint between a_i and b_i, then the degree of closeness (nearness) to β can be used as a ranking criterion. In other words, the closer the β, the higher the rank retrieved documents for a given fuzzy query.

Example 3. Consider the fuzzy query Q_2 in Example 2. In this case, the fuzzy term '*famous*' is represented by a monotonically nondecreasing as in Fig. 4. We assume that Q_2 is A_F by using α-cut. Let A_F be $\{A_F^1, A_F^2, ..., A_F^r\}$ taking values of focal attribute (i.e., no. of visitors) such as $Val(A_F^1) \leq Val(A_F^2) \leq \ldots \leq Val(A_F^r)$, then the search results A_F are ranked as the following order : $A_F^r, ..., A_F^2, A_F^1$.

The other cases in Algorithm 1 can be similarly handled. In the proposed approach, user's search intentions can be explicitly reflected by using the values of focal attributes in the PI. Thus, the proposed approach provides users with the personalized ranking based on user's search intentions (i.e., keywords, preferences).

5 Conclusions

Enhancing the power of voice Web search-GPS-enabled telematics by means of fuzzy query is introduced. Personalized ranking algorithm based on the PI is also introduced. It provides drivers with the personalized ranking based on driver's search intentions while the vehicle is in motion. Using fuzzy query, together with voice Web search, driver's physical location, search distance and search directory, drivers in vehicles can receive more personalized and locally targeted search results.

Acknowledgement

The author wishes to thank Prof. L. A. Zadeh, University of California, Berkeley, for his inspirational address on the perceptual aspects of humans in the BISC (Berkeley Initiative in Soft Computing) seminars and group meetings.

References

1. Andrienko, N., Andrienko, G.: Intelligent Support for Geographic Data Analysis and Decision Making in the Web, Journal of Geographic Information and Decision Analysis 5(2) (2001) 115-128.
2. Carriere, J., Kazman, R.: WebQuery : Searching and Visualizing the Web through Connectivity, Proceedings of the Sixth International Conference on WWW, 1997.
3. Kao, B., Lee, J., Ng, C. Y., Cheung, D.: Anchor Point Indexing in Web Document Retrieval, IEEE trans. on SMC (part C) 30(3) (2000) 364-373.
4. Kraft, D. H., Petry, F. E.: Fuzzy Information Systems : Managing Uncertainty in Databases and Information Retrieval Systems, Fuzzy Sets and Systems 90(2) (1997)183-191.
5. Martin, P., Eklund, P. W.: Knowledge Retrieval and the World Wide Web, IEEE Intelligent Systems, (May/June 2000) 18-25.
6. Yager, R. R.: On Linguistic Summaries of Data, In Knowledge Discovery in Databases, G. Piatetsky-Shapiro and B. Frawley (Eds.), MIT Press, (1991) 347-363.
7. Yager, R. R.: Database Discovery using Fuzzy Sets, Int. J. of. Intelligent Systems 11 (1996) 691-712.
8. http://gpshome.ssc.nasa.gov
9. http://www.SuperPages.com
10. http://www-306.ibm.com/software/pervasive/embedded_viavoice_enterprise

A Method of Recommending Buying Points for Internet Shopping Malls

Eun Sill Jang[1] and Yong Kyu Lee[2]

[1] Department of Computer Engineering, Dongguk University
[2] Corresponding Author, Department of Computer Engineering, Dongguk University
Pil-dong, Jung-gu, Seoul 100-715, Republic of Korea
esjang@dongguk.edu

Abstract. When a customer wants to buy an item in an Internet shopping mall, one of his/her difficulties would be to decide when to buy the item, because its price changes over time. If the shopping mall can recommend appropriate buying points, this would greatly help the customer. Therefore, in this paper, a method of recommending buying points based on time series analysis is proposed using a database of past item prices. The procedure for providing buying points for an item is as follows. First, the past time series patterns are searched for from the database using normalized similarities, which are similar to the current time series pattern of the item. Second, the retrieved past patterns are analyzed and the item's future price pattern is predicted. Third, using the future price pattern, a recommendation on when to buy the item is made.

1 Introduction

With the rising popularity of e-commerce of late, most Internet shopping malls are increasingly providing a variety of services to help customers make the best purchase through active marketing policies [14]. Such services recommend goods that are suitable for the price levels that customers want, in consideration of their preferences. It is difficult, however, to find out what services propose buying points for the goods that customers want to buy. If Internet shopping malls would recommend buying points to allow purchase of goods at moderate prices, along with providing customers information on such goods, customers will have useful information.

Thus, this study analyzes changes in selling prices to identify patterns and help customers purchase goods at moderate prices, and recommends the buying point recommendation approach that gives information on the most appropriate buying points in accordance with changes in selling price patterns. To identify changes in selling price patterns, patterns demonstrating changes similar to current selling price patterns were examined using the selling prices from a database of past sales. Next, future selling price patterns were analyzed on the basis of the examined past price pattern and the buying points were recommended in accordance with the changes in the future selling price patterns. This study also designs a goods recommendation system that includes buying point recommendation agents, using the buying point recommendation approach mentioned above.

The contents of this paper are organized as follows: Section 2 presents related work. In Section 3, we describe how to search similar price change patterns. A buying point

B. Gabrys, R.J. Howlett, and L.C. Jain (Eds.): KES 2006, Part I, LNAI 4251, pp. 992 – 1001, 2006.
© Springer-Verlag Berlin Heidelberg 2006

recommendation method is suggested in Section 4, and we describe the results of the performance experiments in Section 5. Section 6 designs a goods recommendation system and finally, Section 7 derives a conclusion and future study topics.

2 Related Work

In this study, goods recommendations on Internet shopping malls are examined using the similarity measurement utilized in the time series data pattern search and the time series data patterns.

2.1 Item Recommendation

A number of goods have been rapidly introduced through active marketing in recent consumer markets, thus shortening the price change cycle of goods [14]. To reflect such trends, Internet shopping malls must provide buying point information by analyzing the change cycles of selling prices in the past, to help customers purchase goods at more moderate prices. Most researches, however, are about accurate recommendations of goods that customers want. It is difficult to find researches that try to provide information on buying points for purchase of better goods at moderate prices.

2.2 Similarity Measure

Time Series Data comprise real values in series acquired from a certain time cycle. Examples of these are data related to stock prices, exchange rates, sales records, medical measurements, and corporate growth rates [4]. Researches have been performed on similarities of such time series data [9] and time series data inquiry approaches based on similarities [1][4][10][11]. Similar time series data search techniques include Euclidean Distance [4][5][13], Scaling [1], Shifting [1], Normalization [10][13], Moving Average [2][8], Time Warping [10], Landmark Similarity Measurement [9], and Shape-based Search [12].

Several attempts to use only Euclidean Distance to determine the desired time series data have failed. Thus, to effectively search for similar time series data, researches are needed to find out similar time series data using Euclidean Distance by identifying patterns showing significant similarities using k-moving average transformation, which removes noise from total time series data by getting the average of the k data in the series, arranging them in sequence, and normalizing the transformation to facilitate the search of similar patterns by calculating the relative values [1][8][11][12].

Consequently, this study identifies similar selling price patterns by calculating the Euclidean Distance using k-moving average transformation and normalizing transformation, which are widely used to measure similarities.

2.3 Time Series Data Patterns

Time series data patterns are largely classified into Up, Flat, and Down patterns, in consideration of the flow of price changes. Such patterns are defined in stock markets as triangle, flag, and broadening patterns. These patterns are used to estimate the pattern changes over time [7] and to technically estimate the data change from the influence of the independent data flow in the statistics techniques of regression analysis [6].

It is difficult to identify price flows on the basis of the patterns defined in stock markets because of short change cycles and technical limits. Thus, this study adopts the Up, Flat, and Down patterns using regression analysis models.

3 Similar Price Pattern Search

This section examines the normalizing transformation process and similarities in time series data to identify similar price patterns from selling prices of the time series type.

3.1 Definition of Time Series Data

Selling prices on the database of past sale records are used as the total time series data. A set of basic time series data is selected to search for similar partial time series data and each time series data is defined as the following:

- D: Total time series data of the selling prices
- Q: Basic time series data
- S: Partial time series data

Select Q, the length of which from D is n. Compare S (all partial time series data the length of which is n among the remaining time series data of the selling prices) with Q. Next, find out the S of similar selling price patterns satisfying the similarity requirements. The process of identifying similar time series data is summarized below.

① Select Q from D (Q's length is n).
② Compare S with D and Q (S's length is n).
③ Find out the S of similar selling price patterns satisfying the similarity requirements.

3.2 Price Pattern Search

To find out the S the patterns of which are similar to those of Q from D, the normalizing transformation and similarity calculation process below needs to be repeated.

Normalization of Time Series Data

[Step 1]. To eliminate the influence of noise from D, transform the k-moving average for the total time series data using Eq. 1 below.

$$S_k[j] = \frac{1}{k} \times \sum_{i=i}^{j+k-1} S[i] \tag{1}$$

S = Time series data [$0 \le i < Len(S)$]
k = Moving average [$1 \le k < Len(S)$]
$S_k[i]$ = k-moving average transformation time series [$0 \le j < Len(S) - k + 1$]

[Step 2]. To identify similar patterns from the total time series data, carry on the normalizing transformation process using Eq. 2 below.

$$S'[i] = \frac{S_k[i] - \frac{Max(S) + Min(S)}{2}}{\frac{Max(S) - Min(S)}{2}} \quad (2)$$

S[i] = Time series data [$0 \le i < Len(S)$]
S′[i] = Normalizing transformation time series [$0 \le i < Len(S)$]

Similarity of Time Series Data. Calculate the Euclidean Distance of all the S's the length of which is n from the remaining data of D and Q to find out the S that satisfies the similarity requirements.

The Euclidean Distance that will be used to measure the similarity between S and Q is measured using Eq. 3 below.

$$L(S,Q) = \sqrt{\sum_{i=1}^{n} (a_i - b_i)^2} \quad (3)$$

$$S = \{a_1, a_2, \ldots\ldots, a_n\},\ Q = \{b_1, b_2, \ldots\ldots, b_n\}$$

4 Buying Point Recommendation Method

This section examines how to recommend buying points depending on future price changes on the basis of past price patterns.

4.1 Basic Patterns

The basic patterns are classified into three types, as shown in Table 1. The buying points are recommended in accordance with changes in future selling price patterns on the basis of each basic pattern.

Table 1. Types of Basic Patterns

Type of Pattern	Explanation
	Ascending Selling Price
	Maintained Selling Price
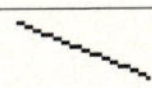	Descending Selling Price

4.2 Generalization of Patterns

Complicated and various patterns of time series data are generalized by applying moving average transformation followed by the simple regression model formula. Fig. 1 shows the generalization example of the ascending pattern. The pattern is generalized as ② by applying the Eq. 4 to pattern ①.

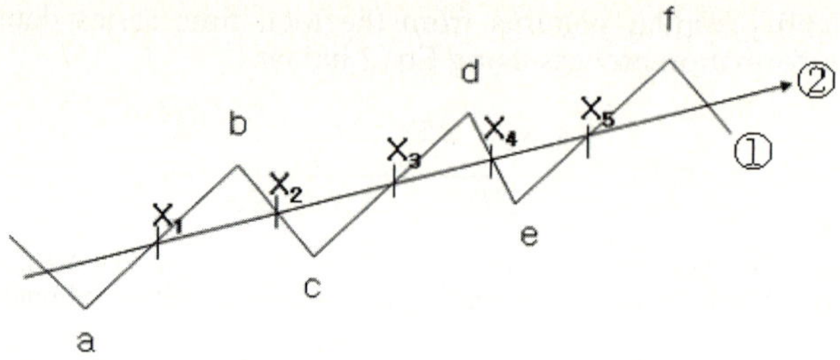

Fig. 1. Generation of an Ascension Pattern

$$Y_i = \alpha + \beta X_i \tag{4}$$

X = Time (Weekly), Y = Selling Price, α = constant, β = slope
$X_i = \{X_1=(a+b)/2, X_2=(b+c)/2, X_3=(c+d)/2, X_4=(d+e)/2, X_5=(e+f)/2\}$

4.3 Past Price Pattern and Future Price Pattern

Fig. 2 illustrates past and future selling price patterns on the basis of the point at which the purchase is requested.

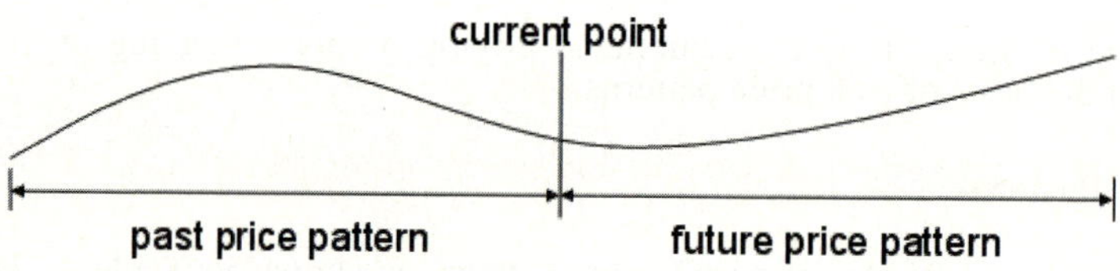

Fig. 2. Past Price Pattern and Future Price Pattern

4.4 Window Size of the Past Price Pattern

The length of time series pattern is called window size. When the length is 5 in a week unit, it refers to 5 weeks. This study applies 3~6 as the window size of past selling price patterns. Then similar price patterns are identified by calculating the basic time series patterns and the Euclidean Distance.

4.5 Window Size of the Future Price Pattern

Future price patterns are analyzed on the basis of similar price patterns found from past selling prices. The window size of the future selling prices is 1~5 weeks in this study.

4.6 Buying Point Recommendation

Table 2 shows the manner of recommending buying points depending on future price patterns analyzed on the basis of past price patterns.

Table 2. Buying Point Recommendation by Future Price Pattern

Type of Future Price Pattern	Explanation
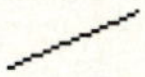	Buying at the Current Point
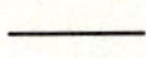	Buying at Any Time
	Waiting for a Purchase

In other words, when a future price pattern tends to ascend, the future selling price will increase. Thus, it is recommended that a purchase be made "at the current point." If a future price pattern tends to be maintained, the future selling price will not change. Therefore, it is recommended that a purchase be made "at any time." When future price patterns show a descending trend, the future selling price will go down. Thus, it is recommended that the right time to purchase "for the time being" be awaited.

4.7 Detailed Recommendation by Pattern Width

Table 3 explains in detail the recommendations on buying points depending on future price pattern widths. The domain refers to the slope of the pattern, and the recommendation rates are determined in accordance with the probability that the selling price slopes are maintained.

Table 3. Detailed Recommendation by Pattern Width

Type of Pattern		Domain	Explanation
Ascending	Steep	1.5 Over	Recommendation Rate 75% Over
	Flat	1 ~ 1.5	Recommendation Rate 50~75%
	Gentle	0.5 ~ 1	Recommendation Rate 25~50%
Maintained		-0.5~ 0.5	Recommendation Rate 100%
Descending	Steep	-1.5 Under	Recommendation Rate 75% Over
	Flat	-1 ~ -1.5	Recommendation Rate 50~75%
	Gentle	-0.5 ~ -1	Recommendation Rate 25~50%

5 Performance Experiments

This section carries on the performance experiments on the similarity and estimation rate of future price patterns to recommend the buying points.

5.1 Experiment Environment

For the experiment, information was collected on sales of used cars that have actually been traded, and a database was formulated. Secured were 554 sales data on the same car models and the database for one year. This study selected 50 time series data and 10 basic time series data by getting the selling price averages on a weekly basis from the sales database for the experiment.

5.2 Similarity of Future Price Patterns

Fig. 3 shows the similarity of the Euclidean Distance of future price patterns analyzed on the basis of past price patterns, by window size. The window size in the example refers to the size of the past price pattern. The shorter the window size of the future price pattern is, the greater its similarity with the Euclidean Distance is. The window size of the past price pattern demonstrated a similar similarity.

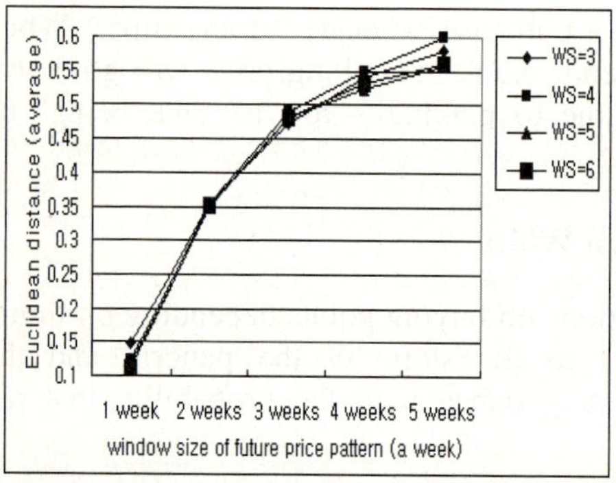

Fig. 3. Similarity of Future Price Patterns

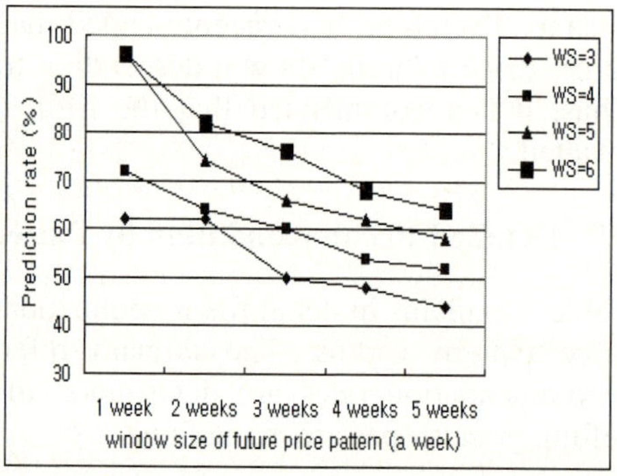

Fig. 4. Prediction Rate of Future Price Patterns

5.3 Prediction Rate of Future Price Patterns

Fig. 4 illustrates the prediction rates of ascending, maintaining and descending changes in future price patterns on the basis of past price patterns by window size. In other words, the shorter the window size of the future price pattern is, the higher the prediction rates are. The prediction rates up to the third week were 60% or higher. The window sizes of past price patterns showed higher prediction rates in the 6th, 5th, 4th and 3rd weeks, in that order.

6 Design of a Goods Recommendation System

This section designs the overall structure and agent model of the buying point recommendation system for the recommendation of buying points.

6.1 Goods Recommendation System

Fig. 5 explains the overall structure of the goods recommendation system, including the agents, to recommend the buying points, as explained earlier. The overall structure comprises the goods registration subsystem, the goods recommendation subsystem, the buying point recommendation agent, and the sales management subsystem.

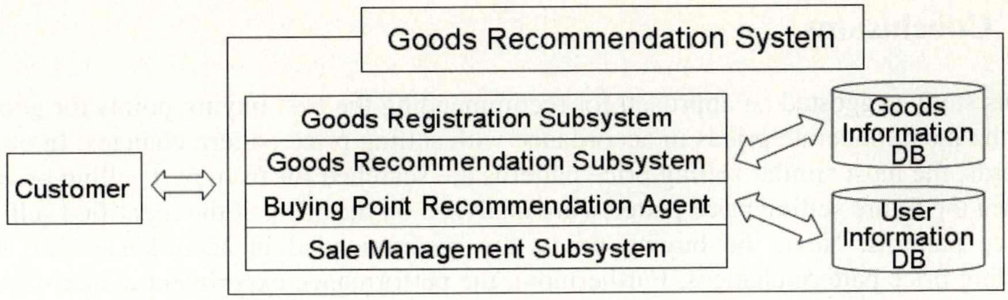

Fig. 5. Structure of the Goods Recommendation System

When a customer wants to purchase goods, the goods recommendation subsystem finds out the goods that the customer wants from the goods registered in the goods recommendation subsystem. The information on the goods from the search is provided to the customer along with the information on the most appropriate buying points through the goods recommendation subsystem and the buying point recommendation agent. All information related to sales of goods provided to customers is managed in the sales management subsystem.

6.2 Buying Point Recommendation Agent Module

Fig. 6 illustrates the agent model design to recommend the buying points using the aforementioned goods recommendation system. In other words, the past price patterns pass through the normalizing transformation process after they are derived. Then the future price patterns are identified on the basis of the past price patterns showing significant similarities. The buying points before the increase in the selling prices are recommended using the future price patterns with significant similarities.

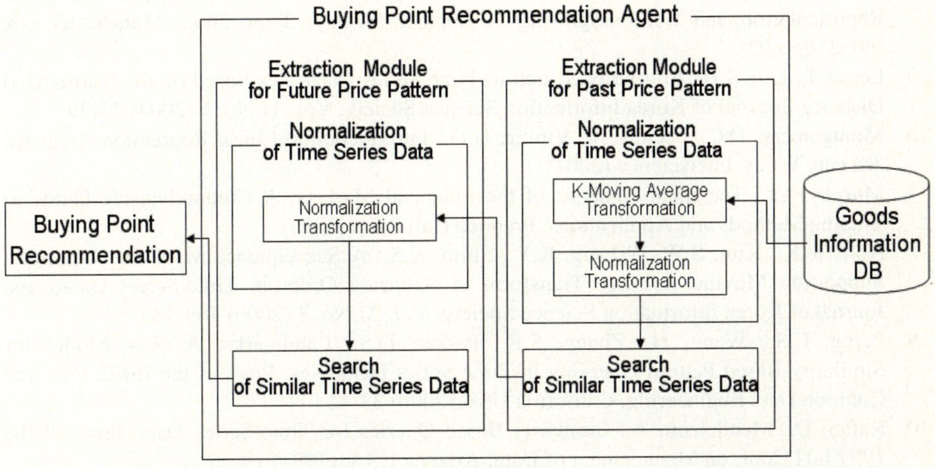

Fig. 6. Structure of the Buying Point Recommendation Agent Module

7 Conclusions

This study suggested an approach for recommending the best buying points for goods at the most moderate prices in accordance with selling price pattern changes. In other words, the most similar selling price patterns are searched for from past selling prices. Then the future selling price patterns are analyzed on the basis of the identified selling price patterns. Next, the buying points are recommended in accordance with the selling price pattern changes. Furthermore, the performance experiment demonstrated the effective prediction of buying points. A goods recommendation subsystem that recommends the buying points was also designed. Consequently, the goods recommendation system suggested in this study will ensure customers' convenient and effective purchase of goods in Internet shopping malls by recommending useful information on buying points through supplementation of previous system functions that recommend goods only in consideration of customers' preferences.

Future studies should concentrate on experiments on the probability that customers will purchase goods the prices of which are cheaper than the actual prices when they purchase goods in accordance with the buying points recommendation system.

References

1. Agrawal, R., Lin, K., Sawhney, H.S., Shim, K.: Fast Similarity Search in the Presence of Noise, Scaling, and Translation in Time-Series Databases. Proc. of the 21st Int'l Conf. on Very Large Databases, Zurich Switzerland (1995) 490-501
2. Chatfield, C.: The Analysis of Time Series: An Introduction. 6th edn. CRC Press (2003)
3. Jang, E.S., Lee, Y.K.: Buying Point Recommendation for E-Commerce Systems. Proc. of the 8th Int'l e-Biz Conf. on Society for e-Business Studies, Seoul Korea (2005) 108-113
4. Kawagoe, K., Ueda, T.: A Similarity Search Method of Time Series Data with Combination of Fourier and Wavelet Transforms. Proc. of the 9th Int'l Symp. on Temporal Representation and Reasoning. IEEE Computer Society, Time 2002. Manchester UK (2002) 86 - 92
5. Lee, S.J., Lee, S.H.: Similarity Search in Time Series Databases Based on the Normalized Distance. Journal of Korea Information Science Society, Vol. 31 No. 1 (2004) 23-29
6. Montgomery, D.C., Peck, E.A., Vining, G.G.: Introduction to Linear Regression Analysis. 3rd edn. Wiley-Interscience (2001)
7. Murphy, J.J.: Technical Analysis of the Financial Markets: A Comprehensive Guide to Trading Methods and Applications. Prentice Hall Press (1999)
8. Noh, W.K., Kim, S.W., Whang, K.Y., Shim, K.S.: A Subsequence Matching Algorithm Supporting Moving Average Transform of Arbitrary Order in Time-Series Databases. Journal of Korea Information Science Society, Vol. 27 No. 3 (2000) 469-485
9. Perng, C.S., Wang, H., Zhang, S.R., Parker, D.S.: Landmarks: A New Model for Similarity-Based Pattern Querying in Time Series Databases. Proc. of the IEEE 16th Int'l Conf. on Data Engineering, California USA (2000) 33-42
10. Rafiei, D., Mendelzon, A.: Similarity-Based Queries for Time-Series Data. Proc. of the 1997 Int'l Conf. on Management of Data, Arizona USA, (1997) 13-25
11. Rafiei, D., Nendelzon, A.O.: Querying Time Series Data Based On Similarity. Journal of IEEE Transactions On Knowledge And Data Engineering, Vol. 12 No. 5 (2000) 675-693

12. Won, J.I., Yoon, J.H., Kim, S.W., Park, S.H.: Shape-Based Subsequence Retrieval Supporting Multiple Models in Time-Series Databases. Journal of Korea Information Processing Society, Vol. 10-D No. 4 (2003) 577-590
13. Yoo, S.K., Lee, S.H.: Effectiveness Evaluations of Subsequence Matching Methods Using KOSPI Data. Journal of Korea Information Processing Society, Vol. 12-D No. 3 (2005) 355-364
14. New Internet Trade, http://search.mk.co.kr/ (2005)

A Sensuous Association Method Using an Association Mechanism for Natural Machine Conversation

Seiji Tsuchiya[1, 2], Hirokazu Watabe[1], and Tsukasa Kawaoka[1]

[1] Dept. of Knowledge Engineering & Computer Sciences, Doshisha University Kyo-Tanabe, Kyoto, 610-0394, Japan
[2] Human Ecology Research Center, R&D H.Q., Sanyo Electric Co,. Ltd. Hirakata, Osaka, 573-8534, Japan

Abstract. Humans manipulate smooth communications by retrieving, understanding and judging the sensuous characteristics from conversations consciously or unconsciously. This paper presents sensuous association methods to judge noun association from combinations of nouns (or adjectives) and adjectives/adjectival equivalents expressing senses as a means of achieving sense judgment similar to the common-sense judgment made by humans. The propose method is based on a mechanism that makes it possible to associate various concepts from a given concept. The precision rate and recall rate for nouns associated from combinations of nouns (or adjectives) and adjectives/adjectival equivalents were approximately 96.3% and 63.6% in reference to human judgment.

1 Introduction

Humans manipulate smooth communications by retrieving, understanding and judging the sensuous characteristics from conversations consciously or unconsciously. For example, the expression "Your face looks like an apple" is interpreted as "You look flushed". A response for "I want to have something warm because it is cold today" would be "How about hot soup?" As indicated in the examples, humans establish conversations by associating sensuously proper things.

To realize such an innovation, the ability to retrieve a concept from a word and associate that concept with various other concepts will play a key role. A number of methods for configuring and realizing the mechanism that makes it possible to associate various concepts from a given concept, such as the concept base [1, 2] and calculation of the degree of association [3], have been proposed to date. A commonsense judgment method that judges the senses that humans retrieve from a noun [4, 5] using this association mechanism has also been suggested.

This paper proposes a sensuous association method to judge nouns that are retrieved from the combination of nouns (or adjectives) and adjectives/adjectival equivalents that express senses.

2 Sensuous Judgment System

Figure 1 shows the structure of the sensuous judgment system. The sensuous judgment system consists of a knowledge base that contains the relationships between

B. Gabrys, R.J. Howlett, and L.C. Jain (Eds.): KES 2006, Part I, LNAI 4251, pp. 1002 – 1010, 2006.
© Springer-Verlag Berlin Heidelberg 2006

the nouns and their characteristic senses (hereinafter referred to as "sensuous judgment knowledge base") and the unknown word processing method that processes unknown words that do not exist in the sensuous judgment knowledge base as known words by relating the unknown words to the known words that exist in the sensuous judgment knowledge base. The unknown word processing realizes the association of the words, processes the words using the concept base [1, 2], which is a large database generated automatically from multiple electronic dictionaries, and the calculates the degree of association method [3], which evaluates the relationships between words (hereinafter these two methods are collectively referred to as the "association mechanism"). The senses mentioned here are those corresponding to the five senses (sight, sound, smell, taste and touch).

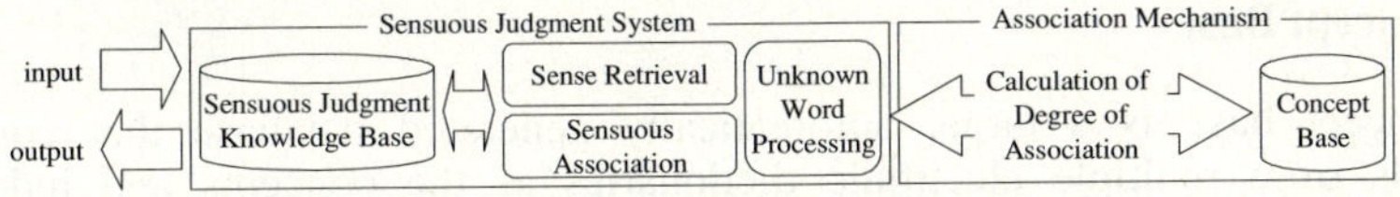

Fig. 1. Structure of Sensuous Judgement System

Two main judgments can be made using the sensuous judgment system: 1) judging in common-sense terms the senses that humans retrieve from a noun (hereinafter referred to as "sense retrieval") [4, 5], and 2) judging the nouns associated from the combination of nouns and adjectives/adjectival equivalents that express senses, or the combination of adjectives and adjectival equivalents that express senses (hereinafter referred to as the "sensuous association"). Sensuous association is proposed in this paper.

3 Sensuous Judgment Knowledge Base

To realize sensuous judgment, knowledge that relates a noun to its characteristic sense is required. However, it is difficult and inefficient to define and register such relations for all nouns. Therefore, the nouns used often in everyday life were selected as representative words, and these words were related to their retrieved senses and registered in the sensuous judgment knowledge base.

The senses retrieved from the nouns can be expressed with adjectives or adjectival equivalents. For example, the noun "apple" retrieves the senses such as "red", "round", or "sweet". The words that are used daily and express senses were manually selected from all the adjectives and adjectival equivalents, which are contained in general dictionaries. Then, by grouping the selected words into synonymous words, 98 words that relate to the senses (hereinafter referred to as "sensuous words") were selected. The sensuous words are used to express the senses retrieved from the nouns. Furthermore, the sensuous words were classified into two levels depending on the types of the five senses. For example, the sensuous word "red" is related to the sense of sight and the "color". Therefore, the sensuous word "red" was classified as "sight: color". Hereinafter, the senses such as "sight" are referred to as "five-sense words", and the factors such as "color" are referred to as "five-sense details". A total of 16 combinations were created from five-sense words and five-sense details.

The sensuous judgment knowledge base was created based on an existing thesaurus using a tree structure to represent its knowledge efficiently. The tree structure of this thesaurus describes the upper/lower level relationship and the whole/part relationship of the semantic attributes (nodes) of 2710 words that represent semantic usages of general nouns. The above-mentioned representative words are the leaves of the sensuous judgment knowledge base (thesaurus). The thesaurus nodes (the upper-level words of the representative words) were related to the retrieved senses, classified, and then registered as classification words. While only the specific expressions using the sensuous words were allowed for the senses related to the representative words, the senses related to the classification words, which are the upper-level words of the representative words, were allowed to have ambiguous expressions with the five-sense words or five-sense details.

4 Concept Base and Calculation of the Degree of Association

4.1 Concept Base

The concept base is a large, automatically generated database that consists of keywords from multiple electronic dictionaries as the concepts and independent words in the descriptions for the keywords as concept attributes. In this study, the concept base [1] was used, which has approximately 90000 concepts after the refining process, that is, after automatic generation, the attributes that were not appropriate as human senses were removed and the necessary attributes were added.

In the Concept Base, a given Concept *A* is expressed with the pair of the Attribute a_i that expresses the semantic characteristics of the concept and the Weight w_i that indicates how important the Attribute a_i is to express the Concept *A*. When the number of attributes of the Concept *A* is *N*, the Concept *A* can be expressed as follows. The Attribute a_i here is called the primary attribute of the Concept *A*.

$$A = \{(a_1, w_1), (a_2, w_2), ..., (a_N, w_N)\}$$

Since the primary Attribute a_i of the Concept *A* is the concept defined in the Concept Base, a_i can also have its own attributes. The a_{ij}, which is an attribute of a_i, is called the secondary attribute of the Concept *A*. The Figure 3 shows the expansion of the Concept "train" to secondary attributes.

	train, 0.36	locomotive, 0.21	railroad, 0.10		a_i, w_i	Primary Attributes
train	train, 0.36	locomotive, 0.21	railroad, 0.10	...	a_{i1}, w_{i1}	Secondary Attributes
	locomotive, 0.21	streetcar, 0.23	subway, 0.25	...	a_{i2}, w_{i2}	
	:	:	:	:	:	
	a_{1j}, w_{1j}	a_{2j}, w_{2j}	a_{3j}, w_{3j}	...	a_{ij}, w_{ij}	

Fig. 2. Example demonstrating the Concept "train" expanded to Secondary Attributes

4.2 Calculation of the Degree of Association

For Concepts *A* and *B* with Primary Attributes a_i and b_i and Weights u_i and v_j, if the numbers of attributes are *L* and *M*, respectively (*L* • *M*), the concepts can be expressed as follows:

$$A = \{(a_1, u_1), (a_2, u_2), ..., (a_L, u_L)\}$$
$$B = \{(b_1, v_1), (b_2, v_2), ..., (b_M, b_M)\}$$

The Degree of Identity *I* (*A*, *B*) between Concepts *A* and *B* is defined as follows (the sum of the weights of the various concepts is normalized to 1):

$$I(A,B) = \sum_{a_i = b_i} \min(u_i, v_j)$$

The Degree of Association is calculated by calculating the Degree of Identity for all of the targeted Primary Attribute combinations and then determining the correspondence between Primary Attributes. Specifically, priority is given to determining the correspondence between matching Primary Attributes. For Primary Attributes that do not match, the correspondence between Primary Attributes is determined so as to maximize the total degree of matching. Using the degree of matching, it is possible to give consideration to the Degree of Association even for Primary Attributes that do not match perfectly.

When the correspondences are thus determined, the Degree of Association $R(A, B)$ between Concepts A and B is as follows:

$$R(A,B) = \sum_{i=1}^{L} I(a_i, b_{xi})(u_i + v_{xi}) \times \{\min(u_i, v_{xi}) / \max(u_i, v_{xi})\} / 2$$

In other words, the Degree of Association is proportional to the Degree of Identity of the corresponding Primary Attributes, and the average of the weights of those attributes and the weight ratios.

5 Simple Sensuous Association Method

Sensuous association is a method to judge associated nouns from the combination of nouns and adjectives/adjectival equivalents that express senses or the combination of adjectives and adjectival equivalents that express senses. Hereinafter, the nouns and adjectives/adjectival equivalents, which are used as the basis for the sensuous association, are referred to as "input nouns" and "input adjectives". Although various patterns of the combinations or more complex combinations are possible, only the two basic combinations, a combination of one noun and one adjective/adjectival equivalent or a combination of one adjective and one adjectival equivalent, are considered in this study.

The relationships between the nouns and adjectives/adjectival equivalents are registered in the sensuous judgment knowledge base. Appropriate nouns can be derived easily for words already registered in the sensuous judgment knowledge base based on their relationships. On the other hand, to derive proper nouns for words not registered in the sensuous judgment knowledge base, the unknown words need to be processed by sense retrieval, which is realized by using the sensuous judgment knowledge base and the unknown word processing method [1, 2], along with the thesaurus and the concept base. Sense retrieval makes it possible to derive nouns from the adjectives and adjectival equivalents, the thesaurus makes it possible to derive nouns from nouns, and the concept base makes it possible to derive nouns from nouns, adjectives and adjectival equivalents. The actual procedures are as follows.

5.1 Simple Sensuous Association Method from Combination of Nouns and Adjectives/Adjectival Equivalents

The following five methods are proposed to realize sensuous association, which associates nouns from the combination of nouns and adjectives/adjectival equivalents.

(1) Select nouns for which a relation between the input nouns and the input adjectives is already registered in the sensuous judgment knowledge base.
(2) Expand the attributes of the input adjective using the concept base. From among the nouns in the expanded attributes, the nouns that are lower than the nodes are selected when the input noun is a node, or the nouns that are lower than the parent nodes are selected when the input noun is a leaf in the thesaurus.
(3) The adjectives/adjectival equivalents were retrieved by sense retrieval from the nouns that are lower than the parent node when the input noun is a leaf or the node when the input noun is a node in the thesaurus. The nouns are selected when the adjectives/adjectival equivalents are the same as input adjective.
(4) Expand the attributes of the input noun using the concept base. The adjectives/adjectival equivalents are retrieved by sense retrieval from the nouns in the expanded attributes. The nouns are selected when the adjectives/adjectival equivalents are the same as input adjective.
(5) Expand the attributes of the input noun and the input adjective using the concept base. The nouns included in the attributes of both the input nouns and the input adjectives are selected.

5.2 Simple Sensuous Association Method from Combination of Adjectives and Adjectival Equivalents

The following three methods are proposed to realize sensuous association by association of nouns through combination of adjectives and adjectival equivalents.
(1) Select the nouns with the relationships of both input adjectives already registered in the sensuous judgment knowledge base.
(2) Expand one input adjective using the concept base. The adjectives/adjectival equivalents were retrieved by the sense retrieval from the nouns in the expanded attributes. The nouns are selected when the adjectives/adjectival equivalents are the same as the other input adjective. This procedure is performed for both input adjectives. The nouns selected from both adjectives are integrated and duplicate nouns are removed.
(3) Expand the attributes of both input adjectives using the concept base. The nouns included in both attributes are selected.

5.3 Evaluation of Simple Sensuous Association Method

To evaluate the methods presented in Sections 5.1 and 5.2, the following experimental data were collected through questionnaire and used to test the methods. For the method in Section 5.1, 50 combinations of nouns and adjectives/adjectival equivalents from which humans can associate concrete nouns were obtained. For the method in Section 5.2, 20 combinations of adjectives and adjectival equivalents from which humans can associate definite nouns were used. The output nouns from each method were judged as follows. When an output noun was the same as the noun that humans associated from each input combinations, that output noun was considered to be a correct answer. When an output noun was different from the noun that humans associated from each input combination, it was considered an incorrect answer. Three people made the judgment by majority vote. Each of the methods was evaluated using the precision rate and the recall rate defined as follows:

Precision rate = number of correct answers / total number of output nouns
Recall rate = average of (number of correct answers for each combination / number of nouns that humans associate from each combination)

Tables 1 and 2 show the results for each method. In this study, the data are evaluated with a focus on the precision rate.

Table 1 shows that method (1) using the sensuous judgment knowledge base gave only correct answers. Among methods (2) to (5), method (3) provided the best precision rate, and method (5) had the worst rate. However, method (3) had the worst recall rate. From these results, it can be deduced that while many nouns can be associated by attribute expansion using the concept base, its characteristics also allow many ill-chosen nouns to be associated, leading to a low precision rate.

Table 2 shows that method (1) using the sensuous judgment knowledge base only gave correct answers. The result that method (3) had a much lower precision rate than method (2) is consistent with the discussion above.

Table 1. Result for methods in Section 5.1

	(1)	(2)	(3)	(4)	(5)
Number of correct answers	234	284	30	69	70
Number of incorrect answers	0	67	5	21	139
Precision rate	100.0%	80.9%	85.7%	76.7%	33.5%
Recall rate	51.2%	52.4%	7.1%	18.2%	19.0%

Table 2. Result for methods in Section 5.2

	(1)	(2)	(3)
Number of correct answers	43	67	59
Number of incorrect answers	0	10	44
Precision rate	100.0%	87.0%	57.3%
Recall rate	57.7%	68.0%	46.7%

6 Sensuous Association Method with a Threshold Value

From the results of Section 5, it is necessary to have a new method to efficiently remove the ill-chosen nouns associated from each method and to avoid eventually associating ill-chosen nouns. The strength of the relationship between the input adjectives and the output nouns for each method was calculated from the degree of association in Section 4.2. The nouns with calculated values smaller than a given threshold value were then processed as not having an output. The method including this process is called the sensuous association method with a threshold value.

6.1 Setting a Threshold Value

The optimal threshold value to use with the sensuous association method was investigated as follows. Data for the research were created based on 50 sets, as used in Section 5.3, of combinations of nouns and adjectives/adjectival equivalents from which humans can associate definite nouns. First, nouns were associated from the 50 sets of experimental data by each of the methods presented in Section 5.1. When the relationships between the output nouns and adjectives/adjectival equivalents in the experimental data were correct, the result was considered as a correct answer. When the relationships were wrong, the answer was considered to be incorrect. Furthermore, since duplicate nouns can be associated from each method in Section 5.1, duplication was removed and all remaining relationships were integrated, resulting 378 correct answers and 71 incorrect answers. The proper threshold value was decided by calculating the degree of association for each relationship and checking the correct answer remaining rate and the incorrect answer removal rate at the given threshold value. The correct answer remaining rate and the incorrect answer removal rate are defined as follows.

Correct answer remaining rate = number of combinations among the correct answers with degree of association above the threshold value / number of correct answers

Incorrect answer removal rate = (number of incorrect answers – number of combinations among the incorrect answers with degree of association above the threshold value) / number of incorrect answers

The variation in the number of correct answers/incorrect answers, the correct answer remaining rate, the incorrect answer removal rate, and the recall rate are examined as the threshold value increases from 0 to 0.25 in increments of 0.01. As the threshold value becomes lower, the correct answer remaining rate and recall rate improve, while the incorrect answer removal rate becomes poorer. Therefore, the threshold value of 0.05, at which the correct answer remaining rate and the incorrect answer removal rate cross, was used.

6.2 Evaluation of Sensuous Association Method with a Threshold Value

Tables 3 and 4 show the results of modifying each method in Section 5 to include sensuous association with a threshold value of 0.05. As method (1) had no incorrect answers, the threshold value was not used for this method.

The precision rate for each method was increased dramatically by processing with the threshold value method. Although the recall rate decreased, the magnitude of decrease was small in comparison with the increase in the precision rate. Therefore, it can be deduced that methods with a threshold value are more effective.

Table 3. Result for methods in Section 5.1 with the threshold value process

	(2)	(3)	(4)	(5)
Number of correct answers	206	27	43	67
Number of incorrect answers	17	1	7	67
Precision rate	92.4%	96.4%	86.0%	50.0%
Recall rate	40.3%	7.1%	12.2%	18.0%

Table 4. Result for methods in Section 5.2 with the threshold value process

	(2)	(3)
Number of correct answers	32	46
Number of incorrect answers	2	8
Precision rate	94.2%	85.2%
Recall rate	40.6%	35.2%

7 Evaluation of Sensuous Association Method

Tables 5 and 6 show the results of combining the methods described in Section 6. Note that the result for method (2) in Section 5.2 without the threshold value is obviously better than the result for method (3) in Section 5.2 with the threshold value. Therefore, method (2) without the threshold value can also be used as a combination method for processing input adjective combinations. Hereinafter, method (2) without the threshold value is referred to as (2)'.

Table 5. Result of combining the methods in Section 5.1 with the threshold value

	(1)	(1)+(2)	(1)+(3)	(1)+(4)	(1)+(2)+(3)	(1)+(2)+(4)	(1)+(3)+(4)	(1)+(2)+(3)+(4)
Number of correct answers	231	496	252	264	512	527	281	547
Number of incorrect answers	0	18	6	22	18	23	9	24
Precision rate	100.0%	96.5%	99.2%	97.1%	96.4%	95.8%	96.9%	95.7%
Recall rate	51.2%	61.9%	54.9%	52.3%	64.5%	61.4%	56.3%	65.3%

Table 6. Result of combining the methods in Section 5.2 with the threshold value

	(1)	(1)+(2)	(1)+(2)'	(1)+(3)	(1)+(2)+(3)	(1)+(2)'+(3)
Number of correct answers	43	51	67	50	56	71
Number of incorrect answers	0	2	10	8	10	18
Precision rate	100.0%	96.2%	87.0%	86.2%	84.8%	79.8%
Recall rate	57.7%	62.6%	72.8%	63.3%	65.5%	74.3%

Table 5 shows remarkably good results (all combinations had over 95% precision rate). When considering the precision rate, the recall rate, and the complexity of the process, the combination of methods (1), (2), and (3) is considered most effective. On the other hand, the combination of methods (1) and (2) is considered most effective in Table 6.

8 Conclusions

This paper presented sensuous association methods to judge noun association from combinations of nouns (or adjectives) and adjectives/adjectival equivalents expressing senses as a means of achieving sense judgment similar to the common-sense judgment made by humans. The proposed method is based on a mechanism that makes it possible to associate various concepts from a given concept.

It is important in the sensuous association method to derive appropriate nouns when processing words not registered in the sensuous judgment knowledge base by the use of sense retrieval, which is realized by using the sensuous judgment knowledge base and the unknown word processing method, the thesaurus, the concept base.

The precision rate and recall rate for nouns associated from combinations of nouns (or adjectives) and adjectives/adjectival equivalents were approximately 96.3% and 63.6% in reference to human judgment. The recall rate is a low level, but it is similar level as other researches. Therefore, the recall rate in this research is an appropriate level. It can be concluded that the method proposed in this paper is effective.

Acknowledgements

This work was supported with the Aid of Doshisha University's Research Promotion Fund.

References

1. Hirose, T., Watabe, H. and Kawaoka, T.: "Automatic Refinement Method of Concept-base Considering the Rule between Concepts and Frequency of Appearance as an Attribute", Technical Report of the Institute of Electronics, Information and Communication Engineers, NLC2001-93, pp.109-116 (2002)
2. Kojima, K., Watabe, H. and Kawaoka, T.: "A Method of a Concept-base Construction for an Association System: Deciding Attribute Weights Based on the Degree of Attribute Reliability", Journal of Natural Language Processing, Vol.9, No.5, pp.93-110 (2002)

3. Watabe, H. and Kawaoka, T.: "Measuring Degree of Association between Concepts for Commonsense Judgements", Journal of Natural Language Processing, Vol.8, No.2, pp.39-54 (2001)
4. Horiguchi, A., Tsuchiya, S., Kojima, K., Watabe, H. and Kawaoka, T.: "Constructing a Sensuous Judgement System Based on Conceptual Processing", Computational Linguistics and Intelligent Text Processing (Proc. of CICLing-2002), Springer-Verlag, pp.86-95 (2002)
5. Watabe, H., Horiguchi, A. and Kawaoka, T.: "A Sense Retrieving Method from a Noun for the Commonsense Feeling Judgement System", Journal of Artificial Intelligence, Vol.19, No.2, pp.73-82 (2004)

Motion-Based Interaction on the Mobile Device for User Interface

Jonghun Baek[1], Ik-Jin Jang[2], Hyun Deok Kim[3], Jae-Soo Cho[4], and Byoung-Ju Yun[3]

[1] Dept. of Information and Communications, Kyungpook National University, Daegu, 702-701, South Korea
[2] Samsung Electro-Mechanics
[3] School of Electrical Engineering and Computer Science, Kyungpook National University, Daegu, 702-701, South Korea
[4] School of Internet Media Engineering, Korea University of Technology and Education, Cheonan, South Korea,
bjisyun@ee.knu.ac.kr

Abstract. As a result of growth of sensor-enabled mobile devices such as PDA, cellular phone and other computing devices, in recent years, users can utilize the diverse digital contents everywhere and anytime. However, the interfaces of mobile applications are often unnatural due to limited resources and miniaturized input/output. Especially, users may feel this problem in some applications such as the mobile game. Therefore, Novel interaction forms have been developed in order to complement the poor user interface (UI) of the mobile device and to increase the interest for the mobile game. In this paper, we describe the demonstration of the gesture and posture input supported by an accelerometer. The application example we created are AM-Bowling game on the mobile device that employs the accelerometer as the main interaction modality. The demos show the usability for the gesture and posture interaction.

1 Introduction

Advances in mobile computing and micro electro mechanical systems (MEMS) enable the development of games which lead about user's interest on the mobile phone, PDA, and other portable devices. Presently, mobile devices have been made popular to everyone, and users always carry with them. In this sense, the mobile game is becoming more popular, especially with the recent release of the portable devices such as Sony's PlayStation Portable or Samsung's SCH-G100/SPH-1000. It shows that this area has huge potential for growth.

Traditional desktop-based UI have been developed on the basis that user's activities are static states. UI design for desktop devices can use all of its visual resources. The representative desktop-based interaction mechanisms are keyboard, mouse and joystick. In general, these are very graphical and still more detailed for desktop-based applications. In contrast, interaction mechanisms of the mobile devices can not utilize all or any of their visual resources by reasons not only that activities of users are

B. Gabrys, R.J. Howlett, and L.C. Jain (Eds.): KES 2006, Part I, LNAI 4251, pp. 1011 – 1020, 2006.
© Springer-Verlag Berlin Heidelberg 2006

dynamic states [1] but also the mobile devices have the limited resource and the small LCD display. Games on the mobile devices rely on key-pad, stylus-pen and input-panel. When using these mechanisms to manipulate mobile devices, it may lead to the time delay and the difficult operation because the small button may be manipulated repeatedly. So, it may make the users to lose their interests in the games [2].

There have been various studies about new input/output methods to solve this problem; voice recognition and gesture recognition are representative cases [1]-[4]. Voice recognition comes hard to apply to it because it is unsuitable for the UI of the mobile game. The other way to design the UI is to recognize the gesture of the user. The user's gesture-based input by interaction between human and computer enable user to approach easily more than desktop PC. The gesture recognition needs additional devices that can detect user's gesture and must be able to embed in mobile devices. The trend of this field is much using the embedded camera [5]. The camera-based UI may be uncomfortable for use because user's gesture must be transmitted to the mobile device under condition that user carries it in hand. Plus, the camera is comparative high cost. Therefore, it needs new method that can detect user's gesture easier and faster in mobile environment. In other words, proper sensors are necessary to detect actions or to recognize gestures of the user in the case of mobile and portable devices, and the processing capacity and mobile convenience should be considered before applying sensors. MEMS accelerometers are suitable sensors that coincide in this purpose. There have been various studies about using acceleration in recognizing gestures [6]-[8]. To use the gesture as input/output interface of the mobile devices, each gesture should be definitely discriminated each other. Otherwise, useless situation or inappropriate gestures can cause malfunction. Therefore, we sort context information in context-aware computing.

In this paper, to supplement lacking the UI of the mobile devices, we propose the novel UI that recognizes user's gestures such as swing and tilt, and estimates the user's posture. The application example we created is bowing game. To improve efficiency that is expected by applying the accelerometer, with considering many restriction elements such as processing capacity, we propose the structure of the system that embedded in the accelerometer and the algorithm of the signal processing of the accelerometer for the mobile devices, which is the motion-based interaction technology and context-aware computing technology.

2 Context Information

Unlike designing UI for traditional desktop applications, mobile devices can sort complex context information. Therefore many kinds of context information in context-aware computing environment are sorted [3][9].

Traditional desktop applications wait until user inputs data through the keyboard or mouse before execute especial action. Context-aware applications with sensors, however, can offer or require information even through non-conscious input of user by estimating motion of user at specific time and in particular place. Context-aware computing analyzes changing contexts in ubiquitous computing environment, and discriminates whether the information is valid or not, and then it generates the control event that requests or delivers information to execute the application.

In this paper, we detect context information called the "gesture" and "behavior", and calculate displacement amount of computer from the acceleration in order to gesture recognition and posture estimation. These user's gestures and behaviors are very important factors in ubiquitous computing environment. For examples, swing can be used to detect user's gait by using "how many swing user's arms?" during walking, tilt can be used to control the TV (e.g. volume and channel control, etc.), posture can be used to automatically turn on the mobile devices or receive any information from the object that attached RFID tag when user puts the mobile devices toward their chest in ubiquitous computing environment. Especially, the user's posture estimation technology can correct the various postures of the user such as walking, running and sitting in daily life as well as be applied to the sport science by the fusion of the sensors.

3 System Description

UI for the mobile device is designed by a general-structure that is available to the sensor's output at the same time in diverse control interface to increase the expected efficiency by applying the sensor, because the mobile device has many restriction elements such as processing capability, limited resources (e.g. memory, battery). Fig. 1 is our system that is designed to use the accelerometer by a general-purpose in mobile device.

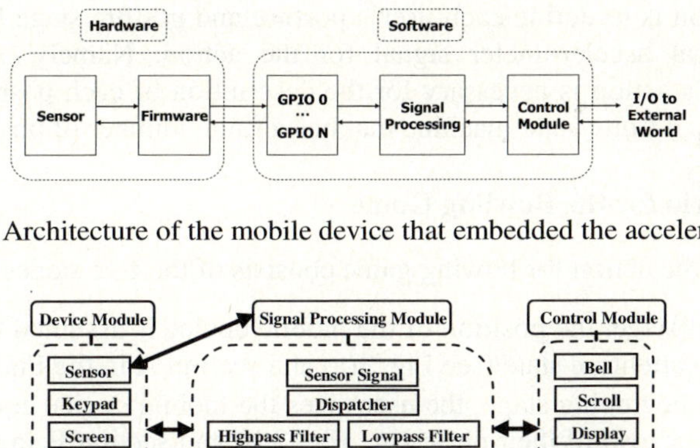

Fig. 1. Architecture of the mobile device that embedded the accelerometer

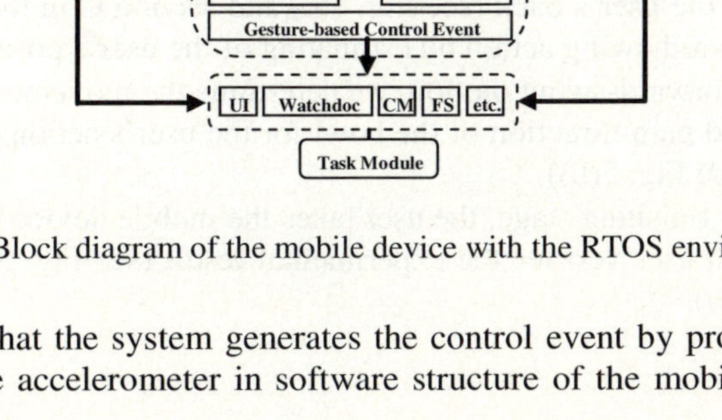

Fig. 2. Block diagram of the mobile device with the RTOS environment.

Fig. 2 shows that the system generates the control event by processing the signal detected from the accelerometer in software structure of the mobile device with the

RTOS (real time operation system) environment. The accelerometer's output is passed through a dispatcher that asks the GPIO for the signal of the accelerometer and passes through two parallel filters; a lowpass filter for a static acceleration such as tilt and gravity and a highpass filter for a dynamic acceleration such as vibration, shock, and impulse. The lowpass and highpass filtered output then goes to a signal processing for each application and then it generates a gesture-based control event and controls or executes the applications. To decrease overheads that are generated by attaching the accelerometer to the mobile device, we designed the structure that control events do not pass to the control module but to a task module.

4 Accelerometer Signal Processing for the Mobile Game

In the game of the existing mobile device, control of the direction and action of a character is done by using the key-pad. In our hand gesture- and posture-based mobile interaction technique, we show the usability for the use of the accelerometer that is embedded in the mobile device and the effect of the input. In the process of the user's motion recognition, actually, there are several cases that the acceleration signal for the user's action often misrecognized. For examples, there are malfunction by the user's hand tremble, the non-necessary gesture, and misrecognition. In addition, in the case that several gestures are occurred almost at the same time in one action as bowling game, how we can distinguish each gesture. To resolve these limitations and problems, we proposed that the signal processing method for the gesture recognition and posture estimation is to define each user's posture and gesture stage by stage as considering a typical accelerometer signal for the action. Namely, a representation scheme for user's action is necessary for the description of each posture and gesture. Therefore we use a finite state machine that has a finite number of possible stages.

4.1 The Scenario for the Bowling Game

We assume that the action for bowing game consists of the five stages as followings:

- In the initial stage, the position of the mobile device is assumed that user is gripping it in an attention state (see Fig. 3(a) and section A in the Fig. 5(b)).
- First, in the beginning stage, the user takes the mobile device equipped with 2-D accelerometer toward their chest (see Fig. 3(b) and section B in the Fig. 5(b)). It informs the beginning of the game by the estimating of the user's posture.
- Second, in the backward-swing stage (preparatory motion), the mobile device is located under the user's back (see Fig. 3(c) and section C in the Fig. 5(b)). It informs the forward-swing action by estimating of the user's posture.
- Third, in the forward-swing motion, we determine the movement distance, velocity, power, and path-direction of the bowl for the user's action (see Fig. 3(d) and section D in the Fig. 5(b)).
- Finally, in the finishing stage, the user takes the mobile device toward their chest again and then user verifies the experimental result (see Fig. 3(e) and section E in the Fig. 5(b)).

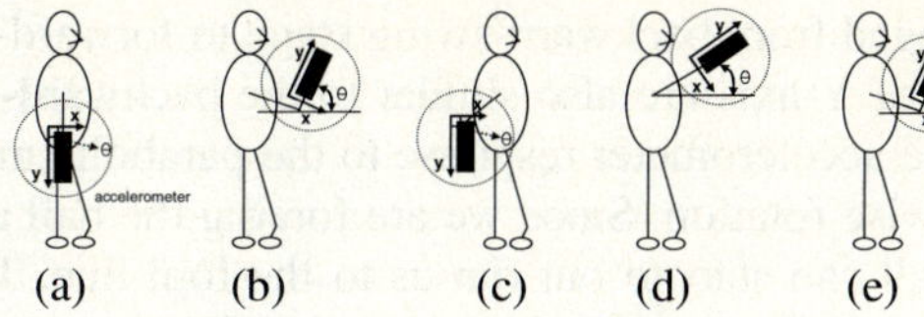

Fig. 3. Examples of the bowling action and the location of the accelerometer in the each stage; (a) initial stage, (b) beginning stage, (c) backward-swing stage, (d) forward-swing stage, and (e) finishing stage. A θ is the angle between the mobile device and the ground.

4.2 Accelerometer Signal Analysis for the Bowling Action

An experiment was conducted to assess the usability of the accelerometer as interaction technique for the mobile games. Thirty subjects (27 males and 3 females, ages 23 to 33, all were colleagues who volunteered for the study) participated in this study. First, we observed the output patterns of the accelerometer for the bowling action as the above scenario (see Fig. 4). Fig. 4 was the result of the experiment for the representative eight subjects among of the thirty subjects. Though the amplitudes are different each other for their action, the all output patterns are almost similar in all the experiments. To analysis the acceleration signal of the gestures for the bowling action, we considered the typical signal type for the bowling action (see Fig. 5(b)).

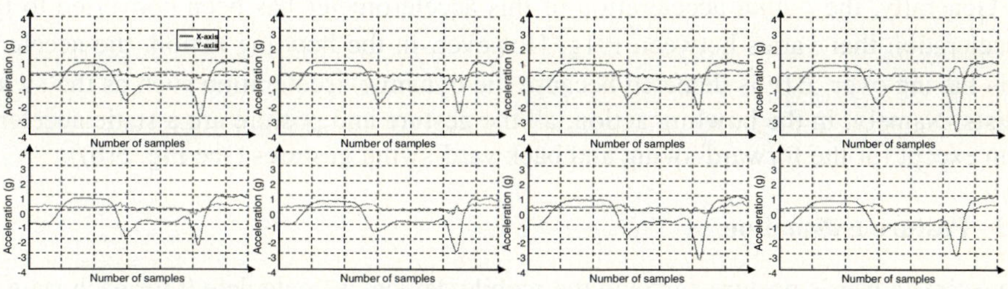

Fig. 4. Signal patterns of the eight participants for the bowling action

In the Fig. 5, the direction of the mobile device is parallel to the Earth's surface, and X- and Y-axis are pointing the right and forward direction, respectively (see Fig. 5(a)). Therefore, the output values of the X- and Y-axis are both 0g. The section A in the Fig. 5(b) indicates that the user's posture is "attention". The direction of the mobile device of this case is that X- and Y-axis are pointing the right and the Earth's gravitational field, respectively, and the output values are 0g (0º) and -1g (90º), respectively. The section B in the Fig. 5(b) indicates that user's posture changed from "attention" to beginning stage. In this case, the output values of Y-axis gradually increase because the Y-axis is changed from direction of the Earth's gravitational field to its reverse. And X-axis nearly remains the same. The section C in the Fig. 5(b) indicates that user's posture changed from beginning stage to backward-swing stage. In this case, the output values of Y-axis are decreasing and then are increasing because the accelerometer undergoes a parabolic motion (up and down direction) by the counterclockwise rotation when user does the backward-swing. The section D in the Fig. 5(b) indicates that user's posture changed from backward-swing stage to forward-swing stage. In this case, the output values of Y-axis are also similar to the backward-swing because, like the backward-swing, the accelerometer response to the parabolic motion (up and down direction) by the clockwise rotation. Since we are forcing the ball after the peak of the backward-swing, the ball can start to out run us to the foul line. The section E in the Fig. 5(b) as the last stage indicates that user ascertains the practice result, and because the accelerometer is located by the user's chest, beginning stage and finishing stage are the same about the output of the accelerometer.

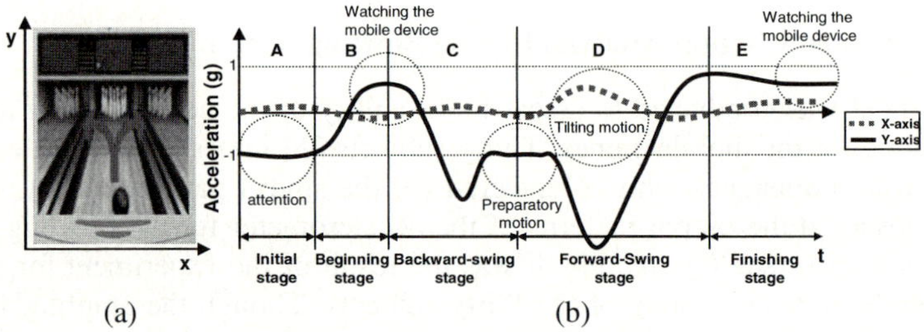

Fig. 5. Typical acceleration of the user; (a) the direction of the accelerometer that is embedded to the mobile device and (b) the typical user's action in bowling game

Generally, the output acceleration of this accelerometer has been converted to the acceleration that varies between ±1g. However, in the bowling action, the acceleration having over ±1g is detected because the dynamic acceleration such as the swing exists. Namely, in the bowling action, all the gesture and posture are a static acceleration except for the forward-swing and backward-swing gestures (see Fig, 5(b)).

4.3 Posture Estimation

To estimate user's posture carrying the mobile device, we calculate θ for each state as shown in Fig. 3. The accelerometer uses the force of gravity as an input vector to determine orientation of an object in space. The X- and Y-axis of the accelerometer are pointing the right and forward direction, respectively. If the accelerometer is ideal state, the accelerometer's digital outputs (X- and Y-axis) are 0g when it is parallel to the earth's surface and a scale factor is 12.5% duty cycle change per g. However, the accelerometer may not always be level when the applications using the accelerometer are executed on the mobile device, and calibration is recommended by the manufacturer. The method to do a more accurate calibration is to slowly rotate the accelerometer 360° and to make measurements at ±90°. The following table 1 shows the calibration results with the changes in the X- and Y-axis as the accelerometer that embedded in the mobile device is tilted ±90° through gravity. These experimental results were recorded at each Y-axis orientation to horizon. If the accelerometer is ideal state, the duty cycle is changed 62.5% into 37.5% both X- and Y-axis when angle is changed ±90°. However, in this experiment, the duty cycles of the X- and Y-axis are (66.2-40.6)/2=12.8%/g and (60.1-36.4)/2=11.9%/g, respectively [9].

To determine a range of the θ for the user's posture, with the various postures of the thirty participants, we collected data a two-dimensional time series for about three seconds at the sampling rate of the 140 samples/s from the outputs of the accelerometer (ADXL202EB). We examined the optimal threshold value to determine the convergence of the user's posture. Here, the convergence of the user posture was determined using the threshold valued and following form,

$$\mu - z \times \frac{\sigma}{\sqrt{n}} < \text{Threshold value} < \mu + z \times \frac{\sigma}{\sqrt{n}} \tag{1}$$

where, μ is mean of the population which is output of the accelerometer, z is normal distribution, σ is variance for the population, and n is number of samples. The 90% (z=1.65) of the total samples is used for the confidence interval. The following table 2 shows the result of the confidence interval.

Table 1. Output of X- and Y-axis respond to changes in tilt

Y-axis	X output		Y output	
Orientation to horizon	PWM (%)	Acceleration (g)	PWM (%)	Acceleration (g)
90	53.40	0.000	60.10	1.000
75	50.43	-0.232	59.69	0.961
60	47.41	-0.468	58.56	0.866
45	44.80	-0.672	56.73	0.713
30	42.77	-0.831	54.31	0.509
15	41.51	-0.929	51.54	0.276
0	40.60	-1.000	48.25	0.000
-15	41.36	-0.941	45.39	-0.240
-30	42.55	-0.848	42.58	-0.476
-45	44.72	-0.678	40.02	-0.692
-60	47.25	-0.481	38.17	-0.847
-75	50.19	-0.251	36.98	-0.947
-90	53.40	-0.000	36.40	-1.000

Through the experiment mentioned above (Table 1), the angle of the mobile device for each user's posture is estimated: it is about -90º for attention, about 40º ~ 50º for watching the mobile device, about -90º ~ -76º for backward-swing (preparatory motion) and about 43º ~ 55º for watching the mobile device after forward-swing gesture.

Table 2. Threshold value for estimating user posture, min/max value, mean and standard deviation of Y-axis for each posture

Posture	Min	Max	Mean	Standard deviation	T(Threshold value)
Attention	-1.0030	-0.9750	-0.9887	0.0043	-0.9958 < T < -0.9816
Watching the mobile device	0.5800	0.7680	0.7015	0.0359	0.6423 < T < 0.7607
Back swing motion	-1.0840	-0.7570	-0.9831	0.0193	-1.0149 < T < -0.9513
Watching after swing	0.6106	0.9130	0.7525	0.0413	0.6844 < T < 0.8207

4.4 Implementation

The gestures used in the bowling game are the forward-swing and tilt; the forward-swing gesture is expressed as the movement distance and the velocity of the bowl by measuring the acceleration for the user's forward-swing gesture, the path-direction of the bowl is expressed by the tilt gesture. We consider the range of the power and the tilt in the forward-swing stage in order to determine the movement distance, the velocity, and the path-direction of the bowl. Because it is difficult to distinguish between the dynamic acceleration like the forward-swing and the static acceleration like the posture, and the noises as the unnecessary user's action and hand tremble have to be removed, the state machine is needed. The state machine for the mobile bowling game recognizes and processes the user's action by the designed scenario by treating only valid gesture or posture in each state. The input of the state machine is the accelerometer signal by lowpass filter and the output is gesture-based control event. Fig. 6 is structure of the state machine.

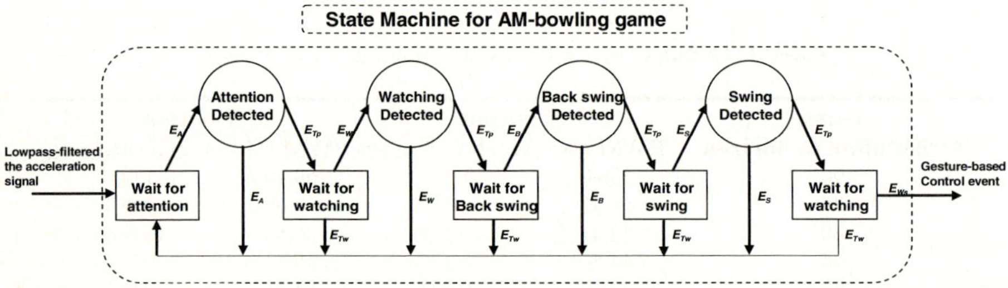

Fig. 6. The finite state machine for the AM-Bowling game; E_A for attention event, E_W for watching event, E_B for back swing event, E_S for swing event, E_{Ws} for watching event after swing, and E_{Tp} and E_{Tw} for timer event

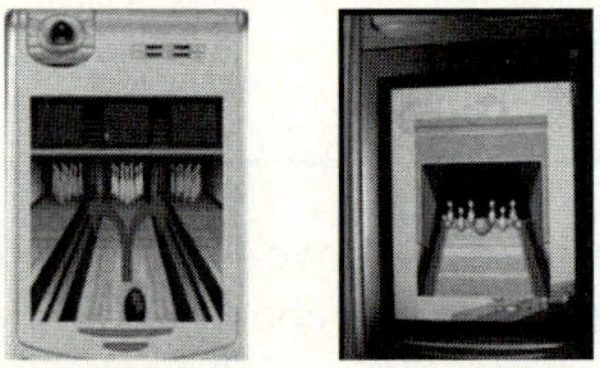

(a) (b)

Fig. 7. AM-Bowling game; (a) initial mode and (b) last mode on the mobile device

To demonstrate our application in games we have developed the mobile game called AM-Bowling game (see Fig. 7). In the game the user has to forward-swing the mobile device with his real arm into a virtual bowling line.

5 Experimental Results

We measured the recognition rate of the gesture and posture for the performance evaluation. The sampling rate is set 140ms. We define practiced man and non-practiced man as the participant and the malfunction includes non-function, mis-recognition, and over-time. The table 3 shows the recognition rate for each gesture and posture.

This result is useful in terms of the gesture- and posture-based control interface and technique for inducing the interest and realism for the mobile game as well as

Table 3. Recognition rate of the gesture and posture in the AM-Bowling game

User	Number of plays	Number of functions	Number of mal-functions	Recognition rate (%)
Practiced man	75	73	2	97.3%
Non-practiced man	75	70	5	93.3%

providing the convenience of the control. For example, in the AM-bowling game, we can improve the realism by using the gesture and posture in the control interface. Also, this enables to induce extra-interest through the dynamic motion for the control of the game. This is because that the interface for controlling the motion of game characters enables user to feel higher realism as we prefer to use joystick instead of keyboard in the personnel computer-based game.

6 Conclusion

We implemented the new interaction by estimating and detecting the context information as user's various gestures and postures with 2-axis accelerometer. We considered the typical bowing action by the output type of the accelerometer for 30 participants. Because it is difficult to distinguish between the dynamic acceleration and the static acceleration, and the noises as the unnecessary user's action and hand tremble have to be removed, the state machine is needed. The state machine for the mobile bowling game recognizes and processes the user's action by the designed scenario by treating only valid gesture or posture in each state. As a result, this will enhance not only the convenience to use but also the interest and realism for the mobile game.

References

[1] M. Ehreumann, T. Lutticke and R. Dillmann, Dynamic Gestures as an Input Device for Direction a Mobile Platform, IEEE International Conference on Robotics and Automation, vol. 3, pp. 2596-2601, 2001.

[2] V. Paelke, C. Reimann and D. Stichling, Foot-based mobile Interaction with Games, ACM SIGCHI International Conference on Advances in Computer Entertainment Technology, pp. 321-324, 2004.

[3] P. Indrajit, S. Togesh, O. Ercan and S. Rajeev, Toward Natural Gesture/Speech HCI: A Case Study of Weather Narration, Workshop on Perceptual User Interfaces, pp. 1-6, 1998.

[4] B. N. Schilit, N. Adams, and R. Want, Context-Aware Computing Applications, In Proceedings of the 1st International Workshop on Mobile Computing Systems and Applications, pp. 85-90, 1994.

[5] L. Gupta and S. Ma, Gesture-Based interaction and Communication: Automated Classification of Hand Gesture Contours, IEEE Transactions on Systems, Man, and Cybernetics, vol. 31, No. 1, pp. 114-120, 2001.

[6] S. Sotiropoulos, K. Papademetriou and A. Dollas, Adaptation of a Low Cost Motion Recognition System for Custom Operation from Shrink-Wrapped Hardware, Proceedings of the 2003 ACM SIGMM workshop on Biometrics methods and applications, pp. 107-114, 2003.

[7] A Wilson and S. Shafer, "XWand: UI for Intelligent Spaces, In Proc. CHI, pp. 545-552, 2003.

[8] K. Partrige, S. Chatterjee, V. Sazawal, G. Borriello and R. Want, TiltType: Accelerometer-Supported Text Entry for Very Small Devices, In UIST02: In Proceedings of the 15th annual ACM Symposium on User Interface Software and Technology, pp. 201-204, 2002.
[9] G. D. Abowd, A. K. Dey, P. J. Brown, N. Davies, M. Smith and P. Steggles, Towards a Better Understanding of Context and Context-Awareness, In HUC'99: In Proceeding of 1st International Symposium on Handheld Ubiquitous Computing, pp. 304-307, 1999.
[10] Low-cost ±2g/±10g Dual Axis iMEMS Accelerometers with Digital Output ADXL202/ADXL210 Data Sheet, Analog Devices Corporation, 1999.

Role-Task Model for Advanced Task-Based Service Navigation System

Yusuke Fukazawa, Takefumi Naganuma, Kunihiro Fujii, and Shoji Kurakake

Network Laboratories, NTT DoCoMo, Inc.
NTT DoCoMo R&D Center, 3-5 Hikari-no-oka, Yokosuka, Kanagawa, 239-8536 Japan
{y-fukazawa, naganuma, kunihi_f, kurakake}@netlab.nttdocomo.co.jp

Abstract. This paper presents a system to navigate a user to appropriate services that help the user to perform real world activities. User activities are modeled as task of solving problems met in the real world. Tasks are decomposed into a set of more concrete tasks, and are finally associated with appropriate services. User tasks are categorized according to the role of the user performing the tasks. A model of changes in appropriate tasks to match changes in the user's role is also proposed. Based on this role-task model, we propose an advanced service navigation system that recommends tasks relevant to the current role of the user in addition to the user's current location. This paper also describes an implemented service navigation prototype that can automatically detect changes in user role by using contactless IC card readers.

1 Introduction

In everyday life people must deal with various kinds of tasks and problems. For example, people feel thirsty, get lost, and need to find a restroom, or miss their train. Fortunately, there are various kinds of contents or services available on the mobile Internet that can solve these problems. Among those mobile services are "Transit-Guide", "Tokyo Restroom Map", and "Gourmet-Navi" [1]. They are useful to some extent, but the user must be prepared to actively find the service desired from among the large number available, and this is very difficult for ordinary users. If appropriate contents and services that could solve the user's problem could be identified and presented to the user, we could perform our daily activities much more comfortably.

Towards the above aim, we have proposed a task-oriented service navigation system[2][3] that supports the user in finding appropriate services. Naganuma et al. proposed a method for constructing a rich task-knowledge-base that represents common sense about typical complex real world activities, which we call tasks. An extraction from task-knowledge-base is shown in Fig.1. The task-knowledge-base holds several kinds of task-models. The task, the white characters, is the top-node of the task-model. The end nodes of the task-model, written in bold font, provide associations to services or contents via their URI.

To use the task-knowledge base for service navigation, Naganuma et al. proposed the **query-based service navigation system**. In this system, the user enters a task-oriented query such as "Go to theme park" and a list of tasks that

B. Gabrys, R.J. Howlett, and L.C. Jain (Eds.): KES 2006, Part I, LNAI 4251, pp. 1021–1028, 2006.
© Springer-Verlag Berlin Heidelberg 2006

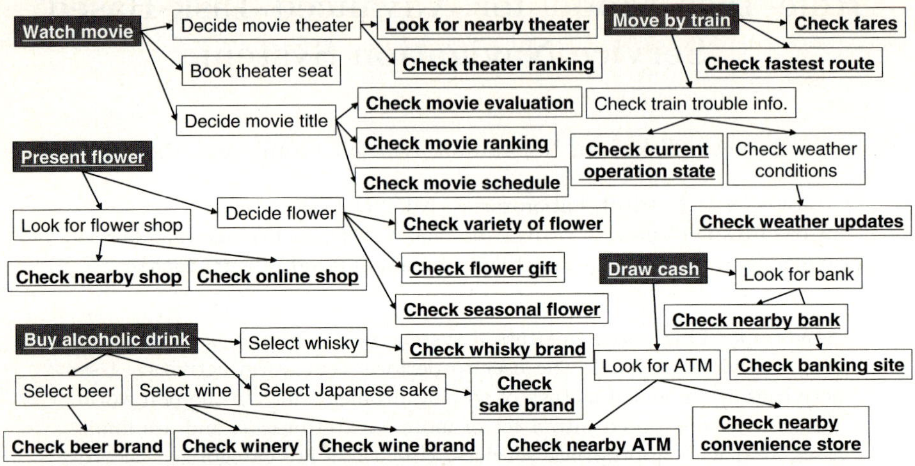

Fig. 1. Extraction from task-knowledge-base[2] that consists of multiple task-models

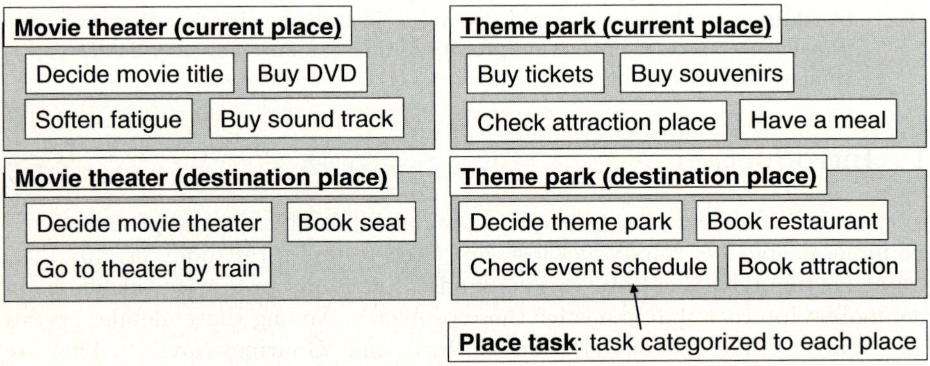

Fig. 2. Example of task categorization based on place-task model

match that query is sent to the mobile device. The most appropriate task can be selected and in turn the corresponding detailed task-model is shown to the user. In a final step, associated services can be invoked by establishing an Internet connection to access actual i-mode services.

In query-based service navigation systems, there is a problem that the step of inputting the user query may bother non-expert users, especially since mobile devices have very restricted UIs. To solve this problem, this paper proposes two variants of the task categorization model that omit the manual query input process and support user's selection of the task. We need to clarify the factors that effect the user's selection of the task from task-knowledge-base, and then categorize the tasks on the basis of the factors.

In Section 2, we focus on the user's current place and current destination is the factors that effect the user's selection of the task. We categorize tasks on the basis of such places. We call this task categorization model the **place-task model**. We also describe a service navigation system based on the place-task model, which we call **place-based service navigation**. In Section 3, we focus on the social role of the user as the factor that effect user's selection of the task and we proposed **role-task model** to categorizing the task by the role assigned to the user and modeled the changes of appropriate tasks according to the changes of the user's role. We also describe a service navigation system based on the role-task model, which we call **role-based service navigation**. Section 4 describes a prototype that can automatically detect changes in the user's role at a station by using contactless IC chip technology.

2 Place-Task Model and Place-Based Service Navigation System

In our developed task-knowledge-base, we have the task user performs only in a specific place such as movie theater, company, station etc. We also have the task that user can perform regardless of any place such as "Have a dinner", "Kill waiting time", "Make a reservation" etc, however, number of such task is few. That is because these task exist only at the top-level in task-knowledge-base and most of the sub-task of these task are the task user perform in a specific place as we firstly described. In this section, we propose the place-task model to categorizing the place-dependent-tasks, which cover most of the tasks in task-knowledge-base.

In the place-task model, we set two kinds of task categories: those associated with the current place and those associated with the destination place. As for current place based tasks, we collect the tasks that are likely to be performed at the current place, i.e. "Check attraction place" for "Theme park" or "Buy sound track" for "Movie theater" from the task-knowledge-base and categorize them as the task category for the current place as shown in the upper part of Fig.2. As for the destination place, we collect tasks associated with "making plans" to go to the destination, such as "Check event schedule" for "Theme park" from the task-knowledge-base and categorize them as the task category associated with the destination place as shown in the lower part of Fig.2.

We describe how to apply the place-task model to a service navigation system. At first, the user is asked to choose the place category, current place or destination place, and then is asked to select one of the places from the place list. The task-list associated with the selected place is then shown to the user. Finally, the user can reach the task-model by selecting the appropriate task from the task-list.

3 Role-Task Model for Advanced Service Navigation

It is obvious that the task is not always clear from the current place or destination. For example, we must take the user's companions such as friend, family, or boss, into consideration in categorizing the task. In this chapter, we consider the social role (role) as a significant factor in determining the user's task in the real world. In other words, we consider that people execute tasks in the real world according to the social role of the user.

To include social role in task categorization, we must deal with changes in the user's social role and determine the user's current social role. This determination is difficult because social role depends on mixture of various contexts such as current place, destination place, accompanies and the task the user is engaged in. In the following sections, we propose the role-task model and a service navigation system based on the model.

3.1 Past Research on Social Role

The social role has been utilized in order to construct ontologies (conceptualization of knowledge domain). Accordingly, past research on the social role concept targeted the role itself such as the definition, properties, categorization and organization of the role. That is because the level at which the role concept is separated from the objective domain impacts the accuracy of the ontology. Kosaki et al. noted that the role-concept represents a role that a thing plays in a specific context and must be defined in association with other concepts[4]. Sunagawa et al. developed Hozo, which can represent role-concepts using two kinds of relation with a context, the part-of relation and the participant-in relation[5]. For example, as for the former relation, "teacher part-of school" represents "a teacher" forms and belongs to "a school (context)" which is an educational institution. As for the latter relation, "brother participant-in brotherhood" represents "a brother participates in a brotherhood". From this point of view, it is challenging to consider the role to support user's daily activities.

3.2 Property of Role Assigned to User

In this section, we describe the property of the role assigned to the user. We consider that user selects and performs the appropriate task according to the social role assigned to the user unconsciously. For example, one user performs those tasks such as "Buy something for kids", "Go to theme park" when the user's role is "Father". The other user performs "Go to business trip", "Send document by mail" when the user's role is "Office worker". In the following, we call these tasks **role-task**.

There is another property that we should consider. In the real world, the role assigned to a user keeps changing; user change their roles dynamically and unconsciously[6]. As the role assigned to a user changes, the actions taken in the real world will change. To allow the role-task model to treat these changes, we consider that people execute tasks that can trigger role changes, which we call **transition-task**. The user's role is changed when the transition-task is completed.

3.3 Proposed Role-Task Model

In this section, we describe the role-task model based on the discussion in the previous section. The concept of the role-task model is shown in Fig.3. The role-task model incorporates role-tasks, which are the tasks associated with each role

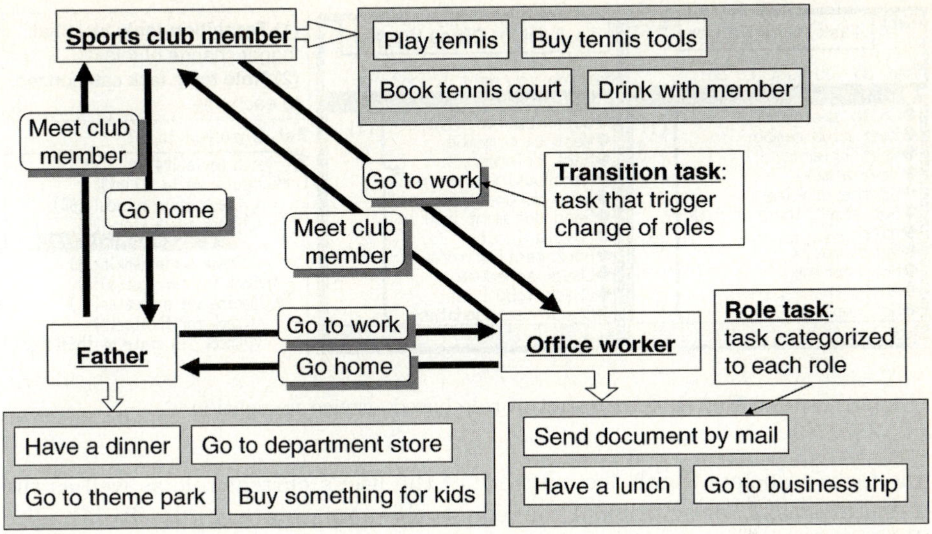

Fig. 3. Role-task model; task categorization model based on user's role

category, and transition-tasks, which can invoke role changes. People execute a role-task while their role remains fixed. On the other hand, when people execute an transition-task, which can be executed regardless of the current role, the user's role is changed.

In the following, we describe how we collected both role-tasks and transition-tasks. As for role-tasks, we extracted the tasks that the user could be expected to execute given his current role from the task-knowledge-base and categorized them using the task category of each role. For example, the role "sports club member" has a set of tasks that include "Play tennis", "Book tennis court" and so on, while the role "Office worker" has a set of tasks that include "Send document by mail", "Have a lunch" and so on.

As for the transition-tasks, we extracted the tasks that are expected to trigger a change in the user's role from the task-knowledge-base. For example, the user's role changes from "Father" to "Office worker" when he executes the transition-task "Go to work", and from "Office worker" to "Sports club member" by executing the transition-task "Meet club member". Each transition-task is associated with the original role and the new role, "Office worker" and "sports club member", respectively. One transition-task may be associated with several pairs of original and new roles. For example, the transition-task "Go to work" can be associated with two pairs of roles, one pair of which is "Father" as original role and "Office worker" as new role, and the pair of "Sports club member " as original role and " Office worker" as new role.

3.4 Flow of Service Navigation Based on Role-Task Model

Flow of the role-based service navigation system is described below. This system always stores the user's current role. When the user opens his handset, it lists the

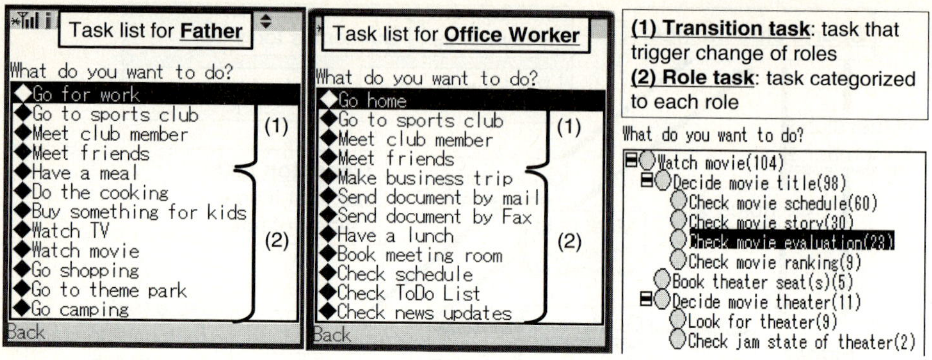

Fig. 4. Screen shot of role-based service navigation

role-tasks associated with the category of the user's current role as well as the transition-tasks whose original role equals the user's current role. To initialize the system, a role must be designated by the user or selected at random. We show an example of the task-list as it appears on a mobile phone in Fig.4(left). In this example, the task-list shows the role-tasks associated with "Father" and transition-tasks whose original role equals "Father".

The user selects the task desired from the task-list. If the user selects a role-task, the task model associated with the selected task appears. We show an example of such a task model as it would appear on a mobile phone in Fig.4(right). This example shows the task model associated with the selected task "Watch movie". By navigating through the task-model, the user finally reaches the service list. If user selects a transition-task, the system replaces the stored user roles with the new roles associated with the selected transition-task. The task-list now shows the role-tasks and transition-tasks appropriate for the user's updated role. For example, the task-list for "Office worker" is shown if the user selected "Go for work" from among the transition-tasks listed for "Office worker" as shown in Fig.4(center).

3.5 Comparison Between Query, Place and Role-Based Service Navigation

The following table provides a comparison of how query-based, place-based and role-based service navigation filters tasks and their usability. In order to compare usability, we categorize the user into two types; (A)the user who has a clear purpose and (B)the user who does not have a clear purpose.

Navigation type	Task filtering basis	User type(A)	User type(B)
Query-based[2][3]	Query input by user	appropriate	
Place-based	Current/destination place		appropriate
Role-based	Social role that depends on place, time, accompanies and task user is engaged in	appropriate	appropriate

Query-based service navigation is useful when the user has a clear purpose. However, the user cannot always express his needs in a query like when user does not have clear purpose. In the case of place-based navigation, user can find the useful task and services for current or destination place by recommending the tasks according to current or destination place user designated. However, the system fails to support the user if user cannot find appropriate place category for the task user wants to do.

Role-based navigation supports both types of user. In role-based service navigation system, user can find appropriate task including the task that is not directly related to the place by filtering the tasks according to user's current role. In addition, user does not need to designate both place and role category; because we acquire the change of user's role by user's selection of transition-task.

4 Acquisition of Change of User's Role at the Station

One weakness of role-based service navigation is that it relies on the user selecting a transition-task. If the user fails to select a transition-task, the system keeps showing the current task-list. There are some locations however, that clearly identify a change in the user's role such as the office, home, and the station used for commuting.

In this section, we focus on roles changes at stations. For example, when the user finishes work and leaves the station on his way home, the station becomes the point at which the user's role changes from "Office worker" to "Father". Moreover, stations are frequently used as a meeting point. If the user leaves home to catch a train to meet tennis club members at a different station, then the station is the point at which the user's role changes from "Father" to "Sports club member".

In order to acquire the exact moment that the user enters or leaves a station, we can use the functions of existing contactless chips mounted on mobile handsets that are now being used for ticketing and fare collection at the gates of railway stations in Japan. This system is currently deployed by Sony and NTT DoCoMo for JR East; it allows commuters to pay their train fares via their mobile phone.

For stations at which the user's role always changes in a predictable manner, the user can select and register his/her role changes. The selected role together with the location of the station used by the user is registered inside the mobile handset. The next time the user passes through a registered station, the user's role is automatically recognized, and the appropriate task-list is shown to the user.

5 Conclusion

We have developed a task-based service navigation system that navigates the user to the appropriate services, those that assist the user in solving the problems currently faced in the real world. In this paper, we categorized user tasks by the role concept of users and modeled changes in user's role. An advanced service navigation system was described that can recommend tasks from the viewpoint of the current role of the user in addition to a consideration of location. We have implemented a new service navigation prototype that can detect changes in the user's role by using contactless IC chips. In future works, we must consider details of changes in the user's role, such as acquisition of role change by recognizing the human relationship with the companion(s), other factors impacting the change and the simultaneous existence of more than one kind of user role.

References

1. A. Serenko and N. Bontis: A model of user adoption of mobile portals, Quarterly Journal of Electronic Commerce 4(1), pp.69-98, 2004.
2. T. Naganuma and S. Kurakake: Task Knowledge Based Retrieval for Service Relevant to Mobile User's Activity, 4th international semantic web conference, Y. Gil et al. (Eds.): ISWC 2005, LNCS 3729, pp.959-973, 2005.
3. Y. Fukazawa, T. Naganuma, K. Fujii and S. Kurakake: A Framework for Task Retrieval in Task-Oriented Service Navigation System, International Workshop on Web Semantics, R. Meersman et al. (Eds.): OTM Workshops 2005, LNCS 3762, pp.876-885, 2005.
4. K. Kozaki, Y. Kitamura, M. Ikeda and R. Mizoguchi: Hozo: An Environment for Building/Using Ontologies Based on a Fundamental Consideration of "Role" and "Relationship", Proc. of the 13th Int. Conf. Knowledge Engineering and Knowledge Management, pp.213-218, 2002.
5. E. Sunagawa, K. Kozaki, Y. Kitamura, and R. Mizoguchi: Organizing Role-concepts in Ontology Development Environment: Hozo, AI Technical Report (Artificial Intelligence Research Group, I. S. I. R., Osaka Univ.), AI-TR-04-1, 2004.
6. C. Masolo, L. Vieu, E. Bottazzi, C, Catenacci, R. Ferrario, A. Gangemi and N. Guarino: Social Roles and their descriptions, Proc. of the 9th Int. Conf. on the Principles of Knowledge Representation and Reasoning, AAAI Press, 2004.

A Discretization Algorithm That Keeps Positive Regions of All the Decision Classes

Feng Honghai[1,2], Liu Baoyan[3], He LiYun[3], Yang Bingru[2], and Li Yueli[1]

[1] Hebei Agricultural University, 071001 Baoding, China
[2] University of Science and Technology Beijing, 100083 Beijing, China
[3] China Academy of Traditional Chinese Medicine, 100700 Beijing, China
liuby@tcmcec.com

Abstract. Most of the existing discretization methods such as k-interval discretization, equal width and equal frequency methods do not take the dependencies of decision attributes on condition attributes into account. In this paper, we propose a discretization algorithm that can keep the dependencies of the decision attribute on condition attributes, or keep the positive regions of the partition of the decision attribute. In the course of inducing classification rules from a data set, keeping these dependencies can achieve getting the set of the least condition attributes and the highest classification precision.

1 Introduction

Discretization has received a great deal of attention in data mining literature and includes a variety of ideas ranging from fixed k-interval discretization [1], fuzzy discretization [2,3], Shannon-entropy discretization due to Fayyad and Irani presented in [4,5], equal width intervals, equal frequency intervals, Naïve Scaler etc [6]. Most of the existing discretization methods such as k-interval discretization, equal width and equal frequency methods do not take the dependencies of the decision attributes on condition attributes into account. The Naïve Scaler method can partially keep the dependencies of one decision class, but it generates too many cuts and too many small width intervals; additionally, it cannot keep the positive regions of two or more decision classes. The Semi Naïve Scaler method sometimes generates no cuts. The goal to handle data sets is to induce the dependencies for the purpose of classification or as a kind of new knowledge. When discretizing continuous values, if the dependencies cannot be kept, the rule induced for classification may influence the classification precision or need more condition attributes to fulfill the classification task.

2 Basic Concepts of Discretization and Rough Set Theory

(1) Discretization

The discretization process can be described generically as follows. Let B be a numerical attribute of a set of objects. The set of values of the components of these

B. Gabrys, R.J. Howlett, and L.C. Jain (Eds.): KES 2006, Part I, LNAI 4251, pp. 1029 – 1036, 2006.
© Springer-Verlag Berlin Heidelberg 2006

objects that correspond to the *B* attribute is the active domain of B and is denoted by adom (*B*).

To discretize *B* we select a sequence of numbers $t_1 < t_2 < \cdots < t_l$ in adom(*B*). Next, the attribute *B* is replaced by the nominal attribute *B* that has l + 1 distinct values in its active domain { $k_0, k_1, \cdots k_l$ }. Each *B* -component b of an object o is replaced by the discretized *B* -component k defined by

$$k = \begin{cases} k_0, & \text{if } b \le t_1 \\ k_1, & \text{if } t_i < b \le t_{i+1} \text{ for } 1 \le i \le l - 1 \\ k_l, & \text{if } t_l < b \end{cases}$$

The numbers $t_1 < t_2 < \cdots < t_l$ define the discretization process and they will be referred to as class separators.

(2) Concept of indiscernibility relation in rough set theory

In a decision table DT=$\langle U, A, V, f \rangle$, where $U = \{x_1, x_2, \cdots, x_n\}$, is a nonempty, finite set called the universe; *A* is a nonempty, finite set of attributes; $A = C \cup D$, in which *C* is a finite set of condition attributes and *D* is a finite set of decision attributes; $V = \bigcup_{a \in A} V_a$, where V_a is a domain (value) of the attribute a, and $f : U \times A \to V$ is called the information function such that $f(x,a) \in V_a$ for every $a \in A, x_i \in U$. To every subset of attributes $B \subseteq A$, a binary relation, denoted by *IND*(*B*), called the B -indiscernibility relation, is associated and defined as follows: two objects, x_i and x_j, are indiscernible by the set of attributes *B* in *A*, if $f(x_i, b) = f(x_j, b)$ for every b$\subset B$. For any element x_i of *U*, the equivalence class of x_i in relation $IND(B)$ is represented as $[x_i]_{IND(B)}$.

(3) Concept of lower approximations in rough set theory

Let *X* denotes the subset of elements of the universe *U* ($X \subset U$). The lower approximation of *X* to *B* ($B \subseteq A$), denoted as $B\_(X)$, is defined as the union of all these elementary sets that are contained in X. More formally:

$$POS_B(X) = B\_(X) = \{x_i \in U \,|\, [x_i]_{IND(B)} \subset X\}$$

(4) Concept of discernibility Matrix in rough set theory

Discernibility matrix of DT, denoted $M_{\mathrm{DT}}(m_{ij})$ a $n \times n$ matrix is defined as

$$m_{ij} = \begin{cases} \{a \in C : f(x_i, a) \ne f(x_j, a) \wedge (d \in D, f(x_j, d) \ne f(x_i, d))\} \\ 0, d \in D, f(x_i, d) = f(x_j, d) \\ \phi, f(x_i, a) = f(x_j, d) \wedge (d \in D, f(x_i, d) \ne f(x_j, d)) \end{cases} \quad (1)$$

Where $i, j = 1, 2\cdots, n$. Thus entry m_{ij} is the set of all attributes that classify objects x_i and x_j into different decision classes in U. From formula (1), all of the distinguishing information for attributes is contained in the above discernibility matrix.

3 Algorithm

Input: Information table.
Output: offer the cuts of every condition attribute.

(1) Sort the attribute values of every condition attribute *C* together with the decision attribute *D* in ascending order.

(2) Calculate every positive region $POS_{C_i}(Y_j)$, where the values of the condition attribute are in the ascending order, and the values of attribute *D* are the same, composing Y_j; Calculate every $POS_{C_i}(\cup Y_j)$, where $\cup Y_j \neq IND(D)$ and the values of decision attribute *D* compose $\cup Y_j$, the values of the condition attribute being in the ascending order.

(3) For every positive region $POS_{C_i}(Y_j)$ or $POS_{C_i}(\cup Y_j), \cup Y_j \neq IND(D)$, set every positive region as an interval.

4 Several Experimental Results

(1) Hand Written Chinese Character Recognition Experiments

1) Discretization results by using the algorithm that keeps all the positive regions *IND* (*Y*) ={Y_1, Y_2, Y_3, Y_4}={{1,2}, {3,4}, {5,6}, {7,8}}. For attribute *B*, we find that the set {5, 6} where the examples' values of attribute *B* belong to [13.0, 14.6] is included in Y_3= {5, 6}; set {7, 8} where the examples' values of attribute *B* is 14.8 is included in Y_4= {7, 8}; set {1, 2} where the examples' values of attribute *B* belong to [15.5, 15.9] is included in Y_1= {1, 2}; set {3, 4} where the examples' values of attribute *B* belong to [16.7, 17.1] is included in Y_3= {3, 4}. So we get the cuts: for intervals [13.0, 14.6] and 14.8, the cut is (14.6 +14.8)/2=14.7, for intervals 14.8 and [15.5, 15.9], the cut is (14.8+15.5)/2=15.15, for intervals [15.5, 15.9] and [16.7, 17.1], the cut is (15.9+16.7)/2=16.3. For attribute *F*, we select [2.58, 2.88] as an interval, whereas [1.78, 3.24] - [2.58, 2.88] = [1.78, 2.58] + [2.88, 3.24] as the other interval. For attribute *G*, we select [1.23, 1.24] as an interval, and select [1.45, 1.65] as another interval, [0.98, 1.68] - [1.23, 1.24] - [1.45, 1.65]=[0.98, 1.23]+ [1.24,1.45]+[1.65,1.68] as the last interval. For attributes *D*, *H*, there is not any positive region of the partition of the *Y*; we can only select all the attribute values into one interval, that is, we cannot get any cut. So we can obtain the discretization results as Table 2.

From Table 2, we find that attribute *B* is enough for classification, since the discretization algorithm keeps all the positive regions of the decision classes. So according to the discretization results we can directly get the classification rules.

Table 1. 4-Hand Written Chinese Character features

U	*A*	*B*	*C*	*D*	*E*	*F*	*G*	*H*	*Y*
1	33.6	15.5	9.3	7.26	2.60	1.83	1.24	0.81	1
2	34.5	15.9	8.9	4.57	2.54	1.78	1.23	0.99	1
3	33.3	17.1	9.9	8.41	3.89	3.24	1.68	1.33	2
4	33.9	16.7	9.8	5.36	3.95	1.85	1.02	0.82	2
5	35.1	13.0	10.2	6.68	3.77	1.82	0.98	0.75	3
6	35.8	14.6	11.7	7.85	4.17	2.17	1.44	1.05	3
7	29.3	14.8	8.2	6.53	5.59	2.58	1.65	0.93	4
8	30.3	14.8	10.0	8.07	5.95	2.88	1.45	1.10	4

Table 2. Discretization results by using the algorithm that keeps all the positive regions

U	*A*	*B*	*C*	*D*	*E*	*F*	*G*	*H*	*Y*
1	1	2	1	0	0	0	0	0	1
2	1	2	1	0	0	0	0	0	1
3	1	3	2	0	1	0	1	0	2
4	1	3	2	0	1	0	1	0	2
5	2	0	4	0	1	0	1	0	3
6	2	0	4	0	1	0	1	0	3
7	0	1	0	0	2	1	2	0	4
8	0	1	3	0	2	1	2	0	4

2) Discretization results by using equal width intervals discretization method

We can obtain the discretization results in Table 3.

From Table 3, some positive regions are mixed up and some of the dependencies of the decision attribute on condition attributes are broken. So in order to discern these examples, we need more condition attributes. In fact, after attribute reduction three condition attributes X_1, X_2, and X_3 are needed for classification.

3) Discretization results by using Naïve Scaler algorithm

The discretization results are shown in Table 4.

From Table 4 we can find that using the Naïve Scaler algorithm may generate many cuts than using our algorithm, especially, with attributes *D* and *H*, the amount of the intervals are equal to the amount of the examples, that is, every example is an interval respectively. Additionally, some of the positive regions of the decision attribute partition have been mixed up. And some of the positive regions of the decision attribute partition have been mixed up.

However, for attribute *B* all the positive regions have been kept and it is enough to classify all the examples, that is, although the Naïve Scaler method may generate more cuts it has the property of keeping most of the positive regions.

(2) Micronutrients in SARS Patients' Body Data Set Experiments

The whole SARS data set includes 60 cases among which examples 31~60 are SARS patients and 61~90 are healthy human beings. Attribute "1", "2", "3", "4", "5", "6", "7" denote micronutrient Zn, Cu, Fe, Ca, Mg, K, Na respectively, and the

Table 3. Discretization results by using equal width intervals

U	X_1	X_2	X_3	X_4	X_5	X_6	Y
1	17	8	5	4	2	1	1
2	18	8	5	3	2	1	1
3	17	9	5	5	2	2	2
4	17	9	5	3	2	1	2
5	18	7	6	4	2	1	3
6	18	8	6	4	3	2	3
7	15	8	5	4	3	2	4
8	16	8	5	5	3	2	4

Table 4. Discretization results by using the Naïve Scaler algorithm

U	A	B	C	D	E	F	G	H	Y
1	2	2	1	4	0	2	2	1	1
2	4	2	1	0	0	0	2	4	1
3	1	3	2	7	2	6	6	7	2
4	3	3	2	1	2	3	1	2	2
5	5	0	4	3	1	1	0	0	3
6	5	0	4	5	3	4	3	5	3
7	0	1	0	2	4	5	5	3	4
8	0	1	3	6	4	5	4	6	4

decision attribute "C" denotes the class "SARS" and "healthy". $V_C = \{0,1\}$, where "0" denotes "SARS", "1" denotes "healthy".For attribute "1" (Zn), [13.5, 132] = $POS_{Zn}(C=0)$, [133, 226] = $BN_{Zn}(C=0) = BN_{Zn}(C=1)$ and [235, 247] = $POS_{Zn}(C=1)$. So the cuts are (132+133)/2=132.5, and 230.5.

Table 5. Comparison results of four different discretization algorithms

Attributes	New algorithm: amount of intervals	New algorithm: amount of intervals that keep dependencies	New algorithm: condition attributes needed for classification	Equal width intervals: amount of intervals	Equal width intervals: amount of intervals that keep dependencies	Equal width intervals: condition attributes needed for classification
1	3	2		3	0	generate 36 inconsistent examples, so all the seven condition attributes is not enough for inducing classification rule
2	3	2		3	0	
3	3	2	y	3	0	
4	3	2	y	3	0	
5	3	2		3	0	
6	3	2	y	3	0	
7	3	2	y	3	0	

Table 5. (*continued*)

Attributes	Naïve Scaler			Semi Naïve Scaler		
	amount of intervals	amount of intervals that keep dependencies	condition attributes needed for classification	amount of intervals	amount of intervals that keep dependencies	condition attributes needed for classification
1	16	12		16	12	
2	21	20		21	20	
3	24	23	y	24	23	y
4	5	5	y	5	5	y
5	11	9		9	9	
6	25	23		23	23	
7	26	25		25	25	

(3) Clinical SARS Data Set Experiment

We obtained the clinical SARS data of 524 SARS patients from Beijing in 2003. And when handling the SARS data, we find the following rules:(1) T<=37.6˚C →C=1 (1 means that the state of illness is slight). (2) 37.7˚C <=T<=38.6˚C →C=1 or 2(2 means that the state of illness is serious). (3) T>=38.7˚C →C=1, 2 or 3 (3 means that the state of illness is critical). (4) C=3 →T>=38.7˚C. Where T denotes the highest body temperature, C denotes the class type. The above rules mean that: The examples (T<=37.6˚C) belong to the positive region of the set in which the examples' states are 1. So we can use this rule to classify the examples directly. The examples (37.7˚C <=T<=38.6˚C) belong to the positive region of the sets in which the examples' states are 1 or 2. So we can use the rule to classify the examples into class 1 and 2. The examples (T>=38.7˚C) belong to the boundary region of the sets in which the example's state are 1, 2 or 3. We cannot classify the examples directly. The four rules have not been found by medical experts, so when discretizing the attribute values this knowledge should be kept well. Using our new algorithm can do that, whereas the other discretization methods cannot.

Table 6. Comparison results of three different discretization algorithms

Attributes	New algorithm		Equal width	
	amount of intervals	amount of intervals that keeps dependencies	amount of intervals	amount of intervals that keeps dependencies
T	3	2	5	0

Table 6. (*continued*)

Attributes	Naïve Scaler		Semi Naïve Scaler	
	amount of intervals	amount of intervals that keep dependencies	amount of intervals	amount of intervals that keep dependencies
T	23	1	0	0

Table 7. Information table of the highest body temperature and class

T	37 … 37.5	37.7	37.7	37.7	37.8	37.8	37.8	37.8 …	38.7 …	38.7…38.7…
C	1 … 1	1	1	2	1	1	2	2 …	1	… 2 … 3 …

In Table 7, the values of decision attribute C start to take 2 when the values of condition attribute T take 37.7°C, and the values of C start to take 3 when the values of T take 38.7°C. Since for every value of T its corresponding values of C mostly take 1, if we adopt the Semi Naïve Scaler discretization method, we cannot get any cut, so the Semi Naïve Scaler method is invalidated in this case.

If we adopt the Naïve Scaler method, we cannot get the cut of (37.5+37.7)/2=37.6, that is, we cannot get the positive region of "C=1" and the positive region of "C=1" as well as "C=2", whereas we get the cut of (37. 7+37. 8)/2=37.75, which breaks the positive region of "C=1" and the positive region of "C=1" as well as " C=2".

5 Discussions

(1) The above three experimental results show that applying our algorithm can generate the least cuts, and keep the dependencies of the decision attribute on condition attributes best, but for some attributes that have not any positive regions of the decision classes, it will generate no cut; The equal width intervals methods create more cuts and few of the intervals can embody the dependencies between two attributes; the Naïve Scaler generates many redundant cuts, and sometimes generates very small width of intervals, in particular, the width may be 0. But it keeps the dependencies better than the equal width method; the Semi Naïve Scaler can get less redundant cuts than the Naïve Scaler does, but in some particular cases it cannot generate any cuts.

(2) Keeping the dependency between condition attributes and the decision attribute can offer fewer attributes for inducing classification rules, and improve the classification precision. If the dependencies cannot be held, more attributes will be added to induce the classification rules, that is, after attribute reduction more condition attributes are needed, as a result, maybe some errors or mistakes are brought in.

(3) Although the Naïve Scaler method keeps most of the positive regions of the partition of the decision attribute it cannot generate the intervals that are the positive regions of two or more equivalence classes. For example in example 3, 37.7°C <=T<=38.6°C →1 or 2, with the Naïve Scaler method the cut of 38.65°C cannot be created where the examples are the positive regions of the class 1 and 2. Additionally, when using the Naïve Scaler method the order of the examples can influence the results of the cuts, so the dependencies may not be held, and the optimum cuts may not be obtained either.

(4) Clustering based discretization algorithms only take the distribution of one attribute values into account, namely, they cannot embody the dependencies of the decision attribute on a condition attribute. So the discretization results of the clustering based discretization algorithms are not changed, though the decision attribute will change. However with our algorithm the discretization results can be different when

the decision attributes change, since the algorithm embodies the dependencies and hold the classification capability of the condition attribute.

(5) Entropy based discretization methods cannot generate the intervals that are the positive regions of two or more equivalence classes either. In particular, it cannot keep the positive regions exactly.

(6) In conclusion, the algorithm proposed in this paper and the Naïve Scaler method can be integrated into one algorithm, that is, when the condition attributes have positive regions of the decision classes, our algorithm should be applied, and when the condition attributes have not any positive regions of the decision classes the Naïve Scaler method should be applied.

Acknowledgements

The paper is supported by the Special Foundation of SARS of the National High-Tech Research and Development Major Program of China (863)- “Clinical Research on the Treatment of Severe Acute Respiratory Syndrome with Integrated Traditional and Western (No.2003AA208101)”.

References

1. Dougherty J, Kohavi R, Sahami M.: Supervised and unsupervised discretization of continuous features. Proc. of the 12th International Conference on Machine Learning. (1995) 194 - 202.
2. Kononenko I.: Naive Bayes classifier and continuous attributes Informatica Vol 16 (1992) 1-8.
3. Kononenko I.: Inductive and Bayesian learning in medical diagnosis. Appl Artif Intell Vol 7 (1993) 317-37
4. Fayyad UM. On the Induction of Decision Trees for Multiple Concept Learning. Ph.D. thesis, University of Michigan (1991).
5. Fayyad UM, Irani K.: Multiinterval discretization of continuous-valued attributes for classification learning. Proc. of the 12th International Joint Conference on Artificial Intelligence (1993) 1022 -27.
6. KerberR.ChiMerge:discretization of numeric attribute [C].In:Proceedings of the 10th National Conference on Artificial Intelligence(AAAI-92). SanJose, CA (1992).123-128

Forecasting Region Search of Moving Objects

Jun Feng[1], Linyan Wu[1], Yuelong Zhu[1], and Toyohide Watanabe[2]

[1] Hohai University, Nanjing, Jiangsu 210098 China
fengjun-cn@vip.sina.com
[2] Nagoya University, Nagoya, Aichi 464-8603 Japan

Abstract. The issue of how to provide location-based service (LBS) is attracted many researchers. In this paper, we focus on an useful searching type in LBS, forecasting search, which provides search result for a future time based on the current information of moving objects. To deal with such kind of searches, a PV-graph is proposed for analyzing the possible situations of the moving objects. By using PV-graph, forecasting queries in LBS applications can be responded efficiently.

1 Introduction

Forecasting search is a kind of spatiotemporal search, which provide region search results or nearest neighbor search results for a future time based on the analysis of the current information. For example, "please list all the possible cars around cross node A in five minutes", or "I am driving on highway h with speed 110 miles per hour, please tell me all the cars within 50 meters from me in 3 minutes", and so on. This kind of queries can be defined as the forecasting region search and forecasting moving region search.

Consider a set of multidimensional moving objects $S = (o_1, o_2, ...o_n)$ and a function D that specifies the distance between the objects. o_i is a moving object on location l_i with speed $\overrightarrow{v_i}$ at current time: $(l_i, \overrightarrow{v_i})$. The forecasting region search is given in a triple (q, r, t), where (q, r) is a region: q is a multidimensional point and is the center of the query region; r is the radius of the query region; t is the future time from now. This query is to find a set $S' \subseteq S$, any object $o' \in S'$ is inside the search region (q, r) at time t. In other words, the distance from q to the possible location of o' at time t is shorter than or equal to r: $D(o', q) \leq r$ on time t. While the forecasting moving region search is (q, r, v, t). The search region (q, r) is moving with speed $\overrightarrow{v_q}$. To process the forecasting moving region search, it is not only the moving situations of all the moving objects, but the moving situations of the search region should be considered.

There are many researches centering on the nearest neighbor search for static or moving query object up to now [1, 2, 3, 4, 5]. However, there is little work for the region queries mentioned here.

In this paper, we propose a PV-graph for analyzing forecasting search and a search method for forecasting search in 1D and 2D space.

This paper is organized as follows. The PV-graph for analyzing forecasting search in 1D and 2D space are depicted in Section 2. Section 3 analyzes our method and makes a conclusion on our work.

B. Gabrys, R.J. Howlett, and L.C. Jain (Eds.): KES 2006, Part I, LNAI 4251, pp. 1037–1042, 2006.
© Springer-Verlag Berlin Heidelberg 2006

2 Forecasting Region Search and PV-Graph

In this section, we propose PV-graph for 1D, 2D forecasting region search and moving region search.

2.1 PV-Graph for 1D Forecasting Search

To deal with the forecasting region search, we first observe an example in 1D space (linear space) given in Fig. 1(a). There are objects moving in two directions with different speed in 1D space. Query region is $[x_1, x_2]$, and the length of x_1x_2 is s. To find all moving objects which are possible inside the query region after time t, it needs to compute all the forecasting locations of moving objects in 1D space, and compare all the computed locations with the query region. This process will be computational extensive when there are a great number of moving objects. In a real system, all the positions and moving speed (direction and moving ratio) of moving objects are managed by a specific index (e.g., spatial index r-tree [6], spatiotemporal index TPR-tree [7] or their extensions and so on [8]). Therefore, if a forecasting query can be changed into a search based on the recorded information directly, by taking the advantages of index structure, the search process can be efficient.

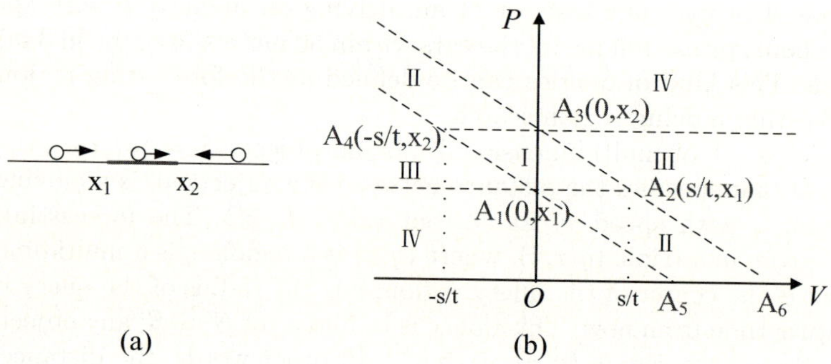

Fig. 1. 1D forecasting search region and PV-graph

PV-graph is a trial of bridging forecasting query and the index of spatiotemporal objects. The PV-graph for the query example of Fig. 1(a) is given in Fig. 1(b). In Fig. 1(b), the horizonal axis is V, which represents the speed of moving objects, and one direction is defined as positive another is negative; the ordinate is P, which represents the location of moving objects. To process the query of $([x_1, x_2], t)$, the PV space is divided into four kinds of regions:

- Region I: Region I is a diamond-shaped figure $A_1A_2A_3A_4$. A_1 and A_3 are decided by the query region limits of x_1, x_2, respectively. A_2 and A_4 are related to the region size and the time t. The coordinate of A_2 is $(s/t, x_1)$, means that a moving point on location x_1 at current time will just reach location x_2 at time t with speed s/t in the direction from x_1 to x_2. While a moving object at location x_2 with speed s/t in the direction from x_2 to x_1 is just depicted by A_4 $(-s/t, x_1)$, which just reaches x_1 at time t. Therefore, all the moving objects inside Region I at current time will still be inside query region $[x_1,\ x_2]$ at time t, because they cannot go out of this region with the current speed until time t.

- Region II: there are two parts of Region II, one is on the upper left of Region I and another is on the lower right of Region I. The objects inside Region II at current time can come into the query region at time t from two directions. For example, the lower part of Region II is a parallelogram $A_1A_2A_6A_5$, $A_5(\frac{sx_1}{t(x_2-x_1)}, 0)$ is the point of intersection between line A_4A_1 and V axis. Moving object on location 0 with speed $\frac{sx_1}{t(x_2-x_1)}$ is just reach location x_1 at time t, while the moving object depicted by A_6 is just reach x_2 at time t, as it is at location 0 with speed $\frac{sx_2}{t(x_2-x_1)}$ at current time. The situation of the upper left part is similar to that of the lower part except that the speed is negative.
- Region III: The moving objects inside Region III will go out of the query region at time t, though they are inside the query region at the current time or go through the query region at a specific time before time t;
- Region IV: the moving objects inside Region IV can never enter the query region from now to time t.

By observing the different situations of moving objects, only those moving objects inside Region I and Region II meet the requirements.

2.2 PV-Graph for 2D Forecasting Search

In 2D space, the forecasting region query is to find all the moving objects inside a circle region at time t, e.g., the example given in Fig. 2(a). In Fig. 2(a), the query region is decided by the center point q and the radius of query region r. To search the moving objects which will be inside this search region at time t is more complex than that in the previous 1D example, because the objects can move on 2D space in all directions. To simplify the analysis, we decompose the speed into two parts, one is with the direction pointing to q and another is in the tangent direction, such kind of a PV-graph is given in Fig. 2(b). In Fig. 2(b), V axis represents the moving ratio decomposed in the direction of pointing to q, X and Y axes represent the 2D space. PV space is also divided into four regions:

- Region I: region I is a cone plus a cylinder (except the cone of Region III) just like the shadow part in Fig. 2(b). The coordinate (x, y, v) of vertex A_1 is $(0, 0, -r/t)$. A moving object related to A_1 is on location q with the moving ratio r/t in contrary direction of pointing to q at current time, which will just reach the edge of the query region (q, r) at time t. The bottom of the shadow region is the edge of circle $x^2 + y^2 = r^2$ and $v = 2r/t$, which means that the object on the edge of query region with speed $2r/t$ in the direction of

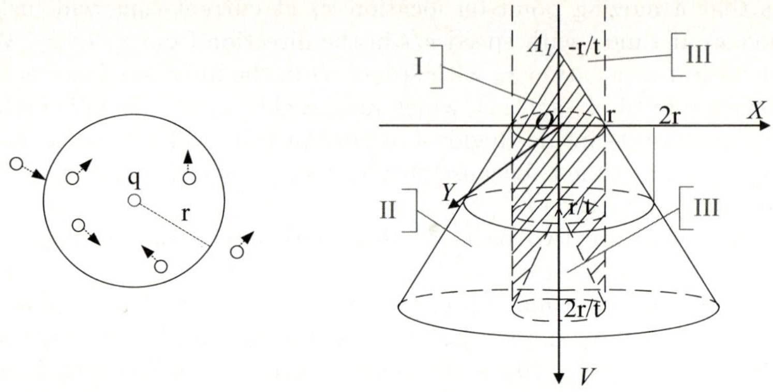

(a) (b)

Fig. 2. 2D forecasting search region and PV-graph

pointing to q will just reach the opposite edge of the query region. Therefore, all the moving object located inside Region I will still be inside the query region at time t;

- Region II: The moving objects inside Region II will go into the query region at time t. Region II is a cone decided by the vertex A_1 and its radial camber through circle ($x^2 + y^2 = r^2$ and $v = 0$) , except the parts of Region I and Region III. Though the object inside Region II is not inside the query region at current time, it is moving with a proper speed pointing to q, at time t it will be inside the query region.
- Region III: There are two parts of Region III with the same meanings: the moving objects inside Region III will go out of the query region at time t, though they are inside the query region at the current time. Simply speaking, the moving object is too speedy, it will go out of the query region before time t.
- Region IV: The moving objects inside Region IV will never go into the query region, because either they are too far from the region or too slow to go toward the region. Region IV is outside the big cone drawn in Fig. 2(b).

Just similar to that in 1D space, only the moving objects inside Region I and Region II need to be considered in the forecasting query.

2.3 PV-Graph for Moving Region Search

When the forecasting query is for a moving object, the current query region will be changed to a new region at time t, the situations are depicted in Fig. 3(a) and (b).

For such kind of query region, the corresponding PV-graph will be changed, too. An example is given in Fig. 4.

The regions in Fig. 4(b) are similar to those in 2D forecasting region search, except that the bottom of cylinder is changed into an ellipse. This is because

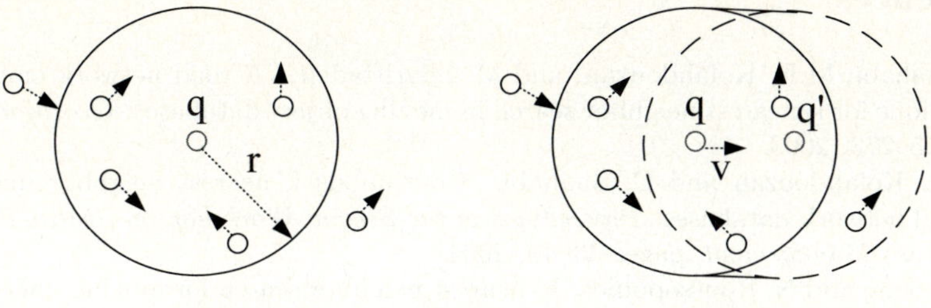

(a) (b)

Fig. 3. Query region and moving query region

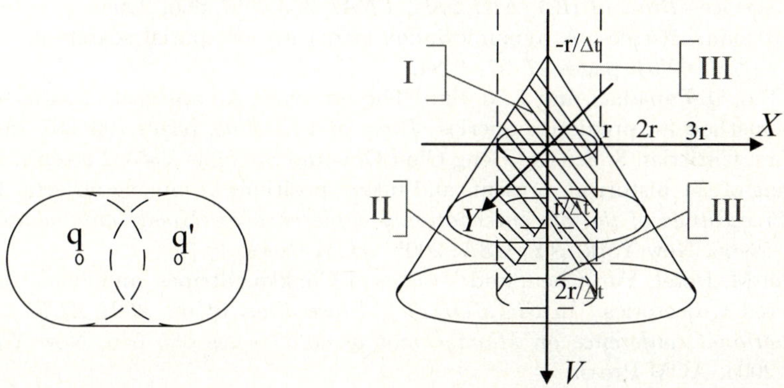

(a) (b)

Fig. 4. 2D moving search region and PV-graph

that the search region is changing all the time, the conservative estimate of the search region is the moving region contrail.

3 Analysis and Conclusion

When the spatiotemporal information of the moving objects are managed by indexing their locations and moving speed (e.g., information indexed by dual transformed space [9]), the forecasting region search can be carried off by using PV-graph efficiently. This is because that there is no need of computing the forecasting location of all the moving objects, only the objects inside Region I and Region II should be tested. Though the size or volume of the regions maybe varied with different time t and the size of query region. Comparing to the total space, Region I and Region II are quite small. Therefore, the testing process or the query process can be more efficient than that without using PV-graph.

In our future work, the performance of our method will be evaluated, and the use of PV-graph in nearest neighbor search and other forecasting search will be considered.

References

1. C. Shahabi, M.R. Kolahdouzan, and M. Sharifzadeh:. A road network embedding technique for k-nearest neighbor search in moving object databases. *GeoInformatica*, (3):255–273, 2003.
2. M. R. Kolahdouzan and C. Shahabi. Continuous k nearest neighbor queries in spatial network databases. *Proceedings of the Second Workshop on Spatio-Temporal Database Management*, pages 33–40, 2004.
3. Z. X. Song and N. Roussopoulos. K-nearest neighbor search for moving query point. *Proc. of SSTD'01*, pages 79–96, 2001.
4. J. Feng, N. Mukai, and T. Watanabe. Stepwise optimization method for k-cnn search for location-based service. *Proc. of SOFSEM 2005, LNCS*, 3381:363–366, 2005.
5. J. Feng, N. Mukai, and T. Watanabe. Search on transportation network for location-based service. *Proc. of IEA/AIE 2005, LNAI*, 3533:657–666, 2005.
6. A. Guttman. R-trees: A dynamic index structure for spatial searching. *Proc. of ACM SIGMOD'84*, pages 47–57, 1984.
7. Y. F. Tao, D. Papadias, and J. M. Sun. The tpr*-tree: An optimized spatio-temporal access method for predictive queries. *Proc. of VLDB'03*, pages 790–801, 2003.
8. Dan Lin, Christian S. Jensen, Beng Chin Ooi, and Simonas Šaltenis. Efficient indexing of the historical, present, and future positions of moving objects. In *MEM '05: Proceedings of the 6th international conference on Mobile data management*, pages 59–66, New York, NY, USA, 2005. ACM Press.
9. Jignesh M. Patel, Yun Chen, and V. Prasad Chakka. Stripes: an efficient index for predicted trajectories. In *SIGMOD '04: Proceedings of the 2004 ACM SIGMOD international conference on Management of data*, pages 635–646, New York, NY, USA, 2004. ACM Press.

An Index-Based Time-Series Subsequence Matching Under Time Warping

Byoungchol Chang[1], Jaehyuk Cha[1], Sang-Wook Kim[1], and Miyoung Shin[2]

[1] Graduate School of Information and Communications, Hanyang University, Korea
bcchang@ihanyang.ac.kr, chajh@hanyang.ac.kr, wook@hanyang.ac.kr
[2] School of Electrical Engineering and Computer Science,
Kyungpook National University, Korea
shinmy@mail.knu.ac.kr

Abstract. This paper addresses efficient processing of time-series subsequence matching under time warping. Time warping enables finding sequences with similar patterns even when they are of different lengths. The *prefix-querying method* is the first index-based approach that performs time-series subsequence matching under time warping without false dismissals. This method originally employs the L_∞ distance metric as a base distance function. This paper extends the prefix-querying method for absorbing L_1 instead of L_∞. We formally prove that the extended prefix-querying method does not incur any false dismissals. The performance results reveal that our method achieves significant performance improvement over the previous methods up to 10.7 times.

1 Introduction

A *time-series database* is a set of data sequences, each of which is an ordered list of elements[1]. Sequences of stock prices, temperature changes, and company growth rates are the typical examples[3]. *Sequence matching* is an operation that finds sequences or subsequences whose changing patterns are similar to that of a given query sequence[4].

In order to measure the similarity of any two sequences of length n, say $X(=[x_1, x_2, \ldots, x_n])$ and $Y(=[y_1, y_2, \ldots, y_n])$, most prior approaches map the sequences into points in n-dimensional space and compute the distance function L_p(X, Y) defined as follows.

$$L_p(X,Y) = \sqrt[p]{\sum_{i=0}^{n} (|x_i - y_i|)^p} \tag{1}$$

The L_p distance function has been widely used to measure the similarity of two sequences X and Y of the same length. L_1 is the *Manhattan distance*, L_2 is the *Euclidean distance*, and L_∞ is the *maximum distance* in any pair of corresponding elements. In real applications, two sequences X and Y are considered similar to each other when their L_p(X, Y) is smaller than a given tolerance ε.

B. Gabrys, R.J. Howlett, and L.C. Jain (Eds.): KES 2006, Part I, LNAI 4251, pp. 1043–1050, 2006.
© Springer-Verlag Berlin Heidelberg 2006

Recent work tends to support transformations such as scaling, shifting, normalization, moving average, and time warping[1, 7, 4, 6, 3] for providing a rich similarity model. Among these, time warping is a transformation that allows any sequence element to replicate itself as many times as needed without extra costs[7]. Given two sequences S and Q, the time warping distance D_{tw}(S, Q) is defined recursively as follows[7]:

Definition 1
(1) $D_{tw}(\langle\rangle, \langle\rangle) = 0$,
(2) $D_{tw}(S, \langle\rangle) = D_{tw}(\langle\rangle, Q) = \infty$,
(3) $D_{tw}(S, Q) = (|L_p(First(S), First(Q))|^p +$
$|min(D_{tw}(S, Rest(Q)), D_{tw}(Rest(S), Q), D_{tw}(Rest(S), Rest(Q)))|^p)^{1/p}$

Here, First(S) is the first element of S, and Rest(S) is a subsequence of S that includes the elements from position 2 to the end. $\langle\rangle$ denotes a null sequence. The min() is a function to take the minimum value amongst the three arguments. L_p can be properly chosen according to applications.

References [4] proposed an index-based sequence matching method under time warping, and reference [6] proposed the *prefix-querying method* for subsequence matching by extending the index-based method. These two methods employ L_∞ as a base distance function for users' convenience in querying. However, through discussions with other researchers working in the same area, we had been recommended to extend those index-based methods to employ L_1, which is most popular for time warping[2]. This paper is a kind of correspondence for this recommendation. This paper addresses an extension of the prefix-querying method to take L_1 as a base distance function. We formally prove that the proposed method does not incur any false dismissals in subsequence matching. To show the superiority of our method, we conduct performance evaluation via extensive experiments.

2 Related Work

2.1 Naive-Scan

It reads all the data sequences from disk and computes the time warping distance of *every subsequence* S[i:j], contained in each data sequence S, to a query sequence Q using the dynamic programming technique[1]. The processing time is too excessive since Naive-Scan does not go through the filtering step[4]. Also, the CPU processing time for computing D_{tw} between the (sub)sequences X and Q is O($|X| * |Q|$), which is quite large.

2.2 LB-Scan

The filtering step is performed using a *lower-bound function* D_{lb} which returns the value always smaller than D_{tw}. After a data sequence S is accessed from disk, D_{lb}(S[i:j], Q) is applied for *every subsequence* S[i:j] contained in S and a query sequence Q. Here, if D_{lb}(S[i:j], Q) returns the value smaller than a tolerance ε, the post-processing step is performed for computing its actual D_{tw}(S, Q) by dynamic programming. Since all the data sequences are accessed from disk in the filtering step, its disk access time is equal to that of Naive-Scan. In the filtering step, D_{lb} is computed between all the (sub)sequences X and a query sequence Q. The CPU processing time for computing D_{lb}(X, Q) is O($|X| + |Q|$), which is much smaller than O($|X| * |Q|$), that for D_{tw}(X, Q).

2.3 ST-Filter

For the filtering step, the elements of data sequences in a database are transformed into symbols, and the symbol sequences are stored into a *suffix tree*. The elements of all the suffixes within every data sequence are converted into symbols, and these symbolized suffixes are stored into the suffix tree. Via the suffix tree traversal, the filtering step finds candidate subsequences S[i:j] whose D_{tw} to a query sequence Q is likely to be smaller than a tolerance ε. For every S[i:j], the post-processing step computes D_{tw}(S[i:j], Q) using dynamic programming. Owing to employing of the suffix tree traversal ST-Filter needs to access only a part of the suffix tree.

2.4 LB-Filter

The subsequence matching with LB-Filter is called *prefix-querying*[6]. It extracts a number of windows of a fixed size from every data sequence, obtains a feature vector consisting of the first, last, greatest, and smallest elements from each window, and then constructs a four-dimensional tree on all the feature vectors. For query processing, *prefixes* are extracted from a query sequence Q, and then a feature vector is obtained from each prefix. The candidate sequences S[i,j] are filtered by the tree traversal with every feature vector. In the post-processing step, for every S[i,j], D_{tw}(S[i,j], Q) is computed by dynamic programming. By using the multi-dimensional index, LB-Filter performs the filtering step much faster than LB-Scan.

3 Extension of the Prefix-Querying Method

This section extends the prefix-querying method to employ L_1 instead of L_∞.

Definition 2. *$D_{tw\_lb}(S, Q) = L_1(Feature(S), Feature(Q))$, where $Feature(S) = \langle First(S), Last(S), Greatest(S), Smallest(S)\rangle$, $Feature(Q) = \langle First(Q), Last(Q), Greatest(Q), Smallest(Q)\rangle$.*

Next, based on Theorems 1 and 2, we prove that the function D_{tw_lb} is not only the lower-bound function of D_{tw} with L_1 but also satisfies the triangle inequality. We employ Assumption 1 for proving Theorems. For cases where the assumption does not hold, Theorem 5 suggests solutions.

Assumption 1. *If the time warped sequences of S and Q are denoted as S' and Q', respectively, a value of s'_1, s'_k, q'_1, q'_k in $S' = \langle s'_1, s'_2, \ldots, s'_k \rangle$ and $Q' = \langle q'_1, q'_2, \ldots, q'_k \rangle$ is neither the maximum nor the minimum within the corresponding sequence.*

Lemma 1. *For two arbitrary sequences $S = \langle s_1, s_2, \ldots, s_n \rangle$ and $Q = \langle q_1, q_2, \ldots, q_m \rangle$, the following is always true:*

$$D_{tw}(S, Q) \geq L_1(\langle First(S), Last(S) \rangle, \langle First(Q), Last(Q) \rangle)$$

Proof: Refers to [8].

Lemma 2. *For two arbitrary sequences $S = \langle s_1, s_2, \ldots, s_n \rangle$ and $Q = \langle q_1, q_2, \ldots, q_m \rangle$, the following is always true:*

$$D_{tw}(S, Q) \geq L_1(\langle Greatest(S), Smallest(S) \rangle, \langle Greatest(Q), Smallest(Q) \rangle)$$

Proof: Refers to [8].

Theorem 1. *For two sequences $S = \langle s_1, s_2, \ldots, s_n \rangle$ and $Q = \langle q_1, q_2, \ldots, q_m \rangle$, the following holds.*

$$D_{tw}(S, Q) \geq D_{tw\_lb}(S, Q)$$

Proof: Refers to [8].

Corollary 1. *For two arbitrary sequences $S = \langle s_1, s_2, \ldots, s_n \rangle$ and $Q = \langle q_1, q_2, \ldots, q_m \rangle$, and for an arbitrary value ε, the following is always true.*

$$D_{tw}(S, Q) \leq \varepsilon \Longrightarrow D_{tw\_lb}(S, Q) \leq \varepsilon$$

Theorem 2. *For three arbitrary sequences X, Y, and Z, the following always holds.*

$$D_{tw\_lb}(X, Z) \leq D_{tw\_lb}(X, Y) + D_{tw\_lb}(Y, Z)$$

Proof: Refers to [8].

Next, it is shown that the prefix-querying method does not incur false dismissals in the case of applying L_1 to D_{tw_lb}.

Theorem 3. *For two arbitrary sequences s and q, and for an arbitrary positive value $w (1 \leq w \leq |s|)$, the following holds in all cases.*

$$D_{tw}(s, q) \leq \varepsilon \Longrightarrow (\exists x)(D_{tw}(s[1:w], q[1:x]) \leq \varepsilon)$$

Proof: Refers to [8].

Since D_{tw_lb} used by LB-Filter is the lower-bound function of the time warping distance D_{tw}, Corollary 2 can be easily induced from Theorem 3.

Corollary 2. *For two arbitrary sequences s and q, and for an arbitrary positive value* $w (1 \leq w \leq |s|)$, *the following holds in all cases.*

$$D_{tw}(s,q) \leq \varepsilon \Longrightarrow (\exists x)(D_{tw\_lb}(s[1:w], q[1:x]) \leq \varepsilon), \text{ where } 1 \leq x \leq |q|$$

Multi-role elements in feature vectors

Until now, we assumed that the values of s'_1, s'_k, q'_1, q'_k are neither the maximum nor the minimum in the time warped sequences $S' = \langle s'_1, s'_2, \ldots, s'_k \rangle$ and $Q' = \langle q'_1, q'_2, \ldots, q'_k \rangle$. In practice, however, they could be the maximum or the minimum, although it is not that frequent. Here, we investigate this issue and propose the solution to it.

If the values of s'_1, s'_k are the maximum or the minimum in a time warped sequence $S' = \langle s'_1, s'_2, \ldots, s'_k \rangle$, it implies that, in Feature(S) = ⟨First(S), Last(S), Greatest(S), Smallest(S)⟩, First(S) = Greatest(S) or Smallest(S), or Last(S) = Greatest(S) or Smallest(S). It is analogous in a query sequence. In this situation, we take two different cases into consideration as follows.

Case 1: One of $s'_2, s'_3, \ldots, s'_{k-1}$ is equal to s'_1 (or s'_k)

This is the case that the first value s'_1 (or the last value s'_k) of a sequence is the maximum or the minimum but one of $s'_2, s'_3, \ldots, s'_{k-1}$ is equal to s'_1(or s'_k). In such cases, although s'_1(or s'_k) is the maximum or the minimum, since there exist other elements which have the same value, the feature vector is to contain four distinct elements in it. Thus, the problem situation does not occur since the following formula used in the proof of Theorem 1 is satisfied.

$$L_1(\langle s'_2, s'_3, \ldots, s'_{k-1} \rangle, \langle q'_2, q'_3, \ldots, q'_{k-1} \rangle) \\ \geq L_1(\langle Greatest(S), Smallest(S) \rangle, \langle Greatest(Q), Smallest(Q) \rangle)$$

Case 2: None of $s'_2, s'_3, \ldots, s'_{k-1}$ is equal to s'_1 (or s'_k)

This is the case that the first value s'_1 (or the last value s'_k) of a sequence is the maximum or the minimum, and does not appear in the remaining part of the time warped sequence. In such cases, if a feature vector is constructed from a sequence, an identical element is represented as two features. We call it *multi-role element.* If there exist multi-role elements in a feature vector, two features in a vector are extracted from the same element in the time warped sequence. In this case, Greatest(S) or Smallest(S) does not appear in $\langle s'_2, s'_3, \ldots, s'_{k-1} \rangle$. Thus, Theorem 1 does not hold in this case.

Example 1. It happens that $D_{tw} < D_{tw\_lb}$ due to a multi-role element in a feature vector.
S = ⟨100, 20, 15, 5, 30⟩, Q = ⟨15, 20, 15, 5, 30⟩
Feature(S) = ⟨100, 30, 100, 5⟩, Feature(Q) = ⟨15, 30, 30, 5⟩
S′ = ⟨100, 20, 15, 5, 30⟩, Q′ = ⟨15, 20, 15, 5, 30⟩
D_{tw}(S, Q) = 85, D_{tw_lb}(S, Q) = 155
Thus, D_{tw}(S, Q) < $D_{tw\_lb}(S, Q)$

Theorem 4. *To make prefix-querying with* L_1 *incur no false dismissals, any two elements of Feature(S) (= ⟨First(S), Last(S), Greatest(S), Smallest(S)⟩)*

should not be extracted from the same element in the time warped sequence $S' = \langle s'_1, s'_2, s'_3, \ldots, s'_{k-1}, s'_k \rangle$.

Proof: Refers to [8].

In this paper, we propose a solution to the case as in ⟨Example 1⟩. This solution is to guarantee no false dismissals by performing index searching in the range of an enlarged ε.

Theorem 5. *If either of S and Q has a multi-role element in its feature vector, index searching with* 2ε *makes prefix-querying with* L_1 *guarantee no false dismissals.*

Proof: Refers to [8].

Prefix-querying with L_1 does not incur false dismissals even for the feature vectors having multi-role elements.

4 Performance Evaluation

For performance evaluation, we used K-Stock-Data, a real-world Korean stock data set that consists of 620 data sequences of length 300. Also, we generated query sequences Q by selecting an arbitrary subsequence of Len(Q) among the sequences chosen from a database. For forming a query, we adjusted ε to make the query meet the specified query *selectivity*.

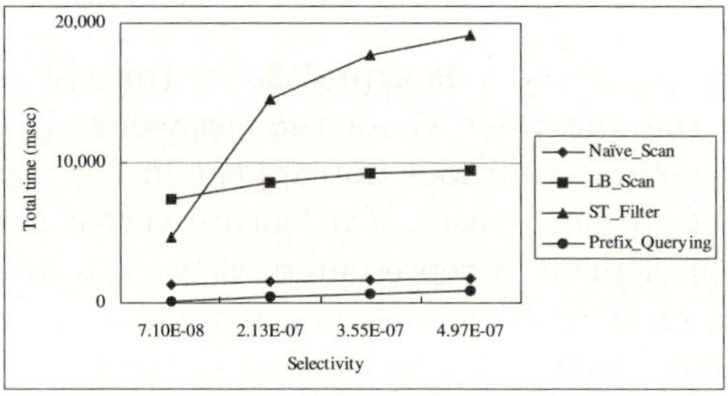

Fig. 1. Performance results with different selectivities

In Experiment 1, we used selectivities of 7.10×10^{-8}, 2.13×10^{-7}, 3.55×10^{-7}, and 4.97×10^{-7}. The length of query sequences used was 110. Figure 1 shows the results. With the increase of selectivities, the total elapsed time increases in all the methods. In particular, ST-Filter shows the great increasing rate. LB-Scan does not show good performance since it needs to access all data sequences in searching for the candidate sequences.

Note that Naive-Scan performs always better than LB-Scan and ST-Filter in all cases. This is because the optimized version of Naive-Scan[5] was used in our experiments for CPU processing. Such Naive-Scan is to maximize CPU performance by removing most redundant calculations. The prefix-querying method shows the superior performance over all other methods, and it shows up to 10.7 times better performance than Naive-Scan, the best one among existing methods.

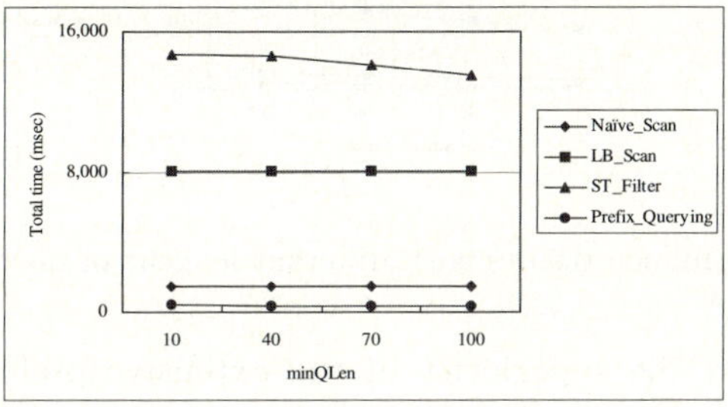

Fig. 2. Performance results with different values of *minQLen*

In Experiment 2, we evaluated the performances with different *minQLen* of 10, 40, 70, and 100. The length of query sequences was 110 and the selectivity was 2.13×10^{-7}. In Figure 2, ST-Filter shows a smaller total elapsed time with a larger *minQLen*. The increase of *minQLen* leads to the reduction of the suffixes in the suffix tree, consequently reducing the computation time in the filtering step. Other methods show almost the same performance even with different values of *minQLen*. The prefix-querying method shows up to 3.8 times better performance than Naive-Scan.

In Experiment 3, we evaluated the performances with changing query sequence lengths of 80, 140, 200, and 260. The query selectivity used is 2.13×10^{-7}. Figure 3 shows the experiment results. As the length of query sequences increases, the cost for computing distances does increase, which leads to the increase of the time spent in the filtering step and the post-processing step. ST-Filter and LB-Scan show the drastic growth of the elapsed time while Naive-Scan and the prefix-querying method do the smooth growth of the elapsed time with increasing lengths of query sequences. In this experiment, the prefix-querying method shows the best performance, and achieves up to 3.6 times speed-up over Naive-Scan.

5 Conclusions

Prefix-querying is the first index-based method to perform time-series subsequence matching under time warping without false dismissals. It employs L_∞ as a base distance function for users' convenience in forming queries. The contributions of this paper are summarized as follows.

- We have extended the prefix-querying method to employ L_1 instead of L_∞ as a base distance function.
- We have formally proven that the prefix-querying method guarantees no false dismissals in subsequence matching under time-warping.

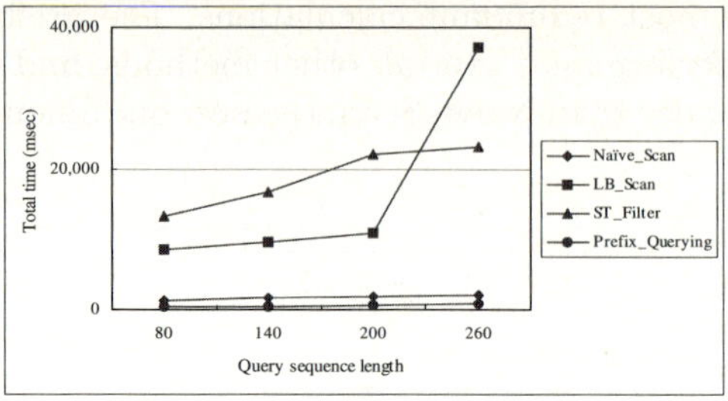

Fig. 3. Performance results with different lengths of query sequences

- We have verified the superiority of the extensive prefix-querying method via experiments. The results reveal that our method achieves significant performance improvement over the previous methods up to 10.7 times.

Acknowledgements

This research was partially supported by the MIC(Ministry of Information and Communications) of Korea under the ITRC(Information Technology Research Center) support program supervised by the IITA(IITA-2005-C1090-0502-0009).

References

[1] D. J. Berndt and J. Clifford: Finding Patterns in Time-Series: A Dynamic Programming Approach, In *Advances in Knowledge Discovery and Data Mining*, pp. 229-248, 1996.
[2] C. Faloutsos, *Private Communication*, 2001.
[3] E. Keogh: Exact Indexing of Dynamic Time Warping, In *Proc. Int'l. Conf. on Very Large Data Bases (VLDB)*, pp. 406-417, 2002.
[4] S. W. Kim, S. H. Park, and W. W. Chu: An Index-Based Approach for Similarity Search Supporting Time Warping in Large Sequence Databases, In *Proc. Int'l. Conf. on Data Engineering*, IEEE ICDE, pp. 1607-614, 2001.
[5] M. S. Kim, S. W. Kim, and M. Y. Shin: Optimization of Subsequence Matching Under Time Warping in Time-Series Databases, In *ACM Symp. on Applied Computing*, pp. 581-586, Apr. 2005.
[6] S. H. Park, S. W. Kim, J. S. Cho, and S. Padmanabhan: Prefix-Querying: An Approach for Effective Subsequence Matching Under Time Warping in Sequence Databases, In *Proc. ACM Intl. Conf. On Information and Knowledge Management*, ACM CIKM, pp. 255-262, 2001.
[7] B. K. Yi, H. V. Jagadish, and C. Faloutsos: Efficient Retrieval of Similar Time Sequences Under Time Warping, In *Proc. Int'l. Conf. on Data Engineering*, IEEE ICDE, pp. 201-208, 1998.
[8] B. C. Chang, S. W. Kim, J. H. Cha and M. Y. Shin: An Index-Based Time-Series Subsequence Matching Under Time Warping, Technical Report, Data and Knowledge Engineering Laboratory, 2005.

An Intrinsic Technique Based on Discrete Wavelet Decomposition for Analysing Phylogeny

Atulya K. Nagar, Dilbag Sokhi, and Hissam Tawfik

Intelligent and Distributed Systems Research Laboratory,
Deanery of Business and Computer Sciences,
Liverpool Hope University,
Hope Park, Liverpool, L16 9JD, U.K.
nagara@hope.ac.uk, dilbag.sokhi@liv.ac.uk,
tawfikh@hope.ac.uk

Abstract. The techniques for sequence analysis can be classified into two categories: sequence matching techniques and intrinsic techniques. Here we present an intrinsic technique of wavelet decomposition which uses the energy contained in the genomic signal for analysing the phylogenetic relationship between species. We apply the Discrete Wavelet transforms to analyze the intron-exon sequences which are represented as numerical series using electron-ion interaction potential. Wavelet transforms take advantage of the periodicity present in the genomic sequences and can adapt to the varying lengths of coding and non-coding sequences. This technique has been found to overcome the shortcomings of other techniques which make use of sequence comparison, since it uses the energy based decomposition to extract the patterns in the genomic data.

Keywords: Wavelet Analysis, Electron-Ion Interaction Potential, Intron, Exon, Phylogeny.

1 Background

Bioinformatics uses the techniques in applied mathematics, informatics, statistics and computer science to solve biological problems. The rate of increase in gene research data has been unprecedented; it is noted that it is doubling every six months [1] and is one of the most abundant data sets available.

The research conducted in Bioinformatics has led the research community towards a very exciting research of finding the relationship between different species in terms of there genetic data. Evidence from various studies conducted in gene sequence analysis suggests the genetic relationship between organisms. The study of there relationships between various organisms represents the phylogeny of organisms. It dictates the history of organismal lineages and the changes occurred between them during the process of evolution. This study is essential for explaining the similarities and differences among various plants, animals and different microorganisms. Many genes in humans and other less complex species are similar in their genomic content,

B. Gabrys, R.J. Howlett, and L.C. Jain (Eds.): KES 2006, Part I, LNAI 4251, pp. 1051 – 1058, 2006.
© Springer-Verlag Berlin Heidelberg 2006

the study of these genes would help in finding the genes which are related to particular diseases.

It is no coincidence that the gene sequences of related species of plants, animals and microorganisms share a complex pattern of similarity between them, which can be considered as the most fascinating aspect of the study of evolution. There is a rich array of computational methods and tools that are available for analysing patterns within this data. One such method discussed here is Wavelet analysis. Wavelet analysis exploits the periodicity present in the genomic sequences. They are also adaptive to the varying lengths of coding and non-coding regions in genomic sequences and does not require any training as compared to other techniques such as Neural network and Hidden Markov models [2].

Recently there have been some successful attempts of applying Wavelet transforms to genomic sequences. A survey of some of these researches has been conducted by Pietro Lio [2]. Trad et al. [3] have published a paper on the wavelet decomposition of protein sequences by discrete wavelet transform and conducted a cross-correlation study. Agarwal et al. [4] used the Euclidean distance for judging the similarity between patterns in sequences.

The current state of biological sequence analysis is based on either comparative methods which use sequence matching for sequence analysis or other methods which use some property of the analysed sequences such as accessible surface area, polarity, electron-ion interaction potential (EIIP). In this paper we demonstrate the use of Wavelet transform for the analysis of genomic data using the electron-ion interaction potential. Traditionally, Wavelet transform has been used as a signal processing tool for efficient multi-resolution analysis of non-stationary signals. The techniques of decomposing the signal into its scale and space components have gained importance in many signal processing applications. It uses smaller window size for analysing data at high frequencies and longer windows for low frequency data.

In this research we have put forward the idea of exploiting the periodicity present in the genomic data for Wavelet analysis. Wavelets provide multi-scale representation of signals by decomposing the signal into several groups of coefficients thus capturing the local as well as global features of the genomic data.

2 Multi-resolution Analysis of Numeric DNA Sequence

The mathematical formulation of the problem has been described in the paper by Krishnan *et al.*[5], please refer to it for further details.

Applying scaling and translation functions on $\psi(t)$, it creates a family of scaled and translated versions of mother wavelet transform.

$$\psi_{a,b} = \frac{1}{\sqrt{|a|}}\psi\left(\frac{t-b}{a}\right). \tag{1}$$

where 'a' is scale and 'b' is translation parameter and the factor $\frac{1}{\sqrt{|a|}}$ is used to ensure that the energy of the scaled and translated versions are the same as that of the mother wavelet.

The discrete wavelet transform is defined as,

$$DWT_f(m,j) = \langle f, \psi_{m,j} \rangle \equiv a_0^{-m/2} \int_{-\infty}^{+\infty} f(t)\psi\left(a_0^{-m}t - jb_0\right)dt, \quad m, j \in z \tag{2}$$

The corresponding family of wavelets is given by,

$$\psi_{m,j}(t) = 2^{-m/2}\psi\left(2^{-m}t - j\right). \tag{3}$$

The corresponding orthonormal wavelet transform for the function $f(t)$ and its reconstruction are:

$$OWT_f(m,j) \equiv \langle f, \psi_{m,j} \rangle = 2^{-m/2} \int_{-\infty}^{+\infty} f(t)\psi(2^{-m}t - j)dt,$$

$$f(t) = \sum_{m,j \in z} \langle f, \psi_{m,j} \rangle \psi_{m,j}. \tag{4}$$

The concept of wavelet transform can be explained by multi-resolution analysis by considering the incremental detail added in obtaining the $(m+1)^{th}$ scale approximation given the mth one. Let $V^{(m+1)}$ represent the space of all functions spanned by the orthogonal set, $\left\{\phi\left(2^{m+1}t - k\right); k \in z\right\}$ corresponding to a finer approximation of functions in $V^{(m)}$, the space of all functions spanned by the orthogonal set, $\left\{\phi\left(2^{m}t - l\right); l \in z\right\}$. Then $V^{(m)} \subset V^{(m+1)}$ and this relation in terms of sets can be stated as:

$$V^{(m+1)} = V^{(m)} \oplus W^{(m)}.$$

and $W^{(m)}$ is orthogonal to $V^{(m)}$ and is the space that contains the information or detail in going from the coarser, $f^{(m)}(t)$, to the finer, $f^{(m+1)}(t)$, representation of the original function $f(t)$. The symbol $\oplus$ is the orthogonal sum.

The basis of space $W^{(m)}$ is spanned by the orthogonal translates of a single wavelet function, $\psi(2^m t)$:

$$f^{(m+1)}(t) = f^{(m)}(t) + \sum \delta f(m, j)\psi(2^m t - j),\ m, j \in z. \tag{5}$$

where $\delta f(m, j)$ are the wavelet coefficients.

The Wavelet function $\psi(2^m t)$ is related to the scaling function $\phi(2^{m+1} t)$ through the relationship,

$$\psi(2^m t) = \sum_k g(j)\phi(2^{m+1} t - j). \tag{6}$$

where $g(j)$ are the wavelet coefficients.

Combining the above equations yields the synthesis form of the wavelet transform,

$$\sum_{j=-\infty}^{+\infty} f(m+1, j)\phi(2^{m+1} t - j) = \sum_{j=-\infty}^{+\infty} f(m, j)\sum_k h(k)\phi(2^{(m+1)} t - j - k)$$

$$+\sum_{j=-\infty}^{+\infty} \delta f(m, j)\sum_k g(k)\phi(2^{m+1} t - j - k), \tag{7}$$

$$m, j \in z.$$

This defines a dynamic relationship between the scaling coefficients $f(m+1, j)$ at one scale and the scaling and wavelet coefficients respectively.

$$f(m, j) = \sum_k h(2j - k) f(m+1, k). \tag{8}$$

$$\delta f(m, j) = \sum_k h(2j - k) f(m+1, k). \tag{9}$$

For further reading about the mathematical proofs please refer to [6, 7].

3 Intrinsic Energy Based Wavelet Decomposition Technique for Genomic Sequences

The data is collected from the NCBI's publicly available database and then pre-processed by zero padding and altering the sequences to make them of equal size for comparison.

Biological processes represent the selective interactions between biomolecules in living organisms. It has been known that certain periodicities within the distribution of energies of delocalized electrons along a protein molecule are critical for protein biological function. In the proposed research we have used the electron-ion interaction potential of nucleotides to convert the genomic sequences into numerical form. The EIIP values describe the average energy status of all valence electrons. The

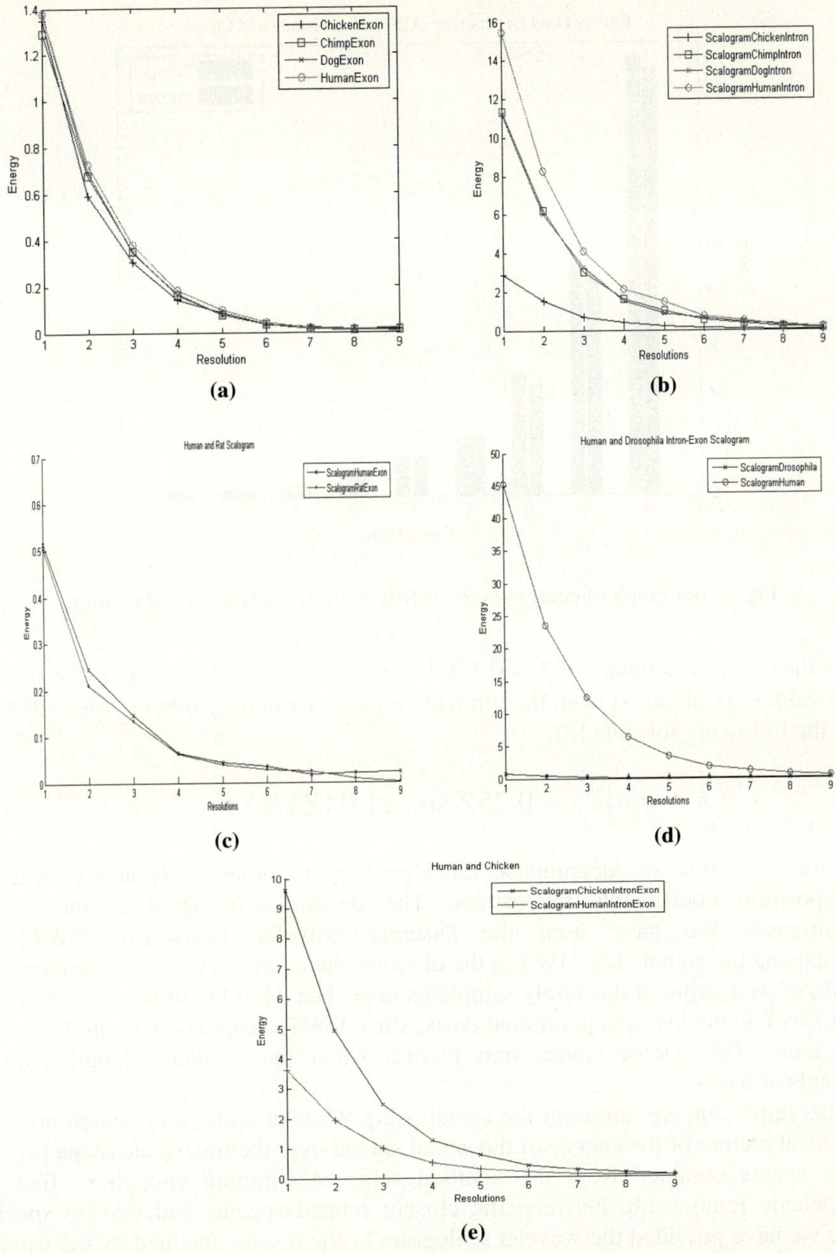

Fig. 1. (a) Exon Scalogram of BRCA1 of Human, Chimp, Dog, and Chicken (b) Intron Scalogram of BRCA1 gene of Human, Chimp, Dog and Chicken (c) Exon Scalogram of Neurod2 gene of Human and Rat (d) Intron-Exon Scalogram of MITF gene of Human and Drosophila (e) Intron-Exon Scalogram of SLC24A5 of Human and Chicken species

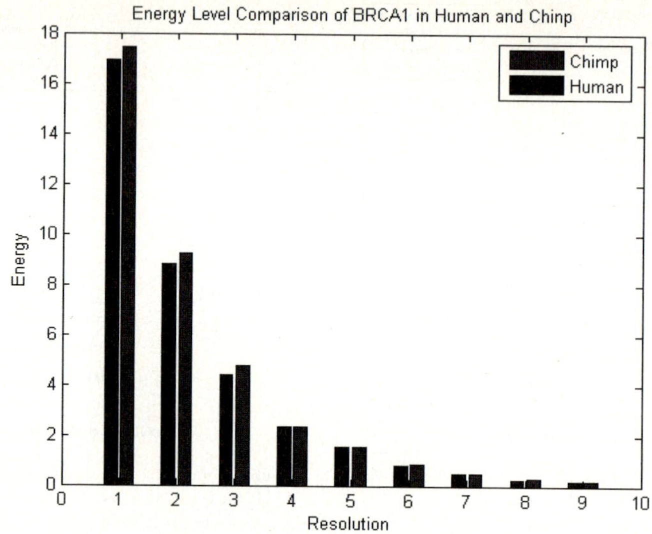

Fig. 2. Bar graph of energy levels of BRCA1 gene in Human and Chimp

EIIP values of nucleotides are A = 0.1260, T = 0.1335, G = 0.0806 and C = 0.1340. These values are obtained from the innovative paper written by Irena Cosic (1994) by using the following formula [8].

$$\left\langle \overrightarrow{k+q} \middle| w \middle| \vec{k} \right\rangle = 0.25Z \sin(\pi 1.04Z)/(2\pi)\,. \tag{10}$$

In the next step we decompose the signal up to nine levels and collect the decomposition coefficients in vectors. The decomposed signal is then again reconstructed. We have used the Discrete Wavelet Transform (DWT) for decomposing the signal. The DWT is the obvious choice as the genomic signal can be considered as a string of discretely sampled signal. The other factor which favours the use of DWT is the less computational costs, since DWT computes only on the dyadic coefficients. This factor comes into picture when the sequence lengths are in thousands of bases.

In the third step, we represent the signal using Wavelet scalogram, which provides a graphical picture of the energy of the signal spread over the time-scale plane [9].

The results obtained from this method were encouraging enough to find the phylogenetic relationship between the closely related species and diverse species. Below we have provided the wavelet scalogram of the results obtained by the wavelet decomposition where X-axis represents the decomposition level and Y-axis represents the energy level at that particular level.

From the figures below we can conclude the effectiveness of Wavelet decomposition in finding out the phylogenetic relationships between different species. In Fig. 1(a) we can see the similarity of scalograms of all the closely related species such as Human and Chimp. On the other hand, Fig. 1(d) clearly rules out any relationship between Humans and Drosophila. One another such comparison of interspecies comparison is done between Human and Chicken by taking into account the SLC24A5 gene which is one of the gene in which a mutation causes the Alzheimer's disease as shown in Fig. 1(e). The difference in the energy content can be easily noticed.

The above figures bring about the effectiveness of the Wavelet technique in finding the similarities as well as dissimilarities between various species.

When a bar graph is drawn of the energy levels at each level of decomposition of BRCA1 gene of Human and Chimp as shown in Fig. 2, it was found to contain about same energy levels thus indicating the strong links between both the species in the ladder of phylogenity.

4 Conclusions and Discussions

This research is a contribution towards the bioinformatics community in analysing the patterns and signals hidden in the DNA coding. We have demonstrated the feasibility of applying Wavelet transform for the analysis of the DNA sequences and found the phylogenetic relationship of some closely related species in the evolutionary cycle. This research has provided the Bioinformatics community with another alternative tool for analysing genomic sequences which uses the energy based decomposition of genomic sequences in finding the phylogenetic relationship between different species. This technique can be used as a benchmark technique for comparing the results obtained from other sequence analysis methods. The proposed research with some biological studies would provide a more robust approach towards gene finding techniques.

The aforementioned research can be taken further by using the amino acid contents of the genomic sequences as compared to the nucleotide sequences, since it has been suggested by some researchers that the amino acids allow distant comparisons as compared to nucleotide sequences. Some other wavelet transforms can be experimented for decomposition of signal. More distant comparisons in terms of species, which are considered to be not much related in evolutionary cycle, will make this approach more robust.

References

1. Kauer G., Blocker H.: Applying Signal Theory to the Analysis of Biomolecules. Bioinformatics. 19 (2003) 2016-2021
2. Lio P.: Wavelets in Bioinformatics and Computational Biology: State of Art and Perspectives. Bioinformatics Review. 9 (2003) 2-9
3. Trad C. H., Fang Q., Cosic I.: Protein Sequence Comparison Based on the Wavelet Transform Approach. Protein Engineering. 15 (2002) 193-203
4. Agrawal R., Faloutsos C., Swami A. N.: Efficient Similarity Search in Sequence Databases, Proc. 4th Int. Conf. Foundations Data Org. Algo., Illinois, USA, (1993) 69-84
5. Krishnan A., Li K. B., Issac P.: Rapid Detection of Conserved Regions in Protein Sequences Using Wavelets. In Silico Biology. 4 (2004) 133-148
6. Chui C. K.: An Introduction to Wavelets. Academic Press, USA (1992)

7. Willsky A. S.: Modeling and Estimation of Multi-Resolution Stochastic Processes, Special issue IEEE Trans. Info. Theory on Wavelet Transforms and Multi-resolution signal anal. 38 (1992) 766-784
8. Cosic I.: Macromolecular Bioactivity: Is it Resonant Interaction Between Macromolecules? – Theory and Applications. IEEE Trans. on Biomedical Eng. 41 (1994) 1101-1114
9. Rioul O., Flandrin P.: Time-Scale Energy Distributions: A General Class Extending Wavelet Transforms. IEEE Trans. on Signal Processing. 40 (1992) 1746-1757.

Analysis of Selfish Behaviors on Door-to-Door Transport System

Naoto Mukai[1], Toyohide Watanabe[1], and Jun Feng[2]

[1] Department of Systems and Social Informatics,
Graduate School of Information Science, Nagoya University
Furo-cho, Chikusa-ku, Nagoya, 464-8603, Japan
naoto@watanabe.ss.is.nagoya-u.ac.jp,
watanabe@is.nagoya-u.ac.jp

[2] Hohai University, Nanjing, Jiangsu 210098, China
fengjun-cn@vip.sina.com

Abstract. The selfishness of people sometime degrades the performance of systems in which resources are shared among people. One of the typical examples is a door-to-door transport system. The door-to-door transport system adopts a flexible style of transportation, different from traditional transport systems such as buses and taxis. In the system, several customers ride one transport vehicle all together, and the share-ride transport vehicle visits from door to door. The selfishness of customers may cause the delay of arrivals and the decrease of benefits. However, it is difficult to control the selfishness of people because of their inherent competitive behaviors. Therefore, in this paper we analyze the selfish decision-making processes in the door-to-door transport system on the basis of game theory. The profit relation among people reaches an equilibrium called Nash Equilibrium. However, the equilibrium relation is not optimal for the system in many cases. Therefore, we also show that the degradation of the system resulted from the selfish behaviors can be avoided by changing parameters of the system such as the number of vehicles.

1 Introduction

We can see selfish behaviors of people in various social systems. The selfishness of people sometime degrades the performance of social systems, especially when people share resources. One of the systems is a packet routing system on Internet. In particular, selfish senders (routers) want to select transfer routes which minimize their latencies, but the result may cause a traffic jam. Such a problem is called "selfish routing" and addressed by several researchers [1,2,3]. In this paper, we focus on the selfishness of people in other social system called "door-to-door transport system (a.k.a. demand bus)".

The door-to-door transport system is recently adopted as a new method of transportation by local governments. In the system, several customers ride one transport vehicle all together, and the share-ride transport vehicle visits from

B. Gabrys, R.J. Howlett, and L.C. Jain (Eds.): KES 2006, Part I, LNAI 4251, pp. 1059–1066, 2006.
© Springer-Verlag Berlin Heidelberg 2006

door to door. Such flexibility of the system can improve both the convenience of customers and the profitability of drivers (companies). However, the flexibility also allows selfish behaviors of customers and drivers. The selfish behaviors may degrade the convenience and profitability. Most of traditional researches related to the door-to-door transport system [4,5,6] focused on mechanical optimality of the system but did not consider the selfishness of people. Therefore, we newly focus on the selfishness of people in the door-to-door transport system.

From the viewpoint of customers, they want to arrive at their destinations as early as possible. However, if many customers share one vehicle, their traveling times (which include waiting times) will be longer. Therefore, in this paper we analyze such selfish decision-making processes on the basis of "game theory". All selfish players (i.e., customers) select the options which maximize their profits, respectively. Such selfish behaviors lead the profit relation among players to an equilibrium relation called Nash Equilibrium (i.e., the profits of all players are the same). The ideal of the system is that the equilibrium relation equals to an optimal relation for the system. However, the equilibrium relation is not optimal in many cases. Therefore, we try to reduce the gap between the equilibrium and optimal relations by changing parameters of the door-to-door transport systems such as the number of vehicles.

The remainder of this paper is as follows: Section 2 describes the traditional selfish routing problem. Section 3 formalizes the model of the door-to-door transport system. Section 4 analyzes the selfishness of people in the system and shows that the profit relation between people can be changed by adjusting parameters. Finally, Section 5 concludes and offers our future work.

2 Selfish Routing

The selfish routing [1,2,3] is a basic idea of our approach to analyze the door-to-door transport system. Here, let consider the packet routing situation shown in Figure 1. There are n selfish senders who want to send their packets from a source S to a destination D. Each selfish sender selects the minimum latency transfer route from two transfer routes (the upper route is f_1, and the lower route is f_2). The cost of senders is defined as Equation (1) where w_i is a volume of packet $i \in n$ and $l(f)$ is a latency function.

$$C(f) = \sum_{i \in n} (w_i \times l(f)) \tag{1}$$

The latency function $l(f)$ represents the latency time of route f caused by a traffic jam. In fact, the latency function of upper route $l(f_1) = x$ means that the latency time is proportional to the number of packets. While, the latency function of lower route $l(f_2) = 1$ means that the latency time is always 1 independently of the number of packets. For simplicity, all volumes of packets w_i are set to 1. Hence, the cost of upper route $C(f_1)$ is $n \times x$, and the cost of lower route $C(f_2)$ is n.

Here, let p be the rate of senders who select f_1, and $(1-p)$ be the rate of senders who select f_2. In fact, the latency function x is replaced by p, and the expected cost of senders $\hat{C}$ is defined as Equation (2).

$$\begin{aligned}\hat{C} &= p \times (n \cdot p) + (1-p) \times (n) \\ &= n(p^2 - p + 1)\end{aligned} \quad (2)$$

The formula $p^2 - p + 1$ is a convex function shown in Figure 2. Hence, the minimum value of the expected cost is $0.75n$ at $p = 0.5$. It means that the optimum of this packet situation is that all senders select f_1 and f_2 evenly as shown in Figure 1(a). However, this optimum situation is not an equilibrium because the cost of upper route $C(f_1)$ is less than the cost of lower route $C(f_2)$ ($C(f_1) = 0.5n$, $C(f_2) = 1.0n$). If there is a difference between the two routes, senders with higher cost envy other senders with lower cost. Consequently, all senders select upper route f_1 (i.e., $p = 1.0$) shown in Figure 1(b), and the expected cost $\hat{C}$ is $1.0 \times n$. Note that there is no difference between the two routes at $p = 1.0$. This equilibrium situation is called "Nash equilibrium" [7] or "Wardrop equilibrium" [8].

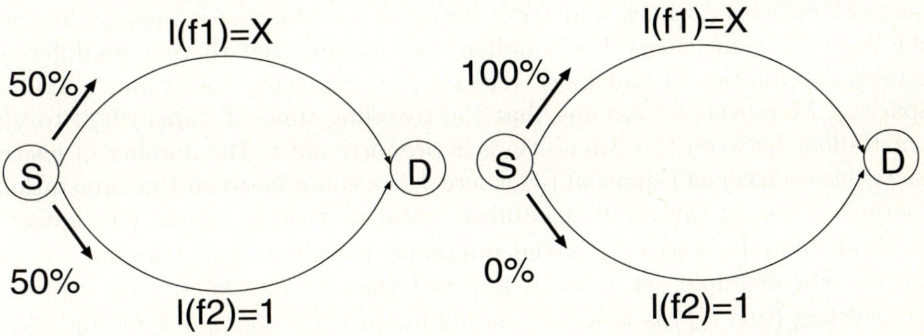

(a) Optimum (b) Nash Equilibrium

Fig. 1. Optimum and Nash Equilibrium on packet routing system

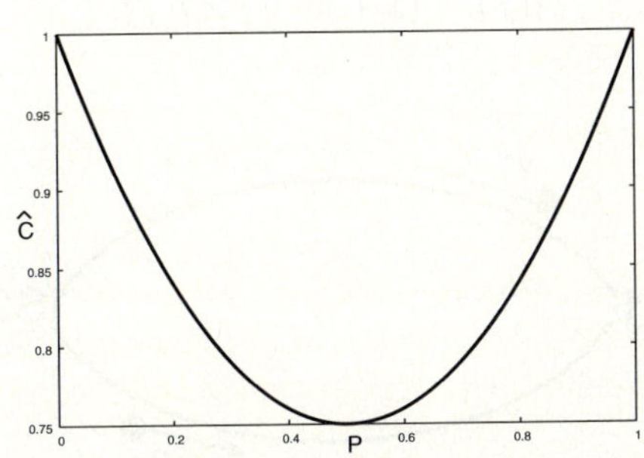

Fig. 2. Expected cost

As described above, selfish behaviors of people degrade the performance of systems in certain situations. This phenomenon can be seen in the door-to-door transport system we focus on. In the following section, we show the selfishness of people in the system from the viewpoint of customers. Moreover, we show the degradation of the system resulted from selfish behaviors can be avoided by changing parameters of the system.

3 Model of Door-to-Door Transportation

In this section, we formalize the model of door-to-door transport system. We can classify the way to visit customers into two types: semi-visit and full-visit. In the semi-visit type, each transport vehicle visits a station to pick up customers, and then visits door-to-door to drop off customers. Alternatively, each transport vehicle visits door-to-door to pick up customers, and then visits a station to drop off customers. In the full-visit type, each transport vehicle always visits door-to-door to pick up and drop off customers. In this paper, we focus on semi-visit types. Figure 3 illustrates the door-to-door transport system of semi-visit type. There are two transport vehicles (v_1 and v_2) and two stations (s_1 and s_2). The transport vehicles go back and forth between two stations in regular intervals (maybe at the same time). For simplicity, we assume that there is no difference between capabilities of transport vehicles (i.e., traveling speed and maximum capacity). Moreover, we assume that the traveling time (distance) $l(f)$ to visit door-to-door between the two stations is proportional to the number of visiting points (customers) as Equation (3) where f is a route between the two stations, w is an increasing rate, x is a number (rate) of visiting points (the range of x is from 0 to 1), and $\acute{l}(f)$ is the minimum traveling time between the two stations. For example, let w be 1, if x is 1 (i.e., all customers ride together), the traveling time $l(f)$ is twice as the minimum traveling time $\acute{l}(f)$, and if x is 0.5 (i.e., half customers ride together), the traveling time $l(f)$ is 1.5 times the minimum traveling time $\acute{l}(f)$.

$$l(f) = (1 + w \cdot x) \times \acute{l}(f). \tag{3}$$

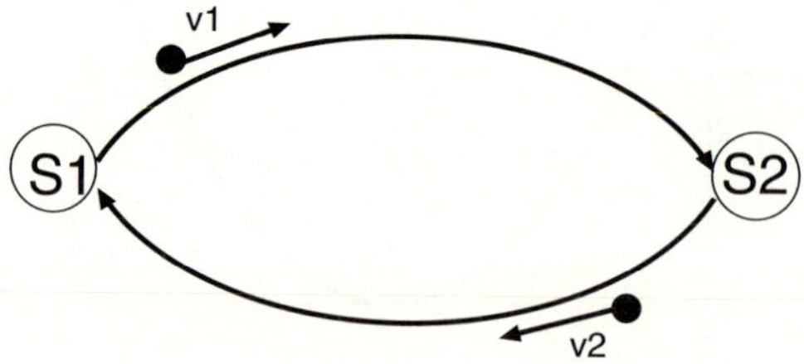

Fig. 3. Semi-visit type

4 Selfishness on Door-to-Door Transportation

In this section, we show the selfish behaviors in the door-to-door transport system from the viewpoint of customers.

Here, let's consider selfish customers who want to minimize their traveling costs (times). Customers have two options to arrive at their destinations. The one is the door-to-door transportation shown in Figure 4(a). The other is the direct transportation such as walking or bicycle shown in Figure 4(b) (i.e., customers go to their destinations, respectively). Each selfish customer selects shorter one from the two options. Figure 5 shows a model of decision-making process related to the two options from the viewpoint of customers. In the model, the route f_1 represents the door-to-door transportation, and the route f_2 represents the direct transportation. The latency function $l(f_2)$ of the direct transportation is always 1 independently of other customers, i.e., the average traveling time of all customers is 1. While, the latency function $l(f_1)$ of the door-to-door transportation is given as follows. Let m be the number of transport vehicles, and v be the minimum traveling time of transport vehicles between two stations. Assuming that the transport vehicles go back and forth between two stations at even intervals, the expected waiting time t_w of customers is given as Equation (4).

$$t_w = \frac{1}{2} \times \frac{2 \cdot v}{m} = \frac{v}{m} \tag{4}$$

Moreover, assuming that the traveling time is independent of the number of visiting points (i.e., customers), the expected riding time t_r of customers is half of the minimum traveling time v as Equation (5).

$$t_r = \frac{v}{2} \tag{5}$$

Hence, the latency function $l(f_1)$ is the sum of the waiting time t_r and riding time t_r as Equation (6).

$$\begin{aligned} l(f_1) &= t_w + t_r \\ &= \frac{v(2+m)}{2m} \end{aligned} \tag{6}$$

However, in actuality, the traveling time increases according to the number of visiting points. Therefore, we replace the latency function $l(f_1)$ with Equation (7), where p is the rate of customers who select the door-to-door transportation, according to the Equation (3).

$$l(f_1) = (1 + w \cdot p) \cdot \frac{v(2+m)}{2m} \tag{7}$$

Consequently, the expected traveling time (i.e., cost) $\hat{C}$ is defined as Equation (8). The formula of $\hat{C}$ is a convex function as well as Equation (2).

$$\hat{C} = p \times (1 + w \cdot p) \cdot \frac{v(2+m)}{2m} + (1-p) \times 1 \tag{8}$$

Figure 6 shows the expected traveling cost $\hat{C}$ from the viewpoint of customers in four parameter patterns ($PT1, \cdots, PT4$) shown in Table 1. The minimum traveling time v of transport vehicles is set to 0.2 in all the patterns. The parameter m is the number of transport vehicles: $PT1$ and $PT3$ adopt 1 vehicle, and $PT2$ and $PT4$ adopt 2 vehicles. The parameter w is the increasing rate of traveling cost (i.e., the relative size of area to visit requested points of customers): $PT1$ and $PT2$ are 2.0, and $PT3$ and $PT4$ are 4.0.

In $PT1$, the optimum (minimum) value is about 0.8 at $p \simeq 0.6$. However, the equilibrium value is about 0.9 at $p = 1.0$. It means that the traveling time of all selfish customers can decrease by only about 10% but it is not optimum. In $PT2$, the optimum and equilibrium values are about 0.6 at $p = 1.0$. It means that the best option is to select the door-to-door system for all customers and the traveling time can decrease by about 40%. In $PT3$, the optimum value is about 0.3 at $p \simeq 0.9$. However, the equilibrium value is 1.0 at $p \simeq 0.6$. It means that the traveling time does not improve at all by using the door-to-door transportation. Furthermore, if all customers select the door-to-door transportation, the traveling time increases by 50% than the direct transportation. In $PT4$, the optimum value is 0.8 at $p = 0.5$. However, the equilibrium value is 1.0 at $p = 1.0$. It means that there is no difference of the traveling time between the door-to-door transportation and the direct transportation at the equilibrium.

From these results, we can say that the degradation caused by selfishness of customers can be avoided by adjusting the number of transport vehicles or the relative size of area to visit requested points. If the parameters of the door-to-door system cannot be changed, customers should take altruistic behaviors to achieve the optimal efficiency of door-to-door system.

Table 1. Parameter setting

Pattern	m	w
$PT1$	1.0	2.0
$PT2$	2.0	2.0
$PT3$	1.0	4.0
$PT4$	2.0	4.0

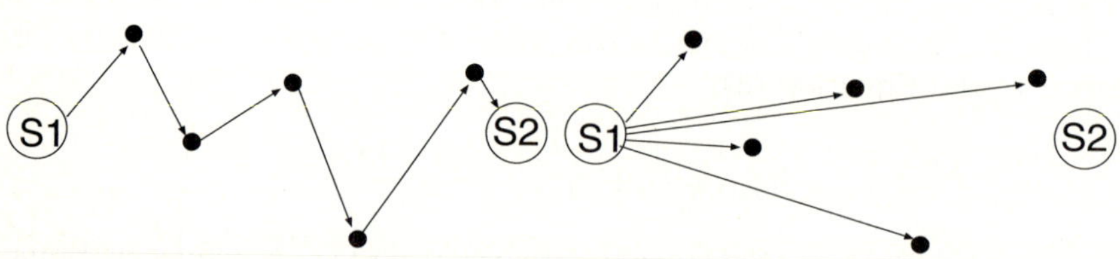

(a) Door-to-door transportation (b) Direct transportation

Fig. 4. Two options to arrive at destinations

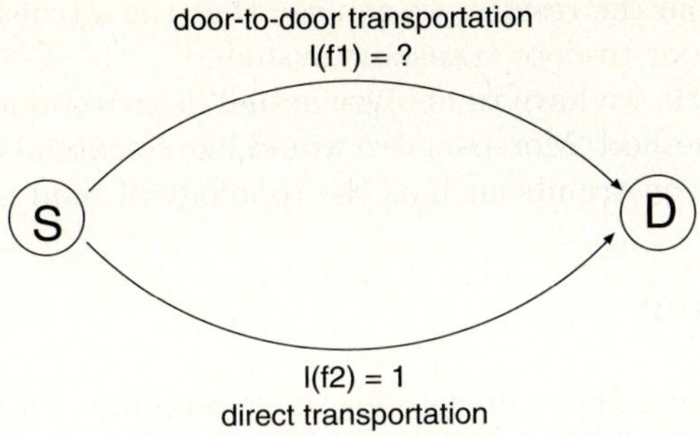

Fig. 5. Model of make-decision process from the viewpoint of customers

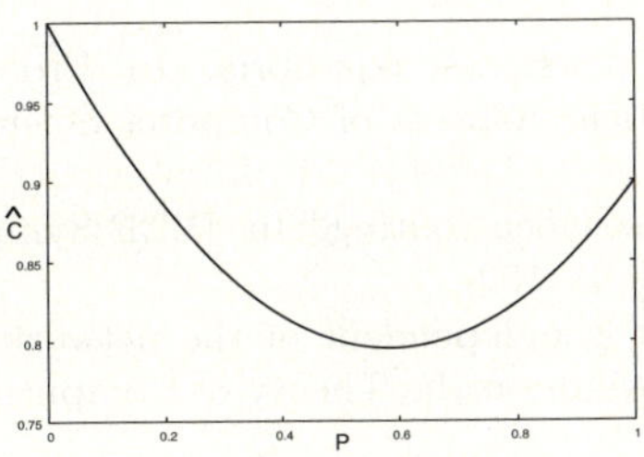

(a) PT1

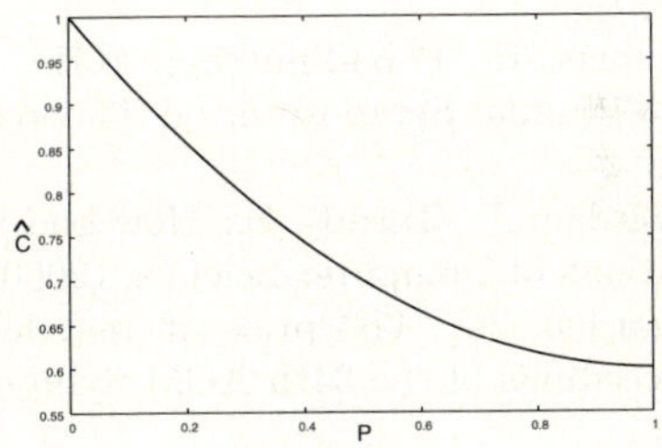

(b) PT2

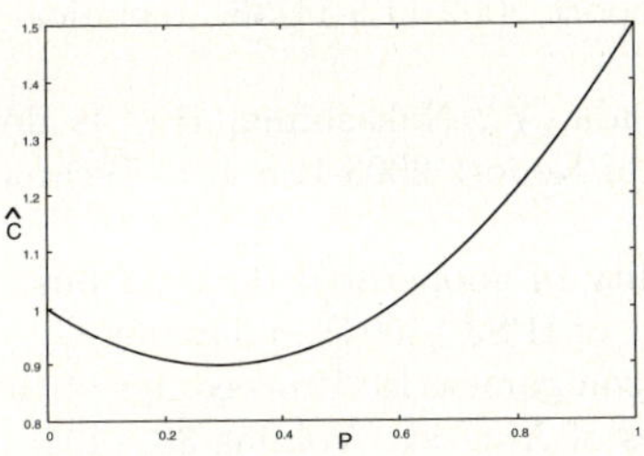

(c) PT3

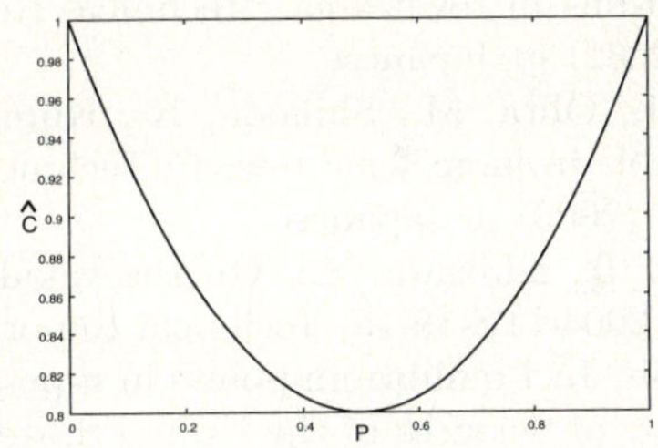

(d) PT4

Fig. 6. Expected traveling cost from the viewpoint of customers

5 Conclusion

In this paper, we focused on selfish behaviors in door-to-door transport system. The flexibility of door-to-door transport system allows selfish behaviors of people. Such selfish behaviors have possibility of degrading the performance of door-to-door system. Therefore, we clarified the make-decision processes of customers and drivers and analyzed the effects of the selfish behaviors on the basis of game theory. We found that the degradation caused by the selfishness can be avoided by adjusting the parameters of the transport system such as the number of cars. From the results, we can say that the altruistic behavior is very important to the door-to-door transport system.

In the future work, we have to analyze actual door-to-door transport systems by our analytical method. Moreover, we would like to extend our method to deal with more specific constraints such as the topology of road networks.

Acknowledgment

We would like to thank Japan Society for the Promotion of Science (JSPS). And, we acknowledge to Prof. Naohiro Ishii of Aichi Institute of Technology.

References

1. Koutsoupias, E., Papadimitriou, C.H.: Worst-case equilibria. In: Proceedings of the 16th Annual Symposium on Theoretical Aspects of Computer Science. (1999) 387–396
2. Roughgarden, T., Tardos, E.: How bad is selfish routing? In: IEEE Symposium on Foundations of Computer Science. (2000) 93–102
3. Roughgarden, T.: The price of anarchy is independent of the network topology. In: Proceedings of the 34th ACM Symposium on the Theory of Computing. (2002) 428–437
4. Ohta, M., Shinoda, K., Noda, I., Kurumatani, K., Nakashima, H.: Usability of demand-bus in town area. Technical Report 2002-ITS-11-33, Technical Report of IPSJ (2002) in Japanese.
5. Noda, I., Ohta, M., Shinoda, K., Kumada, Y., Nakashima, H.: Is demand bus reasonable in large scale towns? Technical Report 2003-ICS-131, Technical Report of IPSJ (2003) in Japanese.
6. Harano, T., Ishikawa, T.: On the vakidity of cooperated demand bus. Technical Report 2004-ITS-19-18, Technical Report of IPSJ (2004) in Japanese.
7. J.F.Nash, J.: Equilibrium points in n-person games. In: Proceedings of the National Academy of Sciences of the United States of America. Volume 36. (1950) 48–49
8. Wardrop, J.: Some theoretic aspects of road traffic research. In: Proceedings of the Institute of Civil Engineering. Volume 1. (1952) 325–378

Fault-Tolerant Cluster Management Tool for Self-managing of Cluster DBMS*

Jae-Woo Chang, Yong-Ki Kim, and Young-Chang Kim

Dept. of Computer Engineering, Chonbuk National University
Chonju, Chonbuk 561-756, South Korea
jwchang@chonbuk.ac.kr, {ykkim, yckim}@dblab.chonbuk.ac.kr

Abstract. In this paper, we design and implement a fault-tolerant cluster management tool which can monitor the status of the nodes in a cluster DBMS (Database Management System) as well as the status of the various DBMS instances at a given node. Based on the monitoring, our cluster DBMS management tool can provide an intelligent fail-over mechanism for dealing with multiple node failures in a cluster DBMS, thus making its fault-tolerance possible when combined with a load balancer. It is shown that our cluster DBMS management tool can support the nonstop and self-managing of cluster DBMS with its intelligent fail-over mechanism even in case of multiple node failures.

1 Introduction

Much research has been conducted regarding the use of cluster DBMSs which are needed to supports 24 hours nonstop database application services, e.g., map database provision in LBS (Location-based Service) and Telematics. They include Oracle 9i Real Application Server, Informix Extended Parallel Server and IBM DB2 Universal Database EEE. As a well-known cluster DBMS management tool, OCMS (Oracle Cluster Management System) [1] is included as a part of the Oracle8i Parallel Server product on Linux. To manage the cluster DBMS efficiently, a cluster DBMS management tool should be designed by considering a couple of requirements. First, a cluster DBMS management tool should provide users with a graphic interface to visualize a cluster system being composed of multiple nodes as a single virtual system, as well as to show the status of all of the nodes in the system and all of the resources (i.e. CPU, memory) at a given node [2]. Next, a cluster DBMS management tool should provide a fail-over mechanism to support its fault-tolerance in case of multiple node failures when combined with a load balancer [3,4]. In this paper, we design and implement a fault-tolerant cluster DBMS management tool which can monitor the status of the nodes in a cluster DBMS as well as the status of the various DBMS instances at a given node. Based on the monitoring, our cluster DBMS management tool can provide an intelligent fail-over mechanism for dealing with multiple node failures in a cluster DBMS, thus making its fault-tolerance possible.

* This work is financially supported by the Ministry of Education and Human Resources Development (MOE), the Ministry of Commerce, Industry and Energy (MOCIE) and the Ministry of Labor (MOLAB) though the fostering project of the Lab of Excellency.

B. Gabrys, R.J. Howlett, and L.C. Jain (Eds.): KES 2006, Part I, LNAI 4251, pp. 1067 – 1074, 2006.
© Springer-Verlag Berlin Heidelberg 2006

2 Fault-Tolerant Cluster Management Tool

A cluster DBMS consists of multiple server nodes and uses a shared disk. All of the server nodes are connected to each other by means of a high-speed gigabit Ethernet, which is called cluster network. A master node receives a service request from clients through a service network and makes job assignment through the cluster network. Each server is connected to a shared disk by means of SAN (Storage Area Network) for data transmission. The Backup node serves as a new master node when it detects the failure of the master node and uses the cluster network for a fail-over mechanism. Figure 1 shows the overall architecture of a cluster DBMS which are needed to supports 24 hours nonstop database application services, e.g., map database provision.

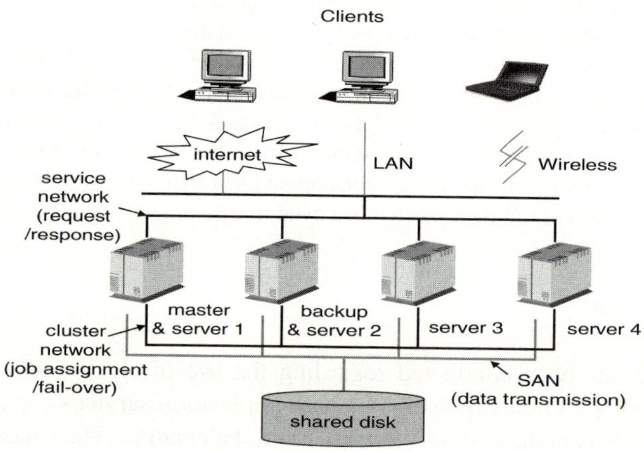

Fig. 1. Overall architecture of a cluster DBMS

By extending the cluster management tool designed for cluster DBMS [5], we proposed a fault-tolerant cluster management tool which can monitor the status of the nodes in a cluster DBMS as well as the status of the various DBMS instances at a given node. Based on the monitoring, our cluster DBMS management tool can provide an intelligent fail-over mechanism for dealing with multiple node failures in a cluster DBMS, thus making its fault-tolerance possible when combined with a load balancer. The cluster DBMS management tool consists of four components; probe, handler, CM (Cluster Manager) and NM (Node Manager). The probe monitors the status of the each node and generates events according to the status of the system. The handler stores the monitored information into a status table and performs a specific procedure for each type of event. The NM performs network communication with the CM at the master node. If the NM detects an error during the communication with the CM, the NM makes a connection with the backup node and transmits all of the available information to it. If the NM does not receive a response from the CM during a defined interval, the NM considers that the CM has failed and makes a connection with a backup node. The CM, running on the master node, perceives the status of the networks by sending a ping message to each node through both the cluster network and the service network. Based on the status of the networks, the CM manages all of the system resources and services. If the CM detects that the NM has encountered an error, it restarts the NM. If the CM detects an error, it generates an event and transmits it to the corresponding service handler, thus requesting it to perform its recovery procedure.

If the master node fails, our fault-tolerant cluster management tool terminates the master node, and then makes the backup node and one of server nodes be a new master and a new backup one, respectively. If the backup node fails, our management tool terminates the backup node and let one of server nodes be a new backup node. If a server node fails, our management tool just terminates the server node. Figure 2(a) shows the status of the system having four nodes before and after a node failure.

(M: master node, B: backup node, S: database server node)

Before node failure				Failure node	After node failure			
1	2	3	4					
M	B	S	S	Master node	failure	M	B	S
				Backup node	M	failure	B	S
				Server node	M	B	failure	S

(a) System having four nodes

Before node failure				Failure node	After node failure			
1	2	3	4					
failure	M	B	S	Master node	failure	failure	M	B
				Backup node	failure	M	failure	B
				Server node	failure	M	B	failure

(b) System having three nodes

Before node failure				Failure node	After node failure			
1	2	3	4					
failure	failure	M	B	Master node	failure	failure	failure	M
				Backup node	failure	failure	M	failure
				Server node	failure	failure	M	failure

(c) System having two nodes

Fig. 2. Status of the system before and after a node failure

At this time, we assume that the master node has failed, and then the second and the third node have been selected as a new master and a new backup one, respectively. Thus, if the new master node fails continuously, our fault-tolerant cluster management tool terminates the master node, and then assigns the backup node and the remaining database server node to the role of a new master and a new backup one, respectively. If the new backup node fails after the first node failure in Figure 2(a), our management tool terminates the backup node (the third one) and let the remaining database server node (the fourth one) be a new backup node. If a database server node fails after the fist node failure in Figure 2(a), our management tool just terminates the database server node (the fourth one). Figure 2(b) shows the status of the system having three nodes before and after a node failure. At this time, we also assume that the new master node has failed, and then the third and the fourth node have been selected as a new master and a new backup one, respectively. Thus, if the new master node fails continuously, our fault-tolerant cluster management tool terminates the master node and just assigns the backup node to the role of a new master node. This is because there is no running node. If the backup node fails after the first two nodes' failures in Figure 2(b), our management tool just terminates the backup node (the fourth one) because there is no running database server node. Figure 2(c) shows the status of the system having two nodes before and after a node failure. The other cases can be dealt with similarly by our fault-tolerant cluster DBMS management tool. Conclusively, it is shown that our cluster management tool is fault-tolerant because it can work well even if one of nodes (database servers) survives in the system.

3 Intelligent Fail-Over Mechanism

The server nodes comprising a cluster system can be classified into the master node, the backup node, and the database server nodes. If there is no failure in both the service and the cluster network, the cluster system runs as a normal situation. If the service network fails, a server node can communicate with the other nodes, but it cannot receive user requests or return the results to the user. If the cluster network fails, a given server node cannot communicate with the others and so the master node cannot distribute user requests to any of the database server nodes. If neither the service nor the cluster network work, the situation is considered as node failure since the nodes cannot function anymore. In order to recover the cluster system from the node failure, we provide an intelligent fail-over mechanism which can deal with multiple node failures in a cluster DBMS. The proposed fail-over mechanism is divided into two procedures; node failure detection and failure recovery. First, in order to detect the node failure, the CM and NM sends a ping message to each database server node in the cluster system and analyze the response of each server node. If it fails to receive any response from any of the database server nodes, it regards the situation as its node failure or its cluster network failure. At that time, there can be an ambiguity as shown in the Figure 3. In the Figure 3(a), all of the database server nodes except the first one have failures (node failures or cluster network failures). In the Figure 3(b), only the first server node has a failure. However, both situations can never be distinguished because they have the same ping status. To solve this ambiguity, we propose an intelligent failure detection procedure which uses the history information of the ping status. Once, our cluster DBMS management tool detects a failure, it compares the current ping status with the past status. If two database server nodes have already failed due to their node failure or their cluster network failures, this situation can be considered as ' the failures of all the nodes except one. Otherwise, this situation can be considered as a failure in only one node. In our intelligent failure detection procedure, it is assumed that more than one node (or cluster network) never fail at the same time. This assumption is very realistic in a cluster DBMS environment. Figure 4 shows the procedure of our intelligent failure detection.

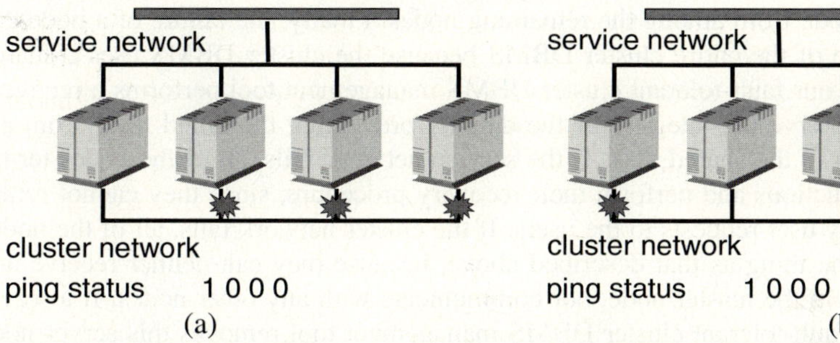

Fig. 3. Ambiguity between two failures

```
Failure_detection_procedure(){
        for(i=0; i<NODE_NUM; i++){
           if (service network of node[i] is down based on old_node_status)
              then increase sudden_death_servicenet
           if (cluster network of node[i] is down based on old_node_status)
              then increase sudden_death_clusternet
           Store status of network into old_node_status }
        if(sudden_death_servicenet >= 2) {
           Generate event ERROR_CLUSTER_NET
           Change mode to TERMINATE;
            Terminate processes in node
        } else if(sudden_death_clusternet >= 2) {
           Generate event ERROR_SERVICE_NET
           Change mode to TERMINATE;
            Terminate processes in node
        } else if(sudden_death_clusternet ==1 && sudden_death_servicenet==1)
             Generate event ERROR_SINGLE } //end of procedure
```

Fig. 4. Intelligent failure detection procedure

After we detect node failures (or network failures), we next perform a failure recovery procedure to bring the cluster DBMS back to a normal situation. First, the master node manages the cluster DBMS by distributing user requests to each server node. The master node sends a ping message to each server node in the cluster system and detects the failure of a node by analyzing the response of each node. If the master node fails to receive any response from the nodes, it regards the situation as cluster network failure. Meanwhile, when the backup node cannot communicate with the master node using a ping message, it regards the situation as the master node failure. The backup node regularly checks for the failure of the master node by sending it a ping message periodically and takes over the role of a new master node if it fails. Secondly, the backup node stores all of the information received from the CM and monitors the master node. If the master node fails, the backup node becomes a new master node and then selects one of the remaining nodes as a new backup node. If the backup node fails, our fault-tolerant cluster DBMS management tool terminates the backup node and terminates its active DB, Sysmon, NM, and backup CM. On the same time, the master node performs its recovery procedure which removes the failed backup node from the available node list and selects a new backup node from among the remaining nodes. Finally, the failure of a node can cause the failure of the entire cluster DBMS because the cluster DBMS uses a shared disk. Therefore, our fault-tolerant cluster DBMS management tool performs a recovery procedure to preserve the integrity of the data by preventing the failed node from accessing any data in the shared disk. If the service network fails, all of the nodes terminates their transactions and perform their recovery procedure, since they cannot return the results of any user requests to the users. If the cluster network fails, all of the nodes perform the same thing as that described above, because they can neither receive any user requests from the master node, nor communicate with any other nodes. If a server node fails, our fault-tolerant cluster DBMS management tool removes this server node from the available node list and informs the other nodes of the failure of this server node. On the same time, the failed node performs its recovery procedure to recovers its transactions and terminate its database. The other nodes recover the data of the failed node. Table 5 shows the failure recovery procedure of the master node due to its importance. The failure recovery procedures of the other nodes can be written similarly.

```
Recovery_procedure_Masternode(int event_num)
    switch(event_num) {
        case ERROR_MASTER :
            Remove VIP from the master node
            Clear the the available server list
            Terminate probe, NM, CM, DBMS
        case ERROR_BACKUP:
            Remove the backupnode from the available server list
            Terminate bcm, DBMS in backupnode
            Select a new backupnode
            break;
        case ERROR_SERVER:
            Remove backupnode from available server list
            Terminate bcm, DBMS in backupnode
        default :
            Print a message of not defining event }} //end of procedure
```

Fig. 5. Failure recovery procedure of the master node

4 Implementation and Performance Analysis

We implement our cluster DBMS management tool using iBASE/Cluster under a cluster system with four nodes, each being 450 Mhz CPU/HDD 30GB/128MB Memory. We make use of Linux virtual server 0.81 as the load balancer [4], the Redhat Linux 7.1 operating system and iBASE/Cluster DBMS [6]. In the normal situation, the service requests of users are transmitted to the master node. The Linux virtual server on the master node distributes a given user request to one of the available server nodes according to its round-robin algorithm. Figure 6 shows the user interface of a normal situation, wherein the four nodes comprising the cluster system are active.

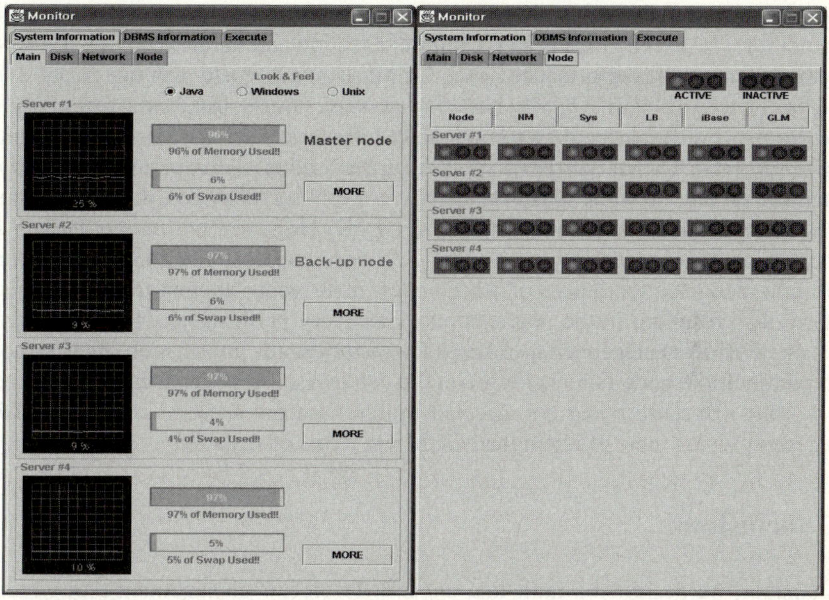

Fig. 6. User interface of a normal situation

The first database server node serves as the master node and the second server node serves as the backup node. It is shown from this figure that the NM, Sys (Sysmon) and iBase (iBASE/Cluster) processes are in the active state and that the Linux virtual server and GLM (global lock manager) of the cluster DBMS are running on the master node. When the master node fails, we disconnect the master node from the cluster network that is used to communicate with the other server nodes. Thus, the second node, which functions as the backup node, takes on the role of a new master node. If the backup node fails, we disconnect the backup node from the cluster network. The second server node, which functions as the backup node, is now in an inactive state, and one of the other server nodes is selected as a new backup node.

For the performance analysis of our fault-tolerant cluster DBMS management tool, we measure failure detection time and failure recovery time for various node failure types. For this, we set the maximum response time for the ping message to 2 seconds. First, the times required for detecting the failure of the master node, the backup node, and the database server node are about 0,9 sec, respectively. It means that the failure detection time is not affected by node failure types because the master or the backup node becomes aware of a node failure through the result of the ping message sent to it. Secondly, the times required for recovering the failure of the master node, the backup node, and the database server node are 0,8, 0.5, 0.7sec, respectively. The time for recovering the master node failure is the longest because there are a lot of work to do. The time for recovering the database sever node failure is relatively long because the master node removes the database server node from the available server list. Finally, if a new master node fails consecutively after one node (or two nodes) has failed, the times required for detecting and recovering the failure in our fault-tolerant cluster management tool are nearly the same as the above failure detection time and failure recovery time. We compare our fault-tolerant cluster DBMS management tool with OCMS, a cluster DBMS management tool for Oracle. The total time for failure detection and failure recovery in OCMS is more than 20 sec. The performance of the OCMS is much worse than our cluster DBMS management tool because OCMS has to wait until Oracle instance's reconfiguration. There are major differences between our fault-tolerant cluster DBMS management tool and OCMS. First, OCMS performs its task on each node and communicate each other through the private network. Assume that there are N server nodes, the total number of messages for communication in OCMS is (N-1)*N. However, our cluster DBMS management tool only needs 2N messages. Secondly, because OCMS uses a quorum partition to recognize the failure of other nodes, more nodes attempt to read the quorum partition, thus resulting in the degradation of recovery performance. However, because our cluster DBMS management tool keeps the history status information of every node, it can detect multiple node failures based on the comparison between a current status and a history status. Because being not affected by the number of nodes, our cluster DBMS management tool is more efficient than OCMS in terms of scalability.

5 Conclusions

In this paper, we designed a fault-tolerant DBMS cluster management tool which can monitor the status of the nodes in a cluster DBMS as well as the status of the various DBMS instances at a given node. Based on the monitoring, our cluster DBMS management tool can provide an intelligent fail-over mechanism for dealing with multiple node failures in a cluster DBMS, thus making its fault-tolerance possible. Finally, we implemented our cluster management tool using iBASE/Cluster DBMS. It was shown that our fault-tolerant cluster DBMS management tool could support nonstop database application service, e.g., map database provision, by using its intelligent fail-over mechanism even in case of multiple node failures.

References

1. Oracle Corporation, "Oracle 8i Administrator's Reference Release3(8.1.7) for Linux Intel", chapter 7, Oracle Cluster Management Software, 2000.
2. Gregory, F.Pfister, In Search of Clusters: 2nd Edition, Prentice-Hall, 1998
3. Linux Clustering, http://dpnm.postech.ac.kr/cluster/index.htm
4. Linux Virtual Server, http://www.linuxvirtualserver.org
5. Y-C. Kim and J-W. Chang, "HACM: High-Availability Cluster Management Tool for Cluster DBMS", Proceedings of the International Whokshop on Database and Expert Systems Applications, 2005.
6. Hong-Yeon Kim, Ki-Sung Jin, June Kim, and Myung-Joon Kim, "iBASE/Cluster: Extending the BADA-IV for a Cluster Environment", Proceeding of the 18th Korea Information Processing Society Fall Conference, Vol. 9, No. 2, pp. 1769-1772, Nov. 2002.

Speed Control of Averaged DC Motor Drive System by Using Neuro-PID Controller

Ali Bekir Yildiz and M. Zeki Bilgin

Kocaeli University, Department of Electrical Engineering,
Vinsan Kampusu, 41300, Kocaeli ,Turkey
{abyildiz, bilgin}@kou.edu.tr
http://www.kou.edu.tr

Abstract. The speed control of a separately excited DC motor driven by DC-DC converter is realized by using Neuro-PID controller. Firstly, a general and unified large-signal averaged circuit model for DC-DC converters is given. This method converts power electronic systems, which are periodic time-variant because of their switching operation, to unified and time independent systems. Therefore, it can be obtained conveniently and straightforwardly various analysis and control processes related to DC motor drive system. Some large-signal variations such as speed, voltage and current relating to DC motor, speed control are easily obtained by using the averaged circuit model. A self-tuning PID neuro-controller is developed for speed control on this model. The PID gains are tuned automatically by the neural network in an on-line way. The controller developed in this work, based on neural network (NN), offers inherent advantages over conventional PID controller for DC motor drive systems.

1 Introduction

The speed control of DC motor with power electronic systems is obtained generally by changing its terminal voltage. When terminal voltage of DC motor is increased between zero and nominal value, the speed can be also controlled from zero to nominal value. Changeable terminal voltage can be obtained by a controlled rectifier or DC-DC converter. Here, speed control of a separately excited DC motor driven by DC-DC converter, which is applied the average circuit technique on, is realized.

Power electronic converters are periodic time-variant systems because of their switching operation. In the analysis of these circuits, circuit equations are separately obtained for every switch-state. During each switch state, the circuit behaves as a linear circuit provided that other circuit elements (R,L,C) are linear. Over a switching period (T_s), the exact solution is obtained by solving sets of circuit equations respectively. But, the design of switching converters, the controller design relating to motor speed control are easily realized by using a circuit model which is valid for the entire switching period (T_s). Therefore, the average circuit approach which contains all switching topologies will be used in driving the DC motor.

B. Gabrys, R.J. Howlett, and L.C. Jain (Eds.): KES 2006, Part I, LNAI 4251, pp. 1075 – 1082, 2006.
© Springer-Verlag Berlin Heidelberg 2006

There are many approaches about the analysis of power electronic converters. Firstly, Middlebrook and Cuk [1] proposed the state-space averaging technique. This study is based on small-signal analysis of circuit variables which have small variations around the operating point. This analysis method has been a reference for many studies, and it has been used in many power converters such as [2]. But the small-signal models can't exactly express the dynamic of systems. A switching converter system may be stable around the operating point, but it may be unstable when the system undergoes to large perturbation or large parameter variations. The large-signal models are essential in order to study the dynamic characteristics of switching converters. Liu and Sen [3] propose a general unified large-signal model for current programmed DC-DC converters. In this study, the current programmed control where the control variable is peak current is used instead of the direct duty ratio (d) control. The paper [4] investigates DC motor control by using Computer-Based Fuzzy Technique. The paper [5] gives an unified large-signal model for DC-DC converters. This model is also essential for this paper. It'll be realized the large-signal analysis and control of DC motor by using this model.

The control performance of the DC servo drive is still influenced by uncertainties, which are usually composed of unpredictable system parameter variations, external load disturbance and nonlinear dynamics of the system. In the past decade, many modern control theories, such as nonlinear control, variable structure system control, adaptive control, optimal control and the robust control have been developed for the motor drive to deal with uncertainties. In the application of such techniques, development of mathematical models is a prior necessity. In recent years, intelligent control has been quite acceptable for real control applications. The back propagation algorithm has been found out as an efficient tool to realize the learning mechanism. The researches on the control of electrical machines based on NN are increased [6].

The purpose of this paper is to develop a self-tuning neuro-PID controlled drive system for DC motor. The analysis, design and simulation of the purposed controller are given. In the simulation, it's used averaged circuit model as a motor model. The following sections are organized as follows. Section 2 gives the electromechanical equations relating to DC motor. In Section 3, it's given the proposed unified large-signal circuit model. Section 4 investigates DC motor drive system driven by the averaged DC-DC converter circuit. Neuro-PID speed controller for averaged DC motor drive system and simulation results are given in Section 5.

2 DC Motor Driven by DC-DC Converter

It's very common application to control the speed of DC motor by changing its terminal voltage. The desired chopped voltage across the load can be obtained by using power semiconductors (Fig.1). The converter is a high-speed switching circuit which includes a diode (D) and a conventional transistor, power mosfet or thyristor (Q) etc.

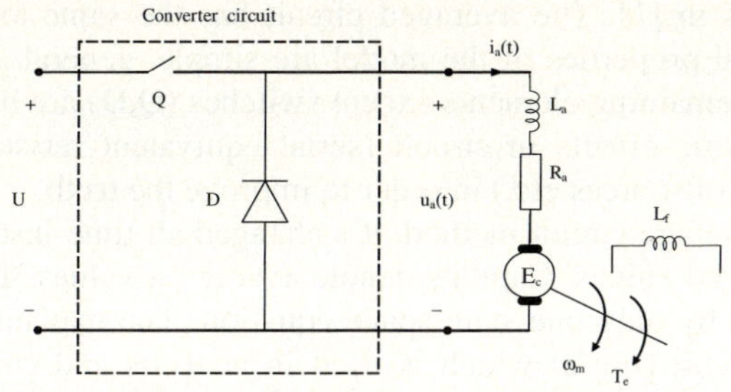

Fig. 1. DC motor drive system

The average of terminal voltage of DC motor is arranged by changing duty cycle (d) of active switch (Q). The continuous-time electromechanical equations relating to a separately excited DC motor circuit in Fig.1 are given as follows [7].

$$E_c + R_a i_a(t) + L_a \frac{di_a}{dt} = u_a(t) \tag{1.a}$$

$$B_1 \omega_m + J \frac{d\omega_m}{dt} = T_e - T_L = T_a \tag{1.b}$$

where, L_a: armature inductance, R_a : armature resistance, i_a: armature current, U: dc voltage source, u_a: terminal voltage of DC motor (chopped voltage), E_c: back emf, T_L: load torque, T_e: electromechanical (air gap) torque, T_a : acceleration torque, J: moment of inertia, B_1 : viscous friction coefficient, ω_m: speed of motor.

In (1.a) and (1.b), the back emf (E_c) is proportional to speed, the produced moment (T_e) is proportional to armature current (2.a and 2.b). At low speeds back emf is small, at high speeds back emf is bigger.

$$E_c = K_b \omega_m \tag{2.a}$$

$$T_e = K_b i_a \tag{2.b}$$

(1.a) and (1.b) constitute the dynamic model of dc motor with load. The circuit in Fig.1 is divided into two different topology relating to state of active switch (Q), with switching period T_s and duty-ratio d. Let's indicate the state with "1" when the active switch is on and the diode is off, $(0 < t < d.T_s)$. And again let's indicate the state with "2" when the active switch is off and the diode is on, $(d.T_s<t< T_s)$. Thus, the continuous-time exact topological analysis is obtained by getting $u_a(t)$=U for state "1" and $u_a(t)$=0 for state "2" in (1.a).

3 The Unified Large-Signal Model for DC-DC Converter

It's proposed an unified model for large-signal analysis of DC-DC converters in [5]. Basically, it's used the state-space averaging technique. But, the averaging process is directly achieved on the switching converter itself instead of taking the average of the circuit equations as in [1]. The averaged circuit has the same topology as the converter. The essential properties of the model are simple, general and unified. It's accepted that the all remaining elements except switches (Q,D) are linear. It's taken into consideration parasitic effects in circuit (serial equivalent resistances of inductors, internal resistances of sources etc.) in order to improve the truth.

Following the average circuit method, it's changed all time-instantaneous variables of circuit by averaged values. Capitals denote averaging values. The averaged circuit model is expressed by only one state-space equation. The explanations are given for Buck converter circuit (Fig.2) which is used in analysis and control of DC motor, here. But, the results can be easily generalized for all DC-DC converters. The waveforms of the current ($i_q(t)$) flowing through the active switch (Q) and the voltage across the diode ($u_d(t)$) are given in Fig.3.a and Fig. 3.b, which include ripples. In Fig.2, when the active switch is on, the inductor current flows through Q and $i_q(t)$ equals $i_L(t)$. The diode voltage, which is off, is found by circuit topology. When the active switch is off, the inductor current flows through diode. $i_q(t)$ and $u_d(t)$ equal 0.

Instead of taking the average of the state-space equations which deal with two different circuit topologies according to state of active switch, as in [1], the averaging process is directly achieved on the circuit itself. Therefore, the active switch (Q) is modelled by a controlled current source (J_Q) in Fig.3.a, the diode (D) is modelled by a controlled voltage source (U_d) in Fig.3.b. The value of controlled current source is obtained by averaging the current flowing through the active switch during a period. Similarly, the value of controlled voltage source is obtained by averaging the voltage across the diode during a period.

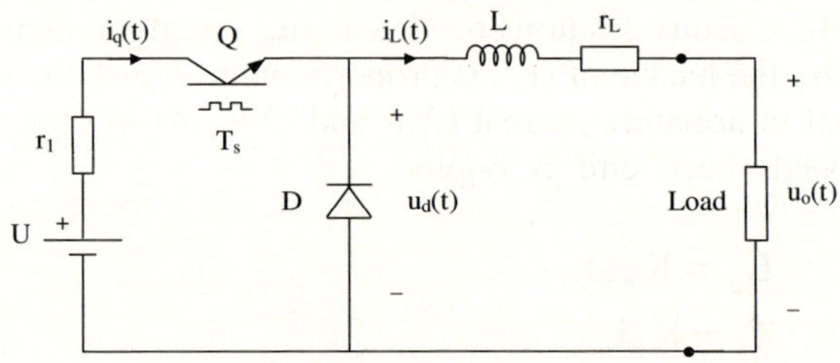

Fig. 2. Buck converter circuit

The average of the current flowing through the active switch is expressed according to Fig. 3.a during a switching period (T_s).

$$J_Q = \frac{1}{T_s}\left[\int_0^{t_{on}}\left(\frac{i_{L2}-i_{L1}}{t_{on}}t + i_{L1}\right)dt\right] = \frac{t_{on}}{T_s}\left(\frac{i_{L1}+i_{L2}}{2}\right) = d\left(\frac{i_{L1}+i_{L2}}{2}\right) \tag{3}$$

The average of the voltage across diode can be also obtained from Fig.3.b.

$$U_d = d.\left(\frac{u_{d1}+u_{d2}}{2}\right) \tag{4}$$

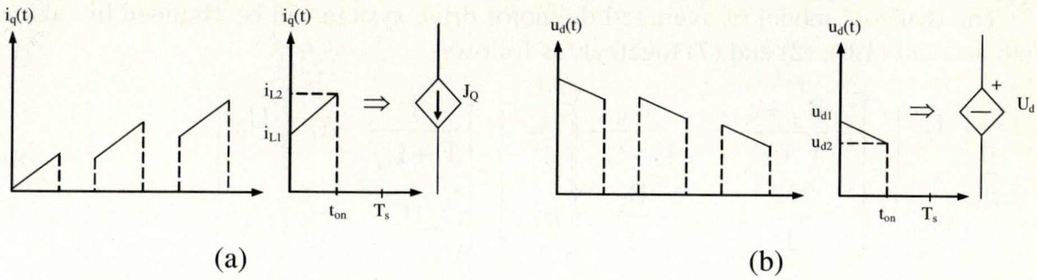

Fig. 3. (a) Current waveform of active switch, (b) Voltage waveform of diode

The values of u_{d1} and u_{d2} are expressed according to circuit topology in Fig.2.

$$u_{d1} = U - i_{L1} r_1, \quad u_{d2} = U - i_{L2} r_1 \tag{5}$$

(4) is rearranged by using (3) and (5).

$$U_d = dU - r_1 J_Q \tag{6}$$

As can be shown from (3) and (6), the averaged values change in every period and are calculated by variables of method during the simulation.

4 The Averaged DC Motor Drive System

After modelling the active switch by controlled current source and the diode by controlled voltage source, it's given an unified DC motor drive system model valid for whole period in Fig.4.

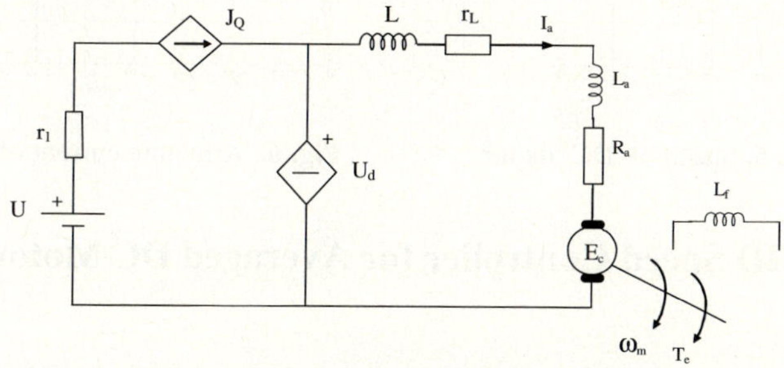

Fig. 4. Averaged DC motor drive system

(1.a) is rearranged according to the averaged circuit in Fig.4,

$$\frac{dI_a}{dt} = -\frac{R_a + r_L}{L + L_a} I_a + \frac{U_d - E_c}{L + L_a} \tag{7}$$

The dynamic model of averaged dc motor drive system can be obtained by taking into account (1.b), (2) and (7) together, as follows.

$$\frac{d}{dt}\begin{bmatrix} I_a \\ \omega_m \end{bmatrix} = \begin{bmatrix} -\dfrac{R_a + r_L}{L+L_a} & -\dfrac{K_b}{L+L_a} \\ \dfrac{K_b}{J} & -\dfrac{B_1}{J} \end{bmatrix} \begin{bmatrix} I_a \\ \omega_m \end{bmatrix} + \begin{bmatrix} \dfrac{1}{L+L_a} & 0 \\ 0 & -\dfrac{1}{J} \end{bmatrix} \begin{bmatrix} U_d \\ T_L \end{bmatrix} \tag{8}$$

The variables in (8) denote the averaged values. This modeling of switches is a good approach to model drive system with an unified equation in the time-independent form. There isn't anymore period (T_s) in the averaged circuit model. The desired analysis and designs are easily realized by the averaged circuit and unified state equations. Moreover, the averaged circuit model given in Fig.4 can be also analyzed with a computer simulation program (PSPICE etc.).

It's given together both the exact topological solutions and the results of averaged circuit for d=0.5 and T_s=0.035s in Fig.5 and Fig.6. (L_a=3mH, R_a=0.5Ω, K_b=0.8 V/rd/s, J=0.0167 kg-m^2, B_1=0.01 Nm/rd/s, T_L=100 Nm, L=20mH, r_L= 0.2Ω, r_l= 1Ω, U=220V). It's seen that the averaging solutions pass from the middle of the exact solutions in Fig.5 and Fig. 6. Especially in the armature current (Fig.6), this situation is very clear. Because the mechanical time constant is much bigger than the electrical time constant, the speed includes very small ripples in Fig.5, and therefore, averaging and exact solutions are almost the same for it.

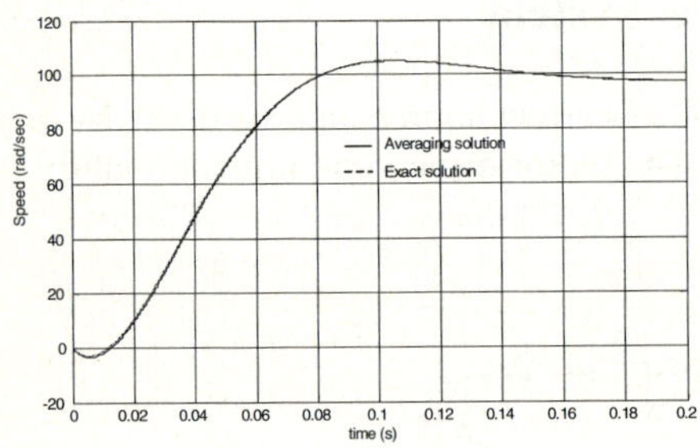

Fig. 5. Speed of DC motor

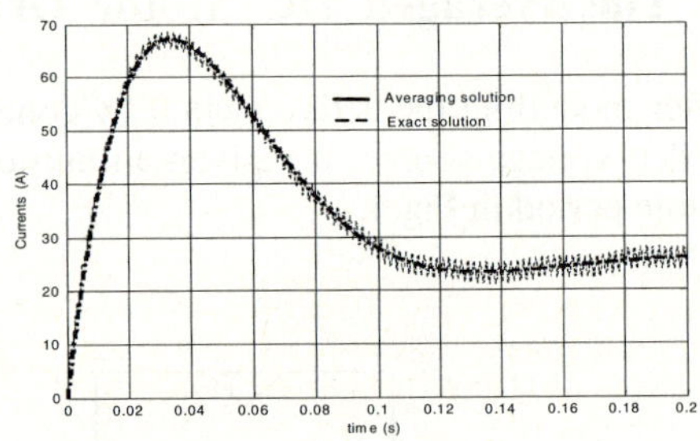

Fig. 6. Armature current of DC motor

5 Neuro-PID Speed Controller for Averaged DC Motor Drive System

For a desired speed and current value (indirectly moment), the controlling of DC motor is realized by using averaged circuit model and changing duty cycle. The closed-loop speed controlled separately-excited DC motor drive system is shown in Fig.7. This block diagram consists of three main parts. These are state-space averaged model containing DC motor and converter, current controller (P), self-tuning neuro PID speed controller (NPID).

Many adaptive control methods have a number of parameters or user defined polynomials. By integrating a neural network into the control scheme, it can be

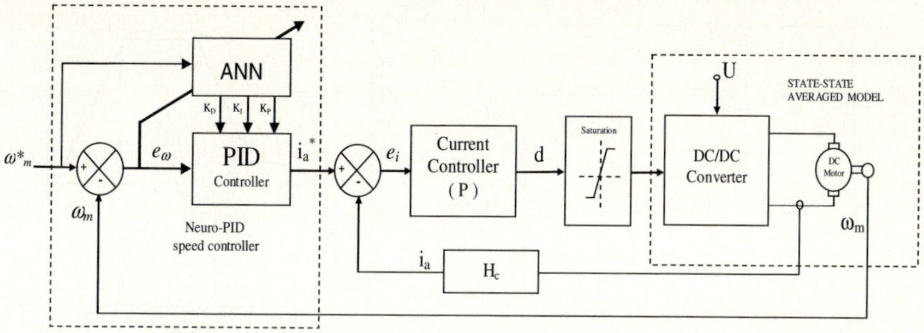

Fig. 7. Speed-control of dc motor drive system with Neuro-PID speed controller

tuned the these parameters in on-line way. In this paper, the neural network is used to tune the parameters of the PID controller. The error function is minimized by adjusting the PID gain, such as K_I, K_P, K_D.

The desired speed value of DC motor is obtained by arranging its averaged terminal voltage. This process is realized by changing duty cycle in the block diagram of Fig.7, as follows: In the block diagram, ω_m^* : reference (desired) speed, ω_m : measuring (actual) speed. The desired value of current (i_a^*) produced by Neuro-PID controller is compared with the actual armature current (i_a). The obtained error is applied to current controller . The current controller's output is duty cycle (d). The current controller generates duty cycle of the converter. d changes between 0 and 1. Therefore, the limiter block is used at the output of current controller. In converter block, the values of J_Q and U_d are calculated by using (3) and (6) respectively. The converter's averaged output voltage is applied to DC motor.

In the simulation of Fig.8, different reference speeds are used by changing its value in 0.1s,0.35s, 0.5s,0.7s and 0.9s,respectively: 100rad/s in 0.1s, a ramp function between 0.35s and 0.5s, 200rad/s in 0.5s, again a ramp function between 0.7s and 0.9s,75rad/s in 0.9s. The reference and actual measuring speeds are shown in Fig.8.

After changing every reference speed, the required duty cycle value is determined by the controllers in Fig. 7. The required duty cycle values (d) relating to different speeds in Fig.8 are given in Fig.9. The motor current and the motor terminal voltage for these different reference speeds and duty cycle values (d) are given in Fig.10, 11.

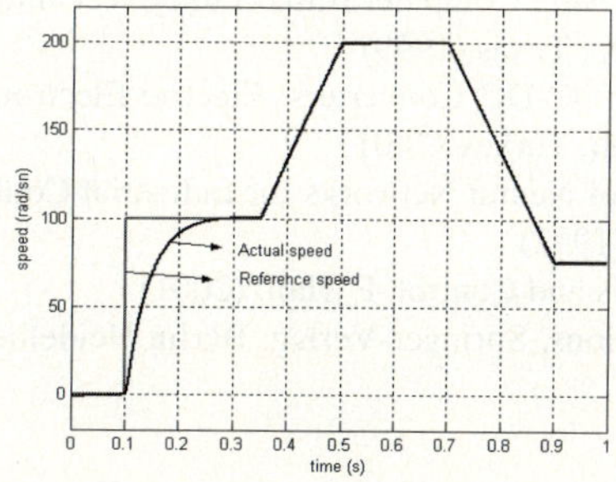

Fig. 8. Reference and actual speeds (ω_m^*, ω_m)

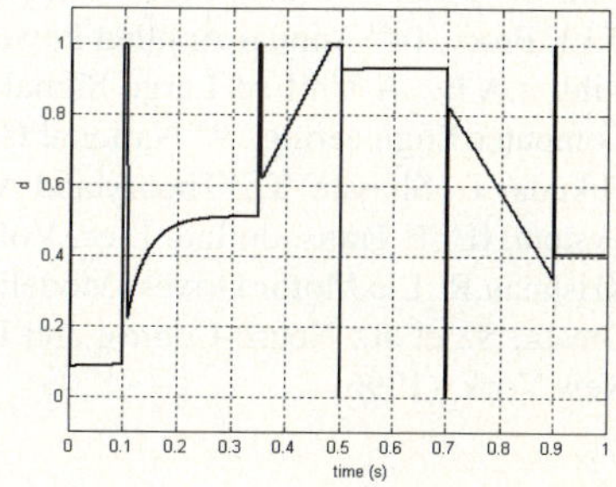

Fig. 9. Required duty cycle values

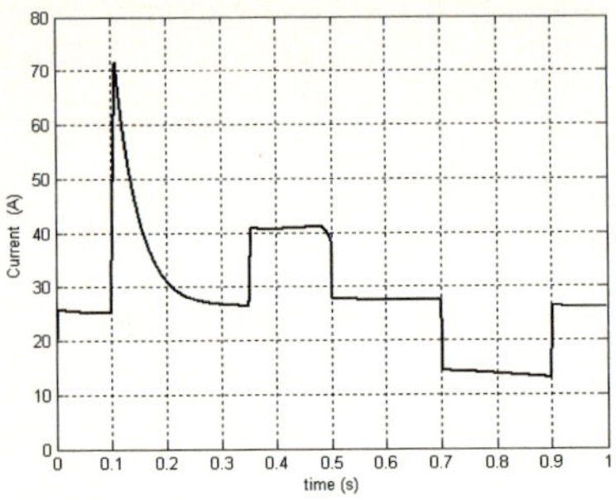

Fig. 10. DC motor current

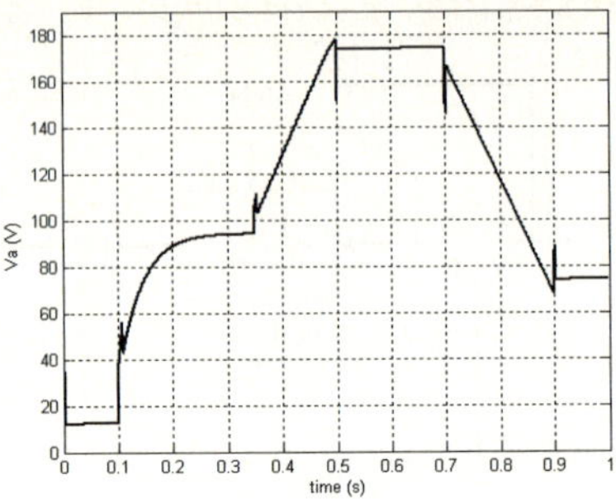

Fig. 11. DC motor terminal voltage

6 Conclusion

In this study, it's shown that it can be easily realized the speed control of DC motor by using the averaged circuit model and neuro PID controller. It's proposed a technique to obtain a general and unified large-signal model for DC-DC converters.

The large-signal averaged DC motor drive system presented in this paper gives a global view of the system dynamics. Since this method can provide a unified time-invariant state-space of DC motor drive system, it enables the designer to achieve easily the analysis and control of DC motor.

The mathematical basis of the proposed controller is obtained by MATLAB/ Simulink software package. The simulation results show that this self-tuning PID type neuro-control strategy achieves effectively the desired dynamic performance of DC motor.

References

1. Middlebrook, R.D.and Cuk, S.: A General Unified Approach to odelling Switching-Converter Power Stages. IEEE Power Elo. Spec. Conf., USA (1976)
2. Mahdavi, J., et al.: Analysis of Power Elo. Converters Using the Generalized State Space Averaging Approach. IEEE Trans. on Cir..and Sys., Vol.44, (1997)
3. Liu,Y. and Sen, P.: A General Unified Large Signal Model for Current Programmed DC-to DC Converters. IEEE Trans. on Power Elo., Vol.9,July (1994)
4. Cakir,B.,Yildiz,A.B et al.: DC Motor Control by Using Computer Based Fuzzy Technique. IEEE Proc., 14th Annual Applied Power Elo. Conf. , Texas (1999)
5. Yildiz, A.B.: A Unified Large-Signal Model for DC-DC Converters. Electric-Electronic-Computer Engineering., 9th National Conf., Kocaeli, Turkey (2001)
6. Fukuda, T., Shibata, T.: Theory and Application of Neural Networks for Industrial Control System, IEEE Trans. on Ind. Elec. Vol.39, No.6, (1992)
7. Krishnan,R.:Elc.Motor Drives, Modeling, Analysis and Control. P..Hall, (2001)
8. Omatu, S., et al.: Neuro-Control and Its Applications, Springer-Verlag, Berlin Heidelberg, New York (1996)

Automatic Detection of SLS Violation Using Knowledge Based Systems

Pedro Alípio, José Neves, and Paulo Carvalho

Universidade do Minho, Departamento de Informática, 4710-057 Braga, Portugal
{pma, jneves, pmc}@di.uminho.pt

Abstract. Self-Management of multiservice network domains must include automatic detection of service quality violations in order to perform appropriate and timely responses, fostering the fulfillment of the service agreements committed with customers, or other ISPs. As knowledge based systems are capable of emulating human knowledge, they are presented here as convenient tools to specify heuristics for QoS violation detection. In this paper, the implementation of a service violation detection agent is described and an illustrative scenario of its use is also presented.

Keywords: Service Level Specification, Knowledge based Systems, Quality of Service, Self-configuration Networks.

1 Introduction

Monitoring Service Level Agreements (SLAs) between Internet Service Providers (ISPs) and customers is becoming a very relevant topic as the number of companies using the Internet to perform their business activities increases. Many of those services are tolerant to some performance degradation, while others are not tolerant at all. As a result, the disruption of the contracted Quality of Service (QoS) may have serious consequences on the customers activities. Under such circumstances, ISPs may be considered responsible and prosecuted. Therefore, in their own interest, ISPs must be sure that the QoS guarantees are being assured as a key component of SLA auditing procedures.

In the context of an SLA, the technical requirements of a service are specified as a Service Level Specification (SLS) which includes, among other information, the expected QoS. The most frequent QoS metrics used to express the service performance are throughput, one-way delay, interpacket delay variation and packet loss ratio. Monitoring SLSs is the process of collecting statistical metrics about service's performance in order to verify if it meets the specified requirements. Our main concern is edge-to-edge SLS monitoring. Therefore, the service traffic has to be monitored at the entry and exit nodes of ISP domains, and then the resulting metrics verified against the information within SLSs. This process may be enhanced when performed autonomously using knowledge based technologies. Such technologies consist of knowledge specification through sets of production rules (**if** antecedents **then** consequents) and an Inference mechanism to perform the deductive process over factual information to reach out for conclusions. These conclusions may be one of the following: to assess the utilization of network resources, to carry out an automatic network reconfiguration process or to re-negotiate SLAs.

B. Gabrys, R.J. Howlett, and L.C. Jain (Eds.): KES 2006, Part I, LNAI 4251, pp. 1083–1089, 2006.
© Springer-Verlag Berlin Heidelberg 2006

This paper has the following structure: related work is presented in Section 2; the proposed SLS violation detection strategy is presented in Section 3; an example of an SLS violation scenario is given in Section 4 and the conclusions and future work are presented in Section 5.

2 Related Work

Detecting SLS violations is crucial for self-management networks pledged in assuring edge-to-edge QoS. The research community has been committed to SLS definition, monitoring and management [1,2,3,4]. Work is currently in progress to establish service classes and their respective configuration in DiffServ networks [5], and on mapping SLS into PDBs (Per Domain Behaviors) [6]. These contributions, as well as the ones resulting from this work aim to assure a consistent forwarding path treatment for the services traffic in a network domain [7].

Most of these approaches use the Common Open Policy Service (COPS) [8] to map SLSs into network configurations. However, the COPS protocol does not include a knowledge base or any form of reasoning mechanisms. On these systems, SLS violations are detected by hard-coded conditions, and they are not flexible enough to support changes in the detection criteria or in the inclusion of new metrics. By using a general purpose rule based system, eg., the C Language Integrated Production System (CLIPS) [9], a wider range of functionalities may be included improving the flexibility and applicability of the solution.

Although, in previous work, XML was used to specify service requirements [3], a XML rule description language such as RuleML [10] was not considered due to the following: (i) XML is very verbose and rules with a large number of conditional elements and consequents tend to be very difficult to read; (ii) CLIPS is easily extendable to support XML parsing and validation, database access, or to include a network configuration protocol to trigger short term actions; (iii) CLIPS is written in C and implements a forward chaining inference engine which is more efficient than XML rule based systems implementations such as Mandarax [11] which is written in Java and uses backward chaining.

3 Using a Knowledge Based System to Detect SLS Violations

To be possible to compare the values measured by the monitoring process with the QoS requirements, SLSs have to be specified in a structured way. An XML based specification was used for that purpose through XML Schema, containng the structure, elements and data types included in SLSs [3]. As XML is an universal language to be used in structured documents, XML SLSs may be used as a common ground for interoperability and heterogeneity within or across ISP domains. Although, many software applications tend to use XML, most of the knowledge based systems shells do not support it yet. Our solution involves transformation of XML SLSs into the public domain CLIPS through XSLT (eXtensible Stylesheet Language Transformations). In addition, knowledge based systems with XML support tend to be slower and the rules are often more verbose and more difficult to read.

In a network domain, services are often mapped into Classes of Service (CoS). In such case, traffic is commonly measured edge-to-edge for each service class. Otherwise, it is measured on a flow basis. In either case, no specific metric collection methods are assumed. Active or passive monitoring may therefore be used by the Measurement Agents (MAs).

Figure 1 illustrates the deployment of MA and SVDA agents in the boundaries of the ISP domain network. As showed, there may be a MA monitoring each node or each MA may collect information form several nodes or interfaces as MA1. MAs inform the SVDA periodically or when a anomalous situation occurs. Whenever a violation is detected, the SVDA informs the self-management network decision agent, i.e., it will decide whether to reconfigure the network domain (short-term actions) or to renegotiate the SLA (long-term actions).

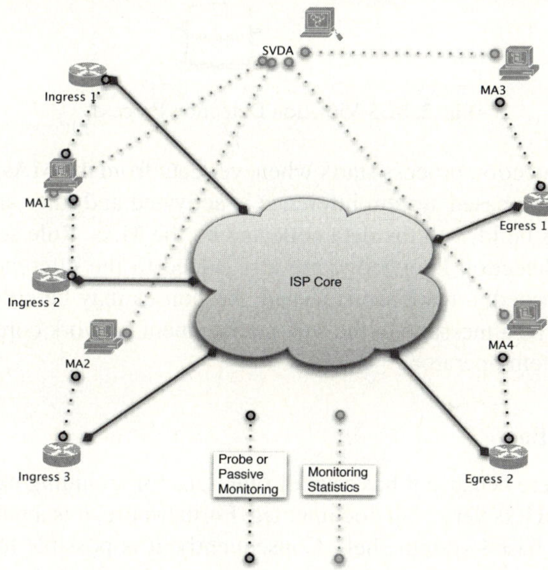

Fig. 1. SLS QoS Monitoring and Violation Detection Deployment

3.1 The Monitoring System Architecture

The architecture of the monitoring system includes the following processing steps: (i) Validation of SLSs; (ii) management of SLSs; (iii) translation of SLSs into the Expert System Language; (iv) getting information from MAs and (v) the inference process. Figure 2 illustrates the whole process of service violation detection starting from the SLSs submission process.

After being submitted to the system, SLSs are validated. This step is accomplished by verifying if the SLS XML structure, elements and data types are written according to the XML Schema. Valid SLSs are passed to the SLS Translator which translates the XML SLS into the knowledge based system language. A record in a database is created by the SLS Management Module saving information about the active SLSs and their respective translation, to be added to the knowledge based system fact list.

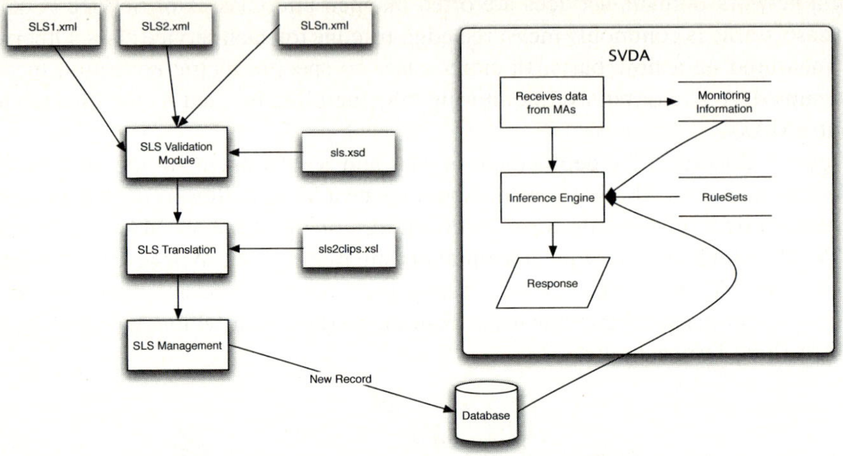

Fig. 2. SLS Violation Detection Process

The violation detection process starts whenever data from the MAs is received. As a result, the knowledge based system inference is activated and rules are fired over SLS facts and over facts build with the data collected by the MAs. Rule sets containing the service violation detection knowledge are also added to the inference engine. When a conclusion is reached, a response is issued. Responses may be either active or passive, i.e, they may be a message to the self-management network core or just an alarm message to the system operator.

3.2 Implementation

CLIPS was used here because it has a set of functions for including new functionalities and extensions, and it is very well documented. Furthermore, it is a public domain open source knowledge based system shell. Consequently, it is possible to modify its code and recompile it.

The SVDA implementation includes several other software modules. The XML validation module was build using *libxml* [12]. This library allows XML validation resorting to a XML Schema document. The transformation into CLIPS facts is accomplished by using *libxst* [13], which allows loading and applying a XSLT stylesheet to the XML document (doc). *SQLite* [14] was used to store SLSs. *SQLite* is a small C library that implements a self-contained, embeddable, without configuration needs. CLIPS system was extended with functions to read, write and modify records in *SQLite* databases.

As the distributed MAs are not yet implemented, in the current version of the SVDA, data is read from a text file resulting from simulated scenarios using the Network Simulator 2 (NS2) [15]. These text files contain the monitored QoS metrics (one way delay, inter packet delay variation and packet drop ratio) for every combination of ingress nodes, egress nodes and CoS. In addition, per flow monitoring may also be used. Under such circumstances, metrics are collected for each combination of source, destination, and flow id. Each line is converted into CLIPS facts as they are read, then a ruleset is applied to check wether the SLSs are violated or not.

4 Example of an SLS Violation Scenario

As an example, a simple scenario is presented. Consider a network domain with three nodes: two edge nodes (E1 and E2) and a core node (Core). In this example, E1 is the ingress node and E2 the egress node. All links have a maximum bandwidth capacity of 10Mbps and a propagation delay of 5ms. The domain accepts two SLAs and their corresponding SLSs, i.e, SLS1 and SLS2 specifying respectively, an Assured Forwarding (AF) and a Best Effort (BE) based service. At the ingress node, service SLS1 traffic is marked with DSCP 20 and SLS2 traffic is marked with DSCP 30 (Figure 3).

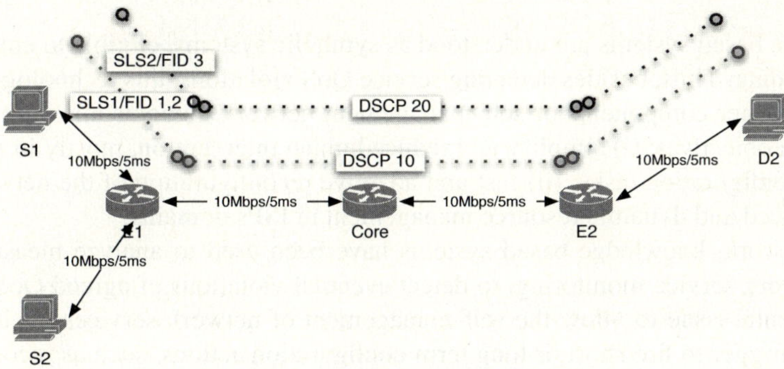

Fig. 3. Test Scenario for the SVDA

Table 1 presents the initial parameters used to test the SVDA. In this case, no congestion is induced on the network. As a result, no alarms are generated by the SVDA, as expected. In order to force an SLS violation, the initial traffic generator parameters are changed in order to force congestion. Source S2 throughput is modified to generate traffic at 10Mbps. Total traffic entering the network is 12Mbps, exceeding the link capacity of 10Mbps.

Table 1. Initial test parameters

Source	Destination	FID	SLS	Expected Delay	Expected Loss	Traffic	Period
S1	D1	1	SLS1	0.01s	0%	CBR 1Mbps	10s
S1	D1	2	SLS1	0.01s	0%	CBR 1Mbps	10s
S2	D1	3	SLS2	1s	10%	CBR 8Mbps	10s

In this case, the SVDA issued the following alerts:

```
SLS2 VIOLATION at 10s: loss(19%) > SLS loss(10%)
SLS2 VIOLATION at 20s: loss(20%) > SLS loss(10%)
SLS2 VIOLATION at 20s: loss(20%) > SLS loss(10%)
SLS2 VIOLATION at 20s: loss(20%) > SLS loss(10%)
```

As packet loss ratio is bigger than 10%, and according to the requirements of the expected QoS metrics in the SLS2, it is considered a violation. If the throughput of SLS1 is also increased to 10Mbps (5Mbps in each flow) than the result will be the following:

```
SLS2 VIOLATION at 10s: loss(51%) > SLS loss(10%)
SLS1 VIOLATION at 10s: loss(0.48%) > SLS loss(0%)
SLS1 VIOLATION at 10s: delay(0.0778) > SLS delay(0.01)
SLS1 VIOLATION at 20s: loss(48%) > SLS loss(0%)
SLS1 VIOLATION at 20s: delay(0.0778) > SLS delay(0.01)
```

SLS1 is more restrictive, it does not tolerate any loss and the maximum one way delay is 0.01s. As SLS2 is more tolerant to loss and delay (1 second), only the expected loss ratio violation is detected.

5 Conclusion

Knowledge based systems are understood as symbolic systems, eligible to emulate human reasoning. Thus, besides detecting service QoS violations, this technology may be applied to other components of self-management network domains allowing to fulfill three main objectives: (i) simplify and reduce human intervention, mostly in repetitive network configuration tasks; (ii) fast and adaptive reconfiguration of the network; and (iii) enhanced and dynamic resource management in ISPs domains.

In this work, knowledge based systems have been used to analyze measurements from network service monitoring, to detect eventual violations of agreed QoS. This is a fundamental issue to allow the self-management of network services, as it may be used as a trigger to fire short or long term configuration actions, such as reconfiguring services at a network level (SLSs) or the renegotiating service requirements with the customer (SLAs).

As future work, several fundamental components required for self-management have to be studied and developed in order to allow building of fully autonomous ISP network domains. An ontology based approach is being considered to specify rules and data in the same model.

Acknowledgments

A PHD grant provided by Fundação para a Ciência e Tecnologia (SFRH/BD/17579/2004) is gratefully acknowledged.

References

1. Morand, P., Boucadair, M., P. Levis, R.E., Asgari, H., Griffin, D., Griem, J., Spencer, J., Trimintzios, P., Howarth, M., Wang, N., Flegkas, P., Ho, K., Georgoulas, S., Pavlou, G., Georgatsos, P., Damilatis, T.: Mescal D1.3 - Final Specification of Protocols and Algorithms for Inter-domain SLS Management and Traffic Engineering for QoS-based IP Service Delivery and their Test Requirements . Mescal Project IST-2001-37961 (2005)
2. Diaconescu, A., Antonio, S., Esposito, M., Romano, S., Potts, M.: Cadenus D2.3 - Resource Management in SLA Networks. Cadenus Project IST-1999-11017 (2003)
3. Alipio, P., Lima, S., Carvalho, P.: XML Service Level Specification and Validation. In: The 10th IEEE Symposium in Computers and Communications (ISCC 2005), IEEE Computer Society (2005) 975–980

4. Goderis, D., T'joens, Y., Jacquenet, C., Memenious, G., Pavlou, G., Egan, R., Griffin, D., Georgatsos, P., Georgiadis, L., Heuven, P.V.: Service Level Specification Semantics, Parameters, and Negotiation Requirements. Internet-Draft, drafttequila-sls-03.txt (work in progress) (2003)
5. Babiarz, J., Chan, K., Baker, F.: Configuration Guidelines for DiffServ Service Classes. Internct Draft (work in progress) (2004)
6. Prieto, A.G., Brunner, M.: SLS to DiffServ configuration mappings. In: 12th International Workshop on Distributed Systems: Operations & Management. (2001)
7. Nichols, K., Carpenter, B.: Definition of Diff. Services Per Domain Behaviors and Rules for their Specification. IETF RFC 3086 (2001)
8. Boyle, J., Cohen, R., Durham, D., Herzog, S., Rajan, R., Sastry, A.: The COPS (Common Open Policy Service) Protocol. IETF RFC2748 (2000)
9. Giarratano, J.C.: CLIPS User's Guide, Volume I - Basic Programming Guide. (2005)
10. : The Rule Markup Initiative. URL: http://www.ruleml.org (2006)
11. : The Mandarax Project. URL: http://mandarax.sourceforge.net (2006)
12. : libxml - The XML C parser and toolkit of Gnome. URL: http://xmlsoft.org (2005)
13. : libxslt - The XML C parser and toolkit of Gnome. URL: http://xmlsoft.org/XSLT (2005)
14. : The SQLite Website. URL: http://www.sqlite.org (2005)
15. : Network Simulator NSv2. URL: http://www.isi.edu/nsnam (2004)

Two-Level Dynamic Programming Hardware Implementation for Real Time Processing

Yong Kim and Hong Jeong

Department of Electronic and Electrical Engineering
POSTECH, Pohang, Kyungbuk, 790-784, South Korea.
ddda@postech.ac.kr, hjeong@postech.ac.kr
http://isp.postech.ac.kr

Abstract. In this paper, we present an efficient architecture for connected speech recognition that can be efficiently implemented with FPGA. The architecture consists of newly derived two-level dynamic programming(TLDP) that use only bit addition and shift operations. The advantages of this architecture are the spatial efficiency to accommodate more words with limited space and the computational speed from avoiding propagation delays in multiplications. The architecture is highly regular, consisting of identical and simple processing elements with only nearest-neighbor communication, and external communication occurs with the end processing elements. In order to verify the proposed architecture, we have also designed and implemented it, prototyping with Xilinx FPGAs running at 33MHz.

1 Introduction

Speech recognition is a process for a computer to map acoustic speech signals to text. That is, speech recognition converts acoustic speech signals provided by a microphone or a telephone into words, a group of words, or sentences. Recognition results may be used as final results by application fields such as instructions, controls, data inputs, and documentation, and may also be used as inputs of language processing to the field of speech understanding. Further, speech recognition is an attractive technique allowing interactive communication between people and computers, making computer usage environments more convenient for people, and thereby enhancing their lifestyle.

For most speech recognition application, it is sufficient to produce results in real time, and software solutions that perform recognition in real time already exist. However, to increase the use of speech recognition in embedded systems, we need a speech recognition chip with low power consumption, small size, and low cost. In previous works, dedicated hardware architectures for hidden Markov model(HMM)-based speech recognition were introduced in [1,2,3,4] The existing architectures are designed for isolated speech recognition. Unfortunately, there is no direct implementation on hardware for connected speech recognition. In this paper, we derived a new architecture based upon bit additions and shift operations only, excluding any integer multiplications for connected word recognition,

B. Gabrys, R.J. Howlett, and L.C. Jain (Eds.): KES 2006, Part I, LNAI 4251, pp. 1090–1097, 2006.
© Springer-Verlag Berlin Heidelberg 2006

using the TLDP [6,7] algorithm. In connected speech recognition, most of the problems arise from the difficulty in reliably determining the word boundaries. TLDP is a well-known speech recognition algorithm that assigns word strings to speech segments.

We introduce an efficient linear systolic array architecture that is appropriate for FPGA implementation. The array is highly regular, consisting of identical and simple processing elements. The design is very scalable, and since these arrays can be concatenated, it is also easily extensible. A scalable technique always provides optimum hardware resources that can cope with variable conditions by making small modifications to the hardware architecture.

2 Background of TLDP Algorithm

When boundaries are unclear (connected speech case), the TLDP algorithm can be used to find them quite well. A very brief outline is given below. For a more detailed description of TLDP theory, see [5]. The notation here is based on [5].

The basic idea of the TLDP is to break up the computation of connected speech recognition into two stages. At the first level the algorithm matches each individual word reference pattern, R_v, against an arbitrary portion of the test string, T. For the range of beginning test frames of the match, s, $1 \leq s \leq M$, and for the range of ending test frames, e, $1 \leq e \leq M$ $(e > s)$, the minimum distance $\hat{D}(v, s, e)$, for every possible vocabulary pattern, R_v, between each possible pair of beginning and ending frames (s, e) is defined as

$$\hat{D}(v, s, e) = \min_{w(m)} \sum_{m=s}^{e} d(\boldsymbol{t}(m) - \boldsymbol{r}_v(w(m))), \tag{1}$$

where $\boldsymbol{t}$ is a test pattern, $\boldsymbol{r}_v(1 \leq v \leq V)$ is a pattern of the vth word from among V reference patterns to be recognized, and $w(m)$ is a window for dividing the total frame input for signal analysis during a very short time that is assumed to be stable. We can eliminate v by finding the best match between s and e for any v, giving

$$\begin{aligned} \tilde{D}(s, e) &= \min_{1 \leq v \leq V} [\hat{D}(v, s, e)] = \text{best score}, \\ \tilde{N}(s, e) &= arg \min_{1 \leq v \leq V} [\hat{D}(v, s, e)] = \text{best reference index}, \end{aligned} \tag{2}$$

thereby significantly reducing the data storage within no loss of optimality.

Given the array of best scores, $\tilde{D}(s, e)$, the second level of the computation pieces together the individual reference pattern scores to minimize the overall accumulated distance over the entire test string. This can be accomplished using dynamic programming as

$$\bar{D}_l(e) = \min_{1 \leq b < e} [\tilde{D}(s, e) + \bar{D}_{l-1}(s - 1)], \tag{3}$$

where $\bar{D}(e)$ is the distance of the best path ending at frame e using a concatenated sequence of l reference patterns. The best path ending at frame e using exactly l reference patterns is the one with minimum distance over all possible beginning frames, s, of the concatenation of the best path ending at frame $s-1$ using exactly $l-1$ reference patterns plus the distance of (2) of the best path from frame s to frame e.

3 The Architecture of Two-Level Dynamic Programming

Fig. 1 is a basic block diagram of a TLDP. As shown in Fig. 1, the system includes a first processing element group, a comparison module, a second processing element group, and a backtracking module. The first processing element group includes a plurality of parallel processing elements that have the same configuration, and calculate matching costs by using the HMM algorithm. The comparison module determines the minimum matching cost from first processing element group, and stores it for later calculation. The second processing element group finds the optimized matching cost with the reference pattern for the total frame by using the minimum value determined by the comparison module, detects the word's end point, and recognizes a connected word. The second processing element group also includes a plurality of parallel processing elements having the same configuration. The backtracking module finds a word arrangement of the reference pattern that corresponds to the speech recognition result based on the calculation result by the second processing element group. In this instance, the first processing element group and the comparison module form the first level dynamic programming(first level DP), the second processing element group and the backtracking module form the second level dynamic programming(second level DP).

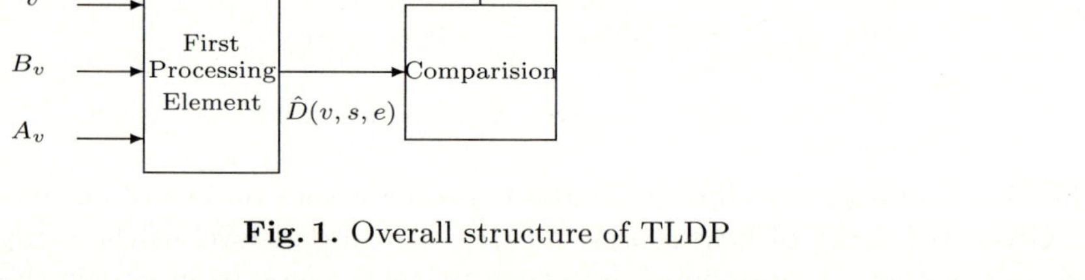

Fig. 1. Overall structure of TLDP

3.1 Architecture of First Level DP

The first group processing elements use the HMM algorithm and the dynamic programming scheme to calculate matching costs. For example, the matching cost $PE_{lev.1}(v,p,m)$ at the pth processing element is:

$$PE_{lev.1}(v, p = s, m = e) = \hat{D}(v, s, e) \qquad 1 \le p,\ m \le M, \tag{4}$$

where M is the dimension of the total frame. (4) shows the matching cost between the test pattern and the reference pattern during the interval of (s,e). Hardware architectures for HMM-based speech recognition were introduced in [1,2,3,4], so it is not presented in this paper.

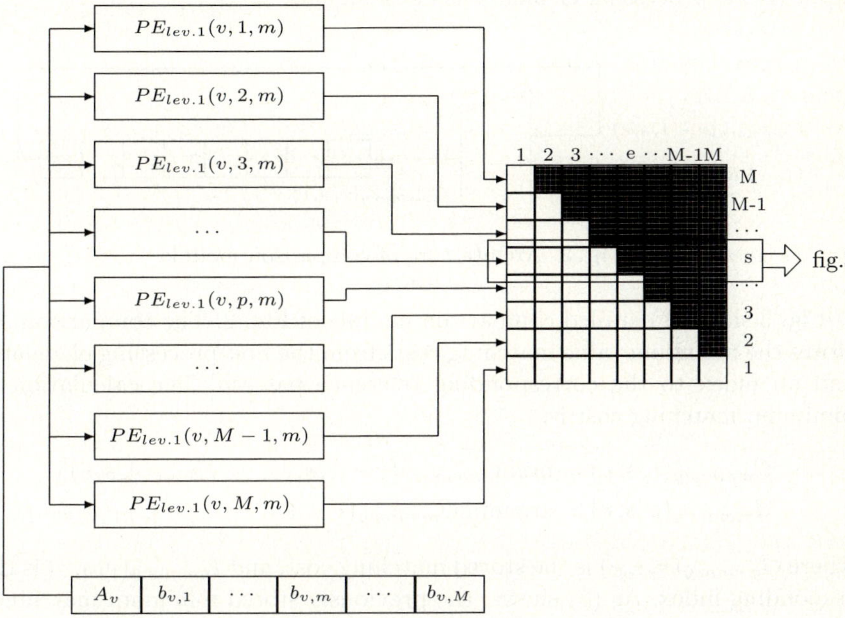

Fig. 2. Architecture of first level dynamic programming

As can be known from (4) and Fig. 2, the pth processing element sequentially calculates matching costs from p to M when the start point is given to be p. The number of functioning processing elements is M; hence, the matching costs from all the start points to all the end points can be calculated. Therefore, realization of the above process through software requires the matching time of M^2 clock signals, but realization through the parallel hardwired configuration of the present architecture generates the same calculation results by using M clock signals, corresponding to the dimension of the total frame.

Fig. 2 shows the systolic array architecture of the first level DP. The first level DP includes the first processing element group and the comparison module. The first processing element group includes a state input unit and a plurality of parallel processing elements that have the same configuration for calculating the matching cost. The first level DP calculates matching costs of a test pattern in comparison with reference patterns at a start point and an end point by using the HMM algorithm and the dynamic programming scheme, determines the minimum matching cost, and extracts an index of the reference pattern corresponding to it.

When the state input unit receives a feature vector of a speech signal from the feature vector generator, the HMM parameters state transition probability distribution, A_v, and observation symbol probability distribution, $b_{v,m}$ are calculated according to the learned probabilistic value, and are provided to the state input unit. To calculate matching costs, the HMM parameters are sequentially input to the processing elements as clock signals.

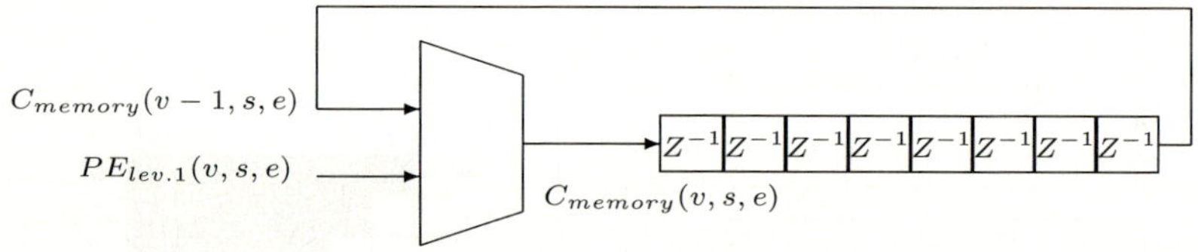

Fig. 3. Architecture of comparison module

Fig. 3 shows a detailed comparison module of Fig. 2. The comparison module stores the minimum value matching costs from the first processing element group and an index to the corresponding reference pattern. The calculation of the minimum matching cost is:

$$\begin{aligned} C_{memory}(v,s,e) &= \min[C_{memory}(v-1,s,e),\quad PE_{lev.1}(v,s,e)], \\ I_{memory}(v,s,e) &= \arg\min[C_{memory}(v-1,s,e),\quad PE_{lev.1}(v,s,e)], \end{aligned} \tag{5}$$

where $C_{memory}(v,s,e)$ is the stored matching cost, and $I_{memory}(v,s,e)$ is the corresponding index. As (5) shows, the previously stored minimum matching cost, $C_{memory}(v-1,s,e)$ is compared to the current input matching cost, $PE_{lev.1}$ (v,s,e), and the lesser one is stored in the memory. That is, the minimum one from among the matching costs that are input up to a specific time is stored in the memory. In this instance, since the values of e in $PE_{lev.1}(v,s,e)$ are sequentially input from 1 to M, the memory in the comparison module is configured to have M first-input first-output(FIFO) memories for sequential comparative calculation, and as shown in Fig. 2, the vertical axis stores the cost of the start point and the horizontal axis stores the cost of the end point. In this instance, since the start point cannot be greater than the end point, the available values correspond to those with slash marks in the comparison module of Fig. 2.

The required information is not only the minimum cost but also the corresponding word index. Therefore, all memory elements store minimum cost and index at $v=V$.

$$\begin{aligned} C_{memory}(V,p=s,m=e) &= \tilde{D}(s,e), \\ I_{memory}(V,p=s,m=e) &= \tilde{N}(s,e). \end{aligned} \tag{6}$$

3.2 Architecture of Second Level DP

At the second level DP, we compute the distance of the best path ending at frame e, $\bar{D}_l(e)$ using a concatenated sequence of l reference patterns as in (3).

Let us define the cost of pth second level DP processing element at l reference patterns as

$$PE_{lev.2}(l, p = e) = \min_{1 \leq s < e} [\tilde{D}(s,e) + \bar{D}_{l-1}(s-1)] = \bar{D}_l(e), \tag{7}$$

As shown in Fig. 4, the second level includes a second processing element group having a plurality of processing elements that have the same configuration, and a register for storing the matching costs calculated by the respective processing elements. The second level is easy to design and modify since all processing elements have the same configuration.

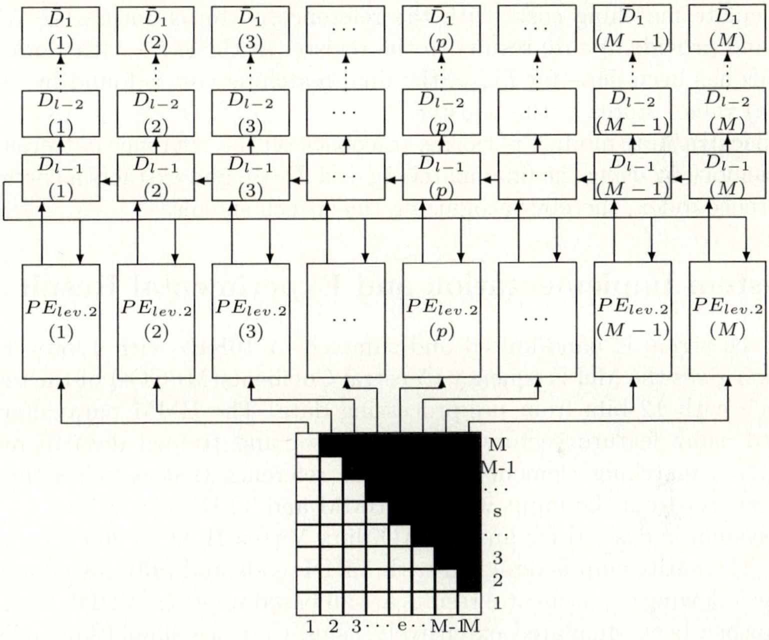

Fig. 4. Architecture of second level dynamic programming

The remaining task is to describe the internal structure of the second level DP processing element. Fig. 5 shows the processing element. The trapezoidal block represent comparators. The block chooses the smaller of $\tilde{D}(s,e) + \bar{D}_{l-1}(s-1)$ and $E_l(s-1)$ for M clock. At $M+1$ clock it updates $\bar{D}_l(s)$ with the register value, $E_l(s)$. Notice that no multiplier is involved in this design, nor in other parts of the system. As shown in Fig. 5, the processing elements of the second processing element group sequentially receive the value of $\tilde{D}(s,e)$ from the memory module, and concurrently receive the value of $\bar{D}_{l-1}(s-1)$ calculated and stored in the register. While M clock signals are applied, these two types of input values $\tilde{D}(s,e)$ and $\bar{D}_{l-1}(s-1)$, are transmitted to the processing elements, the minimum one $\bar{D}_l(e)$ is selected among the sums of the two inputs, and the value of $\bar{D}_l(e)$ is output when the $(M+1)$th clock signal is applied. The output

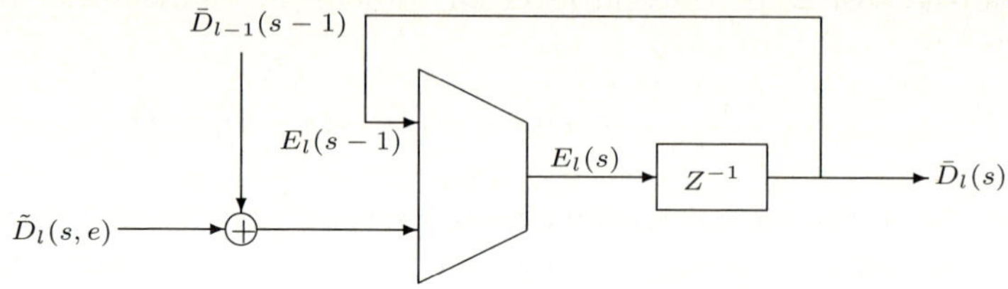

Fig. 5. Architecture of second level processing element

value of $\bar{D}_l(e)$ is stored in the register, and the matching cost thereof with the $(l+1)$th reference pattern is calculated. That is, the processing elements repeatedly calculate matching costs with the reference patterns during the M clock signals and provide update results to the register at the $(M+1)$th clock signal. After this has been done for L_{max}, the final matching cost is found by using all the $\bar{D}_l(e)$ values stores in the register.

The backtracking module performs traceback on the reference patterns stored in the memory by using the final matching cost $\bar{D}_l(e)$, and extracts a corresponding reference index, thereby recognizing the speech signals.

4 System Implementation and Experimental Results

The speech signal is bandlimited and sampled at 16KHz with 12bits. Feature extraction gives the Mel Frequency Cepstral Coeffients(MFCCs) of thirteen vectors, each with 12 bits from pre-processing data. The HMM parameters were extracted using feature vectors in preprocessor and trained data in memory. The pattern matching element chooses the reference that matches the signal parameters set from the input with the HMM and TLDP algorithms.

The system is desiged for an FPGA(Xilinx Virtex-II XC2V8000) running at 33MHz. The entire chip is designed with VHDL code and fully tested and error free. The following experimental results are all based upon the VHDL simulation. The chip has been simulated extensively using Cadence simulation tools. It is designed to interface with the PLX9656 PCI chip which has a maximum clock frequency of 66MHz. The PLX9656 PCI was used within a PC with a Pentium 4, 3.06GHz processor. The full design($M = 40$) occupies 44,027 of the XC2V8000's slices, equal to 94%, requiring 56,063 LUTs and 52,391 FFs.

The speech data used for the testing and training were taken from the database designed by the Speech Technology Research Center of Korea. Both the test and training groups consisted of 10 male and 10 female speakers. We have achieved a very good performance with a 91.4% correctness rate in a vocabulary of more than 500 words.

5 Conclusion

The hardware needs different algorithms for the same application in terms of performance and quality. Because the algorithms used for hardware and software implementation differ significantly, it will be difficult, if not impossible, to migrate software implementations directly to hardware implementations. We have presented a fast and efficient architecture and implementation of a previously presented TLDP algorithm. A systolic TDLP was derived and tested with VHDL code simulation. This scheme is fast and reliable since the architectures are highly regular. In addition, the processing can be done in real time owing to the parallel hardware implementation.

References

1. S. Yoshizawa, Y. Miyanaga, and N. Wada, A Low-power VLSI Design of an HMM Based Speech Recognition System., *Circuits and Systems, 2002. MWSCAS-2002.* 2002; vol.2 pp.II-489 - II-492.
2. Wei Han, Kwok-Wai Hon, Cheong-Fat Chan, Tan Lee, Chiu-Sing Choy, Kong-Pang Pun, P.C. Ching, An HMM-based speech recognition IC., *Circuits and Systems, 2003. ISCAS '03.*, 2003; vol.2, pp.II-744 - II-747.
3. F.A. Elmisery, A.H. Khalil, A.E. Salama, and H.F. Hammed, A FPGA-based HMM for a discrete Arabic speech recognition system., *Microelectronics, 2003. ICM 2003.*, 2003; pp.322-325.
4. Gian Carlo Caradarilli, Alessandro Malatesta, Marco Re, Luigi Arnone, and Sara Bocchio, Hardware Oriented Architectures for Continuous-Speech Speaker-Independent ASR Systems., *Signal Processing and Information Technology*, Dec. 2004; pp.346-352, 18-21.
5. Lawrence Rabiner and Biing-Hwang Juang, Fundamentals of Speech Recognition., *Prentice-Hall*, New Jersey, 1993; pp.321-433.
6. H. Sakoe, Two-Level DP-Matching–A Dynamic Programming-Based Patten Matching Algorithm for Connected Word Recognition., *IEEE Transactions on Acoustics, Speech and Signal Processing*, 1979; vol.27, no.6, pp.588-595.
7. S. Nakagawa, A Connected Spoken Word Recognition Method by O(n) Dynamic Programming Pattern Matching Algorithm., *IEEE International Conference on ASSP*, 1983; vol.8, pp.296-299.
8. Hermann Ney, A Comparative Study of Two Search Strategies for Connected Word Recognition : Dynamic Programming and Heuristic Search., *IEEE Transactions on Pattern Ananysis and Machine Intellinence.*, 1992; vol.14, no.5, pp.586-595.

Towards the Formalization of Innovating Design: The TRIZ Example

Cecilia Zanni and François Rousselot

LGeCo - INSA Strasbourg
24 bd de la Victoire - 67084 Strasbourg - France
{cecilia.zanni, francois.rousselot}@insa-strasbourg.fr

Abstract. The work of our multidisciplinary group at LGeCo is the fine description of the methodology informally applied by the TRIZ experts. TRIZ, innovating design methodology developed in Russia, has given birth to hundreds of patents and is being taught at our institution at graduate and post-graduate levels.

This methodology involves a certain number of similarities with the Artificial Intelligence methods, as it proposes a problem formulation model based on contradictions and heuristics that point out, at an abstract level, to different clues for possible solutions. This paper proposes a CommonKADS modeling of the TRIZ reasoning mechanism and discusses its need.

1 Introduction

The development of methods and tools for the theory of resolution of inventive problems, known under the acronym of TRIZ, was several years associated with a relative success, with the activities of design, in particular of inventive design.

But today, this is not enough. Current problems include supporting and accompanying the use of TRIZ in innovating design, and thus, in every angle of the companies.

From a strategic point of view, knowledge and mastery of TRIZ and of its possible impact in a company make possible to transform the user into a prescriber. It is this last approach, often implicit, which currently tries to emerge and which we propose to facilitate through the formalization of the TRIZ concepts, the clarifying of its operation, the development of specific software and the easing of its systematic utilisation for domain and design knowledge capitalization.

TRIZ is an interesting method from the point of view of Artificial Intelligence. It provides to user with heuristic knowledge which leads him into the search for a solution. In general, design tasks seem to induce a combinatorial research, TRIZ makes possible to strongly reduce the search space [1,2].

The different knowledge sources are situated at different levels of abstraction. Regarding the reasoning, it is very little formalized and the used knowledge is not usually very explicit.

It is really rather difficult to position TRIZ with regards to the artificial intelligence methods. Indeed, there is not ontology of the TRIZ concepts and it is difficult to formally link the different knowledge sources used by the method. A first work which describes the links among these pieces of knowledge was already carried out [3], it shows that they are strongly redundant and inhomogeneous.

B. Gabrys, R.J. Howlett, and L.C. Jain (Eds.): KES 2006, Part I, LNAI 4251, pp. 1098–1105, 2006.
© Springer-Verlag Berlin Heidelberg 2006

The construction of this ontology of the TRIZ concepts is necessary if we want to clarify the knowledge and reasoning mechanism it uses, even if this latter has to be often carried out by a human. A first ontology describing the main TRIZ models has already been defined [2]. It is now important to have a more precise view of the reasoning mechanisms at stake, of the various knowledge sources and of the automatizable tasks.

We are indeed persuaded that the failure of the existing software implementations[1] is due to the fact that they do not model any reasoning and make the user the sole responsible for it.

We have thus decided to study TRIZ within the framework of the CommonKADS [4] methodology, in order to better describe its operation and to have a more precise idea on the elements of the method. CommonKADS separates knowledge, inferences and tasks, a fine analysis of TRIZ in these terms will notably clarify its reasoning mechanisms and the role the different pieces of knowledge play during the reasoning process.

Section 2 of this article gives some notions of TRIZ and section 3 presents our description of TRIZ in the CommonKADS framework. Finally, section 4 exposes our conclusions and perspectives.

2 Notions of TRIZ

TRIZ[2] is primarily about *technical and physical problems*, but is now being used on almost any problem or situation. The key to success in TRIZ is the fact that (technical) systems evolve in similar ways, and by reducing any situation and problem to a functional level, we can apply almost standard solutions and problem solving techniques, even from dissimilar industries.

This section covers the foundation elements of TRIZ along with its problem solving approach.

2.1 The TRIZ Components

The complete list of TRIZ components is very long indeed, and includes laws, methods, and various lists of principles and so on. The following are the main elements from right across TRIZ [8].

- *The ideal final result (IFR)* - The starting point in TRIZ will often be the IFR, where the *ideal solution* provides maximum benefit at zero cost and zero harm. The addition of benefits to a system or solution can stem from many areas, such as increased functionality for the user. Often such additions are at the expense of additional resources, and simply add to the cost and complexity as well as introducing harm such as increased risk of failure. *Idle resources*, such as air, or existing shape and layout, or even knowledge and skill of users, can be added to a system typically at zero cost.

[1] They only use the TRIZ knowledge bases playing the role of data bases.

[2] This section presents a (very) brief introduction to TRIZ. The interested reader is invited to review the pertinent bibliography [5,6,7].

- *Technical and physical contradictions* - The *contradiction* concept is central to TRIZ and inventive problem solving. Inventive solutions solve inherent or imposed contradictions, where two distinct features or one parameter with two required states conflict or oppose each other.
- *Patterns or trends of technological evolution* - Although different fields of technology progress in many different ways, all innovation passes through fairly well defined stages and many *patterns* have been identified and documented. Since almost all future development must also follow these patterns, a study of current position and past evolution will show future direction and potential.
- *The 40 inventive principles and the contradiction matrix* - From a study of the top most inventive patents, G. Altshuller, the creator of TRIZ, identified 40 basic principles that others had used to solve problems. All innovation will use one or more of these principles. The *contradiction matrix* is a grid describing the most popular invention principles to be used to solve each pair of *engineering parameter* conflicts.
- *Separation principles* - Given a conflict between two elements, the best solution is simply to *separate* them. Separation can be by one of *time*, *space*, on a given *condition*, or by *parts/whole*.
- *Substance-field analysis and the inventive standards* - The simplest *system* needed to deliver a *function* contains two *substances* (objects) connected by one *field* or energy source. A *Su-Field analysis* may show a need to modify or add a field or substance, and the *inventive standards* list groups potential solutions that can be used alongside the 40 inventive principles.
- *Use of physical, chemical and geometric effects* - Catalogued lists of scientific *effects* or *knowledge* that can supply answers on how to deliver a particular *function*.
- *ARIZ* - It is an acronym for the Russian phrase "Algorithm for Inventive Problem Solving". It is a logical structured process for working on problems that are not straightforwardly solved by the use of the Contradiction Matrix or the Standards. It incrementally evolves a vaguely defined problem to a point where it is clearly defined and simple to solve.

2.2 The TRIZ Problem Solving Approach

Figure 1 describes the way TRIZ solves problems, showing the interrelations among its different components.

Three different phases are clearly identified:

1. The "reformulation" phase, where the expert uses different tools to express the problem in the form of a contradiction network or another model
2. The "abstract solution finding" phase, where access to different knowledge sources is made to get one or more solution models
3. The "interpretation" phase, where these solution models are instantiated with the help of the scientific-engineering effects knowledge base, to get one or more solutions to be implemented in the real world

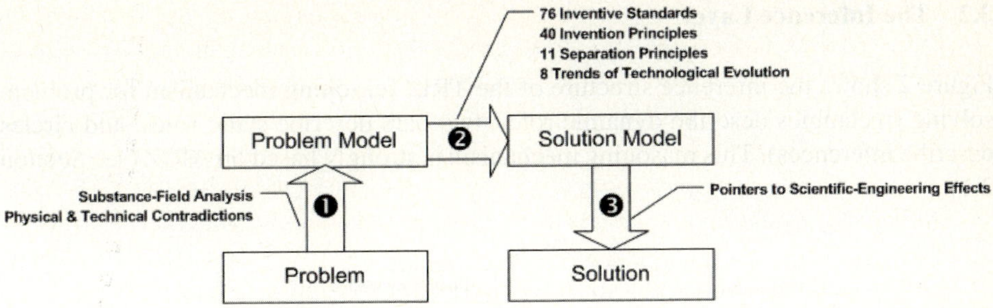

Fig. 1. The TRIZ problem solving approach

3 The CommonKADS Model

The CommonKADS Expertise Model [4] aims at modeling a problem solving behaviour from a knowledge kevel [9] perspective. It is structured in three layers: the domain level and the task and inference levels.

The domain level provides ontology modeling, e.g. conceptualizing and formalizing a domain by means of concepts, expressions and relations.

The task and inference levels aim at modeling reasoning and problem solving. These levels are supposed to be as much as possible independent of the domain: they should be described in a task specific way. Tasks and inferences use a task oriented ontology, the terms of which are called roles.

Tasks describe the goals of the problem solving agent and the actions that can lead to reach the goals, including goals and tasks decomposition.

Basic tasks are implemented by means of inferences. An inference is close to the notion of inference rule in logic. It specifies the basic reasoning steps that are feasible at the domain level. It describes the type of knowledge that it is possible to deduce given inputs and domain models. Inferences also describe the mapping between task oriented roles and domain level entity types.

Due to space constraints, we are not able to develop the full model here. The complete model with examples may be found at [10].

3.1 The Domain Layer

TRIZ contains several knowledge sources which are permanently used at different steps during the resolution process. If we agree that metaknowledge is knowledge about knowledge, it is also true that TRIZ uses different metaknowledge sources.

- *Metaknowledge sources:* The *Trends of Technological Evolution*, the *Inventive Principles*, the *Separation Principles* and the *Inventive Standards*.
- *Knowledge sources:* Apart from the domain knowledge necessary to develop the substance-field models and the technical and physical contradictions, knowledge specific to TRIZ includes the *Physical, Chemical and Geometrical Effects*.

3.2 The Inference Layer

Figure 2 shows the inference structure of the TRIZ reasoning mechanism for problem solving (rectangles describe dynamic roles, two bars describe static roles, and circles describe inferences). This reasoning mechanism is strongly based on ARIZ (see Section 2.1).

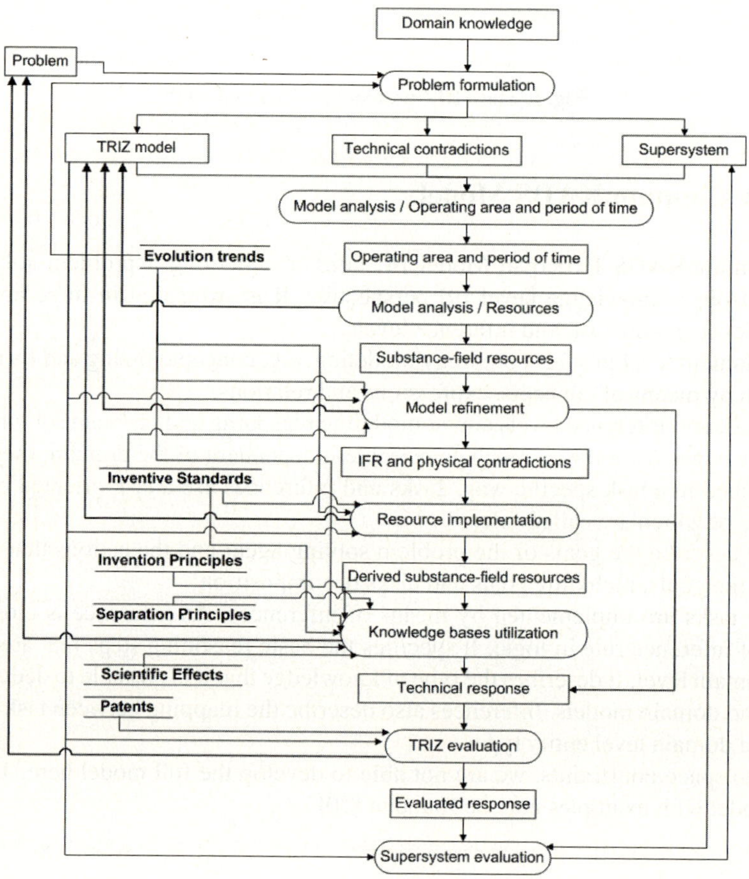

Fig. 2. The TRIZ reasoning mechanism for problem solving

This inference structure includes eight inferences[3]:

– *Problem formulation*: The principal objective of this inference is to evolve from a situation of fuzzy innovation to a clearly built and considerably simplified model, where the key contradictions are identified. This inference provides then the `TRIZ`

[3] We will note in `courier` font the roles involved in each inference.

model, the Technical contradictions and the Supersystem associated to the TRIZ model[4].

- *Model analysis / Operating area and period of time:* The goal of this inference is to precisely determine when and where the contradiction arises.
- *Model analysis / Resources:* The goal here is to list the present resources that are used for the resolution of the problem: substance-field resources. This inference can introduce a modification into the TRIZ model, perhaps because new resources are discovered.
- *Model refinement:* The goal is to formulate the idea of the IFR (ideal final result) and also define the physical contradictions that prevent from obtaining the IFR. It is not always possible to arrive at an ideal situation, however, the IFR directs towards a relevant answer. This inference can also introduce new data into the TRIZ Model. If a solution is obtained using the Inventive Standards, this inference produces a Technical Answer.
- *Resource implementation:* The objective is to methodically increase the substance-field resources already identified in the system. We can obtain a modification of the TRIZ Model, or a Technical Answer if the problem can be solved by means of the Inventive Standards.
- *Knowledge bases utilization:* The objective is to use the experience concentrated in the TRIZ knowledge bases. At this step, the problem is already considerably clarified, its direct resolution using the knowledge bases becomes possible. If the resolution is not possible, it is necessary to reformulate the initial Problem, and to check that it is not an overlapping of several different problems. If a physical answer is possible, this inference gives a Technical Answer, along with the procedure associated for the implementation of it.
- *TRIZ evaluation:* This inference analyzes the found solution, according to various criteria specific to TRIZ. If the evaluation is not satisfactory, the initial Problem has to be reformulated.
- *Supersystem evaluation:* The goal is to specify what it is necessary to modify in the Supersystem which contains the modified system, by evaluating the new and different properties that the system and the Supersystem have.

3.3 The Task Layer

TRIZ involves three distinct stages: an analytical stage, an operative stage and a synthetic stage.

During the analytical stage, the problem is formulated to get a model with some specific information, in particular, the system to analyze, its elements and their roles, the result we expect to obtain and the list of initial available resources that we may use. This stage of construction of the model may lead to the discovery of new resources in its sub-stages, implying that the model may be iteratively modified.

During the operative stage we study the typical methods for solution (with the aid of the TRIZ knowledge bases) and we explore new solution methods by the introduction

[4] The TRIZ model includes the system, its elements and their roles, the ideal final result, the result to reach and the initially identified system resources (that is, the substances and fields already present in the system).

of changes in the system, in its environment or in adjacent systems. If a solution is not found, we may be led to reformulating the initial problem, its definition probably being a set of different problems.

The last stage, the synthetic stage, evaluates the utilization of the method and the found solution. If the evaluation is not satisfactory, we are also led to reformulating the initial problem. This stage also explores the applicability of the solution to other engineering problems.

It is clear that the methodology works in an iterative way, producing successive reformulations or refinements of the initial model or problem.

Figure 3 shows the tasks sequencing on the activity diagram (in the sense of UML) describing the control strategy on the inferences of section 3.2.

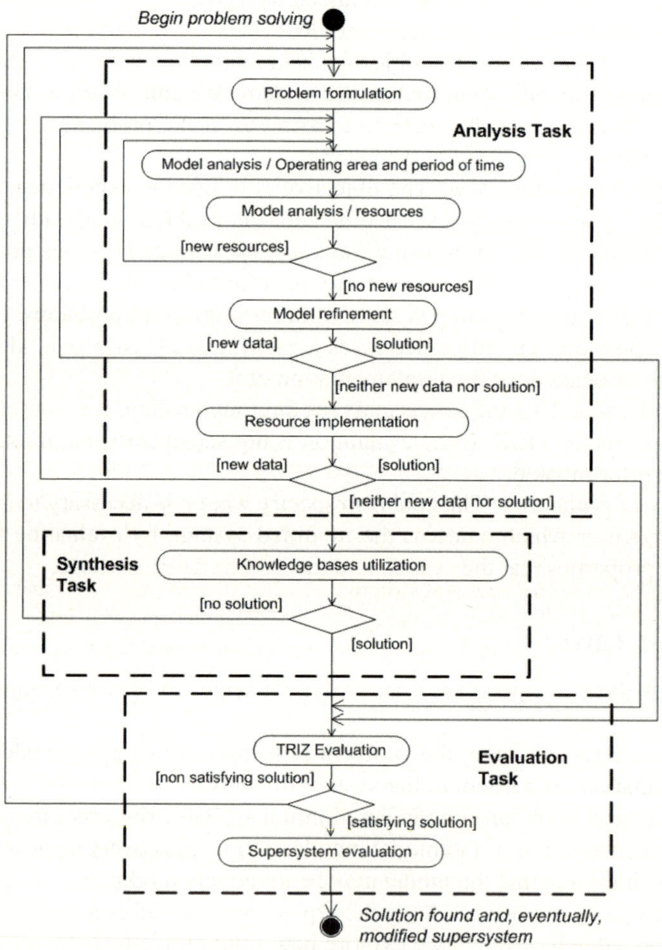

Fig. 3. Tasks sequencing

3.4 Conclusions

The model we have just presented depicts the inferences at stake and the roles of the different TRIZ knowledge sources. The lack of formalization of the method and the non-existence of a terminological normalization, usually make the comprehension of TRIZ a difficult task. The clarification obtained with this CommonKADS model is undeniable.

The foreseeable consequences of this modeling are true improvements in the diffusion and teaching of the method. Also, we can more easily approach the work of formalization and explanation of the TRIZ concepts, with the goal of developing a full ontology. This step of formalization of the TRIZ knowledge bases is important. Certain pieces of knowledge are implicitly used, the goal is to make them explicit. Also, nowadays, the knowledge sources are presented without the context of application. This fact provokes, for example, that certain software implementations of the method advise the segmentation of an object that is not decomposable.

The building of this ontology taking into account the hierarchy of TRIZ concepts along with the application context is our current aim.

Only when all this work will be done, we will be able to provide effective tools for knowledge capitalization and with reasoning capabilities.

References

1. Gartiser, N., Kucharavy, D., Lutz, P.: Le processus convergent de la TRIZ : Une dmarche economiquement efficace de recherche de solutions en conception. In: Colloque IPI. (2002)
2. Dubois, S.: Contribution la formulation de problmes de conception de systmes techniques. PhD thesis, Universit Louis Pasteur, Strasbourg, France (2004)
3. Muller, S.: Vers une mta-structure des bases de connaissances de la TRIZ. Mmoire DEA - LGeCo - INSA Strasbourg (2004)
4. Schreiber, G., Hakkermans, H., Anjewierden, A., de Hoog, R., Shadbolt, N., Van de Velde, W., Wielinga, B.: Knowledge engineering and management - The CommonKADS methodology. MIT Press (2000)
5. Althshuller, G.: The innovation algorithm: TRIZ, systematic innovation and technical creativity. (2000) Translated from russian by Lev Shulyak and Steven Rodman.
6. Altsthuller, G.: Creativity as an Exact Science. Gordon and Breach Scientific Publishers, New York (1984)
7. Altsthuller, G.: TRIZ The innovation algorithm ; systematic innovation and technical creativity. Technical Innovation Center Inc., Worcester, MA (1999)
8. Tennant, G.: Pocket TRIZ for Six Sigma. Mulbury Consulting Limited, Bristol, England (2003)
9. Newell, A.: The knowledge level. Artificial Intelligence **18** (1982) 87–127
10. Rousselot, F., Zanni, C.: La conception innovante : synthse de systmes ou rsolution de problmes ? In: Proceedings of the 17e journes francophones d'Ingnierie des connaissances (IC) 2006, Nantes, France (Jun 28 - 30, 2006)

The "5P" Apprenticeship Which Based the Chinese Traditional Culture

Qin Gu[1] and Liang Liang[2]

[1] School of Management, University of Science and Technology of China
China, An Hui Province, Hefei, the west campus of USTC, 9#724
guqin@mail.ustc.edu.cn

[2] School of Management, University of Science and Technology of China
lliang@ustc.edu.cn

Abstract. The paper sums up five primary characteristics of apprenticeship on the basis of the Chinese traditional culture. Then it analyses the new economic environment and its needs for employees' quality. By expatiating on the effect of the apprenticeship, it demonstrates that the "5P" apprenticeship is accord with the demand of the development of our times, especially in experience management. Finally we conclude that apprenticeship should not be abandoned drastically.

1 Introduction

In ancient times, apprenticeship had rooted in China. Apprenticeship is a training method whose content is teaching production knowledge and skills. Its method is face-to-face explanation and action-imitations [1].

For thousands of years, the Chinese apprenticeship has perfected continually along with the development of productivity, handicraft industries and commodity economy. As far as reform and opening goes, the modern industrialization and scientific management theory resulted in the eclipse of apprenticeship. However, in the face of the trend of the new economy, can apprenticeship readapt to the development of the age?

2 The Characteristics of the Chinese Apprenticeship

The Chinese apprenticeship has formed particular characteristics from its beginning which are patriarchal, practical, progressive, promotive, and pensive. (Figure 1)

If western culture is "wisdom culture", then the Chinese culture is "ethic culture". Different from other countries, China stepped into civilization in the terms of the insufficient disorganization of the kin, therefore formed the particular patriarchal clan system [2]. Generally speaking, the unique technology is only inherited by the eldest son. Even though the expansion of the handcraft workshops' scale had to request recruiting extraneous apprentices in order to guarantee natural production, the key unique technology were hidden by masters. Even now, the connatural tradition isn't

B. Gabrys, R.J. Howlett, and L.C. Jain (Eds.): KES 2006, Part I, LNAI 4251, pp. 1106 – 1114, 2006.
© Springer-Verlag Berlin Heidelberg 2006

broken in some archaic handicraft industries. However, the particular ethic culture of the patriarchal clan society also has positive effects. Fealty is considered the core of all the ethics in China compared with the western self-serving concept [2]. Of course, harmonious relation is also one part of the "ethic culture" for apprenticeship. On one hand, the apprentices respect the masters very much; on the other hand, the masters not only teach them technologies, but also care for their health and life.

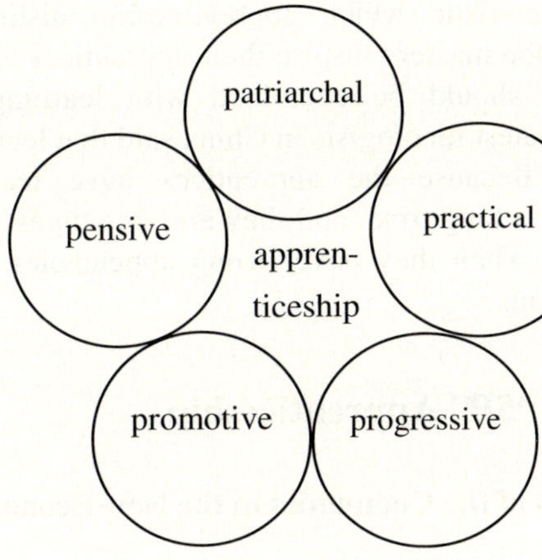

Fig. 1. The "5P" apprenticeship

Practice is an important characteristic of apprenticeship. The essence of western culture is wisdom whose intention is exploring verities. It distinguishes between verity and holy strictly. Oppositely, the Chinese culture considers that verity is namely holiness. It emphasizes individual practical experience, ignoring ecumenical theory principles correspondingly; and it makes much of exploring verities in the self-practice, ignoring general theory analysis and logical reasoning, so it's incapable to develop into formalized and axiom theory system in lots of fields, but accumulates plenty of practical experience [3]. Just as these, apprenticeship is not only a teaching process, but also a process of dividing the work, cooperating and producing together. The practices can feedback the apprentices' study effects in time, and help them to link theory with practice. The biggest advantage is that once the apprentices have finished apprenticeship, they can be independent of producing entirely. Therefore it availably resolves the problems of separation between theory and practice which exit in many other training methods.

At the same time, there is a gradual process of knowledge accumulation and technology increasing. For a long time, the apprenticeship has followed the course of advancing gradually which transfers knowledge from the easy points to the difficult ones. Generally, the apprentices have studied for three to five years following their masters. At first they are only engaging in some basic work and factotums, then taking part in some simple operations till commanding a technology under the directions of the masters.

Teaching is a promotive activity and bidirectional communication. The masters always have taken the fundamental principle of advancing illuminate students step by step. Accordingly, they pay attention to the apprentices' positivity and illuminate them to ponder something. So it does, the heuristic brings up the phenomena that the students surpass the teacher. The masters are not only teaching technology, but also gaining knowledge. They can have new discoveries from seeking the apprentices' doubts, or attain new knowledge directly from their apprentices.

The important characteristic which apprenticeship distinguishes from other training methods is that the masters inspire their apprentices to speculate positively during study. Thinking should be combined with learning in ancient China. Confucius, one of the greatest ideologists in China said that learning without thought results in bafflement. Because the apprentices have learned comprehensive knowledge and skills for a long time, and they are able digest what is propitious to find questions for them. Then they have strong appetencies to think about these questions and resolve them.

3 The Effect of the "5P" Apprenticeship

3.1 The Characteristics of the Companies in the New Economy Times

At the end of the 20th century, the world entered into the new economy times due to the high speed of development. Therefore, the modern companies have some new characteristics. At present the competitions are not established in the traditional viewpoints that a corporation has plenty of manpower or capital. Knowledge is the chief factor in deciding a company's development.

First of all, the rapid development of the information has led to globalization. Therefore, the regional limitation of the traditional competition is broken. So many corporations have begun global management. Second, by way of using the information resources adequately, the information industry has developed consequentially. Accordingly, the running speed of the companies is quickened. If one wants to survive the changing times, it must innovate constantly. Then the concept of soft management is popular for it can adapt to the environment better. At the same time, the organizational structure has changed from pyramid to flat which decreases the administrative levels, but increases the managing extent. Consequently the enterprises are capable to adapt to the exterior environment agility, effectively, elastically and rapidly. Third, the transformation of the organizational structure and the complexity of work have resulted in the cooperation among different departments inevitably. None but the group cooperation can make the item accomplished excellently.

3.2 The Employees' Quality in the New Economy Times

Facing these new tendencies, the companies request much more quality for employees (Table 1) [4].

Table 1. The employees' quality in the new economy times

The first level: Basic Skill	basic knowledge and principle	In order to fulfill the comprehensive needs of the society, ones must be acquainted himself with the general professional principle and other knowledge beyond the professional technology.
The second level: General Skill	professional technology	Having upper professional technology which adapts to the needs of the posts in order to ensure the companies operation.
	communication and cooperation skill	Dealing with the entire relatives well and being capable to cooperate with others. It is important for employees to adapt their changing roles to the situation of group cooperation and global management.
	management skill	Gripping some management knowledge, having strong analysis abilities and decision forces. These management methods adapt to the needs of the flat organization structure, improving the timing decision abilities for replying crisis.
The third level: Core Skill	learning skill	Don't blackball the new things, studying new knowledge and skills rapidly. Although the information world changes quickly, the persons who grip the study skill can fit themselves to the new conditions.
The forth level: Advanced Skill	innovation skill	Having innovation skills, gripping the change. These skills can help the companies to succeed in the market competition.

3.3 The Effect of the "5P" Apprenticeship

The Chinese apprenticeship which is based on the traditional culture is important for training the employees' makings in the new economy times by its own characteristics (Figure 2).

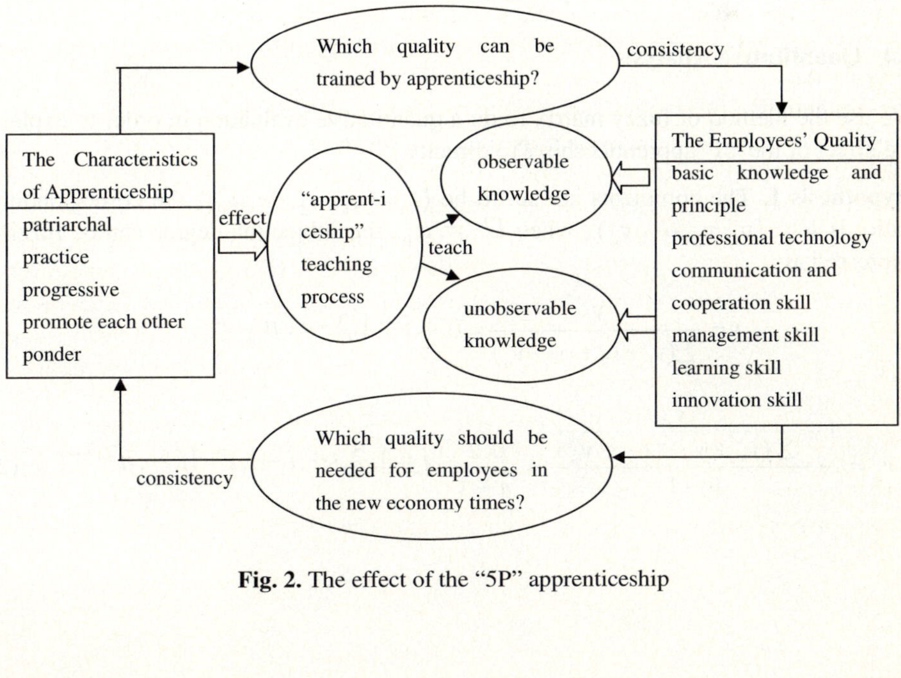

Fig. 2. The effect of the "5P" apprenticeship

First, the patriarchal characteristic emphasizes a harmonious mentality, so it is easy to build harmonious atmosphere during the process of teaching which satisfied the principle of humanity demanded by the modern management. Apprenticeship solves the problem of compulsive study which is belonged to the other training methods, more than that; it is also the basis of communication and cooperation among employees.

Second, the development can't be accomplished in an action. The progressive and practical apprenticeship is beneficial for the apprentices to make their knowledge comprehensively and professionally. On one hand, the process of teaching from the masters is constant; on the other hand, the apprentices are tutored all the while in practice so that they can correct their errors and increase their knowledge. Moreover, because of an abundant of experience, the masters maybe talk about some unobservable knowledge such as particular knack and so on.

Last, the environment brings about many more needs about the employees' abilities nowadays which are ignored in other training methods. However, the training of the abilities is almost the unobservable knowledge; the potential advantages of the apprenticeship appear much more. By the analysis above, we know the promotive characteristic emphasizes on learning from each other among the masters and the apprentices, and during the process there are certainly some ambiguous problems which need their discussions. Therefore, the ability of communication is very important. Of course cooperation is also necessary because it's impossible to communicate effectively among the repellent individuals. Besides these, the comprehensive study among the masters and the apprentices can improve the employees' learning ability. It is beneficial to form a kind of culture so that the enterprises can change to the learning organization easily. Then during this process which is often with the idealistic conflict and the consideration demanded by apprenticeship arouses the employees' concept of innovation to a certain extent and encourages them to innovate by practice.

3.4 Quantitative Analysis

We use the method of fuzzy matrix to do a quantitative evaluation in order to explain the effect of the"5P"apprenticeship (Y) directly [5].

Hypothesis I. The comments are set to be $U=(u_1, u_2,\cdots,u_n)$, the corresponding ballot is $V=(v_1, v_2,\cdots, v_n)$, when $U=u_i$, the subjection degree can be fuzzily expressed as:

$$r_i=\frac{v_i}{\Sigma(v_1+v_2+\cdots+v_n)}=0.8, i=1,2,\cdots, n \tag{1}$$

$$r_j=\frac{1-\dfrac{v_i}{\Sigma(v_1+v_2+\cdots+v_n)}}{n-1}=\frac{0.2}{n-1}, j=1,2,\cdots,i-1,i+1,\cdots,n \tag{2}$$

Table 2. The comparison of training methods.Raymond A.Noe, translated by Fang Xu. (2001). Employee Training & Development. Renmin University of China Press.

	presentation method		hands-on method								group building method		
	lecture	Kines-cope	OJT	self-study	Apprent-iceship	Simula-tion	case study	Busine-ss game	role play	behavior modeling	adventure learning	team training	action learning
study effect（y1）													
language information（a1）	yes	yes	yes	yes	yes	no	yes	yes	no	no	no	no	no
intellect skill（a2）	yes	no	no	yes	yes	yes	yes	yes	no	no	no	yes	no
cognizance policy（a3）	yes	no	yes	yes	no	yes	yes	yes	yes	yes	yes	yes	yes
attitude（a4）	yes	yes	no	no	no	no	no	no	yes	no	yes	yes	yes
sport skill（a5）	no	no	yes	no	yes	yes	no	no	no	yes	no	no	no
study environment（y2）													
clear aim（b1）	middle	low	high	high	high	high	middle	high	middle	high	middle	high	high
practice chance（b2）	low	low	high	high	high	high	middle	middle	middle	high	middle	high	middle
significant content（b3）	middle	middle	high	middle	high	high	middle	middle	middle	middle	low	high	high
feedback（b4）	low	low	high	middle	high	high	middle	high	middle	high	middle	middle	high
observe and communicate with others（b5）	low	middle	high	middle	high	high	high	high	high	high	high	high	high
training transform cost（b6）	low	low	high	middle	high	high	middle	middle	middle	high	low	high	high
exploitation cost（b7）	middle	middle	middle	high	high	high	middle	high	middle	middle	middle	middle	low
management cost（b8）	low	low	low	middle	high	low	low	middle	middle	middle	middle	middle	middle
effect	good for language informati-on	normal	good for organiz-ed OJT	normal	good	good	normal	normal	normal	good	worse	normal	good

Hypothesis II. The comments U_1 (yes, no) is defined as $B_1 = (100,60)$,the comments U_2 (high, middle, low) is defined as $B_2 =$ (100,80,60) .

From table 2:

$$Y = y_1 + y_2 \tag{3}$$

$$y_1 = f(a_1, a_2, a_3, a_4, a_5), y_2 = f(b_1, b_2, b_3, b_4, b_5, b_6, b_7, b_8) \tag{4}$$

$$f = WB^T = PRB^T \tag{5}$$

W is noted the subjection matrix, P is noted the indexes' weight, R is noted the fuzzy subjection matrix.

Then we could set the weight of each index according to the extent of importance to which the era of new economy requires:

$$\Sigma P_1 = 1 \ \Sigma P_2 = 1 \ P_1 = (0.3, 0.25, 0.15, 0.1, 0.2) \ P_2 = (0.1, 0.25, 0.2, 0.2, 0.25, 0, 0, 0) \tag{6}$$

The paper disregards the problem of cost, so the weights of cost are setup 0.

Taking an example of apprenticeship, from the hypothesis I, the fuzzy subjection matrix R is :

$$R_1 = \begin{bmatrix} 0.8 & 0.2 \\ 0.8 & 0.2 \\ 0.2 & 0.8 \\ 0.2 & 0.8 \\ 0.8 & 0.2 \end{bmatrix} \quad R_2 = \begin{bmatrix} 0.8 & 0.1 & 0.1 \\ 0.8 & 0.1 & 0.1 \\ 0.8 & 0.1 & 0.1 \\ 0.8 & 0.1 & 0.1 \\ 0.8 & 0.1 & 0.1 \\ 0.8 & 0.1 & 0.1 \\ 0.8 & 0.1 & 0.1 \\ 0.8 & 0.1 & 0.1 \end{bmatrix} \tag{7}$$

Therefore,

$$W_1 = P_1 R_1 = \begin{bmatrix} 0.3 & 0.25 & 0.15 & 0.1 & 0.2 \end{bmatrix} \begin{bmatrix} 0.8 & 0.2 \\ 0.8 & 0.2 \\ 0.2 & 0.8 \\ 0.2 & 0.8 \\ 0.8 & 0.2 \end{bmatrix} = \begin{bmatrix} 0.65 & 0.35 \end{bmatrix} \tag{8}$$

$$W_2 = P_2 R = [0.1 \ \ 0.25 \ \ 0.2 \ \ 0.2 \ \ 0.25 \ \ 0 \ \ 0 \ \ 0] \begin{bmatrix} 0.8 & 0.1 & 0.1 \\ 0.8 & 0.1 & 0.1 \\ 0.8 & 0.1 & 0.1 \\ 0.8 & 0.1 & 0.1 \\ 0.8 & 0.1 & 0.1 \\ 0.8 & 0.1 & 0.1 \\ 0.8 & 0.1 & 0.1 \\ 0.8 & 0.1 & 0.1 \end{bmatrix} = [0.8 \ \ 0.1 \ \ 0.1] \tag{9}$$

From the hypothesis Π, we can solve for the last scores.

$$Y = [0.65 \ \ 0.35] \begin{bmatrix} 100 \\ 60 \end{bmatrix} + [0.8 \ \ 0.1 \ \ 0.1] \begin{bmatrix} 100 \\ 80 \\ 60 \end{bmatrix} = 180 \tag{10}$$

Following the same methodology, we can compute the scores of all the training methods.(Table3)

The result of quantitative analysis is consistent with that of qualitative analysis. The presentation method is a unilateral channel to communicate. Though knowledge flow could be passed over more rapidly through this channel, it's not advantageous in fostering employees' capabilities. Group-building focuses on the cooperation among employees while the hands-on model combines the virtues of two methods in order to achieve a better training result. However, there are some discrepancies between the quantitative and qualitative analysis. It is justified based on two assumptions which are: (1) whether the various training methods adapt to building up employees' capabilities in the era of new economy; (2) training cost is worthless. Based on the above-mentioned analysis, we conclude that the Apprenticeship mechanism is optimal is this quantitative model.

Table 3. The scores of the training methods

training methods	scores	training methods	scores	training methods	scores
lecture	157.4	simulation	168.3	behavior modeling	167.6
kinescope	149.9	case study	165.4	adventure learning	156.7
OJT	177.6	business game	172.5	team training	166.4
self-study	169.7	role play	157.5	action learning	164.5
Apprentice-ship	180.0				

4 Conclusion

To sum up, we can find that the "5p" apprenticeship which is based on the Chinese traditional culture is advantageous for building up the employees' skills and capabilities after the advent of new economy. Of course, we should also observe its defections. For example, it emphasizes on the narrow kin excessively and so on. Furthermore, a lot of developed countries have actively sought after the modern apprenticeship which can adapt to the situation better, such as the "Dual System" Vocational Education in Germany [6], the CBE (Competency-Based Education) in Canada, the TAFE (Technical and Further Education) in Australia[7], and the MA (the Modern Apprenticeship) in England [8]. Thus, we should overcome our shortcomings after incorporating strengths of other mechanisms. Through increasing the flexibility and discipline within the mechanism, we may promote the "5P" apprenticeship to an ultimate perfection.

Acknowledgements. The authors acknowledge the support of NSFC (70525001).

References

1. Raymond A.Noe, translated by Fang Xu. (2001). Employee Training & Development. Renmin University of China Press.
2. Dainian Zhang, Keli Fang, The Advanced Educational Department of the Ministry of Education. The Conspectus of the Chinese Culture. BeiJing Normal University Press.
3. Ping Li. (2002). The Conspectus of the Chinese Culture. AnHui Normal University Press.
4. Wendong Ma. (2004). The Research of the Method Which Evaluates the Risk of the High-grade Managers in the Companies. Human Resource Development of China 12: 12-14.
5. Pingli Zhu, E'mang Xiao. (2004). The Quantitative Model of Evaluating the Human Resource Managers' Quality. Human Resource Development of China 7: 38-40.
6. Xinshun Dou, Junwei Fu. (2004). Exploring the Teaching Pattern of the "Dual System" Vocational Education in Germany. Metallurgic Education of China 6: 56-59
7. Baizhi Tang, Lixun Wu, (2004). The Foundation and the Actualization of the New Apprenticeship in Australia. Forum of the Vocational Education 12(1): 60-61.
8. Xiaoyu He, (2001). The MA in England. China Training 3: 52-53.

A Unified 2D Representation of Fuzzy Reasoning, CBR, and Experience Based Reasoning

Zhaohao Sun[1] and Gavin Finnie[2]

[1]School of Economics and Information Systems, University of Wollongong
Wollongong NSW 2522 Australia
zsun@uow.edu.au
[2]School of Information Technology, Bond University,
Gold Coast Qld 4229 Australia
gfinnie@staff.bond.edu.au

Abstract. Fuzzy reasoning, case-based reasoning (CBR) and experience-based reasoning (EBR) or natural reasoning have been seriously studied for years. However, these studies essentially can be considered as a 1-dimensional approach, because their reasoning paradigm is 1-dimensional. This paper will propose a 2-dimensional (2D) approach that represents fuzzy reasoning, CBR, and EBR in a unified way. This approach also integrates many reasoning paradigms such as abduction, deduction and similarity-based reasoning into a unified treatment. The proposed approach will facilitate research and development of fuzzy reasoning, knowledge/ experience management, knowledge-based systems, and CBR.

Keywords: Fuzzy reasoning, case-based reasoning, experience-based reasoning, abduction, 2-dimensional (2D) representation, experience management.

1 Introduction

Fuzzy reasoning [21], case-based reasoning (CBR) [13] and experience-based reasoning (EBR) [18] or natural reasoning and also traditional reasoning paradigms such as deduction [9], abduction [17] and similarity-based reasoning (SBR) [11] have been seriously studied in computer science and artificial intelligence (AI) with many applications in engineering [21], commerce [11], business and management [12], to name a few. However, these studies essentially can be considered as a 1-dimensional (1D) approach, because the fundamental inference rule of each of these reasoning paradigms is represented in a 1D way. This 1D representation limits their own development and isolates them from each other, although some interrelationships between fuzzy reasoning and CBR have been studied [13]. This paper will fill this gap by proposing a 2-dimensional (2D) approach that represents fuzzy reasoning, CBR, and EBR in a unified way. This approach also integrates many reasoning paradigms such as abduction, deduction and SBR into a unified treatment. The proposed approach will facilitate research and development of fuzzy reasoning, knowledge/experience management, knowledge-based systems, and CBR.

B. Gabrys, R.J. Howlett, and L.C. Jain (Eds.): KES 2006, Part I, LNAI 4251, pp. 1115 – 1123, 2006.
© Springer-Verlag Berlin Heidelberg 2006

2 Fundamentals

Any intelligent system can only give the answers to problems in a *possible world,* which corresponds to a scenario in the real world [11]. Based on this idea, the possible world of problems, W_p, and the possible world of solutions, W_s, are the whole world of an agent [9] to use an intelligent systems to do everything that he can. If an agent considers a CBR system as a function h from W_p to W_s, it is meaningless to discuss the image of $h(x)$ if $x \notin W_p$. Therefore, the agent can only know and work in the world $W_p \times W_s$. For example, in a CBR e-sale system, the possible world of problems, W_p, might consist of [5]:

- Properties of goods
- Normalized queries of customers
- Knowledge of customer behavior
- General knowledge of business, and so on.

And the possible world of solutions, W_s, consists of:

- Price of goods
- Customized answers to the queries of customers
- General strategies for attracting customers to buy the goods, and so on.

It should be noted that if we assume that P is the subset of problem or antecedent descriptions and Q is the subset of solution or action descriptions [8], then it is obvious that P and Q are subsets of W_p and W_s respectively. A conditional proposition, $p \to q$, can be denoted as an ordered pair (p, q), where $p \in P$ and $q \in Q$. Further, if we assume that W_p is a dimensional world, then W_s is another dimensional world. Therefore, we can use a coordinate system to represent these two dimensional worlds consisting of W_p and W_s.

3 A 2D Representation of Fuzzy Reasoning

In propositional logic, reasoning basically belongs to deductive reasoning. Reasoning is performed by a number of inference rules [11], in which the most commonly used is *modus ponens* (MP):

$$\begin{array}{c} P \to Q \\ \underline{P \qquad} \\ \therefore\ Q \end{array} \tag{1}$$

where P and Q represent compound propositions. One of the most important features of this reasoning is that it satisfies the transitive law, and then this reasoning can be performed as many times or steps as required with the preservation of validity of the result of the inference. This means that the reasoning in traditional logic is a multistep reasoning.

Fuzzy reasoning in fuzzy logic is basically generalized from the deductive reasoning in traditional logic with the exception of its computational process [17]. Its reasoning is based on the following *generalized modus ponens* [13]:

$$\frac{\begin{array}{c} P \to Q \\ P' \end{array}}{\therefore\ Q'} \tag{2}$$

where P and Q represent fuzzy propositions, P' is approximate to P, that is, $P' \sim P$. Model (2) is also commonly represented in the following form in fuzzy logic [21]:

$$\frac{\begin{array}{c} \text{If } x \text{ is } P \text{ Then } y \text{ is } Q \\ x \text{ is } P' \end{array}}{\therefore\ y \text{ is } Q'} \tag{3}$$

For instance,

$$\frac{\begin{array}{c} \text{IF a tomato is red} \quad \text{THEN the tomato is ripe} \\ \text{This tomato is very red} \end{array}}{\text{Conclusion:} \quad \text{This tomato is very ripe}} \tag{4}$$

As we know, inference rules play a central role in reasoning paradigms of AI. For example, one of the main inference rules for fuzzy reasoning is the generalized modus ponens Model (2). Based on the discussion in Section 2, Model (2) can be represented in a 2-dimensional (2D) way as shown in Fig. 1, where the node pointed to by an arrow represents the conclusion of the inference, the small cycle denotes the (fuzzy) proposition p'. Further, the point denoted by (p', q'), $(\neg p', q')$ $(\neg p', \neg q')$ and $(p', \neg q')$ will be in the 1st, 2nd, 3rd and 4th quadrant respectively (Similar explanations are valid for other figures in this paper). Therefore, fuzzy reasoning can be considered as a first quadrant reasoning because it only takes the 1st quadrant of the coordinate system.

For fuzzy reasoning, (p, q) means $p \to q$, and p' is a fuzzy proposition approximate or close to fuzzy proposition p, and then the conclusion inferred from the fuzzy reasoning based on Model (2) is q'.

It should be noted that when $p \equiv p'$, then $q \equiv q'$, the generalized *modus ponens* will degenerate to classic *modus ponens*. The 2D representation of *modus ponens* will be simpler than that in Fig. 1, which we do not discuss any more.

4 A 2D Representation of CBR and SBR

CBR is a reasoning paradigm based on previous experiences or cases that are the operational definition of experiences; that is, a case-based reasoner solves new problems by adapting solutions that were used to solve old problems [13]. Therefore, we call CBR a form of experience-based reasoning [11], briefly,

$$\text{CBR:= Experience-based reasoning} \tag{5}$$

CBR is based on two principles about the nature of the world [6]. The first principle is that the world is regular: similar problems have similar solutions. Consequently, solutions to similar prior problems are a useful starting point for new problem solving. The second principle is that the types of problems that an agent encounter tend to recur. Hence, future problems are likely to be similar to current problems. When the two principles hold, CBR is an effective reasoning paradigm.

The first principle also implies that EBR is based on the experience principle, for example, in business activities it is usually true that "Two cars with similar quality features have similar prices." However, from a logical viewpoint, this is a kind of similarity-based reasoning (SBR) [4]. In other words, SBR can be considered as a special and operational form of EBR. Therefore, CBR can be considered as a kind of SBR from a logical viewpoint [13].

$$\text{CBR:} = \text{Similarity-based reasoning} \tag{6}$$

Similar to the inference engine in expert systems (ESs), one can also use a CBR engine (CBRE) to denote the inference engine in CBR system (CBRS); that is,

$$\text{CBRS} = \text{CB} + \text{CBRE} \tag{7}$$

where CB is the case base. CBRE performs similarity-based reasoning, while the inference engine in ESs performs traditional deductive reasoning.

In CBR, similarity based reasoning is realized by the following model [13].

$$\begin{array}{c} P \to Q \\ P' \\ \hline \therefore\ Q' \end{array} \tag{8}$$

where, $P \in W_p$ and $Q \in W_s$ represent a problem description and the corresponding solution description respectively; that is, P' is the current problem description which is similar to P, $P' \sim P$, $P \to Q$ is a case stored in the CB. The CBR is to find a solution, Q', to the current problem P', and $Q' \sim Q$. Therefore, CBR or SBR based on Model (8) can also be represented in a 2D way as shown in the Fig. 2.

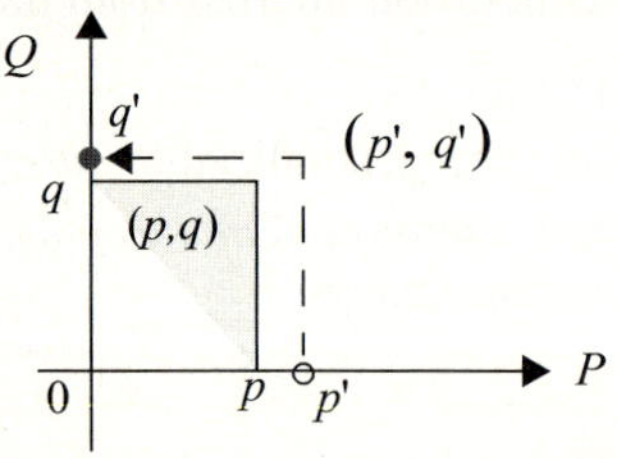

Fig. 1. A 2D representation of fuzzy reasoning

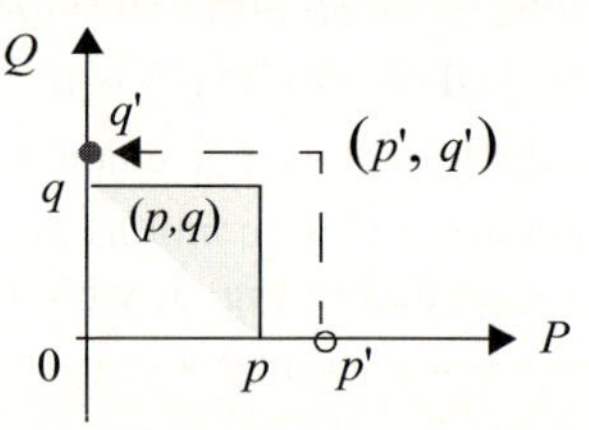

Fig. 2. A 2D representation of CBR and SBR

In this 2D representation, p' and p, q' and q, and (p, q) are the concrete application form of P' and P, Q' and Q, and $P \to Q$ in CBR respectively.

Comparing Fig. 1 with Fig. 2, we find that the 2D representation of fuzzy reasoning and that of CBR and SBR are the same, although they have different semantic interpretations and operational algorithms for performing their own reasoning paradigms based on different real world scenarios. Further, CBR and SBR can also be considered a first quadrant reasoning.

It should be noted that the above 2D representation of CBR is an abstract form of the Leake's model of CBR [4][6].

Based on the above examination, the interesting question arises: Are there other quadrant reasoning paradigms?

5 A Unified 2D Representation of EBR

Experience-based reasoning (EBR) is a reasoning paradigm based on prior experiences. EBR as a technology has been used in many applications [12][14]. Taking into account research and development of CBR, Sun and Finnie [12][18] proposed eight different inference rules for EBR which cover all possibilities of EBR, as shown in Table 1. These inference rules are listed in the first row, and their corresponding general forms with respect to classic logic are shown in the second row respectively [12]. Their corresponding fuzzy logic based forms are shown in the third row respectively [16], which can also be considered as similarity based inference rules respectively taking into account SBR. Because four of them, *modus ponens* (MP), *modus tollens* (MT), *abduction* and *modus ponens with trick* (MPT) are well-known in AI and computer sciences [9][10][11], we do not go into them any more, and focus on reviewing two of the other four inference rules in some detail.

Table 1. Experience-based reasoning: Eight inference rules

MP	MT	abdu- ction	MTT	AT	MPT	IMP	IMPT
P $P \rightarrow Q$ $\therefore Q$	$\neg Q$ $P \rightarrow Q$ $\therefore \neg P$	Q $P \rightarrow Q$ $\therefore P$	$\neg Q$ $P \rightarrow Q$ $\therefore P$	Q $P \rightarrow Q$ $\therefore \neg P$	P $P \rightarrow Q$ $\therefore \neg Q$	$\neg P$ $P \rightarrow Q$ $\therefore \neg Q$	$\neg P$ $P \rightarrow Q$ $\therefore Q$
P' $P \rightarrow Q$ $\therefore Q'$	$\neg Q'$ $P \rightarrow Q$ $\therefore \neg P'$	Q' $P \rightarrow Q$ $\therefore P'$	$\neg Q'$ $P \rightarrow Q$ $\therefore P'$	Q' $P \rightarrow Q$ $\therefore \neg P'$	P' $P \rightarrow Q$ $\therefore \neg Q'$	$\neg P'$ $P \rightarrow Q$ $\therefore \neg Q'$	$\neg P'$ $P \rightarrow Q$ $\therefore Q'$

Inverse modus ponens (IMP) is an inference rule in EBR [12]. For example, if John has enough money, then John will fly to China. Now John does not have sufficient money, then we can conclude that John will not fly to China.

The last inference rule for EBR is *inverse modus ponens with trick* (IMPT) [16]. The difference between IMPT and IMP is "with trick", this is because the reasoning performer tries to use the trick of "make a feint to the east and attack in the west"; that is, he gets Q rather than $\neg Q$ in the *inverse modus ponens*.

It should be noted that the inference rules "with trick" such as MTT, AT, MPT and IMPT are non-traditional inference rules. However, they are really abstractions of some EBR, although few have tried to formalize them. The "with trick" is only a semantic explanation for such models. One can give other explanations for them. For example, one can use "against expectation" instead of "with trick" to explain results of stock trading. Then s/he can use inference rules of "against expectation" to identify and discover the against expectation patterns in the stock trading databases [20].

So far, we have examined fuzzy reasoning, CBR, SBR, and EBR. However, their relationship seems not clear or at least lacks a graphic evidence. Further, the existing studies on them can be essentially considered as one dimensional (1D) approach, because the fundamental inference rule of each of these reasoning paradigms is represented in a 1D way. Based on the discussion in Section 3 and Section 4, we represent the other seven inference rules in a coordinate system, in what follows. We will briefly discuss each of the 2D representations owing to the space limitation.

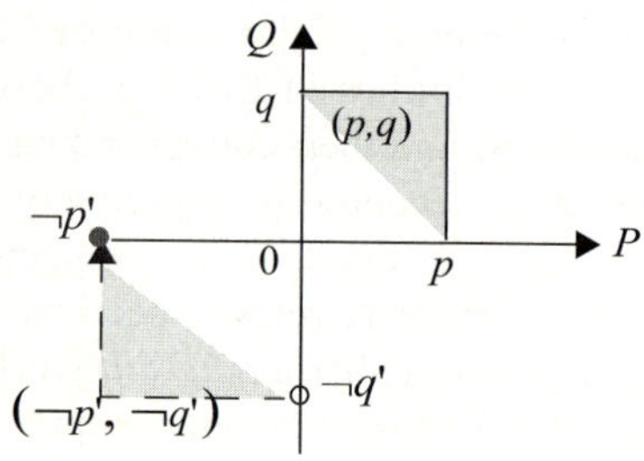

Fig. 3. A 2D representation of fuzzy MT

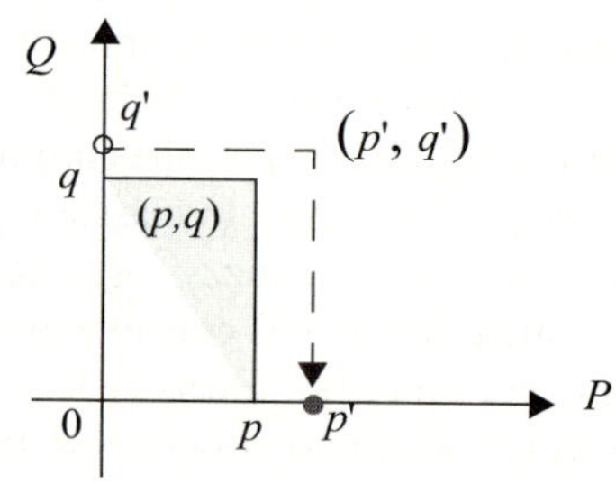

Fig. 4. A 2D representation of fuzzy abduction

Fig. 3 shows the 2D representation of fuzzy *modus tollens* (MT) in EBR, which demonstrates that any reasoning paradigms based on (fuzzy or similarity-based) MT are a non-first quadrant reasoning, because it involves the quantities in the first and third quadrant, we can call it first-third quadrant reasoning.

Fig. 4 shows the 2D representation of fuzzy abduction in EBR, which demonstrates that any reasoning paradigms based on (fuzzy or similarity-based) abduction are a first quadrant reasoning.

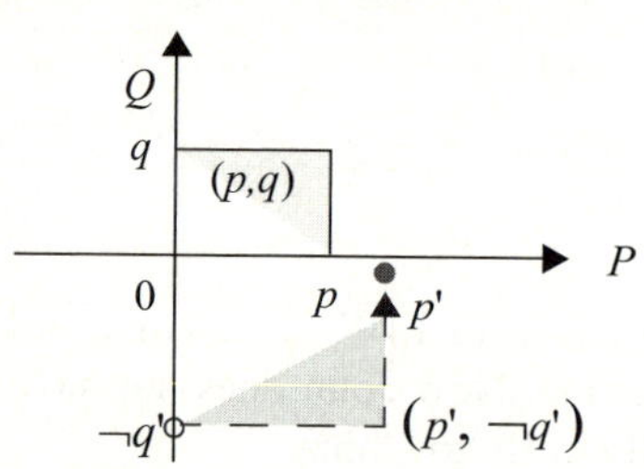

Fig. 5. A 2D representation of fuzzy MTT

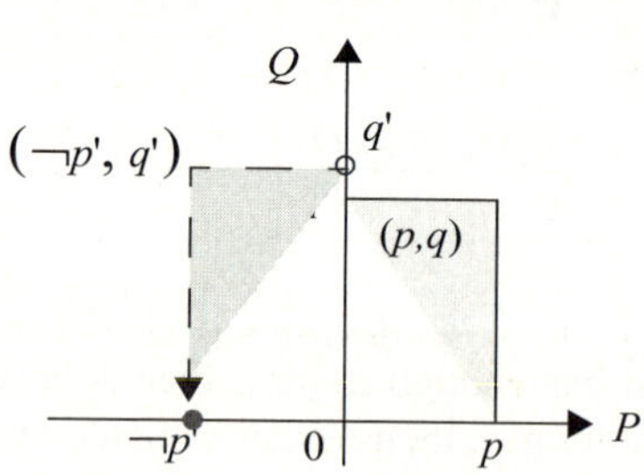

Fig. 6. A 2D representation of fuzzy AT

Fig. 5 shows the 2D representation of (fuzzy or similarity-based) *modus tollens with trick* (MTT) in EBR, which demonstrates that any reasoning paradigms based on (fuzzy or similarity-based) MTT are a non-first quadrant reasoning, but a first-fourth quadrant reasoning.

Fig. 6 shows the 2D representation of (fuzzy or similarity-based) *abduction with trick* (AT) in EBR, which demonstrates that any reasoning paradigms based on (fuzzy or similarity-based) AT are a non-first quadrant reasoning, but a first-second quadrant reasoning.

Fig. 7 shows the 2D representation of (fuzzy or similarity-based) *modus ponens with trick* (MPT) in EBR, which demonstrates that any reasoning paradigms based on (fuzzy or similarity-based) MPT are a non-first quadrant reasoning, but a first-fourth quadrant reasoning.

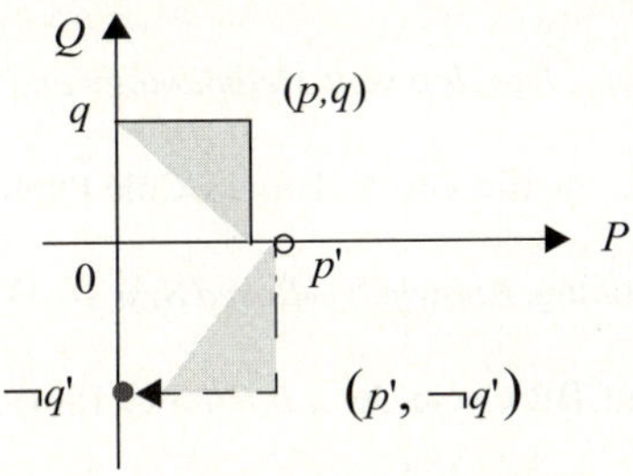

Fig. 7. A 2D representation of fuzzy MPT

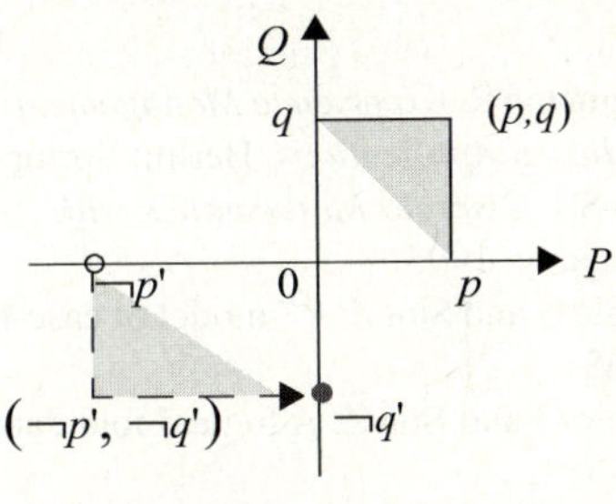

Fig. 8. A 2D representation of fuzzy IMP

Fig. 8 shows the 2D representation of (fuzzy or similarity-based) inverse *modus ponens* (IMP) in EBR, which demonstrates that any reasoning paradigms based on (fuzzy or similarity-based) IMP are a non-first quadrant reasoning, but a first-third quadrant reasoning.

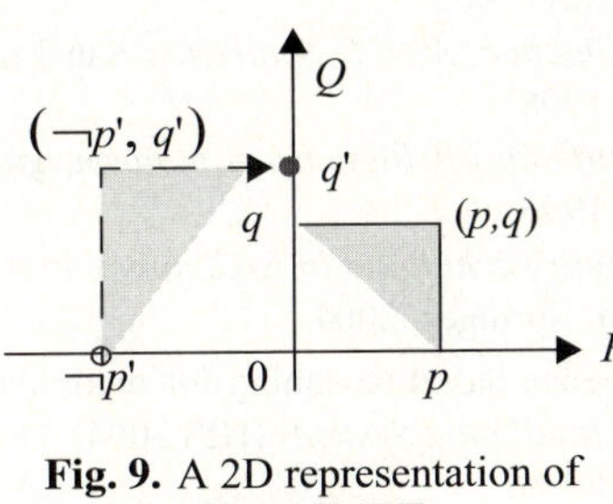

Fig. 9. A 2D representation of fuzzy IMPT

Fig. 9 shows the 2D representation of (fuzzy or similarity-based) *inverse modus ponens* with trick (IMPT) in EBR, which demonstrates that any reasoning paradigms based on (fuzzy or similarity-based) IMPT are a non-first quadrant reasoning, but a first-second quadrant reasoning.

6 Concluding Remarks

This paper examined fuzzy reasoning, CBR, similarity based reasoning and EBR. It then proposed a 2D approach that represents fuzzy reasoning, CBR, and EBR in a unified way. This approach also integrated many reasoning paradigms such as abduction, deduction and similarity-based reasoning into a unified treatment. The proposed approach will facilitate research and development of fuzzy systems, knowledge/experience management, knowledge-based systems, and case-based reasoning.

In future work, we will examine the 2D reasoning properties (such as continuity of reasoning paradigms) of fuzzy reasoning, CBR, similarity based reasoning and EBR and look at the application of the proposed 2D representation of reasoning paradigms in AI, e-commerce and e-services.

References

[1] Bergmann R. *Experience Management: Foundations, Development Methodology and Internet-Based Applications*. Berlin: Springer, 2002

[2] Epp SS. *Discrete Mathematics with Applications*, Pacific Grove: Brooks/Cole Publishing Company, 1995

[3] Finnie G and Sun Z. R^5 model of case-based reasoning. *Knowledge-Based Syst.* 16(1) 2003, 59-65

[4] Finnie G and Sun Z. A logical foundation for the CBR Cycle. *Int J Intell Syst.*18(4) 2003, 367-382

[5] Finnie G and Sun Z. Similarity and metrics in case-based reasoning. *Int J Intell Syst* 17 (3) 2002, 273-287

[6] Leake D. *Case Based Reasoning: Experiences, Lessons & Future Direction.* Menlo Park, California: AAAI Press / MIT Press, 1996

[7] Magnani L. *Abduction, Reason, and Science, Processes of Discovery and Explanation*. NewYork: Kluwer Academic/Plenum Publishers, 2001

[8] Negnevitsky M. *Artificial Intelligence: A Guide to Intelligent Systems* (2nd Edn). Harlow,England: Addison-Wesley, 2002

[9] Nilsson NJ. *Artificial Intelligence. A New Synthesis.* San Francisco, California: Morgan Kaufmann Publishers, Inc. 1998

[10] Russell S and Norvig P. *Artificial Intelligence: A modern approach.* Upper Saddle River, New Jersey: Prentice Hall, 1995

[11] Sun Z and Finnie G. *Intelligent Techniques in E-Commerce: A Case based Reasoning Perspective*, Heidelberg, Berlin: Springer, 2004

[12] Sun Z and Finnie G. Experience based reasoning for recognising fraud and deception. In: *Proc. Inter Conf on Hybrid Intelligent Systems* (HIS 2004), December 6-8, Kitakyushu, Japan, IEEE Press, 2004, pp. 80-85

[13] Sun Z and Finnie G. A unified logical model for CBR-based e-commerce systems. *Int J Intell Syst* 20(1) 2005, 29-46

[14] Sun Z and Finnie G. MEBRS: A multiagent architecture for an experience based reasoning system. *LNAI 3681*, Berlin Heidelberg: Springer-Verlag, 2005, pp. 972-978

[15] Sun Z and Finnie G. Experience management in knowledge management, *LNAI 3681*, Berlin Heidelberg: Springer-Verlag, 2005, pp. 979-986

[16] Sun Z, Finnie G, and Sun J. Four new inference rules for experience-based reasoning, IFSA 2005, Beijing, July 28-31 2005. In: Liu Y, Chen G, and Ying M (eds) *Fuzzy Logic, Soft Computing and Computational Intelligence* (IFSA2005), Beijing: Tsinghua-Springer, 2005,pp.188-193

[17] Sun Z, Finnie G, and Weber K. Abductive case based reasoning. *Int J Intell Syst* 20(9) 2005, 957-983

[18] Sun Z and Finnie G. Experience-based reasoning: A similarity-based perspective. Submitted for publication in *Int J of Intell Syst*, 2006, under review

[19] Torasso P, Console L, Portinale L, and Theseider D. On the role of abduction. *ACM Computing Surveys*, 27 (3) 1995, 353-355

[20] Yang S, You X, Chen F, and Z hang S. Stock-management-based expectation pattern discovery, *Asian Journal of Information Technology* 4 (11) 2005, 1086-1092

[21] Zimmermann HJ. *Fuzzy Set Theory and its Application* (3rd Edn). Boston / Dordrecht / London: Kluwer Academic Publishers, 1996.

MBR Compression in Spatial Databases Using Semi-Approximation Scheme*

Jongwan Kim[1], SeokJin Im[1], Sang-Won Kang[1], SeongHoon Lee[2], and Chong-Sun Hwang[1]

[1] Department of Computer Science and Engineering, Korea University, Seoul, Korea
wany@korea.ac.kr, {seokjin, swkang, hwang}@disys.korea.ac.kr
[2] Division of Information and Communication, Cheonan University, Cheonan, Korea
shlee@cheonan.ac.kr

Abstract. Studies on spatial index, which is used for location-based services in mobile computing or GIS have increased in proportion to the increase in the spatial data. However, these studies were on the indices based on R-tree, and there are a few studies on how to increase the search performance of the spatial data by compressing MBRs. This study was conducted in order to propose a new MBR compression scheme, SA (Semi-approximation), and a SAR-tree that indexes spatial data using R-tree. The basic idea of this paper is the compression of MBRs in a spatial index. Since SA decreases the size of MBR keys, halves QMBR enlargement, and increases node utilization. Therefore, the SAR-tree heightens the overall search performance. The experiments show that the proposed index has increased performance, higher than that of the pre-established schemes on compression of MBRs.

1 Motivation

A spatial database system, which services GIS or LBS information, requires a large memory capacity and a long processing time for management of spatial objects. In order to manage spatial queries effectively, there should be an appropriate spatial index and a technology for query management.

For several years, the increase in spatial data has led to more studies on spatial indexing. The majority of these studies, however, have been based on R-tree, and few have focused on increasing the search performance of spatial data by downsizing the index. This study proposes a semi-approximation R-tree (SAR-tree) that indexes spatial data. The basic concept is to compress the MBRs in the spatial index. By decreasing the size of MBR keys, SA halves QMBR enlargement, increases node utilization, and improves overall search performance. So far, this is the first proposed method that decreases QMBR enlargement.

SAR-tree increases the number of node entries by compressing the MBR keys. The greater number of node entries lowers tree heights, thereby reducing the number of node accesses when processing queries, which leads to faster query results by decreasing disk I/O. The index structure is constructed by including MBR for sub-node details at each node and storing compressed MBR information in the entries.

* This work was supported by the second Brain Korea 21 Project in 2006.

B. Gabrys, R.J. Howlett, and L.C. Jain (Eds.): KES 2006, Part I, LNAI 4251, pp. 1124 – 1130, 2006.
© Springer-Verlag Berlin Heidelberg 2006

This study mathematically analyzes the number of node accesses in a 2-D space, and evaluates the performance of SAR-tree using real location data. The results show that the proposed index performs better than established MBR compression methods.

2 Related Work

A spatial index is used to index the locations of geographical objects. Many studies have evaluated how various index structures handle spatial data, and studies on spatial indexing have improved search performance by setting options in existing indices using R-tree [1]. X-tree [2], which adds the concept of a super node, performs better than R-tree. SR-tree [3] constructs an index using both MBRs and minimum bounding spheres (MBSs).

In R-tree, MBR key values occupy roughly 80% of the entire index. However, compressing MBR keys allows more entries at each node [3], and the greater number of node entries lowers tree heights and thus improves system performance. Methods to downsize an index using MBR compression in a 2-D space include relative representation of MBRs (RMBR); hybrid representation of MBRs (HMBR) [4], which uses relative coordinates; quantized representation of MBRs (QMBR) [5]; and the use of virtual bounding rectangles (VBRs) [6]. Although all of these methods decrease MBR key size, there are differences. RMBR and HMBR calculate MBRs' offset in a search space, while QMBR utilizes quantization to divide the search space into grids in accordance with a quantization level.

The VBR concept was proposed for A-tree [6], which quantizes a region in order to store the keys in fewer units. In other words, a VBR is created by expanding a MBR to the closest quantization unit. However, QMBR-related methods increase the size of the MBR to that of the quantized area, enlarging the search region and causing MBRs to overlap. This increases the number of node accesses, decreasing overall search performance.

RMBR and HMBR calculate the relative distance of keys from the starting coordinates of the search space, thereby storing the keys in less byte. In RMBR, 16 bytes is reduced to 8 bytes, while in HMBR the same keys are stored in 6 bytes. However, HMBR has the disadvantage of requiring two more bytes than QMBR, which decreases the number of entries at each node and thereby increases query times.

3 MBR Compression and Implementation

3.1 Preliminaries

The RMBR compresses the key-values by calculating the offsets of the MBR in each entry in comparison with the search space. Since a axis of MBR is generally stored in 4 bytes, the coordinates occupy 16 bytes as shown by R0 in Fig. 1. It saves 8 bytes by storing the keys of the MBR in each entry as relative representation. The bucket utilization of a node is increased further in decreasing the size of each axis to 2 bytes by calculating the offsets in comparison with the end coordinate of R2. In case of an HMBR, the relative coordinates of the MBR is calculated in its height and width in comparison with the starting point of the same MBR, not in comparison with the entire search space. The lower left corner of HMBR is identical with that of RMBR, but the lengths of MBR represent the distance to upper right corner.

Compression of keys using a space quantized into n-numbers by a given number saves even more key space than an RMBR. This method is referred to as a QMBR. Fig. 2 shows the result of quantizing the x- and y-axes into 16×16, respectively. If a quantized level is smaller than 256, each coordinate is stored in one byte. In Fig. 1 and Fig. 2, although the keys are stored in compression scheme, an RMBR and an HMBR require at least 8 bytes and 6 bytes respectively. QMBR affect the search performance as expanding MBRs, causing overlapping of MBRs.

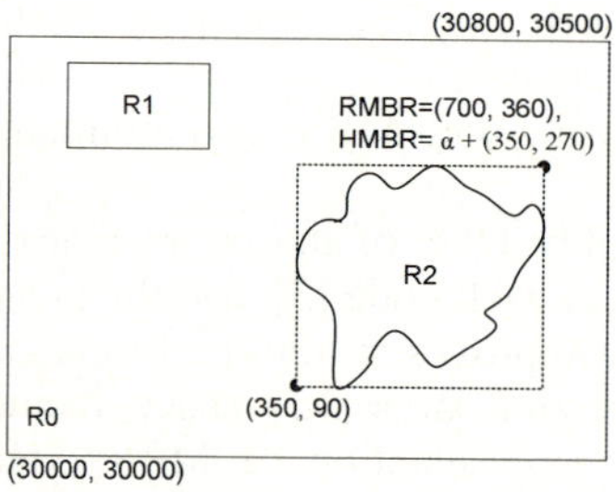

Fig. 1. RMBR and HMBR

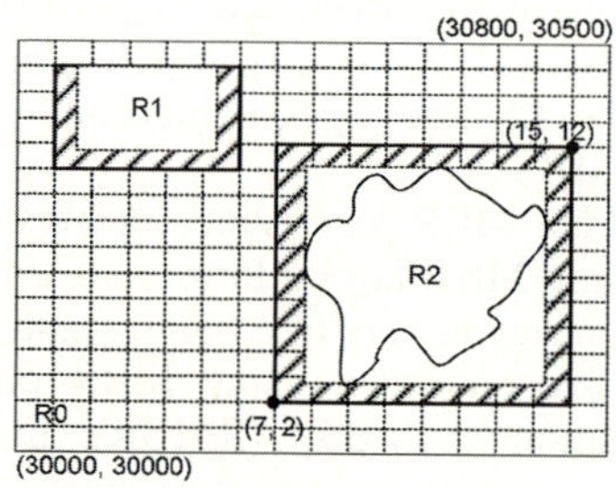

Fig. 2. Quantization MBR

3.2 Semi-Approximation MBR

Since an n-dimensional rectangle such as MBR is viewed as a $2n$-dimensional point [7], the compression of the points increases the search performance and save the space of an index. Four points represent a two-dimensional rectangle. In the semi-approximation (SA) scheme, each node has a MBR, which is composed all entries stored in that node. The MBR of two-dimensional space is represented by two end points (α, β), and each point is represented as $\alpha=(\alpha x, \alpha y)$ and $\beta=(\beta x, \beta y)$ respectively($\alpha \leq \beta$). The idea of the SA scheme is to represent α as relative values and to quantize β in order to decrease the false-search region in half that appears in a QMBR, which quantizes both α and β values. A false-search region refers to the region in which no object is found because the MBRs of real objects are expanded. Thus, decreases of such region heighten the search performance.

The endpoint β is transformed into Q_e according to the quantized level, which minimizes the storage space required in bits. The quantized levels, 2^n ($n = 0, 1, 2, \ldots, 8$), are represented by 1 byte. Because β is determined by the quantized level q, it can be represented by a bit string. In other words, the binary representation is $(Q_e)_2$ and the length of the bit string is $\log_2 q$. The endpoint is stored as $(Q_e - 1)_2$. For example, if $Q_e(\beta.x) = 15$ and $Q_e(\beta.y) = 12$, the bit string is 11101011, the concatenation of the two binary codes (Fig. 3). As a result, the R2 keys require only 5 bytes of storage.

Definition 1. *Representation of the Relative Coordinates of MBR*
Let M be the MBR of an entire search space; then the lower-left and upper-right corners of M are represented by ($M.lx$, $M.ly$) and ($M.rx$, $M.ry$). The entry, R2, is composed of two points (α, β), and the relative coordinate for the starting point α is as follows:

$$R_M(\alpha) = (|M.lx - \alpha.x|, |M.ly - \alpha.y|)$$

Definition 2. *Quantization of MBR*
Let (M_s, M_e) be the two points of M, and q be the quantized level, then β is defined as follows, where Q_e is the endpoint of a quantized MBR:

$$Q_e(\beta) = \begin{cases} 1, & (\beta = M_s) \\ \lceil ((\beta - M_s)/(M_e - M_s)) \times q \rceil, & (otherwise) \end{cases}$$

The SA appears only in the area of deviant crease lines of the expanded space, viz. the false-search region, in Fig. 3. The false-search region is half of a QMBR. The space of coordinates also decreases up to the minimum of 5 bytes. Although high-quantized levels increase the number of bits, thereby increasing the stored bytes of keys, the SA scheme still maintains its advantage over other methods.

In Fig. 4, MBRs are represented by semi-approximation scheme. R0 represents the search region, and the areas of R1 through R7 contain the MBR information (solid line). Each MBR is quantized as it is expanded in the directions of the x- and y-axes. Since R2 is facing the boundary of the parent MBR, the coordinate of its expanded area is the maximum value of its quantized level. R3 and R4 are expanded into the area of R1 (dotted line). In case of R7, the expanded areas in the upper and right sides represent the endpoints.

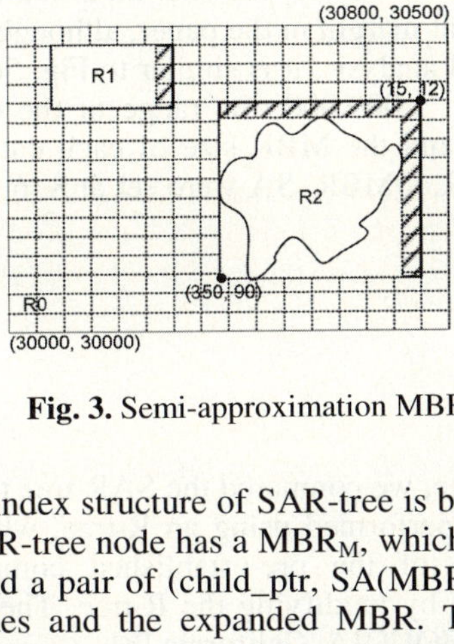

Fig. 3. Semi-approximation MBR

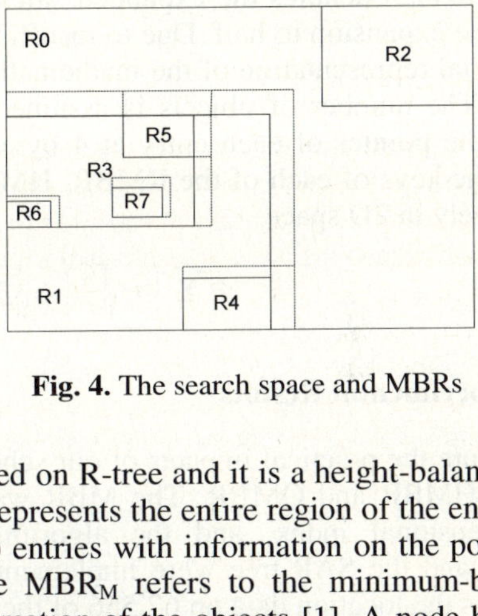

Fig. 4. The search space and MBRs

The index structure of SAR-tree is based on R-tree and it is a height-balanced tree. The SAR-tree node has a MBR_M, which represents the entire region of the entries in a node and a pair of (child_ptr, SA(MBR)) entries with information on the pointers of sub-nodes and the expanded MBR. The MBR_M refers to the minimum-bounding rectangle based on the minimum approximation of the objects [1]. A node has up to the maximum *m* number of entries and has a flag that distinguishes whether the node is an internal node or a leaf one as well as the number of stored entries. The real MBR of an entry is calculated from the SA(MBR), the parent MBR, and the sub-MBR. Accurate information on each of the entries is used to prune the nodes.

4 Performance Evaluations

4.1 Analysis of the Number of Node Accesses

A mathematical analysis of the number of node accesses in R-tree is outlined in [5]. All nodes are assumed to have MBRs of the equal height. Defining the average region of a node as a_h in a tree with height h, a_h of each node is $1/M_h$. Let M_h denote the number of nodes at the height of h. The probability that a node of height h will overlap a given query region is $(\sqrt[d]{s}+\sqrt[d]{a_h})^d$. Let d be a dimension and s be the size of the query region, then the overlapping region of nodes with a height of h and the query region is $M_h(\sqrt[d]{s}+\sqrt[d]{a_h})^d$. The total number of node accesses from the root to the leaf nodes in R-tree is comprised of the summation of nodes at each height, as represented by (1) (*N*: the total number of data; *f*: the average number of fan-out of leaf nodes):

$$1+\sum_{h=1}^{\lceil \log_f N \rceil -1}\left(1+\sqrt[d]{\left\lceil \frac{N}{f^h} \right\rceil}\cdot s\right)^d \tag{1}$$

When the quantized level q is applied, each node has a quantized cell of q^d. Since the access to the nodes in QMBR is first conducted at nodes of height h and then visit the sub-node, the probability is $(\sqrt[d]{s}+\sqrt[d]{a_h}/q+\sqrt[d]{a_{h-1}}+\sqrt[d]{a_h}/q)^d$. Since this applies to all of the nodes, by multiplying by M_h and adding from the root and leaf nodes, the total number of node accesses is shown in (2). The QMBR scheme accesses more nodes than the MBR scheme because the MBRs are bigger than the real MBRs owing to quantization.

$$1+\sum_{h=1}^{\lceil \log_f N \rceil -1}\left(1+\sqrt[d]{\left\lceil \frac{N}{f^h} \right\rceil}\cdot s+\sqrt[d]{\left\lceil \frac{N}{f^{h+1}} \right\rceil}\cdot s/q\right)^d \tag{2}$$

Equation (2) denotes the expanded sides of the MBR, and it is modified as (3) to reduce the expansion to half. Due to insufficient margin in the paper, although there is no pictorial representation of the mathematical analysis, it is similar to Fig. 5(b) in its pattern. The number of objects is assumed at 1,000,000; the range of the query at 0.01%; the pointer of each entry at 4 bytes; and the MBR size of each entry at 16 bytes. The keys of each of the RMBR, HMBR, QMBR, SA were set at 8, 6, 4 and 5 respectively in 2D space.

$$1+\sum_{h=1}^{\lceil \log_f N \rceil -1}\left(\left(1+\sqrt[d]{\left\lceil \frac{N}{f^h} \right\rceil}\cdot s+\sqrt[d]{\left\lceil \frac{N}{f^{h+1}} \right\rceil}\cdot s/q\right)/2\right)^d. \tag{3}$$

4.2 Experimental Results

To measure the practical impacts of our scheme, we compared the SAR-tree to MBR, RMBR, HMBR and QMBR. The MBR was performed using an R-tree, which is a two-dimensional index, and the algorithms of the pre-established compression schemes and the SAR-tree were implemented by modifying the R-tree. The data of this test is the location data on 62,556 of the SEQUOIA, California [8].

The measurement of performance in terms of processing queries was conducted for a range query, and the proportion of the query region in the entire search space was set at a range between 5% and 30%. We generated 10,000 different query rectangles of the same size and the results were accumulated. In Fig. 5(a), the number of node accesses is low than un-compressed MBR in each query region. This is attributable to the raised fan-out of each node due to the lowered MBR keys. As a result, the number of node accesses is decreased more than R-tree's MBR. In the experiments, the RMBR and HMBR have better results than the QMBR, which is due to the false-search region the QMBR has. Changing the size of the nodes reduces the number of node accesses because the other schemes have more entries in a node than MBR. In Fig. 5(b), SA has more entries than the HMBR, whose keys takes up 6 bytes, and thus, SA has lowered number of node accesses.

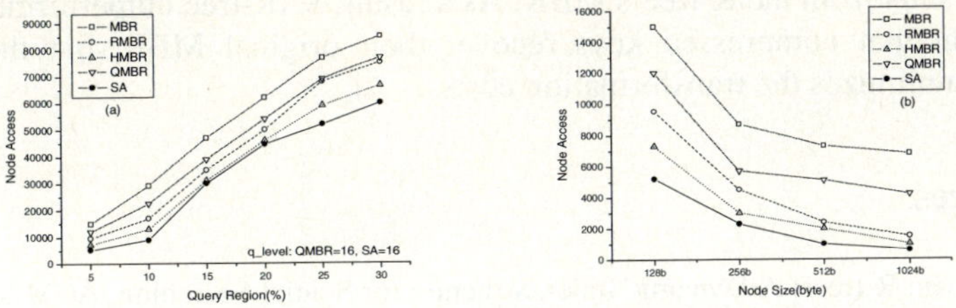

Fig. 5. Number of Node Accesses of Query region (a), Node size (b)

In Fig. 6, the search time is similar to the node access patterns. As the node size grows, the search time approaches quickly to the minimum because the search region of the QMBR increases in proportion to the increased number of entries in each node, it has performance worse than that of other methods.

On the other hand, the entries of the RMBR and HMBR increase in their numbers with the reduced keys, showing better performance. What we should remember is the fact that, even though the keys of the QMBR were reduced to 4 bytes, the backtracking is caused due to the false-search region, which is enlarged by quantization. Thus, the search time increases. Fig. 7 shows the accumulated the number of node accesses in the QMBR and the SA by adjusting the quantized sizes. The QMBR, is changing down slowly but the SA radically decreases at q=16 where the key is stored in 5 bytes, and gradually increases from onward. It represents much less frequency than the QMBR.

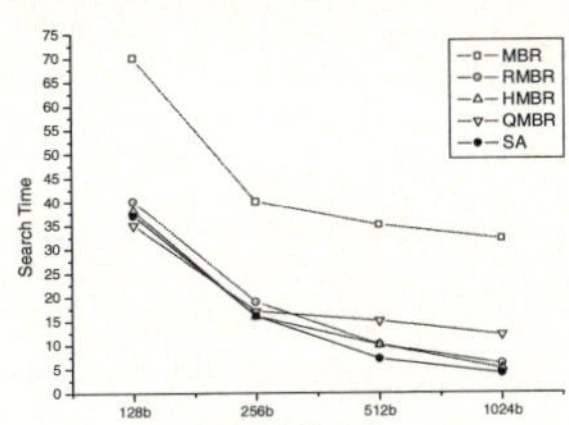

Fig. 6. Search time of nodes

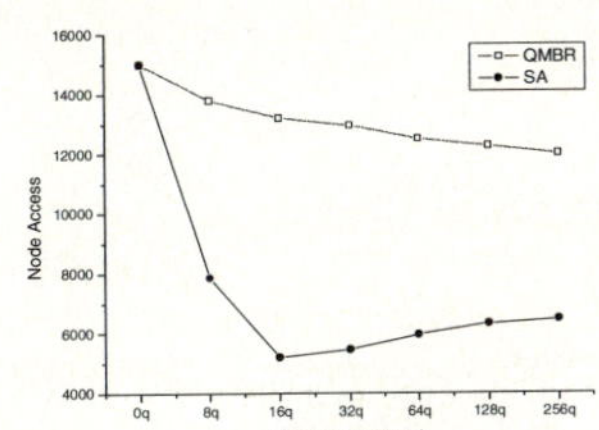

Fig. 7. Number of node accesses of q levels

5 Conclusion

We introduced a new MBR compression scheme, SA, and proposed a SAR-tree, a new spatial index that drastically reduces the MBR size in this paper. Traditional spatial index structures rarely show indices constructed using the MBR compression scheme. The pre-established compression schemes have decreased the MBR keys, but SA scheme decreases the size of MBR keys even further and increasing the search performance by reducing the region of enlargement of the QMBR in half.

The results show that the number of node accesses in the SAR-tree measured by changing the query region, the node sizes, and the quantized levels have distinctively different results from the R-tree's MBR. As a result, SAR-tree outperformed R-tree. It is important that compressed keys recover their original MBR, but the proposed algorithm minimizes the transformation costs.

References

1. A. Guttman: R-trees: A Dynamic Index Structure for Spatial Searching. ACM SIGMOD Int. Conf. on Management of Data (1984) 47-57
2. S. Berchtold, D.A. Keim, H.P. Kriegel: The X-tree: An index structure for high-dimensional data. Proc 22nd Int. Conf. on VLDB (1996) 28-39
3. N. Katayama, S. Satoh: The SR-tree: An index structure for high-dimensional nearest neighbor queries. Proc. ACM SIGMOD Int. Conf. on Management of Data (1997) 396-380
4. J.D. Kim, S.H. Moon, J.O. Choi: A Spatial Index Using MBR Compression and Hashing Technique for Mobile Map Service. DASFA (2005) LNCS3453 625-636
5. K.H. Kim, S.K. Cha, K.J. Kwon: Optimizing Multidimensional Index trees for Main Memory Access. Int. Conf. on ACM SIGMD (2001) 139-150
6. Y. Sakurai, M. Yoshikawa, S. Uemura, H. Kojima: Spatial indexing of high-dimensional data based on relative approximation. VLDB J. (2002) 93-108
7. J. Goldstein, R. Ramakrishnan, U. Shaft: Compressing Relations and Indexes. Proceedings of IEEE Conference on Data Engineering (1998) 370-379
8. The R-tree Portal: http://www.rtreeportal.org
9. H. Schwetman: CSIM19: A Powerful Tool for Building System Models. Proceedings of the 2001 Winter Simulation Conference (2001) 250-255

Annotations: A Way to Capture Experience

Gaëlle Lortal[1], Myriam Lewkowicz[1], and Amalia Todirascu-Courtier[2]

[1] CNRS – ICD Tech-CICO, Université de Technologie de Troyes,
12, rue Marie Curie 10010 Troyes Cedex, France
{lortal, lewkowicz}@utt.fr

[2] Linguistic, Languages and Speech - EA 1339, University Marc Bloch,
22, rue Descartes 67084 Strasbourg Cedex, France
amalia.todirascu@umb.u-strasbg.fr

Abstract. This paper describes a cooperative tool used by a mechanical engineering team carrying out an asynchronous and distributed work. The tool supports communication phases as well as document and product creation phases by means of annotations. Memorizing all the annotations and creating documents from these annotations enable the team members to capture their experience. Annotation is here defined as a continuum fulfilling several purposes from communication and argumentation to indexation.

Keywords: Annotation, Knowledge Management, Experience, Indexation, Communication.

1 Introduction

Distributed collective practices involve multiple actors sharing multiple documents. During a project, members handle documents such as texts, plans, models, specifications. During asynchronous phases of work, they exchange comments, they modify documents and they can negotiate some parts of new documents being created. As they have these activities of interpreting and reviewing documents, they build up their experience. To support these activities, annotation is an interesting tool. Annotation enables communication and argumentation, and might facilitate retrieval of past experiences acquired during a project. We assume that annotation enables the team members to capture the experience developed in asynchronous work. In an experience management framework, we aim at capturing and collecting experience through annotation gathering [27].

Annotations are traces of experience linked to a document. Once captured and structured according to a classification scheme, they can easily be retrieved by project members [22]. Annotations also contain member's expertise and enrich the document with traces of experience. The enriched document is then a matter for Knowledge Management (KM) techniques. In sharing their expertise via annotations, project members create new intellectual capital [8]. This social capital is built by shared interpretations of documents. We can support project members during asynchronous phases of work by a groupware capturing traces of experience.

B. Gabrys, R.J. Howlett, and L.C. Jain (Eds.): KES 2006, Part I, LNAI 4251, pp. 1131 – 1138, 2006.
© Springer-Verlag Berlin Heidelberg 2006

We expose here the behaviour of a mechanical engineering team carrying out an asynchronous and distributed work, from preliminary design phase to production phase. Members of this team discuss design details based on project documents such as description and plans, and they send newly-created documents to the others. On the basis of a distributed collective practices analysis, we describe a scenario used to design a computer-based annotation tool. We classified activities of our scenario, and for each, we describe a function that can support it and how we aim at implementing this function or reuse an existing annotation tools function. Our tool, which is capturing the experience, enables activity awareness [14] and collective sense making [28].

2 Annotation to Capture Experience

Annotations are active elements of document creation and parts of a written document. From [9], we can define several types of annotation: (1) The gloss which is a fragment of text explaining a part of a document, (2) the underlining mark pointing parts of a document, (3) the note paraphrasing the main point of a document, (4) the comment bringing out new ideas, and (5) the discursive comment which is an arguing and organised comment built cooperatively on exchanges among authors. These fragments are firmly linked to a document or to a topic. This relational feature of annotation is the one allowing Semantic Web (SW) [6] to consider annotation as enriching the document in order to improve automatic document indexing and retrieval [16]. Annotation is then used for Electronic Document Management (EDM), document content processing, structuring a document, services interoperability, and for some specific cooperation types (as in [21]'s scenarii of cooperation).

Annotation is also an activity enabling document interpretation and interpretation-sharing with other project members. The document is considered as poor, since it only contains data and information. In our framework, annotating a document means enriching it by the advices and innovative ideas of a member, i.e. a project context [1]. As a member links an annotation to a document, he/she enriches it with traces of his/her interpretation rising from his/her experience.

Annotating also helps forming representation of an event. Individual experience is then built while writing and exchanging. By means of annotation, project members built their experience and let traces of their expertise within project documents. They enrich them with individual knowledge built during the project. This knowledge is built cooperatively and shared by other members through exchanges of interpretation. SW techniques for EDM are not then subtle enough to manage knowledge and social experience.

Capturing annotations and tracing experience through annotation then becomes crucial in memorizing an experience built cooperatively during asynchronous phases of work. As we aim at memorizing this experience, we designed a tool-supporting annotation activity in order to assist collective sense making [28]. This tool is based on the observation of a mechanical engineering project presented in the following section.

3 Scenario-Based Requirements

The scenario is within the context of a project involving an association committee (AC) (mainly plane pilots), and a mechanical engineering team (divided into a Design Team -DT- and a technical team -TT-). The aim is to reuse a car-engine as an aero-engine. Team members participate in the project during their spare time. So, despite members' co-location, the work is asynchronous. Each DT member develops a part of the engine. Mechanical engineering design is a highly collaborative field and consequently, the design team needs perpetual feedback from TT (technicians and suppliers). DT then relays feedbacks from the AC managing expenses. Drawings and documents represent discussion basis between TT and DT about technical feasibility and about tool or material availability. TT mainly manages communication with suppliers. Within the DT, communication is widespread even if information should always be forwarded to the group supervisor. The only medium available in this team to mediate this asynchronous written communication is email. Email is used as a comment coming with a document as in these following descriptions:

1. A member creates a first draft with computer-aided design tools, each member then discusses this draft, asking for complement or verifying calculus. To do so, team members write down their ideas and questions as comments anchored to the draft(s). Each member puts his/her annotation on any documents, but a subtle micro-organization structure within the team involves some ethical rules. After the revision and the updating process, detailed plans are then communicated to the TT in order to begin production phases. TT receives drawing plans and their comments, which explain how the DT made a decision.
2. These annotations are often merged as new documents following an explaining purpose (for example by listing and explaining in an email, all the modifications which are visible in the document).
3. Annotations are identified as elements of solutions, of negotiations and of discussions, helping design process retrieval. As our teams need to share and manage an important number of documents, an adapted classification should be set up, based on several pieces of information: author, date, content, aim, recipient, which part of the document/plan it comments, etc. Since it is an innovative project, they can only partly define an a priori classification for documents to be produced. Their classification is possibly extended as work progresses. They try to organize documents and annotations by storing their mail by date, title and author. Here they need a tool to support their classification and share their storage.

This situation description shows us that indexation and communication functionalities are crucial to understand on which basis a solution has been adopted. It is also crucial for experience building. Annotation is used for communication, indexation and is a part of the experience elaborated during the project. We are now going to look at existing annotation tools which could provide necessary functions to support these different activities.

3.1 Supporting Communication

Supporting interpretation means handling annotations as creating fragments of discourse and enabling discourse by creating annotation threads. Functions, such as document fragments selection (highlighting, circling…), discourse fragments anchoring to documents or other annotations (answering, multi-anchoring…), are then necessary.

Annotation for communication is partly supported by the SoW (Social Web) approach [19], enabling mediated communication by comments. Newsgroup, blog, or wiki [10] enable on-line "polylogue" [24] (dialogue among more than two speakers) by publishing messages related to a discussion thread. Conversation fragments are not yet structured, which leads to topic digression [24] and decay phenomena [18]. Annotation as a comment is supported by annotation clients ([11], [20], [25]) which sort annotation on rudimentary metadata (creation date, author). Annotation could be stored apart on annotation servers, so, differentiated from the document [29] or not [2]. However, these tools do not allow connecting annotations nor structuring exchanges between users about a document. D3E [26] also considers documents as discourse media but does not allow a rich indexation of annotations. These tools do not focus on tracing the design rationale underlying the discussion enabling an *a posteriori* understanding of exchanges which took place.

3.2 Supporting Indexation

Once created, an annotation should be indexed so as to be easily retrieved. Browsing is based on annotation indexing. Indexation allows structuring annotations in a browsable knowledge map as Topic Maps formalism allows [7].

Several annotation clients [23] are available from Semantic Web initiatives (SW). "Semantic annotation" of the SW is "computational annotation". An objective of SW is to index Web pages, and allow search engines a better information recall structuring annotation by underlying ontologies ([17], [12], [13]). These annotations lightly support readers of a page to cooperate or to interact. Magpie [15] uses annotations to support human interpretation and enables multiple viewpoints indexation. These tools enable additional annotation to a document, which permits the DT to explain the TT when and on which document a decision was made. But these annotations only help users to structure or to share her/his understanding of the text. To index subtly these fragments, users should be involved. But to support users in this time-consuming task, we suggest using Natural Language Processing tools to offer users domain specific terms and annotation arguing types.

As we have already claimed, we do not only need annotations to index but also to negotiate. Our purpose is to support document interpretation or recall of an existing interpretation but also to support the creation of new ideas (from collective interpretation). So, annotation is crucial to index and then to capture experience, but we also need a tool enabling exchanges via annotations i.e. enabling discourse around a document.

3.3 Supporting Interpretation

Thirdly, users should be able to create new documents which gather together ideas emerging from collective brainstorming and exchanges around a document. Our tool should contain a gathering functionality allowing creation of a new document to work on.

DT needs to create documents collaboratively i.e. to re-use, re-structure and rewrite existing comments to build a draft on which members can work or to build a *final* document. As we have shown above, two families of annotation tools are available: one focuses on Web pages indexation, while the other focuses on human communication through comments. We can deplore the lack of annotation management or the poverty of cooperative functions in these two families. [26], [15] are the first steps in linking these two points of view. We thus propose to design a tool combining functions supporting SoW and SW activities (i.e. to comment, to answer, to organize).

4 Ant&Cow

AnT&CoW [23], an annotation tool which is still under development, roughly implements these functionalities. It re-uses Annozilla [4], an open-source annotation plugin for the Mozilla-Firefox browser, which follows W3C Annotea protocol [3]. We improved Annozilla in order to facilitate communication and collecting experience. We consider that, in so doing, we are improving activity awareness and expertise-sharing. When a user launches Annozilla, it appears as a frame on the left of the screen. Annotations posted by several members are stored on a server and can be classified according to several viewpoints defined by the group. The user chooses a part of a document to anchor his/her annotations, creating links between several fragments. The link is explained by the argument written by the annotation's author in

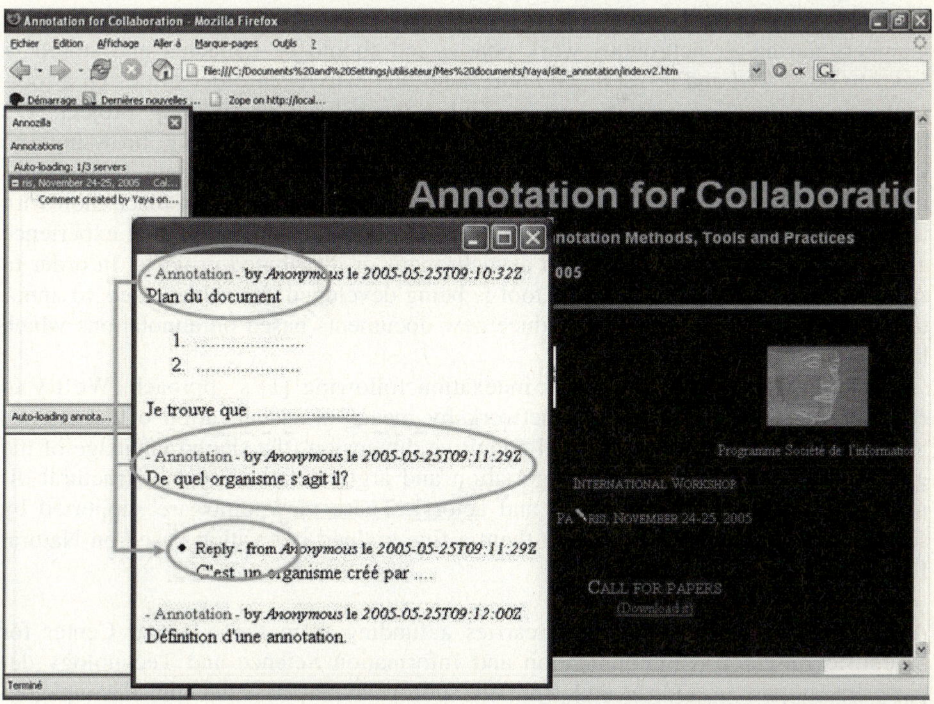

Fig. 3. AnT&CoW generating draft functionality

the annotation body. This body could also be annotated and then it can anchor an annotation. Multi-anchoring and replying could be the genesis of a new document. From the note paraphrasing an external document, new ideas are brought out, and form a discursive comment. The cooperative structuring of this discursive comment during discussion around a document can form a new document. For the moment, our tool enables multi-anchoring only on one document, but we hope to go beyond technical limits in order to enable multi-anchoring among several documents. AnT&CoW allows also the composition of a document representing a collective interpretation. It is called "generate draft" as the new document built is a gathering of linked fragments. Gathering these fragments is done manually by choosing fragments or automatically after a keyword selection. Gathering different fragments written by several authors, means gathering members' viewpoints in one document. This new document is chronologically arranged and represents visualization of experience by showing annotations, their indexes, and their links (see rings in fig.3).

AnT&CoW is still under development but already implements basic collaborative functionalities allowing collective activity awareness, collective sense making and expertise-sharing through communication and indexation means.

5 Conclusion

We have defined an annotation as an interpretation enabling project members to share their expertise. Annotation is an experience contextualisation which enriches the document content with knowledge. This knowledge is built cooperatively among members during asynchronous work phases and should be collected and stored in order to be shared and re-used. To understand this activity in order to design a tool-supporting annotation activity, we studied a mechanical engineering project. We brought to light three main families of functionalities, communicating, browsing and interpreting. Interpreting enables annotation gathering in order to create a new document. This document represents an interpretation, or in other words a discussion with its arguments. Annotation is then a central element to trace the individual experience as well as the project experience in asynchronous or distributed projects. In order to support this activity, an annotation tool is being developed. It enables users to annotate existing documents and to produce new documents based on annotations which have been correctly indexed.

Now, we aim at improving this indexation following [1]'s approach. We try to combine repositories with social network by means of visualization of experience following three viewpoints; a social-cognitive dimension (the shared language of the group), a relational dimension (negotiation and argumentation) and a structural dimension (micro-organisation roles and actors). These viewpoints are supported by several technical means and among them a fine-grained indexation based on Natural Language Processing tools.

Acknowledgments. This research carries a funding from the National Center for Scientific Research /Communication and Information Science and Technology department as part of TCAN and from the European Social Fund and Champagne-Ardenne regional council.

References

1. Ackerman M. and Halverson C., Sharing Expertise: The Next Step for Knowledge Management. In: Huysman M. and Wulf V. (eds): Social Capital and Information Technology (2004)
2. Acrobat PDF, http://www.adobe.com/support/techdocs/ac76.htm (2004)
3. Annotea, http://www.w3.org/2001/Annotea/ (2001)
4. Annozilla, http://annozilla.mozdev.org/ (2004)
5. Bergmann R., Experience Management: Foundations, Development Methodology, and Internet-Based Applications. Lecture Notes in Artificial Intelligence. (2002)
6. Berners-Lee T., Hendler J., et Lassila O. The Semantic Web, Scientific American (2001)
7. Biezunski, M., Bryan, M., et Newcomb, S. R., « Topic Maps », spécification ISO/IEC 13250, 3 Décembre (1999)
8. Bresnan M., Edelman L., Newell S., Scarbrough H. and Swan J., The Impact of Social Capital on Project-Based Learning. In: Huysman M. and Wulf V. (eds): Social Capital and Information Technology. (2004)
9. De Libera A., La philosophie médiévale, Paris, PUF (« Que sais-je ? » 1044), 4e éd. (2000)
10. Delacroix J., Les Wikis M2 éditions (2005)
11. Denoue, L., et Vignollet, L., An annotation tool for Web browsers and its applications to information retrieval, in proceedings of RIAO 2000 (2000)
12. Dingli A., Next Generation Annotation Interfaces for Adaptive Information Extraction. In 6th Annual Computer Linguists UK Colloquium (CLUK 03), January, 2003, Edinburgh, UK (2003)
13. Domingue J.B., Lanzoni M., Motta E., Vargas-Vera M., et Ciravegna F., Mnm: Ontology driven semi-automatic or automatic support for semantic markup. In 13th International Conference on Knowledge Engineering and Knowledge Management (EKAW02), October (2002)
14. Dourish, Paul, & Belotti, Victoria. Awareness and coordination in shared workspaces. Computer Supported Cooperative Work 92 Proceedings, New York: Association for Computing Machinery (1992) 107-115
15. Dzbor, M. - Motta, E. - Domingue, J. B., Opening Up Magpie via Semantic Services. In 3rd ISWC, November 2004, Japan (2004)
16. Handschuh S. and Staab S., Annotating of the Shallow and the Deep Web, in Annotation for the semantic Web, S. Handschuh, S. Staab, IOS Press ,(2003) 25-45
17. Handschuh, S., Staab S., and Ciravegna, F., S-cream - semi-automatic creation of metadata. In 13th International Conference on Knowledge Engineering and Knowledge Management (EKAW02), October (2002)
18. Herring, S. Coherence in CMC. Journal of Computer-Mediated Communication, 4(4). Retrieved January 12, 2005, from http://jcmc.indiana.edu/vol4/issue4/herring.html (1999)
19. Jordan K., Hauser J., Foster S., The augmented Social Network : Building identity and trust into the next-generation Internet, in firstmonday vol.8, N.8, august 4th 2003 http://www.firstmonday.org/issues/issue8_8/jordan/ (2003)
20. Ka-Ping Y. CritLink : Better hyperlinks for the WWW. http://crit.org/ping/ht98.html (1998)
21. Koivunen, M-R, and Swick R.R. Collaboration through annotations in the Semantic Web. In S. Handschuh, S. Staab (eds): Annotation for the semantic Web. IOS Press (2003) 46-60
22. Laclavik M., Gatial E., Balogh Z., Habala O., Nguyen G., Hluchy L., Experience Management Based on Text Notes (EMBET), in Paul Cunnigham and Miriam Cunnigham(eds): Proc. of eChallenges 2005 Conference, Slovenia, Innovation and the Knowledge Economy, Volume 2, Part 1: Issues, Applications, Case Studies IOS Press (2005) 261-268

23. Lortal G., Lewkowicz M., Todirascu-Courtier A. AnT&CoW, a tool supporting collective interpretation of documents through annotation and indexation, In Dieng R., & Matta N.: Proceedings of KMOM-IJCAI Workshop (2005a) 43-54.
24. Marcoccia, Michel. On-line Polylogues: conversation structure and participation framework in Internet Newsgroups. Journal of Pragmatics, vol.36 n°1, (2004) 115-145
25. Price, M., Schilit, B., et Golovchinsky, G. XLibris: The active reading machine. In proceedings of CHI'98 Human factors in computing systems, Los Angeles, California, USA, vol.2 of Demonstrations: Dynamic Documents (1998) 22-23
26. Sumner T., Buckingham Shum S., Wright M., Bonnardel N., Piolat A. & Chevalier A. Redesigning the peer review process: A developmental theory-in-action. In: R. Dieng, A. Giboin, G. De Michelis & L. Karsenty (eds.): Designing cooperative systems: The use of theories and models. Amsterdam: I.O.S. Press (2000) 19-34
27. Sun Z., Finnie G., Experience Management In Knowledge Management, In: Khosla R, Howlett RJ, Jain LC (eds): Knowledge-Based Intelligent Information and Engineering Systems: 9th Intl Conf. KES2005, Melbourne, Australia, September 14-16, 2005, Proc, Part I, LNAI 3681, Springer-Verlag, Berlin Heidelberg 2005 (2005) 979-986
28. Weick, K.E., The Social Psychology of organizing, New York, Random House (1979)
29. Windows Word http://office.microsoft.com/fr-fr/assistance/HA010714941036.aspx (2003)

Cell-Based Distributed Index for Range Query Processing in Wireless Data Broadcast Systems*

SeokJin Im[1], MoonBae Song[1], Jongwan Kim[1], Sang-Won Kang[1], Chong-Sun Hwang[1], and SeongHoon Lee[2]

[1] Department of Computer Science and Engineering, Korea University
1, 5-Ga, Anam-dong, Sungbuk-gu, Seoul, Korea
{seokjin, mbsong, wany, swkang, hwang}@disys.korea.ac.kr
[2] Division of Information and Communication, Cheonan University
Cheonan, Korea
shlee@cheonan.ac.kr

Abstract. In mobile computing, a huge number of users expect various data services based on their location. Wireless data broadcast is suitable for meeting requests from lots of users owing to its scalability. Air indexing techniques have been developed for energy efficient query processing of mobile clients in the wireless data broadcast. Especially for spatial queries related to the user's location, the index based on Hilbert Curve is proposed. However, the index makes clients listen to a lot of data objects in order to process range queries. In this paper, we describe indexing technique for spatial range queries. And we propose the distributed air index based on space partition. For the performance evaluation, it is implemented using a discrete event-driven simulation package, SimJava. A Simulation is conducted with the real dataset that contains 5922 cities of Greece. The results show that the proposed index outperforms the index based on the Hilbert Curve in the energy efficiency and query processing time significantly.

1 Introduction

The most important characteristic of mobile computing is massive clients with unlimited mobility and restricted battery power and the low bandwidth of wireless channels [1, 5-7]. The mobility of clients makes it possible for clients to move and try to access their desired data objects while moving. For example, a great number of clients may want to find hotels within the radius of 1 km or traffic conditions based on their location at the same time.

Data broadcast is fit to data access under the wireless environment thanks to ability of accommodation of a great number of clients simultaneously. Also broadcast scheme is more energy efficient than traditional data access by on-demand scheme because communication through the uplink channel consumes more energy than the downlink channel. Also in order to answer the query based on the

* This work was supported by the Korea Research Foundation Grant funded by the Korea Govenment(MOEHRD).(KRF-2005-041-D00665).

B. Gabrys, R.J. Howlett, and L.C. Jain (Eds.): KES 2006, Part I, LNAI 4251, pp. 1139–1146, 2006.
© Springer-Verlag Berlin Heidelberg 2006

location of clients, data access methods have to support location-dependent spatial queries in the mobile computing environment [5, 6]. Clients can issue various location-dependent queries. A range query is one of the basic queries related to spatial data of clients. The query type is very important type owing to basis of processing of other query types.

Air index [1, 4-6] in the broadcast environment made clients listen selectively the desired data object in the broadcast channel and reduction of client's energy consumption possible. Regarding the performance metrics, tuning time and access latency are considered in the broadcast approach generally [1,6].

In this paper, we will explore spatial data objects broadcast via wireless channel. With the two idea, we propose distributed air index based on cell, more energy efficient than previous works. Our main technical contributions are as follows:

- The proposed index is more energy efficient than the state-of-the-art index based on the Hilbert Curve.
- The proposed index makes it possible more short access latency than the state-of-the-art index.

The rest of this paper is organized as follows. We review previous studies on the range query processing in the broadcast environment in section 2. In section 3, cell-based index is proposed. We evaluate the performance of the proposed index in section 4. Finally, we conclude this paper in section 5.

2 Related Works

In the mobile computing environment, a client that keeps unlimited mobility makes an attempt to access spatial data objects related to its current location. When spatial data objects are broadcasted, air index of the spatial data objects is necessary. R-tree based indexes[2,3] outperform other indexes of spatial data objects in the traditional disk-based spatial database. However, they are not efficient in the mobile spatial broadcast environment, since they do not have time-series characteristics[6]. The recent index in the spatial data broadcast environment is Hilbert-curve index that determines broadcast order of each data object using the Hilbert-curve and points to broadcast time of every data objects using the B+-tree index[5,6]. With the index, a range query is processed by listening to all data objects between the first and last Hilbert number on the query window. However it has also the problems like R-tree based index when the difference between the first and last Hilbert number on the query window is large. The large difference means that the size of dataset for client to search for processing a range query is too large. For example, Fig. 1.(a) shows Hilbert Curve number of order 2 with 16 cells and is assumed that each cell has a data object . For processing the range queries Q1 in the Hilbert Index, a client has to listen to 12 data objects in the shaded region from the Hilbert number 2 to 13. In the Fig. 1.(c), for Q2, the client has to listen to 7 data objects from the

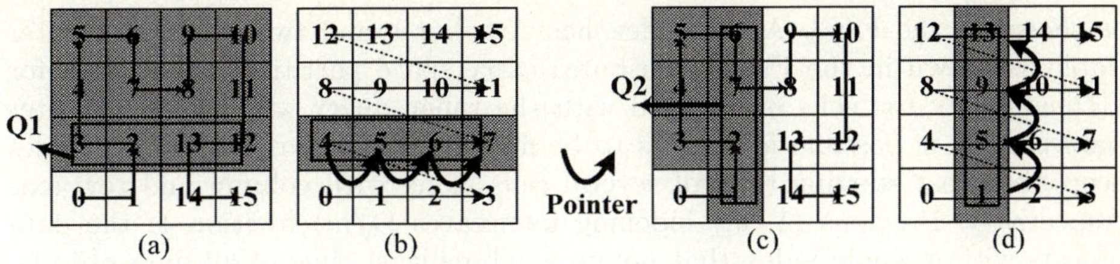

Fig. 1. Running Example

Hilbert number 1 to 7. The dataset for client to search is too large and it cause long tuning time and large energy consumption.

We are strongly motivated by this point. For reducing the size of dataset for client to listen, clients have to listen to only cells overlapped with range query window like Fig 1(b) and (d).

3 Cell-Based Distributed Index

For energy efficient range query processing in wireless data broadcast systems, we propose cell-based distributed index. The index is constructed in units of cell as a result of space partition. Each cell has its own index that contains space partition information and pointers to broadcast time of neighboring cells along with vertical and horizontal direction. Like this, distributed index in each cell makes up of whole index. Clients can make decision of cells to listen for query processing with the space partition information and access the desired cells with the pointers in the index. Index in each cell plays role stepping stone for client to access the desired cells.

3.1 Broadcast Organization and Cell Index Structure

For the broadcast organization, Minimum Bound Rectangle (MBR) enclosing the broadcast service region has to be decided as shown in the Fig.2.(a). Here two coordinates (X_s, Y_s) and (X_e, Y_e) represent left lower point and right upper point each. Then, each side of the MBR along with the column and row axis is partitioned n and m segments respectively. The MBR of the broadcast service region is divided to $n \cdot m$ cells with the length of each side, δ_C and δ_R as shown in the Fig.2.(b). All cells are arranged in row after row from the bottom as shown in the Fig.2.(c).

Broadcast organization is constructed in units of cell that consists of a index segment ,called cell index in this paper, and a data segment as shown in the Fig.3. All data objects in each cell are organized into a data segment and the data segment is consisted of several data buckets as shown in the Fig.3. Every data bucket has a bucket header and data objects of fixed number. The Bucket header of a data bucket keeps bucket identifier and broadcast starting time of the next cell. Then a cell index is added to the every data segment of each cell as shown in the Fig.3. A cell index has Attributes and two kinds of Pointer Table as shown in the Fig.3. Attributes carry space partition information for a client to fix the cells overlapped with the range query window and Pointer Tables make it possible for clients to be able to access the cells. These tables keep broadcast starting time of several cells along with column and row axis respectively. Previous (1, m) indexing interleaves (1/m) fraction of the data objects with a whole index that points out broadcast time of all data objects. However, Pointer Tables in the proposed cell index do not point out broadcast time of all cells but the time of some cells along with column and row axis respectively. Therefore, Pointer Table can make it possible adjusting the size of cell index. Like this, cell-based index scheme can distribute a small size index to all cells. This can make shortening the access latency possible by reducing Probe Wait.

When a client cannot find the broadcast starting time of the cell to access in the tables, the client takes the broadcast starting time of the closest cell to the desired cell. Then the client moves the closest cell index and finds the broadcast starting time of the desired cell. The client can find the broadcast starting time of the desired cell through the procedure like this. Two tables play role of stepping stone for clients to access the desired cells. Attributes that have four items and Pointer Tables are shown in the Table1.

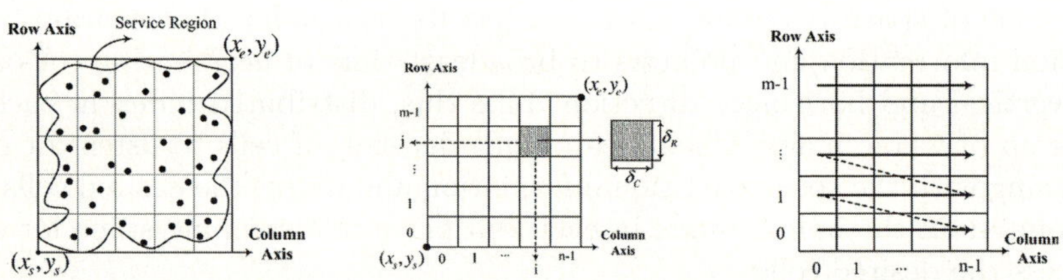

(a) Broadcast region (b) Space Partition (c) Serializing scheme

Fig. 2. Space Partition and Serialization

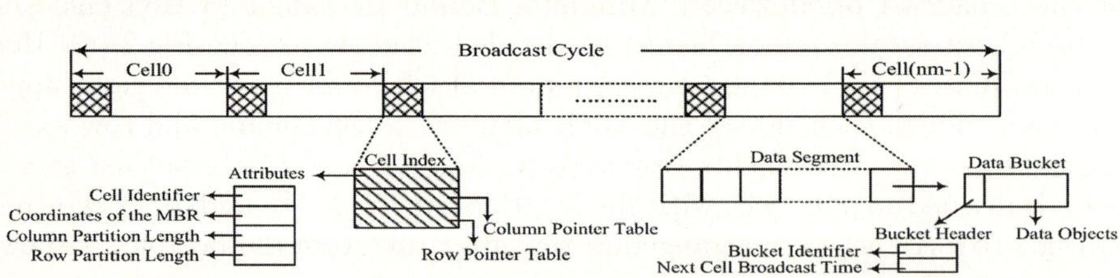

Fig. 3. Structure of the broadcast channel using the cell index

CPT_u and RPT_v of a cell that cell identifier is $C_{ID} = \alpha$ are constructed as follows.

$$CPT_u = \bigcup_{k=1}^{u} T_k \tag{1}$$

Table 1. A list of Attributes and Pointer Tables

Attribute	Description
C_{ID}	identifier of a cell $C_{ID} = i_n j, for\ 0 \leq i \leq n-1, 0 \leq j \leq m-1$
δ_C	Partition Length along with column axis
δ_R	Partition Length along with row axis
CPT_u	**Column Pointer Table**: A table that keeps starting time that u cells neighboring with a specific cell along with column axis are broadcasted
RPT_v	**Row Pointer Table**: A table that keeps starting time that v cells neighboring with a specific cell along with row axis are broadcasted

Here, if $\alpha + k \leq nm - 1$, $T_k = T_{begin}(\alpha + k)$, otherwise $T_k = T_{begin}(\alpha + k - nm) + T_{BCycle}$.

$$RPT_v = \bigcup_{k=1}^{v} T_k \quad (2)$$

Here, if $\alpha + kn \leq nm - 1$, $T_k = T_{begin}(\alpha + kn)$, otherwise $T_k = T_{begin}(\alpha + kn - nm) + T_{BCycle}$.

And $T_{begin}(x)$ means broadcast beginning time of a cell of which the cell identifier is $C_{ID} = x$ and T_{BCycle} means a broadcast cycle time.

3.2 Range Query Processing

Algorithm for clients to process range queries is as shown in Fig.5.

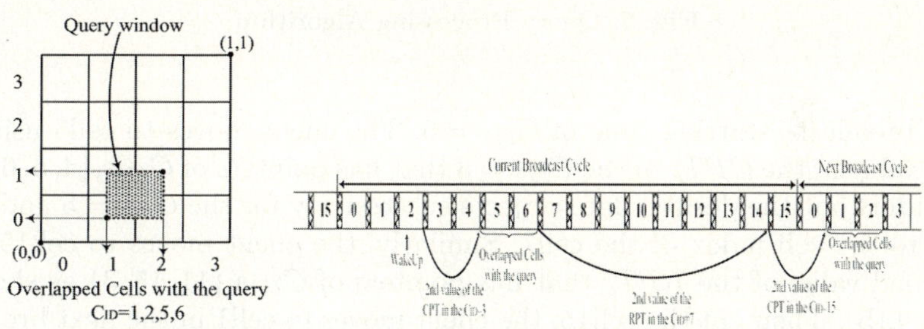

(a) Query window (b) Processing given range query

Fig. 4. Space Partition and Serialization

For example, a client wakes up for processing a range query in the middle of broadcasting the cell2 as shown in the Fig.4.(b). The client listens to the cell index of cell3 and determines the overlapped C_{ID} array, the array that have all C_{ID}s overlapped with the query window, using the Attributes in the cell index of the cell3. It is $\{1, 2, 5, 6\}$ for the given query window as shown in the Fig.4.(a). The client calculates the difference d of each element of the overlapped C_{ID} array in order and its result is $d = \{14, 15, 2, 3\}$. Then, the client sorts all elements of the array by the difference increasingly. The sorted result is $\{5, 6, 1, 2, \}$. In order to access $C_{ID} = 5$, the first cell of the array , the client retrieves CPT_u

```
Range Query Processing Algorithm
Input: a query window, the first cell index after tuning in the channel
Output: a set of data objects contained within the query window

 Result = φ, Pointer = Null

  Initialize current listening C_ID with the C_ID of the given Cell Index as input
  Determine the C_ID array overlapped with the given query window
  Sort the all elements of the array by the difference increasingly
  Initialize the access C_ID with the first value of the array
  Initialize the number of accessed C_ID with zero
  while(the number of accessed C_ID is less than the number of elements of the array)
     if the access C_ID is equal to current listening C_ID then
        Update the access C_ID to the next value of the array
        Add one to the number of accessed C_ID
        for all data objects in the data segment do
           if data object is contained within the query window then
                Result = Result ∪ dataobject
     else
        Calculate d, the difference between the access C_ID and the current listening C_ID
        ifd is no more than n then
           ifd is no more than u then
              Pointer is the d-th value of the CPT
           else Pointer is the u-th value of the CPT
        else
           if ⌊d/n⌋ is no more than v then
              Pointer is the ⌊d/n⌋ -th value of the RPT
           else Pointer is the v-th value of the RPT
        Sleep into the doze mode, the energy saving mode, by the Pointer
        Wake up into the channel
        Listen the Cell Index broadcasting
        Update the current listening C_ID into the C_ID of the Cell Index broadcasting
  end while
```

Fig. 5. Query Processing Algorithm

to find broadcast starting time of $C_{ID} = 5$. The client moves to cell5 using the second value of the CPT_u of the $C_{ID} = 3$ that has pointers of C_{ID}s,$\{4, 5, 6\}$. The client filters out the desired data objects successively for the $C_{ID} = 5$ and 6 and listens to the cell index of the cell7. Similarly, the client moves to cell15 using the second value of the RPT_v that has pointers of C_{ID}s,$\{11, 15, 3\}$ as shown in the Fig.4.(b). Then , at the cell15, the client moves to cell1 in the next broadcast cycle using the second value of the CPT_u that has pointers of C_{ID}s,$\{0, 1, 2\}$ as shown in the Fig.4.(b). The client filters out the desired data objects successively for $C_{ID} = 1, 2$ and completes the processing of the range query.

4 Performance Evaluation

In this section, we present our simulation environment and simulation results for the proposed distributed air index. The testbed for the simulation of the indexing scheme has been developed in order to measure two performance metrics, tuning time and access latency. It has been implemented in java using a discrete event-driven simulation package, SimJava[8]. It consists of a broadcast server, wireless channel and a client. While the one of the characteristics of wireless broadcast environment is massive clients, only a client was implemented since clients are independent each other. The real dataset that contains 5922 cities of Greece[9] is used for the evaluation. In the simulation, 1024 bytes data instance size, 4 bytes coordinate and pointer size, 100000 queries are used. For the control cell size, we define Cell Ratio, that means ratio of side length of a cell to that of the broadcast service region. For the query window size, we define Query Ratio, that means ratio of side length of a query window to that of the broadcast service region.

4.1 Simulation Results

Fig.6.(a) and (b) show the simulation results of tuning time and access latency of the proposed index under three types of query window and varying cell ratio , compared with Hilbert-curve index. The type of query window are square , long rectangle along with horizontal and vertical.

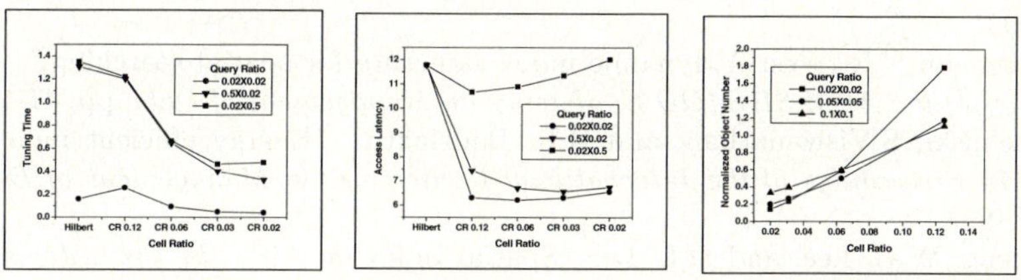

(a) Tuning Time (b) Access Latency (c) Normalized Object Number

Fig. 6. Performance Comparison

Tuning time in the cell index is getting smaller than the Hilbert-curve index from the cell ratio less equal to 0.12. This is due to the fact that the size of data set for client to search is smaller than broadcast using Hilbert-Curve index. Also, the cell index outperforms the Hilbert-curve index for all cell ratios in the side of the access latency. Since cell headers that consist of the index is distributed in every cell, probe wait is very short. Fig.6.(c) shows average data objects number for client to listen for searching answer. The value is normalized by average number in the Hilbert Curve Index. For each queries, when cell ratio is smaller than about 0.12 , average data objects number is much smaller than Hilbert Curve Index. This means that the cell index make clients possible to find the answers in the smaller dataset than the dataset in the Hilbert Curve Index.

5 Conclusion and Future Work

In this paper, we have addressed the problem of range query of the spatial data objects through the wireless broadcast channel and related works. For energy conservation and data waiting time, we proposed a novel index, called cell-based distributed index. The proposed index makes small the size of the data set for client to listen, therefore, power consumption of a client is reduced. Also access latency is decreased thanks to index's being distributed in the broadcast channels. Performance of the proposed index is evaluated through simulation using real dataset. We compare the proposed index with Hilbert-curve index. The proposed index outperforms the Hilbert-Curve index in the tuning time and access latency significantly. As our future work, we are investigating other problem, e.g., nearest neighbor and k nearest-neighbor query.

References

1. T.Imielinski, S.Viswanathan and B.R.Bardrinath, "Data on air: Organization and access," *IEEE TKDE*, 9(3):pp. 353-372, 1997.
2. N.Beckmann, H.Kriegel, R.Schneider, and B.Seeger. "The R*-tree: An efficient and robust access method for points and rectangles," *In Proceedings of the ACM SIGMOD International Conference on Management of Data*, pp. 322-331, May 23-25 1990.
3. A. Guttman, "R-trees: A dynamic index structure for spatial searching," *In Proceedings of the ACM SIGMOD Conference on Management of Data*, pp. 47-54,1984.
4. T.Imielinski, S.Viswanathan, and B.R. Badrinath, "Energy efficient indexing on air," *In Proceedings of the International Conference on Management of Data*, pp. 25-36,1994.
5. B. Zheng, W.C. Lee, and D.L. Lee, Spatial Index on Air", *In Proceedings of. the 1st IEEE International Conference on Pervasive Computing and Communications (PerCom'03)*, Dallas-Fort Worth, Texas, March 23-26, 2003:297-304
6. B. Zheng, W.C. Lee, and D.L. Lee, "Spatial Queries in Wireless Broadcast Systems," *Wireless Network, 10(6)*, pp. 723-736, December, 2004
7. S. Acharya, M.Franklin, and S.Zdonki, "Banancing Push and Pull for Data Broadcast," *In Proceedings of the ACM SIGMOD Conference*, pp 183-194,1997
8. R. McNab and F.W. Howell, "Using Java for Discrete Event Simulation," *In proceedings of the Twelfth UK Computer and Telecommunications Performance Engineering Workshop (UKPEW)*, pp. 219-228. 1996.
9. Real Datasets, "available at http://www.rtreeportal.org"

Genetic-Fuzzy Modeling on High Dimensional Spaces

Joon-Min Gil[1] and Seong-Hoon Lee[2]

[1] Dept. of Computer Science Education, Catholic University of Daegu
330 Geumnak, Hayang-eup, Gyeongsan-si, Gyeongbuk 712-702, Korea
jmgil@cu.ac.kr
[2] Dept. of Computer Science, Cheonan University
115 Anseo-dong, Cheonan 330-180, Korea
shlee@cheonan.ac.kr

Abstract. In this paper, in order to reduce the explosive increase of the search space as the input dimension grows, we present a new representation method for the structure of fuzzy rules, *a graph structured fuzzy system*. The graph structured fuzzy system can flexibly cope with the increase of the input space by selecting these fuzzy rules that significantly affects the input space among the whole set of fuzzy rules. To obtain the optimal structure and parameters of fuzzy systems, an approach to the automatic design of fuzzy systems based on L-systems is also proposed. The proposed method can efficiently construct fuzzy rules without any need for user interaction by using the rewriting mechanism of L-systems.

1 Introduction

When defining an optimal fuzzy system for practical problems, we know intuitively that the whole set of fuzzy rules is impractical, in particular, if the problems are complex; i.e., a fuzzy system with the whole set of fuzzy rules may have a possibility to be included inadequately defined fuzzy rules or conflicted fuzzy rules which are very difficult to identify complex systems. Accordingly, the form of a lookup table expressing the whole set of fuzzy rules can not represent an optimal subset of fuzzy rules. To alleviate this problem, as the first objective of this paper, we propose a new representation method for the structure of fuzzy rules, *a graph structured fuzzy system*. The graph structured fuzzy system can flexibly cope with the increase of the input dimension by selecting these fuzzy rules that significantly affects the input space among the whole set of fuzzy rules.

It is pointed out that an optimal fuzzy system can be extracted from the definition of both the nodes and edges in a graph structured fuzzy system. This is because a good graph structure can generate enough fuzzy rules to precisely represent input-output relation. Genetic algorithms can be utilized to find such the graph structure. However, previous works have mainly focused on the method which directly encodes parameters in fuzzy systems into chromosomes [1,2]. Considering practical problems with high-dimensional input space, as the number of the parameters increase, the complicated search space may still arise from the direct encoding of these parameters. To overcome the problem associated with

B. Gabrys, R.J. Howlett, and L.C. Jain (Eds.): KES 2006, Part I, LNAI 4251, pp. 1147–1154, 2006.
© Springer-Verlag Berlin Heidelberg 2006

the encoding, as the second objective of this paper, we propose the automatic generation method of a fuzzy system, which is is based on L-systems that can define complex objects by successively replacing parts of a simple object using a set of rewriting production rules. The rewriting production rules of L-systems, which can derive the optimal set of fuzzy rules, are encoded into chromosomes in our genetic algorithms. Therefore, our design method can find the optimal structure of a fuzzy system with the reduced search space. Moreover, it can reduce computational requirements for identifying a fuzzy system because the generated system has only the essential fuzzy rules, which are eliminated the inadequate fuzzy terms through evolutionary search.

This paper is organized as follows. In Sec. 2, we provides a brief description of a simplified fuzzy model. Sec. 3 describes a design method of fuzzy systems for solving the explosive increase of search space under high-dimensional input spaces. In this section, we introduce a graph structured fuzzy system. The generation method of the fuzzy system based on L-systems is given in Sec. 4. In Sec. 5, the genetic algorithm used for letting the rewriting production rules of L-systems to be optimized is presented. Sec. 6 shows the simulation with a time-series prediction problem. Finally, our conclusion is given in Sec. 7.

2 Simplified Fuzzy Model

This paper uses the simplified fuzzy model [3] described by the following fuzzy IF-THEN rules: IF x_1 is A_1^i AND, ..., AND, x_d is A_d^i, THEN y is w_i. Here, A_j^i is a fuzzy membership function for the input variable x_j in the ith fuzzy rule ($i = 1, 2, \ldots, n$ and $j = 1, 2, \ldots, d$); w_i is a real value for the output variable y; n and d are the number of fuzzy rules and input variables, respectively.

Given the real-valued input vector $\boldsymbol{x} = [x_1, x_2, \ldots, x_d]$, the real-valued output of the fuzzy model is inferred as follows:

$$f(\boldsymbol{x}) = \frac{\sum_{i=1}^{n} \mu_i \cdot w_i}{\sum_{i=1}^{n} \mu_i}, \quad \mu_i = \prod_{j=1}^{d} \exp\left(-\frac{1}{2} \cdot \left(\frac{x_j - c_j^i}{\sigma_j^i}\right)^2\right) \tag{1}$$

where, μ_i implies the overall truth value of the premise of the ith implication for the input; c_j^i and σ_j^i are the central value and the width of a gaussian fuzzy membership function, respectively.

In order to design a sophisticated fuzzy system, three parameters c_j^i, σ_j^i, and ω_i need to be adjusted by an identification algorithm. This paper uses a self-learning method by the delta rules [3]. The structure identification of the simplified fuzzy model is introduced in next section.

3 Graph Structured Fuzzy System

In the design phase of fuzzy systems, it is useful to select these fuzzy rules that significantly affects an input space among the whole set of fuzzy rules [2]. Let

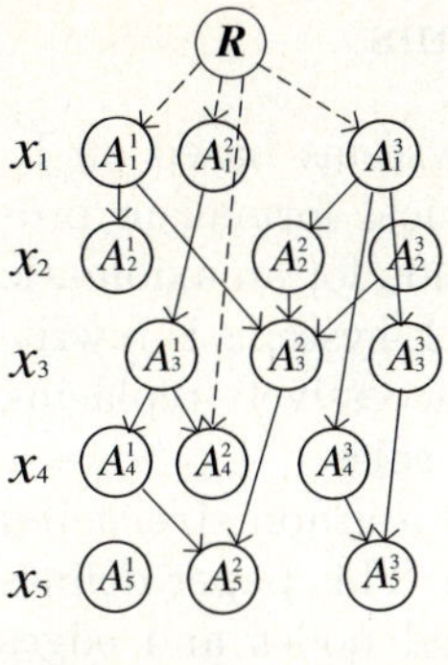

Fig. 1. Graph structure

IF x_1 is A_1^1 AND x_2 is A_2^1 THEN y is w_1
IF x_1 is A_1^1 AND x_3 is A_3^2 AND x_5 is A_5^2 THEN y is w_2
IF x_1 is A_1^2 AND x_3 is A_3^1 AND x_4 is A_4^1 AND x_5 is A_5^2 THEN y is w_3
IF x_1 is A_1^2 AND x_3 is A_3^1 AND x_4 is A_4^2 THEN y is w_4
IF x_4 is A_4^2 THEN y is w_5
IF x_1 is A_1^3 AND x_2 is A_2^2 AND x_3 is A_3^2 AND x_5 is A_5^2 THEN y is w_6
IF x_1 is A_1^3 AND x_4 is A_4^3 AND x_5 is A_5^3 THEN y is w_7
IF x_1 is A_1^3 AND x_3 is A_3^3 AND x_5 is A_5^3 THEN y is w_8

Fig. 2. Fuzzy rules generated by a graph structure

the input space be the cube of r dimension $[-L, L]^r$ and each input domain be partitioned into N fuzzy terms. Then, all the input spaces are partitioned into N^r. In the conventional fuzzy systems, the number of fuzzy rules is N^r. Accordingly, the total number of parameters are $a \cdot r \cdot N + b \cdot N^r$, where a is the number of parameters for fuzzy membership functions and b is the number of parameters for the real parts of fuzzy rules. This means that the fuzzy systems need the order of N^r fuzzy rules in order to approximate a real system. As a result, the fuzzy systems suffer from an exponential rule explosion as the input dimension r grows. As proven in [3,4], there exists a fuzzy system that can approximate a practical problem. However, it is not easy to find such the fuzzy system, in particular, if the problem is complex.

This paper defines fuzzy rules as a graph structure. In the graph structure, nodes consist of a root one and verbal ones. Each verbal node represents a fuzzy membership function for an input variable. The relation between two nodes is expressed as directed edges. Fig. 1 shows an arbitrary graph structure when 5 input values and 3 membership functions are used. In this figure, R and A_j^i represent a root node and a verbal one, respectively. A dashed line represents edges between the root node and a verbal one. The edges between two verbal nodes are depicted as a solid line.

The conversion of the graph structure into fuzzy rules is to search all the paths from the root node to the verbal node which has not any more branch. Fig. 2 shows the fuzzy rules generated by the graph structure shown in Fig. 1. For example, the path $R \dashrightarrow A_1^1 \rightarrow A_2^1$ in Fig. 1 is interpreted as the fuzzy rule whose premise part includes the verbal nodes A_1^1 and A_2^1 but does not exist the input variables x_3, x_4, and x_5. In this way, total 8 fuzzy rules are generated.

Fig. 2 illustrates that fuzzy rules do not necessarily have to involve all the input spaces; i.e., the structure of fuzzy rules defined in this figure can eliminate unnecessary input variables and membership functions from the whole of the input spaces. It should be noted that fuzzy rules can be optimized by the selective combination of membership functions for each input variable.

4 Generation of Fuzzy Systems by L-Systems

In this section, we describe the design method of fuzzy systems based on L-systems, termed DOL-systems. L-systems are a mathematical formalism proposed by biologist Aristid Lindenmayer in 1968 as a foundation for an axiomatic theory of biological development [5]. The principle notion of L-systems is rewriting, where the basic idea is to define complex objects by successively replacing parts of a simple object using a set of rewriting production rules.

It is important to define the optimal nodes and edges of a graph structured fuzzy system in order to obtain an optimal graph structure. This paper defines a set of rewriting production rules from which the optimal nodes and edges of the fuzzy system can be obtained. Let us consider a definition of the nodes that represent fuzzy membership functions. The central value of membership functions on the input domain $[L, R]$ is defined as follows: $c_i = L + i \times \frac{(R-L)}{C}$. Here, C is a precision for the input domain $[L, R]$ $(i = 1, 2, \ldots, C)$.

The edges representing the structure of fuzzy rules are extracted from L-systems defined in the below. The L-systems automatically generate the nodes and edges of a graph structure by the interpretation of the command that can change a graph state successively. The state of a graph is defined by (x, y), where x is the index of an input variable and y is the index of a membership function. Then, the graph responds to the commands defined by

- Fm Move forward a step of length m. The state of the graph changes to (x', y'), where $x' = x + m$ and $y' = y$. A edge between two verbal nodes (x, y) and (x', y') is made.
- fm Move forward a step of length m. The state of the graph changes as above. This command is applied to only between the root node R and a verbal node (x', y'). A edge between two nodes is also made.
- $+n$ Move right a step of length n. The state of the graph changes to (x', y') without making a edge, where $x' = x$ and $y' = y + n$.
- $-n$ Move left a step of length n. The state of the graph changes to (x', y') without making a edge, where $x' = x$ and $y' = y - n$.

In order to represent the graph structure efficiently, the alphabet of L-systems is extended with two new symbols, '[' and ']'. They are interpreted by the graph as follows:

- [Push the current state of the graph onto a pushdown stack.
-] Pop a state from the stack and replace it with the current state of the graph.

Given a string v and an initial state of the graph (x_0, y_0), a graph structure is generated by successively rewriting the graph produced by the interpretation of rewriting production rules. Here, v is composed of an initial symbol S and a set of rewriting production rules $\{p_1, p_2, \ldots\}$. The graph structure shown in Fig. 1 can be obtained by interpreting the strings of the following L-system: $w : S$, $S \rightarrow p_1|p_2|p_3|p_4$, $p_1 \rightarrow +1f1[+0F1]+1F2+0F2$, $p_2 \rightarrow +2f1-1F2[+0F1+1F1]+1F1$, $p_3 \rightarrow +2f4$, $p_4 \rightarrow +3f1[-1F1+0F1+0F2][+0F3+0F1]+0F2+0F2$.

5 Optimization of Production Rules

This section describes genetic algorithms for optimizing the rewriting production rules by which an optimal graph structure can be obtained.

5.1 Encoding

Since an initial symbol is always S, only the rewriting production rules are encoded into a chromosome. The chromosome is subdivided into three subchromosomes: one is for commands, the others are for push and pop operations in a push-down stack. For example, the rewriting production rule p_4 is encoded as follows: $\{\{+3f1-1F1+0F1+0F2+0F3+0F1+0F2+0F2\}, \{010010000\}, \{000010100\}\}$. Here, commands are encoded into first part of the chromosome. For push and pop symbols, the integers which are correspondent with the number of '[' and ']' are encoded into second and third part of the chromosome, respectively.

5.2 Genetic Operations

Since rewriting productions rules have a variable length, each chromosome has also a variable length. Accordingly, to allow a proper derivation and interpretation of genotypes, genetic operations must produce offspring with a valid graph structure. With this consideration, four main operators are designed: two kinds of crossovers and two kinds of mutations.

- **Crossover:** This operator introduces information exchange between two chromosomes to create offsprings. According to the substrings to be exchanged, two kinds of crossovers are designed.
  a. Production Crossover: This is applied to the unit of rewriting production rules. Since a production rule can be interpreted to several fuzzy rules, the information exchanged between two parents is also fuzzy rules. Uniform crossover is used for this crossover.
  b. Symbol Crossover: This is inspired by the Genetic Programming crossover [6]. Koza's Lisp subtrees can be considered analogous to correctly substrings within a L-system. This crossover can search finer fuzzy rules by exchanging the unit of commands.
- **Mutation:** This operator introduces random variations in rewriting production rules.
  a. Symbol Mutation: For the symbols '+' and '−', this mutation changes '+' into '−' or vice verse with a given probability. m and n each is replaced by the value of the randomly selected integer within the given range of values. The value of the genes represented for push and pop operations is also replaced by the value of a randomly selected integer.
  b. Block Mutation: A randomly selected block in a rewriting production rule is replaced by the random string which is syntactically correct.

5.3 Fitness Function

The graph structure generated by the interpretation of L-systems is applied to the evaluation of a chromosome. Accordingly, a fitness function is based on the phenotype representing a graph structure. Generally, as a proper criterion for the verification of fuzzy systems, a learning error has been used [1,2]. Also, the number of fuzzy rules can be used as the design objective of fuzzy systems in order to minimize computational complexity. The fitness function with these objectives is defined by $f(s_i) = \frac{1}{\alpha_e \cdot e + \alpha_r \cdot r}$. Here, $f(s_i)$ is a fitness value for the ith chromosome s_i; e and r denote a learning error and the number of fuzzy rules, respectively; α_e and α_r are nonnegative weights for a learning error and the number of fuzzy rules, respectively.

6 Simulation and Results

To demonstrate the efficiency of the proposed method, we apply the proposed one to the time series prediction problem. The time series used in this paper is generated by the chaotic Mackey-Glass differential delay equation [3] defined as

$$\frac{dx(t)}{dt} = \frac{0.2x(t-\tau)}{1+x^{10}(t-\tau)} - 0.1x(t). \tag{2}$$

The prediction problem is to predict the value at some point in the future $t+P$, $x(t+P)$, using the past values of the time series up to time t, $\{x(t), x(t-\Delta t), \ldots, x(t-(n-1)\Delta t)\}$. According to the selection of Δt and n for P, the number of input dimensions and the property of the problem are determined. In our simulation, the values $P = \Delta t = 6$ and $n = 6$ are used. We use the fourth order Runge-Kutta method to find the numerical solution for (2). Also, the time step used in the simulation is 0.1, initial condition $x(0)$ is 0.8, and τ is 30. In this way, we extract 1000 data from $130 \leq t \leq 1129$. First 500 pairs are used as training data and the remaining 500 pairs used as checking data. The membership functions for all the input variables are defined in the range $[0.15, 1.35]$ and each input domain are partitioned into nine parts; i.e., $C = 9$. We assign 10 rewriting production rules to each chromosome. The probabilities of crossover and mutation are 0.3 and 0.1, respectively. In order to compare the proposed method with other design one, we use the normalized mean squared error (NMSE) defined as the root mean squared error divided by the data standard deviation.

After 100 generations, our genetic algorithm finds the L-system shown in Fig. 3. In the initial phase of the genetic algorithm, 10 rewriting production rules are randomly encoded into chromosomes, but finally only 8 rewriting production rules are used for constructing the graph structure shown in Fig. 4. Since remaining two rewriting production rules generate an invalid graph state, they are excluded from the construction of the graph structure.

By the graph interpretation of the rewriting production rules, we obtained 15 membership functions and 15 fuzzy rules. Initially, each input space is partitioned into nine parts and therefore total 54 membership functions can be used

$$\begin{aligned}
&w : S,\\
&S \rightarrow p_1 \mid p_2 \mid p_3 \mid p_4 \mid p_5 \mid p_6 \mid p_7 \mid p_8,\\
&p_1 \rightarrow +5f4[-2F1] - 0F1[-3F1][-1F1],\\
&p_2 \rightarrow [+4f2] + 4f1[-3f2],\\
&p_3 \rightarrow +3f6,\\
&p_4 \rightarrow +5f5[-3F1][-1F1] + 2F1,\\
&p_5 \rightarrow +1f1 + 0F1,\\
&p_6 \rightarrow +8f1,\\
&p_7 \rightarrow +8f5[-0F1],\\
&p_8 \rightarrow +4f6.
\end{aligned}$$

Fig. 3. L-system obtained by evolutionary search

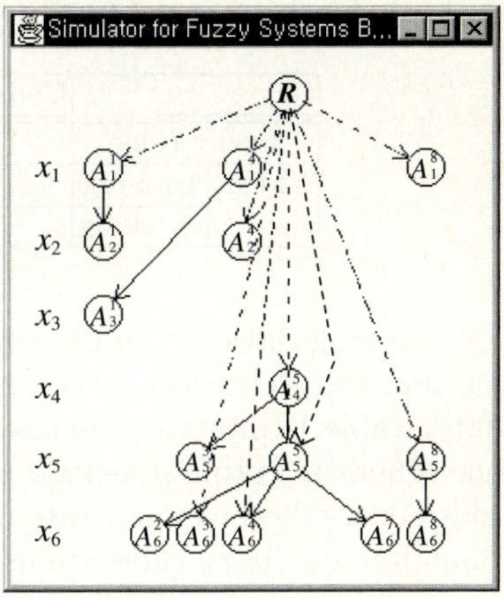

Fig. 4. Graph structure generated by the interpretation of L-system

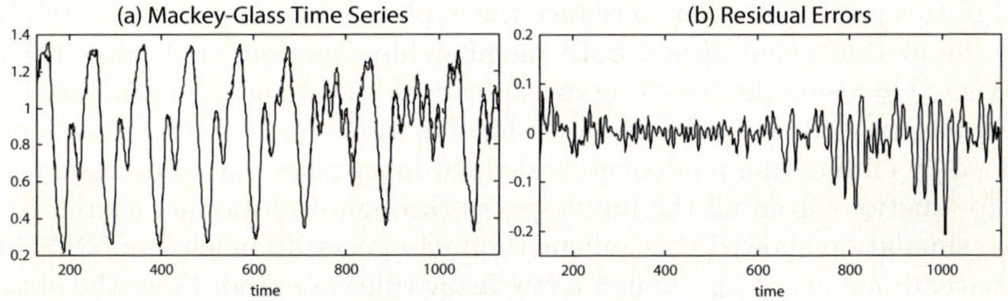

Fig. 5. Prediction results

as verbal nodes. In case of the lookup table structure of fuzzy rules, entire 54 membership functions are used in fuzzy systems without any elimination for the illegal definition of membership functions. Such fuzzy systems will have difficulty in partitioning input domains correctly for a given problem. However, our method selects only 15 membership functions which significantly affect the input space. The fuzzy rules constructed by all the combinations of 15 fuzzy membership functions would be 90 $(3\times2\times1\times1\times3\times5)$, but our method produces only 15 fuzzy rules by the selective combination of the membership functions; i.e., when defining fuzzy rules, it can use only the membership functions that significantly affect the input space through evolutionary search. It should be noted that the fuzzy systems based on L-systems can generate only the fuzzy rules that can correctly express input-output relation for a given problem.

Fig. 5(a) shows the real curve of time series prediction problem and the predicted curve by our method. In this figure, a dashed line represents the real curve and a solid line the predicted curve. Two curves are visually indistinguishable. Fig. 5(b) shows the residual errors that indicate the difference of the real values from the predicted values at each time. Although the time series used

Table 1. Comparison of prediction results

		Conventional method	Proposed method
Fuzzy rules		29	15
NMSE	training	0.0509	0.0385
	checking	0.1637	0.1164

for checking data are more complex than training data, it is obvious that our method can well predict the real curve for checking data as well as for training data. Table 1 compares the results of our method with those of the conventional one where a graph structure is directly encoded into chromosomes. It shows that our method can generate much less NMSE than the conventional one, even though a few fuzzy rules are used.

7 Conclusion

An approach to the automatic design of fuzzy systems by L-systems was proposed in this paper. In order to reduce the explosive increase of the search space as the input dimension grows, both membership functions and fuzzy rules are constructed by using the rewriting mechanism of L-systems. The generated fuzzy system has only the essential membership functions and fuzzy rules, which are obtained by eliminating inadequately defined fuzzy rules and unnecessary membership functions from all the input spaces through evolutionary search.

The simulation showed that our method can generate much less NMSE than the conventional one, even though a few fuzzy rules are used. From the obtained results, we concluded that our method can consistently reduce the computational complexity for designing fuzzy systems by efficiently reducing both the parameters of fuzzy systems and the search space.

We have evaluated with a specific time series prediction to show its feasibility on high-dimensional input spaces. To verify the applicability of our method in various areas, several simulations are being carried out.

References

1. T.C. Chin and X.M. Qi, "Genetic algorithms for learning the rule base of fuzzy logic controller," *Fuzzy Sets and Systems*, vol. 97, pp. 1-7, 1998.
2. K. Shimojima, T. Fukuda, and Y. Hasegawa, "Self-tuning fuzzy modeling with adaptive membership function, rules, and hierarchical structure based on genetic algorithm," *Fuzzy Sets and Systems*, vol. 71, pp. 295-309, 1995.
3. J.-R. Jang, C.-T. Sun, and E.M. Mizutani, *Neuro-Fuzzy and Soft Computing*, Prentice-Hall, 1997.
4. J. Buckley, "Sugeno type controllers are universal controller," *Fuzzy Sets and Systems*, vol. 53, pp. 299-303, 1993.
5. P. Prusinkiewicz and A. Lindenmayer, *The Algorithmic Beauty of Plants*, Springer-Verlag, 1996.
6. M. Mitchell, *An Introduction to Genetic Algorithms*, The MIT Press, 1996.

Considering a Semantic Prefetching Scheme for Cache Management in Location-Based Services*

Sang-Won Kang[1], SeokJin Im[1], Jongwan Kim[1], SeongHoon Lee[2], and Chong-Sun Hwang[1]

[1] Department of Computer Science and Engineering
Korea University
1,5-Ga,Anam-dong,Sungbuk-gu,Seoul, Korea
{swkang, seokjin, wany, hwang}@disys.korea.ac.kr
[2] Division of Information and Communication
Cheonan University
Cheonan, Korea
shlee@cheonan.ac.kr

Abstract. Dissemination techniques of data have been rapidly developed for wireless communication environment. Since the advent of ubiquitous computing, semantic-based researches have been sped-up between service provider and user. We have tried to maintain user information with a semantic cache scheme. In this paper, we propose a *semantic prefetching (SP)* strategy for efficient data management, considering semantic information of user. We concentrate on data prefetching methods to send mass data to high bandwidth channels. In the SP, the prepared semantic descriptions are prefetched from server to client according as the client locates in anywhere. The semantic descriptions are dynamically composed of user patterns, such as the request patterns of user, the size of querying area, the direction of user and the average speed of user. We also propose a new cache management scheme, called to *semantic least recently used (S-LRU)*. The evaluation result shows that the system performance is significantly improved.

1 Introduction

In *Location-Based Services (LBS)*, location is a special feature. Examples of queries for LBS [1, 13] represent "find the nearest gas station from my position," "show me the name of hotels within a radius of 10 kilometers," "which buses will pass by me in the next 10 minutes?", and so forth. We consider the queries issued by clients in these examples, which are the most common kind of queries in LBS. There are many kinds of issues related to the LBS, which include indexing data objects, predicting movements, replacing repositories, processing queries and disseminating the mass data efficiently. In this paper, we focus on the efficient

* This work was supported by the Second Brain Korea 21 Project and This work was supported by the Korea Research Foundation Grant funded by the Korea Government (MOEHRD) (KRF-2005-041-D00665).

B. Gabrys, R.J. Howlett, and L.C. Jain (Eds.): KES 2006, Part I, LNAI 4251, pp. 1155–1162, 2006.
© Springer-Verlag Berlin Heidelberg 2006

cache replacement strategy and, in particular, dissemination technique called semantic prefetching.

Location Dependent Data (LDD) refers to data whose values depend on location. If a client keeps on moving after he or she issues a query, the query result would continue to change in accordance with the client's location. However, it is difficult to get accurate results without requesting additional information. In order to increase the access efficiency of data objects and the validity of cache, we propose the semantic prefetching mechanism with the semantic cache of the LDD queries in LBS. The contents of the semantic prefetching in specific area are based on prefetching description organized by log data. we utilize a range query, called semantic segment [2], which is enclosed by rectangular instead of traditional query. If we maintain contents of data object in local cache, we must last the validity of cache. In semantic cache scheme [4, 5, 7, 8], the meaning of semantics includes not only characteristics of spatial locality but also information of user interest. Contrary to the *page cache* and *tuple cache* [9], the semantic cache manages LDD with a semantic description. Cache replacement researches for LDD have been proposed by manhattan distance strategy of Franklin [2] and furthest away replacement (FAR) strategy of Dunham [4]. In this paper we propose a new cache replacement policy which utilizes the semantic descriptions based on user patterns and other properties. Finally, we investigate a simulation to evaluate the performance of the proposed scheme. The simulation results represent that the proposed scheme outperforms the existing researches.

2 Semantic Prefetching

2.1 Preliminary

A server sends information with prefetching description $S_{d_f} = \{F_r, F_a, F_p, F_l, F_c\}$ to the mobile client, which is expected to use in the future. Here, F_r defines the relation and F_a defines the attribute. F_p represents the predicate, expressing the location range of the mobile client and referencing the F_l which means location visited before. Finally, F_c is contents satisfied to location visited before. This paper describes re-definition of LDD with semantic description using the AND and OR operations. It is applied to *semantic-least recently used (S-LRU)* strategy.

Definition 1. *Given a domain* $G = |X| \times |Y|$, $x \in X$, $y \in Y$, *a prefetching data* D_f *is a set of* $\{D_{ij} | 1 \leq i \leq x, 1 \leq j \leq y\}$.

Definition 2. *Given a Log File* $L_f = \{R_i\}$ *and its attribute set* $A = \bigcup A_{R_i}$, $1 \leq i \leq n$, *a* ***Compare Predicate*** *of* D_f, P, *is of the form* $P = a$ *op* c, *where* $a \in A$, *op* $\in \{ \leq, <, \geq, >, = \}$, *c is a constant in a specific domain, location of* a, $L = \{ L_x, L_y \}$.

Semantic prefetching is composed of a set of items, represented as a prefetching description.

Definition 3. *Given a Log File* $L_f = \{R_i\}$ *and its attribute set* $A = \bigcup A_{R_i}$, $1 \leq i \leq n$, *a* ***Prefetching description***, S_{d_f}, *is a tuple* $< F_r, F_a, F_p, F_l >$, $F_r \in D$, $F_a \subseteq A_{F_r}$, *and* $F_p = P_1 \vee P_2 \vee \ldots \vee P_m$. *where each* P_j *is a conjunctive of compare predicates, i.e.* $P_j = b_{j_1} \wedge b_{j_2} \wedge \ldots \wedge b_{j_l}$, *each* b_{j_t} *is a compare predicate and* $F_l = \bigcup_{i=1}^{n} \bigcup_{j=1}^{m} (L_i, L_j)$.

In definition 4 and 5, we assume that the user selects θ function for the purpose of comparing prefetching description to semantic description in local cache.

Definition 4. *A intersect operation,* ω, *is a intersection of A and B,* $\omega = A \cap B$. *Let A and B be descriptions; The set A is defined as* $A = \{S_{d_{f_i}} | 0 \leq i \leq m\}$, *where* $S_{d_{f_i}}$ *is a prefetching description. The set B is defined as* $B = \{S_{d_j} | 0 \leq j \leq n\}$, *where* S_{d_j} *is a semantic description.*

Definition 5. *A semantic description table* Γ *is defined as*

$$\Gamma = \{\theta, \omega, \omega^c\},$$

where θ *is a sorting function. We suppose that if* Γ *is sorted by S-LRU replacement strategy, the cache follows the recent access of the query: recent data and old data divided by users' semantic description. In S-LRU, the factor of* θ *is time.*

2.2 Cache Management Scheme

We store the semantic description to local cache, but cache must be renewed to the current location, considering a received prefetching description. When a prefetching description is received from server, the contents will be replaced in the cache. In this paper, our scheme is to manage the description-based LDD between server and client. Although the client moves in a new specific area, the validity of cache is kept up with interest of user by the prefetching description.

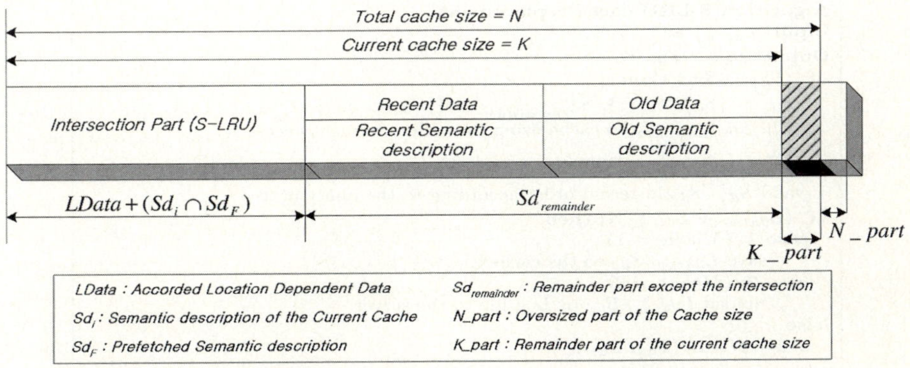

Fig. 1. Contents of S-LRU cache

In the range query, the query is trimmed into two parts: a probe query that retrieves the part of the answer available in the local cache, and a remainder query that must be transmitted to the server through a base station (BS). for transmitting the prefetching description, we define the expanded base station (eBS) that utilize the high bandwidth channel. When caching the received data, we would store the received prefetching description to local cache together. After inserting the received prefetching description into the existing cache contents, the cache is reconstructed to intersection part and union part. The former is a correspondence part and the latter is not a correspondence part at both of descriptions. In this paper, including semantic information, we propose the S-LRU strategy to manage our cache.

In Fig. 1, N defines a total cache size, K/N % notation by the arbitrary K value represents a current cache size, and $L_{Data}+(S_{d_i} \cap S_{d_f})$ denotes a intersection part. In the current cache size, $S_{d_{remainder}}$ denotes a remainder part except the intersection part and K_part means a remainder part except the used cache size. Finally, we denote the N_part that is oversized to the cache size N. Contents of cache maintain data and semantic description.

Fig. 2 shows the cache replacement algorithm. The algorithm adds prefetching contents to the current cache. Comparing contents received by SP to contents of current cache, we make the new contents of client cache. Our scheme determines whether to delete or not, after checking the cache size.

Notations:
L_{data_i} : LDD remained to the current cache
S_{d_i} : semantic description remained to the current cache
L_{data_c} : final LDD organized to the cache
S_{d_c} : final semantic description organized to the cache
$S_{d_{remainder}}$: the discorded remainder semantic description
N_part : an oversized part to the total cache size
K_part : a remainder part of the current cache size
$L_{data_f} + S_{d_f}$: contents received to the server

```
Algorithm S-LRU cache replacement
Input: L_data_f and S_d_f
Ouput: L_data_i and S_d_i
 if (S_d_f ∩ S_d_c) then
    S_d_c = {S_d_f ∩ S_d_c}, S_d_remainder = ({S_d_f ∪ S_d_c} - {S_d_f ∩ S_d_c})
    add Sd_remainder to cache using the order of recent access
    S_d_i = S_d_c + S_d_remainder
 else
    add S_d_f, S_d_i in recent order according to the query access
 if (L_data_c + S_d_c ≤ N) then
   case 1: (Mobile = 1)
       save L_data_i, S_d_i to the cache
   case 2: (Mobile = 0)
       upload L_data_i, S_d_i by Log_file to the server
 else
   delete N_part(S_d_c, L_data_c)
```

Fig. 2. S-LRU cache replacement

3 Performance Evaluation

3.1 System Parameter and Scenario

Cache strategy experiments using SP consider a total cache hit. Our performance compares cache hit ratio with a number of query, distribution, and a query segment size. For our simulation, we basically coordinate one server per eBS. Table 1 represents the parameters used to our simulation model. We suppose that our network utilizes FIFO queue model, having eBS network to use a specific bandwidth. Using 5×5 service area for experiments, total number of queries set 400. We suppose that the number of continuous queries per each area is 10. We implement that 400 LDD including different data type are set in the service area fixed to eBS network. Our eBS area is divided into 100×100 physical coordinates. Query information and previous data used by user are logged to the server.

Table 1. Parameter Setting

Parameter	Description	Setting
CACHE_MAX_SIZE	Size of client cache	10% of server
S_CACHE_MAX_SIZE	Size of server cache	100
TOTAL_QUERY_NUM	No. of query	400
QUERY_NUM_PER_CELL	No. of query per cell area	10
NUM_CELL_eBS	No. of eBS area	5×5
X_Y_eBS	Physical coordinates of x, y about 5×5 eBS area (x,y)	100×100
NUM_eBS	No. of eBS	25
NUM_LDD	No. of distributed LDD	400
TYPE_LDD	No. of data type	30
QUERY_SEGMENT_SIZE	Size of query segment	20, 30

When a user re-enters the service area, the user can receive LDD information. For SP strategy, semantic segment is maintained to the server through a buffer management. Parameter values are determined by the number of queries and the size of query segment. If a user queries a different attribute about same relation, it is assumed to be different data type each other. The semantic predicate including condition_attribute, is organized by the segment area, having a specific distance from the current point. We repeatedly organize the moving pattern to the 20 area that users visit. Our simulation performs the S-LRU strategy for several experiments.

In the processing part of simulation, we assume four scenarios for presentation of real-world moving pattern. Each scenario determines 30 data types and physical location of X, Y coordinates of each data, using the uniform or zipf distribution. In these four scenarios, it is important to determine about moving pattern of clients. We use random work model for real-world moving pattern, and we set specific moving characteristics of the client for the simulation as follows. Examining four scenarios, we consider the different mobility and the query pattern. We compare traditional semantic cache to semantic prefetching, applying to uniform and zipf distribution.

Table 2. Scenario

Scenario	A	B	C	D
Mobility	Low	High	Low	High
Query pattern	Various	Similar	Similar	Various

3.2 Experiments

As shown to Fig. 3 and 4, the point represents that cache hit ratio used to the SP strategy is higher than traditional semantic cache. We try to adjust factors such as moving patterns and query patterns. We observe the traditional semantic cache with SP, using uniform and zipf distribution like Fig. 3 and 4, and our experiments investigate the response time about each scenarios. As started above, we adopt several situations: cache hit ratio of data requested by user, determination of query pattern, definition of locality, space of the local cache, and re-visiting number of the area. The segment size is differently determined to the request type of user. We assume that the query pattern is adequate and it is regularly adjusted from the segment size about each query. Fig. 5 presents the comparison of the number of cache hits, adjusting the querying area. The segment size is determined to the size of a regular square, setting the current location of the user. If the square has a large value, the probability of cache hit about the query more and more increases. In this Fig. 5, each scenario shows that the number of cache hits is remarkably differentiated by the segment size. The increasing rate greatly reduces to the case of the segment size 20. Our paper organizes the

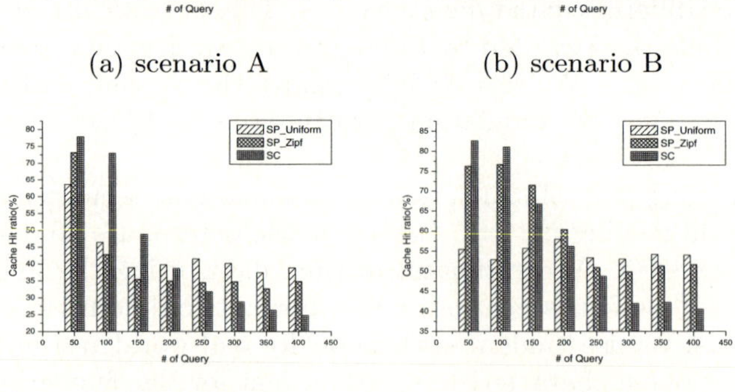

(a) scenario A (b) scenario B

(c) scenario C (d) scenario D

Fig. 3. Comparison of cache hit ratio

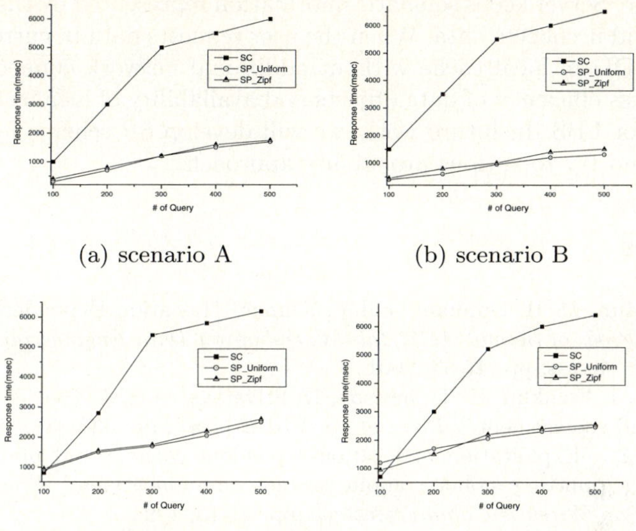

(a) scenario A

(b) scenario B

(c) scenario C

(d) scenario D

Fig. 4. Comparison of response time

segment size, when the simulation executes in proposed model. We can organize the segment area, including the additional elements in real-world implementation. We consider the additional elements in order to include the geographic character, such as the processing of query location and the dynamic organization of the segment area. In this way, we investigate the improvement of the cache hit related to the user request and decide how to organize the segment area.

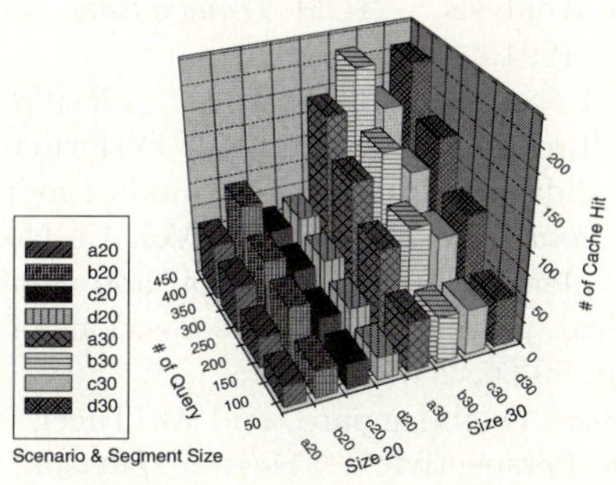

Fig. 5. Comparison of segment size

4 Conclusion and Future Work

In this paper, we propose the semantic prefetching scheme and new cache replacement strategy. By applying the SP through an eBS in LBS, we reduce the network traffic. Server keeps semantic information represented by the prefetching description and it collects data. When the user requests data in current location, we can get LDD in local cache without additional network connection. In the aspect of access efficiency of data objects and availability of local cache, we have best results for LBS. In future work, we will develop SP scheme for real-world application and try to various probability approach.

References

1. A. Y. Seydim, M. H. Dunham, and V. Kumar, "Location Dependent Query Processing," *Proc. of Second ACM Int. Workshop on Data Engineering for Wireless and Mobile Access*, pp. 47-53, 2001.
2. S. Dar, M. J. Franklin, B. T. Jonsson, D. Srivatava, and M. Tan, "Semantic data caching and replacement," *Proc. of the VLDB Conf.*, pp. 330-341, 1996.
3. George Liu, "Exploitation of location-dependent caching and prefetching techniques for supporting mobile computing and communications," *Proc. of the Sixth Int. Conf. on Wireless Communications*, pp. 11-13, 1994.
4. Q. Ren and M. H. Dunham, "Using semantic caching to manage location dependent data in mobile computing," *Proc. of MobiCom'00*, pp. 210-221, 2000.
5. Luo Li, Birgitta K. R., N. Pissinou and K. Makki, "Strategies for Semantic Caching", *DEXA'01*, pp. 284-298, 2001.
6. Daniel Barbara, "Sleepers and Workaholics: Caching Strategies in Mobile Environments," *ACM SIGMOD Int. Conf. on Management of Data*, pp. 1-12, 1994.
7. Ken. C. K. lee, H. V. Leong, and Antonio Si, "Semantic Query Caching in a Mobile Environment," *MC2R*, Vol. 3, No. 2, April 1999.
8. Qun Ren, Margaret H. Dunham, and Vijay Kumar, "Semantic Caching and Query Processing," *IEEE Transactions on Knowledge and Data Engineering*, Vol. 15, No. 1, pp. 192-210, 2003.
9. Nick Roussopoulos, "An Incremental Access Method for ViewCache: Concept, Algorithms, and Cost Analysis," *ACM Transactions on Database Systems*, Vol. 16, No. 3, pp. 535-563, 1991.
10. N. Oren., "A Survey of prefetching techniques," *TR-Wits-CS-2000-10*, July 2000.
11. Jianliang Xu, Xueyan Tang, and Dik Lun Lee, "Performance Analysis of Location-Dependent Cache Invalidation Schemes for Mobile Environments", *IEEE Transactions on Knowledge and Data Engineering*, Vol. 15, No. 2, pp. 474-488, 2003.
12. Dan Lin, Christian S. Jensen, Beng Chin Ooi, and Simonas Saltenis, "Efficient indexing of the historical, present, and future positions of moving objects," *Mobile Data Management*, pp. 59-66, 2005.
13. V. Bellocci, S. Genovese, D. Inuaggiato, and M. Tucci, "Mobile Location-Aware Services: 2002 Market Perspective," *Ericsson, Divison Service Architecture and Interractive Solutions*, 2002.

A Unified Framework for u-Edutainment Development of Using e-Learning and Ubiquitous Technologies

SangHyun Jang[1], Bokyung Kye[1], and YunHee Kang[2]

[1] KERIS: Korea Education Research Information Service
22-1 Ssangnim-dong, Jung-gu, Seoul, 100-400, Korea
{shjang, kye}@keris.or.kr
[2] Department of Computer and Communication Engineering, BaekSeok University
115 Anseo-dong, Cheonan 330-704, Korea
yhkang@bu.ac.kr

Abstract. Edutainment is a type of learning which derives interest within the flow and interaction of learners by adding elements of interest to learning activities and the related contents. This paper introduces the edutainment being applied in current u-learning environment and presents the framework named *Unified Framework* which is used to apply the future direction of the application of ubiquitous technologies. The main characteristic of the Unified Framework is to bolster basic e-learning systems with game theory in ubiquitous computing environments. It can develop four types of applications including m-learning, r-learning, v-learning and c-learning. The basic idea is to exploit the desire, one of the underlying properties of human beings, to be able to learn while playing. Moreover, there is a great expectation to be able to learn in an interesting way in the ubiquitous environment that places human beings at its center.

1 Introduction

Edutainment is a type of learning which derives interest within the flow and interaction of learners by adding elements of interest to learning activities, and the related contents that have been developed for a long time. In particular, the digital-based edutainment combined multimedia and game technology centering on CD-ROM from the early '90s and enjoyed a golden period until 1997, but has rapidly become sluggish after the popular use of the Internet. Meanwhile, "The First Stage of the Comprehensive Development Plan for Educational Information (1997-2000)" from the Korean Ministry of Education and Human Resources Development (MOEHRD) recommended the use of over 10% of Information & Communication Technology (ICT) in the 10 common basic subjects for students. Internet availability was wide spread among all schools, and as a result of the demand for interesting contents of teaching and learning for both teachers and students sharply increased [1,2,3,4]. In addition, as e-learning has expanded and been applied to the elementary and secondary schools from 2004, edutainment is being used extensively not only for the existing active

B. Gabrys, R.J. Howlett, and L.C. Jain (Eds.): KES 2006, Part I, LNAI 4251, pp. 1163 – 1170, 2006.
© Springer-Verlag Berlin Heidelberg 2006

market of infants but also for university education and corporate education. It is also being developed as a strategic industry for export.

Edutainment has been developed using an educational engineering approach such as Game-Based Learning, ARCS Theory, Activity Theory, Recognition Principle and Structuralism. An additional approach has related computer science with 2D or 3D animation/graphics, Virtual or Augmented Reality, voice recognition or mobile technologies. So far, studies and projects have been conducted separately in each field, and so it is difficult to see well-designed contents [5].

This paper introduces edutainment being applied in current the u-learning environments and presents the framework named *Unified Framework* which is used to apply the future direction of the application of ubiquitous technologies. The main characteristic of the Unified Framework is to bolster the basic e-learning system with game in the theory on ubiquitous computing environment. It can develop four types of applications including m-learning, r-learning, v-learning and c-learning.

Thus there is a movement in and outside of the country for a comprehensive study of the teaching/learning design theories and relevant technologies. The government announced a plan to drive the project on a pan-government level with the cooperation of related departments. However, as the change of the relevant development technologies is faster than that of teaching/learning theories, interpreting new technologies in the sense of educational technology can be a realistic strategy for the development of well-designed edutainment.

The current trend of ICT is applying ubiquitous technologies to its main framework. The ubiquitous environment is the fourth revolution after the City Revolution, Industrial Revolution and Information Revolution which have had the most profound influence on the history of mankind in the past and present. It is a concept which integrates the physical space and electronic space by combining the electronic space with different spaces. That is, through the physical space was integrated seamlessly into the computer in the information revolution, in the ubiquitous revolution, computers are planted into the physical space to make man, computer and things combined together with the goal being to escape from the control of machines and establish an environment with humans positioned in the center. Therefore, the teaching/learning of our future education will be performed in a u-learning education with the combination of ubiquitous technologies, and it is self-evident that edutainment will be developed on the basis of ubiquitous technologies [2, 3, 4].

The rest of the paper is organized as follows: Section 2 describes u-Edutainment and its framework. Section 3 describes the applications of the framework. We conclude in section 4.

2 u-Edutainment and Its Framework

The u-learning environment can be realized by the applications of various ubiquitous technologies on the basis of the current e-learning environment. The ubiquitous-related technologies may be summarized through the policy of IT839 announced by the Korea Ministry of Information and Communication for the realization of u-Korea.

Edutainment currently being developed in the e-learning environment has already produced contents for middle school math adopting the game rules of "Challenge! Golden Bell" with the application of Sharable Contents Object Reference Model(SCORM) in each unit of Learning Object based on XML. The standard contents developed in this way are commonly being used for the learning management system(LMS) and the learning contents management system(LCMS), and are planned to be extended to mobile and the Broadband Convergence Network(BcN). In addition, the Advanced Distributed Learning(ADL) of the U.S., which is the developer of SCORM, has recently proposed a method of edutainment development in a larger volume than ever before by suggesting a framework to develop a Massive Multiplayer Online Game(MMOG) in dispersed learning environment[7,8,14].

Fig. 1 shows the unified framework. As shown in Fig. 1, u-learning edutainment will be ultimately be realized as existing edutainment with e-learning system and game technology by adopting ubiquitous technology, and the learning environment that is applied with relevant technologies based on IT839 policy into m(mobile)-learning, r(robot)-learning, v(virtual reality)-learning and c(convergence)-learning. This paper proposes that unified framework provides four learning models into which ubiquitous technology can be applied.

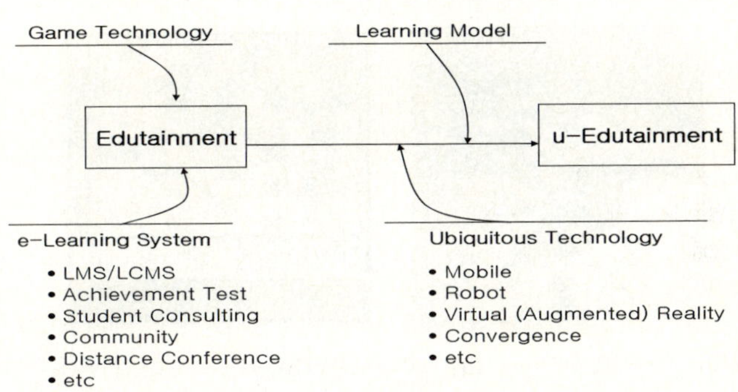

Fig. 1. Unified framework for u-Edutainment

3 Applications of the Unified Framework

In this section, we describe four kinds of applications which are applied the proposed unified framework.

3.1 m-Learning Edutainment

The current ubiquitous technology that we are using is a rudimentary one which is mostly in the stage of m-learning adopting the technologies of WLAN and PIMS. That is, m-learning is an u-learning in a narrow sense. Moreover, the recently commercialized WiBro and DMB are new driving forces that support m-learning.

Fig. 2. Example of learning with wireless terminal

Fig. 2 is a learning environment with PDA and wireless laptops. In m-learning edutainment, the applications of edutainment are developed separately in the case of wireless LAN with slow data transmission speed and portable Internet : a client-server system which stimulates competitions among learners with "Real-time Quiz" or "Summarizing Activities of each Group" and a type of edutainment by learning individually through downloads as in Fig. 3. However the range of development is being extended with the commercialization of WiBro and DMB, as contents with rich capacity such as music, flash, video, 3D and animation are generally used for edutainment.[6]

(a) (b)

Fig. 3. Examples of Mobile Edutainment of BURUXO Ltd. (a) WordWiser (b) Crossword Battle Puzzle

3.2 r-Learning Edutainment

With the increased interest in intelligent robots, there is an active movement for the use of robots for education. Currently, robots are being used as class assistants, museum guides, and for English learning.

Learning with robots is an aspect of edutainment in itself even if without the elements of a game. The development of such intelligent educational robots can be possible with the collection of all state-of-the-art technologies including artificial intelligence, bio engineering, nerve circuits, fuzzy theory, voice and image recognition, micro processors as well as motor control and sensing technology.

Fig. 4. Example of using a robot as class assistant

Recently, the use of robots in life has become easier thanks to the remarkable development of SoC(System-on-Chip) technology which enables a built-in function of image-processing, voice recognition, sensing, etc [9,10].

3.3 v-Learning Edutainment

With robots, contents or tools for Virtual Reality (VR) are also being highlighted a form of an edutainment to draw the interest and flow of learners. They can be divided into conventional contents of simulations or animations that have been previously developed, and the equipments that support the VR like 3D mouse, glasses to show sensor and image, gloves and Head Mounted Display (HMD).

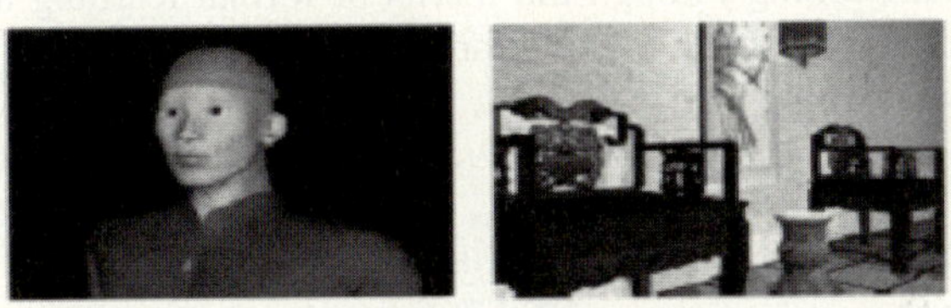

Fig. 5. Virtual Tour Guide and Peranakan Hall

Fig. 5 is an example of the system to guide a cyber museum using VR which proposed the development of edutainment contents and the presentation process with framework [11].

Fig. 6. Example of AR using Marker and PC Camera

Moreover, VR technologies are further developed in the field of augmented reality which is a Hybrid VR system that combines the real world, seen by the viewer, and the virtual world with added information into one image.

Fig. 6 is an example of the development of a realistic edutainment about the phenomenon of raining in which the PC's camera recognizes the marker in the unit of 'water circulation and weather change' in the science subject of the 5th grade in elementary school[12].

3.4 c-Learning Edutainment

It is not an exaggeration to say that the latest technological trend is "*convergence*" and "*semantics*". The ubiquitous learning environment will be established through BcN like the convergence of broadcasting with communication and the convergence of voice with data. In learning too, an environment of a new concept can be created through convergence and semantics.

Fig. 7 shows the convergence learning model for the ubiquitous environment. As shown in Fig. 7, a learning model of the next generation is offered to reinforce the internal motivation and flow of learners through the convergence of the characteristics of psychology, pedagogy and technological environment of the media. This is an attempt for the emergence of formal and informal learning aimed to raise the integrating and creative thinking with the focus on the learner in the aspect of pedagogy. Also reinforces informal learning through the merits of formal learning and provides personalized experiences of formal learning through experiential creativity and the integrity of the informal learning [13].

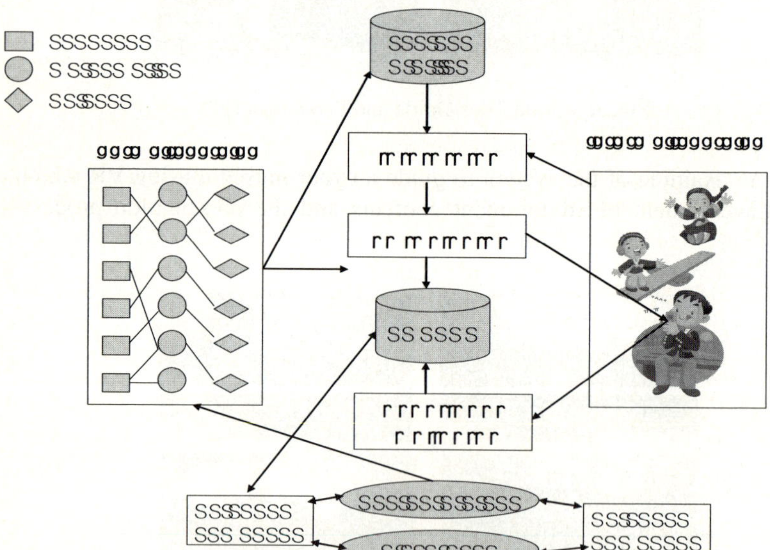

Fig. 7. Convergence Learning Model

The representative case of c-learning edutainment is the "Spungy" which is a TV program broadcast on KBS(Korean Broadcasting System). Internal motivation is accelerated with the critical mind of viewers that occurs from the tremendous knowledge transmitted by various mass media, and the viewers' flow is induced by the transmission of creativity and integrity of informal learning through broadcast media.

4 Conclusion

In this paper, we examined the methods of the development of edutainment based on ICT as applied to the ubiquitous environment as well as their planned application. One of the underlying properties of human beings is the desire to be able to learn while playing, which may be called 'the mentality of a thief'. Moreover, there is a great expectation to be able to do an interesting learning in the ubiquitous environment with human beings positioned in the center. Therefore, if we best utilize this technology well for education or appropriately realize what has been well designed for education, we will be able to liberate our children from the pressure of learning.

The proposed unified framework has assumed that it is feasible to develop u-learning edutainment. However, we have to research the "Roadmap for edutainment using combined approach educational technology, game technology, and ubiquitous technology" in the future.

References

1. Csikszentmihalyi. Flow,*The psychology of optimal experience*. New York: Harper & Row, 1990.
2. YoungGyun Baek, *Understanding and Utilization of Edutainment*, Jung-il, 2005.
3. ShimHo Kang,*Digital Edutainment Storytelling*, Sallim, 2005.
4. SungWoo Choi, JaeGyeong Lee, HwanSik Kim, DaeWon Seo and SangHyun Jang, “A Study on How to Improve EDUNET to Drive the Development Plans for Education Information”, Ministry of Education and Human Resources Development, 2002.
5. Prensky. Marc, *Digital Game-Based Learning*, New York McGraw-Hill, 2001.
6. ByungKil Sohn, BeomSok Koh and BoKyung Kye, “A Study on Development of Edutainment contents Prototype”, Korea Education & Research Information Service(KERIS), 2005.
7. ADL, http://www.adlnet.org.
8. Curtis J. Bonk and Vanessa P. Dennen, "Massive Multiplayer Online Gaming : A Research Framework for Military Training and Education", ADL, 2005.
9. JeongHye Han, DongHo Kim, KyungSeon Lee, SungJu Park and KyungChul Shin, " A Teaching Assistant Robot in Elementary School", 2nd International Conference on Ubiquitous Robots and Ambient Intelligence, 2004.
10. DongSoo Kwon and KiHoon Yang, “Robot and Game”, Korea Game Development and Promotion Institute(KGDI), Report in Forum of Future Game 2003.
11. Meehae Song, Thomas Elias, Ivan Martinovic, Wolfgang Mueller-Witting and Tony K.Y.Chan, "Digital Hertage Application as an Edutainment Tool", ACM, 2004.

12. JungHyun Kim and BoKyung Kye, “A Study on Augmented Reality-based Experiential Learning Contents Development and Field Application”, Korea Education & Research Information Service(KERIS), 2005.
13. GwangSoo Cho, GeunYoung Jang, TaeYeon Jung, SooJin Lee and BoKyung Kye, “A Study on Next Generation e-learning Model and Contents Development Methodology based on the Reinforcement of Learner's Interest, Flow and Motivation”, Korea Education & Research Information Service(KERIS), 2005.
14. InWoo Park, SangHyun Jang, YongSang Cho and ChulGi Hong, "The Research of Design on LMS and LCMS", Korea Education & Research Information Service(KERIS), 2004.

Trace-Based Framework for Experience Management and Engineering

Julien Laflaquière[2], Lotfi S. Settouti[1], Yannick Prié[1], and Alain Mille[1]

[1] LIRIS laboratory, UMR 5205 CNRS, Bat. Nautibus,
Université Claude Bernard Lyon 1,
69622 Villeurbanne CEDEX, France

[2] Institut Charles Delaunay, University of Technology of Troyes, FRE CNRS
2848, Tech-CICO laboratory, 10010 Troyes CEDEX, France

PhD research supported in part by Champage-Ardenne district grant

{jlaflaqu, lsettout, yprie, amille}@liris.cnrs.fr

Abstract. The paper deals with experience management in computer-mediated environments. It particularly focuses on complex tasks whose support relies more on *experience* than on *knowledge*. The presented approach is based on the "use traces" concept to investigate the "activity reflexivity" as a first step in experience management, and experience sharing and reusing as possible applications. The paper also outlines the framework supporting Trace-Based Systems creation. Traces, trace models and trace life cycle are formally defined. The main parts of the framework architecture: collecting, transformation, visualization, and query systems are also detailed.

1 Introduction

It's a common to say that computers are widely used, for more and more various and numerous tasks, usually with a strong documentary dimension. Hence, more and more activities are performed in *computer-mediated environments*, to achieve at least in part these tasks. Such environments are composed of both digital contents *and* the tools that allow manipulating them. Among all the challenges that knowledge management (KM) has to take up, we focus in this paper on *complex mediated tasks support*. As we will show, complex tasks have characteristics that make them hold more onto *experience* than onto simple *knowledge* (in the sense of Sun [1]). Taking into account these characteristics and adopting a specific approach to manage them is a stake for the emerging young experience management and engineering field [1]. In particular, we will focus in this paper on the notion of *interaction trace* for experience management and experience engineering.

This paper has two main parts. Section 2 outlines why and how one can consider *traces of environment use* as a tool to tackle some experience management (EM) objectives, from the activity reflexivity to the experience reuse. Section 3 describes necessary theoretical concepts towards a generic *framework for traces-based systems*.

B. Gabrys, R.J. Howlett, and L.C. Jain (Eds.): KES 2006, Part I, LNAI 4251, pp. 1171 – 1178, 2006.
© Springer-Verlag Berlin Heidelberg 2006

2 Traces for Interaction Experience Management and Engineering: Rationale and Scenarios

We define a complex task in a computer-mediated environment, as a *high level*, *dynamic*, *open-ended*, and strongly *context-dependent* task. Economic Intelligence (EI) on the Web is a good example of this type of task [2]: EI related tasks associate difficulties from process-based classical approach of EI and those from complex Web search tools manipulation (high-level); the tasks are defined step by step during its realization (dynamic); there is neither exact goal, nor identified means to reach this goal (open-ended); and overall, the realization of the tasks depends on the context, here the information found (or not found) during the process (context-dependent). Because of these characteristics, complex tasks are difficult to formalize and to model, and consequently, to integrate in a classical KM process.

As a hint towards a possible solution, we investigate the concept of *activity reflexivity* as a means of making users conscious of their own activity and in a sense conscious of the experience they are living. We consider that an activity is reflexive if its realization provides some information about itself, i.e. leaves an accounted for representation of itself. Several research works underline the role played by activity reflexivity for supporting and enhancing the realization of complex tasks [3]. For instance in Computer Supported Collaborative Learning, reflexivity allows a deeper learning [4] while in Information Retrieval, it is considered as allowing a better activity structuration, avoiding user's disorientation [5].

2.1 Traces for Reflexivity in Computer Mediated Activity

To obtain activity reflexivity in complex mediated tasks, one has to be able to propose a kind of activity representation to the user. This representation has to be constructed from the observation of the mediating environment. Furthermore two opposite general kinds of approaches can be distinguished. *Quantitative* approaches are based on log-files. These log-files, obtained by passive observation, are used to calculate some statistical insights about the user's exploitation of his environment, and/or to profile the user himself [6]. *Qualitative* approaches are mostly proposed in ethnographic and ergonomics research studies [7]. These approaches consist in careful observation, *in situ*, often completed by audio and video recordings, allowing auto-confrontation for observed users. Both quantitative and qualitative approaches remain unsatisfactory to deal with reflexivity: the first are disconnected from miscellaneous contexts in which activity takes place, while these contexts are fundamental elements of the experience; the second are time-expensive and are concerned with a very short period of time. We claim that an intermediate solution is possible, based on *use traces* of computer-mediated environment. This is what our *Trace-Based Systems* aim to do (section 3).

Our main goal is to deal with use traces that "make sense" for the considered user. The structure of a trace and its abstraction level have to be pertinent, accordingly to the tasks the user realizes. In other words, the description power of a trace has to correspond to his reflexivity needs.

With this objective in mind, our approach can be broadly decomposed as follows: a) a modeling of the use is performed, *with* the user participation and *in the context* of a particular task. This allows underlining some salient interaction components called *Objects of interest* (entities, events and relations), which are relevant for him, in the context of *his use* of the environment in the context of his task (there is neither causal nor hierarchical *a priori* constraints and objects of interest definitions belong to a *use model*) and b) an observer agent is built that can accordingly generate traces from the interaction between the user and his environment. Such traces are a graph of objects of interest, where entities and events are linked with relations (see Fig. 1.).

One advantage of this kind of trace is to be "readable" by the user, with an adapted graph visualization tool (which has a critical and complex role in this context). The traces outline the activity structure and its global shape. They can give a continuous representation of the user's activity in the computer-mediated environment as a historic classic view, but it is not necessarily useful. In fact, a trace can be usefully exploited to *contextualize* its proper components. Each object of interest, embedded in the trace, can be contextualized by the context of its use, i.e. by the role it takes relatively to the other components. To be able to replace each object of interest in its context is one dimension of activity reflexivity in the computer-mediated environment.

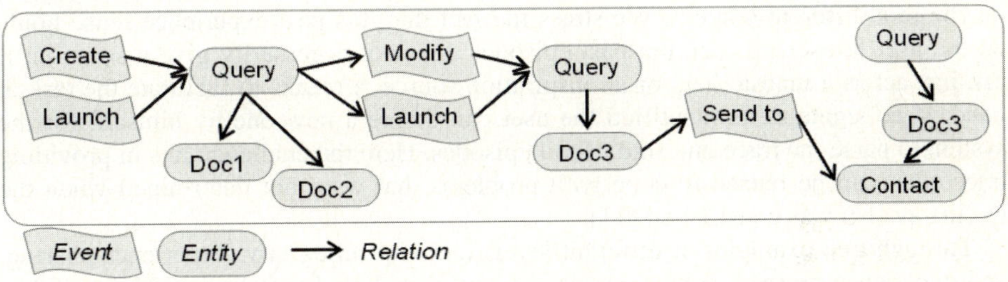

Fig. 1. A part of use trace, as a graph of objects of interest: in this simplistic example, a user of Human-Links® (EI tool), *launches* a *query* with some keywords. He has to *modify* the query to obtain a pertinent result, so he *sends* the *document* to a *contact* (other user of the software). For detailed presentation see [2].

2.2 Traces for Experience Sharing and Reusing

Some applications of our approach have shown that such traces can "make sense" for users as well as for activity analysts [8]. So, one can examine the question of traces exploitation towards the initial issue of activity reflexivity. Two problems seem particularly interesting because they fit with EM purposes: *experience share* and *experience reuse*. Since use traces as the ones we described above can hold a kind of activity reflexivity (e.g. the interpretation of his own trace by a user), it is possible for a user to exploit the same traces as a means to *share* his use/interaction experience. We do not mean that the trace, or part of it, *is* the experience, and that we can exchange it directly (it will probably never be the case). We rather consider the trace as a means, a tool or a *support* for experience sharing.

It can be seen as a "boundary object" [9], useful in several well known situations in KM, when a person has to explain his work face to face with some material supports (analysis of problem situation in a design project or memory project constitution for example). Making such use of a trace needs tackling the trace visualization issue. We do not think as Morse [10] that visualization is the only challenge, but it remains a real and central problem to deal with real trace-based experience sharing.

In KM and EM fields, another recurrent problem is *experience reuse*. In our case we are interested in reusing experience of environment use in a particular task. There are several ways of dealing with this problem, and we want to underline that we are just considering one of them. We can extend the approach detailed in section 2.1: in addition to the objects of interest, we define *tasks signatures* as patterns in the trace graph which characterizes a particular task realization. The goal here is not to describe precisely one pattern for one task, but to be able to know when a particular task occurs. Thanks to tasks signatures, we can extract some significant *episodes* in the trace. One episode at least corresponds to one task signature (a web search for example), and can be considered as reusable piece of use experience. How do we reuse it?

Suppose the user of a computer-mediated environment has a problem and asks his system for help. The system tries to identify in current use the beginning of a known signature. Then it can find and present to the user the past episodes that correspond to this known signature. These episodes, seen as pieces of past use experience, can indicate to the user how in the past he coped with that problem (or not), in a similar task, maybe in a different context. We stress the fact that this past experience reuse holds on the user himself: presenting past episodes does not necessarily give a solution; it can just act as a mnesic help or an inspiration source, a means to facilitate the task at hand. If no signature is identified the user can define a new one by himself, for the system to parse the trace and find useful episodes. Here the challenge lies in providing traces that can be reused to cope with problems that were not determined when the tracing system was constructed [11].

Through tree examples at different levels (activity reflexivity, experience sharing, and experience reuse), we have shown *how* a traces-based system could be useful to support user activity with an EM point of view, and *why* it opens new perspectives to instrument EM approaches for complex computer-mediated tasks. The first part of this paper also gave some clues that suggest the underlying complexity of such systems, and we therefore have to lean on a rigorous theoretical approach and methodological tools. This is why we developed the *traces-based systems framework*. In the following section, we will describe traces-based systems and sketch the idea of how they could be used in order to support the trace exploitation process.

3 Towards a Trace-Based System Framework for EME

In order to tackle the traces management challenge, we present a trace-based systems framework. This framework allows the definition of trace-based systems (TBS), which holds on traces exploitation, e.g. the production, the transformation and the visualization of traces. We will first define our notions of trace and trace lifecycle, before setting up a formal framework for traces-based system.

We introduce traces as *temporal sequences* of *observed items*. "Temporal sequence" indicates the existence of an order-relation that organizes trace data relatively to a time base. "Observed items" indicates that trace data result from an observation. This definition allows us to argue that a trace is a numeric document which reveals temporally located data resulting from an observation. As shown in section 2, we are mainly concerned with use traces of computer-mediated environment. In this context, a trace is considered as a recording of a computer-mediated activity that is potentially constructed from a variety of sources (log-files, videos, transcripts, etc.).

We divide the trace lifecycle into different stages. Firstly, *collecting* deals with deciding what to collect and how to collect it into a basic trace. For example, a basic trace would gather all the events occurring during an interaction as objects of interest. A basic trace is a TBS entry-level trace. Secondly, *transformation* deals with automatically or manually filtering, rearranging or adding information to the basic traces in order to meet the description needs of the user. In our example basic trace, the collected data can be irrelevant or too detailed: *selection transformation* can be used for separating data and sequences of interest from the "noise", while *abstraction transformation* (data aggregation), can be used to provide higher-level objects of interest to the user. Thirdly, *presentation* deals with traces visualization and involves choosing what to present and how to present it to the user. It is important that each of these stages be considered separately to avoid any confusion.

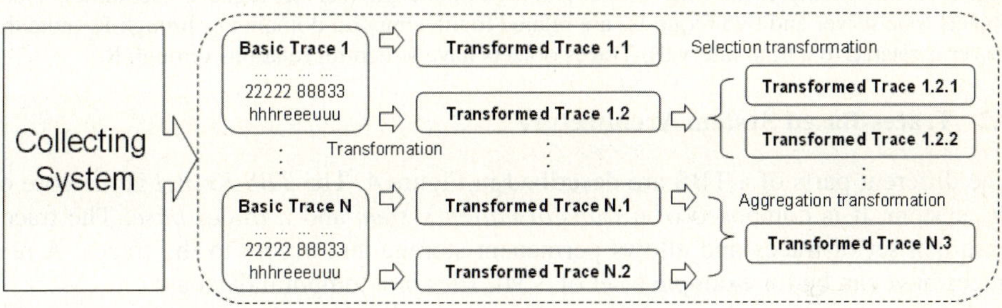

Fig. 2. Traces are collected, transformed and visualized when needed

3.1 Traces and Traces-Models

The trace lifecycle will be supported by traces-based systems. Before describing their various components, it is necessary to present the notion of trace model. In a TBS, each trace must always be associated to an explicit *trace model*, which allows its interpretation. A trace model can be considered as a model describing the vocabulary and the structure of the traces it is associated with. For example, trace log-files often rely on the *Common Logfile Format (CLF)*, traces XML documents are described by DTD trace models or schemas, *etc.* In this work, we will use the ontological model as a general trace model compatible with models that will actually be instantiated in a TBS (simple log models, schema-based models, *etc.*). Formally:

Definition 1. *A Trace Model is an ontology* $M_T = (C; \leq_C; \leq_R; R; T; A; \sigma_A; \sigma_R)$ *consisting of a set of concepts C organized in a hierarchy with an order relation* $\leq_C$*, a set of relations R organized with* $\leq R$*, a relation signature* $\sigma_R : R \rightarrow C \times C$*, a set of data types T, a set of attributes A, and a attribute signature* $\sigma_A : A \rightarrow C \times T$.

The following definition concerns the general formal definition of a trace:

Definition 2. *A trace is a quintuplet* $(M_T, D_p, O_{tr}, R_t, R_s)$ *where* M_T *is the associated trace model;* D_p *is a temporal domain* $(T, <)$ *with* T *a set of time instants and* $<$ *an order on* T*;* O_{tr} *is a set of objects* O*,* $O_{tr}=\{O_0, O_1, \ldots, O_n\}$ *such as* $\forall O_i \in O_{tr}, f(O_i) \in C$*, with* f *a labelling function* $f: O_{tr} \rightarrow C$*.* $R_t \subseteq D_p \times D_p \times O_{tr}$ *is a relation representing the temporal links between objects and time intervals.* $R_s \subseteq O_{tr} \times O_{tr}$ *is a relation representing the structural links between objects, and* $\forall R_{si} \in R_s, g(R_{si}) \in C$*, with* g *a labelling function* $g: R_s \rightarrow C$*.*

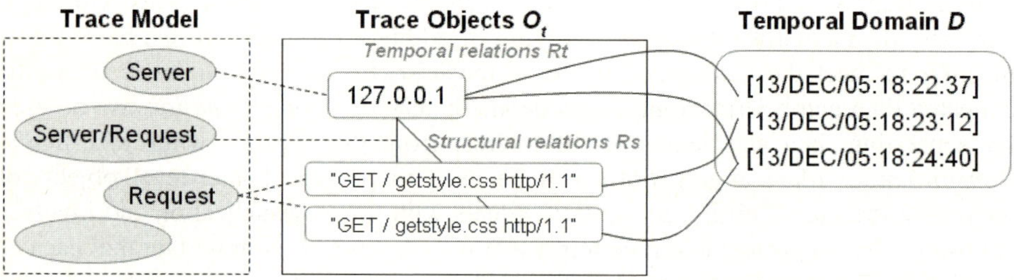

Fig. 3. In this example, the trace model is a set of concepts (server, request, username). Trace objects (one server and two requests) are related to the temporal domain D_p through R_t (note the server is related to a time interval). Traces objects have structural relations through R_s.

3.2 Traces-Based System Architecture

The different parts of a TBS are described in Figure 4. The *TBS Kernel* is the core of the system. It is composed of a *transformation system* and a *traces base.* The traces base is a set of traces and allows permanent storage and access to the traces. A real traces base can be for example a set of XML files or a temporal database.

As seen above, a main part of trace exploitation holds on their transformation. *The transformation system* performs transformations τ during the modification and handling of the traces. The transformation system allows (1) modifying the trace by enriching or filtering its data, (2) modifying the model of trace, (3) updating the traces base or making automatic transformations by using transformation models (sets of formal rules expressed in a rule-based language).

The *query system* allows formulating and resolving queries on the traces base. Queries can be made on traces considering the models they are associated with, their transformations, their objects in their interrelations and time inscription. For instance one can consider all the traces of an individual at a certain transformation level, and query for episodes involving objects in structural and temporal relations (for example two web pages related to the same server, visited in a same hour).

The *collecting system* is a set of processes allowing the conversion of several *tracing sources* into basic traces by appropriate tools. Tracing sources are files or data streams in an unspecified explicit format, from which basic traces will be constructed. The simplest collecting system would do simple data integration and time-based synchronization of the various sources. A more complicated system would incorporate different tracing sources in an intrinsic iterative process for instanced by mixing video annotation and log files use, *etc.*

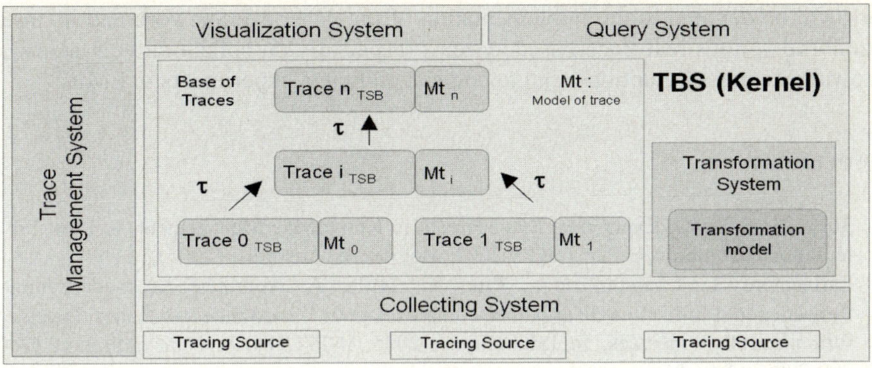

Fig. 4. Trace-based framework architecture

The *visualization system* allows visualizing the traces and thus facilitates analysis and interpretation. The visualization system is a set of techniques for presenting traces (basic traces or transformed traces) in a visual form allowing human's direct exploitation. A well-known visualization involves using a timeline that describes the temporal sequence of data. If video is implied as a tracing source, hence in the temporal domain, visualization can link video and traces. For example, play a video synchronously with trace visualization for better interpretation (as in [8]).

Lastly, the *trace management system* (TMS) allows managing the various models involved in a TBS: trace models, transformation models, queries, *etc*. It allows the archiving of the traces, the granting of rights on traces base and offers also an integrated management of trace lifecycle. Indeed, TMS holds a traceability of transformations and operations performing in TBS. This functionality allows the management of the experience use of TBS and facilitates the trace management and engineering.

4 Conclusion

This paper has presented our approach which suggests to consider use traces, constructed from the interaction between a user and his computer-mediated environment, as a tool for EM approach. In the first part, we have outlined that activity reflexivity is a key step to deal with interaction or use experience of a user. After explanations about the kind of traces involved in our approach, we have proposed examples of use trace exploitation to investigate experience sharing and experience reusing.

In the second part of this paper, we have detailed the framework developed to support use traces exploitation with Trace Based Systems (TBS). We have formally defined the notions of trace and trace model, and have given some details on the different parts of a TBS that are needed to support trace processing.

The general approach we proposed in this short paper is backed by different domain-oriented studies we carried out in our research team. Future work implies developing the general approach both on the theoretical and practical sides. The theoretical work will focus on traces, reflexivity and experience management for individuals and collectives, and formal trace modeling, transforming and querying. The Practical work will be devoted to the implementation of the general framework and its instantiation into various real trace-based systems (especially on collaborative systems), so as to validate our assumptions, and to conduct complete experiments in EME.

References

1. Sun, Z., Finnie, G. Experience Management in Knowledge Management. In Proc. 9th International Conference, Part I, KES 2005, Melbourne, Australia, (2005) p.979.
2. Laflaquière, J., Champin, P.A., Prié, Y., Mille, A. Approche de modélisation de l'expérience d'utilisation de systèmes complexes pour l'assistance aux tâches de veille informatiquement médiées, In ISKO-France'2005, INIST/CNRS, Nancy, France, (2005) 209-230 (in french)
3. Laflaquière, J., Ciaccia, A. : Facilitation de tâches informatiquement médiées : une approche centrée sur la réflexivité de l'utilisation, 6th workshop Colloque des jeunes chercheurs en sciences cognitives 2005, Bordeaux, France (2005), 217(in french)
4. Weerasinghe, A., Mitrovic, A. : Facilitating deep learnig through self-explanation in an open-ended domain, International journal of Knowledge-based and Intelligent Engineering systems, IOS Press, 10 (2006) 3-19
5. Castelli C., Colasso L., Molinari A., Getting lost in Hyperspace: Lessons Learned and Future Directions, in *ED-MEDIA 96/ED-TELECOM 96,* Boston, (1996), 124-130
6. Mobasher, B., Cooley, R., Srivastava, J. 2000a. Automatic personalization based on web usage mining. Commun. ACM, 43 8 (August), 142–151.
7. Dekker S., Nyce, J.M., How can ergonomics influence design? Moving from research findings to future systems. Ergonomics, vol. 47, N. 15 (2004), 1624 - 1639.
8. Georgeon, O., Bellet, T., Mille, A., Letisserand, D., Martine, R.: Driver behaviour modelling and cognitive engineering tools development in order to assess driver situation awareness. Workshop on Modelling Driver Behaviour in Automotive Environments. (2005), Ispra.
9. Star, S. L. 1989. The structure of ill-structured solutions: Boundary objects and heterogeneous distributed problem solving. M. Huhns and L. Gasser, eds. *Readings in Distributed Artificial Intelligence.* Morgan Kaufman, Menlo Park, CA.
10. Morse, E. and Steves, M. CollabLogger: A Tool for Visualizing Groups at Work. Proceedings of WETICE 2000, Workshops on Enabling Technologies:Infrastructure for Collaborative Enterprises, IEEE Computer Society, (2000), 104-109.
11. Champin, P.-A., Prié, Y., & Mille, A. : MUSETTE: a framework for knowledge capture from experience. In Proceedings: Egc'04, (2004), Clermont Ferrand, France (in french)
12. Luotonen A. "The Common Log file Format."
http://www.w3.org/pub/WWW/Daemon/User/Config/Logging.html

Neural Network Classification of Diesel Spray Images

S.D. Walters, S.H. Lee, C. Crua, and R.J. Howlett

Engineering Research Centre, School of Engineering
The University of Brighton, Brighton, U.K.
s.d.walters@brighton.ac.uk
http://www.brighton.ac.uk

Abstract. This paper describes an evaluation of a neural network technique for modelling fuel spray penetration in the cylinder of a diesel internal combustion engine. The model was implemented using a multi-layer perceptron neural network. Two engine operating parameters were used as inputs to the model, namely injection pressure and in-cylinder pressure. Spray penetration length were modelled on the basis of these two inputs. The model was validated using test data that had not been used during training, and it was shown that semi-automated classification of complex diesel spray data is possible. The work lays the foundations for the establishment of an improved neural network paradigm for totally automatic, fast, accurate analysis of such complex data, thus saving many man-hours of tedious manual data analysis.

1 Introduction

In recent years, research has shown that the combustion and emission characteristics of modern diesel engines are dependent on a multitude of parameters, including: fuel atomisation, nozzle geometry, injection pressure, shape of inlet port, and other factors. In order to improve air-fuel mixing and the resulting combustion, it is important to understand the underlying fuel atomisation and spray formation processes. Experimental and theoretical approaches have been adopted by researchers in order to investigate the characteristics of spray behaviour, formation and structure for the high-pressure injector, in order to improve engine efficiency; in practice the goals are: improved combustion, performance and reduced exhaust emissions. However, further detailed studies of the atomisation characteristics and spray development processes of high-pressure diesel sprays are still needed.

Intelligent systems, i.e. software systems incorporating artificial intelligence (AI), have demonstrable benefits for control and modelling of engineering systems. For example, they feature the highly useful ability to rapidly model and learn characteristics of multi-variate complex systems, offering performance advantages over more conventional mathematical techniques. This has resulted in their application in diverse applications within power systems, manufacturing, optimisation, medicine, signal processing, control, robotics, and social/psychological sciences [1-3].

The AI approach is well-suited to the analysis of many combustion-related problems; AI has considerable potential for making faster, more accurate predictions

B. Gabrys, R.J. Howlett, and L.C. Jain (Eds.): KES 2006, Part I, LNAI 4251, pp. 1179 – 1189, 2006.
© Springer-Verlag Berlin Heidelberg 2006

than some traditional methods. Increasingly, the availability of complex sensory and computing systems is resulting in the production of vast quantities of information-rich data. Historically, the analysis of such complex data has required very substantial human effort, often with every image of every video needing to be evaluated by human eye. The aim of this investigation, and the partner investigation featured in [4, 5], has been to apply intelligent systems tools and techniques to the problem of effectively, and semi-automatically, processing and analyzing large complex data sets. The data in question was generated during advanced experimental engine research.

The specific objectives of the experimental work described in this paper were to:

1. Create a semi-automatic diesel spray analysis system in software;
2. Investigate the effect of altering key system parameters, namely:
   - Number of hidden nodes employed, and,
   - Desired output layer threshold;
3. Evaluate the system and assess its future potential.

2 Acquisition of Diesel Spray Penetration Video Clips

2.1 Image Data

A large repository of complex data was produced in the course of an extended investigation using a Ricardo Proteus research engine. These data comprised of '.AVI' video clips depicting the developing spray patterns of diesel injection processes, under selected operating conditions of: fuel injection pressure, in-cylinder air pressure, and injector nozzle size and type [6]. EXCEL files containing additional information about the engine operating conditions were associated with the video clip files, thereby providing a complete set of results for analysis.

It was desired to have the ability to classify images from the videos, using the properties of the images and according to the prevailing engine conditions. In order to classify the images from the video clips, the resulting software program needed to handle vast volumes of data, undertaking image processing, pre-processing and, in future work, subjecting it to neural network analysis by an internal neural network algorithm, before finally the outputting the results. To initially control complexity of the system, it was decided to use a custom-written multi-layer perceptron (MLP) neural network, implemented in the C language [7, 8], as the analysis engine. All other sections of the system were included in specially-written front-end MATLAB ® V.7 based software [9, 10].

The images representing the time-varying spray process under sets of injection and in-cylinder pressure conditions were processed using a specially-written program, coded in MATLAB. The use of MATLAB allowed the data to be relatively quickly visualised and analyzed without the need for writing too many low-level language routines, for example to control file access, perform image processing tasks and display outputs such as graphs. At a convenient juncture in the future, the MATLAB front end could be converted to a C-language addition to the MLP software, or the neural network could be implemented in MATLAB after the other routines, according to the needs of the application.

2.2 Image Processing

In recent years, the field of image processing has advanced apace; MATLAB features an optional Image Processing Toolbox which proved efficacious and convenient [10]. The actual methods chosen were derived from studying the properties of the spray-cone images, coupled with extensive reference to the literature [11-13]. The potentially high degree of spray break-up and the desire to model the spray in a transparent fashion favoured a relatively simple encoding method that consistently produced a fixed data word-length, as opposed to the relatively homogeneous spray shown as an example by Batchelor et al. [14].

The data processing technique was as follows. Each of the 80 .AVI video clips comprised 300 images, each of 320x128 pixels. Key images were automatically selected from each video – in this case those images in which peak spray penetration was attained; Fig. 1 shows examples of the peak penetration images for each of the four classes of data.

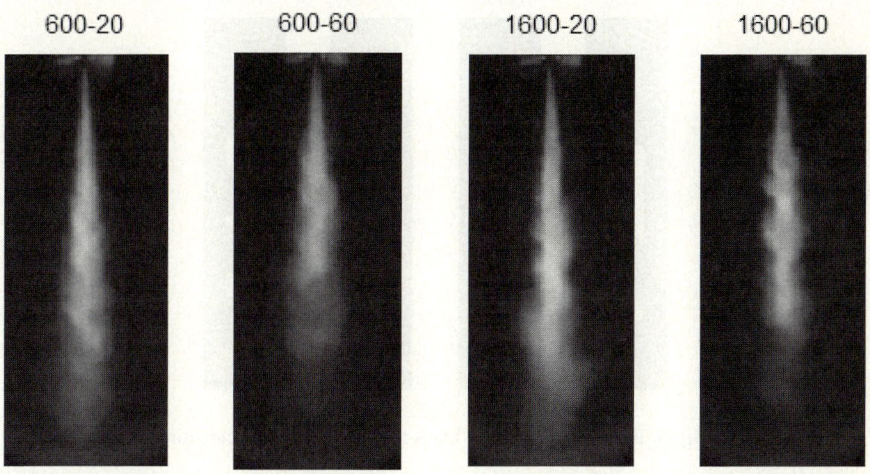

Fig. 1. Peak Penetration Spray Image Examples for each Set of Conditions

The grey-scale images were thresholded [10], yielding binary (black and white) images, as shown in Fig. 2.

A first pass through all of the videos facilitated the automatic selection of the optimum scanning window size to reduce redundant unchanging areas of black pixels, and identified the location and data of the 'peak penetration image' in each video. On the second pass through the data, the desired 80 peak images were selected out and re-sampled accordingly, reducing the window size to 32x7 pixels, hence significantly

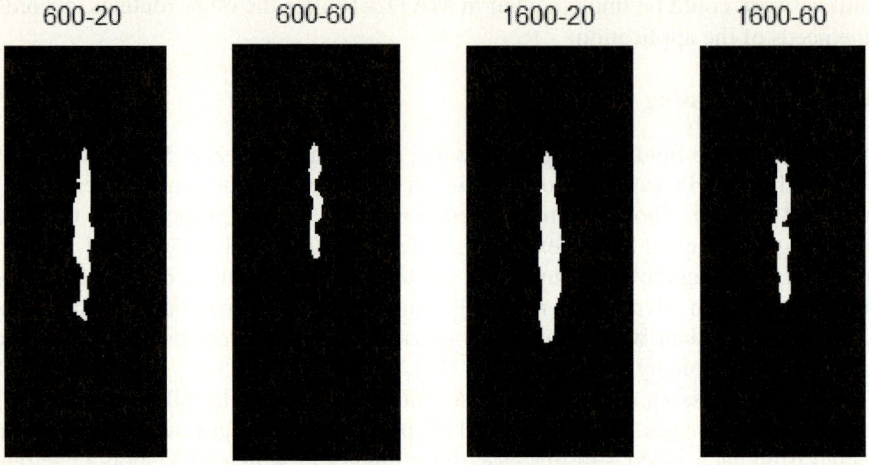

Fig. 2. Thresholded Spray Data for each Set of Conditions

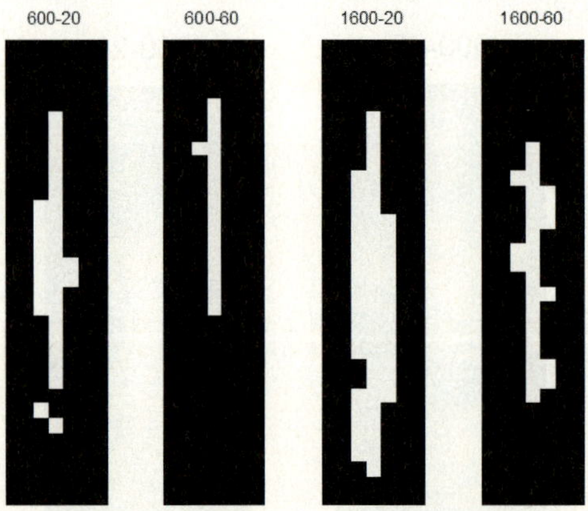

Fig. 3. Re-sampled Spray Data for each Set of Conditions

reducing the number of data-points referred to the neural network. Figs. 3 and 4 indicate the level of detail discernable after re-sampling. The reduced images were scanned line-by-line, producing a total of 80 vectors, each of which featured 224 data-points.

Fig. 5 depicts as vectors the same image examples as shown in previous figures Figs. 1-3. Finally, these vectors were combined with suitable 'Desired Output' information, yielding two data files for evaluation of the neural network in 'Training' and 'Auto-recall' modes. 'Auto-recall' is an automated routine, in which a set of previously unseen data exemplars are subjected to a 'Recall' process; the results yield network performance statistics.

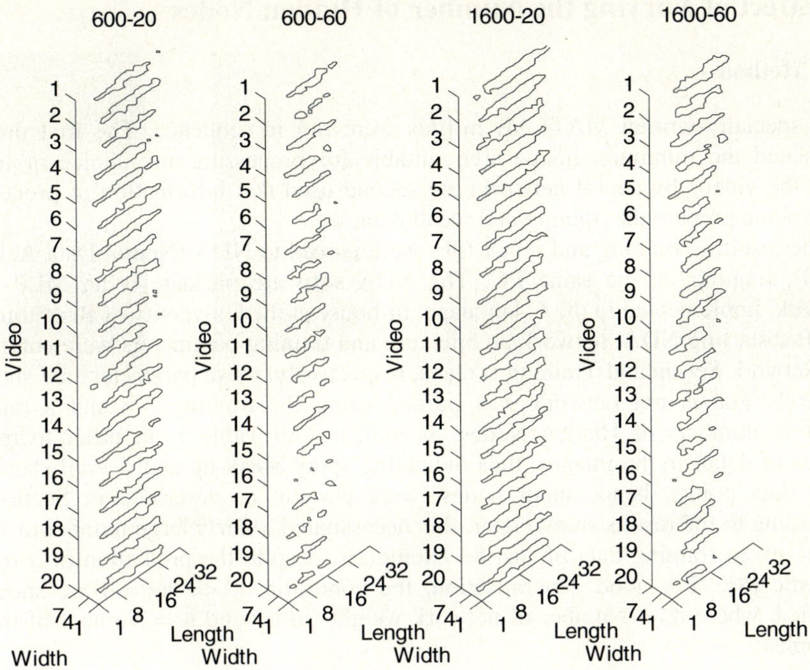

Fig. 4. Visualisation of all the Training (1 – 10) and Recall (11 – 20) Data

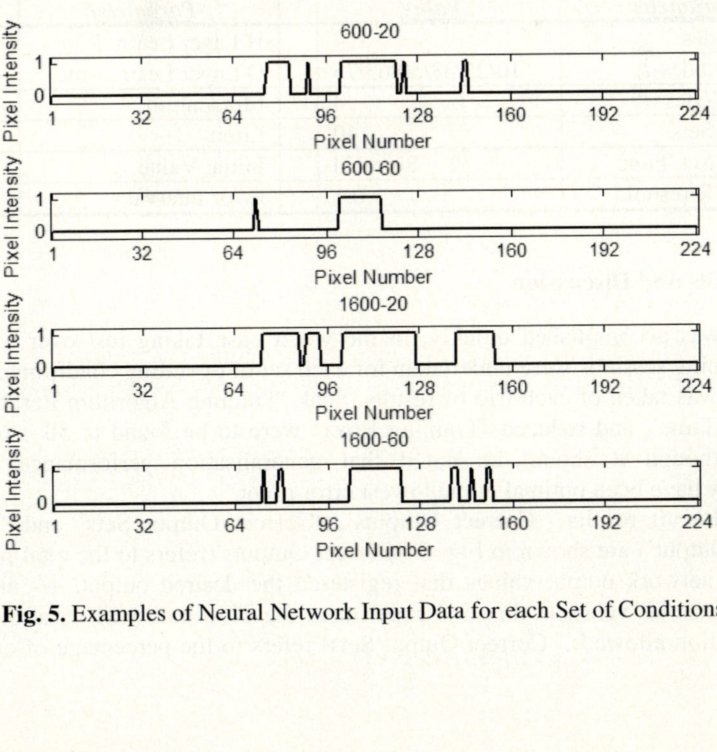

Fig. 5. Examples of Neural Network Input Data for each Set of Conditions

3 Effect of Varying the Number of Hidden Nodes

3.1 Method

Two specially-written MATLAB m-files were run in sequence. The first program calculated the minimum image area suitable for processing any choice of images from the videos by neural network; the second used this information to process the images and prepare the training and recall data.

The resulting training and recall files were passed to NDA (Neural Data Analyzer, V.7.0), running on the same PC. The NDA software package is an MLP neural network, implemented in the C-language, in-house at the University of Brighton.

After starting NDA, network architecture and training parameters were entered, on the Network Set-up and Training screens, respectively; these parameters are shown in Table 1. The neural network was trained using the training file and a range of different numbers of Hidden Nodes, as indicated in Table 1. In order to retain a degree of detail in the images, thus indicating spray break-up and overall shape, 224 input data points, hence input nodes, were chosen, as described in Section 2.2. According to the results shown later, this necessitated a fairly large number of hidden nodes, given training data of just 80 exemplars, despite the prediction of a training heuristic [8]: 'For good generalisation, the condition: n_w<=~n_t<=~10n_w should be satisfied, where n_w = Number of network weight values and n_t = Number of training examples'.

Table 1. Neural Network Configuration and Operating Parameters

Parameter	***Value***		***Parameter***	***Value***
Input Nodes	224		H Layer Learn. Rate	0.080
Hidden Nodes	10/25/38/50/63/75		O Layer Learn. Rate	0.020
Output Nodes	4		Momentum	0.800
Training Sets	40		Limit	0.200
O Layer Act. Func.	Sigmoid		Initial Value	0.500
O Layer Threshold	0.1		Error Interval	10

3.2 Results and Discussion

Training was accomplished quickly, in the worst case taking just over 6 seconds. Three training sessions were undertaken for each set of operating conditions, and then the mean was taken of each trio of results. Peak 'Training Algorithm Iterations' and 'Solution Time', and reduced 'Training Error' were to be found at 50 – 63 hidden nodes, although it should be noted that generalisation performance may not necessarily have been optimal at the lowest error point.

Mean Recall results ('Correct Outputs', 'Correct Output Sets' and 'Matching Greatest Output') are shown in Fig. 6. 'Correct Outputs' refers to the total percentage of neural network output values that registered the desired output, +/- an 'Output Layer Threshold' value; in this first experiment the Threshold was set at 0.1 (i.e. +/- 10% variation allowed). 'Correct Output Sets' refers to the percentage of output sets in which the outputs were all within the 'Output Layer Threshold' tolerance of the desired outputs. 'Matching Greatest Output' refers to the percentage of output sets in which the neural network correctly identified the 'winning' output class by setting its output node to the greatest value within the set.

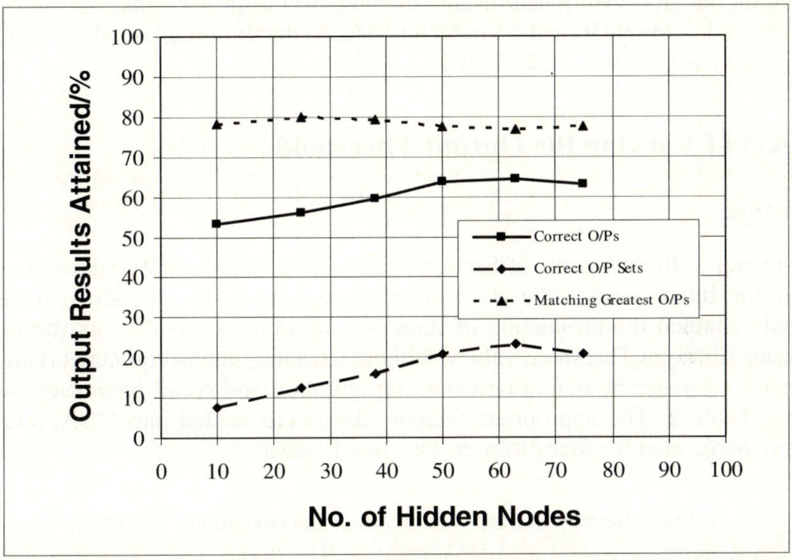

Fig. 6. Variation of Neural Network Performance with Number of Hidden Nodes

The network was best able to maintain +/-10% accuracy of ALL its outputs over 65% of the unseen recall file data, when the number of hidden nodes was between 50 and 75; this is shown by the 'Correct Outputs' plot.

Relatively poor performance was achieved on producing exactly 'Correct Output Sets' that were all within +-10%; the best performance was found to be 23% for 63 hidden nodes. Since the neural network was only being expected to choose the winning output node in this test, this poor performance was not considered very important. In an industrial environment, there would probably have been a different output configuration, for example, two output nodes with linear activation functions, each representing an engine parameter. In addition, the chosen 'Output Layer Threshold' value of 0.1 (or +-10%) was considered to be demanding for this initial experiment.

The 'Matching Greatest Outputs' graph plot was considered important for this experiment, as the network was expected just to choose the greatest output as the correct one. Best performance was 80%, attained between 25 and 38 hidden nodes.

3.3 Conclusions

The MLP neural network was trained quickly and effectively for all values of 10 – 75 hidden nodes.

An output threshold value of 0.1 (or +-10%) was considered strict for this experiment.

The choice of output configuration i.e. four outputs for four classes was not ideal for this application – two linear nodes would be more appropriate and may give better results.

Based on the aforementioned results, the best overall performance was considered to be achieved between 50 and 63 hidden nodes, so the minimum number to give best performance was chosen for future work, i.e.: 50.

4 Effect of Varying the Output Threshold

4.1 Method

In accordance with the results of Section 3, the configuration of 50 hidden nodes was retained for future work. The three training sessions with 50 hidden nodes had previously enabled the production of three sets of weights which could be used for recall using different Threshold values, without retraining the neural network again. For this experiment, the neural network configuration and recall parameters were as shown in Table 2. The appropriate weights files were loaded into NDA, and recall tests were performed for four different Threshold values.

Table 2. Neural Network Configuration and Operating Parameters

Parameter	*Value*		*Parameter*	*Value*
Input Nodes	224		H Layer Learn. Rate	0.080
Hidden Nodes	50		O Layer Learn. Rate	0.020
Output Nodes	4		Momentum	0.800
Training Sets	40		Limit	0.200
O Layer Act. Func.	Sigmoid		Initial Value	0.500
O Layer Threshold	0.1/0.2/0.3/0.4		Error Interval	10

4.2 Results and Discussion

Fig. 7 shows the results obtained with the various ‘Output Layer Threshold’ values, ranging from 0.1, as used in the previous experiment, through 0.2 and 0.3, to the much less- demanding value of 0.4.

It may be readily seen that the number of overall ‘Correct Outputs’ increased from 64% to 87% as the ‘Output Layer Threshold’ was increased to 0.4.

‘Correct Output Sets’ showed a large increase in performance from 23% to 68%.

‘Matching Greatest Outputs’ remained constant at 78% throughout the experiment, showing a robust response regardless of threshold value.

This work relied on the principle of ‘Winner-takes-all’, i.e. the output node exhibiting the greatest value was the winning node in the output set. For this application, the neural network appeared to be relatively undemanding as to the

Number of Hidden Nodes and 'Output Layer Threshold' adopted for training and recall.

Actual output values from the network may attain greater importance in future variations of this work, and methods for increasing the overall performance are currently subject to experimentation. For example, improvements by: (1) removing redundant white area in the image area, in addition to black; (2) re-sampling with even better choice of pixel reduction filtering; and (3) optimizing the number of hidden nodes to the nearest node, rather than to within 10-15 nodes.

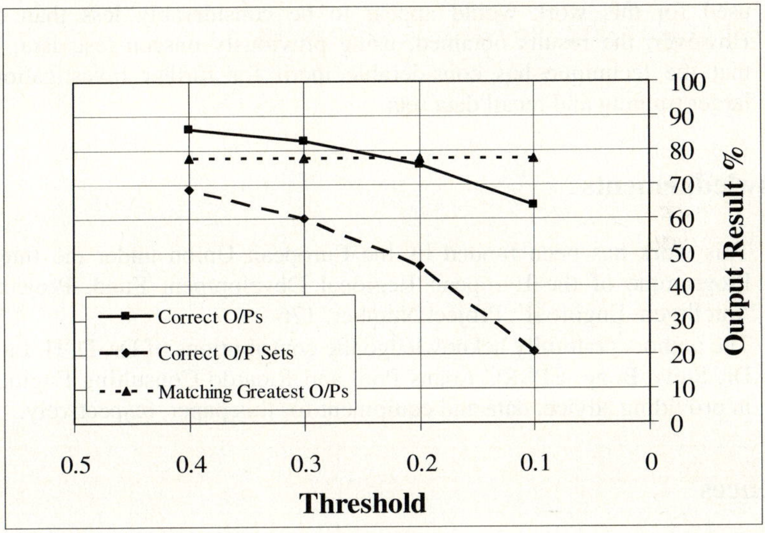

Fig. 7. Variation of Neural Network Performance with Changing Output Threshold

4.3 Conclusions

Given less-demanding 'Output Layer Threshold' values, there was an apparent increase in the performance of the neural network. However, for "winner-takes-all" testing, i.e. using the 'Matching Greatest Outputs' value, the results were exactly the same, at 78%, throughout all the tests.

For selection of classes of diesel spray, the neural network test was robust, but could be refined further, given more time; recommendations have been made. Automatic image processing needs to incorporate safeguards against corrupted images in large data sets.

5 Further Work

The performance of the devised neural network solution was considered satisfactory for an initial experiment. However, future work includes:

- Refinement of the image processing techniques. This will ensure that unchanging pixels (confusing to the neural network) are minimized, assisting with the following analysis.
- Experimentation with the precise number of hidden nodes required.
- Future engine tests to capture more engine parameters to enable more complete modelling of the spray process and evaluation of analysis methods.
- Adoption of new neural network algorithms for elimination of outlying data points and increased processing speed is under consideration.
- Neural network training theory indicates that the amount of training data used for this work would appear to be considerably less than optimal. However, the results obtained, using previously unseen test data, indicate that the technique has considerable merit for further investigation using larger training and recall data sets.

Acknowledgements

- This work has been funded by the European Union under the Interreg III Programme of the European Regional Development Fund, Project name: 'Intelligent Engine II', Project Number: 126.
- The authors gratefully acknowledge the contributions of Dr. D. H. Lawrence, Dr. Steve Begg, EPSRC Loans Pool and Ricardo Consulting Engineers Ltd in providing advice, data and equipment for this paper, respectively.

References

1. Kalogirou S.A.: Applications of artificial neural-networks for energy systems, Applied Energy 67, 2000, pp.17–35.
2. Xu K., Luxmoore A.R., Jones L.M., Deravi F., Integration of neural networks and expert systems for microscopic wear particle analysis, Knowledge-Based Systems 11, 1998, pp.213–227.
3. Walters, S. D., Howson, P. A., Howlett, R. J.: Production Testing of Spark Plugs using a Neural Network, Paper No. kes05-159, Vol. 4, pp. 74-80, Proceedings, KES2005, Melbourne, 14th-16th September 2005, ISSN 0302-9743, ISBN 3-540-28897-X.
4. Lee, S. H., Walters, S. D., Howlett, R. J.: An Adaptive Neuro-fuzzy Modelling of Diesel Spray Penetration, Paper No. SAE-NA 2005-24-64, Proceedings, ICE2005, Capri, 11th-16th September 2005, ISBN 88-900399-2-2.
5. Lee, S. H., Howlett, R. J., Walters, S. D., Crua, C.: Fuzzy Logic and Neuro-fuzzy Modelling of Diesel Spray Penetration, Paper No. kes05-388, Vol. 2, pp. 642-650, Proceedings, KES2005, Melbourne, 14th-16th September 2005, ISSN 0302-9743, ISBN 3-540-28895-3.
6. Crua, C.: Combustion Processes in a Diesel Engine, Ph.D Thesis, University of Brighton, 2002, http://www.crua.net/thesis .
7. Walters, S. D.: Characterization and Analysis of Kettering-type Automotive Ignition Systems and Electrical Spark Profiles, Ph.D Thesis, University of Brighton, in Association with Champion Spark Plug (Europe), 1998, pp. 35-36.

8. Haykin, S.: Neural Networks – A Comprehensive Foundation, 2nd Edition, Chapter 4, pp. 156-255, ISBN 0-13-273350-1, Prentice Hall, Pearson Education, 1999.
9. MATLAB Manual (version 6, Hardcopy), August 2002.
10. MATLAB Manual (version 7.1, Online), August 2005.
11. Forsyth, D. A., Ponce, J.: Computer Vision – A Modern Approach, ISBN 0-13-085198-1, Prentice-Hall, 2003, Ch. 8 pp. 165-189.
12. Rosandich, R. G.: Intelligent Visual Inspection using Artificial Neural Networks, ISBN 0 412 70800 0, Chapman & Hall, 1997, pp. 116-136.
13. Batchelor, B. G., Hill, D. A., Hodgson, D. C.: Automated Visual Inspection, North-Holland (ISBN 0-444-87577-8) / IFS Publications Ltd. (ISBN 0-903608-68-5), 1985, pp. 329-398 and 535-547.
14. Batchelor, B. G., Hill, D. A., Hodgson, D. C.: Automated Visual Inspection, North-Holland (ISBN 0-444-87577-8) / IFS Publications Ltd. (ISBN 0-903608-68-5), 1985, Case Study 22, pp. 501.
15. Lippman, R. P.: An Introduction to Computing with Neural Networks, IEEE ASSP Magazine, Vol. 4, pp. 4-22, 1987.
16. Thompson, S., Fueten, F., Bockus D.: Mineral Identification using Artificial Neural Networks and the Rotating Polarizer Stage, Computers and Geosciences, Vol. 27, pp 1081-1089, 2001.
17. Howlett, R. J., De Zoysa, M. M., Walters S. D.: Monitoring IC Engines using Neural Networks, Paper: SAE-NA 2003-01-09, Procs., 6th International Conference on Engines for Automobiles, Capri, 2003, On CD.
18. Hush, D. R., and Horne, B. G.: Progress in Supervised Neural Networks, IEEE Signal Processing Magazine, pp. 8-39, 1993.
19. Themistoklis, K.: Implementation of an Automated Fuel Spray Image Analysis System, BEng Final Year Report, University of Brighton, 10/5/2001, Ch. 1-2, pp. 2-14.
20. Schalkoff, R.: Pattern Recognition, Statistical, Structural and Neural Approaches, ISBN 0-471-55238-0, J. Wiley & Sons, 1992, Ch. 10, pp. 204-220, and Ch. 12, pp. 236-259.
21. MATLAB Neural Network Toolbox Users' Guide, (version 4, Hardcopy), March 2001.

Molecular Sequence Alignment for Extracting Answers for Where-Typed Questions from Google Snippets

Alejandro Figueroa[1,2] and John Atkinson[3,*]

[1] Deutsches Forschungszentrum für Künstliche Intelligenz - DFKI, Stuhlsatzenhausweg 3, D - 66123, Saarbrücken, Germany
[2] Universität des Saarlandes, Postfach 15 11 50, 66041 Saarbrücken, Germany
alejandro@coli.uni-sb.de
[3] Departamento de Ingeniería Informática, Universidad de Concepción, Concepción, Chile
atkinson@inf.udec.cl

Abstract. This work presents a strategy to answer questions about locations based on a character alignment algorithm that ranks Google snippets according to their similarity with the query. Answer candidates are extracted by using a Name-Entity Recognizer and then ranked according to their position in this new rank. Results concerning questions on location of monuments and cities around the world are discussed.

Keywords: Natural-Language Processing, Question Answering, Semantic Similarity, Internet Applications.

1 Introduction

The increase of the amount of information on the Web has led search engines to deal with a huge amount of data as users have become retrievers of all sorts. Nowadays, search engines are not only focusing on retrieving relevant documents for a user's particular request, they also try to offer other services (i.e., Group Search, News Search, Glossary). Hence the complexity of the request of the users has addressed the research to Question-Answering (QA) systems. QA aims to answer Natural Language (NL) questions prompted by a user by searching the answer in a set of available documents. QA is a challenging task [7] due to the ambiguity of the language and the complexity of the linguistic phenomena that can be found in NL documents.

The main kinds of questions to answer are those that look for name entities (i.e., *locations*, *persons*, *dates*, *organizations*). Nevertheless, QA systems are not

* This research is sponsored by the National Council for Scientific and Technological Research (FONDECYT, Chile) under grant number 1040469 "*Un Modelo Evolucionario de Descubrimiento de Conocimiento Explicativo desde Textos con Base Semntica con Implicaciones para el Anlisis de Inteligencia.*"

B. Gabrys, R.J. Howlett, and L.C. Jain (Eds.): KES 2006, Part I, LNAI 4251, pp. 1190–1197, 2006.
© Springer-Verlag Berlin Heidelberg 2006

restricted to this kind of questions, they try to handle more complex questions that may require reasoning tasks, while the system is looking for the answer. Many approaches have been developed to deal with questions involving a single entity as the answer, in which case the question must be analyzed to determine its type. In recent approaches such as *EagleQA* [7], the question is processed by linguistic tools in order to extract keywords and expansion terms. A statistical analyzer performs Part-of-Speech tagging and bracketing, and answers are identified between 16 categories including currency, location, number, organization and person. Once the question type is identified, the system looks for expansion terms which are usually extracted from Wordnet.

Statistically-motivated methods have also been used in QA tasks such as NER so that the process is faster and more language independent. In [1], some kinds of *who*-questions are handled by an unsupervised algorithm, in which relations are learnt from training data. Definition questions are answered by using WordNet *hyponym-hypernym* relations and machine learning techniques are trained to identify and rank single 250-character snippets containing the answers.

In this work, a statistically-based strategy for extracting answers from Google snippets is proposed for dealing with *where-typed* questions. A where-typed question is a question that expects a location as an answer (i.e., "*Where is Q?*", "*Where is located Q?*"). Our approach ranks Google snippets by measuring their statistical similarity with the query through an alignment algorithm that locates the characters of the query on the snippets. In particular, the *Smith-Waterman* Alignment (SWA) algorithm is applied for measuring molecular common subsequences, showing that higher ranked entities are the most likely answers. Entities are extracted by means of a NER tool (i.e., LingPipe) and then validated by matching the valid locations existing in WordNet [10].

The paper is organized as follows: section 2 describes the proposed molecular sequence alignment, section 3 describes our QA system, and section 4 highlights the main results and conclusions.

2 Molecular Sequence Alignment

Alignment algorithms are one of the most important tools in Biology as they search for homologs sequences of a protein amongst sequences of proteins in a database. This search process provides key information regarding genes and their functions [3]. Like Biology, the only way to identify homologs is by aligning the query sequence against all the sequences in the database, sorting these hits according to the degree of similarity, and assessing their statistical significance that is likely to be indicative of homology. In order to see how the molecular sequence analogy work, consider the query *"where is Sofia?"* sent to Google. One of the retrieved snippets looks like:

Romania covers an area of 110994 sq km (42855 sq mi) with a population of approximately 8.2 million. The capital of Bulgaria is Sofia ...

The query is aligned with the snippet and no distinction between upper and lower case characters are made:

The capital of Bulgaria **is Sofia** .
is Sofia

Here there is a good alignment at the end of the sentence (i.e., molecular sequence). However, it is not always the case that the alignment fits so perfect. Let consider the following snippet for the query *"Where is the leaning tower located ?"* whose alignment is as follows:

Leaning Tower of Pisa by unknown architect, at Pisa, Italy,
Leaning Tower

Location , Pisa, I **t** a **l** y. Dat **e** , 1063 **to** 1350 tim **e** line.
located **t l e to e**

These examples reveal a key issue in the alignment, the existence of *Gaps*. In some NLs, gaps can be thought of the distance between syntactic roles of the words, which for some free-order languages is rather long (i.e., German). In NL texts, many variations of the same text can be found. Thus, it is less desirable that matches occur between large gaps as there is a weaker semantic-syntactic relationship between the matches, or even worst, the alignment might only match characters. In these cases, penalising gaps which are not contributing the alignment, does not seem to be useful. Hence gap penalties should have some influence on the alignment depending on the sintax of the language. This lead us to address the question of which alignments actually reflect homology between two sequences.

In the above example, gaps were required in both sequences. Next, substitutions, deletions and insertions can be seen as a form of inflectional or morphological inflection of one or more terms in the query. Most of the methods use the equation for penalizing gaps given by $G = a + bx$, where a is the base penalty for opening a gap, b is the gap extension unit penalty, and x is the length of the gap. Typically, a is 10, b is 1, and the penalties for substitutions, insertions and deletions are obtained from the penalty matrix [3]. Objective gap penalties can be inferred by analyzing distributions of gaps in structural alignments.

2.1 The Smith-Waterman Alignment Algorithm

Over the last years, many alignment algorithms have been proposed. The most popular one is due to Smith and Waterman [5] and has been used for different purposes. For example, a strategy for expanding a set of candidate answers for list questions is proposed in [12] which is based on the findings that answers for list questions usually occur in the same context. The SWA algorithm is used for matching this natural-language context, which normally consists of long-distance dependencies between the question topic and the answer.

Unlike these approaches, we use the SWA alghorithm for finding specific answers for locations. In order to apply the strategy, key components are considered:

Given two molecular sequences (strings) $A = a_1a_2 \ldots a_n$ and $B = b_1b_2 \ldots b_m$, a similarity measure between segments from both molecules ($s(a_i, b_j)$), and a weight value (W_k) assigned to a deletion gap of length k, the algorithm computes an optimal alignment between a pair of segments of the sequence (a_i and b_j) by using a scoring function from a matrix H_{ij}. This measures the degree of optimal alignment between the two sequences ending at segments a_i and b_j. Initial values of the score matrix H are set according to the following rules:

$$H_{k0} = H_{0l} = 0 \quad 0 \leq k \leq n \text{ and } 0 \leq l \leq m$$

where H_{ij} is the maximum similarity between two segments with endings a_i and b_j respectively. The rest of the values of H is then filled through a dynamic programming strategy:

$$H_{ij} = max\{H_{i-1,j-1} + s(a_i, b_j), max_{1 \leq k}\{H_{i-k,j} - W_k\}, max_{1 \leq l}\{H_{i,j-l} - W_l\}, 0\}$$

which considers the possible endings for two segments at a and b in some of the following cases:

1. There is an association between a_i and b_j: $H_{i-1,j-1} + s(a_i, b_j)$
2. The segment a_i is at the end of a segment of deletion of length k: $max_{1 \leq k}\{H_{i-k,j} - W_k\}$
3. The segment b_j is at the end of a segment of deletion of length l: $max_{1 \leq l}\{H_{i,j-l} - W_l\}$
4. A value of zero avoids maximum negative values, and indicates that there is no similarity between a_i and b_j (i.e., when $s(a, b)$ becomes a negative value).

The maximum value of H has a pair of segments with the higher similarity whereas the other elements are determined by a tracing back procedure ending with a zero value. The second pair of segments with a high similarity is found by tracing back the second highest element in H that has nothing to do with the first [5].

3 Ranking Locations from Google Snippets

Snippets are obtained by sending a query Q to *Google*. We asked Google for a maximum of 30 snippets. The query and the snippets are then normalized so to remove tags and symbols that can introduce noise or are irrelevant to the alignment.

The next task involves sending the snippets to a NER tool so to retrieve the possible entities in the snippets. The tool is applied to the snippets which returns three sets of entities: persons, locations, and organizations. A NER tool does not only identifies single strings entities such as *"Brazil"*, *"IBM"* or *"John"*, but it also identifies sequence of terms that refer to the same entity, for instance, *"New York"*, *"American Airlines"* or *"Tony Blair"*. However, two key problems arise:

1. Some entities are not accurately classified. For instance, frequent organizations can be either in the set of locations or in the set of persons.
2. Since Google's snippets are not always contiguous pieces of text, wrong words which are not entities are considered as entities.

Consequently, stopwords and words belonging to the query are removed. We use stopwords assuming that the answer does not contain a term in the query.

In order to rank snippets, the relationship between the query and each snippet was measured by locating the characters of the query in each snippet. For this, the SWA algorithm was implemented to match different inflections of the words of the query (i.e., the query "where is located Sofia?" will match the stemm "locat" of "location" in the sentence "The location of Sofia is Bulgary"). The algorithm also allows us to match syntactic-semantic variations of the question (i.e., the query "where is Berlin?" will match "The current capital of Germany is Berlin").

As previously explained, alignment strongly depends on the used similarity measure like in molecular sequence alignment. Accordingly, the BLOSUM62 penalty matrix was used [3]. The values in the BLOSUM62 matrix are the log-odds scores of the ratio between two factors:

1. The probability that residuals are corelated (i.e., homologous) which is computed by using a set of trusted alignments.
2. The average frequencies of seing the residuals at any protein sequence.

Whenever two residuals appear more often in homologous sequences than independently, the matrix values are positive (conservative substitution), otherwise these are negative (non-conservative substitution). Similarity is then defined from the BLOSUM62 scores as follows:

Let S_s be the snippet at position s $(1 \leq s \leq N \leq 30)$ and assume that a function sw_sim computes the degree of similarity between a query Q and the snippet S_s using the SWA algorithm: $sw\_sim(S_s, Q) : S \times Q \rightarrow \Re$. Let E be the set of all entities identified by the NER, Υ the number of entities in E, and E_e an entity in E $(1 \leq e \leq \Upsilon)$. The rank of an entity E_e is defined by:

$$rank(E_e) = w(E_e) \sum_{s=1}^{N} sw\_sim(S_s, Q) X_{se} Y_s \tag{1}$$

Where X_{se} is defined as:

$$X_{se} = \begin{cases} 1 \text{ if } E_e \text{ is in } S_s \\ 0 \text{ otherwise} \end{cases}$$

In words, for each instance of an answer candidate, the quality of the alignment of the snippet in which it occurs is considered. Here Y_s is a filtering factor with a value of 1 for the best ranked snippets, and 0 for the others, this is, only the best alignments are taken into account. In addition, $w(E_e)$ is defined as follows:

$$w(E_e) = \begin{cases} 1 \text{ if } E_e \text{ is a location} \\ 0 \text{ otherwise} \end{cases}$$

In our proposed ranking strategy, the value above is obtained from the similarity of the alignments and validated by checking whether the entity -already detected by the NER- is an existing location in WordNet.

4 Results

Our ranking system was assessed with three kinds of (template) questions: the where-typed questions of the *Cross-Language Evaluation Forum* (CLEF) corpus, which refers to answers from 1994/1995 newspaper articles. Some examples of questions provided by CLEF look like the following:

Where was Nick Leeson born?

Where did the 1992 Olympic Games take place?

In addition, all the out-of-date questions were removed (i.e., "*Where does Arafat reside?*"). Next, a set of questions concerning capitals and cities around the world were tested. These have the following template:

where is < city >?

Where < *city* > is replaced with other cities (i.e., *Berlin*). Using this kind of questions is useful to introduce the less amount of semantics as possible so to make the alignment easier. The expected answer is the name of the country, for example, *Germany.* For the third kind of questions, a list of monuments around the world was tested with a template similar to that of the cities:

where is Fontana della Barcaccia?

We accepted either a city or a country as a right answer (i.e., Rome or Italy).

Table 1. Results for location questions

Settings			**Exact Answers**			**Alter. Answers**		
Target	Total	NoAns	1st	2nd	3rd	1st	2nd	3rd
Capitals	193	34	125	6	1	17	7	3
CLEF	53	9	38	-	-	6	-	-
Monuments	50	5	33	8	-	4	-	-
Greece	60	10	34	4	1	8	2	1
UK	56	2	48	4	1	-	1	-
World Heritage Monuments	198	28	142	20	4	4	-	-
Total	615	88 (14%)	430 (70%)	42 (7%)	7 (1%)	39 (6%)	10 (1.9%)	4 (0.06%)

Table 1 shows the results for a set of 615 questions, including the settings used for each run and the best three exact and alternative answers respectively. In addition, the number of question for which no answers were found are highlighted

(*NoAns*). In order to assess the performance of QA systems, the standard metric of *Mean Reciprocal Rank* (MRR) was used. MRR rewards each answer according to its position in the ranking, by assigning each the inverse of its position. Next, the MRR of a system is the average value for a set Qu of N_q questions:

$$MRR = \frac{1}{N_q} \cdot \sum_{\forall q \in Qu} \frac{1}{rank\_answer_q}$$

For the current system, we obtained an MRR value of 0.815 for the top three answers. The approach in [1] scored an MRR of 0.54 for *who*-questions by using an unsupervised learning strategy. The system can also be compared with the approach by [11] which scored MRR values below 0.55 for a given set of locations. Note that this system also uses WordNet and other linguistic resources such as gazetteers.

Overall, the right answer was on the first position of the rank for 430 questions (69.91%), in which 6.34% questions were found to be an alternative valid answer in the first position of the rank. An alternative answer is another city or country that has also a monument or a city with the same name as the requested place. For example, for "*where is Santo domingo?*", the system answered "*Chile*" instead of "*Dominican Republic*" (there is a place named "*Santo domingo*" in Chile). If we consider all the valid answers, the system scored 76.25% of right answers on the first position of the rank.

It is worth highlighting that our system got 83% valid answers from the CLEF corpus, which included complex questions such as "*where did the 1992 Olympics Games take place?*". This high score may be due to the fact that the queries in the CLEF corpus carry many words which are full of semantic context. The accuracy increases as we rank the entities in this snippet which has a high degree of similarity with the query. For the 309 cities, the system got a top one valid answer for 232 questions, that is 75.08% and for 14.87% no answers were obtained. For the set of 248 monuments, a valid top one answer was obtained for 79.73% of the answers, the top two valid answers were obtained for 83.46%. While there are several likely correct answers for cities, countries, etc, our alignment strategy was still capable of capturing correct locations even though the questions were made on places where no much activity is happening on Internet such as ocean islands, African countries, etc.

On the other hand, overall performance of the alignment strategy suggests promise for dealing with several candidate answers. For example, in order to carry out alignments and comparisons based on 2000 samples, the algorithm averaged 1.4 seconds, whereas this took 20 (seconds) in the worst case.

5 Conclusions and Further Work

A strategy for answering questions is proposed to identify locations using an alignment algorithm which is strongly based on analogies with determining sequences of proteins in Biology. This method is used for finding matches in sequences of proteins, showing outstanding results for a set of 615 questions, in which several kind of questions were included. The high accuracy of the proposed alignment strategy suggests that this is an effective way to deal with questions extracted from the CLEF corpus which contain many complex terms.

On the other hand, the approach's main drawback is that the strategy will fail as long as a location does not exist in WordNet. In order to address this issue in a future version, bootstrapping algorithms can be applied to learn phrase structures that contribute a high value for the *rank* function so that the remaining locations can be identified. This would also allow us to deal with scenarios in which the NER is unable to extract the places from the snippets.

The model can be improved by providing algorithms to extract holonyms from WordNet in order to make a more accurate validation of the locations. Another natural extension of this work involves questions that look for a person as an answer. This was left for a future work as new strategies are needed to check whether an answer candidate is a valid person or not.

References

1. L. Lita and J. Carbonell. *Unsupervised Question Answering Data Acquisition From Local Corpora*, Proceedings of the Thirteenth Conference on Information and Knowledge Management (CIKM 2004), Washington, DC, USA, November 813, 2004.
2. A. Savary and C. Jacquemin. *Reducing Information Variation in Text*, ELSNET Summer School, pp. 145-181, 2000.
3. S. Henikoff and J. Henikoff.*Amino acid substitution matrices from protein blocks*, Proceedings of the National Academy of Science USA, 89(2), pp. 10915-10919, 1992.
4. M. Dayhoff, R. Schwartz and B. Orcutt. *A model of evolutionary change in proteins*, In Atlas of Protein Sequence and Structure, Vol 5, Suppl.3, pp. 345-352, Natonal Biomedical Research Foundation, Washington D.C., 1978.
5. M. Waterman. *Estimating statistical of sequence alignments*, Philosophy Transactions of the Royal Society of London, num. 344, pp. 383-390, 1994.
6. Y. Sasaki. *Question Answering as Question-Biased Term Extraction: A New Approach toward Multilingual QA*, Procs. of the 43rd Annual Meeting of the ACL, pp. 215-222, Ann Arbor, USA, 2005.
7. J. Chen, H. Ge, Y. Wu and S. Jiang. *Question Answering Combining Multiple Evidences*, Proceedings of TREC, 2004.
8. N. Needleman and J. Wunsch. *A general method applicable to the search for similarities in the amino acid sequence of two proteins*, Journal of Molecular Biology, num. 48, pp. 443-453, 1970.
9. M. Greenwood. *Using Pertainyms to Improve Passage Retrieval for Questions Requesting Information About a Location*, Information Retrieval for Question Answering: A SIGIR 2004 Workshop, Sheffield, UK, July 2004.
10. C. Fellbaum. *WordNet: An Electronic Lexical Database.* MIT Press, 1998.
11. C. Monz, *From Document Retrieval to Question Answering*, IILC Dissertation Series DS-2003-4, *Institute for Logic, Language and Computation*, University of Amsterdam, 2003.
12. K. Wei, *Improving Answer Precision and Recall of List Questions*, MSc thesis, School of Informatics, University of Edinburgh, UK, 2005.

Temperature Field Estimation for the Pistons of Diesel Engine 4112

Zuoqin Qian[1], Honghai Liu[2,3], Guangde Zhang[3], and David J. Brown[2]

[1] School of Energy and Power Engineering, Wuhan University of Technology Wuhan Hubei, PR China, 430063

[2] Institute of Industrial Research, The University of Portsmouth Portsmouth PO1 3HE, England, UK

[3] Institute of Vehicle Engineering, School of Mechanical Automation Engineering Wuhan University of Science and Technology, Wuhan Hubei, PR China, 430081

Abstract. This paper proposes an approach to estimating the temperature field of the pistons of Diesel Engines 4112 using the combination of experimental measurement and finite element analysis (FEA). Temperatures of the pistons in the testbench of a Diesel Engine 4112 are measured using hardness recovery methods. Then the experimental results are employed as boundary conditions for the pistons FEA, which estimates their temperature field and their thermal loading capability. The FEA result provides effective theoretical evidence for further improving the pistons' performance.

1 Introduction

Due to the increasing requirement of internal combustion engines with heavy loads, the issues related to heat transfer and heat loading capacity have drawn attention to the research communities on engine design and computational analysis [1]. For instance, pistons are the very important component in a diesel engine blast chamber because of their operational conditions [2]. Pistons are operated directly in the conditions of the gas of high temperature and high pressure with heavy heat loading; besides, their reciprocating motions with high velocity in liners are carried out in less lubricating condition. Current relevant open problems generally focus on improving operational life span of the heating parts and their reliability. The piston heat fatigue caused by heat loading and frequent movements is the dominant factor to affect the reliability of the diesel engine [3,4,5], though other factors do make their contributions e.g., the performance of the whole machine, the release indexing data and economical efficiency.

Diesel engine 4112 has been selected as the target platform for the temperature field analysis due to its characteristics. The water coolant in diesel engine 4112 cools the machine body which is connected to the surface of the liners instead of directly cool the surface of the dry liners because of diesel engine 4112 is a compound body of their dry liners and machine body. The extremely high-temperature working condition of the heating parts, e.g., the liners and pistons, makes the heat-loading problem even highlighted. The heat loading of diesel

B. Gabrys, R.J. Howlett, and L.C. Jain (Eds.): KES 2006, Part I, LNAI 4251, pp. 1198–1205, 2006.
© Springer-Verlag Berlin Heidelberg 2006

engines is assessed in terms of many factors, in general, it includes maximum working temperature, temperature gradient from the local surfaces of engine pistons, and their heat stress, thermal strain, the high and low frequency heat loading fatigue. Temperature field is the fundamental parameter used to calculate and assess heat loads, it allows numerically estimate the factors that affect piston's heat loads.

In this work, we focus on estimating the temperature field of diesel engine 4112 pistons. Experimental measurements have been used as the boundary conditions for the FEA theoretical analysis. The rest of the paper is organized as follows: Firstly, Section 2 presents experimental methods for boundary conditions generation. Secondly, Section 3 carries out the research on FEA-based numerical analysis. Finally, Section 4 concludes this paper with detailed discussions and conclusions.

2 Boundary Conditions Generation

An experimental method has proposed in this section to generate temperature boundary conditions for the pistons of diesel engine 4112. Hardness recovery methods are employed to measure the temperature of the diesel engine heating parts in our testbed [6]. The principle of the hardness measurement is the determination of the height of rebound of the spherical indenter of a pin.

2.1 Pistons Measurement Points Selection

The testbed is based on diesel engine YC4112 which is vertical, line hydro cooling, four strokes, direct ejecting and four cylinders. Its major parameters are listed in Table 1. According to our empirical research on diesel engines, we selected the forth cylinder of our testbed as the key component for the collection of sample temperature points and the other cylinders are employed for comparison purpose in higher heat load conditions only. Figs. 1 and 2 show the positions of the selected sampling points on the piston's top and sides. And Fig. 3 illustrates sampling points 1-12 are on the inside of the combustion chamber, sampling point 13 is on the middle point and sampling points 14-16 in the middle of the edge. The rest sampling points are labelled as well. It should note that there is a sampling point labelled as 36 provided on the top of the cylinder's back side.

Table 1. Parameters for the Diesel Engine Testbed

Type	Diameter × Stroke	Volume	Rated Power/velocity	Maximum Torque/Velocity
YC4112	112×132mm	5.202L	105KW/2400rmin	500KW/1400-1600r/min

Fig. 1. Sampling points on the piston top (1)

Fig. 2. Sampling points on the piston (2)

Fig. 3. Sampling points layout

2.2 Experimental Results on the Sampling Points

Diesel engine has been operated in two working conditions, i.e., the 100% and 110% engine rated power. The temperature measurements at the standard experimental condition of the 100% rated power are provided in Table 2, where SP stands for sampling piston, T stands for the temperature of sampling points. The higher values on sampling points from No 5 to 8 indicate that comparison between cylinders have to be taken in order to explore the pistons' heat loads difference. Hence we experimented the four points of No. 2 and 4 cylinders for the comparison, whose values are listed in the Table 3, where PT n stands for temperature of No. n piston's sampling points.

Table 2. Temperature measurements of the fourth piston's samplling points

SP No	1	2	3	4	5	6	7	8	9	10	11	12	13	14	15	16	17	18
T (°C)	301	313	310	318	342	340	355	351	280	268	289	291	292	318	320	321	331	238
SP No	19	20	21	22	23	24	25	26	27	28	29	30	31	32	33	34	35	36
T (°C)	252	259	245	216	231	210	221	180	192	154	136	118	109	230	248	220	241	224

Table 3. The temperatures of measuring points 5.6.7.8 on the second and forth cylinders

SP No	PT 2 (°C)	PT 4 (°C)	SP No	PT 2 (°C)	PT 4 (°C)
5	340	352	7	370	368
6	354	360	8	366	373

3 FEA on Temperature Field

Temperature field is the fundamental tool for the theoretical analysis of tempertuare field in engine community. Finite element analysis on the temperature field of the pistons has been carried out to estimate their temperature field in this section.

3.1 Temperature Field Principle Using FEA

The temperature field of a three-dimensional solid is developed based on the partial differential equations of stationary state of heat transition [7, 8, 9] as follows,

$$\frac{\partial^2 \mathrm{T}}{\partial \mathrm{x}^2} + \frac{\partial^2 \mathrm{T}}{\partial \mathrm{y}^2} + \frac{\partial^2 \mathrm{T}}{\partial \mathrm{z}^2} = 0 \tag{1}$$

its condition constraint is given,

$$-\mathrm{k}\frac{\partial \mathrm{T}}{\partial \mathrm{n}} = \alpha(\mathrm{T} - \mathrm{T_f}) \tag{2}$$

where T stands for temperature distributed function °C, x, y, z stands for the coordinates, k stands for thermal coefficient of conduction W/(m°C), α stands for coefficient of heat transfer W/(m2°C) and T_f stands for medium temperature°C.

Due to the fact that FEA is able to handle complex systems that defy closed-form analytical solutions. Then by applying variational calculus principle to the equation 1, we obtain its optimal function given in the following with a condition $\delta T = 0$,

$$\begin{aligned} J(T) = &\int\int\int \tfrac{1}{2}\lambda \left[\left(\tfrac{\partial T}{\partial x}\right)^2 + \left(\tfrac{\partial T}{\partial y}\right)^2 + \left(\tfrac{\partial T}{\partial z}\right)^2 \right] dxdydz \\ &- \int\int_{\Gamma 3} \alpha \left(T_f - \tfrac{1}{2}T\right) T ds \end{aligned} \tag{3}$$

Further, discreting the solid into finite elements, we obtain its finite element equation,

$$[K]\{T\} + [N]\left\{\frac{\partial T}{\partial t}\right\} = \{P\} \tag{4}$$

3.2 Boundary Condition

The heat transfer in diesel engine cylinders is a complex process due to the fact that the heat transition, heat convection and radiation exist at the same time during the instantaneousness procedure of gas combustion. In this paper we calculate the temperature field using FEA with the experimental results in section 2 as the boundary conditions. In order to achieve higher precision, we compare experimental results and the FEA results, and adjust them to reach a suitable difference before the boundary condition is finalized. We firstly simulate the complex working process of the gas cylinders to work out the instant temperature of the gas in the cylinder T_g and instant coefficient of heat transfer α_g with the Eichelberg equation in the following,

$$\alpha = 1.95\sqrt[3]{\mathrm{C_m}}\sqrt{\mathrm{PT}} \tag{5}$$

where C_m stands for the average speed of a piston (m/s), P stands for the average pressure of inside cylinder (MPa) and T stands for the gas temperature in the cylinder (K).

Then we derive equations 6 and 7 in the following, which allow to calculate the average coefficient of heat transfer α_{gm} and the average temperature T_{gm},

$$\alpha_{gm} = \frac{\int_0^{4\pi} \alpha_g \cdot d\varphi}{4\pi} \tag{6}$$

$$T_{gm} = \frac{\int_0^{4\pi} \alpha_g T_g d\varphi}{\int_0^{4\pi} \alpha_g d\varphi} \tag{7}$$

Further, the coefficients of heat transfer and local temperature of all parts of a piston can be calculated based on empirical formulas provided by Lu's book [1]. On the one hand, the coefficient of local heat release distribution on the top of a piston is determined and estimated by the formula derived by W.J. Seale, The coefficient for the heat transfer of the circle piston rings is estimated by the formula given by [10],

$$\frac{1}{\alpha_1} = \frac{a}{\lambda_1} + \frac{b}{\lambda_2} + \frac{c}{\lambda_3} + \frac{1}{\alpha_w} \tag{8}$$

where a, b, c are the radial thickness of circle piston rings, the thickness of the lubricant film and the wall thickness of cylinder liners, respectively. λ_1, λ_2, λ_3 are the heat-transfer coefficients for piston rings, lubricant film and cylinder liners. and a_w is the coefficient for the heat release of the liner water cavity. Besides, the heat-release coefficient of piston skirts can be estimated by the following equation,

$$\frac{1}{\alpha_2} = \frac{b}{\lambda_2} + \frac{b}{\lambda_3} + \frac{1}{\alpha_w} \tag{9}$$

On the other hand, the coefficient for the heat tranfer of the inside of piston can be estimated by interpolation and empirical data.

3.3 FEA Simulation Results

ANYSYS finite element package and Solid90 are used in this paper to do the FEA of the temperature field of the diesel engine pistons, we divide the piston model in Fig. 4 into a finite element model consisting 32384 units and 51521 nodal points shown in Fig. 5. With the boundary condition in Section 3.2, FEA simulation results are produced as shown in Figs. 6 and 7 corresponding to the two working

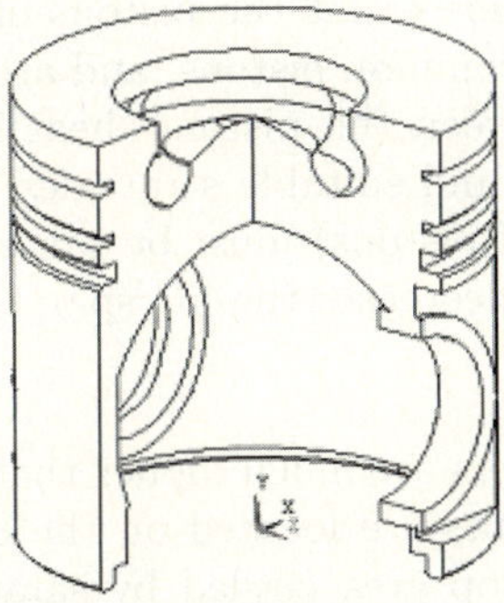

Fig. 4. The solid model of a piston of the Diesel Engine 4112

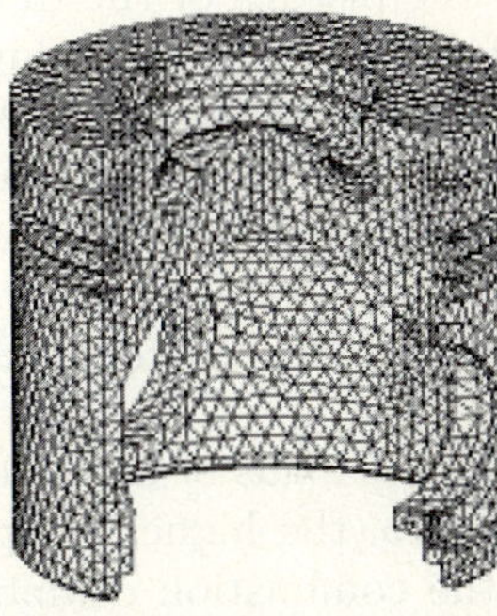

Fig. 5. The finite element model of the model in Fig. 4

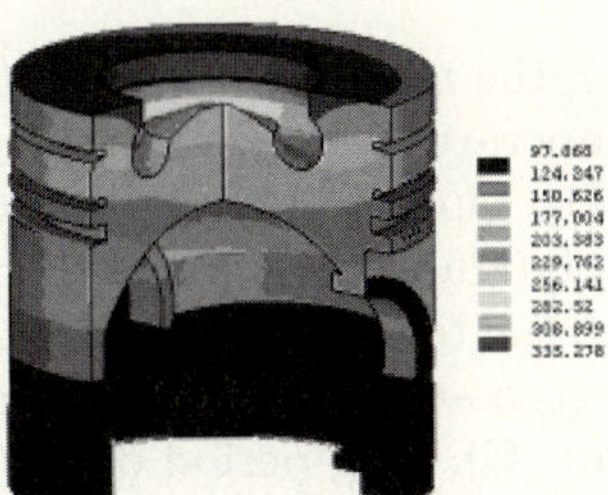

Fig. 6. Temperature field of the piston in Fig. 5under the condition of the 100% rated power

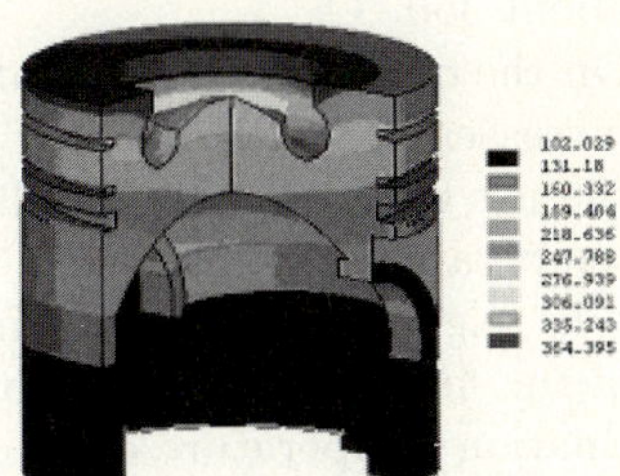

Fig. 7. Temperature field of the piston in Fig. 5 under the condition of the 110% rated power

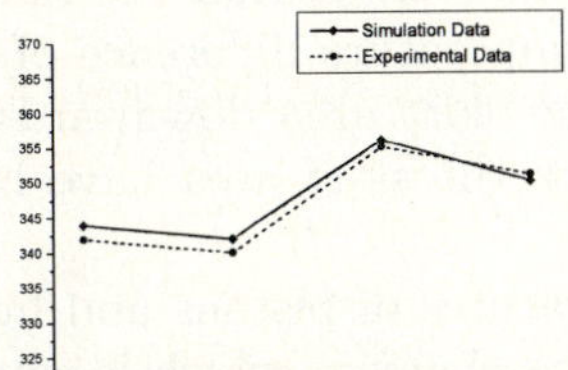

Fig. 8. Comparison of the piston's temperature in Fig. 6 at the 100% rated power

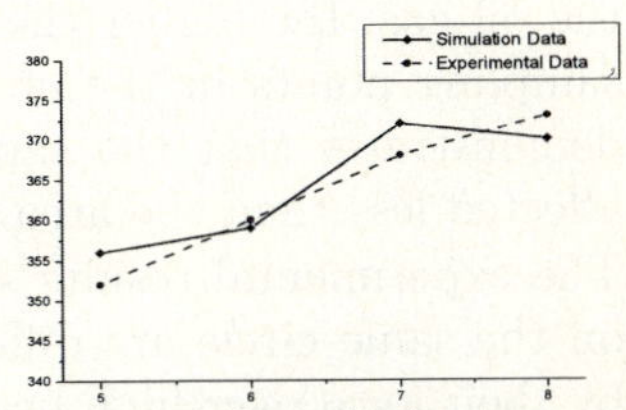

Fig. 9. Comparison of the piston's temperature in Fig. 7 at the 110% rated power

conditions of 100% and 110% rated power with the same measuring rotation velocity as 2400r/min. The comparison of the simulation data and experimental sampling points in Figs. 6 and 7 are given in Figs. 8 and 9.

4 Concluding Remarks

The comparison of the experimental measurements and FEA numerical simulation of the pistons of the dry-liner diesel engine shows that the pistons of this type of diesel engines have higher heat loads than common pistons, and are easier to make heat loads breakdown. Tradeoffs between the pistons' heat loads and their operational life span must be concerned and suitable strategies (e.g., manufacturing cold trap pores in the inside wall of pistons) must be taken due to the proved characteristic of the pistons usually decreases the life span of the pistons. The detailed conclusions are,

1. The heat loads of the second and fourth cylinders are much higher than the rest two, the higher temperature sampling points are located on the edges of the combustion chambers which is on the top area circled by sampling points 5-8 as shown in Fig. 3, in which area, local areas with the maximum heat loads are closed to the sampling points 3,7,11 and 16 when the engine runs at the 100% rated power. The average temperature at the circle area is about 350 °C.
2. On the condition of over rated load(e.g., 110%), the maximum temperature increases up to about 370 °C, which is the temperature threshold value for the use of the aluminum-alloy based pistons [11]. Operating in such higher temperature for a long time will lead to the heat fatigue mutilation of the pistons (e.g., heat check, burning-out). In addition, the temperature in the area of the first piston ring groove (i.e., 266 °C) overpasses the standard agglutination temperature of lubrication oil (i.e., 240 °C), long-period operation in such condition obviously causes the carbon accumulation and scuffing of the cylinder bores.
3. On the same circle area, the temperature difference is more significant on the top of the pistons than the piston skirt area. This phenomena is caused by the uneven gas distribution in the combustion chambers in that the thicker the oil gas, the higher the temperature. The temperature difference of the sampling points in the circle decreases from the skirt area downwards. It demonstrates that the sampling positions below the skirt area have been affected less than the area above it.
4. The experimental results show that the temperatures of pistons and liners on the same circle are different as the circle area changes, which is caused by their local over-high heat loads. Maximum temperatures are close to the sampling areas labelled by 3, 7, 11 and 16. These results can be well explained by the combustion theory, the closer areas that the oil beam of a sprayer can reach with higher speed, the higher the combustion temperatures in the atomisation process of the sprayer oil. It means the areas closer to sprayers

have higher temperatures leading to heavy heat loads. Fig. 3 illustrates that the areas in which the sampling points 3,7,11 and 16 located are closer to the sprayers.

Acknowledgements

The research has been supported partly by the important research fund of Wuhan University of Science and Technology (grant No. 2004XZ11) and partly by Natural Science Foundation of Hubei Province, P. R. China (grant No. 2005ABA017).

References

1. Lu, R.: Thermal transfer and loading of internal combustion engine. China Communication Press, Beijing (1988) 6:1–30
2. Kajiwara, H., Fujioka, Y., Suzuki, T., Negishi, H.: An analytical approach for prediction of piston temperature distribution in diesel engines. JSAE Review (2002) 23:429–434
3. Qiang, Z.: Experiment study on temperature measurement of piston diesel engine yc6108. Journal of Wuhan Technology University (2004) 12(6):814 816
4. Silva, F.: Fatigue on engine pistons- a compendium of case studies. Engineering failure analysis (2006) 13(3):480–492
5. Takiguchi, M., Ando, H., Takimoto, T., Uratsuka, A.: Characteristics of friction and lubrication of two-ring piston. jSAE Review (1996) 17(1):11–16
6. Fitzgerald, L.: Measurement of hardness at very high temperatures. British Journal of Applied Physics (1960)
7. Das, A.: Finite temperature field theory. World Scientific Publishing (1997)
8. Kong, X.: Finite element methods and their applications to heat transfer. The Science Press (1986)
9. Incropera, F., DeWitt, D.: Fundamentals of heat and mass transfer (fourth edition). John Wiley and Sons, New York (1996)
10. Xiong, L.: Design directory of marine diesel. National Defence Industry Press (1982)
11. Huang, Y., Hort, N., Dieringa, H., Kainer, K.: Investigations on thermal fatigue of aluminum- and magesium-alloy based composites. International Journal of Fatigue (in press)

Fuzzy and Neuro-fuzzy Techniques for Modelling and Control

S.H. Lee, R.J. Howlett, and S.D. Walters

Intelligent Systems & Signal Processing Laboratories
Engineering Research Centre, University of Brighton
Moulsecoomb, Brighton, BN2 4GJ, UK
S.H.Lee@Brighton.ac.uk, R.J.Howlett@Brighton.ac.uk,
S.D.Walters@Brighton.ac.uk

Abstract. This paper presents comparative evaluation of two fuzzy-derived techniques for modelling fuel spray penetration in the cylinder of a diesel internal combustion engine and a fuzzy control system for a small internal combustion engine. The first technique used a pure fuzzy paradigm, the parameters of this technique are a collection of intuitively comprehensible rules and fuzzy-set membership functions. While the visual nature of this system facilitates the optimisation of the parameters, the need for this to be accomplished manually is a disadvantage. The second technique used an adaptive neuro-fuzzy inference system (ANFIS), where automatic adjustment of the system parameters was effected by a neural network based on prior knowledge. The ANFIS exhibited improved accuracy compared to a pure fuzzy model. It also has the advantage over the pure fuzzy paradigm that the need for the human operator to tune the system by adjusting the bounds of the membership functions is removed. Future work is concentrating on the establishment of an improved neuro-fuzzy paradigm for adaptive, fast and accurate control of small internal combustion engines.

1 Introduction

In a diesel engine, the combustion and emission characteristics are influenced by fuel atomisation, nozzle geometry, injection pressure, shape of inlet port, and other factors. In order to improve air-fuel mixing, it is important to understand the fuel atomisation and spray formation processes. Researchers have investigated the characteristics of the spray behaviour, formation and structure for the high-pressure injector by experimental and theoretical approaches, in order to improve the combustion performance and reduce exhaust emissions. However, further detailed studies of the atomisation characteristics and spray development processes of high-pressure diesel sprays are still relevant.

Intelligent systems, i.e. software systems incorporating artificial intelligence, have shown many advantages in control and modelling of engineering system. They have the ability to rapidly model and learn characteristics of multi-variant complex systems, exhibiting advantages in performance over more conventional mathematical techniques. This has led to them being applied in diverse applications in power systems, manufacturing, optimisation, medicine, signal processing, control, robotics, and social/psychological sciences [1, 2]. Fuzzy logic is a problem-solving technique

B. Gabrys, R.J. Howlett, and L.C. Jain (Eds.): KES 2006, Part I, LNAI 4251, pp. 1206 – 1215, 2006.
© Springer-Verlag Berlin Heidelberg 2006

that derives its power from its ability to draw conclusions and generate responses based on vague, ambiguous, incomplete and imprecise information. To simulate this process of human reasoning it applies the mathematical theory of fuzzy sets first defined by Zadeh, in 1965 [3]. Fuzzy inference is the process of formulating a mapping from a given input value to an output value using fuzzy logic. The mapping then provides a basis from which decisions can be made, or patterns discerned. It has been proved that the system can effectively express highly non-linear functional relationships [4]. Fuzzy inference systems (FIS) have been successfully applied in fields such as automatic control, data classification, decision analysis, expert systems and computer vision.

The Adaptive Neuro-Fuzzy Inference System (ANFIS), developed in the early 90s by Jang [5], combines the concepts of fuzzy logic and neural networks to form a hybrid intelligent system that enhances the ability to automatically learn and adapt. Hybrid systems have been used by researchers for modelling and predictions in various engineering systems. The basic idea behind these neuro-adaptive learning techniques is to provide a method for the fuzzy modelling procedure to learn information about a data set, in order to automatically compute the membership function parameters that best allow the associated FIS to track the given input/output data. The membership function parameters are tuned using a combination of least squares estimation and backpropagation algorithm for membership function parameter estimation. These parameters associated with the membership functions will change through the learning process similar to that of a neural network. Their adjustment is facilitated by a gradient vector, which provides a measure of how well the FIS is modelling the input/output data for a given set of parameters. Once the gradient vector is obtained, any of several optimisation routines could be applied in order to adjust the parameters, so as to reduce error between the actual and desired outputs. This allows the fuzzy system to learn from the data it is modelling. The ANFIS approach has the advantage over the pure fuzzy paradigm that the need for the human operator to tune the system by adjusting the bounds of the membership functions is removed.

Many combustion and control problems are exactly the types of problems and issues for which a fuzzy logic approach appears to be most applicable and has the potential for making better, quicker and more accurate predictions than traditional methods. For example, fuzzy control been shown to provide a convenient calibration procedure that has the potential to lower costs through reducing the effort required in calibrating an engine. The aim of this investigation was to apply intelligent techniques to achieve improved control and modelling ability. An intelligent paradigm was created based on a fuzzy logic inference system combined with conventional techniques on two separate applications. Basic fuzzy logic model was fairly easy to construct whilst the ANFIS performed well in cases where the input to output relationships become more complex.

2 Techniques

2.1 Pure Fuzzy Logic Model

Fuzzy logic provides a practicable way to understand and manually influence the mapping behaviour. In general, fuzzy logic uses simple rules to describe the system of interest rather than analytical equations, making it easy to implement. Given its advantages, such as robustness and speed, the fuzzy logic method is one of the best solutions for system modelling and control. An FIS contains three main components, the fuzzification stage, the rule base and the defuzzification stage. The fuzzification stage is used to transform the so-called crisp values of the input variables into fuzzy membership values. Then, these membership values are processed within the rule-base using conditional 'if-then' statements. The outputs of the rules are summed and defuzzified into a crisp analogue output value. The effects of variations in the parameters of a FIS can be readily understood and this facilitates calibration of the model.

The system inputs, could be the cylinder pressure and the air density, are called linguistic variables, whereas 'high and 'very high' are linguistic values which are characterised by the membership function. Following the evaluation of the rules, the defuzzification transforms the fuzzy membership values into a crisp output value, for example, the penetration depth. The complexity of a fuzzy logic system with a fixed input-output structure is determined by the number of membership functions used for the fuzzification and defuzzification and by the number of inference levels. A fuzzy system of this kind requires that a knowledgeable human operator initialises the system parameters e.g. the membership function bounds. The operator must then optimise these parameters to achieve a required level of accuracy of mapping of the physical system by the fuzzy system. While the visual nature of a fuzzy system facilitates the optimisation of the parameters, the need for it to be accomplished manually is a disadvantage.

2.2 ANFIS Model

ANFIS largely removes the requirement for manual optimisation of the fuzzy system parameters. A neural network is used to automatically tune the system parameters, i.e. the membership function bounds, leading to improved performance without operator intervention. In addition to a purely fuzzy approach, an ANFIS was also developed for the estimation of spray penetration because the combination of neural network and fuzzy logic enables the system to learn and improve its performance based on past data. The neuro-fuzzy system with the learning capability of neural network and with the advantages of the rule-base fuzzy system can improve the performance significantly and can provide a mechanism to incorporate past observations into the classification process. In a neural network the training essentially builds the system. However using a neuro-fuzzy scheme, the system is built by fuzzy logic definitions and then it is refined using neural network training algorithms.

3 The Case Studies

The aim of the first case study was to demonstrate the effectiveness of an ANFIS for the prediction of diesel spray penetration length in the cylinder of a diesel internal combustion engine. The technique involved extraction of necessary representative features from a collection of diesel engine spray data. A comparative evaluation of two fuzzy-derived techniques for modelling fuel spray penetration was also described [6]. The second case study describes a cost-effective fuzzy control system applied to a small spark-ignition internal-combustion engine to achieve regulation of the fuel injection system. This control system determines the amount of fuel required from a fuzzy algorithm using engine speed and manifold air pressure as input values which has led to improved fuel regulation, and a consequent reduction in exhaust emissions [7].

3.1 Case Study 1: Fuzzy and Neuro-fuzzy Modelling

The first model of this work was implemented using a conventional fuzzy-based paradigm, where human expertise and operator knowledge were used to select the parameters for the system. The second model used an ANFIS, where automatic adjustment of the system parameters was effected by a neural network, based on prior knowledge. Two engine operating parameters were used as inputs to the model; namely in-cylinder pressure and air density. Spray penetration length was modelled on the basis of these two inputs. The models derived using the two techniques were validated using test data that had not been used during training; please see Figure 1.

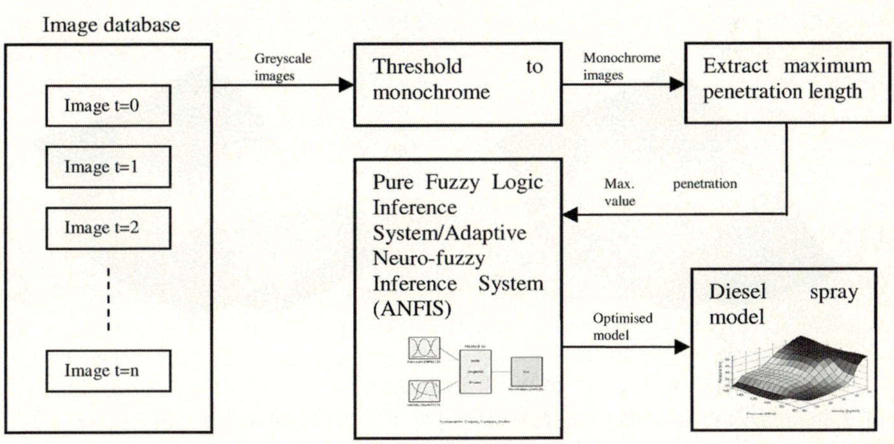

Fig. 1. Schematic diagram of FIS modelling

The generalised bell-shaped membership functions were used for classes: low, medium and high; this was empirically selected, based on the features of all data under consideration, although in many cases membership functions are fixed and somewhat arbitrarily chosen. The process was carried out by examining the ranges of all data sets to determine where the majority of points were located. The functions were also created to have an approximately equal amount of overlap between each membership curve. Experimental adjustment of the limits of the membership classes enabled the response of the model to be tailored to the experimental output from the experimental data. The rule structure was essentially predetermined by the user's interpretation of the characteristics of the input parameters in the model. The contents of these rule-base and membership functions undertake many modifications as part of the process of heuristic optimisation and in many cases it is a continuing process.

The control surface in Figure 2 shows the crisp value of penetration depth at different combinations of in-cylinder pressure and air density, using a pure fuzzy logic model. Each of these intersection points indicates the differing predicted value of spray penetration depth, which is determined by the design of the fuzzy sets, the rule-base and the membership functions. The surface plot acts as a practical means of determining the output needed for each combination of input parameters.

A second model was devised using Matlab® based application, ANFIS. A neuro-adaptive learning technique facilitated the learning of dataset information by the fuzzy modelling procedure; it was then possible to compute the membership function parameters that best allowed the associated FIS to track the given input/output data, rather than choosing the parameters associated with a given membership function arbitrarily.

Raw penetration lengths were plotted against time under each relative pressure and density condition. Polynomial fitting was employed to produce best fitted curves where maximum penetration values can be depicted. These were combined into a vector with which to train the ANFIS. During training in ANFIS, 6 sets of pre-processed data were used to conduct 180 cycles of learning.

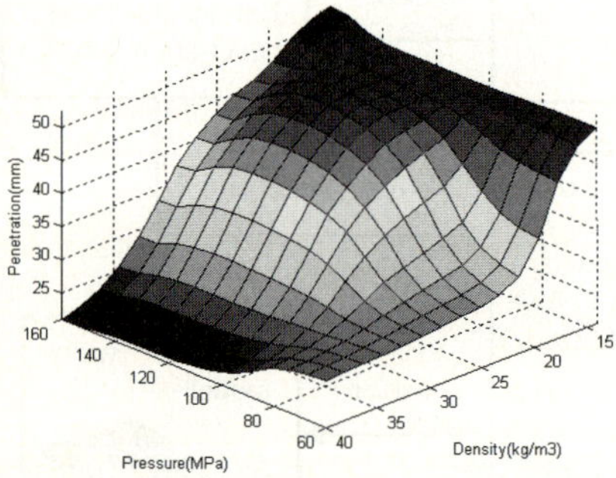

Fig. 2. Pure fuzzy logic model – surface plot

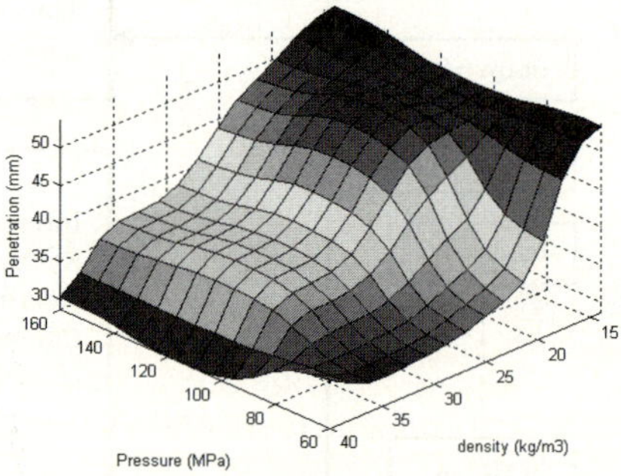

Fig. 3. ANFIS surface plot

Figure 3 depicts a three-dimensional plot that represents the ANFIS mapping from relative pressure and air density to spray penetration length. As the relative pressure and air density increase, the predicted penetration length increases in a non-linear piecewise manner, this being largely due to non-linearity of the characteristic of the input vector matrix derived from the raw image data. This assumes that these raw image data are fully representative of the features of the data that the trained FIS is intended to model.

Figure 4 shows a scatter plot of the measured and FIS-modelled penetration lengths utilising six sets of testing data. These two sets of data demonstrate that the predicted values are close to the experimentally-measured values, as many of the data points fall very close to the diagonal (dotted) line, indicating good correlation. Figure 5 shows similar comparisons between the FIS-modelled and measured values of the penetration length using the same testing data. Clearly the model created by ANFIS has a better agreement than the pure fuzzy logic model.

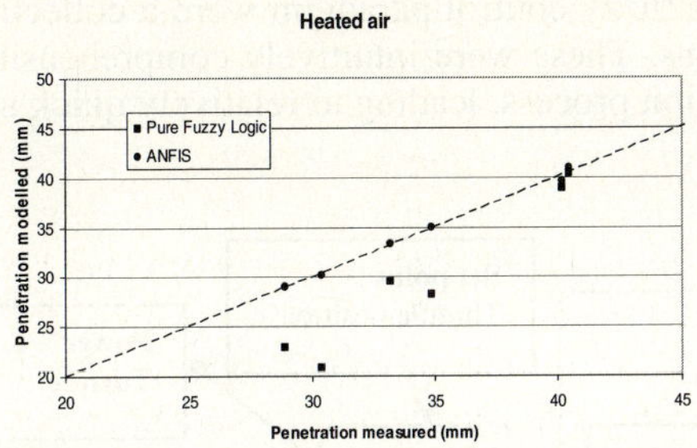

Fig. 4. Scatter plot of measured penetration and predicted penetration

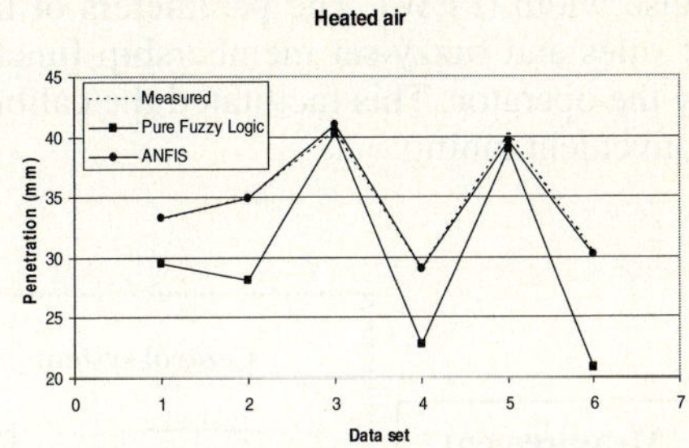

Fig. 5. Comparisons between predicted and measured penetration

This technique ANFIS has several advantages when assigned to applications in which only partial knowledge of the system characteristic exists, as is typically the case with engineering systems. Additionally, the ANFIS can rapidly identify important characteristics of the data, which is an important and useful feature of models used for estimation purposes in internal combustion engines research. In the experiment, an ANFIS was used to predict changes in diesel spray penetration depth as a potential means to evaluate impending changes in combustion chamber and fuel injector design. As an initial step toward modelling and prediction with an ANFIS for this particular application, it has proven very useful for short-term prediction of penetration depth using engine operating parameters as the input. Both models performed fairly well and approximated the output function to a reasonable extent whilst the ANFIS model exhibited improved performance in this respect. Pure fuzzy logic models were conveniently constructed whilst the ANFIS performed well in cases where the input to output relationships became more complex.

3.2 Case Study 2: Engine Fuel Injection Control Using Fuzzy Logic

This work describes a fuzzy control system (FCS) applied to a small engine to achieve regulation of the fuel injection system. It was demonstrated that intelligent systems can be used for the computer control of the fuel supply of a small internal combustion engine. The technique represented a convenient and quick method of achieving ECU calibration; conventionally, ECUs use three-dimensional mappings (3-D maps), in the form of look-up tables, to represent the non-linear behaviour of the engine in real-time. A major disadvantage of the look-up table representation is the time taken to determine the values it should contain for optimal engine operation; the calibration process is an iterative one that requires many cycles of engine measurements and is very time consuming. The fuzzy technique reduced the time and effort required for the calibration process, which is thought to be of considerable interest to engine manufacturers.

Figure 6 shows the block diagram of the test system. The system determines the amount of fuel required from a fuzzy algorithm that uses the engine speed and manifold air pressure (MAP) as input values. These input variables were converted to digital form and the crisp digital signals were then applied to a fuzzy algorithm implemented in the C programming language on a PC. The crisp output from the algorithm was equal to the width of the pulse applied to the fuel injector, the fuel pulse width (FPW). The parameters of this fuzzy control paradigm were a collection of rules and fuzzy-set membership functions. These were intuitively comprehensible by the operator. This facilitated the calibration process, leading to relatively quick and convenient tuning.

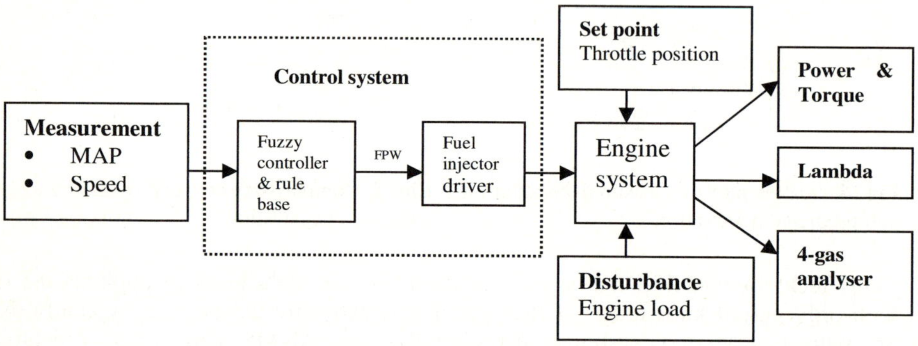

Fig. 6. Block diagram for fuzzy logic control scheme

The control surface in Figure 7 shows the crisp value of FPW at different combinations of speed and vacuum pressure using FCS. Each of these intersection points indicates the differing requirement for fuel, which is determined by the design of fuzzy sets and membership functions. The control surface acts as a means of determining the FPW needed for each combination of speed and MAP value.

Figures 8 and 9 illustrate that the power produced by the engine with the FCS exhibited an increase of between 2% and 21% with an average of approximately 12% compared with the original mechanical fuel delivery system.

A corresponding improvement in output torque also resulted from the use of the fuel injection system with the FCS compared to when the original fuel delivery system was used. Figures 10 to 11 show that the average torque exhibited an increase of between 2% and 20% with an overall average of 12%. These increases in engine performance are partly due to the improvement in charge preparation achieved by the fuel injection process; the improvement in fuel metering also results in improved combustion efficiency hence increased engine power.

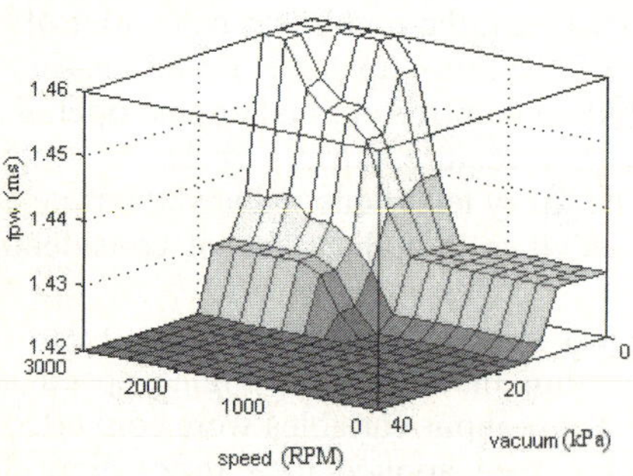

Fig. 7. Three-dimensional FCS map

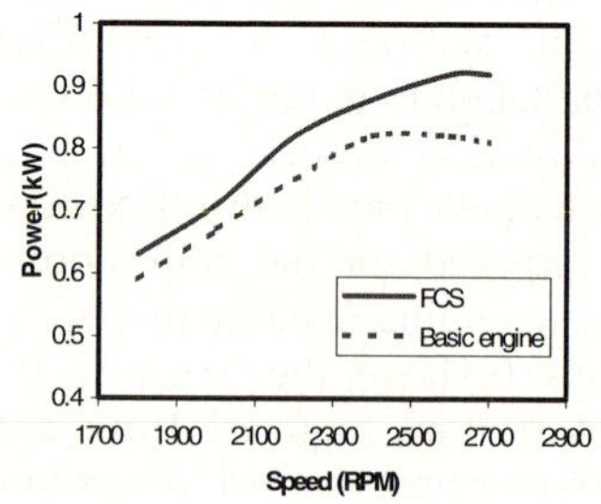

Fig. 8. Engine power when throttle=50%

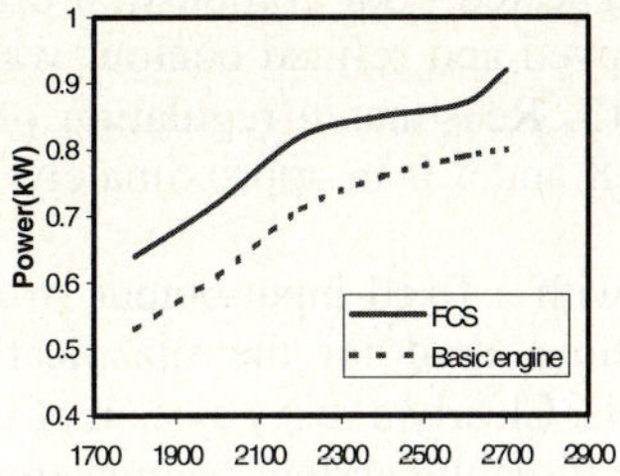

Fig. 9. Engine power when throttle=75%

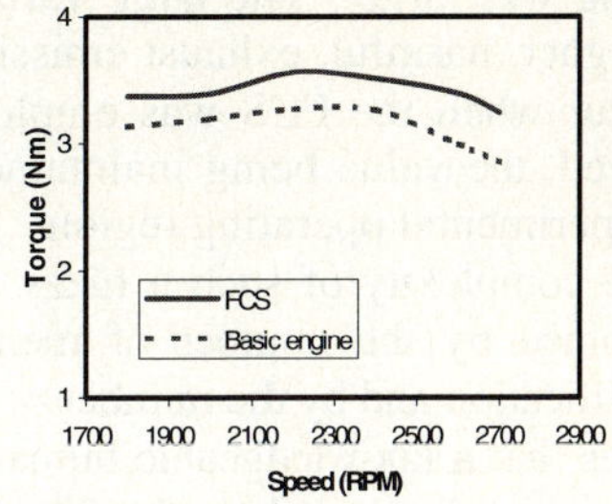

Fig. 10. Engine torque when throttle=50%

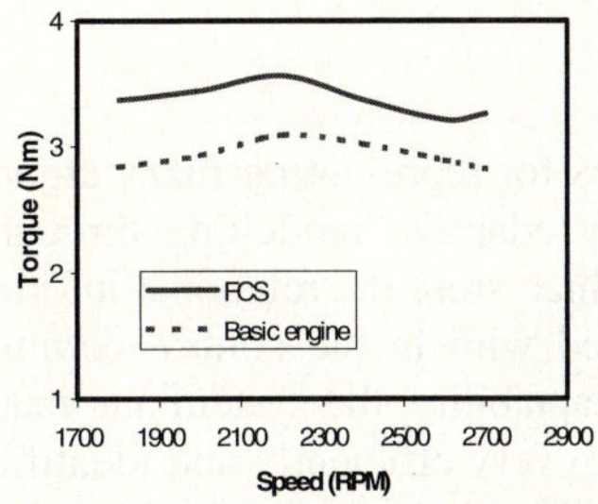

Fig. 11. Engine torque when throttle=75%

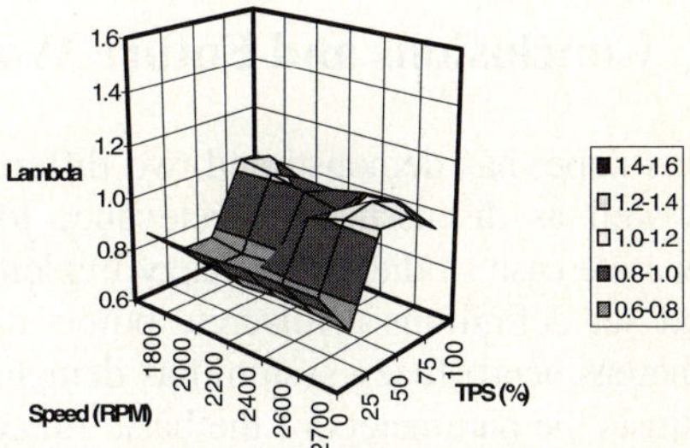

Fig. 12. Variation in lambda with original fuel regulation system

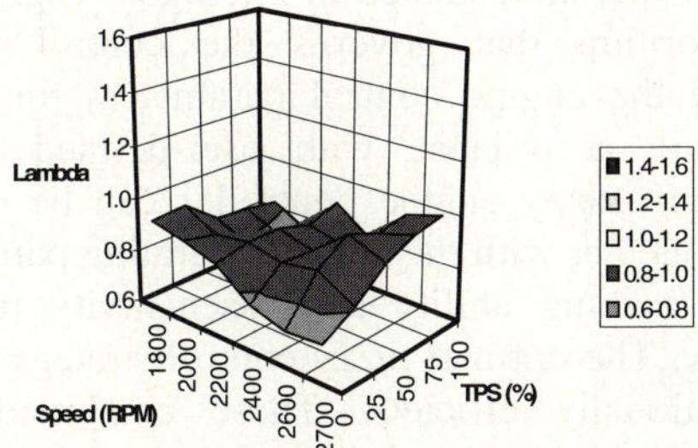

Fig. 13. Variation in lambda with fuzzy-controlled fuel-injection system

Figures 12 and 13 illustrate how the value of λ varied with different combinations of speed and throttle position using the original fuel regulation system and the fuzzy-controlled fuel-injection system, respectively. Figure 12 shows that wide variations in λ occurred when the original fuel regulation system was used, this being due to non-linearities in the characteristic of the carburettor. This resulted in an excessively rich mixture at small throttle openings and an excessively weak mixture when the throttle opening was large. The large variations in λ suggested poor combustion efficiency and higher, harmful, exhaust emissions. An improved and refined contour was found to occur when the FCS was employed, Figure 13. Reasonable regulation of λ was achieved, the value being maintained between 0.8 and 1.0 in approximately 90% of the experimental operating region.

The complexity of such a fuzzy logic system with a fixed input-output structure is determined by the number of membership functions used for the fuzzification and defuzzification and by the number of inference levels. Clearly a fuzzy system of this kind requires that a knowledgeable human operator initialises the system parameters e.g. the membership function bounds. The operator must then optimise these parameters to achieve a required level of accuracy of mapping of the physical system by the fuzzy system. While the visual nature of a fuzzy system facilitates the optimisation of the parameters, the need for it to be accomplished manually is a disadvantage.

4 Conclusions and Future Work

This paper has demonstrated two different techniques for representing fuzzy algorithms as well as discussing the relevance to neuro-fuzzy adaptive modelling through two separate case studies. Pure fuzzy implementations which store the relational information and set definitions at discrete points have been used with in the control community, whereas neuro-fuzzy system has demonstrated its adaptability, the system automatically adjusts the parameters of the basic fuzzy logic system very efficiently and identified the unknown process mapping from input to output data. The latter technique has gained in popularity due to their strong links with neural networks.

The neuro-fuzzy paradigm can be applied in small internal combustion engine control. Future work is concentrating on the establishment of an improved neuro-fuzzy paradigm for adaptive, fast and accurate control of small internal combustion engines drawing from experience gained in intelligent engine control and modelling. The neuro-fuzzy algorithm that governs the control process is designed to automatically optimise the engine control parameters for each operating zone, to achieve performance in accordance with user-defined specifications. The best parameters for the neuro-fuzzy engine controller can be determined by using this ANFIS methodology together with the engine operating parameters. A test rig will be used to validate the tracking ability and insensitivity to engine conditions/load changes experimentally. The optimal fuzzy control strategy should be inexpensive to set up and computationally efficient. These combined or hybrid topological techniques are likely to be highly beneficial in future engine control research.

References

1. Kalogirou S.A. Applications of artificial neural-networks for energy systems, Appl. Energy 67 (2000) pp.17–35
2. Xu K., Luxmoore A.R., Jones L.M., Deravi F., Integration of neural networks and expert systems for microscopic wear particle analysis, Knowledge-Based Systems 11 (1998) pp.213–227
3. Zadeh L.A. Fuzzy sets, Information Control 8 (1965) pp.338–353

4. Wang L.X. Fuzzy systems are universal approximators. Proceedings of the IEEE International conference on Fuzzy Systems, (1992) pp.1163-1170
5. Jang J. ANFIS: Adaptive network-based fuzzy inference systems, IEEE Transactions on Systems, Man, and Cybernetics 23, (1993) pp.665-685
6. Lee, S.H., Howlett, R.J. and Walters, S.D.: An Adaptive Neuro-fuzzy Modelling of Diesel Spray Penetration, Paper No. SAE-NA 2005-24-64, Seventh International Conference on Engines for Automobile (ICE2005), September 2005, Capri, Napoli, Italy, ISBN 88-900399-2-2.
7. Lee, S.H., Howlett, R.J. and Walters, S.D.: Small Engine Control by Fuzzy Logic, International Journal of Intelligent & Fuzzy Systems, Volume 15, Numbers 3, 4, 2004, pp. 207-217, ISSN 1064-1246.
8. Copp, D. G., Burnham, K. J., Locket, F. P.: Model Comparison for Feedforward Air/fuel Ratio Control. KACC International Conference on Control 9, pp. 670-675, (1998).
9. Powers, William F., Nicastri, Paul R.: Automotive vehicle control challenges in the 21st century, Control Engineering Practice 8, pp. 605-618, (2000).

Clustering for Data Matching

Edward Tersoo Apeh[1,2] and Bogdan Gabrys[1]

[1] School of Design, Engineering and Computing, Computational Intelligence Research Group, Bournemouth University, Talbot Campus, Fern Barrow, Poole BH12 5BB, UK
{eapeh, bgabrys}@bournemouth.ac.uk
[2] QGate Software Limited, D2 Fareham Heights, Standard Way, Fareham, Hampshire, UK, PO16 8XT
edward.apeh@qgate.co.uk

Abstract. The problem of matching data has as one of its major bottlenecks the rapid deterioration in performance of time and accuracy, as the amount of data to be processed increases. One reason for this deterioration in performance is the cost incurred by data matching systems when comparing data records to determine their similarity (or dissimilarity). Approaches such as blocking and concatenation of data attributes have been used to minimize the comparison cost. In this paper, we analyse and present Keyword and Digram clustering as alternatives for enhancing the performance of data matching systems. We compare the performance of these clustering techniques in terms of potential savings in performing comparisons and their accuracy in correctly clustering similar data. Our results on a sampled London Stock Exchange listed companies database show that using the clustering techniques can lead to improved accuracy as well as time savings in data matching systems.

1 Introduction

Rapid improvements in database hardware and software technology have resulted in an increased ability to collect and store large quantities of data with varying attributes. This increase in the availability and capacity of data sources/stores has created a number of problems including the occurrence of inconsistent, redundant, and inaccurate structured and unstructured data records.

In the case of structured data, relational database systems in principle, disallow the entry of records containing duplicate primary key values. Unfortunately, due to the high dependence of relational database system on the primary key of data records, any errors on the data that affect the primary attribute would mean that the relational database cannot guarantee the non-existence of redundant records. For example, the consolidation of two databases may result in the occurrence of imprecise primary keys if the databases have different key attributes. Another example is errors that tend to occur during manual data input. A detailed description of the common recurring problems and errors in data can be found in [1, 2].

Data matching of similar entries in databases, also referred to as record linking [3] and data cleansing [4], is the process of detecting multiple, inconsistent, redundant

B. Gabrys, R.J. Howlett, and L.C. Jain (Eds.): KES 2006, Part I, LNAI 4251, pp. 1216 – 1225, 2006.
© Springer-Verlag Berlin Heidelberg 2006

and inaccurate representation of a single entity with the ultimate goal of improving the data quality of collected and stored data.

Efficiently matching data characterized by the problems and errors listed in [1, 2] is the main focus of this paper. We evaluate two promising clustering techniques and present them as potential alternatives for grouping similar records for (potentially) more expensive comparisons in data matching systems.

2 Overview of Related Work

There are currently many approaches to solving the problem of data matching. The approaches differ by the field of specialization and the techniques incorporated in performing the data matching process. These approaches include the following:

Statistical Approach

The problem of data matching was first addressed formally in the foundational work aimed at linking epidemiological records by Newcombe, et al. [5]. Their work gives a clear outline from a statistical perspective of many of the key areas in linking personal records: use of Soundex to handle spelling log-odds and independency assumptions as well as using self-match probabilities to estimate match probabilities. Felligi and Sunter in [6] extended and gave a theoretical foundation to the work done by Newcombe. Winkler and Thibaudeau, under the sponsorship of the US census board further extended the work by Felligi-Sunter in [7]. They addressed the main problem of the Felligi-Sunter model of parameter setting by using the EM algorithm to obtain optimal probability values of record pairs. The statistical approach however, has a high dependence on the structure of the statistical model used and the availability of training data [8].

Data Warehouse Cleansing Approach

Kimball Ralph describes the data warehouse approach to solving the problem of dirty data in [9] as a six step approach involving: (1) elementizing (i.e. parsing), (2) standardizing, (3) verification, (4) matching, (5) house-holding and (6) result documentation. A popular technique for improving performance in data warehouse cleansing is the pre-sorting of the data based on some domain dependent hash function. For example, a hash function can be chosen to be the first four characters of a name attribute or a concatenation of consonants taken from the name, age and address postcode attributes. This approach has a drawback in that the hash function does not guarantee that duplicates will fall next to each other after sorting. In the worst case, the duplicates may occur at opposite ends of the sort. An alternative framework which divides the data cleansing task into a logical and physical level for performing data matching in the data warehousing/database systems genre has been provided by H. Galhardas, et al. in [10]. Choosing the most optimal algorithms to implement the logical and physical level operations of their framework is however, challenging and a wrong choice may result in adverse bottlenecks when processing large databases [11].

Knowledge Based/Expert System Approach

The AI approach to dealing with the problem of data matching is relatively recent. Hernandez and Stoflo in [4] introduced the sorted neighbourhood method which uses the principle of transitive closure to restrict the number of comparisons performed in a sorted database to a fix window size. The sorted neighbourhood method, however, has a heavy dependence on the proximity of the duplicate records after sorting which in turn depend on the key chosen for sorting. Monge and Elkan in [12] describe in detail an alternative performance enhancing algorithm which entails the use of a priority queue algorithm to scan the database sequentially and determine whether each record scanned belongs to a cluster represented in a priority queue. However, this method has the same problem of proximity as the sorted neighbourhood method. Wai Lup Low, et al. in [11] use a knowledge based approach to improve the recall-precision dilemma and show that small changes to the window size of the sorted neighbourhood method has little effect on the results of the sorted neighbourhood method. Their approach however, suffers from a high dependence on domain experts to define business rules used in the data matching process.

In this paper, we will present Keyword and Digram clustering as potential alternatives to dealing with the speed and accuracy performance problem in data matching systems. The canopy clustering technique [13] was recently applied to the problem of medical record linking in [14]. Our approach is however orthogonal to the one in [14] in that we integrate knowledge from the established fields of spell-checking and unsupervised learning to cluster inconsistent business data sampled from the London Stock Exchange with the goal of investigating clustering performance in terms of accuracy and potential reduction in the number of expensive comparisons performed by the data matching process.

In Sect. 2, we more precisely describe the significance of performance in data matching systems and introduce clustering as a performance enhancing measure. We go on in Sect. 3, to give an overview of keyword and Digram clustering methods. Section 4 describes the metrics used to evaluate the performance of the two clustering methods when applied to the problem of data matching of business names. In Sect. 5, we present a side by side evaluation of the clustering methods. We conclude in Sect. 6 with a discussion of the performance evaluation results and concluding remarks on how the clustering approaches offer a better alternative to other approaches to tackling performance bottleneck in data matching systems.

3 Application of Clustering in Data Matching

The data matching process is considered to be a problem of $O(n^2)$ complexity. Performance enhancement measures, therefore, tend to be centred on implementing techniques which perform better than the brute force approach of comparing each record to every other record. The brute force approach while being most reliable (as no record is missed during the comparison process) is also the most time consuming and least efficient in resource usage, as it requires $\frac{n^2 - n}{2}$ comparisons.

In this paper, we analyse clustering as a potential alternative approach to reducing the number of comparisons performed by a data matching system. Clustering methods are mainly divided, based on the structure imposed on the data formed, as hierarchical (divisive and agglomerative) and partitioning clustering methods [15]. Most partitioning algorithms have been reported to fair badly when applied to information retrieval problems [16] which are analogous to the data matching problem [17].

Many agglomerative clustering algorithms (especially those based on the Lance-Williams equation [18]) have a drawback in that the full distance matrix has to be calculated during cluster analysis. This can take considerable computing power and requires high amounts of memory. A remedy to this can be found in the minimal spanning tree (MST), a graph-theory approach which can only be applied to the single linkage clustering. In the case of large data sets, this algorithm can be orders of magnitude faster [19] in comparison to other hierarchical clustering approaches. The efficiency of the MST in terms of space and computing power was the basis of our decision to adopt it for our analysis where we based our implementation according to Prim-Dijkstra [20]. Further details of the MST algorithm can be found in [19, 20].

For any clustering algorithm, the choice of the similarity measure [21-23] is of critical importance as it has a huge influence on the quality of clusters formed. Willet in [24] has suggested that the choice of similarity measure has a greater qualitative impact on clustering results than the choice of clustering algorithm. For our work, we adopted the Dice Coefficient similarity measure to determine the nearest neighbour information. Our decision was based on the need to adequately cluster multilingual data and by the findings in [25] where the dice coefficient was chosen after a detailed analysis of the similarity measures in [21-23], as a better indicator of similarity for translating collocations in bilingual lexicons.

The Dice coefficient is a term based similarity measure (0-1) whereby the similarity is defined as twice the number of terms common to compared entities divided by the total number of terms in both tested entities. That is:

$$s_{ij} = \frac{\left(2 \times c_{ij}\right)}{\left(t_i + t_j\right)} \tag{1}$$

where s_{ij} is the dice coefficient; c_{ij} is the number of common terms in the entities i and j; t_i is the number of terms in the i-th entity; t_j is the number of terms in the j-th entity. The dice coefficient s_{ij} value of 1 indicates identical records whereas 0 indicates records with no common terms. More details about the Dice Coefficient and its use for spelling correction and text searching applications can be found in [21].

The MST algorithm by definition minimizes the sum of the dissimilarities between connected records. To obtain the dissimilarity values for the MST algorithm we used the following relation for converting between dissimilarity and similarity bounded by 0 and 1:

$$d_{ij} = 1 - s_{ij} \tag{2}$$

where s_{ij} is the similarity value computed from (1) above and d_{ij} is their dissimilarity value [26].

4 Overview of Keyword Clustering and Digram Clustering Approaches

Records in the business world tend to have hundreds of attributes describing them. For instance, a business entity might have an SIC code, address, area of specialization, number of employees, etc. as attributes describing it. If during a data matching process each record is considered as a multi-dimensional vector one is potentially left with hundreds of dimensions per record to match on.

Our paper investigates two clustering methods: Keyword and Digram clustering methods, where one attribute is employed to determine records that are most likely to match. Any further expensive comparison which may involve the use of more attributes is then restricted to just records in the same cluster returned by the clustering component as containing potential matches. The goal is to compare one data record against a very large number of others, and to do so without making unnecessary individual comparisons. In our analysis, Keyword clustering was used as a baseline with its performance evaluated against that of the Digram clustering method.

4.1 Keyword Clustering

Keyword clustering is an approach that dates back to the Information Retrieval work by Sparck Jones [27] and Salton [23], in which a thesauri was automatically constructed by clustering words based on the words co-occurrence in documents. The principle behind their work involves the substitution of one word with another if they have the same meaning or refer to a common subject or topic in a document. One way of determining the substitutability of words is by looking at the documents in which they occur. If they tend to co-occur in the same documents, the chances are they have to do with the same subject and so can be substituted for one another [23].

We adopted a similar approach in our research. Using (1), records having the highest number of similar words co-occurring in them were clustered together. For instance, the record First Dental Laboratories was clustered together with 1st Dental Laboratories based on the co-occurrence of the words Dental and Laboratories with First and 1st; with the dice coefficient computed as 0.75.

4.2 Digram Clustering

As discussed in Sect. 1, one of the main goals of a data matching system is to effectively match similar data irrespective of the occurrence of errors in the data. N-grams have been extensively used in spelling correction and text searching applications because of their robustness to errors [21]. An *n*-gram is a subset of characters of length *n* taken from a string of at least length *n*. The characters in the *n*-gram retain the same order as in the source text from which the *n*-gram has been derived.

For our work, we investigated the effect the reported robustness of *n*-grams had on the clusters formed by the Digram clustering method in contrast to those formed by the keyword clustering method which only takes word co-occurrence frequencies into account. To obtain the Digrams for clustering, *n*-grams containing two characters were derived from the words in the records, with spaces padded to ensure that all of the original characters occurred in the *n*-grams. The *n*-grams were generated by moving an *n*-character window across the resulting string, one character at a time and starting at the left-hand side of the string. For instance, the *n*-gram for the word DENTAL will be

D, DE, EN, NT, TA, AL, L

To cluster the records using the derived *n*-grams, the degree of similarity between the records were calculated again by the means of (1).

5 Experiment Methodology

5.1 Evaluation Metrics

Experiments were undertaken to evaluate the performance of the two clustering methods described above on the London Stock Exchange database.

Our experiments involved the evaluation of the clustering methods based on two criteria: accuracy of matching and potential comparison reduction of full records. For accuracy, we were interested in how the clustering methods were accurately able to identify and return data as potential matches for further comparison; while for potential comparison reduction, we were interested in potential reduction in the number of more expensive comparisons which may involve more attributes.

To evaluate for accuracy, we used recall and precision at a specific dissimilarity threshold as the metrics for benchmarking the clustering methods. The first two metrics are established metrics used in benchmarking information retrieval systems. To measure the potential comparison reduction rates at a specific dissimilarity threshold we used the metrics analogous to the one used by M. Elfeky et al. [30].

In information retrieval, recall is defined as the ratio of the number of relevant records retrieved to the total number of relevant records in the database expressed as a percentage; while precision is defined as the ratio of the number of relevant records retrieved to the total number of irrelevant and relevant records retrieved expressed as a percentage.

In our work, we defined the recall performance of a clustering method as the number of records correctly clustered as matches to the total number of known matches in the database at a specific dissimilarity threshold as follows:

$$\text{Clustering Recall} = \frac{\sum_{i-clusters} C_m^i}{N_m} \quad (3)$$

where C_m^i is the number of matching records clustered and N_m is the total number of known matching records in the database.

The precision performance of a clustering method was defined as the ratio of the number of records correctly clustered as matches to the total number of records clustered at each dissimilarity threshold. That is:

$$\text{Clustering Precision} = \frac{\sum_{i-Clusters} C_m^i}{\sum N_C} \quad (4)$$

where C_m^i is the number of matching records clustered and N_c is the total number of records clustered at each dissimilarity threshold.

5.2 Data Sets

For our evaluation, 1000 records were hand-picked from the publicly available database of listed companies on the London Stock Exchange. The records are attributed by their company SIC codes, business names and addresses. We choose to cluster on the business name attribute which we represented as numerical pointers. Subsets of 43 records having the typical data problems and errors reported in [1, 2] were picked as known matches and placed into 20 different clusters. The data sets and matches are available for viewing at www.qgate.co.uk/clustering.html.

Pre-processing activities such as standardization make a big impact on the scores produced by the similarity measure because it eliminates a lot of the (arguably) irrelevant matches on records. Standardisation which involves the pre-processing of terms using look-up tables to convert acronyms and abbreviations to their universally accepted forms (e.g. "IBM" to International Business Machines) as well as removing noisy words such as "to", "Plc", "Ltd". For our experiment, no standardisation or any other pre-processing (e.g. the use of Stop-list, stemming, etc.) was performed on the input data sets.

6 Experimental Results

Experiments were undertaken to evaluate the recall and precision capability of the clustering methods to correctly match records as well as the potential comparison reduction ratio at dissimilarity threshold ranges 0 to 1. The results of the runs are as plotted in Fig. 2 and Fig. 3 for both Keyword and Digram clustering methods.

Figure 2 shows a better recall performance by the Digram method with the optimal recall value occurring at 0.6 dissimilarity threshold when the most matches are identified. Figure 3 shows deterioration in precision as more matches are identified and clustered with records that were not labeled as matches. The plots also show an inverse relationship between the clustering recall and clustering precision that depict a trade-off between the accuracy and time performance of data matching systems, with lower dissimilarity threshold values resulting in low recall values and high precision; and higher dissimilarity threshold values resulting in high recall and low precision values.

Thus, a high recall (low precision) clustering output from the grouping component while providing for more thoroughness will result in a larger number of further comparisons by the comparison component as opposed to a low recall (high precision) which will result in a reduction in the number of further comparisons performed but be less thorough. For instance, a recall of 80% will present for further comparison, clusters with only 20% purity while a clustering purity of 90% will present clusters with only 20% of the total number of known matches/duplicates identified for further comparisons.

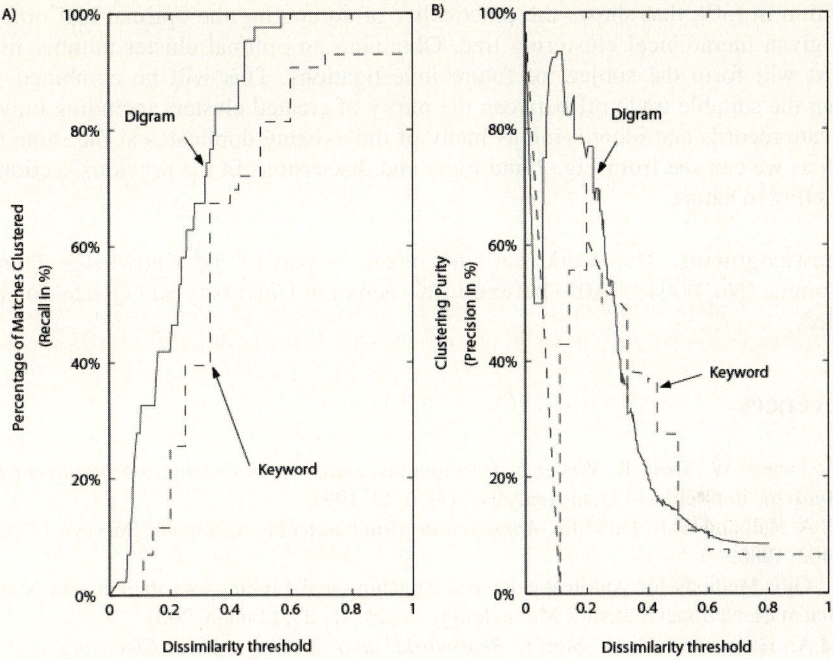

Fig. 1. Plots of recall and precision performance by Keyword and Digram clustering methods at dissimilarity threshold ranges 0 to 1. The recall plots show the matching records clustered while the precision plot show the purity of the clusters formed.

7 Discussion of Results and Conclusions

We have described and presented the implementation and experimental results of our investigation of keyword and Digram clustering as alternatives to dealing with the problem of performance in data matching systems.

The analysis described in this paper is a work in progress. Our findings so far show that the clustering alternatives offer promise in reducing full record comparison costs and improving (recall and precision) accuracy. We are however, particularly concerned about the scalability of traditional clustering algorithms, of which the MST clustering algorithm used in our analysis with $O(n)$ space and $O(n^2)$ time complexity is theoretically the most elegant. Although the implementation of the inverted index algorithm as described in [16, 31] improved performance significantly there is still room for improvement other promising fast clustering approaches for large datasets in the data matching domain such as the clustering approach proposed by Nigram et al in [12] will be further investigated.

Another issue of great interest and importance is obtaining an optimal stopping criterion for the clustering process. Not much work has been carried out in this area in the data matching domain. Eisen et al. [29], while applying the hierarchical clustering to cluster micro-array data, asserted that computing the optimal ordering (clusters) was impractical. However, Z. Bar-Joseph et al. provide an $O(n^4)$ time and $O(n^2)$

algorithm in [30], that shows the practicality of computing the optimal leaf ordering for a given hierarchical clustering tree. Obtaining an optimal cluster number in our context will form the subject of future investigations. This will be combined with finding the suitable trade off between the purity of created clusters including only the duplicate records and identifying as many of the existing duplicates at the same time which as we can see from Fig. 2 and Fig.3,and discussions in the previous section are conflicting in nature.

Acknowledgments. This work was undertaken as part of the Knowledge Transfer Programme (No. 000487-A01) between Bournemouth University and QGate Software Limited.

References

1. A. Famili, W. Shen, R. Weber & E. Simoudis, *Data Preprocessing and Intelligent Data Analysis,* in Intelligent Data Analysis, 1(1), 3-23, 1997.
2. P. A. Hall and G. R. Dowling. *Approximate string matching*. Computer Surveys, (12):381--402, 1980.
3. L. Gill, Methods for Automatic Record Matching and Linking and their use in National Statistics. National Statistics Methodology Series No. 25, London, 2001.
4. M.A. Hernandez, S. J. Stolfo, *Real-world data is dirty: Data Cleansing and The Merge/Purge Problem*, Knowledge Discovery 2 (1) (1998) 9-37.
5. H.B. Newcombe, Kennedy J.M., Axford S.J., and A.P. James. *Automatic linkage of vital records*. Science, 130(3381):954--959, October 1959.
6. I. P. Fellegi and A. B. Sunter. *A theory for record linkage*. J. Amer. Statist. Assoc., 64:1183--1210, 1969.
7. William E. Winkler and Yves Thibaudeau, *An Application of the Fellegi-Sunter Model of Record Linkage to the 1990 U.S. Census*, Number RR91/09.
8. Winkler, W. E. *The State of Record Linkage and Current Research Problems*, Statistical Society of Canada, Proceedings of the Section on Survey Methods, (1999), 73-79.
9. R. Kimball, *Dealing with Dirty Data,* DBMS online, Available at URL: http://www.dbmsmag.com/9609d14.html, September, 1996
10. H. Galhardas, D. Florescu, D. Shasha, E. Simon, and C-A. Saita. *Declarative Data Cleaning: Language, Model, and Algorithms*. In 27th International Conference on Very Large Data Bases, pp. 371--380, 2001
11. Wai Lup Low, Mong-Li Lee, Tok Wang Lin: *A knowledge-based approach for duplicate elimination in data cleaning*. Inf. Syst. 26 (8): 585-606 (2001).
12. A. E. Monge, C. P. Elkan, *An efficient domain-independent algorithm for detecting approximately duplicate database records*, in: Proceedings of the ACM-SIGMOD workshop on Research issues on Knowledge discovery and data mining. AZ. 1997.
13. McCallum AK, Nigam K, Ungar LH: *Efficient clustering of high dimensional datasets with application to reference matching,* in: Sixth International Conference on Knowledge Discovery and Data Mining, *Boston* 2000.
14. Erik A Sauleau, Jean-Philippe Paumier and Antoine Buemi: *Medical record linkage in health information systems byapproximate string matching and clustering*, in BMC Medical Informatics and Decision Making, 5-32, 2005
15. A. K. Jain and R. C. Dubes. *Algorithms for Clustering Data*. Prentice Hall, 1988.
16. *Information retrieval: data structures and algorithms* Pages: 419 - 442 Publication: 1992

17. J. Zobel and P. Dart. *Phonetic string matching: Lessons from information retrieval*, in Proceedings of the Eighteenth ACM SIGIR International Conference on Research and Development in Information Retrieval, Zurich, Switzerland, August 1996, pp. 166-173.
18. Lance, G. and Williams W. *A general theory of classification sorting strategies.* Computer Journal, 9, 373-386, 1967.
19. H. Lohninger: *Teach/Me Data Analysis*, Springer-Verlag, Berlin-New York-Tokyo, 1999.
20. Margaret H. Dunham: *Data Mining: Introductory and Advanced Topics* Prentice-Hall 2002
21. Robertson, A.M. & Willet, P. 1998. *Applications of n-grams in textual information systems.* Journal of Documentation, 54(1), 48-69.
22. C.J. Van-Rijsbergen. *Information Retrieval.* Butterworths, London, England, 2nd edition. Chapter 3, 1979
23. G. Salton and M. J. McGill. *Introduction to Modern Information Retrieval.* McGraw-Hill, 1983.
24. P. Willett. *Recent trends in hierarchic document clustering: A critical review.* Information Processing & Management, 24(4):577--597, 1988.
25. Smadja, Frank A. & Kathleen R. McKeown. 1994. *Translating collocations for use in bilingual lexicons.* In Proceedings of the ARPA Human Language Technology Workshop, Princeton, N.J.
26. Teknomo, Kardi. Similarity Measurement. http://people.revoledu.com/kardi/ tutorial/ Similarity/
27. K. Sparck Jones. *Automatic keyword classification for information retrieval.* Butterworths, London, UK, 1971.
28. M. Elfeky, V. Verykios, and A. Elmagarmid. *TAILOR: A Record Linkage Toolbox.* In Proc. of the 18th Int. Conf. on Data Engineering. IEEE, 2002.
29. Michael B. Eisen, Paul T. Spellman, Patrick O. Browndagger, and David Botstein. *Cluster analysis and display of genome-wide expression patterns.* Proceedings of the National Academy of Sciences of the United States of America (PNAS), 95:25, 1998.
30. Z. Bar-Joseph, D. K. Gi ord, and T. S. Jaakkola. *Fast optimal leaf ordering for hierarchical clustering.* Bioinformatics, 17:22-29, 2001.
31. I. Witten, A. Moffat, and T. Bell. *Managing Gigabytes.* Morgan Kaufmann Publishers, New York, second edition, 1999.

Intelligent Optical Otolith Classification for Species Recognition of Bony Fish

D. Lefkaditis, G.J. Awcock, and R.J. Howlett

[1] Intelligent Systems & Applied Image Processing Research Laboratories
Engineering Research Centre, University of Brighton
Moulsecoomb, Brighton, BN2 4GJ, UK
[2] Fisheries Research Institute, Nea Peramos, 640 07, Kavala, Greece
D.Lefkaditis@brighton.ac.uk, G.J.Awcock@brighton.ac.uk, R.J.Howlett@Brighton.ac.uk

Abstract. The study of otoliths is an established method of age estimation of bony fish. It can also find interesting applications such as dietary studies by conducting species recognition on otoliths found in the stomach contents of marine animals. Moreover, they could even be sourced from geological sediments or pre-Neolithic archaeological excavations, providing useful data for palaeontology research. This paper presents work in progress to develop an alternative method of optical otolith recognition. This methodology is based on the processing and analysis of images acquired using a stereoscopic microscope fitted with a digital camera. Several configurations of neural networks are tested to conduct species recognition of bony fish.

1 Introduction

Otoliths are found in the inner ear of most of osteichthyes (bony fish). There are three pairs of otoliths in every fish. The larger pair is called "sagittae" and the two other smaller pairs are called "asteriscii" and "lapilli". Their physiology is very similar to the otoliths found in the labyrinth organ that exists in mammals. Therefore, they are associated with hearing, the sense of balance and orientation [8].

Otoliths are not bones; they are a kind of sedimentary formation that mainly consists of layers of $CaCO_3$ [5]. Each species creates a unique shape of otolith which is dependent on the size and shape of its brain cavity and balance organs [7]. Furthermore, there is a direct link between the otolith size and the weight/size of the fish.

The sedimentary nature of otoliths makes them more durable than bones, hence they are preserved in harsh environments and they can be recovered from the stomach of many marine fauna species. Therefore, study of recovered otoliths can be used to track their behaviour in terms of their eating habits and population movement. This study can be extended into the field of archaeology and palaeontology, as otoliths are also found preserved in fossils, geological sediments and pre-Neolithic archaeological excavations.

Otolith analysis is a highly complex procedure and a valid study requires hundreds of fish to be sampled. This means hundreds of hours of rigorous and repetitive work in order to accurately characterise otoliths. The aim of this investigation is to develop an automated method of fish species identification from otolith images.

B. Gabrys, R.J. Howlett, and L.C. Jain (Eds.): KES 2006, Part I, LNAI 4251, pp. 1226 – 1233, 2006.
© Springer-Verlag Berlin Heidelberg 2006

2 Materials and Methods

2.1 Sample Collection

This project is carried out in synergy of the "University of Brighton" (UoB) with the "Fisheries Research Institute" (FRI) in Kavala, Greece. Hence the area that this project focuses on is the Northern Aegean Sea in particular. Fish collection is carried out in pelagic (open sea) and demersal (coastal) areas so as to obtain a wide variety of relevant fish species and sizes (ages). The depth of sampling varies from 20 to 600 metres, so both pelagic and benthic (bottom-feeding) species can be fished.

Every fish is measured and weighed. An appropriate technique is utilised to ensure the safe removal of the "sagittae" pair of otoliths, which are then cleaned and stored in special labelled containers.

To date 386 individuals of 31 commercially interesting species have been collected explicitly for this study. Although more would be ideally required for system training, the pragmatic goal set is to collect 60 to 100 individuals from each of 20 to 30 species, because of the difficult nature of the collection work. Further sample collection will be carried out throughout 2006-7 by the FRI personnel.

2.2 Image Acquisition and Scene Constraints

The equipment currently used is a Leica MZ6 stereoscopic microscope fitted with a DFC320 full frame 3.2 MP image sensor. This tool provides the required magnifications (0.63x to 40x), and high quality optics.

The sample is placed on a thin layer of plasticine which sits on the main stage of the microscope. Plasticine provides a simple way of ensuring that the otoliths are held at the correct position and angle, which otherwise is difficult to achieve due to the otolith's irregular 3-D shape. Its colour was selected to be matte black so that it makes high contrast background with the off-white colour of the otoliths.

The lighting of the scene is achieved using two Leica L2 cold light sources that have two fibre optic branches each. These branches are set in such a way that light is cast from four directions minimizing shadow creation.

2.3 Image Processing

The images have to be processed before useful data can be extracted from them. This means that all background noise and unwanted particles have to be removed from the images in a way that provides uniform results, as the images are going to batch processed. Moreover, automatic translation and orientation of the otolith within the image is essential. An unprocessed image can be seen in Figure 1.

A 'intelligent' filtering strategy was designed for batch processing. It opens the set of images as an image "stack". A copy of each image in the stack is thresholded to reveal all the particles. Then a list of the particles is created, sorted according to their size. It is assumed that the largest particle in each image is the otolith. So, the number of the largest particles that equals the number of images in the stack is selected (one otolith per image) and the rest are forced to background colour. This action creates a stack of otolith masks. A logical AND is now performed on the original image stack with the mask stack. This allows only the pixels of the original images that fall within the area of the masks to remain unchanged, while the rest of the pixels are forced to black. This creates an image stack that contains otoliths with unmodified morphology against a clean black background. This strategy gives very satisfactory results.

Next, the otoliths within the image stack have to be translated and rotated so that they are all uniformly aligned in the central region of the images. This is a vital step to ensure that the extracted data in later stages is credible, as some features are sensitive to the position of the otolith in the image.

For this task, an ImageJ plugin called "Turboreg" was used, which was developed by the Biomedical Imaging Group in the Ecole Polytechnique Federale de Lausanne [9]. It can automatically register a "source" image to a "target" image. Hence, a set of "guide" images that was centred and rotated to a specific position was manually created, one for each species. When processing a stack of images, the appropriate guide image is used as a "target" to which source images will be aligned. The plugin was configured to perform a "rigid body" transformation so that the distance between any pair of points is conserved. A single landmark defines the translational component of this transformation, while the rotational component is given by an angle. The mapping is of the form [10]:

$$x = \{\{\cos\theta, -\sin\theta\}, \{\sin\theta, \cos\theta\}\} \cdot u + \Delta u \ . \qquad (1)$$

This means that there is no geometric distortion of the otolith, just translation and rotation. So, the morphology of the otolith is preserved. A typical result of the above processing can be seen in Figure 2.

Fig. 1. Unprocessed otolith image

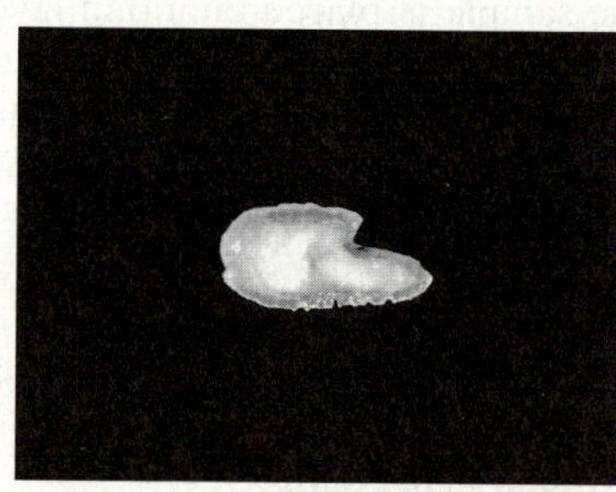

Fig. 2. Otolith image after processing

2.4 Feature Extraction

It would not be realistic, using current technology, to expect that one can avoid the curse of data dimensionality [2]. This is specifically important on this application due to the limited amount of data available and the high resolution of images required for this task.

Reduction of data dimensionality can be achieved by transforming the image data into a set of comparable characteristics of the otoliths that one can use to distinguish different fish species. In this investigation, a set of morphological otolith "features" is used. This set of features forms a four-element vector that aims to describe the 2-D shape of an otolith.

These elements are:

- Circularity.

A shape factor based on the projected area of the otolith and the overall perimeter of the projection according to equation 2:

$$C = \frac{4\pi A}{P^2} \ . \tag{2}$$

where A is the area and P is the perimeter of the otolith [3].

Values range from 1, for a perfect circle, to 0 for a line. It is used to give an impression of elongation of the otolith's shape.

- Maximum Feret's diameter.

This is defined as the greatest distance possible between any two points along the boundary of the otolith. It is used to give a measure of actual size scale of the otolith.

- Skewness.

This is the third order spatial moment of the otoliths's shape distribution around its centre of mass. In other words, it shows the otolith's symmetry with reference to its centre. It takes negative values when it is asymmetric to the left and becomes positive when it is asymmetric to the right.

- Kurtosis.

This represents the fourth spatial moment of the otolith's shape distribution. It shows the "flatness" of the distribution where it equals 0 if it is Gaussian, it becomes negative if it is flatter and takes positive values for a more peaked one.

An ImageJ macro was compiled that automatically carries out the extraction of the above features from an otolith image stack. These feature vectors are then exported in a tab-delimited text file, a file format compatible with most spreadsheet packages enabling further data manipulation.

2.5 Normalisation of Extracted Data

As the input nodes of a neural network accept values in the range 0 to 1, the data extracted has to be normalised. This procedure ought to be robust so that data can qualify for neural network training and recall. It should also remain descriptive in order to resemble the actual otolith characteristics with minimal distortion. So, different normalisation techniques are used for every one of the elements of the feature vector.

Circularity does not need to be normalised as it is already in the expected range and provides highly descriptive data.

Feret's diameter is measured in μm as all the images are calibrated according to the specific objective magnification used on the stereo-microscope at the time of image acquisition. So, it represents the real length of the otoliths. It is then normalised using equation 3:

$$F_{norm} = \frac{F_{real}}{IW_{real}} \ . \tag{3}$$

where IW is the overall image width. F_{real} and IW_{real} are measured in μm.

Skewness and kurtosis are of unknown range. This implies that a more robust normalisation method has to be implemented. Jain *et al.* suggest than the tanh estimator is a both robust and efficient technique for data normalisation [6]. This method can be summarised by the equation 4.

$$s_{norm} = \frac{1}{2}[\tanh(\frac{(s-\mu_G)}{100\sigma_G})+1]\,. \tag{4}$$

where s is the parameter to be normalized and μG and σG are the mean and standard deviation estimates of the parameter's distribution.

2.6 Training the Neural Network

The neural network used to carry out the recognition tasks of the proposed system is a two-layer (one input, one hidden and one output layer) feed-forward network with sigmoid activation functions. It was trained with the back-propagation algorithm.

The network was trained to distinguish between two and then three fish species that could supply the largest amount of data (species with the larger number of collected otoliths at this time). These species were *Sardina pilchardus* (European pilchard), *Trisopterus minutus capelanus* (poor cod) and *Trachurus mediterraneus* (Mediterranean horse mackerel). Images of the inner side of the left otolith were used as training and recall data. The total 75 sets of extracted data were split into 36 sets for training and 39 sets for recall.

The network was trained for two species (*S. pilchardus* and *T. minutus*) and three species (*S. pilchardus*, *T. minutus* and *T. mediterraneus*). Neural network configurations of two input neurons (circularity and Feret's diameter) three input neurons (circularity, Feret's diameter, and skewness) and four input neurons (circularity, Feret's diameter, skewness and kurtosis) were investigated. For each of these, hidden layers of 2, 3, 5, 10, 15 and 18 neurons were tried. The setup of the network, such as learning rates, momentum and initial weights, was kept the same for every training session. So, the network was trained for 36 sessions different architectures in all.

3 Results

3.1 Input Data Analysis

A good way to determine the quality of the extracted data is to plot graphs of features one against the other. These graphs permit clustering to be visualized, which reveals the possible decision boundaries and the existence of outliers.

Figures 3 and 4 show that the extracted data does form a considerable amount of clustering. This means that the features selected should be capable of providing reasonably reliable decision boundaries for the neural network.

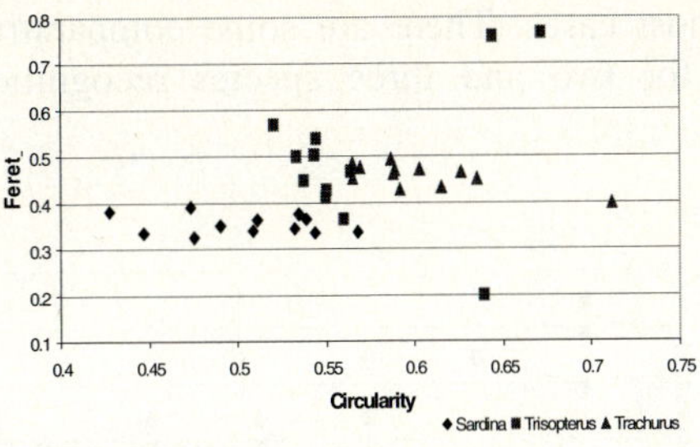

Fig. 3. Data clusters formed by Feret's diameter against circularity

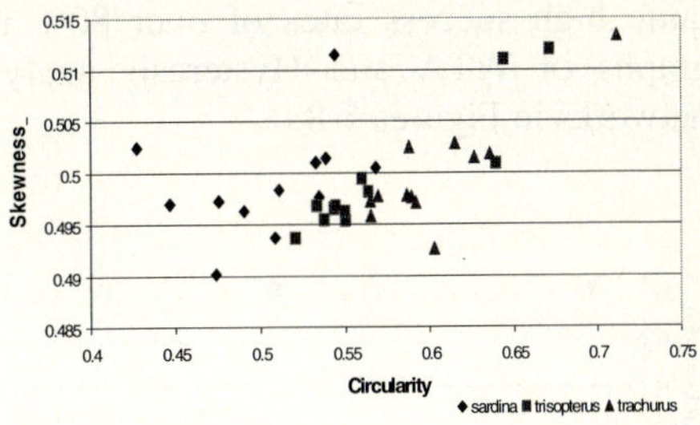

Fig. 4. Data clusters formed by skewness against circularity

Even though the data overlap is not extensive, a few outliers are present. They can potentially decrease the system's performance. There might be a number of possible reasons for this. Potentially otoliths can be subtly damaged during their extraction from the fish, resulting in an altered otolith shape. The age of the fish plays a profound role on the otolith's size and shape. As the fish grows, the otolith turns from almost circular, for juvenile and larval fish, to an elongated shape for the adults [11]. So, both the circularity and Feret's diameter are affected. Another possibility is that some otoliths are not cleaned properly after their extraction, so that blood and other substances can be deposited on them changing their optical properties.

3.2 Neural Output Analysis

In order to assess the success rates, or the measure of confidence in the decisions of the neural network, two types of analysis were conducted, the 'Winner-Takes-All (WTA)' and the 'Hysteresis' analysis.

The Winner-Takes-All analysis treats the output vectors of the neural network as one entity. All neuron outputs are compared to each other. The one with the largest value takes a value of 1 while the rest are forced to 0. This output vector is then contrasted with the desired network output. If these match, then the considered output is successful. For obvious reasons this type of analysis cannot be applied on a single output network.

In Hysteresis analysis two success rates are defined. These are the number of correct outputs and the number of correct recall sets. In the first case, every neuron output is considered. So, the total number of outputs equals the number of output neurons of the network multiplied by the number of recall sets. In the second case, the outputs of the network are considered to be a single output vector, so a successful vector should match the desired one as a whole. In Hysteresis analysis, what defines a successful neuron output is the truth of the following statement:

$$\left| output_{actual} - output_{desired} \right| \leq threshold \; . \tag{5}$$

where the threshold is set by the user. The threshold value used here is 0.4. It has to be mentioned that all the networks configured for two species recognition managed to converge with a limit error value of 0.2. On the other hand, networks configured for three species recognition did not converge. However, they achieved quite high success rates of over 80% in most cases. There are some comparative graphs of WTA and Hysteresis analysis for two and three species recognition networks in Figures 5-8:

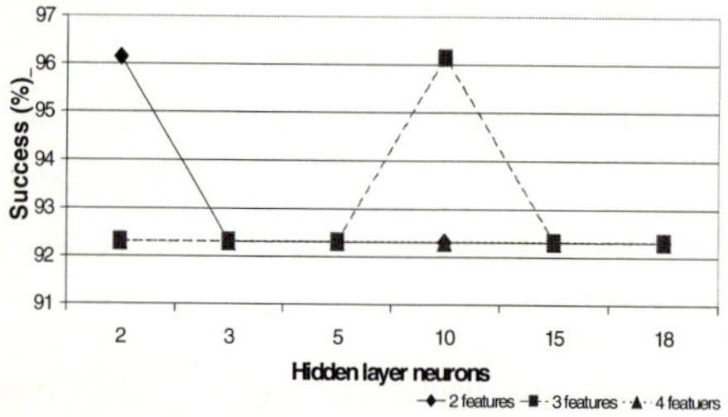

Fig. 5. WTA analysis for 2 species recognition

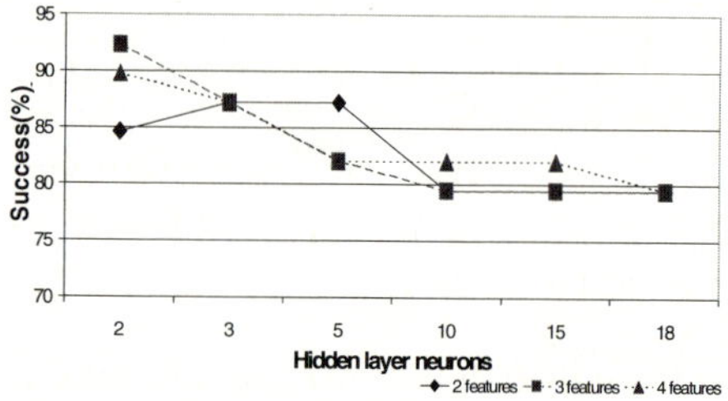

Fig. 6. Hysteresis analysis-No. of correct outputs for 2 species recognition

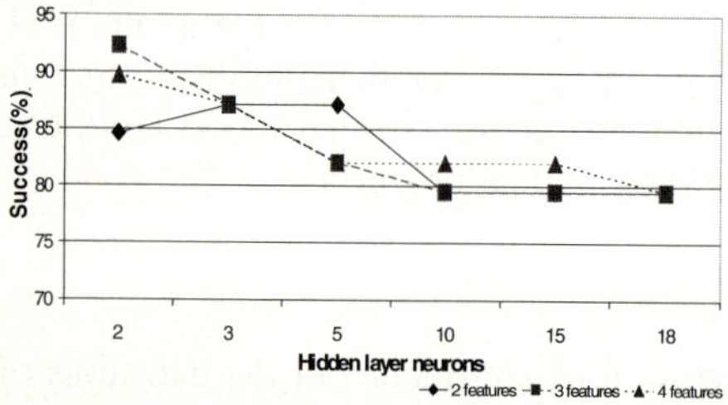

Fig. 7. WTA analysis for 3 species recognition

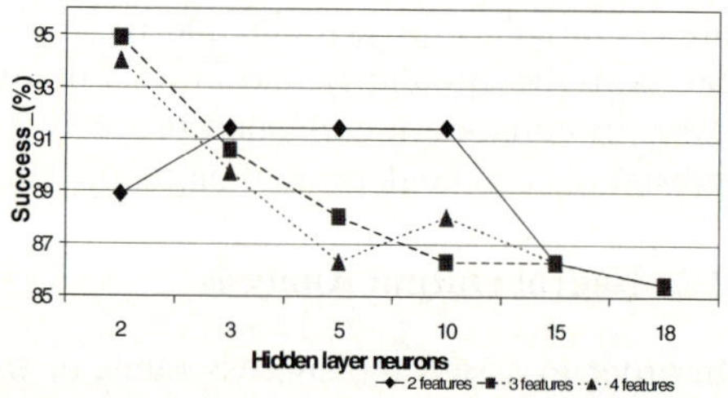

Fig. 8. Hysteresis analysis-No. of correct outputs for 3 species recognition

Looking at the graphs for two species recognition, we can see that the success rates (both WTA and Hysteresis) are virtually constant in respect with the number of hidden layer neurons no matter what the number of features is.

Conversely, for three species recognition (or non-converged networks) the number of neurons has a negative effect on the success of the network.

As expected, three species recognition achieves lower success rates, partly because of the increased complexity of networks used. Even though more hidden neurons means lower success, by adding more features the networks exhibit better behaviour.

The overall performance of the networks is considered to be satisfactory. Nevertheless, more work needs to be done to increase the performance of the system.

3.4 Conclusions

This paper presents an intelligent methodology for fish species classification based on the image analysis of otoliths.

An 'intelligent' image filtering approach is proposed in order to remove unwanted noise and particles from digital images of otoliths. This approach is considered to be superior to traditional neighbourhood operations as it causes minimal shape distortion.

A set of four morphological characteristics are included in the feature vector on which the classification processed is based. These are the Circularity, the maximum Feret's diameter, the Skewness and the Kurtosis. They were found to exhibit reasonable data clustering, which provides clear decision boundaries.

The classification operations were carried out by a feed-forward two-layered neural network. The neural networks were trained to distinguish between two and three bony fish species. A wide variety of configurations were tested, achieving success rates varying from 79.5% to 96%, which is encouraging for future work.

A lot more work is necessary for a clearer image of the subject to be provided. This paper is a review of a set of preliminary results.

References

1. Awcock G.J. & R.Thomas, 1995. Applied Image Processing, pp. 111-118.
2. Bellman, R.E 1961. Adaptive Control Processes. Princeton University Press, Princeton, NJ.
3. Bouwman A.M., J.C. Bosma, P. Vonk, J.A. Wesselingh, H.W. Frijlink, Which Shape Factors Best Describe Granules?, Powder Technology, 2004, vol. 146, pp. 66-72.
4. Campana S.E., Neilson J. D., 1985. Microstructure of fish otoliths. Can. J. Fish. Acquat. Sci. 42: 1014-1032.
5. Degens E.T., W.G. Deuser, R.L. Heandrich, 1969. Molecular structure and composition of fish otoliths. Marine Biology, 2: 105-113.
6. Jain A., K. Nandakumar, A., Ross, Score Normalisation in Multimodal Biometric Systems, Pattern Recognition 38(2005): 2270-2285.
7. Lomparte A., J. Lleonart, 1992. Otolith size changes related with body growth, habitat depth and temperature. Environmental biology of fishes 37: 292-306.
8. Popper A.N., Z. Lu, 2000. Structure-function relationships in fish otolith organs. Fisheries research 46: 15-25.
9. Thenevaz P. , U.E. Ruttimann, M. Unser, A pyramid approach to subpixel registration based on intensity, IEEE Transactions on Image Processing, vol. 7, no. 1, pp. 27-41, January 1998.
10. http://bigwww.epf.ch/thenevaz/turboreg, Turboreg Plugin, last checked on 30/5/06
11. http://www.marine.ie/oto/form+and+structure/otolith+forms/index.htm, Otolith forms, last checked on 30/5/06

General Drawing of the Integrated Framework for Security Governance

Heejun Park[1], Sangkyun Kim[2], and Hong Joo Lee[3]

[1] Department of Information and Industrial Engineering, Yonsei University, 134, Shinchondong, Seodaemoongu, Seoul 129-749, South Korea
h.park@yonsei.ac.kr
[2] Somansa Co., Ltd., 16, Yangpyeongdong 3ga, Yeongdeungpogu, Seoul 150-103, South Korea
saviourkim@somansa.com
[3] The Liberal Arts School, Dankook University, 147, Hannamro, Yongsangu, Seoul 140-714, South Korea
blue1024@dankook.ac.kr

Abstract. To provide the structured approach of the security governance to corporate executives is the purpose of this paper. Previous studies on the governance and security management including international standards, methods for risk analysis, guideline for security policy were reviewed to design the components and requirements of the framework of the security governance. Finally, the framework for the security governance, which consists of four domains and two categories of relationship, is suggested considering the requirements of the framework including three perspectives of an architecture, domain, and presentation. It is believed that, with this framework, corporate executives could create greater productivity gains and cost efficiencies from information security.

1 Literature Review

1.1 Governance

Governance refers to the process whereby elements in society wield power and authority, and influence and enact policies and decisions concerning public life, and economic and social development [1]. Yahoo's dictionary defines governance as "the act, process, or power of governing." Table 1 summarizes previous research and definitions of governance including the enterprise governance, IT governance, and security governance.

The benefits gained from introducing effective governance strategy to an IT department are considerable. The holistic view of IT governance integrates principles, people, processes, and performance into a seamless operational whole [2]. IT governance specifies the decision-making authority and accountability to encourage desirable behaviors in the use of IT. IT governance provides a framework in which the decisions made about IT issues are aligned with the overall business strategy and culture of the enterprise [3].

B. Gabrys, R.J. Howlett, and L.C. Jain (Eds.): KES 2006, Part I, LNAI 4251, pp. 1234–1241, 2006.
© Springer-Verlag Berlin Heidelberg 2006

Table 1. Definitions of governance

Category	Research	Definition
Enterprise governance	IT Governance Institute [9]	Set of responsibilities and practices exercised by the board and executive management with the goal of providing strategic direction, ensuring that objectives are achieved, ascertaining that risks are managed appropriately and verifying that the enterprise's resources are used responsibly.
	OECD [10]	It involves a set of relationships between a company's management, its board, its shareholders, and other stakeholders. It also provides the structure through which the objectives of the company are set, and the means of attaining those objectives and monitoring performance are determined. It should provide proper incentives for the board and management to pursue objectives that are in the interests of the company and its shareholders and should facilitate effective monitoring.
IT governance	Appel [2]	The principles, processes, people, and performance metrics enacted to enable freedom of thinking and action without compromising the overall objectives of the organization.
	Neela and Mahoney [11]	IS governance is the mechanism for assigning decision rights and creating an accountability framework that drives desirable behavior in the use of IT.
	Allen [12]	The actions required to align IT with enterprise objectives and ensure that IT investment decisions and performance measures demonstrate the value of IT toward meeting these.
	IT Governance Institute [13]	The leadership, organizational structures, and processes that ensure that the enterprise's IT sustains and extends the enterprise's strategies and objectives.
Security governance	Moulton and Coles [14]	It is the establishment and maintenance of the control environment to manage the risks relating to the confidentiality, integrity and availability of information and its supporting processes and systems.

Table 2. Limitations of the security management

Success factors of Governance [11, 15, 16]	Related research based on security management	Limitations (Additional requirements of security governance)
Adequate participation by business management (align decision-making authority with the domain)	[17, 18, 19]	Focuses on the provision of decision-making factors; Concentrates on the technology-oriented factors
Clearly defined governance processes (integrate the governance mechanisms and evolve them)	[20, 21, 22, 23]	Focuses on the provision of lifecycle of security controls; Lacks in the integration of the governance mechanism and various participants of enterprise security
Clarify stakeholders' roles	Not available	Does not provide the relationship and responsibility of stakeholders
Measure the effectiveness of governance	[24, 25, 26, 27, 28, 29, 30, 31, 32]	Lacks in the provision of strategic benefits and overall performance
Facilitate the evolution of governance	[33]	Only provides the maturity model of security engineering (doest not provide a virtuous cycle of enterprise security)
Clearly articulated goals	[17, 19, 34, 35, 36, 37]	Lacks in the alignment of security strategy with business strategy
Resolution of cultural issues	Not available	Does not consider the relationship among the participants of enterprise security

1.2 Security Management

Solms stated that 'Information security is a direct corporate governance responsibility and lies squarely on the shoulders of the Board of the company [4].' Conner and Coviello stated that 'Corporate governance consists of the set of policies and internal controls by which organizations, irrespective of size or form, are directed and managed. Information security governance is a subset of organizations' overall (corporate) governance program [5].'

According to Solms, in the process of establishing environments for information security governance, companies are realizing that is it preferable to follow some type of internationally recognized reference framework for establishing an information security governance environment, rather than doing it ad hoc [6]. By using the information security governance framework, CEOs and boards of directors will create a safer business community internally and for their customers and others interconnected throughout the critical infrastructure. In aggregate, such measures serve as an executive call to action that will also help better protect our nation's security. Information security governance will provide organizations far greater benefits than just legal or regulatory compliance. Robust security serves as a catalyst to even greater productivity gains and cost efficiencies for businesses, customers, citizens and governments during times of crisis and normal operations [7].

IT Governance Institute summarized the necessities of security governance: Risks and threats are real and could have significant impact on the enterprise; Effective information security requires coordinated and integrated action from the top down; IT investments can be very substantial and easily misdirected; Cultural and organizational factors are equally important; Rules and priorities need to be established and enforced; Trust needs to be demonstrated toward trading partners while exchanging electronic transactions; Trust in reliability in system security needs to be demonstrated to all stakeholders; Security incidents are likely to be exposed to the public; Reputation damage can be considerable [8].

There is lots of research on the security management. However, research on the security management has some limitations to satisfy the necessities of security governance as described in table 2.

2 Integrated Framework for Security Governance

Table 3 summarizes the requirements of the framework of the security governance considering the success factors of the security governance, the limitations of the security management, and existing concepts of enterprise, IT and security governance.

Based on the requirements of table 3, the framework of the security governance is suggested as illustrated in figure 1.

The security governance framework consists of four domains and two relationship categories. Domains have several objects respectively. Objects consist of components which should be resolved or provided to govern the enterprise security. The relationship among the objects of the security governance framework has two categories of harmonization and flywheel. The harmonization category governs the relationship among the performance, security, and community domain. The harmonization category deals with the problems of social, organizational, and human factors of enterprise security. The flywheel category governs the relationship between the performance domain and security domain. The flywheel category deals with the virtuous cycle of enterprise security.

Table 3. Requirements of the security governance framework

View	Requirements
Architecture	● Domain which adjusts the partitions of every object of enterprise security ● Clear relationship among domains
Domain	● Every participants of enterprise security should be considered ● Characteristics of business information ● Performance management should provide the cost and benefit factors ● Security controls and strategies should be subdivided in further detail
Presentation	● Illustration which shows a bird-eye view of the security governance framework should be provided ● Structured table which describes every object and component of domain, and their characteristics should be provided

3 Conclusion

This paper suggests the framework of the security governance. Following studies are reviewed to design this framework: 1) Definition of enterprise governance, IT governance, and security governance. 2) Previous research on security management. 3) Implications of the governance concept.

Considering the fulfillments and limitations of previous research, the requirements and architectures of the security governance framework are suggested. Finally, the framework of the security governance which consists of four domains including eleven objects and two relative categories including eight relationships is suggested.

Managerial implications of the security governance framework are summarized as follow: 1) It integrates strategies, security controls, performance, and communities into a seamless operational whole. 2) It specifies the decision-making authority and accountability to encourage desirable behaviors. 3) Robust governance of enterprise security creates greater productivity gains and cost efficiencies for businesses.

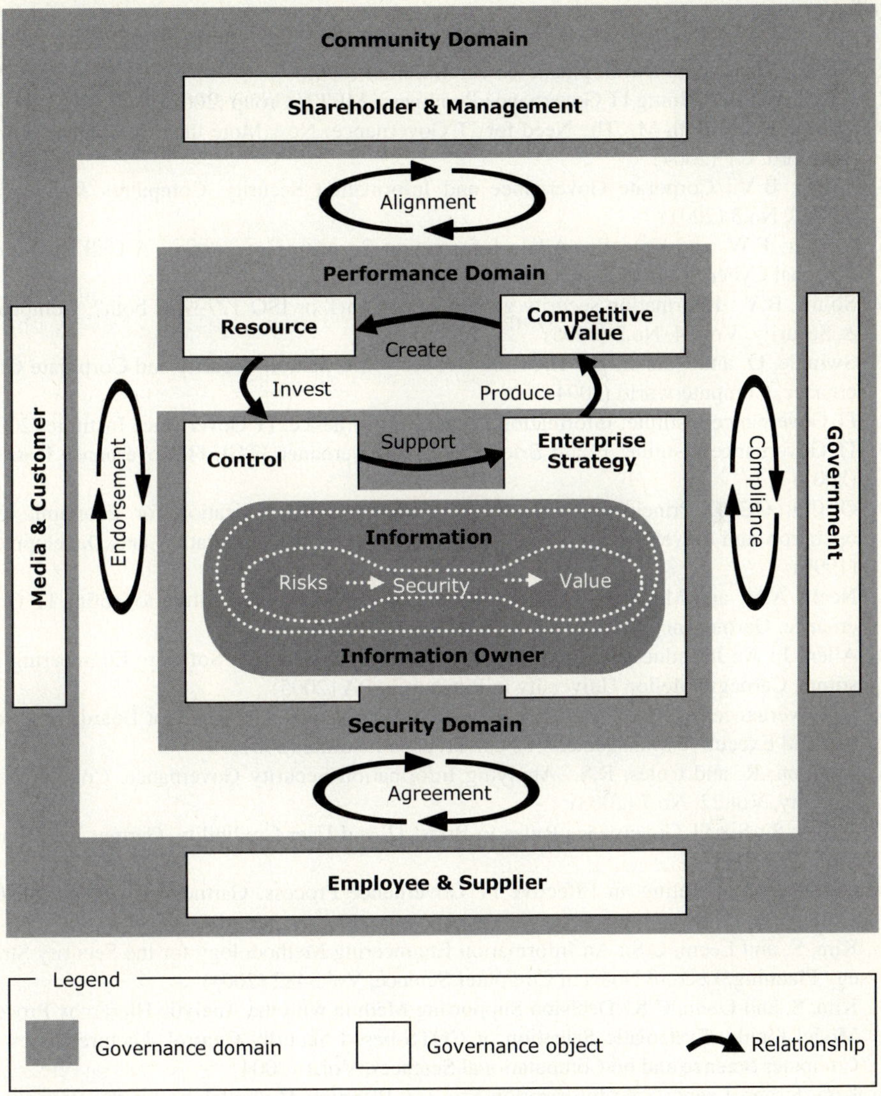

Fig. 1. Security governance framework

Limitations and further research issues are: 1) Detailed descriptions of each component and relationship should be provided. 2) Practical guidelines on the governance of each component and relationship should be provided. 3) Best practices resulted from the operation of this governance framework should be accumulated and influenced to ameliorate this framework. 4) Methods to keep in touch with an enterprise or IT governance strategy should be studied.

References

1. GDRC: The Global Development Research Center (2005)
2. Appel, W.: Redefining IT Governance Readiness. META Group (2005)
3. Dallas, S. and Bell, M.: The Need for IT Governance: Now More than Ever. Gartner Inc., Stamford, CT (2004)
4. Solms, B.V.: Corporate Governance and Information Security. Computers & Security, Vol.20, No.3 (2001)
5. Conner, F.W. and Coviello, A.W.: Information Security Governance: A Call to Action. National Cyber Security Summit Task Force (2004)
6. Solms, B.V.: Information security governance: CobiT or ISO 17799 or both?. Computers & Security, Vol.24, No.2 (2005)
7. Swindle, O. and Conner, B.: The Link Between Information Security and Corporate Governance. Computerworld (2004)
8. IT Governance Institute: Information Security Governance. IT Governance Institute (2004)
9. IT Governance Institute: Board Briefing on IT Governance ITGI. IT Governance Institute (2001)
10. OECD: OECD Principles of Corporate Governance, Organization for Economic Co-operation and Development. Organisation for Economic Co-operation and Development (1999)
11. Neela, A.M. and Mahoney, J.: Work With, Not Against, Your Culture to Refine IT Governance. Gartner Inc., Stamford, CT (2003)
12. Allen, J.: An Introduction to Governing for Enterprise Security. Software Engineering Institute, Carnegie Mellon University in Pittsburgh, PA (2005)
13. IT Governance Institute: Information Security Governance: Guidance for Boards of Directors and Executive Management. IT Governance Institute (2001)
14. Moulton, R. and Coles, R.S.: Applying Information Security Governance. Computers & Security, Vol.22, No.7 (2003)
15. Dallas, S.: Six IT Governance Rules to Boost IT and User Credibility. Gartner Inc., Stamford, CT (2002)
16. Gerrard, M.: Creating an Effective IT Governance Process. Gartner Inc., Stamford, CT (2003)
17. Kim, S. and Leem, C.S.: An Information Engineering Methodology for the Security Strategy Planning. Lecture Notes in Computer Science, Vol.3482 (2004)
18. Kim, S. and Leem, C.S.: Decision Supporting Method with the Analytic Hierarchy Process Model for the Systematic Selection of COTS-based Security Control. Lecture Series on Computer Science and on Computational Sciences, Vol.1 (2004)
19. Kim, S. and Leem, C.S.: Information Strategy Planning Methodology for the Security of Information Systems. ICCIE 2004, Cheju (2004)
20. ISO, ISO13335-1: Information Technology - Guidelines for the Management of IT Security - Part 1: Concepts and Models for IT Security. International Organization for Standardization
21. NIST: An Introduction to Computer Security: The NIST Handbook. NIST, Gaithersburg, MD (1995)
22. Henze, D.: IT Baseline Protection Manual. UK (2000)
23. Kim, S., Choi, S.S. and Leem, C.S.: An Integrated Framework for Secure E-business Models and Their Implementation. INFORMS'99, Seoul (1999)
24. Geer, D.E.: Making Choices to Show ROI. Secure Business Quarterly, Vol.1, No,2 (2001)

25. Scott, D.: Security Investment Justification and Success Factors. Gartner Inc., Stamford, CT (1998)
26. Blakley, B.: Returns on Security Investment: An Imprecise But Necessary Calculation. Secure Business Quarterly, Vol.1, No.2 (2001)
27. Malik, W.: A Security Funding Strategy. Gartner Inc., Stamford, CT (2001)
28. Power, R.: CSI/FBI Computer Crime and Security Survey. Computer Security Issues & Trends (2002)
29. Bates, R.J.: Disaster Recovery Planning. McGraw-Hill, New York, NY (1991)
30. Witty, R.J, Girard, J., Graff, J.W., Hallawell, A., Hildreth, B., MacDonald, N., Malik, W.J., Pescatore, J., Reynolds, M., Russell, K., Wheatman, V., Dubiel, J.P. and Weintraub, A.: The Price of Information Security. Gartner Inc., Stamford, CT (2001)
31. Harris, S.: CISSP All-in-One Exam Guide, Second Edition. McGraw-Hill, New York, NY (2003)
32. Roper, C.A.: Risk Management for Security Professionals. Butterworth-Heinemann, Boston, MA (1999)
33. SEI: A Systems Engineering Capability Maturity Model, Version 2.0. Software Engineering Institute, Carnegie Mellon University in Pittsburgh, PA (1999)
34. Rex, R.K., Charles, S.A. and Houston, C.H.: Risk Analysis for Information Technology. Journal of Management Information Systems, Vol.8, No.1 (1991)
35. Kim, S. and Leem, C.S.: Implementation of the Security System for Instant Messengers. Lecture Notes in Computer Science, Vol.3314 (2004)
36. Kim, S. and Leem, C.S.: Security of the Internet-based Instant Messenger: Risks and Safeguards. Internet Research: Electronic Networking Applications and Policy, Vol.15, No.1 (2005)
37. Ron, W.: EDP Auditing: Conceptual Foundations and Practice. McGraw-Hill, New York, NY (1988)

A Study on the Rain Attenuation Prediction Model for Ubiquitous Computing Environments in Korea

Dong You Choi

#36 Naedok-dong, Sangdang-gu, Cheongju-si 360-764 Korea
Division of Electronics & Information Engineering, Cheongju University
dy_choi@cju.ac.kr

Abstract. In this paper, I present the results of the measurements of rain-induced attenuation in the vertically polarized signal propagating at 12.25GHz during certain rain events, which occurred in the rainy wet season of the year 2001 at Yong-in, Korea (ITU-R rain zone K). The total experimentally measured attenuation over the link path was measured experimentally has been compared with that obtained using the International Telecommunication Union Radio Communication Sector (ITU-R) model. I also propose a prediction model which can predict the rain attenuation in satellite-earth communication links more conveniently and efficiently, by using the 1-h rain rate without conversion.

1 Introduction

In microwave communication systems, vapor, fog, oxygen, rain, and several other gases which make up the air, cause propagation attenuation. Among these different atmospheric constituents, rain-induced attenuation is the most severe, except for the degradations that occur near 22GHz or 60GHz due to vapor or oxygen. In particular, if a high frequency band of 10GHz and over is used, the propagation attenuation due to rainfall is further increased [1][2].

Many rain attenuation prediction models have been developed including the ITU-R model [3][4], in order to permit the design of effect satellite-earth communication links. I have been measuring a satellite system in the 12.25GHz band [5]. However, since these models were based on statistics pertaining to rain attenuation under a rainfall environment native to each country, and thus having regional peculiarities, they are not considered to be adapted to the effects of rain attenuation under the Korean rainfall environment. Furthermore, because the degree of rain attenuation depends on the frequency being used, and a high frequency band is used in Korea, the prediction results obtained using these models are likely to be wide of the mark when it comes to taking into account the effects of rain attenuation under the Korean rainfall environment.

On the other hand, it is recommended in the ITU-R model, that the rainfall be measured at intervals of one minute, in order to determine the rain rate [4]. However, in many countries, it is practically impossible to obtain 1-min rain rate data over period of several years. Moreover, it is inconvenient to convert the available data into a 1-min rain rate and to apply this to a rain attenuation prediction model.

B. Gabrys, R.J. Howlett, and L.C. Jain (Eds.): KES 2006, Part I, LNAI 4251, pp. 1242 – 1249, 2006.
© Springer-Verlag Berlin Heidelberg 2006

For this reason, in this paper I propose a 1-h rain rate-based attenuation prediction model, in order to provide a more convenient and efficient method of predicting rain attenuation.

2 Received Beacon Signal Level and Rain Rate

In this study, I selected Koreasat-3, which uses the Ku-band (14GHz/12GHz) frequency, and analyzed the beacon signal level data according to the rain rate from June to August in 2001 [5].

To measure the rain rate, I used the rain measurement system, which was installed when the Yong-in Satellite Control Office was established. To measure the beacon signal level, which is always received at a certain level from a satellite, the controlling equipment of Koreasat-3 was used. Block diagrams for the two measurement systems are shown Figs. 1 and 2.

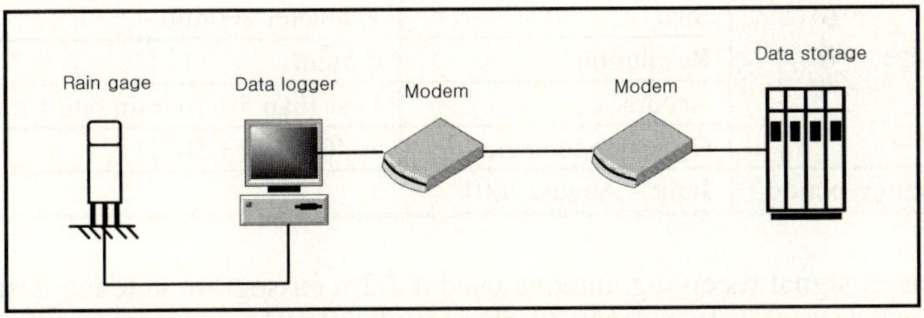

Fig. 1. Experimental system used for measuring the rain rate

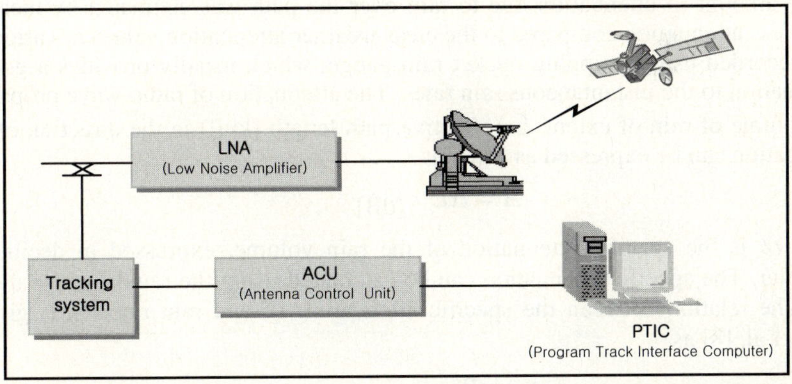

Fig. 2. Experimental system used for measuring the beacon signal level

As shown in Fig. 1, the accumulated rain rate data is first collected in a data collector and is then saved in a computer. The rain rate data may be saved in either 10-min or 10-sec intervals. Since 10-min interval data was not sufficiently accurate for use with the satellite beacon signal level, the 10-sec interval data collection was used. The experimental system saves the received beacon signal level at intervals of 1-min, as shown in Fig. 2.

Table 1. Measurement and Experimental specifications

System location	Latitude(Yong-in)	37.43°N
	Longitude(Koreasat-3)	116°E
	Elevation angle	45.20°
	Azimuth angle	198.1°
	Sea leve	0.142km
Climate zone	ITU-R model	K-zone
Down link	Polarization	Dual liner
	EIRP	34dBW
	Frequency	12.25GHz
Antenna	Type	Cassegrain
	Diameter	7.2m
Rain gage	Type	Tipping bucket
	Size	Diameter 200mm
	Resolution	0.5mm
	Accuracy	Less than 5% (in rain rate 10mm/hr)
	Operative temperature	−40℃ ~ +50℃
Observation period	Jung ~ August 2001	

The beacon signal receiving antenna used a 7.2m cassegrain antenna designed specifically for Koreasat-3. In order to compensate for the changes in power level caused by the perturbation of the transmission from the satellite, which is located in a geostationary orbit, a tracking system using the steptrack tracking method was used.

The amount of attenuation due to rain over the path was estimated by measuring the excess attenuation compared to the clear weather attenuation values at various rain rates recorded using a tipping bucket rain gauge, which usually provides a good approximation to the instantaneous rain rates. The attenuation of radio wave propagation in a volume of rain of extent L (effective path length [km]) in the direction of wave propagation can be expressed as [6], [7]

$$A = \alpha L \quad \text{[dB]} \tag{1}$$

where α is the specific attenuation of the rain volume, expressed in decibels per kilometer. The specific attenuation can be calculated using the raindrop size distribution. The relation between the specific attenuation α and rain rate R, is given by Olsen et al. [8] as

$$\alpha = aR^b \quad \text{[dB/km]} \tag{2}$$

The values of coefficient a and b depend upon the frequency, rain temperature and drop size distribution. In the present work, the coefficients a, b and L were chosen for the ITU-R model at a frequency of 12.25GHz, by comparing the measured values of the rain attenuation over the link path.

Data corresponding to the signal levels and rain rates during the propagation and communication times were collected and analyzed during the period from June to August, 2001.

The rain attenuation is obtained by subtracting the reference level from the measured value of the received beacon signal level. The reference level is obtained by averaging the beacon signal level data obtained during a period in which there is no rain. The relation between the rain attenuation and the rain rate is shown in Table 5. The 1st to 8th set of statistical data represent the mean values of the data collected on rainy days at intervals of 3~4 days after excluding the minimum and maximum values. The total statistical data was obtained by summing all of the mean data values during each measurement period. Fig. 3 shows the log-normal approximated time percentage curves of rain attenuation derived from data measured in Yong-in.

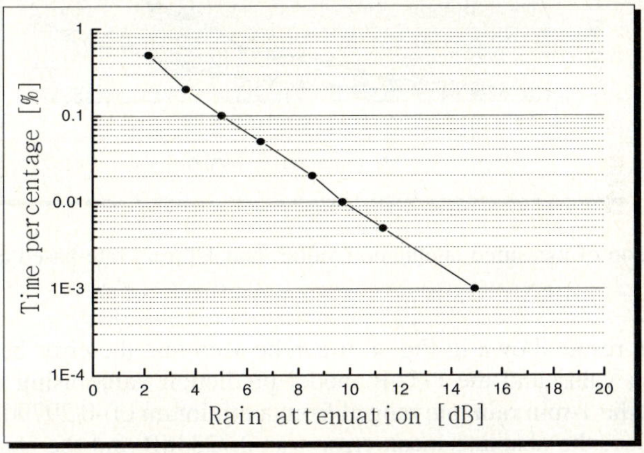

Fig. 3. Measured attenuation data

3 Proposed of Prediction Model

It is recommended in the ITU-R model to measure the rainfall at 1-min intervals, in order to determine or anticipate the rain rate [4]. However, in many countries including Korea, it is very difficult to obtain 1-min rain rate data for a period of several years. Furthermore, the 1-h rain rate data needs to be converted, in order to be able to apply it to the rain attenuation prediction model. Thus, in this paper, I propose to use A_{PHM} as determined by Expression (3) on the basis of the ITU-R model, and which can predict rain attenuation more conveniently and efficiently, by using the 1-h rain rate without conversion. Then, K_{HM} is used as a compensatory factor, in order to approximate the measured results. Also, to convert the 1-min rain rate data into the 1-h rain rate data, Moupfouma-Martin model [9] was used.

$$A_{PHM} = A_{ITU-R\ model} - K_{HM} \quad [\text{dB}] \tag{3}$$

where $A_{ITU-R\ model}$ is rain attenuation prediction value obtained from the ITU-R model, and K_{HM} is a compensatory factor.

To determine the compensatory factor, K_{HM}, the 1-min rain rate data at the time of the measurement was converted into the 1-h rain rate data and then the rain attenuation was predicted using the ITU-R model. The prediction results are shown in Fig. 4.

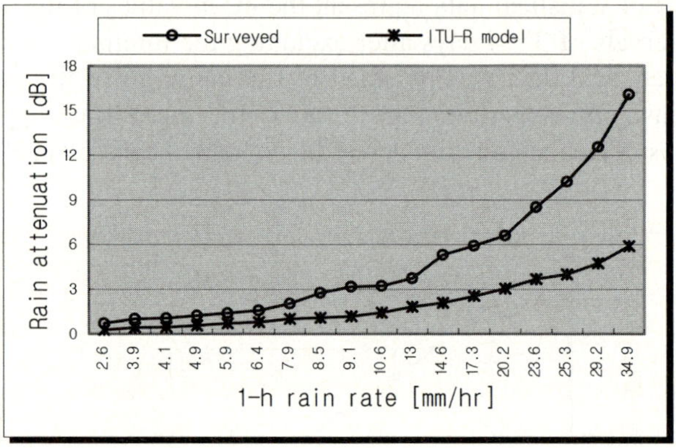

Fig. 4. Comparison of measured attenuation values and 1-h rain rate-based attenuation prediction values

Based on the result shown in Fig. 4, it can be seen that the error between the measured attenuation value and the ITU-R model prediction value using the 1-h rain rate converted from the 1-min rain rate ranged from a minimum of -0.2979dB to a maximum -10.1999dB. Also, the absolute mean error was 2.7234dB and the absolute error standard deviation was 2.7713, suggesting that there was a big error in the predicted value.

Fig. 5 shows the result of the regression analysis of the error between the measured attenuation value and the rain attenuation prediction value of the 1-h rain rate-based ITU-R model in order to determine the value of K_{HM}.

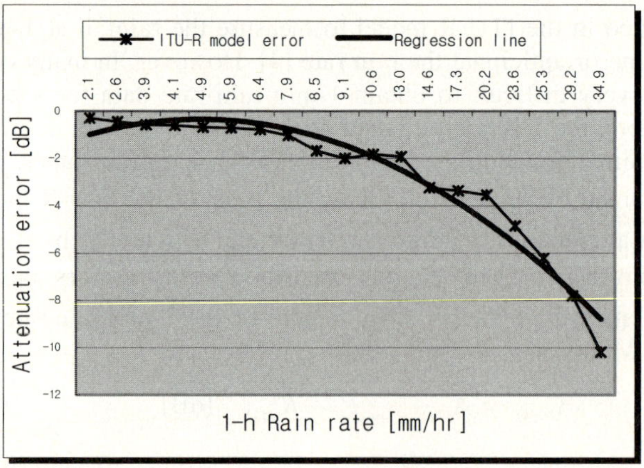

Fig. 5. Regression analysis result of the error of the 1-h rain rate-based ITU-R model prediction values

Equation (4), which represents a regression distribution curve was modeled mathe matically, was decided as a compensation factor K_{HM}.

$$K_{HM} = -0.21359 - (0.077405 \times R_H) - (0.0059611 \times R_H^{\ 2}) \quad [\mathrm{dB}] \tag{4}$$

where R_H is the 1-h rain rate [mm/hr].

In order to predict the rain attenuation in Korea more conveniently and efficiently b y means of the compensatory factor K_{HM}, so determined, this paper proposes a rain at tenuation prediction model, A_{PHM}, as described in Expression (5).

$$A_{PHM} = A_{ITU-R\ \ model} \\ -\{-0.21359 - (0.077405 \times R_H) - (0.0059611 \times R_H^{\ 2})\} \quad [\mathrm{dB}] \tag{5}$$

Fig. 6 shows the simulation result for the rain attenuation prediction model [Expres sion (5)] on the basis of the specifications given in Table 1.

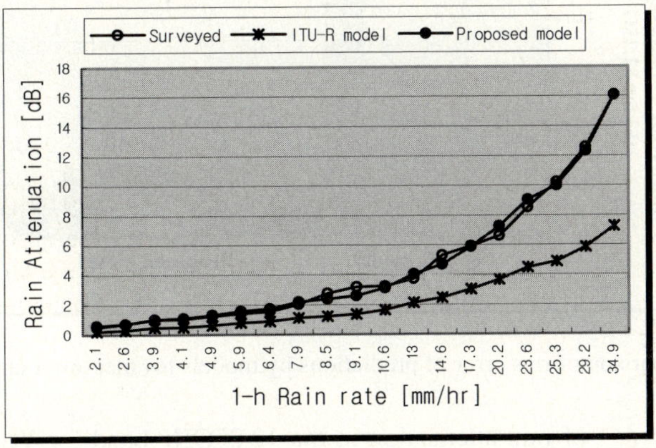

Fig. 6. Simulation result for the proposed rain attenuation prediction model (A_{PHM})

Based on the result shown in Fig. 6, the error in the prediction value based on the measured attenuation value ranged from a minimum of -0.2979dB to a maximum of -10.1999dB for the ITU-R model, whereas that for the proposed rain attenuation prediction model ranged from a minimum of +0.0161dB to a maximum of +0.6673dB, thus demonstrating a considerable decrease in the case of the proposed rain attenuation prediction model. In addition, the absolute mean error decreased from 2.7234dB to 0.2362dB and the absolute error standard deviation decreased from 2.7713 to 0.2161. These results demonstrate reasonableness of the compensatory factory, K_{HM}.

Usually, a method using logarithm errors recommended by ITU-R P.311-11 [10] is used to compare predicted and measured rain attenuation data. Comparison are based on the percentage prediction error defined as

$$e_i = \frac{A_{pi} - A_{mi}}{A_{mi}} \times 100 \tag{6}$$

where A_p is the prediction attenuation and A_m is the measured attenuation (both in decibels) and i represents the percent probability level at which the prediction error is estimated. The root-mean-square (rms) error is used as the metric to judge the overall fit of the prediction with the measurement and is given by

$$e_{rms} = \sqrt{\langle e_i^2 \rangle} \tag{7}$$

where $\langle \; \rangle$ denotes averaging [11].

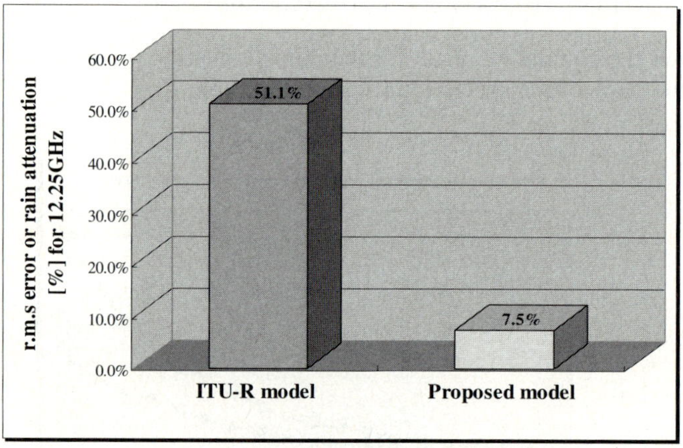

Fig. 7. Comparison of rms error of predictions by models for measured rain attenuation

Fig 7 shows e_{rms} of rain attenuation in the 12.25GHz band for Yong-in. From the figure, it is found that the ITU-R model gives an e_{rms} of 51.1%. In contrast, the proposed model gives an e_{rms} of 7.5%. This results show that the proposed model gives better effectiveness and improvement than ITU-R model from the viewpoint f overall fit.

4 Conclusions

Due to the increasing use of radio wave transmissions, the range of the frequency bands employed is being extended from the existing UHF band to the microwave and millimeter wave bands. However, the higher the frequency band, the higher the rain attenuation. Thus, the effect of rain attenuation should be given serious consideration.

It is recommended in the ITU-R model that a 1-min interval be used to measure rainfall, in order to determine or anticipate the rain rate. However, in practice, it is very difficult to maintain 1-min rain rate data over a period of several years. Furthermore, the 1-h rain rate data needs to be converted into 1-min rain rate data, in order to apply this information to the existing rain attenuation prediction models. In order to overcome thus inconvenience, in this paper, I propose a novel prediction model, which can be used to predict rain attenuation more conveniently and efficiently for in microwave communication systems, using the 1-h rain rate without conversion.

The main results and conclusions of this study can be summarized as fellows.

1) From the comparison of the proposed prediction model with the ITU-R model, based on the measured rain attenuation values, it was found that the proposed model was able to reduce the values of the absolute mean error and the absolute error standard deviation (by 2.4872dB and 2.5552, respectively).
2) An evaluation of rms prediction error of rain attenuation shows that the proposed model gives better effectiveness and improvement than ITU-R model for Korea.

Considering that these measurements were conducted over a short period of time(June ~ August in 2001), further studies need to be done using long-term rain attenuation measurement data for several frequencies, as well as rain attenuation data recorded throughout the country, in order to develop a specific model which is ideally adapted to this environment.

References

1. T. C. Ramadorai: Rain attenuation and prediction in the Satellite-Earth Path. in Proc. Workshop HF VHF and Microwave Communications. New Delhi, India. (Feb. 1987)
2. R. K. Crane: Electromagnetic wave propagation through rain. John Wiley & Sons, Inc. (1996) 1-4
3. ITU-R: Propagation data and prediction methods required for the design of Earth-space Telecommunications systems. Rec. P.618, ITU-R (1997)
4. ITU-R: Specific attenuation model for rain for use in prediction methods. Rec. P.838, ITU-R (1999)
5. Dong You Choi: Measurement of Rain Attenuation of Microwaves at 12.25GHz in Korea. Lecture Notes in Computer Science, Vol. 3462. Springer-Verlag, Berlin Heidelberg New York (2005) 1353-1356
6. CCIR: Propagation data and prediction methods required for terrestrial line of sight systems. Rep. 564-3; Rep. and Recommendations of CCIR. Int. Telecommunication Union (1974-1986)
7. CCIR: Propagation data and prediction methods required for earth-space telecommunication systems. in proc. Plenary Assembly, vol. 5, Geneva, Switzerland, Rep. 564-4 (1990) 447-505
8. R. L. Olsen, D. V. Rogers, and D. B. Hodge: The aR^b relation in the calculation of rain attenuation. IEEE Trans. Antennas propagation, Vol. AP-26 (Mar. 1978) 318-329
9. F. Moupfouma, L. Martin: Model of Rainfall Rate Distribution for Radio System Design. in Proc. IEEE, Vol. 132, Pt. H. No. 1 (Feb. 1985) 39-43
10. Recommendation ITU-R P.311-11 (2003)
11. Asoka Dissanayake, Jeremy Allnutt, and Fatim Haidara: A Prediction Model that Combines Rain Attenuation and Other Propagation Impairments Along Earth-Satellite Paths. IEEE Transactions on Antennas and Propagation, Vol. 45, No. 10 (October 1997) 1546-1558

A Framework for Ensuring Security in Ubiquitous Computing Environment Based on Security Engineering Approach

JooYoung Lee and HoWon Kim

Electronics and Telecommunications Research Institute (ETRI), 161 Gajeong-Dong
YuSeong-Gu, DaeJeon, 305-700, KOREA
{joolee, khw}@etri.re.kr
http://www.etri.re.kr

Abstract. The goal of this paper is to provide a framework for ensuring security in ubiquitous computing environment based on security engineering approach. Security engineering is about building systems to remain dependable in the face of malice, error or mischance. In order to do this, we propose a framework which includes the concept of security state and security flow. The combinations of security state and security flow represent a unique viewpoint and a particular pattern for consideration of security services. We can analyze and identify security issues and threats from these combinations and patterns and provide suitable security services for concerns. For this purpose, we have applied our approach to RFID service network, which is one of the representative examples that are to practically realize ubiquitous computing environment.

1 Introduction

Ubiquitous computing environment[1] is to allow people to use customized services suited their tastes or needs using intelligent and tiny devices through wired or wireless network regardless of the when and the where.

However, as we have pervasive and invisible systems and services for providing convenient ubiquitous lives to us, it essentially involves security-privacy paradox, which it is requisite for collecting information on users, such as identities, preferences, and physical status while requiring security including privacy and data protection. Besides, many ubiquitous computing services are based on the situation awareness for autonomous and automatic services as in [2]. It means that information or applications are not stayed in a specific device, but delivered from one device to another for collecting and sharing information. Such collection and sharing are indispensable for self-growing feature which allows the customized services to sense the varying and evolving environment and to respond it.

Additionally, as the number of devices and systems has increased in ubiquitous computing environment, there would be much room for having difficulties managing them and for laying themselves open to attack. As examples, it can be in danger by changing locations of the managed devices without notice and the exhaustion of device resources paralyzes systems. Such a paralysis can fall society into great confusion.

B. Gabrys, R.J. Howlett, and L.C. Jain (Eds.): KES 2006, Part I, LNAI 4251, pp. 1250 – 1258, 2006.
© Springer-Verlag Berlin Heidelberg 2006

Although a lot of solutions to these problems have been proposed and developed, there is no way to ensure whether the coordinated security solutions for ubiquitous computing environment are properly deployed for providing entire end-to-end security or not. In order to ensure it, we propose a framework which includes the concept of security state and security flow.

This approach is based on the security engineering which is about building systems to remain dependable in the face of malice, error or mischance[3]. The combinations of the security state and the security flow represent a unique viewpoint and a particular pattern for consideration of security services. And then, we analyze and identify security issues and threats from these combinations and patterns and provide suitable security services for these concerns. For this purpose, we have applied our approach to RFID service network, which is one of the representative examples that are to practically realize ubiquitous computing environment.

The paper is organized as follows. A brief description of security engineering and the overview of RFID service network are given in Section 2. A framework for ensuring security is discussed in Section 3 and illustrated via a detailed example in Section 4. Section 5 provides concluding remarks and further works.

2 Related Works

In this section of paper, we describe what security engineering is and introduce RFID service network to apply our framework. The RFID technology is a representative example of ubiquitous computing environment at the present time.

2.1 Security Engineering in Ubiquitous Computing Environment

Technological advances, principally in the field of computers, have now allowed the creation of far more complex systems than before, with new and complex security problems. Because modern systems cut across many areas of human endeavor, security engineers not only need consider the mathematical and physical properties of systems; they also need to consider attacks on the people who use and form parts of those systems using social engineering attacks. Secure systems have to resist not only technical attacks, but also coercion, fraud, and deception by confidence tricksters.

Then, what security engineering is and why it is important in ubiquitous computing environment. According to [3], security engineering is about building systems to remain dependable in the face of malice, error or mischance. As a discipline, it focuses on the tools, processes and methods needed to design, implement and test complete systems, and to adapt existing systems as their environment evolves.

In the ubiquitous ear, new systems including small node and portable devices are often rapidly broken, and the same elementary mistakes are repeated in one application after another. It often takes four or five attempts to get a security design right, and that is far too much. For this reason, we think that security engineering is suitable for analyzing security threats and countermeasures for ubiquitous computing environment. Although security engineering requires multi-disciplinary approach, we will just focus on technical design of the system and its environment in this paper.

2.2 RFID Service Network Environment

Mark Weiser describes a vision of ubiquitous computing in which technology is seamlessly integrated into the environment and provides useful services to humans in their everyday lives[1]. The potential of Radio Frequency Identification (RFID) tags to contribute to the realization of this vision has been demonstrated by many researchers over the years.

RFID refers to the set of technologies that use radio waves for identifying objects or people[4]. A basic RFID system consists of a radio frequency tag, a combination of a microchip and an antenna, and a reader. The reader sends out electromagnetic waves, which are received by the tag's antenna. The tag sends the data, which is usually a serial number stored on the tag, by transmitting radio waves back to the reader.

The EPCglobal network is a set of global technical standards aimed at enabling automatic and instant identification of items in the supply chain and sharing the information throughout the supply chain[6,7]. The set of standards focuses on providing a numbering system for unique identification and define how data is stored and transferred. The EPCglobal network consists of five fundamental elements: the ID System (EPC Tags and Readers), Electronic Product Code (EPC), Object Name Service (ONS), Physical Markup Language (PML), and Savant.

EPC is a product identification code. Like many other product codes that are used in commerce, EPC identifies the manufacturer and product type. EPC is just an information reference; therefore, ONS has been proposed to translate EPC codes directly to IP (Internet Protocol) addresses. The ONS communicates with the computer system and tells the computer where to find information related to a specific EPC. The ONS matches RFID tags to information about the products to which they are attached. Savant is an enterprise software technology developed to provide middleware between RFID reader and databases. Savant sits between tag readers and enterprise applications in order to manage the vast amount of information retrieved from the tags[8].

3 Proposed Model

In this section of paper, we have introduced the concept of security state and security flow to provide security ensuring framework. The security state regards as a component of service network such like a server system, a client device and so on. It can be identified and categorized by its position and role. The security flow corresponding to the business flow is relevant to security requirement of the flow.

To approach the methodology to ensure service network security, we have divided the work into two steps: the preprocessing step and the application step. The preprocessing step is to define the service network environment and to analyze overall security requirements for it. The application step is to define the business flow of a specific application in the pre-defined service network environment, to find security requirement specifically for it, and to determine which security techniques are required.

In the first step of the preprocessing step, we define the service network environment and users who take part in activities in the defined service network. Next, we recognize the security states of service network and categorize users into a few groups according to their activities and roles.

Next, in order to examine how the states are related to others, we identify transitions among states and draw a state diagram to represent them. This state diagram shows interaction among components comprising service network. The directly connected two components in the diagram mean that they have direct communication channel.

We depict a state diagram per a user group because they have different work flow, and analyze security requirement to be needed when transition between any two states of diagram happens. After the process mentioned above is repeated for each user group, respectively, we integrate these diagrams and analyze their security requirement for a specific service network environment. That is a state diagram representing the service network environment and its security requirements analyzed in previous step. The preprocessing step mentioned above is summarized in Fig. 1.

1. Define service network environment and users
2. Identify security state
3. Categorize the users into a few groups according to their activities
4. For each user group
   4.1 depict state diagram
   4.2 analyze security requirement to be needed when transition between any two states of diagram happens
5. Integrate the results from step 4

Fig. 1. Preprocessing step. In this step, we have identified security states and depicted state diagram to analyze security requirements.

Subsequently, we take the application processing step to make the security flow. In the first step of this processing, we derive flows from the state diagram depicted in the preprocessing step and get them together according to the application scenario. The combined flow is a subset of the entire state diagram and models a specific use case.

Secondly, we derive security requirements corresponding to the combined flow from the security state diagram of the preprocessing step. We call the result the security flow for the particular application scenario. The security flow refers the service flow and security requirement of the specific application and is identified as combination of states. The security flow enables us to recognize security requirements and to determine which security technologies would be applied. Fig. 2 shows the summarized processing flow.

1. Define a specific application of service network environment.
2. Derive security flows from the state diagram of the preprocessing step.
3. Determine which security techniques will be deployed.
4. Check if applied security techniques are enough for defense to assumed t hreats.
5. Apply additional security techniques according to the result of step 4.

Fig. 2. Applying step. In this step, we have derived the security flow from the security state according to a particular application scenario.

4 Applying to RFID Service Network

In this section of paper, we have applied the proposed methodology to RFID service network in order to ensure its security. Particularly we have considered mobile RFID network environment in here. The mobile RFID network allows the users to carry their own RFID reader, to read RFID code from the tag, and to get information related to the code from the information server.

On mobile RFID service network environment, we can find 5 states: Tag state, Reader state, Application(EA) state, Information Server(IS) state, Object Naming Service(ONS) state. The tag state represents RFID tags which are attached to products and contain codes such as EPC or ISO. The reader state means a kind of reader getting information from RFID tags and including basic filters. It could be not only wired reader, but also wireless and sensible readers attached to mobile device. The application state regards as application programs that provide particular services using relevant RFID-related information to users and other applications. It would be simple contents services on the mobile phone as well as enterprise application for B2B. The information server state represents information servers which collects and manages not only static data related with RFID attached products, but also dynamic data captured when events occur. Additionally it offers responses to any query about the information. The ONS state represents an ONS server which has mapping table about relation of RFID tags and Information Servers' URL and provides this information according to requests.

Next, we have categorized users into three groups: service customer group, service provider group, authority group, and depicted state diagrams respectively. In the perspective of the service provider group, they require the EA component serving what they want and the IS component keeping static and dynamic information. These two components need direct connection to obtain some information from each other. Also, in the viewpoint of the service consumer group, a reader might need direct communication channel with ONS component to get URIs of IS component, where further information on the object may be found. Fig. 3 shows an example of panels depicting the state diagram for each user group and Fig. 4 presents a state diagram, which is to integrate three state diagrams of users.

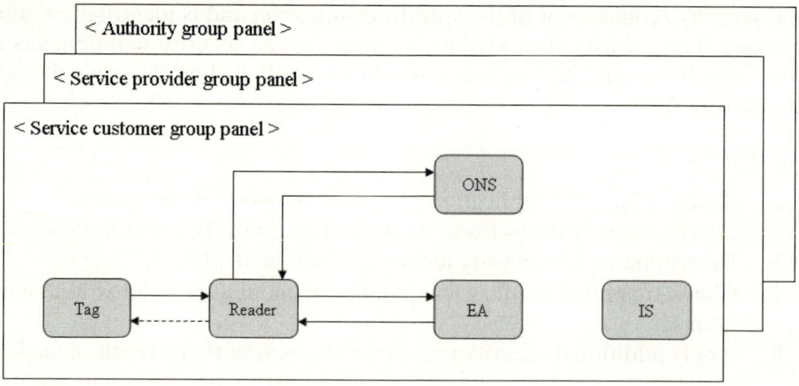

Fig. 3. An example of state diagrams for each user group. In here, we have depicted three panels for three user groups. They are different according to their perspective and role.

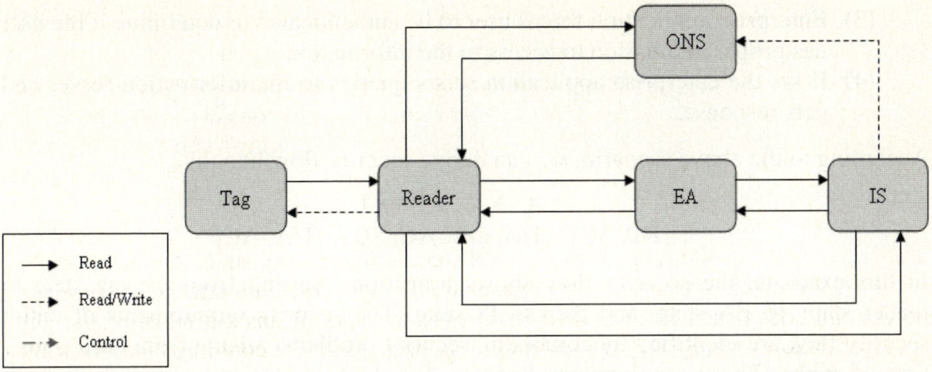

Fig. 4. Integrated security state diagram. We have joined state diagrams from Fig. 3 together. It shows states of mobile RFID service network and their relations.

From the example, we can analyze security requirements. For instance, when state transition from the reader state to EA state occurs, from viewpoint of service provider group, the EA forces the reader to be authenticated, needs access control mechanism, requires security channel between them. Besides, we would define additional security requirement according to application. Such security requirements according to state transition might be represented as shown in Table 1.

Table 1. Security Requirements. It shows security requirement according to state transition.

State Transition	Security Requirements
T→R	DoS Prevention (DP), Secure Channel (SC)
R←T	Device Authentication (DA), Authorization
R→E	User Authentication (UA), Device Authentication, Access Control (AC), Secure Channel
E←R	…
E→I	User Authentication, Access Control
I→E	…
R→I	…
I←R	…
R→O	…
O→R	…
I→O	…

Now, we can define a specific application scenario based on this mobile RFID service network environment. In here, we consider a simple situation, which a user carrying his mobile RFID reader are trying to read a RFID tag attached to a bag that is owned by other person. Then, security flow of this scenario is summarized briefly as follows:

(1) User scans a code saved in RFID tag attached to the bag using his RFID reader.
(2) User sends request with the code to the enterprise application in order to get information on the bag.

(3) Enterprise application forces user to be authenticated to determine if the user has proper permission to access to the information.
(4) If so, the enterprise application sends queries to the information server and gets response.

According to the above scenario, we can derive security flow like this:

$$T \rightarrow R \rightarrow E \rightarrow I$$
$$: \{DP, SC\}, \{UA, DA, AC, SC\}, \{UA, AC\}$$

In this example, the security flow shows transitions starting from the tag state to reader state, to EA state, and then to IS state. The security requirements of entire security flow are identified by combining security problems arising from each transition of states. These requirements become denial of service prevention technique, secure channel, user authentication technique, device authentication, access control, and so on. Therefore, for particular application service having transition of state like T→R→E→I we can find security techniques to meet the security requirements described above and assert whether security techniques are deployed to proper position for protecting assumed security threats or not.

Globally networked RFID service environment is shown in Fig. 5. We have applied our proposed methodology to it. As a result, we can ensure RFID security by addressing security issues and providing suitable security services and techniques to them. Security techniques to deploy are depicted it in Fig. 6.

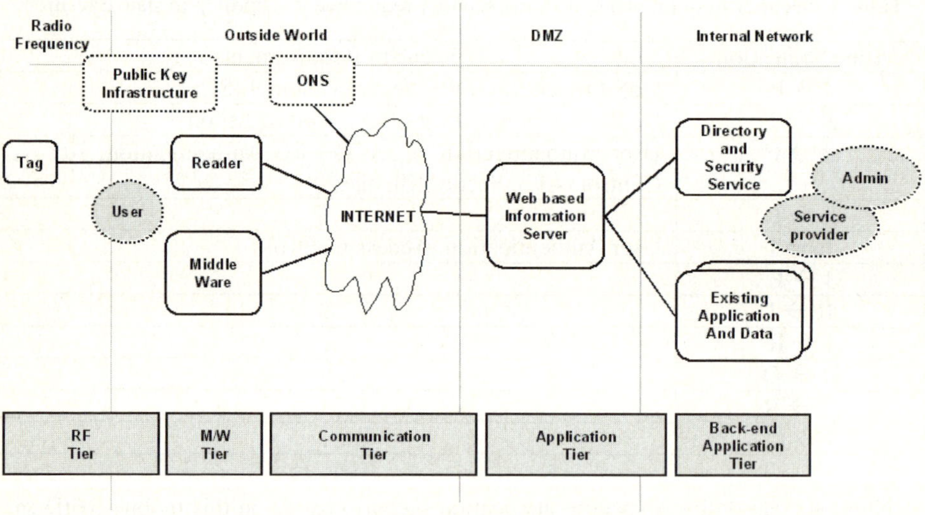

Fig. 5. Globally networked RFID service environment

In this example, we can find what kinds of security requirement are needed for each tier. The RF tier between the tag and the reader requires confidentiality and integrity and we are able to know that availability is important factor and key management scheme is also needed in the middleware tier. Additionally, privacy protection mechanism must be deployed through globally networked RFID service environment.

Through such processes, we can induce security requirements and make decision about what kind of security solutions are deployed for them, and then simply verify if security hole exists or not.

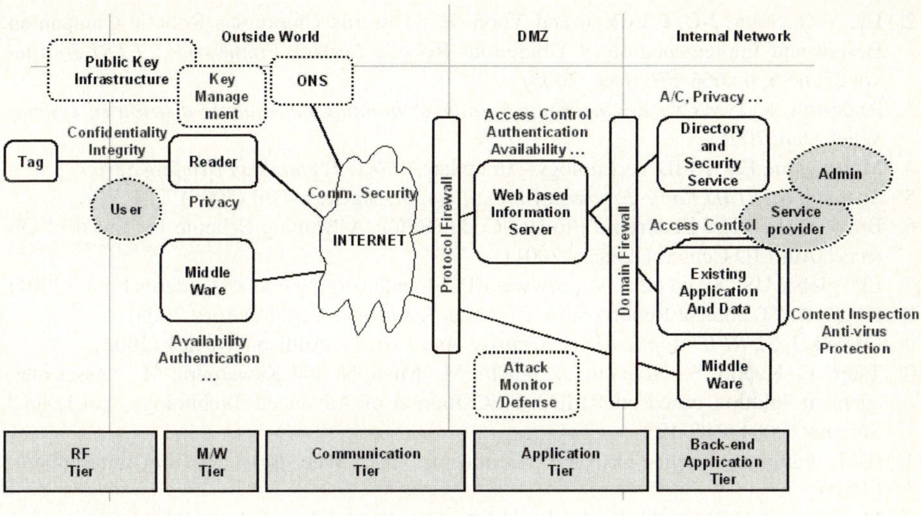

Fig. 6. Applying our approach to address RFID service network security

5 Conclusion and Further Works

In this paper, we have proposed a framework for ensuring security in ubiquitous computing environment. Our approach based on security engineering approach will play important role to ensure security for ubiquitous applications and services because it is difficult to ensure whether the coordinated security solutions are properly deployed for providing entire end-to-end security or not and currently, there is no methodology to be provided for it.

For this, we have introduced the concept of security state and security flow. The combinations of the security state and the security flow represents a unique viewpoint and a particular pattern for consideration of security services and we can analyze and identify security issues and threats from these combinations and patterns and provide suitable security services for concerns. Our approach provides overall viewpoint on RFID service network security. It enables addressing security issues which are requisite for network services, therefore, security services and techniques can be deployed without security holes.

Although it provides convenient and feasible methods that enable to ensure security, there are things to improve our approach. As an example, we might recognize a security flow as a pattern and implement automated system ensuring security using artificial intelligent technologies such as neural networks and genetic algorithm. These artificial intelligent technologies are usually used for developing prediction or classification models from historic data or training samples[12]. Additionally we might use them to analyze security threats and to find an optimal solution for the problems.

References

1. Weiser , M., "The Computer for the Twenty-First Century," *Scientific America* 265(3) (1991) 94-104.
2. Ha, Y-G, Sohn, J-C, Cho Y-J, and Yoon, H, "Towards Ubiquitous Robotic Companion: Design and Implementation of Ubiquitous Robotic Service Framework," *ETRI Journal*, vol.27, no.6, pp.666-676 (Dec. 2005)
3. Anderson, R., *Security Engineering: A guide to building dependable distributed systems*, Wiley (Jan. 2001)
4. Mousavidin, E., "RFID Technology: An update," *ISRC Technology Briefing Series.*
5. Shepard, S., *RFID Radio Frequency Identification*, McGraw-Hill (2005)
6. Brock D. L. "The Electronic Product Code (EPC): A Naming Scheme for Physical Objects", Auto-ID Center, (January 2001)
7. EPCglobal US "Standards", http://www.EPCglobalus.org/Network/standards.html (2004)
8. Goyal, A. "Technical Report: Savant™ Guide", Auto-ID Center, (April 2003)
9. Garfinkel, S., *RFID Applications, Security, And Privacy*, Addison-Wesley (2006)
10. Tsuji, T. Kouno, S. Noguchi, J. Iguchi, M. Misu, N. and Kawamura, M., "Asset management solution based on RFID," NEC Journal of Advanced Technology, vol.1, no.3, Summer (2004) 188-193
11. IBM, Patterns: Service-Oriented Architecture and Web Services, ibm.com/redbooks (2004)
12. M. Shin and C. Park, "A Radial Basis Function Approach to Pattern Recognition and its Applications," ETRI Journal, vol. 22, no. 2, (2000) 1-10.
13. Kaufman, C. Perlman, R. and Speciner, M, *Network Security*, Prentice Hall PTR (2002)
14. Davis, J.F., "Information systems security engineering: a critical component of the systems engineering lifecycle," ACM SIGAda Ada Letters, Volume XXIV, Issue 4 (Dec. 2004)
15. Devanbu, P.T. and Stubblebine, S, "Software engineering for security: a roadmap," *Proceedings of the Conference on The Future of Software Engineering*, ACM Press (May 2000)
16. Mead, N. R. and Stehney, T, "Software Engineering for Secure Systems (SESS) -- Building Trustworthy Applications: Security quality requirement engineering (SQUARE) methodology," ACM SIGSOFT Software Engineering Notes , *Proceedings of the 2005 workshop on Software engineering for secure systems—building trustworthy applications SESS '05*, Volume 30 Issue 4 (2005)
17. Dragovic, B and Crowcroft, J, "Ubiquitous computing: Information exposure control through data manipulation for ubiquitous computing," *Proceeding of the 2004 workshop on New security paradigms*, ACM Press (2004)
18. Seigneur, J. M and Jensen C. D, "Ubiquitous computing (UC): Trust enhanced ubiquitous payment without too much privacy loss," *Proceedings of the 2004 ACM symposium on Applied computing* (2004)
19. Schmandt, C and Ackerman, M, "Personal and Ubiquitous Computing: Issue on privacy and security," *Personal and Ubiquitous Computing*, Volumn 8 Issue 6, Springer-Verlag (Nov. 2004)
20. Eustice, K and Kelinrock, L, Markstrum, S. Popek, G, Ramakrishna, V and Reiher, P, "Ubiquitous comuting/security: Securing nomads: the case for quarantine, examination, and decontamination," *Proceedings of the 2003 workshop on New security paradigms*, (2003)

A Study on Development of Business Factor in Ubiquitous Technology

Hong Joo Lee[1], Choon Seong Leem[2], and Sangkyun Kim[3]

[1] The Liberal Arts School, Dankook University, Korea
blue1024@dankook.ac.kr
[2] Yonsei University, 134, Sinchondong, Seodaemungu, Seoul 120-749, Korea
leem@yonsei.ac.kr
[3] Yonsei University, 134, Sinchondong, Seodaemungu, Seoul 120-749, Korea
saviour@yonsei.ac.kr

Abstract. As 'Ubiquitous Technology' is rapidly growing in recent days, off-line and on-line are being integrated into a single space, creating a totally new concept of space and more business opportunities utilizing it. Yet, ubiquitous technology has rarely been conceptualized, with only superficial studies about it conducted. In this paper, we addressed the problem, derived business factors to be applied to actual business, based on characteristics of technologies and application systems currently developed, and gained specific business factors through it.

1 Introduction

As 'Ubiquitous Technology' is rapidly growing in recent days, off-line and on-line are being integrated into a single space, creating a totally new concept of space and more business opportunities utilizing it. Many countries, including the US, Europe and Japan, are eager to conduct diverse researches and projects in relation to such ubiquitous technology in the hope of utilizing those opportunities, and domestic universities and research institutes also are active to implement relevant researches. Yet, ubiquitous technology has rarely been conceptualized, with only superficial studies about it conducted.[7]

In addition, relevant studies have been excessively focused on its hardware, resulting in only case studies through scenarios. Fragmentary service scenarios applying enabling technologies, application systems, and simple enabling technologies in business can bring technical innovation, but they do not last long.

In this paper, we addressed the problem, derived business factors to be applied to actual business, based on characteristics of technologies and application systems currently developed, and gained specific business factors through it.

New technologies give rise to business opportunities for companies, and companies make every effort to seize those opportunities.

B. Gabrys, R.J. Howlett, and L.C. Jain (Eds.): KES 2006, Part I, LNAI 4251, pp. 1259 – 1266, 2006.
© Springer-Verlag Berlin Heidelberg 2006

2 Literature Review

2.1 Revenue Models

The revenue model is related to how each company can create revenue within the context of a business model [11]. The key to studying business models is discussions on which role the individual entity plays in the business model and which revenue it can create through the process.

Revenue-creating can be defined as seeking a role worthwhile to pay compensation for business entities (customers) [10].

Companies create revenue based on this definition, as seen in [Table 1]. [6]

Table 1. Revenue-creating method

Revenue-creating Method	Definition
Execution-driven revenue	Revenue resulting from companies' business activities (ex: How efficiently companies operate, how much active the corporate culture or business structure is, etc.)
Relationship structure-driven revenue	Revenue through value-creating relationship structure

With the rapidly changing business environment, companies are highly influenced by the business environment in creating revenue, and it, in return, actively interacts with companies, directly influencing their strategies and behaviors.
Thus, the value-creating relationship structure, rather than execution-driven revenue, is more likely to affect companies' strategies and revenue.[10]

2.2 Value-Creating in Technical Aspect

New technology provides enabling effects for diverse business areas, which means advanced technology enables many ideas to come true. Utilizing technical aspects in creating business value is called "Technology Enabling Effects Motives."[10]

The technology enabling effects motive can be seen from two perspectives [4]. First reason is results of technology enabling effects, or what the enabling effects result in. Second reason is what the technology enabling effects can realize. The realization area also can be divided into two categories, transforming and linking.

Table 2. Realization area of technology enabling effects

Area	Meaning
Transforming	Create a new transforming method. The new transforming method is presented in process or product of transforming.
Linking	Promote flow of information and knowledge, and systematically connect all activities

<Source: S.H Jeon, 2001>

Business values that can be gained by applying new technology and [Table 2] are seen in [Table 3].

Table 3. Business values of technology enabling effects motives

		Area of enabling effects	
		Transformation	Linking
Results of enabling effects	Change Eco nomics	Value at Transform Economics	Value at Linking Economics
	Create Innov ation	Value at Transform Innovation	Value at Linking Innovation

<Source: Buttler(1997) & Evans(2000)>

[Table 3] is that value created in new business environment can be examined by reviewing which activities create the value and what composes the value.

Transformation economics value, linking economics value, transformation innovation value and linking innovation value seen in [Table 3] are defined as [Table 4].

Table 4. Definition of business value of technology enabling effects motives

Value	Definition
Value at Transform Economics	Implement conventional activities with lower cost but more rapidly by using a new method.
Value at Transform Innovation	Enable new business by using a new method.
Value at Linking Economics	Remarkably reduce costs of conventional business activities by utilizing technology.
Value at Linking Innovation	Design and execute new business by systematically linking all entities.

After all, creating business models proper for new technology environment begins with exploring new business based on the four types.

2.3 Technology-Driven Business Motives

Technology is utilized not only as a means of business activities. "Technology-driven Business Motive" is triggered by the question of what business can be executed with technology [10]. It means technology provides diverse business opportunities and each business area creates diverse types of value-creating motives.

The question of what business should be executed with technology presents a type of value-creating motives in new business environment, in that it regards technology as the starting point of value creation. There are five key terms in answering to the question in accordance with current state of IT business environment; On, For, With, Around and Through [10].

Table 5. Business application, business motive value and business type

Application	Business motive value	Business type
On	Create new space. Develop new trade based on the n ew space	Commerce Using Network
For	Create technology-related goods a nd services	Develop goods for value-Creating
Around	Promote interactions by engaging related business	Related industry
With	Implement business process more efficiently	Activity means
Through	Create new business methods	Business design

[Table 5] displays 'business' types that can be executed with technology.

3 Business Value-Creating in Technology Environment

<Figure 1> shows 4 business values in [Table 5], value at transform economics, value at linking economics, value at transform innovation, and value at linking innovation, and presents business that can be realized through technology, examined by Seong hyeon Jeon (2002). Seong hyeon Jeon (2002)[10], in his research, described a business measure that can be realized through technology, by using terms of 'On', 'For', 'Around', 'With' and 'Through'. I, however, developed a research framework to derive business factors for the mobile industry, based on the description, because business factors derived from 'On', 'For', 'Around' and 'With' can be explained in a more comprehensive way by using 'Through.'

In the respect, I will suggest 4 aspects of 'On', 'For', 'Around' and 'With' to respond to the question of "what could be done with technology?"

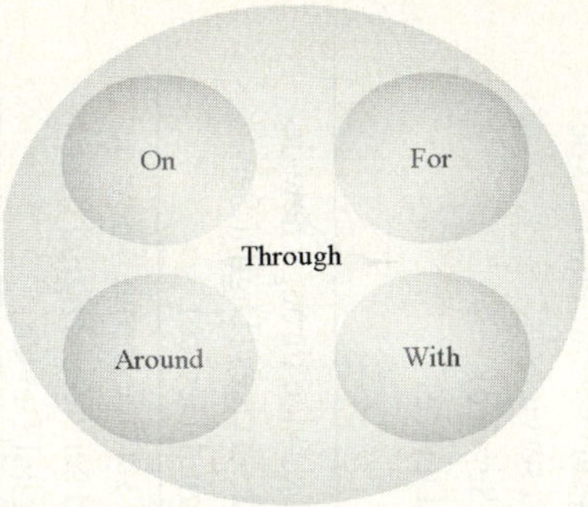

Fig. 1. Business application through technology

Business value and application derived from <Figure 1> can be analyzed as seen in [Table 6].

Table 6. Business value and application of technology enabling effects motives

characteristics		Application	
Value at transform economics	With	On	For
Value at linking economics	With	On	For
Value at transform innovation	On	For	Around
Value at linking innovation	With	For	On

To specifically look at [Table 7], First, Commerce Using Network can produce a business type that can be analyzed when applying business values in previous researches and new trades in new space.

Second, Electronics Device Development for Ubiquitous Business can produce a business type that can be analyzed when applying business values in previously-examined technology and goods development concept for value-creating.

Third, Business Innovation, which is to reduce costs through technology, can produce a business type that can create value with lower cost but more rapidly.

Table 7. Value and application of technology

Application / Value	On	For	With	Around
Value at transform economics	Commerce Using Network	Electronics Device Development For Ubiquitous Business	Business Innovation	Efficient service through new devices and services (Digital Service) (Digital Device)
Value at linking economics			Lower business cost through technology (Digital Service)	Efficient service through new devices and services (Digital Service/(Digital Device)
Value at transform innovation			Sharing of knowledge	Efficient service through new devices and services (Digital Service/(Digital Device)
Value at linking innovation			Efficient business through technology (Digital Service)	Digital Content

Fourth, Lower Business Cost through Technology can produce a business factor called Digital Service, considering value at linking economics to reduce costs through technology, and its specific means.

Fifth, Sharing of Knowledge can produce a business type to develop new business by using a new method.

The sixth, seventh, eighth and ninth all can produce Digital Service/Digital Contents by reducing operative costs, develop new business and apply it to other areas through a new method.

Content Providers can provide Digital Service in order for higher convenience, and this can develop into each business factor.

Table 8. Validity of business factors

Business factors	CSFs derived from previous researches
Commerce Using Network	-Localization (Durlacher, 2000) -Customer Ownership (Durlacher, 2000) -Utility (ARC Group, 2000)
Electronics Device Development for Ubiquitous Business	-Ubiquity (Durlacher, 2000) -Convenience (Durlacher, 2000)
Work innovation through Technology	-Ubiquity (Durlacher, 2000) -Exclusivity(ARC Group, 2000)
Lower business cost through Technology	-Timeliness(ARC Group, 2000) -Cost (Ovum, 2000)
Sharing of knowledge	-Ubiquity (Durlacher, 2000) -Timeliness (Durlacher, 2000) -Content(Oh, 2000)
Digital Contents	-Ubiquity (Durlacher, 2000) -Utility (ARC Group, 2000) -Content(Oh, 2000)
Innovative support for Business	-Utility (ARC Group, 2000) -Timeliness (Durlacher, 2000)
Lower cost thorough Technology (Digital Service)	-Cost (Ovum, 2000) -Ubiquity (Durlacher, 2000)
Digital Service	-Utility (ARC Group, 2000) -Ubiquity (Durlacher, 2000) -Personalization (ARC Group, 2000)
Digital Service Digital Content	-Convenience (Durlacher, 2000) -Personalization (Durlacher, 2000) -Content(Oh, 2000)

Lower Business Cost through Technology can produce a specific business factor called Digital Service, by considering value at linking economics to reduce business costs through technology, and specific means.

Tenth, a business factor that improves conventional work and service through new technology and turns it into business can be produced.

Eleventh, a business factor that reduces business costs through technology and turns it into business can be produced.

Twelveth, a business factor that provides new business with ubiquitous technology can be produced.

Finally, a business factor that systematically links all entities through ubiquitous technology, and turns it into business can be produced.

The factor can be produced when systematic relationship among diverse Contents Providers is required in providing Digital Service or Digital Contents.

In order to verify business factors produced in [Table 29], I analyzed previous researches on CSFs for mobile business.

[Table 8] indicates results of linking those business factors to similar research findings.

4 Conclusion

In this paper, we presented business factors in the emerging ubiquitous technology environment. In order for this research, we analyzed business value that could be produced in technology environment and then, presented previous research for deriving business factors.

References

1. ARC Group.:Contents & Applications for the Wireless Internet: Worldwide Market Analysis & Strategic Outlook 2000-2005 (2000)
2. Buttler, P.,T.W.Hall, A.M. Hanna, L. Mendonca,B.Auguste,J.Manyika, A.Sahay.: A Revolution in Interaction. The McKinsey Quarterly (1997)
3. Durlacher.: UMTS report (2001)
4. Evans & Philips and Thomas S. Wurster.: Strategy and the New Economics of Information, Harvard Business Review(1997)
5. Evans, P. and T.S Wurster.: Blown to Bits: How the New Economics of Information Trans Strategy, Harvard Business Review(2000)
6. H.J Lee.: A Study on Development of Business Factors by Mobile Telecommunication Customer Needs Analysis in Ubiquitous Technology Environment, Doctoral dissertation, The Graduate School Yonsei University (2005)
7. H.J Lee.: Ubiquitous Revolution, EcoBook(2004)
8. J.I Oh and T.W Kim.: The Evolution and CSFs of Mobile Business, KMIS(2000)107-116
9. OVUM.: Mobile E-Commerce: Market Strategies (2000)
10. S.H, Cheon, New Business Model, Jipmoondang, 2001
11. Weill,P. and Vitale, M.R.,Place to Space: migrating to business models, Boston:Harvard Business School Press, 2001

Design of a RFID-Based Ubiquitous Comparison Shopping System

Kyoung Jun Lee and Young Hwan Seo

Kyung Hee University, School of Business,
#1 Hoegi-dong Dongdaemun-Ku, Seoul 130-701, Korea
{klee, seoyn01}@khu.ac.kr

Abstract. With the spread of the so-called always-online environment that allows consumers to be online anytime, anyplace, the next step will be the integration of online and offline markets. Competition will be consequently further intensified and there is a probability that in the process the role of offline retailers will shift from that of a traditional retailer to a displayer role. When this occurs, appropriate technological devices and business models should be explored so that both displayers and retailers can benefit. To this end, this paper proposes an RFID (Radio Frequency Identification) technology-based pervasive comparison shopping business model. RFID will allow consumers to be seamlessly connected to the network, and the advent of a new shopping network will enable a smoothly functioning incentive mechanism between displayers and retailers. Ultimately, a new shopping network will enable consumers to be engaged in seamless commerce.

1 Introduction

The development of the Internet has greatly changed people's shopping patterns. In the traditional commercial transaction, people are required to visit shops to make a purchase. Now, thanks to the Internet, consumers can choose to use the Internet shopping to shop instead of physically visiting stores. In the current shopping environment, both types of consumers are faced with dilemmas: offline consumers cannot be sure whether they are getting a good price, while online consumers cannot be assured of the quality or design of the product they are getting or the credibility of the merchant selling the product. Due to these shortcomings, in brick and mortar commerce, shoppers must visit several stores to compare prices and other terms, while comparison shopping has emerged as a solution in the case of online shopping. Nevertheless, these two markets remain divided, with each constituting a distinct one. The only way to make comparisons between online and offline markets is to learn about the transaction terms of the offline market and then perform a search online or vice versa. The formation of a ubiquitous computing environment will allow these two divided markets to be integrated. This paper presents how the online and offline markets would be integrated in an always-online environment, possible consequential problems that might arise in the market, and business models and methods for overcoming those problems.

B. Gabrys, R.J. Howlett, and L.C. Jain (Eds.): KES 2006, Part I, LNAI 4251, pp. 1267 – 1283, 2006.
© Springer-Verlag Berlin Heidelberg 2006

2 Evolution of Comparison Shopping

In traditional commerce, the product information of each retailer is dispersed and consumers must visit each of them to seek such information. Under these circumstances, the search cost may be greater than the benefit acquired through the search. Now, with the help of the Internet, consumers can easily find product and merchant information. There is no longer a need to pay a visit to physical stores and compare transaction terms. This led to the drastic diminishment of the search cost to be borne by consumers (Bakos 1997), but again the rapidly increasing number of retailers made it difficult for consumers to find what they are looking for (Santos et al 2001). In addition, due to the excessive amount of information available on the Internet, search costs are still incurred. This gave rise to comparison shopping agents to reduce search costs (Krulwich 1996: Kushmerich et al 1997: Yuan et al 2000). Comparison shopping helps consumers make wiser purchase decisions (Yuan 2002) and reduce search costs (Crowston 2001). However, the world-first comparison shopping agent Bargain Finder(Krulwich 1996) was blocked by some sellers feeling threats from it. After the Bargain Finder, various types of comparison shopping services have been developed such as client-based comparison shopping (Jango), collaborative filtering-based recommendation (Firefly.com), and knowledge-based comparison shopping (PersonalLogic). Finally, many commercially successful comparison shopping business came out such as mysimon.com, kelkoo.com, froogle.google.com, and shopping.com. Especially a new type of shopping portal site Become.com provides not only comparison of price but also news or data related shopping. Nevertheless, Internet-enabled comparison shopping only searches among registered online retailers rather than the offline market. In addition, though online comparison shopping services notably curtailed search costs, there remain issues such as the lack of opportunity to actually see and touch products and the question of the credibility of online merchants.

3 Comparison Shopping in Always On-Line Environment

What will happen to the comparison shopping environment when the always-online IT infrastructure and services are widespread, for example through mobile Internet (Wibro) and HSDPA (High Speed Downlink Packet Access)? This section offers a possible scenario.

3.1 Scenario 1

Tom, who owns a digital camera shop downtown, has recently been concerned. The number of consumers visiting his shop has not decreased, but the number of visits leading to actual purchases has dropped sharply. His operating cost is still the same, and reduced sales make it difficult to remain in business. People come to a shop to view products and ask relevant questions. Even those with the intention to make a purchase just browse products at physical shops and use online devices to comparison shop and ultimately purchase from another retailer offering a better deal than Tom. As a consequence, Tom has lost any drive to keep his store afloat and is not making any genuine effort to answer shoppers' questions.

The scenario above described what could happen to physical retailers when the always-online environment becomes prevalent. The commercialization of always-online technologies such as WiBro will make possible the two inherent characteristics of ubiquitousness; that is, anytime and anyplace computing. People will be able to connect to the network anytime, anyplace (Holtjona et al 2004). It will integrate online and offline markets that are currently separate, and, more specifically, enable a market where consumers can use always-online services in physical spaces. Against this backdrop, online and offline markets will no longer be distinct as they are now, and will have to compete fiercely with each other in a single integrated arena. Until now, the retailers in the offline market have had an advantage in terms of information, and consumers have been generally passive in accepting whatever terms they suggest. To overcome this disadvantage, consumers have employed such methods as physically visiting stores, comparing transaction terms, or trying to obtain a discount through negotiation. The tables will be turned in the always-online environment, and consumers will be able to proactively respond to transaction terms through searching and comparison. They will no longer have to accommodate the terms vendors offer without any effort to make them more favorable to themselves. They will be able to purchase products or services as closely in accordance as possible with the terms set by them. With heightening competition, a market equipped with competitiveness will be more active, and by comparison, relatively inferior markets will become lackluster. However these two markets are mutually complementary and both of them must exist for consumers. In this sense, a new market structure is required where markets can be both successful.

4 Pervasive Comparison Shopping Process in Always On-Line with RFID

This section is intended to demonstrate a process wherein the issues introduced in scenario 1 can be resolved using RFID technology and an appropriate business model.

4.1 Scenario 2

Jamie, who loves shopping, wants to buy new fall clothes. She finds an article of clothing she likes in a catalogue. She can buy it online, but wants to go see it for herself, and she enjoys shopping. Accordingly, she goes with a friend to a store located downtown. She is also concerned about costs and wants to buy it at a low price. She likes the product when she sees it, but a slightly high price makes her hesitate. So she uses her terminal to scan the RFID attached to the item to check the prices offered at other stores and to connect to a comparison shopping site. As she is interested only in prices after examining the product herself, she browses a list arranged in order of price. She selects Shopping Mall A, which is reliable and known for quick delivery, and makes a purchase.

This scenario depicts a process from the perspective of consumers. There is not much difference for consumers except that RFID enables seamless comparison shopping. Nevertheless, the relationship between offline stores and online shopping malls changes. If consumers only examine products at offline shops that spend money to display them and then buy elsewhere, offline shops will take countermeasures such as hiding RFIDs or not providing information to prevent consumers from using RFIDs to engage in comparison shopping. This leads to a need for a business model that offsets the costs incurred to offline shops even when consumers use only the information provided at offline stores but make an actual purchase elsewhere.

4.2 Scenario 3

James, who owns a shop in a downtown shopping center, makes a genuine effort to explain about the products in his store to the many consumers who visit his shop on a holiday. Most consumers ask detailed questions, but make their purchase through RFID scan-enabled comparison shopping. Even now, he introduces the new fall products to a female consumer, and she evaluates and buys a product through comparison shopping. After the purchase is completed, the online shopping mall from which the female customer bought the product deposits a payment in James' bank account. Before RFID-based comparison shopping, James put great efforts into acquiring products at a lower price, needed a warehouse in which to store his stock, and bore logistics costs to maintain them. Now he is relieved of all these incidental expenses and all he has to do is to acquire and display products.

These scenarios are centered on offline stores. Offline stores have to deal with significant consumer traffic, which requires them to pay a considerable expense. Offline stores that have to bear such expense should be given incentive to allow the initiation of an RFID-based comparison shopping model. In the model in scenario 3, both physical retailers and displayers can benefit.

4.3 Economic Parties

Consumer: Those who consume products. They are engaged in comparison shopping both in physical spaces and using RFIDs in the always-online environment. Under such circumstances, the cost borne by consumers to research prices and services shrinks. Furthermore, seamless networking is made possible as the search method changes from keyword-based to RFID scanning.

Displayer: Without RFID technology, retailers' sales volumes diminish in contrast to their costs as consumers move away from brick and mortar stores through comparison shopping. When a comparison shopping network based on RFID takes root, they can receive incentives from other retailers, and it will change the focus of their business from sales to display.

Retailer: Once they are registered with a comparison shopping network and post information on their products, they can concentrate their efforts on sales without having to handle marketing. They can be offline retailers that enjoy relatively low maintenance costs thanks to low traffic or they can be online retailers.

Shopping network: Site to support RFID-enabled research and to connect physical product retailers and virtual product vendors displayed on the terminal using RFID.

4.4 Processes in the Scenario

The model in scenario 3 can be divided into two processes. In the first process, a search process, a consumer views the products of a displayer, signs onto a shopping network server, and receives information about retailers registered there. In the second process, the consumer selects a product and makes a payment, and incentives are distributed.

4.4.1 Search Process

Definitions

DID: Displayer Identification that uniquely distinguishes a displayer to whom an incentive is given in a transaction.

RID: Retailer Identification that uniquely distinguishes a retailer for a consumer to communicate with through the shopping network.

PID: Product Identification that discriminates a product such as EPC (Electronic Product Code).

SNID (Shopping Network Identification): The address that consumers access the network for comparison shopping.

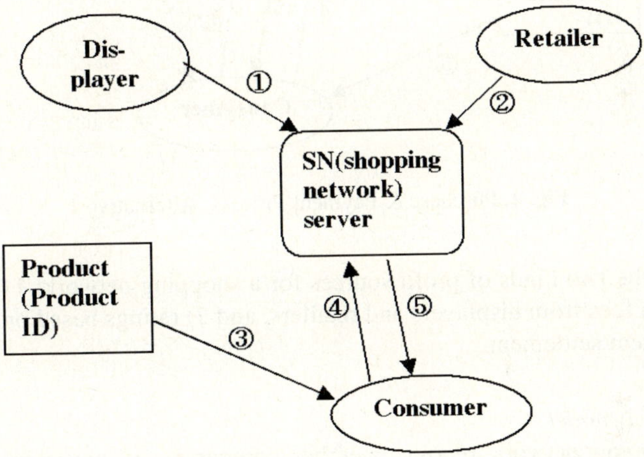

Fig. 1. Search Process of the Pervasive Comparison Shopping

① Displayer ID(DID), Product ID (PID)
② Retailer ID(RID), Product ID(PID), price and transaction condition/update
③ PID, SNID
④ PID
⑤ sorted-list which contains the RFIDs that meet the search criteria

In steps ① and ② of fig. 1, a shopping network recruits member companies before it commences service to ensure the smooth operation of its incentive system. This is because if the network gathers retailers' information through scraps to deliver a sorted list to consumers, retailers can refuse to pay incentives to displayers. However, consumers seek convenience and also prefer to do it their own way. In step ③, a PID ensures automatic and reliable connection to the shopping network. Finally, in steps ④ and ⑤, consumers are logged onto the shopping network and receive a sorted list of retailers that meet the search criteria.

4.4.2 Purchase and Payment Process

The purchase and payment process, which comes after the search process, can have two alternatives. The first alternative is described in fig. 2 where consumers pay through a shopping network.

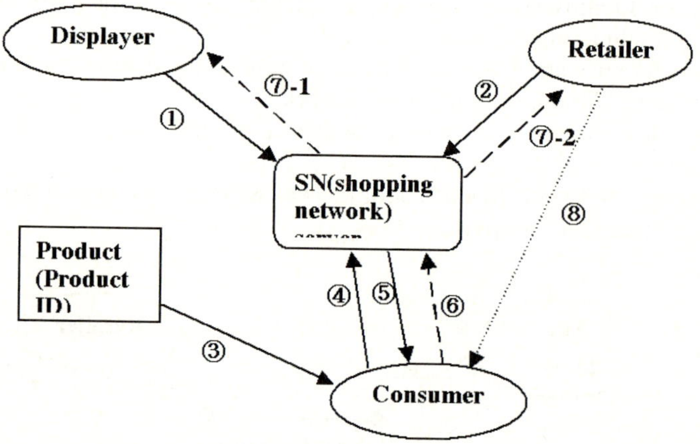

Fig. 2. Purchase & Payment Process: Alternative 1

There are the two kinds of profit sources for a shopping network: 1) the collection of registration fees from displayers and retailers, and 2) ratings based on a CPA model through payment settlement.

Registration fee model

When a shopping network recruits member companies, to participate this service system(① and ②), displayer and retailer pay registration fee to shopping network and they get each identification from shopping network.

CPA (Cost Per Action) model

After receiving full payment from a customer (⑥), the shopping network server pays an incentive to the displayer (⑦-1) and the remainder goes to the retailer (⑦-2) aside from the assigned commission. The retailer sends the corresponding product to the consumer after checking that payment has been remitted (⑧).

The biggest challenge currently faced in online shopping is that consumers' personal and transaction information is stored on the server, resulting in possible loopholes in privacy protection. This is because the entire process is centered on the service of the online shopping server. If consumers pay vendors directly, the shopping network cannot collect individuals' transaction information, thereby securing privacy protection as far as transactions are concerned. This is explained in the following example (Fig. 3).

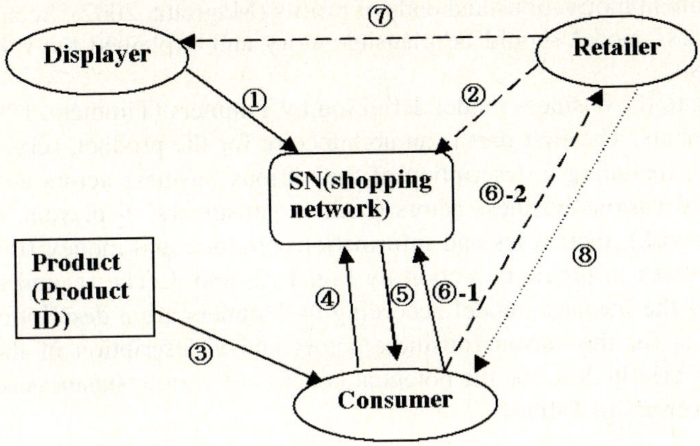

Fig. 3. Purchase & Payment Process: Alternative 2

The consumer decides to buy a product and pays the retailer directly. The displayer's ID is provided to allow the retailer to pay an incentive (⑥-2). In this case, the purchase decision by the consumer is automatically communicated to the shopping network to prevent the retailer from evade paying an incentive to the displayer (⑥-1). After the payment is made, the retailer uses the DID it has received to pay an incentive to the displayer (⑦), and lastly, it delivers the product to the consumer (⑧). Unlike the model in which payment is made through the shopping network server, the CPA method-based rating in this model is difficult to apply. When rating takes place using the CPA method, the retailer is required to pay both an incentive to the displayer and a commission to the shopping network. This may not be a practical model to apply as the retailer's overhead increases.

As seen in the examples above, a new business model is created in the offline market. Though the physical sales volume of products drops, the additional profit source of the commission arises in addition to the existing source of sales revenue. When the comparison shopping becomes more widespread and consumers become even more sensitive to prices, the sales revenue of physical retailers will further diminish and inventory and incidental costs will increase. Against this backdrop, a totally new business model of pervasive comparison shopping will emerge where physical retailers shift their focus from sales to the display of products and commission is the main profit source.

5 Evaluation of Pervasive Comparison Shopping Business Model

5.1 Analysis of the Business Model

Since ubiquitous computing environment is not the Garden of Eden, the service scenario should be based on the proper business model development by each business participants. U-commerce research should be based on the concrete methodology for business model definition, development, representation and evaluation. Creating a business model is a almost same as writing a new story and a good business model starts with human motivations and ends of profits (Magretta, 2002). Scenario, the first step of business model should be plausible story and explained the role of all economic parties.

According to the business model definition by Timmers (Timmers, 1998), there are three components. The first part is an architecture for the product, service and information flows, including a description of the various business actors and their roles. We mentioned various business actors (such as consumers, displayers, retailers, and shopping network), their roles and information, product, and money flow with three parts of processes in previous section by Fig. 1, 2, and 3. The second and the third component of the business model according to Timmers are a description of the potential benefits for the various business actors and a description of the sources of revenues. We briefly describe the potential benefits of various business actors and the sources of revenues in Table 1.

Table 1. The Potential Benefit and Source of Revenue in Pervasive Comparison Shopping

	Potential benefits	Source of revenues
Consumer	- Convenient comparison shopping - Better services from displayer in always online environment	- Diminishing shopping cost
Displayer	- Logistic cost reduction	- Incentive
Retailer	- Reduction in marketing cost - Reduction in display cost	- Sales margin
SN(shopping network)	- A new business opportunity	- Registration fee - Cost-per-purchase revenue

5.2 Conditions for Working the Business Model

Several preconditions are required to work the pervasive comparison shopping business model. In this section, we will find some conditions, for SN-registered sellers competing with SN-unregistered sellers, cooperating with displayers (i.e. offline sellers). SN-registered sellers are those who register the pervasive comparison shopping network (SN) and SN-unregistered sellers are the sellers who do not have the shopping network membership.

Notations

CostOFF: The product unit cost of offline seller reflecting shop operating cost etc.
CostUN: The product unit cost of SN-unregistered seller reflecting marketing cost, web site operating cost and delivery cost etc.
CostSN: The product unit cost of SN-registered seller reflecting SN registration fee and delivery cost etc.
INCEN: The money (incentive) that SN-registered seller pays to Displayer (offline seller) when a transaction occurs by the shopping network.

SCOFF: Shopping cost[1] incurred to a consumer who is currently in front of the displayer when the consumer buys directly from the offline seller including delivery cost etc.
SCSN: Shopping cost incurred to the consumer when the consumer buys from SN-registered seller including delivery delay cost and seller trust cost etc.
SCUN: Shopping cost incurred to the consumer when the consumer buys form SN-unregistered seller including delivery delay cost, seller trust cost, search cost and memory cost etc.

5.2.1 The Condition to Work This Business Model

The minimum price of SN-registered seller is CostSN + INCEN and the minimum price of offline seller is CostOFF + INCEN. If offline sellers set the price less than CostOFF + INCEN, they can sell products but they get less money than INCEN. So they tend to set the price more than (CostOFF + INCEN). If SN-registered sellers set the price less than that of offline seller, offline sellers are hard to sell products, but offline seller can get profit as much as the incentive.

From the consumer side, the minimum price slightly changes due to the shopping cost difference. The minimum price that consumer feels from the SN-registered seller becomes CostSN + INCEN + SCSN and the minimum price that consumer feels from the offline seller becomes CostOFF + INCEN + SCOFF. If we assume that the two parties have the perfect information on each other, then each party will be able to set the maximum price using the other party's information. Therefore, the price that consumer feels will be determined at the MAX(CostSN + INCEN + SCSN, CostOFF + INCEN + SCOFF). If a SN-registered seller wins, its profit becomes (CostOFF - CostSN) + (SCOFF – SCSN) and the offline seller gets the incentive as a displayer. If the offline seller wins, there is no profit for the SN-registered seller is zero and the profit of the offline seller becomes INCEN + (SCSN - SCOFF)+ (CostSN - CostOFF). The analysis is summarized in Table 2.

Table 2. Comparison of the price and profit between SN-registered seller and offline seller

	SN-registered seller	Offline seller
Minimum price of seller	CostSN + INCEN	CostOFF + INCEN
Minimum price that consumer feels	CostSN + INCEN + SCSN	CostOFF + INCEN + SCOFF
Profit of seller	MAX((CostOFF - CostSN) + (SCOFF – SCSN),0)	MAX(INCEN + (CostSN - CostOFF) + (SCSN - SCOFF), INCEN)

From the analysis, we can see that the condition for the SN-registered sellers to participate the pervasive comparison shopping is that CostSN + INCEN + SCSN < CostOFF + INCEN + SCOFF. The condition can be rewritten as (CostOFF - CostSN) + (SCOFF – SCSN) > 0. Especially, the so called Win/Win condition for both the SN-registered seller and offline seller is the point where the profit of each side is the same. There for the Win/Win condition is (CostOFF - CostSN) + (SCOFF – SCSN) = INCEN.

[1] Shopping Cost is invisible cost that consumers always consider when they do shopping.

5.2.2 The Condition for SN-Registered Seller Competing with SN-Unregistered Seller

In the real world, there is a probability that SN-registered sellers have to compete with other online sellers which are not registered to SN. So, we need to consider conditions including other SN-unregistered sellers.

The minimum price of SN-registered seller is CostSN + INCEN while the minimum price of SN-unregistered seller is just CostUN. Since the SN-registered sellers should pay the incentive to offline seller in this model, they should set the price no less than Cost SN + INCEN. However, the SN-unregistered seller does not have to pay anything to offline seller.

From the consumer side, the minimum price slightly changes due to the shopping cost difference as in previous analysis. The minimum price that consumer feels from the SN-registered seller becomes CostSN + INCEN + SCSN and the minimum price that consumer feels from the SN-unregistered seller becomes CostUN + SCUN. If we assume that the two parties have the perfect information on each other, then each party will be able to set the maximum price using the other party's information. Therefore, the price that consumer feels will be determined at the MAX(CostUN + SCUN, CostSN + INCEN + SCSN). If a SN-unregistered seller wins, its profit becomes ((CostSN + INCEN + SCSN) - (CostUN + SCUN). If the SN-registered seller wins, the profit becomes (CostUN + SCUN) - (CostSN + INCEN + SCSN). The analysis is summarized in Table 3.

Table 3. Comparison of the price between SN-unregistered seller and SN-registered seller

	SN-unregistered seller	SN-registered seller
Minimum price of seller	CostUN	CostSN + INCEN
Minimum price that consumer feels	CostUN + SCUN	CostSN + INCEN + SCSN
Profit of seller	MAX ((CostSN + INCEN + SCSN) - (CostUN + SCUN),0)	MAX((CostUN + SCUN) - (CostSN + INCEN + SCSN),0)

For a SN-registered seller to compete with SN-unregistered seller, the profit of the SN-registered seller should be higher than zero. So, the first condition is (CostUN + SCUN) > (CostSN + INCEN + SCSN). In this formula, we can infer the condition for a seller to register SN. We can also infer the range of incentive as INCEN < (CostUN + SCUN) - (CostSN + SCSN). The win-win condition between offline seller and SN-registered seller is INCEN = (CostUN + SCUN) - (CostSN + INCEN + SCSN). So, The second incentive condition is INCEN = [(CostUN + SCUN) - (CostSN + SCSN)]/2. To consider the two Win/Win conditions together, the final incentive can be set as the MIN((CostOFF - CostSN) + (SCOFF – SCSN), [(CostUN + SCUN) - (CostSN + SCSN)]/2).

5.2.3 Example

For the illustration of the above analysis, we assign some values to the variables as follows.

CostOFF = $400
CostUN = $370
CostSN = $350
SCOFF = $10
SCUN = $40
SCSN = $20
INCEN = $20 = [($370 + $40) – ($350 + $20)]/2

Table 4. Example simulation

	SN-unregistered seller	SN-registered seller	Offline seller
The Cost of seller	$370	$350	$400
Minimum price of seller	$370	$370	$400
Price range of seller	Higher than $370	$370~$400	Higher than $400
Minimum price that consumer feels	$410	$390	$430
Profit of seller	0	Up to $20	$20

The simulation result confirms the conditions for working the pervasive comparison shopping business model.

6 Implementation Issue of Pervasive Comparison Shopping

6.1 Why Not Barcode But RFID?

To work this business model, we propose to use RFID as an auto-identification technology instead of barcode. From the point of view that consumers get information from the tag attached to a product through wireless network, barcode can be acceptable as in the applications such as Pocket BargainFinder (Brody et al 1999) and Scanning Objects in the Wild (Brush et al 2004). However, only with the barcode system, we cannot make the mechanism that gives incentive to the participants because there is no way to distinguish each displayer with the barcode system. With the RFID, we can identify each product since there is EPC (Electronic Product Code) system to distinguish each product. In addition, according to many industrial reports (Accenture 2005; Deloitte 2005; Intermec 2005; Symbol 2005), the world trend of the retail industry shows that we can expect RFID-based retail industry instead of barcode. Nowadays major retailers such as Metro AG, Tesco, and Seven-Eleven consider the benefit of RFID (Levinson 2005) and many experts expect that the price of tag will drop down before long even though the current price of RFID is not cost effective for retailers.

6.2 Which RFID Can Be Used?

Nowadays in global market, RFID has been greatly developed and tried to apply to a vast range of business. But it is difficult to define which RFID can be used or required in business area due to the diverse and complex type of features. Type of RFID tag can be divided as fig. 4.

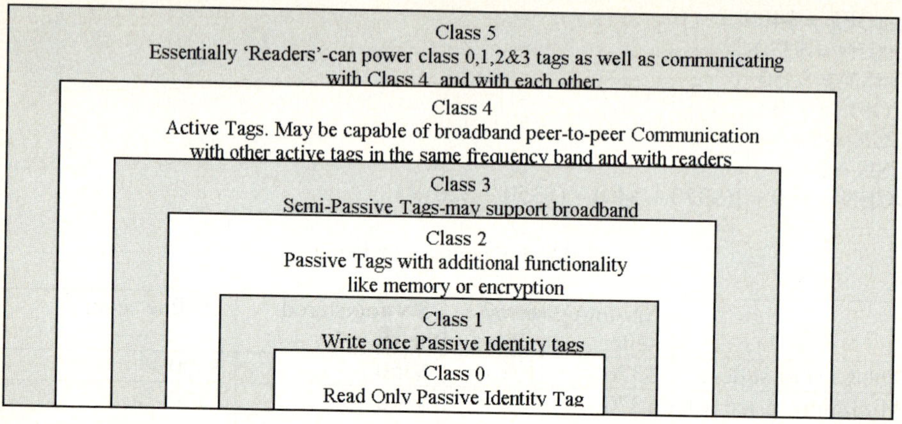

Fig. 4. EPC Class (Source: MIT Auto-ID Center)

In this picture, we can conclude that minimum specification of the tag to work our business model is Class 0. Because the tag does not need to be writable and active.

According to a taxonomy for RFID (Hassan et al 2006), there are four types of signal ranges. Three RFID frequencies such as 135 KHz, 13.56 MHz and 2.45 GHz, are used as a de facto standard in Europe, North America, Japan and Korea. But UHF (Ultra High Frequency) is different depending on countries. The characteristic of each frequency is summarized in table 5.

Table 5. The characteristic of each frequency (Source: IBM Business Consulting Services)

Frequency	Benefits	Drawbacks
Low (125-134 KHz)	- Works Near Metal - In Wide Use Today	- < 1.5m read range - Not in EPC Standards
High (13.56 MHz)	- Works in Most Environments - In Wide Use Today	- < 1.5m read range - Does not work near metal
Ultra High (0.3-1.2 GHz)	- Longer Read-Range Potential - Growing Commercial Use	- Does not work in moist environments
Microwave (2.45 or 5.8 GHz)	- Longer Read-Range Potential, - > 1.5M read range	- Complex Systems Development

Even though all types of frequency range are acceptable to work our business model, High and Ultra High Frequency will be more acceptable than Low Frequency which is relatively high cost. In addition, Ultra High Frequency will be the most appropriate in the future, because the research & development on the UHF has been most active in worldwide. Although the lack of standardization of frequency spectrum was the obstacle to spread of RFID, UHF Gen 2, approved by EPC global in December 2004, is very promising and expected to be improved continuously. In addition, since Gen 2 has the function such as permanent kill capability, it will be useful for privacy protection from scanning after purchase when the user wants.

April 2005, a Korean venture (http://www.myuzone.com) developed a RFID-enabled Mobile Phone. At the end of 2006, Samsung Electronics Co., Ltd.

(http://www.sec.co.kr) and LG Electronics Inc. (http://www.lge.co.kr) are planning to sell a mobile phone embedded a RFID reader-featured chip (900MHz range band). The phone is available to the consumers just with additional cost \$10~20.

6.3 System Architecture

Fig. 5 is the system architecture based on the EPC framework we propose. The flows among the system components are described sequentially in table 6.

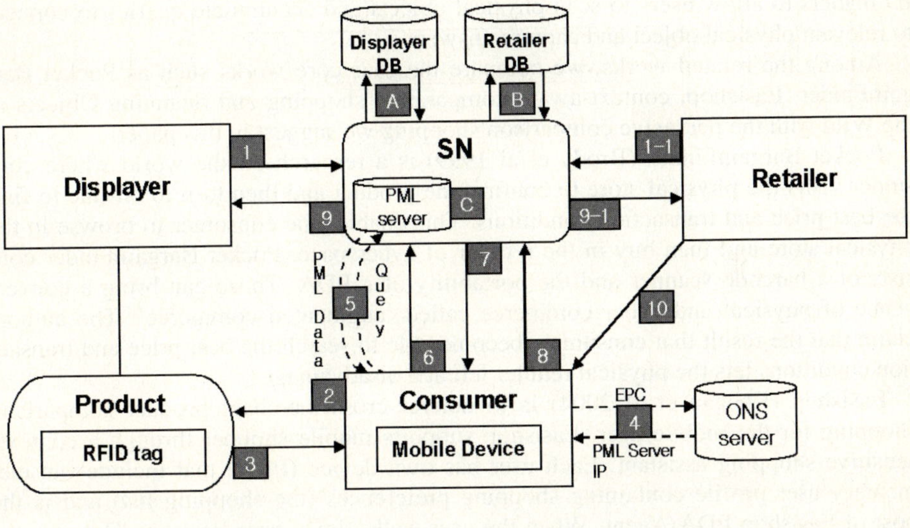

Fig. 5. System Architecture

Table 6. Process description of system architecture

Flow	Description
1	Displayer inputs DID and EPC (Electronic Product Code) to SN
1-1	Retailer inputs RID, EPC and PI (Product Information) to SN
A	SN saves DID into Displayer DB
B	SN saves RID into Retailer DB
C	SN saves PI into PML server
2	Consumer scans RFID tag with Mobile Device
3	Consumer gets EPC
4	Finding IP Address to connect to PML server
5	Getting Product Information
6	Request of Retailers list
B	Searching Retailer DB
7	Getting sorted-list that meets the search criteria
8	Full Payment
A	Find Displayer matched with EPC through searching Displayer DB
9	Giving incentive
9-1	Payment with removed incentive (& commission)
10	Product Delivery

7 Comparison with the Related Works

Up to the present, there are some attempts to integrate online market place with off-line market place. As a initial venture using desktop was Webstickers(Ljungstrand et al 2000) that let the users take advantage in the physical environment when sharing and organizing bookmarks with standard barcode readers and adhesive stickers. With development of mobile devices and wireless networks, QueryLens(Konomi 2002) tried to access ID-based(barcode or RFID) information. QueryLens used PDAs with ID readers to allow users to scan physical objects and accumulate queries to connect to relevant physical object and capture answers.

Among the related works, we compare the four core works such as Pocket BargainFinder, Easishop, context-aware comparative shopping and Scanning Objects in the Wild with the pervasive comparison shopping we suggest in this paper.

Pocket BargainFinder (Brody et al 1999) is a research on the world where consumer visits the physical store to confirm the product and then turn to on-line to find the best price and transaction conditions. This enables the consumer to browse in the physical store and then buy in the retailer of cyberspace. Pocket BargainFinder consists of a barcode scanner and the portability of a PDA. Those can bring a convergence of physical and cyber commerce called 'augmented commerce'. The authors claim that the result that consumers become able to search the best price and transaction conditions lets the physical retailer left at a disadvantage.

Easishop (Keegan et al 2004) is to deliver cross merchant product comparison shopping for the mobile user. Easishop supports mobile shopper through a context-sensitive shopping assistant. Each user has own device (PDA) that includes an elementary user profile containing shopping preferences (the shopping list) and is the host of Easishop PDA Agent. When the user walks down near Easishop Hotspot, the PDA Agent communicates with the Store Agents that represents a multitude of relevant retail outlets and matches the shopping list. PDA Agent and the Store Agent migrate to the Easishop Marketplace providing an auctioneering forum and the Stall Manager Agent is created to make the imminent auction. After completing a purchase, a special Secure Transaction Agent processes payment with ordered the agreed amount. What Easishop differs from our Pervasive Comparison Shopping is user intervention. With user shopping preferences (the shopping list), Easishop PDA Agent purchases the product on behalf of the user through communication with agents. But in Pervasive Comparison Shopping, instead of agent, the user purchases the product directly.

Kwon and Sadeh (2004) proposed context-aware comparative shopping with case-based reasoning and multi-agent intelligent system. In offline environment, this model can let consumer compare online and offline sellers with negotiation between B-agent (buyer agent) and S-agents (seller agent). In addition, they considered contextual information such as location, weather and calendar, and case-based reasoning to find user preference. The goals are aimed fitting buyers' preferences, considering context awareness of external sources and negotiation to provide better transaction conditions. The flow is that after consumer gives a message what he/she wants to buy to B-agent, he/she gets bidding results while Negotiator operates bidding system among S-agents. At this time, S-agent delegated by the own user can negotiate by modifying its own price condition within the allowable price level.

Scanning Objects in the Wild (Brush et al 2005) is the challenge that has been linking physical objects and relevant online information. They conducted experiment with twenty participants over five weeks with a ARUA system that is commercially available pocket computers two parts such as Client Application including barcode scanner and Web Portal to make private and public ratings and comments of the item that user scanned. This study aimed four points, "How do people use the AURA system?", "Do People find AURA system functional and useful?", "Does the privacy model meet users' needs?", "How do people use the sharing features?" What ever the results did not come out as well as expected, this study is well worth enough in view of experiment in the real world. But this study still consider of convenience of consumer not offline seller.

That's why we suggest a so-called win-win(Displayer & Seller) business model to overcome the disadvantage of physical retailer. In addition, rather than using bar code, our business model is based on the RFID technology that can contain and transform information. It maintains the market mechanism where retailers can give incentives to displayers. We briefly describe each characteristic as Table 7.

Table 7. Comparing Pervasive Comparison Shopping with Related Works

	Pocket Bargain-Finder	Easishop	Scanning Objects in the Wild	Context-aware comparative shopping	Pervasive Comparison Shopping
Information Infrastructure	Barcode	HotSpot	Barcode	Query from consumer	RFID
Range of search	Unlimited	Limited location	Unlimited	Unlimited	Unlimited
Personalization	Non-existing	Existing	Non-existing	Existing	Non-existing
Seller Trust	Medium	High	High	Medium	Medium
User intervention	High	Low	High	Medium	High
Role of displayer	Existing	Non-existing	Existing	Existing	Existing
Incentive mechanism	Non-existing	Non-existing	Non-existing	Non-existing	Existing

8 Conclusions

The expansion of the pervasive comparison shopping we suggest will give rise to a new and different marketplace. Currently, high-traffic spaces are occupied by billboards. These ads enable consumers only see products, but do not offer purchase opportunities. They are simply for marketing purposes. However, with the help of RFID, consumers can view advertisements and receive information on the products featured in the ads. If they like what they see, actual purchases can be made. People who have to shop online due to their busy schedule can browse and purchase products shown on billboards in a subway station or on the street during their commute. In addition, people can contain product information by scanning, using this information, they can be connected to the network to purchase the product anytime. Spaces that have been used solely for advertising are turning into unmanned stores where consumers can buy goods on favorable terms through comparison shopping.

Further, displayer does not have to be the store. A woman wearing a dress can be called a displayer because she can serve the role of marketing as a displayer. If we connect retailers and the 'displayer' through a proper incentive mechanism, we are able to make a new market mechanism. Although consumers have been playing a sole role as a buyer so far, with this business model, the consumers can easily become a displayer.

In e-commerce, comparison shopping is divided into online and offline markets. Ubiquitous computing will allow consumers to be connected to the Internet anytime, anyplace, and will consequently lead to the integration of these two markets. The resultant intensification of competition will call for a new business model, and this paper responds to this call with a RFID-based model. Until now, RFID research has focused primarily on supplier management or consumers' simple usage of information provided by RFID. Unlike its predecessors, this paper discusses the changes experienced by each economic player in the always-online environment, uses RFID technology as a solution to the issues brought about by these changes, and offers a new business model that enables seamless networking.

An incentive-based model in which each relevant party can participate is required to allow seamless networking in ubiquitous commerce. It will be a wholly different business model and new spaces can be created through its expansion.

In this paper, we compare the transaction conditions only in terms of price. When consumers make a decision to purchase they consider not only the price but also multi-criteria purchase conditions such as payment options, supplier reputations, a burden of products, multi-item selection, price discount, delivery cost and payment methods that are diverse and complex. To solve this problem, Chang and Lee (2006) suggested a study provides a hybrid model for comparison shopping with "what-if analysis of a Quadratic Integer Programming model". Even though this study explores not only price but multi-criteria purchase conditions, it still remains to study in the online environment. In the future, we need to find a solution including even offline sellers.

References

1. Accenture. Driving High Performance With Silent Commerce in The Supply Chain, White Paper (2005)
2. Bakos, Y. Reducing Buyer Search Costs: Implications for Electronic Marketplaces. Management Science, Vol. 43, No. 12 (1997) 1676-1692

3. Brody, A., Gottsman. E. Pocket BargainFinder: A Handheld Device for Augmented Commerce, Proceedings of the International Symposium on Handheld and Ubiquitous Computing (1999)
4. Brush, A., Turner, T., Smith, M., Gupta, N. Scanning Objects in the Wild: Assessing an Object Triggered Information System, Proceedings of Ubicomp 2005, Springer(2005), 305-322
5. Chang, Y.S., Lee, K.J. A QIP-Integrated Online Comparison Shopping Model and Its Implementation, Submitted to Decision Support Systems (2006)
6. Crowston, K. The Effects of Market-enabling Internet Agents on Competition and Prices: A Model and Empirical Evidence, Journal of Electronic Commerce Research, Vol 2, No. 1 (2001) 1-22
7. Deloitte. Synchronicity: An Emerging Vision of the Retail Future, White Paper (2004)
8. Holtjona, G. J., Fiona, F. H., and Nah. U-commerce: Emerging Trends and Research Issues, Industrial Management & Data Systems, Vol. 104, No. 9 (2004) 744-755
9. Intermec. RFID Technology in Retail, White Paper (2005)
10. Keegan, S., O' Hare, G. EASISHOP: Delivering Cross Merchant Product Comparison Shopping for the Mobile User, Proceedings of the15th IEEE International Symposium on Personal, Indoor and Mobile Radio Communications (PIMRC'04), 5th-8th September Barcalona, Spain (2004) 44-51
11. Konomi, S. QueryLens: Beyond ID-Based Information Access. Proceedings of Ubicomp 2002, Springer (2002), 210-218
12. Krulwich, B.: The BargainFinder agent. Comparison Price Shopping on the Internet. In Williams, J., ed., Bots and Other Internet Beasties. SAMS.NET (1996)
13. Kushmerick, N., Weld, D., Doorenbos, B. Wrapper Induction for Information Extraction. In Proceedings of the International Joint Conference on Artificial intelligence (IJCAI-97), Nagoya (1997)
14. Levinson, M. The RFID Imperative, CIO Magazine (2005)
15. Ljungstrand, P., Redström, J., Holmquist, L.E. Webstickers: Using Physical Tokens to Access, Manage and Share Bookmarks to the Web. Proceedings of DARE 2000 (2000) 23-31
16. Magretta, J. Why Business Models Matter, Harvard Business Review, Vol. 80, No. 5 (2002) 86-89
17. dos Santos, S.C., Angelim, S., Meira, S.R.L. Building Comparison-Shopping Brokers on the Web, Lecture Notes in Computer Science, Vol. 2232 (2001) 26-38
18. Symbol. Business Benefits from Radio Frequency Identification (RFID), White Paper (2005)
19. Taimur Hassan, Samir Chatterjee. A Taxonomy for RFID, Proceedings of the 39th Hawaii International Conference on System Sciences (2006)
20. Timmers, P. Business Models for Electronic Markets, Electronic Markets, Vol. 8, No. 2. (1998) 3-8
21. Yuan, S.-T. and A. Liu. Next-Generation Agent-enabled Comparison Shopping, Expert systems with Applications, Vol. 18, No. 4 (2000) 283-297
22. Yuan, S.-T. A Personalized and Integrative Comparison-Shopping Engine and its Applications, Decision Support Systems, Vol. 43, No. 2 (2002) 139-156
23. Kwon, O. B., Sadeh, N. Applying case-based reasoning and multi-agent intelligent system to context-aware comparative shopping, Decision Support Systems, Vol. 37, No. 2 (2004) 199-213

Author Index

Lecture Notes in Artificial Intelligence (LNAI)

Vol. 4264: J.L. Balcázar, P.M. Long, F. Stephan (Eds.), Algorithmic Learning Theory. XIII, 393 pages. 2006.

Vol. 4253: B. Gabrys, R.J. Howlett, L.C. Jain (Eds.), Knowledge-Based Intelligent Information and Engineering Systems, Part III. XXXII, 1301 pages. 2006.

Vol. 4252: B. Gabrys, R.J. Howlett, L.C. Jain (Eds.), Knowledge-Based Intelligent Information and Engineering Systems, Part II. XXXIII, 1335 pages. 2006.

Vol. 4251: B. Gabrys, R.J. Howlett, L.C. Jain (Eds.), Knowledge-Based Intelligent Information and Engineering Systems, Part I. LXVI, 1297 pages. 2006.

Vol. 4248: S. Staab, V. Svátek (Eds.), Engineering Knowledge in the Age of the Semantic Web. XIV, 400 pages. 2006.

Vol. 4223: L. Wang, L. Jiao, G. Shi, X. Li, J. Liu (Eds.), Fuzzy Systems and Knowledge Discovery. XXVIII, 1335 pages. 2006.

Vol. 4213: J. Fürnkranz, T. Scheffer, M. Spiliopoulou (Eds.), Knowledge Discovery in Databases: PKDD 2006. XXII, 660 pages. 2006.

Vol. 4212: J. Fürnkranz, T. Scheffer, M. Spiliopoulou (Eds.), Machine Learning: ECML 2006. XXIII, 851 pages. 2006.

Vol. 4211: P. Vogt, Y. Sugita, E. Tuci, C. Nehaniv (Eds.), Symbol Grounding and Beyond. VIII, 237 pages. 2006.

Vol. 4203: F. Esposito, Z.W. Raś, D. Malerba, G. Semeraro (Eds.), Foundations of Intelligent Systems. XVIII, 767 pages. 2006.

Vol. 4201: Y. Sakakibara, S. Kobayashi, K. Sato, T. Nishino, E. Tomita (Eds.), Grammatical Inference: Algorithms and Applications. XII, 359 pages. 2006.

Vol. 4198: O. Nasraoui, O. Zaiane, M. Spiliopoulou, B. Mobasher, B. Masand, P. Yu (Eds.), Web Minding and Web Usage Analysis. IX, 177 pages. 2006.

Vol. 4196: K. Fischer, I.J. Timm, E. André, N. Zhong (Eds.), Multiagent System Technologies. X, 185 pages. 2006.

Vol. 4188: P. Sojka, I. Kopeček, K. Pala (Eds.), Text, Speech and Dialogue. XIV, 721 pages. 2006.

Vol. 4183: J. Euzenat, J. Domingue (Eds.), Artificial Intelligence: Methodology, Systems, and Applications. XIII, 291 pages. 2006.

Vol. 4180: M. Kohlhase, OMDoc – An Open Markup Format for Mathematical Documents [version 1.2]. XIX, 428 pages. 2006.

Vol. 4177: R. Marín, E. Onaindía, A. Bugarín, J. Santos (Eds.), Current Topics in Artificial Intelligence. XV, 482 pages. 2006.

Vol. 4160: M. Fisher, W.v.d. Hoek, B. Konev, A. Lisitsa (Eds.), Logics in Artificial Intelligence. XII, 516 pages. 2006.

Vol. 4155: O. Stock, M. Schaerf (Eds.), Reasoning, Action and Interaction in AI Theories and Systems. XVIII, 343 pages. 2006.

Vol. 4149: M. Klusch, M. Rovatsos, T.R. Payne (Eds.), Cooperative Information Agents X. XII, 477 pages. 2006.

Vol. 4140: J.S. Sichman, H. Coelho, S.O. Rezende (Eds.), Advances in Artificial Intelligence - IBERAMIA-SBIA 2006. XXIII, 635 pages. 2006.

Vol. 4139: T. Salakoski, F. Ginter, S. Pyysalo, T. Pahikkala, Advances in Natural Language Processing. XVI, 771 pages. 2006.

Vol. 4133: J. Gratch, M. Young, R. Aylett, D. Ballin, P. Olivier (Eds.), Intelligent Virtual Agents. XIV, 472 pages. 2006.

Vol. 4130: U. Furbach, N. Shankar (Eds.), Automated Reasoning. XV, 680 pages. 2006.

Vol. 4120: J. Calmet, T. Ida, D. Wang (Eds.), Artificial Intelligence and Symbolic Computation. XIII, 269 pages. 2006.

Vol. 4114: D.-S. Huang, K. Li, G.W. Irwin (Eds.), Computational Intelligence, Part II. XXVII, 1337 pages. 2006.

Vol. 4108: J.M. Borwein, W.M. Farmer (Eds.), Mathematical Knowledge Management. VIII, 295 pages. 2006.

Vol. 4106: T.R. Roth-Berghofer, M.H. Göker, H. A. Güvenir (Eds.), Advances in Case-Based Reasoning. XIV, 566 pages. 2006.

Vol. 4099: Q. Yang, G. Webb (Eds.), PRICAI 2006: Trends in Artificial Intelligence. XXVIII, 1263 pages. 2006.

Vol. 4095: S. Nolfi, G. Baldassarre, R. Calabretta, J.C. T. Hallam, D. Marocco, J.-A. Meyer, O. Miglino, D. Parisi (Eds.), From Animals to Animats 9. XV, 869 pages. 2006.

Vol. 4093: X. Li, O.R. Zaïane, Z. Li (Eds.), Advanced Data Mining and Applications. XXI, 1110 pages. 2006.

Vol. 4092: J. Lang, F. Lin, J. Wang (Eds.), Knowledge Science, Engineering and Management. XV, 664 pages. 2006.

Vol. 4088: Z.-Z. Shi, R. Sadananda (Eds.), Agent Computing and Multi-Agent Systems. XVII, 827 pages. 2006.

Vol. 4087: F. Schwenker, S. Marinai (Eds.), Artificial Neural Networks in Pattern Recognition. IX, 299 pages. 2006.

Vol. 4068: H. Schärfe, P. Hitzler, P. Øhrstrøm (Eds.), Conceptual Structures: Inspiration and Application. XI, 455 pages. 2006.

Vol. 4065: P. Perner (Ed.), Advances in Data Mining. XI, 592 pages. 2006.

Vol. 4062: G. Wang, J.F. Peters, A. Skowron, Y. Yao (Eds.), Rough Sets and Knowledge Technology. XX, 810 pages. 2006.

Vol. 4049: S. Parsons, N. Maudet, P. Moraitis, I. Rahwan (Eds.), Argumentation in Multi-Agent Systems. XIV, 313 pages. 2006.

Vol. 4048: L. Goble, J.-J.C.. Meyer (Eds.), Deontic Logic and Artificial Normative Systems. X, 273 pages. 2006.

Vol. 4045: D. Barker-Plummer, R. Cox, N. Swoboda (Eds.), Diagrammatic Representation and Inference. XII, 301 pages. 2006.

Vol. 4031: M. Ali, R. Dapoigny (Eds.), Advances in Applied Artificial Intelligence. XXIII, 1353 pages. 2006.

Vol. 4029: L. Rutkowski, R. Tadeusiewicz, L.A. Zadeh, J.M. Zurada (Eds.), Artificial Intelligence and Soft Computing – ICAISC 2006. XXI, 1235 pages. 2006.

Vol. 4027: H.L. Larsen, G. Pasi, D. Ortiz-Arroyo, T. Andreasen, H. Christiansen (Eds.), Flexible Query Answering Systems. XVIII, 714 pages. 2006.

Vol. 4021: E. André, L. Dybkjær, W. Minker, H. Neumann, M. Weber (Eds.), Perception and Interactive Technologies. XI, 217 pages. 2006.

Vol. 4020: A. Bredenfeld, A. Jacoff, I. Noda, Y. Takahashi (Eds.), RoboCup 2005: Robot Soccer World Cup IX. XVII, 727 pages. 2006.

Vol. 4013: L. Lamontagne, M. Marchand (Eds.), Advances in Artificial Intelligence. XIII, 564 pages. 2006.

Vol. 4012: T. Washio, A. Sakurai, K. Nakajima, H. Takeda, S. Tojo, M. Yokoo (Eds.), New Frontiers in Artificial Intelligence. XIII, 484 pages. 2006.

Vol. 4008: J.C. Augusto, C.D. Nugent (Eds.), Designing Smart Homes. XI, 183 pages. 2006.

Vol. 4005: G. Lugosi, H.U. Simon (Eds.), Learning Theory. XI, 656 pages. 2006.

Vol. 3978: B. Hnich, M. Carlsson, F. Fages, F. Rossi (Eds.), Recent Advances in Constraints. VIII, 179 pages. 2006.

Vol. 3963: O. Dikenelli, M.-P. Gleizes, A. Ricci (Eds.), Engineering Societies in the Agents World VI. XII, 303 pages. 2006.

Vol. 3960: R. Vieira, P. Quaresma, M.d.G.V. Nunes, N.J. Mamede, C. Oliveira, M.C. Dias (Eds.), Computational Processing of the Portuguese Language. XII, 274 pages. 2006.

Vol. 3955: G. Antoniou, G. Potamias, C. Spyropoulos, D. Plexousakis (Eds.), Advances in Artificial Intelligence. XVII, 611 pages. 2006.

Vol. 3949: F. A. Savacı (Ed.), Artificial Intelligence and Neural Networks. IX, 227 pages. 2006.

Vol. 3946: T.R. Roth-Berghofer, S. Schulz, D.B. Leake (Eds.), Modeling and Retrieval of Context. XI, 149 pages. 2006.

Vol. 3944: J. Quiñonero-Candela, I. Dagan, B. Magnini, F. d'Alché-Buc (Eds.), Machine Learning Challenges. XIII, 462 pages. 2006.

Vol. 3930: D.S. Yeung, Z.-Q. Liu, X.-Z. Wang, H. Yan (Eds.), Advances in Machine Learning and Cybernetics. XXI, 1110 pages. 2006.

Vol. 3918: W.K. Ng, M. Kitsuregawa, J. Li, K. Chang (Eds.), Advances in Knowledge Discovery and Data Mining. XXIV, 879 pages. 2006.

Vol. 3913: O. Boissier, J. Padget, V. Dignum, G. Lindemann, E. Matson, S. Ossowski, J.S. Sichman, J. Vázquez-Salceda (Eds.), Coordination, Organizations, Institutions, and Norms in Multi-Agent Systems. XII, 259 pages. 2006.

Vol. 3910: S.A. Brueckner, G.D.M. Serugendo, D. Hales, F. Zambonelli (Eds.), Engineering Self-Organising Systems. XII, 245 pages. 2006.

Vol. 3904: M. Baldoni, U. Endriss, A. Omicini, P. Torroni (Eds.), Declarative Agent Languages and Technologies III. XII, 245 pages. 2006.

Vol. 3900: F. Toni, P. Torroni (Eds.), Computational Logic in Multi-Agent Systems. XVII, 427 pages. 2006.

Vol. 3899: S. Frintrop, VOCUS: A Visual Attention System for Object Detection and Goal-Directed Search. XIV, 216 pages. 2006.

Vol. 3898: K. Tuyls, P.J. 't Hoen, K. Verbeeck, S. Sen (Eds.), Learning and Adaption in Multi-Agent Systems. X, 217 pages. 2006.

Vol. 3891: J.S. Sichman, L. Antunes (Eds.), Multi-Agent-Based Simulation VI. X, 191 pages. 2006.

Vol. 3890: S.G. Thompson, R. Ghanea-Hercock (Eds.), Defence Applications of Multi-Agent Systems. XII, 141 pages. 2006.

Vol. 3885: V. Torra, Y. Narukawa, A. Valls, J. Domingo-Ferrer (Eds.), Modeling Decisions for Artificial Intelligence. XII, 374 pages. 2006.

Vol. 3881: S. Gibet, N. Courty, J.-F. Kamp (Eds.), Gesture in Human-Computer Interaction and Simulation. XIII, 344 pages. 2006.

Vol. 3874: R. Missaoui, J. Schmidt (Eds.), Formal Concept Analysis. X, 309 pages. 2006.

Vol. 3873: L. Maicher, J. Park (Eds.), Charting the Topic Maps Research and Applications Landscape. VIII, 281 pages. 2006.

Vol. 3864: Y. Cai, J. Abascal (Eds.), Ambient Intelligence in Everyday Life. XII, 323 pages. 2006.

Vol. 3863: M. Kohlhase (Ed.), Mathematical Knowledge Management. XI, 405 pages. 2006.

Vol. 3862: R.H. Bordini, M. Dastani, J. Dix, A.E.F. Seghrouchni (Eds.), Programming Multi-Agent Systems. XIV, 267 pages. 2006.

Vol. 3849: I. Bloch, A. Petrosino, A.G.B. Tettamanzi (Eds.), Fuzzy Logic and Applications. XIV, 438 pages. 2006.

Vol. 3848: J.-F. Boulicaut, L. De Raedt, H. Mannila (Eds.), Constraint-Based Mining and Inductive Databases. X, 401 pages. 2006.

Vol. 3847: K.P. Jantke, A. Lunzer, N. Spyratos, Y. Tanaka (Eds.), Federation over the Web. X, 215 pages. 2006.

Vol. 3835: G. Sutcliffe, A. Voronkov (Eds.), Logic for Programming, Artificial Intelligence, and Reasoning. XIV, 744 pages. 2005.

Vol. 3830: D. Weyns, H. V.D. Parunak, F. Michel (Eds.), Environments for Multi-Agent Systems II. VIII, 291 pages. 2006.